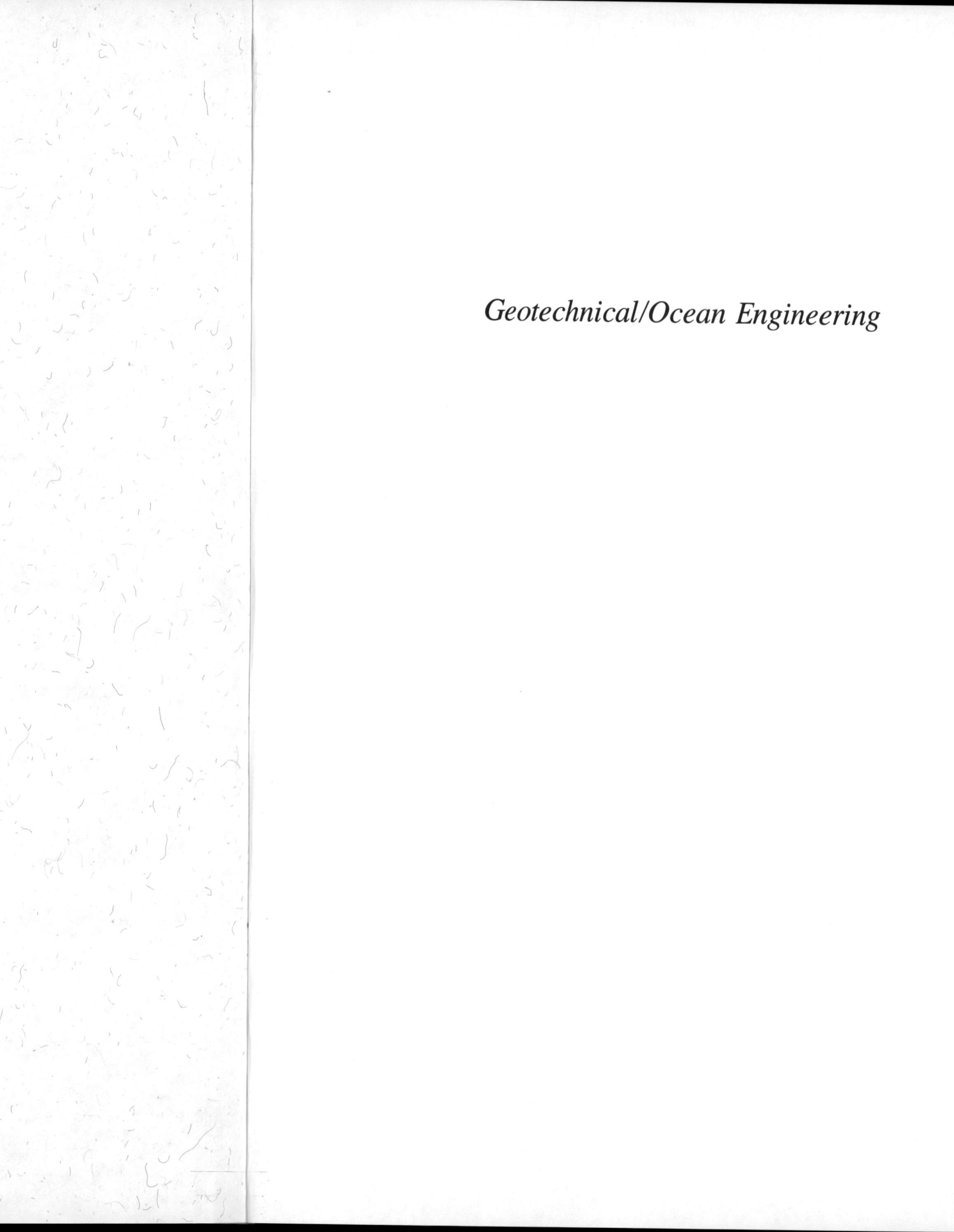

Geotechnical/Ocean Engineering

CIVIL ENGINEERING PRACTICE

3/Geotechnical/Ocean Engineering

EDITED BY

PAUL N. CHEREMISINOFF

NICHOLAS P. CHEREMISINOFF

SU LING CHENG

IN COLLABORATION WITH

S. Ahmed
M. S. S. Almeida
F. Aschieri
C. T. Bishop
G. E. Blight
A. M. Britto
S. K. Chakrabarti
P. N. Cheremisinoff
M. A. Donelan
J. M. Golden
J. Hunt
G. P. Korfiatis
P. C. Kotzias
O. Kusakabe
P. W. Mayne
K. Mizumura
T. S. Nagaraj
A. Nakase
M. Noritake
H. S. Oey
J. A. R. Ortigao
V. M. Paparozzi
S. Prakash
S. Saran
A. Sawicki
U. N. Saran
J. R. Schuring
A. W. S. Smith
B. R. Srinivasa Murthy
A. C. Stamatopoulos
C.-N. Sun
J. Takemura
B. R. Thamm
E. F. Thompson
F. A. Uliana
H.-C. Wu
S. L. Xie

LANCASTER · BASEL

Published in the Western Hemisphere by
Technomic Publishing Company, Inc.
851 New Holland Avenue
Box 3535
Lancaster, Pennsylvania 17604 U.S.A.

Distributed in the Rest of the World by
Technomic Publishing AG

Printed in the United States of America

10 9 8 7 6 5 4 3 2 1

Main entry under title:
Civil Engineering Practice 3—Geotechnical/Ocean Engineering

A Technomic Publishing Company book
Bibliography: p.
Includes index p. 867

Library of Congress Card No. 87-50629
ISBN No. 87762-554-9

TABLE OF CONTENTS

Preface vii

Contributors to Volume 3 ix

SECTION ONE: **SOIL MECHANICS**

1 **Endochronic Description of Sand Response** 3
H.-C. Wu, *University of Iowa, Iowa City, IA*

2 **Consistency Limits of Soils—Principles and Potentials** 23
T. S. Nagaraj and B. R. Srinivasa Murthy, *Indian Institute of Science, Bangalore, India*

3 **Plastic Behavior of Reinforced Earth** 45
A. Sawicki, *Institute of Hydroengineering, Gdansk-Oliwa, Poland*

4 **Footings and Constitutive Laws** 65
S. Prakash, *University of Missouri–Rolla, Rolla, MO*
S. Saran, *University of Roorkie, Roorkie, U.P., India*
U. N. Saran, *Muzuffarpur Institute of Technology, Muzuffarpur Bihar, India*

5 **Stochastic Models of Stress Distributions in Granular Materials** 139
J. M. Golden, *National Institute for Physical Planning and Construction, Dublin, Ireland*

6 **Bearing Capacity of Shallow and Pile Foundations** 175
G. P. Korfiatis, *Stevens Institute of Technology, Hoboken, NJ*
J. R. Schuring, *New Jersey Institute of Technology, Newark, NJ*

SECTION TWO: **STABILITY**

7 **Field Performance of Embankments Over Soft Soil** 223
B. R. Thamm, *Federal Highway Research Institute of Germany, Bergisch-Gladbach, FRG*

8 **Stability and Deformation of Embankments on Soft Clay** 267
J. A. R. Ortigão and M. S. S. Almeida, *Federal University of Rio de Janeiro, Brazil*

9 **Stability of Axisymmetric Excavation in Clay** 337
A. M. Britto, *Cambridge University, Cambridge, England*
O. Kusakabe, *Utsunomiya University, Utsunomiya, Japan*

SECTION THREE: **BEARING CAPACITY**

10 **Preloading: Planning and Predictions** 367
A. C. Stamatopoulos and P. C. Kotzias, *P. Kotzias-A. Stamatopoulos Co., Ltd., Athens, Greece*

11 **Ground Improvement by Dynamic Compaction** 405
P. W. Mayne, *Law Engineering Testing Company, Washington, D.C.*

12 Seasonal Temperature Effects on Foundations **423**
F. Aschieri and F. A. Uliana, *ENEL (Ente Nazionale per l'Energia Elettrica) Milano, Italy*

13 Centrifuge Model Tests on Bearing Capacity of Clay **439**
A. Nakase and J. Takemura, *Tokyo Institute of Technology, Tokyo, Japan*
O. Kusakabe, *University of Utsunomiya, Utsunomiya, Japan*

SECTION FOUR: **BURIED STRUCTURES**

14 Geotechnical Engineering for Pipelines and Tank Farms **459**
S. Ahmed, *Law Engineering Testing Company, Houston, TX*

15 Cement-Mortar Lined Pipes **503**
C.-N. Sun and J. Hunt, *Tennessee Valley Authority, Knoxville, TN*

16 Soil Mechanics Principles in Underground Mining **521**
G. E. Blight, *University of Witwatersrand, Republic of South Africa*

17 Design of Buried Pipelines **555**
H. S. Oey, *University of Texas at El Paso, El Paso, TX*

18 Corrosion of Underground Piping **575**
P. N. Cheremisinoff, *New Jersey Institute of Technology, Newark, NJ*
V. M. Paparozzi, *GHA Lock Joint, Inc., Wharton, NJ*

SECTION FIVE: **WAVES AND WAVE ACTIONS**

19 Waves and Wave Forecasting **597**
E. F. Thompson, *U.S. Army Engineer Waterways Experiment Station, Vicksburg, MS*

20 Waves and Wave Forecasting **653**
C. T. Bishop and M. A. Donelan, *National Water Research Institute, Burlington, Ontario, Canada*

21 Waves on Coastal Structures **697**
A. W. S. Smith, *Queensland Institute of Technology, Brisbane, Queensland, Australia*

22 Waves on Structures **717**
S. L. Xie, *The Ministry of Communications, Tianjin, China*

SECTION SIX: **COASTAL STRUCTURES**

23 Coastal Structure **735**
K. Mizumura, *Kanazawa Institute of Technology, Ishikawa, Japan*

24 Offshore Structures **781**
S. K. Chakrabarti, *CBI Industries, Inc., Plainfield, IL*

25 Ports and Harbors **829**
M. Noritake, *Kansai University, Suita, Osaka, Japan*

Index 867

PREFACE

While the designation civil engineering dates back only two centuries, the profession of civil engineering is as old as civilized life. Through ancient times it formed a broader profession, best described as master builder, which included what is now known as architecture and both civil and military engineering. The field of civil engineering was once defined as including all branches of engineering and has come to include established aspects of construction, structures, and emerging and newer sub-disciplines (e.g., environmental, water resources, etc.). The civil engineer is engaged in planning, design of works connected with transportation, water and air pollution, as well as canals, rivers, piers, harbors, etc. The hydraulic field covers water supply/power, flood control, drainage and irrigation, as well as sewerage and waste disposal.

The civil engineer may also specialize in various stages of projects such as investigation, design, construction, operation, etc. Civil engineers today, as well as engineers in all branches, have become highly specialized, as well as requiring a multiplicity of skills in methods and procedures. Various civil engineering specialties have led to the requirement of a wide array of knowledge.

Civil engineers today find themselves in a broad range of applications, and it was to this end that the concept of putting this series of volumes together was made. The tremendous increase of information and knowledge all over the world has resulted in proliferation of new ideas and concepts, as well as a large increase in available information and data in civil engineering. The treatises presented are divided into five volumes for the convenience of reference and the reader:

VOLUME 1 Structures
VOLUME 2 Hydraulics/Mechanics
VOLUME 3 Geotechnical/Ocean Engineering
VOLUME 4 Surveying/Transportation/Energy/Economics & Government/Computers
VOLUME 5 Water Resources/Environmental

A serious effort has been made by each of the contributing specialists to this series to present information that will have enduring value. The intent is to supply the practitioner with an authoritative reference work in the field of civil engineering. References and citations are given to the extensive literature as well as comprehensive, detailed, up-to-date coverage.

To insure the highest degree of reliability in the selected subject matter presented, the collaboration of a large number of specialists was enlisted, and this book presents their efforts. Heartfelt thanks go to these contributors, each of whom has endeavored to present an up-to-date section in their area of expertise and has given willingly of valuable time and knowledge.

PAUL N. CHEREMISINOFF
NICHOLAS P. CHEREMISINOFF
SU LING CHENG

CONTRIBUTORS TO VOLUME 3

S. AHMED, Law Engineering Testing Co., Houston, TX

M. S. S. ALMEIDA, Federal University of Rio de Janeiro, Brazil

F. ASCHIERI, ENEL (Ente Nazionale per l'Energia Elettrica) DCO, Milano, Italy

C. T. BISHOP, National Water Research Institute, Burlington, Ontario, Canada

G. E. BLIGHT, University of the Witwatersrand, Johannesburg, 2000, Republic of South Africa

A. M. BRITTO, Cambridge University, Cambridge, England

S. K. CHAKRABARTI, CBI Industries Inc., Plainfield, IL

P. N. CHEREMISINOFF, New Jersey Institute of Technology, Newark, NJ

M. A. DONELAN, National Water Research Institute, Burlington, Ontario, Canada

J. M. GOLDEN, National Institute for Physical Planning and Construction, Dublin, Ireland

J. HUNT, Tennessee Valley Authority, Knoxville, TN

G. P. KORFIATIS, Stevens Institute of Technology, Hoboken, NJ

P. C. KOTZIAS, P. Kotzias-A. Stamatopoulos Company, Ltd., Athens 11471, Greece

O. KUSAKABE, Utsunomiya University, Utsunomiya, 1 Shii, Japan

P. W. MAYNE, Law Engineering Testing Company, Washington, D.C.

K. MIZUMURA, Kanazawa Institute of Technology, Ishikawa 921, Japan

T. S. NAGARAJ, Indian Institute of Science, Bangalore, India

A. NAKASE, Tokyo Institute of Technology

M. NORITAKE, Kansai University, Suita, Osaka, Japan

H. S. OEY, University of Texas at El Paso, El Paso, TX

J. A. R. ORTIGÃO, Federal University of Rio de Janeiro, Brazil

V. M. PAPAROZZI, GHA Lock Joint, Inc., Wharton, NJ

S. PRAKASH, University of Missouri–Rolla, Rolla, MO

S. SARAN, University of Roorkie, Roorkie, U.P., India

U. N. SARAN, Muzuffarpur Institute of Technology, Muzuffarpur Bihar, India

A. SAWICKI, Institute of Hydroengineering, Gdansk-Oliwa, Poland

J. R. SCHURING, New Jersey Institute of Technology, Newark, NJ

A. W. S. SMITH, Queensland Institute of Technology, Brisbane, Queensland, Australia

B. R. SRINIVASA MURTHY, Indian Institute of Science, Bangalore, India

A. C. STAMATOPOULOS, P. Kotzias-A. Stamatopoulos Company, Ltd., Athens 11471, Greece

C.-N. SUN, Tennessee Valley Authority, Knoxville, TN

J. TAKEMURA, Tokyo Institute of Technology

B. R. THAMM, Federal Highway Research Institute of Germany, Bergisch-Gladbach, FRG

E. F. THOMPSON, U.S. Army Engineer Waterways Experiment Station, Vicksburg, MS

F. A. ULIANA, ENEL (Ente Nazionale per l'Energia Elettrica) DCO, Milano, Italy

H.-C. WU, University of Iowa, Iowa City, IA

S. L. XIE, The Ministry of Communications, Tianjin, China

CIVIL ENGINEERING PRACTICE

VOLUME 1
Structures

SECTION 1 Reinforced Concrete Structures

SECTION 2 Structural Analysis

SECTION 3 Stability

SECTION 4 Pavement Design

SECTION 5 Wood Structures

SECTION 6 Composites

VOLUME 2
Hydraulics/Mechanics

SECTION 1 Hydraulics/Open Channel Flow

SECTION 2 Flow in Pipes

SECTION 3 Flow With Bed Load

SECTION 4 Mechanics/Solid Mechanics

SECTION 5 Fluid Mechanics

SECTION 6 Solid–Fluid Interaction

VOLUME 3
Geotechnical/Ocean Engineering

SECTION 1 Soil Mechanics

SECTION 2 Stability

SECTION 3 Bearing Capacity

SECTION 4 Buried Structures

SECTION 5 Waves and Wave Actions

SECTION 6 Coastal Structures

VOLUME 4
Surveying/Construction/ Transportation/Energy/ Economics & Goverment/Computers

SECTION 1 Surveying

SECTION 2 Construction

SECTION 3 Transportation

SECTION 4 Energy

SECTION 5 Economics/Government/Data Acquisition

VOLUME 5
Water Resources/Environmental

SECTION 1 Water Supply and Management

SECTION 2 Irrigation

SECTION 3 Environmental

SECTION ONE

Soil Mechanics

CHAPTER 1	Endochronic Description of Sand Response	3
CHAPTER 2	Consistency Limits of Soils – Principles and Potentials	23
CHAPTER 3	Plastic Behavior of Reinforced Earth	45
CHAPTER 4	Footings and Constitutive Laws	65
CHAPTER 5	Stochastic Models of Stress Distributions in Granular Materials	139
CHAPTER 6	Bearing Capacity of Shallow and Pile Foundations	175

Endochronic Description of Sand Response

Han-Chin Wu*

INTRODUCTION

There is a renewed interest among research workers to model the mechanical behavior of cohesionless soil. This new interest arises primarily from the progress of the computational techniques using a large-scale computer and it is now realistic to perform stress analysis and design using a sophisticated model (or constitutive equation) of soil. However, due to the fact that the progress in the computational techniques has out-paced that in the soil modeling, the results of computation are often not usable. Therefore, there is an urgent need for the development of a realistic sophisticated constitutive equation of cohesionless soil.

The need for development of a constitutive equation for cohesionless soil arises also from the research related to earthquake engineering in recent years. Due to the complicated loading condition associated with an earthquake, many an existing model of soil has become unsuitable for application.

The objective of this paper is to develop a constitutive equation which complies with the basic laws of continuum mechanics and satisfactorily describes the complex nature of the mechanical behavior of cohesionless soil under loading. In this development, the following experimental results are observed: (1) effect of void ratio on the mechanical behavior; (2) the volumetric stress–strain curve bending toward the stress axis; (3) the deviatoric (shear) stress–strain curve bending toward the strain axis; and (4) the cross effects which include the deviatoric stress–strain curve, which is strongly affected by confining pressure (there is no shear strength at zero confining pressure), and the densification, which is the volumetric strain due to shear deformation.

*Department of Civil and Environmental Engineering, University of Iowa, Iowa City, IA

Considerable effort has been made by research workers in recent years to model the behavior of sand using the concepts of elasto-plasticity and others. It is not the purpose of this paper to survey the progress made in these other approaches. The reader is referred to literature concerning these works.

The present paper is devoted to the modeling of cohesionless soil using the endochronic theory. It is shown herein that this theory can provide a consistent modeling and satisfy all the requirements previously described. In addition, it is worthwhile to mention that the final form of the constitutive equation is mathematically simple which would mean greater applicability in solving realistic problems. Furthermore, the material constants associated with the theory are easily determined. This latter assertion will be analyzed fully in the text.

The endochronic theory was introduced by Valanis [10] in order to describe the mechanical behavior of metallic materials. Valanis [11,12] and Wu [22,23] have subsequently contributed toward the further development of the theory. The foundation of the theory is built upon the theory of thermodynamics of internal variables with the additional concept of intrinsic (or endochronic) time, which is a monotonically increasing "time-like" parameter measuring the history of deformation.

Only a few years ago, the idea of three dimensional constitutive equation of soil met with skepticism. This is best described by the following excerpt from the preface written by C. P. Wroth and A. C. Palmer for the *Proceedings of the Symposium on the Role of Plasticity in Soil Mechanics* held in Cambridge. Wroth and Palmer wrote in 1973 that "plasticity theory was developed by people who thought in terms of metals, and for about 20 years workers in soil mechanics have been looking at the theory, rather as outsiders, and asking whether it had anything to offer them. Opinions have been very diverse: some have thought that plasticity was highflown mathematics with little to offer the soils engineer, and others have come to the same conclusion in a more rational way, pointing out that materials composed of hard particles slipping and rolling over one another have little in common with polycrystalline metals. Some people, on the other hand, jumped into plasticity theory over hastily, and naively assumed that it could be taken into soil mechanics without any critical assessment of the underlying assumptions."

Today, plasticity has gained broader acceptance among geotechnical engineers. Indeed, the opinion of the geotechnical engineers about the constitutive equations for soils using the concept of continuum mechanics (plasticity is only one of the several approaches) has greatly improved. Nevertheless, some elements of the skepticism mentioned above still remain. The following comments are offered in favor of a continuum mechanics approach and especially from the viewpoint of endochronic theory.

At the microscopic level, materials composed of hard particles slipping and rolling over one another have indeed little in common with polycrystalline metals. However, at the macroscopic level, there exist properties common to both. Due to the small sizes of sand particle and the grains of polycrystalline metals, both materials may be justly treated as continua with density defined. In fact, density, which is closely related to the continuum assumption, has long been used by the geotechnical engineers to describe the state of different soil types.

Gas, liquid, and solid are all treated under one roof in continuum mechanics. The differences among them are expressed through different constitutive equations. Due to different forms of constitutive equations, diversified experimentally observed behaviors of materials may be successfully described. For instance, the observed facts that volume change is important in gas, an invicid fluid cannot take shear stresses, volume change is not important for incompressible fluids, volumetric and shear deformations are usually not coupled in solids, volumetric deformation is recoverable in metals, and others can all be described using different types of constitutive equations.

The macroscopic behaviors of soils and metals are different in many ways, but if both are treated as continua, then the difference between the two is in the constitutive equations.

Plasticity was initially proposed to represent the macroscopic behavior of metals, but its correlation with the microscopic behavior of polycrystalline metals has not yet been fully justified. Therefore, plasticity should be regarded as a macroscopic theory. It is true that yielding is a phenomenon widely recognized in metals. However, in the deviatoric response of soil, a phenomenon similar to yielding is also observed macroscopically. Therefore, the concept of yielding has been conveniently employed in the description of soil. In the volumetric response, great differences exist between soils and metals. This is the area where the concept of yielding is not readily applicable, and many research workers are currently undertaking to solve the problem.

As to the endochronic theory, it is also true that the theory was first developed for metals, and the first applications were related to the description of metallic behaviors. However, if the details of derivation are examined, then one would quickly find that the theory is not made for any particular material, at least not until the very last stage when specific functional forms are chosen. It is at this last stage that the resulting equation is suitable for describing the behavior of metals; otherwise, the volumetric response may be made nonlinear and the resulting equation can then be conveniently used to describe the behavior of geotechnical materials. Therefore, the endochronic theory for metals and the endochronic theory for soils should be viewed as two offsprings under the roof of endochronic theory. They are in parallel positions.

In this summary paper, the endochronic model developed by the writer and his co-workers will be discussed. Emphasis will be given to the loading-unloading phenomenon. The phenomenon of yielding has been found in the deviatoric response of soil as mentioned before. This model has thus been developed to account for yielding in the deviatoric response. On the other hand, yield stress is not apparent in the volumetric stress–strain curve of soils. A nonlinear form of equation has been developed to account for this behavior. The fact that the endochronic theory is able to account for the volumetric response of soil should be viewed as a major advantage of this approach.

Valanis [11] has formulated the theory both in the Helmholtz and the Gibbs free energy representations. Most applications of the endochronic theory used the Helmholtz formulation. The only applications in the Gibbs formulation were due to Wu, et al. [18] and Wei [14]. These writers showed that the Gibbs formulation can be convenient in the description of stress-controlled tests.

The first application of the theory to the description of mechanical behavior of sand was due to Bazant and Krizek [1], using the linear form of the endochronic theory. Through this application the endochronic theory has been recognized by researchers as quite powerful in the investigation of dynamic behavior of sand. However, the linear form of the theory is not adequate in describing a material with noticeable coupling effect between the volumetric and deviatoric responses. Therefore, the theory should be modified; this is accomplished in the present paper.

Valanis [13] has also modified his theory to accommodate the complex behavior of soil. The approach is different than the present one and will not be elaborated here.

The present form of the endochronic theory is based on the Gibbs formulation which is presented in the next section. After the constitutive equations have been derived, they are applied to describe the mechanical behavior of drained sand under isotropic consolidation and shear loading during a constant confining pressure triaxial test (CCT-test). The equations are then applied to describe the effect of densification. In these applications, relatively simple loading paths have been considered. Other loading paths are considered in subsequent sections, which include cyclic shear behavior of sand and loading paths in true triaxial test. Finally, undrained sand is investigated. Correlation between densification and dilation of a sand under drained condition and the pore-water pressure change of the same sand under undrained condition has been achieved. Under undrained condition the concept of effective stress is employed. Furthermore, the volumetric strain is assumed to be unchanged, which implies that the volume change due to compaction and reorientation of the sand grains is counteracted by the volume change caused by the hydrostatic response to the change of the effective mean stress.

GIBBS FREE ENERGY FORMULATION

The details of this model is presented in Reference 19. This development is based on the improved endochronic theory by Valanis [12], which has been proposed for metallic materials.

Within the contest of thermodynamics of internal variables, inelastic deformations of materials are described by r numbers of internal variables, $\underset{\sim}{q}^r$. The values of these variables grow as the internal structures of materials change by deformation. The r internal variables are divided into three groups, each being active only for a specific function. Thus, h number of internal variables q^h_{kk} are active in the representation of hydrostatic response, s internal variables p^s are responsible for deviatoric response, and d number of internal variables q^d_{kk} are related to volume change (densification) due to shearing. Hence, $r = h + s + d$. The last group, q^d_{kk}, distinguishes geotechnical materials from metals in the continuum mechanics sense. The activities of this group of variables are not significant for metals but are very important for geotechnical materials. The group of internal variables, q^h_{kk}, makes the description of the inelastic nonlinear hydrostatic response possible. In the case of metals, the volumetric response is linearly elastic and q^h_{kk} does not play any part at all in the constitutive modelling.

The rate of change (or evolution) of each group of internal variables is a material property and depends on the material that the model is presumed to describe. Therefore, the evolution equation may be linear or nonlinear depending on the material at hand. In Reference 19, it has been found that a nonlinear evolution equation for q^h_{kk}, and linear evolution equations for $\underset{\sim}{p}^s$ and q^d_{kk} are satisfactory for the sands considered.

According to this model, the rate of change takes place with respect to a monotonically increasing time-like parameter called intrinsic (or endochronic) time. The intrinsic time increases only when the material undergoes deformation. Therefore, the intrinsic time serves to record the deformation history.

Different definitions of intrinsic time are possible. However, they do not all lead to the same results. Those leading to controversial results for one material may still give reasonable results for other materials. All definition giving satisfactory results should be further investigated with the purpose of simplifying and improving the mathematical expression. The intrinsic time has been defined in terms of total strain by Valanis [10] which, although convenient in many applications, has led to controversial results for metals during unloading and reloading. Valanis [12] has since proposed a new definition of intrinsic time based on plastic strain, and this has proved to be satisfactory for metals.

In the present model for sand, three intrinsic times are used and denoted by z_H, z_S, and z_D, respectively, for hydrostatic response, deviatoric response, and densification. These intrinsic times are further scaled by factors, so that another set of intrinsic times ζ_H, ζ_S, ζ_D are defined by

$$\frac{d\zeta_H}{dz_H} = h(\zeta_H) > 0, \frac{d\zeta_S}{dz_S} = f(\zeta_s, \sigma) > 0$$

and

$$\frac{d\zeta_D}{dz_D} = g(\zeta_D) > 0 \tag{1}$$

The scaling factor $h(\zeta_H)$ is necessary so that nonlinear volumetric behavior may be described. The factor $f(\zeta_S,\sigma)$, where σ is hydrostatic pressure, is needed to account for isotropic hardening in the deviatoric behavior, and the scale $g(\zeta_D)$ is used so that the phenomenon of densification may be better described. The intrinsic times ζ_H and ζ_S are defined in terms of plastic strain, but the intrinsic time ζ_D is defined in terms of the total shear strain. These points will be further explained later in the text.

The Gibbs free energy formulation is used to derive the constitutive equations. In this formulation, the variables are stress, $\underset{\sim}{\sigma}$, and internal variables, $\underset{\sim}{q}^r$. An important reason for using this formulation is that hydrostatic stress, $\sigma = 1/3\ \sigma_{kk}$, and deviatoric stress, $\underset{\sim}{s}$, may be varied independently, and the procedure for derivation of constitutive equation is thus simplified.

To facilitate the formulation, it is assumed that the Gibbs free energy can be separated into two parts, i.e.:

$$\phi(\underset{\sim}{\sigma},\underset{\sim}{q}^r) = \phi_H(\sigma,q^h,q^d) + \phi_D(\sigma^o,\underset{\sim}{s},\underset{\sim}{p}^s) \tag{2}$$

with $\sigma = 1/3\ \sigma_{kk}$; $q^h = 1/3\ q^h_{kk}$; and $q^d = 1/3\ q^d_{kk}$. The subscripts H and D denote, respectively, the hydrostatic and the deviatoric parts. The stress $\sigma^o = \sigma|_{s\to 0, \underset{\sim}{p}^s \to 0}$ is the hydrostatic stress prior to the application of any deviatoric stress. Thus, σ^o is a constant during deviatoric loading and enters into Equation (2) as a parameter. It should be mentioned that ϕ_H and ϕ_D are separable but dependent due to the presence of q^d in ϕ_H and σ^o in ϕ_D.

It is further assumed the following forms for ϕ_H and ϕ_D are adequate in the description of mechanical behavior of cohesionless soil for loading paths under investigation. The validity of these expressions under other loading paths awaits further investigation. Thus,

$$\phi_H = -\frac{1}{2}A\sigma^2 - \sum_h B^h q^h \sigma + \frac{1}{2}\sum_h M^h q^h q^h - \sum_d C^d q^d \sigma + \frac{1}{2}\sum_{d=2}^{d} D^d q^d q^d + \nu C^1 q^1 \tag{3}$$

$$\phi_D = -\frac{1}{2}\underset{\sim}{s} \cdot \underset{\approx}{E} \cdot \underset{\sim}{s} - \sum_s \underset{\sim}{s} \cdot \underset{\approx}{F}^s \cdot \underset{\sim}{p}^s + \frac{1}{2}\sum_s \underset{\sim}{p}^s \cdot \underset{\approx}{G}^s \cdot \underset{\sim}{p} \tag{4}$$

in which A, B^h, M^h, C^d, and D^d = positive material constants and $D^1 = 0$, υ = constant; $\underset{\approx}{E}$ and $\underset{\approx}{F}^s$ = fourth-order material tensors, and may depend on σ^o; $\underset{\approx}{E}$ = positive semi-definite; and G^s = positive constant material tensors. The dependence of $\underset{\approx}{E}$ and $\underset{\approx}{F}^s$ upon σ^o accounts for the fact that shear response of sand is pressure dependent and that at zero confining pressure the sand does not possess any shear resistance.

In the Gibbs formulation, the strain, $\underset{\sim}{\epsilon}$, is given by

$$\underset{\sim}{\varepsilon} = -\frac{\partial \phi}{\partial \underset{\sim}{\sigma}} \tag{5}$$

Substituting Equations (2), (3), and (4) into Equation (5) and using the following expression for an isotropic fourth-order tensor, M_{ijkl}:

$$M_{ijkl} = M_1 \delta_{ij}\delta_{kl} + M_2 \delta_{ik}\delta_{jl} \tag{6}$$

in which M_1 and M_2 = constants; and δ_{ij} the Kronecker delta. The constitutive equation for cohesionless soil may be given by

$$\underset{\sim}{\varepsilon} = \frac{1}{3}\left(A\sigma + \sum_h B^h q^h + \sum_d C^d q^d\right)\underset{\sim}{\delta} + E_2 \underset{\sim}{s} + \sum_s F_2^s \underset{\sim}{p}^s \tag{7}$$

or, equivalently, by the set of equations

$$\varepsilon_{kk} = A\sigma + \sum_h B^h q^h + \sum_d C^d q^d \tag{8}$$

and

$$\underset{\sim}{e} = E_2 \underset{\sim}{s} + \sum_s F_2^s \underset{\sim}{p}^s \tag{9}$$

It is remarked that in Equation (8) the volumetric strain consists of two parts, i.e., that due to isotropic consolidation, ϵ_{kk}^h, and that due to deviatoric deformation (densification), ϵ_{kk}^d. In Equation (9), $\underset{\sim}{e}$ is the deviatoric part of $\underset{\sim}{\epsilon}$.

In the subsequent sections, Equations (8) and (9) will be shown to describe various aspects of the loading-unloading behavior of sand.

VOLUMETRIC RESPONSE IN ISOTROPIC CONSOLIDATION

The volumetric response in the isotropic consolidation of cohesionless soil is usually nonlinear. When pressure is applied, the soil behaves plastically at the beginning and elastically at a later stage. Moreover, the volumetric stress–strain curve bends toward the stress axis. In this discussion, it is assumed that the hydrostatic stress level is low enough not to cause any crushing of particles during loading.

During isotropic consolidation, no densification (due to shear deformation) takes place. Thus, $d\epsilon_{kk} = d\epsilon_{kk}^h$. The evolution equation for q^h is given by the nonlinear expression

$$\frac{d}{dz_H} q^h = H^h(\sigma) \tag{10}$$

where the dependence of H^h on q^h is suppressed which is equivalent to putting $M^h = 0$. For the sands considered in this paper, Equation (10) is a good approximation. Hence, the following volumetric constitutive equation is obtained:

$$d\varepsilon_{kk}^h = A d\sigma + \sum_h B^h H^h(\sigma) dz_H \tag{11}$$

To account for nonlinear behavior in the volumetric response, an intrinsic time scale is introduced so that intrinsic time ζ_H is defined by Equation (1a) and is further related to deformation by

$$d\zeta_H = |d\theta^h| = \left| d\varepsilon_{kk}^h - \frac{d\sigma}{K_o} \right| \tag{12}$$

in which θ^h is the plastic part of ϵ_{kk}^h. In this treatment, the unloading curve is assumed to be a straight line with constant slope K_o. This is considered to be a good first-order approximation in view of the uncertainty associated with the membrane penetration in an isotropic consolidation test.

The accuracy in the membrane correction directly affects the data of hydrostatic stress–strain curve during loading and unloading. It is popular to use the dummy rod method

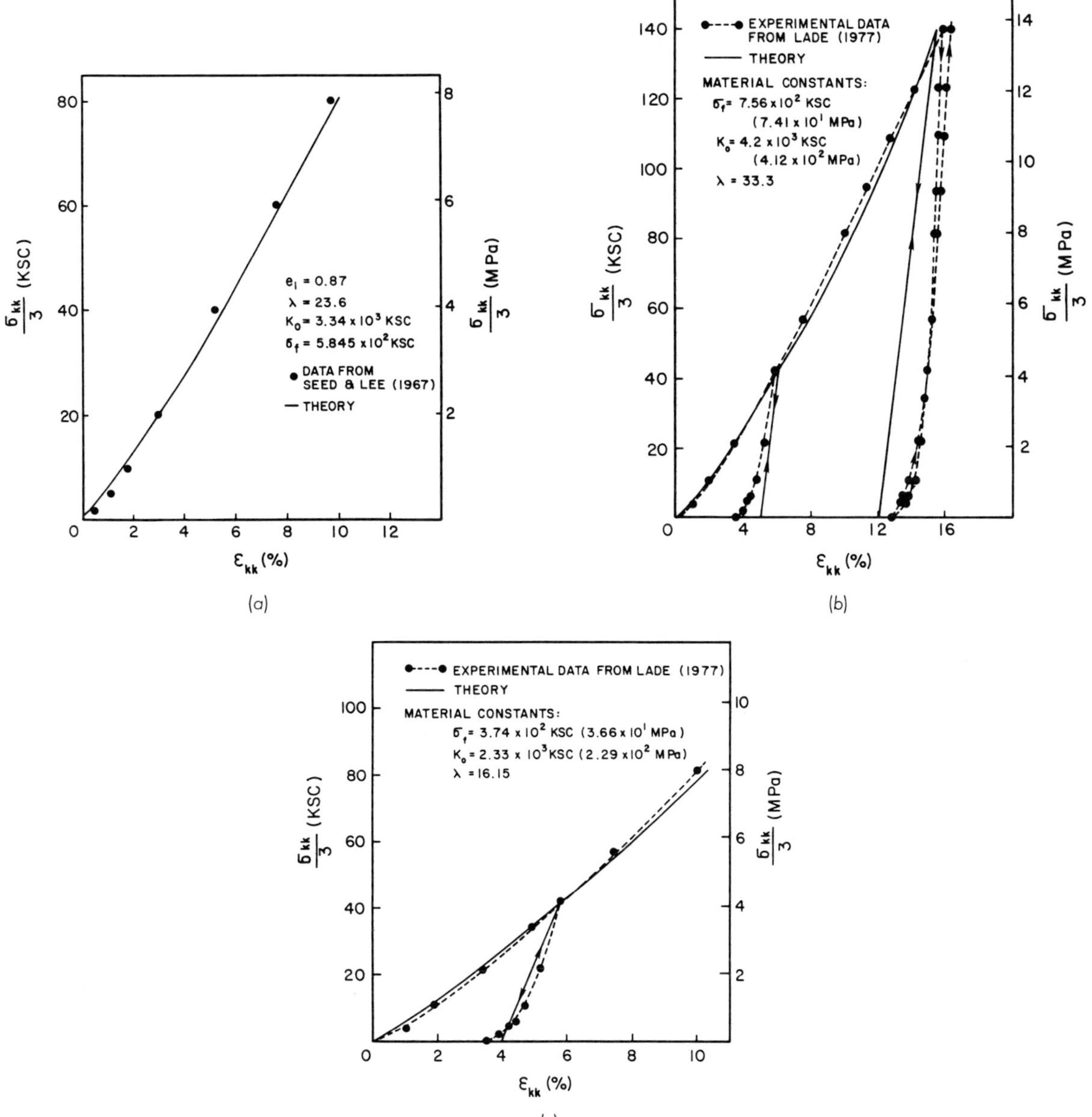

FIGURE 1. Isotropic consolidation test for Sacramento River sand: (a) loading; (b) loading–unloading; (c) linear unloading.

to estimate the correction for membrane penetration. However, the analysis of Wu and Chang [16] shows that the state of stress in the sand specimen with a dummy rod (metal core) is not in an isotropic condition. Hence, the membrane correction obtained by the dummy rod method is questionable. The effect of this finding may be far reaching since the membrane correction constitutes a large fraction of the total measured volume change. In the investigation of volumetric behavior of sand under isotropic compression, unloading–reloading sequence would result in a small hysteresis loop in the stress–strain curve. Due to the uncertainty associated with the membrane penetration, it is reasonable to assume that unloading and reloading follow the same straight line in the volumetric response.

A simple volumetric constitutive equation is obtained if the following special expressions are chosen:

$$H(\zeta_H) = \left(\frac{\sigma_f}{K_o} - \zeta_H\right)^{-1}$$

$$A = \frac{1}{K_o}, \tag{13}$$

$$\sum_h B^h H^h(\sigma) = \frac{K_o}{\sigma_f} e^{(\lambda/K_o)\sigma}$$

in which σ_f and λ are constants. Therefore, Equation (11) reduces to

$$d\theta^h = \frac{K_o}{\sigma_f} e^{(\lambda/K_o)\sigma} \left(\frac{\sigma_f}{K_o} - \zeta_H\right) d\zeta_H \tag{14}$$

In the case of $\zeta_H \leq \sigma_f/K_o$, which will be shown later to be true, the above equation yields $d\theta^h \geq 0$. Therefore, only compression is admissible and this of course is in agreement with the observed behavior of sand. It may then be written that $d\zeta_H = d\theta^h$ for all hydrostatic behaviors.

Thus, Equation (14) may be written as

$$(d\theta^h)\, F = 0 \tag{15}$$

in which

$$F = 1 - \frac{K_o}{\sigma_f} e^{(\lambda/K_o)\sigma} \left(\frac{\sigma_f}{K_o} - \zeta_H\right) \tag{16}$$

This implies either $d\theta^h = d\epsilon_{kk}^h - d\sigma/K_o = 0$, which is the elastic behavior, or $F = 0$ and $d\theta^h \neq 0$. In the latter case, $\zeta_H = \theta^h$ (assuming zero initial values for ζ_H and θ^h) and $F = 0$ resulted in

$$\epsilon_{kk}^h = \frac{\sigma_f}{K_o}\left(\frac{\sigma}{\sigma_f} + 1 - e^{-(\lambda/K_o)\sigma}\right) \tag{17}$$

by use of $\epsilon_{kk}^h = \theta^h + \sigma/K_o$. Equation (17) governs the volumetric behavior when volumetric plastic deformation occurs. By comparing this equation with the nonlinear experimental curve, the constants σ_f and K_o may be determined, and it can easily be shown that

$$\lambda = \frac{K_o}{\sigma_f}\left(\frac{K_o}{K_1} - 1\right) \tag{18}$$

where K_1 is the initial slope of volumetric stress–strain curve.

It should be mentioned that the asympotote of Equation (17) intersects with the ϵ_{kk}^h-axis at σ_f/K_o and that on this line $\zeta_H = \theta^h = \sigma_f/K_o$ (a constant). Hence, the condition

$$\zeta_H \leqq \frac{\sigma_f}{K_o} \tag{19}$$

always holds.

Figure 1a shows the theoretical results obtained from Equation (17). The experimental data were due to Lee and Seed [5] which were obtained for Sacramento River sand with initial void ratio $e_i = 0.87$. The data were originally presented in a confining pressure (σ_r) versus current void ratio (e) relation. This has been replotted in Figure 1a by observing that $1/3\ \sigma_{kk} = \sigma_r$ and $\epsilon_{kk} = (e_i - e)/(1 + e_i)$.

Figure 1b shows the theoretical loading/unloading curves compared with the experimental results by Lade [3] also for Sacramento River sand. Figure 1c shows that by adjusting the constants σ_f, K_o, and λ, the point where unloading begins may be made to coincide with the corresponding experimental point. The agreement in the unloading curve can thus be improved.

DEVIATORIC RESPONSE

The deviatoric response of cohesionless soil is similar to that of metallic material in that the stress–strain curve bends toward the strain axis. However, the response of cohesionless soil is highly sensitive to the confining pressure and the shear modulus is a function of σ^o. In addition, the rate of strain-hardening is also confining pressure dependent.

The deviatoric constitutive equation [Equation (9)] is used in the following discussion. The evolution equation for $\underset{\sim}{p}^s$ is assumed to be

$$\frac{\partial \phi_D}{\partial \underset{\sim}{p}^s} + b_2^s \frac{d\underset{\sim}{p}^s}{dz_S} = 0, \ s \text{ not summed} \tag{20}$$

in which b_2^s are positive constants. It is shown in this section that the above linear evolution equation is adequate in describing the behavior of sand under consideration.

The intrinsic time for deviatoric response is scaled by the

expression [Equation (1b)], in which $\sigma = \sigma_r$ = the confining pressure in the CCT (constant confining pressure triaxial) test. It is further defined that

$$d\zeta_S^2 = d\theta \cdot d\theta \tag{21}$$

where $\underset{\sim}{\theta} = \underset{\sim}{e} - \underset{\sim}{s}/2\mu_o$ is the plastic strain, and $2\mu_o$ the shear modulus.

It has been found that for relatively simple strain-paths the hardening function takes the following simple form:

$$f(\zeta_S, \sigma_r) = 1 + \beta(\sigma_r)\zeta_S \tag{22}$$

where $\beta(\sigma_r)$ is a material function.

Integrating Equation (20) for p^s and substituting the resulting equation into Equation (9), the following deviatoric constitutive equation is obtained:

$$\underset{\sim}{e} = E_2\underset{\sim}{s} + \sum_s G_s \int_o^{z_S} e^{-\alpha_s(z_S - z_S')}\underset{\sim}{s}(z_S')dz_S' \tag{23}$$

where α_s are positive constants and G_s are parameters depending on the hydrostatic stress σ. This equation was derived with the initial condition $p^s(0) = \underset{\sim}{0}$.

It has been shown in Reference 19 that by use of Laplace transformation, Equation (23) can be reduced to

$$\underset{\sim}{s} = s_y \frac{d\underset{\sim}{\theta}}{dz_S} + \int_o^{z_S} \varrho_1(z_S - z_S')\frac{d\underset{\sim}{\theta}}{dz_S'}dz_S' + R_o\underset{\sim}{\theta} \tag{24}$$

where

$$s_y = 1\Big/\sum_s G_s \text{ and } \varrho_1(z_S) = \sum_{r=1}^{s-1} R_r e^{-\beta_r z_S} \tag{25}$$

Equation (24) is the constitutive equation for deviatoric behavior. It should be noted that in the preceding derivation, the condition $\underset{\sim}{\theta}(0) = \underset{\sim}{0}$ was employed. Following other information [12] it may be shown that s_y^o is the initial yield stress at pressure σ^o, and that when $d\zeta_S = 0$, the deviatoric behavior is elastic.

In the following application, only one term of the function $\varrho_1(z_s)$ is employed. In this event, the constitutive equation [Equation (24)] is rewritten as

$$\underset{\sim}{s} = S_y^o \frac{d\underset{\sim}{\theta}}{dz_s} + 2\mu_1\underset{\sim}{\theta} + 2\mu_2 \int_o^{z_S} e^{-\alpha(z_S - z_S')}\frac{d\underset{\sim}{\theta}}{dz_S'}dz_S' \tag{26}$$

with $S_y^o = 1/\Sigma_s G_s = Q_o/\Sigma_s F_2^s F_2^s = 2\mu_o Q_o$, $2\mu_1 = R_o = 1/\Sigma_s\ (G_s/\alpha_s) = Q_1/\Sigma_s F_2^s F_2^s = 2\mu_o Q_1$; $2\mu_2 = R_r = Q_2/\Sigma_s F_2^s F_2^s = 2\mu_o\ Q_2$; Q_o, Q_1, Q_2 and α = constants; and $2\mu_o = \Sigma_s F_2^s F_2^s$ = hydrostatic pressure dependent.

It is demonstrated in this paper that only one term is sufficient in describing the sands considered. However, the theory is capable of describing more complex deviatoric responses by use of more terms of function $\varrho_1(z_S)$, if such a situation is called for.

Equation (26) is now applied to obtain the deviatoric stress–strain curves in a CCT test. In such a test

$$3\sigma = \sigma_a + 2\sigma_r;\ \varepsilon_{kk} = \varepsilon_a + 2\varepsilon_r$$
$$s = \frac{2}{3}(\sigma_a - \sigma_r);\ \eta = \frac{2}{3}(\varepsilon_a - \varepsilon_r) \tag{27}$$

in which σ_a and σ_r = axial and confining stress, respectively; ϵ_a and ϵ_r = axial and radial strain, respectively; and s and η = deviatoric stress and strain, respectively. Under monotonic loading, Equation (26) may be integrated to render [22]:

$$s = S_y^o(1 + \beta\theta) + 2\mu_1\theta + \frac{2\mu_2}{n\beta}[(1 + \beta\theta) - (1 + \beta\theta)^{-n+1}] \tag{28}$$

in which θ = the plastic part of the shear strain η; and $n = \alpha/\beta + 1$ is a constant. The material constants may be determined in the manner used before [23]. The constants involved are S_y^o, S_o (the intercept of the asymptotic straight line of the stress–strain curve with the stress axis), and $2\mu_t$ (the slope of the asymptotic straight line). These may be determined from an experimental curve. The quantity $2\mu_2/n\beta = S_o - S_y^o$ can then be calculated. By choosing a value for $2\mu_1$, which may be zero, the value of $\beta = (2\mu_t^\theta - 2\mu_1)/S_o$ may be calculated. Finally, n is determined by trial and error so that Equation (28) fits the experimental data well.

In order to compare the theoretical results with the experimental data obtained by Lee and Seed [5], a sorting procedure is necessary. Data obtained by Lee and Seed [5] were presented in the form of the principal stress ratio σ_a/σ_r vs. the axial strain measured after shearing, $\bar{\epsilon}_a$. The volumetric strains measured after shearing, $\bar{\epsilon}_{kk}$, were also recorded. Let ϵ_{kk}^o be the volumetric strain measured prior to shearing. Then

$$\epsilon_{kk} = \bar{\epsilon}_{kk} + \epsilon_{kk}^o = \epsilon_a + 2\epsilon_r \tag{29}$$

Thus,

$$\epsilon_r = \frac{1}{2}\left(\bar{\epsilon}_{kk} - \bar{\epsilon}_a + \frac{2}{3}\epsilon_{kk}^o\right) \tag{30}$$

in which

$$\epsilon_a = \bar{\epsilon}_a + \frac{1}{3}\epsilon_{kk}^o \tag{31}$$

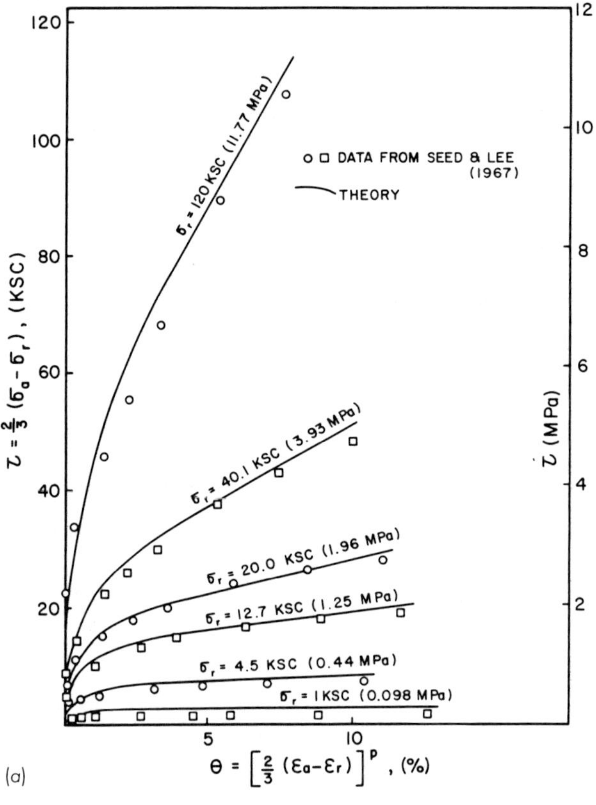

(a)

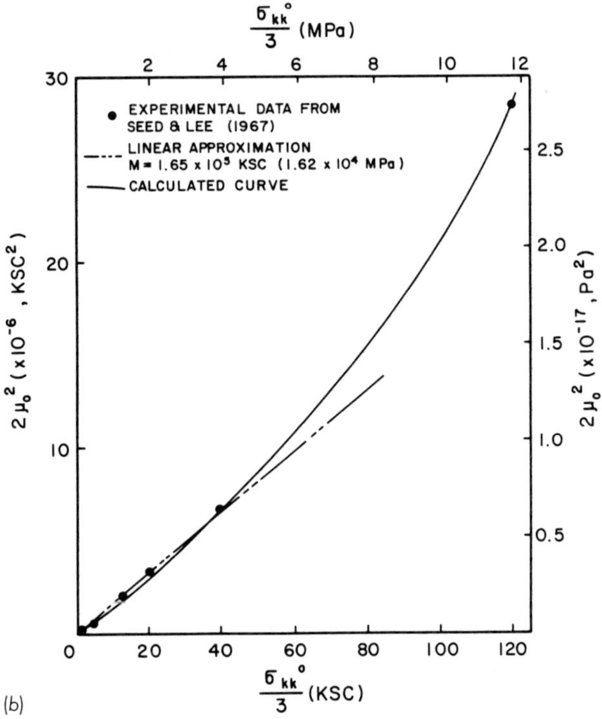

(b)

FIGURE 2. Deviatoric response for Sacramento River Sand: (a) loading at constant confined pressure; (b) shear modulus.

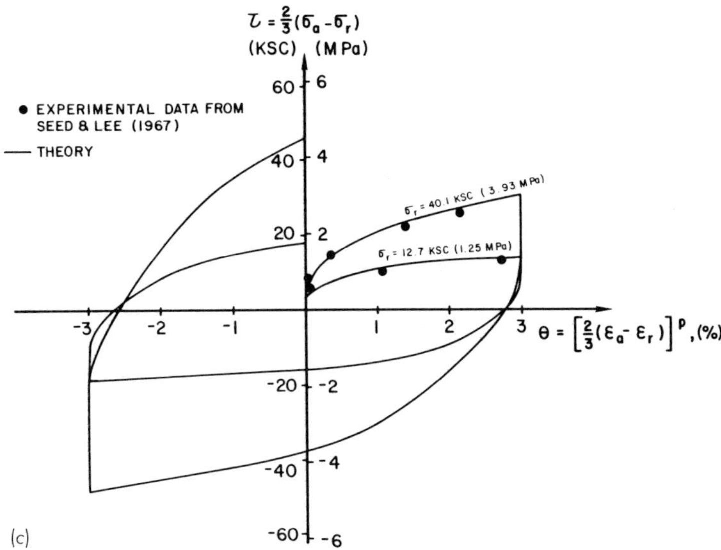

FIGURE 2 (continued). Deviatoric response for Sacramento River Sand: (c) theoretical hysteretic loops.

The shear strain may then be obtained from Equation (27) as

$$\eta = \bar{\varepsilon}_a - \frac{1}{3}\bar{\varepsilon}_{kk} \tag{32}$$

Reconstructing the experimental results using Equations (27) and (32), the data are plotted in Figure 2a. The theoretical results using Equation (28) are also shown. It is seen that reasonable agreement has been achieved. The constants are determined to be $Q_o = 2.87 \times 10^{-3}$, $Q_1 = 0$, $Q_2 = 0.94$, and $\alpha = 140.2$ by fitting the $\sigma_r = 20$ ksc curve. Using the same constants, Equation (28), therefore, predicts the remaining curves.

The shear and tangent moduli are fitted to the experimental results by the following expressions:

$$2\mu_o = K_o\left[5.50\left(\frac{\sigma^o}{K_o}\right)^{1/2} + 1.88 \times 10\left(\frac{\sigma^o}{K_o}\right) - 4.74 \times 10^2\left(\frac{\sigma^o}{K_o}\right)^2 + 1.06 \times 10^4\left(\frac{\sigma^o}{K_o}\right)^3\right] \tag{33}$$

and

$$2\mu_t^\theta = 6.9\sigma_r \tag{34}$$

Equation (33) is plotted in Figure 2b together with the experimental results. A linear approximation popular in the literature is also shown in the figure. It is seen that the linear approximation is reasonable for a small and moderate confining pressure range.

Figure 2c shows the theoretical loading–unloading loops corresponding to two different confining pressures. These curves have been obtained using the material constants previously determined for loading. The method of constructing the unloading–reloading curves is the same as that used by Wu and Yip [23].

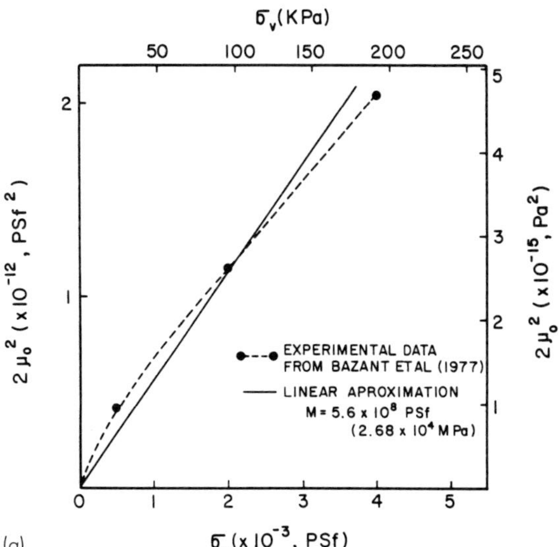

FIGURE 3. Shear response for loose crystal silica no. 20 sand, relative density = 45%: (a) shear modulus.

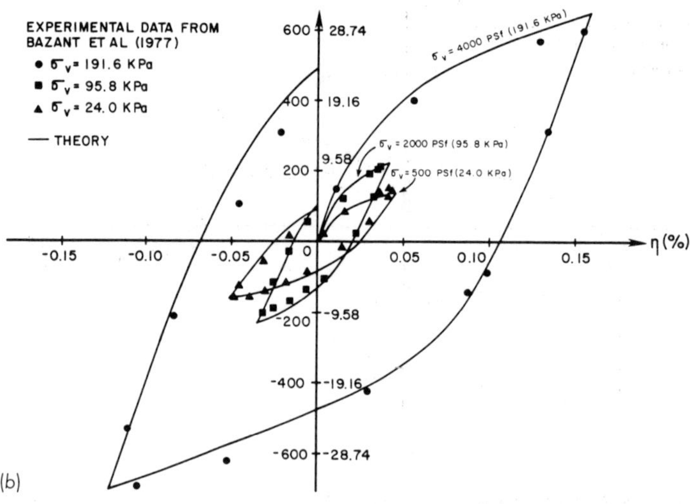

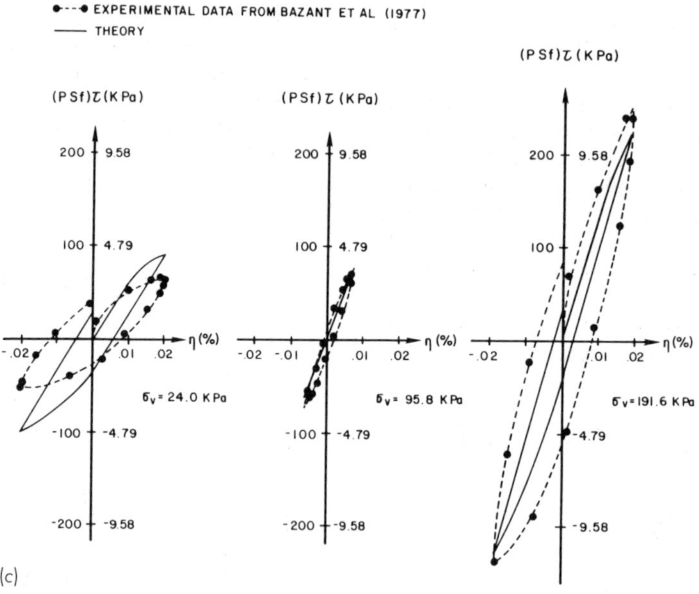

FIGURE 3 (continued). Shear response for loose crystal silica no. 20 sand, relative density = 45%: (b) shear hysteretic loops at moderate strain; (c) shear hysteretic loops at small strain.

Equation (26) is now applied to investigate the behavior of loose crystal silica no. 20 sand with relative density = 45%. The experimental data are reported elsewhere [2]. It is seen from Figure 3a that the linear approximation for the dependence of $2\mu_o^2$ on the vertical stress σ_v is quite reasonable in the range of interest. Thus, this expression is used in the calculation. Figure 3b shows both the theoretical and the experimental results for shear deformations at a moderate range. Figure 3c shows the same for small shear deformations. The constants are determined by fitting the loading curve of $\sigma_v = 95.8$ kPa in Figure 3b. All the remaining curves in Figure 3b and all the curves in Figure 3c have been generated without changing the material constants.

DENSIFICATION AND DILATION OF DRAINED SAND

The effect of densification and dilation is discussed in detail in Reference 21. The volumetric strain has two sources. One source is the "strain coupling" effect between deviatoric and volumetric responses which is called densification and/or dilation and denoted as ϵ_{kk}^d. The other is caused by changes in hydrostatic stress σ, and is denoted by ϵ_{kk}^h. The densification response is controlled by the evolution of the internal variables q^d, whereas the hydrostatic response is a result of the growth of internal variables q^h. Thus,

$$\varepsilon_{kk} = \varepsilon_{kk}^h + \varepsilon_{kk}^d = -\frac{\partial \phi_H}{\partial \sigma} \tag{35}$$

in which ϕ_H is defined by Equation (3), and

$$\varepsilon_{kk}^d = C^1 q^1 + \sum_{d=2}^{d} C^d q^d \tag{36}$$

For this study, the evolution for q^d is written in a linear form:

$$\frac{\partial \phi_H}{\partial q^d} + a_o^d \frac{dq^d}{dz_D} = 0 \tag{37}$$

in which a_o^d = a constant and z_D = densification intrinsic time.

Applying Equation (3) in Equation (37) and integrating the resulting first order differential equation while keeping $\sigma = \sigma_0$ constant, the following expressions for $q^d(z_D)$ are obtained:

$$q^1 = q^1(0) + \frac{C^1}{a_o^1}(\sigma_o - \nu)z_D \qquad \text{for } d = 1 \tag{38}$$

$$C^d q^d = I^d \sigma_o - [I^d \sigma_o - C^d q^d(0)]e^{-\gamma^d z_D} \qquad \text{for } d \neq 1 \tag{39}$$

with initial value $z_D(0) = 0$. In Equation (39), $I^d = (C^d)^2/D^d$ and $\gamma^d = D^d/a_o^d$. The discussion below is for the case of $\sigma = \sigma_o$ = constant, which is true for several stress paths studied in this paper. For the case of $\sigma \neq$ constant such as in the conventional triaxial test, Equation (37) can only be numerically integrated and is discussed later.

Equations (3), (35), (38), and (39) are now used to obtain

$$\begin{aligned} \varepsilon_{kk}^d = {} & I^1\gamma^1(\sigma_o - \nu)z_D + I^2\sigma_o(1 - e^{-\gamma^2 z_D}) \\ & + C^1 q^1(0) + C^2 q^2(0) e^{-\gamma^2 z_D} \end{aligned} \tag{40}$$

in which only two densification internal variables q^1 and q^2 are assumed to be active. Equation (40) is then the equation which governs the densification and dilation behavior of drained sand. The constant ν can now be identified and a relation between the initial values of $q^1(0)$ and $q^2(0)$ can be found. This is done by recognizing that sand would initially densify even for the case that dilation follows. Therefore, there are two parts in Equation (40). The first term on the right-hand side gives densification or dilation according to the sign of the quantity $(\sigma_o - \nu)$. The remaining terms basically describe the behavior of densification and at critical hydrostatic stress σ_{cr}, i.e., at $\sigma_o = \sigma_{cr}$, it is assumed that $\sigma_{cr} - \nu = 0$ for all z_D. Hence, $\nu = \sigma_{cr}$. Furthermore, the initial condition, $\epsilon_{kk}^d = 0$ when $z_D = 0$, should be obeyed, which leads to

$$C^1 q^1(0) + C^2 q^2(0) = 0 \tag{41}$$

It is remarked that σ_{cr} is the minimum confining stress at which, after an initial densification, further shearing produces no dilation. It is a material property depending on the relative density of sand. Applying these conditions in Equation (40) resulted in

$$\epsilon_{kk}^d = I^1\gamma^1(\sigma_o - \sigma_{cr})z_D + (I^2\sigma_o + q_0)(1 - e^{-\gamma^2 z_D}) \tag{42}$$

where $q_0 = C^1 q^1(0)$. The behavior represented by Equation (42) is schematically shown in Figure 4. It is seen that both densification and dilation behavior can be described by Equation (42). The derivative of Equation (42) is

$$\frac{d\varepsilon_{kk}^d}{dz_D} = I^1\gamma^1(\sigma_o - \sigma_{cr}) + \gamma^2[I^2\sigma_o + q_0]e^{-\gamma^2 z_D} \tag{43}$$

In the above derivation, parameter z_D is the densification intrinsic time scaled by

$$dz_D = \frac{d\zeta_D}{1 + \xi\zeta_D} \tag{44}$$

where

$$d\zeta_D = k_{ijkl}^D \, de_{ij} \, de_{kl} \tag{45}$$

(a) DENSIFICATION

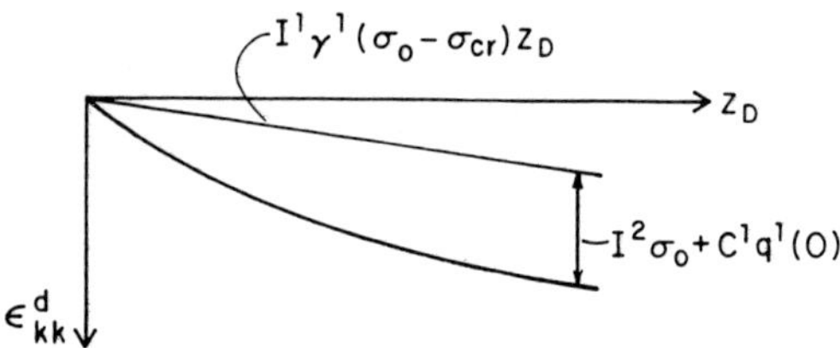

(b) AT σ_{cr}

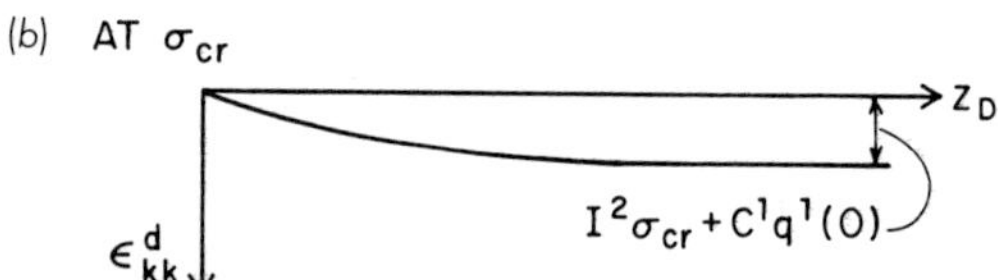

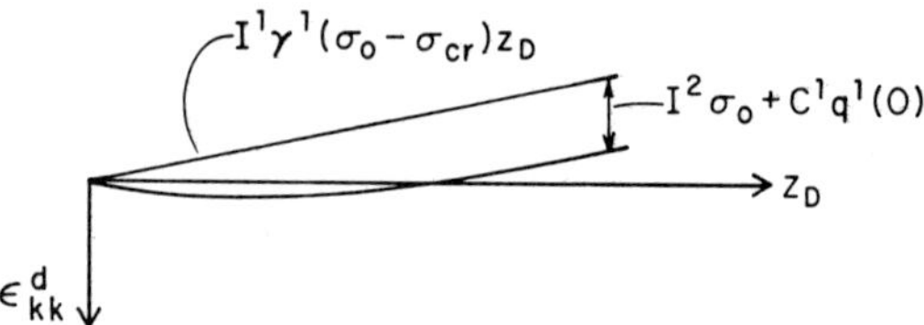

FIGURE 4. Schematic densification-dilation curves for σ = constant.

The choice of the total strain over the plastic strain in Equation (45) is motivated by the fact that densification grows even when shear deformation is small [2]. Furthermore, it has been demonstrated by Oh-Oka [8] that densification also grows when unloading takes place in shear.

The initial slope D_o of the densification curve can be used as a parameter in the description of densification and dilation curves. Currently, Equation (43) is described by six parameters, σ_{cr}, ξ, $I^1\gamma^1$, I^2, γ^2, and q_o, but if D_o is used, then only five parameters are needed. They are σ_{cr}, ξ, $I^1\gamma^1$, γ^2, and D_o.

The slope of densification curve is defined as

$$D_o = \left.\frac{d\varepsilon^d_{kk}}{d\zeta_D}\right|_o \tag{46}$$

By use of Equations (43) and (44), it is found that

$$D_o = I^1\gamma^1(\sigma_o - \sigma_{cr}) + \gamma^2 [I^2\sigma_o + q_o] \tag{47}$$

It would be ideal to define

$$D_o = \left.\frac{d\varepsilon^d_{kk}}{d\zeta_D}\right|_{\zeta=0,\sigma_o=\sigma_{cr}} \tag{48}$$

then

$$D_o = \gamma^2[I^2\sigma_{cr} + q_o] \tag{49}$$

However, no experimental results are available to the writer for this purpose. For Equation (49) to be useful, the sand would have to be tested at $\sigma_o = \sigma_{cr}$.

Substituting Equation (47) into (42) one obtains

$$\epsilon^d_{kk} = I^1\gamma^1(\sigma_o - \sigma_{cr})\, z_D + \frac{1}{\gamma^2} [D_o - I^1\gamma^1 \times (\sigma_o - \sigma_{cr})](1 - e^{-\gamma^2 z_D}) \tag{50}$$

and the differential form is

$$d\epsilon^d_{kk} = I^1\gamma^1 (\sigma_o - \sigma_{cr})\, dz_D + [D_o - I^1\gamma^1(\sigma_o - \sigma_{cr})]e^{-\gamma^2 z_D} dz_D \tag{51}$$

Equations (50) and (51) are convenient to use for computation of densification and dilation.

In the case of non-constant σ, an integration of Equation (37) for q^d and a subsequent substitution into Equation (36) renders

$$\varepsilon^d_{kk} = I^1\gamma^1 \int_0^{z_D} (\sigma - m) dz_D + I^2\gamma^2 \int_0^{z_D} e^{\gamma^2(z'_D - z_D)}\sigma dz'_D - q_o(e^{-\gamma^2 z_D} - 1) \tag{52}$$

Integrating Equation (52) by parts and then differentiating with respect to z_D enables one to write the following incremental form for densification strain:

$$d\varepsilon^d_{kk} = \left[I^1\gamma^1(\sigma - \nu) + I^1\gamma^1\gamma^2 \int_o^{z_D} (\sigma - \nu) dz_D + I^2\gamma^2\sigma + \gamma^2 q^1 - \gamma^2\varepsilon^d_{kk}\right] dz_D \tag{53}$$

For isotropic soil at one-dimensional shearing, Equation (45) reduces to

$$d\zeta_D = k_D\, |d\eta| \text{ with } k_D = k_D\,(\sigma_r) \tag{54}$$

where η is the deviatoric strain. Because the rate of development of densification strain depends on the initial confining stress σ_r, $\xi = \xi(\sigma_r)$. Combining Equations (44) and (54), the following relationship between deviatoric strain, η, and intrinsic time scale z_D is obtained:

$$dz_D = k_D(\sigma_r)e^{-\xi(\sigma_r)z_D}d\eta \tag{55}$$

The above model is used to describe the densification and dilation of drained Sacramento River sand under conventional constant confining stress triaxial (CCT) test. The inital void ratio of this sand is $e_i = 0.87$. Three sets of data for initial confining stresses of 12.65, 4.5, and 1.0 ksc were available. The first two only exhibited densification behavior whereas in the last set densification was followed by dilation. The original data consisted of axial strain ϵ_a and total volumetric strain ϵ_{kk}, which is the sum of hydrostatic and deviatoric strains. From these data the deviatoric strain η was calculated according to

$$\eta = \varepsilon_a - \frac{1}{3}\varepsilon_{kk} \tag{56}$$

and the results are plotted in Figure 5 in $\eta - \epsilon_{kk}$ space. Equations (53) and (17) along with Equation (55) were used to fit the experimental $\eta - \epsilon_{kk}$ data. The algorithm used for this purpose is as follows. An increment of deviatoric strain, $d\eta$, was specified and then from Equation (55), dz_D was calculated. This information along with the value of variables in previous incremental step were then substituted into Equation (53), to solve for $d\epsilon^d_{kk}$. On the other hand, the amount of $d\sigma$ corresponding to $d\eta$ was determined from experimental data and through Equation (17), the amount of $d\epsilon^h_{kk}$ was determined. The increment of total volumetric strain is just the sum of hydrostatic and densified volumetric strains, i.e.,

$$d\epsilon_{kk} = d\epsilon^h_{kk} + d\epsilon^d_{kk} \tag{57}$$

But, note that the value of $d\sigma$ used in evaluation of $d\epsilon^h_{kk}$ can be found through the deviatoric stress–strain response without resorting to experimental data, since in a CCT test where σ_r is constant one can show that

$$d\sigma = 0.5\, ds \tag{58}$$

Thus, if $d\eta$ is specified, one may compute $d\sigma$ through Equations (24) and (58).

The results presented in Figure 5 show that the model developed herein is capable of describing both densification and dilation reasonably well. The material parameters used are as follows. There are four parameters which stem from energy equation and are common to all tests and will only vary with the initial void ratio of the sand specimen. These were determined to have the following values: $I^1\gamma^1 = 2$ ksc^{-1}, $I^2\gamma^2 = 0.0001$ ksc^{-1}, $\gamma^2 = 1{,}000$, and $m = 3$ ksc. The other parameters which depend on the initial confining pressure are presented in Table 1. The constant of Table 1

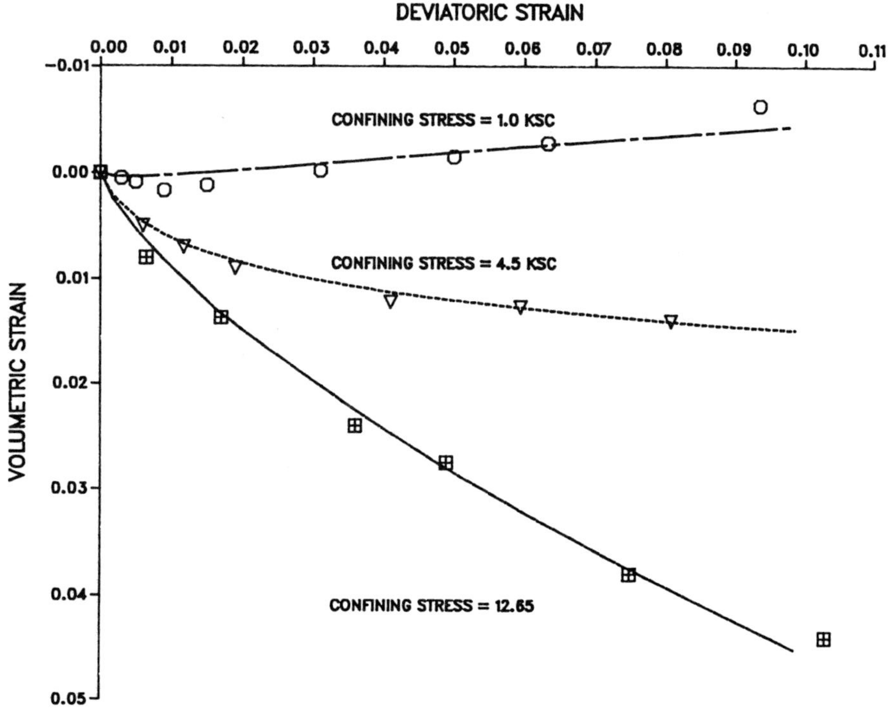

FIGURE 5. Densification and dilation in drained conventional triaxial test.

TABLE 1.

σ_r ksc	k_D	ζ	q_o
1	.001	400	.06
4.5	.18	3000	0
12.65	.015	700	0

shows a trend that suggests that this value peak at the critical confining pressure of $\sigma_{cr} = \nu = 3$ ksc, and therefore it is possible to define them in an analytical form, as functions of σ_r. However, to determine the form of such function, one needs many experimental data at closely spaced intervals of confining pressure, which at present are not available. Nevertheless, if such data are available, the values of the parameters can be found with much more ease.

The constants of the hydrostatic response are those of Reference 19, whose values are $\sigma_f = 3.74 \times 10^2$ ksc, $K_o = 2.33 \times 10^3$ ksc, and $\lambda = 16.15$.

CYCLIC SHEAR BEHAVIOR

For more complex loading, such as cyclic loading, the expression of hardening function given by Equation (22) leads to a prediction that the peak stress of each cycle increases monotonically with the number of cycles. Evidently, this is not true in practice since a steady state should be reached as the number of cycles increases.

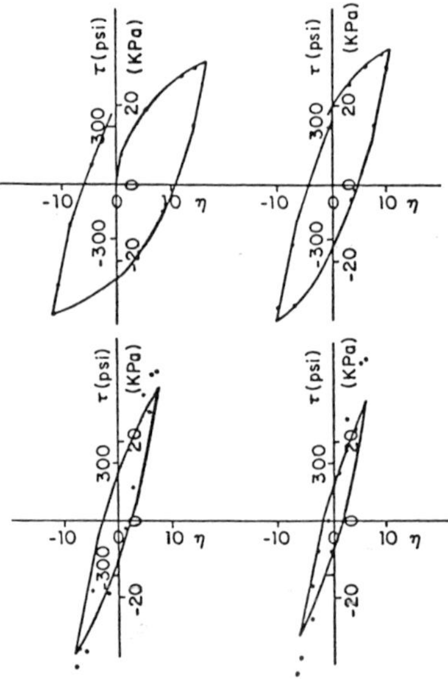

FIGURE 6. Shear hysteresis loops of the 1st, 2nd, 10th and 300th cycles.

To account for the steady state eventually reached by cycling, Wu and Sheu [17] modified the hardening function by writing

$$\frac{d\zeta}{dz} = f(z) = C - (C - 1)e^{-\beta z} \tag{59}$$

in which C, β = material parameters both depending on the confining pressure. It should be mentioned that this form has been previously used by Wu and Yip [24] in their discussion of cyclic hardening of metallic materials.

Equation (59) is now used in conjunction with Equation (26) to obtain the theoretical cyclic stress–strain curves shown in Figure 6. The experimental data due to Silver and Seed [9] are also shown. The material parameters have been determined to match the first 1-1/2 cycles, which are $2\mu_o = 1.43 \times 10^6$ psf (6.85×10^4 kPa), $Q_o = 1.287 \times 10^{-4}$, $Q_1 = 6.997 \times 10^{-2}$, $Q_2 = 0.692$, $\alpha = 3116.5$, $\beta = 169.05$ and $C = 1.76$ for the case of vertical stress = 191.6 kPa and relative density = 45%. Results for all other cycles are predictions by the present model.

There are nine cases all together with details reported in Reference 17. The experimental strain amplitude was varying for all cases. However, only those of the first, second, tenth, and 300th (200th for the case of relative density is 80% and vertical stress is 190 kPa) cycles were given in Reference 9. The authors interpolated the existing data to estimate the unknown strain amplitudes at various cycles. Obviously, the manner of interpolation would slightly affect the computed results.

TRUE TRIAXIAL TEST

The endochronic constitutive equations used in the previous sections were in a form suitable to describing tests with controlled deformation histories. However, there are tests, such as the true triaxial test, which can be stress-controlled and, to describe them, the form of endochronic equations will have to be modified and suitable method for computation will have to be developed. The main objective of this section is to develop specific endochronic equations and procedures which are appropriate for a stress-controlled experiment, especially the true triaxial test. Details of this development may be found in Reference 21.

The constitutive equations for hydrostatic and densification behaviors retain their original form in this consideration. However, the differential form initially developed by Valanis [12] is employed to describe the deviatoric behavior. This differential form is different from the deviatoric constitutive equation described earlier through a constant parameter $\varkappa$ to be explained later. The two equations are equivalent only when $\varkappa = 1$. In this section, by use of the differential form, explicit equations for true triaxial test have been derived and applicability of these equations is explored based on several stress-paths described by Yong and Ko [25] for Ottawa sand. A major advantage of this differential form is that the constitutive equation does not have a built-in discontinuity and thus facilitates numerical computation. The deviatoric constitutive equations are, from References 12 and 22, written as

$$\underset{\sim}{s} = 2\mu_o \int_o^{z_S} \rho(z_S - z_S') \frac{d\underset{\approx}{Q}}{dz_S'} dz_S' \tag{60}$$

in which

$$\underset{\approx}{Q} = \underset{\sim}{e} - \frac{\kappa}{2\mu_o} \underset{\sim}{s} \tag{61}$$

and

$$2\mu_o\rho(z_S) = \frac{2\mu_o}{1-\kappa} e^{-(\alpha_o/1-\kappa)z_S} + 2\mu_1 e^{-\alpha z_S} \tag{62}$$

In Equation (61), $\varkappa$ = constant parameter such that $0 \le \varkappa \le 1$; $\underset{\sim}{Q}$ = strain-like tensor and $\underset{\sim}{Q} = \theta$ = plastic strain when $\varkappa = 1$. The kernel function $\varrho(z_S)$ is represented by a simplified expression, as in Equation (62) in this investigation, in which α_o = constant. Note that the first term on the right-hand side of Equation (62) will become the Dirac delta function when $\varkappa = 1$ and, in this case, Equation (60) reduces to Equation (26).

The tensor $\underset{\sim}{Q}$ and the parameter $\varkappa$ were introduced by Valanis [12] in order to get rid of the singularity of the constitutive equation [Equation (26)] and its counterpart, Equation (60), is thus continuous and differentiable.

By differentiation, Equation (60) may be written in differential form as

$$\frac{\alpha_o}{2\mu_o}(\underset{\sim}{s} - \underset{\sim}{r})dz_S + \frac{1-\kappa}{2\mu_o} d(\underset{\sim}{s} - \underset{\sim}{r}) = d\underset{\approx}{Q} \tag{63}$$

and

$$\frac{1}{2\mu_1}(\alpha \underset{\sim}{r} dz_S + d\underset{\sim}{r}) = d\underset{\approx}{Q} \tag{64}$$

in which

$$\underset{\sim}{r}(z) = 2\mu_1 \int_o^{z_S} e^{-\alpha(z_S - z_S')} \frac{d\underset{\approx}{Q}}{dz_S'} dz_S' \tag{65}$$

This differential form was first obtained in Reference 12 and was later applied in Reference 6 to discuss viscoplastic wave propagation under uniaxial stress. In particular, this form was used to calculate loading–unloading–reloading curves for various constant $\varkappa$ values, and it is felt that if the case of $\varkappa = 1$ is believed to be representative of the deviatoric sand behavior, then by choosing $\varkappa$ to be close to 1, 0.95 in this investigation, an approximate representation may be achieved without having to introduce a discontinuity into the constitutive equation. This aspect is indeed a very important strength of the present approach.

The deviatoric intrinsic time is defined as

$$d\zeta_S^2 = d\underset{\sim}{Q} \cdot d\underset{\sim}{Q} \tag{66}$$

By use of the hardening function, $f(z_S, \sigma)$ of Equation (1b), this equation is further written as

$$[f(z_S, \sigma)\, dz_S]^2 = d\underset{\sim}{Q} \cdot d\underset{\sim}{Q} \tag{67}$$

Equations (63), (64), and (67) constitute a set of thirteen equations with thirteen unknowns dr_{ij}, dQ_{ij}, and dz_S to be solved for every step if the deviatoric stress ds_{ij} is the input. This set of nonlinear algebraic equations may be solved numerically.

In the case of true triaxial test, the three axes are principal axes and denoted by x, y, and z with z vertical. The components for stress $\underset{\sim}{\sigma}$ and strain $\underset{\sim}{\epsilon}$ are σ_x, σ_y, σ_z and ϵ_x, ϵ_y, ϵ_z, respectively. The deviatoric stress components are

$$\begin{aligned} s_x &= \frac{1}{3}(2\sigma_x - \sigma_y - \sigma_z) \\ s_y &= -\frac{1}{3}(\sigma_x - 2\sigma_y + \sigma_z) \\ s_z &= -\frac{1}{3}(\sigma_x + \sigma_y - 2\sigma_z) \end{aligned} \tag{68}$$

and the deviatoric strain components are

$$\begin{aligned} e_x &= \frac{1}{3}(2\varepsilon_x - \varepsilon_y - \varepsilon_z) \\ e_y &= -\frac{1}{3}(\varepsilon_x - 2\varepsilon_y + \varepsilon_z) \\ e_z &= -\frac{1}{3}(\varepsilon_x + \varepsilon_y - 2\varepsilon_z) \end{aligned} \tag{69}$$

The components of $\underset{\sim}{Q}$ are

$$Q_x = \frac{1}{3}\left\{2\left[\varepsilon_x - \frac{\kappa}{2\mu_o}\sigma_x\right] - \left[\varepsilon_y - \frac{\kappa}{2\mu_o}\sigma_y\right] - \left[\varepsilon_z - \frac{\kappa}{2\mu_o}\sigma_z\right]\right\}$$

$$Q_y = -\frac{1}{3}\left\{\left[\varepsilon_x - \frac{\kappa}{2\mu_o}\sigma_x\right] - 2\left[\varepsilon_y - \frac{\kappa}{2\mu_o}\sigma_y\right] + \left[\varepsilon_z - \frac{\kappa}{2\mu_o}\sigma_z\right]\right\} \tag{70}$$

$$Q_z = -\frac{1}{3}\left\{\left[\varepsilon_x - \frac{\kappa}{2\mu_o}\sigma_x\right] + \left[\varepsilon_y - \frac{\kappa}{2\mu_o}\sigma_y\right] - 2\left[\varepsilon_z - \frac{\kappa}{2\mu_o}\sigma_z\right]\right\}$$

From Equations (65) and (70), since $tr\,\underset{\sim}{Q} = 0$, it is obtained that $tr\,\underset{\sim}{r} = 0$, i.e.,

$$r_x + r_y + r_z = 0 \tag{71}$$

and only two of Equations (70) are independent.

Three equations are obtained from each of Equations (63) and (64). However, due to the relation of Equation (70), only two of these three equations are independent. Thus, the following set of four equations is resulted:

$$\frac{\alpha_o}{2\mu_o}(s_x - r_x)dz_S + \frac{1-\kappa}{2\mu_o}d(s_x - r_x) = dQ_x \tag{72}$$

$$\frac{\alpha_o}{2\mu_o}(s_y - r_y)dz_S + \frac{1-\kappa}{2\mu_o}d(s_y - r_y) = dQ_y \tag{73}$$

$$\frac{1}{2\mu_1}(\alpha r_x dz_S + dr_x) = dQ_x \tag{74}$$

$$\frac{1}{2\mu_1}(\alpha r_y dz_S + dr_y) = dQ_y \tag{75}$$

Furthermore, if $f(z_S, \sigma) = 1 + \beta(\sigma)\,\zeta_S = e^{\beta_z s}$ as in Reference 19, Equation (67) is written as

$$e^{2\beta_z s}(dz_S)^2 = dQ_x^2 + dQ_y^2 + dQ_z^2 \tag{76}$$

Substituting Equations (74) and (75) into (76), it is found that

$$e^{2\beta_z s}(dz_S)^2 = \left(\frac{1}{2\mu_1}\right)^2 [(\alpha r_x dz_S + dr_x)^2 + (\alpha r_y dz_S + dr_y)^2 + (\alpha r_z dz + dr_z)^2] \tag{77}$$

By eliminating Q_x and Q_y in Equations (72–75), the following two equations are obtained:

$$\left[\frac{\alpha_o}{2\mu_o}s_x - \left(\frac{\alpha_o}{2\mu_o} + \frac{\alpha}{2\mu_1}\right)r_x\right]dz_S = -\frac{1-\kappa}{2\mu_o}ds_x + \left(\frac{1-\kappa}{2\mu_o} + \frac{1}{2\mu_1}\right)dr_x \tag{78}$$

$$\left[\frac{\alpha_o}{2\mu_o}s_y - \left(\frac{\alpha_o}{2\mu_o} + \frac{\alpha}{2\mu_1}\right)r_y\right]dz_S = -\frac{1-\kappa}{2\mu_o}ds_y + \left(\frac{1-\kappa}{2\mu_o} + \frac{1}{2\mu_1}\right)dr_y \tag{79}$$

In order to evaluate the deviatoric response of sand to different stress paths, an incremental procedure may be applied to solve Equations (77), (78), and (79). These equations may be combined to yield thc following second order algebraic equation:

$$P dz_S^2 + Q\,dz_S + R = 0$$

where P, Q, and R are expressed in terms of the constants of the theory and the quantities s_x, ds_x, s_y, ds_y, r_x, r_y, and z_S which are known for the current incremental iteration. Therefore the above equation may be solved for dz_S which corresponds to an increment ds of the deviatoric stress state. Then Equation (61) is employed to find de_x and de_y.

Triaxial compression test (TC) is used to determine the material constants of the differential form. Other tests considered are the triaxial extension test (TE) and shear tests (SS) with several stress ratios. These are discussed in detail in Reference 21. It has been shown that the theory does lead to reasonable agreement with experimental results of true triaxial test. Both the deviatoric response and the densification–dilation behavior of Ottawa sand have been investigated.

UNDRAINED RESPONSE OF SAND

If drainage is not allowed in monotonic shearing of a saturated sand specimen in conventional triaxial test, the pore water pressure increases. Considerable field and laboratory studies have provided insight into the mechanism of such behavior. The consensus among research workers, as pointed out in Reference 7, is that under shear stresses the tendency of the sand grains to compact and reorient causes the build-up of pore water pressure in the undrained state, resulting in the transfer of load from soil grains to the pore water and in the reduction of effective stress on the soil skeleton, whereupon the soil structure rebounds to counteract the compaction tendencies, so that the condition of constant volume be maintained.

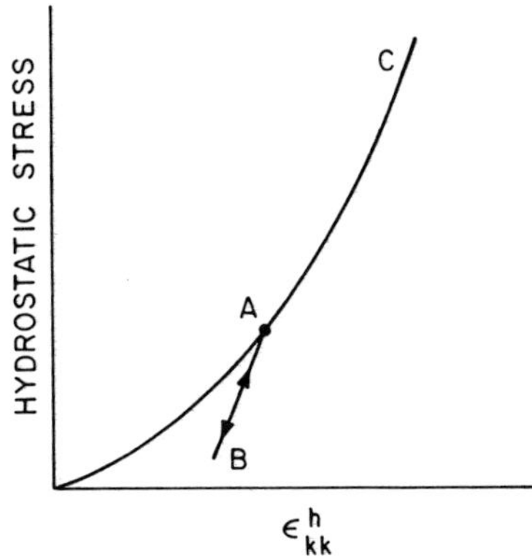

FIGURE 7. Model for hydrostatic loading–unloading–reloading in undrained test.

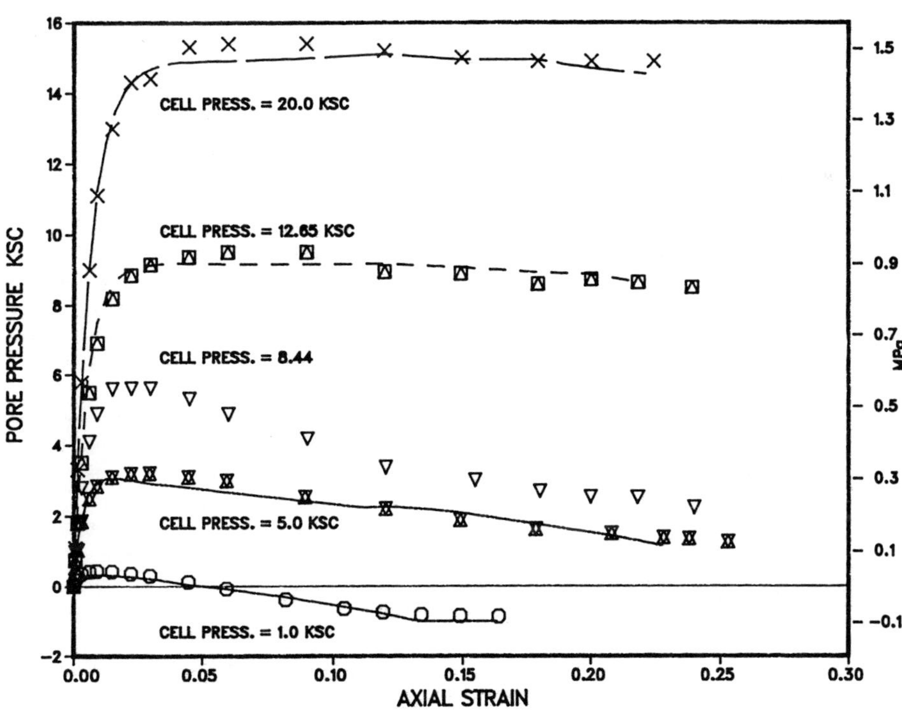

FIGURE 8. Pore–water pressure in undrained test.

It is thus clear that there is a close relationship between densification response in drained condition on the one hand, and the rise in pore water pressure in undrained condition on the other, because the underlying physical mechanisms are identical. Therefore, it is natural to expect that the same equation which explains densification must be also capable of describing pore water pressure rise, if sufficient side conditions are provided. The approximation that the total volumetric strain increment in undrained loading is zero provides such side conditions.

In models developed within the framework of classical theory of plasticity, the condition of zero total volumetric strain increment is satisfied by equating the magnitudes of elastic and plastic components of volumetric strain increment, i.e.,

$$d\epsilon_{kk} = d\epsilon^e_{kk} + d\epsilon^p_{kk} = 0 \qquad (80)$$

In the present endochronic theory the same condition is satisfied by writing

$$d\epsilon_{kk} = d\epsilon^h_{kk} + d\epsilon^d_{kk} = 0 \qquad (81)$$

Focus now on the hydrostatic volumetric strain and refer to Figure 7, which shows the hydrostatic response in a typical undrained test situation. It becomes clear that upon unloading at point A, $d\epsilon^h_{kk}$ is entirely elastic and thus Equations (80) and (81) are in one-to-one correspondence. However, as the test progresses, reloading along path BAC takes place and beyond A, the correspondence ceases to exist. Of course, if in the endochronic theory the hydrostatic unloading is not modeled as a straight line, which would also improve the agreement with experiment, this correspondence would not exist at all. Therefore, Equation (81) has an advantage that beyond point A in the direction of *C*, it represents precisely what actually takes place. In other words, $d\epsilon^d_{kk}$ can be identified with compaction strain due to shear loading, and $d\epsilon^h_{kk}$ with the rebounding of sand structure caused by decrease in effective hydrostatic stress, so that the balance between the two yields zero volumetric strain change.

In the evaluation of pore water pressure rise, the equations of drained condition are directly used with the exception that in undrained condition the stress values are those of effective stresses. In the remainder of this paper effective stresses are shown unprimed, i.e., σ, and the primed quantities correspond to applied chamber stresses. Thus,

$$\sigma = \sigma' - u \qquad (82)$$

in which u represents the pore water pressure.

Again, experimental results for Sacramento River sand with $e_i = 0.87$ [4] are matched against theory. The procedure for calculation of pore water pressure is as follows. An increment of $d\eta$ is specified and then through Equations (53) and (55) an increment of $d\epsilon^d_{kk}$ is found. Since $d\epsilon_{kk} = d\epsilon^d_{kk} + d\epsilon^h_{kk} = 0$, it is possible to evaluate $d\epsilon^h_{kk}$ and then utilizing Equation (17), the increment $d\sigma$ of effective hydrostatic stress is calculated. On the other hand, through experimental data an increment $d\sigma'$ of total hydrostatic stress corresponding to $d\eta$ is determined. Now since $du = d\sigma' - d\sigma$, the increment of pore water pressure is easily determined, and the procedure may be repeated for another increment of du.

The recorded experimental data available consisted of the ratio of effective stresses σ_a/σ_r, the difference between axial and radial applied stresses $\sigma'_a - \sigma'_r$, the axial strain ϵ_a and the pore water pressure u. In the undrained test

$$d\varepsilon_{kk} = d\varepsilon_a + 2d\varepsilon_r = 0; \; d\eta = \frac{2}{3}(d\varepsilon_a - d\varepsilon_r) \qquad (83)$$

and therefore

$$d\eta = d\epsilon_a \qquad (84)$$

It is also very easy to evaluate the total applied hydrostatic stress from the given experimental data. Thus, the data is readily converted to the desired form.

Figure 8 shows the experimental and theoretical results. It is seen that the model does describe the pore water pressure build-up reasonably well for different confining pressures of 1.0, 5.0, 12.65, 20.0. No attempt was made to match the data for confining pressure of 8.44, because this particular set of data does not follow the pattern of other confining pressures, and therefore its validity is questionable.

In the computation, unloading and reloading to initial hydrostatic stress level takes place along a straight line with a constant slope, as shown in Figure 7. Reloading to stress levels even higher than the original unloading stress may occur in an undrained test. In fact the discontinuity in the slope of the theoretical curve for $\sigma_r = 5.0$ ksc at about the strain level of 0.11 is precisely because of this, as shown in Figure 8. Upon unloading from point A (Figure 7) in the hydrostatic response, a straight line is followed, and as reloading continues, the stress level rises and crosses point A into region AC. This abrupt change in the slope at A induces discontinuity in the slope of Figure 8.

Such behavior is not observed in curves of 12.65 and 20.0 ksc, because in these two cases, the effective hydrostatic stress upon unloading at A and reloading at B does not cross into AC. Also, for the curve of $\sigma_r = 1.0$ ksc, the condition that pore pressure cannot go below -1.0 ksc is applied. The details of computation concerning the pore water pressure can be found in Reference 15.

ACKNOWLEDGEMENT

The work was supported by the National Science Foundation. The writer would like to thank Jane Frank for her careful typing.

REFERENCES

1. Bazant, Z. P. and R. J. Krizek, "Endochronic Constitutive Law for Liquefaction of Sand," *Journal of Engineering Mechanics Division,* ASCE, Vol 102, No. EM2, pp. 225-238 (Apr., 1976).
2. Cuellar, V., Z. P. Bazant, R. J. Krizek, and M. L. Silver, "Densification and Hysteresis of Sand Under Cyclic Shear," *Journal of the Geotechnical Engineering Division,* ASCE, Vol. 103, No. GT5, pp. 399–416 (May, 1977).
3. Lade, P. V., "Elasto-Plastic Stress-Strain Theory for Cohesiveless Soil with Curved Yield Surfaces," *International Journal of Solids and Structures,* Vol. 13, pp. 1019–1035 (1977).
4. Lade, P. V., Private communication.
5. Lee, K. L. and H. B. Seed, "Drained Strength Characteristics of Sands," *Journal of the Soil Mechanics and Foundation Division,* ASCE, Vol. 93, No. SM6, pp. 117–141 (Nov., 1967).
6. Lin, H. C. and H. C. Wu, "On the Rate-Dependent Endochronic Theory of Viscoplasticity and its Application to Plastic-Wave Propagation," *International Journal of Solids and Structures,* Vol. 19, No. 7, pp. 587–599 (1983).
7. Mostaghel, N. and K. Habibagahi, "Cyclic Liquefaction Strength of Sands," *Earthquake Engineering and Structural Dynamics,* Vol. 7, pp. 213–233 (1979).
8. Oh-Oka, H., "Drained and Undrained Stress-Strain Behavior of Sand Subjected to Cyclic Shear Stress Under Nearly Plane Strain Condition," *Soils and Foundations,* Vol. 16, No. 3, pp. 19–31 (Sept., 1976).
9. Silver, M. L. and H. B. Seed, "Deformation Characteristic of Sand Under Cyclic Loading," *Journal of the Soil Mechanics and Foundations Division,* ASCE, Vol. 97, No. SM8, pp. 1081–1098 (Aug., 1971).
10. Valanis, K. C., "A Theory of Viscoplasticity Without a Yield Surface, Part I: General Theory; Part II: Application to Mechanical Behavior of Metals," *Archiwun Mechaniki Stosowanej (Archives of Mechanics),* Vol. 23, pp. 517–551 (1971).
11. Valanis, K. C., "On the Foundations of the Endochronic Theory of Viscoplasticity," *Archives of Mechanics,* Vol. 27, pp. 857–868 (1975).
12. Valanis, K. C., "Fundamental Consequences of a New Intrinsic Time Measure—Plasticity as a Limit of the Endochronic Theory," *Archives of Mechanics,* Vol. 32, p. 171 (1980).
13. Valanis, K. C. and H. E. Read, "A New Endochronic Plasticity Model for Soils," in *Soil Mechanics—Transient and Cyclic Loads,* G. N. Pande and O. C. Zienkiewicz, editors, John Wiley (1982).
14. Wei, C. P., "Effects of Temperature, Irradiation and Cold Work on Creep Behavior of Metals," thesis presented to the Chemical and Materials Engineering Department, The University of Iowa, at Iowa City, Iowa, in 1978, in partial fulfillment of the requirements for the degree of Doctor of Philosophy.
15. Wu, H. C. and M. R. Aboutorabi, "Endochronic Description of Undrained Sand," Report G218-84-3, Department of Civil and Environmental Engineering, The University of Iowa, Iowa City, Iowa (1984).
16. Wu, H. C. and G. S. Chang, "Stress Analysis of the Dummy Rod Method for Sand Specimens," *Journal of the Geotechnical Engineering Division,* ASCE, Vol. 108 (September 1982).
17. Wu, H. C. and J. C. Sheu, "Endochronic Modeling for Shear Hysteresis of Sand," *Journal of Geotechnical Engineering,* Vol. 109, No. 12, pp. 1539–1550 (December, 1983).
18. Wu, H. C., K. C. Valanis, and R. F. Yao, "Application of the Endochronic Theory of Plasticity in the Gibbs Free Energy Form," *Letters in Applied and Engineering Sciences,* Vol. 4, pp. 127–136 (1976).
19. Wu, H. C. and T. P. Wang, "Endochronic Description of Sand Response to Static Loading," *Journal of Engineering Mechanics,* ASCE, Vol. 109, No. 4, pp. 970–989 (August 1983).

Wu, H. C. and T. P. Wang, "Loading and Unloading of Drained Sand—An Endochronic Theory," *Proceedings of International Conference on Constitutive Laws for Engineering Materials—Theory and Application,* Desai and Gallagher (ed.), p. 321–329, Tucson, Arizona (January 1983).

21. Wu, H. C., Z. K. Wang, and M. R. Aboutorabi, "Endochronic Modeling of Sand in True Triaxial Test," *Journal of Engineering Mechanics,* ASCE, Vol. 111, No. 10, pp. 1257–1276 (October, 1985).
22. Wu, H. C. and R. J. Yang, "Application of the Improved Endochronic Theory of Plasticity to Loading with Multi-Axial Strain-Path," *International Journal of Nonlinear Mechanics,* Vol. 18, No. 5, pp. 395–408 (1983).
23. Wu, H. C. and M. C. Yip, "Strain Rate and Strain Rate History Effects on the Dynamic Behavior of Metallic Materials," *International Journal of Solids and Structures,* Vol. 16, pp. 515–536 (1980).
24. Wu, H. C. and M. C. Yip, "Endochronic Description of Cyclic Hardening Behavior for Metallic Materials," *Journal of Engineering Materials and Technology,* ASME, Vol. 103, pp. 212–217 (July, 1981).
25. Yong, R. K. and H-Y. Ko, "Pre-Workshop Information Package," *Proceedings of the Workshop on Limit Equilibrium, Plasticity and Generalized Stress-Strain in Geotechnical Engineering,* Mcgill University, Canada, pp. 13–60 (May 1980).

CHAPTER 2

Consistency Limits of Soils—Principles and Potentials

T. S. NAGARAJ* AND B. R. SRINIVASA MURTHY*

INTRODUCTION

Soil is more or less taken for granted by the average person. It makes up the ground on which he lives and it makes him dirty. Apart from this, most people are not overly concerned with the soil. There are, however, certain people deeply concerned, such as civil engineers in general and geotechnical engineers in particular, geologists, mining engineers and hydrologists.

Soil for the geotechnical engineer is the entire thickness of the earth's crust which is accessible and feasible for practical utilization in engineering practice. Soils are formed from the parent material 'rock' through the morphological cycle of weathering, transportation and deposition. In the most general sense, soil refers to the uncemented particulate material with the exception of shale and sensitive soils which possess some degree of cementation bond strength. Since the discrete but rigid particles of soils are not strongly bonded together as the crystal structures of the metal, they are free to have relative movement. Hence the responses are to be characterised by the principles of particulate mechanics. Particulate materials are those composed of solid particles within the liquid or gaseous phase, which exhibit dilatancy and contractancy and are sensitive to hydrostatic stresses [27].

Surfaces of all soil particles, particularly of fine grained particles, adsorb the available moisture forming a thin hull or film. The saturated uncemented soils behave like a dispersed system with solid particles forming the dispersed phase and the water forming the continuous dispersion medium.

It is well demonstrated by Casagrande [19] that for classifying and for defining the specific properties of coarse grained soils, the gradation of particles alone is sufficient, whereas for fine grained soils, additional parameters like liquid limit and group index are required. These additional parameters are intended to account for the physicochemical interactions between the soil and the water.

SOIL CONSISTENCY

Early attention to the various modes by which a soil at different degrees of wetness reacts to the externally imposed forces has resulted in the development of the concept of soil consistency. For quite some time the soil consistency resisted all attempts to quantify and make exact definitions. Russel and Russel [74] define the soil consistency as the one which designates the manifestation of the physical forces of cohesion and adhesion acting within the soil at various moisture contents . . . including the behaviour towards gravity pressure, thrust and pull . . . , the tendency to adhere to foreign bodies, the sensations which are evidenced by the fingers of the observer as "feel." No matter how difficult it was to define the concept of soil consistency, constructive efforts were made from rheological approaches to understand the soil consistency.

Atterberg Limits

About 1911, the Swedish soil scientist Atterberg [2] reported an extensive study on the plasticity of clays. He divided the entire cohesive range from the solid to the liquid state into five stages and set arbitrary limits for these divisions in terms of water contents as follows:

- upper limit of viscous flow above which the soil-water system flows like a liquid
- liquid limit or lower limit of viscous flow or upper limit of plastic flow, above which the soil–water system flows like a viscous fluid and below which the system is plastic

*Department of Civil Engineering, Indian Institute of Science, Bangalore, India

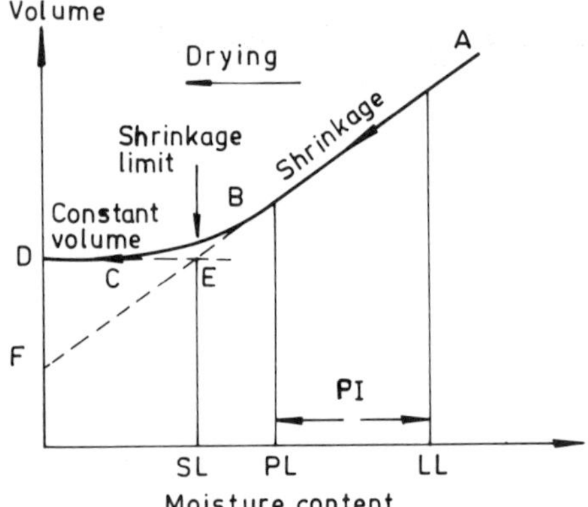

FIGURE 1. Shrinkage curve for clay soil.

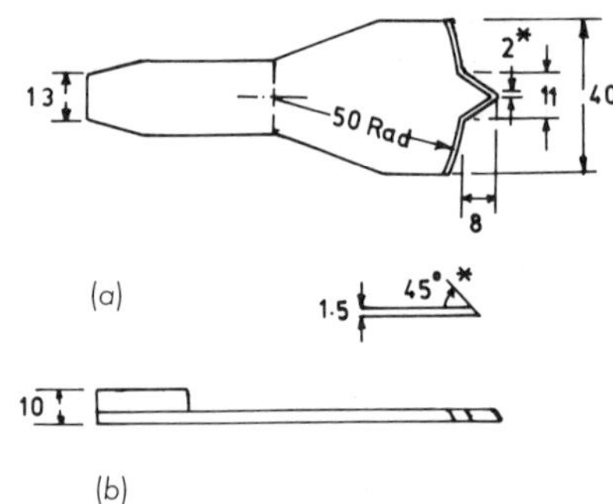

FIGURE 2. Grooving tool for Casagrande apparatus.

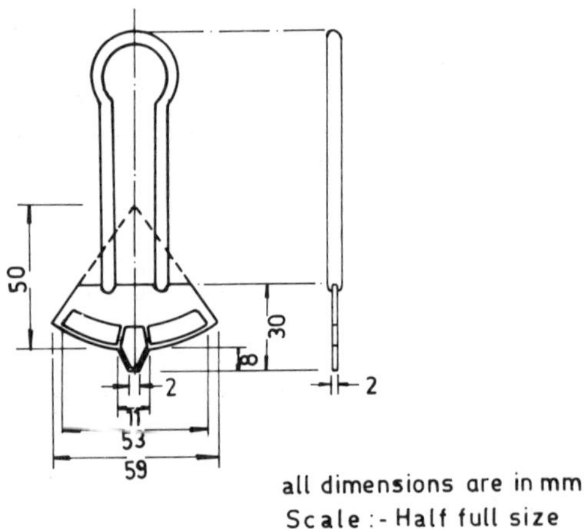

FIGURE 3. Working drawing of new grooving tool in unhardened steel, 2 mm diameter.

- sticky limit above which the mixture of soil and water sticks or adheres to objects that can be wet by water
- cohesion limit, the water content at which crumbs of the soil cease to adhere to each other when placed in contact
- plastic limit or lower limit of plasticity, at which the soil will start to crumble when rolled into a thread under the palm of the hand
- shrinkage limit or lower limit of volume change, at which there is no further decrease in volume as water is evaporated

To Atterberg plasticity meant "capable of being shaped" and the best indication of this was the ability to be rolled into threads. He used the amount of sand that could be mixed with a plastic soil at liquid limit, before the soil completely lost its plasticity, as a measure of the plasticity.

In the progressive transition from liquid state to solid state, the soil undergoes dramatic changes in consistency. Figure 1 schematically indicates the change in volume of the soil with changes in the water content together with the limits of different states which are of engineering significance. In our present context of engineering practice, the upper plastic limit, lower plastic limit and the shrinkage limits are the liquid, plastic and shrinkage limits, respectively. All limits are expressed as the water contents, being the ratio in percent of the weight of water to the weight of soil dried at 105°C till constant dry weight is reached. Realistically, all these transitions in the rheological states are not abrupt but are gradual.

EARLIER EXPERIMENTAL PROCEDURES

Since Atterberg considered that roll-out or sliding ability was the best quantitative characteristic of plasticity he proceeded to study the roll-out and sliding ability of a variety of soils. The soil paste was spread out in a dish with a spatula to a thickness of about one centimeter and divided into two halves. The number of times the dish was vigorously and repeatedly hit, to bring about the closure of the two halves in the lower portion flowing laterally, was recorded. The corresponding water content was determined by drying the sample. The procedure was repeated with varied water contents. The water content corresponding to a definite number of blows was defined as the liquid limit water content. For plastic limit, the same soil was further mixed with more of pulverised soil and rolled out into a thread under the palm of the hand (or fingers) on a base of paper. The thread was folded together and rerolled until it was broken during rolling. The corresponding water content was the plastic limit of the soil. Bauer [5] has provided an historical development of the Atterberg limits. The importance of these limits was never realised in Atterberg's own field of agriculture nor in other fields concerned with clays including soil mechanics, until Terzaghi [90] could see its importance and potential as applicable to soil mechanics.

LIQUID LIMIT

Percussion Testing

Standard device to determine the liquid limit of soils was developed by Casagrande [18]. This device is essentially the same as specified by ASTM (D423) except for the grooving tool (Figure 2). The device has a standard cup which can be lifted through a standard height by a cam and dropped on a base of standard material. The soil paste is transferred into the cup by a spatula and finished to get a smooth surface without air voids. A groove of standard dimensions is cut in the soil using the standard grooving tool. The number of drops required to close the groove by half an inch along the bottom of the groove is recorded. The water content of the sample paste removed from the cup is determined. The procedure is repeated by altering the water content by spreading the soil paste on a glass plate and working with a spatula. The water contents are plotted against their corresponding logarithm of number of blows within a range of 10 to 40 blows to get a linear relationship. The water content corresponding to 25 blows is read out from the graph as the liquid limit of the soil. This number was suggested by Casagrande to obtain about the same value of liquid limit as obtained by Atterberg's technique.

After 30 years of usage, Casagrande [20], based on the observations of use and misuse of his technique, reexamined the factors influencing the test results. He found that his liquid limit determination test was akin to a dynamic shear test. As such, he emphasised that the height of drop, the hardness of the base, which influence the test results, need be standardised. Kenney [45] visualised the liquid limit determination test as a dynamic, stress controlled, strain dependent shear test in which the maximum strength is not involved. Norman [67] reported the results of a series of tests in which he varied the hardness of the base. He found that the American Micarta base was considerably harder than the hardness specified by British standards. The tests with Micarta base resulted in a 3 to 4 percent lesser water content, for the same number of blows, than those obtained with the base of British Standard hardness. Since this variation was considerable, the need for standardisation became more apparent.

Hovayni [37] developed a new grooving tool in the form of a double edged blade (Figure 3). This tool could be used for cutting a clear groove, even in soils of low plasticity, which was otherwise not possible using the other types of grooving tools. By tilting the cup, the excess soil could be removed without disturbing the groove.

Cone Penetrometer Method

During the same period when Atterberg developed the method to determine plasticity characteristics of soils, the Geotechnical Commission of the Swedish State Railways (1912–1914) developed the fall cone test to determine the consistency of clays. John Olsson [39] considered an adaptation of the Swedish Brinell hardness test wherein a ball is pressed into the test material under certain pressure. In the adapted test a cone or cylinder was tried in place of a ball. Eventually the cone was chosen since it provided geometrical similarity of impression independent of the depth. It was also decided to use a free falling cone instead of the one that was pressed slowly into the clay. This eliminated the difficult decision about the rate of loading. A metal cone of certain weight and apex angle is suspended over a horizontally levelled surface with the point barely touching the surface. Then the cone is released for penetration under its own weight. The depth of penetration is related to shear strength. Since there is no fall of the cone through a certain height before it penetrates into the soil, this method can very well be regarded as cone method rather than fall cone method.

To support the use of cone in the prediction of undrained strength, theoretical interpretation of cone method was provided by Hansbo [32] in 1957. Considering the forces acting on the cone element it has been shown that the undrained shear strength τ_f and the depth of penetration h of the cone of weight Q are related in the form

$$\tau_f = K \frac{Q}{h^2} \tag{1}$$

The proportionality factor K was found to be constant for any given value of the cone angle. Its value was found to be influenced by the rate of shear and sensitivity of the soil. The effects of these factors were considered, at that time, less important than those of sampling and transportation disturbance. The values of τ_f obtained by cone method were found to be in close agreement with those from field vane test, provided an appropriate cone angle and sampler factor K were chosen. It has been cited [32] that with a cone of 400 gms weight and 30° apex angle, undrained strengths up to 20 t/m^2 could be estimated satisfactorily.

Casagrande [20], in 1958, realised the serious disadvantages of his experimental device to determine the liquid limit. He even suggested that a simpler direct shear test or an indirect shear test such as static penetration test could possibly eliminate many of the difficulties that one faces in the use of percussion type of liquid limit device. He also realised that none of the other tests had simplified to an extent that could compete in simplicity, acceptance and cost with the percussion type of testing. This became a turning point to look for an alternative method to determine liquid limits of soils. Most of these alternatives were based on some form or other of cone penetration test.

Prior to 1958, Vasilev [94], based on the Swedish cone test, suggested a method to determine the liquid limit. This method specified the water content of the sample in which a cone of 30° apex angle and 76 gms weight penetrated a depth of 10 mm in five seconds as the liquid limit. Since the values obtained from this method differed considerably from those obtained by Casagrande's percussion method, the approach did not catch the attention of soil engineers. Subsequently, the reason for the difference was attributed [85] to high value of shear strength of about 8.5 kN/m^2 reflected in Vasilev's method compared to the range (1 to 3 kN/m^2) of shear strength measured at liquid limit determined by Casagrande's percussion method and a functional relationship between the two was provided. Although Jurgenson [40] appreciated the strong point in Vasilev's procedure in giving results which corresponded to a definite shearing resistance, he pointed out the possible errors in the interpretation and classification of soils.

Karlsson [41] in 1961 realised the potential of the cone method to determine the liquid limit of soils. Instead of developing a method that would enable him to obtain the water contents corresponding to Casagrande's method, he leaned towards the fineness number, earlier suggested by Hansbo [32] to classify the plastic soils. The fineness number is defined as the water content at a certain relative strength of remoulded material determined by cone method. Although the results were examined in relation to Casagrande's percussion method, the advantages of the suggested approach were not strongly evident to replace the approach that had fair universal acceptance. Moore and Richards [58] also attempted to strengthen Karlsson's approach, by developing a graph (Figure 4) for converting the relative strength to conventional units of shear strength, using the functional relation $\tau = 1.5 H_3$, where τ is the shear strength in g/cm^2 and H_3 is the fineness number.

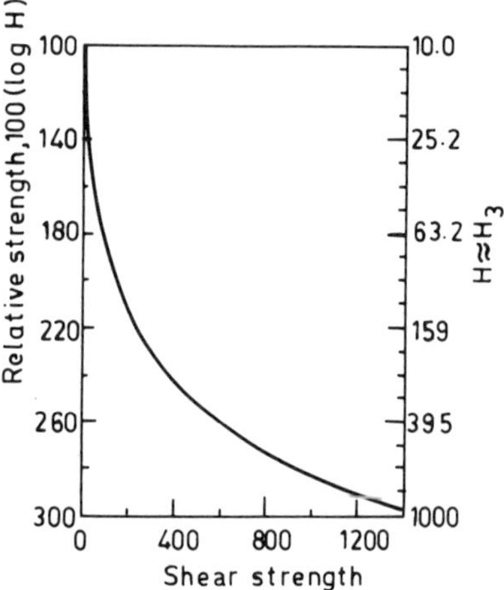

FIGURE 4. Graph for converting "relative strength" fall-cone values to conventional units of shear strength.

It was Sherwood and Ryley [80] who made a detailed assessment of the cone method (Table 1) and directly approached the problem of finding direct equivalence to Casagrande's method. After examination of test results to liquid limit which agreed well with that by Casagrande's percussion method in the range of liquid limit values up to 100%, an alternative procedure was suggested. Liquid limits of soils were defined as the moisture contents corresponding to 20 mm penetration of a 30° apex angle cone of 80 gms weight. It was realised by British Standards Institution that the test could be performed easily, the results of which were not dependent either on the design of the equipment or the manner in which the tests were performed. Campbell [15] reinforced the cone method of liquid limit determination. He demonstrated that the cone method offers worthwhile improvements over the Casagrande's percussion method both in terms of reproducibility and the ease of conduct of the test. The liquid limits of arable top soils covered in the investigation were in the range of 27% to 47%.

In the liquid limit test using the cone penetrometer specified in BS-1377 (1975), a cone of apex angle of 30° ± 1° fixed to a vertically sliding shaft is positioned with its tip just touching the surface of the clay. Liquid limit is defined as the water content of the clay at which the cone penetrates 20 mm in 5 ± 0.5 sec. The stainless steel surface of the cone is required to be smooth and polished. The tip of the cone is allowed to be blunted slightly so that, when the cone is pushed through a hole of 1.5 mm diameter in a plate 1.75 ± 0.1 mm thick, the tip can be felt by a finger brushed across the other face of the plate.

Further concerted research efforts can be traced in the most recent publications to reinforce the above approach and to bring about the distinct advantages of the method. Houlsby [34] has provided a detailed theoretical analysis of the cone method that enabled him to directly compute the undrained shear strength. Subsequently, Houlsby [35] associates the discrepancy in the predicted and measured range of strengths to approximations involved in his analysis. He also indicates that more sophisticated analysis in the future might achieve a closer correspondence with the experimental results. It has been found that the most important factor affecting the liquid limit of soils is the roughness of the cone. It is suggested that the cone could either be as smooth as possible and lightly oiled before use or it could be very rough. However, one condition or the other need be made explicit in the standard procedure. The above study also critically examines the tolerances specified for the liquid limit device, such as cone angle and bluntness of the tip. It has been found by the above analysis that the maximum radius to which deformation is expected to occur while the cone penetrates the soil is 21.4 mm. The cup used to contain the soil for liquid limit test is thus specified approximately 55 mm in diameter.

TABLE 1. Cone-Penetrometers in Use for Determining the Liquid Limit of Soils [80].

Country of origin	References	Angle of cone, degrees	Weight of cone, g	Definition of liquid limit
Sweden	Karlsson (1961 [41]	60	60	Related to the fineness number which is the moisture content corresponding to cone penetration of 10 mm.
U.S.S.R. Bulgaria Yugoslavia East Germany	Vasilev (1949) [94] Stefanov (1958) [89] Bazinovic (1958) [80] Matschak and Rietschel (1965) [80]	30	76	Moisture content corresponding to cone penetration of 10 mm.
India	Uppal and Aggarwal [92]	31	148	Moisture content corresponding to cone penetration of 25 mm.
U.S.A.	Sowers et al. (1960) [86]	30	75	Moisture content corresponding to cone penetration of 10 mm.
France	Laboratoire Central des Ponts et Chaussees (1966) [48]	30	80	Apparatus calibrated against Casagrande apparatus.

An alternative line of support for the use of cone penetrometer to determine liquid limit of soils has been the observation of the same range of shear strength of soils at their liquid limit [84,106]. Wroth and Wood [101,102] appreciate the use of simple strength measuring device such as cone instead of the more elaborate percussion device. It is evident that in making a radical change in the method to determine the liquid limit, the values determined by the new and old methods will not necessarily correspond precisely. Whereas the Casagrande apparatus was an attempt to standardize the procedure described by Atterberg [2] for detecting the moisture contents of soils at which a transition between liquid and plastic states took place, the cone penetrometer method is a more plausible attempt to detect the moisture content at which the soil exhibits a certain strength.

The arguments that are now building up in favour of the cone method to determine the liquid limit of soils might be sufficient to an extent that in the near future this method might be accepted universally as a standard procedure to determine liquid limit of soils.

One other recent significant finding [65] in favour of the cone method is the possibility of determining the liquid limit of soils instantaneously without necessarily waiting for oven drying of the soil for water content determination. It has been found that the amount of soil paste transferred to the cup is a function of both water content and liquid limit of the soil. Hence, instead of water content determination for a particular depth of cone penetration, the bulk density of the soil can be determined just prior to the cone penetration. From the bulk density versus penetration plots, the density associated with 20 mm penetration can be read, or computed by one point method. The water content at liquid limit can be found with assumed value of degree of saturation and specific gravity of soil particles. It has been shown by extensive laboratory study [65] that a degree of saturation of 98% can be assumed. The error associated with the assumption of specific gravity of clay particles within the range of 2.65 to 2.8 are well within the acceptable levels of accuracy of determination. This approach also enables one to independently check the liquid limit determination based on water content determinations.

ONE POINT METHODS

Liquid limit tests do not directly provide water contents at liquid limit but flow lines are obtained between water content on arithmetic scale and number of blows on the logarithmic scale. Functionally,

$$w = -F \log N + c \qquad (2)$$

in which

w = is the percent water content
F = is the flow index constant being the slope of the flow line
N = the number of blows
c = the constant being the intercept at unit number of blows

The observation that the flow curves for the soils in

groups are broadly similar prompted many investigators [25,26,93,69,24] to develop a method to determine liquid limit from a single determination of number of blows in the range of 10 to 40 and the corresponding moisture content, with the use of an algebraic expression for the characteristic flow curve. Norman [67] reinforced these approaches by analysis of his data covering 455 British soils. It was also found that for soils with liquid limits exceeding 120% the order of accuracy was not sufficiently good to justify the use of one point method. With such soils, British Standard Institution provided a subsidiary one point Casagrande liquid limit test (test 2b) in 1967, which was subsequently transferred as test 2c in 1975. Following the introduction of cone penetrometer in Ground Engineering Laboratories, the first 600 points were examined by Clayton and Jukes [22] for possible development of one point method. The analysis has indicated that the cone penetrometer may reasonably be used in one point liquid limit test which will provide a cheap but relatively accurate alternative to British Standard one point Casagrande method.

The following two sets of formulae provide a means to compute w_L from one point data.

$$w/w_L = \frac{1}{A-B\log N} \tag{3}$$

$$w/w_L = \left(\frac{N}{25}\right)^{\alpha} \tag{4}$$

Mohan and Goel [57] observed that, for values of α ranging from 0.068 to 0.121, the error associated with liquid limit water content is less than one percent. No rational basis was provided by the authors.

A scientific base for liquid limit determination by one point method has been provided by Nagaraj and Jayadeva [59]. The details have been discussed later. It is derived from the uniqueness of generalised flow line, which is obtained on the basis of the physico-chemical interactions. Generalised flow line is obtained by normalization of water contents by liquid limit values, thereby nullifying the effects of specific surface (surface area expressed in m^2/gm) and associated physico chemical factors which are otherwise reflected in different responses at macrolevel. Defining the K as the numerical value of the slope of line represented by w/w_L against $\log N$, the Equations (2) and (3) along with (4) can be written as:

$$w/w_L = (N/25)^{\alpha} = \frac{1}{1 - K\log(N/25)} \tag{5}$$

or

$$K = 1 - (25/N)^{\alpha}/\log(N/25) \tag{6}$$

Table 2 shows the values of K for different values of α and N. Assuming a central value for variation in α to be 1.00 and N ranging from 15 to 35 blows, a value of $K = 0.23$ has been found to be appropriate. Taking $K = 0.23$ for $\Delta w_L/w_L = \pm 0.01$, deviations in ΔK values are computed and shown in Table 3.

It has been possible to infer that the equation

$$w_L = \frac{w}{1.3215 - 0.23\log N} \tag{7}$$

would enable one to compute w_L by one point method within an error of one percent water content.

From the data of Sherwood and Ryley [80] a rational one point method to determine liquid limit from cone penetrometer tests has been developed [59]. The functional relationship to compute liquid limit from one point data of water content and depth of cone penetration can be either

$$w_L = \frac{w}{0.77\log D} \tag{8}$$

or

$$w_L = \frac{w}{0.65 + 0.017D} \tag{9}$$

where D is the depth of cone penetration in mm. The above functional relationships arise out of normalising the flow lines of soils in w versus D or w versus $\log D$ form, using their respective liquid limit water contents. These relationships are very simple compared to earlier approaches by Karlsson [41] and Clayton and Jukes [22] which are purely based on statistical correlations.

TABLE 2. Values of K for Different Values of α and N.

N	α=0.068	α=0.092	α=0.100	α=0.109	α=0.114	α=0.121
15	0.1593	0.2169	0.2362	0.2581	0.2703	0.2874
20	0.1578	0.2140	0.2328	0.2541	0.2659	0.2824
30	0.1556	0.2101	0.2282	0.2485	0.2598	0.2756
35	0.1548	0.2086	0.2264	0.2464	0.2575	0.2730

TABLE 3. Variation of ΔK for 1% Variation in Liquid Limit Values.

$\Delta w_L/w_L$	N	ΔK
0.01	15	−0.0479
0.01	20	−0.1066
0.01	30	0.1252
0.01	35	0.0668
−0.01	15	0.0469
−0.01	20	0.1044
−0.01	30	−0.1228
−0.01	35	−0.0655

Indirect Methods

Based on the fact that adsorption of basic dyes on clay minerals depends on the type of clay mineral and of the adsorptate itself, Ramachandran et al. [71] showed that as the liquid limit of the clay increases the adsorption of malchite green and methylene blue increases. Functional relationships were developed between the liquid limit in percent and dye adsorbed in g/g from experimental data on alluvial and black cotton soils. Since liquid limit and dye adsorption are both dependent on specific surface of soils, such correlations are not strange.

The feasibility of using moisture tension as a practical method of determining the liquid limit was studied by Russel and Mickle [73]. Based on extensive laboratory moisture tension curves, in general a total pressure difference of 30 inches of water or 1.09 psi was suggested as the value at which soils equilibriate to their liquid limit water content. It was indicated that obtaining one unique pressure would be possible by taking into consideration the time factor required to reach equilibrium states.

Although the above indirect methods have enabled us to have a better insight into the fundamentals of liquid limit, they cannot be regarded as alternative methods to the conventional method of determination.

PLASTIC LIMIT

In principle, plastic limit of fine grained soils involves both strength and deformability. Hence all methods of determination involve shaping clay pastes of different consistency into balls, threads, cubes and subjecting the same to different types of manipulation. The most commonly used method is by rolling wet clay into threads and determining the limit when it breaks up on rolling to ⅛ inch diameter or less. The definition is based on the test method (rolling and allowing water evaporation), the mode of failure (crumbling) and the diameter of the specimen. According to Kenney [45], the plastic limit test is rather a complicated version of splitting test, in which, during rolling the soil specimen is subjected to rotating applied stresses in addition to reduction in diameter. The test conditions are more strain controlled than stress controlled.

Plastic limit determination is subjected to personal impressions and can very much cause considerable discrepancies even though it is possible to eliminate human factor in reproducibility of test results [4,68]. The other method, subsequently developed, consists of refining the method of identifying the limits of transition from plastic to semisolid state. The cube method [1] consists of moulding the wet soil into a cube of approximately ¾ inch size and pressing the opposite faces between fingers or between spatula and nonabsorptive surface till it develops cracks on being deformed. The moisture content at this stage is plastic limit.

Indirect Methods

Ekman [26] proposed an alternative method to determine plastic limit. It consisted of subjecting clay specimens to different magnitudes of pressure. It was found that clay specimens reach their plastic limit water contents in the range of 15 to 20 kg/cm^2. The uncertainty with respect to equilibrium time has not been examined.

Instead of application of pressure to clay water pastes Russel and Mickle [73] applied suction pressure and developed moisture tension desorption curves for different soils. They inferred that plastic limit value of soils can be approximated to be the water content at 162 psi (1112 kPa) suction pressure within reasonable limits of accuracy. The study was not completely conclusive.

However, the above approaches can be regarded only to provide an inkling to the basic aspects and certainly cannot replace the simple conventional method.

Cone Penetrometer Method

Scherrer [76] analysed the cone penetration test. He considered the case of gradually loading a cone from zero to one kilogram in one minute. The deeper the cone penetrates into the soil the larger becomes the loaded area and lower the specific pressure. This specific pressure decreases until the soil is capable of supporting the load so that an equilibrium pressure is reached. Unlike other types of penetration tests it does not measure the resistance to plastic flow, but measures the equilibrium pressure which is directly proportional to the shear strength of the soil. As shear strength of a saturated clay depends directly on the water content so does the equilibrium pressure. For six soils the relationship between moisture content and equilibrium pressure was developed. The equilibrium pressures corresponding to liquid and plastic limit water contents were examined.

Although some scatter was noticed it indicated the possibility of using cone method to determine both liquid and plastic limits.

Campbell [16] examined the potential of cone method in greater detail. He worked with sixteen arable top soils, one subsoil and one clay mineral whose liquid limits ranged from 27 to 67 percent and plastic limits from 20 to 44 percent. It is claimed that the following procedure, if adopted, has the advantage that the plastic limit test would more closely relate to soil behaviour, be less subjective, and at least as reproducible as the Casagrande's test. The more distinct plus point is that the cone method can be used both for liquid and plastic limit determination.

For the cone method of plastic limit determination, soil is to be wetted to different moisture contents and then packed into a rigid metal cup of 55 mm diameter and 45 mm depth. The penetration of a 30° cone of mass 80 gms is determined after the cone is released for 5 sec from an initial position in which its tip was just in contact with the soil surface. After repeating the procedure for all samples the moisture content versus cone penetration curve is plotted. Typical plot is shown in Figure 5. The plastic limit from the cone method is the moisture content corresponding to the minimum of the curve.

It has been found [16] that the plastic limit moisture contents correlated closely with, but was not equivalent to, those determined by Casagrande's method. It has also been observed that the minima of the curves were in the range of a mean value of 1.36 mm penetration. Wood [103] wonders as to how cone penetration plastic limit gives an indication of the water content at which a soil changes from the plastic to brittle state. Subsequently Campbell et al. [17] have shown that cone penetration plastic limits are better indicators of soil behaviour than those determined by Casagrande's method.

There are two strong views to the reassessment of plastic limit of soils, the one being the possibility of developing a

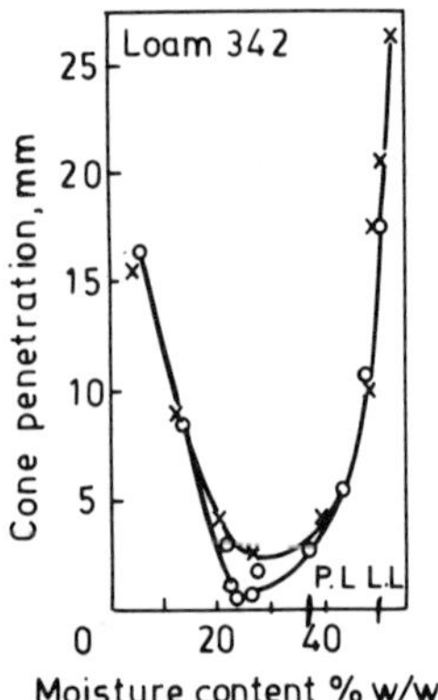

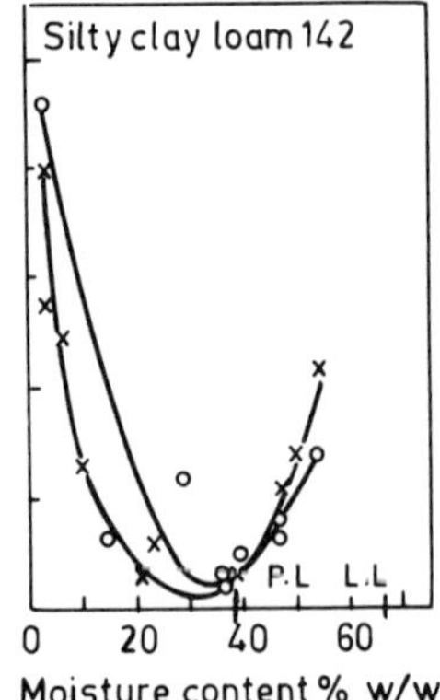

FIGURE 5. Moisture content/cone penetration relations for two soils.

rational method to determine this parameter, such as by use of cone which eliminates the ambiguities of determination, and the second being the possibility of having a unique stress relation with the state of soil at that water content. It might also be necessary to redefine plastic limit. In pursuance of the above lines, Wroth and Wood [101] and Wood [102] attempted to link the plastic limit water content to a level of hundred fold increase in shear strength at liquid limit. This is believed to eventually dilute the concept of plastic limit and it might even be abandoned.

The role of plastic limit in geotechnical engineering needs closer examination. It should be possible in the not too distant future to develop predictive models involving only liquid limit water contents.

SHRINKAGE LIMIT

The phenomenon of shrinking and swelling, which takes place in moist fine grained soils as the water content decreases or increases, has attracted very general observation. This results in undue stresses in structural members due to differential movements and hence progressive damage to structures. Accelerated desiccation such as heat from boilers, ovens and furnaces under inadequately insulated conditions causes rapid and irregular differential movements. Apart from this external source, even growing vegetation can cause desiccation under pavements and shallow foundations founded on soils having potential for volume changes [23,31]. Although the magnitude of volume change experienced by any situation is a direct reflection of soil state and environment, the potential can be assessed by inferential tests such as shrinkage limit and swelling potential tests.

Determination of the cuboidal contraction of clay test pieces has long been a standard test in ceramic laboratories. As water is lost progressively from a saturated sample, there is, at first, a linear relationship between shrinkage and water content. The loss of water would be equal to the volume contraction. When sufficient water is lost, there is an abrupt change in the relationship. Further removal of water is not associated with corresponding volume reduction as the particles would have reached very dense close packing. This transitional water content is the shrinkage limit of soil (Figure 1).

Most of the concepts of the shrinkage of clay soils are based on the work of Haines [30] on shrinkage of remoulded clays. Since physical reconditionining of soils takes place by alternate wetting and drying cycles, detailed studies on apparent specific volume and shrinkage of soils are reported by Lauritzen [50]. In ceramic industry shrinkage of soils has received much attention. Kingery and Franel [46] report a fundamental study of clay drying and plastic properties with surface tension of fluid varying from 16 to 71 dynes/cm². In

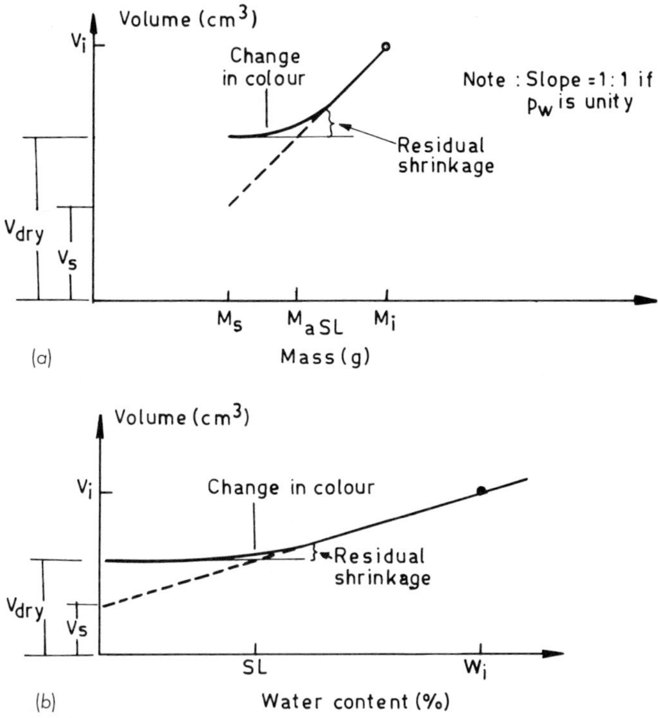

FIGURE 6. Determination of the shrinkage limit, based on (a) total mass, (b) water content.

general it was noticed that bulk density increased with increase in surface tension.

Determination of Shrinkage Limit

Atterberg's [2] original work was with small clay bars which he allowed to dry slowly. He observed the point at which the colour changed and at the same time noticed the constancy of length at that point. Terzaghi figured out that one could just as well measure the dry volume and dry mass and back calculate the water content at the point of minimum volume. This is schematically shown in Figure 6. A small amount of soil of mass w_i is placed in a dish of known volume V and allowed to slowly dry. After the oven dry mass w_{dry} is obtained, the volume V_{dry} is found by the displaced weight of mercury. The shrinkage limit is calculated from

$$SL = \left(\frac{V_{dry}}{w_i} - \frac{1}{G}\right)\gamma_w \times 100\ (\%) \qquad (10)$$

or

$$SL = w_i - \left\{\frac{(V_i - V_{dry})V_w}{w_s}\right\} \times 100\ (\%) \qquad (11)$$

The two equations correspond to the two parts of Figure 6. Unfortunately the test has certain shortcomings arising from air entrapment, cracking and the risk of mercury poisoning. It should be possible to eliminate the use of mercury by more sophisticated methods of volume measurements.

Initial soil fabric pronouncedly influences the shrinkage limit [105]. Shrinkage limits can be higher than plastic limits if the fabric resists the shrinkage. As such, the standard procedure specifies (ASTM Designation D 427) to start the test on remoulded soil with water content close to liquid limit water content.

MICRO-MECHANISTIC INTERPRETATION

A saturated fine grained soil with a water content greater than 37 percent contains a greater volume of water than solids. Generally, water contents greater than 37 percent are more the rule than exception in the case of fine grained soils. This is more so true at liquid limit water contents. Yet more stress has been laid on the study of structure and mineralogy of the solid phase than the study of the properties of the liquid phase and the mutual interactions between the phases.

Since neither water nor soil particle surfaces are chemically inert, they interact with each other. These interactions influence the physical and physico-chemical behaviours of soil. Attempts made to quantify these interactions from the fundamentals of a scientific principle indicate a unique microbehaviour for saturated fine grained soils. The subject of this section has been to discuss the consistency limits from micro-mechanistic considerations.

Soil-Water Interaction

Generally all soil particles, particularly clay particles, carry a net negative electrical charge on their surface. The magnitude of total electrical charge can be directly related to the specific surface of clays which increases with decrease in particle size. The water molecule has a V-shaped arrangement of atomic nuclei with an average H—O—H angle of about 105°, which induces dipolar nature to the water molecule. As a result of uneven charge distribution and dipolar character of water molecules, they are attracted to ions in solution leading to ion hydration. Not all ions hydrate although cations common in soils do.

Interactions between soil particles, adsorbed cations and water arise because there are unbalanced force fields at the interfaces between the constituents. When two clay particles are brought into close proximity their respective force fields begin to overlap and influence the behaviour of the system if the magnitudes of these forces are large relative to the weight of the particles themselves. Fine grained soils, because of their small size and large surface area, are well known to be susceptible to such effects [56].

The effects of surface force interactions and small particle size are manifested by a variety of interparticle attractive and repulsive forces. The principle contributions to the attractive forces come from van der Waal's forces, which can be either ion-dipole or dipole-dipole interactions, in addition to the London dispersion force. Lambe [49] and Rosenquist [72] have suggested interparticle attractive force mechanisms such as Coulombic attraction between negative surfaces and positive edges, particle-cation-particle linkages and hydrogen bonds. Sridharan and Jayadeva [88] have shown that these attractive forces are too small for the range of water contents from liquid limit to plastic limit and do not influence the soil behaviour.

The adsorption of cations by clays and the formation of double layers is responsible for long range repulsive forces between particles. Calculation of interparticle repulsions due to interacting double layers may be done in more than one way; however, the osmotic pressure is convenient. By this approach the pressure that must be applied to a soil system to prevent movement of water either in or out of the system is determined as a function of the solution concentration which can be related to the particle spacing and the corresponding void ratio or water content. The quantification of the repulsive pressure using the Gouy-Chapman diffuse double layer theory for planar surfaces has been made by various investigators [96,11,12,95,3,88]. The magnitude of the repulsive pressure depends on the surface charge density, electrolyte concentration and temperature, valency of the cation and the static dielectric constant of the pore fluid.

Micro Mechanistic Behaviour of Soils

Several investigators [11,9,98,47,79,88] have analysed the compressibility and swelling behaviour of pure clays using the Gouy-Chapman diffuse double layer theory. In these investigations a parallel plate mode, where $(R - A) = p$, (Figure 7) has been assumed and the void ratio, e, versus effective consolidation pressure, p, relation has been predicted from the known physical and physico-chemical properties of the soil and pore fluid. The predicted e vs. p relation has been compared with experimental curve. These

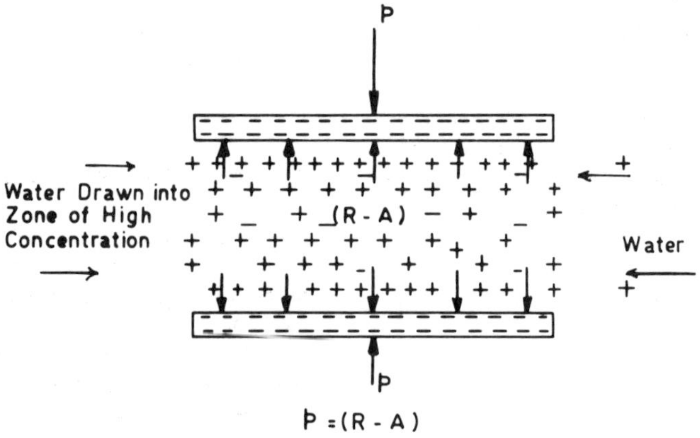

FIGURE 7. Parallel plate model.

indicate that the soils are distinctly different, in their mechanical behaviour.

The micro-mechanistic reexamination of the e versus p relations of the three pure clays (Figure 8) has revealed that the average half space distance, d, versus the effective consolidation pressure, p, relation is unique for a given physico-chemical environment [88]. Nagaraj and Srinivasa Murthy [61,62] have shown that the unique d versus p relation at a physico-chemical environment corresponding to that of natural soils [10] can be linearised in the working stress range of 25 to 800 kPa in the form:

$$d = a - b \log p \qquad (12)$$

Equation (12) micro-mechanistically defines the compressibility behaviour of all soils. In the above formulations the void ratio and half space distance have been linked through the specific surface, S, of the soil [9,59] in the form:

$$e = G \gamma_w S d \qquad (13)$$

where G is the specific gravity of the solids and γ_w is the unit weight of water.

Micro Mechanistic Interpretation of Liquid Limit

From the earlier discussions it is clear that physico-chemical interactions exist between phases in saturated fine grained soils. This is more so true in the range of water contents between liquid limit and plastic limit. Since most of the engineering property correlations are attempted within this range of water contents, micromechanistic interpretation of these two limits is brought out.

States at Microlevel

The liquid and plastic limits represent two states at which the separation distance between particles or their aggregated units are under force field equilibrium [97]. The water holding capacity of soils under the influence of surface forces at liquid and plastic limits has been found to be due to the suction pressures of 5 and 1112 kPa, respectively [73]. The soils at their liquid limit possess small but definite measurable shearing resistance. Casagrande [18] has deduced that the liquid limit corresponds to a water content at which soils have a unique shear strength of 2.5 kPa. Norman [67] has reported a shear strength of 2.0 kPa at liquid limit. Wroth and Wood [101], based on test results of Youssef et al. [106] have concluded that the shear strength at liquid limit is unique and has a value of about 1.7 kPa. Wroth [100], from the critical state concepts has demonstrated the existence of a unique consolidation pressure of 6.3 kPa at liquid limit. Whyte [99], based on extensive test data, has shown the shear strength at liquid limit to be 1.6 kPa.

These discussions testify to the existence of a shear strength of about 1.7 kPa and a corresponding consolidation pressure of about 6.0 kPa at liquid limit of soils. At liquid limit there exists a continuous water phase and a discontinuous solid phase which forbids the stress transfer through effective mineral–mineral contact. Then the effective consolidation pressure of 6 kPa at liquid limit is equal to the net repulsive pressure. Further, Russel and Mickle [73] have shown the existence of a unique consolidation or suction pressure even at plastic limit. Wroth [100] has indicated the existence of a unique ratio of shear strengths at liquid and plastic limits. Thus a unique consolidation pressure can be assumed to exist for all soils at their plastic limits also.

For a given physico-chemical environment, the d versus $(R - A)$ or p relation is unique, which results in two unique values of d, corresponding to liquid and plastic limits of soils. These d values can be termed as d_L and d_p at liquid and plastic limits, respectively. Though soils exhibit a wide range of values for liquid and plastic limit water contents, at macro-level they have same order of d_L and d_p values. The microlevel water content variation is a reflection of the variation in specific surfaces of soils.

Link Between Micro and Macro Behaviours

The unique d versus p relation of Equation (12) can be normalised with the d_L value at $p = 6.0$ kPa. Then Equation (12) can be written as:

$$d/d_L = a - b \log p \qquad (14)$$

From Equation (13) for saturated soils,

$$e = wG = G\gamma_w S d \qquad (13)$$

and

$$e_L = w_L G = G\gamma_w S d_L \qquad (15)$$

Dividing Equation (13) by (15),

$$e/e_L = w/w_L = d/d_L$$

Thus Equation (14) can be written as

$$d/d_L = e/e_L = w/w_L = a - b \log p \qquad (16)$$

Now for the assumed parallel plate model Equation (16) represents the compressibility behaviour of saturated soils in terms of easily measurable parameters. Based on e versus p relations of a number of natural soils in their normally consolidated saturated uncemented state, the authors [64,65] have evaluated the coefficients of Equation (16) and the resulting Equation is in the form:

$$e/e_L = w/w_L = 1.122 - 0.2343 \log p \qquad (17)$$

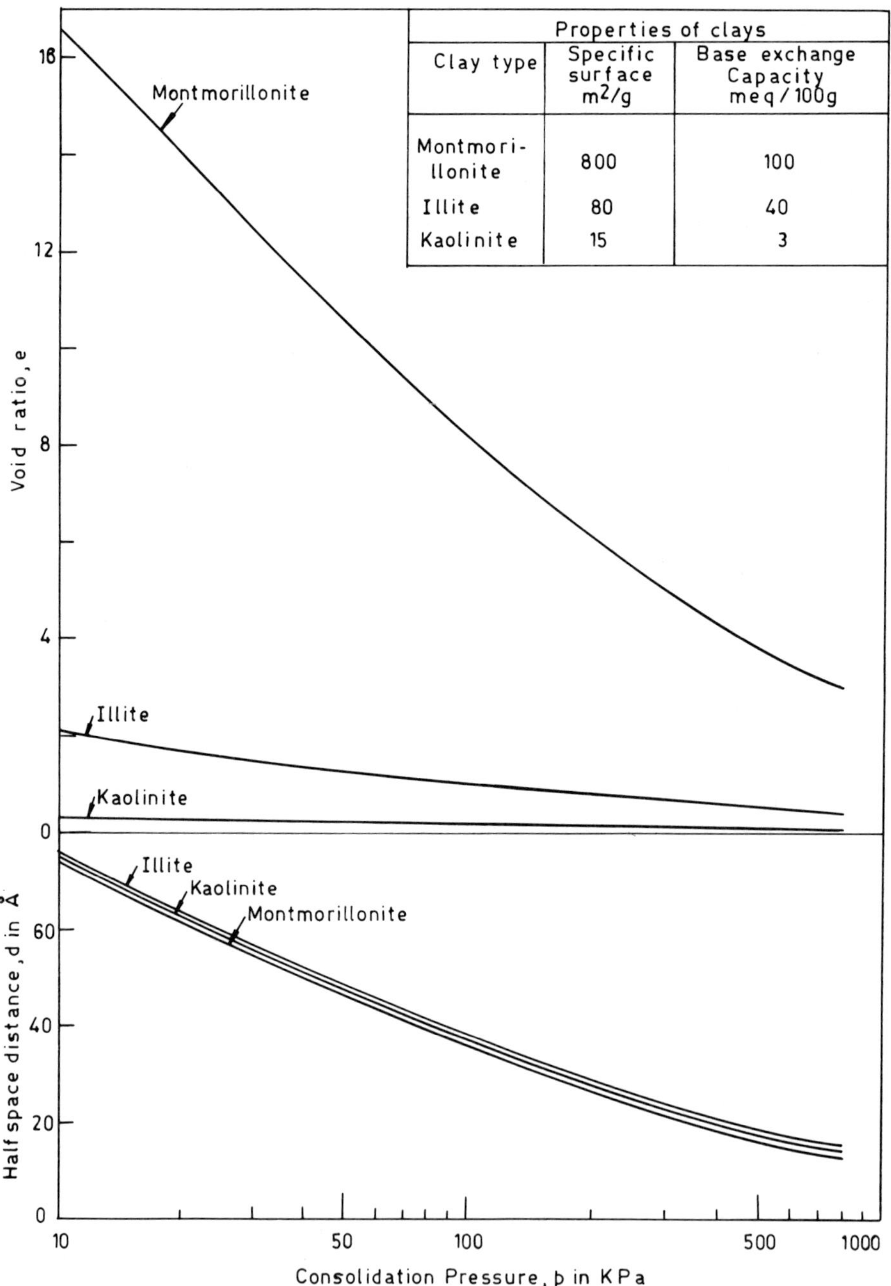

FIGURE 8. Theoretical e-log p and d-log p curves.

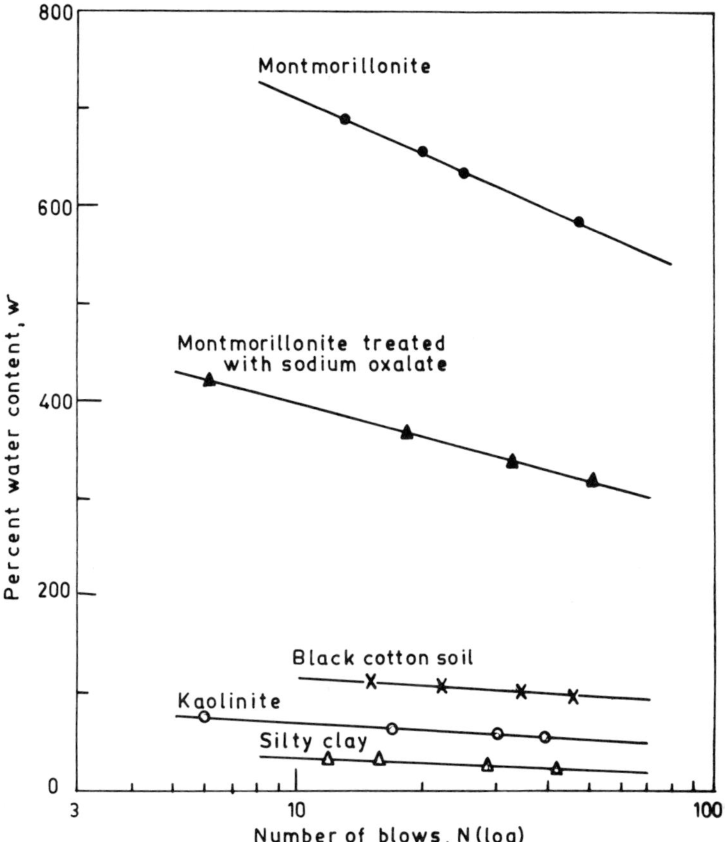

FIGURE 9a. Flow curves of clay water system.

where p is in kPa. This clearly brings out that liquid limit serves as an essential parametric link between micro and macro behaviours.

One Point Method for Liquid Limit Determination

Rutledge [75], Hvorslev [38], Henkel [33], Balasubramaniam and Choudhari [21], Leorouell et al. [53] and others have shown that the shear strength of saturated remoulded clay decreases in a regular manner with increasing water content and the water content versus shear strength relation is a curve nearly parallel to the water content versus consolidation pressure curve. In case of fine grained soils where the surface forces are predominant, the failure surfaces occur through the adsorbed layers [78]. This necessitates the force field equilibrium condition to be in the form:

$$p - u = (R - A) \tag{18}$$

and

$$q = f(p - u) = f(R - A) \tag{19}$$

For $u = O$, $q = f(p)$.

For known linear relations between q and p, Equation (17) can be written in the form:

$$w/w_L = a - b \log q \tag{20}$$

In the test for liquid limit determination, the number of blows, N or the depth of penetration, D, reflect a definite amount of shear stress induced in the soil. Thus q in Equation (20) can be replaced by either N or D of the Casagrande's percussion or cone penetrometer tests, respectively. Then Equation (20) can be rewritten as:

$$w/w_L = a - b \log N \tag{21}$$

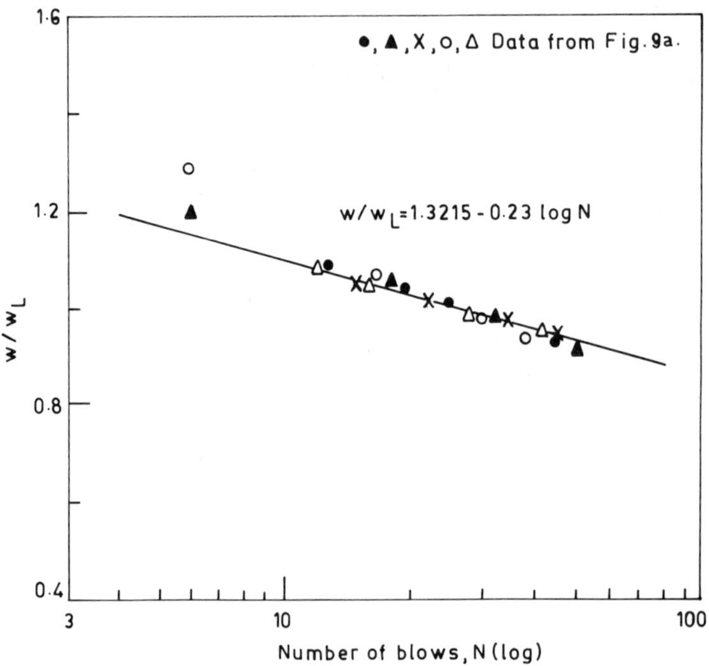

FIGURE 9b. Normalised flow curves.

and

$$w/w_L = a + b \log D \quad (22)$$

For D being in a small range,

$$w/w_L = a + bD \quad (23)$$

With known values of a and b, Equations (21) to (23) can be employed for one point method of liquid limit determination. Figures 9a and 9b indicate flow curves of a number of soils and their collapse into a narrow band upon normalisation with their respective liquid limit. From the normalised plots, a and b values can be evaluated.

DERIVED PARAMETERS

Though in engineering practice the liquid and plastic limits are often used directly, a few other derived parameters have gained greater acceptance.

Plasticity Index

This is defined as the difference in liquid and plastic limit water contents. This has a very wide application in engineering practice. This defines the range of water contents at which the soil behaviour is plastic. Micro-mechanistically this index can be written as:

$$I_p = (w_L - w_p) = \frac{1}{G}(e_L - e_p) = \gamma_w S(d_L - d_p) \quad (24)$$

From the discussions of earlier sections, for assumed unique values of p or q at lipid and plastic limits, $(d_L - d_p)$ represents a unique value. The variation of I_p is due to the variation in specific surface only. Statistical analysis of published data of consistency limits of 520 inorganic soils and the examination of possible relationship between plasticity index, I_p, and liquid limit, w_L, from critical state concepts have shown [60] that plastic limit of soils is a function of liquid limit for the range up to 150 percent liquid limit of soils. The functional relationship obtained is:

$$I_p = 0.74\,(w_L - 8) \quad (25)$$

It should be possible, in the not too distant future, mostly to develop functional relationships to predict soil behavior involving only liquid limit of soils. Since the constants in Equation (25) would be different for clay minerals other than sheet minerals plastic limits are likely to play the role of identifying appropriate physical models for prediction of soil behavior.

Liquidity Index

This is defined as the ratio of the difference of natural water content and plastic limit to the plasticity index, i.e.,

$$LI = \frac{w - w_p}{w_L - w_p} \quad (26)$$

The liquidity index would be unity for a soil whose natural water content is equal to its liquid limit, and zero for a soil having natural water content equal to its plastic limit. Micro-mechanistically the liquidity index can be written as:

$$LI = \frac{d - d_p}{d_L - d_p} \quad (27)$$

For d_L and d_p being unique and constant, liquidity index is a function of the average half space distance only. This also implies that liquidity index is only another form of (w/w_L).

For plastic soils, the value of liquidity index is indicative of the stress history of soil. Normally consolidated natural soils usually have liquidity index nearer to unity while overconsolidated natural soils have liquidity index closer to zero. All intermediate values are possible except in exceptional cases of extra sensitive clays when the liquidity index may be substantially greater than unity or of heavily overconsolidated desiccated soils when the index may even be negative.

Flow Index (I_F)

This is defined as the slope of the flow curve, i.e., the curve of water content versus the log number of blows in the percussion method of liquid limit determination. For convenience it is taken as positive. Thus,

$$I_F = -\frac{(w_1 - w_2)}{\log(N_1/N_2)} \quad (28)$$

Usually it is expressed as the difference in water content over one log cycle of number of blows. The I_F indicates the rate at which a soil loses shearing resistance with increase in water content.

If the water content versus shear strength variation is assumed linear, the flow index can be related to the plasticity index. The flow index represents the change in water content for a ten times increase in shear strength. The ratio of shear strengths at plastic and liquid limits may generally be assumed to be between 100 and 165. Even for such wide variation in the assumed strength values the ratio of I_p to I_F varies marginally from 2.0 to 2.2. Thus a unique relation between I_p and I_F can be written as:

$$I_p = 2.10\ I_F \quad (29)$$

Toughness Index (I_T)

This is defined as the ratio of plasticity index to the flow index. It reflects the ratio of the shearing resistance at liquid and plastic limits. It also serves as a measure of the resistance to deformation or toughness at the plastic limit.

$$I_T = I_p/I_F \quad (30)$$

High toughness at plastic limit indicates a high percentage of colloidal clay particles in the soil. From Equations (29) and (30) it is very clear that this will have a very narrow range of values from 2.0 to 2.2. Though values ranging from 0.0 to 3.0 have been reported, its utility in engineering practice has been diminished.

Activity Number (A)

Activity is a term in reference to the propensity of the soils for undergoing changes in volume in the presence of varying moisture conditions. Skempton [82] proposed the activity of the soil as the ratio of the plasticity index to the percent finer than 2 μm particle size.

$$A = \frac{I_p}{\text{percent} < 0.002\ \text{mm}}$$

Micro-mechanistically I_p is a function of the specific surface and the unique value of $(d_L - d_p)$. The percent weight finer than 2 μm reflects the minimum specific surface of the soil, corresponding to a condition of all particles being of that size. For the range of particles finer than 2 μm, the specific surface is not reflected by the denominator. Thus activity varies with the specific surface of the soil which depends on the geologic origin. However, for soils from a given geologic origin, with increase in clay content, both I_p and percentage finer than 2 μm increase maintaining a constant value of A. Consequently the value of A may be a reliable parameter to ascertain the existence of different geological strata encountered in a bore logging. Clays for which A is less than 0.75 are considered inactive. Normal activity is associated with values of A between 0.75 and 1.5, while values greater than 1.5 indicate clays progressively more active.

PROPERTY CORRELATIONS

Laboratory testing in which mechanical properties are not directly measured but are inferred through empirical correlations, forms the inferential testing programme. Consistency limits and their derived parameters form, to date, the most widely used inferential parameters in assessment and, at times, prediction of soil behaviour. Any correlations with mechanical property would be of proven value so far when there is a clear understanding of the basic mechanisms of the inferential parameters themselves and with respect to the mechanical property they reflect or correlate.

Though there are cautions to the effect that the plasticity limits and indices are determined on remoulded soils and at best are only indicative of other physical properties, many attempts with success have been made to link these parameters with mechanical properties.

Compressibility

For saturated remoulded clays, compressibility is defined in terms of the compression index C_c, which is the slope of the e versus $\log p$ curve. The value of C_c is generally found to increase with increase in liquid limit of the soil. Correlations are in vogue relating the compression index with the liquid limit of the soil [81,91]. Based on test results of random samples from the world over, whose initial water contents were at their liquid limits, Skempton [81] suggested the compression index equation of the form:

$$C_c = 0.007(w_L - 10) \quad (31)$$

Terzaghi and Peck [91], based on Skempton's work and subsequent analysis, observed that for the case of medium to low sensitive soils, Equation (31) can be modified as:

$$C_c = 0.009(w_L - 10) \quad (32)$$

Nagaraj and Srinivasa Murthy [61] have shown that these equations are not just empirical but can be derived from a well established scientific principle, and can be confidently used in predicting the compressibility of normally consolidated saturated uncemented soils. Further differentiating Equation (17),

$$\frac{de}{e_L} = -\ 0.2343\ d\ (\log_{10} p)$$

or

$$C_c = -\ \frac{de}{d(\log_{10} p)} = 0.2343\ e_L \quad (33)$$

For $G = 2.7$ and w_L in percentage,

$$C_c = 0.00633\ w_L \quad (34)$$

Equation (34) could be transformed to the same form of Equation (31), [61]. Schofield and Wroth [77] from the critical state concepts have shown that the slope, λ, of specific volume versus logarithm of effective pressure plot is a function of the plasticity index in the form:

$$\lambda = 0.585\ I_p \quad (35)$$

Nagaraj and Srinivasa Murthy [62] have shown that the slope of the average rebound recompression line which is normally referred to as C_r will also be a function of liquid limit in the form:

$$C_r = 0.0463\ e_L \quad (36)$$

For $G = 2.7$ and w_L in percentage,

$$C_r = 0.00125\ w_L \quad (37)$$

Further it has been shown that the preconsolidation pressure, p_c, and the entire $e - \log p$ curve can be predicted knowing the liquid limit, w_L, insitu overburden pressure, p, and natural water content, w, from the equation

$$e/e_L = \frac{w}{w_L} = 1.122 - 0.188 \log p_c - 0.0463 \log p \quad (38)$$

where p and p_c are in kPa.

Equation (38) has been derived with Gouy-Chapman diffuse double layer theory as the basis. This clearly demonstrates that the liquid limit will serve as a potential parameter in predicting the compressibility of saturated uncemented soils. This is viable from the micromechanistic considerations.

Shear Strengths

Skempton and Northey [84] have observed that the liquidity index of clays and their remoulded shear strengths will follow a unique trend as shown in Figure 10. Wroth and Wood [101] have made a reexamination of the above observation by linear idealisation of the liquidity index versus logarithm of remoulded strength plot as critical state line (Figure 11). In establishing the general relationships between sensitivity, effective stress and water contents for all clays, Houston and Mitchell [36] have used the liquidity index as a useful parameter to provide an appropriate normalization of equilibrium water contents. Figure 12 shows the data of Houston and Mitchell, wherein the liquidity index versus remoulded strengths, as determined by six investigators, are plotted. The test results represent a wide range of liquid limits and different types of shear tests such as vane shear, unconfined compression and triaxial shear, which might have caused the scatter. However, Houston and Mitchell have opined that a plot of liquidity index and remoulded strength follows a specific pattern irrespective of soil type. Examination of additional data (Figure 13) has also indicated that *LI* versus remoulded strength follows the same pattern as observed by Houston and Mitchell. The viability of this observation is micromechanistically reexamined.

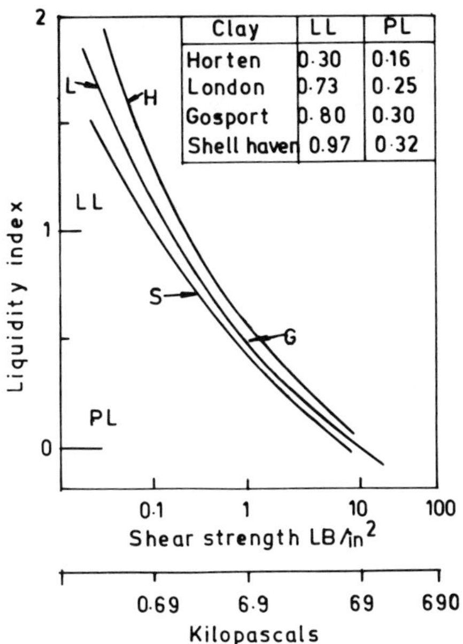

FIGURE 10. LI vs. remoulded shear strength.

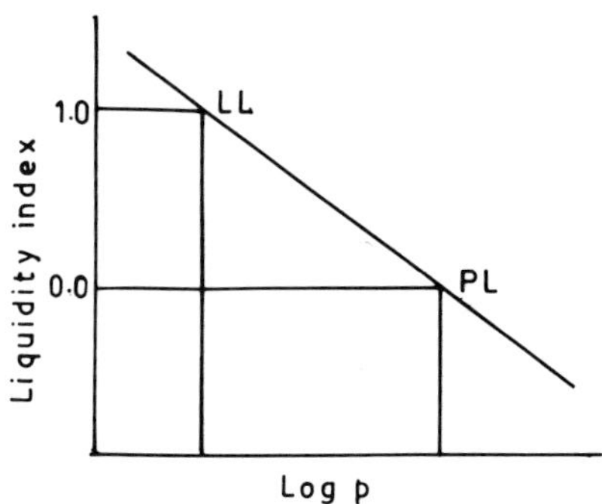

FIGURE 11. Linear idealisation.

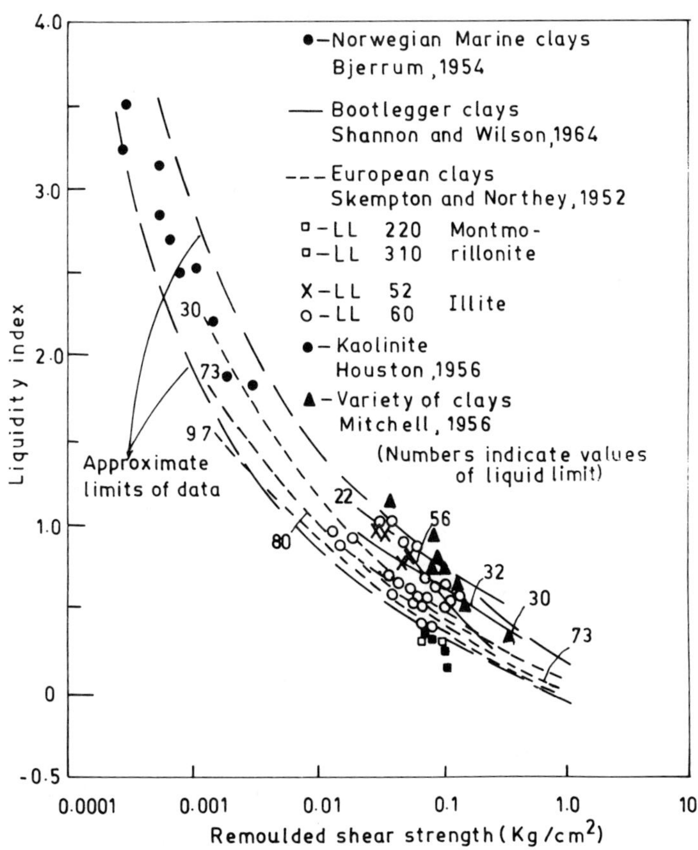

FIGURE 12. LI vs. remoulded strength (Houston and Mitchell, 1969).

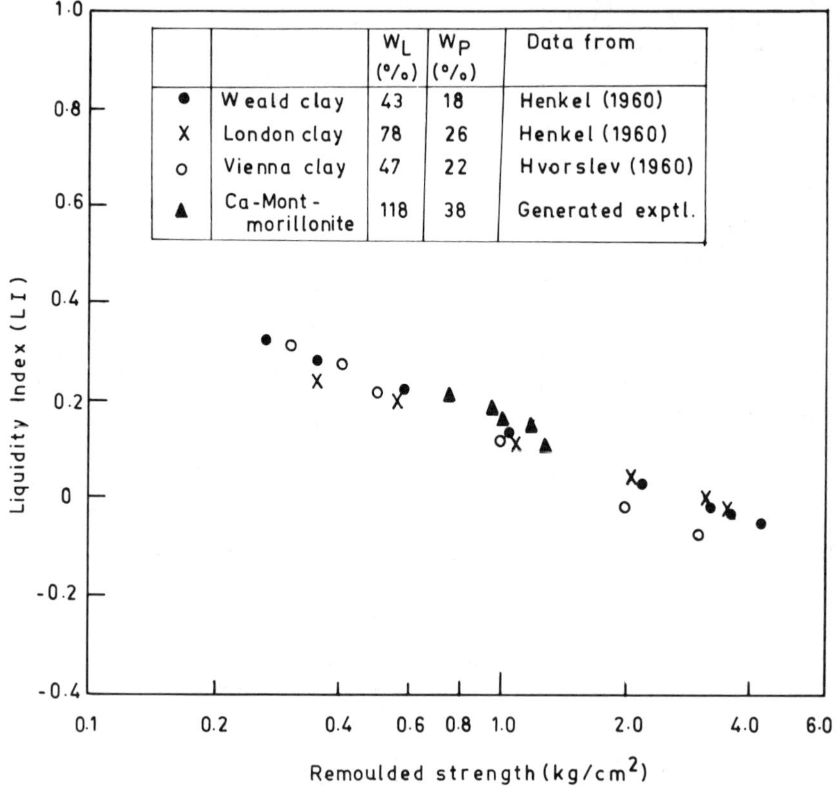

FIGURE 13. Remoulded strength (additional data).

From micro-mechanistic considerations, liquidity index is a function of the average half space distance between the particles [Equation (27)]. It has been possible to show that w/w_L versus the effective consolidation pressure is unique for all soils [Equations (16) and (17)]. Since p and q can be uniquely related, it implies that w/w_L versus shear strength should be a unique relation. This type of possible relation has been brought out by generating experimental relations of the type of Equation (21) in Figure 9. This aspect has been examined in detail by Nagaraj et al. [66] and it has been concluded that the plot of w/w_L versus logarithm of shear strength of all soils lies in a narrow band and within the limits of accuracy at engineering level can be assumed to be a straight line. Extending this concept further, a generalised critical state line and unique $e/e_L - p - q$ surface has been proposed.

Now, since w/w_L can be uniquely related to d, it follows that liquidity index is a function of (w/w_L). This can be further shown as follows:

$$LI = \frac{w - w_p}{w_L - w_p} = \frac{w/w_L - w_p/w_L}{1 - w_p/w_L} \tag{39}$$

Now, for an assumed unique ratio of shear strengths at plastic and liquid limits, w_p/w_L which is equal to d_p/d_L is constant. Thus,

$$LI = f(w/w_L) \tag{40}$$

This establishes that the correlations between liquidity index and remoulded strengths follows a consistant micro-mechanistic behaviour of unique d versus log q relation.

Relation between C_u/p and I_p

The other empirical relation between shear strength and index property is the one suggested by Skempton [83] based on his own and Henkel's work. This relation is of the form:

$$C_u/p = 0.11 - 0.0037\ I_p \tag{41}$$

where C_u is the undrained shear strength, p is the effective overburden pressure and I_p is the plasticity index in percent.

Several investigators [7,29,104,43,44,28,70,54,8,6,51,55, 42,53,36,87] have examined the applicability of the above relationship. The analysis of the above studies indicates that C_u/p could increase or decrease or even remain constant with increase in plasticity index. There is no unanimity amongst various investigators. It is attempted here to examine the validity of such a relation from micro-mechanistic considerations.

It has been shown, from micro-mechanistic considerations that d versus log p and d versus log q are unique relations and are independent of soil type. C_u is a reflection of q and therefore for a given average half space distance, d, which reflects different void ratios at macro-level depending on the soil type, the ratio of C_u/p for uncemented soils is independent of the soil type, i.e.,

$$d = a - b \log p \tag{12}$$

$$d = a' - b \log C_u \tag{42}$$

Subtracting Equation (12) by (42),

$$(a' - a) = b \log (C_u/p) \tag{43}$$

or

$$C_u/p = 10^{(a'-a)/b} = \text{constant} \tag{44}$$

Now the right-hand side of Equation (41) contains I_p which is a variable with S as per Equation (24). This clearly brings out that a correlation of the type of Equation (41) is micro-mechanistically not tenable. It is likely that some nonparticulate strengths in the soil and the fabric effects might have reflected in the variation in Equation (41). This is also the reason for not getting unanimity amongst various investigators.

Similar attempts made [43] to relate the angle of internal friction with plasticity index or liquid limit have not produced any conclusive evidence for the existence of such a relation. This is obvious because ϕ represents only C_u/p.

CONCLUDING REMARKS

It was realised by Terzaghi, as early as 1926, that consistency limits depend precisely on the same physical factors which determine the resistance and permeability of soils only in a far more complex manner. From the standpoint of assessment of soil properties, at the same time, it was recognised that if several soils with similar geologic origins had identical liquid limits, then their mechanical properties too would be identical. The fond hope of realising the predictive capability of consistency limits of soils was re-emphasized by Schofield and Wroth [77]. To quote, "If we have a simple laboratory with only a water supply, a drying oven, a balance and a simple indentation test equipment (such as the falling cone test) we can find the value of λ for silty clay soil".

This chapter is an attempt to detail how far the above objectives can be met with, in a simple way without sacrificing the needed rationality to enhance the credibility. Consistency limits have stood the test of time (1911–1985).

It is possible that some parts of this chapter might be sketchy due to unavoidable constraints. Still, the coverage provided has been to the desired extent to ensure that the important concepts do not become masked or lost.

ACKNOWLEDGEMENTS

The authors wish to place on record their sincere thanks to Prof. A. Sridharan, Chairman, and Smt. A. Vatsala, Scientific Assistant, Department of Civil Engineering, Indian Institute of Science, for helpful suggestions and critical review of the manuscript.

REFERENCES

1. Abdun-Nur, E. A., "Plastic Limit–Comparison of Cube and Standard Thread Test Methods," *ASTM, Spl. Tech. Pub. 254,* pp. 212–216 (1959).
2. Atterberg, A., "The Behaviour of Clays with Water, Their Limits of Plasticity and Their Degrees of Plasticity," *Kungliga lantbruksakademiens Handluigar Och Tidskrift, Vol. 50,* No. 2, pp. 132–158 (1911).
3. Babcock, K. L., "Theory of the Soil Colloidal Systems at Equilibrium," *Hilgardia, Vol. 34,* pp. 417 (1963).
4. Ballard, G. E. H. and W. F. Weeks, "Human Factor in Determining Plastic Limit of Cohesive Soils," *Material Res. and Standards, ASTM, Vol. 3,* No. 9, pp. 726–729 (1963).
5. Bauer, E. E., "History and Development of the Atterberg Limits Tests," *Papers on Soils, 1959 Meetings, ASTM, STP-254,* pp. 160–167 (1959).
6. Bishop, A. W. and D. J. Henkel, *The Measurement of Soil Properties in the Triaxial Test,* Edward Arnold, London, Second Ed.
7. Bjerrum, L., *Theoretical and Experimental Investigation on the Shear Strength of Soils,* Norwegian Geotech. Inst., Pub. No. 5 (1954).
8. Bjerrum, L. and N. E. Simons, "Comparison of Shear Strength Characteristics of Normally Consolidated Clays," *Proc. Conf. on Shear Strength of Cohesive Soils,* Boulder, Colorado, pp. 711–726 (1960).
9. Bolt, G. H., "Physico-Chemical Analysis of the Compressibility of Pure Clays," *Geotechnique, Vol. 6,* pp. 86–93 (1956).
10. Bolt, G. H., *Soil Chemistry A.–Basic Elements,* Ed. G. H. Bolt and M. G. M. Bruggenwert, Elsevier Sci. Pub. Co. NY 11 (1976).
11. Bolt, G. H. and R. D. Miller, "Compression Studies of Illite Suspensions," *Proc. Soil Science Soc. of Amer., Vol. 19,* pp. 285–288 (1955).

12. Bolt, G. H. and R. D. Miller, "Calculation of Total and Component Potentials for Water and Soil," *Trans. Am. Geophysical Union, Vol. 39,* No. 5 (1958).
13. British Standards Institution 1967, "Methods of Testing Soils for Civil Engineering Purposes," Part C, BS 1377, London.
14. British Standards Institution 1975, "Methods of Testing Soils for Civil Engineering Purposes," Part C, BS 1377, London.
15. Campbell, D. J., "Liquid Limit Determination of Arable Top Soils Using a Drop Cone Penetrometer," *J. of Soil Science, Vol. 26,* pp. 234-240 (1975).
16. Campbell, D. J., "Plastic Limit Determination Using a Drop Cone Penetrometer," *J. of Soil Science, Vol. 27,* pp. 295-300 (1976).
17. Campbell, D. J., J. V. Stafford, and P. S. Black, "The Plastic Limit as Determined by the Drop Cone Test, in Relation to the Mechanical Behaviour of Soil," *J. of Soil Science, Vol. 31,* pp. 11-24 (1980).
18. Casagrande, A., "Research on the Atterberg Limits of Soils," *Public Roads, Vol. 12,* No. 8, pp. 121-136 (1932).
19. Casagrande, A., "Classification and Identification of Soils," *Trans. ASCE, Vol. 113,* pp. 901-991 (1948).
20. Casagrande, A., "Notes on the Design of Liquid Limit Device," *Geotechnique, Vol. 8,* pp. 84-91 (1958).
21. Chowdari, A. R. and A. S. Balasubramaniam, "Deformation and Strength Characteristics of Soft Bangkok Clay," *Jl. Geotech. Div. ASCE, Vol. 104,* GT9, pp. 1153-1167 (1978).
22. Clayton, C. R. I. and A. W. Jukes, "A One Point, Cone Penetrometer Liquid Limit Test?" *Geotechnique, Vol. 28,* pp. 469-478 (1978).
23. Cooling, L. F. and W. H. Ward, "Some Examples of Foundation Movements Due to Other Than Structural Loads," *Proc. 2nd Int. Conf. SM & FE, Vol. 2,* pp. 162-167 (1948).
24. Cooper, J. H. and K. A. Johnson, "A Rapid Method of Determining the Liquid Limit of Soils," Lab. Report No. 83, State of Washington, Dept. of Highways (1950).
25. Eden, W. J., "Use of One Point Liquid Limit Procedure," ASTM, *Spl. Tech. Pub. 254,* pp. 168-177 (1959).
26. Ekman, V. E., "Determining the Lower Plastic Limit of Soil by Pressing," *Soils and Foundations (Translations),* No. 2, pp. 74-77.
27. Feda, J., *Mechanics of Particulate Materials—The Principles, Elsevier Scientific Publications,* Amsterdam (1982).
28. Golder, H. G., and R. A. Spence, Author's Reply on "Engineering Properties of the Marine Clay at Port Mann," *Proc. 14th Can. Soil Mech. Conf. Ottawa,* p. 153 (1960).
29. Grace, H. and J. K. M. Henry, Discussion on "The Planning and Design of New Hong Kong Airport," *Proc. Inst. of Engg. London, Vol. 7,* pp. 305 (1957).
30. Haines, W. B., "The Volume Changes Associated with Variations of Water Content in the Soil," *J. of Agricultural Science, Vol. 13,* pp. 296-310 (1923).
31. Hammer, M. J., and O. B. Thompson, "Foundation Clay Shrinkage Caused by Large Trees," *J. of SM & FE Dn. ASCE, Vol. 92,* SM-6, pp. 1-17 (1966).
32. Hansbo, S., "A New Approach to the Determination of the Shear Strength of Clay by Fall Cone Test," *Royal Swedish Geotech. Inst. Proc.,* No. 14, pp. 1-48 (1957).
33. Henkel, D. J., "The Shear Strength of Saturated Remoulded Clays," *Res. Conf. on Shear Strength of Cohesive Soils,* Colorado, pp. 533-554 (1960).
34. Houlsby, G. T., "Theoretical Analysis of the Fall Cone Test," *Geotechnique, Vol. 32,* No. 2, pp. 111-118 (1982).
35. Houlsby, G. T., "Discussion on Cone Penetrometer and Liquid Limit by Wood," *Geotechnique, Vol. 33,* No. 4, p. 463 (1983).
36. Houston, W. N. and J. K. Mitchell, "Property Interrelationships in Sensitive Clays," *Jl. of SM & FE Div. ASCE, Vol. 95,* SM 4, pp. 1037-1062 (1969).
37. Hovayni, P., "A New Grooving Tool," *Geotechnique, Vol. 8,* p. 79 (1958).
38. Hvorslev, M. J., "Physical Components of the Shear Strength of Saturated Clays," *Proc. Res. Conf. on Shear Strength of Cohesive Soils,* Colorado, pp. 169-213 (1960).
39. John Olsson, "Method for Investigating the Strength Properties of Clays as Used by Swedish State Railways," *Tekn. Ukebel, 340,* 41:37:310-311 (1921).
40. Jurgenson, L., Discussion on "Comparison of Liquid Limit Values Determined According to Casagrande and Vasilev," by Skopek and Ter-Stephanian, *Geotechnique, Vol. 26,* p. 539 (1976).
41. Karlsson, R., "Suggested Improvements in the Liquid Limit Test with Reference to Flow Properties of Remoulded Clays," *Proc. 5th Int. Conf. on SM & FE. Parts, Vol. 1/29,* p. 171 (1961).
42. Karlsson, R. and L. Viberg, "Ratio of C/p′ in Relation to Liquid Limit and Plasticity Index with Special Reference to Swedish Clays," *Proc. Geotech. Conf. Oslo, Vol. 1* (1967).
43. Kenney, T. C., Discussion on "Glacial Lake Clays," by T. H. Wu, *Jl. of SM & FE Div. ASCE, Vol. 85,* No. SM 3, pp. 67-69 (1959).
44. Kenney, T. C., "Correspondence on Fundamental Shear Strength Properties of Lilla Edet Clay," *Geotechnique, Vol. 11,* pp. 54-56 (1961m).
45. Kenney, T. C., "Atterberg Limits, Correspondence," *Geotechnique, Vol. 13,* pp. 159-162 (1963).
46. Kingery, W. D. and J. Francl, "Fundamental Study of Clay," XII Drying Behaviour and Plastic Properties, *J. Ame. Ceramic Soc., Vol. 37,* pp. 596-602 (1954).
47. Klausner, Y., Shainberg, I., "Consolidation Properties of Adsorbed Montmorillonites," *Proc. 4th Asian Reg. Conf. on SM & FE, Bangkok,* pp. 371-378 (1971).
48. Laboratoire Central Des Ponts et. Chaussees, *Determination Repride des Limitees d'Atterberg a Laide dun Penetrometre et.d. un Picnometre d'an,* Paris, Dossier SGR/149 (1966).
49. Lambe, T. W., "The Structure of Inorganic Soil," *Proc. ASCE,* Seperate, pp. 315 (1953).
50. Lauritzen, C. W., "Apparent Specific Volume and Shrinkage Characteristics of Soil Materials," *Soil Science, Vol. 65,* pp. 155-179 (1948).
51. Leonards, C. A., "Engineering Properties of Soils," *Founda-*

tion Engineering, G. A. Leonards, Ed., McGraw Hill Book Co.

52. Leroueil, et. al., "Behaviour of Destructed Natural Clays," *Jl. of Geotechn. Engg. Div. ASCE, Vol. 105,* No. GT 6, pp. 759–780 (1979).
53. Lumb, P. and J. K. Holt, "The Undrained Strength of a Soft Marine Clay from Hong Kong," *Geotechnique, Vol. 18,* pp. 25-36 (1968).
54. Metcalf, J. B. and J. B. Townsend, "A Preliminary Study of the Geotechnical Properties of Varved Clays as Reported in Canadian Engineering Case Records," *Proc. 14th Canad. Soil Mech. Conf. Ottawa,* pp. 203 (1960).
55. Milligan, V., L. G. Doderman, and A. Rutka, "Experience with Canadian Varved Clays," *Jl. of SM & FE, Div., ASCE, Vol. 88,* No. SM 4, p. 31 (1962).
56. Mitchell, J. K., *Fundamentals of Soil Behaviour,* John Wiley, New York.
57. Mohan, D., and R. K. Goel, Correspondence, *Geotechnique, Vol. 9,* pp. 144–145 (1959).
58. Moore, D. G. and A. F. Richards, "Conversion of Relative Shear Strength Measurements by Arrhenius on East Pacific Deep Sea Cores, to Conventional Units of Shear Strength," *Geotechnique, Vol. 12,* pp. 55–59 (1962).
59. Nagaraj, T. S. and Jayadeva, M. S., "Reexamination of One Point Methods of Liquid Limit Determination," *Geotechnique, Vol. 31,* No. 3, pp. 413–425 (1981).
60. Nagaraj, T. S. and M. S. Jayadeva, "Critical Reappraisal of Plasticity Index of Soils," *J. of GT, Div. ASCE, Vol. 109,* No. 7, pp. 994–1000 (1983).
61. Nagaraj, T. S. and B. R. Srinivasamurthy, "Rationalisation of Skempton's Compressibility Equation," *Geotechnique, Vol. 33,* No. 4, pp. 433–443 (1983).
62. Nagaraj, T. S. and B. R. Srinivasa Murthy, "Critical Reappraisal of Compression Index Equations," under publication, *Geotechnique.*
63. Nagaraj, T. S. and B. R. Srinivasa Murthy, "Prediction of Compressibility of Over Consolidated Uncemented Soils," *Tech. Note,* under publication, ASCE, GT Div.
64. Nagaraj, T. S. and B. R. Srinivasa Murthy, "Prediction of Preconsolidated Pressure and Recompression Index of Soils," *Geotechnical Testing J., ASTM, Vol. 8,* No. 4, pp. 199–203 (1985).
65. Nagaraj, T. S., B. R. Srinivasa Murthy, and Bindu Madhava, "Liquid Limit Determination–Further Simplified," *Geotechnical Testing Journal, ASTM, Vol. 10,* No. 3, pp. 142–145.
66. Nagaraj, T. S., Srinivasa Murthy, B. R. and A. Vatsala, "Can Cam Clay Model be Generalized?" *J. of Geotechnical Engineering, ASCE, Vol. 114,* No. 5 (1988).
67. Norman, L. E. J., "A Comparison of Values of Liquid Limits Determined with Apparatus Having Bases of Different Hardness," *Geotechnique, Vol. 8,* pp. 79–83 (1958).
68. Nuyens, J. G. E. and R. F. Kockaerts, "Reliable Technique for Determining Plastic Limit," *Materials Research Standards, Vol. 7,* pp. 295–299 (1967).
69. Olmstead, F. R. and C. M. Johnston, "Rapid Methods for Determining Liquid Limits of Soils," *Highway Research Board Bulletin 95,* Washington, pp. 27–35 (1955).
70. Osterman, J., "Notes on the Shearing Resistance of Soft Clays," *Acta Polytechnique,* Scandia, Ser. Ci.2. Stockholm (263/1959).
71. R machandran, V. S., K. P. Kacker, and H. A. B. Rao, "Determination of Liquid Limit of Soils by Dye Adsorption," *Soil Science, Vol. 95,* pp. 414 (1963).
72. Rosenquist, I. Th., "Investigations in the Clay-Electrolyte-Water System," *Norwegian Geotech. Inst. Pub. No. 9,* Oslo, p. 125 (1955).
73. Russel, E. R. and J. L. Mickle, "Liquid Limit Values by Soil Moisture Tension," *J. of SM & FE Div., Vol. 96,* No. SM 3, pp. 967–989 (1970).
74. Russel, E. J. and E. W. Russel, *Soil Conditions and Plant Growth,* Longman, London.
75. Rutledge, P. C., "Review of Cooperative Triaxial Research Programme of the Corps of Engineers," Progress Report on Soil Mech. Fact Finding Survey Waterways Expt. Statn. Mississippi, 1-78.
76. Scherrer, H. V., "Determination of Liquid Limit by Static Cone Penetration Test," *Proc. 5th Int. Conf. Vol. 1,* pp. 319-322 (1961).
77. Schofield, A. N. and C. P. Wroth, *Critical State Soil Mechanics,* McGraw Hill, London (1968).
78. Seed, H. B., J. K. Mitchell, and C. K. Chang, "Swell and Swell Pressure Characteristics of Compacted Clays," *H.R.B. Bulletin,* 313, pp. 12–39 (1962).
79. Shainberg, I., E. Bresler, and Y. Kalusner, "Studies of Na-Ca Montmorillonite Systems. I–The Swelling Pressure," *Soil Science Jl., Vol. 101,* pp. 214–219 (1971).
80. Sherwood, P. T. and M. D. Ryley, "An Investigation of a Cone Penetrometer Method for the Determination of the Liquid Limit," *Geotechnique, Vol. 20,* pp. 203–208 (1970).
81. Skempton, A. W., "Notes on Compressibility of Clays," Qrly. *Jl. of Geo. Soc.,* London, pp. 57–61 (1944).
82. Skempton, A. W., "Colloidal Activity of Clays," *Proc. 3rd Int. Conf. on SM & FE, Vol. 1,* pp. 57–61 (1953).
83. Skempton, A. W., Discussion on "The Structure of Inorganic Soil," *ASCE, GT Div., Vol. 80,* pp. 19–22 (1954).
84. Skempton, A. W. and R. D. Northey, "The Sensitivity of Clays," *Geotechnique, Vol. 3,* pp. 30–53 (1953).
85. Skopek, J. and G. Ter-Stepanian, "Comparison of Liquid Limit Values Determined According to Casagrande and Vasilev," *Geotechnique, Vol. 25,* pp. 135–136 (1975).
86. Sowers, G. G., A. Vesic, and M. Grandolfi, "Penetration Tests for Liquid Limit," *ASTM, STP* 254, pp. 216–226.
87. Sridharan, A. and S. Narasimha Rao, "The Relationship Between Undrained Strength and Plasticity Index," *Geotechnical Engineering, Vol. 4,* No. 1, pp. 41–53 (1973).
88. Sridharan, A. and M. S. Jayadeva, "Double Layer Theory and Compressibility of Clays," *Geotechnique, Vol. 32, No. 2,* pp. 133–144 (1982).
89. Stefanov, G., Discussion on "Liquid Limit," *Proc. 4th Int. Conf. SM & FE, Vol. 3,* p. 97.

90. Terzaghi, K., "Simplified Soil Tests for Subgrades and Their Physical Significance," *Public Road* (1926).
91. Terzaghi, K. and R. B. Peck, *Soil Mechanics in Engineering Practice,* John Wiley and Sons, Inc., NY, pp. 729 (1948).
92. Uppal, H. L. and H. R. Agarwal, "A New Method of Determining the Liquid Limit of Soils," *Road Research Paper 19,* New Delhi.
93. U. S. Waterways Experiment Station, Correlation of Soil Properties with Geologic Information Report No. 1, Simplification of the Liquid Limit Test Procedure, *Tech. Memo.* 3-286, Vicksberg (1949).
94. Vasilev, A. M., "Basic Principles of Methods and Techniques of Laboratory Determination of Physical Soil Properties," *Pochvovedenie,* 11, pp. 675–676 (1949).
95. van Olphen, H., *An Introduction to Clay-Colloid Chemistry,* Inter-Science Publishers, New York.
96. Verwey, E. J. W. and J. Th. G. Overbeek, *Theory of the Stability of Lyophobic Colloids,* Elsevier Pub. Co., NY (1948).
97. Warkentine, B. P., "Interpretation of the Upper Plastic Limit of Clays," *Nature, Vol. 190,* pp. 287–288 (1961).
98. Warkentine, B. P., G. H. Bolt, and R. D. Miller, "Swelling Pressure of Montmorillonite," *Soil Science Soc. of Ame. Proc. Vol. 21,* pp. 495–497 (1957).
99. Whyte, I. L., "Soil Plasticity and Strength—A New Approach Using Extrusion," *Ground Engg. Vol. 15,* No. 1, pp. 16–24 (1982).
100. Wroth, C. P., "Correlations of Some Engineering Properties of Soils," *Second Int. Conf. on Behaviour of Offshore Structures, Vol. 2,* London, pp. 121–132 (1979).
101. Wroth, C. P. and D. M. Wood, "Correlation of Index Properties with Some Basic Engineering Properties of Soils," *Canad. Geotech. Jl., Vol. 15,* No. 2, pp. 137–145 (1978).
102. Wood, D. M., "Cone Penetrometer and Liquid Limit," *Geotechnique, Vol. 32,* No. 2, pp. 152–157 (1982).
103. Wood, D. M., Reply to Discussion on "Cone Penetrometer and Liquid Limit," *Geotechnique, Vol. 33,* No. 1, pp. 80 (1983).
104. Wu, T. H., "Geotechnical Properties of Glacial Lake Clays," *Jl. ASCE, SM & FE Div., Vol. 84,* SM 3, No. 1732 (1958).
105. Yong, R. N. and T. S. Nagaraj, "Investigation of Fabric and Compressibility of Sensitive Clay," *Int. Symp. on Soft Clay,* AIT Bangkok, pp. 327–334 (1977).
106. Youssef, M. S., A. H. El Ramali, and M. El-Demery, "Relations Between Shear Strengths, Consolidation, Liquid Limit and Plastic Limit for Remoulded Clays," *Proc. 6th Int. Conf. on SM & FE, Vol. 1,* pp. 126–129 (1965).

CHAPTER 3

Plastic Behavior of Reinforced Earth

ANDRZEJ SAWICKI*

INTRODUCTION

Reinforced earth was invented by H. Vidal in 1966 as a new composite material formed by the association of a soil (usually sand) and metal strips. Other types of reinforcement like membranes, grids and recently geotextiles are also used in the civil engineering practice. The reinforced earth technique is used today in a variety of applications ranging from retaining structures to the strengthening of a soil beneath footings. The essential feature of reinforced earth behavior is the existence of friction between both constituents. By means of friction, the reinforcement imposes some constraints on the soil deformation. The constrained deformation causes the self-equilibrated stress state in reinforced earth. The reinforcement works in tension, and the soil is subjected to additional compression in the direction of reinforcement. A result of that is the increase in the soil strength. The behavior of reinforced earth can be studied using various approaches which depend on the objectives of the investigation. An extensive review of the literature, as well as an introduction to the engineering analyses of reinforced earth is presented by Ingold [2].

One of the most effective methods in soil mechanics is limit analysis which allows one to study the plastic collapse of earth structures in a relatively simple manner. These methods are based on the rigid-plastic models of earth materials [1]. More advanced methods, but also more complicated, deal with the elasto-plastic behavior of soils.

In the present work these two approaches are applied to the analysis of reinforced earth treated as a two-phase composite material. One of the fundamental problems appearing in the theory of composites is the description of the macro-behavior of the composite knowing the mechanical properties and volume fractions of constituents, and also the geometrical arrangement. The purpose of the approach presented is primarily the macroscopic description of the plastic behavior of reinforced earth and, at this level, it is not necessary to understand problems involving the local interactions between the soil and the reinforcement, even if these effects are important. The approach presented, as any theory, has its limitations, and local effects should be studied using other techniques.

The proposed approaches to the analysis of reinforced earth describe the basic features governing the plastic behavior of reinforced earth. Only one essential feature is not included i.e., a slippage between the reinforcement and the soil; this problem has been studied by Naylor and Richards [6] by a finite element approach. Slippage and edge effects both play an important role in the analysis of reinforced earth.

In the first part of this elaboration the rigid-plastic model of reinforced earth is developed, and then, the applications of the proposed model are given. The latter deals with the analysis of the bearing capacity of a footing on reinforced earth, and with the upper and lower bound estimates of the bearing capacity of a slope. In the second part, the elasto-plastic model of reinforced earth is presented.

RIGID-PLASTIC MODEL

This section deals with the plastic behavior of soil that is unidirectionally reinforced by fibers. The two failure mechanisms of reinforced earth have been determined from experimental data [5,12]. The first mechanism depends on the simultaneous plastic flow of both the soil and the reinforcement. In the second mechanism the reinforcement remains rigid, but the soil becomes plastic. Therefore, either there is slippage on the interfaces between the soil and the

*Institute of Hydroengineering, IBW PAN,ul. Cystersow 11, 80-953 Gdansk-Oliwa, Poland

reinforcement, or the reinforcement works as a rigid constraint to inhibit the plastic flow of the soil. This analysis considers reinforced earth as macroscopically anisotropic and homogeneous, and it describes its gross behavior based on a knowledge of the mechanical properties and interactive contribution of each component.

Basic Relations

The soil is assumed to be a perfectly plastic Coulomb–Mohr material, and the associated flow rule is assumed to be valid. The reinforcement is assumed to work only in one direction. Both components are assumed to work together, i.e., slippage on the interfaces between the soil and the reinforcement is neglected. It is also assumed that both constituents coexist at every point of the material. The thickness of the fibers is neglected, and they may work only in pure tension or compression. Following the above assumptions, three stress tensors are defined at every point; one tensor of macrostresses $\boldsymbol{\sigma}$ and two tensors of microstresses, $\boldsymbol{\sigma}^s$ and $\boldsymbol{\sigma}^r$, in the soil and reinforcement, respectively (Figure 1). These tensors are related by

$$\boldsymbol{\sigma} = \eta_s \boldsymbol{\sigma}^s + \eta_r \boldsymbol{\sigma}^r \tag{1}$$

where η_s and η_r are volumetric fractions of the soil and reinforcement, respectively. In indicial notation Equation (1) takes the form

$$\sigma_{ij} = \eta_s \sigma_{ij}^s + \zeta \eta_r R n_i n_j \tag{2}$$

where R is the plastic locus for the reinforcement in extension, n_k are the components of the vector $\mathbf{n}$ describing the direction of the fibers, and $\zeta \varepsilon \langle -1, \xi \rangle$ is a parameter describing the behavior of the reinforcement. The parameter ξ characterizes the plastic limit of the reinforcement provided that the reinforcement acts in compression. The value of $\xi = 0$ corresponds to the case where the reinforcement does not work in compression. If $\zeta \varepsilon (-1, \xi)$ the reinforcement remains rigid.

The soil is assumed to obey the Coulomb–Mohr yield condition:

$$f^s = (\sigma_{11}^s - \sigma_{22}^s)^2 - (\sigma_{11}^s + \sigma_{22}^s)^2 \sin^2 \phi + 4(\sigma_{12}^s)^2 = 0 \tag{3}$$

where ϕ is the angle of internal friction of the soil. The analysis is performed under the assumption that the stress σ_{33}^s does not influence the plastic flow of the soil. Using Equation (2) one can express the yield condition given by Equation (3) in terms of principal macrostresses and parameters describing both the orientation and the state of the reinforcement:

$$f = (\sigma_1 - \sigma_2)^2 - (\sigma_1 + \sigma_2)^2 \sin^2 \phi + (\zeta \sigma_0 \cos \phi)^2 - 2\zeta\sigma_0[(\sigma_1 - \sigma_2)\cos 2\alpha - (\sigma_1 + \sigma_2)\sin^2 \phi] = 0 \tag{4}$$

where α is the angle between the fiber direction and the direction of the principal microstress σ_1 (Figure 1), and $\sigma_0 = \eta_r R$.

Substitution of $\zeta = -1$ into Equation (4) gives

$$f_{SR} = (\sigma_1 - \sigma_2)^2 - (\sigma_1 + \sigma_2)^2 \sin^2 \phi + \sigma_0^2 \cos^2 \phi + 2\sigma_0[(\sigma_1 - \sigma_2)\cos 2\alpha - (\sigma_1 + \sigma_2)\sin^2 \phi] = 0 \tag{5}$$

which represents a family of hyperbolas in the σ_1, σ_2 plane. The conditions that the energy dissipated during plastic flow should be non-negative, $\sigma_{ij}\dot{\varepsilon}_{ij} \geqslant 0$, takes the form:

$$(\sigma_1^2 + \sigma_2^2)\cos^2 \phi - 2\sigma_1\sigma_2(1 + \sin^2 \phi) + \sigma_0[(\cos 2\alpha - \sin^2 \phi)\sigma_1 - (\cos 2\alpha + \sin^2 \phi)\sigma_2] \geqslant 0 \tag{6}$$

where the strain rates, $\dot{\varepsilon}_{ij}$, were calculated using the associated flow rule:

$$\dot{\varepsilon}_{ij} = \dot{\lambda} \frac{\partial f}{\partial \sigma_{ij}} \tag{7}$$

which is assumed as the most simple case of a flow rule. Equation (6) admits only some branches of hyperbolas (4), and it will be shown that, in regions which do not satisfy Equation (6), the second mechanism of failure occurs.

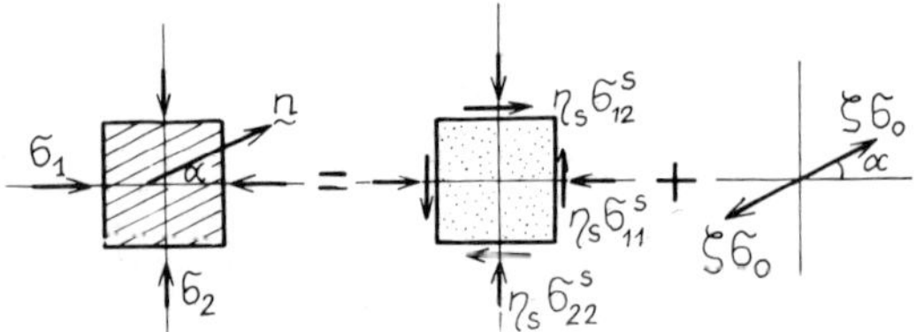

FIGURE 1. Illustration of macro- and microstresses in reinforced earth.

Rigid Reinforcement—Soil Becomes Plastic

Following the alternative assumption, the starting condition is

$$\dot{\varepsilon}_n^s = \dot{\lambda}^s \frac{\partial f^s}{\partial \sigma_{ij}^s} n_i n_j = 0 \tag{8}$$

which means that plastic flow of the soil cannot occur in the fiber direction. Equations (2), (3) and (8) give

$$\zeta = (\sigma_0 \cos^2 \phi)^{-1}[\sigma_1(\cos 2\alpha - \sin^2 \phi) - \sigma_2(\cos 2\alpha + \sin^2 \phi)] \tag{9}$$

Substitution of Equation (9) into Equation (4) gives

$$f_s = \sin^2 \alpha \cos^2 \alpha(\sigma_1 - \sigma_2)^2 - \sin^2 \phi(\sigma_1^2 \sin^2 \alpha + \sigma_2^2 \cos^2 \alpha) = 0 \tag{10}$$

which represents two straight lines in the σ_1, σ_2 plane. These lines are tangent to the hyperbolas given by Equation (5) at the points of intersection of these hyperbolas with the line bounding the region. The global yield condition is the convex combination of the yield conditions given by Equations (5) and (10). The shape of the global yield curve depends on the orientation of the fiber direction with respect to the axes of the principal macrostresses. Figure 2(b) shows a typical yield curve calculated for $\alpha = 30°$, and Figure 2(a) presents the case where $\alpha = 0$, where Equation (5) represents straight lines and Equation (11) reduces to the form $\sigma_2 = 0$.

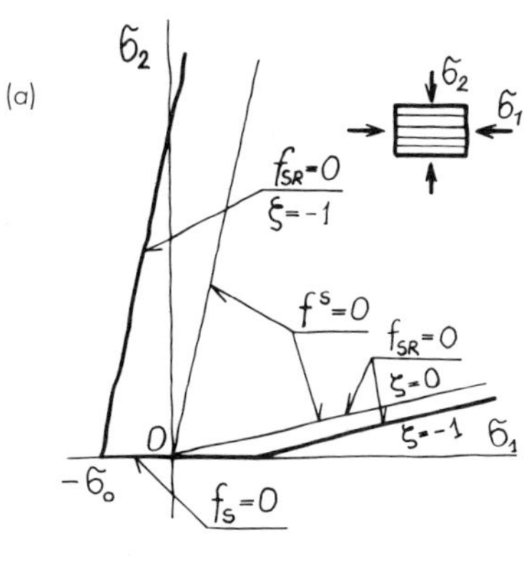

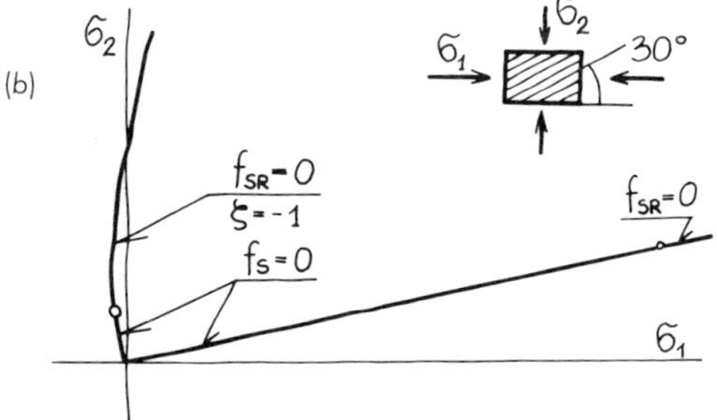

FIGURE 2. Yield conditions.

BEARING CAPACITY OF FOOTING—STATIC SOLUTION

In this Section, an application of the rigid-plastic model of reinforced earth to the analysis of bearing capacity of a strip footing will be shown. Figure 3 shows schematically a long strip foundation pressed by a force P per unit length into a subsoil reinforced by fibers in the x direction. Our problem is the determination of the force P which will cause failure of the subsoil. Solution to this problem will be obtained using the limit analysis theorems, which may be employed to obtain upper and lower bounds of the collapse load, [1].

Simple Lower-Bound Estimate

The statically admissible stress field satisfies:

(a) the equilibrium equations,
(b) stress boundary conditions,
(c) nowhere violates the yield criterion.

The lower-bound theorem can be formulated as follows: If a statically admissible stress field can be found for the problem under consideration, the plastic flow will not occur at a lower load.

There is no restriction regarding the shape of a yield criterion, so the lower-bound theorem is also applicable to anisotropic materials. We shall apply this theorem to obtain a simple estimate of the bearing capacity of a footing. Experimental results show that, during the failure process of such a footing, simultaneous plastic flow of both the soil and the reinforcement occurs in some regions of the subsoil, [3,4,14]. Therefore, the yield condition (5) will be applied, and a simple statically admissible stress field is shown in Figure 4.

This symmetrical "two-legged" system consists of two intersecting strips I and I′, in which uniaxial stress states prevail. There are two unknowns: the stress k and the angle β characterizing the inclination of the strips. Both strips cross under the footing, creating the triangular area II in which a biaxial stress state exists; this is characterized by stress components p and q. It follows from the boundary condition that p is the normal stress under the footing. There

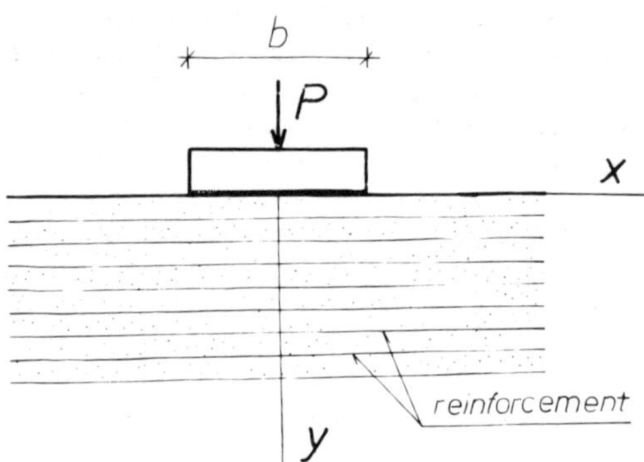

FIGURE 3. Footing on reinforced earth.

are four unknowns (k β, p, q) and four equations to determine these unknowns. Two are equations of equilibrium, and the remaining two are obtained by satisfying the yield condition (5) in regions I and II. Solving this system of equations gives the following simple formula for the limit stress under the footing

$$p = \frac{\sigma_0}{2}(1 + \sqrt{\sin \phi})^2 \frac{1 + \sin \phi}{1 - \sin \phi} \tag{11}$$

On the basis of the lower-bound theorem, the exact value of the limit stress is greater than the value obtained from Equation (11). Comparative values calculated by using the formula given by Stefani and Long [14]

$$p = \frac{\sigma_0}{2}\sqrt{\frac{1 + \sin \phi}{1 - \sin \phi}} \frac{\pi}{1 - \left(\frac{\pi}{2} - \phi\right) \tan \phi} \tag{12}$$

are greater than those obtained from Equation (11). For

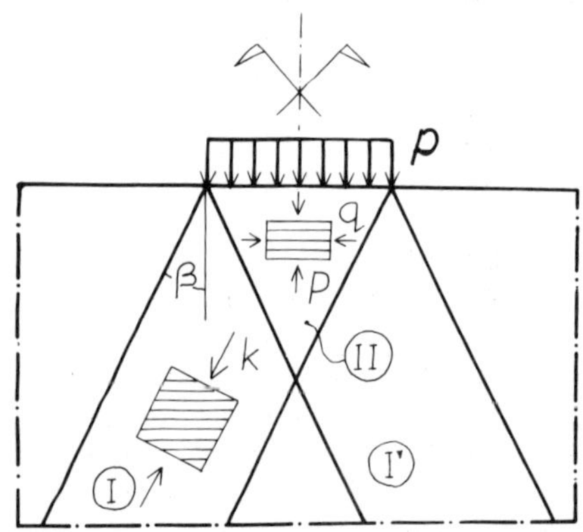

FIGURE 4. Statically admissible stress field under a footing.

example, substituting $\phi = 37°$ gives $p = 6.3\sigma_0$ and $q = 10.4\sigma_0$ for Equations (11) and (12), respectively.

The formula (12) gives values of the limit stress greater than experimental ones, [14]. The solution presented shows only one possibility of the rigid-plastic model of reinforced earth, and serves as an example how to apply the lower-bound theorem in a simple manner. In the next part of this Section we shall present a more elaborated solution to our problem, first given by Kulczykowski [4]. For the sake of simplicity the weight of the soil will be neglected.

General Concept of a Static Solution

It follows from experimental results [3,4] that, at the instant of failure, one can distinguish some regions, in which the reinforcement works in a specific way. These regions are schematically marked in Figure 5.

Region "a": Simultaneous plastic flow of the soil and the reinforcement occurs. In this region, the yield condition (5) is valid: $f_{SR} = 0$.

Region "b": Plastic flow of the soil occurs, but the reinforcement remains rigid and works in tension. The yield condition (10) is valid: $f_S = 0$.

Region "c": Only the plastic flow of the soil occurs. The reinforcement does not work, or works in compression. We can treat this region like a soil without reinforcement: $f = 0$.

The regions "a" and "b" are separated by the line l_1, on which $\zeta = -1$. There is $\zeta = 0$ on the line l_2 separating the regions "b" and "c". We shall write down the respective yield conditions in the x, y co-ordinates, (Figure 3):

$$\begin{aligned} f_{SR} &= (\sigma_x - \sigma_y + \sigma_0)^2 \\ &\quad - (\sigma_x + \sigma_y + \sigma_0)^2 \sin^2 \phi + 4\tau_{xy}^2 = 0 \end{aligned} \tag{13}$$

$$f_S = \tau_{xy}^2 - \sigma_y^2 \tan^2 \phi = 0 \tag{14}$$

$$f = (\sigma_x - \sigma_y)^2 - (\sigma_x + \sigma_y)^2 \sin^2 \phi + 4\tau_{xy}^2 = 0 \tag{15}$$

The equilibrium equations and respective boundary conditions complete the system of static equations:

$$\frac{\partial \sigma_x}{\partial x} + \frac{\partial \tau_{xy}}{\partial y} = 0, \quad \frac{\partial \tau_{xy}}{\partial x} + \frac{\partial \sigma_y}{\partial y} = 0 \tag{16}$$

$$\begin{aligned} 0 \leq x < b/2&: \sigma_x = 0, \quad \sigma_y = p, \quad \tau_{xy} = 0 \\ x > b/2&: \sigma_x = 0, \quad \sigma_y = q, \quad \tau_{xy} = 0 \end{aligned} \tag{17}$$

where q is a surcharge of the subsoil. We assume the solution to be symmetric with respect to the y axis, so further con-

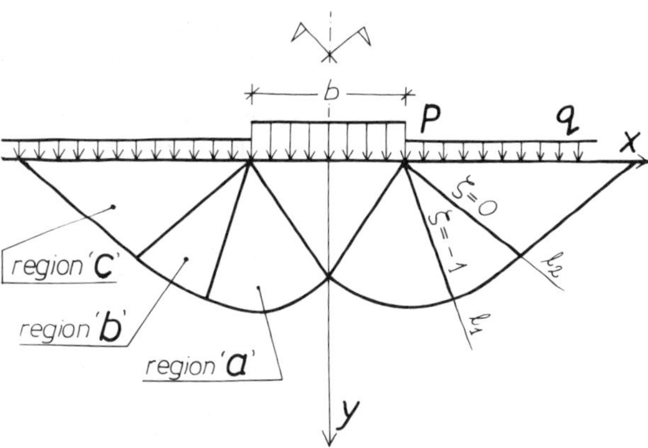

FIGURE 5. Failure zones in reinforced subsoil.

siderations will be performed for $x \geq 0$. The system of Equations (13)–(17) will be solved with the help of the method of characteristics. Applications of the method of characteristics to the soil mechanics problems are presented in [13].

Solution by the Method of Characteristics

We restrict our attention to the regions "a", "b" and "c", in which different yield conditions are valid. Neglecting details, see [4,13], we shall present only the basic results. The following notation is useful:

$$\sigma = \frac{1}{2}(\sigma_x + \sigma_y + \zeta\sigma_0) \tag{18}$$

where $\zeta = -1$ in region "a", $\zeta = 0$ in region "c", and

$$\zeta = \frac{1}{\sigma_0}\left[\sigma_x - \frac{1 + \sin^2\phi}{\cos^2\phi}\sigma_y\right] \quad \text{in region "b"}$$

We denote by β an angle between the x axis and the direction of the bigger principal stress σ_1, (Figure 6). The macrostresses in reinforced earth can be expressed by σ and β. For example, the following relations are valid in the region "c":

$$\begin{aligned} \sigma_x &= \sigma(1 + \sin\phi\cos 2\beta) + \zeta\sigma_0 \\ \sigma_y &= \sigma(1 - \sin\phi\cos 2\beta) \\ \tau_{xy} &= \sigma\sin\phi\sin 2\beta \end{aligned} \tag{19}$$

It is easy to check that relations (19) satisfy the yield condition (15). Substitution of Equations (19) into the equilibrium conditions (16) gives the hyperbolic system of two quasi-linear differential equations with respect to σ and β:

$$\begin{aligned} &(1 + \sin\phi\cos 2\beta)\frac{\partial\sigma}{\partial x} + \sin\phi\sin 2\beta\frac{\partial\sigma}{\partial y} \\ &- 2\sigma\sin\phi\sin 2\beta\frac{\partial\beta}{\partial x} + 2\sigma\sin\phi\cos 2\beta\frac{\partial\beta}{\partial y} = 0 \\ &\sin\phi\sin 2\beta\frac{\partial\sigma}{\partial x} + (1 - \sin\phi\cos 2\beta)\frac{\partial\sigma}{\partial y} \\ &+ 2\sigma\sin\phi\cos 2\beta\frac{\partial\beta}{\partial x} + 2\sigma\sin\phi\sin 2\beta\frac{\partial\beta}{\partial y} = 0 \end{aligned} \tag{20}$$

The solution to Equations (20) is represented by the two families of characteristics given by the following equations:

The first family:

$$\frac{dy}{dx} = \tan(\beta + \varepsilon), \quad d\sigma + 2\sigma\tan\phi d\beta = 0 \tag{21}$$

The second family:

$$\frac{dy}{dx} = \tan(\beta - \varepsilon), \quad d\sigma - 2\sigma\tan\phi d\beta = 0 \tag{22}$$

where

$$\varepsilon = \frac{\pi}{4} - \frac{\phi}{2} \tag{23}$$

The same equations are valid in the region "a". In the region "b" we have the following differential equations for

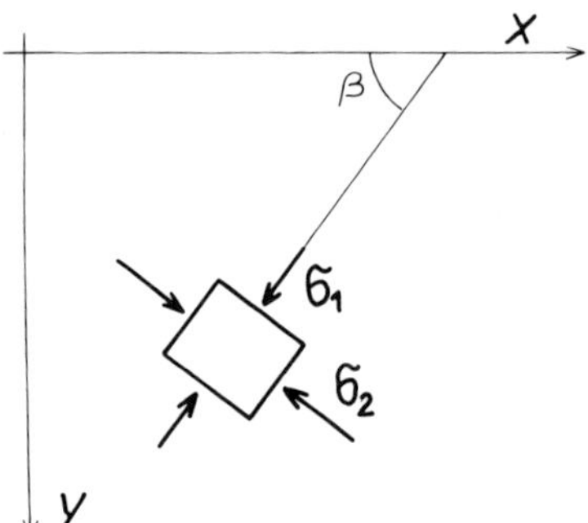

FIGURE 6. Notation.

characteristics:

The first family:

$$\frac{dy}{dx} = \frac{1}{\tan\phi}, \quad \sigma = C_1 \tag{24}$$

The second family:

$$\frac{dy}{dx} = 0, \quad \sigma = -\sigma_0\beta + C \tag{25}$$

where C_1 and C are constants.

Figure 7 presents the net of characteristics described by Equations (21)–(25). In the triangle ABC the state of passive earth pressure exists. Characteristics are straight lines inclined to the x axis by an angle ε. In this region there is

$$\beta = 0 \quad \text{and} \quad \sigma = q/(1 - \sin\phi) \tag{26}$$

which means that

$$\begin{aligned} \tau_{xy} &= 0, \quad \sigma_y = q, \\ \sigma_x &= q(1 + \sin\phi)/(1 - \sin\phi) \end{aligned} \tag{27}$$

In the area ACD, which also belongs to the region ''c'', the first family of characteristics is represented by straight lines, the second one by logarithmic spirals. The parameter β changes along the logspiral characteristics, taking the value 0 in the case of passive earth pressure (AC), and $\beta = (\pi/4) - (\phi/2)$ on the sector AD. There is

$$\sigma = \frac{q}{1 - \sin\phi}\exp(2\beta\tan\phi) \tag{28}$$

The stress components can be computed using Equations (19).

In the region "b" (A′ADD′) we have a simple net of characteristics, which are straight lines. There is $\zeta = -1$ on the A′D′ characteristic, and $\zeta = 0$ on the AD characteristic. The parameter ζ changes along the horizontal characteristics, as well as the parameter σ, which is given by the following formula:

$$\sigma = \frac{q}{1 - \sin\phi}\exp\left[\left(\frac{\pi}{2} - \phi\right)\tan\phi\right] - \zeta\sigma_0 \tag{29}$$

It follows from Equation (29) that the change of σ does not depend on the x and y co-ordinates, so the length L of horizontal characteristics can be optional. In the extreme case, if $L \to 0$, the area ADA′D′ becomes the line, on which:

$$\begin{aligned} \zeta = 0:\ \sigma_x &= \sigma|_{\zeta=0} \times (1 + \sin^2\phi) \\ \sigma_y &= \sigma|_{\zeta=0} \times \cos^2\phi \\ \tau_{xy} &= \sigma|_{\zeta=0} \times \sin\phi\cos\phi \\ \zeta = -1:\ \sigma_x &= (\sigma|_{\zeta=0} + \sigma_0)(1 + \sin^2\phi) - \sigma_0 \\ \sigma_y &= (\sigma|_{\zeta=0} + \sigma_0)\cos^2\phi \\ \tau_{xy} &= (\sigma|_{\zeta=0} + \sigma_0)\sin\phi\cos\phi \end{aligned} \tag{30}$$

The area A′D′E corresponds to the region ''a'' in which both the reinforcement and the soil become plastic. Characteristics, in this region, are the same as those for the Coulomb–Mohr material. The angle β changes from $\beta = (\pi/4) - (\phi/2)$ on the A′D′ characteristics to $\beta = (\pi/2)$ on the A′E sector, where the state of active earth pressure is attained. The parameter σ is given by

$$\sigma = \left\{\frac{q}{1 - \sin\phi} + \frac{\sigma_0}{\exp\left[\left(\frac{\pi}{2} - \phi\right)\tan\phi\right]}\right\}\exp(2\beta\tan\phi) \tag{31}$$

In the triangle A′F′E there is the state of active earth pressure, and we have:

$$\beta = \frac{\pi}{2} \quad \text{and} \quad \sigma = p/(1 + \sin\phi) \tag{32}$$

The sector A′E is a common characteristic for the regions A′F′E and A′D′E, so the substitution of Equation (32) into Equation (31) gives the limit value of vertical stress beneath

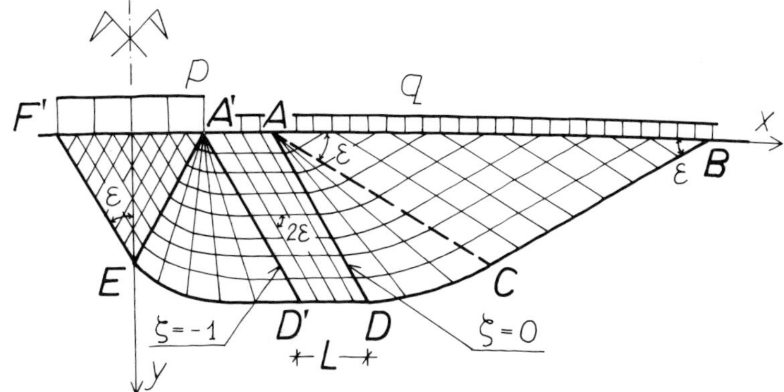

FIGURE 7. Stress characteristics.

a footing:

$$p = (1 + \sin\phi)\left\{\frac{q}{1 - \sin\phi} + \frac{\sigma_0}{\exp\left[\left(\frac{\pi}{2} - \phi\right)\tan\phi\right]}\right\}\exp(\pi\tan\phi) \quad (33)$$

The formula (33), for $q = 0$ and $\phi = 37°$, gives $p = 8.52\sigma_0$, which is a bigger value than that computed using Equation (11).

BEARING CAPACITY OF FOOTING—KINEMATIC SOLUTION

General Remarks

This Section is devoted to the determination of the kinematic solution to the problem of bearing capacity of a footing, [4]. A method of solution is based on the construction of a kinematically admissible velocity field, which must satisfy the assumed flow rule and respective velocity boundary conditions. The dissipation of energy associated with such a field cannot be negative. The collapse load is determined by equating of external to internal rate of work, for assumed failure mechanism. The solution obtained in such a way is an upper-bound estimate of the collapse load, as per the Upper-Bound Theorem: "If a kinematically admissible velocity field can be found, uncontained plastic flow must have taken place previously, [1]."

We shall present a method of solution which is based on the method of characteristics. Let us denote by $F = 0$ a yield condition for reinforced earth, where $F = f_{SR}$, f_S or f, see Equations (13)–(15), depending on the assumed mechanism of plastic flow. We have assumed the associated flow rule to be valid, i.e., plastic strain rates are defined as follows:

$$\dot{\varepsilon}_{ij} = \dot{\lambda}\frac{\partial F(\sigma_{ij})}{\partial \sigma_{ij}} \quad (34)$$

where $\dot{\lambda}$ is a positive scalar function. The strain rates are also given by

$$\dot{\varepsilon}_{ij} = -\frac{1}{2}\left(\frac{\partial v_i}{\partial x_j} + \frac{\partial v_j}{\partial x_i}\right) \quad (35)$$

where i are components of the velocity vector $\mathbf{v}$. Our example deals with the plane strain state, so Equation (35) is equivalent to the following equations:

$$\dot{\varepsilon}_x = -\frac{\partial v_x}{\partial x}$$

$$\dot{\varepsilon}_y = -\frac{\partial v_y}{\partial y} \quad (36)$$

$$\dot{\varepsilon}_{xy} = -\frac{1}{2}\left(\frac{\partial v_x}{\partial y} + \frac{\partial v_y}{\partial x}\right)$$

Substitution of Equation (35) into Equation (34), and elimination of a function $\dot{\lambda}$, gives the following formula:

$$\frac{\left(\frac{\partial v_x}{\partial x}\right)}{\left(\frac{\partial F}{\partial \sigma_x}\right)} = \frac{\left(\frac{\partial v_y}{\partial y}\right)}{\left(\frac{\partial F}{\partial \sigma_y}\right)} = \frac{\frac{1}{2}\left(\frac{\partial v_x}{\partial y} + \frac{\partial v_y}{\partial x}\right)}{\left(\frac{\partial F}{\partial \tau_{xy}}\right)} \quad (37)$$

which represents the system of two partial differential equations with respect to v_x and v_y. This system of equations has been solved by the method of characteristics, for the yield conditions given by Equations (13)–(15).

(a) The case $F = f_{SR}$ and $F = f$:

The first family of characteristics:

$$\frac{dy}{dx} = \tan(\beta + \varepsilon) \qquad dv_x + dv_y \cdot \tan(\beta + \varepsilon) = 0 \tag{38}$$

The second family:

$$\frac{dy}{dx} = \tan(\beta - \varepsilon) \qquad dv_x + dv_y \cdot \tan(\beta - \varepsilon) = 0 \tag{39}$$

(b) The case $F = f_S$:

The first family:

$$\frac{dy}{dx} = \frac{1}{\tan\phi} \qquad dv_y + dv_x \cdot \tan\phi = 0 \tag{40}$$

The second family:

$$\frac{dy}{dx} = 0, \quad dv_x = 0 \tag{41}$$

It follows from comparison of the first of Equations (21), (22), (24), (25) and Equations (38)–(41) that the net of velocity characteristics coincides with that of the stresses.

Velocity Field

Figure 8 shows a strip foundation moving downwards with a velocity v_0. Because of the symmetry with respect to the y axis, our considerations deal with $x \geqslant 0$. The rigid triangular wedge F′EA′ moves together with a foundation, with the same velocity v_0. In the area A′ED′ the plastic flow of both the soil and the reinforcement occurs. In the area ADCB the reinforcement does not work, but the soil becomes plastic. The area ADA′D′ represents a region in which the reinforcement remains rigid, and works in tension, but plastic flow of the soil occurs. The region ADA′D′ has been reduced to the sector. The characteristics BCDD′EF′ and A′E are the velocity discontinuity lines. The semi-infinite region below the characteristic BCDD′E remains at rest.

If the velocity discontinuity line belongs to the family of characteristics given by $dy/dx = \tan(\beta \pm \varepsilon)$, then

$$v = C \exp(\pm\beta \tan\phi) \tag{42}$$

On the characteristic A′E we have:

$$v = \frac{v_0}{2 \sin\varepsilon} \tag{43}$$

The angle between the direction of v and A′E is ϕ. On the characteristic ED′ there is:

$$v = \frac{v_0}{2\sin\varepsilon} \exp\left[\left(\frac{\pi}{2} - \beta\right)\tan\phi\right] \tag{44}$$

The sectors A′D′ and AD are not the velocity discontinuity lines. There is:

$$v = \frac{v_0}{2\sin\varepsilon} \exp\left[\left(\frac{\pi}{4} + \frac{\phi}{2}\right)\tan\phi\right] \tag{45}$$

and

$$v_x = v\cos\phi, \quad v_y = -v\sin\phi \tag{46}$$

In the area ADC velocity vectors are inclined at the angle ϕ against the logspiral characteristics. The velocity field is given by

$$v = \frac{v_0}{2\sin\varepsilon} \exp\left[\left(\frac{\pi}{2} - \beta\right)\tan\phi\right] \tag{47}$$

The triangular area ABC moves as a rigid body with the velocity

$$v = \frac{v_0}{2\sin\varepsilon} \exp\left(\frac{\pi}{2}\tan\phi\right) \tag{48}$$

which is inclined at the angle ϕ against the CB sector.

The obtained velocity field is kinematically admissible, and the rate of the energy dissipation is non-negative in the area of the plastic flow of reinforced earth.

Determination of the Collapse Load

The collapse load will be determined by equating of external to internal rate of work dissipated during the plastic flow. This balance equation takes the following form in the case of plane strain state:

$$\int_S \sigma_{ij}\dot{\varepsilon}_{ij} dS + \int_{l_v} \sigma_{ij} n_i [V_i] dl_v = \int_B T_i V_i dB \tag{49}$$

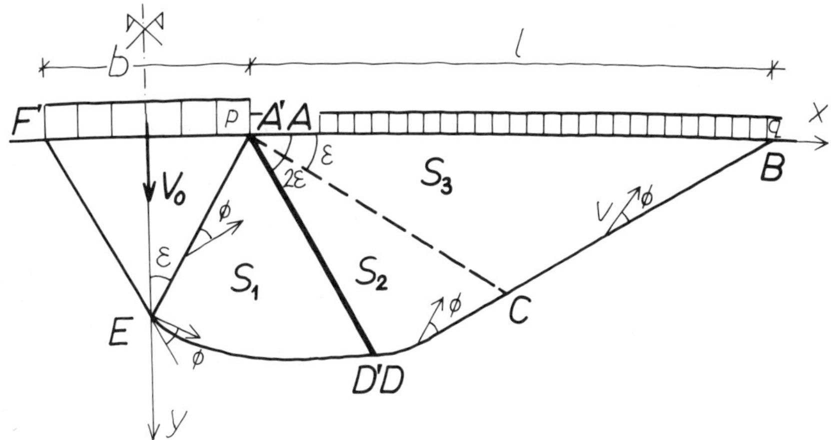

FIGURE 8. Kinematic solution.

where S is a region in which the plastic flow of subsoil occurs, n_i are components of a unit vector normal to the velocity discontinuity line, $[V_i]$ is the velocity jump on a discontinuity line, B denotes the boundary, T_i are the boundary forces, and V_i is the velocity field which satisfies the boundary conditions, l_v denotes the velocity discontinuity line.

The first integral on the left-hand side of Equation (49) denotes the rate of work dissipated in the subsoil during the plastic flow. The second integral represents the rate of work dissipated on the velocity discontinuity lines. The right-hand side of Equation (49) represents the rate of work done by external loads.

Equation (49) takes the following form, in the case studied:

$$pbv_0 + 2qlv_y|_{\beta=0} = 2\int_{S_1+S_2+S_3} \sigma_{ij}\dot{\varepsilon}_{ij}dS + 2\int_{\mathrm{F'ED'DCB}} \sigma_{ij}n_i[V_i]dl_v \tag{50}$$

where $v_y|_{\beta=0}$ is a vertical velocity of the edge AB. S_1, S_2, S_3 and 1 are marked in Figure 8. The second member on the left-hand side of Equation (50) is

$$2qlv_y|_{\beta=0} = -[qbv_0 \exp(\pi \tan\phi)]/\tan^2\varepsilon \tag{51}$$

Contribution of S_2 and S_3 into the first integral on the right-hand side of Equation (50) is zero, and the integral takes the following form:

$$\int_{S_1} \sigma_{ij}\dot{\varepsilon}_{ij}dS = \frac{\sigma_0 v_0 b}{8\sin^2\varepsilon} \times \left\{(\sin^2 2\varepsilon)\exp\left[\left(\frac{\pi}{2}+\phi\right)\tan\phi\right] - \sin^2\left(\frac{\pi}{2}+\varepsilon\right)\right\} \tag{52}$$

The second integral on the right-hand side of Equation (50) is following:

$$\int_{\mathrm{A'E+ED'}} \sigma_{ij}n_i[V_i]dl_v = v_0 b\sigma_0/4\tan^2\varepsilon + \frac{v_0 b\sigma_0}{8\sin^2\varepsilon} \times \left\{(\sin^2 2\varepsilon)\cdot\exp\left[\left(\frac{\pi}{2}+\phi\right)\tan\phi\right] - \sin^2\left(\frac{\pi}{2}+\varepsilon\right)\right\} \tag{53}$$

Substitution of Equations (51)–(53) into Equation (50) gives, after simple manipulations, the formula for the collapse load, which is identical to that given by Equation (33). The obtained result means that the formula (33) represents the exact solution to the problem of bearing capacity of a footing on reinforced earth.

The formula (33) is of a practical importance. Figure 9

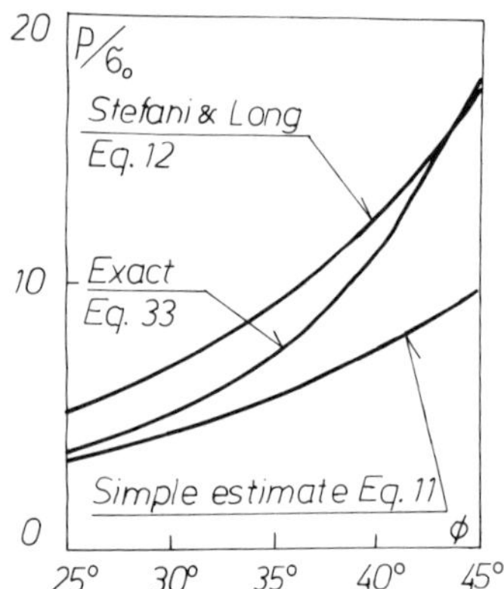

FIGURE 9. Bearing capacity of footing as a function of ϕ.

shows a comparison of results obtained using Equation (33) with those computed using Equations (11) and (12).

BEARING CAPACITY OF A SLOPE—SIMPLE ESTIMATES

From the view point of engineering practice it is not always necessary to know the exact solution to the problem of bearing capacity of earth structures, and the lower and upper bounds are often sufficient to estimate the collapse load. In order to obtain an exact solution one has to solve a system of partial differential equations, which is a rather laborious task. Mathematical methods employed to obtain the estimates are usually based on a simple algebra, which is an advantage in comparison with more advanced approaches.

The aim of this Section is to present some applications of simple techniques to obtain the lower and upper bound estimates of the bearing capacity of the slope (vertical cut) horizontally reinforced by fibers.

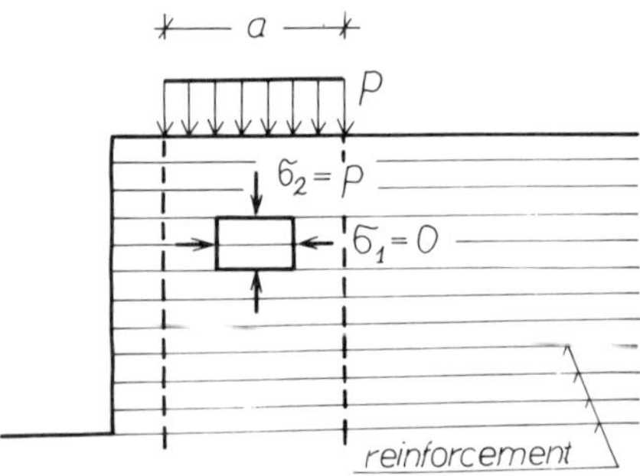

FIGURE 10. Simple statically admissible stress field.

Lower-Bound Estimates

Let us consider a weightless vertical cut of the horizontally reinforced earth, which is loaded by the pressure p uniformly distributed on the length a, Figure 10. The most simple statically admissible stress field is represented by a vertical column, in which a uniaxial stress state exists. The vertical stress is $\sigma_2 = p$, the horizontal stress component is zero. The assumed stress field satisfies the equilibrium conditions and stress boundary conditions. Let us assume that the soil and the reinforcement become plastic in this column, so the yield condition (5) is valid, and takes the following form for $\alpha = 0$:

$$(p^2 + \sigma_0^2)\cos^2\phi - 2\sigma_0 p(1 + \sin^2\phi) = 0 \quad (54)$$

There are two solutions to Equation (54):

$$p_{1,2} = \sigma_0\left(\frac{1 \pm \sin\phi}{\cos\phi}\right)^2 \quad (55)$$

The reinforcement has been assumed to work in tension, so the following condition must be fulfilled:

$$\dot{\varepsilon}_2 = \dot{\lambda}\frac{\partial f_{SR}}{\partial \sigma_2} < 0 \quad (56)$$

Substitution of Equation (5) into Equation (56), for $\sigma_2 = p$, gives the inequality:

$$\frac{1 + \sin^2\phi}{\cos^2\phi}p - \sigma_0 > 0 \quad (57)$$

which admits only one root of Equation (54), namely

$$p = \sigma_0\left(\frac{1 + \sin\phi}{\cos\phi}\right)^2 = \sigma_0\tan^2\left(\frac{\pi}{4} + \frac{\phi}{2}\right) = p_l \quad (58)$$

The simple formula (58) gives us a lower-bound estimate of the collapse load.

The second example of a lower-bound estimate deals with the determination of the critical height H_l of a vertical cut presented in Figure 11, where the statically admissible stress field is also shown. The critical height is defined here as the height at which the vertical cut will collapse due its own weight. The unit weight of reinforced earth is denoted by γ. The critical height H_l will be determined from the condition that the yield criterion (5) is fulfilled at the stress discontinuity line l. Substitution of $p = \gamma H_l$ into Equation (54), after simple computations, gives the following formula for a lower-bound of the critical height:

$$H_l = \frac{\sigma_0}{\gamma}\tan^2\left(\frac{\pi}{4} + \frac{\phi}{2}\right) \quad (59)$$

It can be easily shown that the yield condition (5) is not exceeded in the zones II and III, (Figure 11).

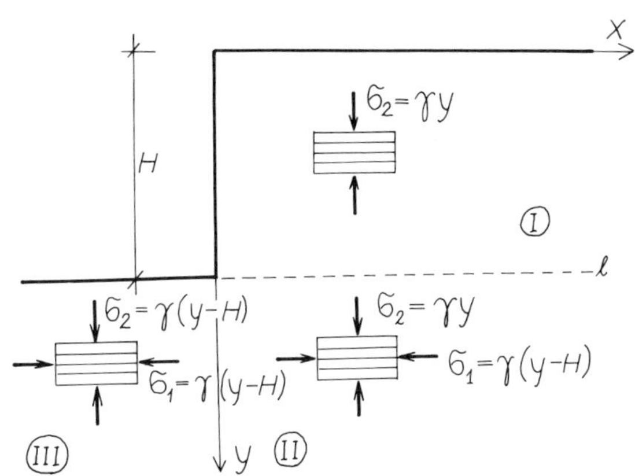

FIGURE 11. Stability of a vertical cut—statically admissible stress field.

Upper-Bound Estimate

Figure 12 shows a possible mechanism of failure of a vertical cut loaded by the pressure p. The assumed mechanism of failure appears in practice, [2]. The failure occurs by sliding of the rigid triangular chunk ABC along a plane making an angle β with the horizontal. A limiting condition is reached when the rate of work of the external load is equal to the rate of energy dissipation along the surface of sliding AB. The velocity of sliding is denoted by v. An angle between directions of the velocity and the sliding surface is ϕ. The rate of work done by the external load can be expressed as follows:

$$M = vpa \sin(\beta - \phi) \tag{60}$$

where β is an unknown angle.

The rate of energy dissipated along the velocity discontinuity line is equal to zero in the case of ideal (cohesionless) Coulomb–Mohr material. In the case of reinforced earth, the dissipation of energy along the sector AB depends on the state of reinforcement. Figure 13 shows a situation, when the reinforcement becomes plastic. In this case, the rate of energy dissipation is given by the formula:

$$D = \sigma_0 vb \cos(\beta - \phi)\tan \phi \tag{61}$$

The following formula for the collapse load is obtained by equating the right-hand sides of Equations (60) and (61):

$$p = \sigma_0 \frac{b}{a} \frac{\tan \beta}{\tan(\beta - \phi)} \tag{62}$$

There is $p = p_{min}$ for

$$\beta = \frac{\pi}{4} + \frac{\phi}{2} \tag{63}$$

Substitution of Equation (63) into Equation (62) gives an upper-bound for the collapse load:

$$p_u = \sigma_0 \frac{b}{a} \tan^2\left(\frac{\pi}{4} + \frac{\phi}{2}\right) \tag{64}$$

If there is $b = a$, then Equation (64) is identical to Equation (58), which is, in this case, the exact solution to the bearing capacity problem of a slope.

Applying the same procedure to the problem of a critical height of a slope, Figure 14, we have

$$H_u = \frac{2\sigma_0}{\gamma} \tan^2\left(\frac{\pi}{4} + \frac{\phi}{2}\right) = 2H_l \tag{65}$$

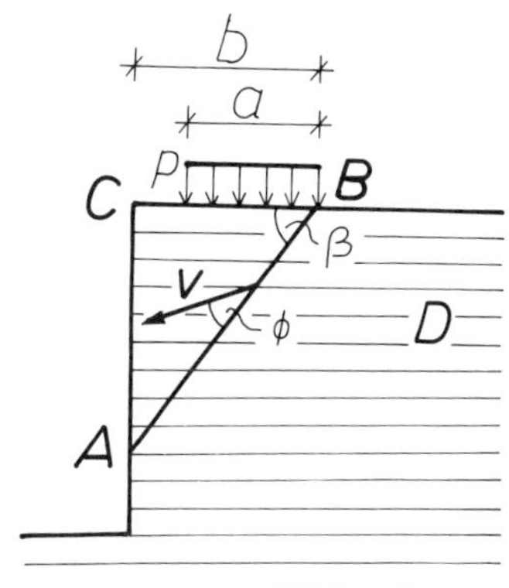

FIGURE 12. Kinematically admissible mechanism of failure.

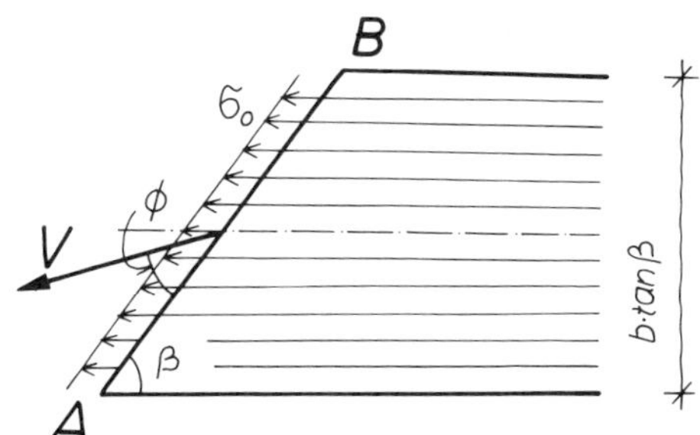

FIGURE 13. Failure along the sector AB.

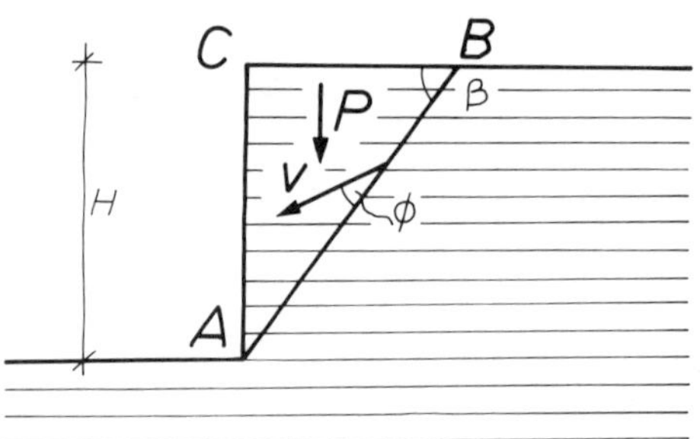

FIGURE 14. Stability of a vertical cut—kinematically admissible mechanism.

Example

The present example deals with the calculation of the critical height of a vertical cut horizontally reinforced by steel strips, these being 3.0 mm thick. The plastic limit of the strips in a simple extension test is $R = 353.16$ MPa. The width of a strip is 6.0 cm in the first version, and 8.0 cm in the second one. The strips are connected to the facing unit at 0.75 m horizontal and vertical centres. There is $\eta_r = 3.17 \times 10^{-4}$ in the first version, and $\eta_r = 4.23 \times 10^{-4}$ in the second one. σ_0 is 0.11 MPa and 0.15 MPa, respectively. The unit weight of reinforced earth is $\gamma = 16.58$ kN/m^3. The lower-bound estimate of the critical height [Equation (59)], for $\phi = 37°$, is 27.14 m and 36.18 m, in the first and the second version respectively.

AXISYMMETRICAL ELASTO-PLASTIC BEHAVIOR

Assumptions

The aim of the present Section is a macroscopic description of the behavior of an axisymmetrical specimen of soil, reinforced by membranes, through elastic and elasto-plastic states up to the limit state, [11].

The macrobehavior of the specimen is described in terms of the mechanical properties of the soil and the reinforcing membranes and of their volume fractions. The description is based on the continuum theory of composites proposed by the author [8,10] which does not distinguish between particles of the reinforcement and of the soil. The reinforced earth is treated as a macroscopically homogeneous composite, the behavior of which is described using macroscopic quantities such as macrostresses $\boldsymbol{\sigma}$ and macrostrains $\boldsymbol{\epsilon}$. These quantities are defined as respective averages over the representative elementary volume of the composite, as is widely used in the mechanics of composites. There appear also microscopic quantities, such as microstresses $\boldsymbol{\sigma}^{\alpha}$ and microstrains $\boldsymbol{\epsilon}^{\alpha}$, which are the averages over the volumes of respective constituents. The following assumptions are made:

(a) Soil and reinforcement exist simultaneously at every point of the composite.
(b) There is no relative motion between the constituents.
(c) Each constituent is an ideal elasto-plastic material and the associated flow rule is valid.

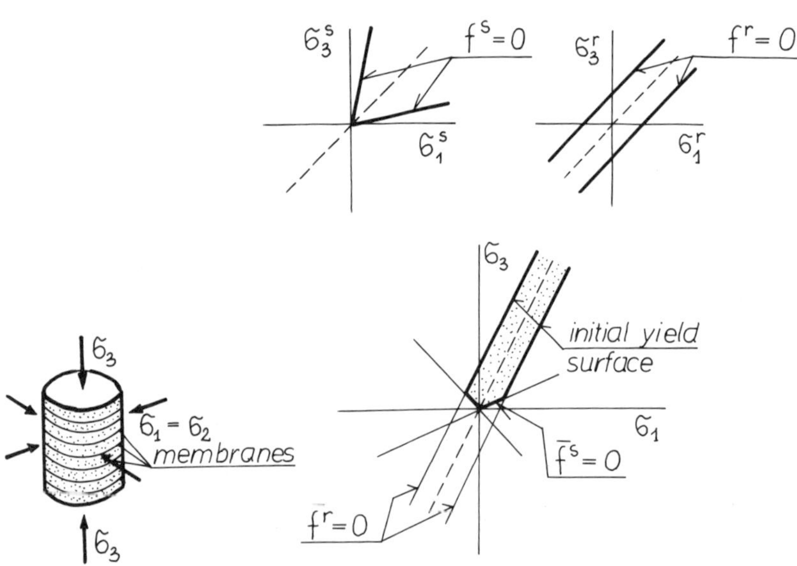

FIGURE 15. Transformation of yield conditions. Initial yield surface.

The example analysed is important from the view point of experimental verification, which may be performed using the standard techniques applied in geotechnical laboratories. The results obtained are compared with experimental data and a good agreement is obtained.

Elastic Behavior and Initial Yield Surface

The example analysed is presented in Figure 15. The specimen is subjected to the macrostress state $\boldsymbol{\sigma} = \{\sigma_1, \sigma_3\}^T$ which causes the respective microstress states in soil and reinforcement: $\boldsymbol{\sigma}^\alpha = \{\sigma_1^\alpha, \sigma_3^\alpha\}^T$. There is a corresponding macrostrain vector $\boldsymbol{\epsilon} = \{\varepsilon_1, \varepsilon_3\}^T$ with respective microstrain vectors $\boldsymbol{\epsilon}^\alpha = \{\varepsilon_1^\alpha, \varepsilon_3^\alpha\}^T$. Macrostrains describe the macrobehavior of the reinforced earth, which behaves as a homogeneous and anisotropic material. The following relations follow from the equilibrium and compatibility conditions respectively:

$$\sigma_3^s = \sigma_3^r = \sigma_3 \tag{66}$$

$$\varepsilon_1^s = \varepsilon_1^r = \varepsilon_1 \tag{67}$$

where superscripts r and s refer to the reinforcement and the soil respectively. The same basic relations are valid for stress and strain increments.

The basic formulae are the structural relations for stresses:

$$\boldsymbol{\sigma}^\alpha = \mathbf{B}^\alpha \boldsymbol{\sigma} = \left[\begin{array}{c|c} b_1^\alpha & b_2^\alpha \\ \hline 0 & 1 \end{array}\right] \begin{Bmatrix} \sigma_1 \\ \sigma_3 \end{Bmatrix} \tag{68}$$

where $\mathbf{B}^\alpha$ are matrices depending on the elastic moduli of constituents and of their volume fractions in reinforced earth. The detailed derivation of these matrices for layered composites is presented in [7]. The final form of the coefficients is

$$\begin{aligned} b_1^s &= \frac{E_s(1 - \nu_r)[\eta_s E_s(1 + \nu_r) + \eta_r E_r(1 + \nu_s)]}{\eta_s^2 E_s^2(1 - \nu_r^2) + \eta_r^2 E_r^2(1 - \nu_s^2) + 2\eta_s\eta_r E_s E_r(1 - \nu_s\nu_r)} \\ b_2^s &= \frac{\eta_r(\nu_s E_r - \nu_r E_s)[\eta_s E_s(1 + \nu_r) + \eta_r E_r(1 + \nu_s)]}{\eta_s^2 E_s^2(1 - \nu_r^2) + \eta_r^2 E_r^2(1 - \nu_s^2) + 2\eta_s\eta_r E_s E_r(1 - \nu_s\nu_r)} \\ b_1^r &= \frac{1}{\eta_r}(1 - \eta_s b_1^s) \\ b_2^r &= -\frac{\eta_s}{\eta_r} b_2^s \end{aligned} \tag{69}$$

where E_α, ν_α are Young's modulus and Poisson's ratio for the α constituent and η_α is the volume fraction of the α constituent in reinforced earth (recall $\eta_r + \eta_s = 1$).

The reinforcement is assumed to obey the Huber–Mises yield condition

$$f^r = (\sigma_1^r - \sigma_3^r)^2 - \sigma_0^2 = 0 \tag{70}$$

which represents two parallel straight lines in the plane of microstresses for the reinforcement. Here, σ_0 denotes the plastic limit in uniaxial extension. The soil is assumed to obey the Coulomb–Mohr yield condition

$$f^s = (\sigma_1^s - \sigma_3^s)^2 - (\sigma_1^s + \sigma_3^s)^2 \sin^2\phi = 0 \tag{71}$$

representing two straight lines intersecting at the zero point in the respective microstress plane, and where ϕ denotes the angle of internal friction. These yield conditions may be transformed from the respective microstress spaces into the space of macrostresses with the help of the structural relations (68), hence

$$\bar{f}^r = [b_1^r\sigma_1 + (b_2^r - 1)\sigma_3]^2 - \sigma_0^2 = 0 \tag{72}$$

$$\begin{aligned} \bar{f}^s = {} & [b_1^s\sigma_1 + (b_2^s - 1)\sigma_3]^2 \\ & - [b_1^s\sigma_1 + (b_2^s + 1)\sigma_3]^2\sin^2\phi = 0 \end{aligned} \tag{73}$$

These transformations are schematically presented in Figure 15. The initial yield surface of reinforced earth is composed of two sectors corresponding to the plastic state of soil and two parallel straight half-lines corresponding to the plastic state of reinforcement. The dotted region corresponds to the elastic states of both constituents.

Elasto-Plastic Behavior and Limit Surface

When the initial yield condition is attained, one of the constituents, or both of them, becomes plastic. If further loading proceeds the structural relations (68) do not hold. Let it be assumed that the initial yield condition (72) is reached; this means that the reinforcement becomes plastic first.

An arbitrary infinitesimal macrostress increment $d\boldsymbol{\sigma}$ is considered. If $d\boldsymbol{\sigma}$ is directed into the interior of the initial yield surface the process of unloading takes place. If $d\boldsymbol{\sigma}$ is tangent to the curve (72) a regrouping of elastic states in soil and reinforcement appears. The macrostress increment $d\boldsymbol{\sigma}$ outward to the initial yield surface causes the regrouping of elastic states in both constituents, as well as the appearance of plastic strains in the reinforcement. In this case, each microstress increment vector is tangent to the yield surface (70), i.e.

$$df^r = \frac{\partial f^r}{\partial \boldsymbol{\sigma}^r} \cdot d\boldsymbol{\sigma}^r = 0 \tag{74}$$

The strain increment in the reinforcement is expressed as follows:

$$d\boldsymbol{\epsilon}^r = \mathbf{M}^r d\boldsymbol{\sigma}^r + d\lambda^r \frac{\partial f^r}{\partial \boldsymbol{\sigma}^r} \tag{75}$$

In the soil, only the elastic strain increment appears:

$$d\boldsymbol{\epsilon}^s = \mathbf{M}^s\, d\boldsymbol{\sigma}^s \tag{76}$$

Here $\mathbf{M}^\alpha$ are matrices of elastic compliances, $d\lambda^r$ is a scalar function appearing in the associated flow rule. Using relations (66), (67) and (74)–(76) the incremental relations describing the elasto-plastic macrobehavior of reinforced earth are

$$\begin{Bmatrix} d\sigma_1^r \\ d\sigma_3^r \end{Bmatrix} = \begin{bmatrix} 0 & 1 \\ 0 & 1 \end{bmatrix} \begin{Bmatrix} d\sigma_1 \\ d\sigma_3 \end{Bmatrix} \tag{77}$$

$$\begin{Bmatrix} d\sigma_1^s \\ d\sigma_3^s \end{Bmatrix} = \begin{bmatrix} 1/\eta_s & -\eta_r/\eta_s \\ 0 & 1 \end{bmatrix} \begin{Bmatrix} d\sigma_1 \\ d\sigma_3 \end{Bmatrix} \tag{78}$$

$$\begin{Bmatrix} d\varepsilon_1 \\ d\varepsilon_3 \end{Bmatrix} = \frac{1}{\eta_s E_s} \begin{bmatrix} 1 - \nu_s & -\eta_r(1 - \nu_s) - \eta_s \nu_s \\ -\eta_r(1 - \nu_s) - 2\eta_s \nu_s & 3\eta_r\eta_s\nu_s + \eta_r^2(1 - \nu_s) + 2\eta_r\eta_s(E_s/E_r)(1 - 2\nu_r) + \eta_s^2 \end{bmatrix} \begin{Bmatrix} d\sigma_1 \\ d\sigma_3 \end{Bmatrix} \tag{79}$$

Analysis of the unloading process down to the zero macrostress state followed by further loading, leads to the equation of the actual yield surface

$$\hat{f}^r = [b_1^r(\sigma_1 - \Delta\sigma_1) + (b_2^r - 1) \times (\sigma_3 - \Delta\sigma_3)]^2 - \sigma_0^2 = 0 \tag{80}$$

$$\hat{f}^s = \left\{ b_1^s\sigma_1 + (b_2^s - 1)\sigma_3 + \frac{\eta_r}{\eta_s} \times [b_1^r\Delta\sigma_1 - (1 - b_2^r)\Delta\sigma_3] \right\}^2 - \left\{ b_1^s\sigma_1 + (b_2^s + 1)\sigma_3 + \frac{\eta_r}{\eta_s} \times [b_1^r\Delta\sigma_1 - (1 - b_2^r)\Delta\sigma_3] \right\}^2 \sin^2\phi = 0 \tag{81}$$

where $\Delta\sigma_i = \int_\Sigma d\sigma_1$ and Σ denotes the loading path in the macrostress space. The results are illustrated in Figure 16. It may be easily proved that the crossing points of the straight lines (80) and (81) lie on the following straight lines:

$$\mathfrak{F}_1 = \sigma_3 - \frac{1 + \sin\phi}{1 - (1 - 2\eta_r)\sin\phi} \times (\sigma_1 + \eta_r\sigma_0) = 0 \tag{82}$$

$$\mathfrak{F}_2 = \sigma_3 - \frac{1 - \sin\phi}{1 + (1 - 2\eta_r)\sin\phi} \times (\sigma_1 - \eta_r\sigma_0) = 0 \tag{83}$$

The lines (82) and (83) form the limit surface for reinforced earth. They form the envelope of stress states for which both the soil and the reinforcement become plastic, and plastic flow appears which satisfies the associated flow rule.

A similar analysis may be performed in the case when the soil first becomes plastic. In this case only the incremental relations differ from (77)–(79), but Equations (80)–(83) are still valid.

Experimental Results and Conclusions

The specimens of the soil reinforced by horizontal membranes were investigated in the triaxial compression test by Long *et al.* [5]. They and Schlosser [12] state that there is a good agreement between the experimental results and the following formula,

$$\sigma_3 - \frac{1 + \sin\phi}{1 - \sin\phi}(\sigma_1 + \eta_r\sigma_0) = 0 \tag{84}$$

which has been derived on the basis of the specimen's limit behavior [12], expressed with the notation $R_T/\Delta H$ instead of $\eta_r\sigma_0$. The formula (84) corresponds to Equation (82), and it is obtained from Equation (82) by assuming:

$$1 - (1 - 2\eta_r)\sin\phi \cong 1 - \sin\phi \tag{85}$$

The approximate Equation (85) follows from the fact that the parameter η_r is usually considerably less than unity, so Equation (84) is a special case of the more general Equation (82). Substituting $\eta_r = 1$ into Equation (82) gives the Huber–Mises yield condition. Substitution of $\eta_r = 0$ gives the Coulomb–Mohr yield condition.

The result obtained confirms the accuracy of the simple Schlosser–Long formula from the view point of continuum mechanics. The theory presented is more general, however, than the Schlosser–Long approach because it makes possible the analysis of reinforced earth in stages preceding the limit state. It seems that even in situations where the present approach is inadequate, it may serve as the first approximation or suggest alternative methods of analysis. It may be also a starting point for the problem of how to design the required macro-properties of reinforced earth.

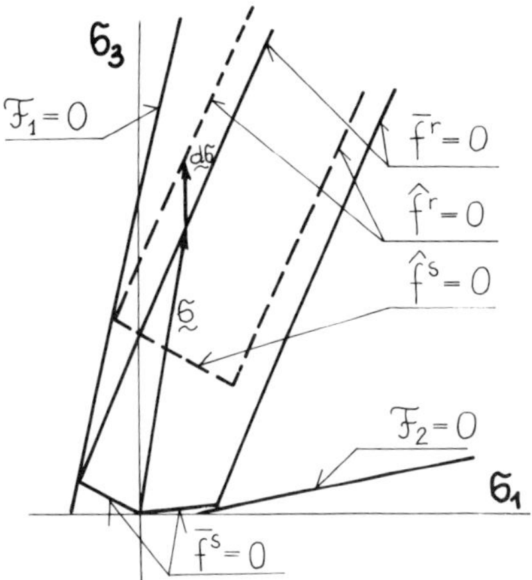

FIGURE 16. Axisymmetrical case—yield conditions.

ELASTO-PLASTIC BEHAVIOR: PLANE STRAIN STATE

Elastic Behavior

The present Section deals with an example of application of the continuum theory of composites [8,10] to the elasto-plastic analysis of fiber reinforced earth in a plane strain state. For simplification it is assumed that principal macrostresses coincide with the principal axes of macroscopic anisotropy. The assumptions made in the previous Section are valid. Consider the special case of reinforced earth behavior presented in Figure 17. The fiber direction coincides with the x_3 axis. From the equilibrium and compatibility conditions it follows:

$$\varepsilon^r_{33} = \varepsilon^s_{33} \tag{86}$$

$$\sigma^r_{11} = \sigma^s_{11}, \quad \sigma^r_{22} = \sigma^s_{22} \tag{87}$$

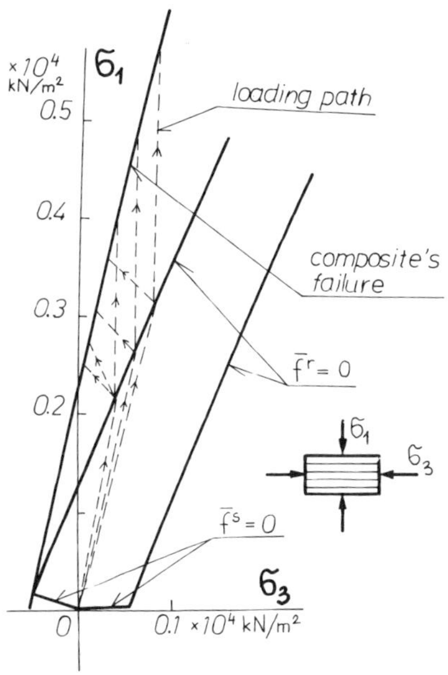

FIGURE 17. Plane strain—yield conditions.

The microstresses are determined by macrostresses in a unique way, with the help of structural relations

$$\begin{Bmatrix} \sigma_1^\alpha \\ \sigma_2^\alpha \\ \sigma_3^\alpha \end{Bmatrix} = \begin{bmatrix} 1 & & \\ & 1 & \\ b_1^\alpha & b_1^\alpha & b_2^\alpha \end{bmatrix} \begin{Bmatrix} \sigma_1 \\ \sigma_2 \\ \sigma_3 \end{Bmatrix} \tag{88}$$

where

$$b_1^r = \frac{v_s(\eta_r E_s/E_r - v_s)}{\eta_r + \eta_s E_s/E_r}$$

$$b_1^s = \frac{-\eta_r(v_r E_s/E_r - v_s)}{\eta_r + \eta_s E_s/E_r}$$

$$b_2^r = \frac{1}{\eta_r + \eta_s E_s/E_r} \tag{89}$$

$$b_2^s = \frac{E_s/E_r}{\eta_r + \eta_s E_s/E_r}$$

where E_α and v_α ($\alpha = r, s$) denote respectively the Young's modulus and the Poisson ratio for the α-th constituent. A plane strain condition is equivalent to the following relation:

$$\sigma_2 = m\sigma_1 + n\sigma_3 \tag{90}$$

where

$$m = \frac{v_s - \eta_r(v_s - v_r E_s/E_r)(1 + b_1^r)}{1 + \eta_r[E_s/E_r - 1 + b_1^r(v_s - v_r E_s/E_r)]}$$

$$n = \frac{v_2 - \eta_r(v_s - v_r E_s/E_r)b_2^r}{1 + \eta_r[E_s/E_r - 1 + b_1^r(v_s - v_r E_s/E_r)]} \tag{91}$$

Substitution of Equation (90) into Equation (88) yields

$$\begin{Bmatrix} \sigma_1^\alpha \\ \sigma_2^\alpha \\ \sigma_3^\alpha \end{Bmatrix} = \begin{bmatrix} 1 & 0 \\ m & n \\ b_1^\alpha(1 + m) & nb_1^\alpha + b_2^\alpha \end{bmatrix} \begin{Bmatrix} \sigma_1 \\ \sigma_3 \end{Bmatrix} \tag{92}$$

Initial Yield Surface

It is assumed that the reinforcement obeys the Huber–Mises yield condition:

$$f^r = (\sigma_1^r - \sigma_2^r)^2 + (\sigma_1^r - \sigma_3^r)^2 + (\sigma_2^r - \sigma_3^r)^2 - 2\sigma_0^2 = 0 \tag{93}$$

Substitution of Equation (92) into Equation (93), and after making some simplifications [10], gives the following formula

$$\bar{f}^r = \left\{\left(b_1^r - \frac{1}{2}\right)(1 + m)\sigma_1 + \left[\left(b_1^r - \frac{1}{2}\right) n + b_2^r\right]\sigma_3\right\}^2 - \sigma_0^2 = 0 \tag{94}$$

which represents two straight lines in the space of macrostresses. The soil is assumed to obey the Coulomb–Mohr yield condition (71), which takes the following form in the space of macrostresses:

$$\bar{f}^s = \{\sigma_1[1 - b_1^s(1 + m)] - \sigma_3[nb_1^s + b_2^s]\}^2 - \{\sigma_1[1 + b_1^s(1 + m)] + \sigma_3 \times [nb_1^s + b_2^s]\}^2 \sin^2\phi = 0 \tag{95}$$

representing the straight lines in this space. Equations (94) and (95) define the initial yield surface for reinforced earth, which is illustrated in Figure 17 for the following data:

$$E_r = 706.1 \times 10^5 \text{ kN/m}^2, \quad v_r = 0.34 \text{ (aluminum)}$$

$$E_s = 0.44132 \times 10^5 \text{ kN/m}^2, \quad v_s = 0.3$$

$$\eta_r = 0.01, \quad \eta_s = 0.99$$

The respective coefficients are

$$b_1^r = -27.95, \quad b_2^r = 94.172$$

$$b_1^s = 0.2823, \quad b_2^s = 0.0589$$

The initial yield surface shown in Figure 17 corresponds to macroscopic compression stresses. A reasonable engineer-

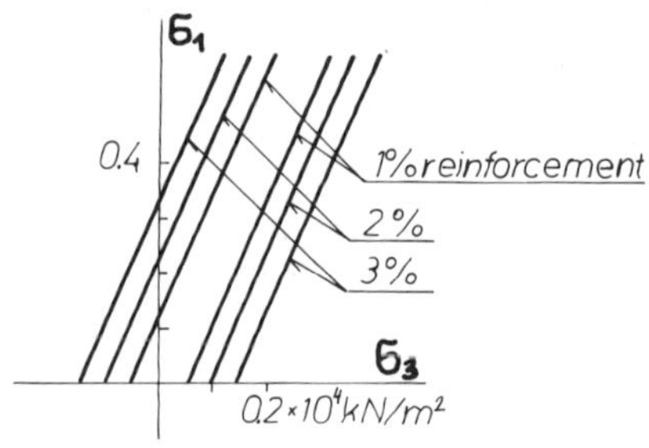

FIGURE 18. Influence of reinforcement on the initial yield surface.

ing approximation may be obtained by assuming that the quantity E_s/E_r is small in comparison with unity. In this case Equation (95) reduces to

$$\sigma_3 = 0 \tag{96}$$

In practice this result means that for almost all loading paths the reinforcement first becomes plastic. This conclusion is supported by experimental results [5]. In Figure 18 the influence of the amount of reinforcement on simplified initial yield surface is shown.

Elasto-Plastic Behavior

When the initial yield condition (94) is attained the reinforcement becomes plastic. It is useful to introduce a parameterization of the yield condition (93), such that

$$\begin{aligned} \sigma_1^r &= \frac{2}{3}\sigma_0 \cos \omega + \sigma_m \\ \sigma_2^r &= \frac{2}{3}\sigma_0 \cos\left(\omega - \frac{2}{3}\pi\right) + \sigma_m \\ \sigma_3^r &= \frac{2}{3}\sigma_0 \cos\left(\omega + \frac{2}{3}\pi\right) + \sigma_m \end{aligned} \tag{97}$$

where $\sigma_m = ⅓(\sigma_1^r + \sigma_2^r + \sigma_3^r)$ and ω is a parameter.

The following relations are valid in the case of elasto-plastic behavior:

and

$$\begin{Bmatrix} d\varepsilon_1 \\ d\varepsilon_3 \end{Bmatrix} = \left[\begin{array}{c|c} D_{11} & D_{12} \\ \hline D_{21} & D_{22} \end{array}\right] \begin{Bmatrix} d\sigma_1 \\ d\sigma_3 \end{Bmatrix} \tag{100}$$

where

$$D_{11} = \frac{\eta_r}{E_r}(1 - v_r M) + \frac{\eta_s}{E_s}(1 - v_s M) + \eta_r(\Omega_1 + M\Omega_2)\left(\frac{v_s}{E_s} - \frac{v_r}{E_r}\right) + \eta_r\Omega_1(X + YM)$$

$$D_{12} = -\frac{\eta_r v_r}{E_r}N(1 + \Omega_2) - \frac{v_s}{E_s} \times (\eta_s N + 1 - \eta_r\Omega_2 N) + \eta_r\Omega_1(Z + YN)$$

$$D_{21} = -\frac{1}{E_s}\left[v_s(1 + M) + \frac{\eta_r}{\eta_s} \times (\Omega_1 + M\Omega_2)\right]$$

$$\begin{Bmatrix} d\sigma_1^r \\ d\sigma_2^r \\ d\sigma_3^r \end{Bmatrix} = \left[\begin{array}{c|c} 1 & 0 \\ \hline M & N \\ \hline \Omega_1 + M\Omega_2 & \Omega_2 N \end{array}\right] \begin{Bmatrix} d\sigma_1 \\ d\sigma_3 \end{Bmatrix} \tag{98}$$

$$\begin{Bmatrix} d\sigma_1^s \\ d\sigma_2^s \\ d\sigma_3^s \end{Bmatrix} = \left[\begin{array}{c|c} 1 & 0 \\ \hline M & N \\ \hline -(\eta_r/\eta_s)(\Omega_1 + M\Omega_2) & (1/\eta_s)(1 - \eta_r N\Omega_2) \end{array}\right] \begin{Bmatrix} d\sigma_1 \\ d\sigma_3 \end{Bmatrix} \tag{99}$$

where

$$M = \frac{-\eta_r(\eta_r + \eta_s E_s/E_r)\Omega_1\Omega_2 - \eta_s v_s}{\eta_r\Omega_2[2\eta_s(v_s - v_r E_s/E_r) + \Omega_2(\eta_r + \eta_s E_s/E_r)] + \eta_s(\eta_s + \eta_r E_s/E_r)}$$

$$N = \frac{\eta_r\Omega_2 + \eta_s v_s}{\eta_r\Omega_2[2\eta_s(v_s - v_r E_s/E_r) + \Omega_2(\eta_r + \eta_s E_s/E_r)] + \eta_s(\eta_s + \eta_r E_s/E_r)}$$

$$\Omega_1 = \frac{2\cos\omega}{\cos\omega + \sqrt{3}\sin\omega}, \quad \Omega_2 = \frac{-\cos\omega + \sqrt{3}\sin\omega}{\cos\omega + \sqrt{3}\sin\omega}$$

$$D_{22} = \frac{1}{E_s}\left[-\nu_s N + \frac{1}{\eta_s}(1 - \eta_r \Omega_2 N)\right]$$

$$X = \Omega_1(\eta_r E_r + \eta_s E_s)/(\eta_s E_r E_s) + (E_r \nu_s - E_s \nu_r)/(E_r E_s)$$

$$Y = \Omega_2(\eta_r E_r + \eta_s E_s)/(\eta_s E_r E_s) + (E_r \nu_s - E_s \nu_r)/(E_r E_s)$$

$$Z = -1/(\eta_s E_s)$$

The incremental equation for $d\omega$ is

$$\frac{1}{3}\sqrt{3}\sigma_0(\cos\omega + \sqrt{3}\sin\omega)d\omega = (M - 1)d\sigma_1 + Nd\sigma_3 \tag{101}$$

Limit Surface

The simplest method to determine the limit surface is by numerical integration of the incremental Equations (98) and (99) for a given loading path, up to the state when the second constituent (soil) becomes plastic, [10]. Numerical analysis shows that the soil becomes plastic on a straight line which may be treated as a limit surface. Figure 17 shows results of some calculations performed for data presented in the previous Subsection, for different loading paths.

A simplified form of a limit surface can be obtained in an analytic way, [10]. This approximate equation of the limit surface is of the form:

$$\sigma_3 + \left[\eta_s\psi - \frac{1}{2}\eta_r(1 + m)\right]\sigma_1 \pm \eta_r\sigma_0 = 0 \tag{102}$$

where

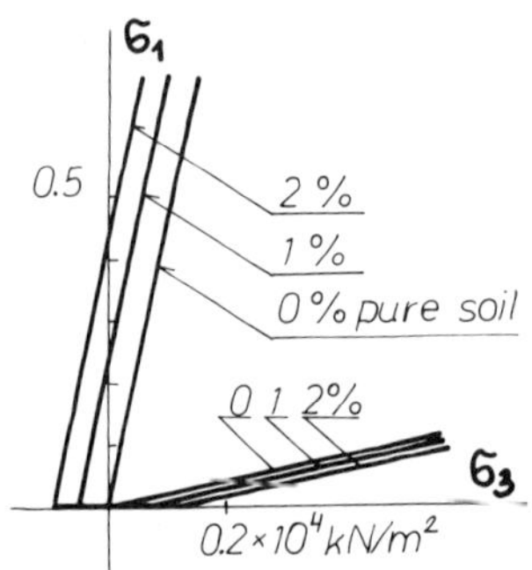

FIGURE 19. Influence of reinforcement on the global yield surface.

$$\psi = -\frac{1 + \sin\phi}{1 - \sin\phi} \quad \text{or} \quad \psi = -\frac{1 - \sin\phi}{1 + \sin\phi}$$

depending on the straight line on which the yield condition (71) is reached. In Figure 19 the influence of the reinforcement on a limit surface is shown. Equation (102) may be also compared with Equation (5), which takes the following form for $\alpha = 0$:

$$\sigma_3 + \psi\sigma_1 \pm \eta_r\sigma_0 = 0 \tag{103}$$

Equation (103) may be treated as a further approximation to Equation (102). For the data given in the previous Subsection the coefficients of σ_1 are respectively -4.56 and -4.6.

CONCLUSIONS

The present elaboration is devoted to the description of plastic behavior of reinforced earth, from the view-point of continuum mechanics. The work divides roughly into two parts. In the first part, the rigid-plastic model and its application to the analysis of some problems of practical importance are shown. In the second part, a more elaborated, elasto-plastic theory of reinforced earth is presented. Both approaches provide a theoretical framework to an engineering analysis of reinforced earth, treated as a macroscopically homogeneous composite. The results presented may also serve as a theoretical tool to the analysis of modern geotextiles and geomembranes.

The mathematical structure of the rigid-plastic model is much more simple than that of the elasto-plastic one. This latter, however, gives more informations about the reinforced earth behavior and allows for better understanding of this material. For example, a practical conclusion which follows from the elasto-plastic analysis is that the reinforcement, almost always, first becomes plastic if an ideal friction between both constituents is assumed. This conclusion is supported by experimental observations [2,5,12]. The elasto-plastic model enables us to study the reinforced earth behavior preceding the plastic collapse of the composite. To apply this model to the analysis of boundary value problems one requires the use of numerical techniques, for example the finite element method. The advantage of the rigid-plastic approach is a possibility to obtain the estimates of a collapse load, usually in a simple way.

Some of the obtained results are of a practical importance. For example, Equation (33) is a closed form solution to the problem of a bearing capacity of a footing on reinforced earth. Also Equations (58), (59), (64) and (65) may serve as engineering formulae helpful in practice. Equation (84) derived on the basis of elasto-plastic approach confirms the accuracy of some simplified approaches to the analysis of reinforced earth, as well as Equation (102) shows that the rigid-plastic model [Equation (5)] is quite good from the engineering point of view.

The practical importance of the presented models follows from a fact that they can be applied to the analysis of various geotechnical problems. In this work only a limited range of possible applications has been shown. An engineer has a choice between a relatively simple rigid-plastic model or a more advanced elasto-plastic one. For details see References.

NOTATION

The following symbols are used in this paper:

a = length
b = width of a footing, length
$\mathbf{B}^\alpha$ = structural matrix for stresses
b_i^α = component of a structural matrix
D_{ij} = coefficients
$d(\ )$ = increment of a quantity ( )
E_α = Young's modulus of α-th constituent
f^α = yield condition for α-th constituent
$\hat{f}^\alpha$ = actual yield surface
$\underline{f}$ = yield condition
$\bar{f}$ = initial yield condition
$F, \mathfrak{F}$ = yield conditions
H = height of a vertical cut
l = length
M = parameter
m = parameter
N = parameter
n = parameter
$\mathbf{n}$ = unit vector indicating the direction of reinforcement
n_i = components of a unit vector
p = vertical pressure
q = horizontal pressure
R = plastic locus for the reinforcement in extension
v = velocity
v_i, V_i = components of the velocity vector
α = angle between directions of the reinforcement and the principal stress σ_1
β = angle, parameter
γ = own weight
ε = parameter
$\boldsymbol{\epsilon}$ = macrostrain tensor
$\boldsymbol{\epsilon}^\alpha$ = microstrain tensor
ζ = parameter
η_α = volume fraction of α-th constituent
λ = scalar function
ν_α = Poisson ratio
ξ = parameter
$\boldsymbol{\sigma}$ = macrostress tensor
$\boldsymbol{\sigma}^\alpha$ = microstress tensor
σ = parameter
σ_m = mean pressure
σ_x, σ_y = stresses
σ_0 = plastic limit of the fibers
τ_{xy} = shear stress
ϕ = angle of internal friction of the soil
ψ = parameter
Ω, ω = parameters

Subscripts

l = lower bound estimate
u = upper bound estimate
α = deals with the α-th constituent
S = soil in a plastic state
SR = both soil and reinforcement in a plastic state

Superscripts

α = deals with the α-th constituent
r = reinforcement
s = soil

REFERENCES

1. Chen, W.-F., *Limit Analysis and Soil Plasticity*, Elsevier Scientific Publishing Company, Amsterdam-Oxford-New York, 1975.
2. Ingold, T. S., *Reinforced Earth*, Thomas Telford Ltd., London, 1982.
3. Kulczykowski, M. and Sawicki, A., ''Bearing Capacity of Footing on Reinforced Earth,'' *Archiwum Hydrotechniki XXXI*, 3, 1984, pp. 275–281.
4. Kulczykowski, M., ''Analysis of the Bearing Capacity of Reinforced Earth Loaded by a Footing,'' thesis presented to the Institute of Hydroengineering of the Polish Academy of Sciences in Gdansk, in 1985, in partial fulfillment of the requirements for the degree of Doctor of Philosophy.
5. Long, N. T., Guégan, Y., and Legeay, G., ''Etude de la Terre Armèe à l'appareil Triaxial,'' Lab. des Ponts et Chaussèes, Rapp. de recherche No. 17, 1972.
6. Naylor, D. J. and Richards, H., ''Slipping Strip Analysis of Reinforced Earth,'' *International Journal for Numerical and Analytical Methods in Geomechanics*, Vol. 2, 1978, pp. 343–366.
7. Sawicki, A., ''Yield Conditions for Layered Composites,'' *International Journal of Solids and Structures*, 17, 1981, pp. 969–979.
8. Sawicki, A., ''Elasto-Plastic Theory of Composites with Regular Internal Structure,'' *Comportement Mécanique des Solides Anisotropes*, J. P. Boehler, ed., CNRS and Martinus Nijhoff, 1982, pp. 409–422.
9. Sawicki, A., ''Plastic Limit Behavior of Reinforced Earth,''

Journal of Geotechnical Engineering, ASCE, Vol. 109, No. 7, Paper No. 18075, July 1983, pp. 1000–1005.

10. Sawicki, A., ''Engineering Mechanics of Elasto-Plastic Composites,'' *Mechanics of Materials*, 2, 1983, pp. 217–231.
11. Sawicki, A., ''Axisymmetrical Elasto-Plastic Behavior of Reinforced Earth,'' *International Journal for Numerical and Analytical Methods in Geomechanics*, Vol. 7, 1983, pp. 493–498.
12. Schlosser, F., ''La Terre Armèe. Recherches et Réalisations,'' Bull. Liaison Lab. des Ponts et Chaussées, 62, Nov.–Dec. 1972, pp. 79–92.
13. Sokolovskii, V. V., *Statics of Granular Media*, Pergamon Press, New York, 1965.
14. Stefani, C. and Long, N. T., ''Comportement de Semelles sur un Massif Armèe Semi-Infinit,'' *Proc. Colloque International Renforcement des Sols*, Vol. I, Paris, 1979.

CHAPTER 4

Footings and Constitutive Laws

Shamsher Prakash,* Swami Saran and Uma Nath Saran†**

CONSTITUTIVE LAWS OF SOILS

Constitutive laws of soils define the mechanical behaviour of soils. They are of prime importance for analysing almost all applied non-linear problems of soil mechanics, e.g. stability analysis of slopes, dams and embankments, analysis of rafts, piles and wells. Constitutive laws are known to be functions of the (a) physical properties of soils, (b) moisture content and relative density, (c) confining pressure, (d) intermediate principal stress, (e) strain rate, and (f) stress history of soil. A review of all pertinent literatures available regarding the influence of the above factors on the mechanical behaviour of soils are summarised in the next paragraphs.

FACTORS AFFECTING CONSTITUTIVE LAWS

Influence of Particle Physical Properties

The influence of physical properties of granular soil particles characterized by shape, size and grading have been studied to quite a limited extent. It has been found that shear strength decreases with increase of particle size and increases with increase in angularity of particle shape. But grading has little effect on shear strength as long as coefficient of uniformity remains unaltered (Vallerga, 1957; Leslic, 1963; Kolbuszewski, 1963; Kirkpatric, 1965; Marshall, 1965; Leussink, 1965; Lee and Seed, 1967; Fumagalli, 1969 and Koerner, 1970). The behaviour of soils composed of flaky (platey) mineral particles such as mica, talcum, montmorillonite, kaolonite, allumina and chlorite have been found to be quite different from those composed of bulky particles such as feldspar, iron ores, halocytes and calcite or rod like particles such as hornblende. The basic change in stress-strain curve, the tremendous volumetric decrease and terminal orientation of particles were noticed in soils composed of flaky particles (Horn and Deere, 1962; Taylor, 1948; Jumikis, 1965; and Koerner, 1970).

Effect of Moisture Content and Relative Density

The strength characteristics of clay are dependent upon moisture content, structural formation i.e., whether it is flocculated, dispersed, honeycombed and also upon the properties of the pore fluid. The moisture content has, however, no significant effect on the strength characteristics of granular soils (Kirkpatric, 1965). Higher initial tangent modulus and strength are obtained for denser soils. The increase of density to the extent that there is no structural breakage, causes an increase in the strength.

Effect of Confining Pressure

The significant effect of confining pressure on the behaviour of soils has been well recognized. It is generally found that the strength and compressibility of soil varies with confining pressures. At higher confining pressures structural breakdown and increased compressibility are observed. Mohr's rupture envelope has been found to be curved with more pronounced curvature for dense sand and gravelly samples in higher pressure range. In case of partially saturated clays, both the tangent modulus and the compressive strength have been found to be affected by confining pressure, but in case of saturated clays, they remain

*Department of Civil Engineering, University of Missouri–Rolla, Rolla, MO

**Department of Civil Engineering, University of Roorkie, Roorkie, U.P., India

†Department of Civil Engineering, Muzuffarpur Institute of Technology, Muzuffarpur Bihar, India

unaffected (Holtz and Gibbs, 1967; Hall and Gordon, 1963; Leslie, 1963; Vesic and Barkdale, 1963; Bishop et al., 1965; Lee and Seed, 1967; Vesic and Clough, 1968; and Pounce and Bells, 1971).

Effect of Intermediate Principal Stress

Limited literatures are available on the effect of intermediate principal stress on the strength characteristics of soils. The strength and initial tangent modulus of soil under plane strain condition has been found to be higher but strain at peak stress lower than in the triaxial compression and extension tests (Cornforth, 1964; Ko and Scot, 1967). The failure stress has been found to be minimum when $\sigma_2 = \sigma_3$ and increased with increase in σ_2 keeping σ_3 constant (Bishop, 1953; Southerland and Mcsdary, 1969, Duncan, 1970). In general this behaviour may be shown graphically in Figure 1. The intermediate principal stress has the effect of stiffening and increasing the strength of soil when it goes from condition of triaxial compression to plane strain condition. The maximum possible shear stress is given by one half of the difference in these stresses,

$$\tau_{max} = -1/2(\sigma_1 - \sigma_3) \qquad (1)$$

The octahedral shear stress is given by

$$\tau_{oct} = \frac{1}{3}\sqrt{(\sigma_1 - \sigma_2)^2 + (\sigma_2 - \sigma_3)^2 + (\sigma_3 - \sigma_1)^2} \qquad (2)$$

It can be shown from the above expression that τ_{oct} can vary only between limits of 0.816 (when $\sigma_2 = \sigma_1$ or σ_3) to 0.943 [when $\sigma_2 = (\sigma_1 + \sigma_3/2)$] times the maximum shear stress. There is consequently a change of about 15 percent only in the relationship between τ_{oct} and τ_{max} over the entire range of possible values. In many cases, therefore, one may be substituted for the other (Newmark, 1960).

Effect of Rate of Strain

The effect of strain rate on small samples of sand has been investigated by Casagrande and Shannon (1947), Skempton and Bishop (1951), Whitman (1967), Prakash and Venkatesan (1960), Whitman and Healey (1962). It has been found that there is no significant effect of rate of strain on the strength characteristics of granular materials. The strength characteristics of clay are affected to a significant extent by the rate of strain. Higher tangent modulus and strength have been observed at faster rate of loading. The increase in strength from the percent strain of 1 per minute to 10^{-3} per minute has been found to increase the strength by 20 percent (Taylor, 1948).

The Effect of Stress History

The shear strength parameters have been found to be independent of the stress history and the use of anisotropic consolidation in case of sand (Bishop and Eldin, 1953). But the shear strength characteristics soft clay are greatly influenced by consolidation history and appear to depend upon the over consolidation ratio (Simon, 1960; Henkel, 1966).

Discussion

The factors influencing the stress-strain characteristics of soils are quite large in number and it is not possible to establish any general law which can take into account all the above factors. Therefore, laboratory testing under simulated field condition is done to establish constitutive laws for a given foundation soil.

The constitutive relationships for simplified practical cases only may be studied with the help of rheological and mathematical models. A review of literatures on rheological and mathematical models suitable for soils has been given subsequently.

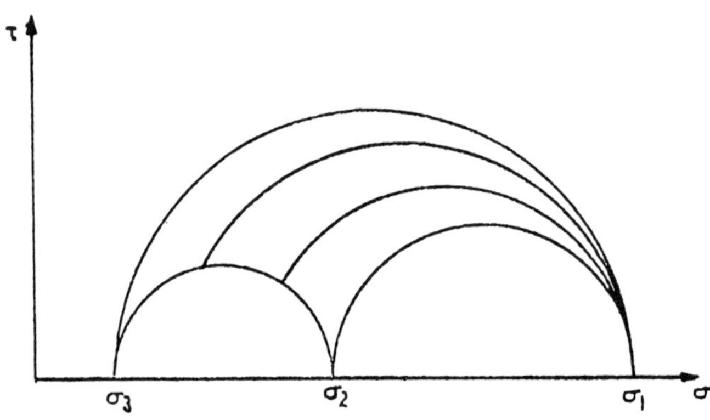

FIGURE 1. Principal stresses on Mohr's diagram.

MODELS FOR CONSTITUTIVE RELATIONSHIPS

Rheological Models

Soils are generally a three phase system consisting of solids, liquid and gaseous materials. The solid part is composed of numerous grains of various minerological composition and is porous. The pore is generally filled with water and air. Soils resist the effects of external forces in a manner different from a simple solid continua. In non-cohesive soils, the external force is resisted by the intergranular friction at the contact surfaces. In cohesive soils, composed of clay minerals, the strength of the films of absorbed water surrounding the grains accounts for the resistance of the soil to deformation (Suklje, 1969).

Soils exhibit elasticity as well as creep under constant stress. Creep occurs at a rate which remains either constant or varies with time. Stress relaxation under constant applied strain is also observed in soils. This behaviour of soil can be described by visco-elastic models consisting of rheological elements namely Hookean elastic body, Newtonian viscous liquid, Saint Venant plastic body and Pascal's liquid. The relationships between stress and strain of these bodies are given by Scott (1963). The rheological models are constructed in an intuitive way and the corresponding relationships between stresses and strains are deduced and compared with the experimental observations. This comparison controls the applicability of the assumed rheological models. Various rheological models which may be suitable for soils have been proposed by Rao (1967).

The model discussed here and given in Figure 2 is constructed to represent elasto-plastic and viscous behaviours of soils and found to better represent soil behaviours (Chakarvorty, 1970; Prakash, Saran and Sharan, 1975). The model consists of spring (H) and viscous dash-pot (N) on the left branch of the model and on the right branch of the model, gas or air (G) and incompressible liquid (IP) represented by a cylinder and a tight fitting piston filled with air and a cylinder with water and tight fitting piston respectively.

The relation between stress and strain is given by the following expression (Prakash, Saran and Sharan, 1975).

$$\sigma = \dot{\epsilon}\lambda(1 - \bar{e}^{E/\lambda T}) + \frac{u_0\epsilon}{1 - \epsilon} \qquad (3)$$

where

σ = applied stress
$\dot{\epsilon}$ = rate of strain
λ = Trouton's coefficient of viscosity
E = spring constant (modulus of elasticity)
T = time
ϵ = strain
u_o = initial pore pressure

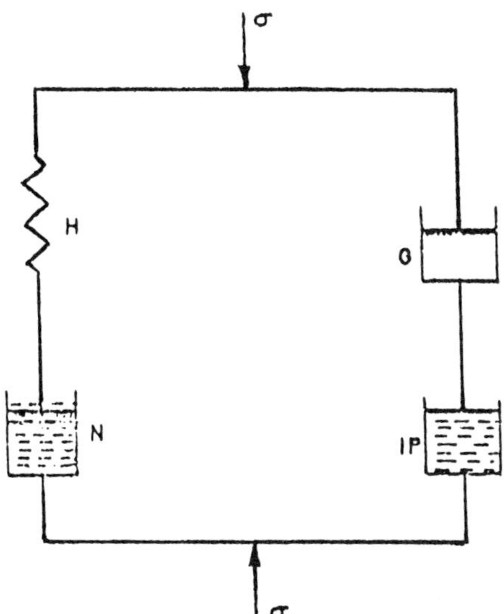

FIGURE 2. Rheological model.

At large strains, time T in the above equation becomes large and it reduces to:

$$\sigma = \dot{\epsilon}\lambda + \frac{u_o\epsilon}{1 - \epsilon} \qquad (4)$$

Therefore, for a given stress, strain at failure and rate of strain, the value of λ can be obtained.

The values of E at various strains for a given curve can also be calculated. The theoretical stress-strain curves for various values of E and λ for this model are given in Figure 3.

The rheological model discussed herein has been found to predict well the stress-strain curves for cohesive as well as non-cohesive soils by taking E at 1 percent strain (Chakravarty, 1970; Prakash, Saran and Sharan, 1975). The great drawback of this method is the correct evaluation of the elastic constant E which is done arbitrarily at 1 percent strain. Soils exhibit non-linearity in stress-strain relationship even at low strain level and these curves are sensitive to magnitude of E as shown in Figure 3.

Mathematical Models

The behaviour of soil over a wide range of stresses is nonlinear, inelastic and dependent upon the magnitude of the confining pressure employed in the test. A review of literatures on simplified practical mathematical models which takes into account this non-linearity, stress dependency and inelasticity of soil behaviour is given in the following paragraphs.

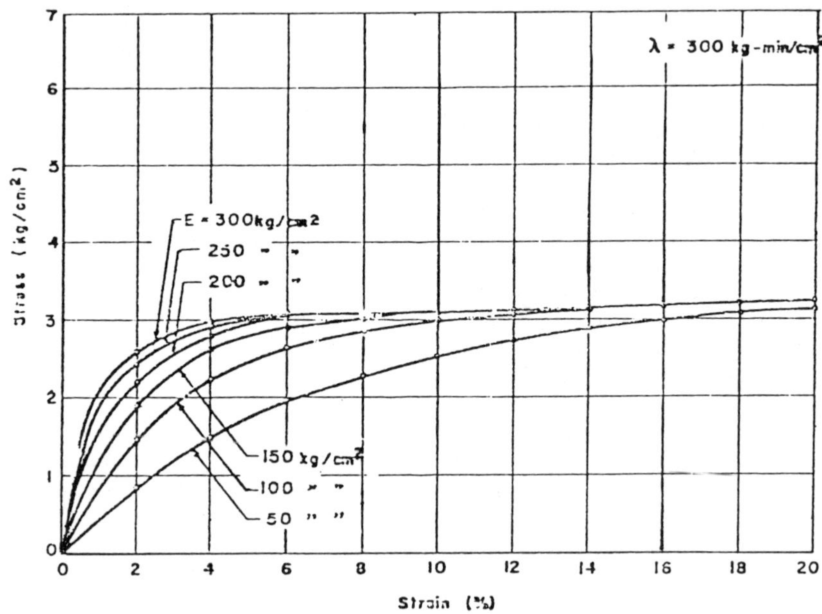

FIGURE 3. Stress-strain curves (rheological model).

HYPERBOLIC FUNCTIONS

Kondner (1963) and Kondner and Zelasko (1963) have shown that the non-linear stress-strain curves of both clay and sand may be approximated by hyperbolae of the following form with high degree of accuracy:

$$\frac{\epsilon}{\sigma_1 - \sigma_3} = a + b\epsilon$$

or,

$$\epsilon = \frac{a(\sigma_1 - \sigma_3)}{1 - b(\sigma_1 - \sigma_3)} \tag{5}$$

where

ϵ = axial strain
a,b = constants of hyperbola

The plot $\epsilon/(\sigma_1 - \sigma_3)$ vs. $a + b\epsilon$ gives a straight line where a is the intercept on Y-axis and b is the slope of the line. Working in limit $\epsilon \rightarrow \infty$ with Equation (5), the reciprocal of b represents the ultimate compressive strength of the soil which is larger than the failure compressive strength. This is expected because the hyperbola remains below the asymptote at all finite values of strain. The ratio R_f of compressive strength $(\sigma_1 - \sigma_3)_f$ to ultimate compressive value σ_u varies from 0.75 to 1.0 for different soils independent of confining pressure (Kondner, 1963).

A differentiation of the Equation (5) with respect to strain ϵ gives the inverse of a as the initial tangent modulus.

Duncan and Chang (1970) have expressed the Kondner's expression in terms of the shear strength and initial tangent modulus as given below:

$$(\sigma_1 - \sigma_3) = \epsilon \left[1 - \frac{R_f(1 - \sin\phi)(\sigma_1 - \sigma_3)}{2\,c\cos\phi + 2\sigma_3\sin\phi} \right] E_i \tag{6}$$

where

c = cohesion
ϕ = angle of internal friction
E_i = initial tangent modulus
$R_f = (\sigma_1 - \sigma_3)_f$

Janbu's (1963) experimental studies have shown that the relationship between initial tangent modulus and confining pressure may be expressed as follows:

$$Ei = K\,pa \left(\frac{\sigma_3}{p_a}\right)^n \tag{7}$$

where P_a = atmospheric pressure, and K and n are pure numbers.

The values of K and n can be readily determined from the results by plotting E_i against σ_3 on log-log scales and fitting a straight line to the data.

PARABOLIC FUNCTIONS

Hansen (1963) proposed two additional functional representation of stress-strain relationships:

$$(\sigma_1 - \sigma_3) = \left[\frac{\epsilon}{a + b}\right]^{1/2} \tag{8}$$

$$(\sigma_1 - \sigma_3) = \frac{\epsilon^{1/2}}{a + b} \tag{9}$$

The first equation accounts for the possibility of parabolic variation of stress-strain curves at small strains. The second equation is an alternative form to account for the parabolic variation and possesses the property of giving a maximum value of $\sigma_1 - \sigma_3$ for finite strain, i.e., it is suitable when the curve shows a decrease after the peak stress. Hansen (1963) used one of the data from Kondner and compared the stresses obtained from the above equations. He observed that if the stress-strain curve is initially parabolic, Equation (8) will probably be better and if it is work softening, the Equation (9) should be attempted.

SPLINE FUNCTION

Desai (1971) compared the above results with that of his spline functions which approximate the given non-linear stress-strain relations by a number of polynomials of a given degree spanning a number of data points. Details of splines and their mathematical properties are described by Schoenburg (1964), Ahlberg et al. (1967) and Graville (1967). Kondner's (1963) work compared very well with that of Desai's (1971) spline functions.

Discussion

The mathematical models discussed above are purely empirical in nature; therefore, it is likely that one relation may represent the actual stress-strain to a better degree than the other. A comparative study for predicting bearing capacity and settlement was carried out by Desai (1971) for circular footing in clay using the spline functions and hyperbolic equations. His study showed that ultimate bearing capacity computed using spline function differed by about 0.52 percent and that using hyperbola differed by 1.58 percent from the experimental results. The settlements, predicted at 60 percent of the ultimate load differed 3 to 6 percent and 20 to 25 percent on computation by spline function and hyperbolic equations respectively.

Basavanna (1975) has discussed the cause of this large variation in case of hyperbolic equation of Kondner (1963). Desai had computed the values of a and b constants of hyperbola by using the criteria suggested by Duncan and Chang (1970) which takes into account of straight line passing through the points of 70 to 95 percent of mobilized strength. Basavanna (1975) suggests that if due weightage is given to the experimental points in the initial portion of the curve, different values of a and b parameters are obtained and then the predicted pressure settlement curve is close to the experimental one.

The above discussions lead to the conclusion that the stress-strain relationship represented by mathematical model suggested by Kondner is simple and acceptable and therefore, may be adopted for investigation of constitutive laws of soils.

PRESSURE-SETTLEMENT CHARACTERISTICS OF FOOTINGS

General

The rational method to compute bearing capacity and settlement of an actual footing is from the pressure settlement curve. It is difficult to determine experimentally the pressure-settlement curve of an actual footing. Therefore, experiments are performed on a smaller size of plate and results are extrapolated for actual size of footing by formulas which are at best semi-empirical. Analytical methods based on the soil properties have also been evolved for the computation of pressure settlement curve of an actual footing. A brief critical review of the experimental and analytical methods is given in the following paragraphs.

Experimental Method

Plate load test is performed on in situ soil by the method described in I.S. Code, 1971. The data are plotted in the form of a pressure-settlement curve. The failure stress is determined by the point of inter-section of tangents drawn at the initial and final portion of the pressure-settlement curve. The settlement of actual footing is then calculated by formulae which are at best semi-empirical relations. The expressions show that under same intensity of pressure, the settlement increases with the increase in the size of footing.

There are many limitations to the experimental method of plate load test. They are enumerated below:

1. A plate load test is essentially a test of short duration. The elastic and compression settlement for foundations on cohesionless soil can be estimated from plate load test. But it does not give the ultimate settlement in case of cohesive soils.
2. The plate load test is an expensive and time consuming test and is difficult to perform under difficult site and weather conditions.
3. The extrapolation of the results of plate load test to other larger sizes of footings may give erroneous values, if the soil properties change.

Analytical Approach

Pressure-settlement characteristics of a soil-foundation system for a given size of footing depend upon the stress-strain relation of the soil mass loaded at the surface on finite area. Analytical approaches for determining the pressure settlement curve using linear stress-strain relation have been given by Steinbrenner (1934) and Terzaghi (1943) and others. The pressure settlement curve works out to be a straight line.

The evaluation of pressure settlement characteristics using non-linear stress-strain relation for soil presents a difficult task. It is the availability of the advanced numerical technique and fast computer nowadays that has made it possible to compute the pressure-settlement characteristics using non-linear soil property. The works of Duncan and Chang (1970), Desai (1971) and Basavanna (1975) in this direction are discussed in the following paragraphs.

Duncan and Chang (1970) used hyperbolic stress-strain relation in finite element method to obtain the pressure settlement curve for strip footing of 0.062 m × 0.316 m placed at a depth of 0.508 m from the surface. The soil was Chatahoochee river sand. The computed curve was compared with the experimental plate load test results. The above two curves compared well up to a bearing pressure of 1.06 kg/cm². The ultimate bearing pressure from plate load test could not be computed as the soil did not show failure up to a bearing pressure of 2.12 kg/cm². Ultimate bearing pressure from Terzaghi's formula gave a value of 3.8 kg/cm². The computed pressure settlement curve deviated to a large extent from the experimental curve after a bearing pressure of 1.06 kg/cm².

Desai (1971) used non-linear stress-strain property represented by spline function and hyperbolic function in his finite element technique to predict load deformation behaviour of a model circular footing made of steel 7.62 cm (3.0 inch) and 1.27 cm (0.5 inch) thick resting on cohesive soil (= 0.48). He found that the ultimate bearing pressures were about the same, i.e., experimental value was 0.697 kg/cm² (9.95 psi), spline function gave 0.7 kg/cm² (10.0 psi) and hyperbolic function − 6.87 kg/cm² (9.8 psi). However, the settlement behaviour at lower and intermediate ranges of pressure were significantly different. For example, at about 60 percent of ultimate load, the absolute difference in displacement with respect to experimental value using spline and hyperbolic functions were about 3 to 6 percent and 20 to 25 percent respectively.

Basavanna (1975) used hyperbolic stress-strain relation for soil in his finite element method to compute pressure-settlement curves for square footings resting on sand and clay. He used a footing of 25.4 cm × 25.4 cm resting on Buckshot clay (Carrol, 1963). The computed (Basavanna 1975) and experimental (Carrol, 1963) pressure-settlement curves tallied very well. The ultimate bearing pressures from experimental results and finite element method were 1.48 kg/cm² and 1.50 kg/cm² respectively (a difference of about 3 percent only). He analysed a square footing, 10 cm × 10 cm, resting on Ranipur sand. The pressure-settlement curves tallied only up to a pressure of 0.13 kg/cm², beyond which there was a large variation. The computed ultimate bearing pressure was 0.16 kg/cm² as against 1.45 kg/cm² obtained from plate load test results. He has explained that the large variation in case of sand is due to use of constant value of Poisson's ratio in the finite element programme, which may be greater than 0.50 in the failure range and stress dependent anisotropic deformation property which is more pronounced in sand.

Conclusions

The computation of load settlement characteristics of soil by finite element technique is a very costly procedure. It requires availability of large electronic computer and a lengthy computer programme making it a costly operation. Thus simplified procedures are needed which may give adequate degree of precision in error.

BEHAVIOUR OF FOOTINGS USING CONSTITUTIVE LAWS

This aspect has been divided into two parts:

A. Footings in clay
B. Footings in sand

The part (A) deals with the assumptions and procedure adopted for the establishment of analytical methods for evaluation of pressure settlement characteristics of footings using constitutive laws of clay. Since contact pressure plays an important part in transmitting stresses to the soil supporting a footing, the latter has been taken as an important criterion to evaluate the stresses in soil masses. The general procedure of analysis of pressure-settlement characteristics for uniformly loaded strip and square footings as well as eccentrically loaded strip footing resting on clay has been dealt with in various steps. The procedure of analysis of all particular cases, mentioned below, will follow.

a. Smooth flexible strip footing
b. Smooth rigid strip footing
c. Rough flexible strip footing
d. Rough rigid strip footing
e. Smooth flexible square footing
f. Smooth rigid square footing
g. Eccentrically loaded smooth rigid strip footing

The effect of flexibility and roughness of base of footing on pressure settlement characteristics has been studied from cases (a) to (d). The contact pressure distribution has also been studied from the above cases in two dimensional space. The contact pressure distribution in three dimensional space has been studied for cases (e) and (f). No rational method is available for determining contact pressure distribution in two dimensional and three dimensional spaces.

Case (g) has been set up for evaluating pressure-settlement characteristics for eccentrically loaded footing. No method is available for evaluating pressure-settlement curves of eccentrically loaded footing.

Part (B) discusses the difficulties faced by previous investigators in obtaining pressure settlement characteristics for footings in sand. In case of sands, the problem becomes more involving as the stresses and settlement equations become dependent on confirming pressure. A new procedure is described for computing pressure settlement characteristics of uniformly loaded strip footings in sand.

The constitutive relations of soil represented by hyperbola were established from triaxial compression tests. Drainage conditions in the tests were controlled according to the field conditions for obtaining the constitutive relations. Usually in case of footings in clay, the settlement occurs slowly and sufficient time is available for pore pressure dissipation; therefore the constitutive relations obtained by performing drained tests are used. In case of footing in sand settlement occurs instantaneously; therefore constitutive laws for consolidated undrained tests are used. The parameters of constitutive laws obtained from triaxial tests have been suitably modified while using them in two dimensional stress system.

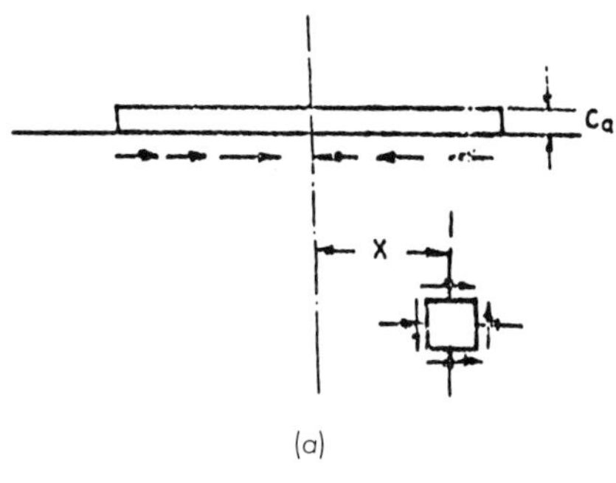

(a)

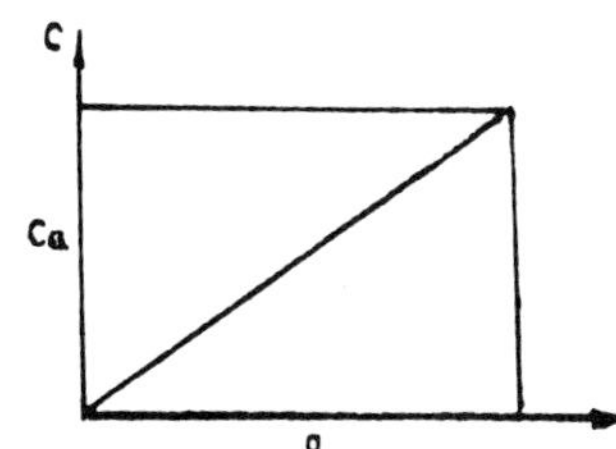

q = Intensity of uniform surface load
C = Cohesion
Ca = Tangential force at the base of footing.

(b)

FIGURE 4. (a) Tangential stresses at contact surface due to roughness of base of footing; (b) Variation of tangential stresses on contact surface for footing on cohesive soil.

Footings in Clay

ASSUMPTIONS

1. The soil mass has been assumed as semi-infinite and isotropic medium.
2. The footing base has been assumed as fully flexible or fully rigid.
3. The roughness of footing has been assumed to generate tangential forces (Figures 4a and 4b) at the contact surface, which follows the relationship

$$C_a = C\left(\frac{q}{CNc}\right)$$

for cohesive soils, where C_a is the tangential force at the contact surface.
4. The contact pressure distribution in case of smooth rigid strip footing has been assumed as linear from centre to edges (Figure 5).
5. The contact pressure distribution in case of eccentrically loaded smooth rigid footing has been assumed as linear (Figure 6).
6. The whole soil mass supporting a footing has been divided into a large number of thin horizontal strips (Figure 7) in which stresses and strains have been assumed as uniform along any vertical section.
7. (a) The stresses in each layer have been computed using Boussinesq's theory as the stress equations for various types of loads are available.
(b) The strains have been computed from the known stress condition using constitutive laws.
8. There is no slippage at the interface of layers of the soil mass.

GENERAL PROCEDURE

The general procedure for evaluation of pressure settlement characteristics of footings has been described in the following steps:

Step 1—For a given load q_1 on the footing, the contact pressure distribution at the interface of footing base and supporting soil may be assumed as shown in Figure 7. The contact pressure distribution is the loading pattern for the soil at the surface below the footing and the stresses in the soil mass will be induced according to this loading pattern. The correctness of the assumption will be tested in Step 11.

Step 2—The soil mass supporting the footing has been divided into n layers as shown in Figure 7.

Step 3—Taking any vertical section, the normal and shear stresses at the centre of a layer at depth z below the footing have been computed using equations of the theory of elasticity for appropriate pattern of loading. The stress compo-

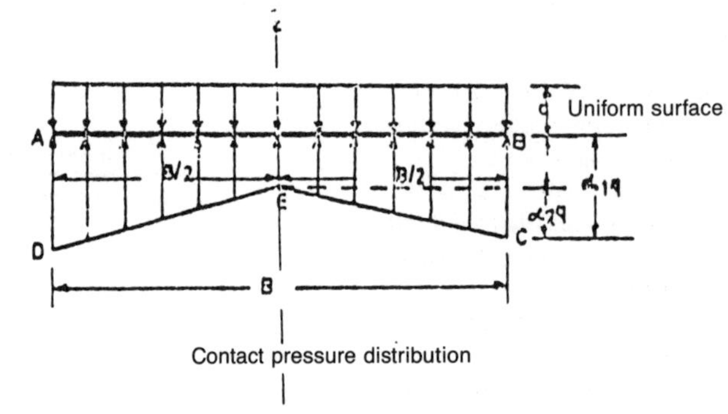

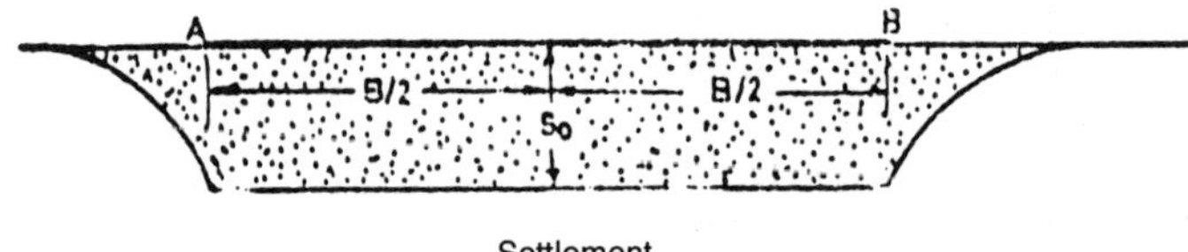

FIGURE 5. Contact pressure distribution and settlement of centrally loaded footing with rigid base.

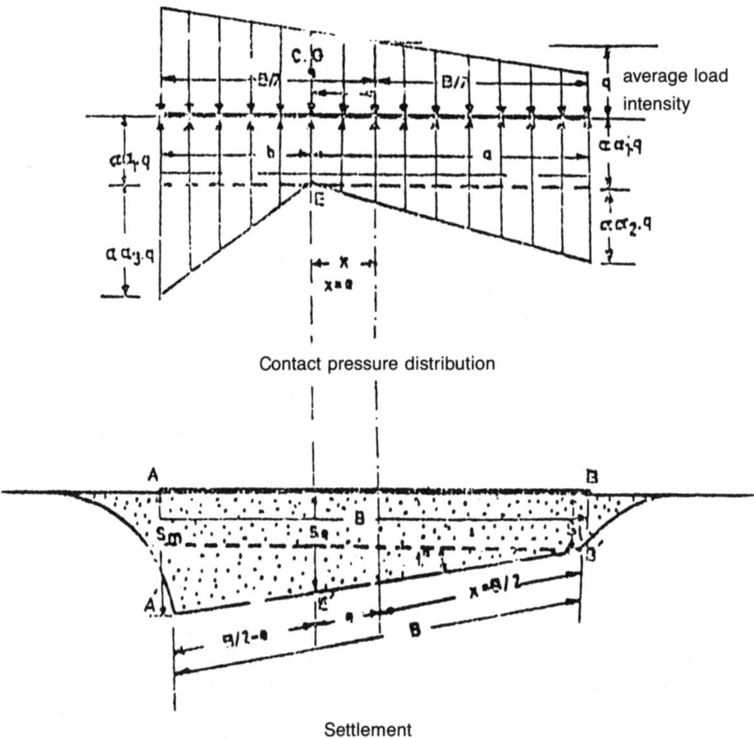

FIGURE 6. Contact pressure distribution and settlement of eccentrically loaded smooth rigid footing.

nents at a point are shown in Figure 8(a) for three dimensional space and 8(b) for two dimensional space.

Step 4—The principal stresses at a point in the soil mass and their directions with respect to the vertical z-axis have been computed using the equations of the theory of elasticity. The principal stresses are shown in Figures 8(a) and 8(b) for three dimensional and two dimensional stress system and their directions with respect to z-axis.

Step 5—The constitutive equations for soils have been obtained from triaxial compression test results using Kondner's (1963) two constant hyperbola.

Step 6—Evaluation of Strains: The strains in the direction of principal stresses are evaluated from the constitutive relationship as follows: In three dimensional cases,

$$\epsilon_1 = \frac{1}{E}[\sigma_1 - \mu(\sigma_2 + \sigma_3)] \tag{10}$$

$$\epsilon_2 = \frac{1}{E}[\sigma_2 - \mu(\sigma_3 + \sigma_1)] \tag{11}$$

$$\epsilon_3 = \frac{1}{E}[\sigma_3 - \mu(\sigma_1 + \sigma_2)] \tag{12}$$

where ϵ_1, ϵ_2 and ϵ_3 are the principal strains in the direction of the principal stresses σ_1, σ_2 and σ_3, respectively; μ is the Poisson's ratio; and E is the modulus of elasticity.

The strain ϵ_1 in the direction of major principal stress σ_1 is determined from constitutive relationship given by Equation (5), i.e.,

$$\epsilon_1 = \frac{a(\sigma_1 - \sigma_3)}{1 - b(\sigma_1 - \sigma_3)}$$

The strains ϵ_2 and ϵ_3 are computed from the following relationship:

$$\epsilon_2 = \frac{\sigma_2 - \mu(\sigma_1 + \sigma_3)}{\sigma_1 - \mu(\sigma_3 + \sigma_2)} \times \epsilon_1 \tag{13}$$

$$\epsilon_3 = \frac{\sigma_3 - \mu(\sigma_1 + \sigma_2)}{\sigma_1 - \mu(\sigma_3 + \sigma_2)} \times \epsilon_1 \tag{14}$$

The strain in the vertical direction ϵ_z for each layer is calculated from the following equations:

$$\epsilon_z = \epsilon_1\cos^2\theta_1 + \epsilon_2\cos^2\theta_2 + \epsilon_3\cos^2\theta_3 \tag{15}$$

where θ_1, θ_2 and θ_3 are the directions of the principal strains with respect to the vertical axis (Figure 8).

Step 7—The vertical settlement s of any layer is computed by multiplying the strain ϵ_z with the thickness of each layer (δ_z).

$$s = \epsilon_z \cdot \delta_z \tag{16}$$

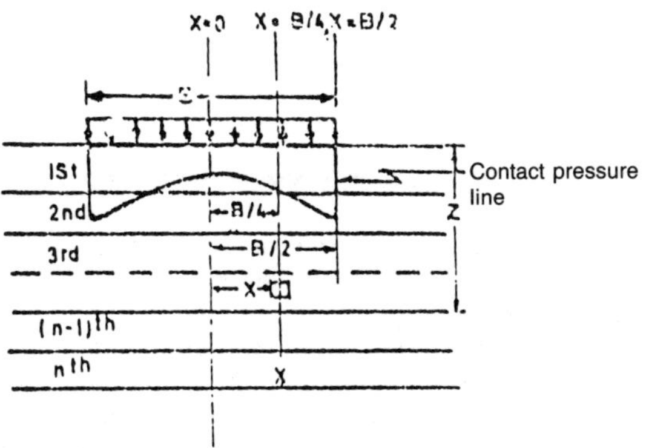

FIGURE 7. Contact pressure profile and soil below footing divided in n layers.

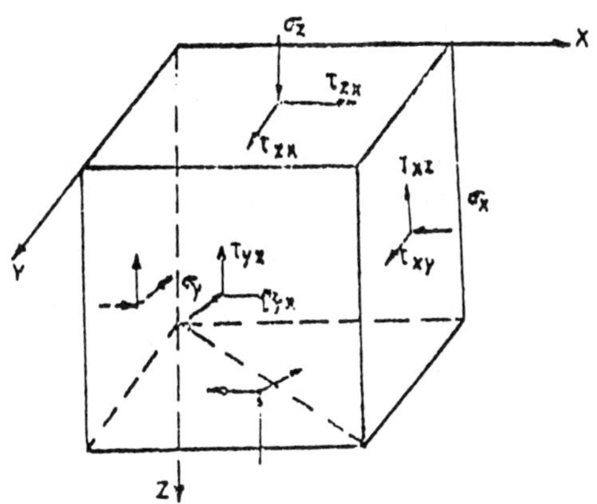

(a) Three dimensional

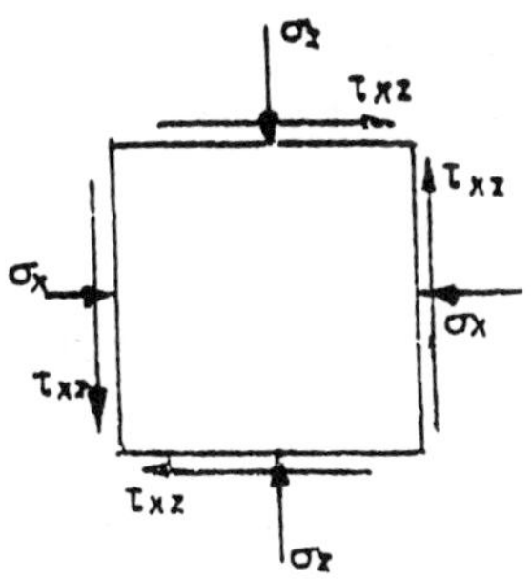

(b) Two dimensional

FIGURE 8. Stresses in soil mass.

Step 8—The total settlement (S) along any vertical axis is computed by numerically integrating the expression

$$S = \int_{o}^{n} \epsilon_z . dz \tag{17}$$

The numerical integration is done with the help of a computer.

Step 9—The total settlement is computed at three vertical sections, namely, at $x = 0$, $x = B/4$ and $X = B/2$ for the given pressure.

Step 10—For rigid footings, the settlements at the centre, $B/4$ and $B/2$ sections are compared. If the contact pressure distribution assumed in step 1 is correct, then settlement at all three points shall be the same. If the settlements of all the above three points do not fall within a close range, then new pressure distribution is assumed and steps 1 to 10 are repeated.

For flexible footings, the contact pressure distribution is taken as uniform and the average of the settlement computed in step 10 is calculated by dividing the area of settlement diagram by width of footing.

Step 11—The footing load is varied and step 1 to 10 are repeated. The pressure versus settlement is plotted for the case studied.

Step 12—The ultimate bearing capacity (qu) is determined by intersection tangent method as shown in Figure 10.

SOLUTION OF PARTICULAR CASES

The procedure adopted for the analysis of the pressure-settlement characteristics of various footings are given in the following paragraphs:

Smooth Flexible Strip Footing in Clay

Step 1—Evaluation of Contact Pressure—The contact pressures for a smooth footing are the normal stresses. The distribution of contact pressures is uniform for a flexible footing.

Step 2—Evaluation of Stresses—The stresses at the centre of each layer (Figure 7) along the vertical section passing through $x = 0$, $B/4$ and $B/2$ are computed using Equations (18) to (20) given below:

$$\sigma_x = \frac{q}{\pi} (\alpha^1 - \sin\alpha^1 \cdot \cos 2\beta) \tag{18}$$

$$\sigma_z = \frac{q}{\pi} (\alpha^1 + \sin\alpha^1 \cdot \cos 2\beta) \tag{19}$$

$$\tau_{xz} = \frac{q}{\pi} (\sin\alpha^1 \cdot \cos 2\beta) \tag{20}$$

Step 3—Evaluation of Principal Stresses—The principal stresses and their directions with vertical axis shown in

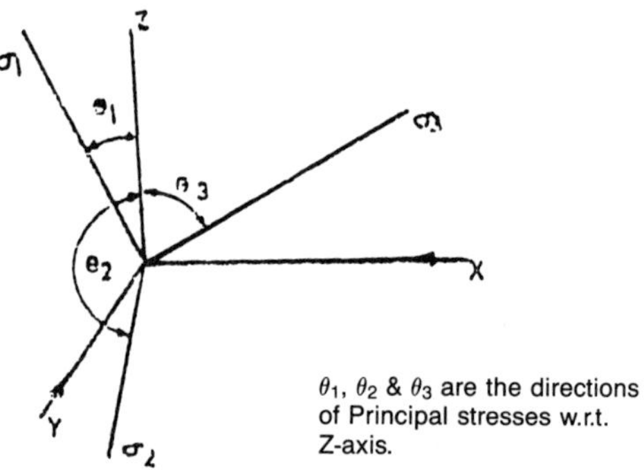

(a) Three dimensional

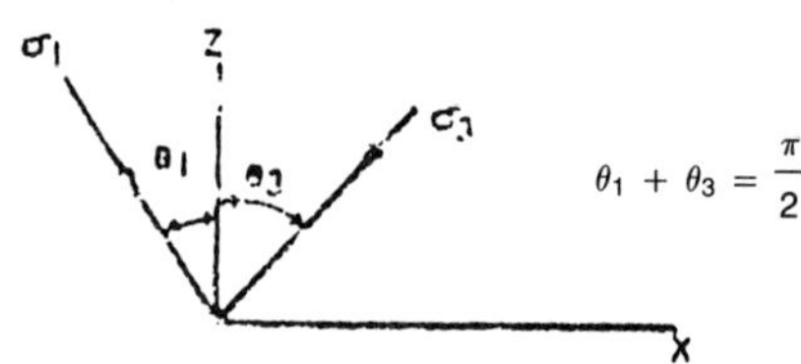

(b) Two dimensional

FIGURE 9. Principal stresses at a point and their directions.

Figure 9 are computed using the following equations:

$$\sigma_1 = \frac{\sigma_z + \sigma_x}{2} + \sqrt{\left(\frac{\sigma_z - \sigma_x}{2}\right)^2 + \tau^2 xz} \tag{21}$$

$$\sigma_3 = \sqrt{\frac{\sigma_z + \sigma_x}{2} - \left(\frac{\sigma_z - \sigma_x}{2}\right)^2 + \tau^2 xz} \tag{22}$$

$$\tan 2\theta = \frac{2\tau_{xz}}{\sigma_z - \sigma_x} \tag{23}$$

The positive value of θ is measured counterclockwise with direction of σ_z.

Step 4—Evaluation of Lateral Strain—Strip footing is a case of plane strain. The strain ϵ_2 in the direction of intermediate principal stress σ_2 equals zero in plane strain. Therefore, from Equations (10), (11), and (12), we have:

$$\epsilon_2 = 0 = \frac{1}{E} [\sigma_2 - \mu(\sigma_1 + \sigma_3)]$$

or

$$\sigma_2 = \mu(\sigma_1 + \sigma_3)$$

And

$$\epsilon_1 = \frac{1 - \mu^2}{E}\left[\sigma_1 - \frac{\mu\sigma_3}{1 - \mu}\right]$$

$$\epsilon_3 = \frac{1 - \mu^2}{E}\left[\sigma_3 - \frac{\mu}{1 - \mu}\sigma_1\right]$$

Assuming

$$\mu_1 = \frac{\mu}{1 - \mu}$$

and

$$E_1 = \frac{E}{1 - \mu^2}$$

We have

$$\epsilon_1 = \frac{1}{E_1}(\sigma_1 - \mu_1\sigma_3)$$

$$\epsilon_3 = \frac{1}{E_1}(\sigma_3 - \mu_1\sigma_1)$$

and

$$\epsilon_3/\epsilon_1 = \frac{\sigma_3 - \mu_1\sigma_1}{\sigma_1 - \mu_1\sigma_3} \tag{24}$$

Let

$$\epsilon_3/\epsilon_1 = -\mu_2 \tag{25}$$

Then

$$\mu_2 = \frac{-\sigma_3 + \mu_1\sigma_1}{\sigma_1 - \mu_1\sigma_3} \tag{26}$$

Step 5—Evaluation of Principal Strains—The strain in the direction of major principal stress is computed from constitutive relations as given below:

$$1 = \frac{a'\ (\sigma_1 - \sigma_3)}{1 - b'\ (\sigma_1 - \sigma_3)}$$

where $a' = a\ (1 - \mu^2)$, $b' = 1.1 \times b$, and a and b are parameters of Equation (5).

From Equation (25),

$$\epsilon_3 = -\mu_2\ \epsilon_1 \tag{27}$$

Step 6—Evaluation of Vertical Strain—The strain in the vertical direction (ϵ_z) for each layer is calculated using the following expression:

$$\epsilon_z = \epsilon_1\cos^2\theta_1 + \epsilon_3\cos^2\theta_3 \tag{28}$$

since $\epsilon_2 = 0$ in plain strain.

Step 7—Evaluation of Settlement—The evaluation of the total settlement along any vertical section is done by numerically integrating the quantity:

$$S = \int_o^n \epsilon_z\, dz \tag{29}$$

The total settlement was computed along vertical section passing through centre of the footing, at $B/4$ and $B/2$ from centre. The average settlement is computed by dividing the area of settlement diagram by width of the footing.

Step 8—Pressure Settlement Curve—The total settlement for various pressure intensity on the surface of the footing is computed according to the method described above and the average surface load intensity versus average settlement curve is drawn. The curve of average surface load intensity versus maximum settlement at centre is also drawn and is recommended for use in design.

Smooth Rigid Strip Footing in Clay

Step 1—Evaluation of Contact Pressure—The contact pressure distribution for centrally loaded rigid footing is taken as shown in Figure 5. It is linear from the centre to the edges and is described by two coefficients: α_1 and α_2. These coefficients have been evaluated on the following assumptions:

1. The total surface load equals the area of contact pressure diagram.

$$q \cdot B = q \cdot \alpha_1 \cdot B - q \cdot \alpha_2 \cdot \frac{B}{2}$$

Therefore,

$$\alpha_2 = 2(\alpha_1 - 1) \tag{30}$$

2. The settlements at the centre of the footing (*So*), at quarter point and at edges are all equal since the footing base is rigid.

Step 2—Evaluation of Stresses—The stresses at the centre of each layer along any vertical section have been computed by taking the difference of stresses due to rectangular load area ABCD and triangular loaded area DEC (Figure 5), using Equations (18) to (20) for uniformly distributed load and following Equations (31) to (33) for triangular loads:

$$\sigma_x = \frac{q}{\pi}\left[\alpha_1^1 + \alpha_2^1 + \frac{2x}{B}\left(\alpha_1^1 - \alpha_2^1 \right) - \frac{4z}{B}\,\ell_n\,\frac{\gamma_1\,\gamma_2}{\gamma_0^2}\right] \tag{31}$$

$$\sigma_z = \frac{q}{\pi}\,\alpha_1^1 + \alpha_2^1\,\frac{2x}{B}\,(\alpha_1^1 - \alpha_2^1) \tag{32}$$

$$\tau_{xz} = \frac{q}{\pi}\,\frac{2z}{B}\,(\alpha_1^1 - \alpha_2^1) \tag{33}$$

The principal stresses, constitutive equations, vertical strain and settlement along any vertical section are computed as in case (a) of smooth flexible strip footing.

Step 3—Pressure-Settlement Curve—The settlements at the edges and centre of a footing have been computed as described in Step (7) of case (a) for various values of contact pressure coefficient (α_1) for a given surface load intensity. That α_1 which gave equal settlements at centre quarter point and edges is adopted for determining the contact pressure distribution.

The surface load intensity is then varied and corresponding values of α_1 which fulfilled the above condition of uniform settlement are determined. Thus a set of values of surface load intensity, contact pressure coefficients and corresponding settlements are obtained. The pressure settlement curve is now plotted.

Rough Flexible Strip Footing in Clay

Roughness of base induces frictional resistance to deformation at the interface of soil and base of footing. Therefore, the base of footing is subjected to two types of stresses, namely, normal stresses due to surface load and horizontal stresses due to frictional force. The distribution of normal stresses is uniform for flexible footing, i.e., the contact pressure coefficient $\alpha_1 = 1$ (Figure 5).

Step 1—Evaluation of Contact Pressure—The horizontal frictional force has been assumed to be uniformly distributed along the width of the footing as shown in Figure 4a. The intensity of horizontal stress (C_a) has been assumed to vary as per following relationship.

$$C_a = \left[\frac{q}{C\,N_c}\right] \times C \text{ for cohesive soils} \tag{34}$$

where

q = intensity of surface load
C = cohesion
N_c = bearing capacity factor

The above relationships have been shown graphically in Figure 4b.

Step 2—Evaluation of Stresses—The stresses at the centre of each layer along a vertical section have been evaluated by considering uniform vertical and horizontal intensity of loads of magnitude (q) and (C_a) respectively. The Equations (18) to (20) have been used for vertical loads and following Equations (35) to (37) for shear loads:

$$\sigma_x = \frac{Ca}{\pi}\left[\ell_n\,\frac{(x + A)^2 + z^2}{(x - A)^2 + z^2} - \frac{4A\,xz^2}{(x^2 + z^2 - A^2)^2 + 4A^2z^2}\right] \tag{35}$$

$$\sigma_z = \frac{Ca}{\pi}\left[\frac{4A\,xz^2}{(x^2 + z^2 - A^2)^2 + 4A^2z^2}\right] \tag{36}$$

$$\tau_{xz} = \frac{Ca}{A}\left[\tan^{-1}\frac{z}{x - A} - \tan^{-1}\frac{z}{x + A} - \frac{2Az(x^2 - z^2 - A^2)}{(z^2 + z^2 - A^2)^2 + 4A^2z^2}\right] \tag{37}$$

All other steps followed are the same as in case of smooth strips footing with flexible base for computation of pressure-settlement relationships.

Rough Rigid Strip Footing in Clay

Step 1—Evaluation of Contact Pressure—In addition to the contact pressure distribution assumed in case (b), horizontal stress (C_a) as in case (c) is also assumed to act at the footing-soil interface.

Step 2—Evaluation of Stresses—The stresses at the centre of each layer along any vertical section have been computed by summing up the stresses in each layer due to normal contact pressure computed in case (b) and stresses due to horizontal stress (C_a) computed in case (c).

Steps 1 to 3 of case (b) are followed for the computation of contact pressure distribution coefficients, settlements and pressure-settlement curves.

Smooth Flexible Square Footing in Clay

It was found from the study of the above four cases, namely (a), (b), (c) and (d), that the roughness has no significant effect on pressure-settlement characteristics. Therefore, case of smooth footing has only been studied in square footings. The contact pressure distribution for centrally loaded square footing is three dimensional. All stress equations are not available for such a case so as to use constitutive laws in the analysis. All stress equations for point load are available. Therefore, stresses in soil mass due to centrally loaded square footing have been computed by dividing the whole base area into n equal squares as shown in Figure 13. The squares are very small; therefore, the total load in each square has been taken as point load.

Step 1—Evaluation of Contact Pressure—The contact pressure has been assumed as uniform as the footing is flexible.

Step 2—Evaluation of Stresses—The stresses at the centre of each layer along a vertical section have been computed using Equations (31) to (37).

Step 3—Evaluation of Principal Stresses—The Principal stresses and their directions with respect to vertical Z-axis have been computed using equations of the theory of elasticity (Durelli, 1958; Sokolnikof, 1956 and Selby, 1972) and are given below:

The stress invariant:

$$I_1 = \sigma_x + \sigma_y + \sigma_z \qquad (38)$$

$$I_2 = \sigma_x \cdot \sigma_y + \sigma_y \cdot \sigma_z + \sigma_x\sigma_z - \tau^2_{xy} - \tau^2_{yx} - \tau^2_{zx} \qquad (39)$$

$$I_3 = \sigma_x \cdot \sigma_y \cdot \sigma_z - \sigma_x\tau^2_{yz} - \sigma_y\tau^2_{zx} - \sigma_z\tau^2_{xy} + 2\tau_{xy} \cdot \tau_{yz} \cdot \tau_{zx} \qquad (40)$$

Also,

$$I_1 = \sigma_1 + \sigma_2 + \sigma_3 \qquad (41)$$

$$I_2 = \sigma_1\sigma_2 + \sigma_2\sigma_3 + \sigma_3\sigma_1 \qquad (42)$$

$$I_3 = \sigma_1\sigma_2\sigma_3 \qquad (43)$$

Solving for σ_1, σ_2 and σ_3 we have

$$\sigma_1^3 - I_1\sigma_1^2 + I_2\sigma_1 - I_3 = 0$$

$$\sigma_2^3 - I_1\sigma_2^2 + I_2\sigma_2 - I_3 = 0$$

$$\sigma_3^3 - I_1\sigma_3^2 + I_2\sigma_3 - I_3 = 0$$

Thus, σ_1, σ_2 and σ_3 are the roots of equation

$$y^3 + Py^2 + qy + r = 0 \qquad (44)$$

where

$$P = -I_1,\ q = I_2 \text{ and } \gamma = -I_3$$

Putting

$$y - x - P/3 = x + I_1/3$$

$$a = 1/3\,(3q - P^2) = 1/3\,(3I_2 - I_1^2) \qquad (45)$$

and

$$b = 1/27\,(2P^3 - 9Pq + 27r)$$

$$= 1/27\,(-2I_1^3 + 9I_1 I_2 - 27I_3) \qquad (46)$$

Equation (44) reduces to

$$x^3 + ax + b = 0 \qquad (47)$$

which may be compared with

$$\cos^3\frac{\theta}{3} - 3/4\cos\theta/3 - 1/4\cos\theta = 0 \qquad (48)$$

$$\text{Let } x = \alpha \cdot z$$

Then Equation (47) becomes

$$\alpha^3 z^3 + a\,\alpha\, z + b = 0$$

$$z^3 + (a/\alpha^2)z + (b/\alpha^3) = 0$$

or

$$z^3 - 3/4z + \frac{b\,3\sqrt{3}}{8(-a)\sqrt{-a}} = 0 \qquad (49)$$

Comparing equals of Equations (48) and (49)

$$a/\alpha^2 = -3/4$$

or

$$\alpha = 2\sqrt{-a/3}$$

Therefore,

$$1/4\cos\theta = \frac{b\,3\sqrt{3}}{8\,a\sqrt{-a}} \qquad (50)$$

$$\cos\theta = \frac{b\,3\sqrt{3}}{2a\sqrt{-a}} = \left(-\frac{27b^2}{4a^3}\right)^{1/2} \qquad (51)$$

$$\tan\theta/2 = \sqrt{\frac{1-\cos\theta}{1+\cos\theta}}$$

$$\therefore\ \theta = 2\tan^{-1}\sqrt{\frac{1-\cos\theta}{1+\cos\theta}} \qquad (52)$$

Solution is cos $\theta/3$ which will correspond to z in Equation (49). Then the roots of Equation (49) are as follows:

$$x_1 = 2\sqrt{-a/3}\ \cos\theta/3$$

$$x_3 = 2\sqrt{-a/3}\ \cos(2\pi/3 + \theta/2)$$

$$x_2 = 2\sqrt{-a/3}\ \cos(4\pi/3 + \theta/3)$$

The principal stresses are as follows:

$$\sigma_1 = x_1 + I_1/3 \quad (53)$$

$$\sigma_2 = x_2 + I_1/3 \quad (54)$$

$$\sigma_3 = x_3 + I_1/3 \quad (55)$$

Solving for direction cosines with respect to z axis, we have (Sokolnikoff, 1956)

$$A_1 = (\sigma_y - \sigma_1)(\sigma_2 - \sigma_1) - \tau_{yz} \cdot \tau_{zy}$$

$$B_1 = -\tau_{xy}(\sigma_z - \sigma_1) + \tau_{xy} \cdot \tau_{xz}$$

$$C_1 = \tau_{xy} \cdot \tau_{yz} - (\sigma_y - \sigma_1)\tau_{xy}$$

$$A_2 = (\sigma_y - \sigma_2)(\sigma_2 - \sigma_2) - \tau_{yz} \cdot \tau_{zy}$$

$$B_2 = -\tau_{xy}(\sigma_z - \sigma_2) + \tau_{zy} \cdot \tau_{xz}$$

$$C_2 = \tau_{xy} \cdot \tau_{yz} - (\sigma_y - \sigma_2)_{xy}$$

$$A_3 = (\sigma_y - \sigma_3)(\sigma_z - \sigma_3) - \tau_{yz} \cdot \tau_{zy}$$

$$B_3 = -\tau_{xy}(\sigma_z - \sigma_3) + \tau_{zy} \cdot \tau_{xz}$$

$$C_3 = \tau_{xy} \cdot \tau_{yz} - (\sigma_y - \sigma_3)\,\tau_{xz}$$

and

$$\cos\theta_1 = \frac{C_1}{\sqrt{A_1^2 + B_1^2 + C_1^2}} \quad (56)$$

$$\cos\theta_2 = \frac{C_2}{\sqrt{A_2^2 + B_2^2 + C_2^2}} \quad (57)$$

$$\cos\theta_3 = \frac{C_3}{\sqrt{A_3^2 + B_3^2 + C_3^2}} \quad (58)$$

Step 4—Evaluation of Vertical Strain

$$\epsilon_z = \epsilon_1 \cos^2\theta_1 + \epsilon_2 \cos^2\theta_2 + \epsilon_3 \cos^2\theta_3 \quad (59)$$

The principal strains ϵ_1, ϵ_2, and ϵ_3 were evaluated from Equations (5), (13) and (14) respectively.

The evaluation of constitutive equations settlement and pressure-settlement curves is done as in case (a) of smooth flexible strip footing.

Smooth Rigid Square Footing in Clay

The analysis of rigid square footing on clay has also been done by dividing the whole area of foundation in n equal parts as shown in Figure 11. The loads in each small square have been taken as point loads.

The contact pressure distribution has been assumed as shown in Figure 12 in three dimensional space and has been defined by two coefficients, α_1 and α_2. The coefficients have been evaluated by the following assumptions.

1. The total surcharge load equals the volume of the contact pressure diagram:

$$q \cdot A = \alpha_1 \cdot q \cdot A - 1/3\alpha_2 \cdot q \cdot A$$
$$\alpha_2 = 3(\alpha_1 - 1) \quad (60)$$

when $\alpha_1 = 1$, $\alpha_2 = 0$ and when $\alpha_1 = 1.5$, $\alpha_2 = 1.5$.

2. The settlement at the centre quarter point and edges have been taken equal since the footing is rigid.

The evaluation of stresses at the centre of each layer, the principal stresses and vertical strains have been done as for flexible square footing. The constitutive equations, strains, settlement and pressure-settlement curves are evaluated as in case (b) of rigid strip footing.

Eccentrically Loaded Smooth Rigid Strip Footing in Clay

Step 1—Evaluation of Contact Pressure—The contact pressure distribution for eccentrically loaded rigid footing has been shown in Figure 6 and is described by three coefficients: $\alpha\alpha_1$, $\alpha\alpha_2$ and $\alpha\alpha_3$. These coefficients have been evaluated on the following assumptions:

1. The total surcharge load equals the area of the contact pressure diagram:

$$q(\alpha + b) = \alpha\alpha_1 \cdot q(a + b)$$
$$+ 1/2\alpha\alpha_2\, q \cdot a + 1/2\alpha\alpha_3\, q \cdot b$$

2. Moment of areas about point E are equal:

$$\alpha\alpha_1 \cdot qb \cdot b/2 + 1/2\alpha\alpha_3 q \cdot b \cdot 2/3b$$
$$= \alpha\alpha_1 \cdot q \cdot a \cdot B/2 + 1/2\alpha\alpha_2 \cdot q \cdot a \cdot 2/3a$$

Solving above two equations in terms of $\alpha\ \alpha_1$, we get

$$\alpha\alpha_2 = \frac{4b(a + b) - (3a^2 + 4ab + b^2)\alpha\alpha_1}{2(a^2 + ab)} \quad (61)$$

$$\alpha\alpha_3 = \frac{4a(a + b) - (a^2 + 4ab + 3b^2)\alpha\alpha_1}{2(b^2 + ab)} \quad (62)$$

3. The slope of the line $A\ E\ B$ has been assumed as uniform

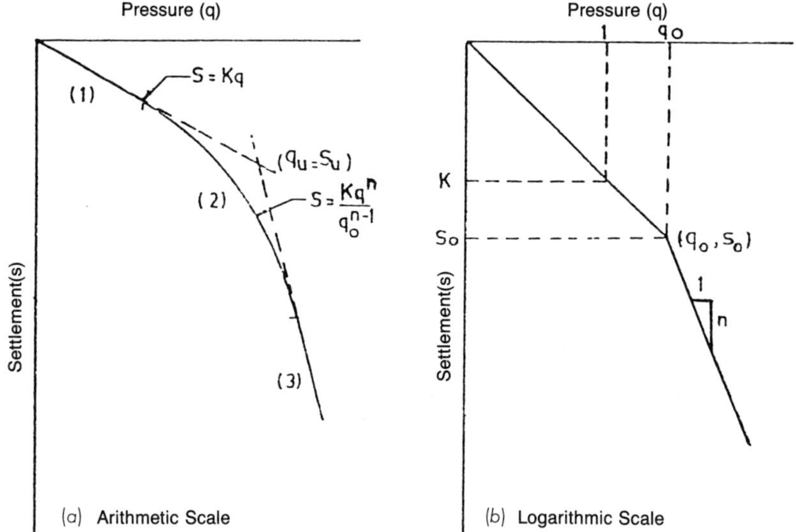

FIGURE 10. Pressure vs. settlement curves on arithmetic and logarithmic scale.

for evaluation of the value of $\alpha\ \alpha_1$ for any surface load, i.e.,

$$\frac{S_m - S_c}{\frac{B}{2} - e} = \frac{S_e - S_L}{\frac{B}{2} - e} \tag{63}$$

where e = eccentricity of load.

Step 2—Evaluation of Stresses—The contact pressure diagram Figure 6 has been divided into three parts: (1) a rectangular loaded area *ABCG* over width *B*, (2) a triangular loaded area *CDE* over width *a*, and (3) another triangular loaded area *EFG* over width *b*. The stresses in the soil mass at the centre of a layer have been computed by summing up the stresses due to all the above three surface loads.

The evaluation of principal stresses, constitutive laws, strains and settlements were done as for smooth flexible strip footing.

Step 3—Pressure Settlement Curve—The settlements S_e at the section *E*, and settlements S_m and S_L at edges have been computed as described above for various values of $\alpha\alpha_1$. For a given eccentricity the value of $\alpha\alpha_1$, which satisfied the condition (3) in step 1 of this section for a given surface load intensity, is taken for determining contact pressure distribution. The corresponding values of S_e, S_m and S_L are noted. The surface load is varied and a set of values of $\alpha\alpha_1$, S_e, S_m and S_L are computed.

The eccentricity of load is changed to various values and the values of $\alpha\alpha_1$, S_e, S_m and S_L at various surcharge loads are computed which fulfilled the condition (3) in step 1 of this section.

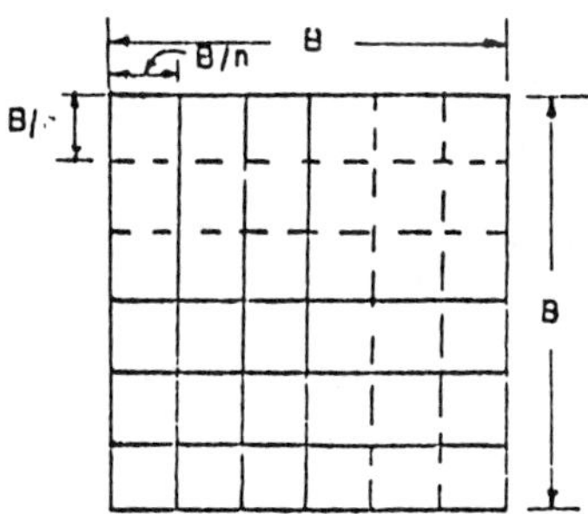

FIGURE 11. Division of square footing in small squares.

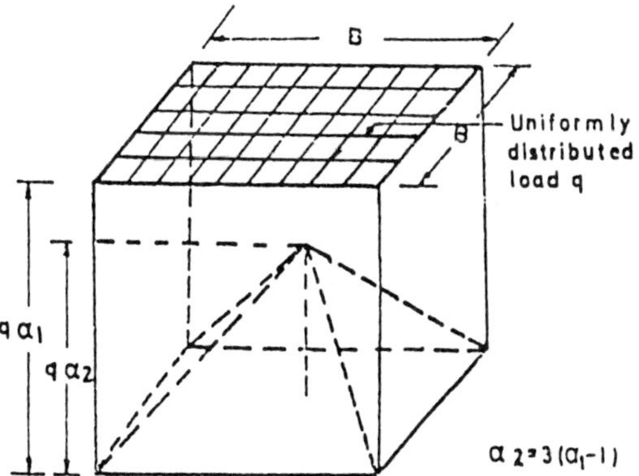

FIGURE 12. Contact pressure distribution for uniformly loaded square footing.

Curves of average pressure versus S_e and S_m for various eccentricities are then plotted.

Step 4—Tilt of Footing—The tilt (t) of footing for various load eccentricities is calculated from the following relationship:

$$\sin t = \frac{S_m - S_e}{\left(\frac{B}{2} - e\right)} \tag{64}$$

as shown in Figure 6. The tilt at various eccentricities is plotted against various intensities of surcharge loads to obtain pressures versus tilt relationship.

Step 5—Contact Width—Footing base tends to lose contact with the supporting soil surface when the load eccentricity increases. Since soil takes no tension, the contact pressure at the point of loss of contact will be zero. The point of zero contact pressure is found from the contact pressure distribution diagram. The ratio of contact width to actual width of footing has been represented by X_1.

Footings in Sand

INTRODUCTION

The pressure-settlement characteristics for sand using non-linear stress-strain relationship have been computed by Duncan and Chang (1970) and Basavanna (1975) using Finite Element Technique and non-linear constitutive relationships.

Duncan and Chang worked with Chatahookhee river sand and strip footing of 0.06 m × 0.316 m, placed at a depth of 0.508 m from surface. The load settlement curve computed by them was found to be in good agreement with that of plate load test up to a bearing pressure of 1.06 kg/cm² and gave ultimate bearing capacity of 0.882 kg/cm² as against the value of 3.80 kg/cm² from Terzaghi's formula. Within the range of pressures, their plate load test did not show the failure of soil.

Basavanna considered 10 cm × 10 cm square footing resting on the surface of Ranipur sand and computed the pressure-settlement curve. His results, shown in Figure 13, compare with experimental curve only up to the pressure intensity of 0.13 kg/cm². Beyond 0.13 kg/cm² his computed values deviate very largely. He found the ultimate bearing capacity as 0.16 kg/cm² as against 1.45 kg/cm² obtained by plate load test.

From the above investigations, it is evident that the pressure-settlement curve of sand obtained by F.E.M./. Technique deviates to a large extent from that of plate load test.

The pressure-settlement curve for strip footing of 5 cm width resting on Ranipur sand was investigated by the same method used in this investigation for the analysis of strip footing in clay. The results are plotted in Figure 14. The experimental curve is also partly shown in the above figure. It is observed that the rate of settlement decreases at higher intensity of surface loads. It is contrary to the physical behaviour of footing under load. The amount of settlement is also very small. The reason of these results is that the stress equations used in the formulation are valid for isotropic, elastic and homogenous mass. The value of youngs modulus E or ($1/a$) is taken as constant. Sand, on the contrary, is an anisotropic material. The Young's modulus is dependent upon confining pressure and varies in both vertical and horizontal directions. No stress equations are available where variation of E in both vertical and horizontal directions have been considered.

ASSUMPTIONS

In the present investigation an empirical approach has been developed which has been found to yield satisfactory results. The following assumptions have been made:

1. The effect of the weight of soil mass has been considered in determination of stresses in sand. Vertical stress due to weight of soil has been taken equal to γz where γ is the density of soil and z is the depth. The horizontal stress due to weight of soil has been taken equal to $K_o \gamma_z$ where K_o is the coefficient of earth pressure at rest. The soil-state varies from at rest condition at centre to active condition at the edges. However, the pattern of variation of earth pressure coefficient from centre to edges is not known.
2. The ultimate bearing capacity (q_u) has been computed from Terzaghi's formula.
3. A coefficient F has been introduced such that at all stress levels the following relationship is satisfied:

$$\frac{q_u}{q} = \frac{\sigma_u}{\sigma_1' - \sigma_3'} = F \tag{65}$$

where q is the intensity of surface load, σ_u is the ultimate stress from hyperbolic relationship and is equal to ($1/b$), σ_1', σ_3' are the major and minor principal stresses in the soil mass due to surface load q and weight of soil.
4. Constitutive law described by hyperbola has been taken. Therefore, σ_u is taken equal to ($1/b$).
5. The modulus of elasticity (E_s) has been calculated from the Figure 15 at stress level of σ_u/F. The strain at σ_u/F

$$= \frac{a(\sigma_u/F)}{1 - b(\sigma_u/F)} \tag{66}$$

Therefore,

$$E_s = \frac{\text{stress}}{\text{strain}} = \frac{1 - b(\sigma_u/F)}{a} \tag{67}$$

σ_u of ($1/b$) is a function of confining pressure which

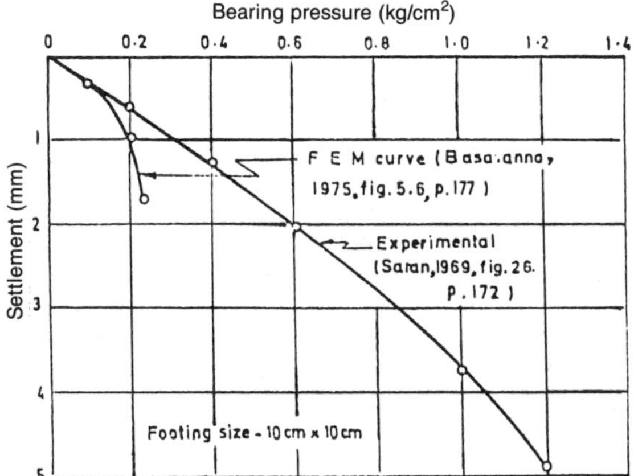

FIGURE 13. Pressure vs. settlement curve for square footing resting on Ranipur sand (D_R = 84%) computed by F.E. method.

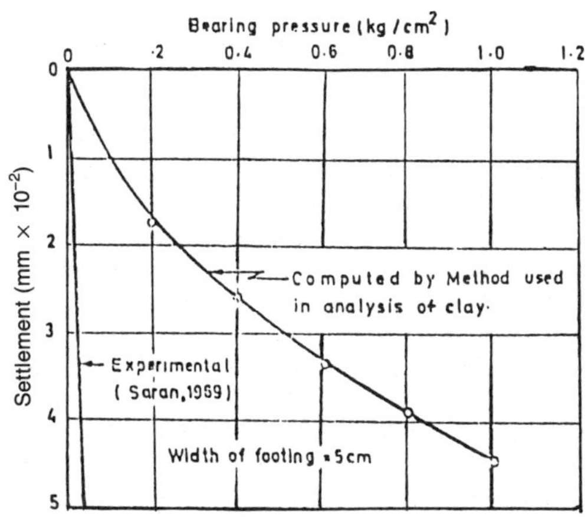

FIGURE 14. Pressure vs. settlement curve for strip footing on Ranipur sand (D_R = 84%).

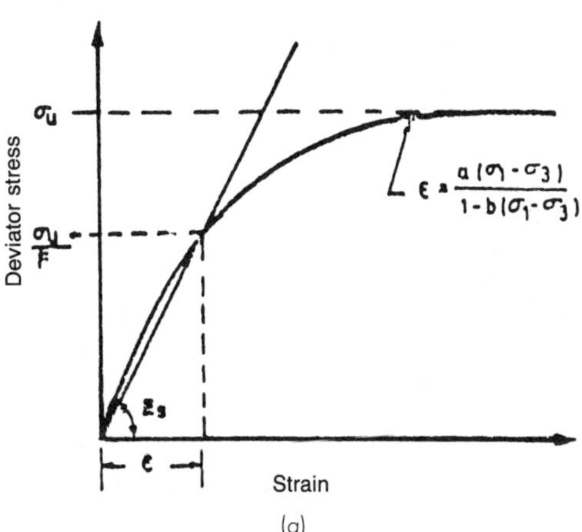

(a)

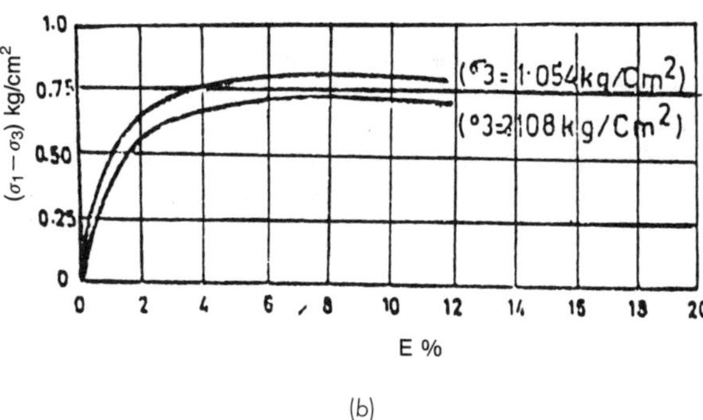

(b)

FIGURE 15. (a) Hyperbolic stress-strain relationship; (b) U.U. stress-strain curves for Buckshot clay (Carrol, 1963).

varies both in vertical and horizontal directions in sand. Therefore, E_s also varies in both vertical and horizontal directions.

6. It has been found in case of clay that pressure versus average settlement curve for flexible footing assuming uniform contact pressure distribution and pressure versus uniform settlement curve for rigid footings were very close to each other. Therefore, for the sake of computation, uniform contact pressure distribution has been assumed for computing pressure settlement curves. The average settlements have been computed for various pressure intensities. The pressure-settlement curve computed by the present empirical approach may be taken as for rigid strip footing.

PROCEDURE FOR ANALYSIS

The procedure followed in the analysis of uniformly loaded strip footing resting on sand is described in the following steps.

Step 1—The stresses at the centre of each layer along the vertical axis passing through centre, quarter point and edges shown in Figure 7 using Equations (18) to (20) for a given surface load intensity q are computed.

Step 2—The principal stresses σ_1 and σ_3 and their directions with vertical axis are computed using Equations (21) to (23). The effect of the weight of sand has been incorporated by adding $(\gamma \cdot z)$ to σ_1 and $K_o\,\gamma \cdot z$ to σ_3 where $K_o = 1 - \sin\Phi$. The value of Φ is determined from triaxial tests.

Step 3—The ultimate shear strength (σ_u) of sand for a given confining pressure σ_3' is computed from constitutive laws.

Step 4—The ultimate bearing pressure (q_u) is calculated from conventional bearing capacity equations (Terzaghi, 1943; Meyerhof, 1951).

Step 5—A coefficient F for a given surface load intensity (q) is computed from the following relationship:

$$\frac{q_u}{q} = F \qquad (65)$$

Step 6—The modulus of elasticity (E_s) is calculated by the following equation:

$$E_s = \frac{1 - b(\sigma_u/F)}{a} \qquad (67)$$

where a and b are the constants of hyperbola whose values depend upon confining pressure.

Step 7—The strain in each layer in the direction of major principal stress is calculated from the equation

$$\epsilon_1 - (\sigma_1' - \sigma_3')/E_s \qquad (68)$$

where $\sigma_1' = \sigma_1 + \gamma z$ and $\sigma_3' = \sigma_3 + K_o\,\gamma z$.

The strain in the direction of minor principal stress is calculated from the following relationship:

$$\epsilon_3 = -\mu_2\epsilon_1 \qquad (69)$$

Step 8—The strain in the vertical direction is calculated using Equation (28), namely,

$$\epsilon_z = \epsilon_1\cos^2\theta_1 + \epsilon_3\cos^2\theta_3 \qquad (28)$$

Step 9—The settlement of each layer and the total settlement along any vertical section are calculated using Equations (16) and (17) respectively.

Step 10—The average settlement is computed by dividing the area of settlement diagram by width of footing.

Step 11—The surface load intensity is varied and the steps 1 to 10 are repeated. The pressure settlement curve is obtained by plotting settlements obtained in step 10 against corresponding surface load in tensity.

APPLICATION OF METHODOLOGY

General

A footing resting on clay may fail immediately after construction due to shear failure of the supporting soil. This condition is similar to undrained triaxial compression test condition. If the footing does not fail immediately after construction then it undergoes long-term settlement. This condition is similar to drained triaxial compression test condition.

The pressure settlement characteristics of footings resting on Buckshot clay have been predicted using stress-strain relationship for undrained shear strength tests as this was used by Carrol (1963) in his experiments to determine the pressure-settlement curves for the footings. His rate of loading in plate load test was fast as he was mainly concerned with the determination of dynamic bearing capacity of footings. The soil in his plate load test was nearly saturated (90 percent) and b/a ratio was 142 where b and a are the constant of transformed hyperbolic stress-strain plot.

The pressure-settlement characteristics of footings resting on Dhanori clay have been predicted using the stress-strain relation of drained shear test.

Dhanori clay was used by Prakash (1975) in conducting plate load test at a very slow rate so as to allow pore-pressure dissipation. The pressure-settlement curves using undrained shear strength test data were also computed but were found to give very small values of settlements compared to the experimental values of Prakash (1977). The soil used was nearly saturated (saturation = 91.15 percent) and b/a ratio was 15 where b and a are the constants of transformed hyperbolic stress-strain plot.

The stress-strains for Ranipur sand was taken from the

consolidated undrained shear strength test to predict the pressure-settlement curves of footings. The sand had a coefficient of uniformity of 1.75 and relative densities were 84 percent and 46 percent.

The pressure-settlement characteristics for the following cases have been obtained by the method described earlier.

Footing Resting on Clay

a. Smooth flexible strip footing
b. Smooth rigid strip footing
c. Rough flexible strip footing
d. Rough rigid strip footing
e. Smooth flexible square footing
f. Smooth rigid square footing
g. Eccentrically loaded smooth rigid strip footing

Footings Resting on Sand

a. Smooth rigid strip footing

In this article the results of computations have been analysed for evaluating pressure-settlement curves, ultimate bearing capacity, settlements, tilts and contact pressure distribution. The results have been presented in the form of graphs, tables and equations. Comparison with the existing theories and results of previous investigators have also been presented.

FOOTING RESTING ON CLAY

Smooth Flexible Strip Footing

Flexible footings are generally not encountered in practice. But they are easy to solve and are, therefore, solved in developing new theories. It has been found later that the average settlement of flexible footing is almost the same as the settlement of a rigid footing for equal pressure intensities.

A flexible footing produces uniform pressure at the soil-footing interface. The contact pressure coefficient α_1, defined earlier, will be unity. Soil offers no lateral resistance to deformation when the footing base is smooth. Therefore, the frictional force at the soil-footing interface has been taken as zero.

PRESSURE-SETTLEMENT CURVE

The settlements for flexible strip footings of widths 25 cm, 60 cm, 100 cm, 200 cm and 400 cm resting on Buckshot clay were computed for various surface load intensities varying from 0.25 kg/cm^2 to 1.75 kg/cm^2 at three sections, namely, central, quarter point and edges. It was found that the maximum settlement occured at the central section. The pressure versus maximum settlement curves are plotted in Figure 16a. The curves show that for a particular footing, the settlement increases with load intensity at the surface and the rate of increase increases at higher load intensities. The values of pressures and the corresponding maximum settlements are given in column 2 of Table 1 for footing width of 25 cm.

TABLE 1. Maximum and Average Settlements of Smooth Flexible Strip Footing Resting on Buckshot Clay, B = 25 cm.

Pressure kg/cm^2	Maximum Settlement (mm)	Average Settlement (mm)	Percent Difference (%)
1.	2.	3.	4.
0.25	1.06	1.05	1.0
0.50	2.40	2.30	4.16
0.75	4.21	3.95	6.17
1.00	7.09	6.50	9.08
1.25	14.70	13.02	11.43
1.50	61.77	49.50	19.86
1.75	139.96	102.00	27.12

Average settlements of the footings mentioned above were computed by dividing the area of settlement diagrams by the width of the footing for 25 cm wide footing. Diagrams for determining settlements are shown in Figure 16b. For various pressure intensities these are listed in Table 1 for footing width of 25 cm. The pressure versus average settlements are plotted in Figure 17 for all the footings under consideration. The trends are the same as pressure versus maximum settlement curves.

Table 1 shows that the maximum settlements are always higher than the average settlements and their differences increase with increase of surface load intensity. Maximum settlements are needed for design purposes.

ULTIMATE BEARING CAPACITY

The ultimate bearing capacity was computed by intersection tangent method, and for 25 cm wide footing the ultimate bearing capacity values came out as 1.22 kg/cm^2 and 1.26 kg/cm^2 from pressure versus maximum settlement and pressure versus average settlement curves respectively. The ultimate bearing capacity values were also determined by intersection tangent method (Figures 16, 17) and they also give the above values.

The pressure-settlement curves of Figure 16 were further analysed by plotting a curve between pressure versus S_f/B, where S_f is the settlement of footing of width B for a given pressure and is shown in Figure 18a. It is found that a single curve is obtained for all footing sizes. The ultimate bearing pressure obtained by intersection tangent method from the above curves of Figure 18a equals 1.22 kg/cm^2.

A single curve was also obtained for pressure versus aver-

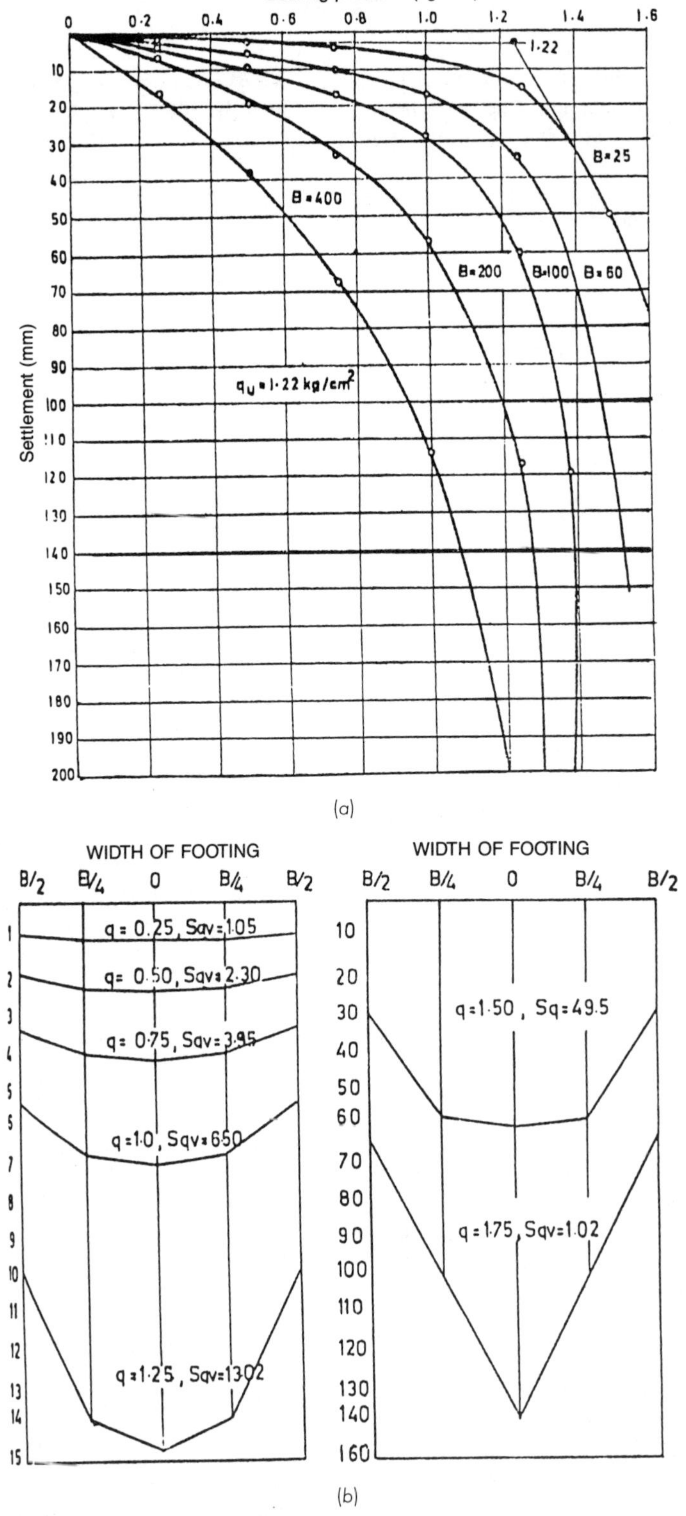

FIGURE 16. (a) Pressure vs. max. settlement of smooth flexible strip footing retin on Buckshot clay; (b) Average settlement calculation for flexible strip footing with smooth base on clay.

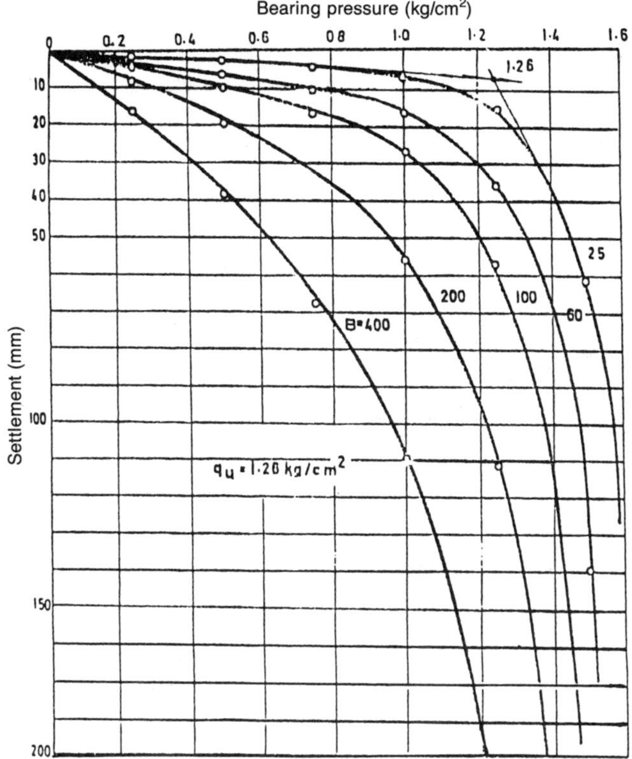

FIGURE 17. Pressure vs. average settlement of smooth flexible strip footing resting on Buckshot clay.

age settlement curves of Figure 17 for pressure versus S_f/B relationship and is shown in Figure 18b. The ultimate bearing pressure was found to be 1.26 kg/cm². From the curves of Figure 18a and 18b, it may be concluded that the ultimate bearing pressure for flexible strip footing in clay is independent of the size of the footing and is practically the same from the pressure versus maximum settlement and pressure versus average settlement considerations.

SETTLEMENT AT FAILURE

The settlement corresponding to the ultimate bearing capacity has been defined as settlement at failure (S_u). The value of S_u has been computed from curves of Figure 16 and Figure 17. The ratio of S_u versus B, where B is the width of footing, has been plotted in Figures 19a and 19b for pressure versus maximum settlement curves and pressure versus average settlement curves respectively. This plot is a straight line parallel to horizontal axis indicating that the ratio (S_u/B) is found to be 5.0 percent of the width of footing from pressure versus maximum settlement and 4.8 percent from pressure versus average settlement considerations. It is quite expected since the maximum settlement will be always more than the average settlement.

S_f/S_p VERSUS S_f/B_p RELATIONSHIP

The S_f/S_p versus B_f/B_p relationship were plotted in Figure 20, where S_f and S_p are the settlements of footings of widths B_f and B_p respectively. This plot has been made from pressure versus maximum settlement and pressure versus average settlement considerations. B_p has been taken as 60 cm and B_f = 25 cm, 100 cm, 200 cm and 400 cm. It is found that this gives a unique straight line with equal inclination to both the axes and therefore is governed by the equation

$$S_f/S_p = B_f/B_p \tag{70}$$

It also confirms that the relationship of the above quantities given in Taylor (1948) is true for footings in clay.

Smooth Rigid Strip Footing

A rigid footing settles uniformly. Since a uniform pressure distribution causes a dish shaped pattern of settlement, in order to produce a uniform settlement, the contact pressure must increase on the edges and decrease near the centre line of the footing. A simple pattern of distribution of

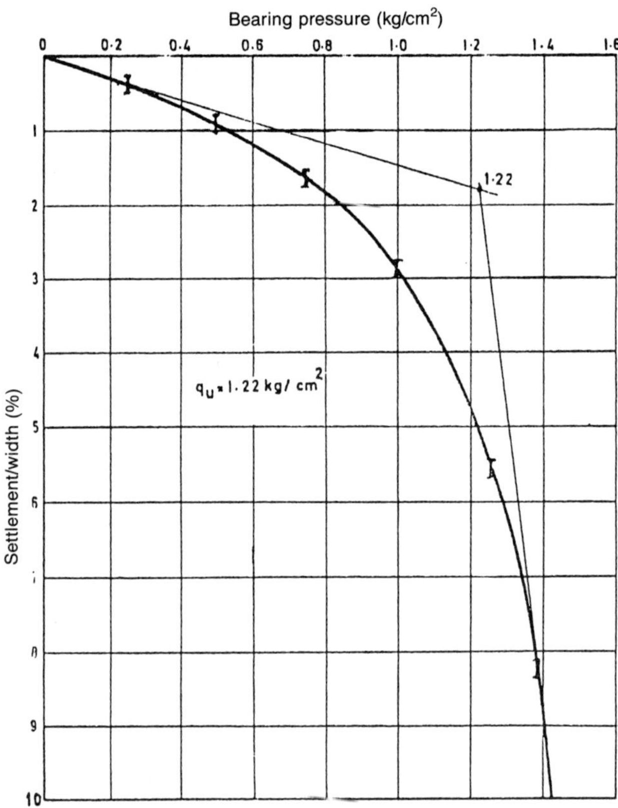

FIGURE 18a. Pressure vs. S_{max}/B for smooth flexible strip footing resting on Buckshot clay.

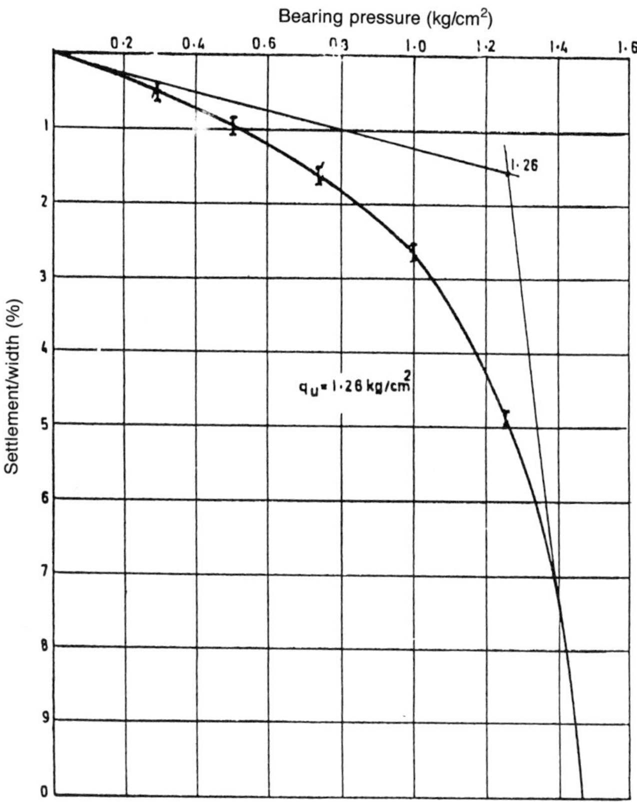

FIGURE 18b. Pressure vs. S_{av}/B for smooth flexible strip footing resting on Buckshot clay.

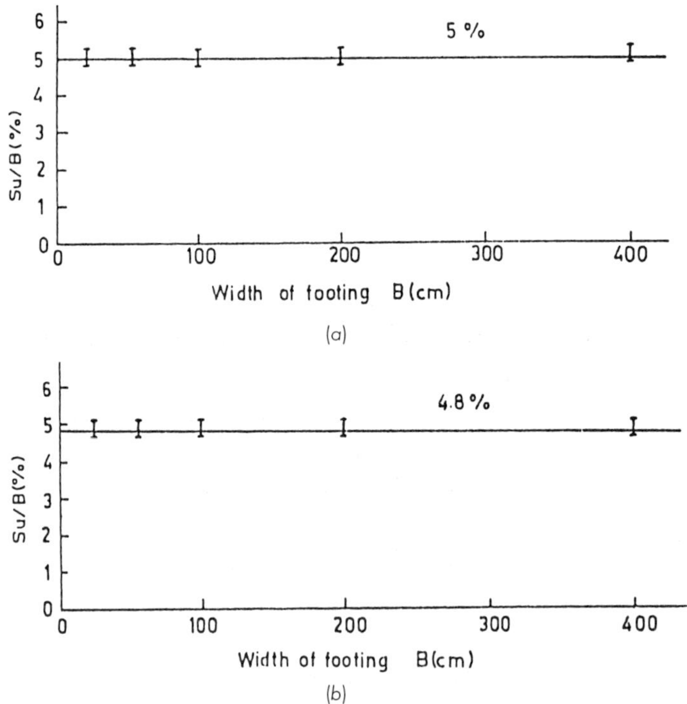

FIGURE 19. (a) S_u/B vs. B for smooth flexible strip footing resting on Buckshot clay using maximum settlement criteria; (b) S_u/B for smooth flexible strip footing resting on Buckshot clay using average settlement criteria.

contact pressure has been assumed (Figure 5). The coefficient of contact pressure α_1 has been evaluated in the analysis.

CONTACT PRESSURE DISTRIBUTION

The contact pressure distribution was determined by evaluating settlements at three sections, namely, $x = 0$, $x = B/4$ and $x = B/2$ for one pressure intensity and different values of contact pressure coefficient α_1. For example, for a pressure intensity of 1.00 kg/cm², the value of α_1 was varied from 1.0 to 1.25 at an internal of 0.05 and the corresponding values of settlements were obtained at the above-mentioned three sections. These values for footing width of 25 cm have been plotted in Figure 21.

The above figure shows that for $\alpha_1 = 1.07$, the settlements at $x = 0$ and $x = B/4$ sections are equal. Similarly, for $\alpha_1 = 1.09$, the settlements at $x = 0$ and $x = B/2$ sections are equal. In the present investigation, that value of α_1 which gave the minimum variation of settlements at $x = 0$, $B/4$ and $B/2$ was computed from the graphs which works out to be 1.080. Similar analysis was done for other footing sizes and for other pressure intensities. The values of α_1 computed as above and the corresponding settlements are given in Table 2.

The Table 2 shows that for a given pressure, the contact pressure coefficient α_1 is the same for all footing sizes. The pattern of distribution of contact pressure is, therefore, the same for all footing sizes in the two-dimensional problem of smooth rigid strip footing.

The contact pressure distribution has been plotted in Figure 22. It is found that the contact pressure curve becomes flatter with increase of pressure at surface and finally becomes uniform at failure. Soil remains in the elastic state before loading. As the load at the surface is increased, plastic flow starts at the edges of the footing while the soil in the central region remains in elastic range. As the failure load is reached, plastic failure occurs all over and therefore, contact pressure distribution becomes uniform.

The contact pressure distribution is further studied by plotting α_1 versus non-dimensional quantity (q/c_u) in Figure 23 where c_u is the compressive strength of soil in triaxial compression test. The curve shows that α_1 varies linearly with surface load intensity and is governed by the following equation.

$$\alpha_1 = 1.36 - 0.778\ (q/c_u) \tag{71}$$

within the range of $q = 0.25$ kg/cm² to q_u.

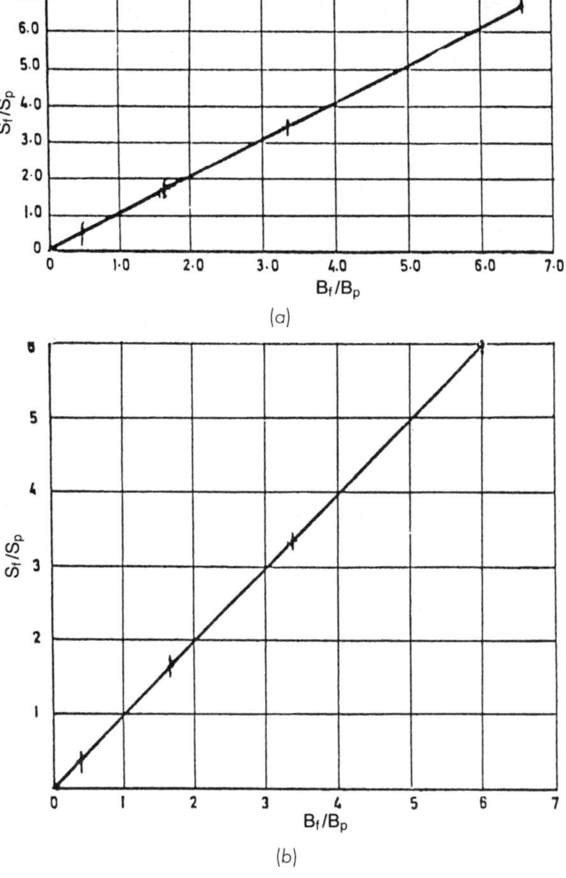

FIGURE 20. (a) S_f/S_p vs. B_f/B_p for smooth flexible strip footing resting on Buckshot clay using max. settlement criteria; (b) S_f/S_p vs. B_f/B_p for smooth flexible strip footing resting on Buckshot clay.

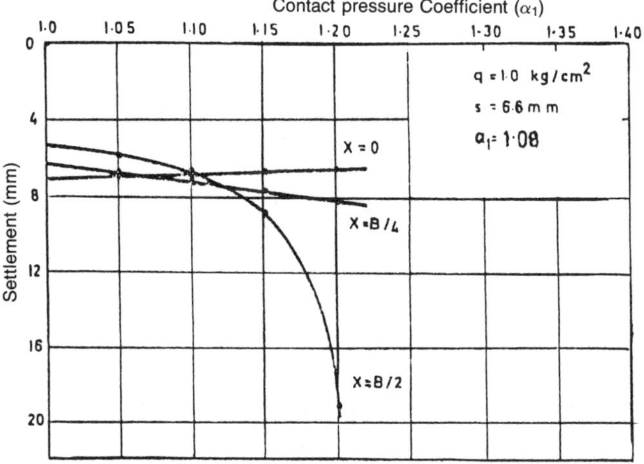

FIGURE 21. Contact pressure coefficient vs. settlement of smooth rigid strip footing on Buckshot clay of 25.0 cm width.

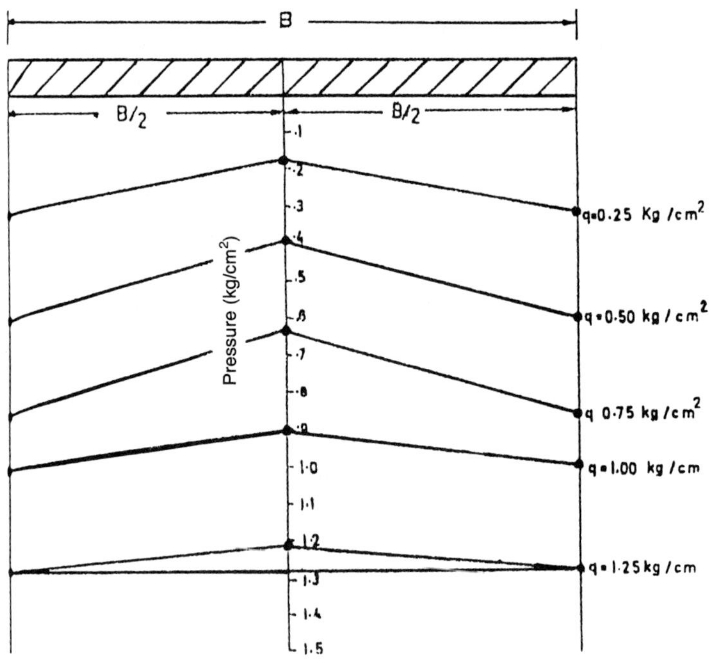

FIGURE 22. Contact pressure distribution below smooth rigid strip footing resting on Buckshot clay.

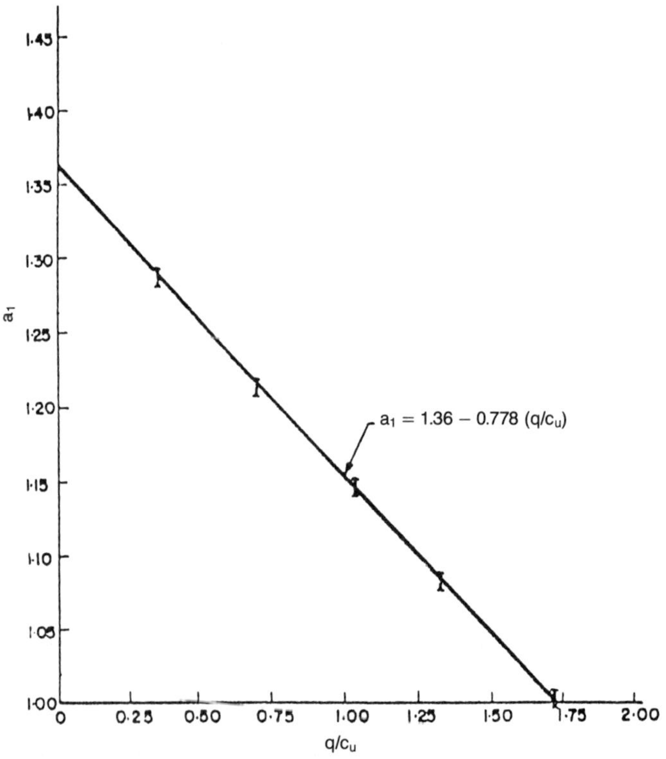

FIGURE 23. a_l vs. q/c_u for smooth rigid strip footing resting on Buckshot clay.

TABLE 2. Contact Pressure Coefficients and Settlements for Smooth Rigid Strip Footing Resting on Buckshot Clay.

	B = 25 cm		B = 60 cm		B = 100 cm		B = 200 cm		B = 400 cm			
Pressure q kg/cm²	α_1	S (mm)	α_1	S (mm)	α_1	S (mm)	α_1	S (mm)	α_1	S (mm)	Cu kg/cm²	q/Cu
1.	2.	3.	4.	5.	6.	7.	8.	9.	10.	11.	12.	13.
0.25	1.258	1.006	1.285	2.41	1.281	4.02	1.28	8.05	1.285	16.0	0.72	0.347
0.50	1.215	2.28	1.215	5.45	1.215	9.1	1.215	18.2	1.215	36.3		0.694
0.75	1.55	4.0	1.155	9.60	1.155	16.0	1.155	32.0	1.155	64.0		1.044
1.00	1.080	6.4	1.095	16.1	1.095	26.8	1.095	54.0	1.095	107.5		1.388
1.25	1.00	14.5	1.025	35.0	1.00	56.0	1.025	112.0	1.005	225.0		1.735
1.50	1.00	29.0	1.00	70.0	1.00	84.0	1.000	208.0	1.00	425.0		2.082

The above equation gives a realistic contact pressure distribution for various surface load intensities and is based upon the compressive strength of soil. No simple equation is available for determining contact pressure distribution which gives the true physical picture.

PRESSURE-SETTLEMENT CURVES

Pressure-settlement curves of smooth rigid strip footings of different widths are shown in Figure 24. These curves are analysed in the same way as in the previous case to obtain the pressure verses S_f/B (Figure 25), S_u/B versus B (Figure 26) and S_f/S_p (Figure 27) relationships. It is found that the trend of the above relationships are exactly the same as for smooth flexible footings. The ultimate bearing pressure was found to be 1.26 kg/cm² from both pressure versus settlement curves and pressure versus S_f/B curve. The intersection tangent method gave nearly the same value. The corresponding value from Terzaghi's formula works out to 1.39 kg/cm². The settlement at failure was 5.1 percent of the width of footing. The equation

$$S_f/S_p = B_f/B_p \qquad (70)$$

is also confirmed in case of smooth rigid strip footing.

EFFECT OF RIGIDITY

The effect of rigidity on ultimate bearing capacity and settlement has been shown in Figure 28. Curve 1 shows the pressure-settlement curve of smooth flexible strip footing of width equal to 25 cm taking average settlement as criterion and curve 2 for smooth rigid strip footing of the same width. It is found that the settlements at failure are 13.0 mm and 12.5 mm and ultimate bearing pressures are 1.26 kg/cm² for both rigid and flexible cases respectively. Thus, there is practically no significant difference in the values of ultimate bearing capacity, and settlement of rigid and average settlement of flexible smooth strip footings at failure.

Rough Flexible Strip Footing in Clay

This case was studied for the sake of testing the validity of the programme for its use in the practical case of rough rigid strip footing. Roughness of footing base induces frictional force proportional to surface load at the interface of footing base and soil. The intensity of frictional force has been assumed as uniform from centre to the edges of the footing but of opposite sign (Figure 4a). Frictional force at the interface causes stress in the soil mass which has been considered while solving this problem.

As discussed in case (a), the contact pressure for flexible footing is uniform. Therefore, the stresses in soil mass have also been considered due to uniform pressure. The total stress at a point in soil mass has been computed from the above two considerations.

PRESSURE-SETTLEMENT CURVES

The pressure settlement curves for footing widths of 25 cm, 60 cm, 100 cm, 200 cm and 400 cm were computed and are plotted in Figure 29. The settlements for given surface load and the corresponding frictional force were computed at three sections, namely central, quarter points and edges according to the procedure described earlier, case (c). The average settlement was computed by dividing the area of settlement diagram by width of footing. These values are shown in Table 3 for surface load intensities of 0.25 kg/cm² to 1.75 kg/cm² and footing width of 25 cm.

The values of maximum settlements at the centre for the smooth flexible case (Table 1, column 2) and rough flexible case (Table 3, column 2) are found to be the same. The frictional forces due to roughness of base on either side of the central axis of footing are of opposite sign. Therefore, the sum total of stresses at the centre due to frictional force remains the same. Hence, there is no change in the maximum settlement of the uniformly loaded smooth flexible and rough flexible footings.

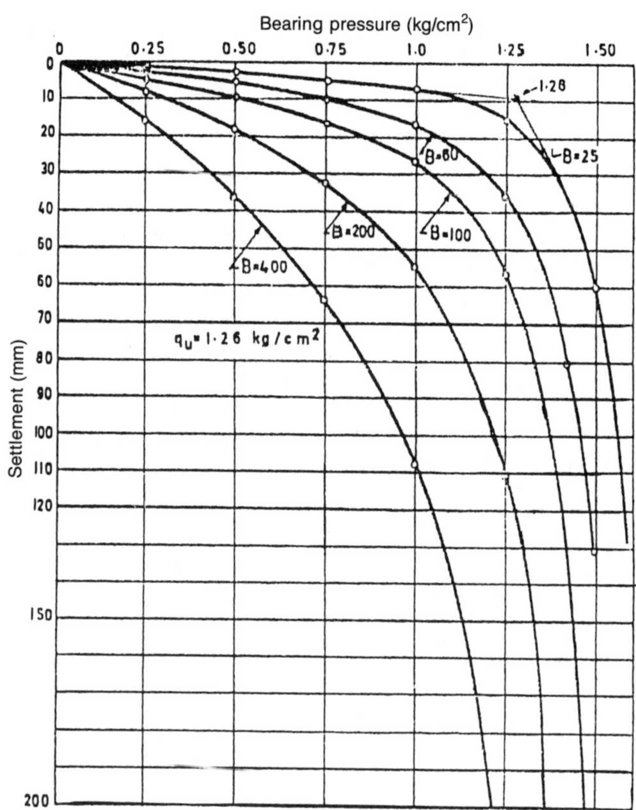

FIGURE 24. Pressure vs. settlement curves for smooth rigid strip footing on Buckshot clay.

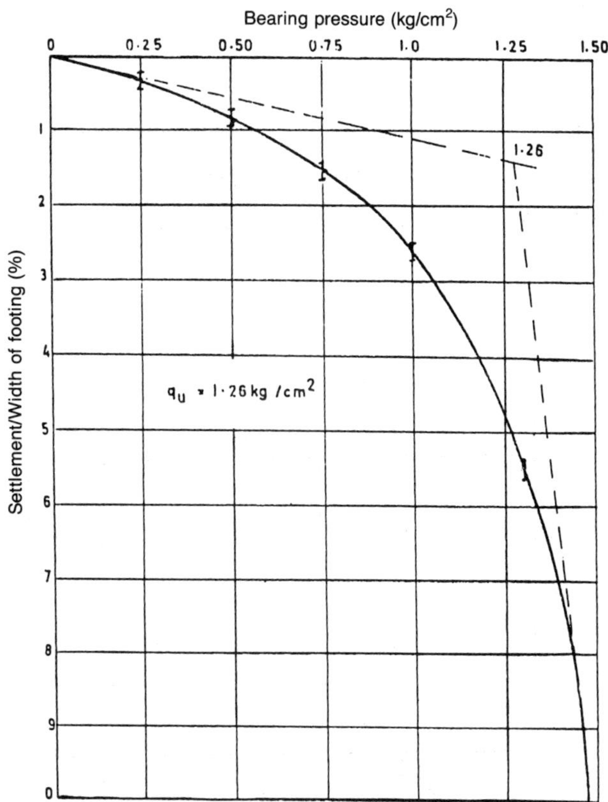

FIGURE 25. Pressure vs. S_f/B for smooth rigid strip footing resting on Buckshot clay.

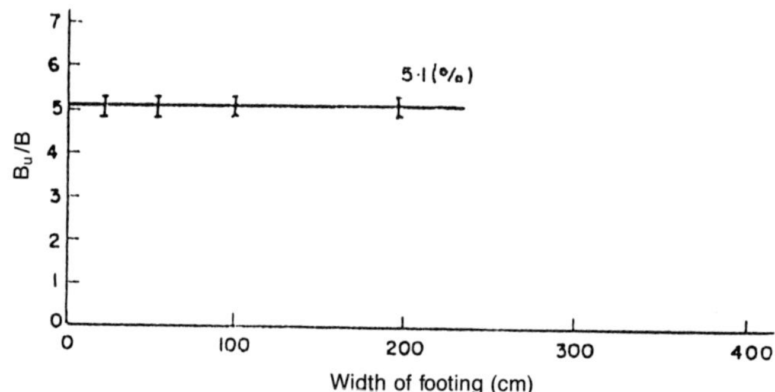

FIGURE 26. S_u/B vs. B for smooth rigid strip footing resting on Buckshot clay.

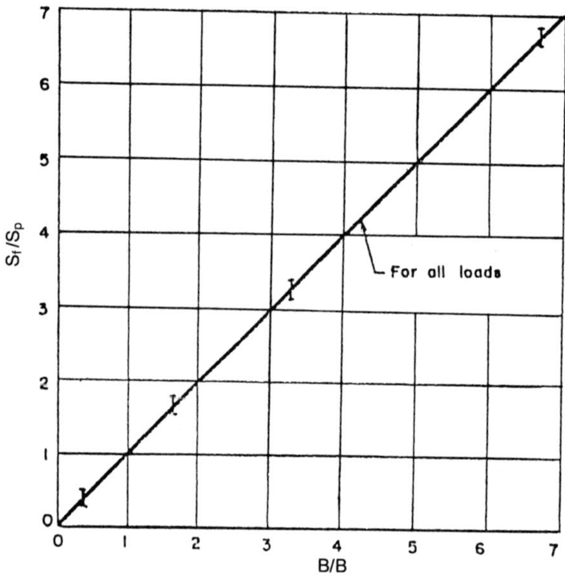

FIGURE 27. S_f/S_p vs. B_f/B_p curves for smooth rigid strip footing resting on Buckshot clay.

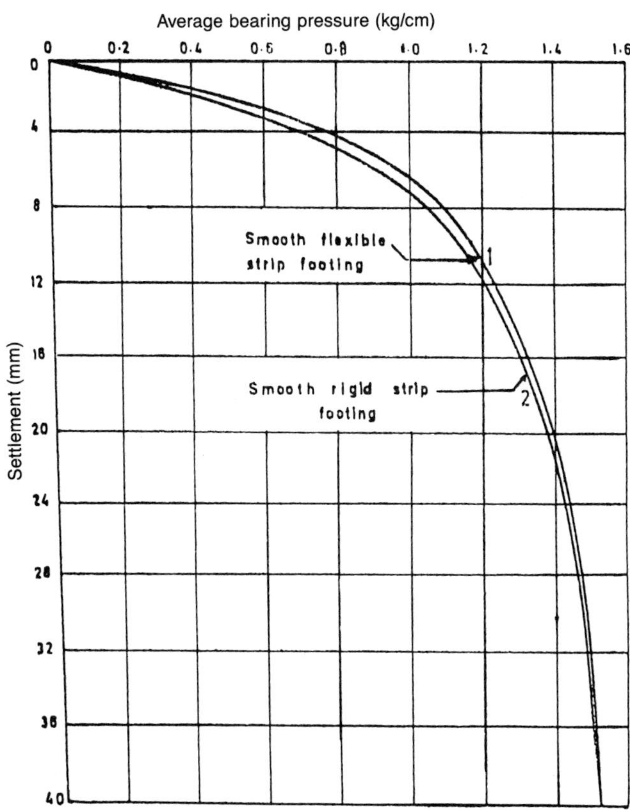

FIGURE 28. Pressure vs. settlement curves for smooth strip footing with flexible and rigid base resting on Buckshot clay (B = 25 cm).

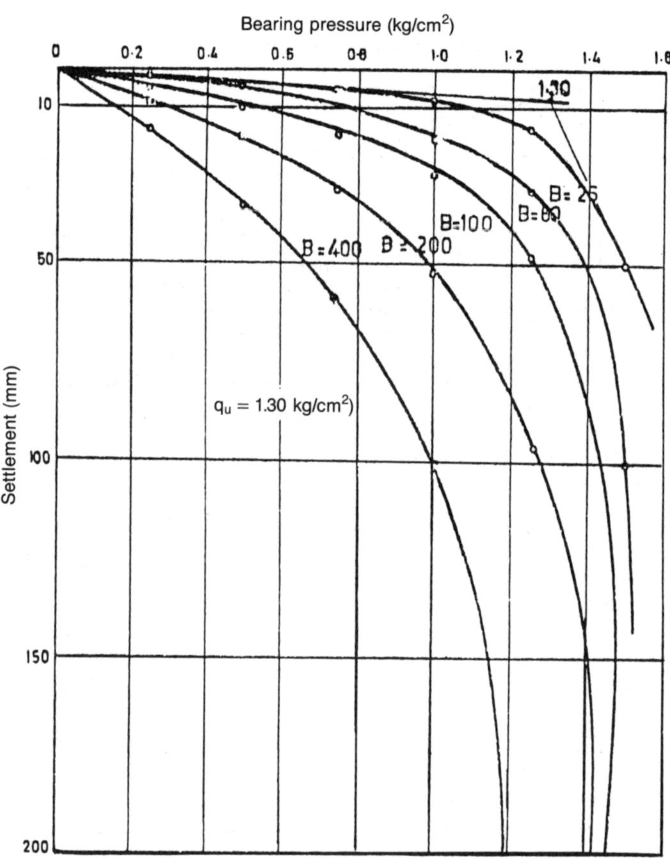

FIGURE 29. Pressure vs. average settlement of rough flexible strip footing resting on Buckshot clay.

TABLE 3. Maximum and Average Settlements of Rough Flexible Strip Footing in Buckshot Clay, B = 25 cm.

Q	Maximum in Settlement	Average Settlement	Difference (percent)
1.	2.	3.	4.
0.25	1.06	0.96	9.43
0.50	2.40	2.03	11.2
0.75	4.21	3.43	18.5
1.00	7.09	5.65	20.3
1.25	14.70	10.50	29.43
1.50	61.77	42.00	32.0
1.75	139.96	82.5	41.05

The above settlement of rough flexible footing has been shown in column 3, Table 3. Compared to the average settlement of smooth footing (column 3, Table 1) the above values are less. The difference between the two increases with the pressure intensities and the difference is of the order of 2.7 mm at $q = 1.25$ kg/cm². Thus friction force reduces the average settlements of footings by restraining the deformation of the footing base.

Ultimate Bearing Capacity

The ultimate bearing capacity was found to be 1.30 kg/cm² for all footings. The intersection tangent method also gave the same value. This value is about 3 percent more compared to the corresponding case with smooth base. Thus roughness of base increases the bearing capacity of flexible strip footing.

PRESSURE VERSUS S_f/B

The pressure versus S_f/B was computed and the values are plotted in Figure 30. A unique curve is found as in the previous cases. The ultimate bearing capacity works out to 1.30 kg/cm².

S_u/B VERSUS B RELATIONSHIP

The above is plotted in Figure 31. It is found that a single horizontal line is generated showing that the settlement at ultimate failure for all footing sizes is the same and is equal to 5 percent. The corresponding value for smooth case was 4.8 percent.

S_f/S_p VERSUS B_f/B_p RELATIONSHIP

The above relationship is plotted in Figure 32. The plot is a straight line and is found to have equal inclination to both axes, showing that

$$S_f/S_p = B_f/B_p \tag{70}$$

Rough Rigid Strip Footing

This case has been analysed by assuming uniform tangential force intensity as in case (c) and non-uniform contact pressure distribution as in case (b) at the interface of soil and footing base. The total stress at every point has been computed by algebraically adding the stresses due to the above two pressures at the interface of soil and footing base. The computation of uniform settlement and the corresponding value of contact pressure coefficient has been done as in case (b). Footings of widths 25 cm, 60 cm, 100 cm, 200 cm and 400 cm have been analysed.

CONTACT PRESSURE COEFFICIENTS

The contact pressure distribution has been determined in the same way as in case (b). The values are given in Table 4. The settlements are also shown for all the footings analysed.

The Table 4 shows that for a given pressure, the contact pressure coefficient α_1 is almost the same for all footing sizes. Thus the pattern of distribution of contact pressure is, therefore, the same for all sizes of rough rigid strip footing.

The contact pressure distribution has been plotted in Figure 33. It is found that the contact pressure curve is concave downward. As the pressure at the surface increases, the concavity tends to disappear as the failure load is reached.

TABLE 4. Contact Pressure Coefficients and Settlements for Rough Rigid Strip Footing Resting on Buckshot Clay.

q	B = 25 cm		B = 60 cm		B = 100 cm		B = 200 cm		B = 400 cm		c_u	
kg/cm²	α_1	S(mm)	α_1	S(mm)	α_1	S(mm)	α_1	S(mm)	α_1	S(mm)	kg/cm²	q/c_u
0.25	1.60	0.94	1.595	2.25	1.60	3.75	1.60	7.25	1.65	14.80	0.72	0.347
0.50	1.34	2.18	1.46	5.10	1.45	8.50	1.43	17.20	1.515	34.00		0.694
0.75	1.32	3.90	1.30	9.20	1.30	15.30	1.27	31.00	1.38	59.00		1.044
1.00	1.215	5.80	1.215	15.2	1.22	26.0	1.22	51.00	1.21	100.00		1.388
1.25	1.145	12.00	1.155	27.5	1.155	47.0	1.095	100.00	1.15	180.00		1.735
1.50	1.00	45.00	1.00	140.0	1.00	200.0	1.00	450.00	1.00	910.00		2.082

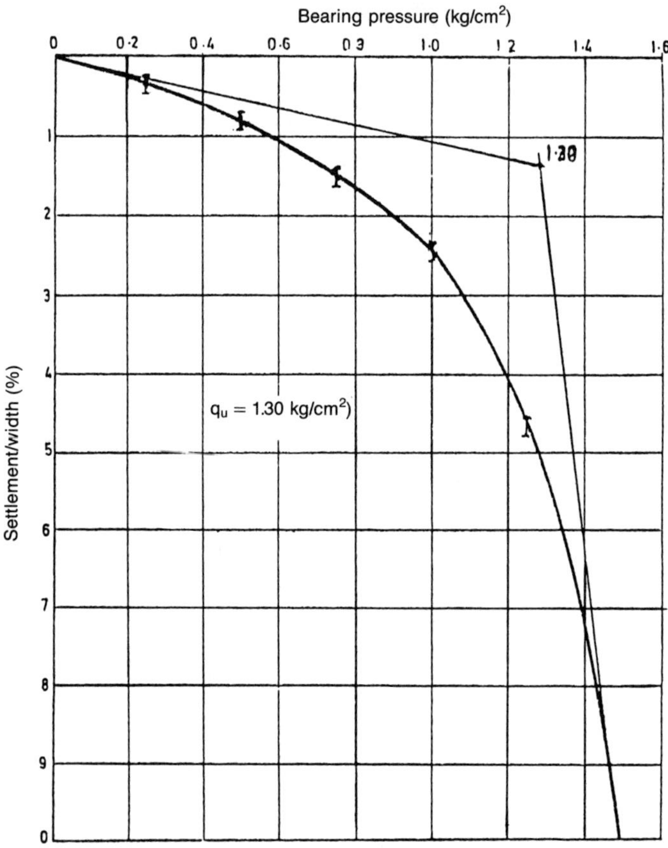

FIGURE 30. Pressure vs. $S_{av.}/B$ for rough flexible strip footing resting on Buckshot clay.

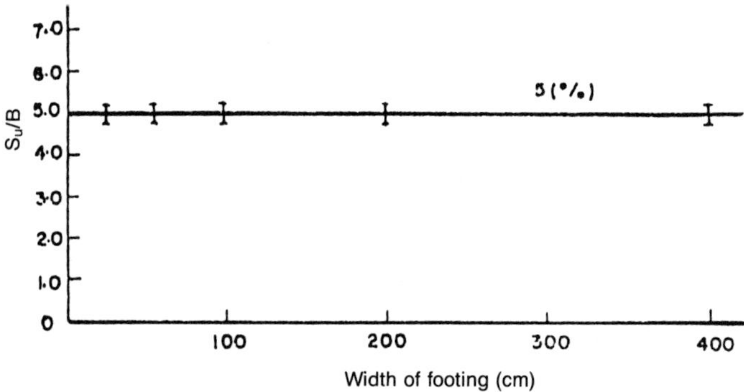

FIGURE 31. S_u/B for rough flexible strip footing resting on Buckshot clay.

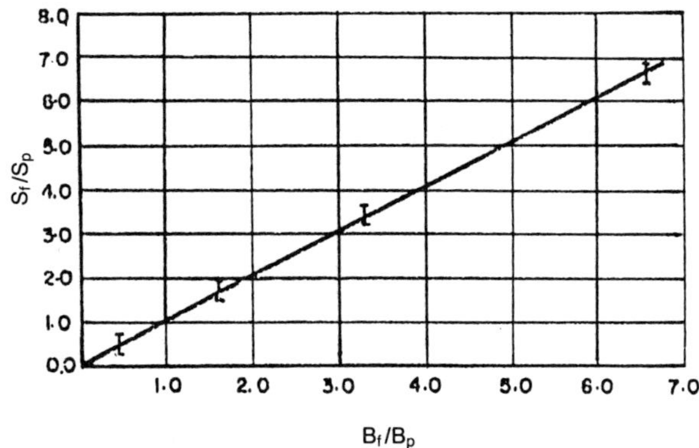

FIGURE 32. S_f/S_p vs. B_f/B_p for rough flexible strip footing resting on Buckshot clay.

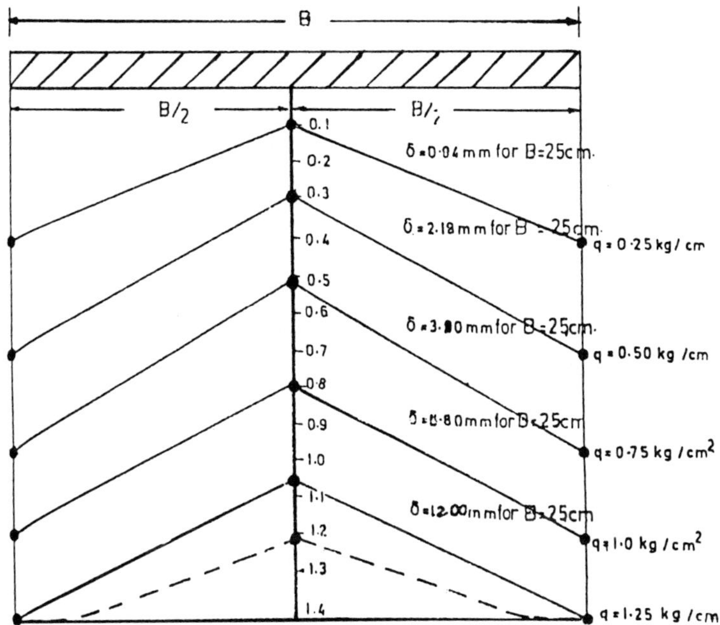

FIGURE 33. Contact pressure distribution below rough rigid strip footing resting on Buckshot clay.

The contact pressure distribution coefficient has been correlated with non-dimensional quantity q/c_u, where c_u is the compressive strength of soil and is plotted in Figure 34 on arithmetic scale. The equation of the curve is following:

$$\alpha_1 = 1.72 - 0.86\,(q/c_u) \qquad (72)$$

with the range of pressures of 0.25 kg/cm² to failure. This equation gives a realistic picture of contact pressure distribution from centre to edges for which no simple equation is available.

PRESSURE-SETTLEMENT CHARACTERISTICS

The average settlement shown in Table 3 has been plotted in Figure 35 for all footing sizes. These curves are analysed in the same way as in the case (a). The pressure versus S_f/B, S_u/B versus B and S_f/S_p versus B_f/B_p relationships have been plotted in Figures 36, 37 and 38 respectively. The trends of the above curves have been found to be the same as in cases (a) to (c). The ultimate bearing capacity computed from curves of Figures 35 and 36 is found to be 1.32 kg/cm². Terzaghi's formula gives ultimate bearing capacity as 1.54 kg/cm². The settlement at failure is equal to 5.5 percent of the width of footing.

The relationship

$$S_f/S_p = B_f/B_p \qquad (70)$$

is found to hold good in this case also.

EFFECT OF ROUGHNESS

The effect of roughness on ultimate bearing pressure and settlement has been plotted in Figure 39 for strip footing of 25 cm width. Curve 1 is for rough base and curve 2 is for smooth base. The value of ultimate bearing pressure is found to be 1.32 kg/cm² and 1.26 kg/cm² for rough and smooth bases respectively. The rough footing gives approximately 5 percent higher ultimate bearing pressure than the smooth footing. The settlements at failures are 13 mm and 20 mm for smooth and rough bases respectively. Thus failure occurs at higher settlement for rough rigid case.

PREDICTION OF PRESSURE SETTLEMENT CHARACTERISTICS OF FOOTINGS OF DHANORI CLAY

The pressure settlement curve for a 10 cm wide strip footing resting on Dhanori clay was predicted by the present method. The parameters (a) and (b) of constitutive equation for Dhanori clay were obtained from drained triaxial test. The coefficient of contact pressure distribution for a given pressure at surface was calculated from Equation (72) developed for the present case.

The pressure-settlement curve obtained by Prakash (1975) from tests on 10 cm wide model strip footing resting on Dhanori clay in two dimensional tank is shown in Figure 40. The pressure-settlement curve predicted by the present method has also been shown in the above figure. It is found that the two curves compare very well. The curves show that the soil yields to local shear failure. The predicted and experimental values of ultimate bearing pressures are 4.25 kg/cm² and 4.50 kg/cm² respectively.

Terzaghi's theory gives a value of 5.63 kg/cm² for this soil considering local shear failure. Thus Terzaghi's theory overestimates the ultimate bearing pressure by 36 percent compared to the experimental value.

The pressure settlement curve, using undrained shear strength parameter obtained by Prakash (1975) from undrained triaxial compression test, was also predicted for Dhanori clay and is shown in the Figure 40. It is found that this curve lies very high above the experimental curve, i.e., the settlements are very small for a given surface load compared to experimental value.

Smooth Flexible Square Footing

A uniformly loaded square footing induces stresses in the supporting soil such that $\sigma_2 = \sigma_3$ at the central section and $\sigma_2 \neq \sigma_3$ at the edges. Thus in analysing square footing and three-dimensional stress, distribution in the soil mass has been considered. The contact pressure in flexible footings is uniform or $\alpha_1 = 1$. The method of computing pressure settlement characteristics has been discussed in previous sections.

Pressure Settlement Characteristics

The pressure settlement curves for square footings on flexible base have been computed for footing-widths of 25 cm, 60 cm, 100 cm and 200 cm and are given in Figure 41. It is found that the settlements increase with increase of surface load and width of footing. The ultimate bearing capacity for all widths comes to 1.35 kg/cm² as against 1.26 kg/cm² for smooth flexible strip footing. The maximum and average settlement for load intensities of 0.25 kg/cm² to 1.25 kg/cm² has been given in columns 2 and 3 of Table 5 for footing width of 25 cm. It is noticed that percent difference of maximum and average settlements increases with increase of load intensity at the surface. Compared to the case of smooth flexible strip footings (column 4, Table 1) the above differences are less.

The ultimate bearing pressure of smooth square footing is more and settlement less compared to smooth strip footing. A strip footing is a two-dimensional problem in which the failure planes are generated only on two sides, whereas a square footing is a three-dimensional problem and the failure planes are formed on all sides of the footing. Therefore, the resistance to deformation is more in case of square footing than strip footing.

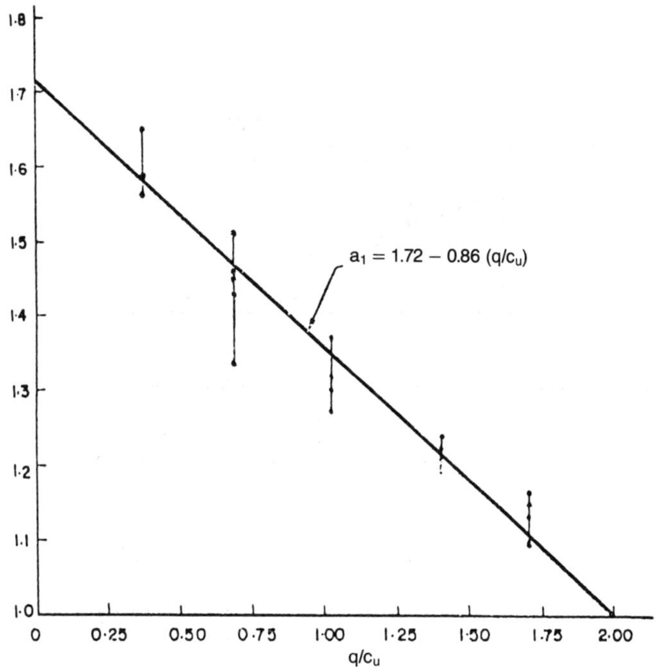

FIGURE 34. a_l vs. q/c_u for rough rigid strip footing resting on Buckshot clay.

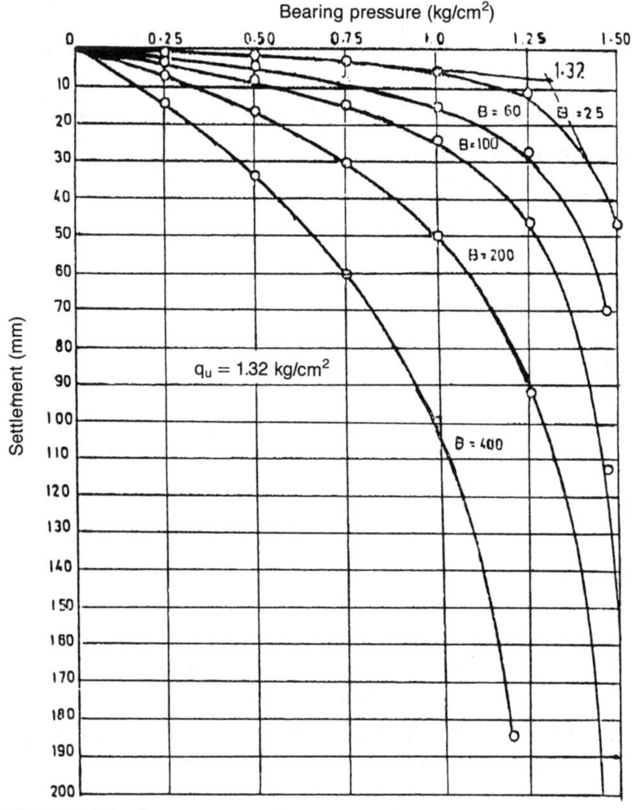

FIGURE 35. Pressure vs. settlement curve for rough rigid strip footing resting on Buckshot clay.

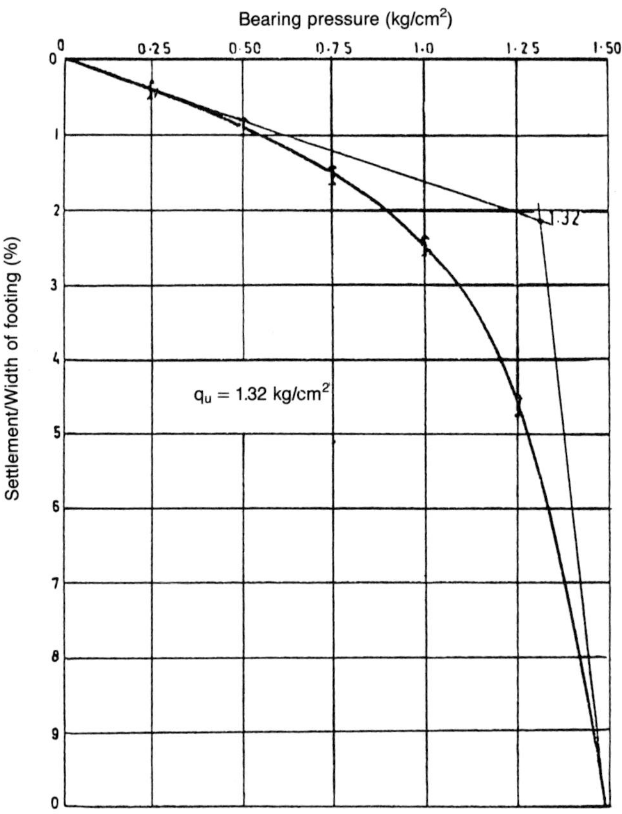

FIGURE 36. Pressure vs. S_f/B curve for rough rigid strip footing resting on Buckshot clay.

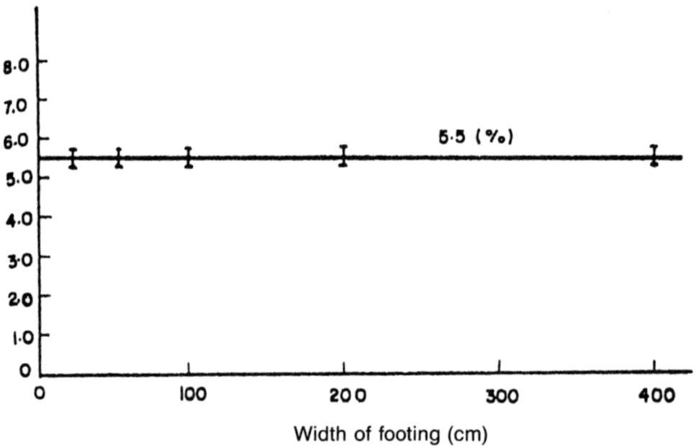

FIGURE 37. S_u/B vs. width of footing for rough rigid strip footing resting on Buckshot clay.

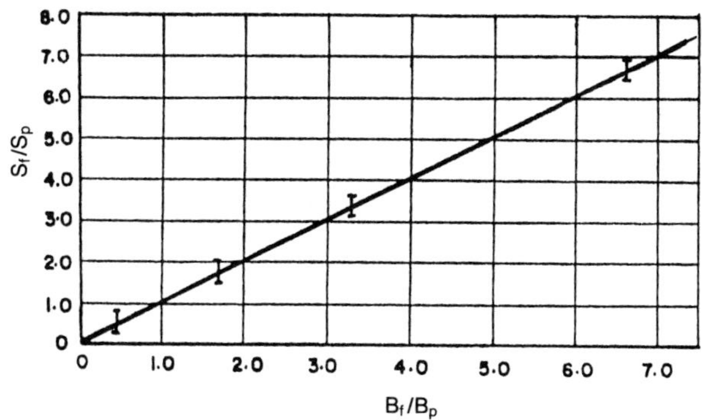

FIGURE 38. S_f/S_p vs. B_f/B_p curve for rough rigid strip footing resting on Buckshot clay.

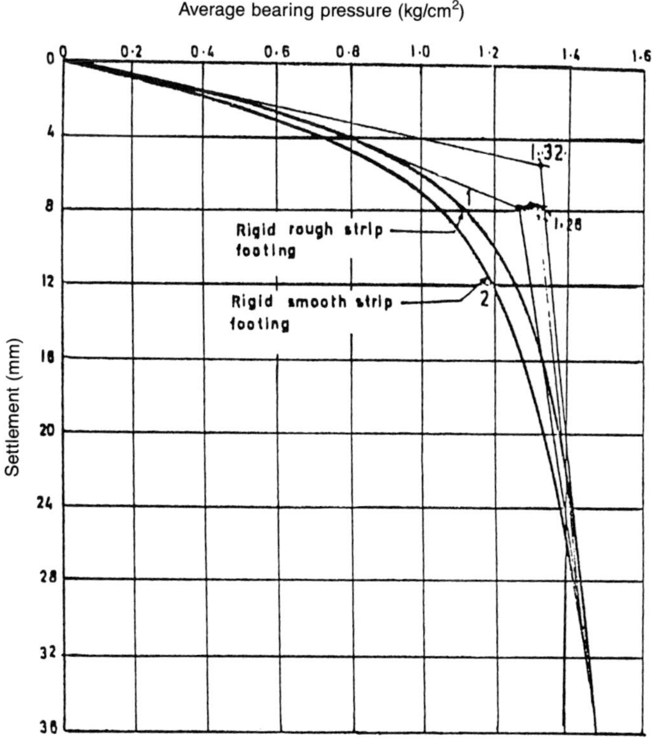

FIGURE 39. Pressure vs. settlement curves for rigid strip footing with smooth and rough base resting on Buckshot clay (B = 25 cm).

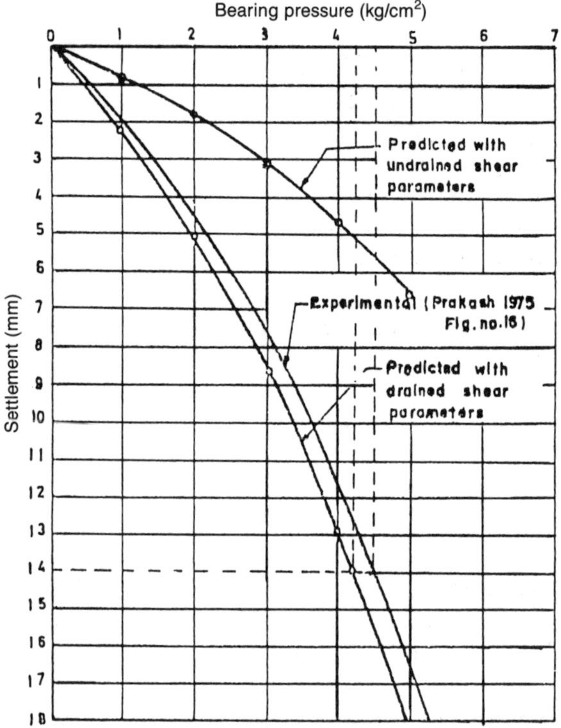

FIGURE 40. Comparison of pressure vs. settlement curve of 10 cm wide rigid strip footing resting on Dhanori clay.

The pressure settlement curves were further analysed by plotting (1) pressure versus S_f/B, (2) S_u/B versus B, and (3) S_f/S_p versus B_f/B_p relationship and are plotted in Figures 42, 43 and 44 respectively. A unique curve is obtained for all footing sizes for pressure versus S_f/B curve (Figure 42). The

TABLE 5. Maximum and Average Settlement of Square Flexible Footing Resting on Buckshot Clay, B = 25 cm.

Pressure kg/cm²	Maximum Settlement (mm)	Average Settlement (mm)	Percent Difference
1.	2.	3.	4.
0.25	0.88	0.87	0.8
0.50	1.99	1.92	3.5
0.75	3.55	3.33	6.0
1.00	6.29	5.82	8.0
1.25	19.12	17.30	10.0
1.50	71.4	58.5	21.0

TABLE 6. Contact Pressure Coefficient for Smooth Rigid Square Footing.

Sl. No.	Load Intensity q(kg/cm²)	Shear Strength c_u'(kg/cm²)	q/c_u	α_1	$\alpha_2 = 3(\alpha_1 - 1)$
1.	2.	3.	4.	5.	6.
1.	0.25	0.72	.347	1.36	1.05
2.	0.50		.694	1.28	.84
3.	0.75		1.044	1.20	.60
4.	1.00		1.388	1.12	.36
5.	1.25		1.735	1.05	.15

ultimate bearing capacity worked out on 1.35 kg/cm². S_u/B versus B relationship (Figure 43) gives a failure settlement of 10 percent of the width of footing and the plot of S_f/S_p versus B_f/B_p relationship (Figure 44) confirms that

$$S_f/S_p = B_f/B_p \tag{70}$$

Smooth Rigid Square Footing

The pressure settlement curves for smooth rigid square footings resting on Buckshot clay have been computed. The effect of roughness has not been considered as it was found not to affect the bearing capacity significantly. The computed curves have been compared with the curves obtained experimentally by Carrol (1963) and by finite element method by Basavanna (1975). The results are discussed below.

CONTACT PRESSURE DISTRIBUTION

The contact pressure distribution in case of rigid square footing has been taken as shown in Figure 12 and has been described by coefficients α_1 and α_2. For a given intensity of surface load, the value of α_1 has been varied and settlements at central axis quarter point and edges have been computed. The value of α_1 which gave uniform settlement at centre, quarter point and edge was computed as in case (b). The surface load intensity was varied and the corresponding values of α_1 which gave uniform settlements were computed. The settlements thus computed were taken as settlements of a rigid square footing. The values of α_1 obtained for all footing sizes were found to be the same and are given in Table 6 for various load intensities.

It is noted that as the load intensity approaches failure load the value of α_1 approaches unity and α_2 approaches zero. In other words the contact pressure distribution tends to be uniform at failure.

The contact pressures given in Table 6 have also been plotted in Figure 45.

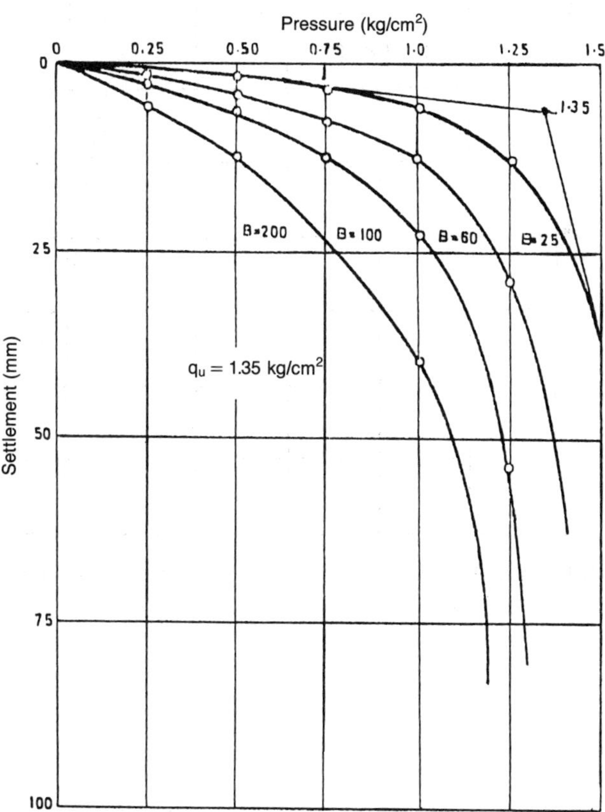

FIGURE 41. Pressure-settlement curves for smooth flexible square footing resting on Buckshot clay.

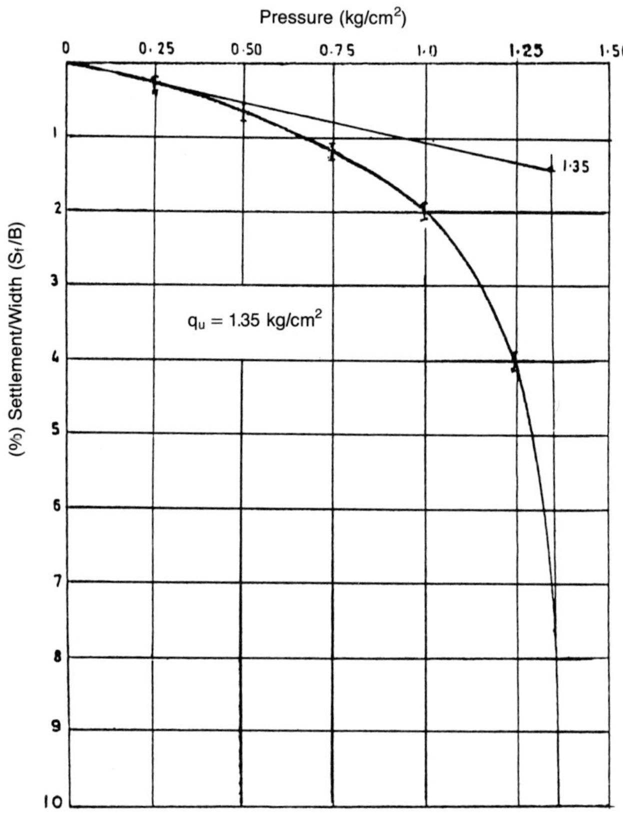

FIGURE 42. Pressure vs. S_f/B for a smooth flexible square footing resting on Buckshot clay.

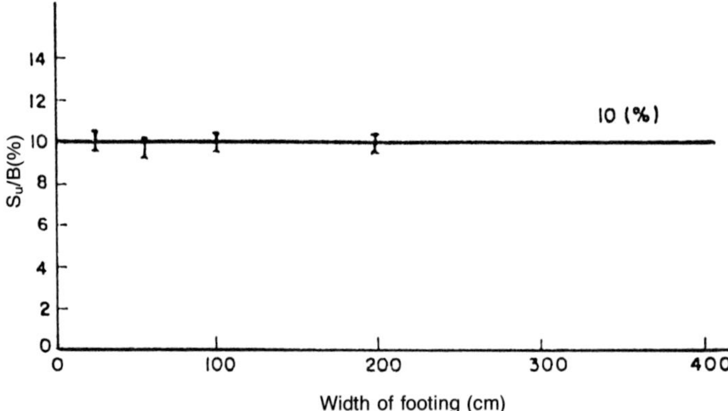

FIGURE 43. S_u/B vs. width of footing for smooth flexible square footing resting on Buckshot clay.

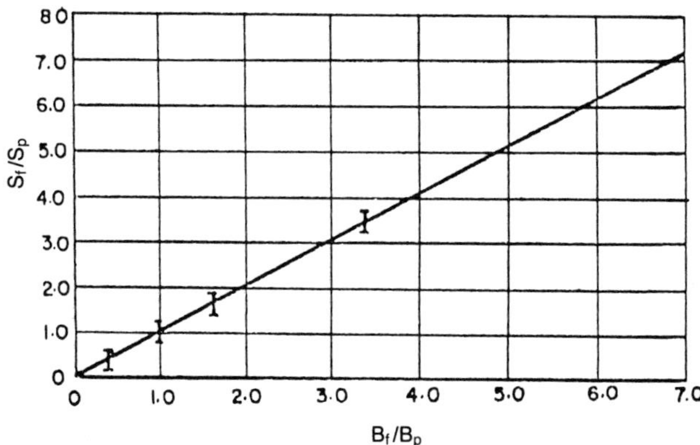

FIGURE 44. S_f/S_p vs. B_f/B_p for smooth flexible square footing resting on Buckshot clay.

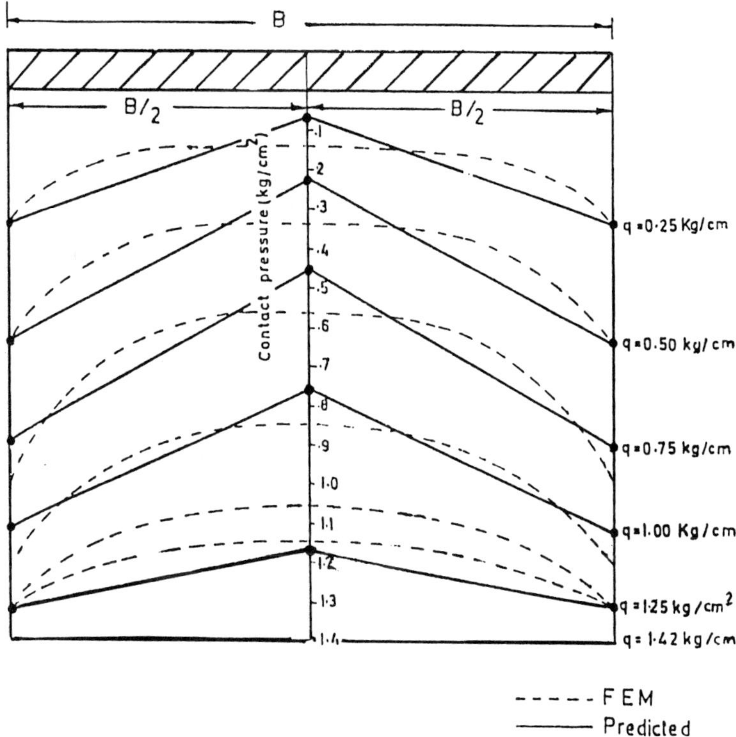

FIGURE 45. Contact pressure distribution below smooth rigid square footing resting on Buckshot clay.

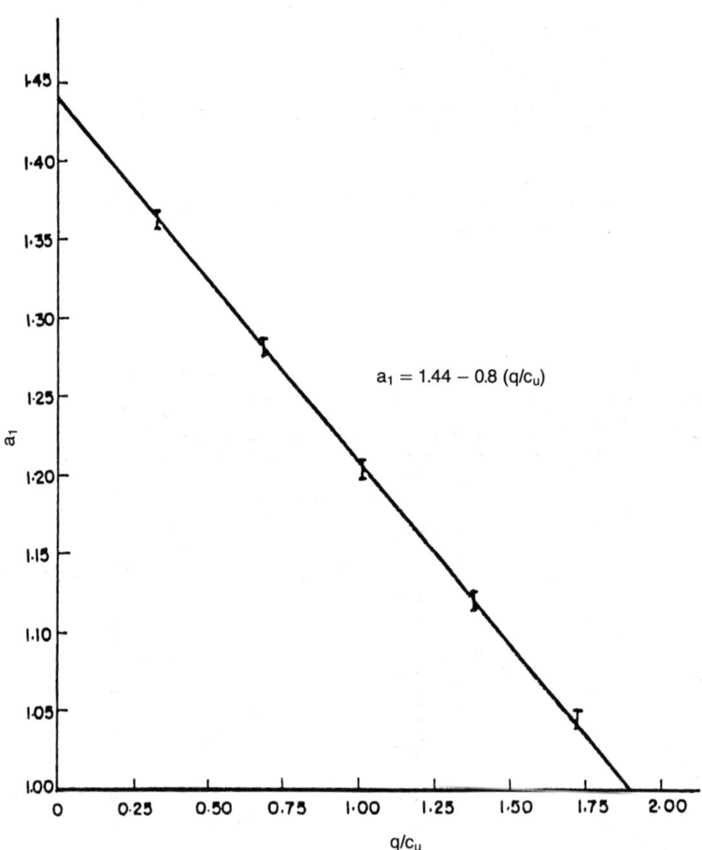

FIGURE 46. a_l vs. q/c_u curve for smooth rigid square footing resting on Buckshot clay.

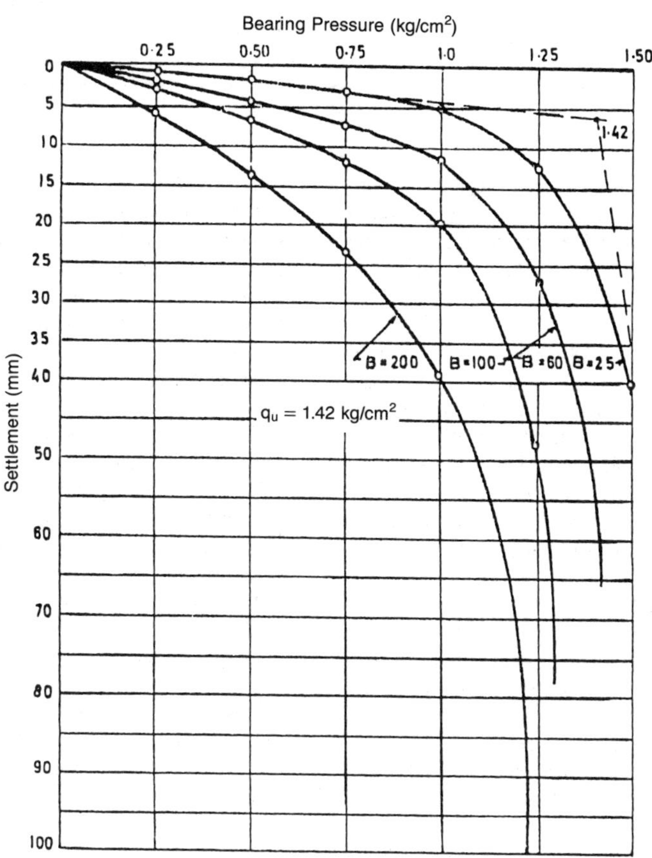

FIGURE 47. Pressure settlement curves for smooth rigid square footings resting on Buckshot clay.

TABLE 7. Settlement of Smooth Rigid Square Footing on Buckshot Clay.

q	α_1	S(mm) B = 25 cm	S(mm) B = 60 cm	S(mm) B = 100 cm	S(mm) B = 200 cm
1.	2.	3.	4.	5.	6.
0.25	1.36	0.75	1.82	3.10	6.20
0.50	1.28	1.70	4.15	6.80	13.60
0.75	1.20	2.92	7.10	11.70	23.50
1.00	1.12	4.85	11.75	19.50	39.00
1.25	1.05	12.10	28.50	48.00	97.00
1.50	1.00	57.00	128.00	200.00	450.00

The results have further been analysed by plotting non-dimensional relation of α_1 against q/c_u and is shown in Figure 46. A straight line is obtained having the following equation between load intensity of 0.25 kg/cm² and failure load:

$$\alpha_1 = 1.44 - 0.8(q/c_u) \quad (73)$$

The contact pressure for rigid square footing on clay was computed by Basavanna by finite elements method and is plotted in Figure 45. The two plots differ because Basavanna has assumed a different pattern of the contact pressure distribution. However, no equation has been developed by him for readily computing the contact pressure distribution.

PRESSURE-SETTLEMENT CURVES

The settlements computed above for uniform settlement at centre, $B/4$ and edges of footing have been plotted against the corresponding pressure intensities and shown in Figure 47 for footing widths of 25 cm, 60 cm, 100 cm and 200 cm. The trend of the curve shows that the rate of settlements increases with increase of surface load intensity and wider footings give larger settlements. The values of settlements are also given in Table 7.

The pressure settlement curve for footing width of 25 cm has been compared with hose obtained experimentally by Carrol (1963) and by F.E. method by Basavanna (1975) by plotting them simultaneously in Figure 62. The curve computed by the present method compares very well with the experimental and F.E.M. up to about 50 percent of the failure load of 1.42 kg/cm². At the 1.42 kg/cm² the computed curves deviate from experimental curve. The settlement computed at failure load differs by about 25 percent compared to experimental curve.

The pressure settlement curves have been further analysed by plotting pressure versus S_f/B for all footing sizes in Figure 49. A unique curve for all footing sizes has been obtained. This shows that pressure versus S_f/B relationship is independent of footing sizes for smooth rigid square footings on clay.

ULTIMATE BEARING CAPACITY

The ultimate bearing capacity was found to be 1.42 kg/cm². The ultimate bearing capacity was also found from pressure vs S_f/B curve of Figure 49 and the value comes to again 1.42 kg/cm². The ultimate bearing capacity was also found by intersection tangent method from the above curves and was found to be 1.42 kg/cm². It shows that the bearing capacity for square footings in clay is independent of footing widths.

The ultimate bearing capacity has also been found from the curves of Figure 48 by intersection tangent method. The values are 1.50 kg/cm² for experimental curve and 1.48 kg/cm² from F.E.M. curve. The bearing capacity computed by Terzaghi's formula comes to 1.81 kg/cm². The values are given in Table 8.

The ultimate bearing pressure predicted by the present method is about 5 percent on conservative side compared to the experimental value and about 4 percent on the conservative side compared to F.E.M. value. Terzaghi's method overestimates bearing capacity by 17.25 percent compared to the experimental value. Thus the present method can predict reasonably well the bearing capacity of smooth rigid square footings.

TABLE 8. Ultimate Bearing Capacity and Settlement of Smooth Square Footing on Buckshot Clay.

Sl. No.	Footing Size (cm)	Method Used	Ultimate Bearing pressure (kg/cm²)	Settlement/ Width (S_u/B) Percent
1.	2.	3.	4.	5.
1.	25	Experimental Carrol (1963)	1.50	15.0
2.	25	F.E.M.	1.48	15.0
3.	25	Present	1.42	12.5
4.	25	Terzaghi	1.81	—

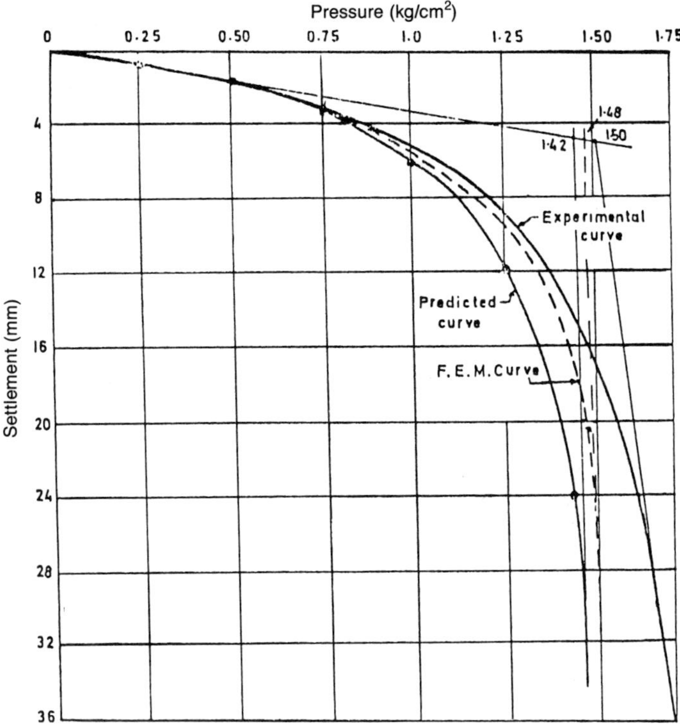

FIGURE 48. Comparison of computed and experimental pressure-settlement curves for 25 cm square footing on Buckshot clay.

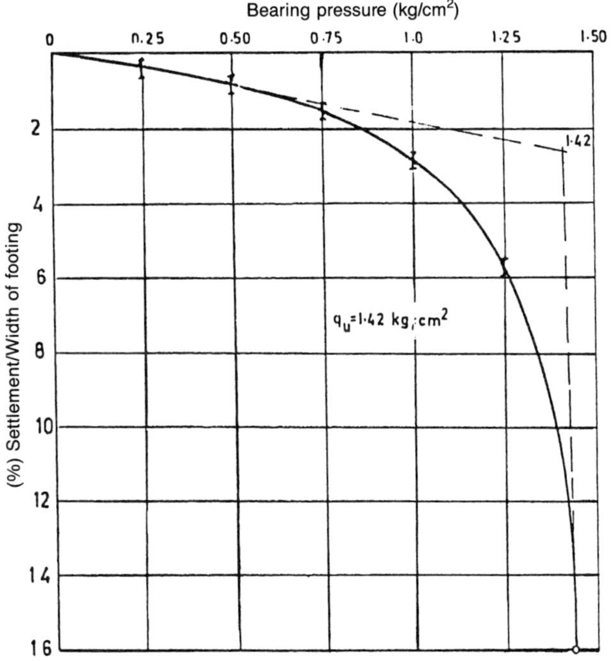

FIGURE 49. Pressure vs. S_f/B for smooth rigid square footing resting on Buckshot clay.

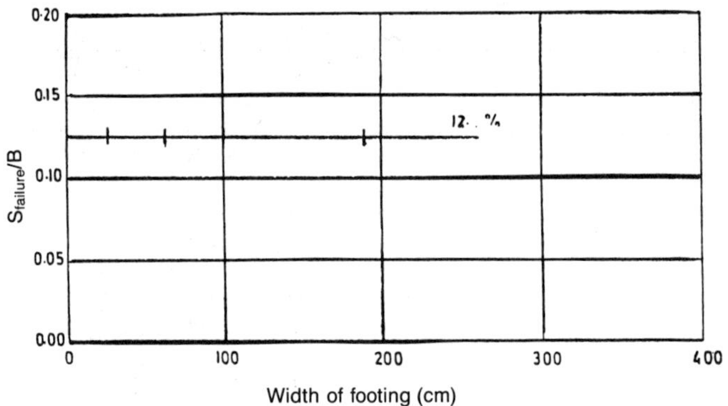

FIGURE 50. S_{fail}/B vs. B for smooth rigid square footing resting on Buckshot clay.

SETTLEMENT AT FAILURE

The settlement at ultimate bearing pressure for all footing sizes was obtained from Figure 47. The values of S_u/B versus B, where B is the footing width, are plotted in Figure 50. It is found that a line parallel to horizontal axis is obtained. The value of S_u/B works out to 12.5 percent. The settlements at failure for experimental and F.E.M. curves were also worked out from Figure 48 and they are found to be 15 percent of the footing widths. The above values are also tabulated in column 5, Table 8.

S_f/S_p VERSUS B_f/B_p RELATIONSHIP

The above relationship has been plotted as in case (a) and shown in Figure 51. A straight line with equal inclination to both axes is obtained showing the following relationship.

$$S_f/S_p = B_f/B_p \tag{70}$$

This confirms that the relationship given in Taylor (1942) is true for smooth square footings on clay also.

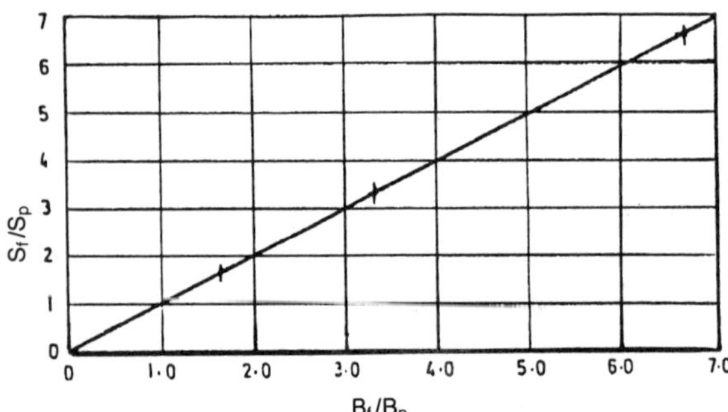

FIGURE 51. S_f/S_p vs. B_f/B_p for rigid square footing resting on Buckshot clay.

SUMMARY OF RESULTS OF UNIFORMLY LOADED FOOTINGS

The results of analysis of pressure-settlement curves are summarized in Table 9. The following observations are listed for uniformly loaded footing:

1. It is found that the bearing capacity of flexible strip footings from pressure versus average settlement curves and for rigid footings from pressure versus uniform settlement curve are practically the same for both smooth and rough bases. Similarly the bearing capacities of smooth flexible and smooth rigid square footings are also nearly the same.
2. The bearing capacity of rough rigid strip footing is 5 percent higher than the bearing capacity of smooth rigid strip footing.
3. The settlement of failure has been found to be approximately 5 percent of the width of the footing for all cases of strip footing and 12.5 percent for rigid square footing.
4. $S_f/S_p = B_f/B_p$ relationship holds good in all cases.
5. A simple equation for coefficient of contact pressure distribution has been derived in terms of non-dimensional parameter (q/c_u) for smooth and rough rigid strip footing and square footing cases.
6. The shape factors for square footings have been worked out from the bearing capacity values. It comes to 1.08 for flexible square footing and 1.27 for rigid square footing. The Terzaghi's shape factor is 1.2 for square footing.
7. Drained shear strength parameters give better prediction of pressure settlement curves for Dhanori clay (Prakash 1975) and undrained shear strength parameters for better prediction for Buckshot clay (Carrol, 1963).

Eccentrically Loaded Rigid Strip Footing in Clay

GENERAL

A footing foundation is often subjected to eccentric loads. The procedure to evaluate pressure-settlement curves for

TABLE 9. Summary of Results of Uniformly Loaded Footings in Clay.

Sl. No.	Description of Cases Studied	Contact Pressure Coefficient	Ultimate Bearing Capacity (kg/cm²)	Terzaghi's Value (kg/cm²)
1.	2.	3.	4.	5.
1.	Smooth Flexible Strip Footing	1.0	1.26	
2.	Smooth Rigid Strip Footing	$1.36 - 0.778(q/c_u)$	1.26	1.39
3.	Rough Flexible Strip Footing	1.0	1.30	
4.	Rough Rigid Strip Footing	$1.72 - 0.86(q/c_u)$	1.32	1.54
5.	Smooth Flexible Square Footing	1.0	1.35	
6.	Smooth Rigid Square Footing	$1.44 - 0.8(q/c_u)$	1.42	1.81
	Flexible —	1.08		
	Rigid —	1.27		
	Terzaghi —	1.2		

TABLE 10. Computation of Contact Pressure Coefficient for Eccentrically Loaded Rigid Strip Footing.

e/B	q (kg/cm²)	$\alpha\alpha_1$	S_m (mm)	S_o (mm)	S_L (mm)	$\frac{S_m - S_e}{B/2 - e}$	$\frac{S_e - S_L}{B/2 + e}$
1.	2.	3.	4.	5.	6.	7.	8.
0.1	1.5	0.80	100.40	49.60	9.320	12.60	6.71
		0.81	99.22	51.84	9.308	11.83	7.09
		0.82	98.44	54.18	9.296	11.07	7.31
		0.83	97.66	56.47	9.284	10.29	7.86
		0.84	96.88	58.76	9.272	9.53	8.24
		0.85	96.10	61.05	9.260	8.76	8.63
		0.86	95.32	63.34	9.248	7.99	9.01
		0.87	94.54	65.63	9.236	7.23	9.40
		0.88	93.76	67.92	9.224	6.46	9.78
		0.89	92.98	70.21	9.212	5.69	10.17
		0.90	92.20	72.50	9.200	4.42	10.55

TABLE 11. Contact Pressure Coefficients for Eccentrically Loaded Rigid Strip Footing Resting on Buckshot Clay.

e/B	q kg/cm²	c_u kg/cm²	q/c_u	$\alpha\alpha_1$
1.	2.	3.	4.	5.
0.1	0.25	0.72	0.347	0.30
	0.50		0.694	0.60
	0.75		1.044	1.00
	1.00		1.388	1.35
	1.25		1.735	1.05
	1.50		2.082	
0.2	0.25	0.72	0.347	0.60
	0.50		0.694	1.70
	0.75		1.044	1.70
	1.00		1.388	1.50
	1.25		1.735	1.15
	1.50		2.082	1.05
0.3	0.25	0.72	0.347	1.4
	0.50		0.694	1.8
	0.75		1.044	2.0
	1.00		1.388	1.7
	1.25		1.735	1.5
	1.50		2.082	1.3

such footings is not known. The pressure-settlement characteristics of eccentrically loaded rigid strip footing of widths 10 cm, 25 cm, 100 cm and 200 cm resting on Buckshot clay have been computed. For comparison with experimental results, the pressure-settlement curves of rigid strip footing resting on Dhanori clay have also been predicted.

The base of footing has been taken as smooth as it was found in the analysis of strip footing cases (b) and (d) that the roughness of base has insignificant effect on the pressure settlement curves. The contact pressure distribution shown in Figure 12 has been taken. Load eccentricities of $0.1B$, $0.2B$ and $0.3B$ have been considered.

The contact pressure distribution, ultimate bearing pressure and tilt of footings were computed from the pressure-settlement curves. Non-dimensional factors S_e/S_o and S_m/S_o at various arbitrary factors of safety of 1, 2 and 3 were computed and plotted against e/B ratio for all footing sizes. The analysis of results is discussed in the following paragraphs.

CONTACT PRESSURE COEFFICIENTS

The contact pressure distribution as shown in Figure 12 was taken for computation of pressure-settlement curves. For a given eccentricity and load intensity, the settlements S_e at E, S_m at A and S_L at B were computed for various values of contact pressure coefficient $\alpha\alpha_1$. For example, for an eccentricity of $e/B = 0.1$ and pressure intensity of 1.50 kg/cm², the value of $\alpha\alpha_1$ has been varied at intervals of 0.1 within suitable range and the values of settlement were obtained at the above three sections. These values have been plotted in Figure 66 for $\alpha\alpha_1 = 0.8$ and 0.9 and tabulated in Table 10. The slopes of line AE and EB have been computed by linear interpolation of settlement values for different values of $\alpha\alpha_1$.

It is found from the Tables 6–10 and also from Figure 66 that for $\alpha\alpha_1 = 0.85$, the line joining A, E and B is a straight line. The contact pressure defined by this value of $\alpha\alpha_1$ corresponds to the eccentrically loaded rigid footing case at bearing pressure intensity of 1.50 kg/cm². Similar analysis was done for other pressure intensities. The values of $\alpha\alpha_1$ which gave a straight line joining the settlements at the sections A, E and B are given in Table 11. The eccentricity of load was varied to $e/B = 0.2$ and 0.3 and the above procedure of evaluating $\alpha\alpha_1$ and settlements S_e, S_m and S_L were repeated. Thus, a set of values of pressure intensity q, $\alpha\alpha_1$, and S_e, S_m and S_L were obtained. The values of $\alpha\alpha_1$ for $e/B = 0.2$ and 0.3 are also given in Table 11.

The coefficient $\alpha\alpha_1$ for Buckshot clay has been correlated with the non-dimensional quantity q/c and has been plotted in Figure 53 for various eccentricities. The plot is a straight line up to approximately the ultimate bearing pressure and forms a peak and then slopes down. The peak is formed approximately at the ultimate bearing pressure. The contact pressure distribution changes when failure occurs and no more remains the same as assumed. Similar things happen at large eccentricity of load. Therefore, $\alpha\alpha_1$ starts falling after failure load is reached or eccentricity becomes large. This non-dimensional plot was used for predicting settlements of rigid footing on Dhanori Clay.

PRESSURE-SETTLEMENT CURVES

Average pressure versus settlements (S_e) for eccentricity $e/B = 0.1$, 0.2 and 0.3 for footing sizes of 10 cm, 25 cm, 100 cm and 200 cm have been plotted in Figures 54, 55, 56 and 57 respectively. The pressure versus settlement (S_m) for the above sizes are plotted in Figures 58, 59, 60 and 61 respectively. The figures show that for a particular footing, the settlement increases with increase of pressure at surface. For the same pressure, say 0.75 kg/cm², the settlements S_e and (S_m), increase with increase of footing sizes and eccentricity.

The pressure-settlement characteristics of eccentrically loaded rigid strip footing resting on Dhanori clay have also been computed, using the constitutive relationship for Dhanori clay. The values of contact pressure coefficient for a given surface load intensity was computed from pressure versus q/c_u relationship plotted in Figure 53 by the method described herein. The pressure versus settlement (S_e) for $e/B = 0.1$ and 0.2 has been plotted in Figure 62, for a footing of width 10 cm. The results obtained by Prakash (1975) for the same footing on laboratory model footing resting on

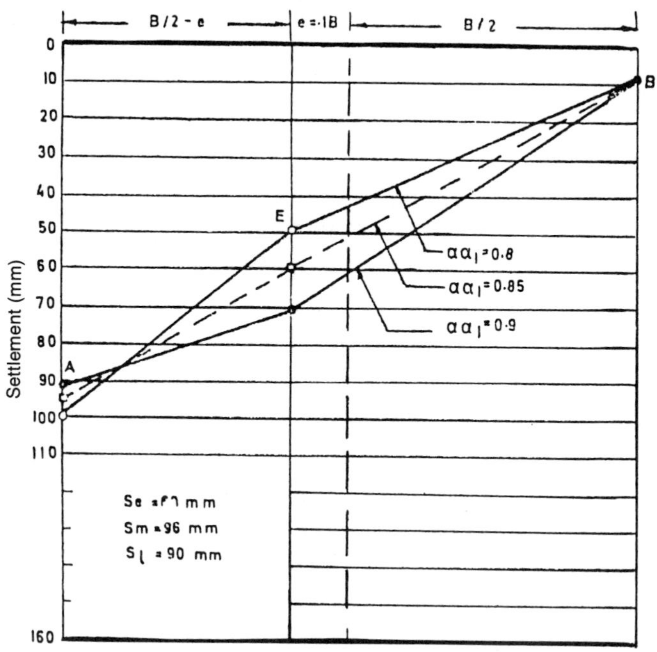

FIGURE 52. Settlement for e/B = 0.1 and B = 25 cm for Ecc. loaded rigid strip footing resting on Buckshot clay.

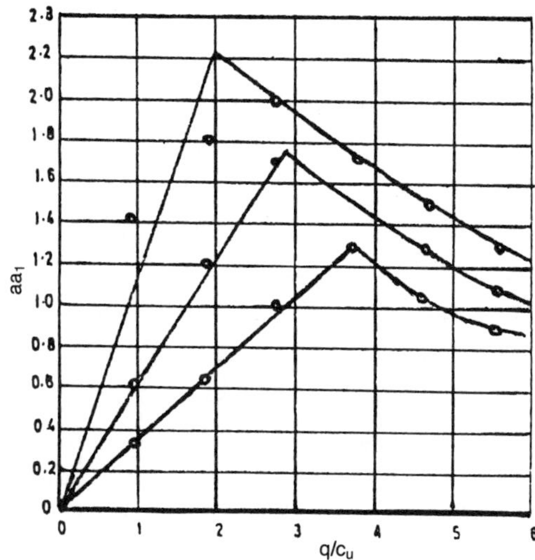

FIGURE 53. aa_l vs. q/c_u for eccentrically loaded rigid strip footing.

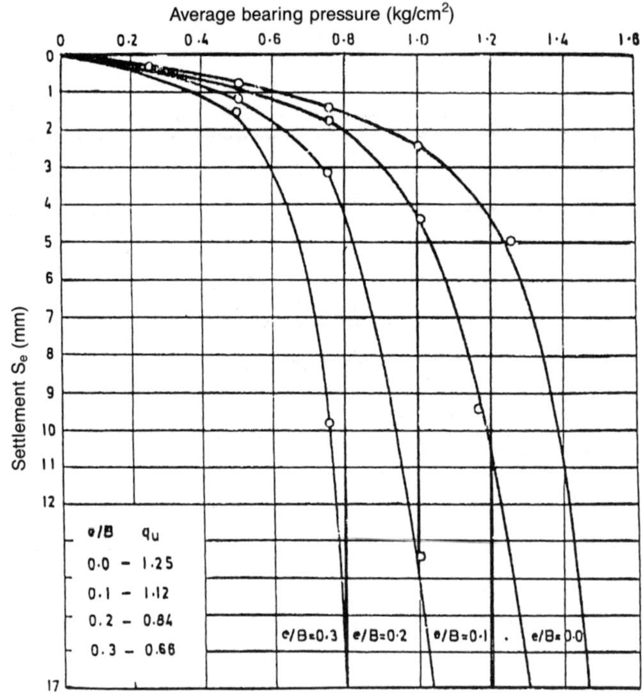

FIGURE 54. Pressure vs. S_e for an eccentrically loaded rigid strip footing resting on Buckshot clay (width 10 cm).

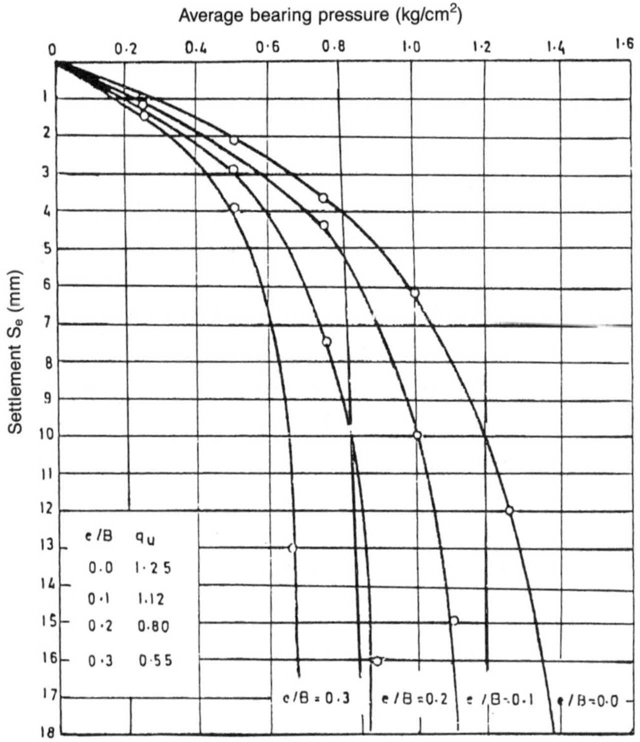

FIGURE 55. Average bearing pressure vs. S_e for eccentrically loaded rigid strip footing resting on Buckshot clay (width = 25 cm).

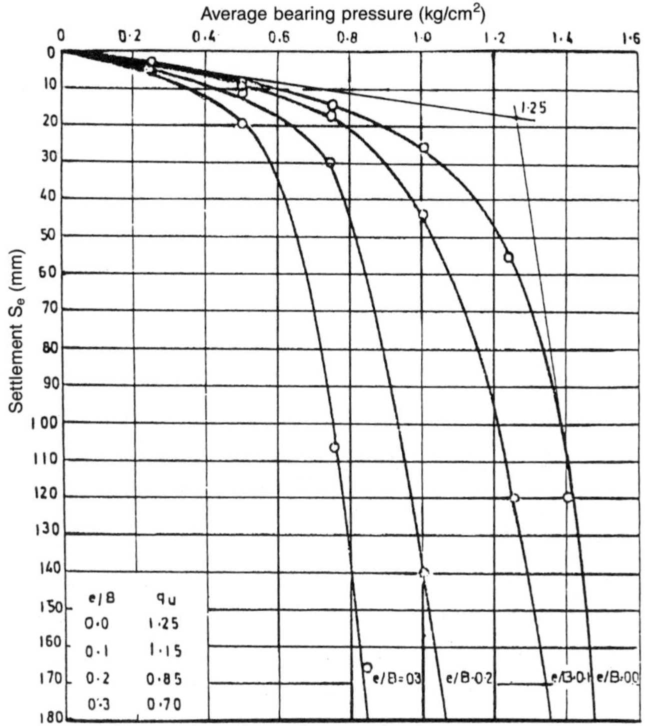

FIGURE 56. Pressure vs. S_e for an eccentrically loaded rigid strip footing resting on Buckshot clay (B = 100 cm).

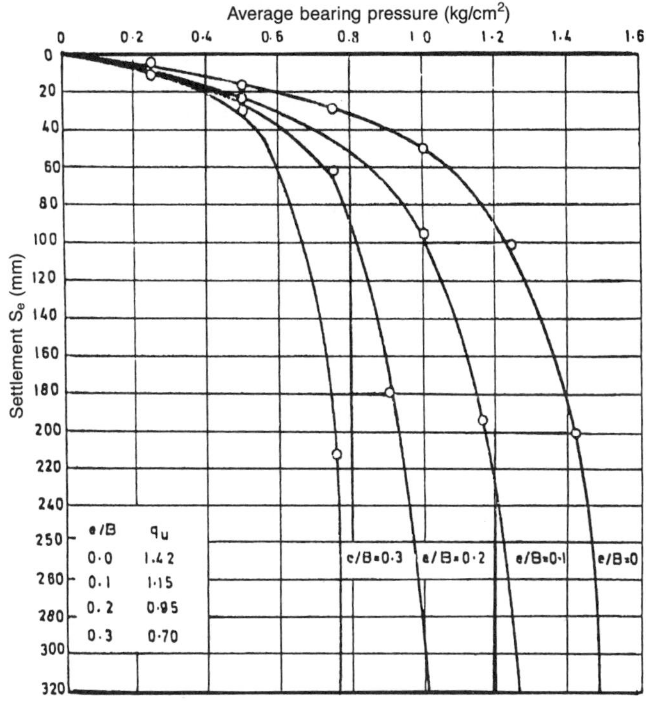

FIGURE 57. Average bearing pressure vs. S_e for an eccentrically loaded rigid strip footing resting on Buckshot clay (B = 200 cm).

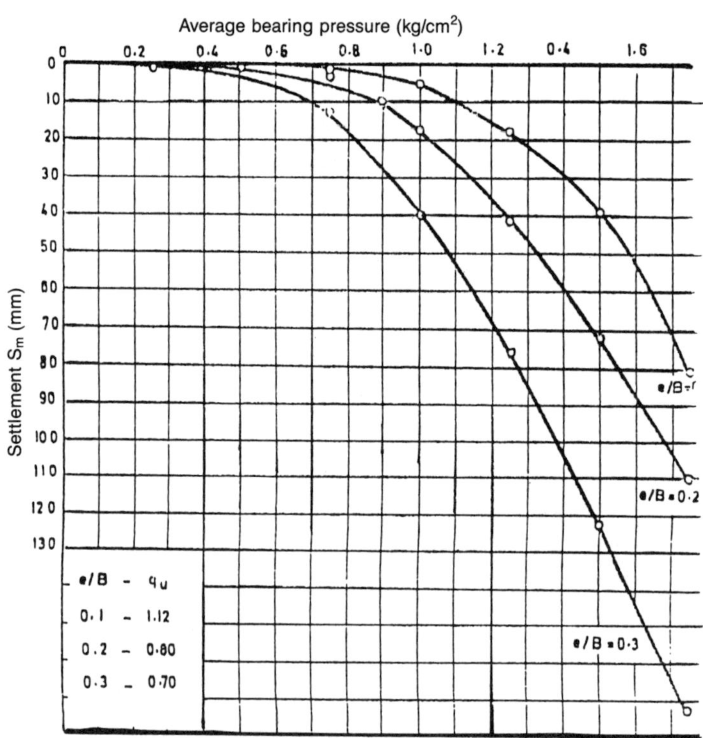

FIGURE 58. Pressure vs. S_m for eccentrically loaded rigid strip footing on Buckshot clay of width B = 10 cm.

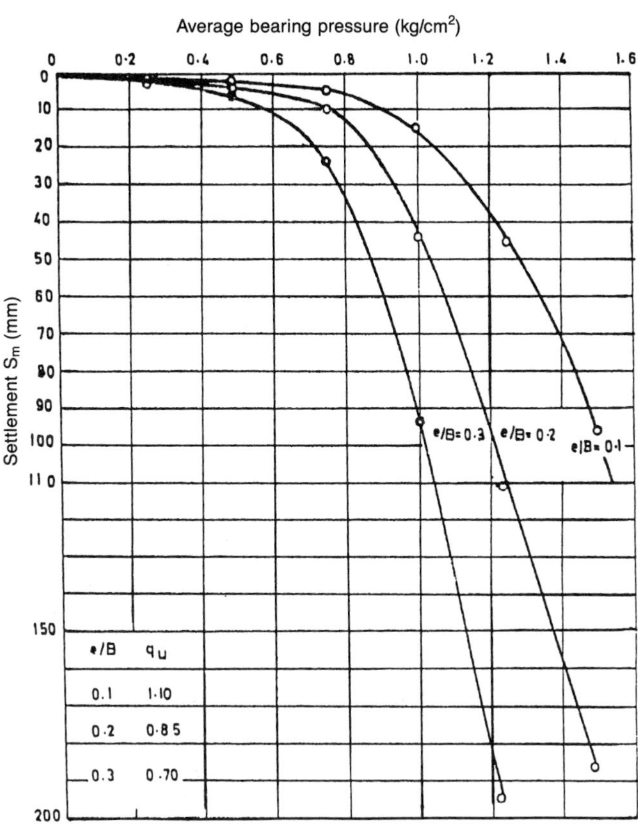

FIGURE 59. Pressure vs. S_m curve for an eccentrically loaded rigid strip footing resting on Buckshot clay (B = 25 cm).

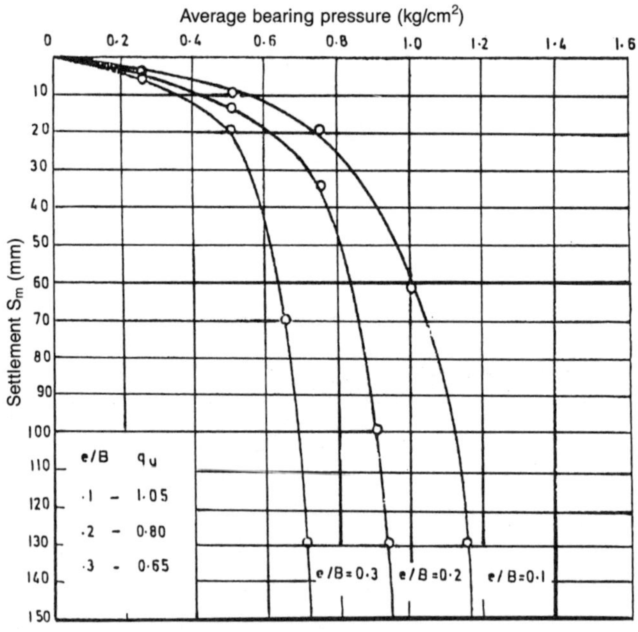

FIGURE 60. Average bearing pressure vs. S_m for an eccentrically loaded rigid strip footing on Buckshot clay (B = 100 cm).

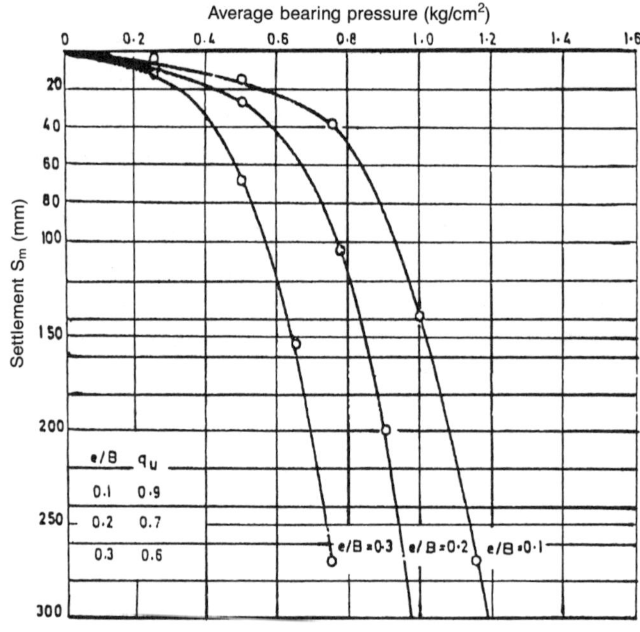

FIGURE 61. Average bearing pressure vs. S_m for eccentrically loaded rigid strip footing resting on Buckshot clay (B = 200 cm).

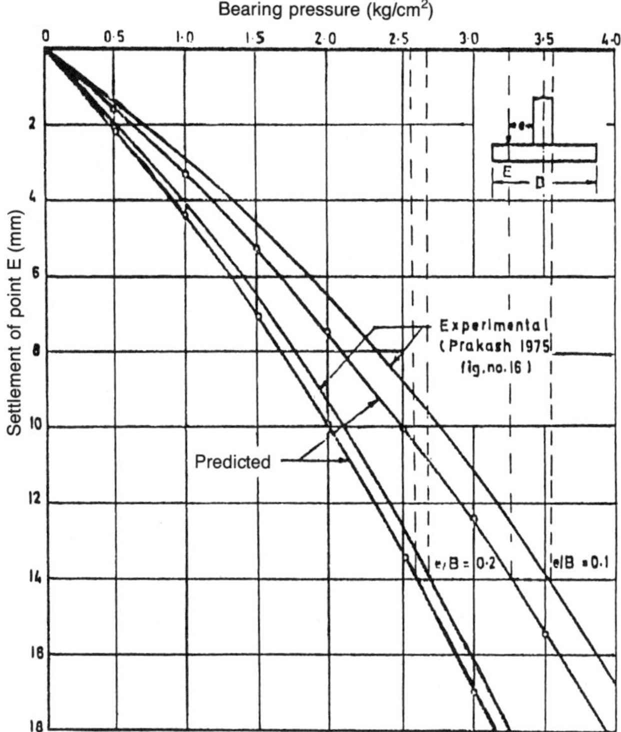

FIGURE 62. Comparison of pressure vs. settlement curves of eccentrically loaded 10 cm wide rigid strip footing resting on Dhanori clay.

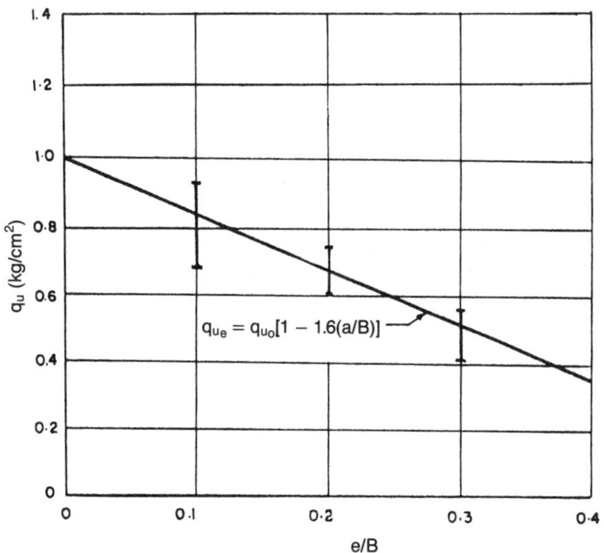

FIGURE 63. Ultimate bearing capacity vs. e/B ratio for eccentrically loaded rigid strip footing resting on Buckshot clay.

TABLE 12. Settlements S_e and S_m for Rigid Strip Footing Resting on Buckshot Clay.

e/B	q kg/cm²	B (cm)	S_e (mm)	S_m (mm)
0.1	0.75	10	1.768	1.974
		25	4.300	4.800
		100	17.700	19.75
		200	35.370	39.48
0.2	0.75	10	3.0	3.7
		25	7.5	9.5
		100	31.22	39.48
		200	62.45	69.18
0.3	0.75	10	10.65	13.78
		25	24.0	30.0
		100	99.4	174.0
		200	213.21	275.62

TABLE 13. Ultimate Bearing Capacity of Eccentrically Loaded Footings on Buckshot Clay.

Width	e/B = 0 Computed Terzaghi		e/B = 0.1				e/B = 0.2				e/B = 0.3			
			a	b	c	d	a	b	c	d	a	b	c	d
10	1.25	1.39	1.12	1.12	1.0	1.11	0.84	0.80	0.80	0.834	0.66	0.70	0.70	0.556
25	1.25	1.39	1.12	1.10	1.0	1.11	0.80	0.85	0.80	0.834	0.55	0.70	0.70	0.556
100	1.25	1.39	1.15	1.10	1.0	1.11	0.85	0.80	0.80	0.834	0.70	0.65	0.70	0.556
200	1.30	1.39	1.15	1.0	1.0	1.11	0.85	0.80	0.80	0.834	0.70	0.60	0.70	0.556

a = q vs. S_e curve b = q vs. S_m c = q vs. tilt d = computed

TABLE 14. Ultimate Bearing Capacity of Eccentrically Loaded Footing on Dhanori Clay.

Width (cm)	e/B	Ultimate Bearing Pressure kg/cm²: Computed	Observed	Meyerhof (1953)
10	0.0	4.26	4.50	5.7
10	0.1	3.25	3.50	4.56
10	0.2	2.55	2.65	3.42

Dhanori clay are also plotted. It is found that the computed and experimental curves compare very well. The curves show that the type of failure is similar to Terzaghi's local shear failure.

ULTIMATE BEARING PRESSURE

The ultimate bearing pressures were computed from pressure versus S_e and pressure versus S_m curves for Buckshot clay. The ultimate bearing pressure was also evaluated from pressure versus tilt curves shown in Figure 68. The ultimate bearing pressure for footings on Dhanori clay was computed using Terzaghi's (1943) method considering local shear failure. The values are tabulated in Tables 13 and 14 for footings in Buckshot clay and Dhanori clay respectively.

The ultimate bearing pressures computed in Tables 13 and 14 further analysed by plotting non-dimensional plot of q_{ue}/q_{uo} versus e/B ratio. This plot is shown in Figure 63 and is found to be a straight line having the following equation:

$$q_{ue} = q_{uo}\,[1 - 1.6(e/B)] \qquad (74)$$

where

q_{ue} = ultimate bearing pressure of eccentrically loaded footing, kg/cm²

q_{uo} = ultimate bearing pressure of centrally loaded footing, kg/cm²

The equation suggested by Meyerhof (1953) for the above case is following:

$$q_{ue} = q_{uo}[1 - 2(e/B)] \text{ for surface footings on } \Phi = 0 \text{ soil}$$

Thus Meyerhof equation under estimate the bearing capacity of eccentrically loaded footings.

NON-DIMENSIONAL CORRELATIONS

For various bearing pressures, the values of settlements S_o at $X = O$ S_e at $x = O$ and S_m at $X = B/2$ (Figure 12) have been computed for footing sizes of 10 cm, 25 cm, 100 cm and 200 cm and e/B = 0.0, 0.1, 0.2 and 0.3 and plotted in Figures 54 to 61. The values of S_e/S_o and S_m/S_o at arbitrary factor of safety of 1, 2 and 3, i.e., at q_u, $q_u/2$ and $q_u/3$ from the above curves were computed and are given in Table 15.

The ratios of S_e/S_o and S_m/S_o given in Table 15 were analysed by plotting them against e/B in Figures 64 and 65 respectively. It is found that a unique curve is obtained for both the above relationships for all factors of safeties. The relationship S_e/S_o vs. e/B can be expressed by the following equation:

$$S_e/S_o = 1.0 - 1.5\,(e/B) \qquad (75)$$

The curve S_m/S_o versus e/B follows a horizontal straight line from e/B = 0.0 to e/B = 0.066 and then from e/B = 0.066 to e/B = 0.3m; it is also represented by a straight line

TABLE 15. S_e/S_o and S_m/S_o for Factors of Safety of 1, 2 and 3.

Factor of Safety	e/B	Width	S_o (mm)	S_e (mm)	S_m (mm)	S_e/S_o	S_m/S_o
1.	2.	3.	4.	5.	6.	7.	8.
1.0	0.1	10	13.0	9.5	13.0	0.732	1.00
		25	33.0	24.0	32.0	0.727	0.97
		100	130.0	96.0	130.0	0.728	1.00
		200	260.0	195.0	270.0	0.750	1.04
1.0	0.2	10	13.0	9.0	10.0	0.692	0.77
		25	33.0	23.0	25.0	0.700	0.758
		100	130.0	88.0	100.0	0.678	0.77
		200	260.0	180.0	200.0	0.693	0.77
1.0	0.3	10	13.0	6.5	7.0	0.500	0.536
		25	33.0	17.0	17.0	0.515	0.515
		100	130.0	65.0	70.0	0.500	0.535
		200	260.0	130.0	155.0	0.500	0.597
2.0	0.1	10	1.3	1.1	1.2	0.848	0.922
		25	3.4	2.9	3.0	0.850	0.882
		100	13.0	11.0	12.5	0.846	0.96
		200	23.0	20.0	22.0	0.870	0.955
2.0	0.2	10	1.3	0.9	1.0	0.692	0.77
		25	3.4	2.4	2.6	0.705	0.764
		100	13.0	9.0	10.0	0.692	0.77
		200	23.0	16.0	18.0	0.696	0.782
2.0	0.3	10	1.3	0.7	0.78	0.538	0.60
		25	3.4	2.0	2.40	0.588	0.705
		100	13.0	8.0	9.1	0.615	0.700
		200	23.0	15.0	16.0	0.654	0.695
3.0	0.1	10	0.8	0.68	0.8	0.85	1.00
		25	1.9	1.6	2.0	0.842	1.055
		100	8.0	7.0	8.0	0.875	1.00
		200	17.0	14.5	15.0	0.85	0.88
3.0	0.2	10	0.8	0.6	0.6	0.75	0.75
		25	1.9	1.4	1.5	0.735	0.79
		100	8.0	6.0	6.2	0.75	0.776
		200	17.0	12.0	15.0	0.706	0.765
3.0	0.3	10	0.8	0.4	0.45	0.50	0.56
		25	1.9	1.0	11.0	0.526	0.58
		100	8.0	4.0	5.0	0.52	0.625
		200	17.0	10.0	10.5	0.590	0.618

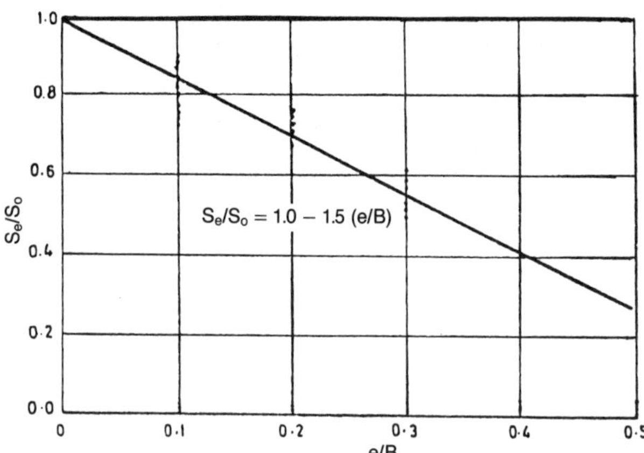

FIGURE 64. S_e/S_o vs. e/B and S_m/S_o vs. e/B for eccentrically loaded rigid strip footing resting on Buckshot clay.

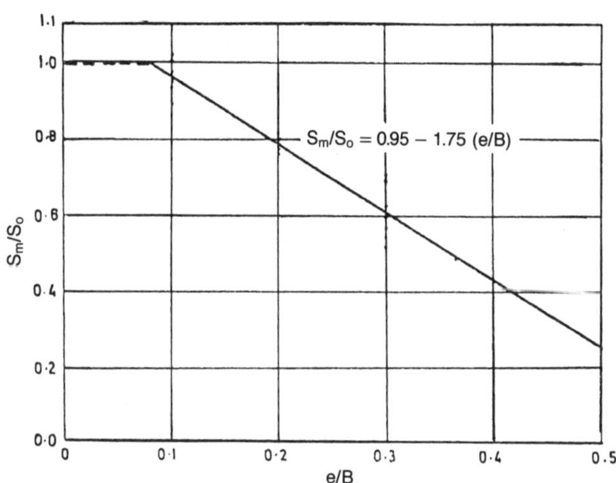

FIGURE 65. S_e/S_o vs. e/B and S_m/S_o vs. e/B for eccentrically loaded rigid strip footing resting on Buckshot clay.

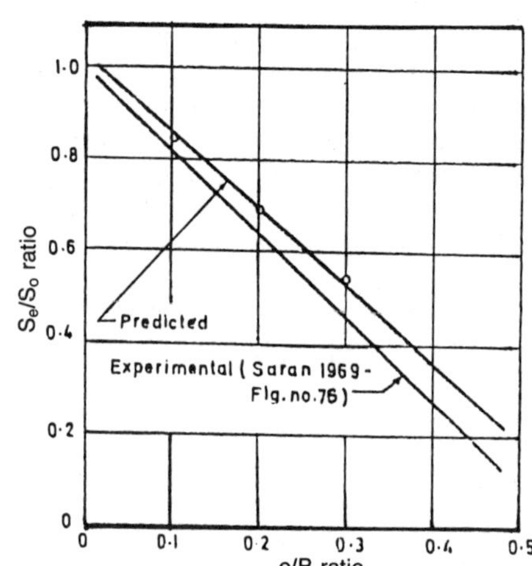

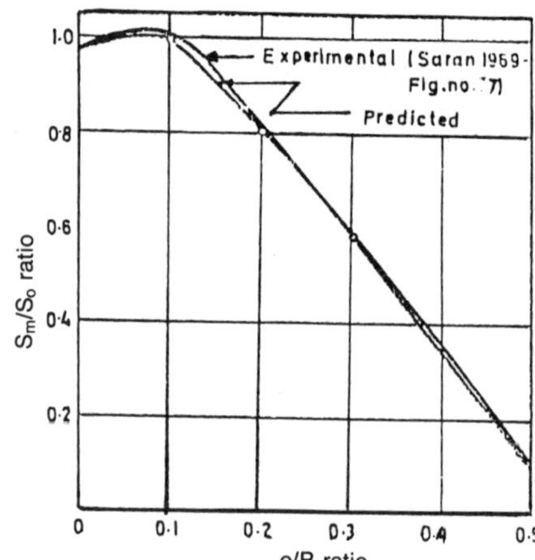

FIGURE 66. Comparison of S_e/S_o and S_m/S_o vs. e/B with Saran (1969) results.

with the following equation:

$$S_m/S_o = 0.95 - 1.75\ (e/B) \qquad (76)$$

The equations for the above relationships have been determined from experimental results of footing tests on Ranipur sand by Sarn (1969) and are given below:

$$\frac{S_e}{S_o} = 1.0 - 1.63\ (e/B) - 2.63(e/B)^2 + 5.83(e/B)^3 \qquad (77)$$

$$\frac{S_m}{S_o} = 1.0 + 2.31\ (e/B) - 22.61(e/B)^2 + 31.64(e/B)^3 \qquad (78)$$

The equations are in third degree power fractions. For comparison of the values of two equations they are plotted together in Figure 66. The two plots compare very well. The equation suggested in the present case is simpler than that of Saran (1969).

The non-dimensional plots S_e/S_o versus e/B and S_m/S_o vs. e/B for Dhanori clay are shown in Figure 67. The corresponding values obtained experimentally by Prakash (1975) are also shown on the above figure. It is found that the plots compare well, showing that equations derived in this case can give satisfactory results.

From the above comparisons, it may be concluded that the relationships S_e/S_o versus e/B and S_m/S_o versus e/B are unique for rigid strip footing and are independent of the type of soil. The plots of S_e/S_o and S_m/S_o versus e/B can be used to predict the settlement and tilt of an eccentrically loaded rigid strip footing if the settlement (S_o) of a centrally loaded footing is known. The settlement (S_o) of strip and square footings can be predicted by the method described herein using the constitutive relationships.

TILT

The tilts of the footings (10 cm, 25 cm, 100 cm and 200 cm) have been computed from the values of S_e and S_m plotted in Figures 54 to 60 respectively for e/B = 0.1, 0.2 and 0.3 by the following equations:

$$\sin t = \frac{S_m - S_e}{\frac{B}{2} - e} \qquad (64)$$

The pressure versus tilt relationship computed from above formula were plotted in Figure 68 for all footing sizes and eccentricity. It is found that angular tilt increases with increase of eccentricity and bearing pressure but is independent of footing size. The ultimate bearing pressure can be computed from the pressure versus tilt curves.

The settlements S_e and S_m and the corresponding tilt for rigid strip footing 10 cm wide resting on Dhanori clay were computed for the sake of comparison with the experimental results of Prakash (1975). The tilts for e/B = 0.1 and 0.2 were plotted in Figure 69. The tilts observed by Prakash (1975) are also shown in the above figure. The plots compare very well showing that the procedure followed in this investigation can yield satisfactory results.

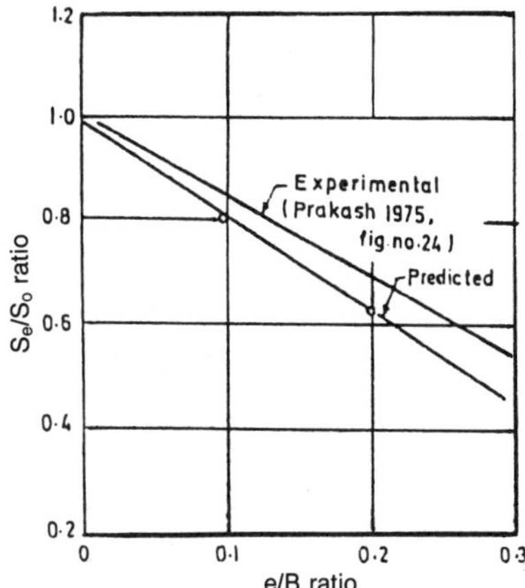

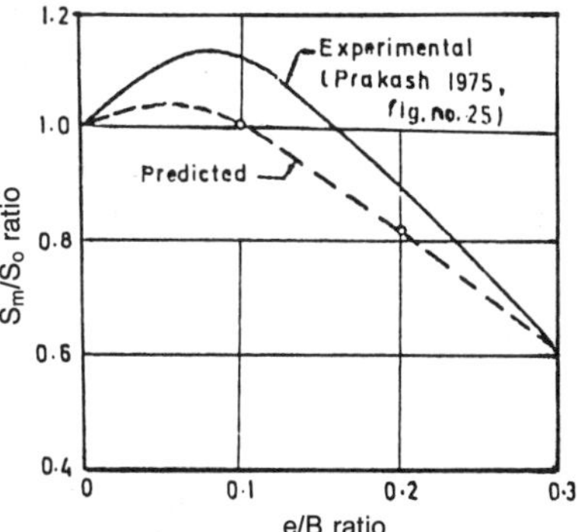

FIGURE 67. Comparison of S_e/S_o and S_m/S_o vs. e/B of 10 cm wide strip footings resting on Dhanori clay.

CONTACT PRESSURE DISTRIBUTION

The values of contact pressure at *A*, *E* and *B* (Figure 14) for various load intensities were computed for e/B = 0.1, 0.2 and 0.3. The values are plotted in Figures 70, 71 and 72. It is observed that the curve tends to be flatter and then straight at failure load for all e/B ratios.

It is further observed that the contact pressure at section *B* starts becoming negative for e/B = 0.2 and q = 0.75 kg/cm^2 as shown in Figure 72. Since there cannot be negative contact pressure at the soil–footing interface, this phenomenon shows that there is loss of contact at e/B = 0.2 and 0.3 from q = 0.75 kg/cm^2 and 0.50 kg/cm^2 respectively.

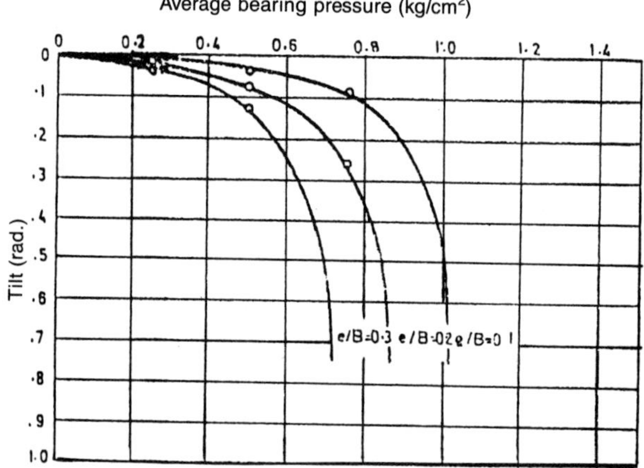

FIGURE 68. Average bearing pressure vs. tilt for eccentrically loaded rigid strip footing resting on Buckshot clay.

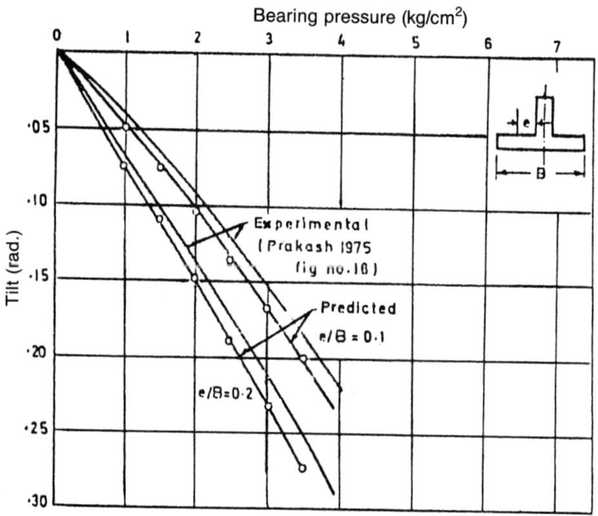

FIGURE 69. Comparison of pressure vs. tilt of eccentrically loaded rigid strip footing resting on Dhanori clay.

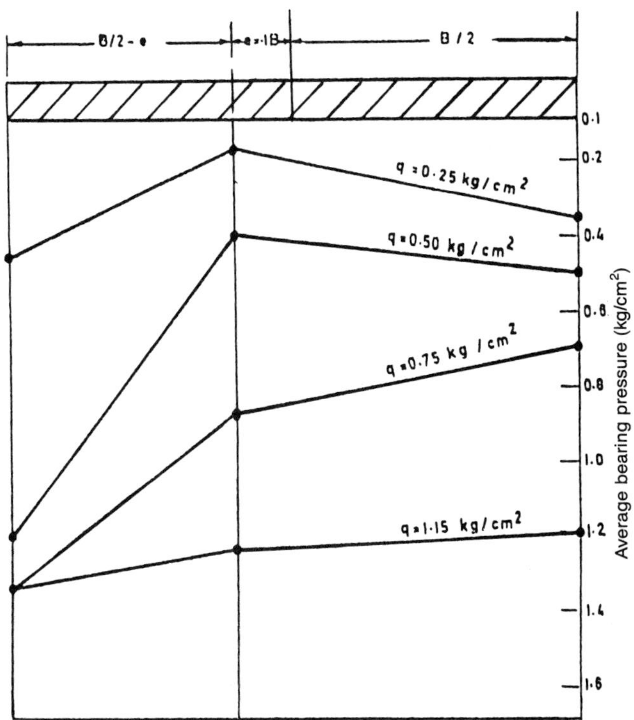

FIGURE 70. Contact pressure distribution for eccentrically loaded rigid strip footing resting on Buckshot clay (e/B = 0.1).

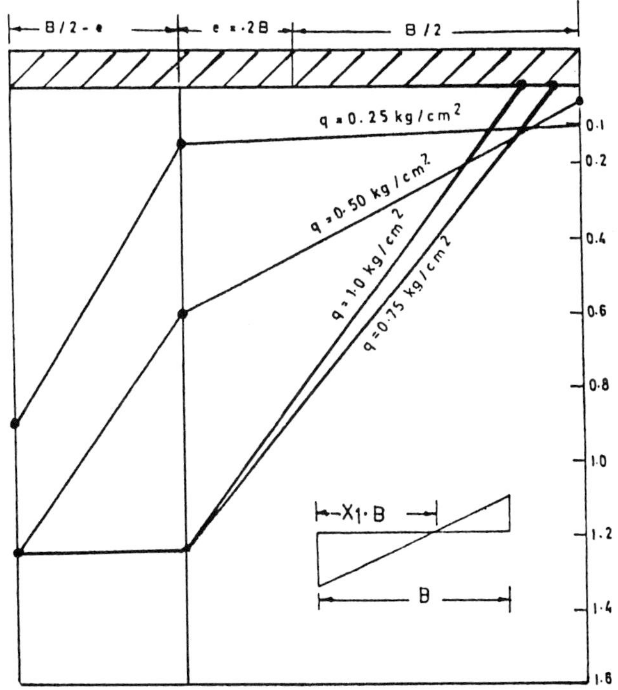

FIGURE 71. Contact pressure distribution for eccentrically loaded rigid strip footing resting on Buckshot clay (e/B = 0.2).

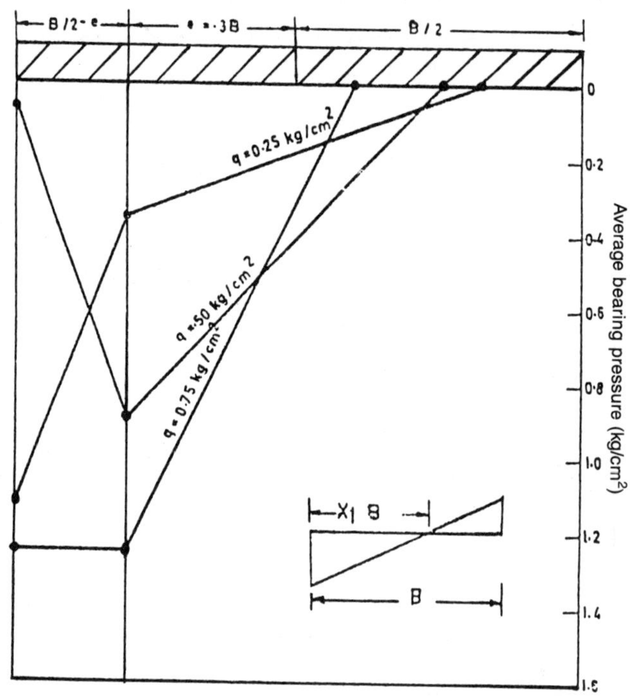

FIGURE 72. Contact pressure distribution for eccentrically loaded rigid strip footing resting on Buckshot clay (e/B = 0.3).

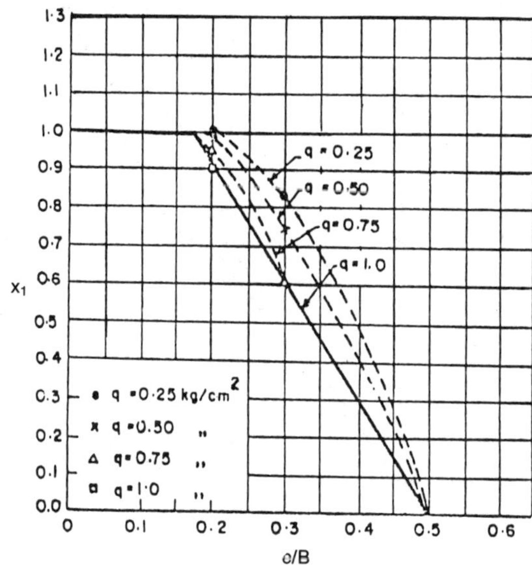

FIGURE 73. Contact with x_1 vs. e/B for eccentrically loaded rigid strip footing resting on Buckshot clay.

TABLE 16. Contact Width x_1.

e/B	Load kg/cm²	x_1 (computed)	x_1 (conventional)	Ultimate Bearing Pressure
1.	2.	3.	4.	5.
0.2	0.25	1.00		
	0.50	1.00	0.90	0.90 kg/cm²
	0.75	0.95		
	1.00	0.90		
0.3	0.25	0.845		
	0.50	0.75	0.60	0.70 kg/cm²
	0.75	0.60		

CONTACT WIDTH

There is loss of contact at the soil–footing interface with increase of eccentricity and bearing pressures (Figures 71 and 72). The ratio of contact width and footing width B denoted by x_1 has been computed for $e/B = 0.2$ and 0.3 at various bearing pressures and are given in column 3 of Table 16. The value of x_1 from conventional method has also been given in column 4 of Table 16. The values of x_1 from computation and from conventional methods have been plotted in Figure 73.

It is observed that computed values tend to approach the conventional values at failure loads. The computed values of x_1 are higher at lower pressure levels compared to the conventional values. In case of sand, Saran (1969) has reported that the contact width follows the conventional rule. In the present approach, the guiding factor for determining contact width is surface load, eccentricity and soil parameters. In conventional method eccentricity is the only guiding criterion. Hence there is difference in values.

SUMMARY

The following observations may be listed from the results of eccentrically loaded rigid footings.

1. The contact pressure coefficient $\alpha\alpha_1$ is a function of load intensity (q) and load eccentricity (e/B) and is independent of footing width. A co-relation of α_1 with non-dimensional quantity (q/c_u) has been plotted in Figure 53 for various load eccentricity e/B. α_1 reaches maximum value near failure load and then falls down.

2. The ultimate bearing capacity can be found from q vs. S_e curve, q vs. S_m curves and q vs. tilt curves. The equation for determining the ultimate bearing pressure is as follows:

$$q_{ue} = q_{uo}\,[1 - 1.6(e/B)] \tag{74}$$

3. Non-dimensional correlations for S_e/S_o vs. e/B and S_m/S_o vs. e/B have been obtained and are given by the following equations:

$$\frac{S_e}{S_o} = 1.0 - 1.5(e/B) \tag{75}$$

$$\frac{S_m}{S_o} = 0.95 - 1.75(e/B) \tag{76}$$

These relationships have compared well with the values obtained by Prakash (1975) and Saran (1969). Prakash (1975) had used Dhanori clay and Saran (1969) Ranipur sand. It may thus be inferred that the above relationships are unique and independent of the type of soil, size and shape of the footing.

4. The tilt of a footing has been found from the following equation:

$$\sin t = \frac{S_m - S_e}{\frac{B}{2} - e} \tag{64}$$

The pressure versus tilt has been plotted in Figure 68 for various load eccentricities (e/B). The curves have been found to be independent of the footing width.

5. The footing base loses contact with the supporting soil with increase of eccentricity (e/B) and load intensity (q). The contact width factor x_1 has been found to be a function of eccentricity and load intensity. The values of x_1 tend to equal the corresponding values obtained from conventional method as the surface load intensity reaches ultimate value (Figure 73).

6. The pressure settlement curves predicted by the analytical procedure have been found to compare well with the experimental curves of Prakash (1975).

FOOTINGS IN SAND

General

The pressure settlement characteristics of smooth strip footings resting on Ranipur sand have been investigated for footing widths of 5 cm, 10 cm, 15 cm, 60 cm and 200 cm. The computations were done for sand at two relative densities, namely, 84 percent and 46 percent. Settlements at central, quarter and edge sections for a given pressure intensity were computed by the method described earlier and constitutive relations were taken accordingly. The average settlements were calculated by dividing the area of settlement diagram by width of footing.

Pressure-Settlement Curves

Pressure-settlement curves were drawn by plotting average settlements against bearing pressures. These curves for footings width of 5 cm, 10 cm, and 15 cm are plotted in

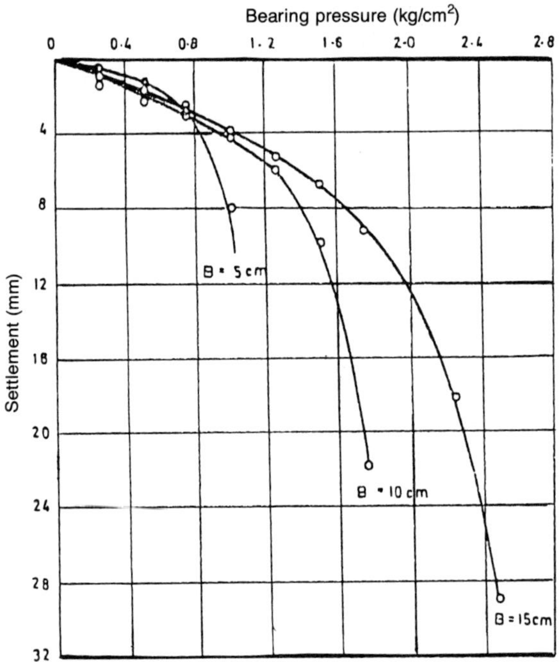

FIGURE 74. Pressure vs. settlement for smooth strip footing resting on Ranipur sand (D_R = 84%, B = 5 cm, 10 cm and 15 cm).

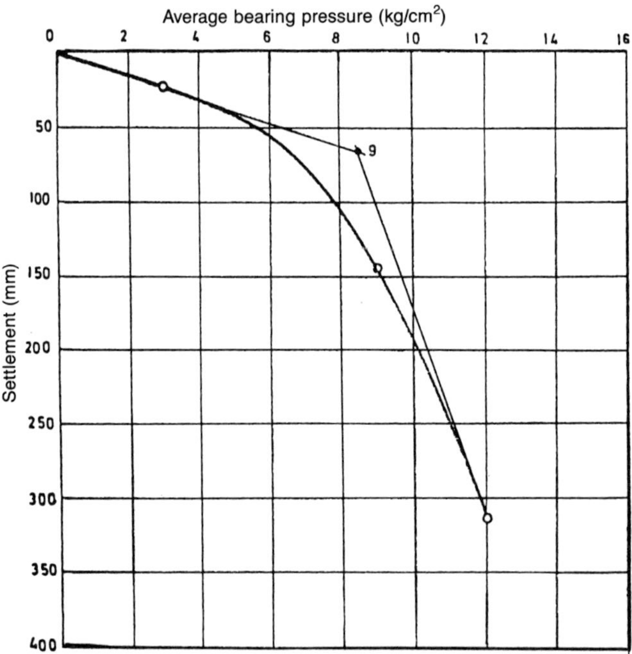

FIGURE 75. Pressure vs. settlement curve for smooth strip footing resting on Ranipur sand (D_R = 84%, B = 60 cm).

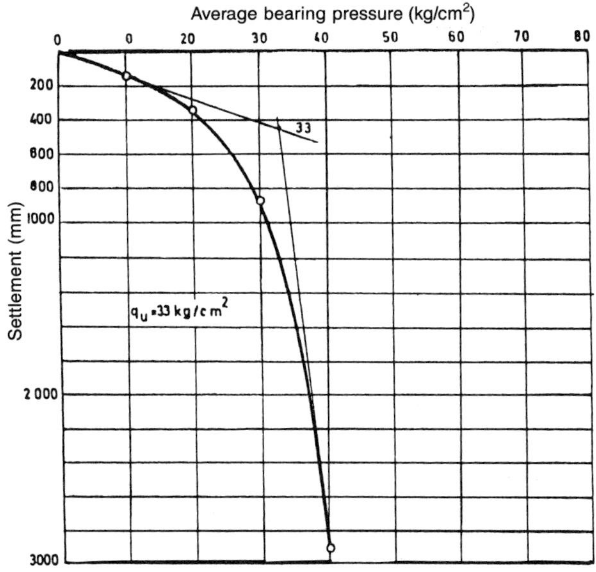

FIGURE 76. Pressure vs. settlement for smooth strip footing on Ranipur sand (D_R = 84%, B = 200 cm).

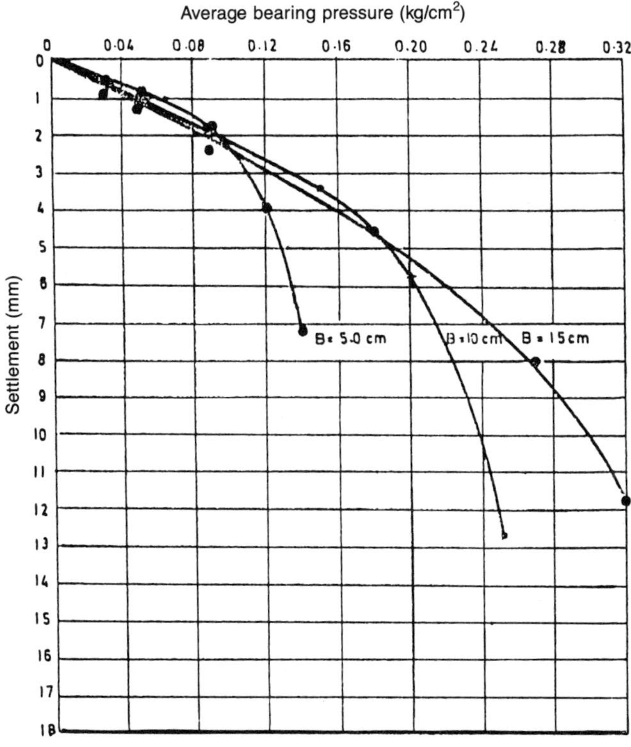

FIGURE 77. Pressure vs. settlement curve for footings resting on Ranipur sand (D_R = 46%, B = 5.0 cm, 10 cm and 15 cm, smooth strip).

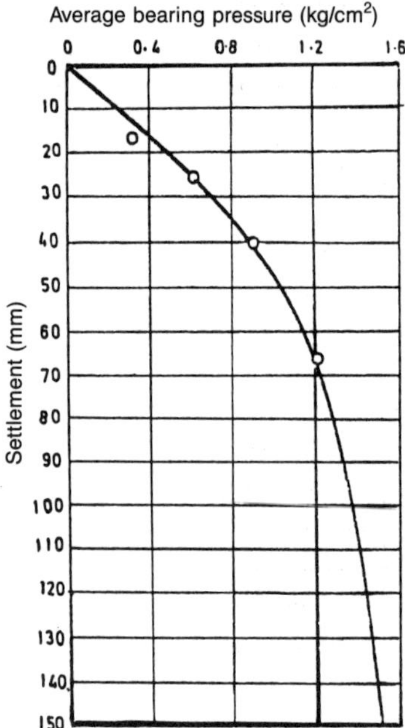

FIGURE 78. Pressure vs. settlement curve for smooth strip footing resting on Ranipur sand (D_R = 46%, B = 60 cm).

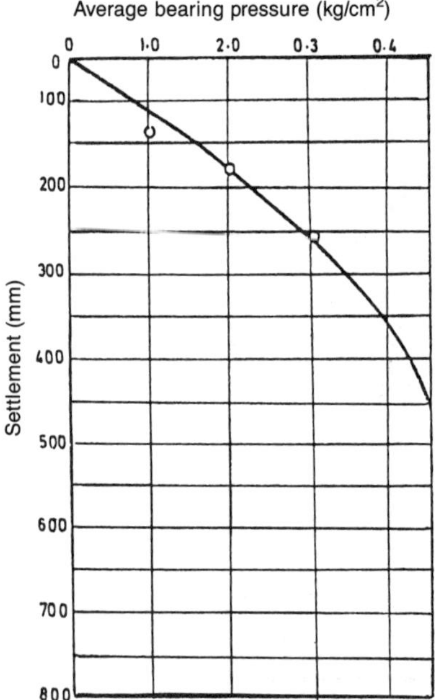

FIGURE 79. Pressure vs. settlement curve for smooth strip footing resting on Ranipur sand (D_R = 46%, B = 200 cm).

TABLE 17. Settlements of Strip Footing in Sand.

D_R (%)	q kg/cm²	Footing Width(B) (mm)	Average Settl.(s) (mm)	B_f/B_p	S_f/S_p Taking $B_p = 11$
84	0.50	5	1.0	0.5	0.667
		10	1.5	1.0	1.0
		15	2.0	1.5	1.335
		60	3.0	6.0	2.00
		200	4.5	20.0	3.0
46	0.50	5	0.8	0.5	0.726
		10	1.1	1.0	1.0
		15	1.4	1.5	1.27
		60	2.1	6.0	1.91
		200	3.0	20.0	2.73

Figure 74, for 60 cm in Figure 75, and for 200 cm in Figure 76 for sand at relative density of 84 percent. Similar curves are drawn for footing widths of 5 cm, 10 cm, 15 cm, 60 cm and 200 cm in Figures 77, 78 and 79 respectively for relative density of 46 percent. The results show that for the same density and pressure, the settlements increase with footing size. The settlements for same pressure and footing size increase with decrease of relative density. The values are shown in Table 17.

Ultimate Bearing Pressure

The ultimate bearing pressures have been computed by intersection tangent method from pressure-settlement curves and the values are shown in column 2, Table 18 for D_R = 84 percent and in the column 4 for D_R = 46 percent. The corresponding values computed by Terzaghi's formula are given in columns 3 and 4 of Table 18. It is observed that ultimate bearing pressures increase with increase of footing size and relative density.

Size Effect

To study the footing size effect on the pressure-settlement curves for footings in sand, the S_f/S_p versus B_f/B_p relation has been computed for pressure of 0.50 kg/cm² taking B_p = 10 cm and B_f = 5 cm, 15 cm, 60 cm and 200 cm and shown in Figure 80 for relative densities of 84 percent and 46 percent. The above relationship computed from conventional formula is also shown in the above figure. It is observed that the curve becomes asymptotic with increase of footing size (B_f). After B_f/B_p is greater than 6.0 the curve is practically horizontal.

TABLE 18. Ultimate Bearing Pressures for Footings in Sand.

Footing Width (cm)	D_R = 84 percent, ϕ = 42° Computed From Terzaghi's Method kg/cm²	kg/cm²	Observed kg/cm²	D_R = 46 percent, ϕ = 36.5° Computed From Terzaghi's Method kg/cm²	kg/cm²	Observed kg/cm²
5	0.95	1.0	1.08	0.11	0.075	0.07
10	1.50	2.0		0.22	0.150	
15	2.10	3.0		0.31	0.225	
60	9.0	12.0		1.25	0.90	
100	33.0	40.0		4.25	3.00	

Prediction of Pressure-Settlement Curves for Strip Footing in Sand

The pressure-settlement curves of rigid strip footings computed by the method described herein for a 5 cm width have been plotted in Figure 81 and Figure 82 for relative density of 84 percent and 46 percent respectively. The pressure settlement curves for the above footings, obtained by Saran (1969) by performing tests on model footing in Ranipur sand, have also been shown in the above figures. It is found that the experimental and theoretical curves compare very well. The ultimate bearing capacity for 84 percent and 46 percent relative densities have been found to be 0.95 kg/cm² and 0.11 kg/cm² respectively. The corresponding experimental values are 1.0 kg/cm² and 0.70 kg/cm²; the difference in the latter case is largely due to the fact that it is difficult to determine ultimate bearing pressure by intersection tangent method in the latter case. Terzaghi's formula gives a value of 1.08 and 0.075 kg/cm² for D_R = 84 percent and 46 percent respectively. Thus it is found that the

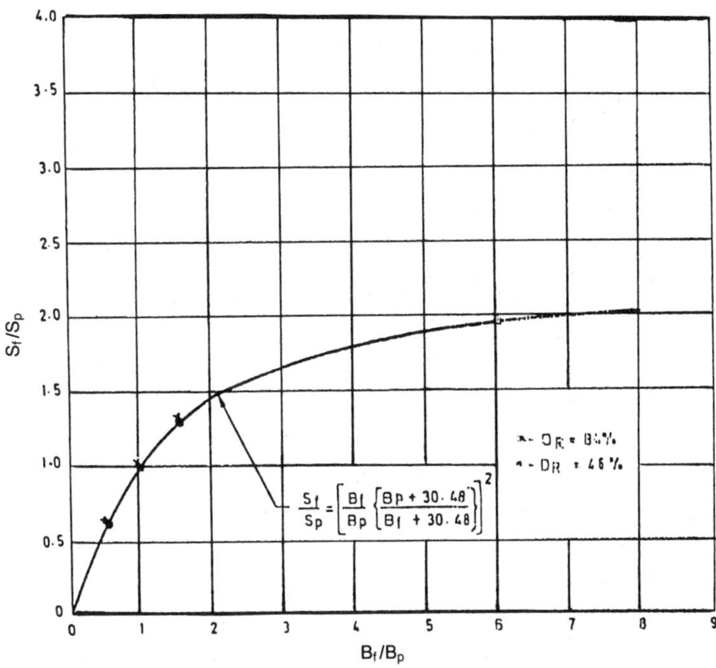

FIGURE 80. S_f/S_p vs. B_f/B_p relationship for Ranipur sand (D_R = 84% and 46%).

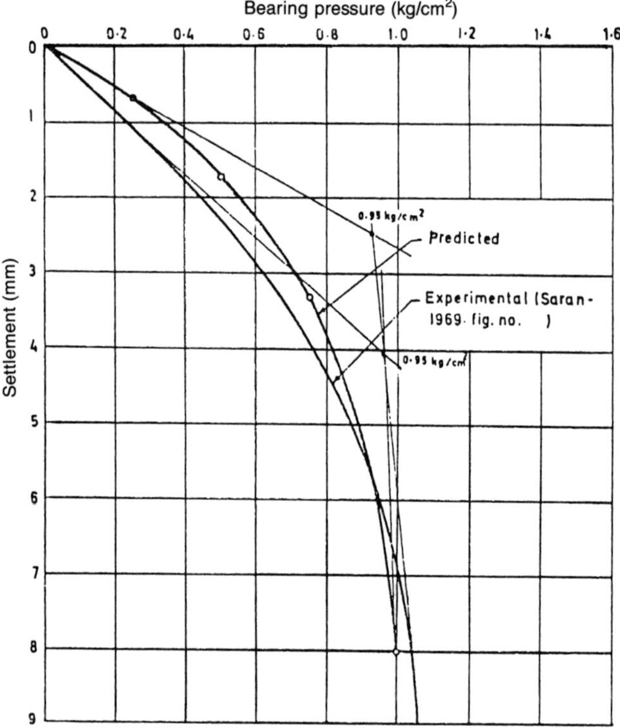

FIGURE 81. Comparison of pressure vs. settlement curves of 5 cm wide strip footing on Ranipur sand (D_R = 84%).

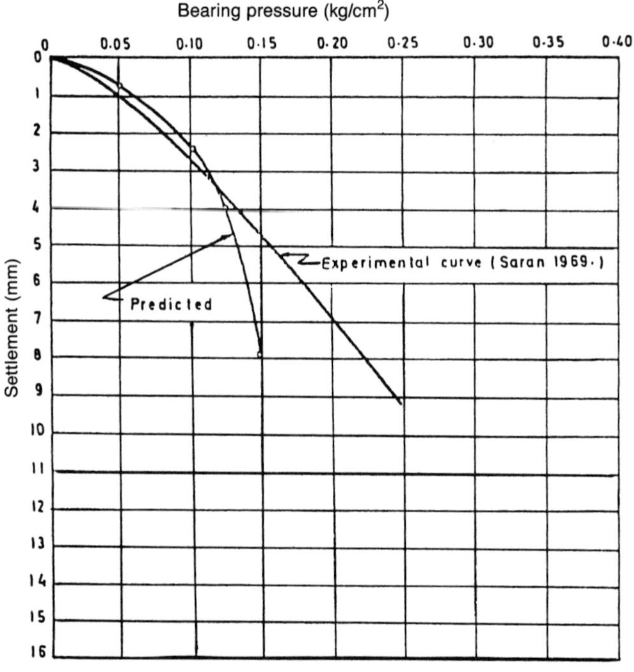

FIGURE 82. Comparison of pressure vs. settlement curves of 5 cm wide strip footing resting on Ranipur sand (D_R = 46%).

empirical method described earlier for pressure settlement curves gives satisfactory results in case of sand for which no method is available to date.

Summary

The prediction of pressure settlement characteristics using non-linear stress-strain relationship for footings resting on sand was attempted by Duncan and Chang (1970) and Basavanna (1975) using finite element technique. Their results show large variation with the experimental values. A semi-empirical approach using constitutive relationship has been given. The pressure settlement characteristic of strip footing in sand at relative density of 84 percent and 46 percent have been predicted. The following observations may be listed from the results obtained for strip footing on sand:

1. The ultimate bearing capacity predicted by the present method compares with the experimental values (Table 18).
2. The size effects on settlement has been studied by drawing curves between S_f/S_p versus B_f/B_p (Figure 80) and has been found that they confirm the established relationship

$$\frac{S_f}{S_p} = \left[\frac{B_f}{B_p}\left(\frac{B_p + 30.48}{D_f + 30.48}\right)\right]^2$$

3. The pressure-settlement curves predicted by this method compare well with the experimental curves of Saran (1969).

REFERENCES

1. Ahlberg, J. H., F. W. Nilson, and J. L. Walsh, *The Theory of Splines and their Applications*, Academic Press, New York (1967).
2. Awojobi, A. O. and R. E. Gibson, "Plane Strain and Axially Symmetric Problems of a Linearly Non-homogeneous Elastic Half Space," *Quarterly Jl. Mech. Appl. Math. 26*, p. 285-302 (1973).
3. Barden, L., "Distribution of Contact Pressure under Foundation" *Geotechnique Vol. XII, No. 3*, pp. 181-198 (1962).
4. Basavanna, B. M., "Bearing Capacity of Soil under Static and Transient Load," Ph.D. Thesis, School of Research and Training in Earthquake Engineering, University of Roorkee, Roorkee (India) (1975).
5. Bishop, A. W. and A. K. G. Eldin, "The Effect of Stress History on the Relation Between Angle of Shearing Resistance and Porosity in Sand," *3rd International Conference on Soil Mechanics and Foundation Engineering, Vol. I*, pp. 100-105 (1953).
6. Bishop, A. W. and L. Bjerrum, "The Relevance of the Triaxial Test on the Solution of Stability Problems," *Proc. Research Conf. Shear Strength of Cohesive Soils*, ASCE. pp. 437-501, Colorado, June (1960).
7. Bishop, A. W. et al., "Triaxial Tests on Soil at Elevated Cell Pressure," *Proc. of 6th International Conference on Soil Mechanics and Foundation Engineering, Vol. I*, pp. 170-174 (1965).
8. Bishop, A. W. and D. J. Henkel, "The Measurement of Soil Properties in Triaxial Test," English Language Book Society 1 (1969).
9. Bhushan, K. and S. C. Haley, "Stress Distribution for Heavy Embedded Structures," *Proc. of ASCE. Journal of Geotechnical Engineering Division, Vol. 102*, No. GT 7, July (1976).
10. Boussinesq, J., "Application des Potentiels a l'Etude de *l'Equilibre et du Mouvement des Solides Elastiques*, Paris, Gauthier-Villard (1985).
11. Chakarborty, S. K., "Bearing Capacity in Relation to Stress-Strain Characteristics of Soils," Ph.D. Thesis, Ranchi University, Ranchi (1970).
12. Carrol, W. F., "Dynamic Bearing Capacity of Soil," Technical Report No. 3-599, U. S. Army Engineers, Waterways Experiment Stations, Corps of Engineers, Vicksburg, Mississippi (1963).
13. Casagrande, A. and W. L. Shannon, "Research on Stress Deformation and Strength Characteristics of Soil and Soft Rocks under Transient Loadings," Publication, Harvard University, Grad. School of Engineering, No. 447, p. 132 (1947).
14. Carothers, S. D., "Direct Determination of Stresses," *Proceedings of the Royal Society of London, Series A, Vol. XCVII*, p. 110 (1920).
15. Carothers, S. D., "Elastic Equivalence of Statically Equipollent Loads," *Proceedings of International Mathematical Congress,—Vol. 11*, Toronto University Press, p. 518.
16. Cornforth, D. H., "Some Experiments on the Influence of Strain Conditions on the Strength of Sands," *Geotechnique, Vol. 14*, No. 2, pp. 143-167 (1964).
17. Detye, K. R., "Shallow Foundation in Clay: Suggested Methodology for Investigation and Design," *Proc. of Symp. on Shallow Foundation, National Society of Soil Mechanics and Foundation Engineering, Bombay, Vol. I*, pp. 1 to 10 (1970).
18. Davis, E. H. and H. G. Poulos, "Use of Elastic Theory for Settlement Prediction under Three-Dimensional Conditions," *Geotechnique, London, Vol. 18*, pp. 67-91 (1968).
19. Desai, M. D., G. R. S. Jain, S. Saran, and P. K. Jain, "Penetration Testing in India," *Proceedings of European Conference on Penetration Testing held in Bulgaria, Vol. I*, pp. 167-184 (1974).
20. Desai, C. K., "Non-Linear Analysis Using Spline Functions," *Proceedings American Society of Civil Engineer, SM 10:* 1461:8462, Oct. (1971).
21. Dhillon, G. S., "The Settlement, Tilt and Bearing Capacity of Footings on Sand under Central & Inclined Loads," *Journal of National Buildings Organization, 6*, p. 66 (1961).

22. Durelli, A. J. and TSA cH Phillips, *Introduction to the Theoretical and Experimental Analysis of Stress and Strain,* McGraw Hill Book Company, Inc. (1958).
23. Duncan, J. M. and C. Y. Chang, "Non-linear Analysis of Stress-Strain in Soils," *Proceedings of ASCE, SM5,* 1629-1651, 7513 Sept. (1970).
24. Fumagalli, E., "Tests on Cohesionless Materials for Rockfill Dams," *American Society of Civil Engineers, SM I,* pp. 313-332 January (1969).
25. Frohilich, O. K., "On the Settlement of Buildings Combined with Deviation from their Originally Vertical Position," *Proceedings of Third International Conference on Soil Mechanics and Foundation Engineering, Vol. I.* (1963).
26. Gibbs, H. J. and W. G. Holtz, "Research on Determining the Density of Sands by Spoon Penetration Testing," *Proc. 4th International Conference on Soil Mechanics and Foundation Engineering, Vol. I,* pp. 3539, 1557 (1957).
27. Gibson, R. E. and G. C. Sills, "Settlement of Strip Load on a Non-Homogeneous Orthotropic Incompressible Elastic Half Space," *Qr. Jl. Mech. Appl. Math., 28, Part 2,* 233-243 May (1975).
28. Gibson, R. E., "Some Results Concerning Displacement and Stresses in a Non-Homogeneous Elastic Half Space," *Geotechnique Vol. XVII,* pp. 58-67 (1967).
29. Gibson, R. E. and G. C. Sills, "Some Results Concerning the Plane Deformation of a Non-Homogeneous Elastic Half Space," *Proceedings of Roscoe Memorial Symposium,* Cambridge University, p. 564-572 March (1971).
30. Green, G. C. and A. W. Bishop, "A Role on the Drained Strength of Sand under Generalized Strain Condition," *Geotechnique, London, Vol. 19,* No. 1, p. 144-49 (1969).
31. Greville, T. T. N., "Spline Functions," Interpolation and Numerical Quandrature," *Mathematic Methods of Digital Computers Vol. II A.* Ralston and H. S. Wilfeds, John Wiley and Sons (1967).
32. Hall, E. B. and B. B. Gordon, "Triaxial Testing with Large Scale High Pressure Equipment, Laboratory Shear Testing of Soils, *ASTM., STP 62,* pp. 315-328 (1963).
33. Hansen, J. B., "Discussion of Hyperbolic Stress-Strain Response of Cohesive Soils," *Proceedings American Society of Civil Engineer, SM 4:* p. 241-242 July (1963).
34. Haythornthwaite, R. M., "Mechanics of the Triaxial Test on Soils," *Proceedings American Society of Civil Engineers, Vol. 86,* SM 5, pp. 35-61 (1960).
35. Hankel, D. J., "The Effect of Over Consolidation on the Behaviour of Clays under Shear," *Geotechnique Vol. 6,* p. 139-150 (1956).
36. Hankel, D. J. and N. H. Wade, "Plane Strain Test on a Saturated Remoulded Clay," *Proc. American Society of Civil Engineer, Vol. 92,* SM 6, 67-80 (1966).
37. Holtz, W. G. and H. J. Gibbs, "Triaxial Shear Tests on Pervious Gravelly Soils," American Society of Civil Engineer, *Journal of Soil Mechanics and Foundation Engineering, No. I,* pp. 1-22 Jan. (1956).
38. Horn, M. R. and D. U. Deere, "Frictional Characteristics of Minerals," *Geotechnique, London, Vol. 12,* pp. 319-335 (1962).
39. Harr, M. E., *Foundations of Theoretical Soil Mechanics,* McGraw Hill Company, N.Y. (1966).
40. I.S.I., I.S.1498—Indian Standard Classification and Identification of Soils for General Engineering Purposes, I.S.I., New Delhi (1970).
41. Janbu, N., "Soil Compressibility as Determined by Oedometer and Triaxial Tests," *European Conference on Soil Mechanics and Foundation Engineering Wiesbaden, Germany, Vol. I,* p. 19-25 (1963).
42. Jumikis, R. A., *Soil Mechanics,* D. Van Nostrand Company, Inc., N.Y. (1965).
43. Jurgenson, Leo, "The Application of Theories of Elasticity and Plasticity to Foundation Problems," *Contribution of Soil Mechanics,* 1925-40, Published by Boston Society of Civil Engineers, 20 Pemberton Square, Boston, Massachusetts, p. 148-198 (1925-40).
44. Kirpatric, W. M., "Effect of Grain Size and Grading on the Shearing Behaviour of Granular Materials," *Proceedings of 6th International Conference on Soil Mechanics and Foundation Engineering, Vol. I,* pp. 273-277 (1965).
45. Kirpatric, W. K. and J. Y. Scott, "Strain Conditions in Compression Cylinders," *Proceedings of American Society of Civil Engineers, Journal of Soil Mechanics and Foundation Engineering Vol. 96,* No. SM 5, pp. 1683-1698 (1970).
46. Kjellman, W., "Report on an Apparatus for Consummate Investigation of Mechanical Properties of Soil," *Proceedings of 1st International Conference on Soil Mechanics and Foundation Engineering,* Cambridge, *Vol. II,* pp. 16-20 (1936).
47. Koeglar, F. and A. Scheidig, "Druckverteilung in Baugrande" *Bautechnik,* (1927-29).
48. Koerner, R. M., "Behaviour of Single Mineral Soils in Triaxial Shear," *American Society of Civil Engineers, SM 4,* pp. 1373-1390—7433 July (1970).
49. Kolbuszewski, J. and M. R. Fredrick, "The Significance of Particle Shape and Size on the Mechanical Behaviour of Granular Materials," *European Conference on Soil Mechanics and Foundation Engineering, Vol. I,* Wiesbaden, pp. 253-263 (1963).
50. Kolosov, G. B., "Application of Complex Diagrams and the Theory of Functions of Complex Variables to the Theory of Elasticity," *ONTI,* (1935).
51. Ko, H. Y. and R. F. Scott, "A New Soil Testing Apparatus," *Geotechnique, London, Vol. 17,* pp. 40-57 (1967).
52. Kondner, R. L. and S. Zelasko, "A Hyperbolic Stress-Strain Formulation for Sands," *Proceedings of 2nd Pan American Conference on Soil Mechanics and Foundation Engineering, Brazil, Vol. I.* p. 289-324 (1963).
53. Kondner, R. L., "Hyperbolic Stress-Strain Response of Cohesive Soils," *Proceedings of American Society of Civil Engineers, Journal of Soil Mechanics and Foundation Engineering, Vol. 89,* No. SM 3, pp. 115-143 (1963).
54. Lade, P. V. and J. M. Duncan, "Cubical Triaxial Tests on Cohesionless Soils," *Proceedings of American Society of*

Civil Engineers, Journal of Soil Mechanics and Foundation Engineering, SM 10, Vol. 99, pp. 793–811 (1973).

55. Lambe, T. W., "Methods of Estimating Settlement," *Proceedings of American Society of Civil Engineers, Journal of Soil Mechanics and Foundation Engineerings Vol. 90,* SM 3, Proc. Paper 4060 pp. 47–71 Sept. (1964).
56. Lambe, T. W., *Soil Testing for Engineers,* John Wiley and Sons, New York (1967).
57. Lee, K. L. and H. B. Seed, "Drained Strength Characteristics of Sand," *Proceedings of American Society of Civil Engineers,* SM 6, 117, 5561 Nov. (1967).
58. Lee, K. L. and H. B. Seed, "Undrained Strength Characteristics of Cohesionless Soils," *Proceedings of American Society of Civil Engineers, Journal of Soil Mechanics and Foundation Engineering,* No. 6, pp. 333–360 Nov. (1967).
59. Leussink, H., Discussion, "Effect of Specimen Size on the Shear Strength of Granular Materials," *Proceedings of 6th International Conference on Soil Mechanics and Foundation Engineering, Vol. III,* p. 316 (1965).
60. Leslie, D. E., "Large Scale Triaxial Tests on Gravelley Soils," *Proceeding of 2nd Pan American Conference, Vol. I,* pp. 181–202 (1963).
61. Lundgren, H. and K. Mortenson, "Determination by Theory of Plasticity of the Bearing Capacity of Continuous Footings on Sand," *Proceedings of 3rd International Conf. on Soil Mechanics and Foundation Eng., Vol. I,* p. 409 (1963).
62. Marshall, R. J., "Discussion," *Proceedings of 6th International Conference on Soil Mechanics and Foundation Engineering, Vol. III,* pp. 310–316 (1965).
63. Melan, E., Die Druckverteilung durch eine elastishe Schieht, *Beton u. Eisen, Vol. 18,* pp. 83–85 (1919).
64. Meyerhof, G. G., "Ultimate Bearing Capacity of Foundation," *Geotechnique, Vol. 2,* p. 301 (1951).
65. Meyerhof, G. G., "The Bearing Capacity of Footings under Eccentric and Inclined Loads," *Proceeding of 3rd International Conference on Soil Mechanics and Foundation Engineering, Vol. I,* p. 384 (1953).
66. Meyerhof, G. G., "Influence of Roughness of Base and Ground Water Conditions on the Ultimate Bearing Capacity of Foundation," *Geotechnique, Vol. 5,* No. 3 (1955).
67. Mindlin, R. D., Force at a point in the Interior of a Semi-Infinite Solid, *Physics, Vol. 7,* May (1936).
68. Nadai, A., *Theory of Flow and Fractor of Solids,* McGraw Hill Book Company, N.Y. (1962).
69. Newmark, N. W., "Stress Distribution in Soil," *Proceedings, Purdue Conference on Soil Mechanics and its Application,* Purdue University, Lafayette, Indiana, p. 295-303 (1940).
70. Newmark, N. W., "Influence Charts for Computation of Stresses in Elastic Foundation," Bulletin, University of Illinois, Urbana, Illinois, No. 338, Vol. 40 Nov. 10 (1942).
71. Newmark, N. M., "Failure Hypothesis for Soils," *Proceedings of the Research Conference on Shear Strength of Cohesive Soils,* American Society of Civil Engineers, University of Colorado, Boulder, Colorado, June (1960).
72. Ponce, V. M. and J. M. Bell, "Shear Strength of Sand at Extremely Low Pressure," *Proceedings of American Society of Civil Engineers, Journal of Soil Mechanics and Foundation Engineering,* No. 4, April 1971, pp. 625 (1977).
73. Prakash, C., "Behaviour of Eccentrically Loaded Footings in Clay," M. E. Thesis, Department of Civil Engineering, University of Roorkee (India) (1975).
74. Prakash, S. and S. Venkatesan, "Shearing Characteristics of Dense Sands at Different Rates of Strain," *Journal of Scientific and Industrial Research, Vol. 19A,* No. 5, pp. 219–223 (1960).
75. Prakash, S. and G. Ranjan, "Effect of Rate of Strain on Strength of Obra Sand," *Journal of Indian Society of Soil Mechanics and Foundation Engineering, Vol. 5,* No. 4, pp. 423 (1966).
76. Prakash, S. and G. Ranjan, "Effect of Rate of Strain on Strength Characteristics of Non-Cohesive Soils," *Proceedings of Third South East Asian Conference on Soil Engineering,* Hong-Kong, pp. 288–295 November (1972).
77. Prakash, S. and S. Saran, "Bearing Capacity of Eccentrically Loaded Footings," *Proceedings of American Society of Civil Engineers,* SMI, 95:7814 Jan. (1971).
78. Prakash, S., S. Saran, and U. N. Sharan, "Prediction of Constitutive Laws for Granular Soils", *Proceedings of Istembul Conference on Soil Mechanics and Foundation Engineering,* pp. 120–128 (1975).
79. Prandtl, L., Uber die Harte Plastischer Korper, Nachr. Kgl, Ges., Wiss, Gottingen (1970).
80. Ramamurthy, T., "A Universal Triaxial Apparatus," *Journal of Soil Mechanics and Foundation Engineering,* Indian National Society of Soil Mechanics and Foundation Engineering, Vol. 9, No. 3, pp. 251–269 (1970).
81. Rao, N. V. R. L. N., Rheological Models and their Use in Soil Mechanics," *Journal of Indian National Society of Soil Mechanics and Foundation Engineering,* pp. 225–267 (1967).
82. Rao, N. V. R. L. N. and S. K. Chakarvarty, "Rheological Parameters of Soils," Symposium on Modern Trends in Civil Engineering, Department of Civil Engineering, University of Roorkee, p. 15 (1972).
83. Reissaner, H., "Zum Erddruck Problem," *Proceedings, 1st International Congress of Applied Mechanics* (Delft) pp. 295–311 (1924).
84. Saran, S., "Bearing Capacity of Footings Subjected to Moments," Ph.D. Thesis, Geotechnical Engineering Section, Civil Engineering Department, University of Roorkee, Roorkee (India) (1969).
85. Saran, S., P. K. Jain, and C. Prakash, "Behaviour of Eccentrically Loaded Footings in Clay," *Proceedings of the Symposium on Foundations and Excavations in Weak Soil, Vol. I,* Calcutta, India (1976).
86. Schoenburg, I. J., "On Interpolation by Spline Functions and Its Minimal Properties," International Series of Numerical Analysis, Vol. 5, Academic Press (1964).
87. Scott, R. F., *Principles of Soil Mechanics,* Addision Wesley Pub. Co. Inc. Readin Mao. p. 280 (1963).

88. Schuttze, E., "Distribution of Stress Beneath a Rigid Foundation," *5th International Conference on Soil Mechanics and Foundation Engineering, Vol. I,* p. 807 (1961).
89. Seed, H. B., J. K. Mitchel, and C. K. Chan, "The Strength of Compacted Cohesive Soils," Research Conference on Cohesive Soil, University of Colorado, Boulder, Colorado, p. 923 (1960).
90. Selby, Samuel, M., "Standard Mathematical Tables," The Chemical Rubber Co. Cleveland, Ohio, pp. 103 (1972).
91. Shibata, T. and D. Karube, "Influence of the Variation of Intermediate Principal Stress on the Mechanical Properties of Normally Consolidated Clays," *Proceedings of the 6th International Conference on Soil Mechanics and Foundation Engineering, Vol. I,* p. 359-363 (1965).
92. Simon, N. E., "Effect of Over Consolidation of Shear Strength Characteristics of Undisturbed Oslo Clay," *Reseema Conference on Cohesive Soils, University of Colorado,* Boulder, Colorado, p. 747–763 (1960).
93. Skempton, A. W. and A. W. Bishop, "Measurement of Shear Strength of Soils," *Geotechnique, Vol. II,* 1950–51, p. 90–108 (1951).
94. Skempton and Bjerrum, "A Contribution to the Settlement Analysis of Foundations in Clay," *Geotechnique,* London, *Vol. 17,* pp. 168–178 (1957).
95. Skupek, J., "The Influence of Foundation Depth on Stress Distribution," *5th International Conference on Soil Mechanics and Foundation Engineering, Vol. I,* p. 815 (1961).
96. Southerland, H. S. and M. S. Mesdary, "The Influence of Intermediate Principal Stress on Strength of Sand," *Proceedings of the 7th International Conference on Soil Mechanics and Foundation Engineering, Vol. I,* Mexico, p. 391–394 (1969).
97. Sokolnikoff, *Mathematical Theory of Elasticity,* McGraw Hill Book Company (1956).
98. Sovinic, I., "Calculus of Contact Pressure, Displacements and Inclinations of Centrally and Eccentrically Loaded Rectangular Shaped Footings," *Proc. Yogoslav. Soc. on Soil Mechanics and Foundation Engineering, No. 10* (1955).
99. Stamatopoulos, A. C., "Linearly Variable Load Distribution on Rectangular Foundations," *Proceedings American Society of Civil Engineers,* SM 6, p. 137 (1959).
100. Steinbrenner, W., "Tafeln Zur Sotzungsberechnung, p. 21, No. 4, Schriftenreihe der Strasse 1, Strasse (1934).
101. Sukelje, L., *Rheological Aspects of Soil Mechanics,* Wiley Interscience, N.Y., p. 4 (1969).
102. Taylor, D. W., "Research on Consolidation of Clays," Department of Civil and Sanitary Engineering, Massachusetts Institute of Technology, U.S.A. (1942).
103. Taylor, D. W., *Fundamentals of Soil Mechanics,* Asia Publishing House, pp. 53, 259, 377 (1948).
104. Terzaghi, K., "Opening Discussions on Settlement of Structures," *Proceedings of 1st International Conference on Soil* Mechanics and Foundation Engineering, Cambridge, Mass., Vol. 3, p. 79–87 (1956).
105. Terzaghi, K., *Theoretical Soil Mechanics,* John Wiley and Sons, Inc. N.Y. (1943).
106. Terzaghi, K. and R. B. Peck, *Soil Mechanics in Engineering Practice,* John Wiley and Sons Inc., New York, p. 495 (1967).
107. Tittnek, W. and F. Matl, "A Contribution to Calculating the Inclination of Eccentrically Loaded Foundation," *Proceedings 3rd International Conference on Soil Mechanics and Foundation Engineering, Vol. 1,* p. 46 (1953).
108. Vallerga, B. A., et al., "Effect of Shape, Size and Surface Roughness of Aggregates Particles on the Strength of Granular Materials," *ASTM, STP* 212 (1957).
109. Vesic, A. J. and R. D. Barkdale, "On Shear Strength of Sand at High Pressure," Laboratory Shear Testing of Soils," *ASTM, STP, 361,* pp. 301–305 (1963).
110. Westergaard, H. M., "A Problem of Elasticity Suggested by a Problem in Soil Mechanics, Soft Material Reinforced by Numerous Strong Horizontal Sheets," *Contribution to the Mechanics of Solids,* S. Timoshenko, 60th Anniversary Volume, Macmillan Co., N.Y. (1938).
111. Whitman, R. V. and K. A. Healey, "Shear Strength of Sand During Rapid Loading," *Proceedings of ASCE, Journal of S.M.F.D., Vol. 88,* SM 2, pp. 99–132 (1962).
112. Whitman, R. V., "The Behaviour of Soils under Transient Loading," *Proceedings of 4th International Conference on Soil Mechanics and Foundation Engineering, Vol. II,* p. 717 (1967).
113. Weissmann, G. F., "Rigid Foundation Subjected to Overturning Moments," *Proceedings Vth Symposium of Civil and Hydraulics Engineering Department,* Indian Institute of Science, Bangalore (1965).

CHAPTER 5

Stochastic Models of Stress Distributions in Granular Materials

J. M. GOLDEN*

INTRODUCTION

A granular material is particulate in nature, the individual particles exhibiting little or no cohesion other than that induced by capillary moisture. It generally consists of a random assortment of particles of various sizes and shapes, exchanging stresses through isolated contact points of unpredictable position and orientation. Thus, it has long been realized that reasoning of a stochastic or probabilistic nature should prove useful in understanding the properties of such materials. The objective of this chapter is to use these techniques to obtain information on stress distributions which prevail in layers of granular material which are subjected to specified surface loading.

Background

The traditional method of estimating stress distributions in granular materials and other soils subject to surface loading has been to use the results of the theory of Linear Elasticity, as given by Boussinesq [6] (see Appendix II). These formulae have the advantage of being simple and explicit, particularly for the case of a point load and on the main axis for a single, uniformly distributed load. Also, the vertical compressive and shear stresses are actually independent of the material characteristics of the medium.

The Boussinesq formulae apply to a single homogeneous half-space. More recently, however, and largely in the context of structural pavement design, multi-layer elastic theory has become a standard tool for analysing layered media, where some of the layers are particulate; see Reference 57 for discussion of this topic and Reference 21 for an example of its application to pavement design.

*Roads Division, National Institute for Physical Planning and Construction, Dublin, Ireland

From an early date, however, experimental evidence pointed to the inadequacy of linear elastic theory for describing the stress distribution and displacement characteristics of granular materials under load [15,31,55]. In particular, measurement of vertical stresses indicated that the distribution is more "bell-shaped" than given by the Boussinesq formula. In other words, it falls off more sharply in a lateral direction, at a given depth [31,55]. Morgan and Gerrard [36] discuss, in a unified manner, stress, strain and displacement measurements in sands, reported by various authors. Their analysis highlights the extent and nature of the discrepancies between linear elastic theory and measurement.

There are fundamental reasons for expecting that the predictions of Linear Elasticity will not be valid for granular materials. Such materials can withstand little or no tensile stresses while a basic assumption of Elasticity Theory is that tensile and compressive responses are identical; and indeed some linear elastic stress components for the half-space problem do become tensile in certain regions. Also, granular media exhibit strong stress dependence in their response to applied loads. This stress dependence is equivalent to non-linear elastic behaviour. Provided such behaviour can be characterized in mathematical form, standard techniques, notably the finite element method [58] can be used to calculate stress distributions and indeed strains and displacements as well. This is a complex procedure, however, involving significant laboratory and computer resources.

An approximate method of taking account of the stress dependence would be to regard the granular half-space as inhomogeneous in the sense that its modulus depends on depth. Explicit formulae for the stresses may be given for such a structure ([16,44]; for more recent references, see Reference 46), though only for very restricted types of variation of the modulus with depth.

Another option is to model the material as an anisotropic elastic medium, using available theoretical and numerical solution techniques [18,19,46]. Holden [25] successfully fits his data to such a theory. The complexity of the theory and the number of free parameters to be evaluated render it a somewhat clumsy tool. We remark that Holden [25] discusses in some detail the various approaches to modelling stress distributions in granular media, and refers extensively to the earlier literature.

An alternative approach is to develop a fundamental particulate model, using probabilistic concepts. In its most basic sense, the construction of such a model would be very difficult, involving the use of the inter-relationship between the size and shape distribution of particles, the nature of the packing, the position of contact points and their orientation. Little progress has been made at this level. It turns out, however, that, on the basis of relatively simple, general reasoning, which makes no reference to the detailed structure of the medium, it is possible to deduce a considerable amount about the stress distributions in a loaded granular half-space. This information is not complete, however, and in order to obtain formulae which are finally useful, it is necessary to add certain assumptions, guided by criteria, such as the fact that tensile stresses are forbidden and, of course, by the need for simplicity. The results obtained are at least as simple and as explicit as the Boussinesq formulae. The validity of such a theory must finally rest, as always, on whether it agrees with observed stress distributions or not. The available evidence indicates that there is considerable agreement, as we shall show in the section Comparison with Measured Stresses, below. In this chapter, we shall outline the content of this theory, provide a range of formulae and numerical data to facilitate its practical application, and also indicate the extent to which it reproduces measured stresses. Background probabilistic concepts are discussed in an appendix.

The suggestion that probabilistic reasoning might be useful in explaining stress distributions was made by Kögler and Scheidig [31], though a little earlier, Pokrovsky [45] had obtained a Gaussian-type distribution without explicit use of probabilistic concepts. This approach was developed in the work of Kandaurov [27–29] and Muller [37,38]. Later Sergeev [49], using detailed probabilistic reasoning, formulated the theory in terms of diffusion or Markov processes. Independently, several western authors developed a similar approach. Smoltczyk [52], on the basis of general rather than detailed reasoning, proposed formulae for the stress distribution, which were in effect the same as those emerging from the Russian work. Harr was also developing the probabilistic approach at this time, which was published later [23]. This book, which contains a thoroughgoing formulation of the stochastic nature of stress transmission in granular materials, is the central reference on the subject. Chikwendu and Alimba [8] give a derivation of the distribution formulae which casts light on the physical significance of the material constant, the coefficient of lateral stress, which is introduced below. They calculate its value for an idealized material. A somewhat different approach adopted by Endley and Peyrot [13] is interesting in that it introduces the concept of stiffness of a stress path, which determines its ability to transmit stress. Golden [20] presented a general formulation of the stochastic theory of stress distributions, and more recently proposed a generalization to take account of the presence of horizontal surface forces [22]. Application of the theory to pavement design has recently been reported [4,5].

Other Applications of Stochastic Methods

This chapter deals only with the determination of stress distributions. However, we mention briefly other properties of granular materials which have been usefully modelled by means of probabilistic techniques. There is a large and developing literature on several topics, of which a small sample is referred to here.

At the most basic descriptive level, McAdams [34] applies relatively sophisticated mathematical techniques to the problem of characterizing particle shape and size. The porosity of randomly packed spheres has been studied theoretically and experimentally [3,48,50].

Particulate models have long been used to provide insight into the mechanical response of granular materials. Early work in this area [10–12,54] did not make use of stochastic methods. However, more recently, Davis and Deresiewicz [9] have applied discrete probabilistic methods to predict the compressibility of a random packing of identical spheres, as well as coordination number (number of contacts per particle) and the angular distribution of contacts, which are compared with experimental measurements. Data on the distribution of contact forces is also presented.

The angular distribution of contacts is a central parameter in much recent work on granular materials. Oda et al. [43] explore in depth the properties of this parameter and its relation to the observed properties of the granular medium. Murayama and Kitamura [42] and Kitamura [30] model the mechanical response of the material as a Markov process (Appendix I), using angle of contact as a random variable.

Finally, there is extensive work on the flow of granular materials, using stochastic methods, notably that of Litwiniszyn and co-workers (for example [33,26]) and Mullins [39–41]. A failure model for granular materials is proposed by Shimbo [51]. His approach is to regard the coefficient of particle friction as a random variable distributed over the particle surface.

Definition of Stress

Only stresses due to interparticle contacts are considered. We leave aside consideration of stresses transmitted by the liquid phase, if present.

The definition of stress in a granular medium which is of primary engineering interest is the force per unit area over an area that is small from the macroscopic viewpoint but large compared with average particle size. Let s be a component of the stress tensor defined in this manner, which might be termed macro-stress to distinguish it from the micro-stresses on individual particles. Then s is the average of micro-stresses in the vicinity, individual values of which may vary considerably from the mean value. Harr [23] emphasizes this fact by using the notation $\bar{s}$, a procedure which for notational simplicity will not be followed here.

The micro-stresses in a given small region will be distributed about the macro-stress. It is difficult to determine the nature of this distribution in realistic materials. Measurements have been reported on a related matter, the distribution of forces between particles [9,35]. The results are derived from observations of two-dimensional models where the particles are spheres or discs, so that the materials are somewhat idealized. The frequency distribution of contact forces obtained in both cases are qualitatively similar in that they are characterized by an early maximum followed by a slow decline. Mendoza [35] points out that his results seem to fit the Gamma distribution (Appendix I). This distribution contains a parameter λ which must be evaluated from the data. Mendoza's data suggest a value $\lambda \doteq 3$.

An unsuccessful attempt by Auvinet and Marsal [2] to model the phenomenon of particle breakage leads the authors to argue that the distribution of contact forces may possess a more complex structure than the work discussed previously would indicate. They suggest that the contact forces may, in fact, depend on the size of the particles involved, with the larger particles carrying greater loads.

The relationship between contact forces and micro-stresses may be expected to depend upon the sizes of the particles in contact, and the nature of this dependence will be influenced by whether the contact is purely elastic or generates some plastic flow. It will, therefore, be affected by the grading of the material and the mechanical properties of the particles.

THEORETICAL DEVELOPMENTS

Consider a thick granular layer, subject to surface loading. This will be idealized as a half-space occupying the region $z > 0$ (Figure 1). The surface is the xy plane passing through the origin. The z direction will be regarded as vertical and the xy directions as horizontal, though in fact gravity is neglected in our considerations. We will confine the discussion to isotropic and also, mainly speaking, homogeneous materials—in a macroscopic sense. Clearly a granular material is never microscopically isotropic or homogeneous. The theory in its general form can, however, easily handle the case of a medium the properties of which vary with depth. In particular, this includes the case of a layered medium.

Our object is to determine how the various stress components are distributed throughout the half-space. We use stress now to mean averaged or macroscopic stress as defined earlier.

The convention that compressive stresses are positive will be adopted.

Surface Loading

For purposes of the initial theoretical discussion, two special loadings will be used. These are (1) a point loading at the origin and (2) an infinite uniform line loading along the y axis. For case (2), the problem reduces from three dimensions to two dimensions, since there can be no variation along the y direction and thus none of the stresses can depend on that coordinate. In other words, this loading generates plain strain conditions.

Later, we will determine the stress components under distributed loads, which are more relevant for practical purposes and in particular for comparison with experimental results.

Physical Constraints

There are two properties of the stress distributions which are necessary in a realistic theory of granular materials: (a) Tensile stresses cannot occur; (b) The horizontal fall-off of the vertical compressive stress should be stronger than that predicted by Linear Elasticity.

The first property excludes cohesive materials. In fact, the model developed does apply, in its general form, to cohesive particulate media, and they will be discussed briefly. However, the main emphasis of the discussion will be on the non-cohesive case.

The second point is based on experimental observation, as discussed in the section Comparison with Measured Stresses, below. It may be seen as an expression of the fact that granular materials are less effective at spreading load than materials with cohesion.

The theory presented below is incomplete in certain aspects. These requirements are useful in helping to reduce the indeterminacy and provide explicit formulae for the stress distributions.

The Diffusion Analogy

Consider a point vertical load L acting at the origin. This will be transmitted in a random manner, by inter-particle contacts, downwards through the medium. At any given depth, the vertical forces will be spread out over the xy plane and their sum will be equal to L. As depth increases, they will spread further and become weaker. This is quite analogous to a diffusion process, such as occurs in a thermal conductor from a heat source. The vertical force is dissipated with depth, which is the phenomenon of load spreading. We now give a quantitative description of this effect.

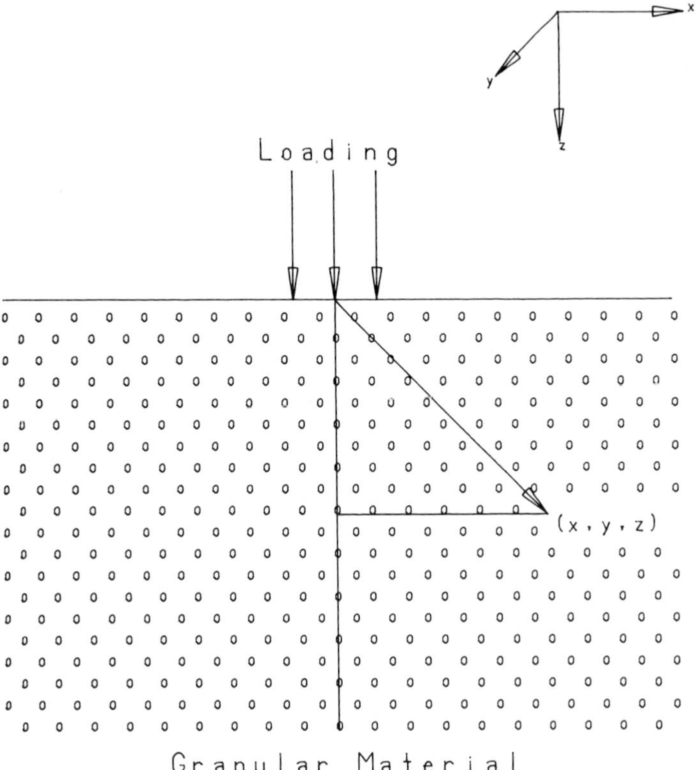

FIGURE 1. Schematic representation of the Granular Half-Space.

The Fundamental Equation

Let there be a vertical point load L and a horizontal point force along the x-axis of magnitude qL, both acting on the surface. Consider the vertical compressive stress $s_z(\underline{R})$ at a point $\underline{R} = (x,y,z)$. This vector notation will be used occasionally, for brevity. It is argued in Appendix I on the basis of certain assumptions, that $s_z(\underline{R})$ obeys Equation (A39), which is the Fokker-Planck equation, given by

$$\frac{\partial}{\partial z} s_z(\underline{R}) = -\frac{\partial}{\partial x}[b(\underline{R})s_z(\underline{R})] + \left\{\frac{\partial^2}{\partial x^2} + \frac{\partial^2}{\partial y^2}\right\}[a(\underline{R})s_z(\underline{R})] \qquad (1)$$

This equation expresses the fact that the diffusion of the vertical stress with depth may be regarded as a Markov process. In physical terms, this means that the stress distribution at any level z is a consequence only of the distribution prevailing at a level slightly above z, say $z - \Delta z$. The stress at each point at level $z - \Delta z$ is distributed stochastically at the lower level z. The function $a(\underline{R})$ may be interpreted from the observation that $2a(\underline{R})\Delta z$ is the variance of this distribution, so that $a(\underline{R})$ is a measure of the load spreading capability of the medium. It must always be positive. We will refer to it as the influence function. The physical meaning of $b(\underline{R})$ may be deduced from the fact that $b(\underline{R})\Delta z$ is the shift in the x direction of the mean point of the distribution at level z from the source point giving rise to that distribution at level $z - \Delta z$. It will be zero unless a horizontal surface force is present. Further properties of these functions and discussion of the background theory and its physical interpretation may be found in Appendix I.

The influence function $a(\underline{R})$ is a fundamental characteristic of the medium, a reflection of the degree of interlocking that exists between the particles. It has dimensions of length. The quantity $b(\underline{R})$ is dimensionless.

The physical assumption underlying Equation (1) described above would seem reasonable for a granular material. However, if this equation is to be useful, we must be able to assign simple forms to $a(\underline{R})$ and $b(\underline{R})$. It is in doing this that we introduce the really strong assumptions.

It follows from Equation (1) that the integral of $s_z(\underline{R})$ over the xy plane at any depth z is a constant, which, according to the boundary conditions, must equal L. Therefore

$$L = \int_{-\infty}^{\infty} \int_{-\infty}^{\infty} dxdy\ s_z(x, y, z) \tag{2}$$

In showing this, it is necessary to use the fact that $s_z(\underline{R})$ vanishes at large horizontal distances.

The Case of a Uniform Line Force

For an infinite line force along the y-axis, Equation (1) reduces to [see Equation (A29)]

$$\frac{\partial}{\partial z} s_z(x, z) = -\frac{\partial}{\partial x} [b(x, z)s_z(x, z)] + \frac{\partial^2}{\partial x^2} [a(x, z)s_z(x, z)] \tag{3}$$

where $b(x,z)$ is zero if no horizontal surface force is present. We shall denote the vertical load per unit length l and the horizontal load in the x direction ql.

Application to a Linear Elastic Continuum

It is interesting to observe that we can find a form of $a(\underline{R})$ and $b(\underline{R})$ such that the linear elastic form for s_z obeys Equation (1). Similarly, for the two-dimensional case, we can find an $a(x,z)$ and $b(x,z)$ such that the plane strain linear elastic form of s_z obeys Equation (3). The elastic results are given in Appendix II on Tables A1, A2. It may be checked that these satisfy Equations (1) and (3) respectively, if

$$\begin{aligned} a(\underline{R}) &= (x^2 + y^2 + z^2)/(3z) \\ b(\underline{R}) &= qa(\underline{R})/(z + qx) \end{aligned} \quad \text{(3-D)} \tag{4}$$

and

$$\begin{aligned} a(x,z) &= (x^2 + z^2)/(2z) \\ b(x,z) &= qa(x,z)/(z + qx) \end{aligned} \quad \text{(2-D)} \tag{5}$$

Observe that $a(\underline{R})$ increases horizontally as $(x^2 + y^2)/(3z)$ and $a(x,z)$ as $x^2/(2z)$. This is interesting when one recalls the physical significance of $a(\underline{R})$. Note also that at large z,

$$\begin{aligned} a(\underline{R}) &\sim z/3 \quad \text{(3-D)} \\ a(x,z) &\sim z/2 \quad \text{(2-D)} \end{aligned} \tag{6}$$

Hill and Harr [24] propose a generalisation of Equation (1) and (3), based on rotational symmetry arguments, which includes both the elastic and stochastic theories as special cases. They use an influence function given by Equation (7) below. An underlying theoretical motivation for this direction of development is the possibility of eliminating from the theory the preferred space direction, defined by the normal surface force.

Choice of $a(\underline{R})$ and $b(\underline{R})$

It is intuitively reasonable that by choosing the horizontal increase of the influence function less than in the elastic case, we obtain narrower vertical stress distributions, in line with requirement (b) above. The simplest such choice is to take $a = a(z)$, a function of depth only, with no horizontal increase. It emerges that the resulting stress distributions exhibit very strong horizontal fall-off. Also, it is possible to ensure the absence of tensile stresses with such a form. The simplest choice of $a(z)$ consistent with these requirements is

$$a(z) = \nu z \tag{7}$$

for all z, where ν is a constant, named by Harr [23] the *coefficient of lateral stress.* Equation (7) has been adopted by various authors [28,37,49,23,8,52,20,22] to describe a homogeneous granular material, and will be used as such henceforth in this chapter. The justification for this choice is that it leads to stress distributions that have the desired properties and gives reasonable agreement with experiment, as we shall show in Comparison with Measured Stresses, below.

The constant ν characterizes the medium, not unlike a modulus. If a material is inhomogeneous with depth, we would expect ν to be a function of z, changing in a discontinuous manner in the case of uniform layers of different materials. These different layers could, of course, be composed of the same material but compacted to different degrees. Therefore, a general choice of $a(z)$ would correspond to a material, the properties of which vary with depth.

Comparing the form $a(\underline{R}) = a(z)$, which can generate distributions free of tensile stresses, with the elastic forms given by Equations (4) and (5), it is tempting to conclude that a horizontal increase in $a(\underline{R})$ corresponds to the presence of cohesiveness in the material. This must be regarded as a conjecture, however, at the present stage of understanding of the theory.

A simple and acceptable choice of $b(\underline{R})$ is to take it to be constant [22]. Clearly, it has the advantage of simplicity. For a point or line source, it may be shown that, if this assumption is valid, then

$$b(\underline{R}) = q \tag{8}$$

where q is the ratio of the horizontal to the vertical load. We will adopt this choice for the remainder of the chapter.

Other Stress Components

The medium is assumed to be in static equilibrium. Therefore, at every point, the components of the stress tensor must obey the equilibrium equations. Let us consider the two-dimensional problem first. We have

$$\frac{\partial}{\partial x} s_x + \frac{\partial}{\partial z} s_{xz} = 0$$
$$\frac{\partial}{\partial x} s_{zx} + \frac{\partial}{\partial z} s_z = 0 \qquad (9)$$

These conditions, together with Equation (3) and the fact that all components must vanish at large x, give that (22)

$$s_{zx} = q s_z - a(z) \frac{\partial}{\partial x} s_z$$
$$s_x = q s_{zx} + \frac{\partial}{\partial z} [a(z) s_z] \qquad (10)$$

Putting q to zero gives the results for a purely vertical force.

In the three-dimensional case, an analagous argument does not determine all the components of the stress tensor uniquely, which exposes the present incompleteness of the theory. This is not surprising, given that no constitutive assumption has been made for the material. However, Muller [38], Smoltczyk [52] and Harr [23] have proposed a particular choice of stress tensor components which exhibit a natural symmetry and which, as we will see later in this chapter, describes fairly well what is observed experimentally. This form of the stress tensor for a point source will be given explicitly in the next section. Here, we are concerned with writing down a general differential form of these results, similar to Equation (10). The advantage of this form is that it can be applied in situations where the surface loading is other than a point load. In the presence of a horizontal load, the relations in question are

$$s_{xz} = q s_z - a(z) \frac{\partial}{\partial x} s_z$$

$$s_{yz} = -a(z) \frac{\partial}{\partial y} s_z$$

$$s_x = \left[q^2 - 2qa(z) \frac{\partial}{\partial x} + a'(z) + a^2(z) \frac{\partial^2}{\partial x^2} \right] s_z \qquad (11)$$

$$s_y = \left[a'(z) + a^2(z) \frac{\partial^2}{\partial y^2} \right] s_z$$

$$s_{xy} = \left[-qa(z) \frac{\partial}{\partial y} + a^2(z) \frac{\partial^2}{\partial x \partial y} \right] s_z$$

where $a'(z)$ is the derivative of $a(z)$ and q is the ratio of horizontal to vertical surface load. This ratio is assumed to be the same at all surface points. In the absence of horizontal stresses, it is convenient to use cylindrical polar coordinates r, θ, z, where r, θ are two-dimensional polar coordinates in the xy plane, if the applied load has circular symmetry. In terms of these coordinates and putting $q = 0$, Equation (11) becomes [20].

$$s_{rz} = -a(z) \frac{\partial}{\partial r} s_z$$
$$s_r = \left[a'(z) + a^2(z) \frac{\partial^2}{\partial r^2} \right] s_z$$
$$s_\theta = \left[a'(z) + a^2(z) \frac{1}{r} \frac{\partial}{\partial r} \right] s_z \qquad (12)$$
$$s_{r\theta} = s_{\theta z} = 0$$

FORMULAE FOR THE STRESS DISTRIBUTIONS

Point and Line Load Solutions

The solutions of Equations (1) and (3) corresponding to a line or a point load are given by Equations (A38) and (A40) of Appendix I. The remaining components may be deduced from Equations (10–12). The quantity $w(z)$ is related to the influence function by Equation (A37), which for a homogeneous medium, becomes

$$w(z) = \nu z^2/2 \qquad (13)$$

by virtue of Equation (7). The stress components in this case are given on Table 1.

The formulae applicable in the presence of a horizontal force will not be discussed further in view of the dearth of controlled measurements available. However, before leaving this topic, we present a comparison of the stochastic and linear elastic (Appendix II) behaviour under a line load on Figure 2. The most apparent contrasting feature is the fact that no tensile stresses occur in the stochastic theory. The positions of the maxima are different.

Note that in the limit of a purely horizontal load (q very large) the theory predicts that all stresses go to zero, which is reasonable for a granular material. Muller [37] proposes

TABLE 1. Stress Components in a Homogeneous Medium for a Point Load (L) at the Origin, and a Line Load (1 per Unit Length) Along the y-Axis, with Horizontal Load Acting Along the x-Axis of qL and ql Respectively.*

Stress Component	Line Load	Point Load
s_z	$\frac{1}{(2\pi\nu)^{1/2}z}\exp[-(x-qz)^2/(2\nu z^2)]$	$\frac{L}{2\pi\nu z^2}\exp\{-[(x-qz)^2+y^2]/(2\nu z^2)\}$
s_{xz}	$\frac{x}{z}s_z$	$\frac{x}{z}s_z$
s_{yz}		$\frac{y}{z}s_z$
s_{xy}		$\frac{xy}{z^2}s_z$
s_x	$\frac{x^2}{z^2}s_z$	$\frac{x^2}{z^2}s_z$
s_y		$\frac{y^2}{z^2}s_z$

*Purely vertical load: $q = 0$.

a quite different method of describing the effect of a horizontal load, which is also discussed by Harr [23]. This theory predicts that a purely horizontal load would generate compressive horizontal stresses over the entire surface of the half-space, which seems unlikely.

It is interesting to note that the relationship between s_z and the other components, for a line load, is the same as for the linear elastic solutions, given in Appendix II. A similar remark applies to the point load configuration if the elastic material is incompressible.

Distributed Loads

The form of s_z under a distributed load is given by the formula

$$s_z = \begin{cases} \int_C dx'dy'\, p(x',y')s_{pz}(\underline{R}-\underline{R}') & \text{(3-D)} \\ \int_C dx'\, p(x')s_{lz}(x-x',z) & \text{(2-D)} \end{cases} \tag{14}$$

where

C = contact region

$p(x,y)$, $p(x)$ = surface vertical pressure distribution, which is independent of y in the 2-D case

$s_{pz}(\underline{R}), s_{lz}(x,z)$ = vertical stress at $\underline{R}$ [or (x,z)] due to a vertical point (line) load at the origin

The other components are then determined with the aid of Equation (10) for the two-dimensional configuration and Equations (11) and (12) for the three-dimensional configuration.

An alternative approach to obtaining expressions for s_z under a distributed surface load is to apply standard integral transform techniques for solving the partial differential equation which it obeys. In the case of Equation (1), with $a(\underline{R})$ given by Equation (7) and $b(\underline{R})$ equal to zero, this procedure leads to solutions expressed as infinite integrals over Bessel Functions. These forms, though less explicit than obtained from Equation (14), and mainly speaking less suitable for calculation, are more compact and convenient for theoretical manipulation, and will also be given.

On Table 2[1] are given the stress components under a uniform vertical pressure over an infinite strip centred over the y-axis, for a homogeneous medium. Also given are the stress components under a uniform vertical rectangular load. Some of these results have been given by Kandaurov [29], Smoltczyk [52] and Harr [23]. The quantity $\Psi(u)$ is introduced in Appendix I and plotted on Figure A1. The functions $f(x,a)$, $g(x,a)$, and $h(x,a)$ are plotted on Figures 3a, 3b, and 3c for various values of the combination $(z\sqrt{\nu})/a$.

[1]We have used the conventional symbols a,b as the size parameters of the surface loads in these tables, which should not be confused with the functions $a(\underline{R})$ and $b(\underline{R})$.

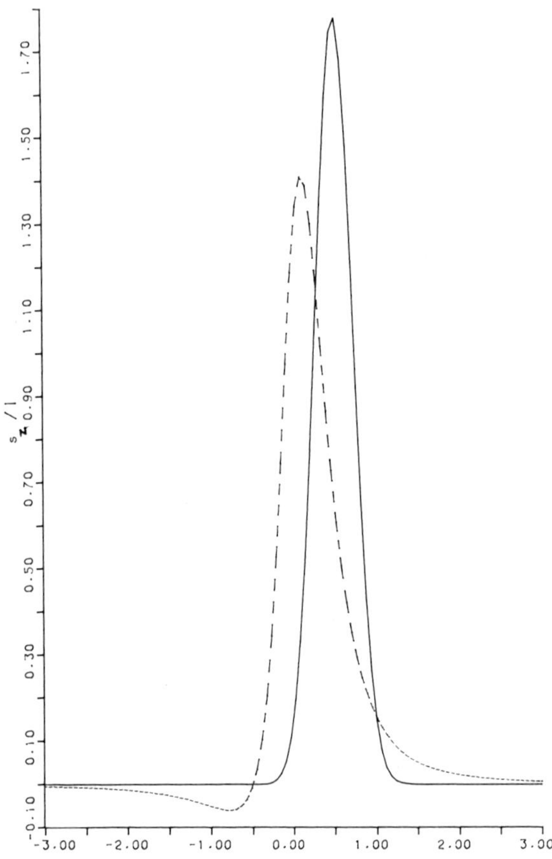

FIGURE 2. Comparison of the behaviour of s_z predicted by the stochastic and linear elastic models under a line load 1 with horizontal component 1 at depth $z = 0.5$ units and $\nu = 0.2$.

Only the positive x region is shown, since the function $f(x,a)$ is even, while $g(x,a)$ is odd and $h(x,a)$ is even. Once these are known, all the stress components may be determined.

On Table 3, stress components in cylindrical polar coordinates are given under a vertical point loading at the origin and under a uniform vertical pressure centred on the origin. This latter loading would exist, at least approximately, under a flexible plate. Several of these formulae, given originally by Kandaurov (27), are discussed by Harr (23). Examples of s_z, s_{rz}, and s_r, under a uniform loading are plotted on Figures 4a, 4b, and 4c. The component s_θ may be evaluated from s_z and s_{rz}.

Stress components under a parabolically distributed vertical pressure over a circular area are given on Table 4[27,23]. Examples of the form of s_z, s_{rz}, and s_r are plotted on Figures 5a, 5b, and 5c.

The curves for circular loads were obtained from the first form of the stress components on Tables 3 and 4, which are generally easier to evaluate than the integrals over products of Bessel Functions. The oscillatory behaviour of the functions $J_n(u)$ at large u causes problems in the evaluation of the latter formulae. However, they have certain advantages near the surface.

Layered Media

We have given formulae on Tables 1–4 only for homogeneous materials, for the sake of simplicity and explicitness. However, the generalization of these results to take account of more general forms of the influence function $a(z)$ is relatively straightforward. In the expressions for s_z, it is a matter of replacing the combination νz^2 (or $\sqrt{\nu}\, z$) by $2w(z)$ (or $[2w(z)]^{1/2}$) where it occurs. In particular, the line and point load solutions are given by Equations (A38) and (A40). The remaining stress components can be evaluated with the aid of Equations (10–12), using the general form of $a(z)$. Note that in a layered system, s_x, s_y, s_r, and s_θ will be discontinuous at the layer boundaries. Consider the two layer case, for example. Let the coefficients of lateral stress for the top and lower layer be ν_1, and ν_2, respectively, and denote the thickness of the top layer by h.. Then

$$w(z) \begin{cases} = \nu_1 z^2/2, \ z \leq h \\ = \{h^2(\nu_1 - \nu_2) + \nu_2 z^2\}/2, \ z \geq h \end{cases} \tag{15}$$

Observe that if $\nu_1 > \nu_2$ then $w(z)$, $z \geq h$, is increased by the presence of the upper layer.

Example—In a homogeneous medium with coefficient of lateral stress $\nu_2 = 0.2$, the vertical stress at a depth of 1m(3.28ft.) directly below purely vertical point load of 1kN (224.81bf) is (Table 1):

$$s_z = (2\pi\nu_2)^{-1} = 0.8\text{kN/m}^2(0.12 \text{ lb/in}^2)$$

If, however, the top 0.5m(1.64ft) is replaced by a material with coefficient of lateral stress $\nu_1 = 0.3$, then from Equation (15) and Table 1,

$$s_z = \{2\pi[(0.5)^2(\nu_1 - \nu_2) + \nu_2]\}^{-1}$$

$$= 0.7\text{kN/m}^2 \ (0.1 \text{ lb/in}^2)$$

Formulae for more than two layers may be written down without difficulty from Equation (A37). In fact, a significant positive feature of this theory is the simplicity with which it can be extended to the multi-layer case, in contrast to linear elastic theory. Kandaurov [29] (see also Harr [23]) has de-

TABLE 2. Stress Components in a Homogeneous Medium Under a Uniform Vertical Load Acting Over (a) an Infinite Strip Centred on the y-Axis, of Width 2a; and (b) a Rectangle Covering [– a,a] and [– b,b] in the x and y Directions Respectively. The Functions f(x,a), g(x,a) and h(x,a) all Depend on z Which is Omitted for the Sake of Notational Compactness. The Surface Pressure is p.

Stress Component	Ininite Strip	Rectangle
s_z	$pf(x,a)$	$pf(x,a)\ f(y,b)$
s_{xz}	$pg(x,a)$	$pg(x,a)\ f(y,b)$
s_{yz}		$pf(x,a)\ g(y,b)$
s_{xy}		$pg(x,a)\ g(y,b)$
s_x	$\nu[s_z + ph(x,a)]$	$\nu[s_z + ph(x,a)\ f(y,b)]$
s_y		$\nu[s_z + pf(x,a)\ h(y,b)]$
$s_z(0)$*	$2p\psi(A)$	$4p\psi(A)\ \psi(B)$
$s_x(0)$*	$\nu[s_z(0) - 2Ap(2\pi)^{-1/2}\exp(-A^2/2)]$	$\nu[s_z(0) - 4Ap(2\pi)^{-1/2}\exp(-A^2/2)\psi(B)]$
$s_y(0)$*		$\nu[s_z(0) - 4Bp(2\pi)^{-1/2}\psi(A)\exp(-B^2/2)]$

$f(x,a) = \psi(Z_+) - \psi(Z_-);\ \psi(u) = (2\pi)^{-1/2}\int_0^u dv\ \exp(-v^2/2)$
$g(x,a) = (\nu/(2\pi))^{1/2}[\exp(-Z_-{}^2/2) - \exp(-Z_+{}^2/2)]$
$h(x,a) = [(x-a)\exp(-Z_-{}^2/2) - (x+a)\exp(-Z_+{}^2/2)]/[z(2\pi\nu)^{1/2}]$
$Z_\pm = (x \pm a)/(z\nu^{1/2});\ A = a/(z\nu^{1/2});\ B = b/(z\nu^{1/2})$
*at x = 0 (2 – D) and under the origin (3 – D).

veloped an approximate "equivalent thickness" approach to the multi-layer problem, which gives results similar, though not identical to Equation (A37). Their method has been applied to pavement design [4,5].

COMPARISON WITH MEASURED STRESSES

We will now discuss the question of how accurately the stochastic theory represents measured stress distributions in granular media and the related question of assigning appropriate values to the coefficient of lateral stress ν.

Comparing with Linear Elastic Formula

Comparison of the line and point load expressions for s_z from Table 1 for a purely vertical load with the corresponding elastic results given in Appendix II, directly under the load, gives

$$\begin{aligned} \nu &= \pi/8 = 0.39 \qquad (2\text{-D}) \\ \nu &= 1/3 \qquad (3\text{-D}) \end{aligned} \tag{16}$$

Alternatively, comparison of Equations (6) and (7) would suggest values of ½(2-D) and ⅓(3-D). A theoretical argument [8], based on an idealized particulate model, suggests a value of $\nu = 0.27$ for loosely packed materials. While there is no reason to attach fundamental significance to the values obtained by these arguments they at least indicate the probable order of magnitude of the parameter ν.

In the case of a distributed load, these values of ν give approximate agreement between elastic and stochastic predictions for $s_z(r = 0)$ at large depths ($z/a > 2$, typically), as one may check from the formula on Table 3 and the elastic stresses under a uniform circular pressure, given on Table A3 of Appendix II. However, near the surface, it is necessary to take increasingly larger values of ν to obtain agreement. The stochastic theory predicts a much faster approach to the surface stress as z decreases, than elastic theory. This is a consequence of the exponential behaviour in the former.

Relation of ν to Soil Mechanics Parameters

In Soil Mechanics, *the coefficient of lateral pressure K* at a point in a medium is defined by the relation

$$s_x = K\,s_z \tag{17}$$

between the horizontal and vertical diagonal components of the stress tensor. Consideration of Tables 2–4 (using s_r or s_θ as the horizontal component) shows that near the surface, we have the approximate relation

$$s_x = \nu\,s_z \tag{18}$$

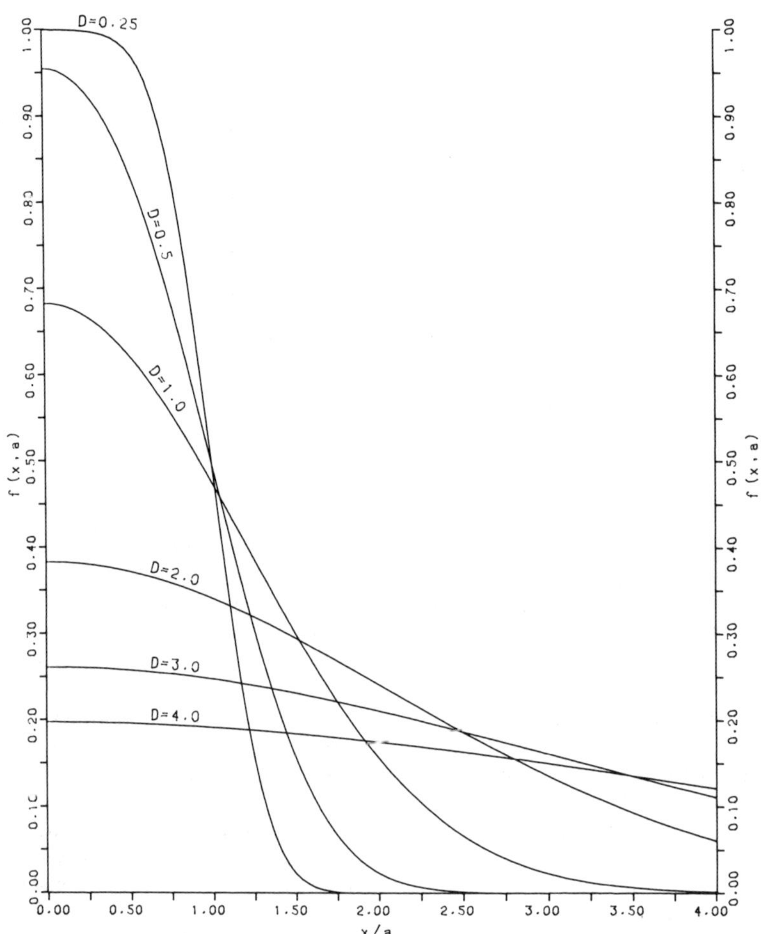

FIGURE 3a. The form of the function f(x,a) used on Table 2, for various values of $D = \sqrt{\nu}\, z/a$.

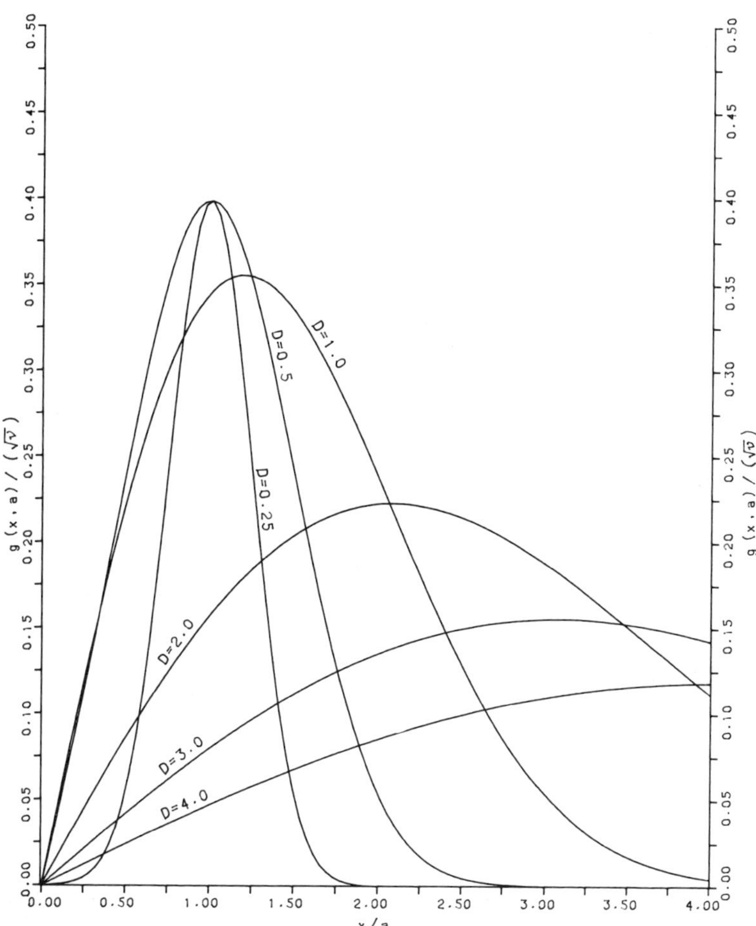

FIGURE 3b. The form of the function $g(x,a)/\sqrt{\nu}$ used on Table 2, for various values of $D = \sqrt{\nu}\, z/a$.

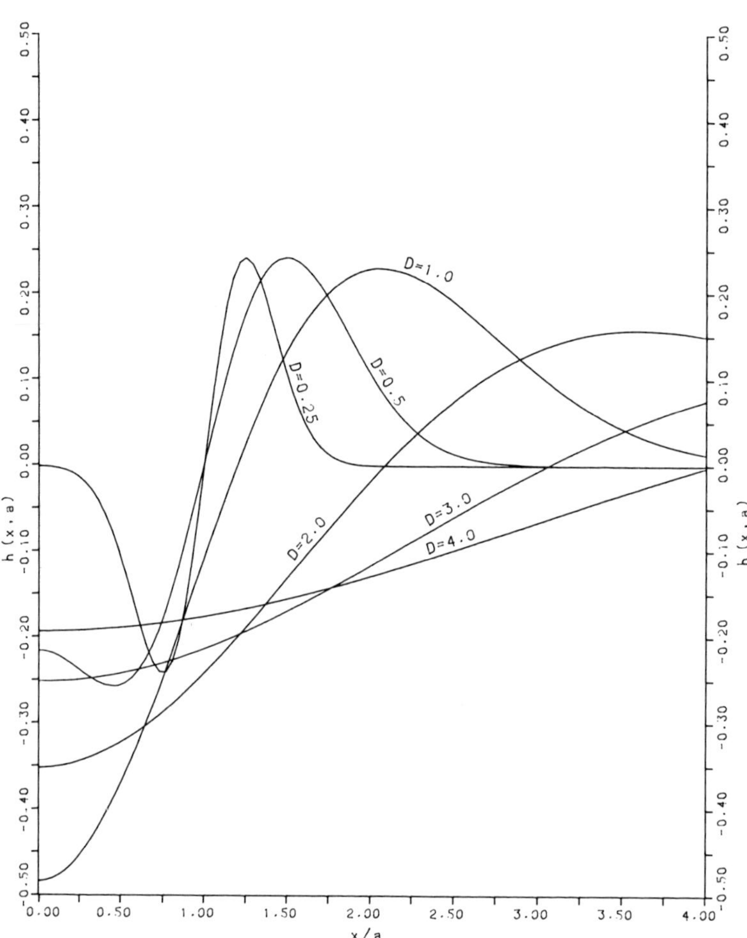

FIGURE 3c. The form of the function h(x,a) used on Table 2, for various values of $D = \sqrt{\nu}\, z/a$.

TABLE 3. Non-Zero Stress Components in Cylindrical Coordinates Under a Point Load and Under a Uniformly Distributed Circular Load. In Both Cases, the Loading is Purely Vertical. The Circle is Centred at the Origin with Radius a. Two Alternative Forms are Given in the Distributed Load Case.

Stress Component	Point Loading (L = surface load)	Uniform Loading of Intensity p per Unit Area*
s_z	$\frac{L}{2\pi\nu z^2}\exp[-r^2/(2\nu z^2)]$	$\frac{p}{\nu z^2}\, e^{-[r^2/(2\nu z^2)]}\int_0^a dr'\, r'\, I_0(\alpha)\, e^{-[r'^2/(2\nu z^2)]} = paI(0,0,1)$
s_{rz}	$\frac{r}{z}s_z$	$\frac{pa}{z}\exp[-(a^2+r^2)/(2\nu z^2)]\, I_1(\beta) = zpa\nu I(1,1,1)$
s_r	$\frac{r^2}{z^2}s_z$	$\nu\{s_z + p\beta\exp[-(a^2+r^2)/(2\nu z^2)][2rI_1(\beta) - a(I_0(\beta) + I_2(\beta))]/(2r)\}$ $= \nu\{s_z + (\nu z^2 ap/2)[I(2,2,1) - I(2,0,1)]\}$
s_θ	0	$\nu(s_z - (z/r)s_{rz})$
$s_z(r=0)$	$L/(2\pi\nu z^2)$	$p\{1 - \exp[-a^2/(2\nu z^2)]\}$
$s_r(r=0)$	0	$p\nu\{1 - [1 + (a^2/(2\nu z^2))]\exp[-a^2/(2\nu z^2)]\}$
$s_\theta(r=0)$	0	$s_r(r=0)$

*$\alpha = rr'/(\nu z^2)$; $\beta = ar/(\nu z^2)$; $I_n(\alpha)$ = Bessel Function of Imaginary Argument.
$I(n,m,r) = \int_0^\infty dk\, k^n J_m(kr)\, J_r(ka)\exp(-\nu z^2k^2/2)$; $J_m(u)$ = Bessel Function of the First Kind.

TABLE 4. Non-Zero Stress Components in Cylindrical Coordinates Under a Parabolically Distributed Vertical Load Over a Circular Area of Radius a Centered on the Origin. Two Alternative Forms are Given.*

Stress Component	Parabolic Loading Over a Circular Area Surface Pressure = $p(1-(r/a)^2)$; p = maximum pressure.
s_z	$(p/\nu z^2)\, e^{-r^2/(2\nu z^2)}\int_0^a r'dr'\,(1-(r'/a)^2)\, e^{-r'^2/(2\nu z^2)}\, I_0(\alpha) = 2pI(-1,0,2)$
s_{rz}	$(2p/za^2)\, e^{-r^2/(2\nu z^2)}\int_0^a r'^2dr'\, e^{-r'^2/(2\nu z^2)}\, I_1(\alpha) = 2\nu zpI(0,1,2)$
s_r	$\nu\{s_z - (p/(\nu z^2a^2))\, e^{-r^2/(2\nu z^2)}\int_0^a r'^2dr'\, e^{-r'^2/(2\nu z^2)}[r'(I_0(\alpha) + I_2(\alpha)) - 2rI_1(\alpha)]\} = \nu\{s_z + \nu z^2p[I(1,2,2) - I(1,0,2)]\}$
s_θ	$\nu(s_z - (z/r)s_{rz})$
$s_z(r=0)$	$p\{1 - (2\nu z^2/a^2)[1 - \exp(-a^2/(2\nu z^2))]\}$
$s_r(r=0)$ $s_\theta(r=0)$	$p\nu\{1 + \exp[-a^2/(2\nu z^2)] - (4\nu z^2/a^2)[1 - \exp(-a^2/(2\nu z^2))]\}$ $s_r(r=0)$

*See Table 3 for definitions of various quantities.

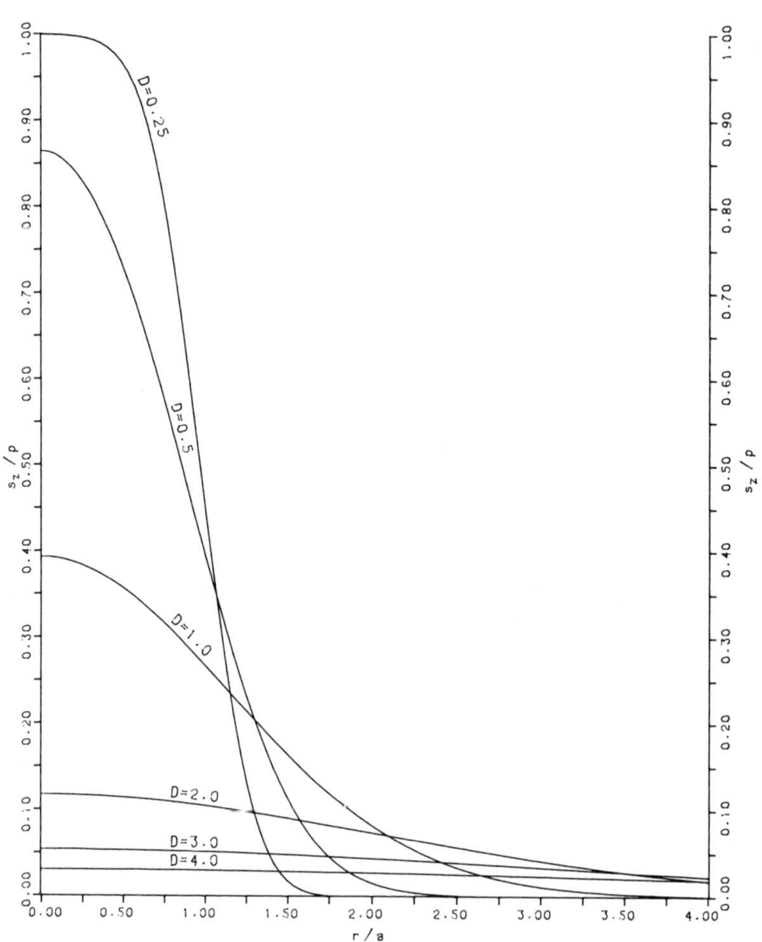

FIGURE 4a. The forms of s_z under a uniform circular load of radius a, for various values of D = $\sqrt{\nu}$ z/a.

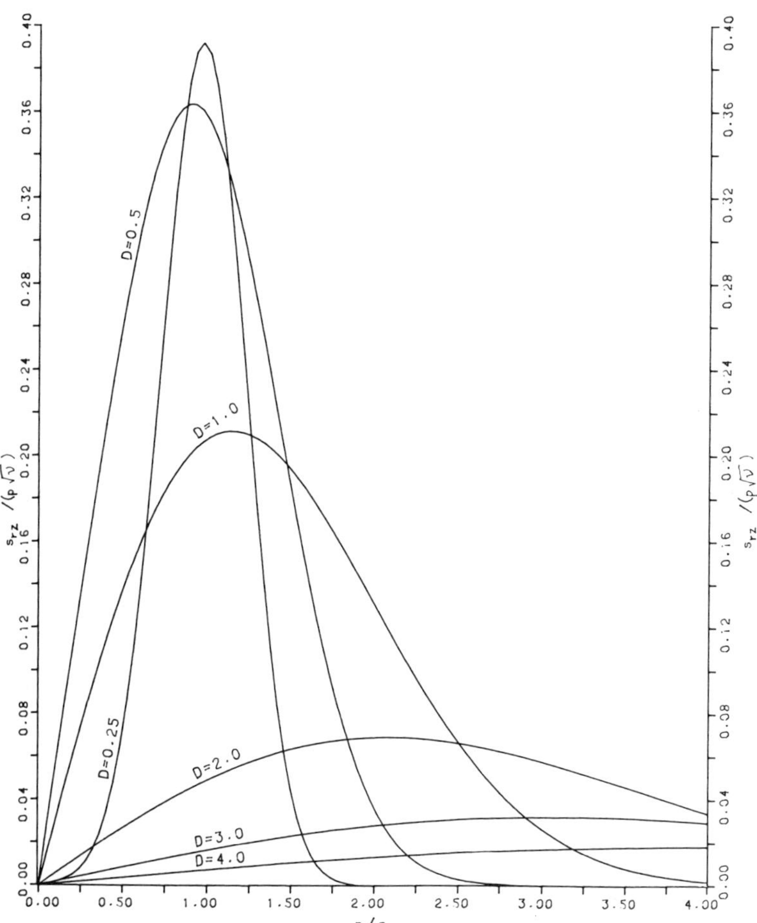

FIGURE 4b. The forms of $s_{rz}/\sqrt{\nu}$ under a uniform circular load of radius a; for various values of $D = \sqrt{\nu}\ z/a$.

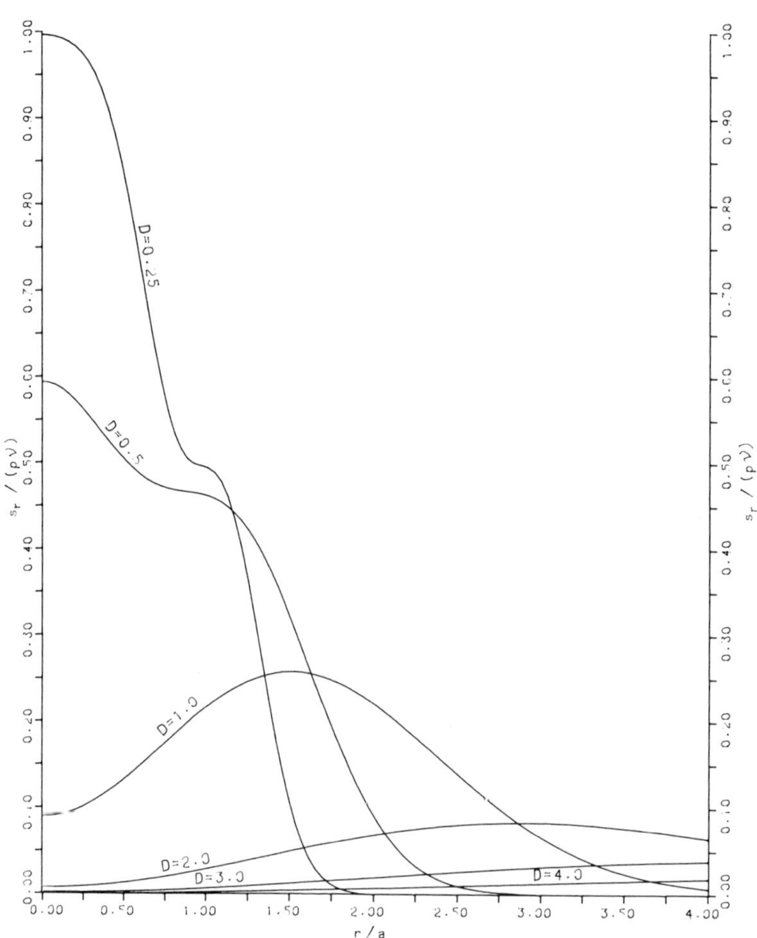

FIGURE 4c. The forms of s_r/ν under a uniform circular load of radius a, for various values of $D = \sqrt{\nu}\, z/a$.

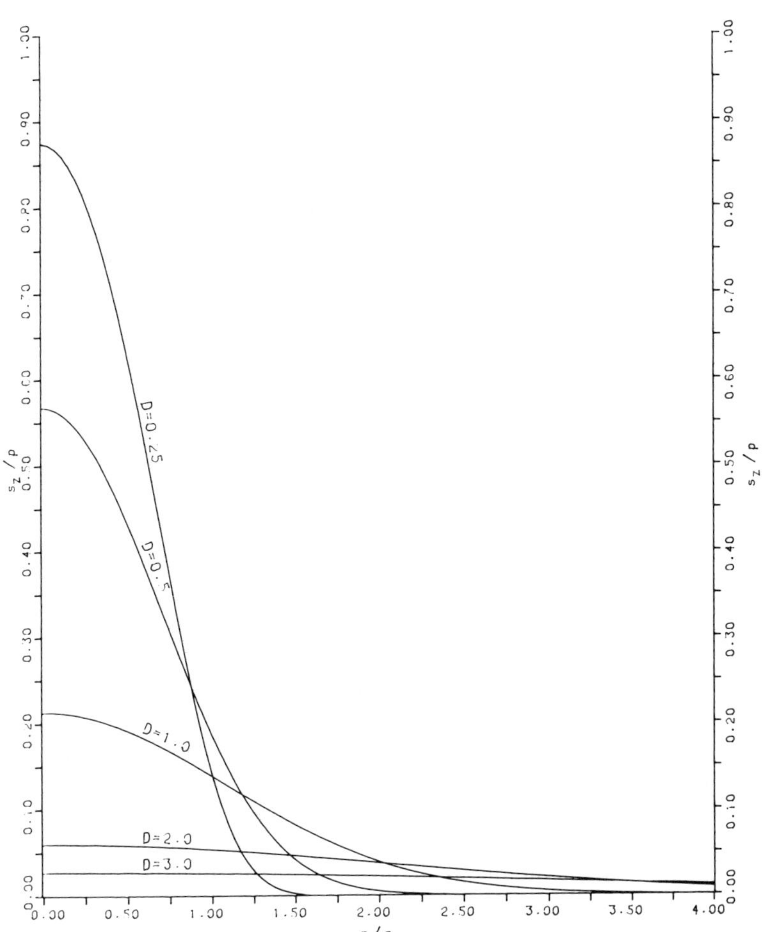

FIGURE 5a. The forms of s_z under a parabolic circular load of radius a, for various values of $D = \sqrt{\nu}\, z/a$.

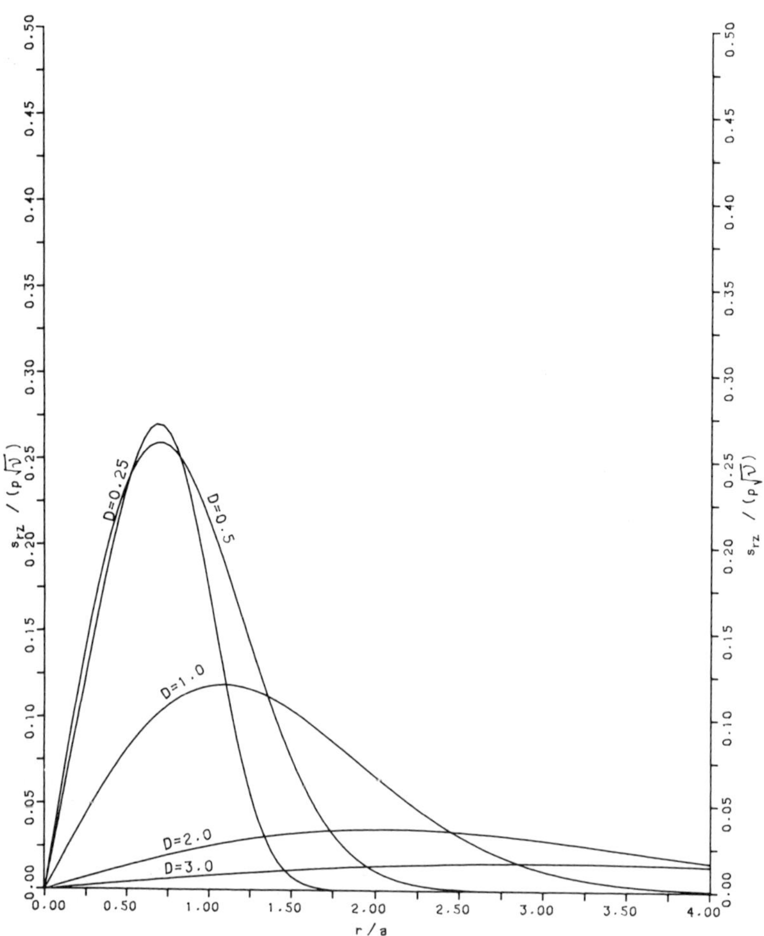

FIGURE 5b. The forms of $s_{rz}/\sqrt{\nu}$ under a parabolic circular load of radius a, for various values of $D = \sqrt{\nu}\ z/a$.

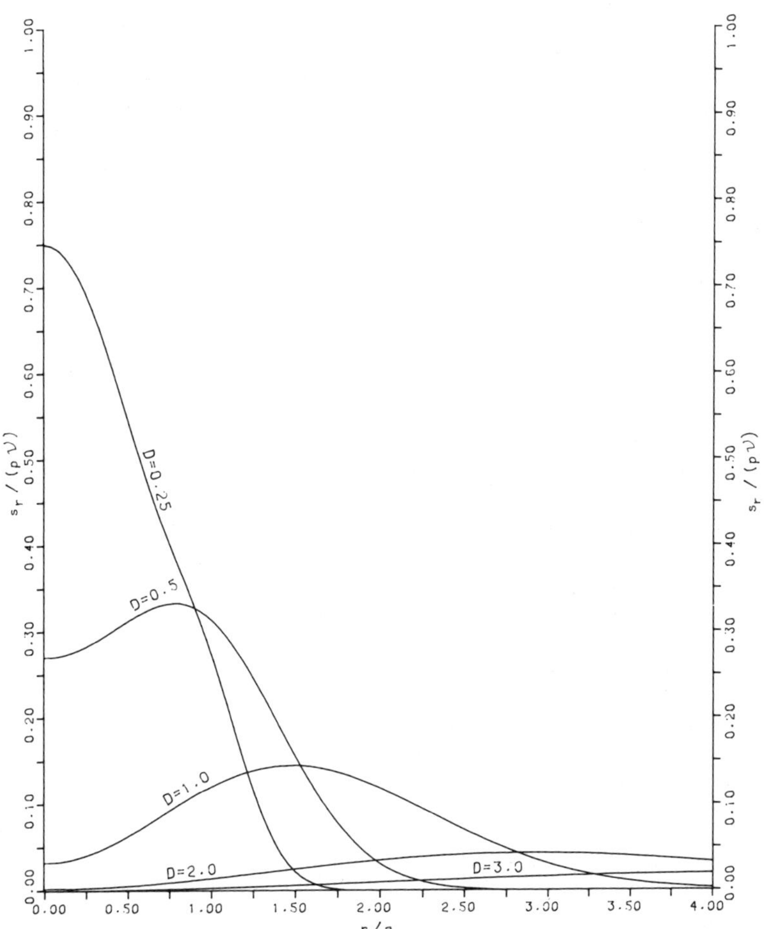

FIGURE 5c. The forms of s_r/ν under a parabolic circular load of radius a, for various values of $D = \sqrt{\nu}\ z/a$.

Therefore ν may be identified approximately with K where the latter is evaluated near the surface, which motivates the name coefficient of lateral stress for ν. Reimbert and Reimbert [47] have reported that for a variety of particulate materials, K lies in the range

$$\left\{\frac{45 - \gamma/2}{45 + \gamma/2}\right\}^2 \leq K \leq \left\{\frac{45 + \gamma/2}{45 - \gamma/2}\right\} \tag{19}$$

where γ is the angle of repose of the material.

Detailed Comparison with Data

Harr [23] using data from the Waterways Experimental Station (WES) reported in Reference 55 for sand and clayey silt and from Reference 31 for sand, concluded that ν is reasonably approximated by the Reimberts' minimum (active thrust) value for K in Equation (19). For the clayey silt data of Reference 55 and the sand data of Reference 31, γ was taken to be the friction angle from triaxial and direct shear tests respectively.

The values of ν suggested by this analysis for sand were in the region 0.19–0.24, except that from WES data [55], a higher value of 0.29 emerged at shallow depths. Harr argues that this may be a consequence of the way the experiment was carried out in that the material used to give results from shallow depths was subjected to more loading than for other depths. A higher level of compaction should lead to better load spreading and therefore a higher value of ν.

The analysis of the WES clayey silt data by Harr [23] indicated large variation in the values of ν obtained (0.29–0.71). The values were higher at shallower depths (material which received more loadings) and declined with increasing surface contact pressure over the range 15 lb/in^2 (103 kPa) to 60 lb/in^2 (414 kPa). The experiment involved significant disturbance of the material and it might be argued [23] that the values of ν for the highest loads best reflect undisturbed conditions. They certainly showed least variation (0.29–0.37) which includes the elastic value of ⅓ given by Equation (16). The measured stresses were, in fact, in good agreement with those predicted by the Theory of Elasticity for a homogeneous, isotropic medium.

The values of ν obtained for clayey silt are higher than those obtained for sand. This is to be expected in that the latter is a cohesive material which should lead to more effective load spreading properties. It will be recalled, however, from the discussion of the section Theoretical Developments, above, that a choice of $a(\underline{R})$ depending only on z may not be appropriate for cohesive materials.

This analysis of Harr [23] was based on comparing $s_z(r = 0)$ from Tables 3 and 4 with measured stresses under the centre of the circular loading.

Morgan and Gerrard [36] have presented a unified digest of stress measurements in sands reported by a number of workers, including [55,25,1,32,17]. A least squares fit of s_z from Table 3 at a depth $z/a = 1.5$ to the curves in Reference 36, extracted from [55] and what the authors refer to as the Melbourne Series 1 data, give a value of $\nu = 0.21$. On Figure 6, the form of s_z predicted by the stochastic theory for this value of ν is compared with the experimental data and the Boussinesq result, as given in [36]. The stochastic prediction is closer to the experimental curves. Other data presented by Morgan and Gerrard [36] derived from [25,1,32] can also be well fitted by the stochastic theory. This better agreement is not surprising since the stochastic theory contains a free parameter, which the Boussinesq result does not. The point is that experiment would indicate that, given the divergence of the experimental curves, this free parameter is necessary. Observe that the stronger fall-

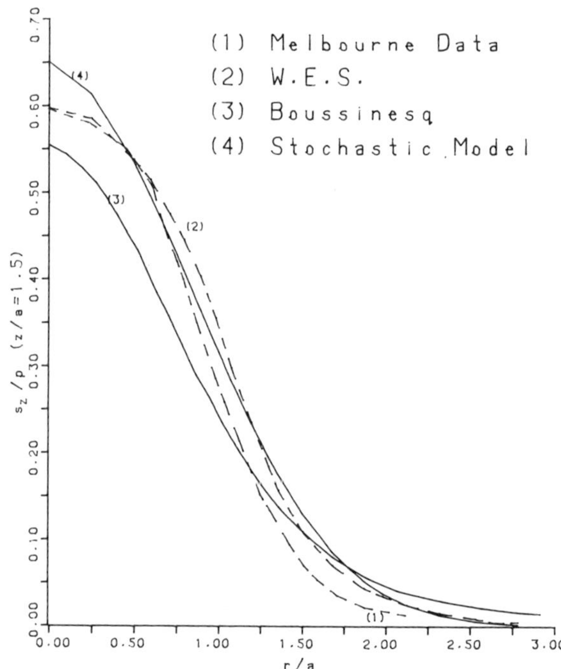

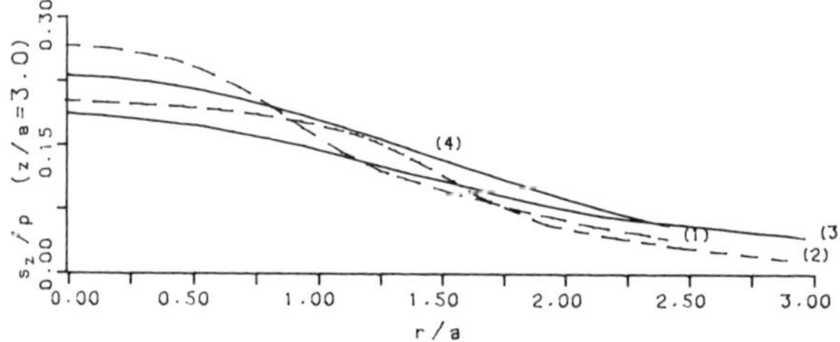

FIGURE 6. Comparison of s_z under a loaded flexible circular plate (approximated as a uniform distributed load) as predicted by the stochastic and linear elastic models with data from two sources, at $z/a = 1.5$ and 3.0.

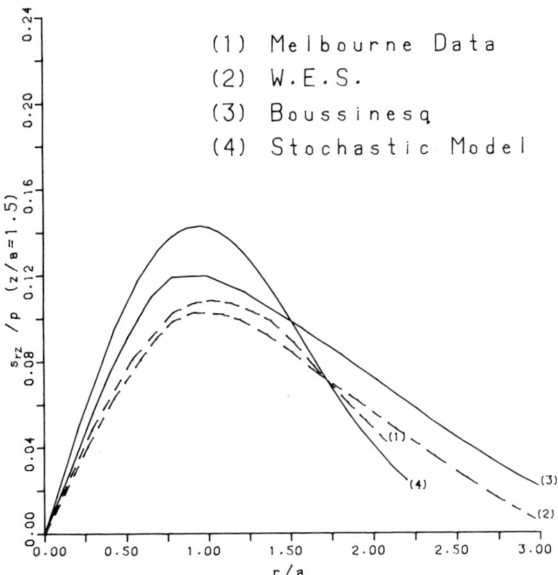

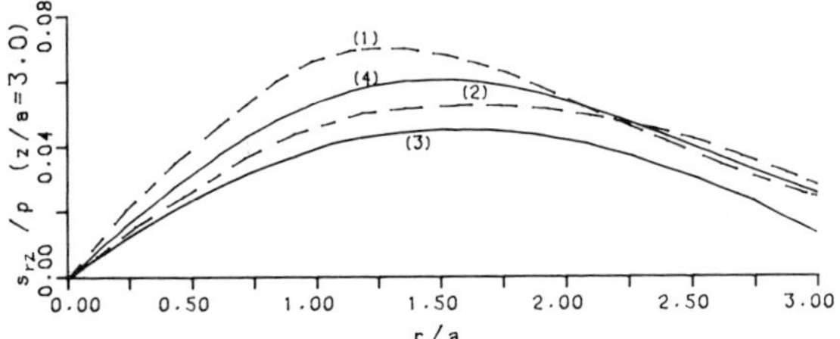

FIGURE 7. Comparison of s_{rz} under a loaded flexible circular plate (approximated as a uniform distributed load) as predicted by the stochastic and linear elastic models with data from two sources, at z/a = 1.5 and 3.0.

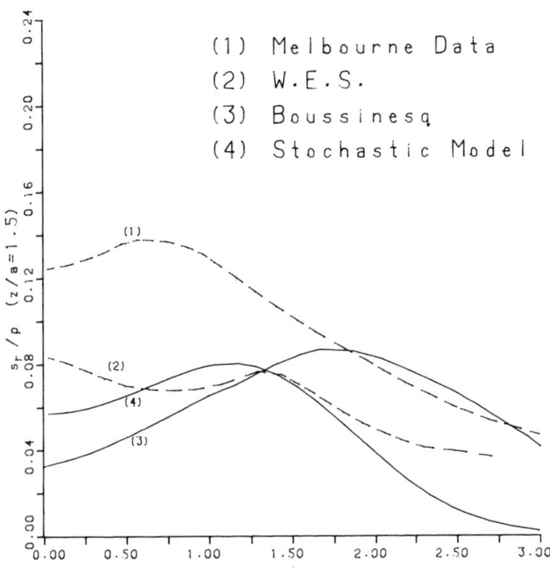

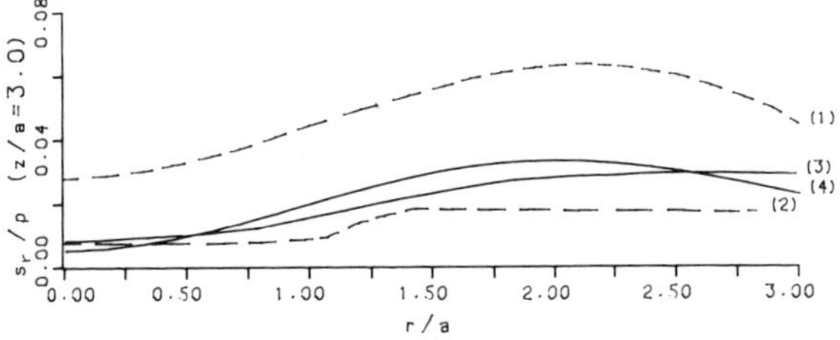

FIGURE 8. Comparison of s_r under a loaded flexible circular plate (approximated as a uniform distributed load) as predicted by the stochastic and linear elastic models with data from two sources, at z/a = 1.5 and 3.0.

off at large r of the experimental curves is better approximated by the stochastic model than by the Boussinesq curve.

Fitting in the same manner to the curves presented in [36] for $z/a = 3.0$ gives higher values of ν. However, using $\nu = 0.21$, deduced for $z/a = 1.5$, gives agreement slightly better than the Boussinesq curve (Figure 6).

Other components of the stress tensor given on Table 3, for $\nu = 0.21$, were compared with the experimental results and the Boussinesq result as presented in [36]. Where the Boussinesq formula depends on Poisson's ratio η (it does in s_r, s_θ) this was taken, following Reference 36, to be 0.4, as giving the most appropriate fit to s_r.

On Figure 7 the measured and predicted curves for s_{rz} at $z/a = 1.5$, 3.0 are plotted. At $z/a = 3.0$, the stochastic curve fits the data better than that derived from the Boussinesq model. However, at $z/a = 1.5$, the Boussinesq curve is in somewhat better agreement with the data, particularly at small r. At large r, the experimental curves are intermediate between the two predictions.

On Figures 8 and 9, the theoretical and experimental curves for s_r, s_θ at $z/a = 1.5$ and 3.0 are presented. Neither the stochastic nor the elastic predictions are particularly good. The Boussinesq prediction fits s_r, better at large r. However, the reverse is true at small r, and in particular, the position of the maximum is more accurately approximated by the stochastic theory. For s_θ (no Melbourne data) at $z/a = 1.5$, the stochastic prediction is somewhat better than the elastic prediction. At $z/a = 3$, neither are good at small r, though at large r, the stochastic model falls to zero with the data.

The curves for s_z on the central axis are given on Figure 10. The Melbourne curve is that labelled Series 2 in Reference 36 which is more consistent with Figure 6 than the curve labelled Series 1. The experimental data is intermediate between the two theoretical predictions. Recall however that the value of ν used in the stochastic case was determined at one depth from the data on Figure 6, so that it does not take account of any variation of material properties with depth. Also, even on Figure 6 the agreement is not particularly good near the central axis between the stochastic curve and the data at $z/a = 1.5$.

The curves for the component $s_r(r = 0)$ are given on Figure 11. An interesting feature here is that the Boussinesq result predicts that near the surface $s_r(r = 0)$ approaches the value $p(\frac{1}{2} + \eta)$ while the stochastic theory gives that it approaches νp, a quite different prediction. The experimental curves shown on Figure 11 and others in [36] qualitatively support the stochastic theory. The Melbourne data near the surface agrees quite well with the stochastic curve for $\nu = 0.21$, which value was deduced for a depth $z/a = 1.5$. Clearly, the WES data does not agree closely with the theoretical curve. However, we recall from Harr's [23] analysis that the top layer in the WES experiment tended to exhibit large values of ν. Below the depth $z = a$, the situation is reversed. Agreement between the stochastic result and the WES data is quite good, better than the Boussinesq prediction, while neither theory agrees with the Melbourne data.

Holden [45] presents data on the central axis under a circular load and compares his result with various theories, including the stochastic model discussed here. He finds reasonable agreement with $s_z(r = 0)$ for $\nu = 0.25$, while in the case of $s_r(r = 0)$, agreement is adequate above $z = a$ but below this value, both the stochastic and Boussinesq predictions are too low.

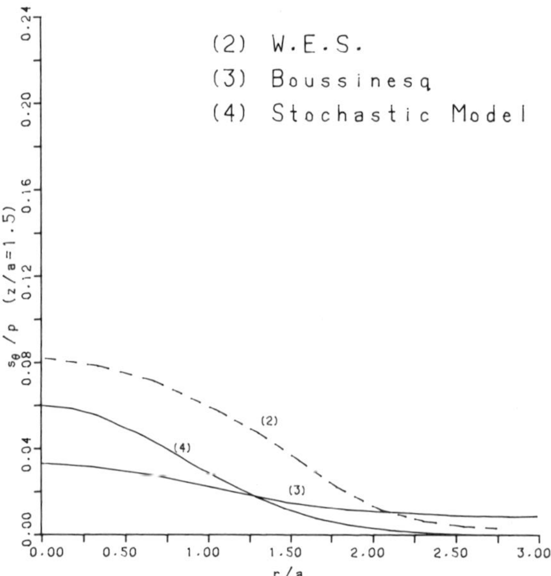

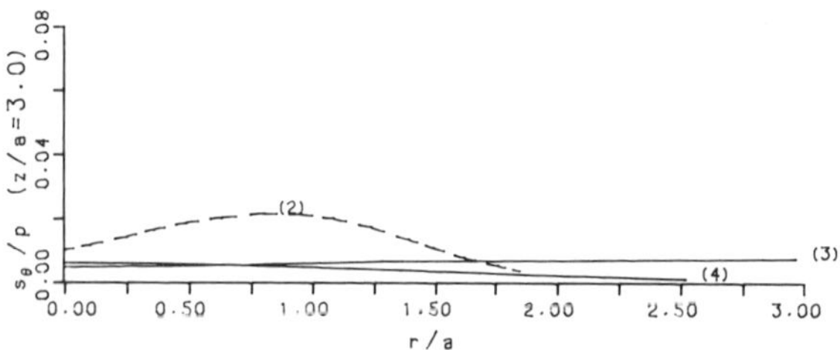

FIGURE 9. Comparison of s_θ under a loaded flexible circular plate (approximated as a uniform distributed load) as predicted by the stochastic and linear elastic models with data from two sources, at $z/a = 1.5$ and 3.0.

CONCLUSIONS

The stress formulae given on Table 3 provide a description of stresses in uniformly loaded granular half-spaces that

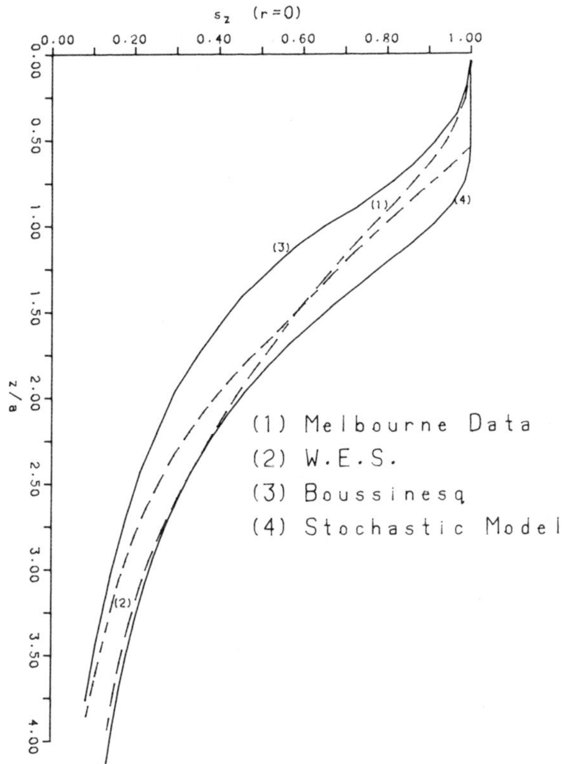

FIGURE 10. Comparison of s_z on the central axis under a loaded flexible circular plate (approximated as a uniform distributed load) as predicted by the stochastic and linear elastic models with data from two sources.

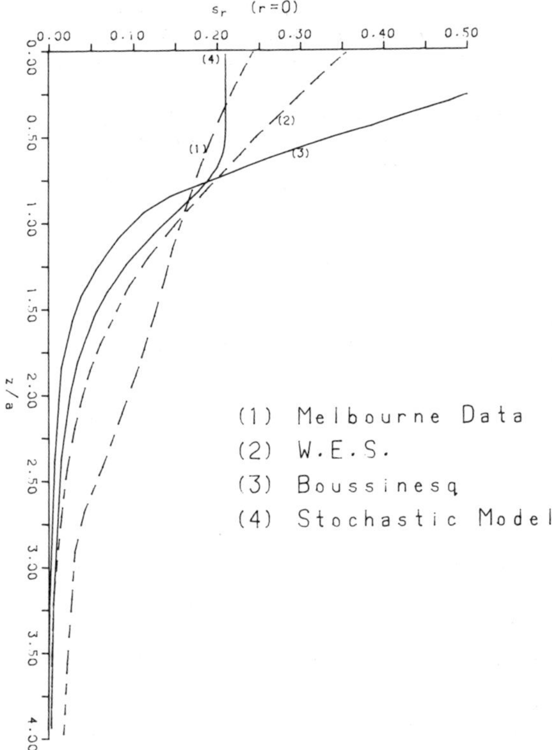

FIGURE 11. Comparison of s_r on the central axis under a loaded flexible circular plate (approximated as a uniform distributed load) as predicted by the stochastic and linear elastic models with data from two sources.

is as good as and in several respects somewhat better than the predictions of homogeneous isotropic linear elastic theory. Also, the stochastic theory is mathematically simpler. This is particularly true when one considers the application of the theory to multi-layered media, which is almost as simple as the single-layered case, if the stochastic model is used, and significantly more complicated, if Linear Elasticity is adopted as the descriptive theory. The formula for s_z on the central axis from Table 4 has also been found to compare reasonably well with experimental results under rigid plates. It must be admitted that comparisons with experiment rely totally on data for sands. Further work is required on other types of granular media.

No doubt, in particular circumstances, better agreement than that of the stochastic theory could be obtained by using an anisotropic or inhomogeneous linear elastic theory or alternatively, non-linear models with the aid of finite element techniques. However, these options are all mathematically far more cumbersome than the stochastic theory.

Perhaps the strongest argument in favour of the stochastic theory is that, for an influence function depending only on depth, the stress distributions are free of tensile stresses which may occur in the linear elastic formulae. For example, under a uniform vertical pressure, distributed over a circle, the horizontal stress on the central axis becomes tensile below a certain depth, provided that the material is compressible (Table A3 of Appendix II). These tensile stresses are not often encountered in practice, under a purely vertical force. However, if a horizontal surface component is present, they are more widespread. This is clear from the formulae for point and line surface loads in Appendix II, and from Figure 2. In this context it should be noted that the absence of tensile stresses in the stochastic theory has been conclusively demonstrated only for the point and line load solutions. It may be checked also along the central axis under circular distributed loads.

The stochastic theory where the influence function depends only on depth should be approximately applicable to media which manifest some cohesiveness. However, as the cohesiveness increases, it may be necessary to include a de-

TABLE 5. Stress Values on the Central Axis Under a Uniform Circular Pressure of Radius a as Predicted by the Stochastic Theory for the Design Value ν = 0.2, Appropriate to the Granular Material, and by Linear Elastic Theory with Poisson's Ratio η = 0.4, Respectively.

		Stochastic Theory $\nu = 0.2$	Elastic Theory $\eta = 0.4$
s_z/p	z = a	0.92	0.65
	z = 2a	0.46	0.28
s_r/p	z = 0	0.2	0.9
	z = a	0.14	0.087
	z = 2a	0.026	0.006

pendence on horizontal coordinates in the influence function, which would add significantly to the complexity of the theory. Also, the mechanical behaviour of such materials is strongly affected by moisture, which further complicates matters as far as the applicability of the stochastic model is concerned.

We suggest that a value of $\nu = 0.2$ may be appropriate for design purposes. On Table 5, the predictions of the stochastic theory, using this value, are compared with those of elastic theory, on the central axis under a uniform circular load, using the formulae of Tables 3 and A3.

NOTATION

Note: The notation of Appendix I is not included in this list, nor are certain symbols used and defined on the tables of formulae and Equation (14).

a = size parameter of applied distributed loads
$a(\underline{R}),a(x,z),a(z)$ = influence function
b = size parameter of applied distributed loads
$b(\underline{R}),b(x,z)$ = function taking account of horizontal forces
K = coefficient of lateral pressure
l = vertical load per unit length
L = vertical load
q = ratio of horizontal to vertical load
$\underline{R}$ = position vector (x,y,z)
$\underline{R}'$ = integration variable
r = cylindrical polar coordinate $(x^2 + y^2)^{1/2}$
s = a component of the stress tensor
s_z,s_x,s_{xz}, etc. = components of the stress tensor
s_r,s_θ,s_{rz}, etc. = components of the stress tensor in cylindrical polar coordinates
$w(z)$ = integral with respect to z of $a(z)$
x,y = components of the position vector
x',y' = integration variables
z = depth coordinate
γ = angle of repose
η = Poisson's Ratio
θ = cylindrical polar coordinate
λ = parameter of Gamma Distribution
ν, ν_1, ν_2 = coefficient of lateral stress

REFERENCES

Note: The references in the subsequent appendices are included below.

1. Allwood, R. T., "An Experimental Investigation of the Distribution of Pressure in Sand," thesis presented to the University of Birmingham at Birmingham, England, in partial fulfillment of the requirements for the Degree of Doctor of Philosophy (1956).
2. Auvinet, G. and R. J. Marsal, "Statistical Model of Grain Breakage," *Proceedings of the Fifth Panamerican Conference on Soil Mechanics and Foundation Engineering,* Vol. 1, Buenos Aires, Argentina, pp. 193–204 (Nov. 17–22, 1975).
3. Bernal, J. D. and J. Mason, "Coordination of Randomly Packed Spheres," *Nature,* Vol. 188, pp. 910–911 (1960).
4. Bourdeau, P. L., M. E. Harr, and R. D. Holtz, "Soil–Fabric Interaction—An Analytical Model," *Proceedings of the Second International Conference on Geotextiles,* Las Vegas, pp. 387–391 (1982).
5. Bourdeau, P. L. and E. Recordon, "Dimensionnement des Chausses Souples," *Annales de L'Institut Technique du Batiment et des Travaux Publics*, Serie-Travaux Public 198, No. 420 (December 1983).
6. Boussinesq, J., *Application des Potentials a l'Etude de l'Equilibre et du Mouvement des Solides Elastiques,* Gauthier-Villard, Paris, France (1885).
7. Chandrasekhar, S., "Stochastic Problems in Physics and Astronomy," *Reviews of Modern Physics,* Vol. 15, No. 1, pp. 1–87 (1943).
8. Chikwendu, S. C. and M. Alimba, "Diffusion Analogy for some Stress Computations," *Journal of the Geotechnical Engineering Division,* ASCE, Vol. 105, No. GT11, Proc. Paper 1111, pp. 1337–1342 (Nov., 1979).
9. Davis, R. A. and H. Deresiewicz, "A Discrete Probabilistic Model for Mechanical Response of a Granular Medium," *Acta Mechanica,* Vol. 27, pp. 69–89 (1977).
10. Deresiewicz, H., "Stress–Strain Relations for a Simple Model of a Granular Medium," *Journal of Applied Mechanics,* Vol. 25 pp. 402–406 (1958).
11. Duffy, J., "A Differential Stress–Strain Relation for the Hexag-

onal Close-Packed Array of Elastic Spheres," *Journal of Applied Mechanics,* Vol. 26, pp. 88–94 (1959).

12. Duffy, J. and R. D. Mindlin, "Stress-Strain Relations and Vibrations of a Granular Medium," *Journal of Applied Mechanics,* Vol. 24, pp. 585–593 (1957).
13. Endley, S. N. and A. H. Peyrot, "Load Distribution in Granular Media," *Journal of the Engineering Mechanics Division,* ASCE, Vol. 103, No. EM1, Proc. Paper 12719, pp. 99–111 (Feb., 1977).
14. Feller, W., *An Introduction to Probability Theory and its Applications,* John Wiley & Sons, Inc., New York, N.Y., Vol. 1 (1950); and Vol. 2 (1971).
15. Foppl, A., *Die wichtigsten Lehren der hoheren Elastizitatstheorien,* 4th ed., Verlag Teubner, Leipzig, Germany (1922).
16. Fröhlich, O. K., *Druckverteilung in Baugrund,* Springer Verlag, Vienna, Austria (1934).
17. Gerrard, C. M., "Theoretical and Experimental Investigations of Model Pavement Structures," thesis presented to the University of Melbourne, Melbourne, Australia, in partial fulfillment of the requirements for the degree of Doctor of Philosophy (1969).
18. Gerrard, C. M. and W. Jill Harrison, "Circular Loads Applied to a Cross-anisotropic Half Space," Commonwealth Scientific Industrial Research Organisation, Division of Applied Geomechanics, Technical Paper No. 8 (1970).
19. Gerrard, C. M. and P. Mulholland, "Stress, Strain and Displacement Distributions in Cross-Anisotropic and Two-Layer Isotropic Elastic Systems," Proceedings of the Third Conference of the Australian Road Research Board (1966).
20. Golden, J. M., "Stochastic Models of Granular Media," *Journal of the Engineering Mechanics Division,* ASCE, Vol. 110, No. EM11, Proc. Paper 19256, pp. 1610–1626 (Nov., 1984).
21. Golden, J. M., "Rational Thickness Design of Flexible Pavements for Irish Conditions," National Institute for Physical Planning and Construction Research, Report No. RC.243 (1984).
22. Golden, J. M., "Stochastic Stress Models with Horizontal Forces," *Journal of the Engineering Mechanics Division,* ASCE, Technical note, Vol. 112, No. EM5, Proc. Paper 20568, pp. 517–522 (May 1986).
23. Harr, M. E., *Mechanics of Particulate Media,* McGraw-Hill, New York, N.Y. (1977).
24. Hill, J. M. and M. E. Harr, "Elastic and Particulate Media," *Journal of the Engineering Mechanics Division,* ASCE, Vol. 108, No. EM4, Proc. Paper 17255, pp. 596–604 (Aug., 1982).
25. Holden, J. C., "Stresses and Strains in a Sand Mass Subjected to a Uniform Circular Load," thesis presented to the University of Melbourne, Australia, in partial fulfillment of the requirements for the degree of Doctor of Philosophy (1967).
26. Iordache, O. and S. Corbu, "A Discrete Stochastic Model of Particles Dynamics in Granular Media," *Bulletin de L'Academie Polonaise des Sciences, Serie des Sciences Techniques,* Vol. 19, No. 11–12, pp. 45–51 (1981).
27. Kandaurov, I. I., "Theory of Discrete Distribution of Stress and Deformation," *Izd. VATT.* (in Russian) (1959).
28. Kandaurov, I. I., "Theory of Stress Distributions in a Granular Medium," *Bases, Foundations and Soil Mechanics,* Vol. 2, No. 4, pp. 6–9 (in Russian) (1960).
29. Kandaurov, I. I., "Mechanics of Discrete Media and its Application to Construction," *Izd. Liter Po Stroitel'stvu* (in Russian) (1966).
30. Kitamura, R., "A Mechanical Model of Particulate Material Based on Stochastic Process," *Soils and Foundations,* Vol. 21, No. 2, pp. 63–72 (June 1982).
31. Kögler, F. and A. Scheidig, *Baugrund und Bauwerk,* Ernst & Sons, Berlin, Germany (1938).
32. Kolbuszewski, J. and G. C. Y. Hu, "An Interim Research Report on Pressure Distribution and Measurements in Sands," *Proceedings, Midland Soil Mechanics and Foundation Engineering Society,* UK, Vol. 4, pp. 73–93 (1961).
33. Litwiniszyn, J., "The Model of a Random Walk of Particles Adapted to Researches on Problems of Mechanics of Loose Media," *Bulletin de L'Academie Polonaise des Sciences, Serie des Sciences Techniques,* Vol. 11, No. 10, pp. 593–602 (1963).
34. McAdams, H. T., "Probability Foundations of Particle Statistics," *Powder Technology,* Vol. 2, pp. 260–268 (1968/69).
35. Mendoza, M. L., "Measurement of Contact Forces in Photoelastic Models," *Proceedings of the Fifth Panamerican Conference on Soil Mechanics and Foundation Engineering,* Vol. 1, Buenos Aires, Argentina, pp. 1–10 (Nov. 17–22, 1975).
36. Morgan, J. R. and C. M. Gerrard, "Behaviour of Sands Under Surface Loads," *Journal of the Engineering Mechanics and Foundations Division,* ASCE, Vol. 97, No. SM12, Proc. Paper 8577, pp. 1675–1699 (Dec., 1971).
37. Muller, R. A., "Statistical Theory of Stress Distribution in a Granular Soil Foundation Bed," *Bases, Foundations and Soil Mechanics,* Vol. 4, No. 4, pp. 4–6 (in Russian) (1962).
38. Muller, R. A., "Deformation of Granular Soil," *Proceedings of the All Union Scientific Research Mining Institute,* Leningrad (in Russian) (1963).
39. Mullins, W. W., "Stochastic Theory of Particle Flow Under Gravity," *Journal of Applied Physics,* Vol. 43, pp. 665–678 (1972).
40. Mullins, W. W., "Experimental Evidence for the Stochastic Theory of Particle Flow Under Gravity," *Powder Technology,* Vol. 9, pp. 29–37 (1974).
41. Mullins, W. W., "Critique and Comparison of Two Stochastic Theories of Gravity-Induced Particle Flow," *Powder Technology,* Vol. 23, pp. 115–119 (1979).
42. Murayama, S. and R. Kitamura, "Mechanical Model of Particulate Material Based on Markov Process," *Bulletin of the Disaster Prevention Research Institute, Kyoto University,* Vol. 28, No. 254, pp. 9–24 (April 1978).
43. Oda M., J. Konishi, and S. Nemat-Nasser, "Some Experimentally Based Fundamental Results on the Mechanical Behaviour of Granular Materials," *Geotechnique* Vol. 30, No. 4, pp. 479–495 (1980).
44. Ohde, J., "Zur Theorie der Druckverteilung in Baugrund," *Der Bauingenieur,* p. 451 (1939).

45. Pokrovsky, G. I., "Investigation on the Physics of Soils," *Vodgeo*, ONTI (in Russian) (1937).
46. Poulos, H. G. and E. H. Davis, *Elastic Solutions for Soil and Rock Mechanics*, John Wiley & Sons, New York, N.Y. (1974).
47. Reimbert, M. L. and A. M. Reimbert, *Retaining Walls*, Trans Tech Publishing, Bay Village, Ohio (1974).
48. Scott, D. G., "Packing of Equal Spheres," *Nature*, Vol. 188, pp. 908–909 (1960).
49. Sergeev, I. T., "The Application of Probability-Processes Equations to the Theory of Stress Distribution in Non-Cohesive Soil Foundation Beds," *Soil Mechanics and Foundation Engineering*, No. 2, pp. 84–88 (Mar.-Apr., 1969).
50. Shahinpoor M., "Statistical Mechanical Considerations on the Random Packing of Granular Materials," *Powder Technology*, Vol. 25, pp. 163–176 (1980).
51. Shimbo, M., "Stochastic Processes of Fracture in Granular Materials," *International Journal of Engineering Science*, Vol. 12, pp. 1635–1638 (1981).
52. Smoltczyk, H. U., "Stress Computation in Soil Media," *Journal of the Soil Mechanics and Foundations Division*, ASCE, Vol. 93, No. SM2, Proc. Paper 5142, pp. 101–124 (Mar., 1967).
53. Spiegel, M. R., *Theory and Problems of Probability and Statistics*, McGraw-Hill Book Company, New York, N.Y. (1975).
54. Thurston, C. W., "Discussion of Stress-Strain Relations and Vibrations of a Granular Medium," *Journal of Applied Mechanics*, Vol. 25, pp. 310–311 (1958).
55. Turnbull, E. J., A. A. Maxwell, and R. G. Ahlvin, "Stresses and Deflections in Homogeneous Soil Masses," *Proceedings of the Fifth International Conference on Soil Mechanics and Foundations Engineering*, Paris, pp. 337–345 (1961).
56. Wang, M. C. and G. E. Uhlenbeck, "On the Theory of the Brownian Motion II," *Reviews of Modern Physics*, Vol. 17, No. 2, 3, pp. 323–342 (1945).
57. Yoder, E. J. and M. W. Witczak, *Principles of Pavement Design*, John Wiley and Sons, New York, N.Y. (1975).
58. Zienkiewicz, O. C. and Y. K. Cheung, *The Finite Element Method in Structural and Continuum Mechanics*, McGraw-Hill Book Co. Inc., New York, N.Y. (1968).

APPENDIX I—PROBABILITY AND STOCHASTIC PROCESSES

Introduction

The object of this appendix is to summarize some basic concepts from the Theory of Probability and Stochastic Processes which are relevant to the developments of the present chapter. For a more complete treatment of these topics, we refer to the many books on Probability Theory, for example [53]. Harr [23] also deals extensively with this topic in an engineering context. There are fewer texts dealing with Stochastic Processes, particularly those with a continuous independent variable. We note the classic papers of Chandrasekhar [7] and Wang and Uhlenbeck [56]. Also, there is the two volume work of Feller [14], the second volume dealing more deeply with continuous stochastic processes, though mainly in general and abstract terms.

This appendix is intended for the reader who wishes to gain a deeper insight into the mathematical concepts which are the basis of the theory discussed in the main text. Mathematical arguments, at times of a somewhat abstract nature, are included, but only to the extent which is felt necessary to achieve this aim. At the end of the appendix, there is some discussion of the physical interpretation of the concepts introduced, in the context of stress transmission in granular materials.

The fundamental notion we begin with is that of an experiment and its possible outcomes. The word experiment is used here in a very broad sense to mean the observation of any physical or engineering process—for example, the transmission of stress from one level in a granular material to a slightly lower level. The number of possible different outcomes may be finite or infinite. We next distinguish between a *deterministic* experiment and a non-deterministic or *random* experiment. The former is where the outcome is entirely predictable on the basis of information initially provided and a theoretical model of the process. The latter is where this is not so, either because of insufficient initial data or an incomplete theoretical understanding of the process, or both. An example of a deterministic experiment is where a particle is projected under gravity in a vacuum from a known position with known velocity. The outcome of the experiment is its place of landing, and this is entirely predictable. A simple example of a random experiment is the classic one of tossing a coin, which we will suppose, for purposes of the discussion, to take place in a vacuum also. There are two possible outcomes, heads or tails, and we cannot predict which will occur, though it will certainly be one or the other. It is instructive to compare these two experiments since in a physical sense, they are similar, both involving motion under gravity. Thus, there is a complete theory of the motion of the coin. What is lacking is complete information about its initial conditions.

A more relevant example of a random experiment would be to load a granular half-space with a specified boundary pressure and regard the outcome as the stresses exerted on a given particle at a specified position within the medium. The surface loading may be regarded here as initial data (strictly it is boundary data). A complete theory of the process would involve precise knowledge of how the stresses are transmitted from particle to particle, which is in principle possible but in practice totally out of the question. Thus, in this case, the theoretical model of the process is incomplete. Also, in practical situations, the precise surface loading on individual particles is also indeterminate so the initial data is also incomplete.

More generally, it may be remarked that engineering materials are almost invariably very complex in their structure, so that even if their response to applied loads is understood in principle, it is usually impossible to apply this understanding in a complete manner. Thus, the outcome of most engineering experiments is unpredictable, which is essentially a reflection of the complexity of the real world. In fact, the notion of a deterministic experiment in any field of study is an idealized one, since even if a precise theory can be applied, experimental error will cause uncertainty in the initial and final data.

Definition of Probability and Its Properties

Let a given random experiment be repeated many times, under conditions which are not distinguishably different (though of course they will usually be different in actuality since all features are not controlled). Let the experiment for simplicity have two possible separate outcomes A and B. It is frequently observed that the proportion of times A occurs as an outcome is a fraction that approaches a stable value as the experiment is repeated. Let this value be p. It follows that the proportion of times that B occurs approaches a stable fraction $q = 1 - p$ since one or other must occur. It is the stability of p and q which is the key property. Clearly, these two numbers are telling us something about the system we are dealing with, something that is useful if the experiment or process is in fact repeated many times in real situations of interest.

We are led in this manner to define the probability of an outcome or *event,* this latter term being often used to mean one or more specific outcomes of a random experiment. The probability of an indeterminate event A is empirically defined as follows. Let a large number N of trials or repetitions of the random experiment of interest be carried out and let the outcome in n of these be the event A. Then, the probability of A, conventionally denoted by $P(A)$; is given to a close approximation by n/N. Note that $P(A)$ is always in the interval [0,1], zero indicating the impossibility of A and unity implying that A is certain to occur.

We need to define certain terms. Two events A and B are said to be *mutually exclusive* if the occurrance of A excludes the possibility of B and vice versa. For example, let the experiment be the loading of a granular material and the outcome of interest be the stresses on a given particle. Then if A is that the vertical compressive stress is less than 10 units and B is that it is greater than or equal to this value, the events A and B are mutually exclusive.

The concept of *conditional probability* is important also, for our purposes. Imagine that the experiment or process of interest consists of several stages. Let A be a possible outcome at one stage and B a possible outcome at a later stage. Then $P(B|A)$ is the probability that B will occur, given that A has definitely occurred. If the conditional probability of B, given that A has occurred, is equal to the absolute probability of B, or in symbols, if

$$P(B|A) = P(B) \tag{A1}$$

then A and B are said to be *independent events.* There is no causal relationship between them. If this is true of all possible outcomes at the two stages, then the stages constitute independent experiments.

A sequence of stages of this kind, whether dependent or independent of each other, is essentially a stochastic process, a concept discussed in detail later. Within this context, mutually exclusive events generally arise as different outcomes at a given stage, while independence or dependence arises between events at different stages.

From the empirical definition of Probability, one may prove the fundamental laws on which the Theory of Probability is based. We state them here without proof. If A and B are mutually exclusive events then

$$(1)\ P(A \text{ or } B) = P(A) + P(B) \tag{A2}$$

where the event (A or B) means the occurrence of either A or B. This law is easily extended to give that the probability of (A or B or C or . . .) is the sum of the individual probabilities of A, B, C, etc., where these are mutually exclusive events. Also, if $P(A,B)$ is the probability of both A and B occurring then

$$(2)\ P(A,B) = P(B|A)P(A) \tag{A3}$$

If A and B are independent events then

$$P(A,B) = P(A)P(B) \tag{A4}$$

by virtue of Equation (A1). The quantity $P(A,B)$ is referred to as the joint probability of A and B.

Random Variables

The results of an engineering experiment is usually to assign values to a number of physical variables which were previously unknown. Let X be one such variable and let its range of possible values be $[a,b]$. If we are dealing with a random experiment then repetitions of the experiment will give a sequence of values $x_1, x_2, \ldots$ in the interval $[a,b]$. A variable of this kind, the value of which is unpredictable, is termed a random variable. It is conventional to represent the variable by an upper case letter and values of it, resulting from a particular experiment, by a lower case letter. Its possible values may be discrete or continuous. We will consider only the case where they are continuous.

Consider again the example where the experiment consists of a determination of the stresses at a specified position in a half-space subject to known surface loading. Ideally, repetition of this experiment would involve filling several containers with indistinguishable samples of the granular material and subjecting them to identical loadings. Alternatively, if the material is homogeneous, one could move the loading and, correspondingly, the location of measurement.

The Probability Density Function

Let us divide $[a,b]$ into a large number of small intervals and carry out many repetitions of the experiment. As each value of x is determined we tick off the interval into which it falls. Eventually we can construct a histogram with very thin columns, the area of each column giving the frequency or more conveniently, the fraction, of values falling in this interval. According to our previous discussion, these fractions will generally tend to stable values, these being the probability of X falling in the corresponding intervals. Consider a typical interval $[x,x + dx]$ and let the height of this column be $f(x)$. Then the probability of the random variable X having a value in $[x,x + dx]$ is the area of this column, namely $f(x)dx$. In the limit as the intervals go to zero, $f(x)$ usually becomes a continuous, smooth function. It is termed the *probability density function* or *probability distribution* of the random variable X. This function summarizes all that we can expect to know about X.

Consider some interval $[c,d]$ in $[a,b]$. We ask what the probability of X taking a value in $[c,d]$ is. Let us divide it up into a series of very small intervals. The probability of X falling in any one of these is a mutually exclusive event. Therefore, from Equation (A2) we have that

$$P(X \text{ in } [c, d]) = \int_c^d dx\, f(x) \quad \text{(A5)}$$

Since the value of X resulting from a given experiment must fall in $[a,b]$, we have

$$\int_a^b f(x)dx = 1 \quad \text{(A6)}$$

The function $F(x)$ given by

$$F(x) = \int_a^x dx'\, f(x') \quad \text{(A7)}$$

is known as the *cumulative distribution function* of the random variable X. It is the probability that the value of X is less than x. Note that the derivative of $F(x)$ with respect to x gives the density function $f(x)$.

Consider the average of N repetitions or trials of the experiment, given by

$$m_N = [x_1 + x_2 + \ldots + x_N]/N \quad \text{(A8)}$$

As N becomes large, this can be shown to tend to the true mean of X, namely

$$m = \int_a^b xf(x)dx \quad \text{(A9)}$$

Similarly, the variance

$$\sigma_N^2 = [(x_1 - m_N)^2 + (x_2 - m_N)^2 + \cdots + (x_N - m_N)^2]/N \quad \text{(A10)}$$

tends to the true variance

$$\sigma^2 = \int_a^b (x - m)^2 f(x)dx \quad \text{(A11)}$$

as N becomes large.

We will need to consider *conditional* probability density functions later. A function $f(x|B)$ is a conditional probability density function if $f(x|B)dx$ is the probability that X lies in $[x,x + dx]$ given that the event B has occurred. All the above observations on density functions apply to $f(x|B)$ also.

These concepts can be generalized without difficulty to the case where there are two or more random variables. Let (X,Y) be two random variables, which may be thought of as a two-dimensional vector. The joint probability density function $f(x,y)$ of X and Y is that function with the property that $f(x,y)dxdy$ is the probability that the values of X and Y, resulting from a given experiment, lies in $[x,x + dx]$ and $[y,y + dy]$, respectively. The mean of (X,Y) is given by

$$(m_x, m_y) = \int_a^b dx \int_b^b dy(x, y)f(x, y) \quad \text{(A12)}$$

where we have assumed that the possible range of values of both X and Y is $[a,b]$. Similarly, the variances of X and Y can be defined:

$$\sigma_x^2 = \int_a^b dx \int_a^b dy(x - m_x)^2 f(x, y)$$
$$\sigma_y^2 = \int_a^b dx \int_a^b dy(y - m_y)^2 f(x, y) \quad \text{(A13)}$$

Also, the covariance of X and Y is defined as

$$\sigma_{xy}^2 = \int_a^b dx \int_a^b dy(x - m_x)(y - m_y)f(x, y) \quad \text{(A14)}$$

If X and Y are independent random variables then it follows from Equation (A4) that $f(x,y)$ factorizes into $f(x)g(y)$. In this case, one may show that the covariance of X and Y is zero.

We will now briefly discuss two examples of density functions which are relevant in the context of granular materials.

The Normal Distribution

The density function of a random variable X given by

$$f(x) = \exp[-(x - m)^2/2\sigma^2]/[\sigma\sqrt{2\pi}] \qquad -\infty < x < +\infty \tag{A15}$$

is centrally important in any application of statistical and probabilistic concepts. The quantity m is the mean value of X while σ^2 is its variance. This is the normal or Gaussian distribution. It is symmetrical about m, exhibiting the well-known bell shape the positive half of which is shown in Figure A1. Its cumulative distribution function is given by

$$F(x) = (2\pi\sigma^2)^{-1/2} \int_{-\infty}^{x} dx' \exp[-(x' - m)^2/2\sigma^2] = 1/2 + \psi[(x - m)/\sigma] \tag{A16}$$

where

$$\psi(u) = (1/\sqrt{2\pi}) \int_0^u dv \exp[-v^2/2] \tag{A17}$$

The function $\Psi(u)$ is important in the section Formulae for the Stress Distributions, above, and is also plotted on Figure A1. It is closely related to a function known as the Error Function $Erf(u)[= 2\Psi(u\sqrt{2})]$, so called because of the importance of the normal distribution in the analysis of experimental errors. In fact, the average of several measurements of a physical parameter is distributed about the true value according to Equation (A15). This is a consequence of the Central Limit Theorem, which is now discussed.

The Central Limit Theorem

The enormous importance of the normal distribution derives largely from a classic theorem, the content of which will now be sketched without proof. Let $X_1, X_2, \ldots X_N$ be N independent random variables obeying unspecified probability distributions, with means $m_1, m_2, \ldots, m_N$ and variances $\sigma_1^2, \sigma_2^2, \ldots, \sigma_N^2$. As N becomes large, the random variable

$$S_N = X_1 + X_2 + \ldots + X_N \tag{A18}$$

tends towards a normal variable with mean

$$M_N = m_1 + m_2 + \ldots + m_N \tag{A19}$$

and variance

$$\Sigma_N^2 = \sigma_1^2 + \sigma_2^2 + \ldots + \sigma_N^2 \tag{A20}$$

The power of this theorem derives from the remarkable fact that the density functions of the individual X_i need not be given. One may see it intuitively as an expression of the fact that biases or unsymmetrical attributes in the original distributions cancel out on taking the sum of the variables, to give the simple, symmetrical form of the normal distribution.

If the X_i all have the same mean m and variance σ^2, then the mean and variance of S_N is Nm and $N\sigma^2$. It follows that the average S_N/N is normally distributed with mean m and variance σ^2/N. Thus, we see that the variance of the average becomes small as N increases. In the particular case that the randomness of the X_i is a result of experimental error, we see that by repeating the experiment many times and taking the average, the result is distributed normally about the true value m with variance σ^2/N, which, as mentioned earlier, motivates the title given to the Error Function. In this context the Central Limit Theorem may be seen as a precise form of the intuitive statement that on taking the average, the errors tend to cancel out.

Taking a broader viewpoint, we see that the Central Limit Theorem guarantees the stability of mean values of a large number of measurements of a random variable. In particular, in the context of granular materials, it tells us that the macro-stress at a point, as defined in the Introduction, above, should be stable in the sense that if it is measured at the same location in different but identical media, the result should be the same. This is because it is defined as the average of a large number of microstresses in the vicinity of the point of interest and one may regard, at least to a good approximation, the measurement of each of these microstresses as being a repetition of a random experiment.

The Gamma Distribution

In the context of granular materials, random variables arise with density functions, which are not symmetrical and thus cannot be approximated well by the normal form. One such variable, namely the contact force between particles, was discussed in the Introduction, above. Experimental results suggest the Gamma distribution given by

$$f(x) = (x/x_0)^{\lambda-1} \exp[-(x/x_0)]/(x_0\Gamma(\lambda)) \qquad 0 \le x < \infty;\ \lambda, x_0 > 0 \tag{A21}$$

where $\Gamma(\lambda)$ is the standard Gamma function. The quantity λ is a parameter which must be assigned a value based on the data. The quantity x_0 is a scaling parameter of the same dimensions as x. The mean and variance of the distribution are given by

$$m = \lambda x_0 \qquad \sigma^2 = \lambda x_0^2 \tag{A22}$$

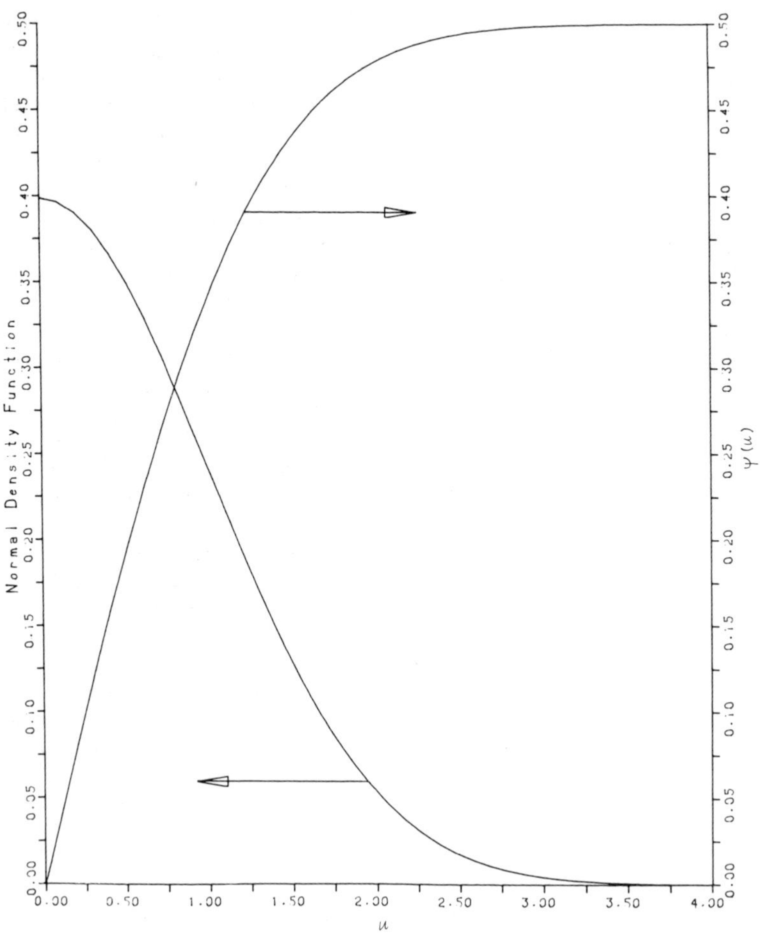

FIGURE A1. The normal curve (Eq.A16) and the function $\Psi(u)$, $u = (x - m)/\sigma$ for positive u. These functions are even and odd, respectively.

If λ is greater than unity, $f(x)$ has the characteristic early maximum followed by a long tail. Some examples of this density function are plotted on Figure A2. In general, the maximum point, or the *mode,* occurs at $x = (\lambda - 1)x_0$. If the mean m and standard deviation σ of the distribution is known, then $\lambda = (m/\sigma)^2$ can be determined and the value of x_0 (in a particular set of units) is fixed. A simple initial check on whether a given set of data obeys the Gamma distribution is to determine the position of the mode from λ, x_0 as fixed above. If it does not agree at least approximately with the position of maximum frequency in the data, then the Gamma distribution may be rejected.

Stochastic Processes

Many processes can be viewed as a chain of random experiments, the outcome of one experiment generally influencing those that follow. We will refer to the experiments as stages, numbered 1, 2, . . ., N, where N may be infinite. This type of structure is referred to as a discrete stochastic process. Our application requires us to consider continuous stochastic processes. However, it is enlightening to start with the discrete case and then go over to the continuous case.

Consider a two stage process. Let there be only two possible outcomes of each stage, namely $A_1^{(1)}$, $A_2^{(1)}$ for the first and $A_1^{(2)}$, $A_2^{(2)}$ for the second. These are assumed to be mutually exclusive at each stage. We ask what the probability of a final outcome $A_1^{(2)}$ is, for example. Applying Equations (A2) and (A3) gives

$$P(A_1^{(2)}) = P(A_1^{(2)}, A_1^{(1)} \text{ or } A_2^{(1)}) = P(A_1^{(2)}, A_1^{(1)})$$

$$= + P(A_1^{(2)}, A_2^{(1)}) = P(A_1^{(2)} | A_1^{(1)}) P(A_1^{(1)})$$

$$+ P(A_1^{(2)}) | A_2^{(1)}) P(A_2^{(1)}) \tag{A23}$$

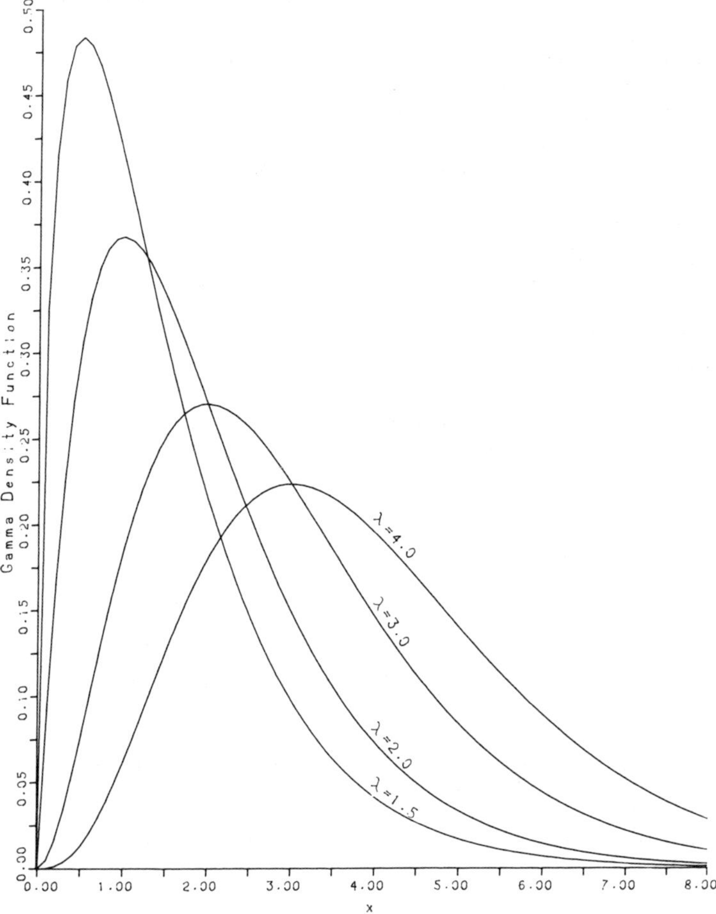

FIGURE A2. Examples of the gamma density function ($x_0 = 1$).

where the probabilities at the first stage and the conditional probabilities are assumed to be known. The conditional probabilities take account of the connection between the two stages. If the stages are independent, Equation (A23) reduces to an identity by virtue of Equation (A1) and the fact that $P(A_1^{(1)}) + P(A_2^{(1)})$ must be unity. In this trivial special case therefore, Equation (A23) is without content. An example would be a series of independent trials of the same experiment. Formula A23 can be generalised without difficulty to the case where there are more than two possible outcomes at each stage.

Let us now consider processes with more than two stages. A given stage can in principle be influenced by the outcomes of all previous stages. However, we confine the discussion to the simplest non-trivial case, namely that where the outcome of a given stage can be influenced only by what happens at the previous stage. Such stochastic processes are *Markov processes*. Let $A_i^{(k)}$, $i = 1,2, \ldots, n$ be the n outcomes at stage k. The probabilities of these can be expressed in terms of those at stage $k - 1$ by a generalized form of Equation (A23), namely

$$P(A_i^{(k)}) = \sum_{j=1}^{n} P_k(A_i^{(k)}|A_j^{(k-1)})P(A_j^{(k-1)}) \qquad \text{(A24)}$$

The conditional probabilities are referred to as *transition probabilities,* which enable one to determine the probabilities of the various outcomes at stage k in terms of those at stage $(k - 1)$. Note that we have allowed for the fact that they may be different at each stage. We can express the probabilities at stage $(k - 1)$ in terms of those at stage $(k - 2)$ and so on back to stage 1. Thus, if the initial conditions are known, and all the transition probabilities, then everything can be determined. This procedure is rendered much more transparent by the use of matrix notation. Note that the transition probabilities really determine the nature of the Markov process.

Let the outcomes of interest at each stage be the values of a random variable X, representing some physical quantity, which takes continuous values. There are therefore a continuously infinite number of possible outcomes at each stage so that Equation (A24) cannot apply, in its present form. We must represent the probabilities of the various outcomes by a probability density function, which may vary depending upon the stage. We can regard the stochastic process therefore, as a sequence of random variables X_k, $k = 0, 1, \ldots,$ each representing the same physical quantity, but with differing probability density functions.

Let the conditional probability density function of X_k be $P_k(x|x')$. In other words, the probability that X_k has a value in the interval $[x, x + dx]$, given that the outcome of stage $(k - 1)$ was x', is $P_k(x|x')dx$. Equation (A24) is replaced by the integral equation

$$f_k(x) = \int_{-\infty}^{\infty} dx'\, P_k(x|x') f_{k-1}(x') \qquad \text{(A25)}$$

where $f_k(x)$, $f_{k-1}(x)$ are the density functions at stages k and $k - 1$. We have assumed that the range of possible values of every X_k is $[-\infty, \infty]$.

Let us now go over to the case where the stages are continuous rather than discrete. They will be labelled by the variable z, ultimately interpreted as the vertical coordinate in a granular half-space. We approximate this situation by supposing that the distance from the origin to z is made up of $(k - 1)$ stages, each of thickness Δz, so that $z = (k - 1)\Delta z$. We put $f_{k-1}(x) = f(x,z)$ and $P_k(x|x') = P(x,x'; z,z + \Delta z)$. Equation (A25) becomes

$$f(x, z + \Delta z) = \int_{-\infty}^{\infty} dx'\, P(x, x'; z, z + \Delta z) f(x', z) \qquad \text{(A26)}$$

where Δz is assumed to be small. What this equation is telling us is that if the transition density functions are known at each z, we can determine the absolute probability density functions $f(x,z)$ at successively greater values of z by carrying out successive integrations. This is a cumbersome procedure, quite apart from the enormous practical difficulty of determining the transition density functions, in a real situation. We will show, however, that it is possible to convert Equation (A26) into a partial differential equation for $f(x,z)$ and that furthermore, this equation depends only on two quantities which are related in a certain sense, made precise below, to the mean and variance of the transition densities. Thus, we do not need complete knowledge of the functions $P(x,x';z,z + \Delta z)$.

The Fokker-Planck Equation

Let us put

$$x = x' + u \qquad \text{(A27)}$$

and change the variables x' and x to x' and u, writing

$$K(x', u; z, z + \Delta z) = P(x, x'; z, z + \Delta z) \qquad \text{(A28)}$$

Expanding the quantity $K(x', u; z, z + \Delta z) f(x', z)$ in a Taylor expansion about x and changing the variable of integration in Equation (A26) to u gives the following equation for $f(x,z)$:

$$\frac{\partial}{\partial z} f(x, z) = -\frac{\partial}{\partial x}[b(x, z) f(x, z)] + \frac{\partial^2}{\partial x^2}[a(x, z) f(x, z)] \qquad \text{(A29)}$$

where

$$\lim_{\Delta z \to 0} \frac{1}{\Delta z} \int_{-\infty}^{\infty} du\, u K(x, u; z, z + \Delta z) = b(x, z)$$

$$\lim_{\Delta z \to 0} \frac{1}{2\Delta z} \int_{-\infty}^{\infty} du\, u^2 K(x, u; z, z + \Delta z) = a(x, z) \qquad \text{(A30)}$$

and provided that

$$\lim_{\Delta z \to 0} \int_{-\infty}^{\infty} du\, K(x, u; z, z + \Delta z) = 1 \qquad \text{(A31)}$$

In order to justify these relations, we first observe that $K(x,u;z,z+\Delta z)$ is the transition probability $P(x+u,x; z,z+\Delta z)$. This follows from Equation (A28). Thus Equation (A31) is really a form of Equation (A6). The first question to be decided in respect of Equation (A30) is whether the limits are finite. The limit of $P(x+u,x;z,z+\Delta z)$ as $\Delta z \to 0$ will be zero everywhere except at $u = 0$. This is clear on physical grounds. But the factors u and u^2 ensure that the product will be zero everywhere. Therefore, one can see, at least intuitively, that the ratio of these integrals with Δz should approach a finite limit. Let us now move on to the question of the physical meaning of $a(x,z)$ and $b(x,z)$. We see that $b(x,z)\Delta z$ for small Δz, is the mean value of u which is the shift in x in passing from z to $z + \Delta z$. Interpreting x and z as horizontal and vertical coordinates, we see that $b(x,z)\Delta z$ is the average horizontal shift in the random variable X, passing from z to $z + \Delta z$, given that, at level z, the value of X was x. If the transition probability is symmetrical about the source point x then $b(x,z)$ is zero. This is where there is no preferential drift in a positive or negative direction. In the context of granular materials, it corresponds to the absence of horizontal surface forces.

In order to interprete $a(x,z)$ we first write it in the form

$$a(x, z) = \lim_{\Delta z \to 0} \frac{1}{2\Delta z} \int_{-\infty}^{\infty} du(u - b(x, z)\Delta z)^2 \times K(x, u; z, z + \Delta z) \quad \text{(A32)}$$

which follows by expanding the brackets and using the first line of Equation (A30). Remember that terms proportional to $(\Delta z)^2$ do not contribute. Thus, on referring to Equation (A10), we see that $2a(x,z)\Delta z$ is the variance of the transition probability distribution, so that $a(x,z)$ is a measure of the horizontal spread of the process originating at the point x, at a slightly lower level of z. In the context of granular materials, it is related to the extent of load spreading at a given position.

Let us now look for solutions of Equation (A29). Consider first the case where $b(x,z)$ is zero. Also, the discussion will be confined to situations where the transition probability from x' to x depends only on $x - x'$ and z. It does not depend on x' directly. This means that $K(x',u;z,z + \Delta z)$ is independent of x'. In particular, by virtue of Equation (A30), it implies that $a(x,z) = a(z)$, in other words, this function depends on z only. Such a form is the most relevant in the context of granular materials.

We consider processes which start at the origin $x = 0$. Let us return for a moment to labelling the consecutive stages by an integer variable instead of the continuous variable z. The solution of Equation (A29) is the probability distribution of X at some level N. Let us write

$$X_N = U_1 + U_2 + \ldots + U_N \quad \text{(A33)}$$

where

$$U_r = X_r - X_{r-1} \quad \text{(A34)}$$

since we have put $X_0 = 0$. It follows from the assumption that the transition probabilities depend only on $(x - x')$ that the U_r are independent of each other. This is intuitively plausible but difficult to demonstrate conclusively. It is left as a (challenging) task for the reader. The mean value of each U_r is zero as a consequence of the assumption that there is no average shift in the X_r at each transition ($b(x,z) = 0$). A stochastic process X_N, $N = 0,1, \ldots$, defined by Equation (A33), is known as a random walk.

Provided that N is large, we can say, by virtue of the Central Limit Theorem, discussed earlier, that X_N is normally distributed, without knowing the detailed shape of the transition probability distributions of the individual U_r. Thus, putting $m = 0$ by virtue of Equation (A19), and $\sigma^2 = 2w(z)$ in Equation (A15), we can deduce that a solution of Equation (A29) is

$$f(x,z) = [4\pi w(z)]^{-1/2} \exp[-x^2/(4w(z))] \quad \text{(A35)}$$

Direct substitution verifies this and identifies $w(z)$ as being related to $a(z)$ by

$$a(z) = \frac{dw}{dz}(z) \quad \text{(A36)}$$

Let us take $w(0) = 0$, so that $f(x,z)$ has the required boundary behaviour at $z = 0$, namely, it is zero everywhere except at the origin and infinite there. This is in fact the line load solution in the context of granular materials. Then Equation (A36) gives

$$w(z) = \int_0^z dz'\, a(z') \quad \text{(A37)}$$

which is really the continuous limit of Equation (A20).

This solution and its three-dimensional counterpart, which is given later, play a central role in the theory developed in the main body of the chapter. It is interesting to observe that it is an immediate consequence of the Central Limit Theorem. This point is emphasised by Harr [23].

Consider the case where $b(x,z)$ is non-zero but is a constant b_0. One may show that if $f_0(x,z)$ is a solution of Equation (A29) with $b_0 = 0$, then $f_0(x - b_0z,z)$ is a solution for b_0 non-zero. Therefore, Equation (A35) is replaced by

$$f(x,z) = [4\pi w(z)]^{-1/2} \exp[-(x-b_0z)^2/(4w(z))] \quad \text{(A38)}$$

The theory of Markov processes and the Fokker-Planck Equation may be developed without difficulty also where X is replaced by a two-dimensional random variable (X,Y). The details will not be given. Instead, we refer to the discussion in Reference 20,22. Let X and Y be statistically uncorrelated in the sense that their covariance, given by Equation (A14), is zero. Also, let the mean change in Y for a small change in z be zero. In other words, the process is not biased in one Y direction or the other. Then, Equation (A29) generalizes to

$$\frac{\partial}{\partial z} f(\underline{r}, z) = -\frac{\partial}{\partial x}[b(\underline{r}, z)f(\underline{r}, z)] + \nabla^2[a(\underline{r}, z)f(\underline{r}, z)] \quad \text{(A39)}$$

$$\nabla^2 = \frac{\partial^2}{\partial x^2} + \frac{\partial^2}{\partial y^2}$$

where $f(\underline{r},z)(= f(x,y,z))$ is the joint probability density of X

and Y at stage z. The quantity $b(r,z)\Delta z$ is the mean shift in the X variable resulting from a small change Δz in z, as before. Also, $2a(r,z)\Delta z$ is again the variance of the transition probability density. It is the variance of changes in both the X and Y variables (see Equation (A13)) resulting from a small change Δz in z. The fact that these are equal is an expression of material isotropy. In this context, $a(r,z)$ is, as before, a measure of the ability of the material to spread load.

The solution to Equation (A39) where X and Y are definitely zero at $z = 0$, corresponding to a point load on a granular half-space, with $a(r,z) = a(z)$ and $b(r,z) = b_0$, is

$$f(x,y,z) = [4\pi w(z)]^{-1}\exp\{-[(x-b_0z)^2 + y^2]/(4w(z))\} \tag{A40}$$

This form may be deduced from the Central Limit Theorem, or alternatively [23] by recalling Equations (A4), (A35) and the fact that X and Y are independent. That it is a solution may be confirmed by direct substitution in Equation (A39) and using Equation (A36).

Physical Interpretation

We have discussed stochastic processes in relatively abstract terms with occasional references to granular materials. It is worthwhile to finish with a more detailed discussion of how a stochastic process describes stress transmission through a granular medium.

Equations of the form of Equation (A39) crop up in the context of diffusion or dissipation phenomena, where $a(r,z)$ is a constant, referred to as the diffusion coefficient and $b(r,z)$, also usually constant, is associated with an overall drift superimposed on the random diffusion effect. The variable z is generally identified as time. Also, there are usually three space dimensions. Examples of applications are heat dissipation through a solid or the diffusion of pollution through water or air. Equation (A39) is often referred to as a diffusion equation.

In the present context however, what we are interested in is the transmission of stress through a granular medium. At first sight, this would seem to be an entirely different type of phenomenon. However, when one considers the mechanism of stress transmission, which takes place through large numbers of randomly positioned and oriented inter-particle contacts, and the fact that as depth increases, the stresses become more widely diffused and less intense, the picture begins to resemble a physical diffusion or dissipation process of the conventional type, with z playing the role of the time variable.

The detailed question which must be considered is how a stochastic process of the type discussed in this appendix corresponds to the stress transmission process. For simplicity, let us use as illustration the one-dimensional process $X(z)$, which we claim describes a granular half-space subjected to infinite line loading. The first picture we will propose is essentially artificial, from the point of view of granular materials, but does resemble more familiar diffusion processes. Let the vertical loading on the surface be thought of as consisting of a very large number of tiny separate units of vertical force, each of which may be transmitted independently. We are ascribing an independent existence to these units, treating them like a particle of pollution in air for example, which is of course an artificial procedure.

Each of these units is transmitted down through the medium in a very erratic manner, reflecting the position and orientation of interparticle contacts. The stochastic process $X(z)$ represents the x coordinate occupied by the unit of force at level z. A given trial of this stochastic process would map out a particular path $x(z)$ of a particular force unit. The loaded medium can be viewed as a large number of trials of the process, one for each force unit. The fraction of the units accumulating in a given interval $[x,x + dx]$ at level z of the medium (in fact, the quantity $f(x,z)dx$) is a measure of the vertical load per unit length at that point, so that $f(x,z)$ is proportional to the stress. Recall from the discussion in the main body of the chapter that the vertical stress is an averaged quantity over an area large compared with the particle size. It is this quantity that we claim is proportional to $f(x,z)$.

An alternative and probably better procedure is to adopt this last observation as the starting point. The vertical stress $s_z(x,z)$ is proportional to $f(x,z)$, the probability distribution of a certain Markov process $X(z)$ at level z, where $f(x,z)$ is defined incrementally at different levels by Equation (A26). We then ask if this is reasonable. The vertical stress at a point (x',z) will be transmitted in part to the vicinity of a point x at a slightly lower level $z + \Delta z$. The fraction transferred will depend upon the number and the nature of the paths through contact points joining small areas around (x',z) and $(x,z + \Delta z)$. This is summarized in the function $P(x,x';z,z + \Delta z)$, which we have termed the transition probability. Such a model is entirely consistent with the physical mechanism of stress transmission in a particulate medium. It is not a detailed particulate model in the sense of starting from individual particles and their contact points and, by enumeration procedures, building up to a macroscopic picture. Nor is it a macroscopic model such as Elasticity Theory. It is intermediate between these two extremes in that it takes account of the particulate nature of the medium in a summarized manner. What would render the theory more complete would be the development of insight into the relationship between the transition probability $P(x, x';z,z + \Delta z)$ and the detailed structure of the granular material. This would allow one to relate the form of the influence function to the properties of the granular material and thereby transform the model from a partly empirical one, as it is at present, into a truly fundamental theory.

Accepting that detailed understanding along these lines

TABLE A1. Three-Dimensional Elastic Solutions Under a Vertical Point Load L.

Stress Component	Point Load Solution (purely vertical load)	S_z for Point Load with Horizontal Load qL along x-axis
S_z	$\frac{3Lz^3}{2\pi R^5}$	$\frac{3Lz^2}{2\pi R^5}[z+qx]$
S_{rz}	$\frac{r}{z} S_z$	
S_r	$\frac{L}{2\pi R^2}\left[\frac{3zr^2}{R^3} - \frac{(1-2\eta)R}{R+z}\right] = \frac{r^2}{z^2} S_z$ (incompressible case)	
S_θ	$\frac{(1-2\eta)L}{2\pi R^2}\left[\frac{R}{R+z} - \frac{z}{R}\right] = 0$ (incompressible case)	

$R^2 = x^2 + y^2 + z^2$
η = Poisson's Ratio

TABLE A2. Two-Dimensional (Plane Strain) Solutions Under a Surface Line Load Along the y-Axis with Components (ql,l).

Stress Component	Line Load Solution
S_z	$\frac{21z^2(z+qz)}{\pi(x^2+z^2)^2}$
S_x	$\frac{x^2}{z^2} S_z$
S_{xz}	$\frac{x}{z} S_z$

TABLE A3. Stress Components on the Central Axis Under a Uniform Vertical Pressure of Magnitude p, Over a Circle of Radius a.

Stress Component	Formula
$S_z(r=0)$	$p(1 - u^3)$
$S_r(r=0)$	$\frac{p}{2}[(1+2\eta) - 2(1+\eta)u + u^3]$
$S_\theta(r=0)$	$S_r(r=0)$
$S_{rz}(r=0)$	0

$u = [1 + (a/z)^2]^{-1/2}$
η = Poisson's Ratio

must await future developments, it is still of interest to seek intuitive insight into why the influence function $a(z)$ increases with depth. It is a consequence of the need to keep horizontal stresses in compression [20,22]. With increasing depth, the contact forces must retain a sufficient horizontal component to maintain compression, thereby giving rise to ever greater spread of vertical forces. This is, however, a statement of what must happen rather than why it happens.

Future work in this field should attempt to come not only from below in the sense of looking more deeply at the particulate nature of the material, using probabilistic methods, but also from above, in the sense of attempting to link the stochastic theory with Elasticity Theory, more particularly with the nonlinear version. Before this can be done at a general level, it is necessary to generalize the existing stochastic theory so that it applies to arbitrary body shapes. Establishment of these links should help to eliminate the indeterminateness in the existing theory.

One final remark is of interest. We have assumed that stress transmission in a granular medium is described by a Markov process. The physical content of this assumption is that the vertical stress distribution at a given level, $z + \Delta z$ depends only on the distribution prevailing at a slightly higher level z. This assumption is intuitively reasonable for a truly granular material. It would not, however, be valid if, for example, vertical reinforcing mesh were present in the material.

APPENDIX II – LINEAR ELASTIC SOLUTIONS

We list on Table A1 the linear elastic solutions for a purely vertical point load L, acting at the origin. This is the well-known Boussinesq solution [6]. Also, we give the vertical stress in the presence of an added horizontal surface force along the x axis, of magnitude qL. More complete information on the stress tensor in this case may be found in Reference 46. On Table A2 the stress components are given for a line load of intensity 1 along the y-axis and acting vertically and a horizontal line load also along the y-axis of intensity ql, acting along the x-axis. These are referred to in the section on Theoretical Developments in the main text.

Formulae for the stress components on the central axis under a uniform vertical pressure over a circular area are given on Table A3. More general formulae and tables for stresses in elastic media under distributed loads may be found in Reference 46.

CHAPTER 6

Bearing Capacity of Shallow and Pile Foundations

GEORGE P. KORFIATIS* AND JOHN R. SCHURING**

SHALLOW FOUNDATIONS

Introduction

The basic function of a foundation is to transmit loads imposed by a structure to the underlying soil or rock. They are usually classified into two broad categories:

1. Shallow Foundations, where the depth of embedment is approximately the same as the foundation width.
2. Deep Foundations, where the depth of embedment is at least two to five times larger than the foundation width.

The topic of shallow foundations is presented in the first part of the chapter, followed by deep foundations in the second part.

Three types of shallow foundations are commonly used.

1. Individual or column footings: used to support individual columns.
2. Continuous wall or strip footings: used to support load bearing walls.
3. Combined footings or mat foundations: used to support more than one column or large distributed loads.

The selection of the appropriate shallow foundation type depends on several factors, such as type and magnitude of loading, soil bearing capacity, anticipated settlements and space constraints.

The designer of shallow foundations must ensure that the foundation system meets several requirements associated with safety and functional performance. These requirements include:

1. Safety requirements. Foundation systems must be designed with:
   a. Adequate safety against shear strength failure of the supporting soil which could result in rotation, sliding or overturning of the footing.
   b. Adequate safety against excessive post-construction settlement of the foundation, especially differential movements which are harmful to the superstructure.
   c. Adequate safety against structural failure of the foundation, i.e., compression, tension or shear failure of the footing.
2. Depth requirements. The footings must be installed deep enough to:
   a. Prevent movement due to soil volume changes caused by seasonal freezing and thawing of the ground. Approximate frost depth contours for United States are shown in Figure 1.
   b. Prevent wind and/or water erosion of bearing soils which could result in extensive footing undermining and subsequent failure.
   c. Prevent footing movement or destruction by plant and tree root growth.
   d. Bypass unsuitable soil layers such as peat and muck, expansive clays, soft unconsolidated deposits, and old topsoil layers.
3. Spacing requirements. The foundations must be spaced appropriately in order to prevent distress in adjacent foundations or other underground structures by overstressing the underlying soils.
4. Economic and functional requirements. In general, a foundation system must be designed so that it will be economical to construct, and will perform satisfactorily its intended function.

*Department of Civil and Ocean Engineering, Stevens Institute of Technology, Hoboken, NJ

**Department of Civil and Environmental Engineering, New Jersey Institute of Technology, Newark, NJ

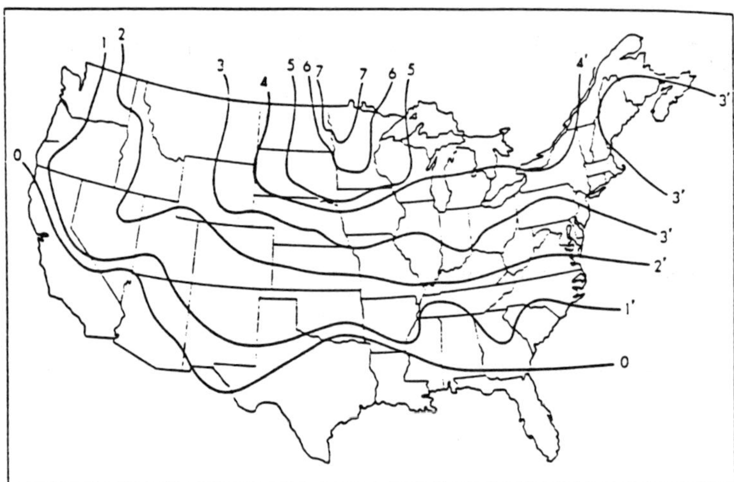

FIGURE 1. Frost depth contours for United States [from Bowles (1982), reprinted with permission].

Behavior of Shallow Foundations

Bearing capacity is the ability of the soil to sustain the imposed loads without shear failure. Consider the square footing of width B, resting on the surface of dense, sandy soil and subjected to an increasing load, Q (Figure 2a). The footing is undergoing a continuous settlement, s. In the plot of the vertical movement of the footing versus the load Q, (Figure 2b), one can observe that an ultimate load Q_{ult} will be reached beyond which the soil fails in shear. The contact pressure corresponding to the ultimate load or $q_u = Q_{ult}/B^2$ is the ultimate bearing capacity of the soil.

Vesic (1963) reported three modes of soil shear failure (Figure 3):

1. General shear failure: characterized by a well defined failure pattern consisting of a wedge and slip surface and bulging of the soil surface adjacent to the footing. This type of failure results in a sudden collapse, accompanied by tilting of the footing.
2. Local shear failure: This failure pattern consists of a wedge and slip failure surface as in the case of general shear but is well defined only under the footing. Soil bulging may take place, but the slip surfaces are not noticeable in the soil surface.
3. Punching shear failure: This failure pattern is not easy to observe. It takes place only in the compression zone immediately below the footing and the surrounding soil mass remains relatively unaffected.

The type of failure mode depends on several factors, including the shear strength and relative compressibility of the soil. In general, it is expected that footings resting on dense incompressible soils will exhibit a general shear failure mode and footings resting on very compressible soils will fail in punching mode. From laboratory model footings in sand, Vesic (1973) has shown that the mode of failure depends on the relative density of the sand and the depth of embedment of the footing (Figure 4).

The ultimate bearing capacity of soils is therefore a very important parameter in the design of shallow foundations. There are four methods commonly used for evaluation of the bearing capacity of soils:

1. Analytical Methods (using the bearing capacity equations)
2. Correlation with field test data (such as the Standard Penetration Resistance Test and the Dutch Cone Penetrometer Test)
3. Plate load tests (on-site determinations of the ultimate load)
4. Presumptive Bearing Capacity (recommended bearing capacities in various codes)

These methods will be described in the following sections of this chapter along with several factors that influence the bearing capacity of soils.

The designer must realize that evaluation of the soil bearing capacity using any of the above methods does not yield an exact or definitive answer. Limitations associated with simplifying assumptions, scale effects, boundary conditions, and uncertainties in soil properties require that caution and engineering judgement be exercised before adopting computed bearing capacity values in the final design.

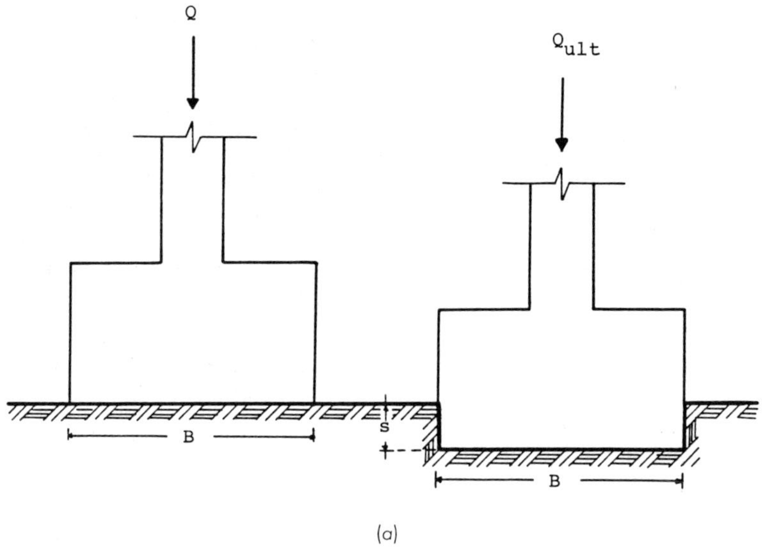

(a)

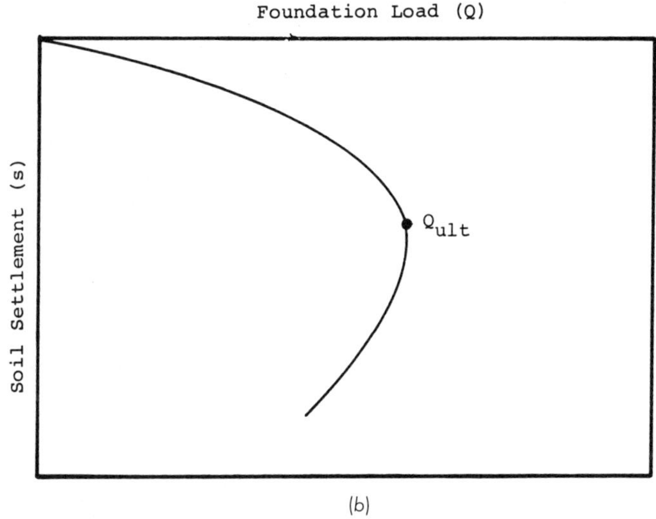

(b)

FIGURE 2. (a) Rectangular footing subjected to increasing load Q; (b) plot of vertical footing movement versus load Q.

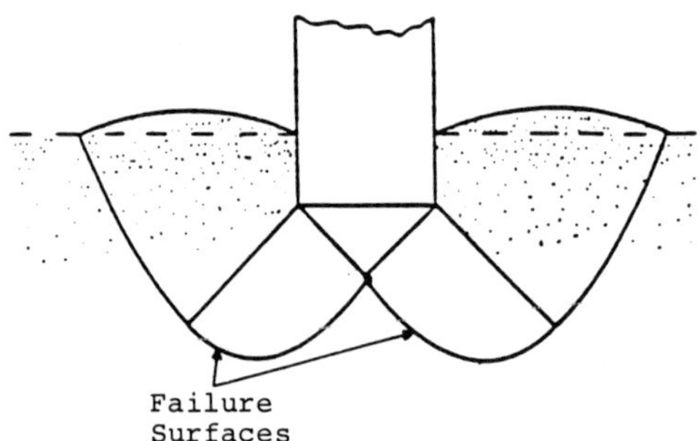

(a) General Shear Failure Pattern

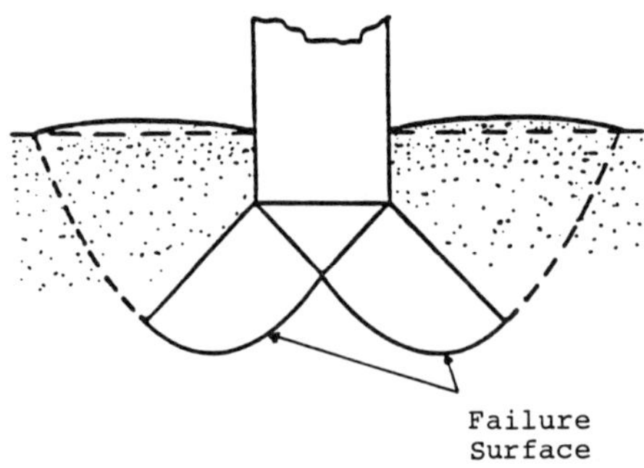

(b) Local Shear Failure Pattern

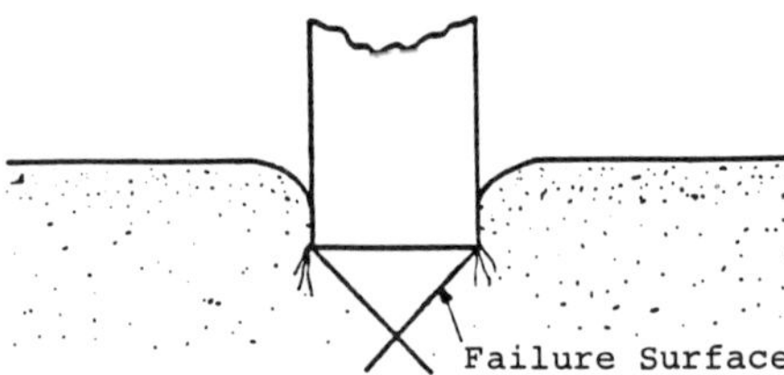

(c) Punching Shear Failure Pattern

FIGURE 3. Modes of shear failure: (a) general shear failure pattern; (b) local shear failure pattern; (c) punching shear failure pattern (after Vesic 1973).

Bearing Capacity Equations

Terzaghi (1943) presented the first set of equations for evaluation of the ultimate bearing capacity of shallow foundations. The equations are based on formulations presented by Prandtl in the early 1900s, derived by using the Theory of Plasticity. The bearing capacity equation was derived for a continuous (strip), rough-bottom foundation of infinite length for general shear failure conditions and was subsequently modified for estimation of the capacity of square and circular foundations.

Terzaghi's bearing capacity equations are based on the soil foundation system depicted in Figure 5.

The failure area below the foundation consists of three distinct zones:

1. Zones I (Triangular Rankine Passive Zones)
2. Zones II (Radial Shear Zones)
3. Zone III (Triangular Shear Zone)

The equations are as follows:

$$q_u = cN_c + \bar{q}N_q + 0.5\gamma BN_\gamma \quad \text{(Strip Foundation)} \tag{1}$$

$$q_u = 1.3cN_c + \bar{q}N_q + 0.4\gamma BN_\gamma \quad \text{(Square Foundation)} \tag{2}$$

$$q_u = 1.3cN_c + \bar{q}N_q + 0.3\gamma DN_\gamma \quad \text{(Circular Foundation)} \tag{3}$$

where

q_u = ultimate bearing capacity
c = soil cohesion
γ = unit weight of soil
$\bar{q}$ = overburden pressure = $D_e\gamma$.
B = foundation width
D = foundation diameter
N_c, N_q, N_γ = bearing capacity factors

$$N_c = \cot\phi\left[\frac{e^{2(3\pi/4-\phi/2)\tan\phi}}{2\cos^2(45+\phi/2)} - 1\right] \tag{4}$$

$$N_q = \frac{e^{2(3\pi/4-\phi/2)\tan\phi}}{2\cos^2(45+\phi/2)} \tag{5}$$

$$N_\gamma = \frac{1}{2}\left[\frac{K_{p\gamma}}{\cos^2\phi} - 1\right]\tan\phi \tag{6}$$

where

$K_{p\gamma}$ = passive earth pressure coefficient
ϕ = angle of internal friction of soil

More details on the derivation of Equations (1) through (6) can be found in Terzaghi (1943).

The bearing capacity factors depend on the angle of internal friction of the soil and are tabulated in Table 1, for $\phi = 0°$ to $50°$. For the local shear failure mode, Terzaghi

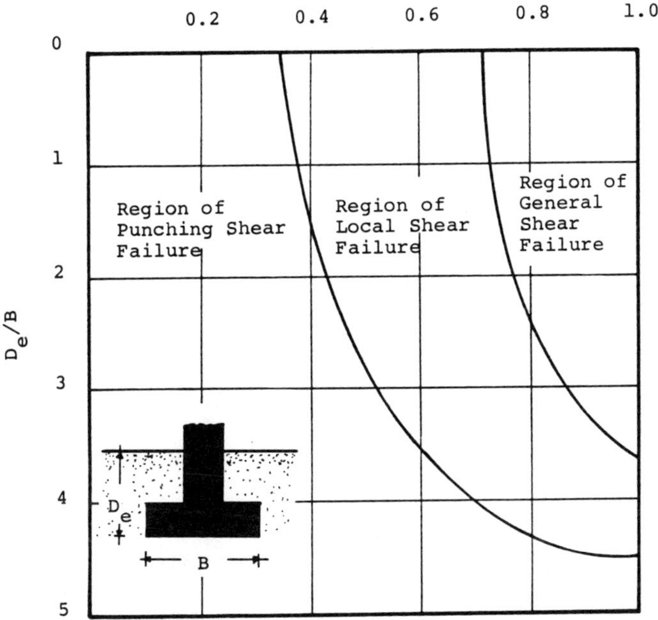

FIGURE 4. Modes of shear failure in relation to depth of embedment and relative density for footings in sand [after Vesic (1973)].

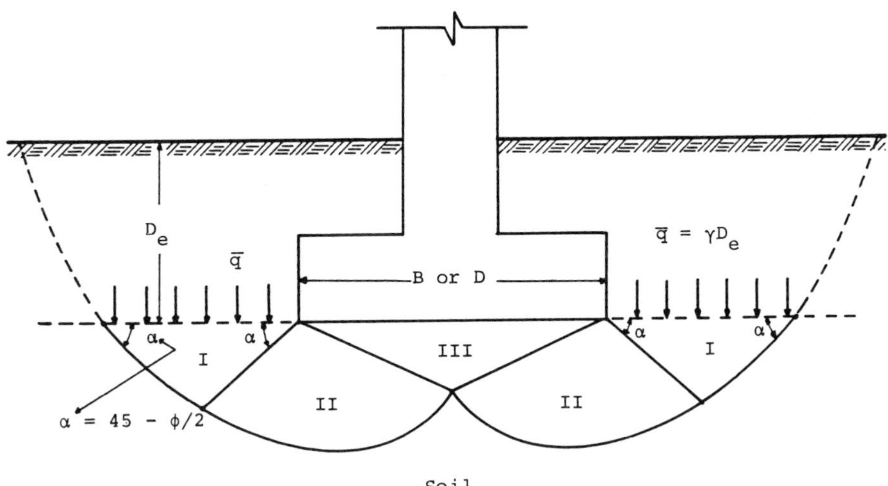

FIGURE 5. General shear bearing capacity failure under a rough rigid foundation.

TABLE 1. Bearing Capacity Factors for Use With Terzaghi's Bearing Capacity Equations.

ϕ	N_q	N_c	N_γ
0.00	0.00	5.70	0.00
1.00	1.10	5.95	0.09
2.00	1.22	6.27	0.17
3.00	1.35	6.60	0.27
4.00	1.49	6.95	0.38
5.00	1.64	7.32	0.49
6.00	1.81	7.71	0.62
7.00	2.00	8.14	0.76
8.00	2.21	8.59	0.91
9.00	2.44	9.07	1.07
10.00	2.69	9.59	1.25
11.00	2.92	10.15	1.45
12.00	3.28	10.75	1.67
13.00	3.63	11.39	1.92
14.00	4.01	12.09	2.21
15.00	4.44	12.84	2.54
16.00	4.92	13.66	2.91
17.00	5.45	14.54	3.33
18.00	6.04	15.50	3.81
19.00	6.69	16.54	4.35
20.00	7.43	17.67	4.97
21.00	8.25	18.90	5.68
22.00	9.18	20.25	6.48
23.00	10.22	21.72	7.41
24.00	11.39	23.33	8.47
25.00	12.70	25.10	9.70
26.00	14.19	27.05	11.13
27.00	15.88	29.20	12.80
28.00	17.78	31.57	14.75
29.00	19.95	34.19	16.04
30.00	22.42	37.11	19.73
31.00	25.25	40.35	22.89
32.00	28.48	43.97	26.61
33.00	32.18	48.02	31.01
34.00	36.45	52.56	36.23
35.00	41.38	57.66	42.43
36.00	47.08	63.43	49.87
37.00	53.71	69.95	58.86
38.00	61.45	77.37	69.84
39.00	70.50	85.82	83.40
40.00	81.13	95.50	100.39
41.00	93.68	106.62	121.97
42.00	108.56	119.45	149.75
43.00	126.27	134.33	186.00
44.00	147.46	151.66	233.82
45.00	172.15	171.95	297.51
46.00	203.79	195.83	382.96
47.00	241.31	224.10	498.21
48.00	287.26	257.75	654.23
49.00	343.91	298.08	865.73
50.00	414.25	346.75	1153.18

modified the equations as follows:

$$q_u = 0.667cN_c^\circ + qN_q^\circ + 0.5\gamma BN_\gamma^\circ \quad (7)$$

(Strip Foundations)

$$q_u = 0.867cN_c^\circ + qN_q^\circ + 0.4\gamma BN_\gamma^\circ \quad (8)$$

(Square Foundations)

$$q_u = 0.867cN_c^\circ + qN_q^\circ + 0.3\gamma DN_\gamma \quad (9)$$

(Circular Foundations)

The modified bearing capacity factors N_c°, N_q°, and N_γ° are computed by replacing ϕ with ϕ° = arctan (2/3 tanϕ), in Equations (4), (5) and (6).

Das (1984) suggested that the modified bearing capacity factor N_q° is better estimated by using the expression presented by Vesic (1963):

$$N_q^\circ = (e^{3.8\ \tan\phi}) \tan^2 (45 + \phi/2) \quad (10)$$

The modified bearing capacity factors have been plotted versus ϕ as shown in Figure 6.

It is usually not necessary to design foundations for local shear failure, since settlement almost always controls the design for soils where local shear failure is critical.

The Terzaghi bearing capacity equations have been extensively used in the design of shallow foundations. They are considered to be very conservative and they have not taken into account all of the factors that influence the bearing capacity. Such factors include the shear resistance of the overburden soils, shape effects for rectangular footings, and inclined loading conditions. A general bearing capacity equation was suggested by Meyerhof (1963), and several other authors have contributed in further developing various terms of the equation (Hansen, 1970; DeBeer, 1970; Vesic, 1973). The general bearing capacity equation is of the form:

$$q_u = cN_c'S_cD_cI_cG_cB_c + \bar{q}N_q'S_qD_qI_qG_qB_q + \tfrac{1}{2}\gamma BN_\gamma'S_\gamma D_\gamma I\gamma G_\gamma B_\gamma \quad (11)$$

where

c = soil cohesion
N_c', N_q', N_γ' = bearing capacity factors
S_c, S_q, S_γ = shape factors
D_c, D_q, D_γ = depth factors
I_c, I_q, I_η = inclination factors
G_c, G_q, G_γ = ground factors
B_c, B_q, B_γ = base factors
$\bar{q}$ = effective overburden pressure

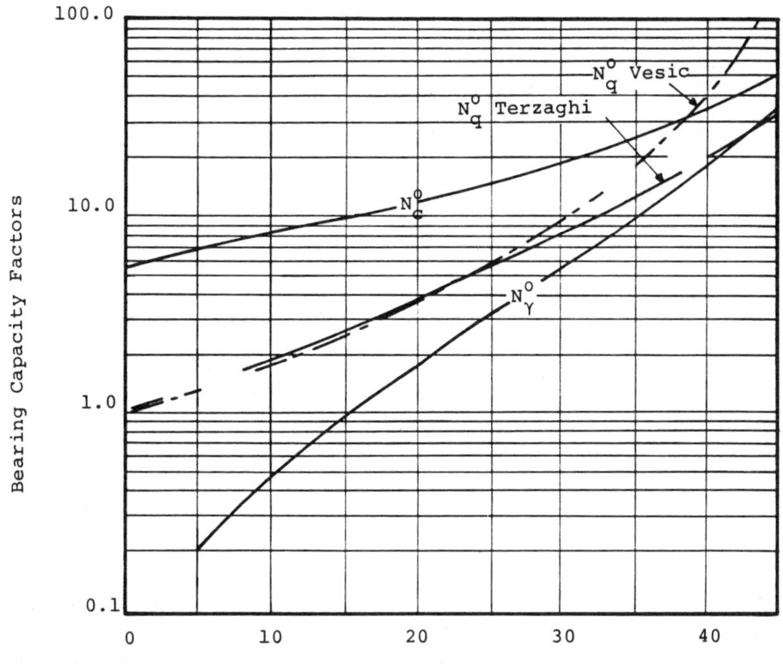

FIGURE 6. Bearing capacity factors for modified bearing capacity equations [redrawn from Das (1984)].

BEARING CAPACITY FACTORS

The bearing capacity factors for the general bearing capacity equations are different from those used in Terzaghi's equations. The following expressions are most widely used (Vesic 1973):

$$N_q' = \tan^2 (45 + \phi/2)\, e^{\pi \tan \phi} \quad (12)$$

$$N_c' = (N_q' - 1) \cot \phi \quad (13)$$

$$N_\gamma' = 2(N_q' + 1) \tan \phi \quad (14)$$

The values of the bearing capacity factors based on the above equations are given in Table 2.

Two other expressions for N_γ' found in the literature are:

$$N_\gamma' = (N_q - 1) \tan (1.4\phi) \quad (15)$$

$$N_\gamma' = 1.5(N_q - 1) \tan \phi \quad (16)$$

Equation (15) is based on Meyerhof's (1951) analysis, and Equation (16) is based on Hansen's (1970) analysis.

SHAPE FACTORS

The effect of the foundation shape on the bearing capacity has been studied in laboratory tests. The following expressions are recommended by DeBeer (1970):

$$S_c = 1 + (B/L)(N_q'/N_c') \quad (17)$$

$$S_q = 1 + (B/L) \tan \phi \quad (18)$$

$$S_\gamma = 1 - 0.4(B/L) \quad (19)$$

where

B = foundation width
L = foundation length ($L > B$)

The values of the shape factors are tabulated in Tables 3, 4 and 5.

DEPTH FACTORS

The depth factors are used in order to take into account the shearing resistance along the failure surface in the soil located above the base of the footing.

Meyerhof has suggested the following expressions (Bowles, 1982):

$$D_c = 1 + 0.2\, K_p\, (D_e/B) \quad (20)$$

$$D_q = D_\gamma = 1 \text{ for } \phi < 10° \quad (21)$$

$$D_q = D_\gamma = 1 + 0.1 \sqrt{K_p}\, (D_e/B) \quad (22)$$

TABLE 2. Bearing Capacity Factors for the General Bearing Capacity Equation [After Vesic (1973)].

ϕ	N_c'	N_q'	N_γ'
0	5.14	1.00	0.00
1	5.38	1.09	0.07
2	5.63	1.20	0.15
3	5.90	1.31	0.24
4	6.19	1.43	0.34
5	6.49	1.57	0.45
6	6.81	1.72	0.57
7	7.16	1.88	0.71
8	7.53	2.06	0.86
9	7.92	2.25	1.03
10	8.35	2.47	1.22
11	8.80	2.71	1.44
12	9.28	2.97	1.69
13	9.81	3.26	1.97
14	10.37	3.59	2.29
15	10.98	3.94	2.65
16	11.63	4.34	3.06
17	12.34	4.77	3.53
18	13.10	5.26	4.07
19	13.93	5.80	4.68
20	14.83	6.40	5.39
21	15.82	7.07	6.20
22	16.88	7.82	7.13
23	18.05	8.66	8.20
24	19.32	9.60	9.44
25	20.72	10.66	10.88
26	22.25	11.85	12.54
27	23.94	13.20	14.47
28	25.80	14.72	16.72
29	27.86	16.44	19.34
30	30.14	18.40	22.40
31	32.67	20.63	25.99
32	35.49	23.18	30.22
33	38.64	26.09	35.19
34	42.16	29.44	41.06
35	46.12	33.30	48.03
36	50.59	37.75	56.31
37	55.63	42.92	66.19
38	61.35	48.93	78.03
39	67.87	55.96	92.25
40	75.31	64.20	109.41
41	83.86	73.90	130.22
42	93.71	85.38	155.55
43	105.11	99.02	186.54
44	118.37	115.31	224.64
45	133.88	134.88	271.76
46	152.10	158.51	330.35
47	173.64	187.21	403.67
48	199.26	222.31	496.01
49	229.93	265.51	613.16
50	266.89	319.07	762.89

where

K_p = $\tan^2 (45 + \phi/2)$ = coefficient of passive resistance
D_e = depth of embedment of the footing

Meyerhof also suggested that in cases where the value of the angle of internal friction is obtained from triaxial tests, it must be modified to account for plane strain conditions as follows:

$$\phi_{ps} = (1.1 - 0.1\, B/L)\, \phi_{tr} \quad (23)$$

Hansen (1970) proposed the following expressions for depth factors:

$$D_c = 1 + 0.4\,(D_e/B), \text{ for } D_e/B \leq 1 \quad (24)$$

$$D_c = 1 + 0.4 \tan^{-1}(D_e/B), \text{ for } D_e/B > 1 \quad (25)$$

$$D_q = 1 + 2 \tan\phi\,(1 - \sin\phi)^2\,(D_e/B), \text{ for } D_e/B \leq 1 \quad (26)$$

$$D_q = 1 + 2 \tan\phi\,(1 - \sin\phi)^2 \tan^{-1}(D_e/B) \quad \text{for } D_e/B > 1 \quad (27)$$

$$D_\gamma = 1 \text{ for all } D_e/B \text{ and all } \phi \quad (28)$$

For $\phi < 0$ and by using the appropriate D_q, D_c can be computed as:

$$D_c = D_q - \frac{1 - D_q}{N_q' \tan\phi} \quad (29)$$

INCLINATION FACTORS

In many engineering applications, the Q load applied on the footing is inclined, as shown in Figure 7. Laboratory experiments have shown that load inclination reduces the ultimate bearing capacity of the soil. In order to compensate for this reduction, inclination factors are introduced to the general bearing capacity equation [Equation (11)].

Meyerhof (1951, 1953) suggested the following expressions:

$$I_q = I_c = 1 - (\alpha/90°) \quad (30)$$

$$I_\gamma = (1 - \alpha/\phi)^2 \quad (31)$$

where α is the inclination angle of the load in degrees with respect to the vertical. Hansen (1970), reporting work done at the Danish Geotechnical Institute, presented the following equations for the inclination factors:

$$I_q = \left[1 - \frac{Q \sin\alpha}{2(Q\cos\alpha + BLc\cot\phi)}\right]^5 \quad (32)$$

$$I_c = I_q - \frac{(1 - I_q)}{(N_q' - 1)} \quad (33)$$

$$I_\gamma = \left[1 - \frac{0.7\, Q\sin\alpha}{Q\cos\alpha + BLc\cot\phi} \right]^5 \quad (34)$$

where L is the length of the footing.

Vesic (1975) suggested that the orientation of the load inclination and the shape of the footing affect the inclination factors. Based on the results of experimental investigations, he modified Equations (32) and (34) as follows:

$$I_q = \left[1 - \frac{Q\sin\alpha}{Q\cos\alpha + BLc\cot\phi} \right]^m \quad (35)$$

$$I = \left[1 - \frac{Q\sin\alpha}{Q\cos\alpha + BLc\cot\phi} \right]^{m+1} \quad (36)$$

where the exponent m is given by either:

$$m_B = (2 + B/L)/(1 + B/L) \quad (37)$$

for cases where the inclination is along the direction of the footing width, B, or:

$$m_L = (2 + L/B)/(1 + L/B) \quad (38)$$

for cases where the inclination is along the direction j for the footing length, L.

In cases where the inclination is along a direction other than the above (see Figure 8), Vesic suggests the use of an interpolated value for the exponent m as follows:

$$m_j = m_L \cos^2\alpha_j + m_B \sin^2\alpha_j \quad (39)$$

where α_j is the angle between the load inclination direction, j, and the direction of the footing length L (see Figure 8).

In case of a foundation shape other than rectangular, an equivalent rectangle may be constructed to compute the effective area as suggested by Hansen (1961).

GROUND AND BASE INCLINATION FACTORS

In many practical engineering situations the base of the foundation may be designed to be inclined at a certain angle ψ, with respect to the horizontal as shown in Figure 9. This condition may be desirable when large horizontal loads are

TABLE 3. Values of S_c Recommended by DeBeer, 1970.

B/L ϕ	0.1	0.2	0.3	0.4	0.5	0.6	0.7	0.8	0.9	1.0
0.0	1.019	1.039	1.058	1.078	1.097	1.117	1.136	1.156	1.175	1.195
2.0	1.021	1.043	1.064	1.086	1.107	1.129	1.150	1.171	1.193	1.214
4.0	1.023	1.046	1.070	1.093	1.116	1.139	1.163	1.186	1.209	1.232
6.0	1.025	1.050	1.076	1.101	1.126	1.151	1.177	1.202	1.227	1.252
8.0	1.027	1.055	1.082	1.110	1.137	1.164	1.192	1.219	1.246	1.274
10.0	1.030	1.059	1.089	1.119	1.148	1.178	1.208	1.237	1.267	1.296
12.0	1.032	1.064	1.096	1.128	1.160	1.192	1.224	1.256	1.288	1.321
14.0	1.035	1.069	1.104	1.138	1.173	1.208	1.242	1.277	1.311	1.346
16.0	1.037	1.075	1.112	1.149	1.186	1.224	1.261	1.298	1.336	1.373
18.0	1.040	1.080	1.120	1.161	1.201	1.241	1.281	1.321	1.361	1.401
20.0	1.043	1.086	1.129	1.173	1.216	1.259	1.302	1.345	1.388	1.432
22.0	1.046	1.093	1.139	1.185	1.232	1.278	1.324	1.371	1.417	1.463
24.0	1.050	1.099	1.149	1.199	1.249	1.298	1.348	1.398	1.447	1.497
26.0	1.053	1.107	1.160	1.213	1.266	1.320	1.373	1.426	1.479	1.533
28.0	1.057	1.114	1.171	1.228	1.285	1.342	1.399	1.456	1.513	1.571
30.0	1.061	1.122	1.183	1.244	1.305	1.366	1.427	1.488	1.550	1.611
32.0	1.065	1.131	1.196	1.261	1.327	1.392	1.457	1.522	1.588	1.653
34.0	1.070	1.140	1.209	1.279	1.349	1.419	1.489	1.559	1.628	1.698
36.0	1.075	1.149	1.224	1.299	1.373	1.448	1.522	1.597	1.672	1.746
38.0	1.080	1.160	1.239	1.319	1.399	1.479	1.558	1.638	1.718	1.798
40.0	1.085	1.170	1.256	1.341	1.426	1.511	1.597	1.682	1.767	1.852
42.0	1.091	1.182	1.273	1.364	1.456	1.547	1.638	1.729	1.820	1.911
44.0	1.097	1.195	1.292	1.390	1.487	1.584	1.682	1.779	1.877	1.974
46.0	1.104	1.208	1.313	1.417	1.521	1.625	1.729	1.834	1.938	2.042
48.0	1.112	1.223	1.335	1.446	1.558	1.669	1.781	1.893	2.004	2.116
50.0	1.120	1.239	1.359	1.478	1.598	1.717	1.837	1.956	2.076	2.196

TABLE 4. Values of S_q Recommended by DeBeer, 1970.

B/L ϕ	0.1	0.2	0.3	0.4	0.5	0.6	0.7	0.8	0.9	1.0
0.0	1.000	1.000	1.000	1.000	1.000	1.000	1.000	1.000	1.000	1.000
2.0	1.003	1.007	1.010	1.014	1.017	1.021	1.024	1.028	1.031	1.035
4.0	1.007	1.014	1.021	1.028	1.035	1.042	1.049	1.056	1.063	1.070
6.0	1.011	1.021	1.032	1.042	1.053	1.063	1.074	1.084	1.095	1.105
8.0	1.014	1.028	1.042	1.056	1.070	1.084	1.098	1.112	1.126	1.141
10.0	1.018	1.035	1.053	1.071	1.088	1.106	1.123	1.141	1.159	1.176
12.0	1.021	1.043	1.064	1.085	1.106	1.128	1.149	1.170	1.191	1.213
14.0	1.025	1.050	1.075	1.100	1.125	1.150	1.175	1.199	1.224	1.249
16.0	1.029	1.057	1.086	1.115	1.143	1.172	1.201	1.229	1.258	1.287
18.0	1.032	1.065	1.097	1.130	1.162	1.195	1.227	1.260	1.292	1.325
20.0	1.036	1.073	1.109	1.146	1.182	1.218	1.255	1.291	1.328	1.364
22.0	1.040	1.081	1.121	1.162	1.202	1.242	1.283	1.323	1.364	1.404
24.0	1.045	1.089	1.134	1.178	1.223	1.267	1.312	1.356	1.401	1.445
26.0	1.049	1.098	1.146	1.195	1.244	1.293	1.341	1.390	1.439	1.488
28.0	1.053	1.106	1.160	1.213	1.266	1.319	1.372	1.425	1.479	1.532
30.0	1.058	1.115	1.173	1.231	1.289	1.346	1.404	1.462	1.520	1.577
32.0	1.062	1.125	1.187	1.250	1.312	1.375	1.437	1.500	1.562	1.625
34.0	1.067	1.135	1.202	1.270	1.337	1.405	1.472	1.540	1.607	1.675
36.0	1.073	1.145	1.218	1.291	1.363	1.436	1.509	1.581	1.654	1.727
38.0	1.078	1.156	1.234	1.313	1.391	1.469	1.547	1.625	1.703	1.781
40.0	1.084	1.168	1.252	1.336	1.420	1.503	1.587	1.671	1.755	1.839
42.0	1.090	1.180	1.270	1.360	1.450	1.540	1.639	1.720	1.810	1.900
44.0	1.097	1.193	1.290	1.386	1.483	1.579	1.676	1.773	1.869	1.966
46.0	1.104	1.207	1.311	1.414	1.518	1.621	1.725	1.828	1.932	2.036
48.0	1.111	1.222	1.333	1.444	1.555	1.666	1.777	1.888	2.000	2.111
50.0	1.119	1.238	1.358	1.477	1.596	1.715	1.834	1.953	2.073	2.190

TABLE 5. Values of S_γ Recommended by Debeer (1970).

B/L	S_γ
0.1	0.96
0.2	0.92
0.3	0.88
0.4	0.84
0.5	0.80
0.6	0.76
0.7	0.72
0.8	0.68
0.9	0.64
1.0	0.60

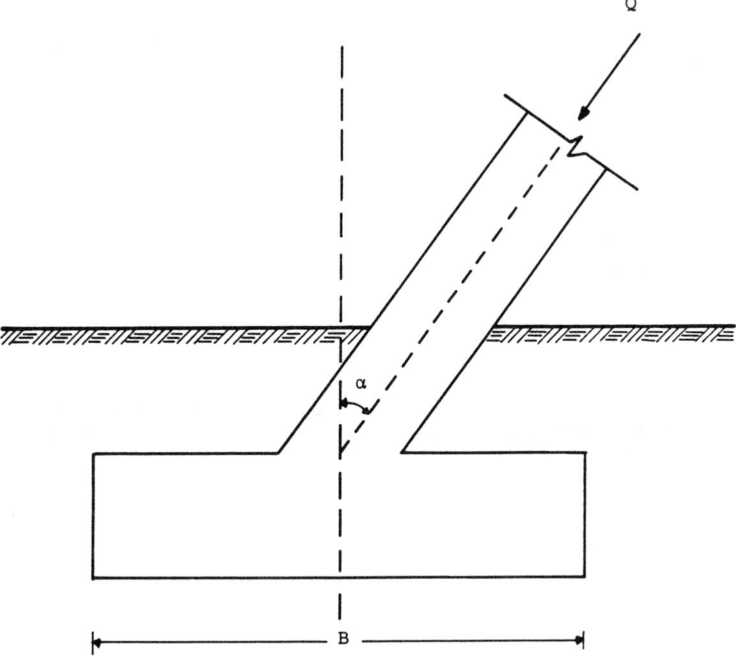

FIGURE 7. Typical foundation with inclined load.

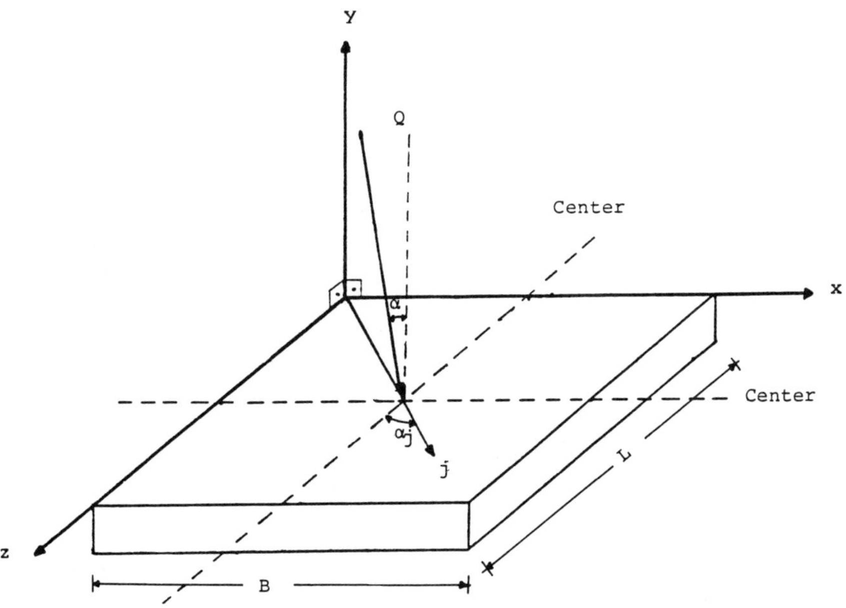

FIGURE 8. Definition sketch for Equation (39).

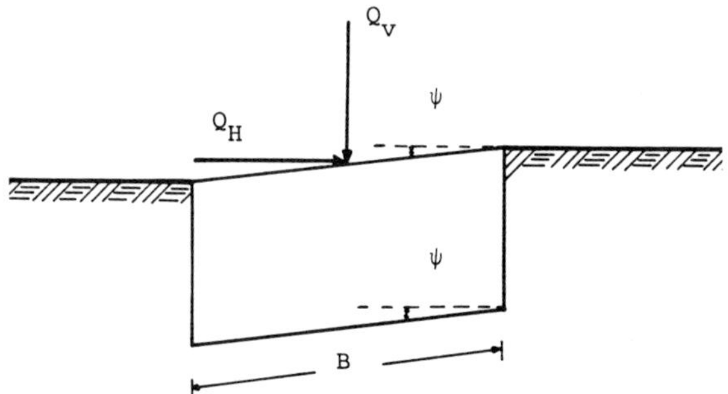

FIGURE 9. Definition sketch for a footing with inclined base.

to be carried by the footing. The following base inclination factors have been proposed by Vesic (1970) to be used in the general bearing capacity equation:

$$B_q = B_\gamma = (1 - \psi \tan\phi)^2 \quad (40)$$

and

$$B_c = 1 - [2\psi/(\pi + 2)] \quad (41)$$

where ψ is in radians.

Equations (40) and (41) can be used if $\psi < 45°$. Other expressions that have been suggested are (Bowles 1982):

$$B_q = B_\gamma = e^{(-2\psi \tan\phi)} \quad (42)$$

and

$$B_c = 1 - \psi°/147° \quad (43)$$

In certain cases the surface of the ground below which the foundation is placed may be sloping at an angle θ as shown in Figure 10.

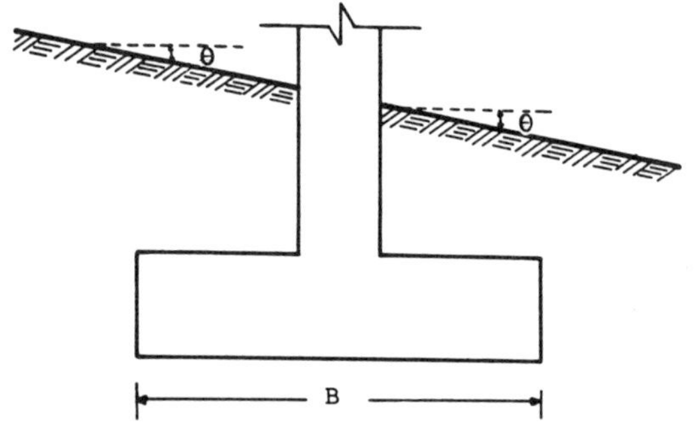

FIGURE 10. Definition sketch of a footing placed on an inclined ground surface.

Hansen (1970) proposed the following ground factors be used in the general bearing capacity equation to account for ground slope.

$$G_q = G_\gamma = (1 - \tan\theta)^2 \quad (44)$$

and

$$G_c = 1 - [2\theta/(\pi + 2)] \quad (45)$$

where θ is in radians.

Equations (44) and (45) are applicable when $\theta < 45°$ and $\theta < \phi$. In cases where the ground inclination angle θ is larger than $\phi/2$, a slope stability analysis must be performed to assess the overall slope condition.

Other expressions for ground factors found in the literature are (Bowles 1982):

$$G_q = G_\gamma = (1 - 0.5 \tan\theta)^5 \quad (46)$$

and

$$G_c = 1 - (\theta°/147°) \quad (47)$$

When the conditions shown in Figures 9 and 10 co-exist, both the ground factors and the foundation base inclination factors must be used in the general bearing capacity equation.

It must be pointed out that the equations for ground and base inclination factors have been derived for plane strain conditions and are applicable strictly speaking only for strip foundations. In the absence of equations for other foundations shapes, however, these equations can be used adequately.

OTHER FACTORS INFLUENCING BEARING CAPACITY

Besides the factors discussed in the preceding sections, several investigations have shown that several additional parameters may effect the bearing capacity. These investigations are not general enough to be included in this text, but these factors will be mentioned for the sake of completeness. They are:

a. Compressibility of underlying soils
b. Shape and roughness of footing bottom
c. Rate of loading
d. Rigidity of footing
e. Influence of adjacent footings

Eccentrically Loaded Foundations

The equations described, up to this point, estimate the bearing capacity of foundations subjected to centric loads. In many cases of practical interest, however, the loads are applied eccentrically on the footing, thus introducing a moment in addition to the vertical and horizontal loads that may exist. As a result, the footing contact pressure distribution is not uniform (Figure 11). In this case, the maximum and minimum contact pressure can be computed from:

$$q_{max,\ min} = \frac{Q}{BL}\left[1 \pm \frac{6e_B}{B}\right] \tag{48}$$

where e_B is the eccentricity along the direction of the footing width, B.

The eccentricity can be one way, i.e., along the direction of either B or L (Figure 12a), or two way, i.e., along both directions (Figure 12b).

It can be seen from Equation (48) that when $e > B/6$, the minimum contact pressure becomes negative. Since soil cannot take tension, uplift of the footing may occur. Therefore, in most applications, it is recommended that footing be designed so that $e \leq B/6$.

Theoretical computations and experiments have shown that the effect of eccentricity can be accounted for by introducing the effective footing dimensions in the computation of the ultimate bearing capacity, such as:

$$B_{eff} = B - 2e_B \tag{49}$$

$$L_{eff} = L - 2e_L \tag{50}$$

$$A_{eff} = L_{eff}\, B_{eff} \tag{51}$$

In order to compute the ultimate bearing capacity of eccentrically loaded rectangular foundations, the effective width, B_{eff}, is used in the third term of the general bearing capacity equation [Equation (11)]. In addition, all the bearing capacity equation factors are computed based on the effective dimensions, rather than the actual dimensions of the footing. Note that the smallest of the two effective dimensions becomes the effective width of the foundation.

If the load is inclined, in addition to being eccentric, the inclination factors are computed based on the effective dimensions. Note that Equations (37) and (38) are computed using the actual footing dimensions.

Effects of the Groundwater Table

The location of the groundwater table relative to the foundation may have a significant effect on the bearing capacity of the soil. This effect stems from two conditions. First, the reduction of the unit weight of the soil due to the submergence and second, the loss of the apparent cohesion of the soil due to the elimination of the capillary stresses.

The general bearing capacity equation [Equation (11)] was derived assuming dry soil conditions. The terms of the equation must be therefore corrected for the effect of submergence. There are two cases where correction for submergence is required. In the first case, the groundwater surface is above the bottom of the foundation (Figure 13a) For this condition, the effective overburden, $\bar{q}$, in the second term of the equation is computed as follows:

$$\bar{q} = D_1\gamma + D_2\gamma_{sub} \tag{52}$$

where

γ_{sub} = submerged unit weight = $\gamma_{sat} - \gamma_w$
γ_{sat} = saturated unit weight
γ_w = unit weight of water

In the second case, the groundwater surface is located at distance z_w below the bottom of the foundation (Figure 13b). For this condition, the overburden is computed as:

$$\bar{q} = \gamma D_e \tag{53}$$

and the unit weight (γ) in the third term of the equation is replaced by:

$$\gamma' = \gamma_{sub} + z_w(\gamma - \gamma_{sub})/B \tag{54}$$

It is apparent that the bearing capacity need only be reduced when the groundwater table is in the proximity of the footing. If the groundwater table is located a distance larger than the depth of the failure wedge zone below the footing, the effect of submergence can be neglected and no correction is required. In cases where foundations are located in areas of fluctuating groundwater tables, the highest groundwater elevation should be considered for the computation of the bearing capacity.

In the preceding discussion, the groundwater has been considered static and no seepage forces have been taken into account. If the groundwater is moving, the resulting seepage forces will add an additional force component in the direction of the flow. The magnitude of the seepage force component is $\gamma_w i$, where i is the hydraulic gradient causing the flow.

Bearing Capacity of Layered Soils

The bearing capacity equation [Equation (11)] has been derived for homogeneous and isotropic soils. In cases where shallow foundations are placed on layered deposits, the properties of all soils present within the shearing failure surface must be considered in the computation of the ultimate bearing capacity.

In this section, some two-layered soil profiles frequently encountered in practice will be addressed.

CASE I: Cohesive Soils

a. Bearing stratum is soft clay and underlying stratum is stiff clay (Figure 14a).
b. Bearing stratum is stiff clay and underlying stratum is soft clay (Figure 14b).

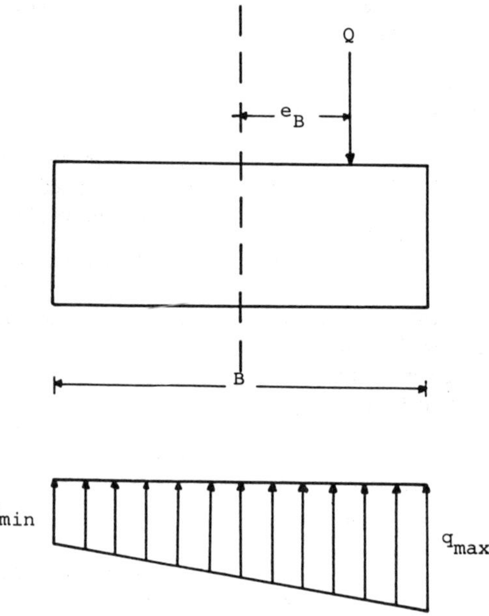

FIGURE 11. Pressure distribution below an eccentrically loaded footing.

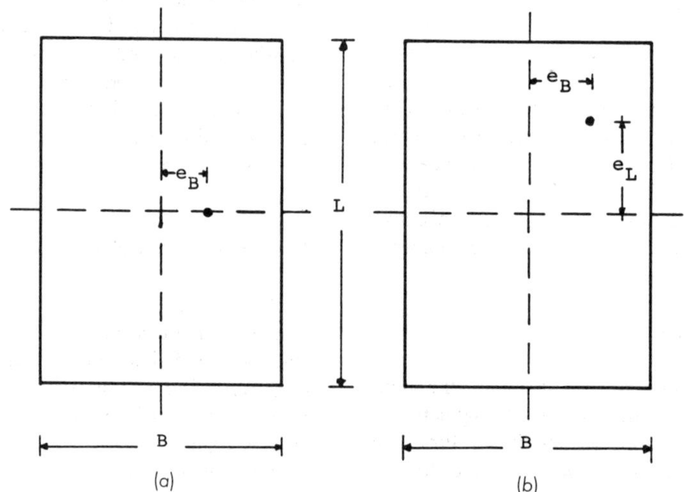

FIGURE 12. (a) One way eccentricity; (b) Two way eccentricity.

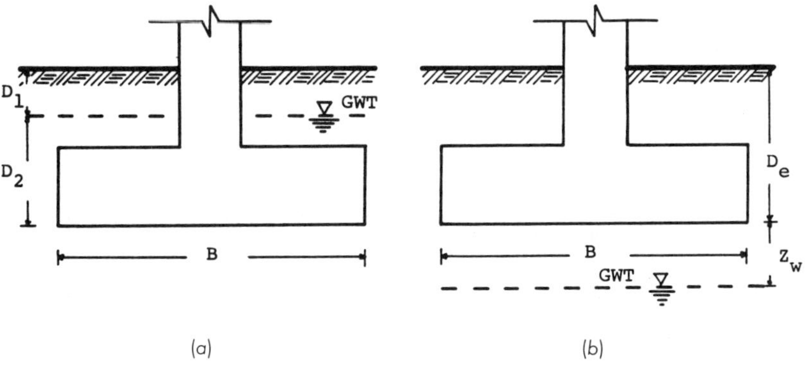

FIGURE 13. Typical locations of the groundwater table relative to foundations.

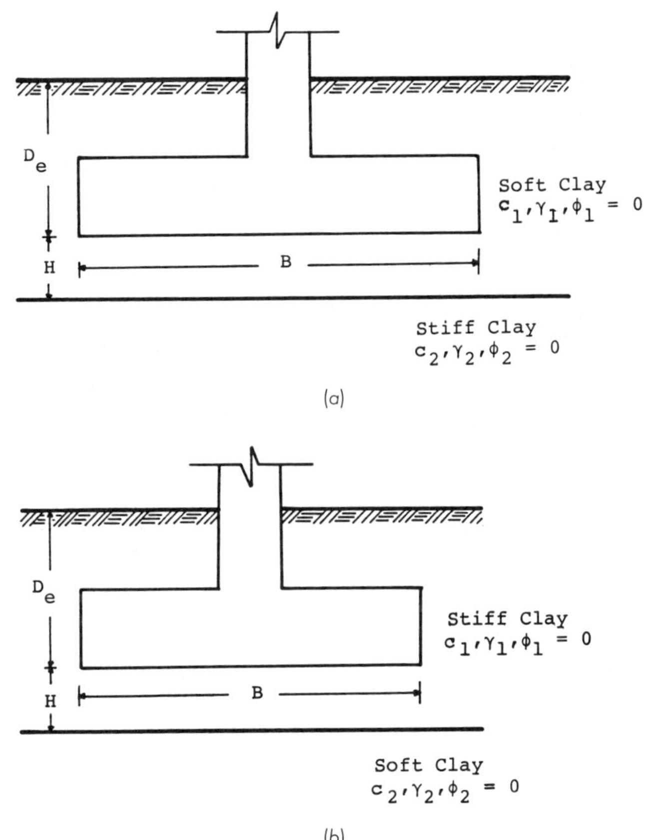

FIGURE 14. Foundation on a two layer clay soil.

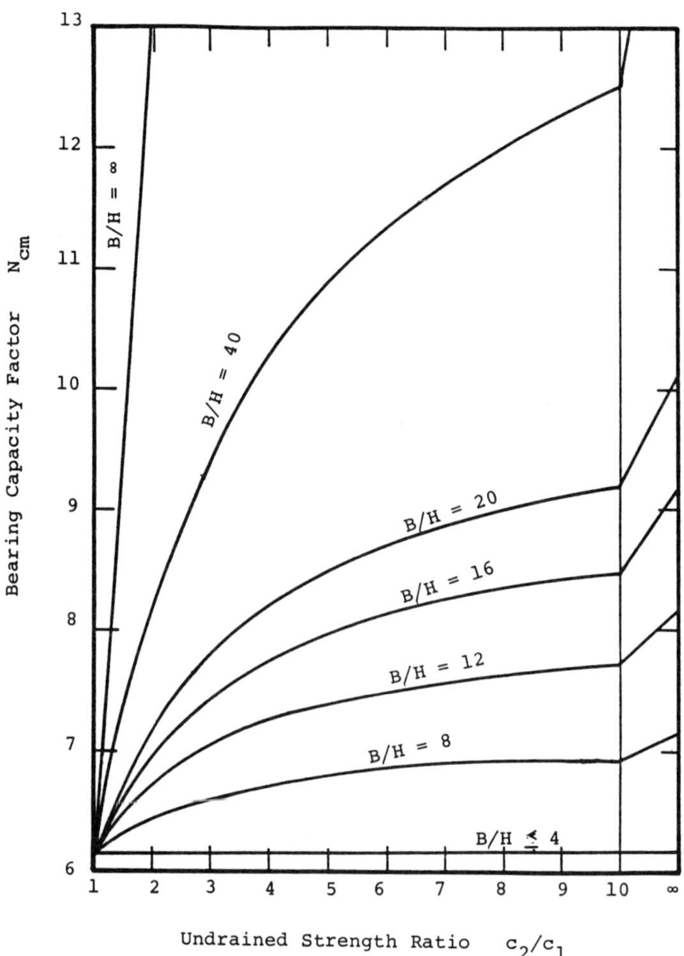

FIGURE 15. Modified bearing capacity factor N_{cm} for square and circular footings [after Vesic (1975)].

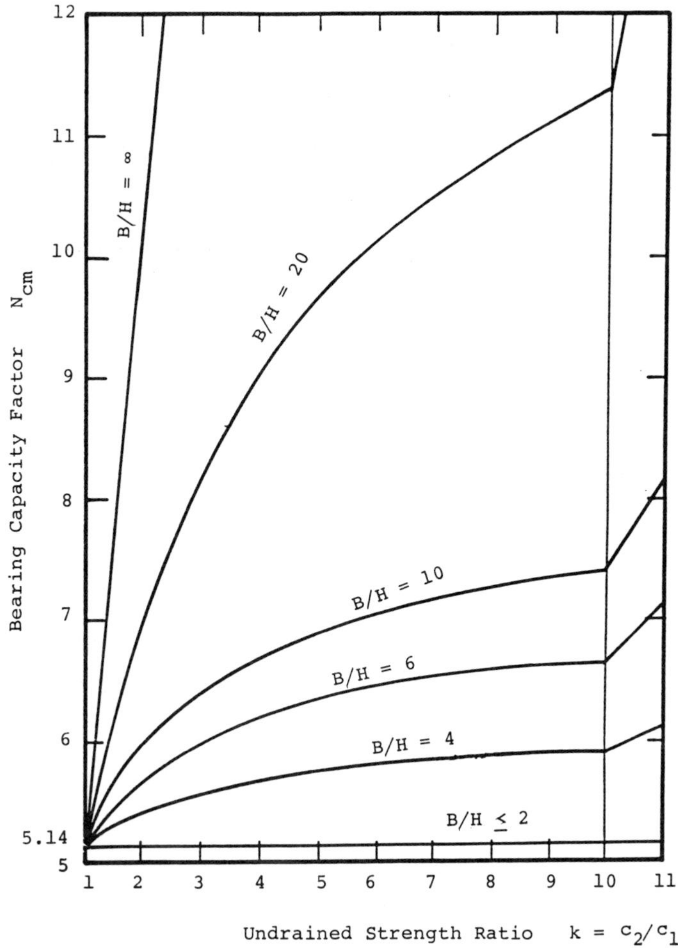

FIGURE 16. Modified bearing capacity factor N_{cm} for continuous rectangular footings [after Vesic (1975)].

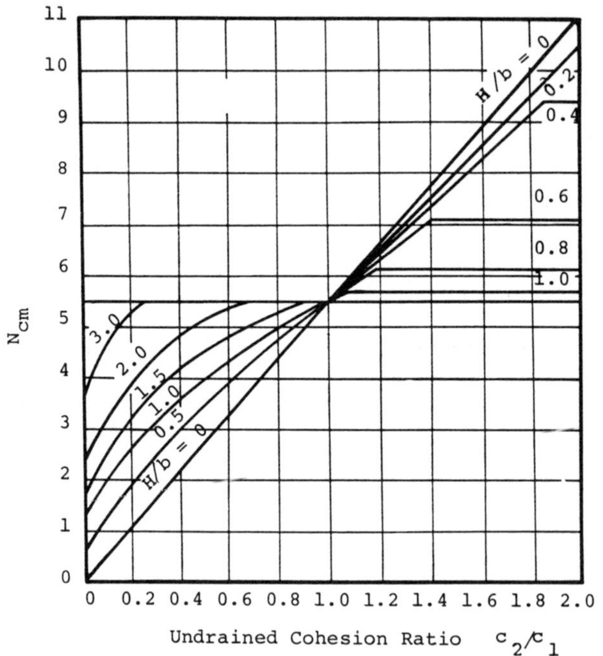

FIGURE 17. Modified bearing capacity factor N_{cm} [after Reddy and Srinivasan (1967)].

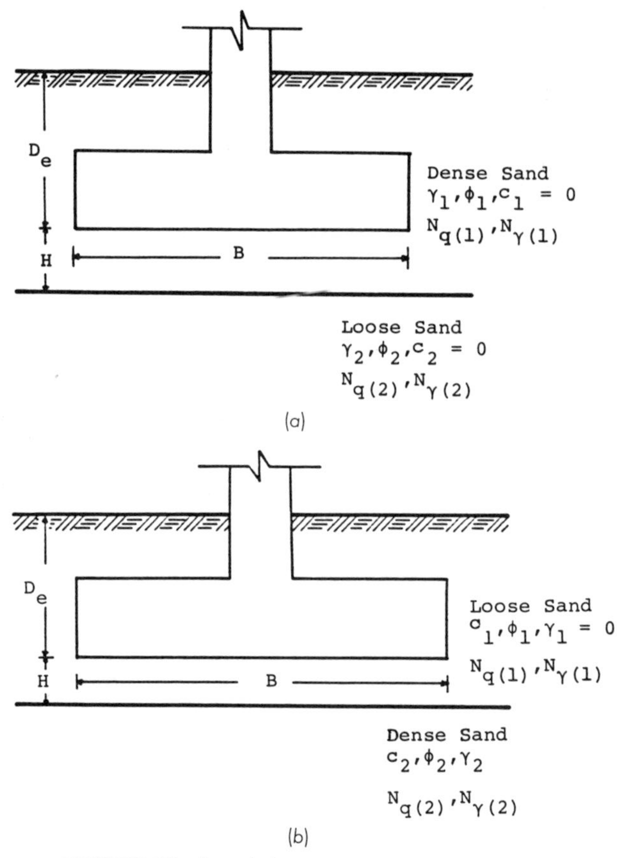

FIGURE 18. Foundation on a two layer sandy soil.

The bearing capacity for both conditions of Case I is computed by:

$$q_u = c_1 N_{cm} + \bar{q} \tag{55}$$

where

c_1 = undrained cohesion of upper layer
N_{cm} = a modified bearing capacity factor
$\bar{q}$ = effective overburden

Values for the modified bearing capacity factor N_{cm}, have been reported by Reddy and Srinivasan (1967), Brown and Meyerhof (1969), and Vesic (1975). This factor depends on the relative undrained cohesion of the two layers, the relative depth of the layer and the shape of the footing.

For the condition of Figure 14a, (soft clay over stiff clay), Vesic (1975) has suggested the modified bearing capacity factor, N_{cm}, as shown in Figures 15 and 16 for square and strip footings, respectively.

Values of N_{cm} for both conditions (a) and (b) can be obtained from the chart given by Reddy and Srinivasan (1967) shown in Figure 17. Note that $b = B/2$ in this chart. This graph can be also used in the case of anisotropic soils using the average shearing strength $q_u = (q_1 + q_2)/2$ and the corresponding coehsion where q_1 and q_2 are the vertical and horizontal shearing strengths, respectively.

The depth of the rupture surface (shear surface) is approximately $(B/2)\tan(45 + \phi/2)$ and if the second layer is deeper than this value the system is considered as a one layer system.

CASE II: Cohesionless Soils
a. Bearing stratum is dense sand and underlying stratum is loose sand (Figure 18a).

For this condition Meyerhof and Hanna (1978) proposed the following equations for the ultimate bearing capacity.

1. For $H \geq B$

$$q_u = \gamma_1 D_e N_{q(1)} + 0.5\gamma_1 B N_{\gamma(1)} \tag{56}$$

Continuous footings

$$q_u = \gamma_1 D_e N_{q(1)} + 0.3\gamma_1 B N_{\gamma(1)} \tag{57}$$

Circular or Square footings

$$q_u = \gamma_1 D_e N_{q(1)} + 0.5\,[1 - 0.4\,(B/L)]\,\gamma_1 B N_{\gamma(1)} \tag{58}$$

Rectangular footing

2. For $H \leq B$

$$q_u = \gamma_1(D_e + H)N_{q(2)} + 0.5\gamma_2 B N_{\gamma(2)} + \gamma_1 H^2\,(1 + 2D_e/H)\,k_s(\tan\phi_1)/B - \gamma_1 H \tag{59}$$

Continuous footings

$$q_u = \gamma_1(D_e + H)N_{q(2)} + 0.3\,\gamma_2 B N_{\gamma(2)} + 2\gamma_1 H^2\,(1 + 2D_e/H)(k_s/B)(\tan\phi_1)\,\lambda_s' - \gamma_1 H \tag{60}$$

Square or Circular Footings

$$q_u = \gamma_1(D_e + H)N_{q(2)} + 0.5[1 - 0.4(B/L)]\,\gamma_2 B N_{\gamma(2)} + (1 + B/L)\gamma_1 H^2\,(1 + 2D_e/H)(k_s/B)(\tan\phi_1)\lambda_s' - \gamma_1 H \tag{61}$$

Rectangular Footings

where k_s = a punching shear coefficient shown in Figure 19 and λ_s' = shape factor ≈ 1.

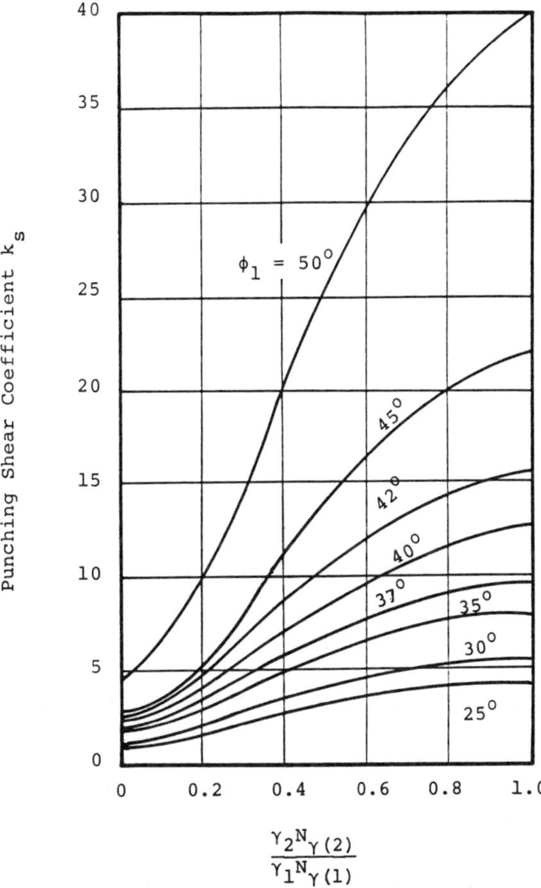

FIGURE 19. Punching shear coefficient k_s [after Meyerhof and Hanna (1978)].

If the values obtained by Equations (59), (60), and (61) exceed the values obtained by Equations (56), (57) and (58), respectively, then the later equations control.

b. Bearing stratum is loose sand and underlying stratum is dense sand (Figure 18b).

For the condition of $H > 2B$, Equations (55), (57) and (58) can be used to compute the ultimate bearing capacity of the upper layer $q_u(u)$. In the case of $H \leq 2B$, Meyerhof and Hanna (1978) suggested:

$$q_u = q_u(u) + [q_u(\ell) - q_u(u)](1 - H/H_f)^2 \quad (62)$$

where

$q_u(u)$ = the ultimate bearing capacity of the upper layer (computed from Equations (56), (57), and (58)

$q_u(\ell)$ = bearing capacity of the lower layer (computed from Equations (56), (57), and (58) using γ_2 and ϕ_2.

H_f = depth of shearing failure below the base of the footing $\approx 2B$

Note that Equation (62) has a limiting condition $q_u(u) \leq q_u \leq q_u(\ell)$.

CASE III: $c - \phi$ Soils
Bearing and underlying strata are $c - \phi$ soils (Figure 20) (with upper layer stronger than the lower layer).

For this condition Vesic (1975) suggested that for rectangular footings the bearing capacity can be computed as:

$$q_u = [q_u(\ell)(1/k)c_1 \cot \phi_1] \exp \{2 [1 + (B/L)]k(H/B) \tan \phi_1\} - (1/k)c_1 \cot \phi_1 \quad (63)$$

where $q_u(\ell)$ is the bearing capacity of a fictitious footing of the same size and shape, resting on the top of the lower layer and

$$k = (1 - \sin^2 \phi_1)/(1 + \sin^2 \phi_1)$$

The limiting depth of the upper layer beyond which the lower layer does not influence the bearing capacity can be computed from

$$(H/B)_{limit} = \frac{3\ell n(q_{u(u)}/q_{u(\ell)})}{2(1 + B/L)} \quad (64)$$

In the case where the footing is underlaid by thin layers of $c - \phi$ soils Bowles (1982) suggests that the average values of c and ϕ can be obtained as follows:

$$c_{av} = \frac{c_1H_1 + c_2H_2 + c_3H_3 + \ldots c_nH_n}{\Sigma Hi} \quad (65)$$

$$\phi_{av} = \tan^{-1}\left[\frac{H_1\tan \phi_1 + H_2\tan \phi_2 + \ldots H_n\tan \phi_n}{\Sigma Hi}\right] \quad (66)$$

These average values can then be used to obtain the bearing capacity factors and the system can be treated as one layer.

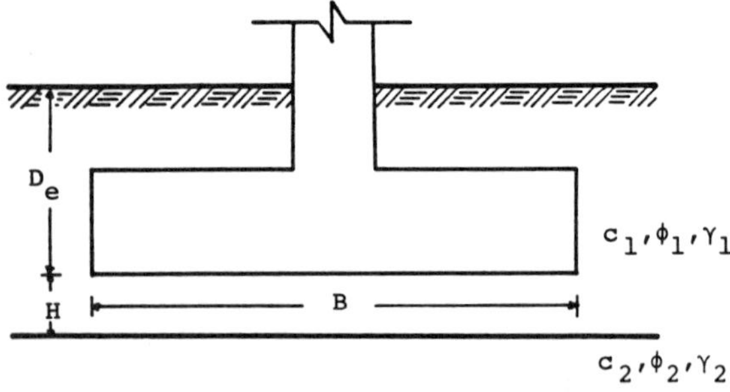

FIGURE 20. Foundation on a two layered $c - \phi$ soil system.

Bearing Capacity Evaluation From Field Tests

PLATE LOAD TESTS

Often it may be desirable to conduct field load tests in order to determine the bearing capacity of a soil. A standard method of conducting plate load tests is given by the ASTM Designation D-1194. The designer should refer to this designation for step by step guidance of conducting and interpreting the results of the test. A very short description will be included here.

The test consists of incremental loading of a standard size (usually 1ft × 1ft) steel plate and monitoring of the plate settlement for each load increment. The results are plotted on a load–settlement curve and settlement is used as the criterion for approximating the ultimate bearing capacity as follows:

$$q_{u(footing)} = q_{u(plate)} \text{ for clayey soils} \quad (67)$$

$$q_{u(footing)} = q_{u(plate)} B_{(footing)}/B_{(plate)} \quad (68)$$

where B = is the width.

If the settlement is used as a criterion, the soil bearing capacity of cohesive soils can be computed for a given settlement by a method proposed by Housel (1929). This method requires that two tests are performed with two plates of different size. The following two equations must be solved simultaneously.

$$Q_1 = A_1q + P_1s \quad (69)$$

and

$$Q_2 = A_2q + P_2s \quad (70)$$

where

$Q_{1,2}$ = the loads at plates 1 and 2, respectively, required to produce the same settlement
$A_{1,2}$ = the areas of plates 1 and 2
$P_{1,2}$ = the perimeter of plates 1 and 2
q = compression stress below the footing
s = unit shear stress at the plate perimeter

The two equations can be solved simultaneously for q and s.

BEARING CAPACITY FROM THE STANDARD PENETRATION RESISTANCE TEST (SPT)

Practicing geotechnical engineers use the SPT widely in estimating the bearing capacity of soils. Expressions relating the standard penetration number N with the allowable bearing capacity have been presented by Terzaghi and Peck (1967), Meyerhof (1956, 1974) and Peck, Hansen and Thornburn (1974).

The expressions presented by Terzaghi and Peck (1967) and Meyerhof (1956, 1974) are considered to be conservative. Bowles (1982) modified Meyerhof's equations by increasing the allowable bearing capacity by 50% and presented the following equations.

$$q_{all} = \frac{N}{2.5} k_d \qquad \text{for } B \leq 4 \text{ ft} \qquad (71)$$

$$q_{all} = \frac{N}{4.0}\left(\frac{B+1}{B}\right)^2 \text{ for } B > 4 \text{ ft} \qquad (72)$$

where

q_{all} = the allowable bearing capacity for a settlement of one inch in kips per square foot (ksf)
K_d = $1 + 0.33D_e/B \leq 1.33$
N = statistical average value of the standard penetration resistance number for the zone of about $0.5D_e$ above the footing base to at least $2B$ below the base

The allowable bearing capacity is the ultimate bearing capacity reduced by a factor of safety (FS) or

$$q_{all} = q_u/FS \qquad (73)$$

The settlement of a footing subjected to ultimate load will increase as the width increases. Therefore, for footings with relatively small widths, the ultimate load may be reached before the limiting condition for one inch settlement is reached. When this occurs, another limiting condition must be prescribed in terms of a safety factor. On the other hand, for footings of large width, the limiting settlement of one inch may be far exceeded before the ultimate load conditions are reached. For this situation the controlling factor is the allowable settlement of the structure. The one inch settlement criterion has been decided from experience and study of the structural behavior of various types of buildings.

Peck, Hansen and Thornburn (1974) produced design charts for the allowable bearing capacity of foundations of relatively small width in sand, for a safety factor $SF = 2$ and for settlement up to one inch. The charts for De/B ratios 0.25, 0.5 and 1.0 are shown in Figures 21, 22 and 23 respectively.

In these charts N' is the standard penetration resistance number, corrected for the effective overburden as:

$$N' = 0.77 \log (20/\bar{q})N_F \quad \text{for } \bar{q} > 0.25 \text{ tons/ft}^2 \qquad (74)$$

where

N_F = field standard penetration number
$\bar{q}$ = effective overburden

If the allowable bearing pressure at any settlement other than 1 inch is desired, it can be obtained from the following

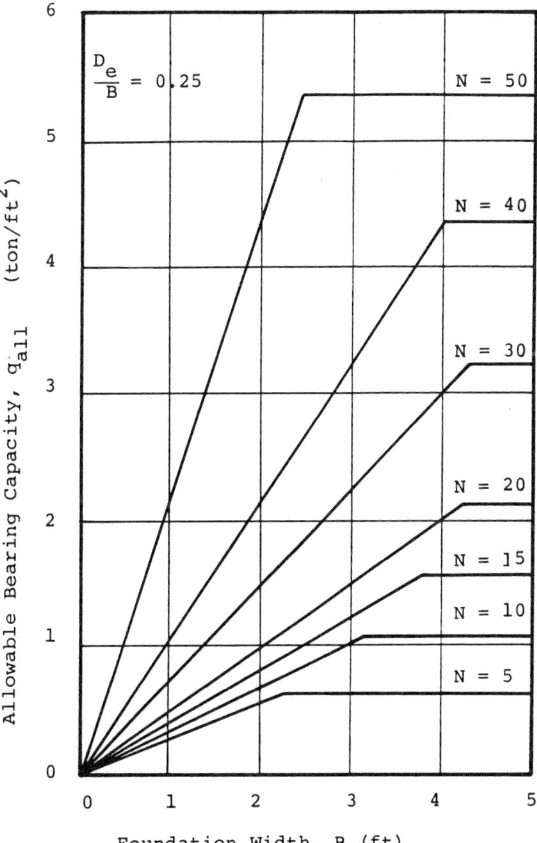

FIGURE 21. Correlation of q_{all} with N for D_e/B = 0.25 [after Peck et al. (1974)].

relationship:

$$q_{all(s)} = S\, q_{all(1)} \tag{75}$$

where

$q_{all(s)}$ = the allowable bearing presure for a settlement s (ksf)

$q_{all(1)}$ = the allowable bearing pressure for a settlement of one inch (ksf)

The preceding equations and charts are applicable for sandy and gravelly soils. Correlations of SPT results with bearing capacity of cohesive soils are not reliable and should be used with great caution. In these soils, in many cases, the controlling design factor is settlement rather than shearing failure.

BEARING CAPACITY FROM DUTCH CONE PENETRATION TEST DATA

Field data from cone penetration resistance tests can be correlated to the allowable bearing capacity as follows (Meyerhof 1965).

$$q_{all}\ (\text{ksf}) = \frac{q_c}{30} \qquad \text{for } B \leq 4 \text{ ft} \tag{76}$$

$$q_{all}\ (\text{ksf}) = \frac{q_c}{50}\left(\frac{B+1}{B}\right)^2 \qquad \text{for } B > 4 \text{ ft} \tag{77}$$

where q_c is the cone point resistance (ksf). These equations are expected to produced conservative results.

Presumptive Bearing Capacity

Most building codes contain a table of suggested allowable bearing pressures for shallow foundations. These are often called "presumptive" bearing capacities since they are based on a visual textural description of the soil. Table 6 contains presumptive bearing values from three different building codes.

Presumptive bearing values are based on years of cumulative experience in foundation construction, and they tend to be conservative. Most building codes allow the use of higher

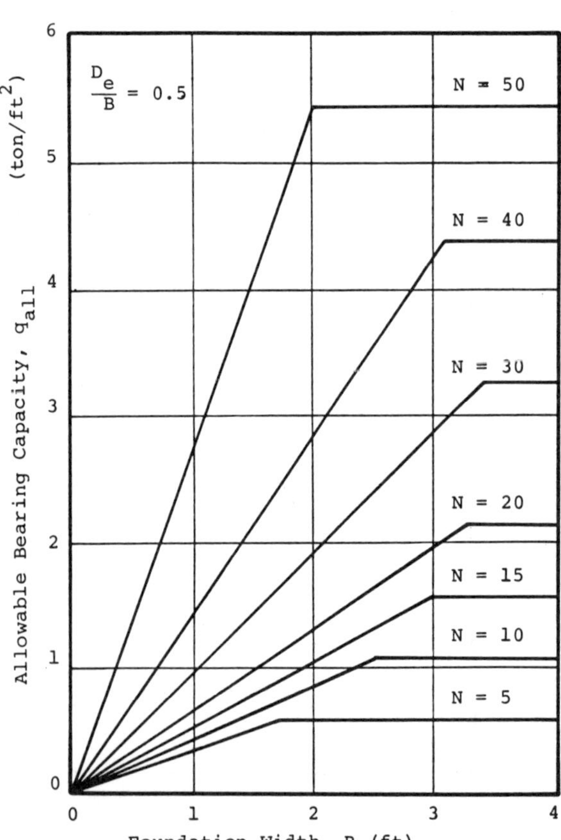

FIGURE 22. Correlation of q_{all} with N for D_e/B = 0.5 [after Peck et al. (1974)].

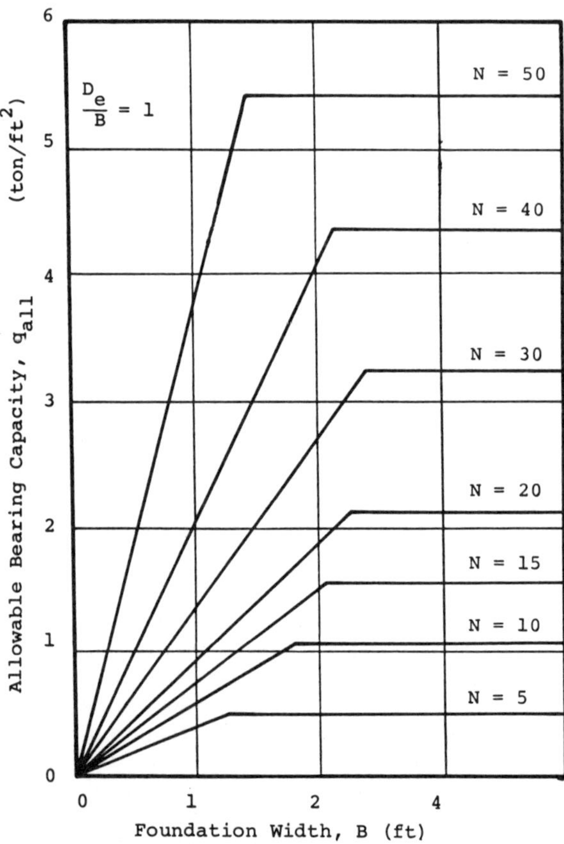

FIGURE 23. Correlation of q_{all} with N for D_e/B = 1.0 [after Peck et al. (1974)].

TABLE 6. Presumptive Bearing Pressures From Selected Building Codes (Pounds Per Square Foot).*

Soil Description	New York City 1968	BOCA 1984	Uniform Building Code 1976
Soft Clay		3,000	
Medium Clay	4,000		
Stiff Clay		4,000	
Hard Clay	10,000		
Clay			3,000
Inorganic Silt, Medium Dense	3,000		
Inorganic Silt, Dense	6,000		
Silty Clay and Clayey Silt			3,000
Medium to Coarse Sand	6,000–12,000	4,000–8,000	
Fine Sand	4,000–8,000	6,000	
Sand, Silty Sand, and Clayey Sand			4,500
Sand and Gravel Mixtures	8,000–20,000	12,000	6,000
Hardpan, Cemented Sand and Cemented Gravels	16,000–24,000	20,000	
Soft Rock	16,000	20,000	
Bedrock of Various Descriptions	40,000–120,000	50,000–200,000	6,000–12,000

*Maximum values. Code considers footing width and depth of embedment.

bearing pressures than the presumptive values only if a detailed soils investigation and analysis is performed.

Presumptive bearing values already contain a safety factor so they may be used directly. The technical disadvantage of presumptive bearing values is that they typically do not take into consideration a number of important factors associated with shallow foundation design including embedment of footing, size of footing, amount of settlement, and location of water table.

Safety Factors

Before using ultimate bearing capacities in design situations, it is first necessary to reduce them by a suitable factor of safety. The resulting "allowable bearing pressure" is defined by Equation (73).

Owing to the higher level of uncertainty associated with the nature of soils, higher safety factors are typically specified in foundations design compared with superstructure design. When selecting a factor of safety, the following should be considered:

1. The variability of the soil or rock conditions at the site
2. The certainty to which the properties of the soil or rock are known
3. The method used to predict the ultimate bearing capacity
4. Experience of the designer under similar conditions
5. The structure's tolerance to settlement
6. The probability that the actual loads will exceed the estimated loads
7. Construction tolerances and procedures
8. The relative economy of increasing or decreasing the safety factor

The safety factor applied to shallow foundation typically ranges from 1.5 to 3.0. As with superstructure design, critical combinations of live and dead loads should be checked. Load combinations which either have a shorter duration or a lower probability, e.g., wind and earthquake loads, require that the safety factor be selected from the lower end of the range, say 1.5–2.0. For the more permanent load combination, sometimes called "service load," the safety factor should be towards the upper end of the range, say 2.5–3.0.

Some investigators have proposed that a different safety factor be applied to c and ϕ (Hansen, 1967). Others have applied probability to the selection of the safety factor based on the statistical distribution of the soil parameters and other related factors (Meyerhof, 1970 and Wu and Kraft, 1967). These alternate approaches to safety factor selection are more cumbersome to apply and are not widely used at present.

PILE FOUNDATIONS

Introduction

Pile foundations are used to support structures at sites where the bearing capacity of the shallower soils is not sufficient to carry the foundation loads. They function by transmitting the load through the weak strata to deeper soil or rock, which has a higher bearing capacity. Pile foundations are also occasionally called upon to perform other functions such as scour protection, uplift anchorage, and soil compaction.

There is a large variety of deep foundation types presently in use of which pile foundations are a part. Three generic types of deep foundations are generally distinguished: piles, caissons, and piers. Piles are slender structural members which are installed by driving, drilling, vibrating, jacking, or some combination of these processes. Caissons are similar to piles although they are larger in diameter and always involve a structural outer shell used to exclude the soil from an interior working space. The 1981 New York City Building Code establishes the lower limit of a "caisson pile" as 18 inches, and the diameter of caissons can range into the hundreds of feet as for a bridge caisson. The third generic type of deep foundation, piers, are distinguished from the other two in that they are generally constructed in deep, unsupported open excavations.

The scope of this section is limited to a discussion of methods for establishing the load capacity of driven pile foundations, which are by far the most common type of deep foundation. Methods to establish the capacity of piles installed by means other than described in this chapter are similar, although some modifications may be warranted to account for the differences in installation procedure.

The determination of vertical and horizontal load capacity is the principal concern in designing pile foundations. Three general methods are available to establish load capacity: 1) static analysis; 2) dynamic analysis; and 3) load testing. The three methods are not exclusive of one another, and ideally the foundation engineer should use a combination of two, or possibly all three of the methods to establish load capacity if the project budget permits. It should be noted that the methods may give significantly different results. Therefore judgement, tempered with experience, is required to establish safe and economical pile foundations. This fact is traditionally a source of frustration to engineers becoming acquainted with the design of pile foundations.

A consideration in the design of pile foundations, quite separate from load capacity, is settlement. As with shallow foundations, pile foundations which experience excessive uniform or differential settlement may be considered as "failed," even though they are safe from a bearing capacity point of view. Although methods of estimating settlement of pile foundations will not be covered in this chapter, it should always be examined as part of a complete design. Settlement is most often of concern in larger pile groups which are underlain by soft, compressible soils.

Behavior of Pile Foundations Under Load

Prior to any discussion of methods to estimate load capacity, it is useful to examine the behavior of pile foundations when loads are applied. Two very distinct considerations control the behavior of pile foundations:

1. The inherent structural properties of the foundation element
2. The ability of the soil or rock to support the foundation element

The first consideration, the so-called "structural capacity," is determined in a straightforward manner by considering the deep foundation element to behave as a columnar structural element. When piles are completely embedded in all but the very softest of soils, simple column behavior is assumed as depicted in Figure 24. For partially embedded piles, as in the case of bridge bents, combined compression and bending must be analyzed as depicted in Figure 25.

The allowable structural stresses which can be sustained by the pile vary with the type of material, and are usually controlled very explicitly by the prevailing building code. Table 7 provides a summary of the allowable material stresses indicating the usual range of values. It will be noted that the allowable stresses on piles are generally more conservative than those allowed in above-ground structural applications owing to the inherent uncertainties of subsurface work.

The second consideration in determining the load capacity of pile foundations is the ability of the soil or rock to support the foundation element. This is by far the more complex question associated with the design of pile foundations. Structural failure of piles is very rare, while failure of the soil or rock supporting the pile is a more common occurrence. This fact has been substantiated by the numerous field load tests which have been carried out to failure. Prediction of the failure threshold of the soil or rock supporting the pile will be the focus of the remainder of this chapter section.

The support afforded to pile foundations by the soil and/or rock in which they are embedded can be divided into two categories: 1) skin resistance along the pile shaft and 2)

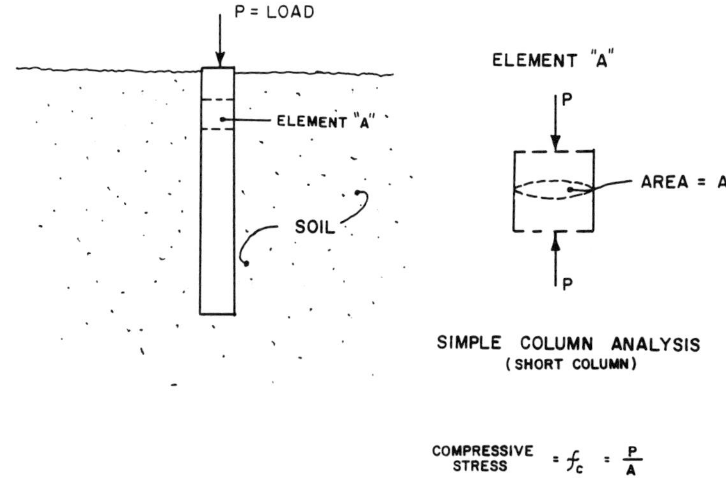

FIGURE 24. Structural capacity of a pile—full embedment.

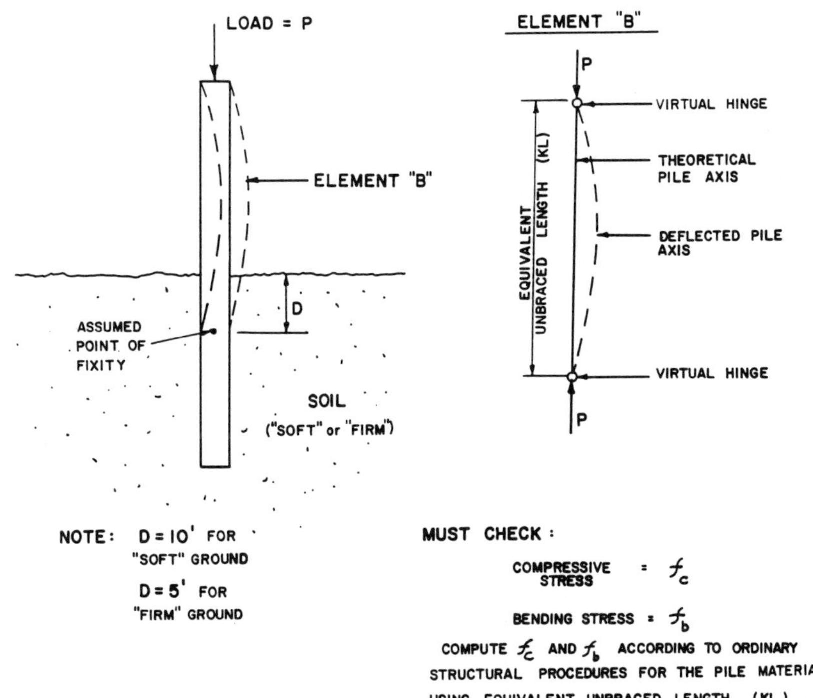

FIGURE 25. Structural capacity of a pile—partial embedment.

TABLE 7. Allowable Material Stresses for Pile.

Pile Type	Allowable Unit Stress
Timber	800–1200 psi depending on the type and grade of the wood.
Cast-in-Place Concrete	33 to 40% of f'_c
Pre-Cast Concrete	concrete: 22 to 33% of f'_c steel: 33 to 40% of F_y
Steel H-piles and Pipe Piles	9,000–12,000 psi

point resistance below the pile tip. These two supporting resistances are depicted in Figure 26. The support of pile foundations differs significantly from that of shallow foundations. Shallow foundations, owing to their comparatively shortened depth of embedment, do not consider skin resistance in the determination of bearing capacity.

These two types of load resistance suggest that deep foundations may behave in one of three possible ways depending on the geologic conditions at the site. The first possibility is that the load may be taken entirely by skin resistance along the shaft, as in the case of a "friction pile." The second possibility is that the load may be taken entirely by point resistance, as in the case of an "end-bearing pile." The third possibility is the load may be taken by a combination of both skin and point resistances. This combined action is realistically to be expected in most cases, although singular behavior is frequently assumed to provide a degree of conservatism and to simplify analysis.

Expressing these two resistances in a mathematical form, the ultimate capacity of a deep foundation is:

$$Q_{ult} = Q_{point} + Q_{skin} \tag{78}$$

Various methods exist for evaluating the point and skin resistance terms, depending on the type of soil or rock present, and some will be discussed in future sections.

An examination of Equation (78) raises an important question: Can skin and point resistances be relied upon to be mobilized simultaneously? Work by various investigators (Whitaker and Cooke, 1966, Vesic, 1970 and O'Neill and Reese, 1972) has revealed that skin resistance is fully mobilized at relatively small displacements on the order of less than ½ inch. In contrast, point resistance requires more displacement to become fully mobilized, ranging from approximately 8% of the pile diameter for driven piles, up to 30% of the pile diameter for bored piles. A load–displacement relationship for a typical pile indicating the different rates of load transfer is shown in Figure 27.

This discrepancy between the rate of development of skin and point resistances prompted Skempton (1966) to recommend that different factors of safety be applied to each in determining working capacity. For large diameter bored piles, he suggested a value of 1.5 be applied to the skin resistance and a value of 3.0 be applied to the point resistance. As a practical matter, however, for most design applications, a single factor of safety is typically applied against the combined ultimate load capacity, and no attempt is made to distinguish between the two components. In view of the many uncertainties involved with pile foundation analysis, this approach seems reasonable.

Static Analysis of Vertical Capacity—Single Piles

Computation of static pile capacity is performed during the design phase of most piling projects, even if other

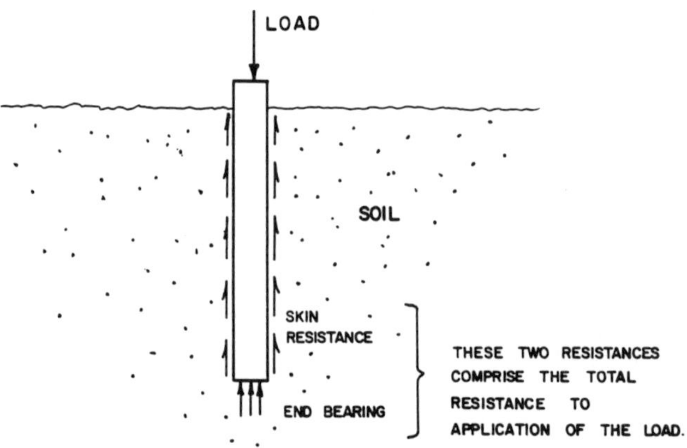

FIGURE 26. Load resistance provided by soil or rock.

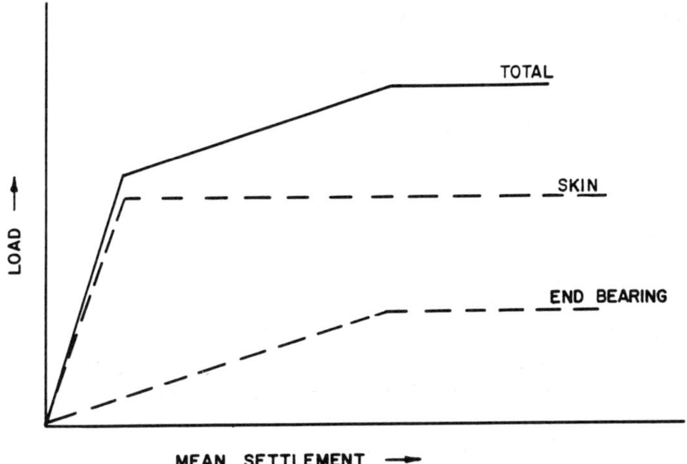

FIGURE 27. Idealized mobilization of point and skin resistances.

methods will be used during construction to establish final load capacity. Static analyses provide estimates of the number and length of piles necessary during the design phase. The computational procedures for static analysis are different for different kinds of soils and rock. In the sections which follow the more commonly used procedures for estimating the static capacity of piles will be presented for: cohesive soils; cohesionless soils; layered and mixed soils; and rock.

COHESIVE SOILS

For piles embedded in saturated cohesive soils, static load capacity is estimated using undrained soil parameters ($\phi = 0$ analysis). A general expression for the ultimate bearing capacity of a pile in a saturated cohesive soil is:

$$Q_{ult} = Q_{point} + Q_{skin}$$
$$Q_{ult} = c_u N_c A_p + c_a P L \quad (79)$$

where

c_u = undrained shear strength
N_c = bearing capacity factor
A_p = area of pile point
c_a = undrained adhesion between pile and soil
P = pile perimeter
L = length of embedment

The bearing capacity factor, N_c, in the point resistance term is usually taken to be 9, if the embedded pile length is at least 5 times more than the pile diameter (Meyerhof, 1951). For most applications in cohesive soils, the point resistance contributes a relatively small portion of the total load capacity and may be neglected in medium to soft clays. Thus, piles driven into clay soils are traditionally called "friction" piles.

Estimation of the undrained adhesion, c_a, involves the highest degree of uncertainty. The usual approach is to correlate the adhesion with the undrained shear strength, c_u.

Tomlinson (1977) has studied the correlation between undrained adhesion and shear strength extensively and has provided the series of design curves shown in Figure 28. These curves correlate the undrained shear strength with an "adhesion factor," which is related to the undrained adhesion in Equation (79) as follows:

$$c_a = (\text{Adhesion Factor}) \times c_u \quad (80)$$

The following trends and limitations of the design curves in Figure 28 should be noted:

1. The adhesion factor depends on the cohesive strength of the soil. It is higher than 1.0 for very soft clays and generally decreases to as low as 0.20 for very stiff clays. The explanation for the decreasing trend is the stiffer soil's inability to remold around the pile after termination of driving.
2. The adhesion factor depends on the nature of the soil which overlies the clay bearing stratum. Piles driven through soft clay overburden will "dragdown" a soft clay skin into the stiffer clay bearing stratum. This reduces the adhesion factor for the stiffer clay thus reducing the anticipated pile capacity. In contrast, pile driven through sand overburden will "dragdown" a sand skin into the clay bearing stratum which tends to increase the adhesion factor for the clay. These sand and clay "dragdowns" skins have been observed to extend approximately three pile diameters into the clay bearing stratum and ob-

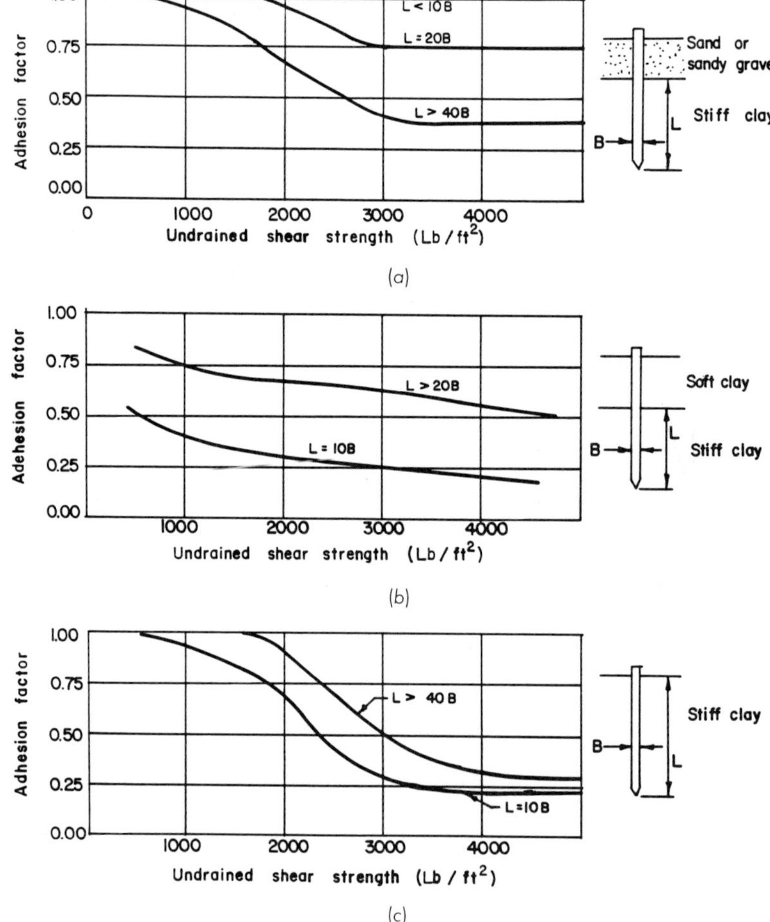

FIGURE 28. Adhesion factors for piles driven into clay [after Tomlinson (1977)].

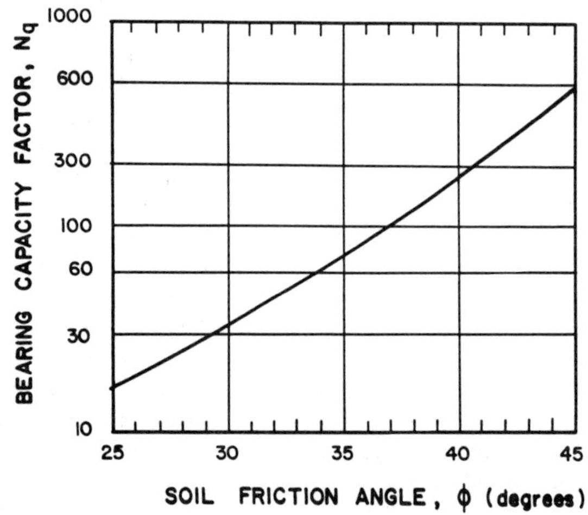

FIGURE 29. Bearing capacity factor, N_q [after Berezantzev (1961)].

viously have the most pronounced effects on piles with short penetrations.

3. Piles driven directly into stiff clay with no overburden soil will tend to form an annular gap between the pile and the soil which extends down several pile diameters from the ground surface. This reduces the average adhesion along the pile shaft and in effect reduces the overall adhesion factor for this case.
4. The design curves apply to straight, tubular piles which do not have projecting base plates or stiffening devices, which would tend to ream an oversize hole and reduce adhesion. Tests with tapered piles indicate that there is no significant increase in adhesion compared with straight piles, and are therefore not distinguished on the design curves.
5. The design curves do not apply to *H*-piles which are generally not recommended for use as friction piles in cohesive soils. When they are used, Tomlinson recommends calculating the skin resistance on the outer flange surfaces only, although point resistance may be calculated on the gross cross-sectional area on account of the beneficial effects of soil plugging at the tip.

Alternative methods to the adhesion factor approach are available to estimate pile load capacity in cohesive soil. Vijayvergiya and Focht (1972) presented a method using a λ coefficient which is based on the assumption the soil displaced during pile driving causes passive lateral earth pressure on the sides of the pile. Another method, known as the β method, was first described by Chandler (1968), and it assumes that the bond between the pile and soil is purely frictional in nature, similar to piles installed in cohesionless soils.

Since the piles installed in cohesive soils derive their capacity primarily from skin friction, it follows that the longer the pile length, the higher the allowable load capacity. It is interesting to note, however, that many construction codes limit the allowable capacity of "friction" piles. For example, the 1981 New York City Building Code sets 60 tons as the maximum allowable load on any friction pile, regardless of length.

COHESIONLESS SOILS

For piles embedded in cohesionless soils, static load capacities are estimated using cohesionless soil parameters ($c_u = 0$). A general expression for the ultimate load capacity of a pile in a cohesionless soil is:

$$Q_{ult} = Q_{point} + Q_{skin}$$
$$Q_{ult} = f_p A_p + f_s A_s \tag{81}$$

where

f_p = unit end-bearing resistance under pile tip
f_s = average unit skin resistance along the pile shaft
A_p = area of pile point
A_s = area of pile shaft

Expanding f_p and f_s we have:

$$Q_{ult} = \bar{\sigma}_v N_q A_p + (K_s \bar{\sigma}_{va} \tan \delta) A_s \tag{82}$$

where

$\bar{\sigma}_v$ = the effective overburden pressure at the pile tip ($\bar{\sigma}_v = \bar{\gamma} L$)
N_q = bearing capacity factor
L = total length of embedment
$\bar{\gamma}$ = effective unit weight of soil
K_s = lateral earth pressure coefficient
$\bar{\sigma}_{va}$ = average effective overburden pressure over the length of the pile
δ = angle of friction between pile and soil

The first term of Equation (82) may be evaluated using the bearing capacity factor, N_q. There is a wide range of values for N_q proposed by various investigators for deep foundations as summarized by Vesic (1967). One of the more commonly used relationships in design practice was proposed by Berezantzev, et al. (1961) and is shown in Figure 29. The selection of the bearing capacity factor requires an estimate of the angle of friction, ϕ, which is typically obtained from standard correlations with field tests such as the Standard Penetration test or the Cone Penetrometer Test.

Research by Vesic (1973) has shown that f_p does not increase indefinitely but reaches a limiting value at a depth of between 10 to 20 pile diameters. The cause of this limiting behavior is not certain and has been variously attributed to soil arching, pulverization of soil beneath the point, and variation of soil friction angle with confining pressure. Generally, the peak value of f_p is taken at a depth of 20 diameters, although Tomlinson (1977) recommends that the peak value should never exceed 100 tons/ft^2.

Unlike piles embedded in cohesive soils, point bearing typically contributes the majority of load capacity to piles embedded in coehsionless soils. For this reason, shorter piles tend to be the most efficient and economical in cohesionless soils.

The general form of the skin resistance term in Equation (82) is generally accepted by most authorities. The average unit skin resistance, f_s, is computed as the product of the average effective lateral earth pressure and the friction coefficient developed between the pile and the soil, $\tan \delta$.

Different methods exist for evaluating the lateral earth pressure coefficient, K_s. Its value is dependent on the initial density of the soil and the method of installation. Values of K_s which have been porposed by Broms (1966) and Mansur and Hunter (1970) are presented in Table 8. Nordlund (1963) proposed a more elaborate procedure for determining K_s using the curves shown in Figure 30. Nordlung's notation and procedure may be summarized as follows:

1. Divide the pile into discrete segments according to the soil characteristics. Each segment shall be analyzed separately.
2. Calculate the displaced soil volume per unit length, V,

TABLE 8. Typical Values of K_s and δ.

		Broms (1966) K_s		
Pile Type	δ	Low Relative Density	High Relative Density	Mansur and Hunter (1970) K_s
Steel	20°	0.5	1.0	H-Piles: 1.4–1.9 Pipe: 1.0–1.3
Concrete	¾ ϕ	1.0	2.0	Precast: 1.45–1.6
Wood	⅔ ϕ	1.5	4.0	—

for use in Figure 30b (for H piles use the gross cross section and for tapered piles use the average cross section in the segment).

3. Select the value of δ/ϕ from Figure 30b.
4. Obtain K_s from Figure 30c, where ω = angle of pile taper (for a straight pile, $\omega = 0$).
5. Apply the correction factor to K_s from Figure 30d depending on δ/ϕ and ϕ. This is necessary since Figure 30c assumes that $\delta/\phi = 1.0$.
6. Evaluate the skin resistance for the segment under consideration using the second term of Equation (82).
7. Repeat for each segment and add together to obtain the Q_{skin}.

Research by Vesic (1970) has shown that like point resistance, skin resistance does not increase indefinitely as the depth of embedment increases, but reaches some limiting value. He reported, based on a series of load tests in sand, that the unit skin resistance reaches a maximum value when the ratio of the penetration depth to pile width exceeds 10 to 20. He proposed the following maximum value of average unit skin resistance, f_s:

$$f_s = 0.08\ (10)^{1.5(D_r)^4} \tag{83}$$

where

f_s = unit skin resistance in tons/ft²
D_r = relative density

It should be noted that in actual practice, the value of the average unit skin resistance should never exceed 1.0 tons/ft², regardless of the computations.

LAYERED AND MIXED SOILS

When installing piles into layered soil deposits, the subsurface profile must be established with a high degree of reliability. The total capacity is estimated by summing the skin resistances provided by each layer and the point resistance in the stratum in which the pile terminates. In soils which contain layers of both cohesive and cohesionless soils, it is usually desirable to terminate the pile in a cohesionless soil layer owing to the much higher point capacity which can be developed compared with cohesive soils.

Where it is desired to terminate a pile in a firm stratum which overlies a weak stratum, e.g., sand over clay, the piles should not penetrate too deeply into the firm stratum. Otherwise "punch-through" of the pile tip may occur with the result that the point resistance will be controlled by the weaker soil. In order to avoid this reduction in point capacity, Meyerhof (1976) recommends that the pile tip be situated at least 10 pile diameters above the interface of the two strata.

When driving piles into soils which are mixtures of both cohesive and cohesionless soils, e.g., sandy clays and clayey sands, two different approaches may be used:

1. If one or the other soil component is dominant, the pile capacity may be analyzed assuming either $c = 0$ if the soil is predominantly cohesionless, or $\phi = 0$ if the soil is predominantly cohesive.
2. If the soil contains appreciable amounts of both cohesive and cohesionless components, then the soil may be analyzed as a $\phi - c$ soil. Total resistance would then be the summation of the ultimate capacities determined in accordance with the previous two sections.

Friction piles driven into silts are often the most difficult to analyze statically. If the silt is clearly plastic in nature, e.g., clayey silt, and the cohesion is measurable, the pile may be designed as in a cohesive soil. If the silt is nonplastic and saturated, the pore pressures generated by pile driving may transfer the soil into a viscous liquid resulting in disturbance of previously driven piles. Such piles will obviously have negligible capacity immediately after driving but will develop skin resistance as the pore pressure dissipates and "set-up" occurs. Static analyses in such soils are of questionable value and capacity should be verified by field load tests.

Another possible behavior in a dense, non-plastic silt is dilatancy. This phenomenon is caused by development of excess pore pressures beneath the tip of the pile during driving which give false high penetration resistances. Then, sometime after the determination of driving, the pore

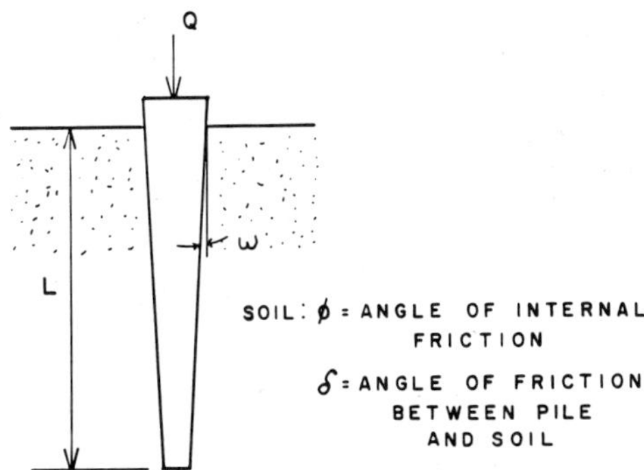

FIGURE 30a. Nordlung's procedure to determine K_s [after Nordlung (1963)]: assumed pile configuration.

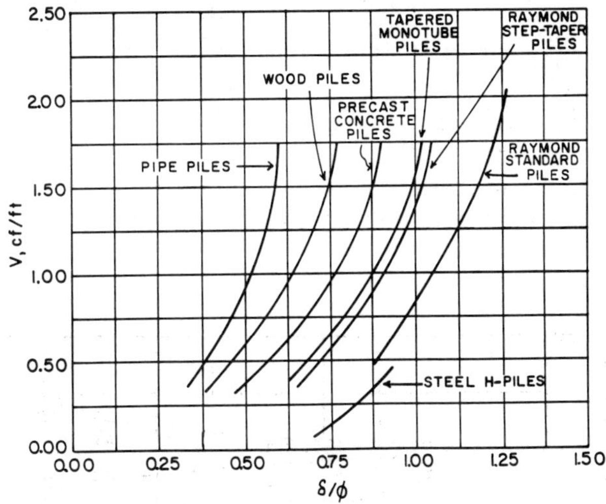

FIGURE 30b. Nordlung's procedure to determine K_s [after Nordlung (1963)]: selection of volume displacement, v.

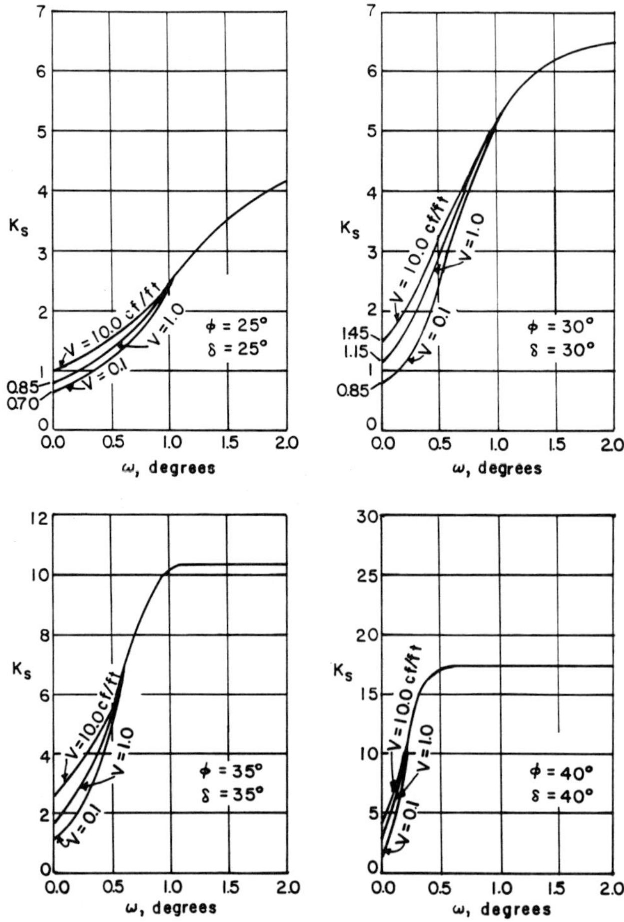

FIGURE 30c. Nordlung's procedure to determine K_s [after Nordlung (1963)]: values of lateral earth pressure coefficient, K_s.

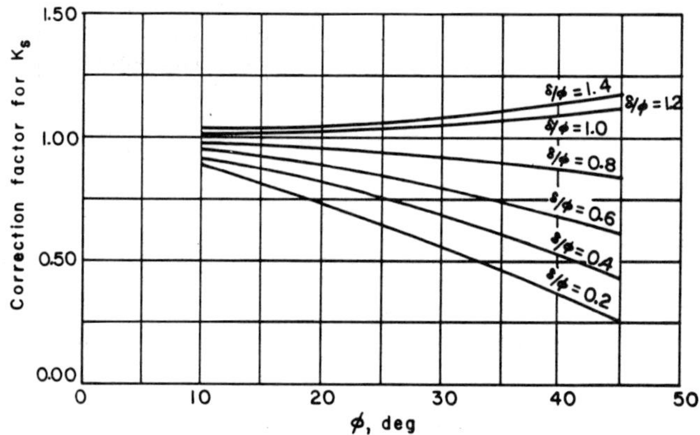

FIGURE 30d. Nordlung's procedure to determine K_s [after Nordlung (1963)]: correction factor for K_s when $\delta \neq \phi$.

pressures dissipate and "relaxation" of the pile resistance occurs.

PILES DRIVEN TO ROCK

As with soils, the ultimate capacity of a pile driven to bear on rock is dependent on the nature of the rock formation. It is a misconception that piles driven to rock are foolproof and require minimal design and field control. This belief probably stems from the fact that most rock formations are much more competent than soil formations. Correspondingly, though, we expect higher capacities from piles driven to rock, thus reducing or eliminating their advantage. It is significant to note that the 1981 New York City Building Code considers the hazard to be sufficient so as to require a registered professional engineer to personally inspect the driving of timber piles and steel piles to bear on hardpan and bedrock. This stringent level of field inspection is not specified by the New York City Code for piles bearing in soil.

The satisfactory installation of driven piles to end bearing on rock requires the development of an intimate contact between the pile tip and the bearing surface. The most common hazards of installing piles to rock include:

1. Damage to the pile tip or shaft by overdriving
2. Lifting of the piles off the bearing stratum due to friction heave of the surrounding soil
3. Sliding of piles driven on a steeply sloping rock surface
4. Fracturing of the bedrock stratum

All of the hazards listed above are usually considered to be construction problems, but they should also be addressed during the development of the design details and specifications. The last item on the list, bedrock fracturing, must be carefully analyzed during the design phase when estimating static capacity, especially when driving piles to rock which is not hard and sound.

The capacity of piles driven to hard, sound rock is controlled by the structural strength of the pile itself. As described earlier in the chapter and as summarized in Table 7, the allowable material stresses will control the allowable load for such piles. Even when installing piles with low cross-sectional areas such as steel H-piles, research has shown that the local rock crushing caused by the high contact stresses beneath the pile tip do not significantly effect the pile capacity. The greatest hazard when driving piles to hard sound rock is usually overdriving.

In contrast, the capacity of piles driven to weak rock are dependent on the ability of the rock to support the pile just as with piles installed to soil. Rock may be classified as "weak" due to a variety of reasons including:

1. Rocks with low intact compressive strength
2. Intensely jointed or fractured rock
3. Jointed rock with open joints
4. Weathered rock
5. Any combination of the above

Piles driven to weak rock will typically penetrate into the formation so that the static capacity includes contributions from both skin and end bearing resistances. Poulos and Davis (1980), based on a review of available and empirical data, recommended the following relationships be used to estimate the static capacity of bored or driven piles in rock:

Allowable End Bearing Pressure $\approx$ 30% of uniaxial compressive strength

Allowable Skin Resistance (Adhesion) $\approx$ 5% of uniaxial compressive strength

or

5% of f'_c of the concrete, whichever is less

Note: For highly fractured rocks, the allowable value of skin resistance should not exceed 10 to 20 psi.

The evaluation of bearing capacity of piles driven to weak rock becomes more complicated when they are driven in closely spaced groups. As each successive pile is driven, it tends to disturb the bearing of the previously driven piles. Some relaxation of the skin and end bearing resistances is inevitable and one or more redrives may be necessary to seat the piles. The tendency for this type of relaxation can be reduced by increasing the pile spacing, and also by not overdriving the pile beyond the point at which a sustained, near refusal resistance is obtained.

Static Analysis of Vertical Capacity–Pile Groups

The load capacity of a group of piles does not necessarily equal the sum of the individual load capacities of the piles in the group. It may be less than or greater than the sum of the individual capacities depending on the pile type, the pile spacing, and the subsurface conditions. This aspect of pile group behavior is usually expressed quantitatively using a Group Efficiency Ratio which is defined as:

$$\eta = \frac{\text{ultimate load capacity of pile group}}{\text{sum of ultimate load capacity of individual piles}} \quad (84)$$

The behavior of pile groups has been studied using both full-scale load tests and model pile tests. Data from full-scale load tests is relatively scarce, however, in view of the practical difficulties and expense of performing such high capacity tests.

For piles installed in granular soils, both models and full-scale load tests have established that the Group Efficiency Ratio is generally greater than one and may be as high as two. This greater than unity value may be attributed to the increased compaction and lateral compression caused by driving the piles. For design purposes, however, it is cus-

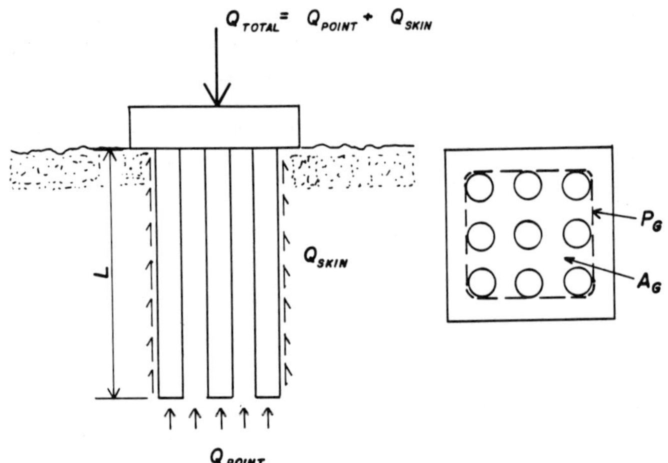

FIGURE 31. Combined capacity of pile group, cohesive soil.

tomary in granular soil to conservatively use a value of efficiency ratio equal to 1.0.

For piles driven into cohesive soils, both model and full-scale load tests have shown that the Group Efficiency Ratio is frequently less than 1.0. A popular method of determining group efficiency for piles in cohesive soils was originally suggested by Terzaghi and Peck (1948) in which the bearing capacity of the group was computed by treating it as a block foundation defined by scribing an imaginary line around the perimeter of the group as shown in Figure 31. The actual pile group capacity may then be computed by:

$$Q_{GROUP} = Q_{POINT} + Q_{SKIN}$$
$$Q_{GROUP} = c_u N_c A_G + L\, c_{av} P_G \tag{85}$$

where

c_u = undrained cohesion at the base of the group
N_c = bearing capacity factor
A_G = area of the pile group
L = length of piles
c_{av} = average undrained cohesion over the length of the piles
P_G = perimeter of the pile group

Inspection of Equation (85) reveals that the group capacity and the Group Efficiency Ratio increase as the pile spacing increases. The minimum spacing at which the full capacity of each individual pile is realized ($\eta = 1.0$) can be found by trial and error, and usually occurs at a spacing of about three pile diameters. As with granular soils, the use of a Group Efficiency Ratio greater than unity is not recommended.

A second approach for estimating group efficiency in cohesive soils is to apply the graphical relationship between pile spacing and group efficiency shown in Figure 32. This figure was developed by Whitaker (1957) by performing tests on model pile groups.

Some sources recognize the additional bearing capacity contributed by the concrete pile caps which are in contact with the soil and extend beyond the perimeter of the group. This issue is treated in considerable detail in Poulos and Davis (1980). In actual design practice, however, owing to the possibility of soil erosion and or settlement away from the pile cap, this additional capacity is not normally considered.

Dynamic Analysis of Vertical Capacity

Dynamic analysis of pile load capacity is based upon the seemingly sound assumption that a pile's resistance to driving is related to its ultimate capacity after driving stops. All things being equal (i.e., hammer, pile, soil), the higher the driving resistance, the higher the ultimate capacity of the pile. Although the overall approach is valid, it turns out that the determination of pile capacity using dynamic analysis is highly complex and has met with varying degrees of success. After more than 100 years of use, the dynamic approach can still be considered under development.

Two general methods of dynamic analysis can be distinguished: 1) dynamic formulas and 2) wave equation. Dynamic formulas are probably the oldest method of quantitative pile load capacity determination, and were first used in the early 1800s. The wave equation is recent by comparison, having been first put into a usable form by Smith in 1960.

DYNAMIC FORMULAS

Dynamic pile driving formulas attempt to relate the ultimate capacity of a pile through the dynamic parameter which is simplest to measure: the penetration resistance, also known as pile "set" or "blowcount." Literally hundreds of formulas have been developed and are generally derived from an energy balance between the driving system and the pile. The following derivation of the well-known Engineering News Formula illustrates the fundamental principle on which most of the formulas are based.

$$\text{Work of Falling Hammer} = \text{Work of Soil Resistance} \qquad (86)$$

$$WH = QS$$

where

W = weight of free-falling hammer
H = height of hammer fall
S = average pile penetration
Q = soil resistance against hammer impact

Recognizing that Q also represents ultimate pile capacity, we get:

$$\text{Ultimate Pile Capacity} = Q = \frac{WH}{S} \qquad (87)$$

This formula does not account for the energy losses in the driving system and pile due to heat and elastic deformation. Wellington (1888), the developer of the formula, accounted for these losses by adding a constant, C, to the denominator. This represents the additional pile penetration which would have occurred if no losses were present. We have then:

$$Q = \frac{WH}{S + C} \qquad (88)$$

C is determined empirically and Wellington proposed a value of 0.1 in. for single-acting steam hammers and 1.0 in. for drop hammers. The final step in the formula's development involves the application of a safety factor of 6 to ac-

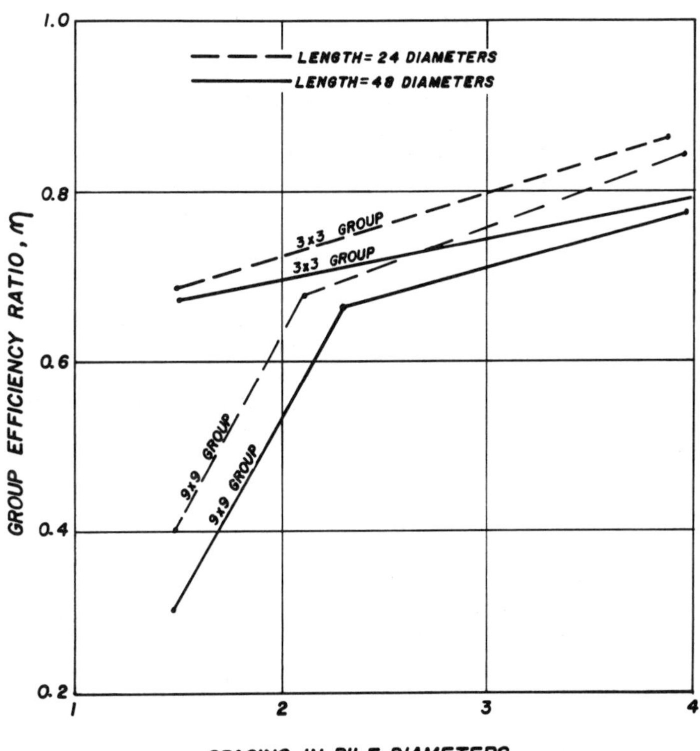

FIGURE 32. Group efficiencies for pile groups in cohesive soils [after Whitaker (1957)].

count for the many uncertainties:

$$\text{Allowable Pile Capacity} = Q_a = \frac{Q}{6} = \frac{WH}{6(S + C)} \tag{89}$$

By expressing H in feet, S in inches, and W and Q_a in pounds, the formula becomes the recognizable Engineering News Formula for single-acting air or steam hammers:

$$Q_a = \frac{2WH}{S + 0.1} \tag{90}$$

This fundamental dynamic formula is probably the most widely used. Many of the other formulas are actually refinements of the Engineering News Formulas, which include additional driving system parameters and energy losses. Despite repeated attempts to develop a highly reliable driving formula, the application of all dynamic formulas should be accompanied with caution.

Dynamic formulas are probably most useful when calibrated with static load test results for a particular pile type at specific sites or in local geologic regions. As an example, the Michigan Highway Department conducted a comprehensive pile load test program in 1961 consisting of 88 piles driven with five different pile driving hammers. All of the piles were load tested statically to failure and their ultimate capacities were compared with predictions from dynamic formulas. Upon analyzing test results some changes were recommended to the Engineering News Formula resulting in the development of the Modified Engineering News Formula shown below:

$$R = \frac{2.5\,E\,(W_r + e^2\,W_p)}{(S + 0.1)(W_r + W_p)} \tag{91}$$

where

Q_a = allowable pile load capacity (lb)
E = rated hammer energy (ft–lb)
S = final average penetration of pile per blow (in)
W_r = weight of ram (lb)
W_p = weight of pile, including driving appurtenances (lb)
e = coefficient of restitution (dimensionless)

Another dynamic formula worthy of note is the Hiley Formula, which reportedly provides one of the highest correlations with static load capacity. The extended form of the equation is:

$$Q_a = \frac{e_h\,E}{s + \frac{1}{2}\,(c_1 + c_2 + c_3)}\;\frac{W_r + e^2\,W_p}{W_r + W_p} \tag{92}$$

where

Q_a, E, s, W_p, W_r, and e are as defined for Equation (91)

e_h = hammer efficiency (dimensionless)
c_1 = elastic compression of pile head assembly (in)
c_2 = elastic compression of the pile (in)
c_3 = elastic compression of soil, also known as "quake" (in)

An inspection of the Hiley Formula shows that it is an expansion of the basic Engineering News Formula, but considers additional energy losses as summarized below:

1. The energy at impact is reduced by the coefficient e_h to account for the frictional losses in the hammer system.
2. The energy during impact is reduced by the coefficient of restitution, e, depending on the condition at the pile head assembly.
3. Additional energy losses occur due to the elastic compression of the pile head assembly (c_1), the pile (c_2), and soil (c_3).

WAVE EQUATION

The primary shortcoming of the dynamic formulas is their assumption that the pile is perfectly rigid resulting in instantaneous transmission of the driving forces from the top to the bottom of the pile. In contrast, the wave equation method recognizes that each hammer blow produces a stress wave which travels down the length of the pile causing non-uniform stresses along the pile at any instant. As with dynamic formulas, the wave equation seeks to establish a relationship between the ultimate pile load and the penetration resistance, but it provides a more rational analysis of the various factors in the driving process including the characteristics of the pile, hammer, cushion, and soil. The method also includes a computation of the stresses in the pile during driving which is useful in selecting the optimum combination of hammer and pile as well as checking whether or not a particular driving system will damage a pile.

The theoretical basis for the wave equation is the partial differential equation which describes one-dimensional wave propagation in a freely suspended bar subjected to an impact. The basic equation includes an additional term for soil resistance and may be written as follows:

$$\frac{\partial^2 D}{\partial t^2} = \frac{E}{s}\frac{\partial^2 D}{\partial x^2} \pm R_s \tag{93}$$

where

D = longitudinal displacement of the bar from its original position at a point
t = time
E = modulus of elasticity of the bar
s = mass density of bar
x = distance along longitudinal axis
R_s = soil resistance term

Equation (93) is typically solved numerically using the finite-difference method. The actual pile, hammer, and soil system are idealized using the mechanical model shown in Figure 33. Computer programs are available for solving the wave equation to establish a relationship between the ultimate capacity and the penetration resistance. These analyses are typically performed prior to driving any piles in the field and are used to interpret field data.

Recent developments over the last 20 years in electronics instrumentation have made it feasible to measure dynamic parameters other than penetration resistance during actual pile driving. The Case Method developed by Goble, et al. (1975) involves the field measurement of force and acceleration at the top of a test pile during driving. These measurements can then be correlated by various means with a prediction of static pile capacity. Since actual pile capacity in the soil may change with time due to set-up or relaxation effects, the Case Method is often applied by restriking a pile some period of time after the termination of initial driving.

Pile Load Tests

The most reliable way to determine the load capacity of a pile is to load test it in the field. "Static" load tests, as they are called, should ideally be performed on all piling projects. However, they are considerably more expensive than the other methods used to determine pile capacity, and economic considerations sometimes preclude their use on projects.

Similitude is the key to performing pile load tests. It is essential that the piles and the driving systems used in the load test be identical with those used in the production operations. Also, while driving of the test piles, the penetration resistance must be carefully recorded so that if the test is successful, the same driving criterion can be applied to the production piles. Even slight differences between the test and production piles, e.g., blowcounts, hammer operating pressure, or capblock material, will invalidate the similitude and may result in unsatisfactory foundation performance.

Load tests should be correlated with nearby boring logs to aid in the interpretation of results. If the general character of the subsurface conditions changes throughout the project site, at least one pile load test should be performed in each area. The actual procedure for performing pile load tests is often dictated by the local building code. A general method is available from the American Society of Testing Materials (ASTM) in their Standard D 1143. Most procedures require that the test piles be loaded to at least twice the design capacity proposed for use in the final construction. Loading is normally applied to the test piles with hydraulic jacks, and load reaction is provided by one of the

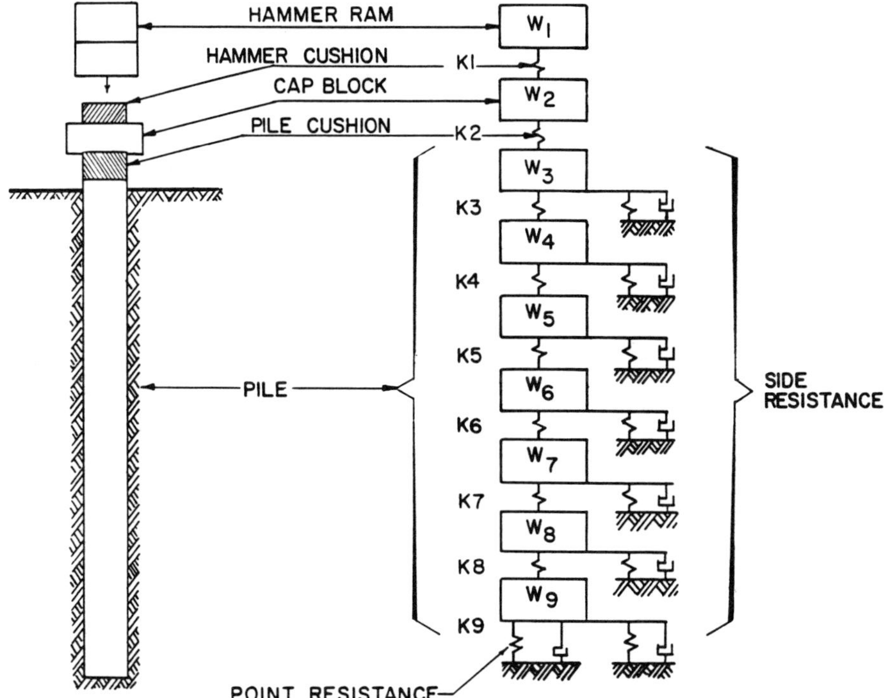

FIGURE 33. Idealized mechanical model of pile for wave equation analysis.

two following methods: 1) dead load and 2) anchor piles. The load to the test pile is applied in increments with 25% of test load being typical.

Typical load settlement curves for two different pile load tests are shown in Figure 34. Curve (a) represents a pile which derives its support primarily from skin resistance as with a friction pile in a cohesive soil. It displays a distinct ultimate capacity beyond which it "plunges." Curve (b) represents a pile which is primarily end-bearing. Unlike the friction pile, it does not show a well defined ultimate capacity, but settles with a gradually increasing rate. The ultimate capacity of a pile which behaves like curve (b) obviously requires some amount of interpretation. It should be apparent that piles which derive their support from both friction and end-bearing will display a load settlement behavior which is intermediate between curves (a) and (b).

In view of the fact that many load tests do not display distinct failures, or if they do, the failure occurs at very large settlements, it is necessary to establish a criterion to define the load capacity. For example, Terzaghi (1942) defined the ultimate capacity as that which causes a penetration equal to 10% of the pile diameter. A number of other rules for establishing load capacity have been developed based on settlement observations during load tests. A sampling of the criteria for "working" or allowable capacity from various sources is provided below:

1. *1981 New York City Building Code:* The allowable pile load is the lesser of two values computed as follows:
   a. "Fifty (50) percent of the applied load causing a net settlement of the pile of not more than 0.01 inches per ton of applied load. Net settlement in this paragraph means gross settlement due to the total test load minus the rebound after removing 100% of the test load.
   b. Fifty (50) percent of the applied load causing a net settlement of the pile of 0.75 inches. Net settlement in this paragraph means the gross settlement as defined in 1a. above less the amount of elastic shortening in the pile section due to total test load."
2. *AASHTO:* The allowable pile load is that which causes a net settlement (gross settlement minus rebound) of the test pile of 0.25 inches.
3. *Raymond Concrete Pile Company:* The allowable pile load is that which causes a gross settlement of 0.05 inches per ton of additional load, or at which the plastic settlement exceeds 0.03 inches per ton of additional load.

Test piles are usually driven in advance of the production piles to establish a driving criterion. When load testing piles in cohesive or other soils sensitive to disturbance by driving, it is necessary to wait up to several days to permit dissipation of pore pressures, otherwise the load test results would not be representative of the long-term carrying capacity of the pile.

Most soils and rock formations exhibit a certain amount of creep under test loads so that the load test procedure must include a limitation on the maximum rate of settlement permitted before the load can be increased to the next increment. Frequently, a holding period is specified at the total test load to check for stable behavior of the test pile. Depending on the specified holding period, a pile load test typically takes a few days or more to perform.

Butt settlement is always monitored during a load test using dial extensometers and/or optical surveying. Installation of settlement "telltales" at the tip or at other intermediate points along the pile shaft are useful in interpreting the load test results.

Driving tests are sometimes specified in addition to, or instead of, pile load tests. Driving tests may be justified as a

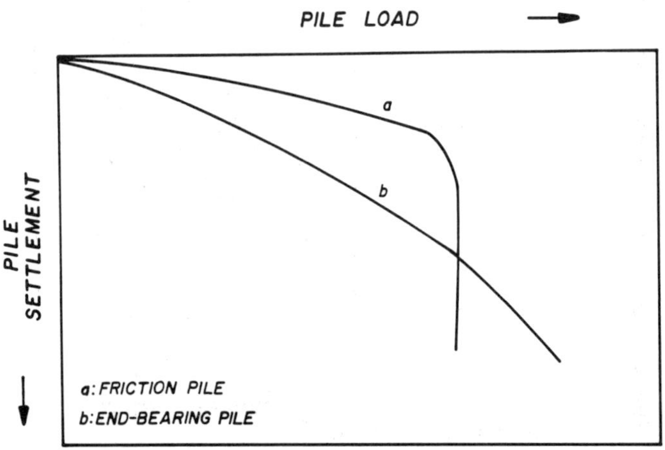

FIGURE 34. Typical load settlement curves for pile load tests.

less expensive alternative to load testing if extensive driving experience and load test data is available from previous piling projects in the same vicinity. Driving tests are performed by carefully observing the penetration resistance of selected test piles and correlating this data with adjacent boring logs. The piles are frequently overdriven well beyond the tip elevation and penetration resistance proposed for the production piles in an attempt to assess ultimate pile behavior. This overdriving also provides information on the soil or rock strata below the production piles.

As pointed out during the discussion of pile group behavior, the load capacity measured for a single test pile may not be representative of a group of test piles. The previously described considerations of group efficiency and settlement must be investigated.

Factor of Safety

For pile foundations, factor of safety may be defined as:

$$FS = \frac{\text{ultimate failure load of pile}}{\text{working load on pile}} \tag{94}$$

When selecting a suitable safety factor for design, the following must be considered:

1. The variability of the soil or rock conditions at the site
2. The certainty to which the properties of the soil or rock are known
3. The method used to predict the ultimate capacity
4. Experience of the designer under similar conditions
5. The structure's tolerance to settlement
6. The probability that the actual loads will exceed the estimated loads
7. Construction tolerances and procedures
8. The relative economy of increasing or decreasing the safety factor

The safety factor applied to driven pile foundations is usually 1.5 or more, depending on the designer's assessment of the factors listed above. For field load tests, most building codes require that a safety factor of 2.0 be applied to the total test load whether or not the test was carried out to failure. Tomlinson (1977) recommends safety factor of 2.5 for piles, and indicates that cumulative experience has indicated that this value will generally result in pile settlements of less than 0.4 inches at the working load. A safety factor of 3.0 is used by many designers when static analysis is the only method used to establish load capacity. Many dynamic formulas already have a safety factor built-in to them such as the Engineering News Formula, which presumably contains a safety factor of 6.0.

Negative Skin Friction

When end-bearing piles are installed through cohesive soils which are undergoing consolidation, the pile is subject to drag-down loads as the soil slips by the surface of the shaft. This drag force imposed on the pile is known as "negative skin friction." The two most common site conditions where negative skin friction should be anticipated are shown in Figure 35.

The downward load caused by negative skin friction is additive to the loads imposed on the pile by the strucutre. This additional load may result in a total pile load which exceeds the design capacity. Structural collapse of the pile shaft can result.

Vesic (1977) reports that full negative skin friction may be expected when there is a relative movement of 0.6 inches between the soil and the pile shaft. It is important to note negative skin friction only develops along that portion of the pile shaft where the soil settlement exceeds the downward displacement of the pile shaft. That point along the pile shaft where the soil settlement and the pile displacement are equal is known as the "neutral point." The location of the "neutral point" varies and is dependent on the pile characteristics and the site conditions. It will depend, for example, on the strength of the bearing stratum and the tendency of the pile tip to settle in it. Vesic (1977), in a review of related literature, reported that the "neutral point" for three separate investigations occurred approximately at 75% of the pile length as measured from the surface, and this value is frequently used as a general estimate.

A conservative approach to estimate the magnitude of negative skin friction is to assume that it will be equal to or less than the mangitude of the positive skin friction. Therefore, methods discussed previously are applicable and the total negative skin friction force for individual piles would be:

$$F_n = \sum_{o}^{z} \tau_a P \tag{95}$$

where

F_n = total negative skin friction force
τ_a = shear stress between pile and soil
P = perimeter of pile

The friction of all soil layers above the neutral point must be included when evaluating this expression, including any overlying fill soil. The shear stress, τ_a, must be evaluated using drained soil parameters and may be calculated as follows:

$$\tau_a = c_a + K_s\bar{\sigma}_v \tan \delta' \tag{96}$$

c_a = drained adhesion between pile and soil
K_s = lateral earth pressure coefficient
$\bar{\sigma}_v$ = effective vertical overburden pressure
δ' = drained angle of friction between pile and soil

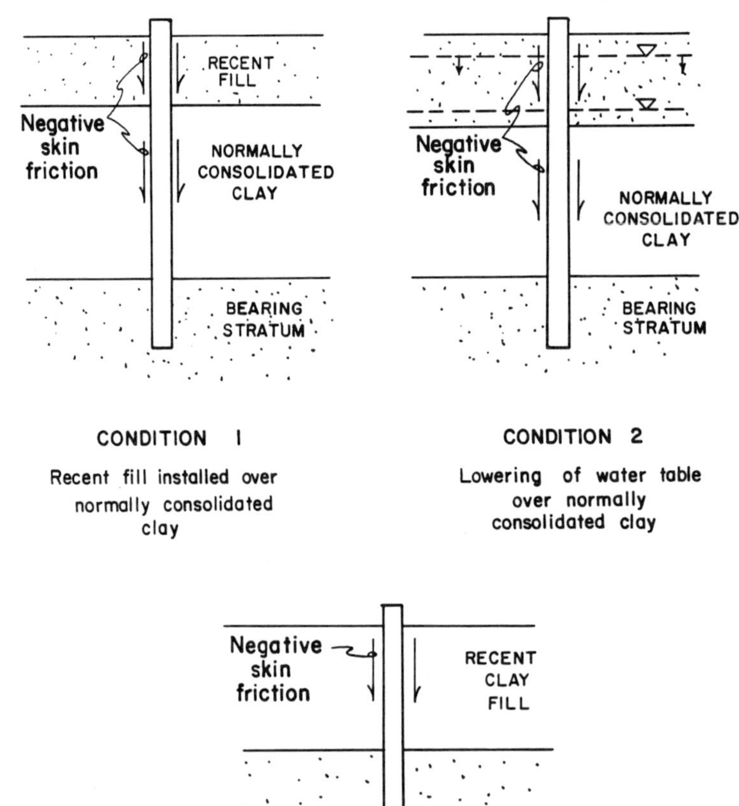

FIGURE 35. Common site conditions for negative skin friction.

It is recommended that the contribution of any positive friction to the load capacity of the pile be neglected in the zone where negative skin friction occurs. It is also interesting to note that negative skin friction may sometimes reduce the point-bearing capacity of piles in addition to adding an additional load. Zeeveart (1959) shows that negative skin friction reduces the effective vertical overburden pressure, so that if the piles are bearing in sand, the load resistance will also be reduced.

Batter piles should be avoided in soil deposits where negative skin friction occurs, since drag-down loads on the upper side of the pile will be even higher than those estimated with Equation (95) above. This condition is aggravated by the settlement of the soil away from the underside of the battered pile. Under these conditions, substantial pile bending and horizontal thrust will occur.

When estimating negative skin friction for pile groups, it is often assumed that the total negative skin friction force will not exceed the total weight of the soil enclosed by the group. This should include the weights of both the compressible soil and overlying fill, and effective soil unit weights should be used.

Some field procedures have been developed to reduce the magnitude of negative skin friction. They include:

1. Application of a bituminous coating on the pile shaft
2. Predrilling an oversized hole and filling around the pile with a bentonite slurry
3. Installation of a casing sleeve around the pile

Of these procedures the most effective at present appears to be the application on the bituminous coating. Reductions in negative skin friction of up to 90% have been attained.

Lateral Capacity of Piles

In many applications, piles are subjected to lateral loading such as from wind or earth pressure. The resistance of pile to lateral load depends on: 1) its structural bending characteristics and 2) the passive support provided by the soil. If the magnitude of the lateral load is small, no formal analysis is performed since many governing construction codes specify a maximum allowable lateral load. Some typical values for allowable lateral forces on piles are given in Table 9.

When either the lateral loads exceed those given by the governing code, or in special situations where even small lateral deflections of the foundation cannot be tolerated, more elaborate methods of analysis are justified. Generally, two different kinds of lateral loading problems are encountered:

1. Determination of ultimate lateral load capacity
2. Determination of lateral pile deflections under loads less than ultimate

The approach to solving each of these problems is different, and they will be treated separately.

TABLE 9. Safe Allowable Lateral Load on Vertical Piles, in Pounds [McNulty (1956)].

Pile Type	Medium Sand	Fine Clay	Medium Clay
Free-head Timber, 12 in. Diameter	1500	1500	1500
Fixed-head Timber, 12 in. Diameter	5000	4500	4000
Free-head Concrete, 16 in. Diameter	7000	5500	5000
Fixed-head Concrete, 16 in. Diameter	7000	5500	5000

ULTIMATE LATERAL LOAD CAPACITY

A widely used method to analyze the ultimate behavior of pile was developed by Broms for cohesive soils (1964a) and cohesionless soils (1964b). This approach distinguishes between short and long piles, as well as free-headed and fixed-headed piles. The assumed ultimate behavior for the various combinations of length and head fixity is shown in Figure 36.

It is apparent from Figure 36 that the ultimate lateral capacity of short piles is controlled by the failure of the soil along the length of the pile. In contrast, the lateral capacity on a long pile depends on the moment capacity of the pile. An intermediate condition was recognized by Broms as indicated for the fixed-head piles only.

Solution diagrams which relate the ultimate lateral load, P_u, to the other pertinent parameters are given in Figure 37 for cohesive soils and Figure 38 for cohesionless soils. For long piles, moment on the pile shaft at the eventual point of yielding, M_{yield}, may also be determined from these figures.

LATERAL DEFLECTION UNDER LOAD

Matlock and Reese (1957, 1961) have presented a set of solutions for estimating the deflection and moment caused by the application of a lateral load to the pile head at loads less than ultimate. The method assumes that the modulus of subgrade reaction, k_h, increases linearly with depth along the pile. This assumption is reasonably valid in most soils with the exception of heavily overconsolidated clays where the modulus is relatively constant with depth. A correlation procedure to account for effects of overconsolidation is available (Carter, 1983).

The method begins with a determination of n_h, the coefficient of subgrade reaction which is related to k_h by:

$$n_h = k_h \left(\frac{D}{z}\right) \tag{97}$$

where

D = pile width
z = depth

Guidance for selecting n_h is presented in Figure 39.

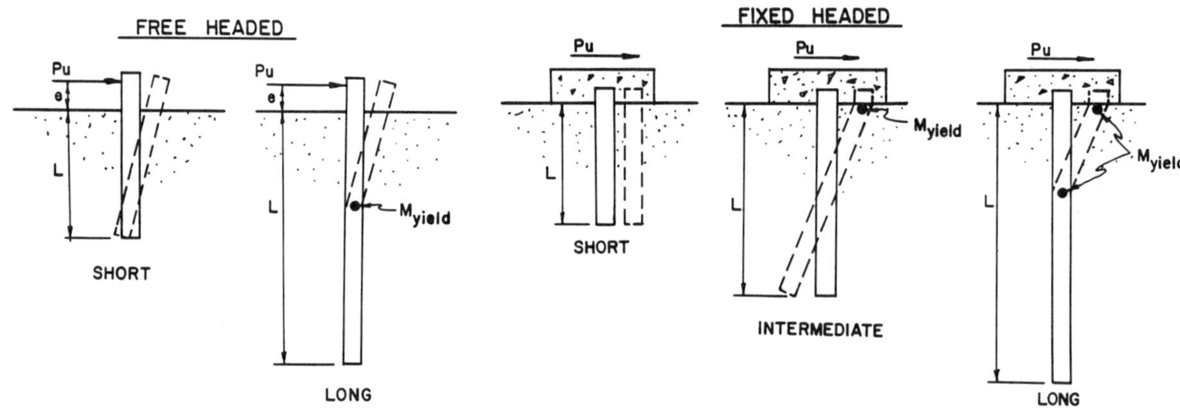

FIGURE 36. Ultimate behavior of laterally loaded piles.

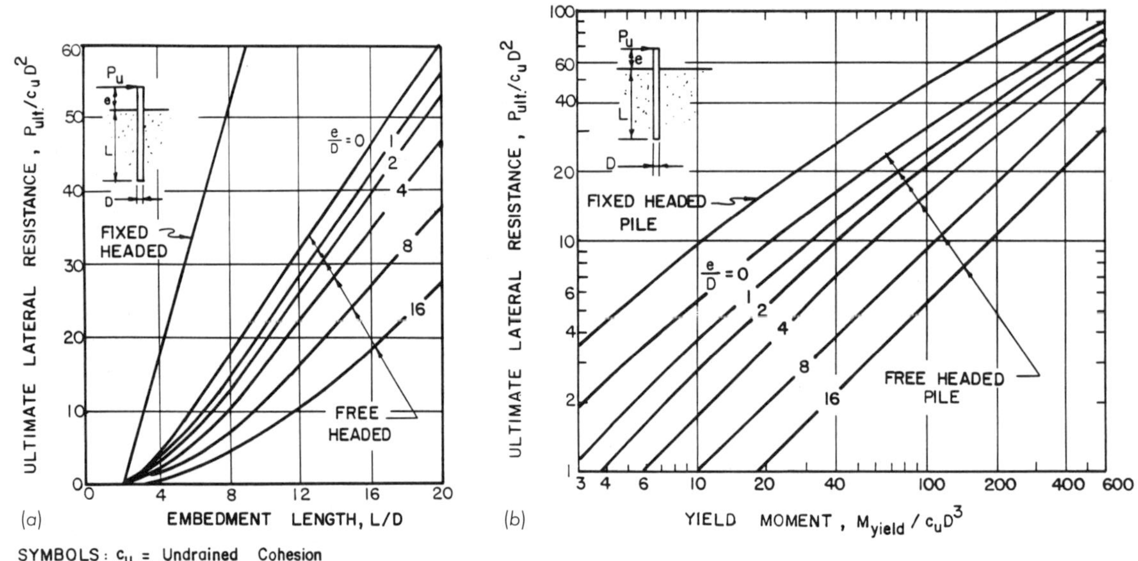

FIGURE 37. Ultimate lateral resistance in cohesive soils [after Broms (1964a)]: (a) short piles; (b) long piles.

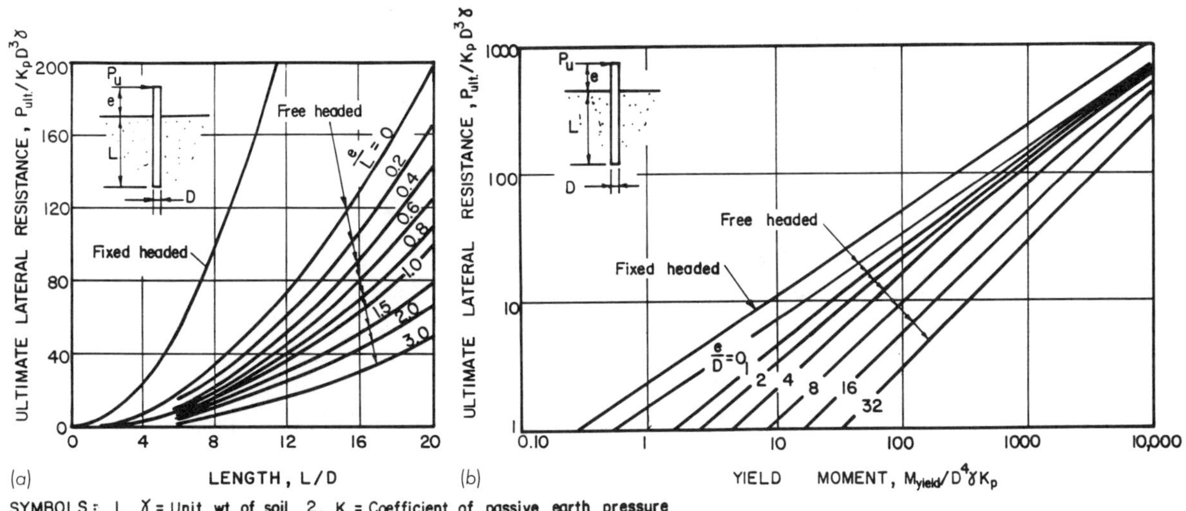

FIGURE 38. Ultimate lateral resistance in cohesionlesss soils [after Broms (1964b)]: (a) short piles; (b) long piles.

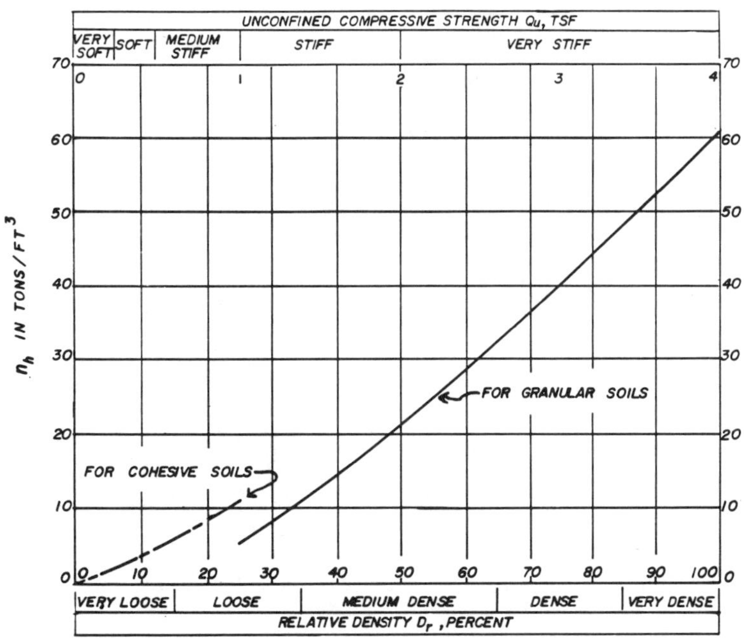

FIGURE 39. Coefficient of subgrade reaction, n_h (after U.S. Navy).

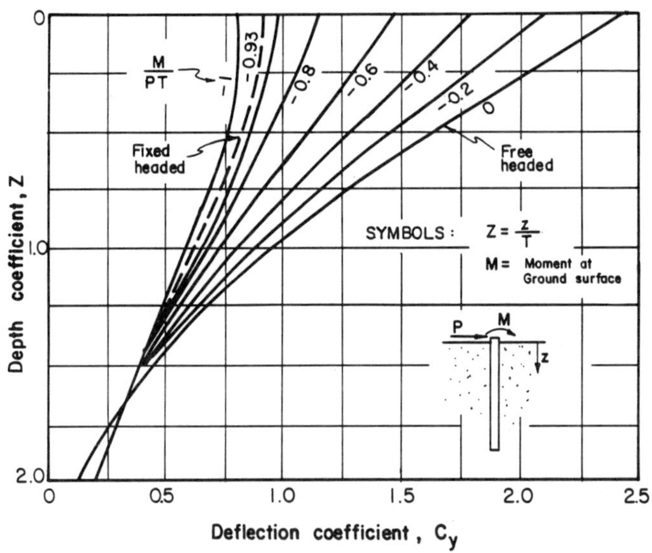

FIGURE 40. Deflection coefficients for laterally loaded piles [after Matlock and Reese (1961)].

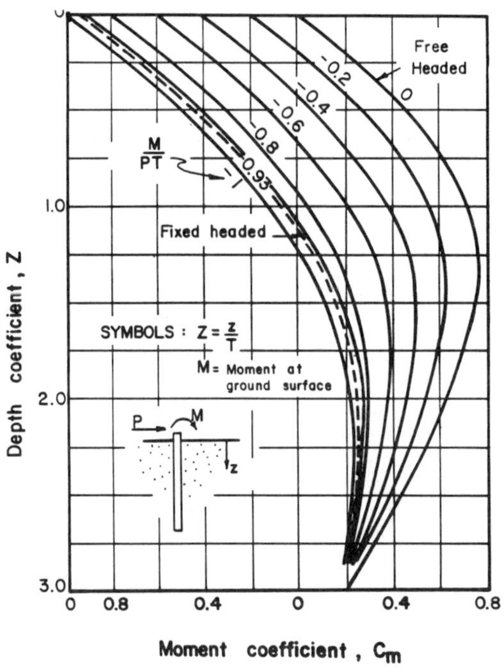

FIGURE 41. Moment coefficients for laterally loaded piles [after Matlock and Reese (1961)].

Once n_h is obtained, a relative stiffness factor, T, is computed by:

$$T = \left(\frac{EI}{n_h}\right)^{1/5} \tag{98}$$

where

T = relative stiffness factor, in
E = modulus of elasticity of pile, lb/in^2
I = moment of inertia of pile, in^4
n_h = coefficient of subgrade reaction expressed in lb/in^3

The lateral deflection of the pile is then given by:

$$y = \frac{C_y P T^3}{EI} \tag{99}$$

where

y = lateral deflection, in
P = applied horizontal force at ground level, lb
C_y = deflection coefficient from Figure 40

The moment on the pile at any depth may also be determined by:

$$M_z = C_m PT$$

where

M_z = maximum moment on pile shaft, in-lb
C_m = moment coefficient from Figure 41

REFERENCES

ASTM D 1143-81, "Piles Under Static Axial Compressive Load," American Society for Testing and Materials, Philadelphia, 16 pp. (1981).

Berezantzev, V. G., V. Khristorforov and V. Golubkov, "Load Bearing Capacity and Deformation of Piled Foundations," *Proc. 5th Int. Conf. Soil Mechanics and Foundation Engineering*, Vol. 2, pp. 11-15 (1961).

Bowles, J. E., *Foundation Analysis and Design*, 3rd Ed., New York:McGraw Hill Book Co. (1982).

Broms, B. B., "Lateral Resistance of Piles in Cohesive Soils," *Journal Soil Mechanics and Foundation Division*, ASCE, Vol. 90, SM. 2, pp. 27-63 (1964a).

Broms, B. B., "Lateral Resistance of Piles in Cohesionless Soils," *Journal Soil Mechanics and Foundation Division*, ASCE, Vol. 90, SM. 3, pp. 123-156 (1964b).

Broms, B. B., "Methods of Calculating the Ultimate Bearing Capacity of Piles, A Summary," *Sols-Soils*, Vol. 5, No. 18-19, pp. 21-31 (1966).

Brown, J. D. and G. G. Meyerhof, "Experimental Study of Bearing Capacity in Layered Clays," *Proc. 7th Int. Conf. Soil Mech. and Found. Eng.*, Mexico City, Vol. 2, pp. 45-51 (1969).

Chandler, R. J., "The Shaft Friction of Piles in Cohesive Soils in Terms of Effective Stresses," *Civil Engineering Public Works Review*, Vol. 63, pp. 48-51 (1968).

Das, B. M., *Principles of Foundation Engineering*, Monterey, CA:Brooks/Cole Engineering Division Co. (1984).

DeBeer, E. E., "Experimental Determination of the Shape Factor and the Bearing Capacity Factors for Sand," *Geotechnique*, Vol. 20, No. 4, pp. 387-411 (1970).

Goble, G. G., G. E. Likins and F. Rausche, "Bearing Capacity of Piles From Dynamic Measurements," Final Report, Dept. of Civil Engineering, Case Western Reserve Univ., Cleveland, OH (March 1975).

Hansen, J. B., "A Revised and Extended Formula for Bearing Capacity," Danish Geotechnological Institute, Bulletin 28, Copenhagen (1970).

Hansen, J. B., "The Philosophy of Foundation Design: Design Criteria, Safety Factors, and Settlement Limits," *Proceedings Symposium on Bearing Capacity and Settlement of Foundations*, Durham, NC:Duke University, pp. 9-13 (1967).

Hansen, J. B., "A General Formula for Bearing Capacity," Danish Geotechnical Institute, Bulletin No. 11, Copenhagen (1961).

Housel, W. S., "A Practical Method for the Selection of Foundations Based on Fundamental Research in Soil Mechanics," Univ. of Michigan, *Eng. Res. Bulletin 13*, Ann Arbor, MI (1979).

Mansur, C. I. and A. H. Hunter, "Pile Tests—Arkansas River Project," *Journal Soil Mechanics and Foundation Division*, ASCE, Vol. 96 SM. 5, pp. 1545-1582 (1970).

Matlock, H. and L. C. Reese, "Foundation Analysis of Offshore Pile Supported Structures," *Proceedings 5th Int. Conf. Soil Mechanics and Foundation Engineering*, Vol. 2, pp. 91-97 (1961).

McNulty, J. F., "Thrust Loading on Piles," *Journal Soil Mechanics and Foundation Division*, ASCE, Vol. 82, No. SM 4, Paper 1081 (1956).

Meyerhof, G. G., "Bearing Capacity and Settlement of Pile Foundations," *Journal Geotechnical Engineering Division*, ASCE, Vol. 102, No. GT 3, pp. 195-228 (1976).

Meyerhof, G. G., "General Report: Outside Europe," *Proc. Conf. on Penetration Testing*, Stockholm, Vol. 2, pp. 40-48 (1974).

Meyerhof, G. G., "Safety Factors in Soil Mechanics," *Canadian Technical Journal*, Vol. 7, No. 4, pp. 349-355 (1970).

Meyerhof, G. G., "Shallow Foundations," *Journal Soil Mechanics and Foundation Division*, ASCE, Vol. 91, SM 2, pp. 21-31 (1965).

Meyerhof, G. G., "Some Recent Research on the Bearing Capacity of Foundations," *Canadian Geotechnical Journal*, Vol. 1, No. 1, pp. 16-26 (1963).

Meyerhof, G. G., "Penetration Tests and Bearing Capacity of Cohesionless Soils," *Journal Soil Mechanics and Foundation*, ASCE, Vol. 82, No. SM 1, pp. 1-19 (1956).

Meyerhof, G. G., "The Bearing Capacity of Foundations Under Eccentric and Inclined Loads," 3rd ICSMFE, Vol. 1, pp. 440-445 (1953).

Meyerhof, G. G., "The Ultimate Bearing Capacity of Foundations," *Geotechnique,* Vol. 2, No. 4, pp. 301-332 (1951).

Meyerhof, G. G. and A. M. Hanna, "Ultimate Bearing Capacity of Foundations in Layered Soil Under Inclined Loads," *Canadian Geotechnical Journal*, Vol. 15, No. 4, pp. 565-572 (1978).

Michigan State Highway Commission, "A Performance Investigation of Pile Driving Hammers and Piles," Lansing, Michigan, 338 pages (1965).

New York City, *Building Code of the City of New York,* Dept. of Buildings, Chap. 26, Title C, Part II, Article 11 (1981).

Nordlung, R. L., "Bearing Capacity of Piles in Cohesionless Soils," *Journal Soil Mechanics and Foundation Division,* ASCE, Vol. 89, SM 3, pp. 1-36 (1963).

O'Neill, M. W. and L. C. Reese, "Behavior of Bored Piles in Beaumont Clay," *Journal Soil Mechanics and Foundation Division,* ASCE, Vol. 98, No. SM 2, pp. 195-213 (1972).

Peck, R. B., W. E. Hansen and T. H. Thornburn, *Foundation Engineering*, 2nd Ed., New York:Wiley (1974).

Poulos, H. G. and E. H. Davis, *Pile Foundation Analysis and Design,* Chap 3, New York:Wiley (1980).

Reddy, A. S. and R. J. Srinivasan, "Bearing Capacity on Layered Clay," *Journal of the Soil Mechanics and Foundations Division,* ASCE, Vol. 93, No. SM 2, pp. 83-99 (1967).

Skempton, A. W., "Summing Up," Symposium on Large Bored Piles, London, p. 155 (1966).

Smith, E. A. L., "Pile Driving Analysis by the Wave Equation," *Journal Soil Mechanics and Foundation Division,* ASCE, Vol. 86, SM 4, pp. 35-61 (1960).

Terzaghi, K., *Theoretical Soil Mechanics,* New York:Wiley (1943).

Terzaghi, K., "Discussion of the Progress Report of the Committee on the Bearing Value of Pile Foundations," *Proceedings, ASCE,* Vol. 68, pp. 311-323 (1942).

Terzaghi, K. and R. B. Peck, *Soil Mechanics in Engineering Practice,* 2nd Ed., New York:McGraw Hill Book Co. (1967).

Tomlinson, M. J., *Pile Design and Construction Practice,* Viewpoint Publications, Chap. 4 (1977).

U.S. Navy, "Design Manual 7.2, Foundations and Earth Structures," Naval Facilities Engineering Command, U.S. Government Printing Office, p 7.2-236 (1982).

Vesic, A. S., "Design of Pile Foundations," Transportation Research Board, Washington, DC, Synthesis 42, pp. 25-26 (1977).

Vesic, A. S., "Bearing Capacity of Shallow Foundations," Chapter 3 in *Foundation Engineering Handbook*, H. F. Winterkorn and H. Y. Fang, eds., New York:Van Nostrand, Reinhold Co. (1975).

Vesic, A. S., "Analysis of Ultimate Loads of Shallow Foundations," *Journal of the Soil Mechanics and Foundations Division,* ASCE, Vol. 99, No. SM 1, pp. 45-73 (1973).

Vesic, A. S., "Tests on Instrumented Piles, Ogeechee River Site," *Journal Soil Mechanics and Foundation Division,* ASCE, Vol. 96, SM 2, pp. 561-584 (1970).

Vesic, A. S., "A Study of Bearing Capacity of Deep Foundations," Final Report Project B-189, School of Civil Engineering, Georgia Inst. of Tech., Atlanta, GA (1967).

Vesic, A. S., "Bearing Capacity of Deep Foundations in Sand," Highway Research Record No. 39, National Academy of Sciences, pp. 112-153 (1963).

Vijayvergiya, V. N. and J. A. Focht, "A New Way to Predict the Capacity of Piles in Clay," *Proceedings 4th Annual Offshore Technical Conference,* Vol. 2, Houston, pp. 865-874 (1972).

Wellington, A. M., "Formulae for Safe Loads of Bearing Piles," *Engineering News,* Vol. 20, pp. 509-512 (1888).

Whitaker, T., "Experiments with Model Piles in Groups," *Geotechnique*, Vol. 7, pp. 147-167 (1957).

Whitaker, T. and R. W. Cooke, "An Investigation of the Shaft and Base Resistances of Large Bored Piles in London Clay," *Proceedings Symposium of Large Bored Piles*, Institute of Civil Engineers, London, pp. 7-49 (1966).

Wu, T. H. and L. M. Kraft, "The Probability of Foundation Safety," *Journal Soil Mechanics and Foundation Division*, ASCE, Vol. 93, SM. 5, pp. 213-231 (1967).

Zeevaert, L., "Reduction of Point Bearing Capacity Because of Negative Skin Friction," *Proceedings 1st Pan American Conference Soil Mechanics and Foundation Engineering,* Vol. 3, p. 1145 (1959).

SECTION TWO

Stability

CHAPTER 7 Field Performance of Embankments Over Soft Soil 223

CHAPTER 8 Stability and Deformation of Embankments on Soft Clay 267

CHAPTER 9 Stability of Axisymmetric Excavation in Clay 337

CHAPTER 7

Field Performance of Embankments Over Soft Soil

BERND R. THAMM*

INTRODUCTION

Economic and environmental constraints of modern road construction have led to an increased need to site new roads and especially by-passes of towns on poor or more marginal areas of land. Many towns are situated near estuaries, deltas, lakes or on alluvial flood plains. These areas are nearly always locations of soft and compressible subsoils. They are characterised by low strength and high voids content. Other areas are peat bogs, swamps and marshes which are soils with high organic content. To allow roads on those soft soils to be economically designed reliable assessments of the stability and of the rate and magnitude of settlement of embankments are needed. Without this information the design can be costly since more elaborate construction techniques than actually needed are often used and because unforeseen problems can arise during and after construction.

There are two limiting approaches to the design of embankments on soft soil:

- At one extreme the difficulties arising from the soft ground are avoided by their complete excavation.
- At the other extreme the engineering behaviour of the soft soil is fully taken account of.

The first approach may lead to very high costs. The latter will usually involve a much greater engineering appreciation of the problem and will generally produce very large savings, when compared to the first method.

Between these two extremes there is a range of possible measures and expedients with costs between the above limiting approaches. Less costly expedients may involve special construction procedures such as preloading the subsoil by surcharges, installing vertical drainage systems to accelerate the consolidation process and ground reinforcement with geotextiles or geomembranes.

In the past there has been an understandable tendency to avoid building on soft soil because of aforementioned problems. Recently, however, particularly due to environmental reasons, there has been a reversal of these trends leading to an increase in the number of construction projects in difficult terrain including soft areas. New construction methods and technologies are being developed at an increasing pace to meet the needs of a rapidly changing transportation program. These will require field verification before widespread implementation can be recommended. Therefore field measurements of embankment on soft soil play an important role as input to the design, to ensure safety, reduce construction costs and control construction procedures.

Measurements can also be used to ensure long-term satisfactory performance, to provide legal protection to an owner responsible for construction and to advance the state-of-the-art of geotechnical engineering.

Field performance monitoring commonly includes routine instrumentation for measurement of settlement of embankment and ground, pore pressure in the soil and horizontal spreading of the toe of the embankment. Specialized instrumentation includes geotechnical instruments for measurements of settlements at different depths, horizontal displacements of subsoil and embankment fill, location of possible failure plane, in situ stresses and strains as well as permeability.

A properly planned and specified instrumentation program can make a large contribution toward increasing safety, reducing cost and potential litigation, and advancing the state-of-the-art.

With improved technology and a better understanding of the behaviour and characteristics of compressible soils, it is

*Federal Highway Research Institute of Germany, Bergisch-Gladbach, FRG

certain that the useful life of embankments founded on such soils could be extended and their maintenance costs reduced.

Before dealing with several case histories planned and constructed mainly in the northern part of the Federal Republic of Germany (FRG) some comments on the analysis and design considerations and on the construction expedients commonly used in the FRG will be given.

A main part of this report is the general background and need for instrumentation of embankments and its function. The commonly used field instrumentation in the FRG will briefly be explained and the results obtained with this instrumentation discussed during the description of four different case studies performed on soft soils in the northern part of the FRG.

ANALYSIS AND DESIGN CONSIDERATIONS

Embankment design and analysis can be divided into stages as shown in Figure 1. Scanning through all stages shows that a great deal of modeling the reality has to be done to accomodate complex geological situations, geometries, loading conditions and material behaviour.

The numerical analysis of embankments over soft soils includes the stability analysis and the analysis of settlements and plastic deformations.

DESIGN AND ANALYSES OF AN EMBANKMENT

- interpretation of i.e. geological maps
- geotechnical investigation
- collecting data and preliminary design
- evaluation of a model

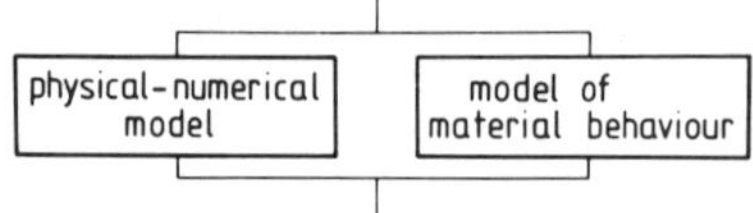

- numerical analyses
- interpretation of numerical results
- comparison with in-situ measurements and if possible correction of the model (observational method)
- design optimization process
- final design
- embankment construction
- investigation of long-term in-situ behaviour of final embankment

FIGURE 1. Stages of design and analyses of an embankment.

Stability Analysis

The method of analysis most commonly employed in practice is based on the "limit equilibrium" approach. Extensive studies have indicated that this method generally produces results which are adequate for most practical situations. The application of more complex methods do not necessarily improve the reliability of assessments of stability as the most important factor in the analysis is the quality of the soils data.

For "undrained" analysis the most important soil property is the undrained shear strength c_u. The undrained shear strength obtained by using the field vane test is found to be between the unconfined compressive strength of the soil in undisturbed and remolded conditions. In most cases the field vane shear strength was less than that obtained by laboratory undrained shear tests with pore pressure measurements.

Until quite recently the vane shear test was assumed to give a direct and accurate measure of in situ undrained shear strength. However, recent indications are that the undrained shear strength of field vane test may be in some cases unrealistically high. Bjerrum [16] attributed the problem mainly to rate of strain effects and suggested a correction vane factor to account for these effects as well as for anisotropy and progressive failure. Nevertheless engineering judgement and local experience is required to use adequate values for the analysis. The results of field vane tests can be used to compute the initial "undrained" stability of an embankment, which is in most cases the critical, and can help in selecting the appropriate construction method.

The analysis of stability can be carried out by two fundamentally different approaches. In one method the analysis considers only the conditions at the point of failure. In this method it is necessary to make some assumptions concerning the stress distributions at the point of failure. Also this method is less sound from the theoretical point of view it is simple to perform, requires less information and thus is widely used [1,2,3].

An alternative approach to analysing stability takes the form of a detailed study of the deformations occuring at a large number of locations in the soil mass. This method allows a continuous assessment to be made of soil behaviour both prior to and at the point of failure and is usually defined as the "stress–strain" method. It would normally be carried out by the finite element method on the basis of a very detailed and complete knowledge of the soil characteristics. The author has worked with this method since 1970 [4].

Settlement Analysis

The calculation of one-dimensional settlement behaviour of soft soils can be divided into two major parts: (1) initial (or undrained) settlements; and (2) final settlements due to primary and secondary consolidation. The initial settlements may be associated with the undrained, elastic deformation of the foundation. They may be computed from

$$s_i = q^*b^*[(1-\nu^2)/E_u]^*I \quad (1)$$

in which q = applied embankment load; b = width of the embankment; E_u and ν = elastic parameters of the soil; and I = stress influence factor [5].

In routine practice the total final settlement s of a layer of thickness H, with one-dimensional settlement behaviour (Figure 2) assumed is computed from

$$s = [H/(1 + e_o)]^*\{C_s^*\log(\sigma'_p/\sigma'_{vo}) + C_c^*\log[(\sigma'_{vo} + \delta\sigma'_v)/\sigma'_p] + C_\alpha^*\log(t)\} \quad (2)$$

in which σ'_p = maximum past pressure; C_s = recompression index; C_c = compression index; and C_α = coefficient of secondary consolidation.

According to Ohde [24], one-dimensional compression data can be expressed by

$$1/m_v = v^*[(\sigma'_{vo} + \delta\sigma'_v)/\sigma_{at}]^w \quad (3)$$

in which m_v = coefficient of volume compressibility; σ'_{vo} = in situ vertical effective stress; $\delta\sigma'_v$ = vertical effective stress change; σ_{at} = atmospheric pressure; and v,w = empirical soil constants derived from compression curves. The empirical soil constant, w, gives indication of the nature of soil behaviour. Values close to 0.5, e.g., confirm elastoplastic behaviour of the soil.

The coefficient of compressibility a_v can be computed from

$$a_v = m_v^*(1 + e_o) \quad (4)$$

with e_o = initial void ratio.

Depending on the magnitude of the stresses and the thickness and the compressibility of the subsoil, the embankment loads can produce settlements of the ground surface from a few centimetres to several meters. In some cases the duration of these movements may be only a few months and will be largely completed within the construction period. However in other cases it can produce settlements lasting for many years. Thus the rate of settlement is of importance to the engineer, since if these movements are large they may seriously affect the riding quality of a road and also result in its gradual deterioration.

The one-dimensional consolidation theory by Terzaghi [6] is still the basis of all methods currently employed for predicting the rate of primary settlement of compressible soils. Despite intensive research and development into the analysis of settlement, the method of calculation most widely employed is still largely based on the above approach. This simple method is often justified in practice because the variation in the soil profile and the limitations of the soils data would not warrant more involved calculations. The most important soil constant is with respect to settlement rate the coefficient of consolidation c_v, which can be computed according to Terzaghi from:

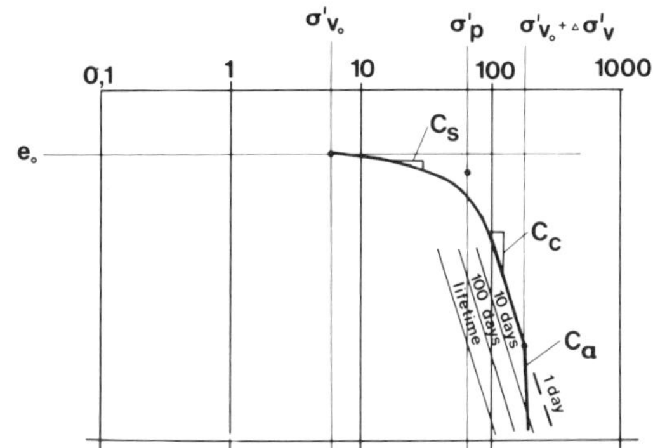

FIGURE 2. One-dimensional settlement behaviour.

$$c_v = (T_v/t)^*H^2 \quad (5)$$

in which T_v = time factor depending on the conditions of drainage; t = consolidation time; and H = thickness of layer. For important works, however, especially if two-dimensional conditions are dominant, consideration should be given to use more realistic methods of analysis which can be carried out by readily available computer programs [4].

CONSTRUCTION EXPEDIENTS

Embankments which are frequently built on layers of soft soils are confronted with geotechnical problems such as:

- failure
- settlements
- horizontal movements (Figure 3)

The engineer may have several options as to how to deal with such problems. These options are e.g.:

- to accept the in situ soil properties and vary the loading applied to the ground surface or to transfer the load to a more stabel stratum
- to improve the soil properties through prior treatment
- to replace totally or partially the compressible layer by a better material

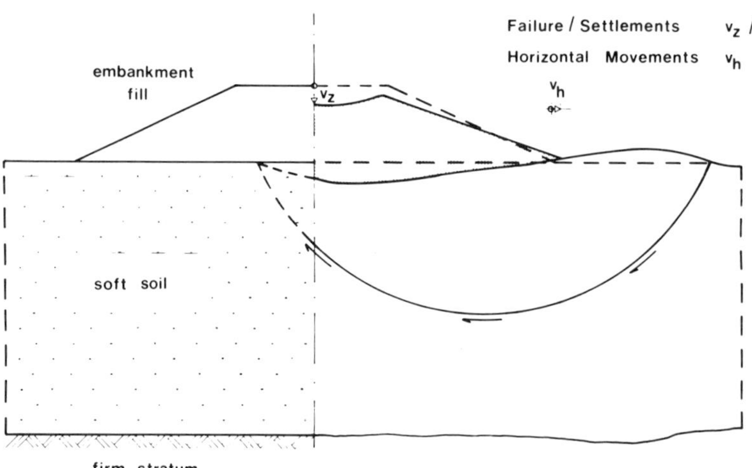

FIGURE 3. Geotechnical problems with embankments.

It is of course possible to combine those options for specific problems.

To ensure embankment stability without prior treatment of the subsoil, the following measures are commonly used in the FRG:

- modify the geometry of the embankment by reducing the embankment slopes and/or constructing loading berms
- construct the embankment by stages to derive benefit from the increase in the soil resistance during consolidation
- insert reinforcements, e.g., geotextiles or geomembranes at the embankment base to improve stability

The improvement of soil properties through prior treatment will be reached commonly by:

- the installation of vertical plastic drains
- preloading with surcharges
- compaction with heavy impact loads

Recently the partial replacement of compressible soil combined with above described measures is common practice. In some cases there has been a reversal of these trends, particularly due to environmental reasons and the impossibility of finding deposits for the replaced compressible often high organic soils. In these cases therefore reinforcements at the base of embankments in form of geotextiles or geomembranes are placed directly on the original ground surface.

INSTRUMENTATION

Embankments are often built as full-scale loading tests carried out to investigate the engineering properties of the underlying soil.

In addition to its obvious applications in these tests, instrumentation has a major role to play in a wide range of activities associated with the construction of roads on compressible soils. Observation of the settlement of embankments for purpose of assessing the total quantities of fill provides an example of a very simple and common application of the use of instrumentation. A much more complex application relates to its use for making observations on which design decisions are based. The latter approach is usually referred to as the "observational method." The method is of particular value where the compressibility of the subsoil conditions could lead to over-conservative design.

Another well established application of instrumentation relates to its use for controlling embankment construction such that the filling operations proceed at a rate consistent with the embankment stability.

To be successful an embankment must be designed and constructed such that neither the embankment material nor the underlying soil are overstressed to the point where the resulting deformations exceed the limit that is acceptable for the intended use of the embankment. Instrumentation is required, therefore, to monitor deformations as well as the factors that can be correlated to deformation and stability.

Another need for instrumentation is to document that improved stability and a reduction in post construction settlements can be achieved by means of special construction procedures such as vertical plastic drains, ground reinforcement with geotextiles or geomembranes, berms and preloading.

An example of the use of instrumentation for embankments is illustrated in Figure 4.

The usual data obtained with field instrumentation for embankments are:

- embankment settlements

- lateral and vertical movement of the natural ground at the toe of the embankment
- pore water pressures within the soft soil

The settlement data reflects the overall performance of an embankment. The initial settlements associated with the undrained elastic deformation caused by each increment of loading as the fill is placed, provide a basis for an appraisal of stability during construction.

Both magnitude and rate of settlement are important to an evaluation of post construction performance. Therefore systematic observations of the ground surface adjacent to and beneath the embankment are necessary. Additional deep settlement points are useful, in particular if the subsoil is stratified.

Lateral yielding of the ground is an important evaluation and control parameter. Therefore the measurement of horizontal displacements at serval distances from the center of the embankment and monitoring these displacements frequently during the placement of the fill is necessary. The best possible installations for these measurements are inclinometer casings.

A warning of impending failure of an embankment can be disclosed from a careful study of measured settlements, horizontal movements and pore pressures. Measured pore pressures often give a better first indication of failure conditions. Thus, if the factor of safety is low or questionable pore pressures should be monitored. In order to be certain of having instruments located in the zones of initial yielding, a relative large number of piezometer may be required to locate the critical zones.

Pore pressure measurements are also essential to evaluation of progress of consolidation settlements and to aid in establishing the rate at which an embankment can be constructed.

Test embankments which are to be deliberately failed should be instrumented with mechanical and/or electrical devices that will permit location of the rupture surface for post-failure analysis.

The range of available instrumentation is large and the choice of appropriate equipment difficult. In the selection of instruments, the overriding desirable feature is reliability. Inherent in reliability is maximum simplicity, maximum durability in the installed environment, minimum sensitivity to climatic conditions and good past performance record [7].

In the following, measurement devices will be presented which are used in the FRG for monitoring field performances of embankments.

Settlement Gauges

Settlement gauges are installed to measure the progress of settlement, especially at surcharged areas, to determine when final paving can be carried out.

The main type used are surface settlement plates, because of simplicity, very good reliability and an accuracy of ± 2mm, which is usually adequate. In peaty soils they are often the only acceptable type.

Settlement plates should be placed on a platform on the original ground before construction begins (see Figure 5). Their chief disadvantage is the presence of vertical pipes on the construction grade which tends to obstruct construction equipment.

Deep settlement points are often useful, especially if the subsoil is stratified. For measuring these points remote single or multipoint gauges are used.

Remote gauges require the leads to be placed in a trench under the proposed fill (Figure 6). These leads should be

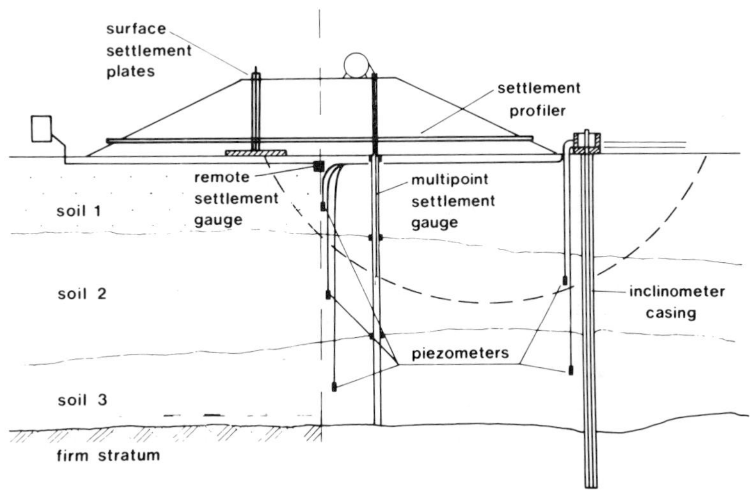

FIGURE 4. Instrumentation for embankments.

FIGURE 5. Surface settlement plates installed at ground surface.

FIGURE 6. Remote gauges with their leads placed prior to embankment filling.

FIGURE 7. Reading station for remote gauges.

buried with sufficient slack to allow for movement due to settlement. Their advantage is that they overcome the need to extend a riser pipe through the embankment. The disadvantage may be that the leads are liable to rupture under severe settlements. The remote gauges mainly used are with cells below reading station, which is accomplished by installing a pressure transducer at the bottom of a liquid column. The accuracy depends on the transducer used (±5 to 20 mm). It decreases with increased difference between elevation of cell and reading station. Due to its good reliability, robustness, stability, accuracy and simplicity pneumatic pressure transducers are very often used in the FRG. Therefore most remote gauges are read with pneumatic pressure transducers. A reading station for remote gauges is shown on Figure 7 and the pneumatic readout is performed with a compressed gas arrangement (Figure 8).

If deep settlement points are within the subsoil then multipoint gauges are used. These gauges require the installation of a plastic pipe in a vertical borehole. The space between pipe and borehole wall is grouted with a material (mostly a bentonite–cement mixture) having modulus and undrained shear strength as similar to the soil as possible.

Three different types of multipoint gauges are common in the FRG:

- a readout probe with an inductive coil and pipes having stainless steel rings at certain intervals
- a magnetic senior probe, which can detect magnetic O-type rings placed at certain intervals around the plastic pipes
- a telescoping inclinometer alu-casing (see under lateral movement gauges) with magnetic O-type rings

The probe readouts are somewhat subjective and the accuracy can be reduced by stray electric currents. But nevertheless the results obtained are of good reliability and the accuracy for most applications is adequate (±5 mm).

Figure 9 gives a view of a field application with a multipoint settlement gauge with an inductive coil and accoustic probe readout. Figure 10 shows a magnetic senior probe with cable reel and calibrated cable.

If a multipoint settlement gauge system (a so-called profiler) is installed horizontally, profiles of settlement data under or within the embankments can be obtained.

A common method is to use a horizontally placed inclinometer casing prior to filling (Figure 11) and take zero readings with a torpedo-like inclinometer at installation level. Successive readings are taken during fill placement.

Figure 12 shows a view on an embankment shoulder with one end of the casing and the inclinometer placed for measurement purposes.

Lateral Movement Gauges

Movement at the toe of an embankment is monitored primarily to detect possible impending failure. Sometimes it could be necessary also to determine the magnitude of lateral subsoil deformation, especially when piles or culverts are embedded in the embankment or when sensitive structures exist near the toe.

Two main types of lateral movement gauges are used: alignement stakes and inclinometers.

Because of simplicity and ease of operation alignement stakes are used widely to monitor lateral movements. They are usually embedded about 2 m (6.6 ft) and placed roughly on a line of sight parallel to the toe and about 2 m (6.6 ft) from it. Alignement stakes are rapidly read by measuring offsets from a line of sight using ordinary surveying techniques, e.g., a theodolite. The reference points of this line must be chosen to be on stable ground.

Inclinometers are used to measure horizontal movement also below the ground surface. Inclinometer casings with internal longitudinal guide grooves (Figure 13) are installed in boreholes, which are, whenever it is possible, drilled to a firm stratum. If firm bearing cannot be reached, determination of movement will have to be made by referencing the top by surveying techniques. The latter increases the data collection effort and decreases the data accuracy. The top of tube with installed inclinometer casing protected against deterioration in the upper regions is shown on Figure 14.

Care must be taken to ensure that the annular area between hole and casing is adequately filled (see also comments on vertical pipe installation for multipoint settlement gauges). The casings are available with rigid or telescoping couplings. They could be of aluminium, and if this reacts with some soils, they should be of plastic.

The inclinometer preferred in the FRG uses a closed-loop servo-accelerometer as a tilt sensor. It requires only 2 insertions in a casing for measuring movements in two axes. A magnetic tape readout is available (Figure 15).

The results of inclinometer measurements are of very good reliability. The accuracy is about 0.02 mm to 0.04 mm ($8 * 10^{-4}$ to $1.6 * 10^{-3}$ in) per m casing.

Inclinometers are time consuming to read and to plot the results. Data reduction may be readily carried out by using a microcomputer, which can use a magnetic tape as input. Developed software may help to analyse the results by graphical plots on the screen directly on site. A typical arrangement which helps in analyzing the enormous amount of inclinometer data is shown on Figure 16.

The automatic recording was mainly developed to control construction on the basis of allowable movements. The maximum lateral movement which is permissible before stopping construction is based largely on engineering judgement. It has been found that acceleration of movement is a much better indicator of impending instability.

Porewater Pressure Gauges

For measuring the development and dissipation of porewater pressures piezometers are installed beneath the embankments within the soft soil layers.

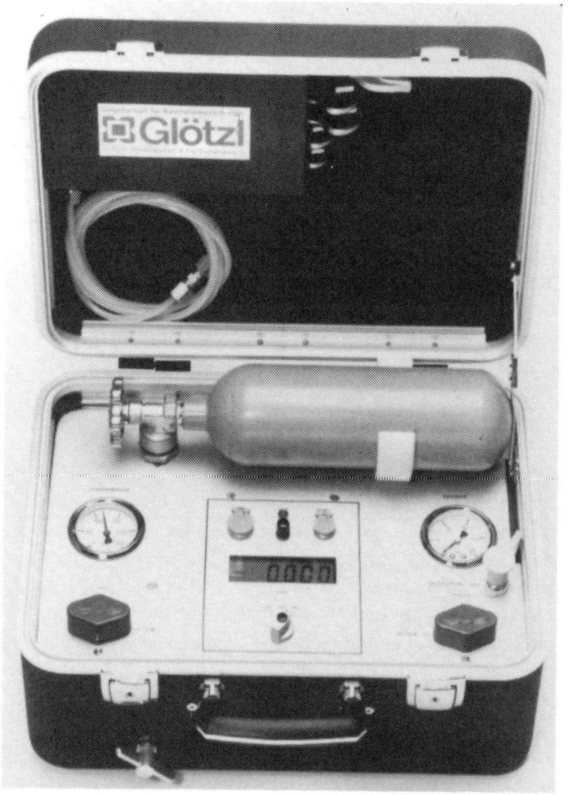

FIGURE 8. Pneumatic transducer readout arrangement.

FIGURE 9. Multipoint settlement gauge with inductive coil.

FIGURE 10. Magnetic senior probe with cable reel and calibrated cable.

FIGURE 11. View of horizontally placed inclinometer casing.

FIGURE 12. View of profiler with inclinometer torpedo.

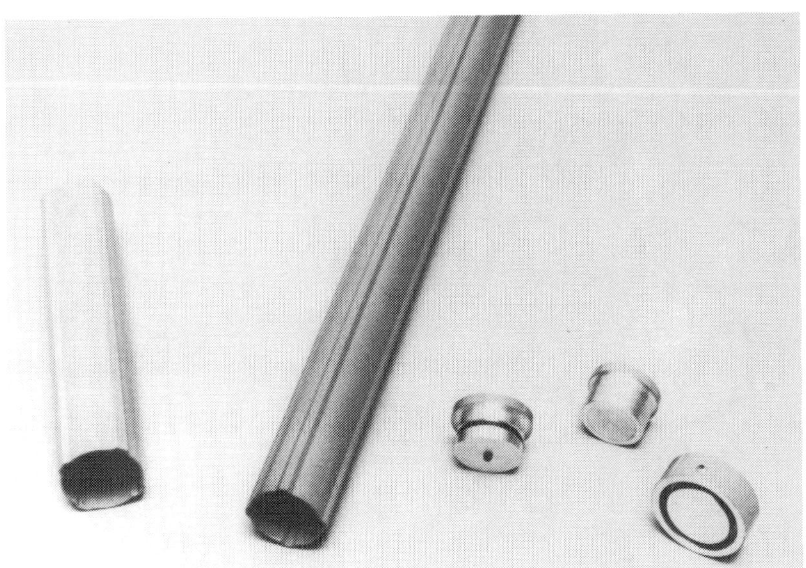

FIGURE 13. Inclinometer casing with couplings and end plates.

FIGURE 14. Top of a tube with inclinometer casing.

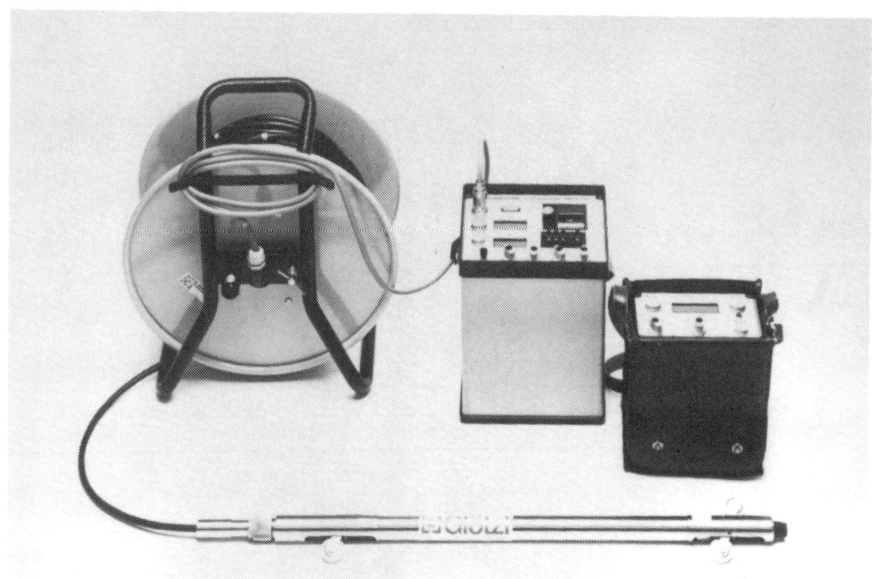

FIGURE 15. Inclinometer torpedo with cable reel, readout and magnetic tape arrangement.

FIGURE 16. Microcomputer equipment used for analyzing inclinometer data on site.

These measurements are particularly necessary when effective stress analysis controls the stability of the embankment. They are also performed to determine the in situ soil permeability. More reliable estimates of consolidation can be made with results of porewater pressure measurements, especially the degree of consolidation U_z, and thus deformations due to primary consolidation can be controlled during fill stages.

The main type of piezometers used in the FRG are:

- pneumatic piezometer
- vibrating-wire strain gauge piezometer (Figure 17)

A pneumatic piezometer consists of a porous stone (a_1), a sensitive check valve with a diaphragm (b) separating the porewater from the measuring system.

In a two tube-version, pressurized gas applied to the inlet tube (c) causes the check valve to open and vent from the outlet tube (d) when the applied gas pressure equals the porewater pressure. The pressure is read at the pressure gauge (see also Figure 8).

Figure 18 shows a pneumatic piezometer with a free porous stone at the top for the placement in a borehole. They have to be sealed carefully with a good bentonite.

A pneumatic drive type piezometer prepared with a cone for penetration into soft subsoil layers can be seen on Figure 19. Drive type piezometers are self-sealing, but the piezometers must not be rotated or displaced during driving.

Pneumatic piezometers are stable and generally preferable because of their good track records and they have a short time lag. There is a minimum interference to construction and no freezing problems are to occur.

The vibrating-wire piezometer, which also has a good track record, may be preferable where negative pore pressures could develop or where automatic recording or transmission of data over long distances is required. They are more expensive than pneumatic piezometers.

Assuming that all piezometers are installed correctly, that they are functioning and that no time lag remains, the accuracy of all piezometers can be within 0.25 m (10 in) of water head and they are working satisfactorily for short terms and uncertain for long terms. If gas is present, erroneously high readings can be taken if the tip is not vented to atmosphere.

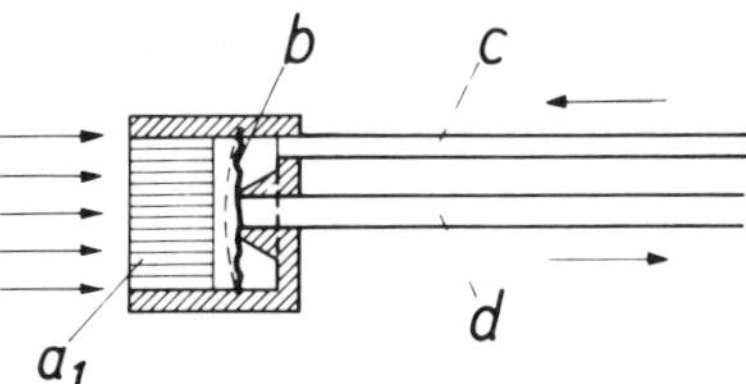

FIGURE 17. Function of a typical pneumatic piezometer.

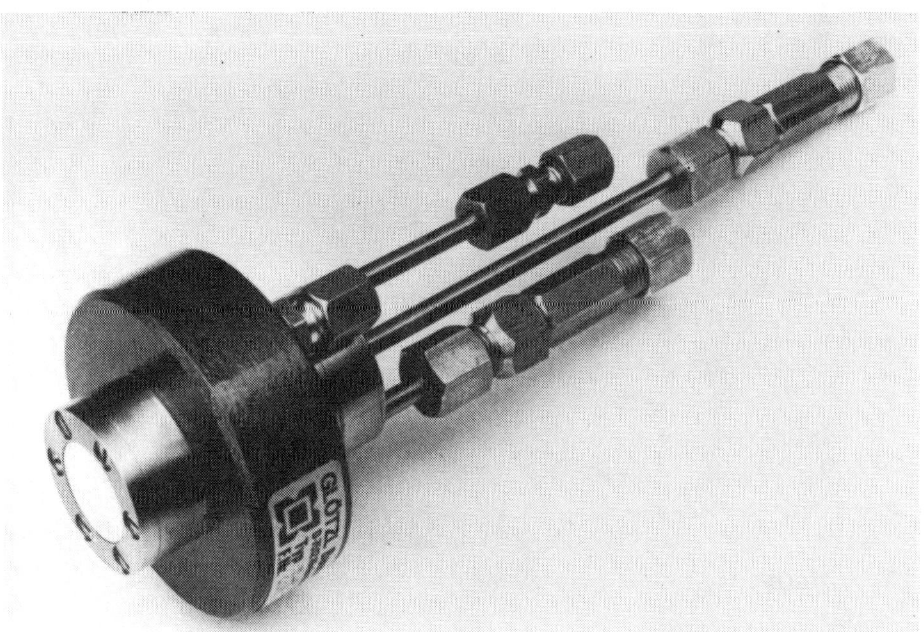

FIGURE 18. View of a pneumatic piezometer with a free porous stone.

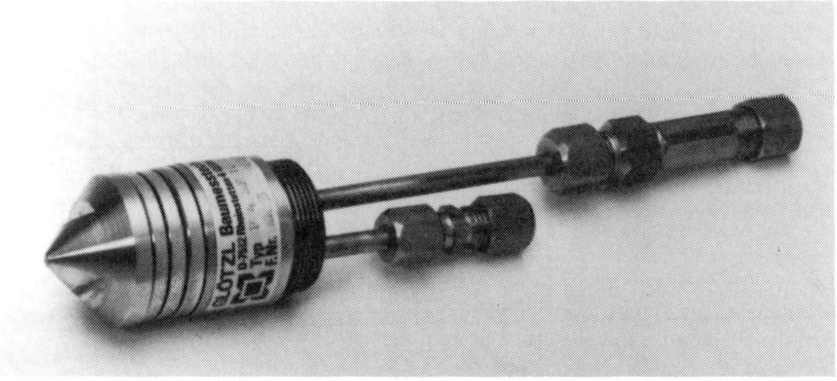

FIGURE 19. View of a pneumatic piezometer with a cone for penetration into soft soils (drive type).

It is relatively expensive to install piezometers and to monitor and evaluate the data. On routes underlain by heterogeneous materials over significant lengths, it is however impossible to instrument all the strata. A practical approach is to provide monitoring in a few cross sections in critical areas. The number of piezometers installed must be greater than required to allow for instrument malfunction or damage.

Case Histories

Since the early seventies there has been a trend leading to an increase in the number of construction projects in the FRG in difficult terrain including soft areas, particularly due to environmental reasons.

The needs of a rapidly changing transportation system in the northern part of the FRG have led to site new roads and especially by-passes of towns on poor or more marginal areas of land. These areas are nearly always locations of soft and compressible subsoils.

New construction methods and technologies were developed to met these needs, but they will require field verification before widespread implementation can be recommended. Therefore field measurements of embankments on soft soil were undertaken at several construction sites in the northern part of the FRG in order to get input to the design, to ensure safety, to reduce construction costs and to control construction procedures. The measurements were and are also used to ensure long-term satisfactory performance of roads.

The design, construction and performance of 4 different embankments constructed on soft soil are reviewed. These embankments are situated near estuaries and on alluvial flood plains in the north of the FRG. Due to the poor subsoil conditions the design included a combination of surcharge fills, stage construction or spoiling in one stage, compaction with impact loads, installation of vertical drains and reinforcements with geotextiles or geomembranes.

Predictions of the behaviour of such combined systems are extremely difficult. Therefore field measurements of movements of and within the subsoil and pore pressures in the soil were undertaken at all construction sites, not only to ensure the already above-mentioned points but also to establish design rules and to advance the state-of-the-art in this field.

CUXHAVEN EMBANKMENT

In the northern part of Germany, close to the town of Cuxhaven the newly built highway A 27 had to be constructed on 2 to 35 m (6 to 114 ft) thick deposits of highly compressible soft silty clay and peat. A 2.9 km long part of this highway has to cross a deep deposit of alluvial flood plains, consisting of different layers of normally consolidated soft silty clay, peat, silty sand, clayey silt, partially with organic matter, with different thickness and different consolidation behaviour. Floss, Brueggemann, and Heinisch [8] present an overall view of the construction site and the field measurements. Special reports on different topics were presented by Heinisch and Floss [9] and Heinisch and Blume [10] elsewhere. The subsoil conditions are illustrated in Figure 20.

To allow an embankment of 2.0 m (6 ft) height to be economically designed reliable assessments of the stability and of the rate and magnitude of settlement of embankments are needed. Therefore it was decided to arrange for 8 test sections of different length from 200 m to 1400 m (600 to 4200 ft) with different approaches as to the construction techniques used for the final embankment placement, for the improvement of soil properties through special treatments and for the replacement of compressible layers. The standard

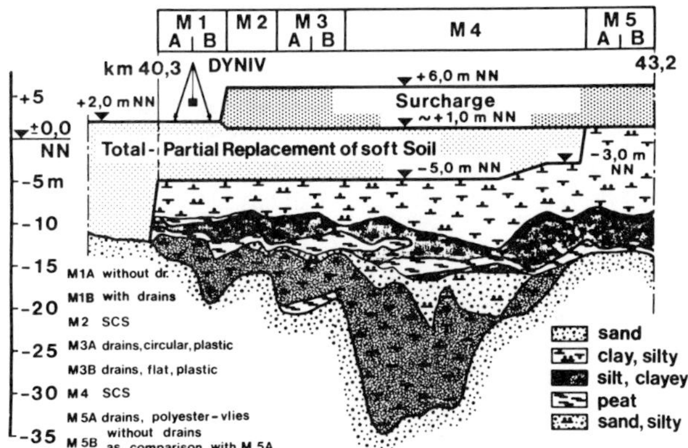

FIGURE 20. Subsoil conditions, construction expedients and test sections.

construction scheme (SCS) was chosen to consist of:

- partial replacement of soft soil up to 6.0 m (18 ft) beneath ground surface by excavating the highly compressible soils with barge-dredges and backfilling sand by spoiling techniques from a closed-by deposit (see Figure 21).
- construction of a surcharge embankment up to a height of 5.0 m (15 ft) with a sand fill rapidly spoiled in one stage [100 m (300 ft) of embankment within 24 h] between dikes of sand (see Figure 22).

A standard cross section of the Cuxhaven embankment is illustrated on Figure 23. The rule for determining the width of excavation of compressible soils beneath road embankments was chosen on the basis of experience and according to specifications (EspE-NS77; [11]) outlined by the county of Niedersachsen, the region where all sites are situated. The result of this is illustrated on the right-hand side of Figure 23.

In order to compare different types of construction expedients a great testing program with 8 test sections was established which is illustrated in Figure 20.

The eight different test sections were as follow:

M1A Partial Replacement up to 6 m (18 ft) depth and heavy compaction with impact loads (DYNIV) without drains; length l = 200 m (600 ft)
M1B Similar as M1A but with vertical plastic drains, flat; b = 95 mm (3.7 in); d = 2 mm (.08 in)
M2 Standard Construction Scheme (SCS) as described above; l = 300 m (900 ft)
M3A SCS; l = 200 m (600 ft) with vertical plastic drains, perforated and circular; diameter 50 mm (2 in)
M3B Similar as M3A but with vertical plastic drains, flat; b = 92 mm (3.6 in); d = 2 mm (.08 in)
M4 SCS as M2; l = 1400 m (4200 ft)
M5A Surcharge embankment without soil replacement, l = 200 m (600 ft) but with reinforcement of a geotextile at the base and flat drains (polyester-vlies-type); b = 300 mm (11.8 in); d = 4 mm (.16 in)
M5B Similar as M5A but without drains

Comparable test sections were chosen to be mainly in regions with almost equal layers and layer thicknesses (see Figure 20). Each of the test sections were equipped with measuring devices for:

- settlement measurements with settlement plates (Figure 5), multipoint gauges with an inductive coil probe (Figure 9) and remote single gauges (Figures 6 and 7)
- lateral movements measurements with inclinometer (SGI-type)
- porewater pressure measurements with vibrating-wire piezometers and pneumatic piezometers (Figure 19) (drive types)

The large test program enabled the use of different measuring systems and to compare results of relatively new developed gauges with reliable already existing ones.

Due to instrumentation it was possible to control the construction procedures in such a way that no deformations occurred which were not acceptable for the intended use of the embankment.

Before placement of the road pavement the following guidelines had to be fulfilled:

- 85% degree of consolidation after one year of surcharging for all construction schemes with drains
- differential settlements between neighbouring test sections less than 10 to 20 cm (4 to 8 in)

For a long-term satisfactory behaviour of the road the differential settlements of the base course should be not more than 5 cm (2 in) along a basis length of 50 m (150 ft) and the elastic deformation of the pavement under a wheel-load of 8.15t should be less than 1 mm (.04 in).

During the construction period from late 1978 to mid 1980 an enormous amount of field data was read. These field data were already presented in the aforementioned papers and reports [8,10]. The results of field section M1 were presented by Heinisch/Floss [9] elsewhere.

It is far beyond the scope of this article to review all data. Therefore it is necessary to concentrate on some topics, e.g., the overall settlement behaviour of test sections M2 to M5 and settlement and pore pressure data of test sections M2 and M5.

Overall settlement data for the different field sections for 10, 100, and 500 days respectively from settlement plates situated at the surface of the partial replacement is shown on Figure 24.

From the results it is obvious that relative high initial (undrained) settlements occur after 10 days (around 37% of the total settlements to expect), which reach about 17 to 47 cm (6.7 to 18.5 in). The settlement distribution curve of the almost initial settlements after 10 days dominates the form of the following curves as can be seen when comparing with similar curves for 100 and 500 days respectively. The differential settlements between neighbouring field sections are as high as 6 to 20 cm (2.4 to 7.9 in). Thus they are within the anticipated range after 500 days of surcharge. The settlements and also the differential settlements in field sections with drains are generally higher than those without drains.

For the extrapolation of the final settlements from field data a hyperbolic time–settlement relationship was used as proposed by Christow [12] in order to obtain the degree of consolidation

$$U_t = s_t/s_e * 100(\%) \tag{6}$$

in which s_t = settlements at time t and s_e = final settlement at the end of primary consolidation.

FIGURE 21. Air view of construction site during partial replacement of compressible layers.

FIGURE 22. Spoiling of surcharge embankment.

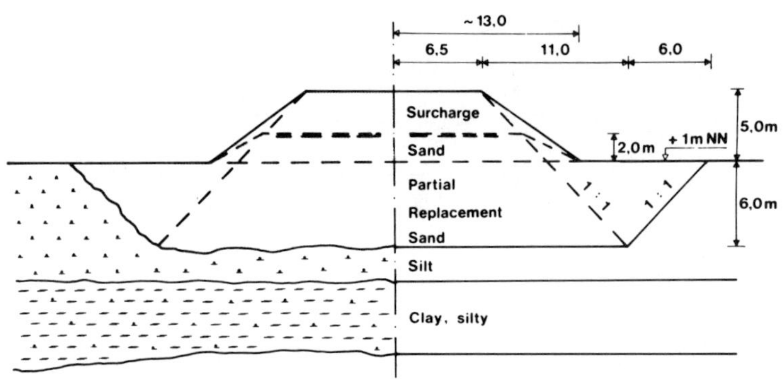

FIGURE 23. Standard cross section of Cuxhaven embankment.

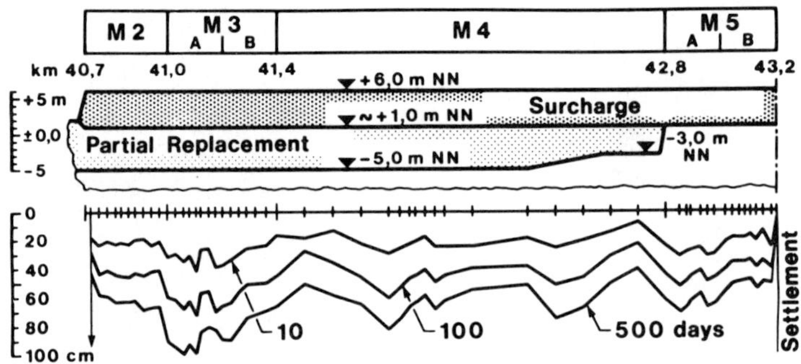

FIGURE 24. Overall settlement behaviour of Cuxhaven embankment.

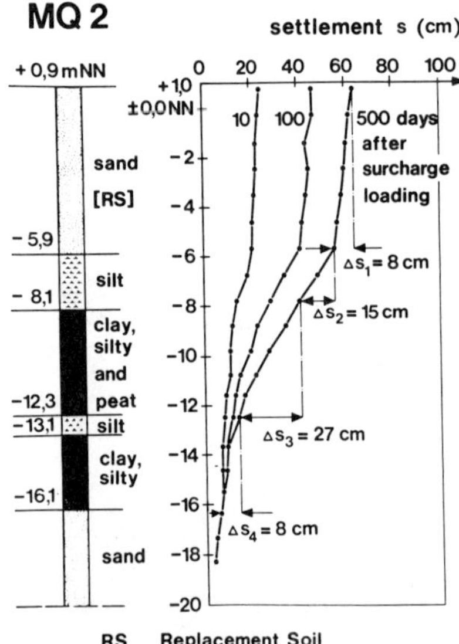

FIGURE 25. Settlement with depth for M2 with multipoint settlement gauge.

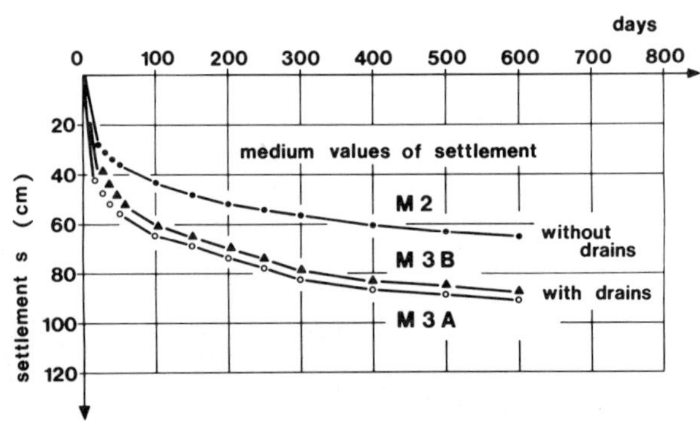

FIGURE 26. Medium values of settlement for field sections M2, M3A, M3B versus time.

TABLE 1. Degree of Consolidation U_t After 500 Days of Surcharge.

Test Section	U_t %	H m	s_R cm
M2	77–85	10	10–20
M3A	83–88	13–19	12–20
M3B	84–87	13–16	10–13
M4	71–87	20–30	12–17
M5A	84–86	15	11–12
M5B	78–84	15	10–14

Due to the preliminary field data the final settlements s_e were calculated to be within the range of 84 cm to 131 cm (33 to 51.5 in) for test sections M2 to M5.

In Table 1 the results of the degree of consolidation after 500 days of surcharge and the still to expect settlements

$$s_R = s_e - s_t \tag{7}$$

together with the whole thickness of the soft soil layers are presented for the test sections M2 to M5.

For all test sections with drains M3A/B and M5A the reached degree of consolidation fulfill the criteria established for the placement of the road pavement. The sections without drains M2, M4, M5B have only little less values. The calculated settlements s_R are in the order of 10 cm to 20 cm (4 to 8 in). These values are theoretical, since they are only to expect if the surcharge is left in place.

As an example of multipoint settlement gauge results the illustration of settlement with depth is plotted on Figure 25 for M2 together with a soil profile. The differences in settlement behaviour of the different soft soil layers can thus be clearly shown.

From the illustration it is obvious that the upper silt layer [H = 2.2m (7.3 ft)] and the combined silty clay and peat layer [H = 4.2 m (14.3 ft)] have a major influence on the overall settlement behaviour. After 500 days of surcharge load around 66% of the overall settlements (24% for the silt layer and 42% for the silty clay/peat layer) were measured. Thus 2/3 of all settlements are due to these two upper layers. To compare field sections with and without drains medium values of settlement for M2, M3A, M3B were plotted versus time in Figure 26.

The difference in settlement between field section M2 (without drains) and M3 reached values from 19 cm to 31 cm (7.5 to 12.2 in) after 500 days of surcharge load. This difference may be due to the difference of soft soil layer thickness (see Table 1) and thus direct comparison is not possible with respect to settlement behaviour. The time settlement behaviour of both test sections with different drains is almost identical and M3A has only around 3 cm (1.2 in) more settlement as M3B. Pore pressure measurements results of the upper silt layer were used to compute the degree of consolidation U_z of this layer:

$$U_z = \frac{\delta u_{max} - \delta u_t}{\delta u_{max}} * 100(\%) \tag{8}$$

with δu_{max} = maximum porewater pressure developed directly after load application and δu_t = porewater pressures measured at time t.

Analyzing the results in Figure 27 leads to the trend expected, namely that sections with drains have a greater acceleration of porewater pressure reduction than sections without drains. But the porewater pressure δu_t of all sections after a time of 3 months of surcharge are within the same magnitude. The degree of consolidation U_z versus time shows little difference between sections M2 and M3 (Figure 28).

The distribution of maximal porewater pressure δu_{max} versus depth for field section M3B is plotted together with a dotted line for σ_z from the surcharge in Figure 29.

From the distribution of the maximal porewater pressures the overall thickness of the soft soil layer for consolidation calculations can be read to be around 11.0 m (37 ft) from −5 m to −16 m under NN (−16 to 53 ft u.NN).

For field sections with drains the overall degree of consolidation $U_{z,r}$ can be calculated from:

$$U_{z,r} = [1 - (1 - U_r) * (1 - U_z)] * 100(\%) \tag{9}$$

in which U_r = degree of consolidation for horizontal drainage and U_z = degree of consolidation for vertical drainage. Theoretical curves for different layer thickness H, drain distance D = 2.10 m (7 ft) and circular drains with a diameter of 0.05 m (2 in) are shown in Figure 30.

For medium thickness of the soft soil layers H = 11.0 m (37 ft) the overall degree of consolidation for t = 365 days is around 86% (U_r = 79%; U_z = 34%). With increasing thickness U_z decreases rapidly. Since the major influence in changing U is the relatively high values for U_r, a change in thickness and thus also in U_z leads only to an overall decrease in the order of 3%.

In order to use design charts developed for average consolidation rates for stabilisation by vertical circular drains (e.g., [13]) an equivalent diameter d_{eq} for flat drains:

$$d_{eq} = \frac{2 * b}{\pi} * f \tag{10}$$

has to be used, in which b = width of the flat drains, π = 3.1416 and f = factor between $\pi/4$ and 1.0, as to account mainly for the hydraulic differences between flat and circular drains.

The equivalent diameter for the flat plastic drains of sec-

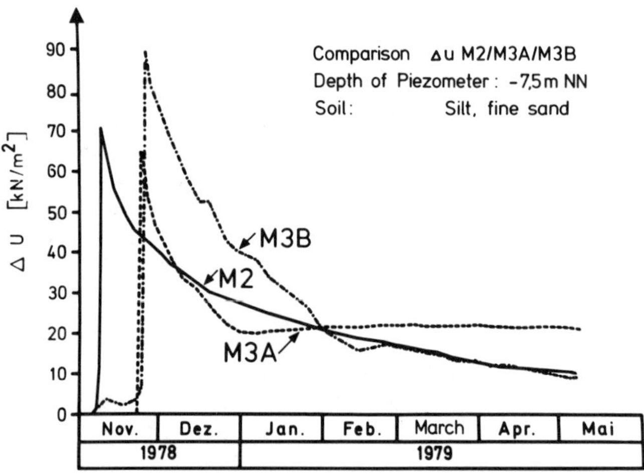

FIGURE 27. Porewater pressure results versus time for field sections M2, M3A, M3B.

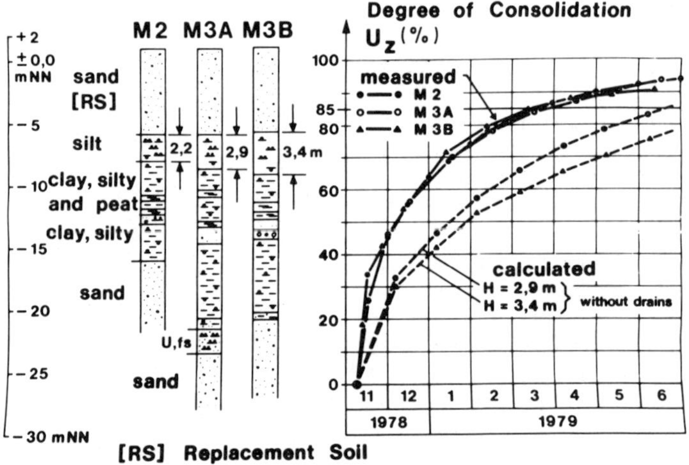

FIGURE 28. Degree of consolidation U_z versus time for upper silt layer and the test sections M2, M3A, M3B.

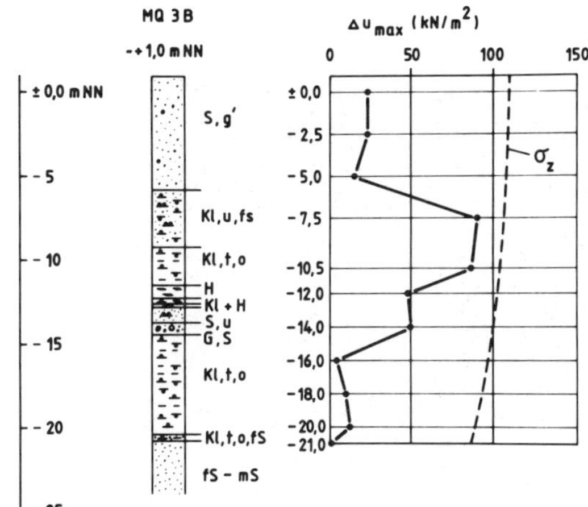

FIGURE 29. Maximal porewater pressures δu_{max} versus depth for field section M3B.

tion M3B was calculated to be in the order of 4.5 to 5.7 cm (1.8 to 2.2 in) and is thus comparable to the circular drains used in section M3A with 5 cm (2 in) diameter.

Where as field sections M3A and M3B are comparable to a certain amount, comparisons between field section M2 without drains and M3 with drains are not possible, because the difference in thickness of the soft layers plays a major role (Table 1). For field sections M5A with drains and M5B without drains this comparison is possible, since the soft soil thickness of both sections is almost equal. As can be seen from the soil profile, the soil conditions at section M5 are slightly better than in all other field sections. Therefore a different construction scheme was used, namely, instead of partial replacement of compressible soil the placement of a geotextile directly at the base of the surcharge embankment (Figure 31).

In Figure 32 the consolidation behaviour of the three involved soft soil layers of field section M5A is illustrated.

From porewater pressure measurements within the three involved layers: silt (*u*), silty sand (*fS,u*) and silty clay (*Clay,t*) degrees of consolidation U_z were plotted with time. The difference between the silt layers, which reach the anticipated degree of 85% within 2 months, and the underlain clay layer, which reaches $U_z = 80\%$ only after one year, is obvious.

In Figure 33 ranges of degree of consolidation U_z are plotted for both field sections M5A with drains and M5B without drains. The effect of the drains is clearly visible, since section M5A reaches the desired value of 85% after 500 days of surcharge, whereas the section M5B without drains at that time has only a degree of consolidation between 35–55%. On the other hand absolute settlements are higher for section M5A with drains than for section M5B without drains and the settlement distribution in section M5B is much smoother than in section M5A (see also Figure 24).

Finally the specifications outlined before construction of the Cuxhaven embankment were fulfilled. At the end of surcharging time the final embankment for the placement of the pavement (see Figure 23) was constructed.

Meanwhile A27 is under traffic for many years and the post-settlement behaviour is still within the anticipated range.

STOTEL EMBANKMENT

The Stotel embankment is a case study in which laboratory and field data was developed in order to get input data for monitoring a stage construction procedure with a versa-

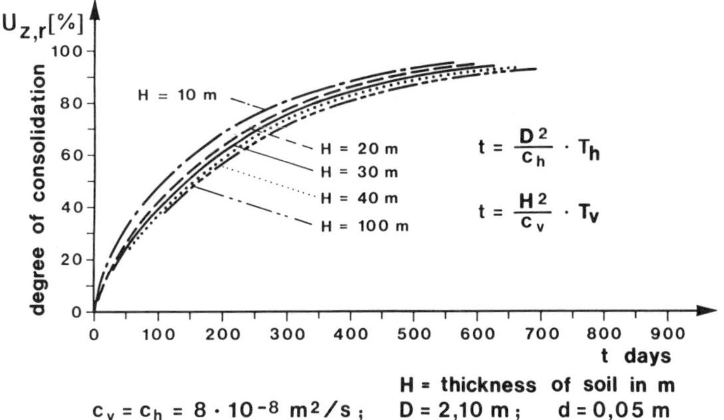

FIGURE 30. Overall degree of consolidation $U_{z,r}$ for soft soil layers with drains.

FIGURE 31. Placement of geotextile at the base of a surcharge embankment.

Degree of Consolidation

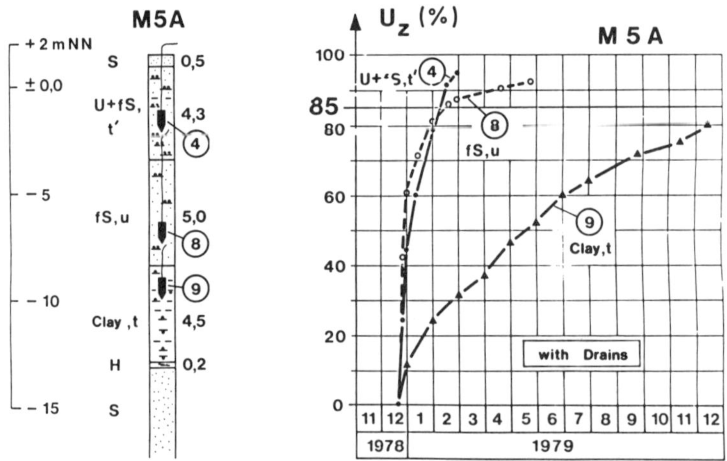

FIGURE 32. Degree of consolidation versus time for field section M5A.

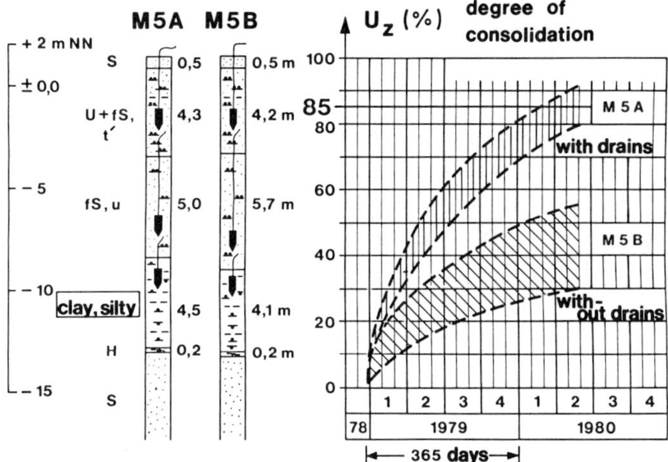

FIGURE 33. Comparison of degree of consolidation versus time for test section M5 with and without drains.

tile numerical analysis using an FE-program with features as to account for consolidation and creep.

An embankment approximately 400 m (1200 ft) long had to be constructed over soft subsoil to connect the existing Federal Road B6 with the new built Highway Route A12 Bremen-Cuxhaven near Stotel/Bremerhaven in Northern Germany. A cut through the whole section and an overall view of the embankment is shown on Figure 34.

A technical solution had to be found to minimize differential settlements under traffic for the intersection of the highway founded on a so-called "moor-bridge" and the existing road. Furthermore a construction time less than 8 months had to be considered.

The existing subsoil consists of about 11 m (37 ft) of normally consolidated soft organic clay. Underlying the clay is a layer of medium dense sand. Several tests were run:

in situ: vane shear and permeability tests
in the laboratory: consolidation-, creep- and triaxial shear tests

to obtain soil data as input for the design of the embankment. More details about these tests and their results are given by the author (Thamm [14]) elsewhere. The soil profile and results of the undrained shear strength with depth are shown on Figure 35.

The distribution of the undrained shear strength of the clay with depth as found by vane shear tests and triaxial tests can be simplified as shown by the straight lines drawn in Figure 35.

It was found [15,16] that the undrained Young's modulus E_u is almost linearly dependent upon the undrained shear strength c_u:

$$E_u = C * c_u \tag{11}$$

where C is a number of approximately 400 for the in situ soils of the Stotel embankment.

This simplified assumption was used as input for undrained Young's modulus into the numerical analyses.

After preliminary studies it was decided to preload the soft clay with an embankment of about 3 m (10 ft) height (Figure 34) using a fine to medium sand. Due to this surcharge the stress levels in the soft clay:

$$S = (\sigma_1 - \sigma_3)/2 * c_u \tag{12}$$

were low (0.3 to 0.5). Thus a linear elastic analysis may therefore be justified for practical reasons.

An important part of soil data had to be established for the consolidation and creep phase. The consolidation data was simply obtained from several oedometer tests with samples of different layers of the soft clay.

The following material constants were derived from the tests:

- The coefficient of volume change m_v:

$$m_v = 1/E_s \tag{13}$$

- The coefficient of consolidation c_v:

$$c_v = (k * E_s)/\gamma_w \tag{14}$$

- The pore pressure parameter at the beginning of loading a:

$$a = u/\sigma_m \tag{15}$$

In Table 2 results from samples taken at 6 different borings

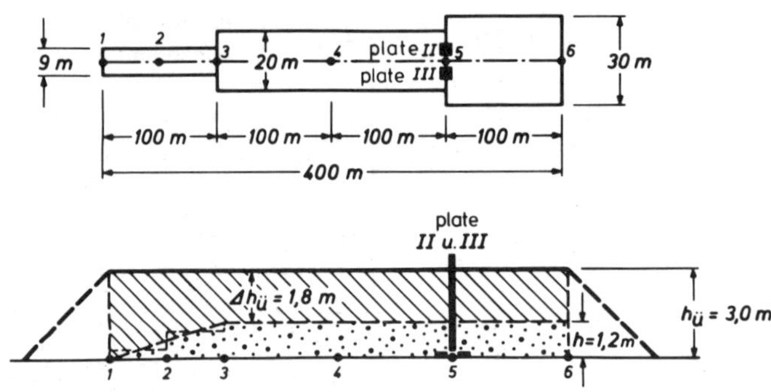

FIGURE 34. Stotel embankment cut through section and overall view.

and different depths are presented. The results were used to establish the above-mentioned material constants.

The creep data was obtained from several creep tests with steps of up to 14 days of loading stages.

The empirical creep function (creep strains as function of the state of stress and time) can be derived as:

$$\epsilon = k_1 * \sigma^{k2} * \sigma^{k3} \quad (16)$$

where k_1, k_2, and k_3 are creep constants obtained by plotting creep data as shown in Figures 36 and 37.

Numerical analyses of the Stotel embankment for different cross sections were performed using the soil data derived from all tests and a finite element program SSOILS which can account for consolidation and creep [17,18]. The discretization of the Stotel embankment is illustrated in Figure 38.

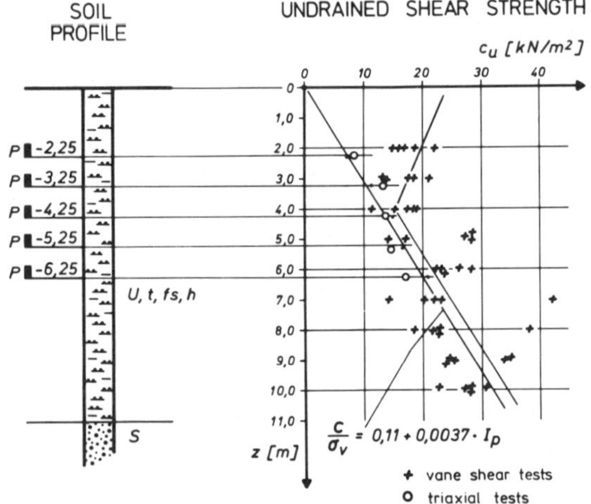

FIGURE 35. Soil profile and undrained shear strength with depth.

The construction sequence of the embankment was simulated with four lifts of placement and one lift of replacement to the final desired height of the embankment.

The appropriate time scale for the whole construction time (8 months) was divided into time steps of 30, 46, 74, 52, and again 30 days respectively, using time increments of 48 h. During construction of the embankment settlements were monitored by 5 settlements plates in different cross sections. Figure 39 shows the displacement of the subsoil surface after one month for the cross section at km 10 + 520 as calculated by the finite element program. At this time about 30% of the total settlement had already taken place. Comparison with settlement plate 3 about 5 m right from the centre line of the cross section showed good agreement between measured and calculated settlements at that time.

Time loading and time settlement curves are plotted on Figure 40 again for cross section at km 10 + 520. The calculated time-settlement curve is thereby plotted as dotted line. Again comparison between settlement plate 3 and the calculated values showed good agreement with respect to magnitude and timing of the settlements.

Comparison with settlement plate 2 about 5 m left from the centre line of the same cross section led to differences in settlement of about 25%. This difference is probably due to preconsolidation effects of an old nearby railroad embankment.

In general the analysis showed that after a construction time of about 8 months no significant differential settlements will occur.

RUEBKE EMBANKMENT

The Ruebke embankment is a case study to construct road embankments on peaty soils without any soil replacement and with reinforcement at the base using a geotextile. In order to establish data for the design of such embankments two full-scale test embankments (test sections A + B) were planned along the new highway A26 Horneburg-Hamburg in Northern Germany close to the vicinity of Ruebke (Figure 41).

TABLE 2. Soil Property Values from Laboratory Tests.

boring Nr.	depth z	γ_f	σ_v	c_u	φ_{cu}	τ_{cu}	E_u	E_s	m_v $*10^{-3}$	k $*10^{-6}$	c_v $*10^{-5}$	f
—	m	kN/m³	kN/m²	kN/m²	[°]	kN/m²	kN/m²	kN/m²	m²/kN	m/s	m²/s	—
I	3,075	15,7	48,3	2,0	6,5	13,5	5400	465	2,15	0,115	0,53	0,99
II	6,125	14,7	90,0	7,0	6,6	17,4	6960	253	3,95	0,115	0,29	0,56
III	4,125	15,4	63,5	6,5	6,0	13,7	5480	669	1,49	2,60	17,45	0,94
IV	2,125	14,4	30,6	6,0	5,0	8,7	3480	—	—	—	—	—
V	5,125	13,7	70,2	6,0	6,8	14,4	5760	582	1,72	1,70	9,88	0,65

$\tau_{cu} = c_u + \sigma_v \cdot \tan \varphi_{cu}$ $\quad$ $C \sim 400$ $\quad$ $c_v = k \cdot E_s/\gamma_w$

$E_u = C \cdot \tau_{cu}$ $\quad$ $m_v = 1/E_s$ $\quad$ $f = u_o/\sigma_v$

Field vane shear tests have proved to be useful for soil investigation on saturated peats in different states of decomposition. It was found that the shear strength obtained using the vane in the field was between unconfined compressive tests in the undisturbed state and the remolded state for peat. The results can be used to calculate the stability of embankments formed on such soils and can help in selecting the construction method, e.g., the appropriate fill stages to be applied.

From the borings on both test sections it was found that an approximately 4 m (13 ft) thick soft layer mainly consists of peat. This layer is underlain by a fine to medium sand. Ten field vane shear tests were performed at section A prior to fill placement.

Faust, Moritz and Stiefken [19] have reported in detail on the results of these tests elsewhere. Field vane shear tests were conducted at test section A prior to fill placement and during stage construction of the test embankment, in order to measure the change in undrained shear strength of the peat under the fill load. A view of the test section during fill placement is shown on Figure 42.

The geometrie of half of the embankment together with the 3 fill stages used and the already encountered settlements are shown on Figure 43.

Results of the undrained shear strength versus depth for mid-embankment are plotted for the three fill stages with fill heights 0.7, 2.0 and 3.5 (2.3, 6.6 and 11.5 ft) respectively and prior to fill placement in Figure 44.

The results revealed that the undrained shear strength of the peat increased directly after applying the load. The minimum value of the undrained shear strength at mid-depth of the peat layer prior to the placement of the fill was increased almost by 50% from 8 kN/m² to around 12 kN/m² (1.2 to 1.8 psi) after fill stage 1. After fill stage 3 at a height of 3.5 m (11.5 ft) surcharge (see Figure 43) the undrained shear strength climbed to almost four times of its initial value.

All tests were run directly after completion of each of the three fill stages. Test repetitions carried out after a period of five month after the placement of fill stage 3 did not reveal any significant changes.

From the results of different tests undertaken from the

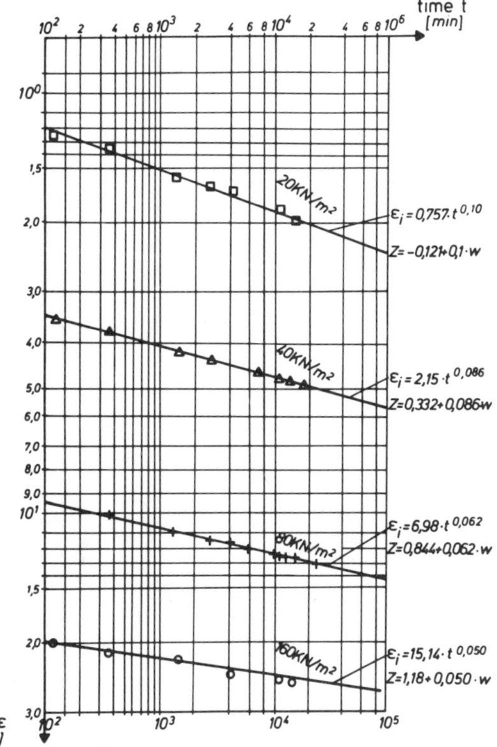

FIGURE 36. Determination of creep constants k_1, k_2 due to overburden pressure.

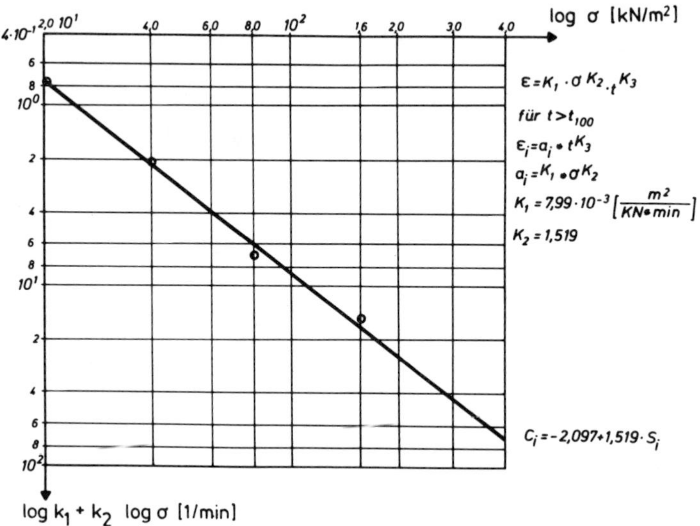

FIGURE 37. Determination of creep constant k_3 due to overburden pressure for boring Nr. III, depth 4 m (13ft).

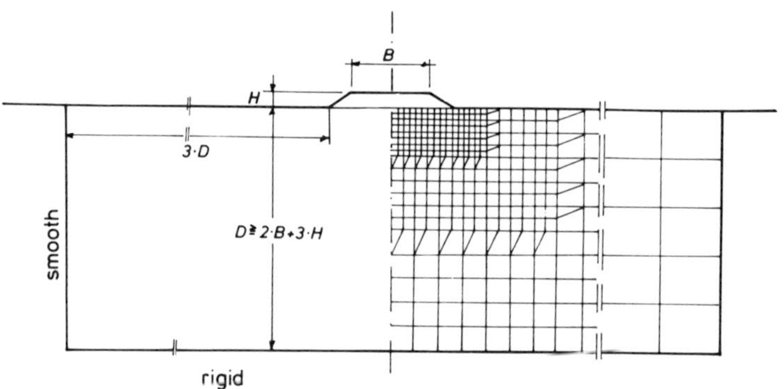

FIGURE 38. Discretization of an embankment with finite elements.

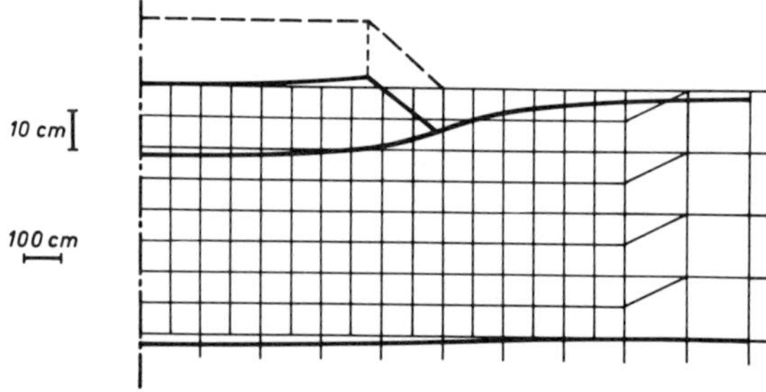

FIGURE 39. Displacements of the surface of the subsoil after 1 month of embankment for cross section at km 10 + 520.

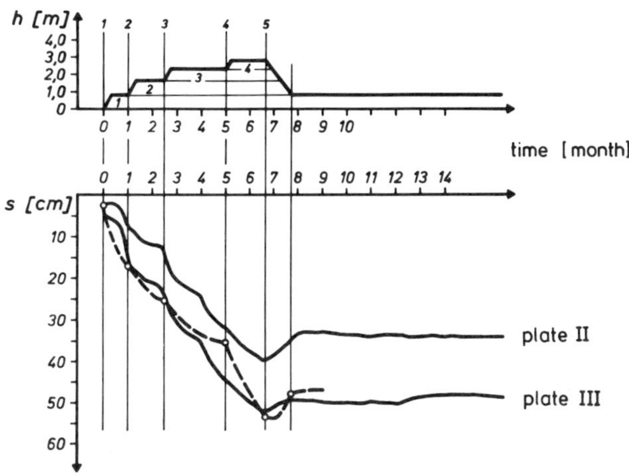

FIGURE 40. Comparison of measured and calculated settlements with time for cross section at 10 + 520.

FIGURE 41. Test embankment near Ruebke along highway A 26.

FIGURE 42. View of Ruebke test embankment during fill placement.

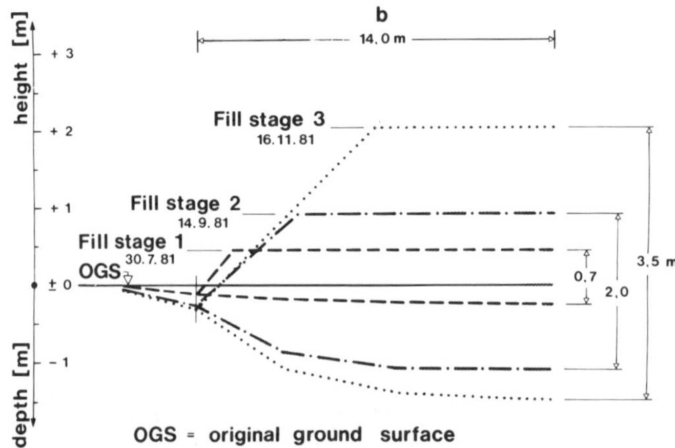

FIGURE 43. Cross section of the test embankment: geometry, fill stages and already encountered settlements.

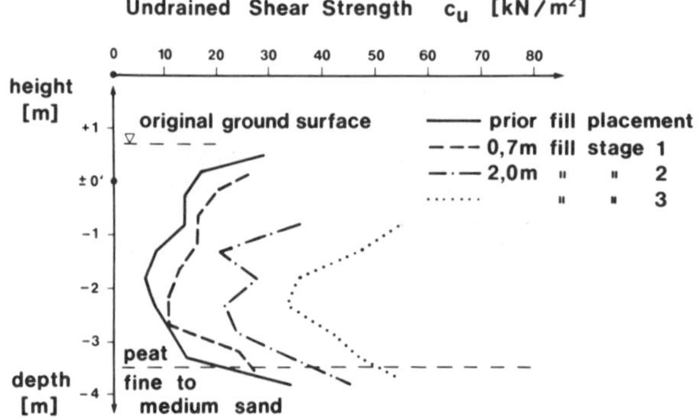

FIGURE 44. Undrained shear strength c_u from field vane tests versus depth prior fill placement and for fill stages.

crest of fill a range of undrained shear strength c_u versus vertical stress at the base could be obtained. This range is shown on Figure 45.

The undrained shear strength c_u almost increases linearly with the vertical stress at the base as can be seen from the line within the range which represents the mean values for the whole peat layer. Results of field vane shear tests performed outside of the test embankment [around 3 m (10 ft) from the toe] revealed that this increase is restricted clearly to the region directly under the embankment.

The stability analysis was performed using the aforementioned results. For the calculations the soft peat layer was divided into 6 layers each having a different undrained shear strength according to the findings in Figure 44. For the region outside of the embankment the undrained shear strength according to the situation prior to fill placement (see also Figure 44) was chosen. The results for the factor of safety (F.S.) computed for the three fill stages 1, 2 and 3 shown on Figure 43 were 3.42, 1.2, and 1.41 respectively.

The form of stage construction, especially for stage 1, was seen to be somewhat conservative. Thus the design of a second test embankment (test section B) was planned, where the whole surcharge embankment should be placed in one lift up to the final height of 3.5 m (11.5 ft) [20]. The stability analyses were performed for the four different time-steps already shown on Figure 43.

The results of these analyses are found in Table 3 in which t = time after beginning of fill placement, H_f = height of fill and H_e = height of embankment above original ground surface (OGS).

The results show that the placement of the surcharge embankment with one lift up to the full height leads to stability problems within the first two months.

To overcome these difficulties it was decided to use a geotextile reinforcement [tension strength 400 kN/m; (27500 lbf/ft) strain failure ~10%] at the base of the embankment. From Figure 46 the placement of a geotextile at the base of the embankment can be seen.

For the analysis the tension strength of the geotextile was chosen to be only 100 kN/m. This will mobilise a strain of about 2% in the geotextile. With this assessment the factor of safety F.S. after initial loading was computed to increase from 0.72 to almost 0.99. After these preliminary investigations it was decided to construct a full-scale test embankment [40 times 29 m (132 times 95 ft)] using two lifts : 2.3 and 1.3 m (7.5 and 4.3 ft) respectively, directly on the original ground surface, leaving the topsoil in place and placing a geotextile at the base prior to filling operations. In Figure 47 the rapid placement of fill on the geotextile can be seen.

Together with the design the instrumentation for monitoring the behaviour of the embankment was planned. Figure 48 illustrates a quart of the test section with locations of different measuring devices for settlement, lateral movement and porewater pressure. The following instrumentation was used for the test section Ruebke:

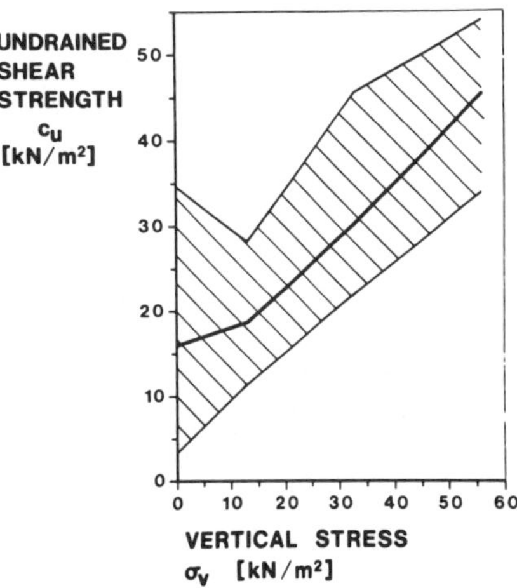

FIGURE 45. Undrained shear strength from field vane shear test versus vertical stress at the base of the embankment.

- settlement plates (Figure 5), remote settlement gauges (Figures 6,7), multipoint settlement gauges (Figure 9) and a profiler (Figures 11,12)
- lateral movement with inclinometer (Figure 15)
- porewater pressure with pneumatic piezometer (Figure 19) and a reference well point

From the enormous amount of field data obtained since summer 1980 only few results can be reviewed. They were already presented elsewhere [20].

During the filling operations no significant stability problem occurred.

Profiles of settlement results from the settlement plates under the test embankment are plotted after the placement of the first lift of around 2.0 m (6.5 ft) height, after leaving stage 1 for about two months in place and after reaching the final height of the surcharge (Figure 49).

From the settlement-time results soil property values can be backfigured for the two stages (see Table 4).

Due to the surcharge the coefficient of volume compressi-

TABLE 3. Results of Stability Analysis of Ruebke Test Section.

Nr.	t	H_f	H_e	F.S.
—	days	m	m	—
1	0	3.5	3.5	0.72
2	10	3.5	3.2	0.84
3	55	3.5	2.4	1.41
4	120	3.5	2.0	2.00

FIGURE 46. Placement of a geotextile as reinforcement at the base.

FIGURE 47. Placement of fill on a geotextile at the test section.

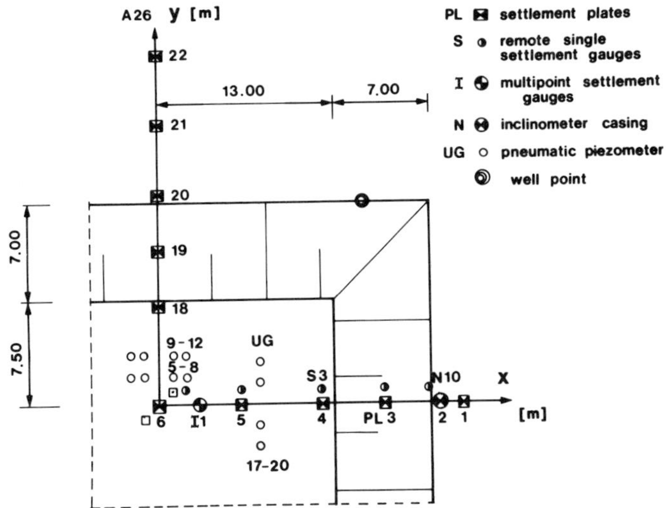

FIGURE 48. Ruebke test section—quart of the test embankment.

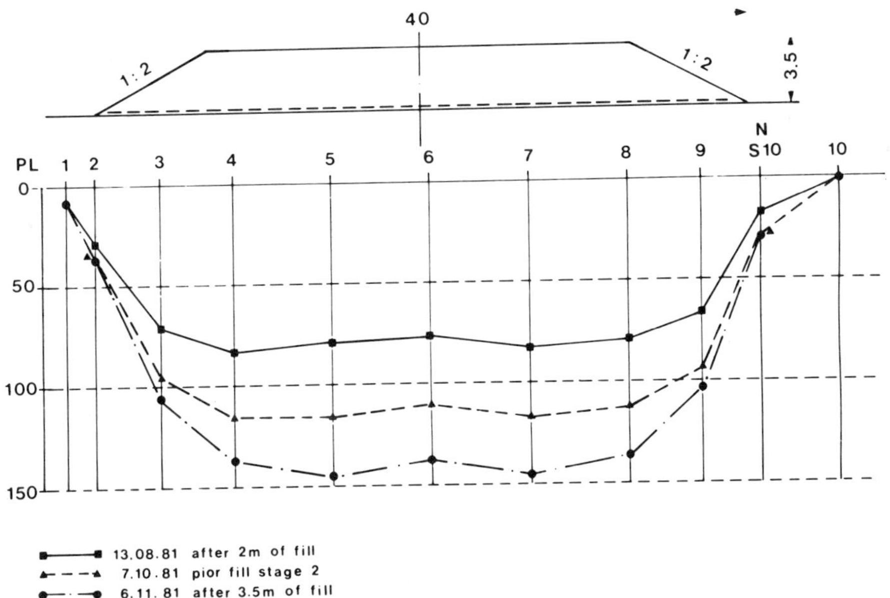

FIGURE 49. Settlement profiles under test section Ruebke.

TABLE 4. Soil Property Values Backfigured from Field Data of Ruebke Test Section.

Stage	m_v	k	c_v
—	m²/kN	m/s	m²/day
1	$5.50*10^{-3}$	$3*10^{-8}$	0.047
2	$1.66*10^{-3}$	$5*10^{-9}$	0.026

bility m_v increased to three times of its initial value. The permeability k and the coefficient of consolidation c_v decreased at the same time.

The initial (undrained) settlements were around 35 cm (14 in) with an undrained Young's modulus E_u of 3950 kN/m² (573 psi) (E_u/c_u from 303 to 492). Shortly after the placement of fill stage 1 the degree of consolidation reached already a value of 37% with settlements around 80 cm (31.5 in). Two months later the degree of consolidation climbed to a value of 87% and it was decided to place the second fill stage up to the final height. At that stage a total settlement of 1.20 m (4 ft) was reached.

Only one month later the degree of consolidation of the fill stage 2 was around 48% with total settlements close to 1.50 m (5 ft). A proposed degree of consolidation of U_z = 90% for the final surcharge was found to exist already

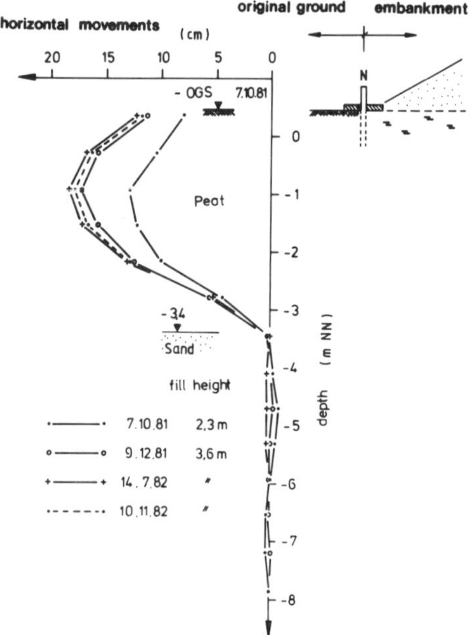

FIGURE 50. Horizontal movements at the toe of the test embankment versus depth.

in February 1982, eight months after beginning of the fill. The total settlements then were in the order of 1.70 m (5.6 ft) (see Figure 52).

The horizontal movements measured with an inclinometer at the toe of the embankment are plotted versus depth in Figure 50. The maximum horizontal movements were encountered in the middle of the peat layer. Prior to fill stage 2 these maximal movements are about 12.5 cm. This is around 10% of the vertical maximal settlement encountered at the same time. From experience with embankments on similar soft soils but without a geotextile at the base a rate of 10 to 15% is usual [21]. The low rate is probably due to the geotextile.

After placement of fill stage 2 the horizontal movements developed similar to fill stage 1 with a maximum lateral movement of 17 cm (6.7 in) at mid-depth of the peat layer. Measurements taken a year later confirmed only slight increases in these movements.

In order to reduce post pavement settlements and because the creep behaviour was not yet fully understood the surcharge was left in place for almost two years, before the surcharge fill was partially removed and the road pavement was brought in place.

The whole construction sequence is illustrated in Figure 51, where 1 = the surcharge fill, 2 = partial removal of surcharge and 3 = the placement of base course and road pavement. The construction sequences were monitored with instrumentation and the locations of the three different devices for monitoring the pressures at the base (EV2), the porewater pressures (UG6) and the settlements (PL6) are shown on Figure 51 as well.

To monitor the field behaviour of an embankment with heavily installed measuring equipment automatic recording and data handling with microcomputers is necessary [20]. Especially plots of different results versus construction time are necessary to control field operations. Examples of these plots are shown on Figure 52 for the development of pressures at the base (EV2), porewater pressures (UG6) and settlements (PL6) with time.

Due to the results of both test sections at Ruebke a safe and economical design of a road embankment on soft peaty soils of Northern Germany could be established. The construction scheme of the embankments consists only of surcharging combined with reinforcement at the base using geotextiles. The surcharge fill can be placed by rapid spoiling techniques [500 m³ (17600 cu ft) within 24 h].

BRUNSBUETTEL EMBANKMENT

In the northern part of Germany several bridge abutments had to be constructed along highway B5 Brunsbuettel-Itzehoe together with appropriate approach embankments on 12 to 20 m (40 to 65 ft) thick deposits of highly compressible soft silty clay and peat. Since the abutments are generally founded on large diameter piles, horizontal forces are induced in piles by unsymmetrical surcharges on the soil. These forces cause more or less severe bending moments within the piles.

The economical design of abutment foundation and embankment and the prediction of its behaviour is therefore of general interest to the profession. To verify existing simplified assumptions and empirical approaches to the design of such soil-structure interaction systems a large field testing program was started in May 1981 on a bridge abutment. Part of the testing program was to study the behaviour of an approach embankment [base length 38 m (125 ft), slopes 1:4, final height 4.25 m (14 ft)] on that soil (Figure 53).

Due to the poor subsoil conditions the embankment design included a combination of surcharge fills, stage construction, the installation of vertical drains and reinforcements at the base with geotextiles. Before embankment construction 10 combined settlement-inclinometer casings (Figure 54) were installed to a depth of 15 m (50 ft) to control horizontal and vertical movements of and within the subsoil (see Figure 14). The design, construction and performance of this embankment during the first 10 months after placement was already presented elsewhere [22].

A detailed boring program was undertaken. In fact, each location of the combined settlement-inclinometer casing was a borehole, in order to determine the extent of the soft soil layers and to obtain samples for laboratory testing.

A typical subsoil profile B8 is shown in Figure 55. Directly beneath a thin organic layer is about 3 m (10 ft) of very soft silty clay.

The top of the clay has been weathered and is quite firm while the remainder of the deposit is very weak and extremely sensitive. A 3.6 m (12 ft) thick layer of peat follows, and underneath are another 5 to 6 m (16.5 to 20 ft) of very soft to soft silty clay. At El. −12 m(−40 ft) a loose to medium dense fine sand deposit was found.

The laboratory testing program consisted of the usual classification tests, one-dimensional consolidation tests, and direct shear tests. In addition, field vane tests were performed with a Geonor apparatus. Atterberg limits and natural water contents versus elevation are plotted in Figure 55.

The soft silty clay has the following typical properties:

Natural water content: $w_n = 119 \pm 27(\%)$
Liquid limit: $w_L = 129 \pm 26(\%)$
Plastic limit: $w_P = 47 \pm 10(\%)$
Plasticity Index: $P_I = 82 \pm 16(\%)$
Clay size: minus 2 microns = 39(%)
Total unit weight: $\gamma = 14$ kN/m³ (89 pcf)
Specific gravity: $\varrho_s = 2.3$ t/m³ (142 lbm/cu ft)

The peat has the following typical properties:

Natural water content: $w_n = 550 \pm 150(\%)$
Total unit weight: $\gamma = 10.2$ kN/m³ (65 pcf)
Specific gravity: $\varrho_s = 1.5$ t/m³ (92.5 lbm/cu ft)

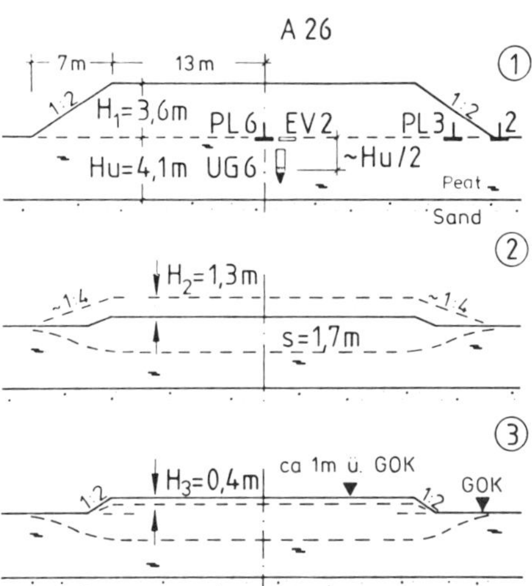

FIGURE 51. Construction sequence of Ruebke embankment.

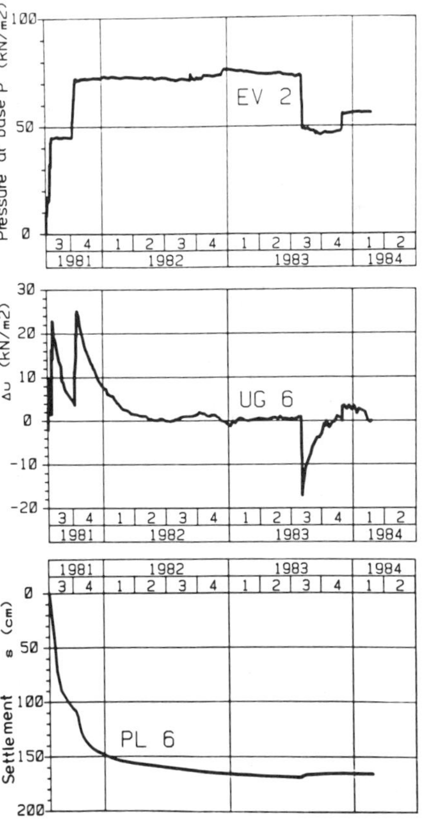

FIGURE 52. Pressures at the base, porewater pressures and settlements with construction time.

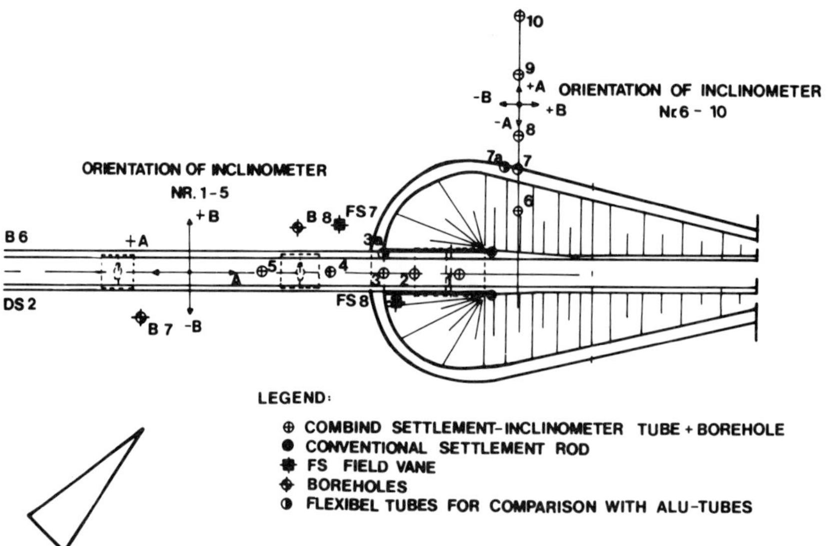

FIGURE 53. Location plan of field instrumentation.

For the upper very soft silty clay the natural water content was greater than the liquid limit below El. -1 m (3 ft), indicating that the material is very sensitive. The peat layer shows natural water contents in the range from 400–700%. The lower very soft to soft silty clay gives some indication of decreasing water content and plasticity with increasing depth below El. -8 m (-26 ft).

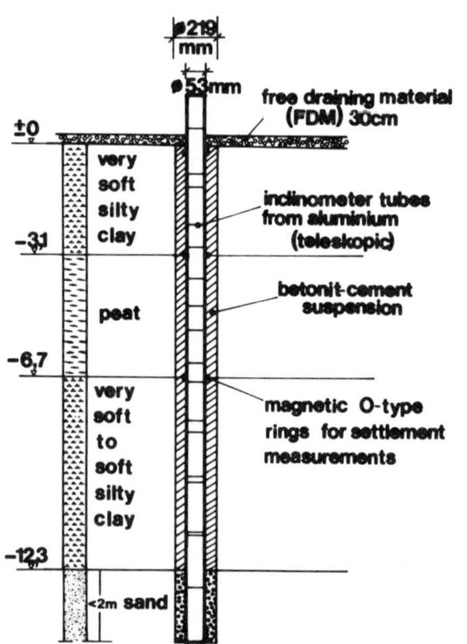

FIGURE 54. Combined settlement-inclinometer casing.

Results of field vane tests performed prior to stage construction from two locations FS 7 and FS 8 are plotted with depth in Figure 55. For stability analysis a vane correction factor was chosen [23] to account for the effects of strain rate, anisotrophy and progressive failure. The corrected values of the vane tests are also shown with depth in Figure 55.

A series of one-dimensional consolidation tests were performed on "undisturbed" samples of soft clay and peat. Typical compression curves from two samples are shown in Figure 56. The vertical strain, ϵ_v, is plotted against the vertical consolidation stress, σ'_{vc}, to a log scale. The sensitive nature of the clay is shown by the very steep slope of the compression curves at stresses just beyond the maximum past pressure, σ'_p. The curves shown in Figure 56 were plotted using the strain at the end of 24 hr load increments. The maximum past pressures, $\sigma'_{vm} = \sigma'_p$ were determined by the well known procedure by Casagrande and plotted with depth as stress history data in Figure 55. According to Ohde [24], one-dimensional compression data can be expressed by Equation (3) and Equation (4).

Values for the empirical soil constants v, w and the initial void ratio e derived from compression data are also shown in Figure 56. The empirical soil constant, w, gives indication of the nature of soil behaviour. Values close to 0.5 confirm elastoplastic behaviour of the soil.

For the present case, the upper soft clay layer is supposed to behave almost elastoplastically. The peat layer will behave almost elastically and the lower soft clay layer elastically to elastoplastically.

For the calculation of consolidation, it was decided to use data from field measurements rather than from laboratory

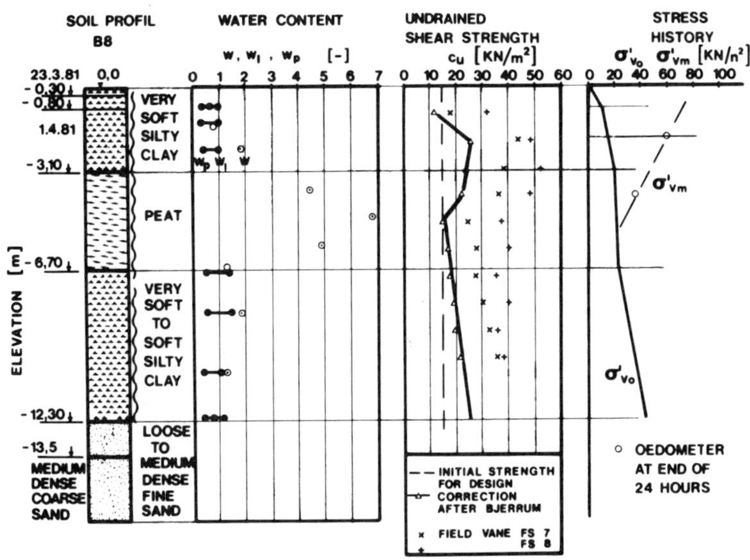

FIGURE 55. Soil profile, classification strength and stress history data.

tests, in order to give more reasonable predictions of time–settlement behaviour of the approach embankments. Similar reasons also lead to the calculation of the possible rate of secondary compression, C_α, from field measurements only. The proposed bridge dictated the height of the approach embankments. Embankments as high as 3.5–4 m (11.5–13 ft) are common. Together with the use of 1.5–2 m (5 to 6.5 ft) of surcharge this meant that a total of 5–6 m (16.5 to 19.5 ft) fill had to be placed above the existing subsurface. According to regional experiences with embankments on similar subsoil [25], a combined system of stage construction, vertical drains, and a geotextile at the base was adopted to construct the embankments safely and to achieve sufficient consolidation of the soft soils so as to minimize post pavement settlements. Before selecting the number of stages and the required vertical drain spacings, estimates of the total settlement of the clay under the embankments were made based on the stress history shown in Figure 55 and the compression data obtained from laboratory tests (Figure 56).

In the present case, assuming the embankment to be placed in one lift, and using Equation (1) together with the material constants $E_u/c_u = 400$ and $\nu = 0.50$, the initial settlements would reach a value of almost 0.30 m (1 ft).

In routine practice the total final settlement s of a layer of thickness H, with one-dimensional settlement behavior (Figure 2) assumed, is computed from Equation (2).

In the present case, neglecting secondary compression in the first place, the total final settlement can be estimated from stress history data (see Figure 55) and with the soil property values in Table 5.

Assuming the embankment will be placed in one lift to the desired height of +4.25 m (74 ft), the final settlement due to primary consolidation will reach 1.88 m (6.16 ft) within almost 16 months. That means a total of 6.13 m (20 ft) of fill has to be placed above the existing subsurface.

The stability analysis using the initial undrained strength of Figure 55 leads to a factor of safety lower than 0.7. Additionally experience with organic soils shows that secondary

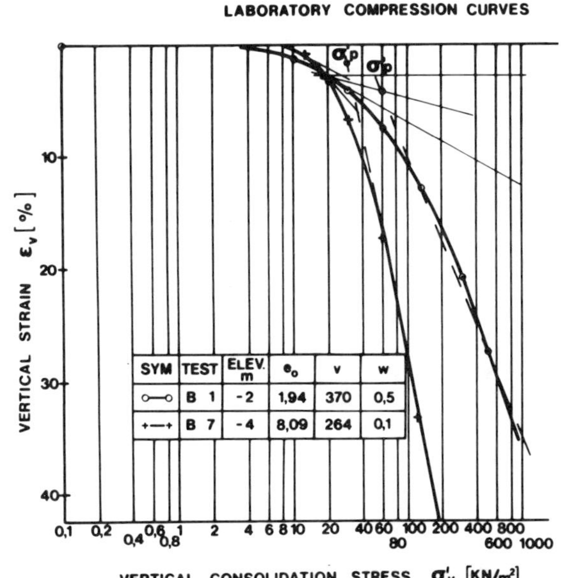

FIGURE 56. Typical laboratory compression curves.

TABLE 5. Soil Property Values from Laboratory Tests (1m = 3.278 ft; 1m² = 10.753 sq ft; 1kN = 224.72 lbf).

El.	m_v	k	C_c	C_R	c_v
m	m^2/kN	m/s	—	—	m^2/day
0 to −3.1	$8.9*10^{-4}$	10^{-9}	0.20	0.068	0.010
−3.1 to −6.7	$3.0*10^{-3}$	10^{-7}	2.70	0.297	0.288
−6.7 to −12.3	$8.9*10^{-4}$	10^{-9}	0.36	0.122	0.010

compression plays an important role. Secondary compression somctimes accounts for half or even more of the final settlement due to primary consolidation and is characterized by a straight line relationship between settlement and the logarithm of time. In Table 5 m_v = coefficient of volume compressibility in m^2/kN; k = permeability in m/s; C_c = compression index; and c_v = coefficient of consolidation in m^2/day [see Equation (14)].

$$C_R = C_c/(1 + e_o) \qquad (17)$$

with e_o = initial void ratio.

Especially for the peat layer [El. −3.10 to 6.70 m(−10 to −22 ft)] primary consolidation will occur rapidly followed by creep or long-term secondary consolidation. Estimation of the effects of this layer shows that even more fill [up to about 1 m (3.3 ft)] has to be placed to reach the desired height of the embankment. To achieve this task safely and economically and to minimize the consolidation time, differential settlements and horizontal forces on the piles of the abutment, a system of stage construction, combined with vertical drains and a geotextile placed at the base of the embankment had to be adopted. Thc stage fill was placed to 1.05 m (3.45 ft) above the existing unprepared subsurface as a working platform for the installation of vertical plastic drains and the geotextile. The factor of safety was at least 2.75 without considering any strength increase with consolidation during the placement of the fill. The stage 1 fill was to remain in place only for less than 2 months prior to stage 2 fill. During stage 1 fill field data was to be collected from the field measurement instrumentation installed prior to filling in order to correct, if necessary, laboratory data and to select the required drain spacings. The selection had to be done on the basis of two more lifts (stages 2 and 3) within less than 8 months prior to the construction of the abutment piles. The degree of consolidation at that time was decided to be greater than 80% to minimize lateral pressures on the piles.

In the following the collection of field data for the decision-making process and the predictions for stages 2 and 3 will be described.

An extensive program of field instrumentation was particularly important for the following reasons:

- Embankment settlements exceeding 2 m (6.5 ft) were expected in many areas.
- The design required at least 80% consolidation under the surcharge fills in order to meet design criteria for the abutment piles within less than 10 months.
- The rate of consolidation in the field was required to provide warning signals of impending stability problems.
- The combination of vertical plastic drains with a geotextile should be investigated.
- The forces on and within the abutment piles should be recorded in order to get realistic design criteria against horizontal forces on such piles.

In this review only the field instrumentation for monitoring the actual performance of the surcharge fill will be described. The used system is a combined settlement-inclinometer-tube (Figure 54), which is of telescopic nature and contains at several depths magnetic O-type rings for measuring settlements within the subsoil. From the inclinometer measurements horizontal movements within the soil can be computed.

The location of the combined tubes is shown in Figure 53 together with the orientation for the inclinometer measurements. A view on the top of a tube with inclinometer casing is shown in Figure 14. In addition, standard surface settlement platforms were placed and two flexible tubes for comparison with the aforementioned combined tubes were located at the foot of the surcharge fill. All of the instruments were installed prior to fill and drain placement. The results of the field data obtained during the placement of the stage 1 fill were extensively analyzed. In particular, the in situ compressibility and rate of consolidation of the soil layers were determined.

The settlements were read and horizontal movements were calculated from inclinometer measurements three times during the construction of stage 1 (after 13, 32, and 45 days, respectively) along the 10 combined inclinometer settlements tubes. The settlement data in Figure 57 for tube 6 shows that the peat layer was extremely compressible, while the clay crust and the underlying soft silty clay show less settlement.

From the settlement data and computed vertical stress changes, elasticity modulii were derived for almost undrained (immediate settlements) condition and drained condition (mainly primary consolidation); the results of which are also shown, e.g., for tube 6 in Figure 57. Analyzing the undrained condition immediately after the placement of stage 1 fill values of E_u/c_u were found to range from 330–420 for the present soil layers.

Backfiguring settlement data after 45 days of stage 1 fill leads to the soil property values shown in Table 6. Comparing these values with data derived from laboratory tests shows that the upper silty clay crust is less stiff than before,

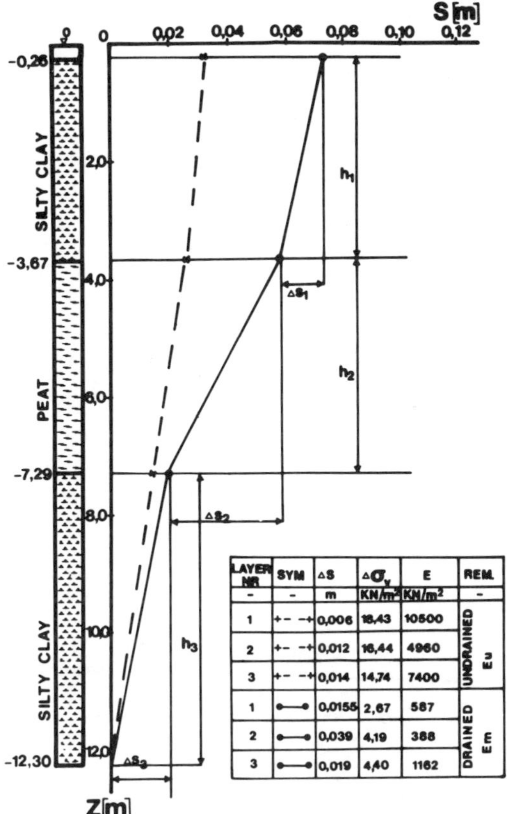

LAYER NR	SYM	ΔS	$\Delta\sigma_v$	E	REM.
-	-	m	KN/m²	KN/m²	-
1	+- -+	0,006	18,43	10500	UNDRAINED Eu
2	+- -+	0,012	16,44	4960	
3	+- -+	0,014	14,74	7400	
1	•—•	0,0155	2,67	587	DRAINED Em
2	•—•	0,039	4,19	388	
3	•—•	0,019	4,40	1162	

FIGURE 57. Settlements at tube 6.

but that consolidation will take place almost six times faster in situ. One reason for the discrepancies in stiffness may be the difficulties of reconsolidation of the upper clay to realistic values in the laboratory tests. For the peat layer, the in situ compressibility is as expected before; the in situ rate of consolidation is also faster.

The lower silty clay has soil property values almost as expected from laboratory tests. Using the corrected soil property values, an FE-analysis was performed with an FE-program described by the writer elsewhere [17]. In the analysis, creep was neglected in the first place. Some of the typical results are shown in Figures 58–60. In Figure 58 the surface settlements due to stage 1 fill are shown for two different times compared with measurements taken at 4 different tubes along half of the cross section.

It is obvious that the program does not meet the requirements of upheave prediction close to the toe of the fill. Nevertheless, the settlement prediction does not seem bad, keeping in mind the simplified assumptions of three almost homogeneous layers.

The settlements along tube 6 are shown in Figure 59. The prediction of settlement with depth is not as expected for the peat layer. These discrepancies are obviously due to creep effects which are common in those soils. From time-settlement data of peat it is extremely difficult to decide when primary consolidation ends and when secondary consolidation starts. Whether or not secondary consolidation takes place already before primary consolidation ends is one of the unsolved questions of the behaviour of organic soils.

The prediction of settlements along the silty clay layers is reasonably well done. Pore pressure parameters, $\delta u/\delta\sigma v$, are shown in Figure 60 along tube 6 for two different times of stage 1 fill.

The pore pressures were computed in the following manner:

1. The FE-program was started using estimated pore pressures and coefficients of consolidation from laboratory tests.
2. Computed pore pressures with the preceding program were used to estimate the rate of consolidation.
3. New coefficients of consolidation were computed for actual time steps.
4. The FE-program was used again to compute pore pressures.

The aforementioned iterative procedure was necessary until the requirements of appropriate pore pressure to coefficients of consolidation were met. The final pore pressures along tube 6 show that on average almost 40% of consolidation has been reached after 45 days of stage 1 fill. During stage 1 fill lateral deformations of less than 2 cm (0.8 in) were recorded (Figure 61).

Prior to fill stages 2 and 3 vertical plastic drains and a geotextile (tension strength 200 kN/m²) had to be installed from and on the working platform described as stage 1 fill [El. + 1.05 m (3.45 ft)] (see Figure 46). For the design of the spacings of vertical drains, experience from sand drains [13] and regional experience from other case histories [25] were used. This led to center-to-center triangular spacings between a = 1.5–2.0 m (5–6.5 ft) with about 10–20 cm (8–16 in) diameter of the drains. The most important drain design parameter is the coefficient of consolidation for horizontal drainage, c_h, which was assumed to be equal to the

TABLE 6. Soil Property Values Backfigured After 45 Days of Stage I Fill (1m = 3.278 ft; 1m² = 10.753 sq ft; 1kN = 224.72 lbf).

El.	m_v	k	C_c	C_R	c_v
m	m²/kN	m/s	—	—	m²/day
0 to −3.1	1.70*10⁻³	1.33*10⁻⁸	0.25	0.085	0.068
−3.1 to −6.7	2.58*10⁻³	1.45*10⁻⁷	7.27	0.800	0.486
−6.7 to −12.3	8.46*10⁻⁴	1.38*10⁻⁹	0.44	0.150	0.014

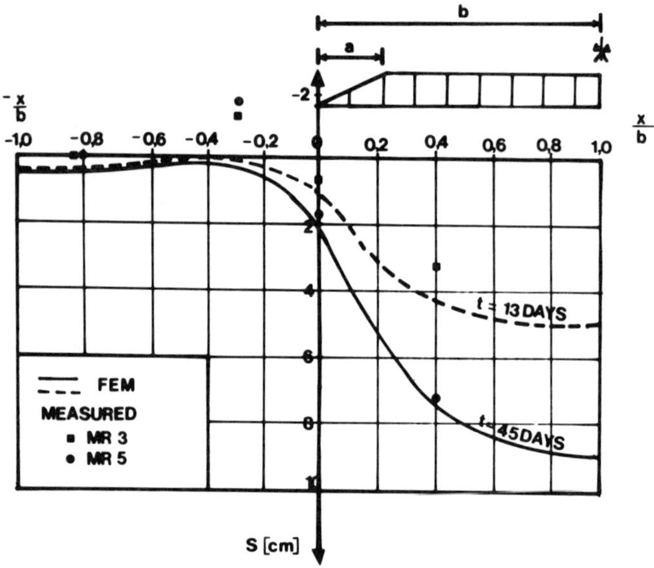

FIGURE 58. Surface settlements of stage 1 fill for two different times.

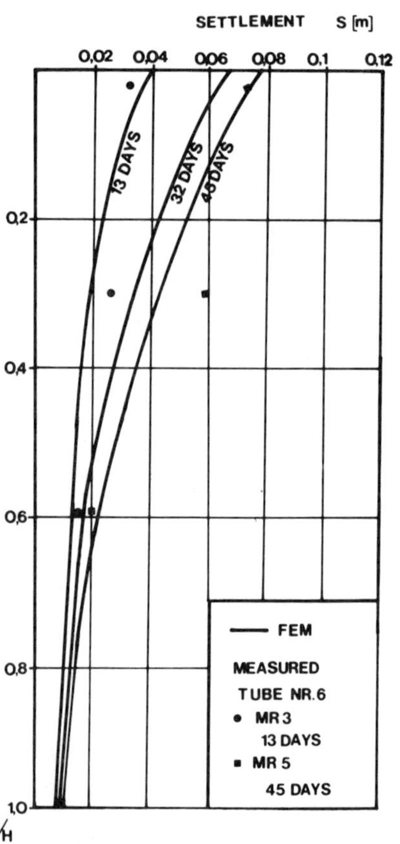

FIGURE 59. Settlements along tube 6 for three different times of stage 1 fill.

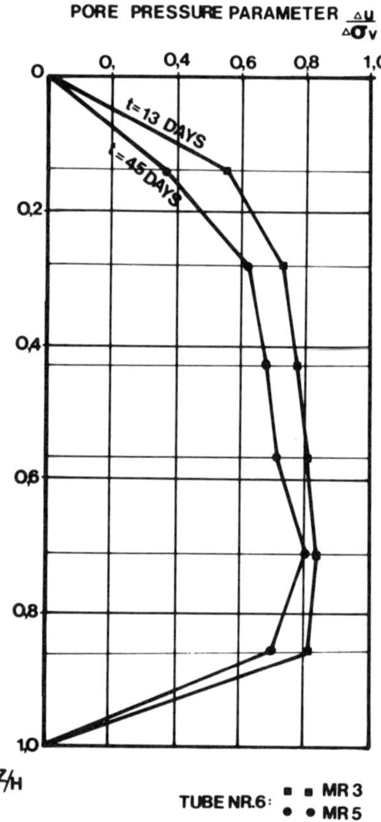

FIGURE 60. Pore pressure parameters, $\delta u/\delta\sigma$, along tube 6 for different times of stage 1 fill.

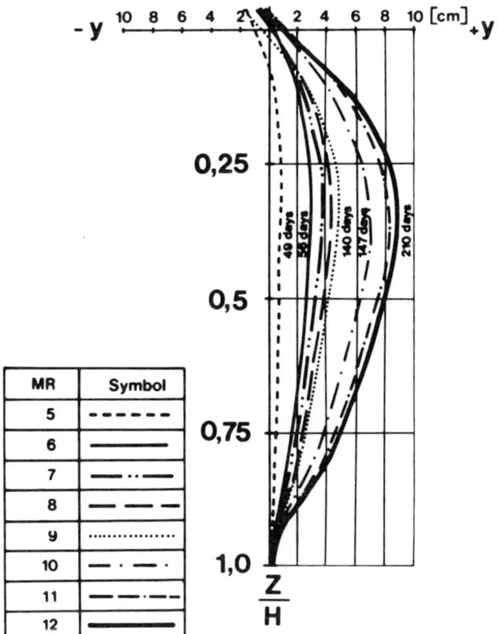

FIGURE 61. Horizontal movements versus normalized height of subsoil for tube 7.

coefficient of consolidation for vertical drainage, c_v. The thickness of the clay was reduced by a factor of almost 3 in determining the drainage height for two-way drainage. Using average consolidation rates from [13], estimates of an equivalent coefficient of consolidation, c_v, including drains and preceding assumptions can be calculated as follows:

$$c^{+}_{vi} = \frac{T_v}{T_h} * \frac{H_i^2}{a^2} * c_{vi} \tag{18}$$

For a required degree of consolidation of 80% the following values can be found for c_v^+ (Table 7). Different drain distances and diameter account for the preceding differences in equivalent coefficients of consolidation.

An average equivalent coefficient of consolidation, c^{+}_{vm}, of 0.44 ± 0.15 m²/day can be chosen, which leads to a time of about 40 days (compared to 106 days without drains) for 80% of consolidation. For stage 2 fill a lift height of 1.70 m (5.6 ft) was chosen [from El. +1.05 − 2.75 m (3 or 5–9 ft)] which fulfilled the stability requirements (factor of safety 1.46).

The maximum settlement was predicted to reach, e.g., at tube T1 0.84 m (2.744 ft) for 80% of consolidation. For stage 3 a lift height of 1.50 m (5 ft) [from El. +2.75 − 4.25 m (9–14 ft)] was chosen. The stability analysis based on initial data leads to a factor of safety of 1.16. The calculated factor is on the safe side because the actual undrained shear strength is supposed to be higher than initially assumed.

The maximum settlement after the placement of stage 3 fill was predicted to reach, e.g., at tube T1 a value of 1.79 m (5.867 ft), including immediate settlement for 100% consolidation.

The primary consolidation was estimated to take place within almost five months from the beginning of placement. This is about three times faster than expected in first place without drains.

The real construction time for the approach embankment was:

- stage 1 fill 1½ months
- stage 2 fill 3 months
- stage 3 fill 3½ months

The whole approach embankment prior abutment construction is shown in Figure 62.

This accounts for secondary compression effects which were neglected in the present design. These effects have yet to be studied in more detail.

During stage 2 and 3 fills, readings of settlements and horizontal and vertical displacements within the subsoil were made at certain time intervals. Some of the settlement and vertical displacement results plotted versus time are shown in Figures 63–65.

The time is plotted to a normal scale in Figure 63. The settlement of the surfaces are shown for different tubes, 1, 3, and 6. It can be seen that after a time of almost 10 months there are still more deformations to expect for tube 1 and 6.

The time can also be plotted to a log scale as shown, e.g., in Figure 64 for the vertical deformations along the tube 1. Most of the vertical deformations take place in the upper third of the upper soft soil. The rather steep slope of vertical deformations just below the surface is a sign of secondary compression effects as expected for organic soils.

To compute the time at which primary consolidation ends, and where secondary consolidation starts, is hardly possible from this diagram.

Therefore, another possible scale for the time is common in practice, namely, the square root scale as shown in Figure

TABLE 7. Coefficients of Consolidation for a Required Degree of Consolidation of 80% (1m = 3.278 ft; 1m² = 10.753 sq ft).

E1.	$c_v = k/(m^* {}_w)$	$c_v^+ = T_v/T_h * H_i^2/a^2 * c_v$
m	m²/day	m²/day
0 to −3.1	0.068	0.35 ± 0.12
−3.1 to −6.7	0.486	0.85 ± 0.30
−6.7 to −12.3	0.014	0.23 ± 0.06

FIGURE 62. Approach embankment prior abutment construction.

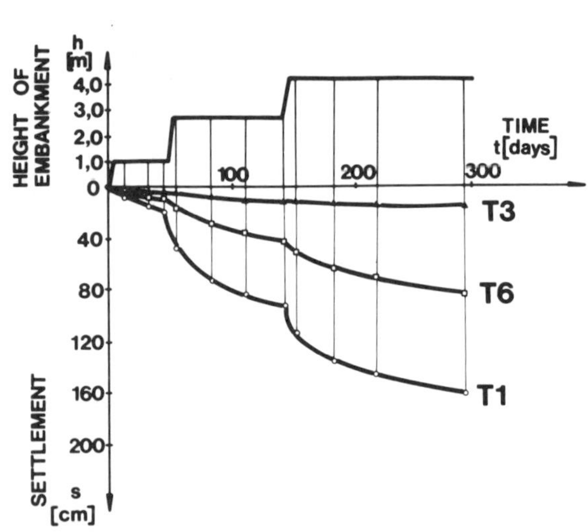

FIGURE 63. Settlements versus time for tube 1, 3 and 6.

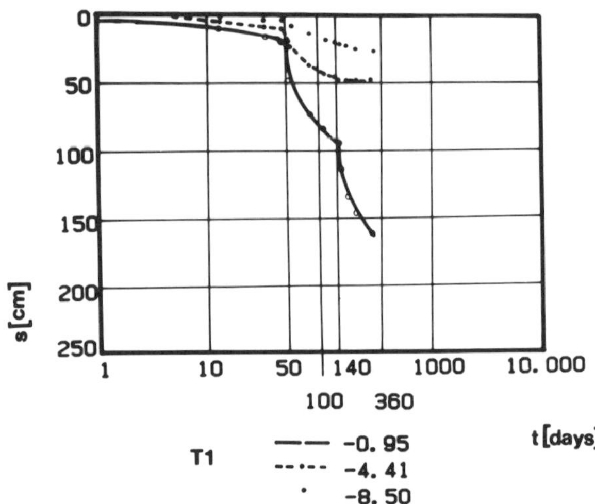

FIGURE 64. Vertical deformations at different depth of tube 1.

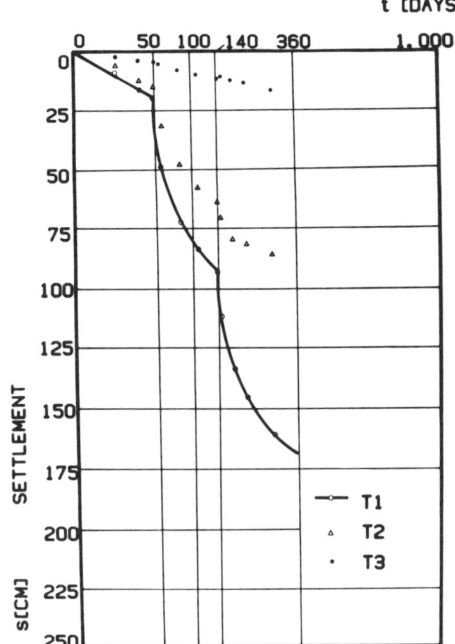

FIGURE 65. Settlement versus square root of time for tubes 1–3.

65 for tubes 1, 2 and 3. From the diagramm in Figure 65 the computation of the end of primary consolidation according to Taylor's method is possible. Thus, estimates of the time of 100% consolidation can be made from such diagrams, coefficients of consolidation can be computed, and estimates of coefficients of secondary consolidation, C_α, can be established for further computations.

For tube 1 an average value, c_v, of 0.637 m²/day was obtained, which is close to the upper bound of the predicted value $c^+_{vm} = 0.44 + 0.15\text{m}^2/\text{day}$.

The coefficient of secondary consolidation, C_α, for the three almost homogeneous layers of the present problems are shown in Table 8. The maximum rate of secondary compression was 8.05% per log cycle of time for the peat layer. Results of the horizontal movements were plotted versus the normalized height of the soft subsoil, e.g., for tube 7 in Figure 61. The maximum horizontal deformation was found to be at a depth about one-third of the full height of the soft soil layers and to reach at tube 7 almost 9 cm (3.5 in) after 210 days, less than expected before.

In Figure 66 the maximal horizontal displacements along tubes 1, 2, and 3 are plotted versus time. Secondary compression effects can be found for horizontal displacements only for tube 1. From horizontal displacement versus time diagrams, coefficients of consolidation, c_h, for horizontal directions were computed. For tube 1 a value of 0.762 m/day was found for c_h. Comparing the result with the c_v-value for the same tube leads to a factor c_h/c_v of 1.20. Thus the assumption that c_h-values are almost identical with c_v-values seems to be reasonable for the present soil.

The maximal horizontal displacement was registered along tube 2 with about 25 cm (10 in). This is almost 15% of the vertical maximal displacement at the same time.

At the toe of the fill (close to tube 3) the horizontal displacements are only about 8% of the vertical maximal displacement. The horizontal displacements due to embankment loading after the placement of stage 3 fill are plotted for different distances from the fill in Figure 67. The results show that about 15 m (50 ft) from the toe of embankments almost no horizontal deformation takes place. The maximum horizontal deformation was registered within the peat layer.

Displacement vectors due to embankment loading are shown in Figure 68 for stage 3 fill (210 days after placement). A slight upheave was registered within the peat layer at tube 8. The main direction of vectors is as expected showing an outward movement of the subsoil due to the embankment loading.

The following conclusions from the measurements of the Brunsbuettel embankment can be drawn:

- Early measurements of vertical and horizontal movements can assist in correcting laboratory data.
- Control of impending stability problems was possible; there were no stability problems during construction.
- The use of vertical drains reduced the time for reaching 80% of consolidation by a factor of almost 3 (40 days compared to 106 days without drains).
- The expected time of all the placement was (including stage 1 without drains) 10 months. The time needed to reach 80% of consolidation for the whole placement was reduced by factor of 2.
- The ratio between horizontal and vertical coefficients of consolidation c_h/c_v was found to be approximately 1.2 for the present soil layers.
- The maximum rate of secondary compression was 8.05% per log cycle of time for the peat layer (compared to 1.64–1.85% for the clay).

TABLE 8. Coefficients of Secondary Consolidation C_α (1m = 3.278 ft).

E1.	C_α
m	—
0 to −3.1	0.0185
−3.1 to −6.7	0.0805
−6.7 to −12.3	0.0164

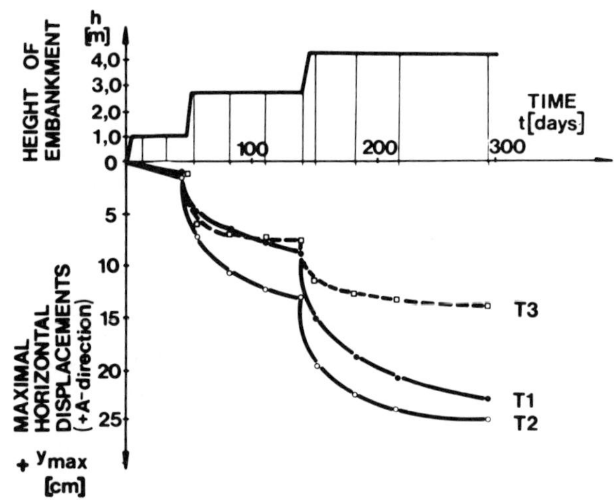

FIGURE 66. Maximal horizontal displacements versus time for tubes 1–3.

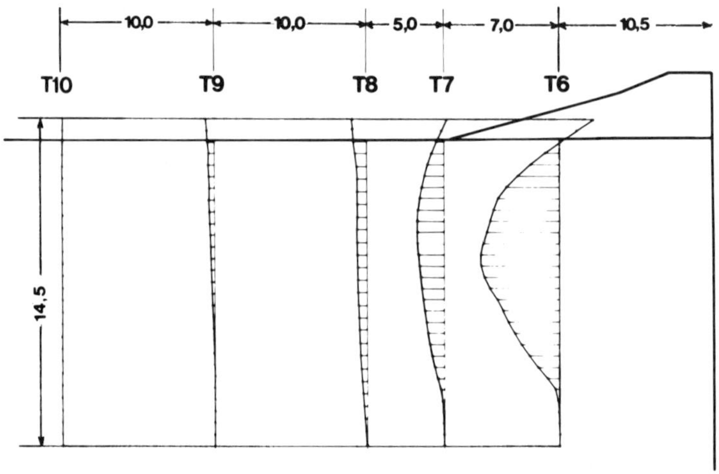

FIGURE 67. Horizontal displacements due to embankment loading for stage 3 fill (210 days after placement).

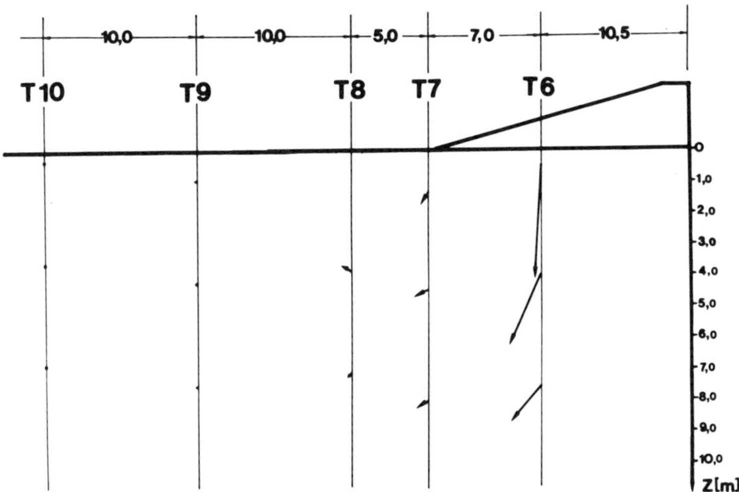

FIGURE 68. Displacement vectors due to embankment loading for stage 3 fill (210 days after placement).

- Horizontal movements of the subsoil were found to be less than expected (less than 15% of the vertical maximal displacement at the same time).
- Due to the construction procedure the supposed horizontal forces on the later placed piles of the abutment could be reduced by almost 30%; thus, a more economical design of the pile foundation could be used.

CONCLUSIONS

The construction of roads on compressible soils is of general concern to the profession, since environmental constraints and limitations on the use of good quality land are likely to occur today and in the future.

Although problems of failure and deformation of embankments on soft soils can be avoided by complete excavation of such soils, such an approach may lead to significant increases in construction costs. For economic reasons, therefore, the normal practice is to design embankments by taking into account the behaviour of the compressible layers. In this case different construction expedients are needed to overcome the problems of instability and settlement. They require in most cases field verification, before widespread implementation can be recommended. Recent research has demonstrated that the use of laboratory data for assessing the rate of consolidation is not always reliable. The use of large-scale loading tests which attempt to simulate the stress conditions likely to occur in the subsoil by the proposed embankment, is now more commonplace.

This may be the only feasible method to obtaining design parameters when conditions become too complex. Such full-scale tests are on the other hand very costly. They should be carried out well in advance of the main contract to permit an adequate period of consolidation to take place and for the results to be analysed. In order to obtain maximum benefit from these tests, as many factors as possible should be investigated.

The value of safety factor to employ in a particular situation is governed by a number of considerations, such as the method of calculation, the reliability of soil parameters and the variability of the subsoil. The selection of appropriate values is mainly based on engineering judgement which takes account of these factors as well as available local experience of similar situations.

The simple approach to analysing settlement originally proposed by Terzaghi [6] is still the most widely used method despite intensive research and development over several years. Although more advanced theories now exist, the use of these complex methods is often unwarranted because of lack of information on soil properties. However, for important works or case studies where two-dimensional conditions predominate consideration should be given to a more realistic analysis by employing readily available software in association with detailed soil information based on in situ measurements or on results of large-scale tests.

The use of instrumentation on site is now recognised also in the FRG as a valuable engineering tool. In the last years field measurements of embankments have played a major role during the design of embankments on soft subsoils in the northern part of the FRG. In this review four case studies of embankments founded mainly on silty clays and peat in the county of Niedersachsen around the cities of Bremen and Hamburg were presented. Case histories can provide invaluable information on the behaviour of roads constructed on such soft soils. Such information can lead to significant cost savings by allowing the design assumptions to be checked and predictive methods to be validated as well as providing data on measures for stability control.

During these studies experience was gained using different types of instrumentation for the measurement of settlement, lateral deformation and pore pressures within the subsoil. In order to check the reliability of new gauges results were compared with reliable already existing ones. The given numbers of accuracy are mainly based on this experience.

With respect to construction expedients, there has been a reversal of trend from partial replacement of compressible soil with drains to preloading with surcharges directly on the ground using geotextiles or geomembranes at the base for improving resistance to instability. This recent trend is particularly due to the impossibility of finding adequate deposits for the replaced compressible, often highly organic, soils.

Despite intensive research during the past years a number of aspects of road construction on soft soil are still not fully understood. Moreover, technological developments introduce new materials and techniques, e.g., geotextiles and geomembranes, to civil engineering. Future research is therefore required to assist in understanding, e.g., the behaviour of a geotextile at the base of an embankment and how reliable assessments for stability calculations of such combined systems can be made. But a number of aspects of the settlement behaviour of roads are also still unclear. The assessment of secondary consolidation, e.g., requires further research as there is evidence to suggest that the methods currently employed are not always reliable, particularly with high organic soils such as peat. On the other hand there is a need of data on acceptable levels of differential settlement, taking into account also accident effects due to surface uneveness and pavement deterioration. With improved technology and a better understanding of the behaviour and characteristics of compressible soils, it is certain that the useful life of embankments founded on such soils could be extended and their maintenance costs reduced.

ACKNOWLEDGEMENTS

The writer would like to thank Mr. Heinisch, Mr. Haberland, and Mr. Faust for the kindness of using some of the original drawings, which were already presented in German papers, for translation and for some of the pictures from the sites.

REFERENCES

1. Fröhlich, O. K., "Sicherheit gegen Rutschung einer Erdmasse auf kreiszylindrischer Gleitfläche mit Berücksichtigung der Spannungsverteilung in dieser Fläche," *Federhofer-Girkmann-Gedenkschrift,* Wien, p. 181 (1950).
2. Krey, H. D., *Erddruck, Erdwiderstand und Tragfähigkeit des Baugrundes,* Verlag von Wilhem Ernst & Sohn, Berlin (1926).
3. Bishop, A. W., "The Use of Slip Circle in the Stability Analysis of Slopes," *Proc. Europ. Conf. Stability of Earth Slopes," Stockholm, Vol. 1,* 1–13 (1954).
4. Thamm, B. R., "Numerical Analysis of Embankments over Soft Subsoils," *Proc. of the IIIrd Int. Conf. on Numerical Methods in Geomechanics,* Aachen, pp. 725–731 (1979).
5. Osterberg, J. O., "Influence Value for Vertical Stress in a Semi-infinite Mass Due to an Embankment Loading," *Proc. Fourth I CSMFE,* London, England, *Vol.I,* pp. 393–394 (1957).
6. Terzaghi, K. V. and O. K. Fröhlich, *Theorie der Setzungen von Tonschichten,* Verlag F. Denticke, Wien (1936).
7. Dunnicliff, J., "Geotechnical Instrumentation for Monitoring Field Performance," Transportation Research Board, NRC, Washington, DC (1982).
8. Floss, R., K. Brüggemann and H. Heinisch, "Porenwasserdruck- und Setzungsmessungen in normal konsolidiertem geschichtetem Untergrund," *Vorträge Baugrundtagung,* Mainz, pp. 281–302 (1980).
9. Floss, R. and H. Heinisch, "Dynamische und statische Vorkonsolidation organischer Schluffe und Tone im Untergrund einer Dammschüttung," 6.Donau-Europäische Konferenz für Bodenmechanik und Grundbau, Varna (1980).
10. Heinisch, H. and K. H. Blume, "Zur Wirksamkeit von Vertikaldräns," 22.*Erfahrungsaustausch über Erdarbeiten im Strassenbau,* Nürnberg, pp. 64–82 (1982).
11. EspE-NS77- *Ergänzungen der ZtVE-StB76 für das Spülverfahren bei Erdarbeiten im Strassenbau in der Niedersächsischen Strassenbauverwaltung,* Niedersächsisches Landesverwaltungsamt, Abt. Strassenbau, Hannover (1978).
12. Christow, K., "Beitrag zur praktischen Setzungsberechnung und Auswertung von Zeit-Setzungsmessungen," Donau-Europ. Konferenz für Bodenmechanik im Strassenbau, Wien (1968).
13. Johnson, S. J., "Foundation Precompression with Vertical Sand Drains," *ISMFD, ASCE, Vol.96,* No.SM1, pp. 145–175 (1970).
14. Thamm, B. R., "Numerische Analyse von einem Damm auf weichem Untergrund," *Geotechnik, I.* 1, pp. 84–92 (1978).
15. D'Appolonia, P. J., H. G. Poulos and C. C. Ladd, "Initial Settlement of Structures on Clay," *J. of the SMFD, ASCE, Vol.97,* No.SM10, pp. 1359–1377 (1971).
16. Bjerrum, L., "General State of the Art Report: Problems of Soil Mechanics and Construction on Soft Clays and Structurally Unstable Soils," *Proc. VIII, ICSMFE, Vol. 3,* Moskau, pp. 111–159 (1973).
17. Thamm, B. R., "Numerische Berechnung von Dämmen auf

weichem Untergrund," "Finite Elemente in der Baupraxis," *Vorträge anlässlich einer Tagung an der TU Hannover,* Verlag Wilh. Ernst & Sohn, Berlin, pp. 196–211 (1978).

18. Thoms, R. L., Pequet and Ara Arman, "Numerical Analyses of Embankments over Soft Soils," *Proc. 2nd. Int. Conf. on Numerical Methods in Geomechanics, Blackburg, Vol.II,* pp. 623–638.
19. Faust, J., K. Moritz, H. Stiefken, "Vane Shear Tests in Peat" *Int. Symp. on In-Situ Testing, Vol. 2,* Paris, pp. 283–286 (1983).
20. Haberland, J., "Einsatz von Tischrechnersystemen in der Messtechnik," *24.Erfahrungsaustausch uber Erdarbeiten im Strassenbau,* Bergisch-Gladbach, pp. 83 (1984).
21. Floss, R., "Verformungen erstkonsolidierter weicher Böden unter Dämmen," *ZTVE- Zusätzliche technische Vorschriften und Richtlinien für Erdarbeiten im Strassenbau,* Kirschbaum-Verlag, Bonn, pp. 347–355 (1979).
22. Thamm, B. R., "Field Performance of Embankment over Soft Soil," *Journal of Geotechn. Eng., Vol.110,* No.8, August, pp. 1126–1146 (1984).
23. Bjerrum, L., "Embankments on Soft Ground," *Proceedings, ASCE Speciality Conference on Earth and Earth-Supported Structures, Purdue University, West Lafayette, Ind., Vol. II,* pp. 1–54 (1972).
24. Ohde, J., "Grundbaumechanik," *Hütte,* Bd.III,27.Aufl., pp. 886 ff (1951).
25. Quast, P., "Probleme beim Bau von Dämmen auf weichem Untergrund," *Vorträge der Baugrundtagung,* Mainz, pp. 303–325 (1980).

CHAPTER 8

Stability and Deformation of Embankments on Soft Clay

J. A. R. ORTIGÃO* AND M. S. S. ALMEIDA**

1. INTRODUCTION

When soft soils are encountered below a proposed earth embankment, the geotechnical engineer has four alternatives:

- Bypass the soft soil by relocating the embankment structure.
- Remove the soft soil and replace it with a good one.
- Design the embankment structure for the soft soil.
- Treat the soil to improve its properties.

As good sites and materials get scarce, the last two alternatives increase in importance.

If stability requirements prevent construction of an embankment in a single lift, a number of methods can be used for the design of the embankment-foundation system. Some common methods of construction are the use of stabilizing berms or light weight fills, preloading, surcharge loading, construction in stages, and dynamic compaction. Sand drains or prefabricated drains associated with stage construction or surcharge loading are commonly used when the acceleration of settlements and rapid increase of the foundation strength are important requirements.

Alternatively the weight of the embankment can be transferred to more competent layers by stone columns, lime columns or embankment piles, or redistributed to the clay foundation by the use of reinforcement under the embankment. Thermal stabilization such as heating and freezing, grouting and eletro-osmosis are less common procedures used for improvement of the foundation soil under embankments.

The methods for construction and treatment of foundation soils under embankments, as outlined above, have been described in detail by Broms (1979), Pilot (1981), ASCE (1978), and Mitchell and Katti (1981), Magnan (1983), among others, and it is beyond the scope of this chapter to discuss them further. Factors which lead to the choice of the most suitable method for construction or treatment of the foundation soil for a particular problem will depend on technical, economical and political aspects.

This chapter examines methods of analysis of embankment on untreated ground in terms of stability and deformation. An outline of the most important steps in site and laboratory tests is presented and procedures are recommended. Instruments and techniques used to control stability and deformations are reviewed.

This chapter is mainly aimed at the non-specialist civil engineer and, therefore, the brief coverage of certain aspects of basic soil mechanics can be found tedious by the specialist. Nevertheless, the writers bore in mind the need of covering the most up-to-date equipment and techniques of analysis in a effort to be comprehensive, but not exhaustive.

2. SITE INVESTIGATION

2.1 Introduction

In this item an outline of recommended procedures for site investigation is presented. For a detailed discussion see, for instance, Clayton et al. (1982), Hvorslev (1949), Andresen (1981), Lowe and Zacchev (1975), OCDE (1979).

Site investigation works can be grouped in the following steps: preliminary study, initial and detailed geotechnical investigation. Each of them will be summarized as follows.

*School of Engineering, Federal University of Rio de Janeiro, Brazil

**COPPE (Post-Graduation School), Federal University of Rio de Janeiro, Brazil

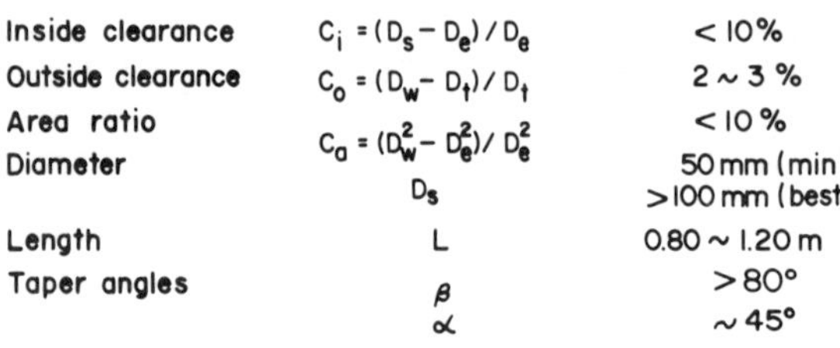

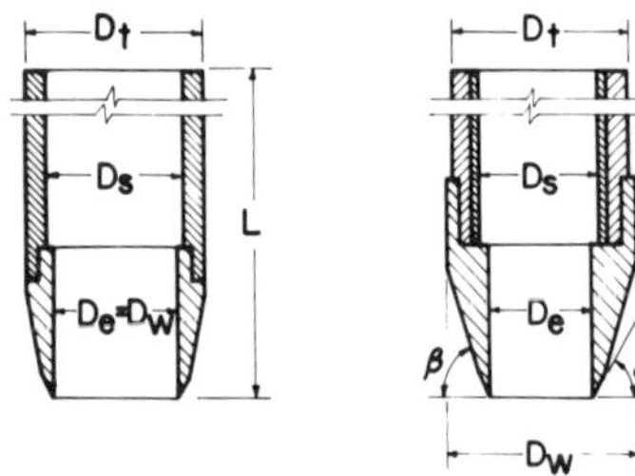

FIGURE 1. Recommendations for samplers design (from Hvorslev, 1949).

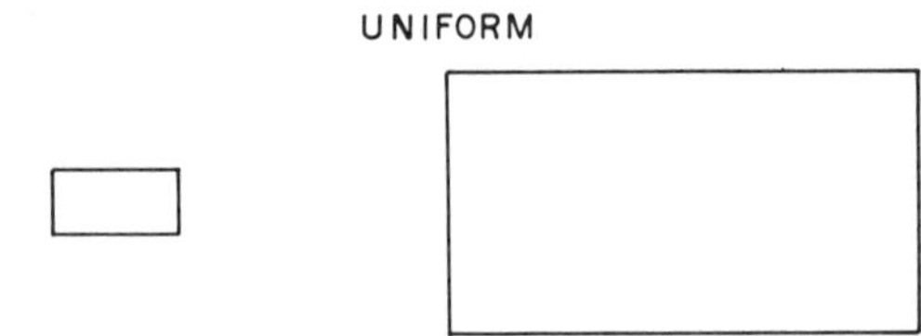

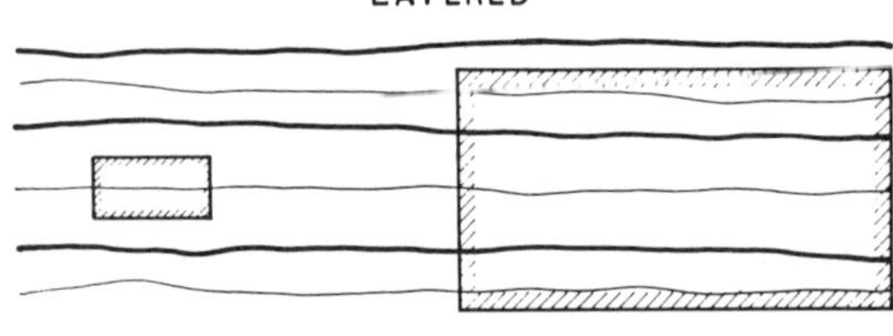

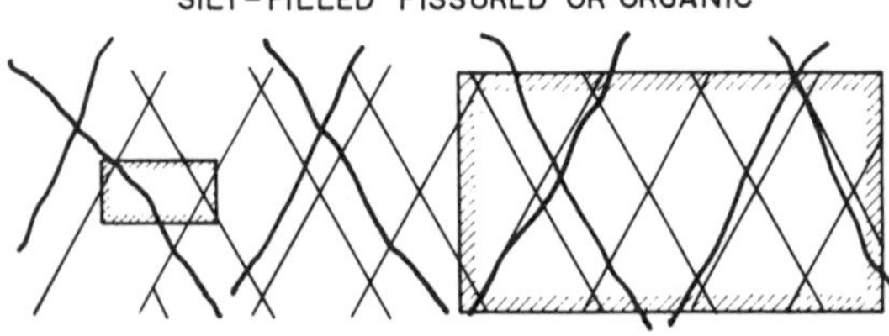

FIGURE 2. Influence of size on fabric representation (from Rowe, 1971).

2.2 Preliminary Investigation on Geological and Soils Conditions

These initial works comprise an investigation on existing documentation, including:

- topographical and geological charts
- aerial photographs
- literature review and pre-existing experience in this area, etc.

During this phase the possibility of existence of soft soils deposits near the proposed structure is investigated. In addition, the pre-existing foundation experience in the area would recommend foundation solutions previously employed with success.

2.3 Initial Geotechnical Investigation

The extent and depth of soil layers and groundwater close to the proposed structure should be investigated in this phase. This can be accomplished by drilling strategic located boreholes, usually having a diameter varying between 60 to 100mm. This provides a detailed sequence and description of soil layers (the borehole log) and at least one sample from each layer.

Laboratory tests to be carried out on these samples include index properties (water content, liquid and plastic limit), sieve and sedimentation analysis, unit weight and crude strength tests (e.g., torvane, pocket penetrometer, fallcone). This information enables a preliminary estimative of embankment settlement and stability to be performed for each type of foundation solution under consideration. This, in turn, will lead to a decision such as:

- The solution is simple and further studies are unnecessary.
- Relocation of the embankment.
- Embankment will be replaced by a structure.
- A further and detailed site investigation is necessary for the embankment design.

TABLE 1. Sampler Quality Classification.

Quality Index	Type	Sample Usual Dimensions
1	Block samples	300 mm side length cube
2	Specially large diameter sample	diameter = 300 mm 1 m length
3	Stationary piston samplers	diameter: 75 to 200 mm 1 m length
4	Thin-walled samplers	diameter: 50 to 100 mm 1 m length

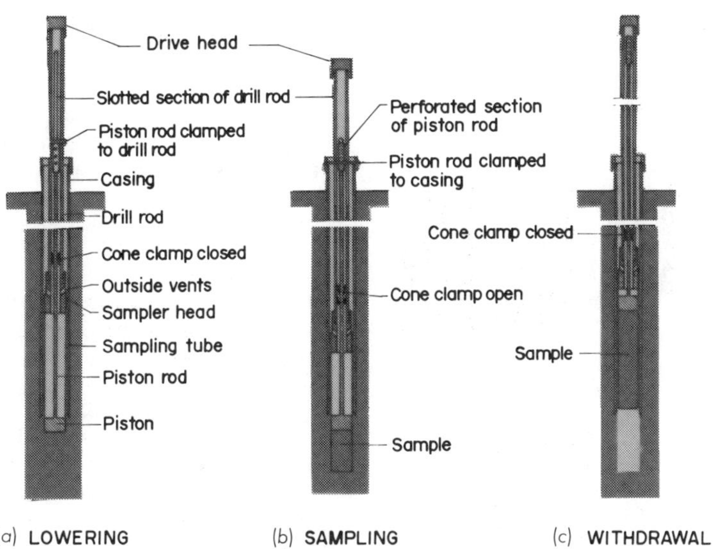

FIGURE 3. Stationary piston sampler (from Hvorslev, 1949).

2.4 Detailed Geotechnical Investigation

This phase may include high quality samples for special laboratory work and in situ tests. The effort spent in this phase will, of course, depend on the importance of the project, the nature of soil and the risks involved.

Sampling techniques can vary according to the soil type and the necessary sample quality which, in turn, depend on the soil parameters to be measured. The more is the effort towards preserving the in situ water content and stress conditions, the higher the sample quality requirement.

Table 1 presents a classification of sampling techniques listed in decreasing quality. Block samples are regarded to present the highest quality (e.g., Milovic, 1971) and can be easily obtained in situations where an excavation is needed for the construction of the structure (e.g., Soares, 1983). Another favorable situation occurs in slope stability studies in residual soils when ground water conditions, depth to be reached and soil type enable an unbraced low cost shaft to be excavated and block samples are obtained at a reasonable cost. This, however, is seldom the case of soft soils in which the costs involved generally make the choice of block samples impossible. In such situations tube samplers, corresponding to quality no. 2, 3 and 4 in Table 1 are obtained. A relationship between wall thickness and tube diameters for these samplers has been recommended by Hvorslev (1949) in order to minimize sample disturbance. These criteria are presented in Figure 1.

There are cases in which specially large diameter samples have been recommended (Rowe, 1971). These situations arise when determining permeability and consolidation properties of soils presenting structural nonhomogeneity, i.e., fissures, slickensides and fine layers of sand (Figure 2). In these cases large (300mm) diameter samples tested in hydraulic oedometers have produced good results.

Nevertheless, in most cases and specially in soft soils, only stationary piston thin-walled samplers, as shown in Figures 3 and 4, are commonly employed. The stationary

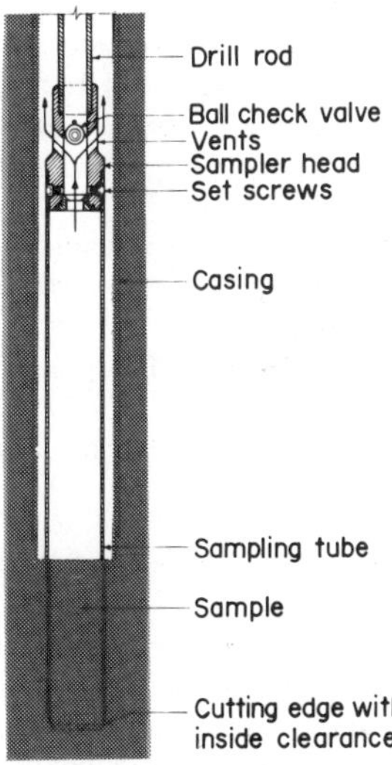

FIGURE 4. Thin-walled sampler (from Hvorslev, 1949).

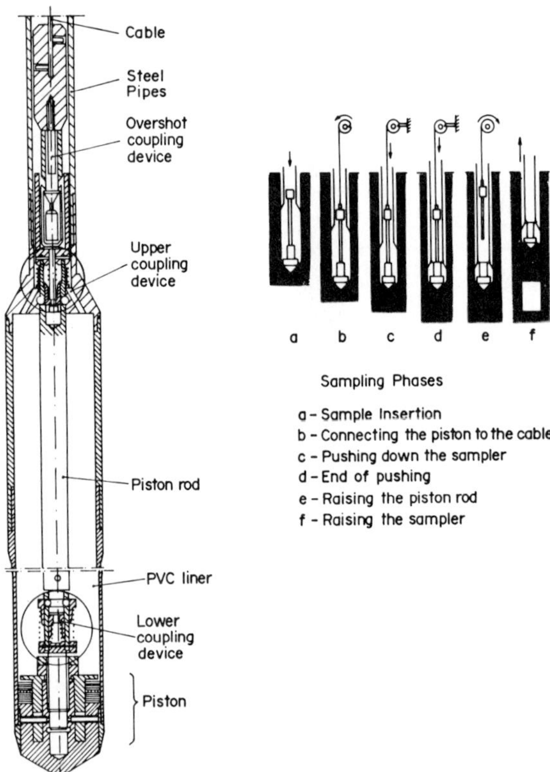

FIGURE 5. LPC stationary piston sampler (from Lemasson, 1973).

piston allows vacuum to be formed on the top of the sample and thus prevents downward movements when extracting it from the ground. This feature has been regarded (e.g., Hvorslev, 1949) to produce a better quality sample. Other precautions to improve sample quality are:

1. To use a thick drilling mud to minimize the stress relief
2. To provide, after sampling, adequate sealing against water content variation
3. To avoid shock and vibration during transportation to the laboratory
4. To provide adequate ambient temperature and moisture content storage facilities

Additionally, some more elaborate samplers, such as the one employed by the French Laboratories des Ponts et Chaussées (Lemasson, 1973), are provided with a plastic internal liner which accommodates the soil until the start of the testing programme. For releasing the sample, instead of extrusion, the liner is longitudinally cut preventing further stressing to the soil.

3. LABORATORY TESTS

3.1 Introduction

A laboratory testing programme on soft foundation soil for the design of an embankment would certainly include:

1. Index properties tests (liquid limit, plastic limit, water content, unit weight and sieve or sedimentation analysis)
2. Oedometer tests
3. Unconsolidated-undrained triaxial tests
4. Consolidated-undrained triaxial tests with pore pressure measurements

The index properties tests are well covered by many textbooks and will not be discussed here. Also, special laboratory tests such as plane strain or simple shear tests will not be described in this section, as they are not commonly used in engineering practice. The remaining tests will be treated as follows.

3.2 Oedometer Tests

Oedometer tests are performed to obtain soil compressibility and soil consolidation parameters. A soil specimen is tested under a zero lateral displacement condition in a steel ring (Figure 6). Porous stones are employed on the top and bottom of the specimen to allow drainage during testing. Among the oedometer techniques, the incremental load test is the most common one in present day practice. The test starts with a small value of pressure (e.g., 10 kPa) applied to the specimen and remains constant during 24 hours. Vertical displacements are recorded during this period. The pressure increment is, then, duplicated and the same sequence of operations is performed. The usual number of increments is 8 to 10 and the test continues until a desired maximum vertical pressure is reached. The pressure is, then, decreased to zero in 3 or 4 stages, time being allowed for the specimen to swell. The readings obtained during a pressure stage enable a plot of specimen height versus log of elapsed time (Figure 7). From this curve, according to the Casagrande or t_{50} method, a value for the coefficient of consolidation c_v is obtained through the equation:

$$c_v = \frac{T_{50}\ (0.5\ d_{50})^2}{t_{50}} \tag{3.1}$$

where

T_{50} = time factor at 50% consolidation, which according to Terzaghi's theory is 0.198
d_{50} = specimen height at 50% of consolidation (Figure 7)
t_{50} = time correspondent to 50% of consolidation obtained through Casagrande's method (Figure 7)

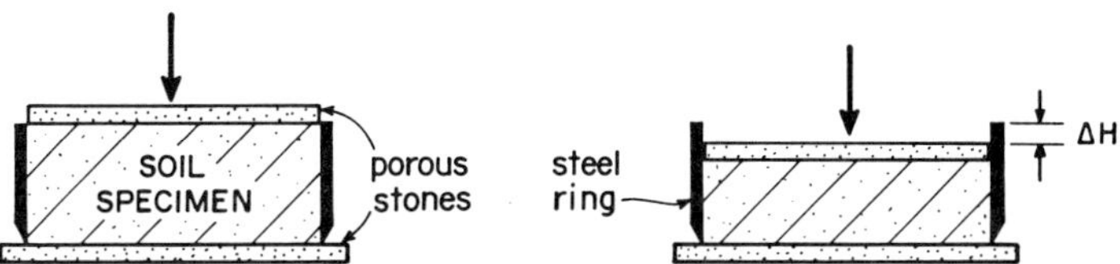

FIGURE 6. Oedometer test conditions.

Alternatively the coefficient of consolidation can be obtained through the t_{90} method from a $\sqrt{t}$ versus specimen height (Figure 8), where t_{90} is the elapsed time during a pressure stage. The equation for computing c_v is:

$$c_v = \frac{T_{90}\,(0.5\,d_{90})^2}{t_{90}} \tag{3.2}$$

in which T_{90} takes the value of 0.848 and other terms are analogous to the previous equation. The writers' experience in applying these methods shows that:

1. c_v from the t_{90} method is slightly different from the one given by t_{50} method. However, due to the scattering of c_v, this difference is not significant.
2. The t_{90} method can usually be applied to all load increments of a test, whereas Casagrande's method is not applicable to the first stages of loading. Indeed, as shown in Figure 9, test data below the vertical pressure of 300 kPa do not follow the pattern shown in Figure 7, which prevents the application of the t_{50} method.

A set of compressibility parameters is obtained from the log of vertical pressure versus void ratio relationship at the end of each loading stage. Figure 10 presents the standard form of obtaining these parameters. The initial flat part of the curve is terminated by a sharp bend, called "virgin" consolidation line. At the end of this line follows the unloading phase of the test or swelling phase. The slopes of these lines, as shown in the figure are the coefficient of compressibility C_c and coefficient of swelling C_s. Compression curves of good quality clay specimens usually have a curved virgin line. In this case it is suggested to take values of C_c for the stress range relevant for the field situation.

The maximum past or preconsolidation pressure σ'_{vm}, defined as a yield pressure below which the strains are approximately recoverable, can be obtained from the void ratio or vertical strain versus log σ'_v relationship as shown in Figure 10. Small loading increments at the beginning of the test, until the sharp bend is reached, are usually recommended for a better definition of σ'_{vm} in very soft and sensitive clays. On the other hand, it is also recommended (e.g., Jamiolkowski et al., 1985) that σ'_{vm} is obtained in a compression curve (Figure 10) drawn from the end of primary consolidation data, as mentioned in Figure 7. The so obtained σ'_{vm} will: (1) be independent of sample thickness and

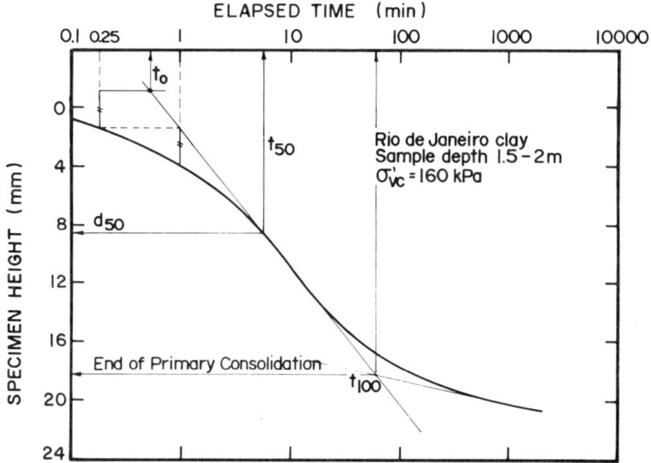

FIGURE 7. The t_{50} method for computing the coefficient of consolidation.

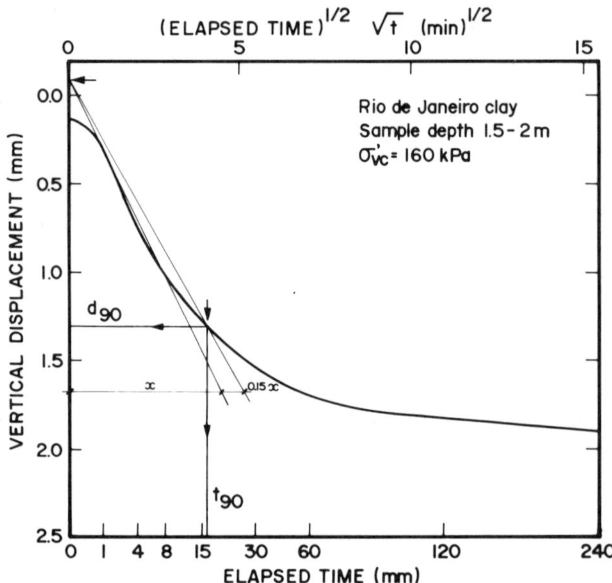

FIGURE 8. The t_{90} method for computing the coefficient of consolidation.

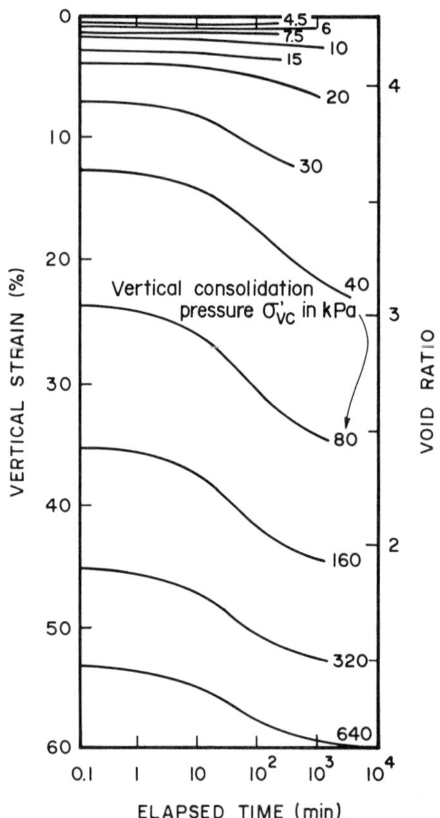

FIGURE 9. Consolidation test data of Rio de Janeiro soft clay.

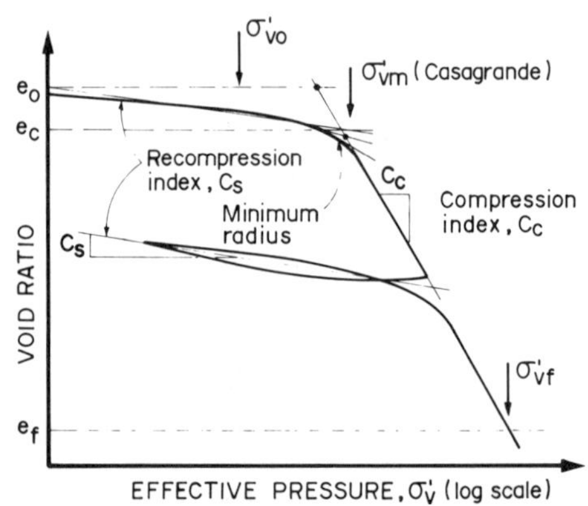

FIGURE 10. Notation and terminology for oedometer compression curves.

drainage path length: (2) be slightly greater, for it doesn't include secondary compression effects. The relationship between the maximum vertical past pressure and in situ overburden pressure gives the overconsolidation ratio ($OCR = \sigma'_{vm}/\sigma'_{vo}$).

As an example of treatment and presentation of oedometer test results, data on Rio de Janeiro soft clay (Ortigao et al., 1983) is shown in Figures 11 and 12. These plots summarize test data from over 60 oedometer tests. Stress history is presented in Figure 11(a). Since σ'_{vm} is greater than in situ effective vertical overburden stress σ'_{vo}, this clay presents a slight overconsolidation. Figures 11(b) and 11(c) present the ratios $C_c/(1+e_o)$ and $C_s/(1+e_o)$ versus depth, in which e_o is the initial void ratio. This ratio was preferred instead of C_c and C_s alone, for it presents a more or less constant value versus depth, being then closer to a material property (Ladd, 1973). Results of the coefficient of consolidation versus log of vertical stress are presented in Figure 12 where the wide scattering of data makes it difficult to evaluate a design value for c_v from oedometer tests.

Oedometer techniques other than the standard incremental load test have been proposed. The main features of the available oedometer tests are summarized in Table 2. The controlling factors of these tests are shown diagramatically in Figure 13. These improved testing techniques produce a continuous compression curve and enable a more accurate determination of compressibility and consolidation properties. Their main advantage, however, is the greatly reduced testing time compared to the conventional incremental test. On the other hand, more sophisticated equipment is needed, and the continuous readings to be taken require an automated data acquisition system.

3.3 UU Triaxial Tests

In an unconsolidated-undrained triaxial test or simply UU triaxial test the soil specimen is sheared at the same water content occurring in situ. No consolidation or drainage is allowed before or during the test.

A scheme of the triaxial cell used for this purpose is shown in Figure 14. The test consists of two phases. Firstly a confining stress σ_c is applied, and then the specimen is sheared by increasing the deviator stress $(\sigma_1 - \sigma_3)$, while vertical displacements are recorded. A plot of deviator stress against vertical strain (Figure 15) enables the value of $(\sigma_1 - \sigma_3)_{max}$, corresponding to failure conditions, to be obtained.

When the value of σ_c is set to zero, the test is called unconfined compression test, and does not require a triaxial cell, which makes the test simpler. However, the soil sample is exposed during testing and its water content may vary, yielding higher strength values than the same sample tested in a triaxial chamber. Therefore, a UU test is usually preferred.

A UU test is assumed to represent a common field condition in which the loading is quick enough to avoid any consolidation taking place at the clay foundation. In this situation, the soil strength during loading is the same of that occurring before construction. Therefore, the "$\phi = 0$ method" is applicable (Figure 16) and the undrained strength c_u is taken equal to $(\sigma_1 - \sigma_3)/2$.

A useful way for exploiting UU test results is to plot c_u data against depth, as shown in Figure 17, and try to obtain a c_u profile for use in the design. Data scattering, however, may make the choice difficult. Computation of the un-

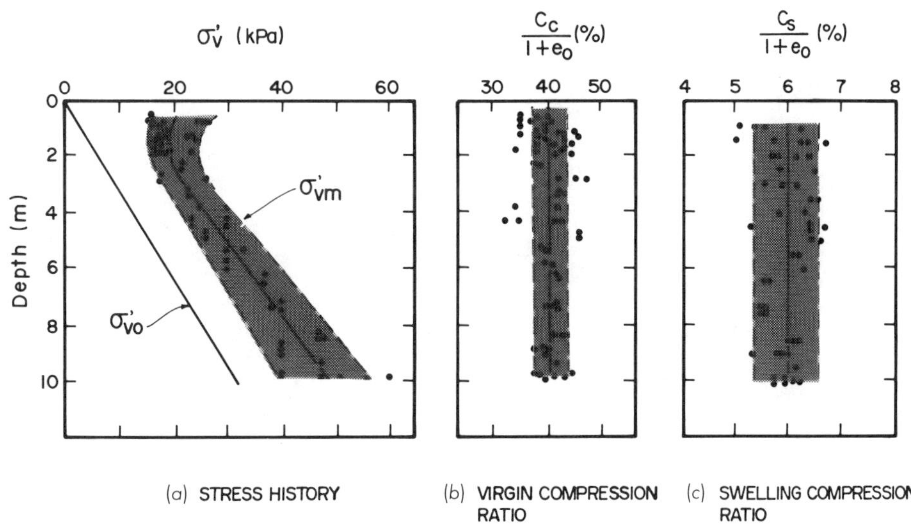

FIGURE 11. Oedometer test data of Rio de Janeiro soft clay.

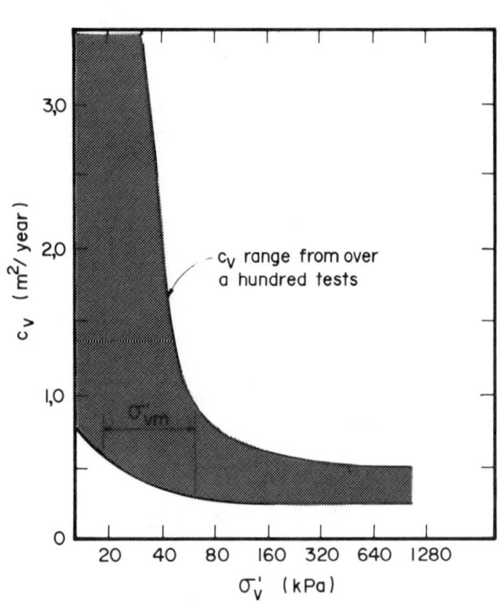

FIGURE 12. Coefficient of consolidation versus effective vertical stress from oedometer tests on Rio de Janeiro soft clay.

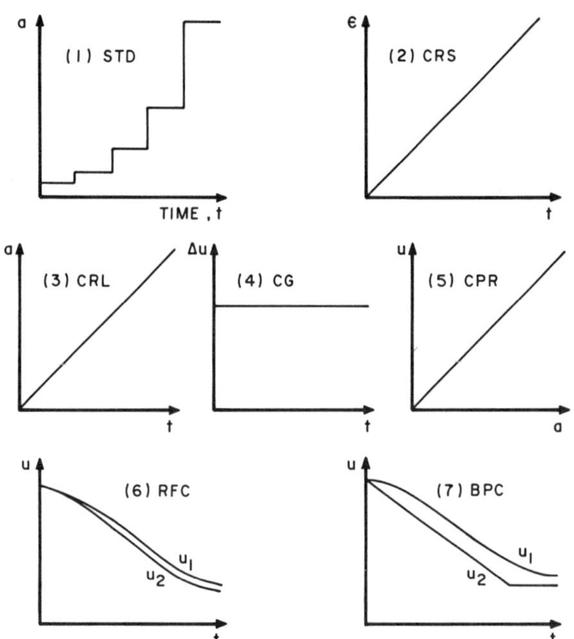

FIGURE 13. Controlling factors for consolidation tests.

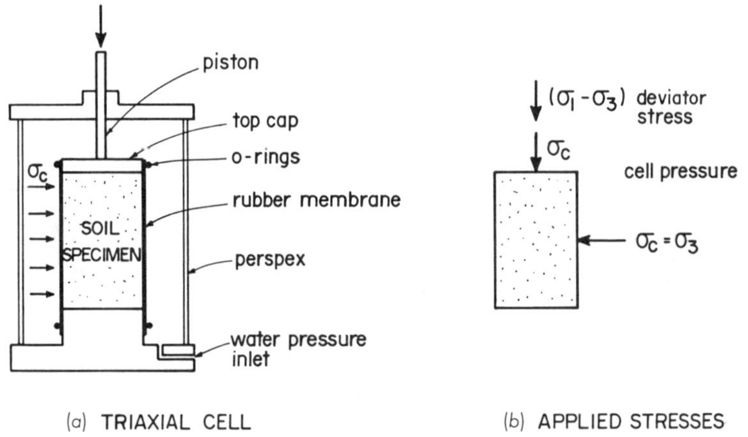

FIGURE 14. Conditions during UU triaxial test.

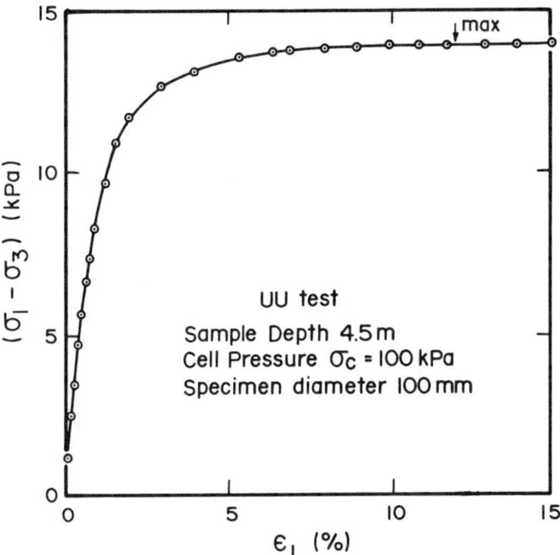

FIGURE 15. UU test results of Rio de Janeiro soft clay.

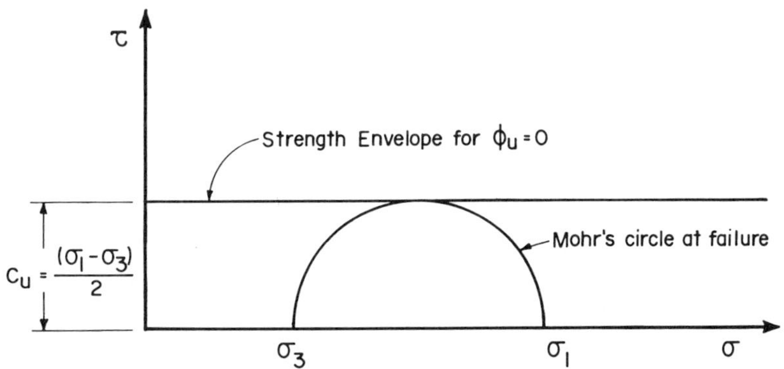

FIGURE 16. Strength envelope under UU conditions.

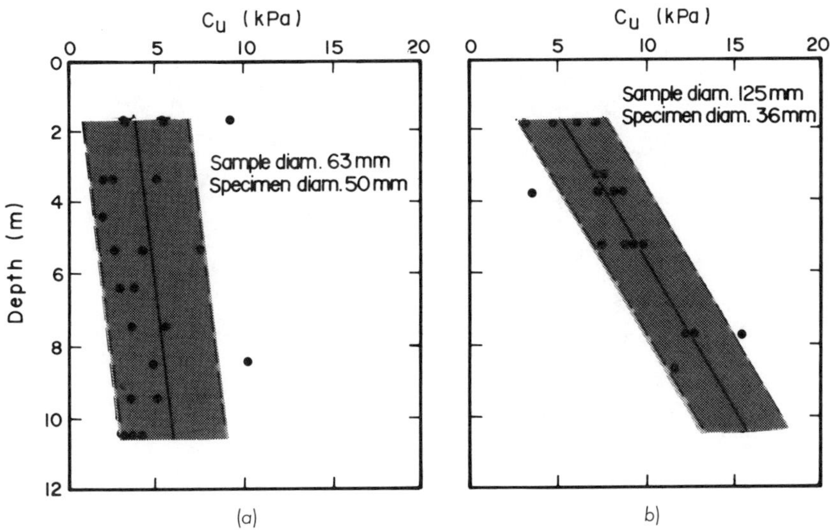

FIGURE 17. UU triaxial test results of Rio de Janeiro soft clay.

TABLE 2. Features of Consolidation Tests (Head, 1983).

Test type	References	Advantages	Disadvantages	Special requirements
1. Standard incremental loading (STD)	BS 1377:1975 Head	Simple apparatus Easy to do Procedures and interpretation well established Direct determination of c_v	Very slow (about 2 weeks) Separate points obtained Curve fitting analysis needed	Constant attention Manual analysis
2. Constant rate of strain (CRS)	Gorman et al. Smith & Wahls Wissa et al. Sällfors	Easy to run Steady state condition reached at slow strain rate if $p < p_c'$ Back pressure not essential Unloading simple by same principle	Rate of strain must be decided No correlation yet with STD unloading curves Various interpretation methods suggested—no standard yet	Load frame Special load piston Very low speed may be needed
3. Constant rate of loading (CRL)	Aboshi et al. Wissa et al. Burghignoli	Straightforward interpretation	Rate of loading must be decided and controlled No correlation with STD unloading	Pacer control or Feed-back control
4. Constant pore pressure gradient (CG)	Gorman et al. Lowe et al. Sällfors	Simple test criterion Steady state condition for $p > p_c$ Lag in pwp response not critical Simple interpretation Choice of gradient not critical	Servo mechanism may be sensitive to trimming No agreement with STD unloading	Null sensor with feed-back or Computer control
5. Constant pressure ratio (CPR)	Janbu et al. ('CL' test)	Testing time < ½ day Ratio not critical if < 0.8 p_c can be determined from several graphical plots	Computer control essential No correlation with STD unloading	Computer control
6. Restricted flow (RFC)	Hoare	One load application only Testing time about ½ day	Reliability of flow restrictor to be proven May not be applicable to unloading	Flow restrictor Differential pressure transducer
7. Back pressure control (BPC)	Head	One load application Quick Can be reversed for unloading Control of back pressure could be related to parameters other than time	Rate of back pressure change must be decided	Feed-back control Moisturised pressure application system

drained strength using the methods described in section 5.3.2 will help with the choice of the c_u profile to be adopted for design.

The purpose of Figure 17 is to show the effect of sample disturbance on c_u data. In fact, the only differences between the tests in Figures 17(a) and 17(b) are the sample and specimen dimensions and the sampler type employed. The data on Figure 17(a) are from specimens trimmed from 63mm diameter samples obtained in thin-walled sample tubes, whereas data on Figure 17(b) are from specimens from 125mm diameter stationary piston sampler. The higher resistances shown on this latter figure are due to the better quality of the samples tested.

The mean value of c_u against depth from UU tests on specimens with a varying degree of disturbance is compared in Figure 18. Again, it can be seen that the less the disturbance the higher the strength.

UU tests have been criticized (Ladd and Foot, 1974; Jamiolkowski et al., 1985) because their results are seriously affected by sample disturbance. Deformation parameters (e.g., Young's Modulus, E_u), in particular, are not considered reliable also because the tests do not start from the correct initial stress state, and c_u data are usually highly scattered.

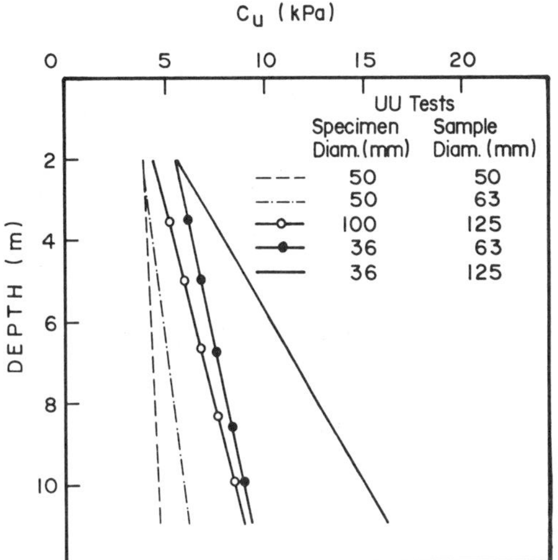

FIGURE 18. The effect of sample disturbance mean c_u profiles from UU triaxial tests on Rio de Janeiro clay.

3.4 CU Triaxial Tests with Pore Pressure Measurements

CU stands for consolidated undrained test in which the sample is consolidated and followed by undrained shear. In most cases the sample is isotropically consolidated (CIU test), i.e., the all around consolidation pressure σ_c is kept constant, while time and drainage are allowed. During the shear phase, the drainage values are closed and the pore water pressure buildup is monitored by means of an electric pressure transducer (Figure 19). The phases of the test are shown on Figure 20. Figures 21 and 22 show typical treatment and presentation of triaxial test data. Figure 21 presents the stress-strain curve and pore pressures monitored during a test on a normally consolidated clay sample. Figure 22(a) shows how the effective strength envelope can be obtained from Mohr's circles drawn from at least 2 specimens consolidated at different σ_c. Alternatively, a $p' \times q$ diagram [Figure 22(b)] yields the parameters a' and α' which can be converted into the traditional c' and ϕ' through the following equations (Lambe and Whitman, 1968):

$$\phi' = \text{arc sin } \tan\alpha' \tag{3.3}$$

$$c' = \frac{a'}{\cos \phi'} \tag{3.4}$$

The advantage of the $p' \times q$ diagram is that it enables the strength envelope to be interpolated among the points correspondent to failure. Interpolation of a straight line between points is easier than drawing a line tangent to circles.

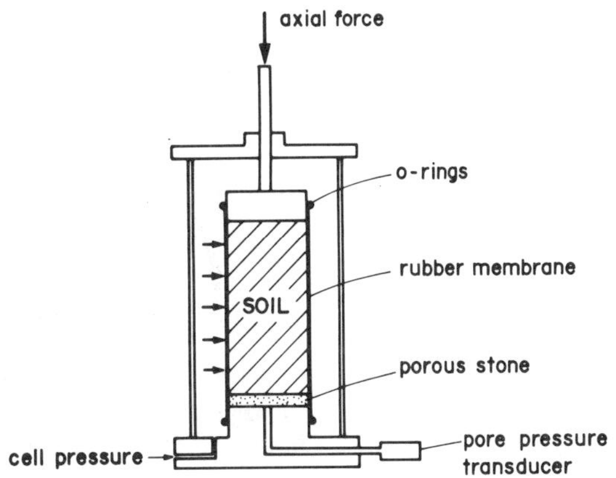

FIGURE 19. Test set up for a CU triaxial test with pore pressure measurements.

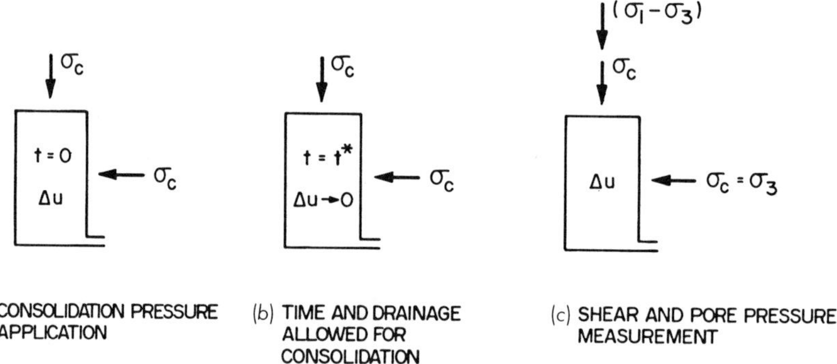

FIGURE 20. Conditions during a CU triaxial test with pore pressure measurements.

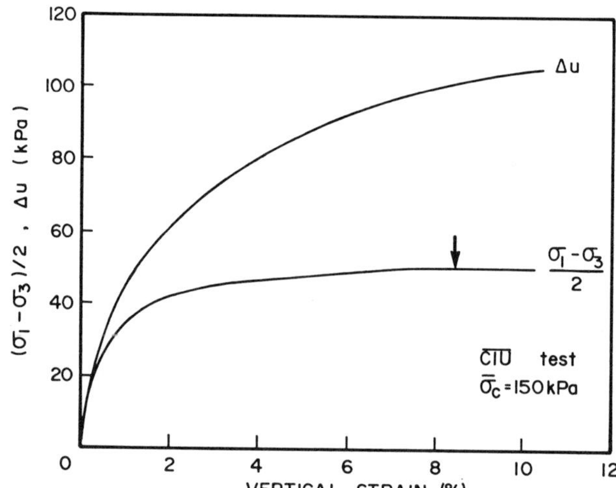

FIGURE 21. Results from a CU test on an isotropically and normally consolidated specimen of Rio de Janeiro soft clay.

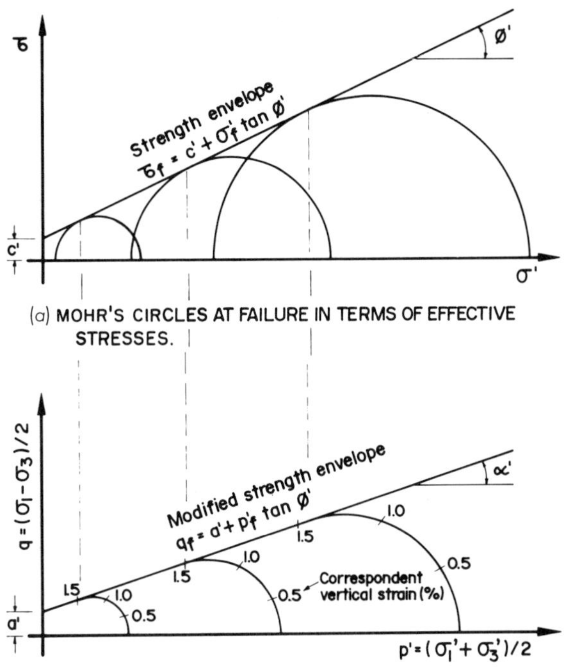

FIGURE 22. Methods for obtaining strength envelopes.

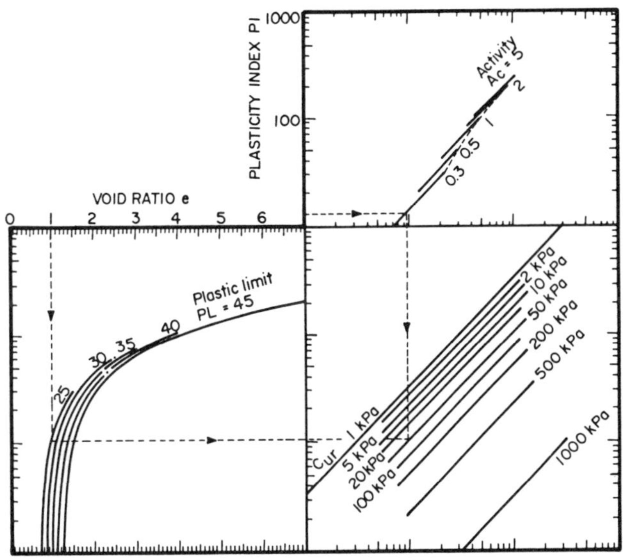

FIGURE 23. Chart for estimating the remoulded undrained strength c_{ur} (from Carrier and Beckeman, 1984).

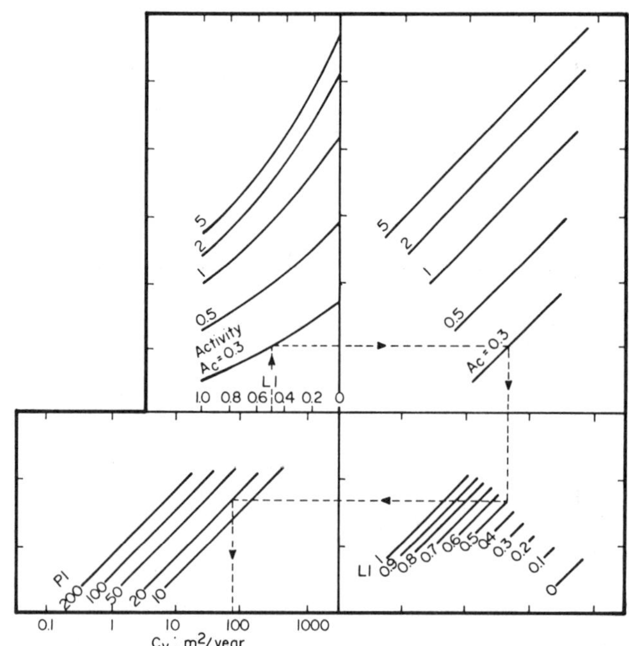

FIGURE 24. Chart for estimating the coefficient of consolidation c_v (from Carrier, 1985).

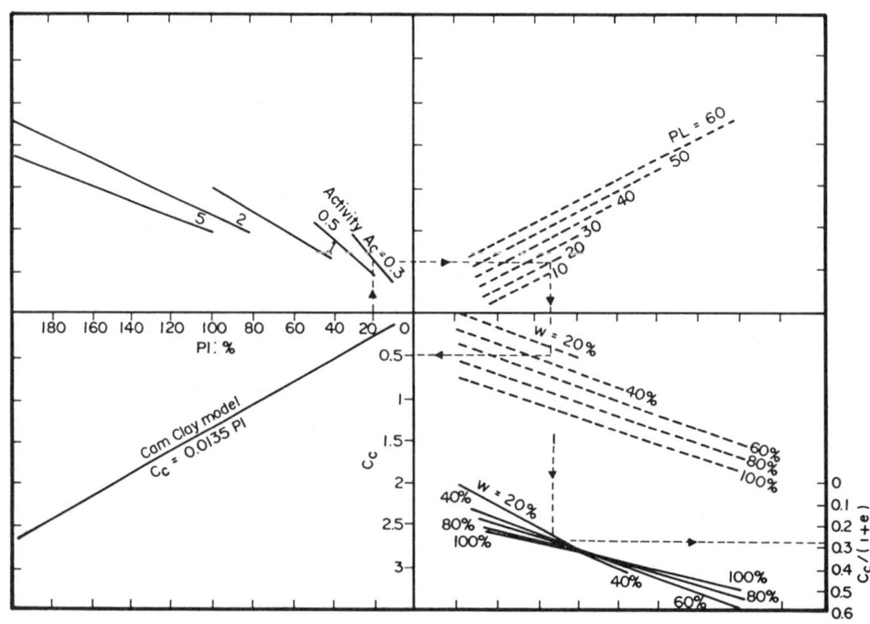

FIGURE 25. Chart for estimating the compression index C_c and the normalized compression index $C_c/(1 + e)$, (from Carrier, 1985).

TABLE 3. Correlations Between Soil Properties.

Correlation	Reference
$c_{ur} = P_{atm}\left(\dfrac{0.166}{0.163 + \dfrac{37.1e - PL}{PI\,[4.14 + A_c^{-1}]}}\right)^{6.33}$	(Carrier and Beckman, 1984)
$c_u/\sigma'_{vo}\ (NC) = 0.652 \sin\phi' + 0.031$	(Mayne, 1980)
$c_u/\sigma'_{vo}\ (OC) = c_u/\sigma'_{vo}\ (NC)\ OCR^{0.8}$	(Ladd et al., 1977)
$\dfrac{C_c}{1 + e_o} = 0.329\left[1 + \dfrac{0.0133\,PI\,(1.192 + A_c^{-1}) - 0.027\,PL - 1}{1 + 0.027\,w}\right]$	(Carrier, 1985)
$C_c = \dfrac{1}{2}\left[\dfrac{\gamma_\omega}{\gamma_d}\right]^{12/5}$	(Herrero, 1980)
$C_v = \dfrac{28.67}{PI} \times \dfrac{(1.192 + A_c^{-1})^{6.993}\,(4.135\,LI + 1)^{4.29}}{(2.03\,LI + 1.192 + A_c^{-1})^{7.993}}\ m^2/year$	(Carrier, 1985)
$\sin\phi' = 0.656 - 0.409\,(PI/LL)$	(Mayne, 1980)
$K_o = (1 - \sin\phi')\ OCR^{\sin\phi'}$	(Mayne and Kulhawy, 1982)

where P_{atm} = atmospheric pressure (100 kPa)
c_{ur} = remoulded undrained strength
PL = plastic limit
LL = liquid limit
LI = liquidity index LI = (LL − W)/PI
A_c = activity, defined as PI divided by the percentage of clay fraction, i.e., less than 2μm in diameter
w = water content
e_o = initial or in situ void ratio
C_c = compression index
γ_w = unit weight of water
γ_d = dry unit weight of soil $\gamma_d = \dfrac{\gamma_t}{(1 + w)}$

Moreover, the stress path representation is definitively an important information on the soil behaviour. Figure 21(b) shows the effective stress path followed by two normally consolidated clay specimens. A valuable feature can be added on the $p' \times q$ diagram to provide information on the strain behaviour of soils. The stress paths shown in Figure 21(b), include values of vertical strain reached by the specimens during loading.

CU tests are sometimes performed under anisotropic consolidated conditions and, in this case the test is called CK_oU. This type of test is usually preferred when using SHANSEP or recompression techniques described in section 5.3.2. Extension CU tests, which are tests in which the specimen is sheared under increasing confining stress and constant vertical stress, are not commonly performed for the design of embankments on soft clays.

3.5 Correlations Between Soil Parameters from Laboratory Tests

In many practical engineering situations important design decisions may have to be based on poor quality or very little soil data. In these circumstances available correlations between soil parameters may be useful for assessing soil parameter values. Also, in early stages of design such correlations may enable the assessment of important soil properties based on Atterberg limits and clay activity.

A few and recent published correlations for clay are summarized in Table 3. They are applicable to normally consolidated clays, with the exception of the equations to compute c_u/σ'_{vo} suggested by Ladd et al. (1977) and the equation to compute the coefficient of earth pressure at rest K_o proposed by Mayne and Kulhawy (1982), which are applicable overconsolidated clays. Charts for estimating the remoulded undrained strength, c_{ur}, the coefficient of consolidation, c_v, and the ratio $C_c/(1 + e)$ are shown in Figures 23, 24, and 25.

4. IN SITU TESTS

4.1 Introduction

In situ tests have several advantages, as compared to laboratory tests, for determining geotechnical properties of

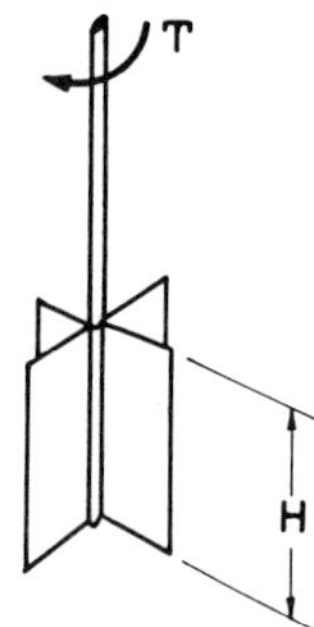

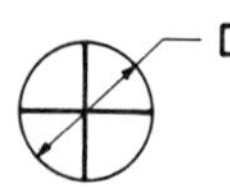

FIGURE 26. The vane blades.

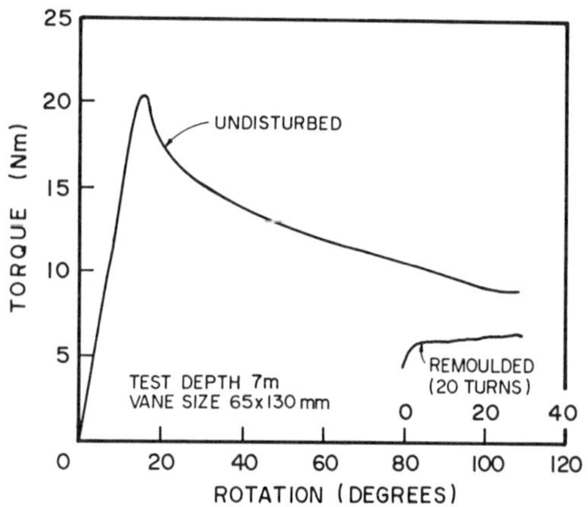

FIGURE 27. Typical torque x rotation curve from FVT on Rio de Janeiro soft clay.

soft clays, such as:

1. The soil is tested in situ, avoiding disturbance due to sampling, handling, and transportation.
2. Continuous profiling is possible, if certain types of in situ testing are chosen, giving a better insight on the stratification and enabling thin sandy or silty layers or seams to be detected.
3. In situ test results can show influence of soil structure and texture which, in some cases, do not appear in laboratory tests.

Nevertheless, there are also disadvantages in in situ tests. The main one is the fact that drainage and boundary conditions cannot be controlled as it can be in the laboratory.

Among the in situ tests types described in the literature (e.g., Andresen, 1981, Clayton et al., 1982 and OCDE, 1979) the most suitable tests for the design of embankments on soft clay are:

- the field vane test (FVT)
- the cone penetration test (CPT) and, recently, the piezocone test (PCPT)
- the pressuremeter test (PMT)

4.2 The Field Vane Test

4.2.1 INTRODUCTION

Early attempts to determine the shear strength of clays through the FVT occurred in Sweden and Germany and go back as far as 1920–30. However, FVT only became fully operational and its use widespread after the work of Cadling and Odenstad (1950).

The tests consists in inserting vertically into the soil a four blade vane (Figure 26) and rotating it according to a standard rate of rotation of 6 degrees per minute. The necessary torque is recorded (Figure 27) and, through the assumption that the soil shear strength is constant over the cylinder which contains the vane and ratio height to diameter of the vane is 2, the undrained shear strength can be computed by:

$$c_u = 0.86 \frac{T_{max}}{\pi D^3} \tag{4.1}$$

where

T_{max} = maximum measured torque
D = vane diameter

4.2.2 EQUIPMENT AND TEST METHODS

Commercially available FV equipments can be classified according to the method of installation (Figure 28):

(a) *Unprotected vane through a prebored hole.* In this case the measured torque includes soil friction on the rods

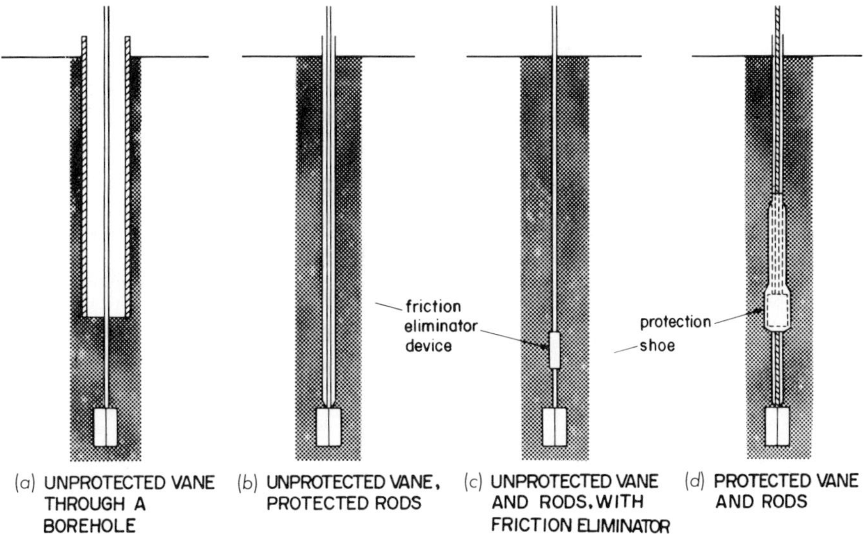

FIGURE 28. Types of FV equipments, classified according to the vane insertion method.

just above the vane. Although friction influence can be evaluated through a dummy vane test, only with the rods, and no vane, this is time consuming and subject to error especially in very soft clays, as discussed in detail by Collet (1978).

(b) *Protected rods and unprotected vane.* An equipment of this type has been in use by the Laboratoire des Ponts et Chaussées (Lemasson, 1973).

(c) *Unprotected vane with friction eliminator device.* This is the case of the Nilcon vaneborer which has a friction eliminator device, allowing the rods to rotate 45° freely before actuating the vane. This device enables rod friction to be accounted for and eliminated in the computations.

(d) *Protected vane and rods.* The vane is installed recoiled in a protection shoe to avoid damage when encountering hard materials. The vane rods are fully protected against contact with the soil, avoiding any undesired friction. At the required depth, the vane rods are released and the vane is, then, pushed some 0.50m in the ground. Details of this apparatus are presented in Figure 29.

In most cases a 65 mm diameter vane with a 1:2 (diameter to height) ratio is used.

Vane dimensions and other test characteristics have been normalized in some countries, e.g.:

- in USA ASTM-D2573–1972
- in Britain BS 1377 : 1975
- in France LPC (1980)
- in Germany DIN 4096

4.2.3 INTERPRETATION OF THE RESULTS

Stress strain conditions during a FVT are very difficult to analyse. Attempts to perform more sophisticated analyses such as those suggested by Aas (1967) and Wiesel (1973) may lead to inconsistent results, as observed by Costa-Filho et al. (1977) and Collet (1978). In addition, test results were found (Aas, 1965 and 1967) to depend on vane shape, dimensions, rate of rotation and delay between vane insertion and rotation. Therefore, it is recommended that the FVT should be standardized and its results regarded rather strength indexes than clay properties.

As an example of FV data interpretation, Figure 30 shows the undrained shear strength profile plotted against depth for the Rio de Janeiro soft clay. These data were obtained from an FV test carried out in 0.50 m intervals, employing the equipment shown in Figure 29. In addition to the test performed in an undisturbed clay condition, another test was carried out after rotating the vane some 20 times. This yielded the remoulded strength which, in comparison with the test result in the undisturbed condition, is an indication of the sensitivity of the clay.

Many years of experience in using FV data for analysing the stability of embankments on soft ground have demonstrated that FV data may have to be corrected. These corrections were originally proposed by Bjerrum (1972) but recently, however, Azzouz et al. (1983) reanalyzed existing data and proposed an improved correction curve as shown in Figure 31. Vane correction factors are discussed in detail in section 5.3.1.

In the writers' point of view, the FV tests are the best way for determining the undrained strength for the design of em-

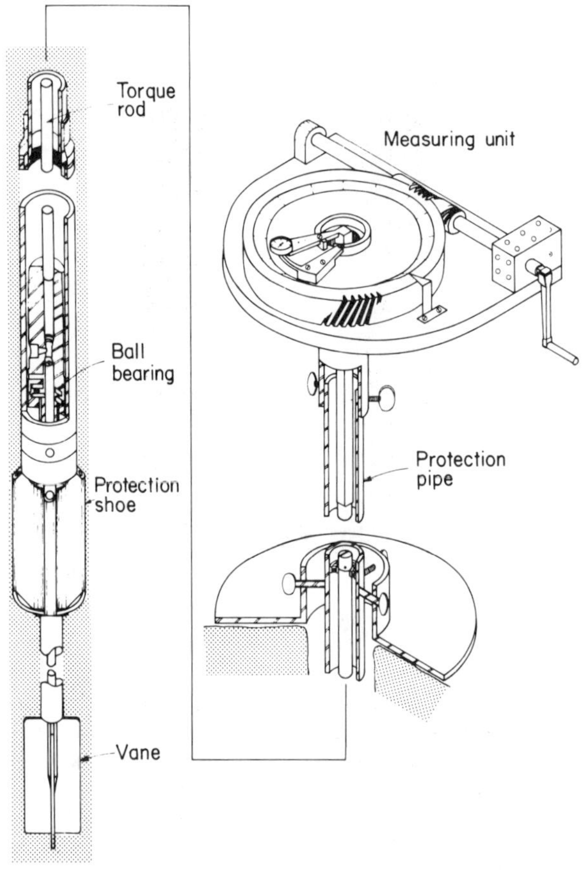

FIGURE 29. Details of a field vane borer apparatus.

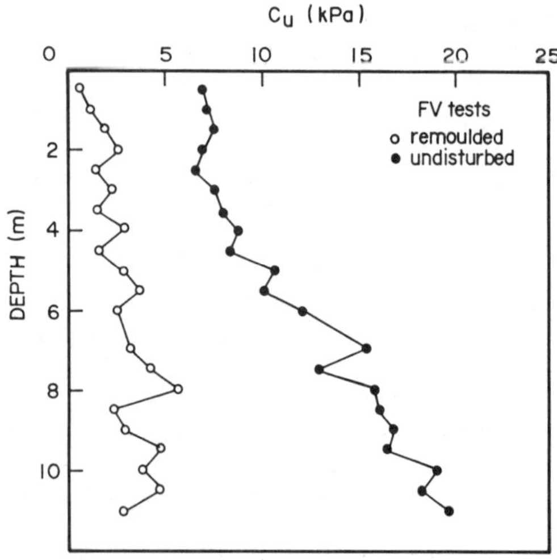

FIGURE 30. FVT results on Rio de Janeiro soft clay.

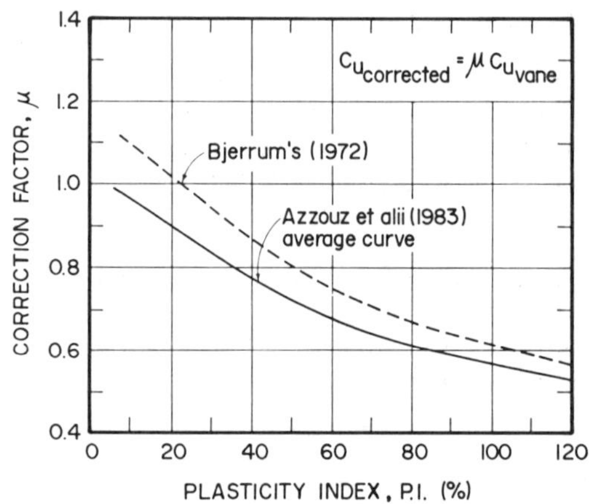

FIGURE 31. Field vane correction factors.

bankments on soft ground. The main reasons for this are:

1. The necessary equipment is simple, rugged and relatively inexpensive and, therefore, appropriate for field use.
2. Only two persons are necessary to operate it and, by using the FV borer equipment shown in Figure 29, a 20m borehole with intervals of 1m tests can be completed in one day's work.
3. The total cost of the strength profile obtained through FV test is certainly the least one.
4. Provided that the necessary corrections are made to the FV data and a total stress analysis is performed, reliability in the design of embankments on soft soils can be achieved.

Nevertheless, it is suggested to undertake at least one additional determination of the clay strength, in order to compare it with the vane strength. This point is further discussed in section 5.3.2.

4.3 The Cone Penetration Test

4.3.1 INTRODUCTION

Originally developed in Holland about 1930, the Cone Penetration Test (CPT) has had in the last decade increasing interest in its application. A comprehensive treatise on CPTs which summarizes developments up to the early 1970s was presented by Sanglerat (1972). Recently an enormous amount of information and research have been published on CPTs and most of it can be found in the following conference proceedings:

- ASCE Symposium on Cone Penetration Testing and Experience, St. Louis, 1981
- 2nd ESOPT, the European Symposium on Penetration Test, Amsterdam, 1982
- ASTM Symposium, San Diego, 1984

The test consists in penetrating into the soil a steel cone having a 10 cm² base area (diameter 35.6mm) and a 60° apex angle. The rate of penetration is kept at 2 cm/s. Cone shape, dimensions and rate of penetration, which influence test results, have been standardized both in Europe and North America (ISSMFE, 1977 and ASTM D3441, 1979).

4.3.2 CONE DEVELOPMENT

Early mechanical cone design is shown in Figure 32. In these devices, by advancing point and pipes independently, the point and friction loads could be measured through hydraulic load cells attached to the penetration rig. Problems due to soil entering the space between the point and the friction rods have led to improve the mechanical design as shown in Figure 33(a); whereas the cone shown in Figure 33(b) has an additional device: a 150cm² skin area friction sleeve which allows the measurement of local friction.

Although still in use today, mechanical cones present the following disadvantages: reduced sensitivity and accuracy, ineffectiveness in soft soils, tedious data processing and result presentation. These drawbacks have led to a significant improvement in the cone design: the electrical cone (Figure 34). Details concerning the cone design have been discussed by Schaap and Zuidberg (1982).

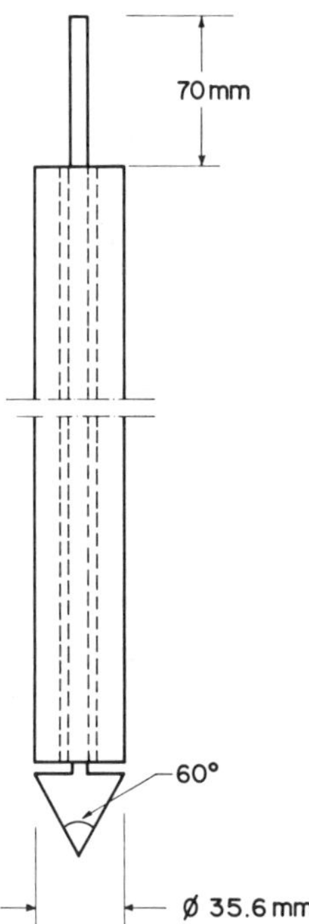

FIGURE 32. Original mechanical cone (from Vermeiden, 1948).

During penetration, the point resistance and the local friction are measured as standard parameters. However, more recently, much other information has been obtained through the incorporation of different transducer types within the cone, such as (de Ruiter, 1981): pore pressure, temperature, acoustic data, resistivity, inclination and gamma-densimetry. Among these new additions, the pore pressure measurements have been one of the most valuable developments in the CPT technique. Details of a piezocone device used by Robertson et al. (1983) are presented in Figure 34.

The use of electric cones enables continuous profiling with improved accuracy and sensitivity to be obtained, but needs a more complex recording system as shown in Figure 35. Land CPT equipment has been fitted in a van (de Ruiter, 1981) resulting in an improved overall productivity. Magnetic tapes containing all field data can be later used to process data through a computer and produce CPT plots as shown in Figure 36. Typical results from piezocone (PCPT) tests are shown in Figure 37.

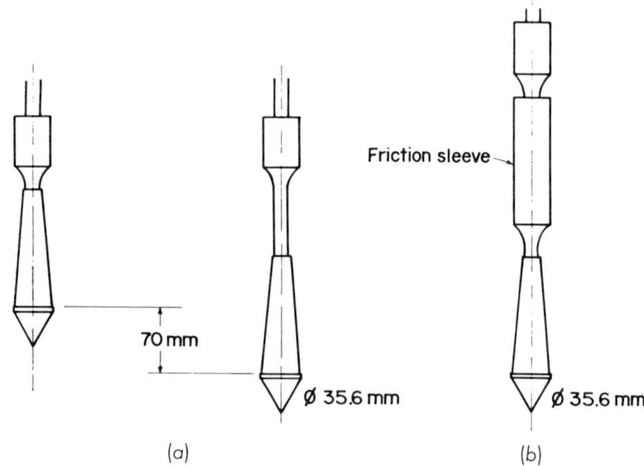

FIGURE 33. Present form of mechanical cones: (a) standard cone; (b) cone with friction sleeve (Begemann, 1965).

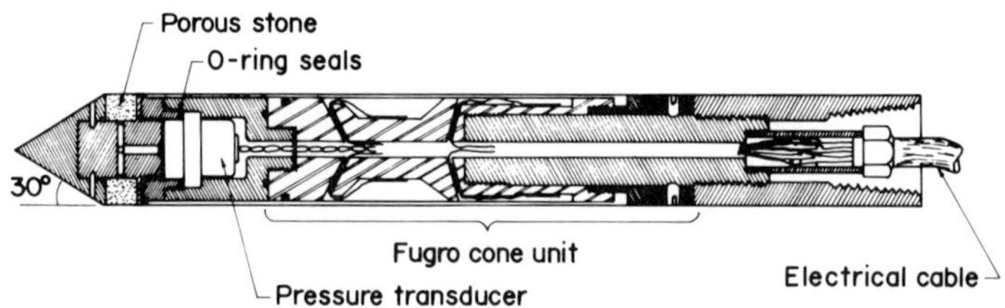

FIGURE 34. Piezocone details (adapted from Campanella et al., 1983).

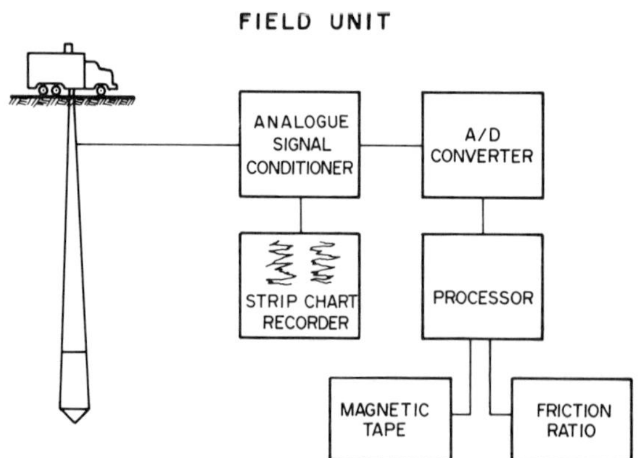

FIGURE 35. Field data acquisition system (from de Ruiter, 1981).

4.3.3 CPT AND PCPT INTERPRETATION

CPT and PCPT interpretation techniques have been summarized and discussed in the state-of-art papers by Robertson and Campanella (1983a,b) and Jamiolkovski et al. (1985). Another important reference on interpretation was presented by Senneset and Janbu (1984). The reader should refer to these papers for a more detailed discussion, however, a brief review on cone interpretation in soft clays will be presented here.

4.3.4 DATA CORRECTION

Corrections to the recorded point resistance and local friction were found to be necessary for the following reasons (see Figure 38):

(a) Unequal end areas in the friction sleeve
(b) The pore pressure built up at the cone base decreases the measured point resistance, as shown by the dead load and chamber calibration tests shown in Figure 39.

The correction mentioned in (a) can be avoided by an improved cone design, yielding equal end areas, whereas the one mentioned in (b) cannot be avoided and has to be accounted for when dealing with soft fine grained soils. The reasons for this are threefold:

1. Soil displacement during cone penetration in a fine grained soil will produce high pore-pressure gradients, as can be observed in Figure 40.
2. More consistency and less scatter in interpreted cone data are obtained if the following correction is applied (Rocha-Filho, 1982, Robertson and Campanella, 1983a and Senneset and Janbu, 1984):

$$q_c = q_r + u\,(1 - \alpha) \qquad (4.2)$$

where

q_r = recorded point resistance
u = recorded pore pressure
α = area ratio (A_n/A_q) (Figure 38)
q_c = corrected point resistance

3. The location of the porous element along the piezocone has not yet been standardized, although there is a trend to locate it just behind the point (Campanella et al. 1983). Thus, the recorded pore pressure in different piezocones cannot be compared unless it refers to the same measurement position. It is possible, however, to use data from Levadoux and Baligh (1980) to correct this problem (Figure 41).

4.3.5 SOIL STRATIGRAPHY

One important application from either CPT and PCPT has been the evaluation of soil stratigraphy. A considerable experience has been accumulated on the use of q_c and the

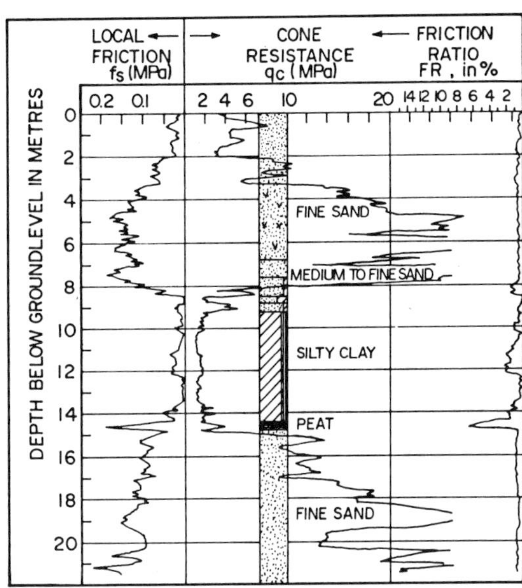

FIGURE 36. Typical CPT results, showing: local friction, point resistance and friction ratio (from de Ruiter, 1981).

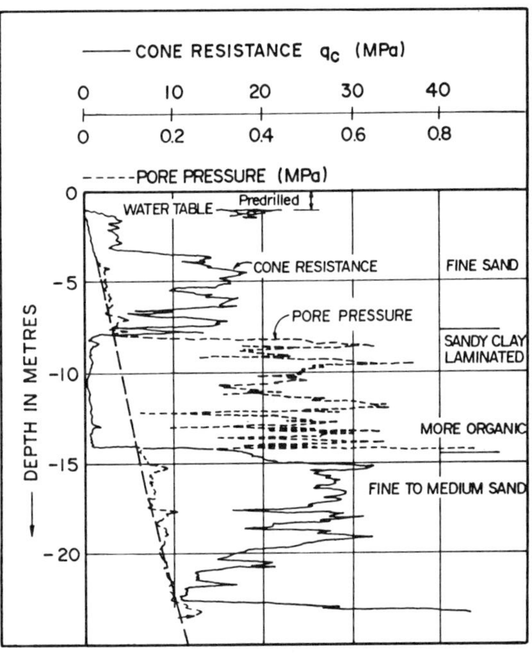

FIGURE 37. Typical PCPT (piezocone) results, showing: pore pressure and cone resistance (from de Ruiter, 1981).

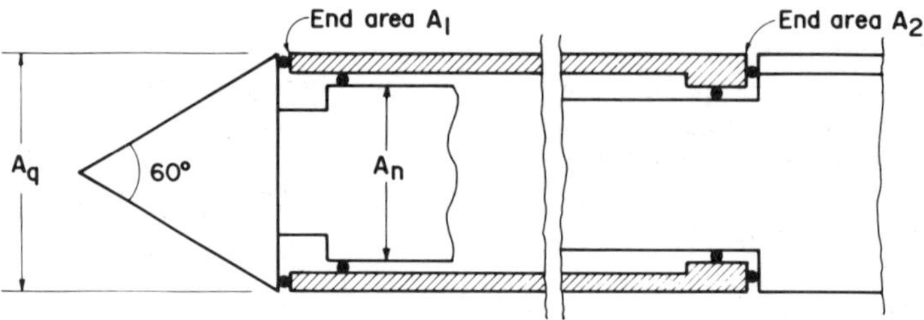

FIGURE 38. Cone unequal end areas.

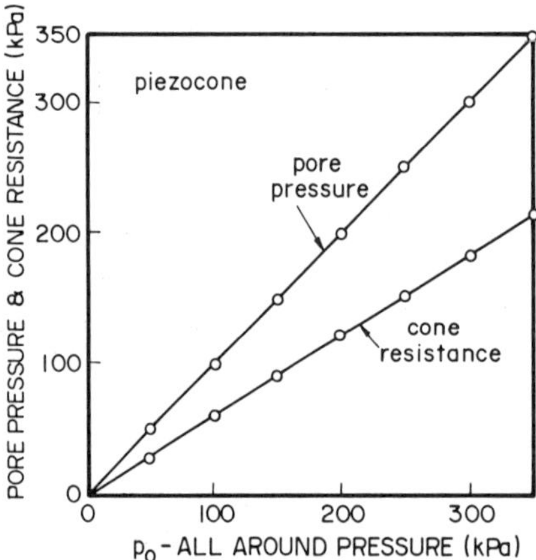

FIGURE 39. Calibration of a piezocone under all around pressure (from Almeida and Parry, 1985).

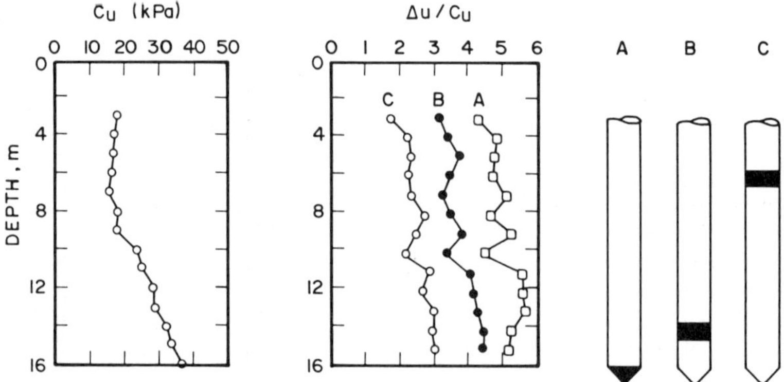

FIGURE 40. Piezocone tests in soft clay: influence of the pore pressure measuring position (from Torstensson, 1977).

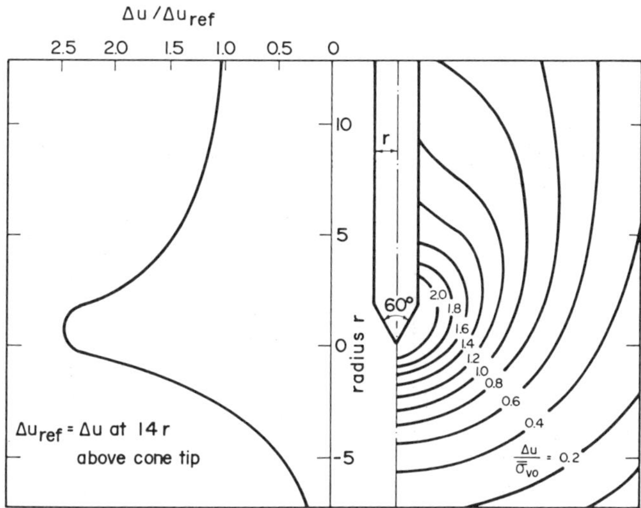

FIGURE 41. Theoretical distribution of initial normalized pore pressures (from Levadoux and Baligh, 1980).

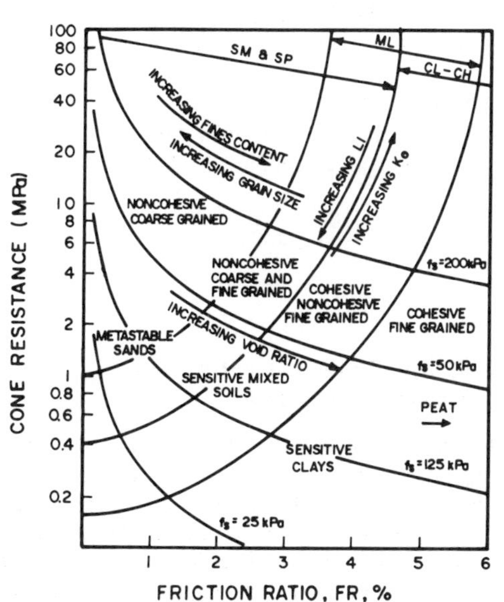

FIGURE 42. Soil identification chart for standard electric cone (from Douglas and Olsen, 1981).

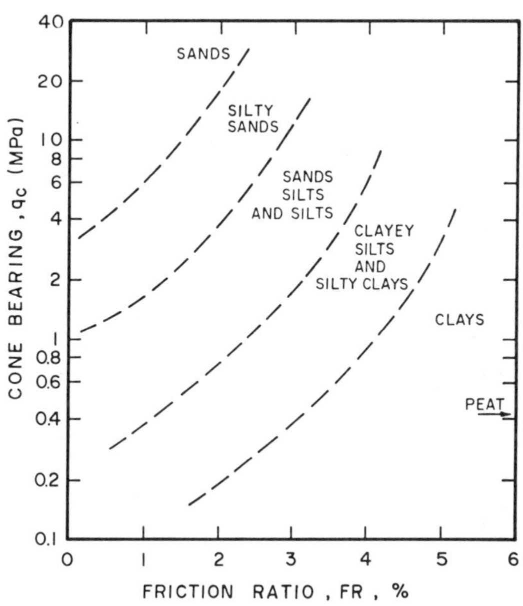

FIGURE 43. Simplified soil classification chart for standard electric friction cone (from Robertson and Campanella, 1983).

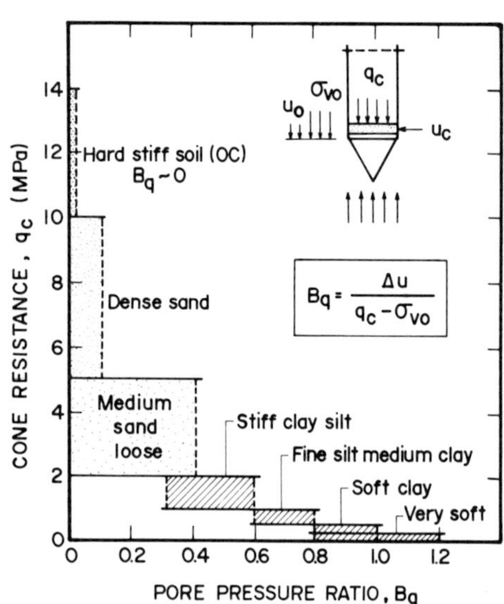

FIGURE 44. Tentative soil classification chart for PCPT (from Senneset and Janbu, 1984).

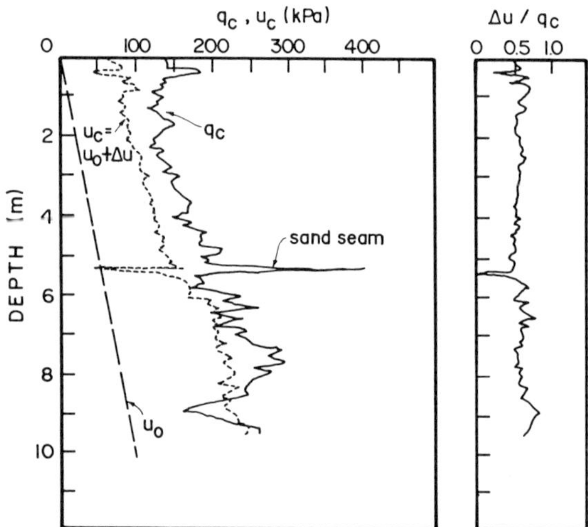

FIGURE 45. Detection of a sand seam in Rio de Janeiro soft clay through PCPT (from Rocha-Filho et al., 1983).

friction ratio (FR) for soil type identification defined as:

$$FR = \frac{f_s}{q_c} (\%) \tag{4.3}$$

where

q_c = point resistance
f_s = local friction

Figures 42 and 43 present classification charts which can be used for this purpose.

The outcome of the PCPT has provided new capabilities in the cone technique. Soil classification will depend more on soil permeability, which influences considerably the measured excess pore pressures. A tentative chart for this purpose is shown in Figure 44, where B_q is defined by (Senneset and Janbu, 1984)

$$B_q = \frac{u_{max} - u_o}{q_c - \sigma_{vo}} \tag{4.4}$$

where

u_{max} = measured pore pressure
u_o = hydrostatic pore pressure
σ_{vo} = total overburden pressure

Additionally, the PCPT is able to detect thin pervious layers and seams in a clay layer as shown in Figure 45.

4.3.6 STRENGTH PARAMETERS

Estimates of undrained strength from CPT can be accomplished through the simple equation:

$$c_u = \frac{q_c - \sigma_{vo}}{N_k} \tag{4.5}$$

where

σ_{vo} = the total vertical overburden pressure
N_k = an empirical cone factor

Robertson and Campanella (1983b) suggest the use of N_k equal to 15 for preliminary assessment of c_u. For sensitive clays the N_k value, according to the authors cited, should be reduced to about 10 or less, depending on the degree of sensitivity. There are indications that, for a given clay, values of N_k increase with the overconsolidation ratio (Almeida and Parry, 1985). It is suggested to use values of N_k calibrated regionally.

Effective stress parameters from PCPT results can be obtained by the method suggested by Senneset et al. (1982 and 1984). Very promising results have so far been obtained from this method (e.g., Rocha-Filho and Alencar, 1985).

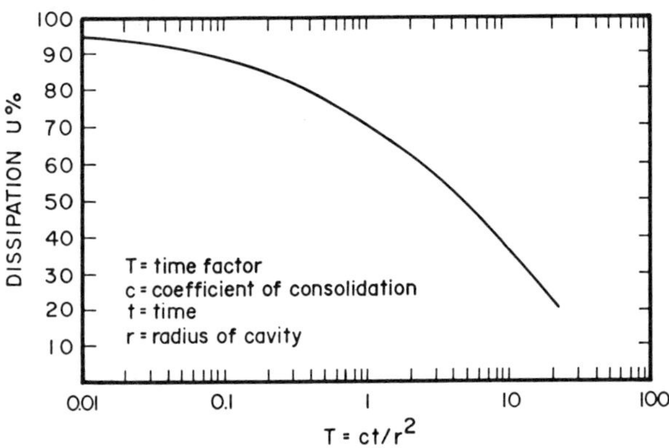

FIGURE 46. Consolidation around a cylindrical cavity (from Torstensson, 1977).

4.3.7 OTHER PARAMETERS FROM PIEZOCONE INTERPRETATION

As discussed comprehensively by Robertson and Campanella (1983b) the following information can be obtained from PCPT data:

- soil type and stress history
- coefficient of consolidation (c_h) in the horizontal direction
- permeability (indirectly from c_h)

One of the most important capabilities of the PCPT is the evaluation of the coefficient of consolidation. This can be accomplished by monitoring the pore pressure dissipation that occurs if the penetration is stopped. Robertson and Campanella (1983b) suggested the use of Torstensson (1977) theoretical solution for consolidation around a cylindrical cavity, as shown in Figure 46.

A more rigorous procedure for the evaluation of c_h was recommended by Baligh and Levadoux (1980), based on an uncoupled numerical analysis of the diffusion around a cone. An application of this solution to dissipation tests around a 60° cone in laboratory is shown in Table 4 (Almeida and Parry, 1985), where theoretical time factors are also presented. Values of c_h have been computed for each of the time factors, and good consistency of results was obtained. Using values of c_h obtained for 50% of consolidation, theoretical dissipation curves were compared with experimental curves as shown in Figure 47. The agreement is good and improves as OCR increases.

The parameter B_q (see Equation (4.4) can also be useful to give indications about the soil stress history (OCR or σ'_{vm}), as suggested by Baligh et al. (1980). Field indications that B_q reflects OCR changes within a soil deposit were given by Jamiolkowski et al. (1985) for a number of cases. However, these authors believe that more data are necessary to increase the reliability of the $B_q = f$(OCR) relations.

4.4 The Pressuremeter Test

The pressuremeter test (PMT) was developed in France by L. Menard between 1950–60 and since then it has had an intense application in France, but less in other countries. It consists in inflating a cylindrical cavity installed in a bore-

TABLE 4. Dissipation Tests for Gault Clay (from Almeida and Parry, 1985).

				OCR = 1		OCR = 1.9		OCR = 7.0	
U %	u	T_h	R^2T (mm²)	t (s)	c_h m²/year	t (s)	c_h m²/year	t (s)	c_h m²/year
20	0.8	0.44	17.74	8	70	9.5	45	6.7	83
40	0.6	1.90	76.61	90	27	68	35	40	60
50	0.5	3.65	147.18	190	24	140	33	80	58
60	0.4	6.50	262.10	340	24	270	62	150	55
80	0.2	27.0	1089.0	1150	30	1000	34	680	50

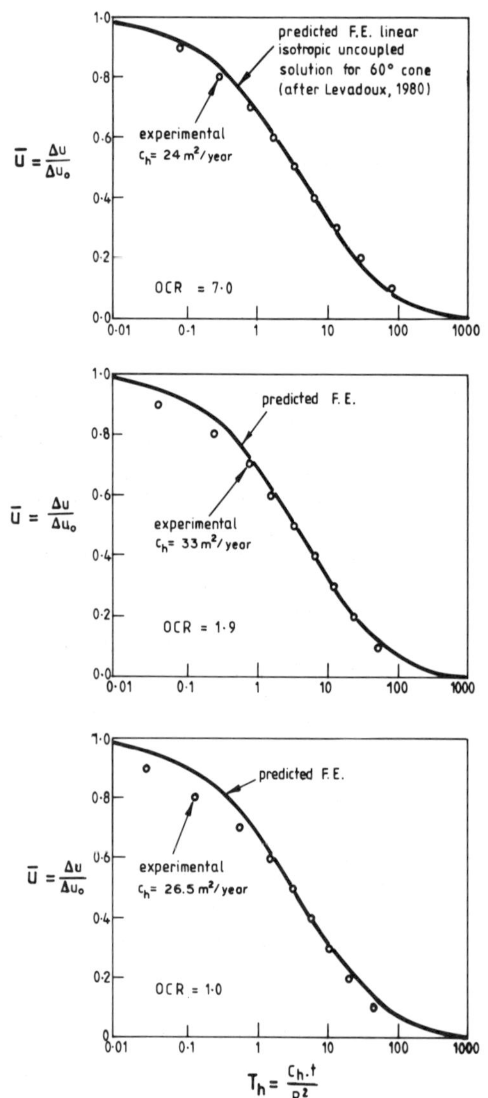

FIGURE 47. Measured and predicted pore pressure during a piezocone dissipation test (Almeida and Parry, 1985).

hole and measuring the corresponding pressure and volume change. Soil properties derived from this pressure–volume relationship have been used in the design of many soil structures. A comprehensive review covering many aspects of equipment, test procedures, theoretical analysis and applications was published by Baguelin et al. (1978), which the reader should refer to further information. A basic scheme of the test apparatus is shown on Figure 48 and a typical pressure–volume curve is shown in Figure 49. This curves comprises three distinct phases: (1) an initial phase in which the soil is assumed to return to the at rest condition (point A) and the applied inflation pressure should correspond to the in situ total horizontal stress σ_{ho}; (2) an elastic phase (AB) in which full recovery of strain is assumed if the soil is unloaded; (3) a plastic phase (BC) in which the limit horizontal pressure p_l can be determined, and as will be shown later, p_l is related to the undrained strength c_u.

There are several approaches for exploiting the pressuremeter test results (Baguelin et al., 1978). The simplest of all considers an elasto-plastic behaviour, in which the shear modulus G is given by:

$$G = V_m \frac{p}{v} \tag{4.6}$$

where p and v are increments of pressure and the corresponding volume changes in the elastic region (AB in Figure 49) and V_m is the average volume of the cavity over the pressure increment p. By making assumptions on the value of the Poisson ratio ν, one can obtain the Young modulus E through the equation:

$$E = \frac{G}{2\,(1 + \nu)} \tag{4.7}$$

which is valid for linear-elastic materials.

Following the elasto-plastic approach, the undrained strength c_u can be obtained through:

$$p_l - \sigma_{ho} = \ln\,(G/c_u + 1)\; c_u \tag{4.8}$$

where p_l and σ_{ho} are the limit and in situ total horizontal pressures, as shown in Figure 49. In the above equation, G has been obtained as shown previously and, since c_u is not explicit, an interative solution will yield a value for c_u.

The borehole drilling and probe insertion significantly affects the pressuremeter results. This effect is predominant in σ_{ho} determined from pressuremeter, which leads to incorrect values of K_o. On the other hand, the limit pressure p_l is less affected by disturbance.

In order to avoid the problems caused by soil disturbance, specially in soft soils, an improved equipment called selfboring pressuremeter was developed in France

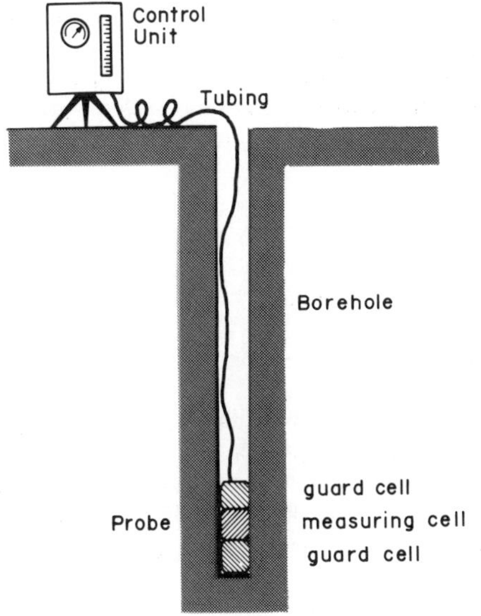

FIGURE 48. Basic principles of the pressuremeter (from Baguelin et al., 1978).

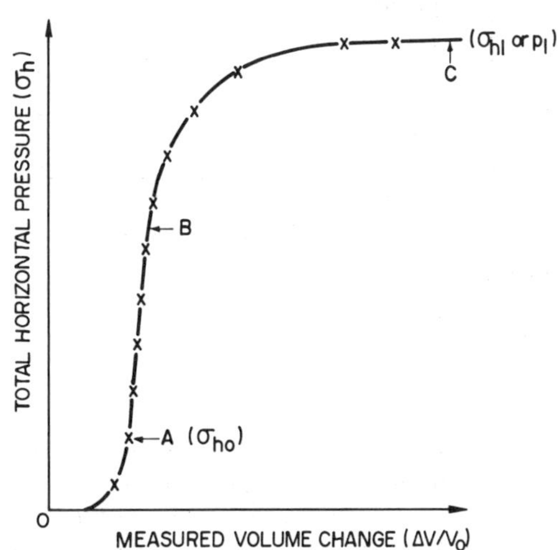

FIGURE 49. Typical pressuremeter test results.

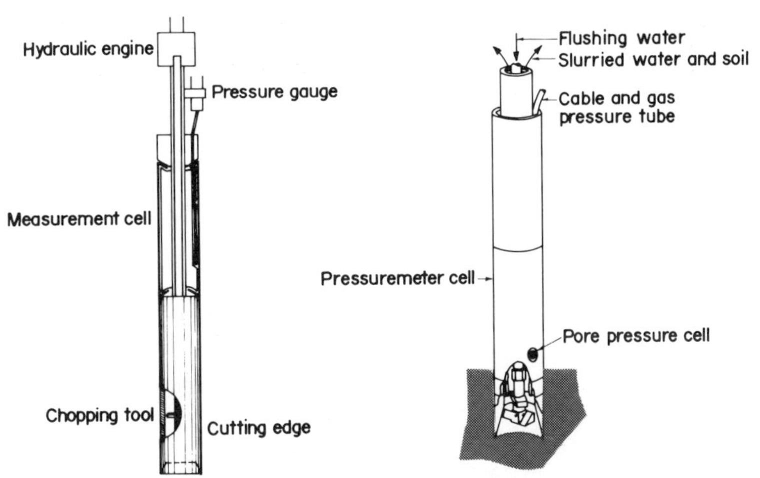

FIGURE 50. Selfboring pressuremeter types (Baguelin et al. 1972, Wroth and Hughes, 1973).

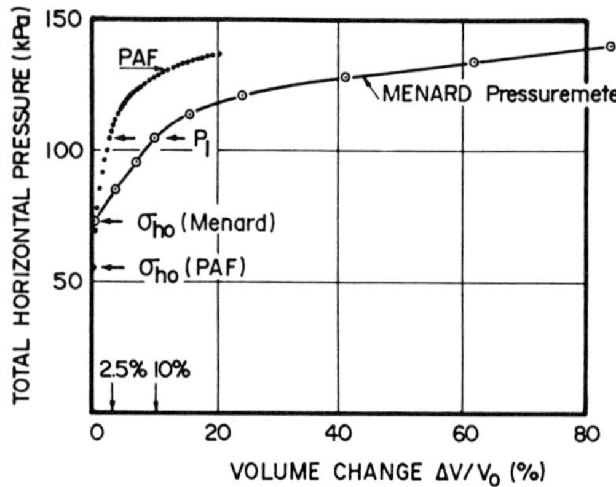

FIGURE 51. Comparison results between a Menard and PAF (pressiometre autoforeur) equipment in soft clay (from Baguelin et al., 1978).

(Baguelin et al., 1972) and in Britain (Wroth and Hughes, 1973). This new pressuremeter generation includes a rotary chopping tool at the lower end of the apparatus (Figure 50) which disintegrates the soil. Mud circulation is used to bring the soil waste up to the surface. Carefully designed sharp edges at the lower end of the equipment, help in the probe insertion avoiding significant disturbance.

The advantages of the selfboring probe are shown in Figure 51 which compares its results with data from a standard Menard pressuremeter. The increase in quality yields: (a) a steeper stress-strain curve from which higher G values can be obtained; (b) consistent values of in situ horizontal stress. As expected, the limit pressure p_l has approximately a unique value from either equipments since it is little affected by disturbance.

In summary, the selfboring pressuremeter provides reliable means of determining the at rest horizontal pressure and a sound assessment of the stress-strain behaviour of soils. On the other hand, costs considerations would, in many cases, prevent its use, unless in major civil engineering projects.

5. STABILITY ANALYSIS

5.1 Introduction

The evaluation of the stability of embankments on soft clays is a most important step in the design of such structures. This section discusses the methods currently used to perform stability analysis. Some examples illustrate the applications of these methods to real cases.

5.2 Methods of Analysis

A number of computational solutions have been developed to perform the stability analysis of embankments on soft clays. The simplified Bishop method presents several advantages over more sophisticated methods (Fredlund and Krahn, 1977) and this is the reason why it is widely used in cases where a circular surface is likely to occur.

A common feature of all approximate methods of slices (Figure 52) is the definition of the factor of safety in terms

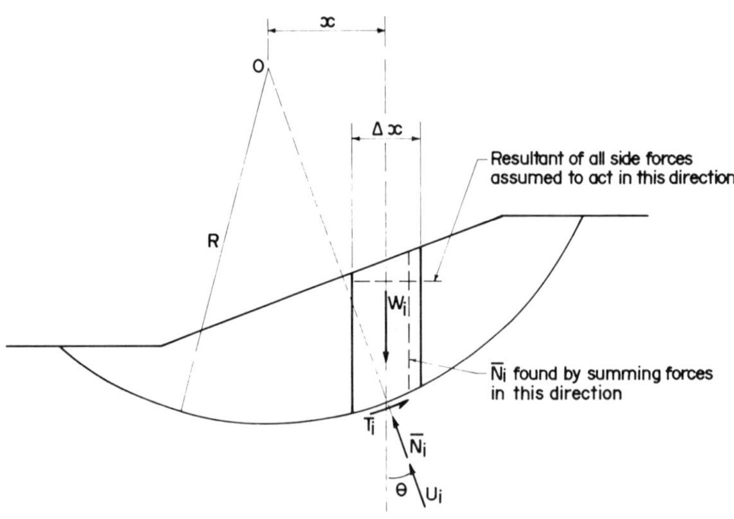

FIGURE 52. Circular failure surface in method of slices.

of moments about the centre of the failure arc:

$$F_s = \frac{\text{resisting moment}}{\text{disturbing moment}} \quad (5.1)$$

The simplified Bishop method assumes that the forces acting on the sides of any slice have zero resultant in the vertical direction, as shown in Figure 52. The normal forces acting at the base of any slice are found by considering the equilibrium of the forces in the vertical direction. A value of factor of safety must be used to express the shear forces T_i and it is assumed that this factor of safety equals F_s defined by Equation (5.1), then:

$$N_i = \frac{W_i - u_i\Delta x_i - c'\,\Delta x_i tg\alpha_i/F_s}{\cos\alpha_i[1 + (tg\alpha_i \cdot tg\phi)/F_s]} \quad (5.2)$$

Developing the equations of resisting and driving moments and substituting these as well as Equation (5.2) into Equation (5.1) gives

$$F_s = \frac{\Sigma[c \cdot \Delta x_i + (W_i u_i \Delta x_i)tg\phi]}{\Sigma W_i \cdot tg\alpha_i \dfrac{(1 + (tg\phi \cdot tg\alpha)/F_s)}{1 + tg^2\alpha}} \quad (5.3)$$

Equation (5.3) requires a trial and error solution since F_s appears on both sides of the equation, which is usually performed by a computer.

Factors of safety adopted in practice are of the order of 1.5. Smaller factors of safety, as low as 1.2, are sometimes adopted when soil conditions are well defined. Factors of safety higher than 1.5, of the order of 2, can be employed in connection with peat or very soft soils, to avoid excessive deformation.

Problems sometimes associated with the mathematical solution of the simplified Bishop method (Whitman and Bayley, 1967; Duncan and Poulos, 1981; Ching and Fredlung, 1983) have to be properly considered in the program utilized.

Other more rigorous solutions for slope stability analysis have been developed. These solutions make certain assumptions about the distribution of the interslice forces and can handle non-circular failure surfaces. In these instances, the methods by Janbu (1957), Morgenstern and Price (1965) and Spencer (1967) are more commonly used. For practical purposes, it has been suggested (Fredlund and Krahn, 1977; Whitman and Bayley, 1967) that Janbu's simplified method could be used with reasonable accuracy in cases where a non-circular surface is likely to occur.

Stability analyses can be performed by either total stress $\phi = 0$ analysis or by effective stress analysis. Both types of analyses are discussed below.

5.3 Total Stress Analysis (TSA)

The assessment of the short term embankment stability is generally made by means of the TSA, the undrained strength being measured by means of vane tests. However, as a number of embankments have failed with factors of safety in excess of unity (Parry, 1971; Bjerrum, 1972), correction factors to vane strength as a function of the plasticity index (Bjerrum, 1972) or liquid limit (Pilot, 1972) have been proposed.

5.3.1 VANE CORRECTION FACTORS

Vane corrections are made on the basis of strain rate effects and influences of anisotropy and progressive failure. Briefly, strain rate effects arise when the difference between the vane testing rate and the rate of embankment construction is large. This appears to be the case in most practical situations. Regarding the influence of anisotropy, the undrained strength of clays varies with the direction along which the clay is sheared (e.g., Duncan and Seed, 1966), amongst other factors. Also, different modes of failure intervene along a slip surface, a compression-active mode of failure under the embankment crest, a simple shear type of failure under the embankment slope and an extension-passive type of failure developing outside the embankment toe, as shown in Figure 53.

The reduction of the shear strength due to the progressive failure occurs because of nonuniform straining along the potential failure surface. Hence, failure starts in the more severely stressed zones beneath the loaded area, gradually extending into the less stressed zones. When sliding finally occurs over the full length of the sliding surface, the clay beneath the loaded area has, in general, been strained beyond the peak. In a strain-softening clay, the average shear strength along the surface of sliding will be smaller than the peak strength, thus the peak vane strength cannot be expected to represent the field strength at the moment of failure.

The vane correction proposed by Bjerrum (1972) is shown in Figure 31. Recently Azzouz et al. (1983) presented a revised vane correction factor that considers the influence of "end effects." These "end effects" provide additional resistance due to the actual three-dimensional mode of failure which is ignored in plane strain analyses. Both Bjerrum's correction and Azzouz et al. correction are shown in Figure 31. It is seen that the revised correction gives strength values about 10% lower than Bjerrum's correction.

Total stress stability calculations of a trial embankment on Rio de Janeiro clay using uncorrected vane strength have produced a factor of safety close to unity (Ortigão et al., 1983, 1985), as seen in Figure 54. The fact that the Rio de Janeiro clay has an average plasticity index of 80%, for which the vane correction is about 0.6–0.7, Figure 31, highlights the type of uncertainty still involved in total stress

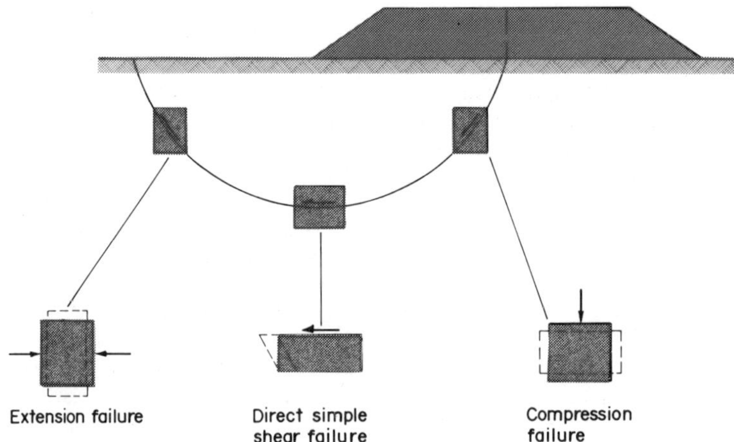

FIGURE 53. Modes of failure developing along a failure surface.

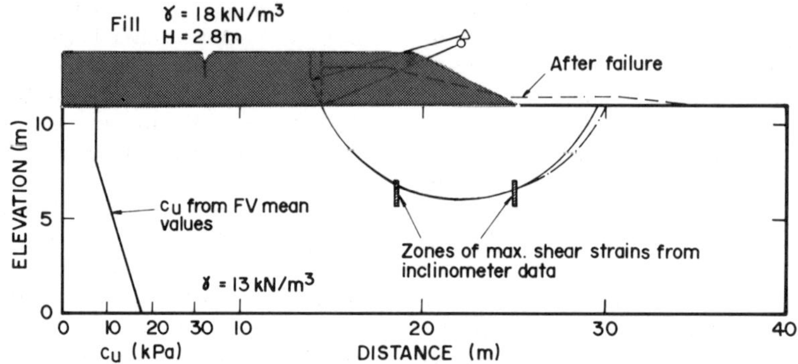

FIGURE 54. Total stress stability calculation of the Rio de Janeiro trial embankment employing FV strength (adapted from Ortigão et al., 1983 and 1985).

stability analyses. Nevertheless, in general the stability of single lift embankments can presently be carried out with a reasonable degree of reliability by the use of total stress approaches calibrated regionally.

5.3.2 THE UNDRAINED STRENGTH OF CLAY

Alternative approaches to the vane strength have been proposed and these are discussed below.

5.3.2.1 SHANSEP Method

This approach recommended by Ladd and Foot (1974) attempts to minimize sample disturbance and is the so-called SHANSEP method, which consists in the following steps;

1. Obtain good undisturbed samples and define the vertical profiles of the effective overburden pressure σ'_{vo} and the preconsolidation pressure σ'_{vm} from which the range of OCR is determined.
2. Select the type of test best representing the field loading conditions. A combination of compression, extension and direct shear tests has been suggested as the best representative of the stress conditions under embankments.
3. Determine the normalized strength parameters by first consolidating the sample to stresses σ'_{vc} in excess of 1.5–2σ'_{vm} and then unloading them to initial consolidated stress σ'_{vo} corresponding to OCR of about 2, 4, and 8. The normalized strength parameters c_u/σ'_{vc} are then plotted against OCR.
4. Determine the undrained shear strength profile from the σ'_{vo} profile and the OCR by directing reading on the c_u/σ'_{vo} versus OCR curves.

The undrained strength profile determined by SHANSEP has been applied to the total stress analysis of a number of failed embankments (e.g., Lacasse et al., 1980).

One of the most contentious aspects of this method is the procedure of consolidating specimens beyond the preconsolidation pressure, so the method has been criticized (Mesri, 1975, Leroueil et al., 1979) to be used for natural clays in general, as they possess some intrinsic structure as a result of the geologic process, which is destroyed by the application of high consolidation pressures.

A practical difficulty involved with SHANSEP method is the suggested use of the simple shear test, since the equipment to perform such a test is not commonly available in practical situations. An alternative for this is to perform CIU or CK_0U test as adopted by Ortigão et al. (1983).

Trak et al. (1980) have proposed a semi-empirical approach based on an interpretation of the Bjerrum's data made by Mesri (1975), which led to the observation that the available strength at failure under an embankment is nearly independent of the plasticity index and is a function of the preconsolidation pressure σ'_{vm}. The equation $c_u = 0.22\sigma'_{vm}$ has been shown to be applicable to sensitive Canadian Clays.

5.3.2.2 Recompression Technique

The recompression technique was suggested by Bjerrum (1973) and consists in reconsolidating the soil specimen at exactly the same stresses it carried in the ground. The beneficial effects of this technique according to Bjerrum are: firstly, to replace the field stresses with an identical set of effective stresses in the laboratory; and secondly, to squeeze out of the specimen the water content which is absorbed during sampling. This technique was regarded by Bjerrum as the most reliable to obtain stress–strain data of the soft clay. Jamiolkowski et al., concluded that the recompression technique is superior to the SHANSEP method for highly structured (high sensitivity and liquidity index) deposits, as well as for testing weathered and highly overconsolidated deposits where SHANSEP is difficult to apply.

5.3.2.3 Critical State Soil Mechanics (CSSM) Theory

Critical state soil mechanics (CSSM) can be a useful tool to provide estimates of the clay undrained strength. The concepts of this theory have been presented by a number of authors (e.g., Schofield and Wroth, 1968; Atkinson and Bransby, 1978) and it is out of the scope of this document to review those concepts.

For the estimate of the undrained strength c_u only three fundamental and previously defined soil parameters are necessary:

C_c = the compression index
C_s = the swelling index
ϕ' = the effective angle of friction for the soft clay, at the normally consolidated state

For the calculation of the undrained strength ratio of normally consolidated clays the following simple equation based on the modified Cam-clay model can be used:

$$\frac{(c_u)}{\sigma'_{vo}} = \frac{3\ \sin\phi' \cdot (0.5)^{\Lambda}}{3 - \sin\phi} \tag{5.4a}$$

where σ'_{vo} is the in situ effective stress and Λ, the plastic volumetric strain ratio defined as:

$$\Lambda = 1 - \frac{C_s}{C_c} \tag{5.4b}$$

as Λ does not vary significantly for a wide range of clays, it is reasonable, for practical purposes, to adopt the average value $\Lambda = 0.8$.

In nature, soil deposits become consolidated under one-dimensional conditions. As the behaviour under anisotropic and isotropic conditions differs, this must be accounted for. The undrained strength of an anisotropic clay depends on the value of K_0. Assuming that K_0 for a normally consolidated clay is given by:

$$K_0 = 1 - \sin\phi'$$

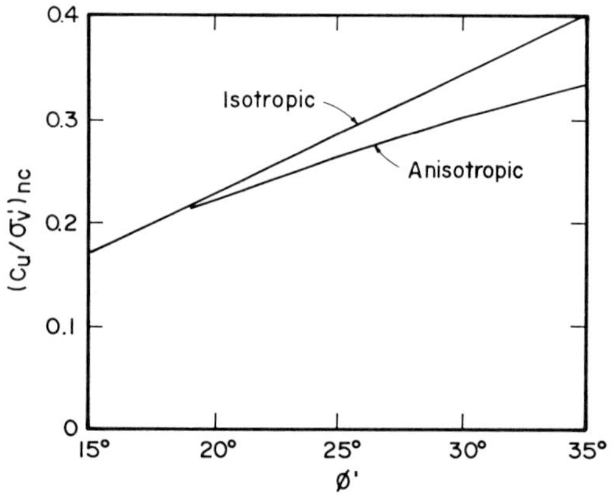

FIGURE 55. Variation in the undrained strength ratio against the angle of friction for normally consolidated clays (from Wroth, 1984).

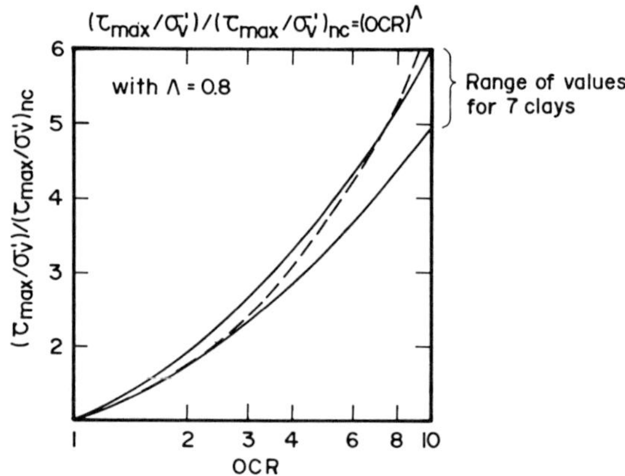

FIGURE 56. Variation in the normalized undrained strength ratio with the overconsolidation ratio for various clays in simple shear (Ladd et al., 1977).

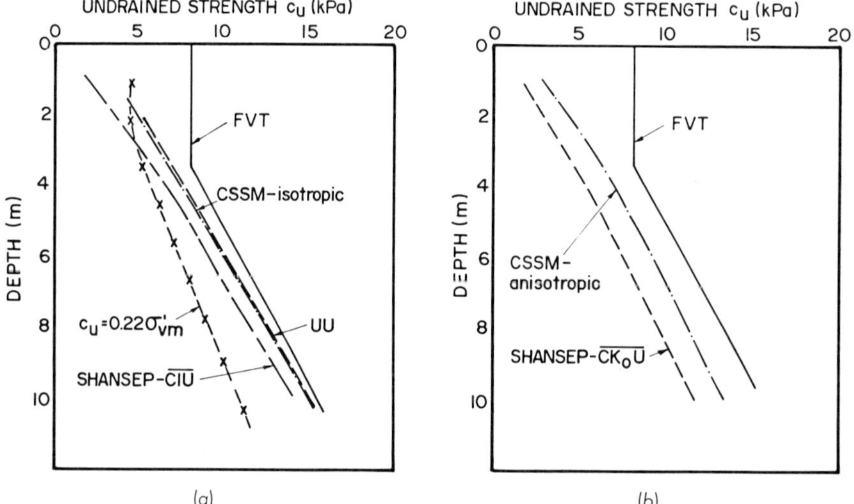

FIGURE 57. Undrained strength profiles of Rio de Janeiro clay (Almeida, 1985).

The following equation can be used, as suggested by Wroth (1984)

$$\frac{c_u}{\sigma'_{vo}} = \frac{\sin\phi' \cdot (a^2 + 1)}{4a} \qquad (5.5a)$$

where

$$a = \frac{3 - \sin\phi'}{2(3 - 2\sin\phi')} \qquad (5.5b)$$

The undrained strength ratio given by isotropic and anisotropic normally consolidated conditions are plotted against the angle of friction in Figure 55. It is seen that there are substantial differences between the two curves, specially for higher friction angles.

Estimates of the undrained strength of overconsolidated clays can be also made based on CSSM theory. The equations are written in terms of the constants ϕ' plus values of OCR, and K_0, which in turn is also dependent on the OCR. However, these equations are cumbersome and a more straight-forward approach is required for practical applications, as explained below.

Ladd et al. (1977) analyzed the data of seven different clays tested under conditions of simple shear, Figure 56, and suggested to relate undrained strength ratios, at overconsolidated and normally consolidated conditions, using the following expression

$$\frac{(c_u/\sigma'_{vo})oc}{(c_u/\sigma'_{vo})nc} = (\mathrm{OCR})^m \qquad (5.6)$$

where m lies in the narrow range 0.68–0.86 and is typically 0.8. Equation (5.5), which was solely based on empirism is, confirmed by CSSM theory. Moreover, CSSM theory relates the exponent m with well recognized physical properties of the clay in question, by stipulating that $m = \Lambda$, the plastic volumetric strain ratio.

The importance of the relationship in Equation (5.6) is that if the undrained strength ratio of a clay can be measured or estimated for a normally consolidated specimen, then its value can be predicted for different overconsolidation ratios, provided these are known.

Considering the three different modes of failure intervening along a failure surface: compression, extension and direct simple shear, it is not theoretically correct to use values of c_u obtained just from triaxial compression tests. Jamiolkowski et al. (1985) have pointed out that the above practice might lead to significantly unsafe results for clays of low to moderate OCR. Jamiolkowski et al. suggest that if the project cannot afford to run laboratory tests with different types of loading modes, use could be made of Equation (5.6) adopting $c_u/\sigma'_{vo} = 0.23$ and $m = 0.8$.

A comparison between undrained strength profiles for the Rio de Janeiro soft clay, obtained by different methods (Almeida, 1985, Ortigão et al., 1983 and 1985) is presented in Figure 57. All methods, except that proposed by Trak et al., have produced, for this particular case, fairly close strengths below the crust. However, the field vane test was the only one to detect the high strength at the crust. Stability analyses employing uncorrected field vane data produced a factor of safety very near one, as shown in Figure 54.

5.3.3 DESIGN USING STABILITY CHARTS

A rough estimate of the critical embankment height can be made using bearing capacity theory. The corresponding rule of thumb is that the critical height of an embankment on soft clay is about 5 c_u/γ_{emb}, where c_u is the average soft clay strength and γ_{emb} the embankment specific weight. This rule neglects the effect of side slopes, embankment strength and strength variation with depth, but can be useful for many practical purposes.

For preliminary design purposes, the stability charts proposed by Pilot and Moreau (1973) may be used. These charts, which consider the embankment resistance and assume a constant profile of the undrained clay strength, can be particularly useful for the preliminary definition of the geometry of embankments with berms.

For clay strength increasing with depth, which is a characteristic of nearly all soft clays, the charts developed by Pinto (1974), Figures 58 and 59, for shallow and deep soft clay deposits can be used.

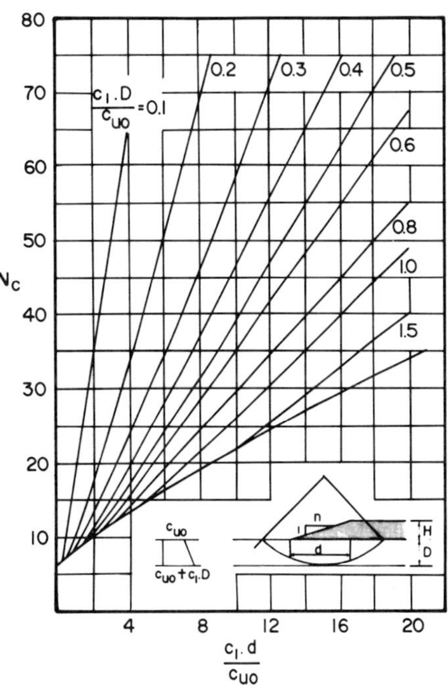

FIGURE 58. Stability charts for embankments on shallow clay deposits (from Pinto, 1974).

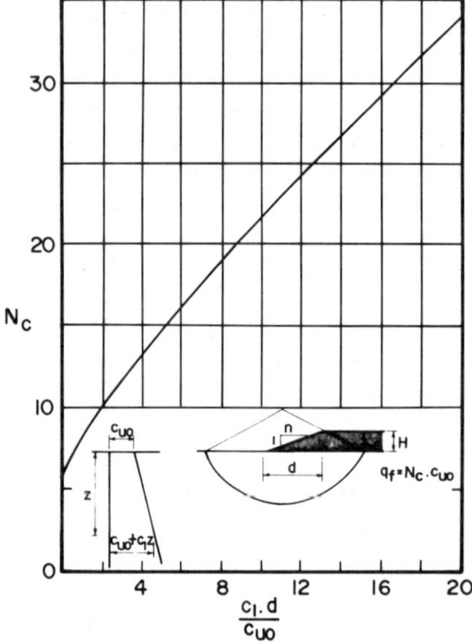

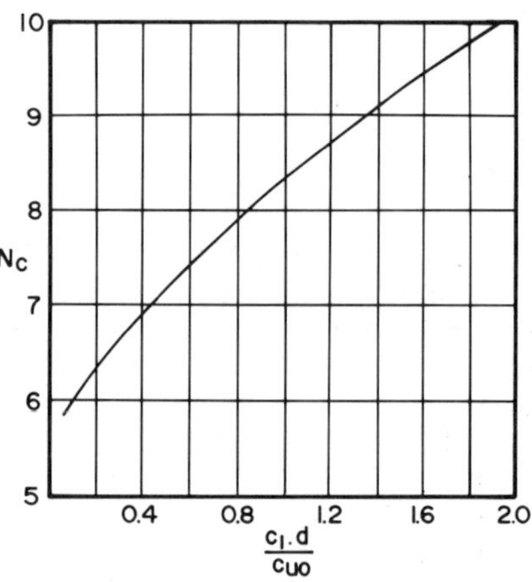

FIGURE 59. Stability charts for embankments on deep soft clay deposits (from Pinto, 1974).

Examples of application of these charts are given below:

Example 5.1

Given data: Embankment height : $H = 3$ metres
unit weight of clay : $\gamma = 19 \text{kN/m}^3$
depth of clay layer : $D = 1.6$ metres
slope data (see Figure 58) : $n = 5$
$d = 15$ metres
clay strength data: $c_{uo} = 4$ kPa
$c_1 = 1.5$ kPa/m

Solution: $$\frac{c_1 \cdot d}{c_{uo}} = \frac{1.5 \times 15}{4} = 5.62$$

$$\frac{c_1 \cdot D}{c_{uo}} = \frac{1.5 \times 1.6}{4} = 0.6$$

from Figure 58 one obtains $N_c = 20.5$, then

$$\sigma_f = N_c \times c_{uo} = 82 \text{ kPa}$$

$$\sigma = 19 \times 3 = 57 \text{ kPa}$$

$$F_s = \frac{82}{57} = 1.44$$

Example 5.2

Given data: $H = 3$ metres
$\gamma = 18 \text{ kN/m}^3$
$n = 4$
$d = 12$ metres
$c_{uo} = 8$ kPa
$c_1 = 1$ kPa/m

Solution: $$\frac{c_1 \, d}{c_{uo}} = \frac{1 \times 12}{8} = 1.5$$

from Figure 59 one obtains $N_c = 9.25$, then

$$\sigma_f = 9.25 \times 8 = 74 \text{ kPa}$$

$$\sigma = 18 \times 3 = 54 \text{ kPa}$$

$$F_s = \frac{74}{54} = 1.37$$

As seen in Figures 58 and 59, these charts do not consider the influence of the embankment resistance, but can be very useful in cases of low embankments, as the resistance component provided by the embankment will be relatively small compared with that provided by the clay mass.

For low cohesive embankments, it is suggested to assume a full cracked embankment, thus making direct use of Pinto's (1974) charts. In this case, straightforward analysis, using a computer program, might grossly underestimate the factor of safety. This is because the driving moment provided by the wedge ABD, Figure 60, if erroneously considered, will lower considerably the computed factor of safety. An example of this is shown in Figure 61, for the stability analyses of a 2.8 metres embankment in which the

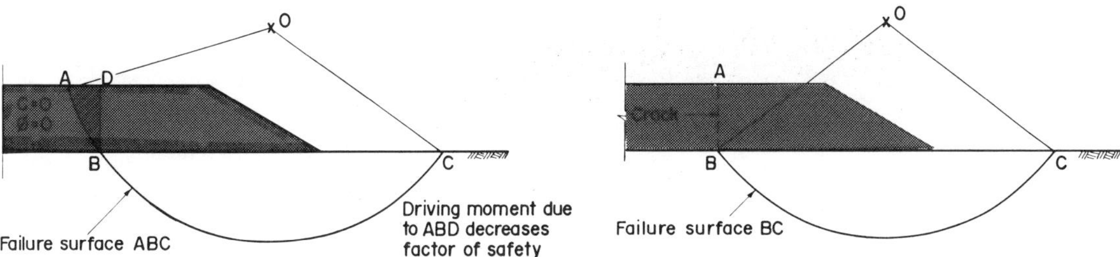

FIGURE 60. Stability analysis of fissured embankments.

embankment resistance was neglected. Whereas Pinto's charts produced a factor of safety $F_s = 0.93$, straightforward analysis (with the simplified Bishop method), using a failure arc going through the embankment, produced $F_s = 0.63$. However, the correct considerations of a composite failure surface (still using the simplified Bishop method) produced $F_s = 0.933$, which is virtually the same value obtained using Pinto's charts.

5.4 Effective Stress Analysis (ESA)

5.4.1 SINGLE LIFT EMBANKMENTS

Effective stress analyses of single lift embankments are sometimes performed with estimated pore pressures and with c' and ϕ' determined from triaxial tests. Predictions of pore pressures generated during construction are made by means of stress computations by elasticity theory and use of pore pressure parameters (Skempton, 1954; Henkel, 1960) obtained from triaxial tests. Other methods for pore pressure calculation have been proposed by Hoeg et al. (1969), Burland (1971) and Leroueil et al. (1978), but these methods have not yet found widespread use. Parry (1971) suggested that the ESA can be more reliable, provided realistic values of pore pressures can be obtained. Parry argued that some of the factors affecting the TSA have a smaller effect on the ESA, in particular strain rate effects. In the light of ESA performed by Parry and McLeod (1967), it was suggested that c' values measured in the laboratory should be ignored in calculating stability, unless definitive evidence to the contrary is available.

The use of ESA for the stability of embankments on soft soils has been recently reviewed by Pilot et al. (1982). These authors concluded that analyses using measured pore pressures and the final geometry of the embankment, at the moment of failure, produced satisfactory results in general. However, analyses were very sensitive to values of c' used in the calculations. It was suggested that the parameters c' and ϕ' should preferably be measured by triaxial tests carried out to large strains and at confining pressures corresponding to the overconsolidated range.

Effective stress analysis of the trial embankment on the Rio de Janeiro clay carried out by Ortigão et al. (1983) using the simplified Bishop method produced a factor of safety well below one ($F_s = 0.6$). Costa-Filho et al. (1985) conducted stability analyses using Sarma's method and factors of safety very close to unity were obtained, as seen in Figure 62. These writers attributed the discrepancy between the rigorous Sarma method and the simplified Bishop method to the rather low effective normal stresses operating on the section of the failure surface beyond the toe of the embankment.

The above example illustrates the difficulties involved in

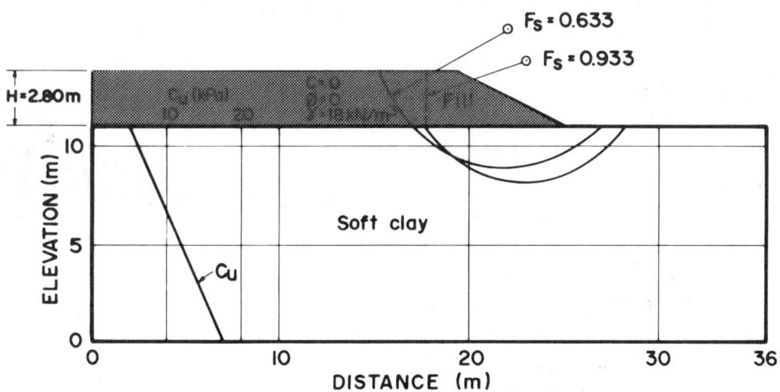

FIGURE 61. Stability analysis of low embankments.

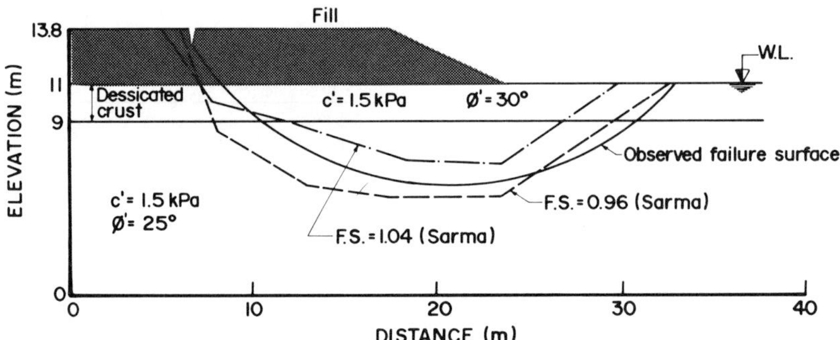

FIGURE 62. Effective stress analyses of the Rio de Janeiro trial embankment (from Costa-Filho et al., 1985).

ESA and, despite its sound theoretical superiority over TSA, the latter is recommended as a design procedure for single lift embankments. Pilot et al. (1982) also pointed out that ESA remains more difficult to apply and requires more measurements, more elaborate testing and a more thorough interpretation of the available data. However, ESA is necessary in two important situations:

(a) Construction of embankments in stages, discussed below, in which estimates of the increase of undrained strength with consolidation is difficult

(b) Control of construction of high embankments on soft compressible soils by means of in situ measurements of pore pressures, also discussed below

5.4.2 STAGE CONSTRUCTED EMBANKMENTS

Incremental or stage construction is a common method employed for embankments on soft ground. The consolidation of the foundation following each stage of construction produces a progressive increase in the shear strength of the foundation. Assessment of the stability for each stage of construction is an important aspect in the design of such structures.

Tavenas et al. (1978) suggested that the only way of obtaining an estimate of the true stability condition, at any stage of construction, is by means of effective stress analysis using the actual pore pressures observed in the foundation and c' and ϕ' obtained at large strains. However, pore pressures are not always measured and reliable methods of pore pressure prediction are not always available (Almeida et al., 1985) for the application ESA.

An example of the stability analysis of a model embankment constructed in stages during a centrifuge test (Almeida et al., 1985) is discussed below. Centrifuge modelling allows the behaviour of on corresponding prototype structure to be accurately simulated using the similitude laws (Schofield, 1980). The model embankment, Figure 63, was constructed in stages from lift 1 to lift 3 and was then taken quickly to failure. This occurred immediately after lift 5 was constructed, at an average prototype height of 11.6 metres.

Results of stability analyses using measured pore pressures are shown in Figure 64, together with the soil parameters adopted in the calculation. Theoretical and observed slip surfaces are very close and the computed factor of safety was 0.91.

Factors of safety computed before and after each construction lift using measured pore water pressures are shown in Figure 65. Upward arrows for constant height of embankment denote the increase of factor of safety during each stage.

5.4.3 CONTROL OF CONSTRUCTION USING ESA

The example above illustrates that effective stress analysis is a useful way of assessing at any moment the stability of stage constructed embankments, provided actual pore pressures are measured. This approach has been successfully used (Margason and Symons, 1969, Tavenas et al., 1978) to accelerate the construction schedule of road embankments. Margason and Symons (1969) developed stability charts relating minimum factor of safety for different em-

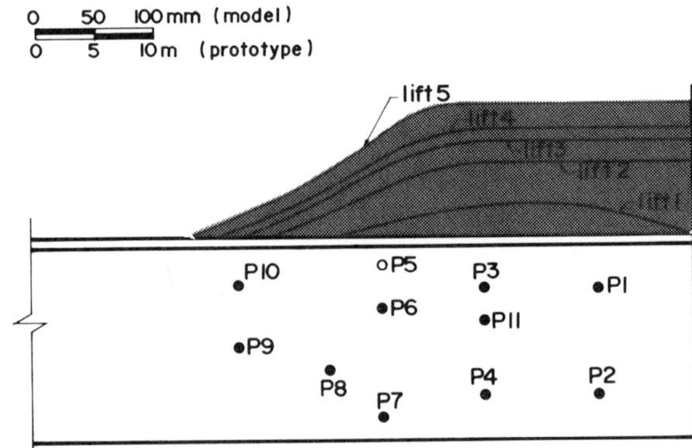

FIGURE 63. Model embankment constructed in stages during a centrifuge test (Almeida et al., 1985).

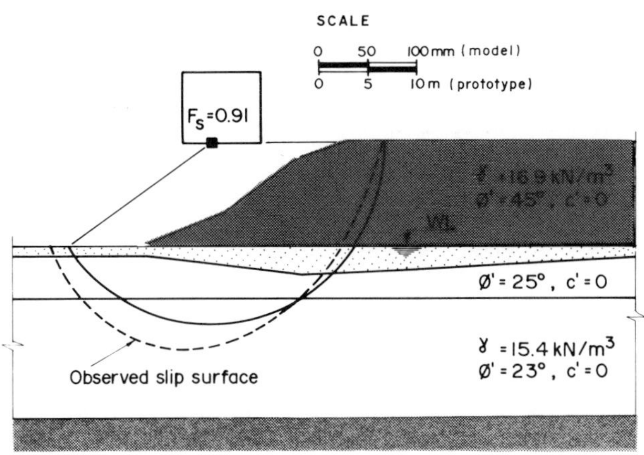

FIGURE 64. Failure of the model embankment constructed in stages (Almeida et al., 1985).

bankments, heights to the pore pressure condition in the subsoil. These charts were prepared as part of the design procedure and provided a convenient method for site control of the rate of construction. Tavenas et al. (1978) suggested that factors of safety computed by effective stress analyses should be kept in excess of 1.3, or preferably 1.4, to avoid the initiation of local failure. This suggestion appears to be supported by the results of the model test embankment reported above.

6. DEFORMATION ANALYSIS

6.1 Introduction

The deformation analysis of single lift constructed embankments usually requires the prediction of short and long term settlements and their variation with time. Estimates of lateral deformations are sometimes required and more rarely a more complete picture of the deformation and pore pressure pattern is necessary, in which case a finite element (F.E.) analysis is performed.

6.2 One-dimensional Deformation

6.2.1 INTRODUCTION

If the embankment width is large compared to the thickness of the compressible layer, strains are essentially one-dimensional and methods of settlement analysis based on one-dimensional models should then give adequate results. A useful reference for applied settlement analysis is Perloff (1974) and a comprehensive review of the available methods to performed settlement analysis was presented by Balasubramanian and Brenner (1981). This section is restricted to the methods usually adopted for practical estimates of settlements of embankments on soft clays.

The necessary information required for a settlement analysis is summarized in Table 5. The three necessary compo-

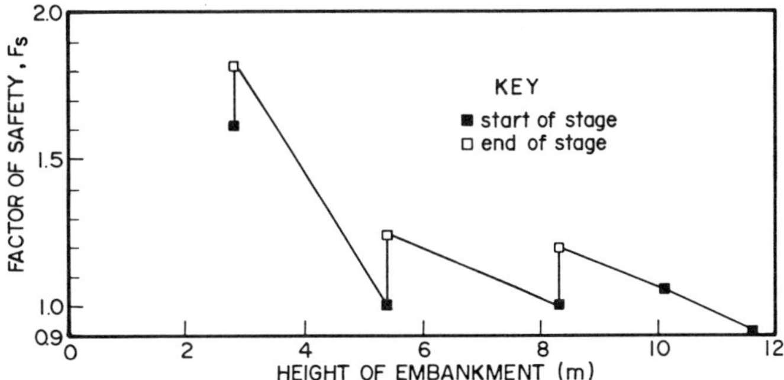

FIGURE 65. Variation of the factor of safety during stage construction (Almeida et al., 1985).

TABLE 5. Components of a Settlement Analysis (Lambe, 1964).

Determination of subsoil section
1. Vertical and lateral extent of soils; location of compressible soils, drainage surfaces and any special boundary conditions
2. Variation of initial pore pressure with depth
Stress analysis
1. Initial effective stress versus depth
2. Magnitude, distribution and time rate of application of surface load, including any shear stress between ground surface and applied load
3. Stress distribution theory compatible with boundary conditions; effect of rigid boundaries or layers
4. Variation of σ_1, σ_2 and σ_3 with consolidation; influence of arching, change in Poisson's ratio
Selection of soil parameters ($m_v, C_c, C_\alpha, \sigma'_{vc}, k, E_u, E, \nu', K_o, A, c_v$)
1. Representatives of samples tested
2. Sample disturbance
3. Environmental factors
4. Testing technique
Estimation of settlement and pore pressures
1. Method of analysis
2. Rotation of principal planes
3. Variation of m_v, k, c_v with consolidation
4. Secondary compression

nents for any successful deformation analysis are (Ladd et al., 1977):

(a) A model to describe soil behaviour
(b) Suitable method to evaluate the required parameters
(c) Computational procedure for applying the model to practical problems

This section discusses (a) and (c) above, as (b) has been discussed in Section 3 and 4.

Total settlements are a sum of initial and long term settlements. These two components of settlements are discussed below.

6.2.2 INITIAL SETTLEMENTS

Initial settlements, also called immediate settlements or undrained settlements, are the settlements which take place immediately after load application and are associated with undrained elastic shear deformation. Initial settlements may be computed from

$$S_i = \Delta\sigma \cdot b \frac{(1 - \nu^2)}{E} \cdot I \tag{6.1}$$

where $\Delta\sigma$ is the applied embankment load, b is the width of the embankment, E and ν are the elastic parameters and I is the stress influence factor, which depends on the geometry of the problem, obtained for instance from Poulos and Davis (1974).

If the soil is saturated, deformations are of undrained type, i.e., at constant volume. In this case, it is suggested to assume a Young's modulus E_u, determined from CU triaxial tests, and a Poisson's ratio $\nu_u = 0.5$, consistent with undrained behaviour. However, the proper selection of E_u is

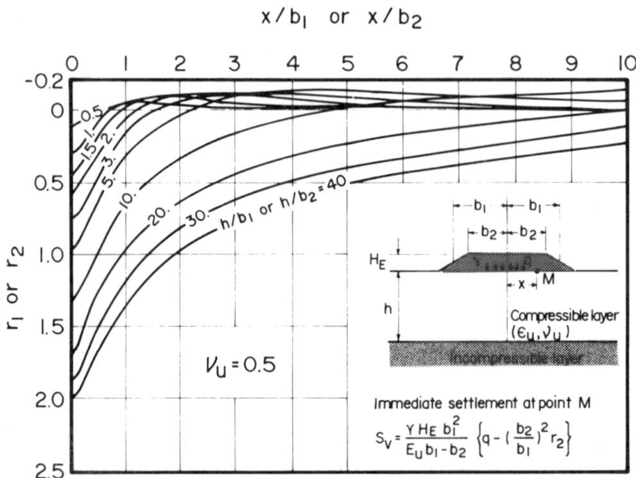

FIGURE 66. Chart for calculating immediate settlement under embankment (after Giroud, 1973).

made difficult by its dependence on stress level and stress path. The standard practice is to assume the secant E_u corresponding to 50% of the maximum deviator stress. The immediate settlement under any point of an embankment can be readily obtained using the chart of Figure 66 developed by Giroud (1973).

Corrections for the non-linear soil behaviour can also be applied, using for instance the charts developed by D'Appolonia et al. (1971). These corrections become important when the fill height exceeds 50% of the height at which failure occurs.

6.2.3 LONG TERM SETTLEMENTS

Long term settlements can be divided into consolidation or primary settlements S_c and secondary settlements S_s. Consolidation settlements are due to the dissipation of the excess pore water pressure and, consequently, produce an increase in the soil effective stresses. When complete dissipation of excess pore water pressure is obtained, primary settlements cease to take place. Secondary settlements occur essentially after complete dissipation of excess pore pressure, i.e., at practically constant effective stresses.

For a soil profile consisting of n layers, the primary settlement can be computed from

$$S_c = \sum_{i=1}^{n} (m_v \cdot \sigma_v' \cdot \Delta z)i \qquad (6.2)$$

vertical stress due to the embankment load, computed at the centre of the layer, and m_v is the coefficient of volume compressibility, computed from a oedometer test and given by

$$m_v = \frac{\epsilon_v}{\sigma_v'} = \frac{\Delta e}{\sigma_v'(1 + e_0)} \qquad (6.3)$$

where ϵ_v is the increment in vertical strain, Δe is the change in void ratio and e_o, the initial void ratio.

In practice, the primary settlement involving both recompression and virgin compression, i.e., the loading of a soil from the overconsolidated to the normally consolidated state, can be computed using an alternative expression for Equation (6.2), as

$$S_c = \sum_{i=1}^{n} \left[\frac{C_s}{1 + e_0} \Delta z \cdot \log \frac{\sigma'_{vm}}{\sigma'_{vo}} + \frac{C_c}{1 + e_0} \Delta z \log \frac{\sigma'_{vf}}{\sigma'_{vc}} \right] \qquad (6.4)$$

where C_s and C_c are the coefficients of swelling and compressibility as defined previously, and σ'_{vo} is the initial vertical overburden pressure, σ'_{vf} is the final vertical effective stress, σ'_{vm} is the preconsolidation pressure, and e_0 is the void ratio corresponding to the intersection of the recompression and virgin lines. The terms used in Equation (6.4) are explained in Figure 10. It should be pointed out that this method of analysis is one-dimensional only with respect to the integration of strains, as the vertical stress increment is usually obtained from three-dimensional stress distribution theory.

The secondary settlement is sometimes taken into account. It is usually computed from the coefficient of secondary compression C_α, using the expression

$$S_s = \frac{h}{1 + e_o} C_\alpha \log \frac{t_s}{t_p} \qquad (6.5)$$

where C_α, measured in long term oedometer tests, is defined as

$$C_\alpha = \Delta e / \Delta \log t \qquad (6.6)$$

t_p is the time of the end of the primary consolidation and t_s is the time for which secondary consolidation is to be computed. The initial time, from which all time values have to be measured, should be, in the case of instantaneous load application, the instant at which the load is applied. However, for loads which increase linearly with time, the initial time should be that at which one-half of the total load is applied.

6.2.4 DEVELOPMENT OF SETTLEMENTS WITH TIME

The classical one-dimensional Terzaghi theory is based on the assumptions:

(a) One-dimensional deformation of clay layer
(b) One-dimensional drainage following Darcy's law for any hydraulic gradient
(c) Homogeneous fully saturated soil
(d) Incompressibility of soil grains and pore fluid
(e) Linear relationship between effective stress and void ratio
(f) Infinitesimal one-dimensional strains and flow velocities
(g) No structural viscosity or secondary compression of the soil

By combining the relationship for the continuity of the fluid flow with the linear relationship effective stress-void ratio and assuming constant total stress, Terzaghi's classical consolidation equation was derived. Solutions of the consolidation equation are commonly presented in the form of graphs or tables of degree of consolidation U versus time factor T_v. The degree of consolidation settlement U_s is given by

$$U_s = \frac{S_c(t)}{S_c} \qquad (6.7)$$

TABLE 6. One Dimensional Consolidation Theory Solutions for Four Cases of Initial Excess Pore Water Pressure Distribution in Double Drained Stratum (Perloff, 1974).

T	Average Degree of Consolidation, U(%) Case 1	Case 2	Case 3	Case 4
0.004	7.14	6.49	0.98	0.80
0.008	10.09	8.62	1.95	1.60
0.012	12.36	10.49	2.92	2.40
0.020	15.96	13.67	4.81	4.00
0.028	18.88	16.38	6.67	5.60
0.036	21.40	18.76	8.50	7.20
0.048	24.72	21.96	11.17	9.60
0.060	27.64	24.81	13.76	11.99
0.072	30.28	27.43	16.28	14.36
0.083	32.51	29.67	18.52	16.51
0.100	35.68	32.88	21.87	19.77
0.125	39.89	36.54	26.54	24.42
0.150	43.70	41.12	30.93	28.68
0.175	47.18	44.73	35.07	33.06
0.200	50.41	48.09	38.95	37.04
0.250	56.22	54.17	46.03	44.32
0.300	61.32	59.50	52.30	50.78
0.350	65.82	64.21	57.83	56.49
0.400	69.79	68.36	62.73	61.54
0.500	76.40	76.28	70.88	69.95
0.600	81.56	80.69	77.25	76.52
0.700	85.59	84.91	82.22	81.65
0.800	88.74	88.21	86.11	85.66
0.900	91.20	90.79	89.15	88.80
1.000	93.13	92.80	91.52	91.25
1.500	98.00	97.70	97.53	97.45
2.000	99.42	99.39	99.28	99.26

and the time factor is a dimensionless variable given by

$$T_v = \frac{c_v \cdot t}{h^2} \tag{6.8}$$

where c_v is the coefficient of one-dimensional consolidation, t is the time and h is the length of the drainage path. From Table 6 and Figure 67, the settlement $S_{c(t)}$ at any time t can be conveniently obtained, using Equations (6.7) and (6.8). The value of c_v is determined from oedometer tests for the range of stresses corresponding to the embankment loading and is considered to be a constant. The first main difficulty for application of the consolidation theory is that c_v is not a soil constant, as shown in Figure 12. Also c_v varies with depth, which limits the straightforward application of the Terzaghi's consolidation theory to stratified deposits.

The second main difficulty is related to the definition of drainage boundaries and thin sand layers, as discussed by Rowe (1972). Because of this, rates of settlements using c_v values obtained from standard oedometer tests are usually underpredicted. Field observations applied regionally might be a useful tool for deriving backfigured values of c_v to be used in future projects. This simple observational method has been applied to a number of projects in Brazil (Cortes, 1981), where c_v values adopted in design range from 4 to 10 times the values measured in oedometer tests. In situ tests, such as piezocone tests might produce more relevant values of c_v than laboratory tests. This trend has been observed by Lewis et al. (1975), where c_v obtained by in situ permeability tests produced much better agreement with settlements observations than those obtained by laboratory tests.

6.2.5 IMPROVED METHODS FOR ONE-DIMENSIONAL CONSOLIDATION

Most of Terzaghi's assumptions listed in the previous section are unrealistic or are not satisfied in practical applications. As a matter of fact, real loadings are time dependent, soils are inhomogeneous, soil properties vary nonlinearly with respect to effective stress, strains are occasionally large and pore fluid is compressible, amongst other aspects. Suggestions of how to tackle some of these problems in practical terms are summarized below. An alternative and more rigorous approach is to use elaborate one-dimensional consolidation models (e.g., Tavenas et al., 1979; Mesri and Choi, 1985) in finite difference programs, which take into account some of the features described above.

6.2.5.1 Time Dependent Loading

An approximate method for the prediction of settlements for linear time dependent loading is given by the so called Terzaghi-Gilboy method. This method assumes that the set-

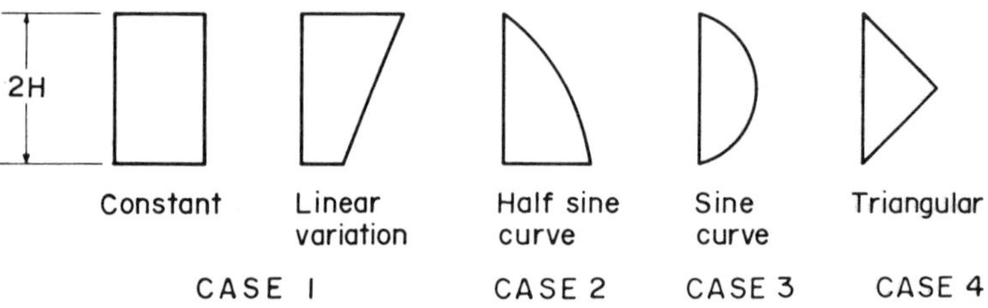

FIGURE 67. Degree of consolidation U against time factor T.

tlement at the end of construction t_c is equal to the settlement at time $\frac{1}{2}t_c$ of the instantaneous settlement curve. The time settlement curve for hypothetical instantaneous loading is assumed to be line OCD of Figure 68. Thus, from point C on the instantaneous curve, point E is obtained on the corrected curve. At any smaller time $t < t_c$, the method was extended as follows: the instantaneous curve at time $t/2$ shows the settlement KF. At time t the load acting is t/t_c times the total load, and the settlement KF must be multiplied by this ratio. This may be done graphically as indicated in Figure 68.

Line OF intersects time t at point H, giving a settlement at H which equals $(t/t_c) \times KF$. Thus H is a point on the settlement curve, and as many points as desired may be obtained by similar procedure. Beyond point E, the curve is assumed to be the instantaneous curve CD, offset to the right by one half of the loading period; for example, DJ equals CE. Thus, after construction is completed, the elapsed time from the start of loading, until any given settlement is reached is greater than it would be under instantaneous loading by one half of the loading period.

A more rigorous approach (Olson, 1977) of considering a linear time dependent loading is by using the graph of Figure 69, in which the time factor t_c for construction is defined by

$$T_c = \frac{c_v\, t_c}{h^2} \tag{6.9}$$

in which h is the drainage distance and c_v and h are as defined previously.

An example of how to use the graph in Figure 69 is given in Table 7. The following data were used: consolidation of a 3 m thick, doubly-drained layer with $c_v = 0.0045$ m²/day, and an ultimate settlement, $S_c = 254$ mm, subjected to a single ramp loading with a construction time of 30 days. This gives a construction time factor $T_c = 0.060$. Interpolation between the curves in Figure 69 gives a degree of consolidation after, for example, 100 days ($T_v = 0.200$) of 47%. If now a second ramp loading of 6 m is added, say between days 30 and 100, then, according to Olson, the combined solution can be obtained by superposition of each loading step. Table 7 shows all the calculations and also includes a step of instantaneous unloading of 3 m of fill, at day 200. The construction time factors, T_c(0.060, 0.100 and 0.000) and the ultimate settlements S_c (254,508 and -254 mm) are also indicated.

6.2.5.2 Finite Strains

In the conventional Terzaghi theory, strains are assumed to be small or, in the mathematical sense, infinitesimal. This means that the drainage distance h remains constant during consolidation, which might lead to considerable errors in the case of embankments on highly compressible soils.

A sufficiently accurate approach to take into account finite

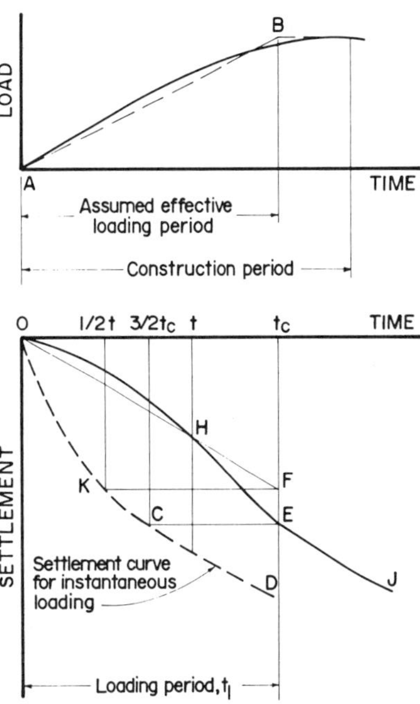

FIGURE 68. Graphical method for adjusting the settlement curve to account initial ramp loading.

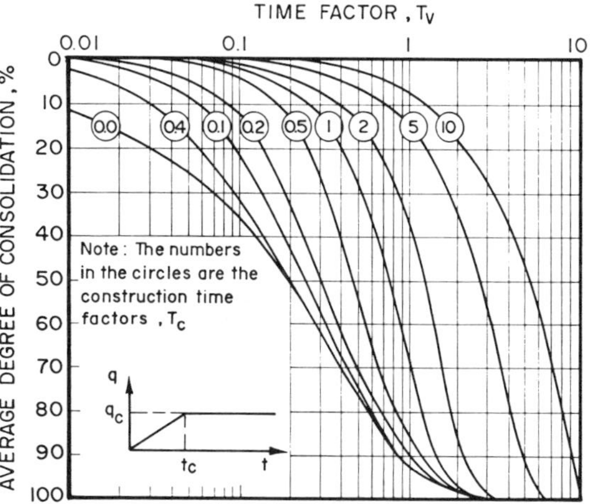

FIGURE 69. Degree of consolidation against time factor curves for single ramp loading.

TABLE 7. Example of Consolidation Under Construction Loading (Olson, 1977).

First ramp loading (T_c = 0.060; S_c = 254 mm)				Second ramp loading (T_c = 0.100; S_c = 508 mm)				Instantaneous unloading (T_c = 0.000, S_c = −254 mm)				
t_1 (days)	T_1	U_1 (%)	S_{c1} (t) (mm)	t_2 (days)	T_2	U_2 (%)	S_{c2} (t) (mm)	t_3 (days)	T_3	U_3 (%)	S_{c3} (t) (mm)	S_{ctot}(t) (mm)
10	0.020	4	10.2								10.2	
30	0.060	17	43.2									43.2
50	0.100	27	68.6	0	0	0	0					68.6
100	0.200	47	119.4	50	0.100	21	106.7					226.1
200	0.400	68	172.7	150	0.300	57	289.6	0	0	0	0	462.3
400	0.800	89	226.1	350	0.700	84	426.7	200	0.400	70	−177.8	475.0

strain was proposed by Olson and Ladd (1979). These authors suggested using a constant average value of h in the classical small theory defined as

$$h = h_o - \frac{S_t}{2N_D} \tag{6.10}$$

where S_t is the ultimate settlement and N_D is the number of drainage boundaries.

A review of the finite strain consolidation theory with applications was presented recently by Schiffman et al. (1984).

6.2.6. GRAPHICAL EVALUATION OF SETTLEMENTS

A new and practical approach to estimate final total settlement and settlements rates from settlements data obtained during a certain time period was proposed by Asaoka (1978). It is not the purpose of this chapter to discuss the basis of the method, but rather to describe its application.

Asaoka suggested a graphical procedure which has been applied to a number of settlement records of embankments on soft clays (Magnan and Deroy, 1980; Magnan and Mieussens, 1980). The steps in the graphical procedure used by Magnan and Mieussens (1980) are as follows:

1. The observed time settlement curve plotted to an arithmetic scale is divided into equal intervals, Δt, (usually Δt is between 30 and 100 days). The settlements S_1, S_2, . . . corresponding to the times t_1, t_2, . . . are read off and tabulated [see Figure 70(a)].
2. The settlements values S_1, S_2, . . . are plotted as points (S_{i-1}, S_i) in a coordinate system with axes S_{i-1} and $S_{i'}$ as shown in Figure 70(b). The 45° line is also drawn.
3. A straight line (l) is fitted through the points. The point where this line intersects the 45° line gives the final consolidation settlement. The slope β_1 is related to the coefficient of consolidation, c_v, by:

$$c_v = \frac{5\,h^2 \ln \beta_1}{12\,t} \tag{6.11}$$

and therefore indicates the rate of settlement. The slope, β_1, depends on the time step, Δt, selected and decreases when Δt increases.

In some cases, it occurs that the data points are better fitted by two straight lines (I) and (II), as shown in Figure 71(a). The second straight line (II) corresponds to the tail of the settlement curve and, thus, represents secondary compression. In the case of multi-stage loading, a straight line corresponding to each loading stage can be fitted. If the time between the loading increments is large, a secondary compression line can also be drawn. Figure 71(b) shows a case of a load applied in three stages with secondary compression at the end of the third stage. In some rare cases, according to Magnan and Deroy (1980), it is difficult to define a straight line through the points.

The graphical procedure in the form presented here is limited to single layers with one-way or two way drainage. An extension of the general theory to multilayer systems has recently been accomplished (Asaoka and Matsuo, 1980).

An important feature of Asaoka's method, as pointed out by Balasubramanian and Brenner (1981) is that it allows the separate determination of the coefficient of consolidation for vertical drainage, c_v, and the final value of settlement. In the conventional method, both the final settlement and c_v must be estimated by fitting the theoretical curve to the observed one, and final settlements cannot be determined accurately if the settlement record is not known from the beginning of consolidation. Furthermore, the sharp bend, characteristic for most settlement records, occurs around 90% of the primary consolidation, and when Terzaghi's solution is employed, no conclusions regarding c_v can be made unless the observation time is sufficiently long. With Asaoka's method, good predictions are usually possible after 60% consolidation has been achieved (Magnan and Deroy, 1980).

Examples of applications of Asaoka's method to embankment settlements have also been given by Magnan and Deroy (1980). They first determined c_v and S_∞, and then the settlement relationship was obtained by means of the

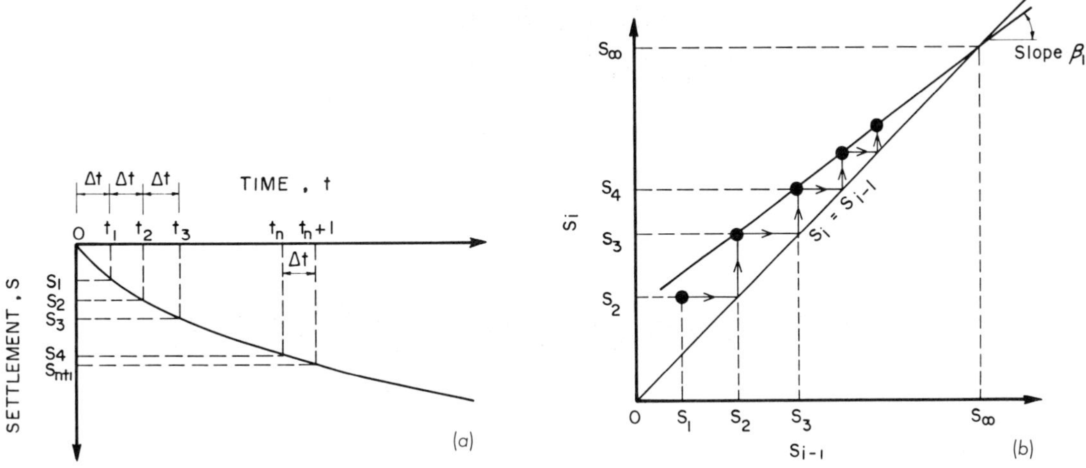

FIGURE 70. Asaoka's graphical method to determine final settlement and coefficient of consolidation.

degree of consolidation values U_s, calculated from the one-dimensional Terzaghi's theory.

6.3 Lateral Displacement Calculation

Estimates of lateral displacements under embankments are sometimes required. The need for such calculations arises owing to the detrimental effect of lateral displacements on the behaviour of adjacent structures, particularly pile bridge abutments.

An empirical method for the calculation of lateral displacements was proposed by Bourges and Mieussens (1979) based on field observations. This method was subsequently extended by Tavenas et al. (1979) and is recommended for practical applications. The method correlates maximum settlement S_m under the embankment centre line with maximum lateral displacements δ_{hm} at the embankment toe, as shown in Figure 72.

Tavenas et al. (1979) have shown that during construction lateral displacements of test embankments on a variety of clays were initially small. This was attributed to the partial drained behaviour of the initially overconsolidated clay foundation. It was observed that in cases of embankments with slopes of the order of 1.5–2.5 horizontal to 1 vertical, initial lateral displacements could be in average correlated with settlements using the equation

$$\delta_{hm} = 0.18\ S_m \qquad (6.12)$$

Tavenas et al., also observed that at a certain point that

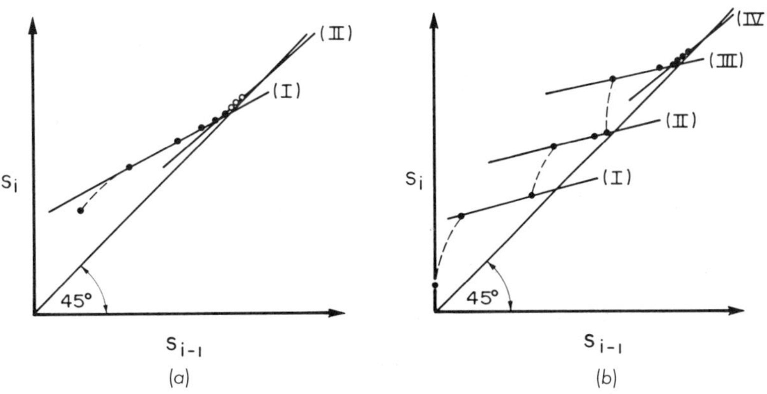

FIGURE 71. Asaoka's method for single and multi-stage loading.

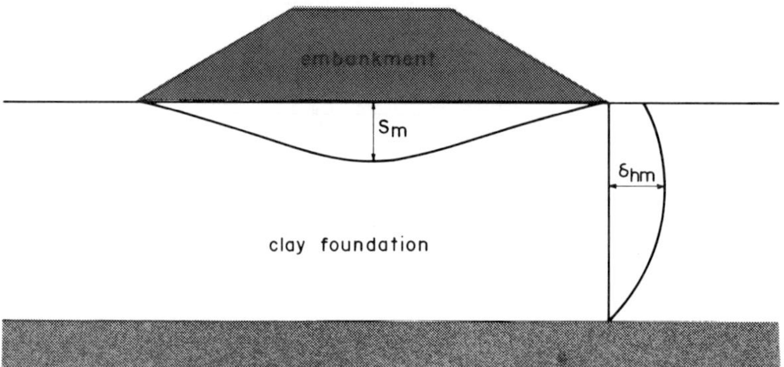

FIGURE 72. Maximum settlement and maximum lateral displacements at the embankment toe.

corresponded to the passage of parts of the foundation to a normally consolidated state, lateral displacements and settlements increased at about the same rate up to the end of construction. However, the point corresponding to the change of slopes from 0.18 to 1.0 depends on a number of factors, as recognized by Tavenas et al., making it very difficult to be predicted.

Tavenas et al. (1979) compiled available field observations of long term lateral displacements and observed that these correlated with settlements developing after the end of construction, as shown in Figure 73. This relationship may be expressed as

$$\delta_{hm} = 0.16\ S_m \tag{6.13}$$

for the geometry and stability conditions noted in Figure 73. The relationship given by Equation (6.13) does not seem to be applicable to the entire duration of the primary consolidation, and the ratio δ_{hm}/S_m may decrease to less than a third of the value observed at the beginning of the consolidation.

The similarity between Equations (6.12) and (6.13) seems to suggest the drained nature of foundation response at the beginning of construction, as pointed out by Tavenas and Leroueil (1980).

6.4 Two Dimensional Consolidation

6.4.1 *TERZAGHI-RENDULIC AND BIOT THEORIES*

In case of deep soil layer, embankment loading might produce substantial lateral drainage which might increase the rate of consolidation. In these situations, analyses employing two or three-dimensional consolidation theories are more appropriate, as the one-dimensional theory may underestimate the rate of settlement.

Two basic approaches are commonly used for analysing consolidation problems in two and three dimensions. The first was developed from the Terzaghi one-dimensional

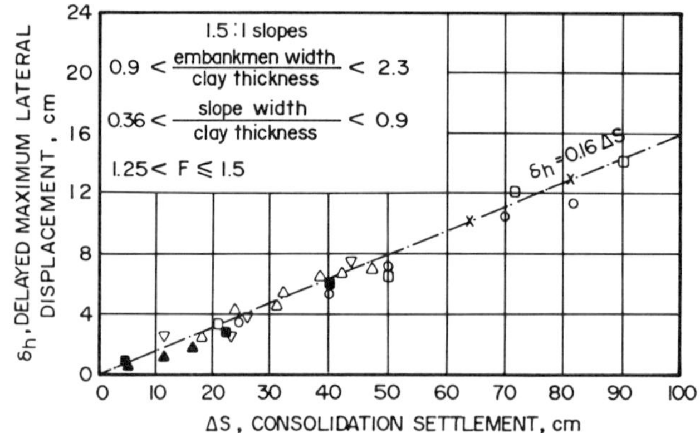

FIGURE 73. Maximum lateral displacements and settlements during long term consolidation of embankment foundations (after Tavenas et al., 1979).

diffusion theory as proposed by Rendulic and is often called the Terzaghi-Rendulic theory or diffusion theory. The second theory was derived directly from the theory of elasticity by Biot (1941) and is known as Biot theory. The mathematical bases of both theories are discussed in detail by Murray (1978) and only the main features of the theories are discussed below.

The basic differences between the theories is that the diffusion theory uncorrectly assumes that total stresses remain constant during consolidation. This is not the case as during consolidation, pore pressure dissipation rates vary depending on the distance from the drainage boundaries and, since Young's modulus and Poisson's ratio change from their undrained to their drained values, differential strains occur, which in turn require an adjustment of the total stress to satisfy the stress–strain law. Unlike the diffusion theory, Biot theory takes into account the change in total stress during consolidation.

Biot theory, which is mathematically more complex than the Terzaghi-Rendulic theory, provides a coupling between the magnitude and progress of the settlement and, at any point in the consolidating layer, there is continuous interaction between the dissipating excess pore pressure and changing total stress.

6.4.2 RATE OF SETTLEMENTS PREDICTED BY TWO-DIMENSIONAL THEORIES

The two-dimensional Terzaghi-Rendulic theory has been applied to the study of consolidation beneath embankments by Dunn and Razouki (1974). Their analysis included the influence of embankment geometry, depth of compressible strata and anisotropy of the soil and are presented in a form suitable for use as design charts. Figure 74 shows the influence of drainage conditions (permeable or impermeable base), embankment shape (defined as the ratio of the width at the top of the embankment to that at the base) and layer thickness, on two degrees of consolidation. The shape of the embankment does not seem to have a great influence at any case. Since a road engineer often assesses his design according to the time required to achieve a degree of consolidation of 90%, the influence of the embankment shape could be ignored for the conditions described.

In order to evaluate the differences between the Terzaghi-Rendulic theory and the Biot theory, Murray (1978) presented results of these two solutions for the case of a strip footing on a finite layer, as shown in Figure 75. The results of the Biot theory are for values of Poisson's ratio varying between 0 and 0.45. It can be seen that the curve based on the simpler theory lies generally near the centre of the range and agrees best with the curves based on ν' of 0.2, at least for the ratio depth of clay to width of the footing $h/b = 1.0$ presented. Murray (1978) showed that for other values of h/b, the maximum ratio between the time factors given by the two theories was 2. Thus, it appears that for practical cases, particularly in view of the uncertainties in the soil parameters selected in an analysis, the simple diffusion theory is a sufficient approximation to obtain the settlement versus time relationship under embankments on soft clays.

6.5 Finite Element Analyses

6.5.1 INTRODUCTION

When a more general picture of vertical and horizontal deformations as well as pore pressures are required, numerical analyses using the finite element method are performed.

The basis of the finite element method is the representation of a body or a structure by an assemblage called "finite elements." These elements are interconnected at nodal points, and the solutions are obtained in terms of displacements at these nodal points and average stresses at the elements. It is out of the scope of this chapter to describe the basis of the finite element method and for this purpose the reader is referred to Zienkiewicz (1977) and Desai and Abel (1972). Such a method can take into account various types of stress–strain behaviour, non-homogeneity, irregular geometry and complex boundary conditions, as well as time-dependent loading (e.g., simulation of construction sequence).

6.5.2 MATERIAL MODELS

The material models commonly used in finite element analyses are:

(a) Elastic linear models
(b) Elastic non-linear models, using bilinear or hyperbolic stress–strain relationships
(c) Elasto perfectly plastic models, using Tresca or Von Mises criteria for the soft clay foundation
(d) Elasto plastic strain hardening models such as the models embodied into critical state soil mechanics (CSSM) theory

Finite element analyses can be performed to predict either short undrained-behaviour or long term drained behaviour, or even time dependent deformations or pore pressures, in which case a coupled consolidation analysis is necessary. The common practice is to use different soil models and parameters for short and long term analysis. The only single theory that allows analyses of both undrained and drained problems, using the same set of soil parameters obtained from standard laboratory tests, is the critical state soil mechanics (CSSM) theory. This theory can also handle partial drained problems by means of the finite element method coupled with Biot two or three-dimensional consolidation theory. The modified Cam-clay model is the critical state model more extensively used in CSSM finite element computations. A brief description of this model follows.

6.5.2 THE MODIFIED CAM-CLAY MODEL

The modified Cam-clay model is applicable to normally and lightly overconsolidated clays which are treated as iso-

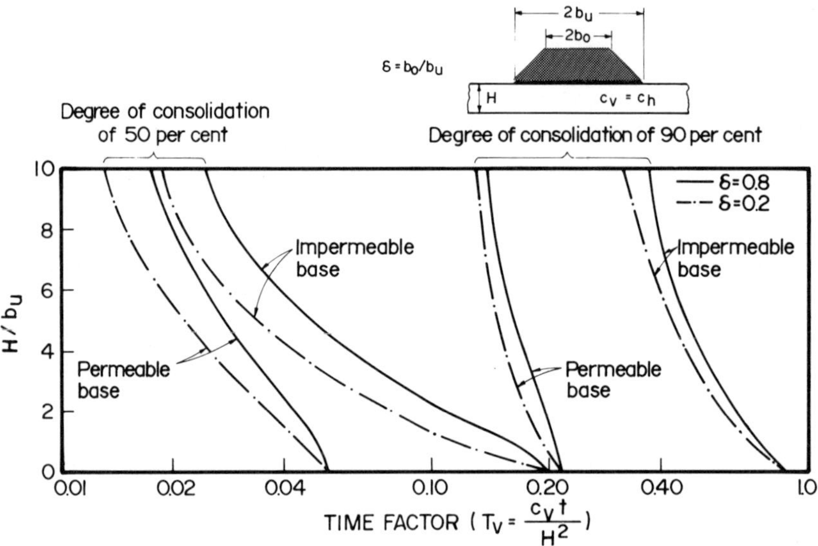

FIGURE 74. Influence of the embankment width on the progress of consolidation (after Dunn and Razouki, 1974).

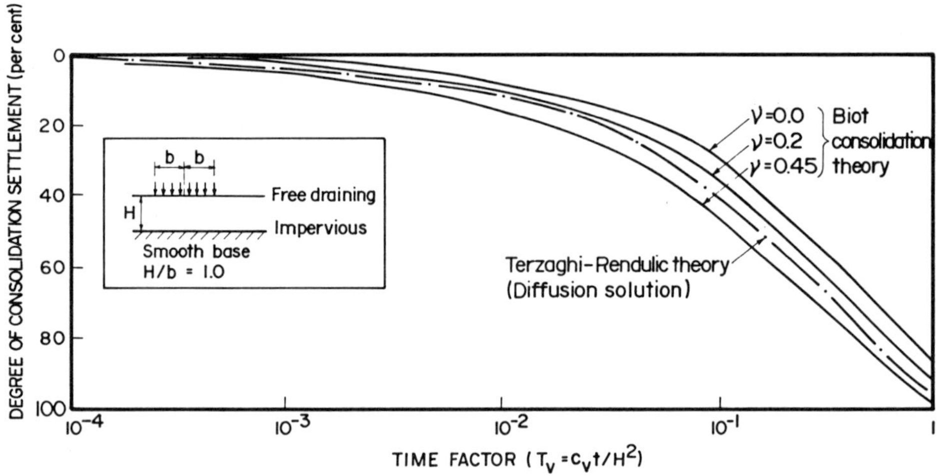

FIGURE 75. Comparison between Terzaghi-Rendulic and Biot theories (after Murray, 1978).

tropic, elasto-plastic, strain-hardening materials. The soil is assumed to possess a yield locus and to have a flow rule which satisfies the normality condition of the plasticity theory. For details about the CSSM theory and Cam-clay models the reader is referred to Schofield and Wroth (1968), Atkinson and Bransby (1978) and Roscoe and Burland (1968).

The soil parameters that are required to specify completely the modified Cam-clay Model are: C_c, C_s, ϕ', and e_{cs}, the void ratio for which $p^* = 1$kPa at the critical state line, where $p^* = (\sigma_1' + \sigma_2' + \sigma_3')/3$.

6.5.3 NUMERICAL ANALYSES OF THE EMBANKMENT ON THE RIO DE JANEIRO CLAY

Numerical analyses of the embankment on the Rio de Janeiro clay were performed using two distinct soil mechanics approaches: CSSM effective stress theory using the CRISP program (Gunn and Britto, 1981) and total stress non-linear elastic method using the FEECON program (Simon et al. 1974). The analyses carried out were: undrained, drained and coupled consolidation using the CRISP program (the modified Cam-clay model was adopted for both embankment and clay foundation), and the elastic non-linear undrained using the FEECON program as described in Table 8. A summary of the results of displacements and pore pressures from numerical analyses compared with observed values during the embankment construction is presented below. Further details about the analyses can be seen in Almeida and Ortigão (1982).

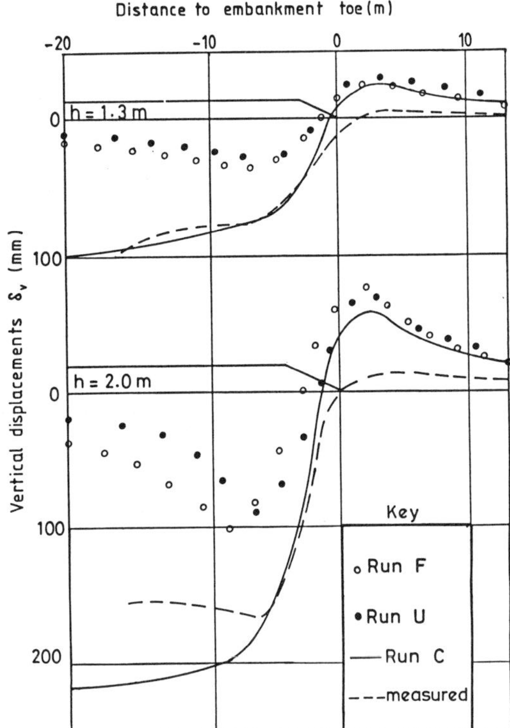

FIGURE 76. Vertical displacements at embankment base (after Almeida and Ortigão, 1982).

6.5.3.1 Vertical Displacements

Figure 76 shows computed and measured vertical displacements at the embankment base at the end of three stages of construction. The most important points to highlight from this figure are:

(a) Measured values below the embankment lie in between undrained (Run U and Run F) and coupled consolidation (Run C) analyses, but only Run C produced settlement curves with the same trend as observed values.

(b) All computations overpredicted the measured heave in front of the embankment.

(c) Run U and Run F produced similar results for the early stages of construction, but the differences between the two analyses increase towards the end of construction.

TABLE 8. Summary of Finite Element Analyses

Program	Analysis	Name
CRISP	Undrained	Run U
	Drained	Run D
	Coupled Consolidation	Run C
FEECON	Undrained	Run F

Figure 77 presents long term settlements (Run D) which, as expected, are greater than all the other analyses. The same figure shows settlements obtained from standard oedometer test computations which agree very well with Run D. However, the results of both computations may be unrealistic due to submerging effects and finite strain consolidation, which were not accounted in either computation.

Measured and computed (Run C) settlements with time during embankment construction are plotted in Figure 78. It is seen that the agreement between both sets of values is quite good mainly with respect to embankment heights up to 2.0 m. Another feature of these results is that the settlements which occurred during embankment loading have a more important contribution than settlements under subsequent load conditions. Discrepancies for the later stages of construction can be due to the reasons mentioned in the last paragraph.

6.5.3.2 Horizontal Displacements

Computed results and measured values of inclinometer displacements are presented in Figure 79. The most important points shown in this figure are: (a) all computations resulted in values higher than those measured; (b) as for vertical displacements, Run F and Run U produced results

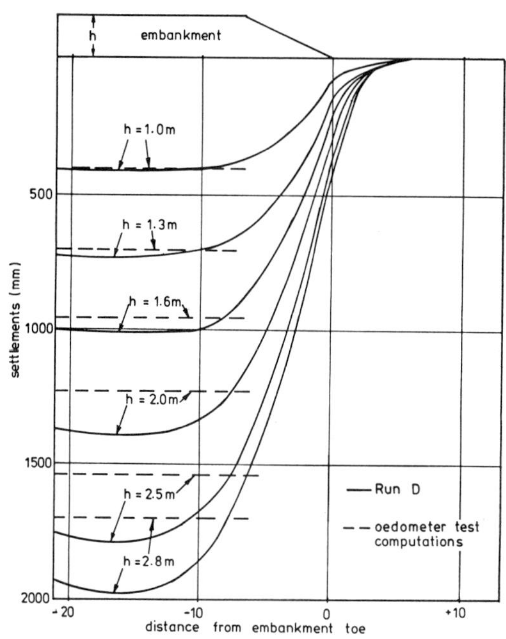

FIGURE 77. Computed long term settlements (after Almeida and Ortigao 1982).

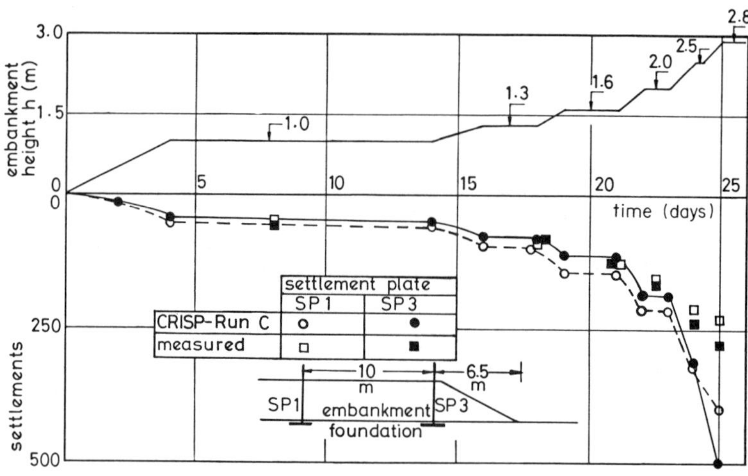

FIGURE 78. Settlements versus time.

which were close for the early stages of the construction, but not for the later stages.

Figure 80 presents measured and computed horizontal versus vertical displacements during embankment construction. It can be seen that computed values by Run C are much closer to measured than the ones resulting from undrained analysis.

6.5.3.3 Pore Pressures

Figure 81 presents computed (Run C) and measured pore pressures for three piezometers during embankment construction. It can be seen that the agreement between observations and predictions is particularly good for P4 and P13. One very interesting feature of these plots is the continuous increase of pore pressures after load application, which is very clear after the first layer. This fact, which was typical for piezometers at mid depth, could be due to the Mandel-Cryer effect (Gibson et al., 1963).

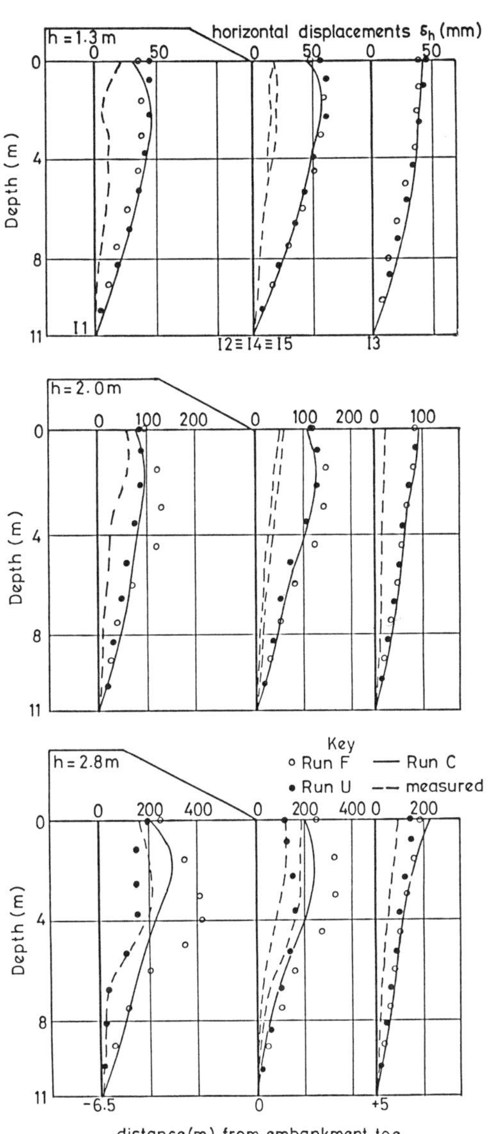

FIGURE 79. Horizontal displacements at inclinometers.

7. INSTRUMENTATION

7.1 Introduction

Instrumentation can be defined as the set of techniques employed for the observation of a structure. It covers the selection of the type and number of instruments, their location, installation, data acquisition, analysis, and interpretation.

In recent years, a great deal of information has been published on this subject, which demonstrates the importance and the development in this field. Two international symposia (BGS, 1973 and Kovari, 1983) were held on this subject and a comprehensive treaty (Hanna, 1973) is available. Additional useful information on instrumentation for embankments can be found in: Dunniclif (1971), Gould and Dunniclif (1971), Di Biaggio and Myrvoll (1981), OCDE (1979), Magnan and Mieussens (1980 and 1981), among others.

This section is aimed at an introduction to the subject and, therefore, will cover briefly the following items: purpose of instrumentation, instruments types and selection, instrumentation location, and data analysis. A few case histories will be presented and discussed.

7.2 Embankment Monitoring Programme

The several phases of a monitoring programme are listed in Figure 82 and will be discussed in the next items. Only by careful planning of each phase of the work, the investments in instrumentation can yield the expected revenue.

An outline of the relative instrumentation costs is presented below:

Instrumentation acquisition	10%
Installation	20%
Data acquisition	20%
Analysis	50%
	100%

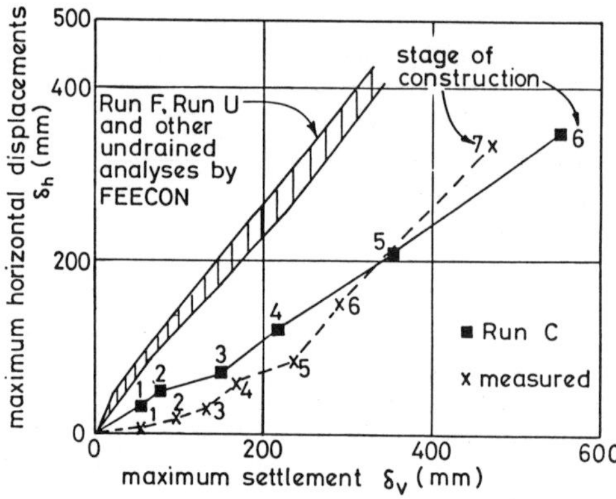

FIGURE 80. Horizontal displacements versus vertical displacements.

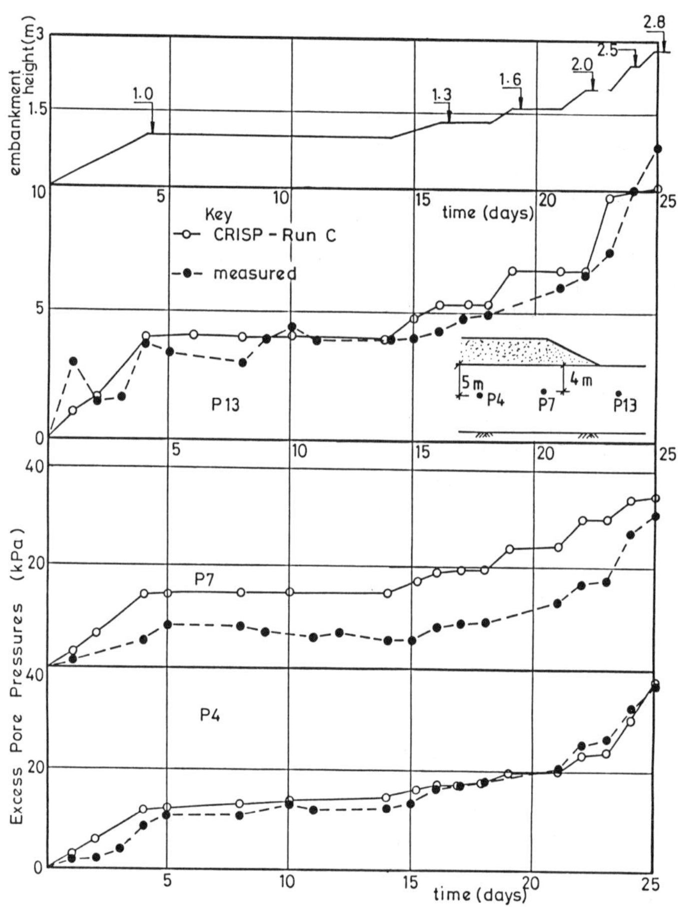

FIGURE 81. Pore pressure versus time.

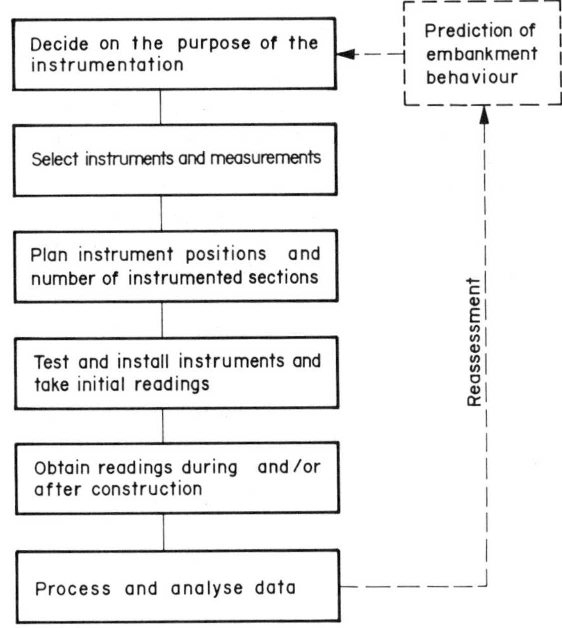

FIGURE 82. Phases of a monitoring programme.

It is important to note that the relative effort needed for the instruments follow-up and data analysis is frequently greater than the previous phases. As a matter of fact, many monitoring programmes fail due to insufficient effort in data acquisition and analysis, rather than the quality of the instruments or their installation.

7.3 Purposes of Instrumentation

Several purposes for instrumenting a civil engineering structure have been summarized by Peck (1969) in his Rankine Lecture (see also Terzaghi and Peck 1967):

1. Observation for detecting imminent danger
2. Observation for obtaining vital information during construction
3. Observation for information for corrective measurement (i.e., underpinning, etc.)
4. Observation as a means of improving construction methods
5. Observation for accumulating local experience
6. Observation to provide evidence in legal actions
7. Observation for the assessment of theories and behaviour mechanisms

In embankments on soft ground, stability and settlement control through instrumentation are certainly of primary importance, as covered by items (1) and (2) above.

7.4 Guidelines for Instrument Selection

In most cases an embankment behaviour is monitored with regard to:

- pore pressures
- vertical and horizontal displacements
- earth pressures

In relation to embankments on soft ground, the interest of monitoring earth pressures is very limited and, therefore, this section will only cover the first two variables above.

The following factors may affect the selection of an instrument:

- physical principle
- ruggedness
- reliability
- response time
- accuracy
- long term stability
- pointwise versus linewise instrument
- method of obtaining readings
- cost

Concerning the physical principle, instruments can be mechanic, hydraulic, pneumatic, or electric. Examples of instruments employing these principles are, respectively, Bourdon pressure gauges, mercury manometers, pneumatic pressures gauges (as will be described later) and electric pressure transducers.

One important lesson from experience in field instrumentation says: keep simple. The simpler the instrument, the lesser the probability of malfunctioning. Therefore, mechanical or hydraulic instruments should generally be preferred. On the other hand, electrical instruments allow automatic data acquisition, and in some cases considerable savings can be done by avoiding excessive labour costs of recording manually the readings.

Instrument ruggedness should match severe field ambient conditions like temperature and humidity, moisture, corrosion, etc. Instrument reliability will depend on many factors, among which simplicity of operation and the experience with the instrument may be of utmost importance.

In quasi-static phenomena, such as the case of embankment construction, response time is not of major concern except for pore pressure observations. Due to the low permeability of the soft foundation soils, a piezometer system must be properly designed, as will be described later, to achieve response times compatible with the rate of construction.

Accuracy should comply with the expected range of field measurements values. In general, an accurate field instrument will yield an overall accuracy of 1 to 2% of the full scale output. A comprehensive discussion about this point can be found in Gould and Dunnicliff (1971).

Permanent embedded instruments should be designed for

stability during the period of measurements. Electrical instruments are, in general, more susceptible to malfunctioning and frequent recalibration than simple mechanical devices. Even properly designed and installed piezometers in unsaturated soils may show read-out errors due to ingress of air in the porous tip (Vaughan, 1973).

Pointwise versus linewise instrumentation schemes are presented in Figure 83. The linewise instrumentation concept (Kovari and Amstad, 1983) has significant advantages for displacement monitoring in quasi-static phenomena and has been in use for several years (e.g., inclinometer). Linewise instrumentation allows continuous recording and, since the probe may not be permanently embedded, retrieving and recalibrating it can be performed easily.

The selection of the method for obtaining readings will significantly affect the overall cost of a monitoring programme. The following aspects should be considered:

- number and relative distance between instrumented sections
- possibility of conveying all instruments leads and cables to one instrument house
- time length of the monitoring programme
- number of instruments and frequency of readings
- local labour costs (for manual data acquisition)
- remoteness of the construction site
- possibility and cost of automatic data acquisition and transmission (via telephone or satellite)

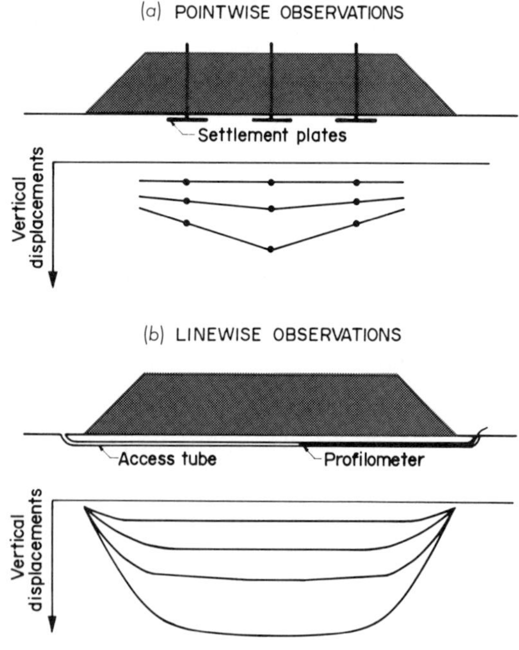

FIGURE 83. Pointwise versus linewise observations.

7.5 Pore Pressure Measurement

Piezometers are used to monitor pore pressures within the soil mass. They can be classified according to their operation principle as hydraulic, pneumatic, or electric.

The basics of hydraulic piezometers are shown in Figure 84. It consists of a porous tip installed in a 50 to 100mm diameter borehole in a sand pocket [Figure 84(a)] or partially driven by a 0.5 m push from the end of the borehole [Figure 84(b)]. The two main types of hydraulic piezometers are shown in the same figure: the open standpipe, also known as Casagrande type, and the closed circuit or twin-tubing type. In the first type, the pore pressure acting in the soil around the porous tip is measured through water level observations in the vertical pipe. In the closed circuit type, the leads, consisting in flexible, 2 to 4mm in diameter, nylon coated polyethylene tubings, are conveyed to a measuring post or cabin that houses the pressure measuring devices. These coated tubings avoid ingress of air and are not previous to water. The whole system is initially saturated with deaired water.

The read-out units used for closed circuit hydraulic piezometers are:

- individual mercury manometers
- electric pressure transducer unit
- individual Bourdon gauges

The main characteristics of these systems are summarized in Table 9. Figure 85 portrays schemes of the first two systems which are more appropriate for the low pore pressure range which are generated in soft foundations under low embankments. These readout systems allow pressure measurement in both leads of the circuit and, therefore, a check on the observed pressures is performed. In the writers' experience, the use of the pressure transducer read-out type is recommended for its simplicity and easiness of operation.

One of the main advantages of the hydraulic piezometers is the tip deairing facility which is provided by the twin-tubing. The necessary equipment and recommended procedures to prepare and circulated deaired water are described elsewhere (e.g., Penman, 1972; Hanna, 1973).

Electric (vibrating-wire) and pneumatic piezometers are shown in Figure 86. In both types the pore water pressure actuates on one side of a diaphragm while its value is sensed at the other. They differ on the physical principle used in the pressure sensing element.

The electric type usually employs a vibrating wire gauge consisting of a thin steel wire stretched between a fixed point at the middle of a steel diaphragm. The water pressure deflects the diaphragm. If water pressure changes, a resultant strain variation in the wire occurs. This phenomenon modifies the wire acoustic properties: the change in strain is proportional to the square of the vibrating frequency which can be remotely sensed.

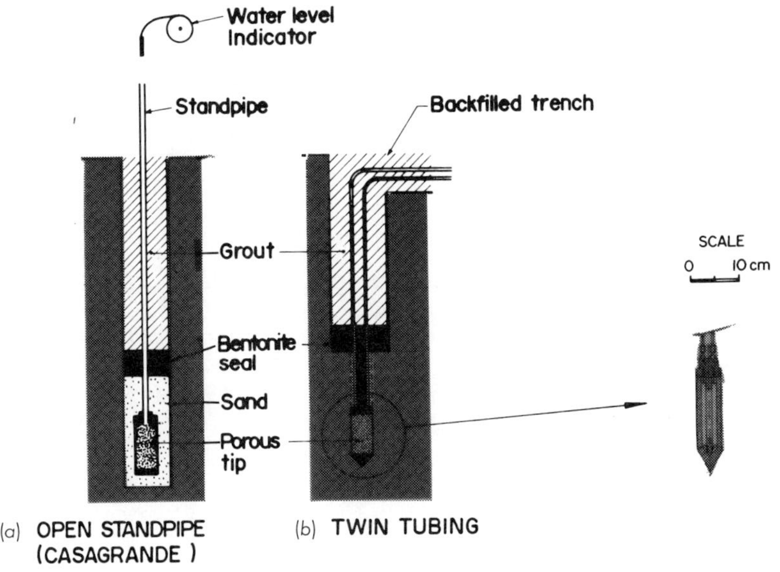

FIGURE 84. Hydraulic piezometer types.

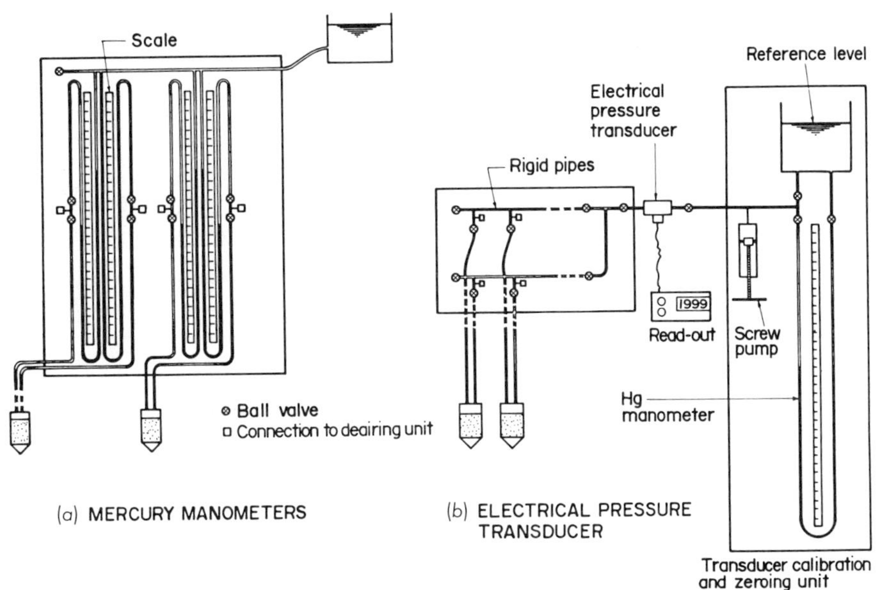

FIGURE 85. Read out units for hydraulic piezometers.

TABLE 9. Characteristics of Read-out Units for Hydraulic Twin Tubing Piezometers.

Read out unit type	Disadvantages	Advantages
Individual mercury manometers	• individual mercury manometers • large instrument houses are needed to accommodate read-out panels • tedious to obtain readings • frequent deairing of the read-out unit may be necessary when negative pressures are read at read-out level • limited pressure range (due to panel height limitation)	• simple • fault-proof • appropriate to low pore pressure range (embankments on soft soil) • very reliable
Pressure transducer gauges	• relatively high cost of electric transducer and read-out • a delicate electric instrument is necessary	• easy to read • readings directly in engineering units • transducer unit can be portable or exchangeable • any pressure range • lower cost for increasing number of piezometers • small instrument houses
Individual Bourdon gauges	• individual gauges are necessary • high cost of accurate Bourdon gauges • not recommended for low pore pressure range (as in the case of embankments on soft soils)	• usually appropriate for high pressure range (embankment dams) • simple • fault-proof

TABLE 10. Advantages and Disadvantages of Piezometer Types.

Piezometer	Advantages	Disadvantages
Open Stand pipe	• simple, low cost, easy to install • high reliability • self-deairing • allows field permeability tests	• long response time in soils of low permeability • vertical standpipe interface with embankment construction
Hydraulic twin tubing	• allows porous tip deairing • very reliable and simple (no calibration necessary) • advantages for long term installations • allows field permeability tests • easy to install • low tip cost	• system deairing is a time consuming operation • pressure measuring level maximum 6 metres above piezometric head to avoid cavitation
Pneumatic	• small and portable read out unit • pore pressure value is read directly (no correction necessary for difference in level • no limitation of piezometric level measurement difference (due to cavitation) • very low response time	• usually no deairing facility for the porous tip • no possibility of performing soil permeability tests • needs very careful installation to ensure tip saturation • diaphragm may present failure in long term installations • medium to high tip cost
Electric	• easier to perform automatic data acquisition • installation in any level difference between tip and read out • very low response time	• high cost • no tip deairing facility usually provided • no possibility of performing soil permeability through porous tip • possibility of electric faults mainly in long term installations

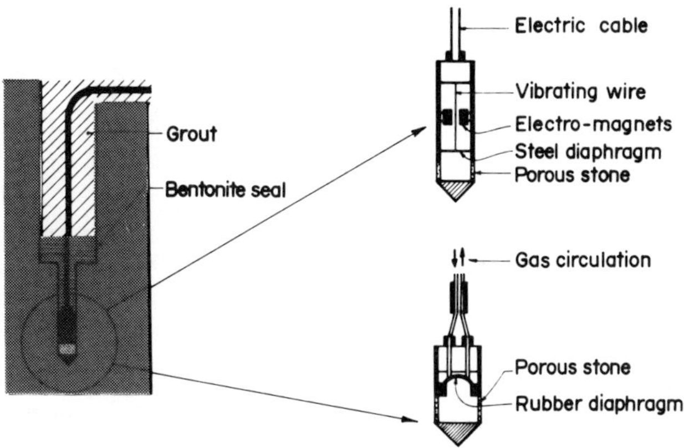

FIGURE 86. Electric (vibrating wire) and pneumatic piezometers.

The pneumatic piezometer operation is shown in Figure 87. The regulated pressure is slowly increased until gas pressure deflects the diaphragm and finds its way to the other side of the tubing, increasing the main pressure gauge. At this moment, the vent valve is opened to atmosphere, releasing pressure on one side of the system. The main Bourdon gauge pressure then falls until equilibrium with the pore pressure is reached.

The advantages and disadvantages of piezometer types are compared in Table 10. One of the main points of concern when selecting a piezometer system for installation in low permeability soils is the response time. This can be a problem when rapid variation of pore pressure within this soil is expected. A certain quantity of water should enter or exit the system. Since this may take some time, in low permeability soils, the piezometer indication may also be delayed. This phenomenon is studied in detail elsewhere (Hvorslev, 1951; Hanna, 1973; Vaughan, 1973). Table 11 portrays typical response times of piezometer installations.

7.6 Vertical Displacement Measurement

Observations of vertical displacements along the base or in the soft foundation of embankments can be performed with the instruments shown in Figures 88 and 89. Pointwise observations can be obtained by levelling: (1) the top end rod of settlement plates, (2) deep anchor plates; and (3) surface marks. The use of vertical magnetic extensometer provide a low cost mean of monitoring several points along the same vertical access tube. This instrument operates by location of magnet targets through a sensor which activates a buzzer when it reaches the magnetic field.

Linewise observations of settlements along the base of an embankment can be performed by a profilometer or simply by a row of steel plates. A profilometer employing a simple air-water principle (Palmeira and Ortigao, 1981) is shown on Figure 89. Other more sophisticated types, such as those employing an inclinometer, are commercially available. A simple and fault-proof alternative method was described by Rozsa and Vidacs (1983), consisting in layering a row of steel plates along the base of the embankment. The embankment settlement profile is obtained from monitoring the plates' position through boreholes in the embankment.

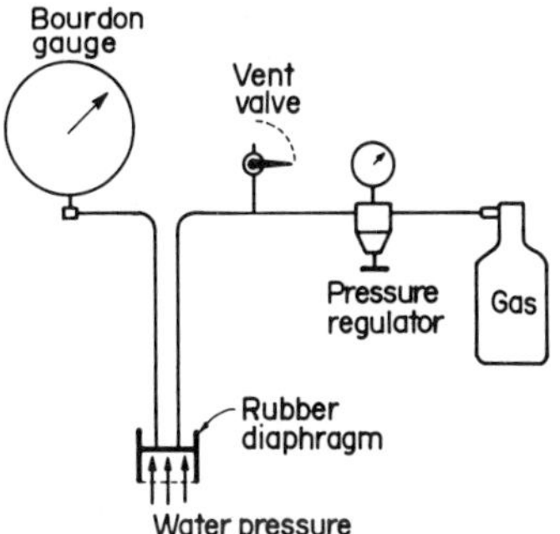

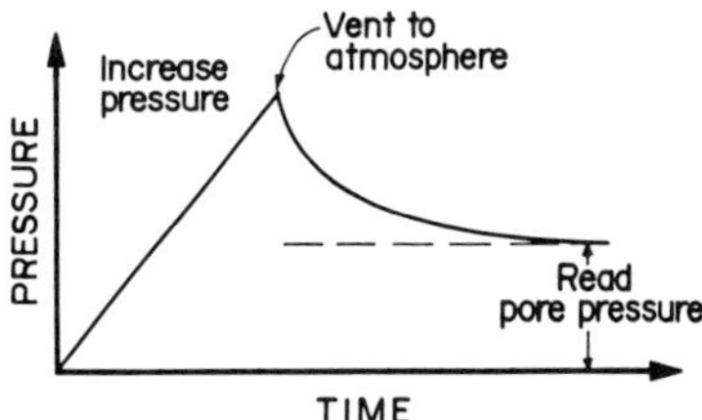

FIGURE 87. Pneumatic piezometer principles.

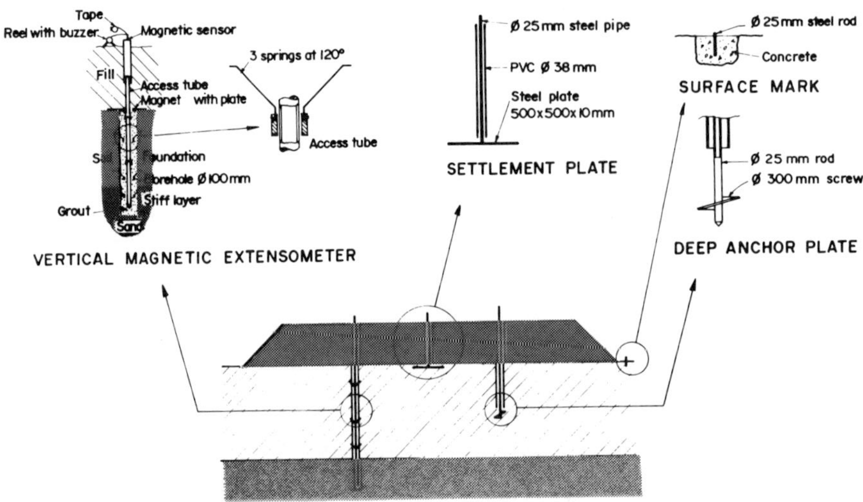

FIGURE 88. Pointwise instruments for settlement observations.

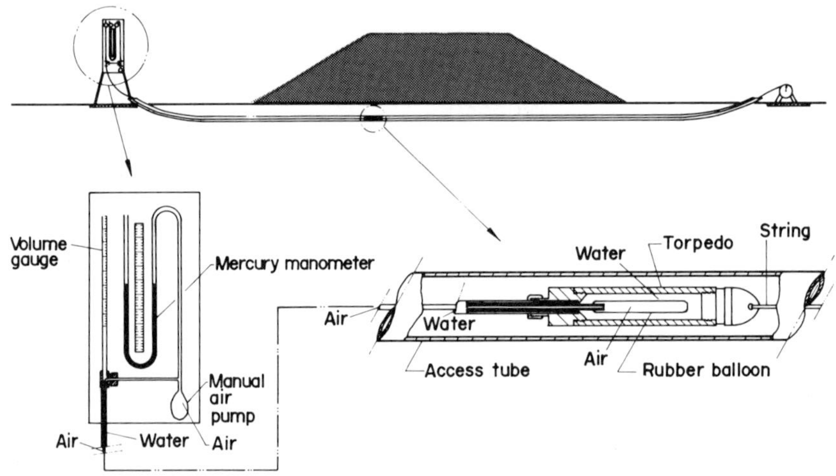

FIGURE 89. The profilometer (from Palmeira and Ortigao, 1981).

TABLE 11. Piezometer Response Times.

Piezometer	Piezometer Characteristics	Order of Magnitude of Calculated Response Time for 95% Equalization
Open Standpipe (Casagrande)	• sand pocket: L = 140 cm D = 6.4 cm • sandpipe diam. 13mm	• 1000 hours
Hydraulic Twin Tubing	• 150m long leads, 3mm ID nylon tubings, and 3mm mercury manometers • Idem, but with Electric transducer read-out unit	• 2 hours • 30 minutes
Electric or Pneumatic		• 10 seconds

Soil permeability: 10^{-10} m/s

7.7 Horizontal Displacement Measurements

Observations of horizontal displacements can be performed by:

1. Surveillance of surface marks at the slope or at the foot of the embankment
2. Magnetic or mechanic extensometers, as shown in Figure 90
3. An inclinometer, as shown in Figure 91

The magnetic extensometer is a low cost instrument used generally in conjunction with a profilometer through the same access tube.

An inclinometer provides a linewise mean of observing horizontal displacements along depth. It plays a major role in many monitoring programmes, such as those aimed at the control of stability of embankments.

7.8 Slip Surface Indicators

Two instrument types can be employed for detecting the depth of occurrence of a shear slip. An inclinometer, as shown in Figure 91, is certainly the most appropriate one for locating a shear surface, for it enables a step by step picture of deep ground displacements.

There are, however, simpler instruments that perform the task of locating the occurrence of a shear slip at a lower cost. The simplest of all consists of a brittle plastic pipe installed in a borehole (Figure 92). A lead weight attached to a string is lowered inside and left there until the slip occurs and the pipe is broken. By pulling the string up, the lower level of the slip can be located, for the lead weight will not pass at the broken level. On the other hand, by lowering another lead weight until it stops in the pipe, an indication of an upper level of the shear slip is obtained.

7.9 Instrumentation Lay-Out and Results from Field Observations

The choice of the number and location of instruments depends on several factors such as the purpose of the instrumentation, equipments and labour available, costs, etc. There are no fixed rules. Nevertheless, certain hints on the design of the instrumentation lay out and on the use of the

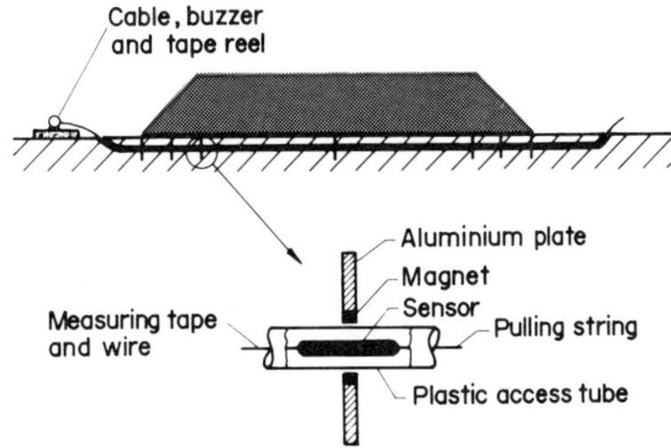

(a) MAGNETIC HORIZONTAL EXTENSOMETER

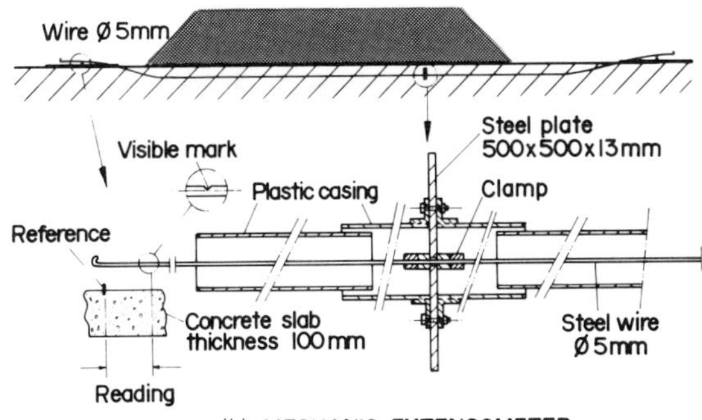

(b) MECHANIC EXTENSOMETER

FIGURE 90. Details of horizontal extensometers.

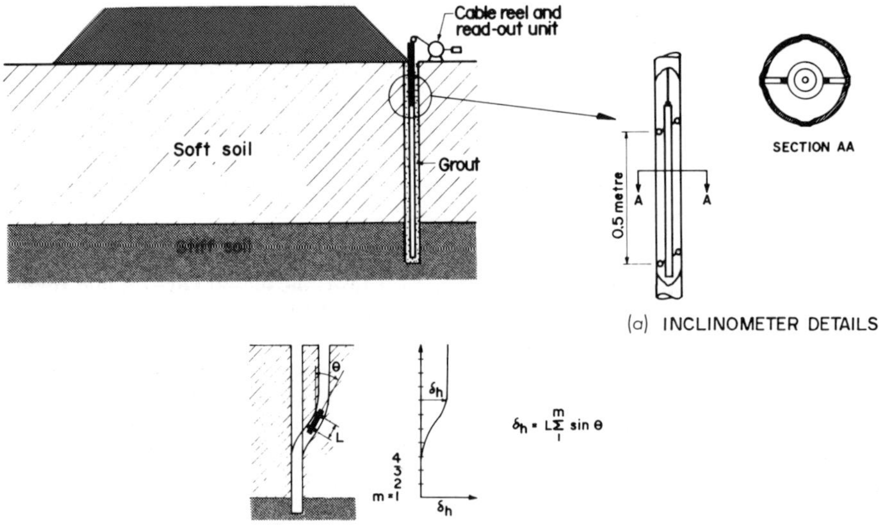

FIGURE 91. The inclinometer.

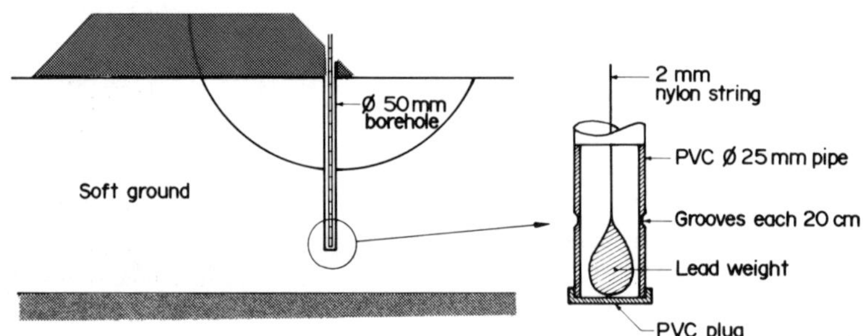

FIGURE 92. The slip surface indicator.

results will be given herein by means of a few examples from the writers' experience.

7.9.1 TRIAL EMBANKMENT BUILT UP TO FAILURE ON RIO DE JANEIRO CLAY

Figure 93 shows a research trial embankment built up to failure on Rio de Janeiro soft clay and described in detail by Ortigão et al. (1983). One of the purposes of this field trial was to test the instrumentation itself and to gain experience with locally designed instruments. Therefore, several different instruments types were used to monitor the same variable.

The purposes of the instrumentation of the main section were:

(a) To monitor pore pressures close to the failure surface
(b) To observe settlements and horizontal displacements along the base of the embankment
(c) To observe deep horizontal displacements
(d) To locate the failure surface

Typical results from field observations are shown in Figures 94 to 97. These data plus suggestions for the use of field observations are as follows:

1. *Pore Pressures*—Pore pressure response to loading is plotted in Figure 94, for the instruments under the middle of the embankment. The lines drawn through the experimental points show three different slopes for the instruments close to the clay surface. This pattern agrees with Leroueil et al. (1978) observations: an initial flatter slope is followed by a steeper one, due to clay foundation passing from an overconsolidated to a normally consolidated state. At failure, as local yield occurs close to the porous tips, a further increase in the rate of pore pressure buildup is observed.

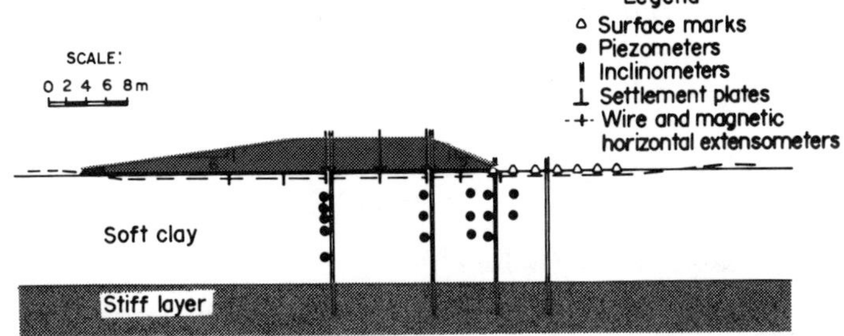

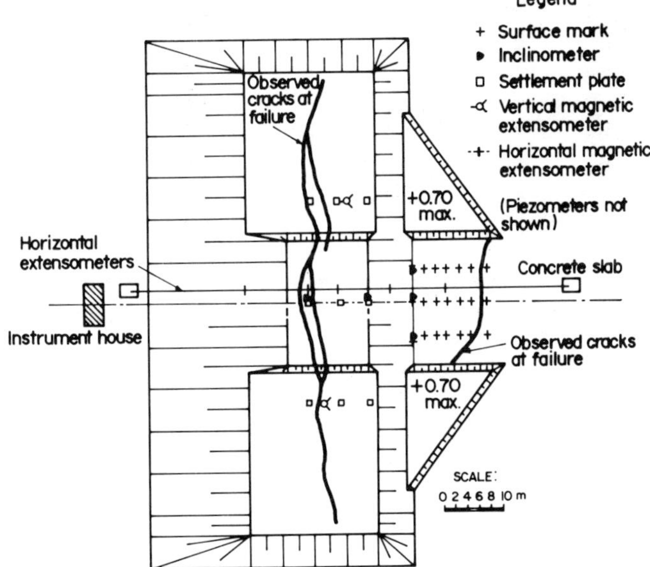

FIGURE 93. Instrumentation of the trial embankment I built up to failure on Rio de Janeiro clay (from Ortigão, Werneck and Lacerda, 1983).

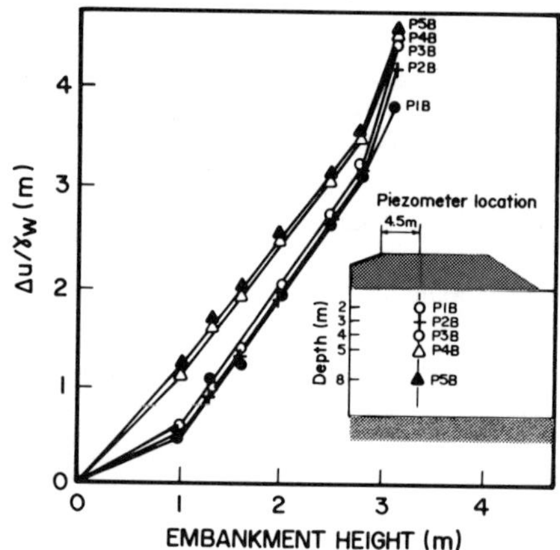

FIGURE 94. Construction pore pressures at an embankment on Rio de Janeiro clay (from Ortigão et al., 1983).

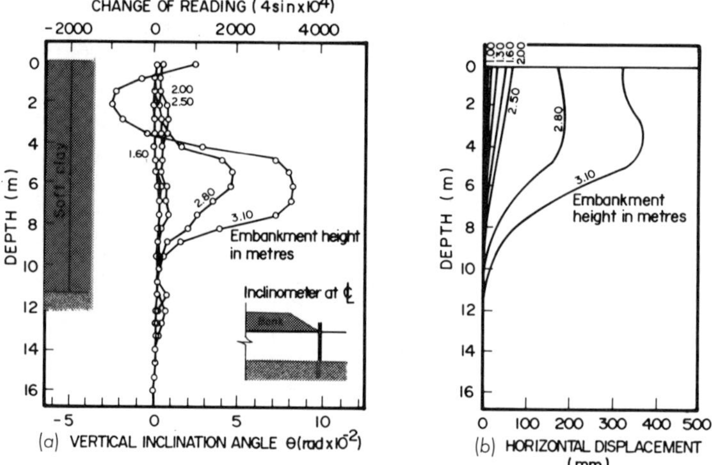

FIGURE 95. Inclinometer observations at an embankment on Rio de Janeiro clay (from Ortigão et al., 1983).

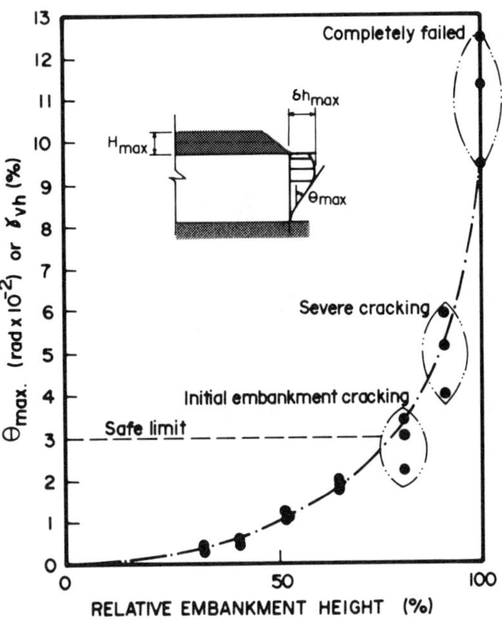

FIGURE 96. Stability control chart through inclinometer observations at the embankment toe (from Ortigão et al., 1983).

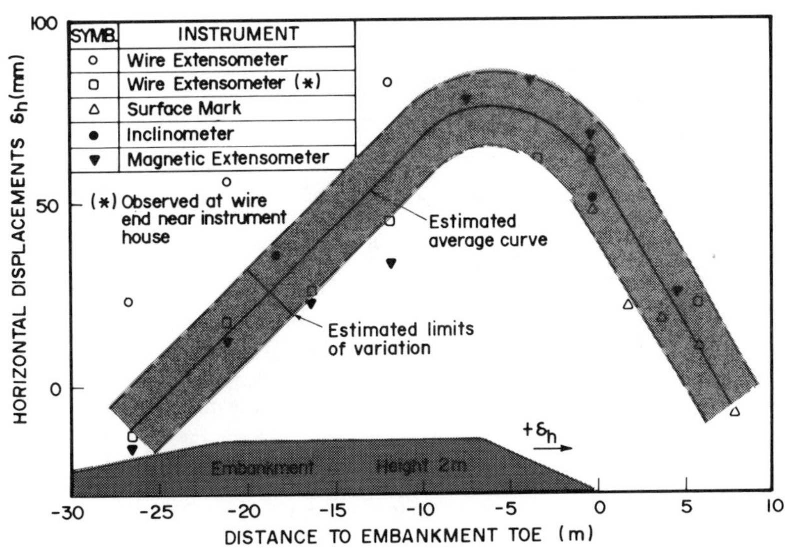

FIGURE 97. Horizontal displacements at the base of the embankment on Rio de Janeiro clay (from Ortigão et al., 1983).

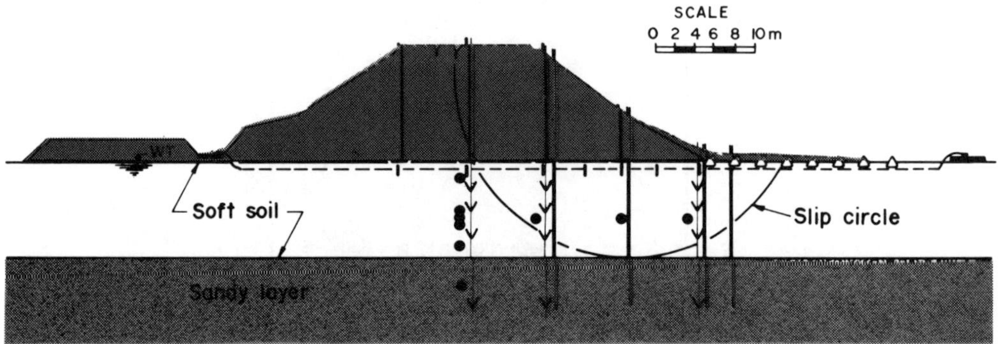

FIGURE 98. Instrumentation of Juturnaiba trial embankment (from Coutinho, 1985).

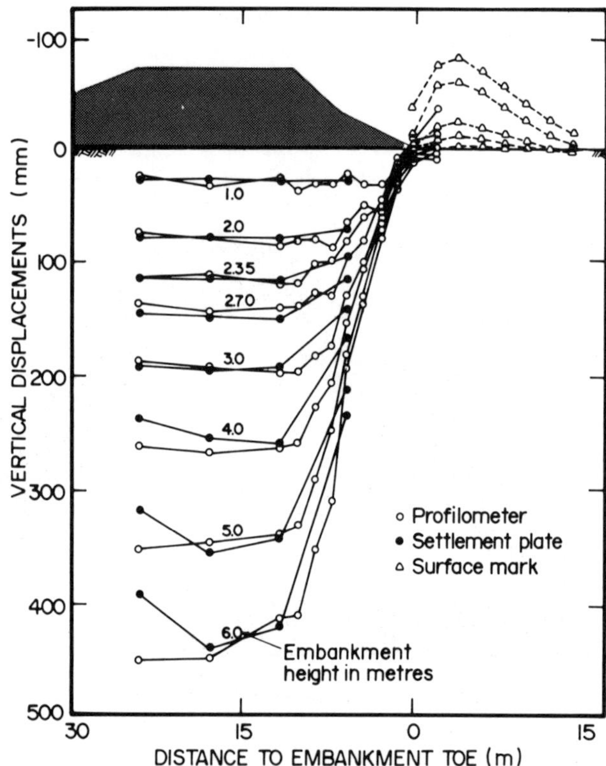

FIGURE 99. Observations of vertical displacements at the Juturnaiba trial embankment (from Palmeira and Ortigão, 1981).

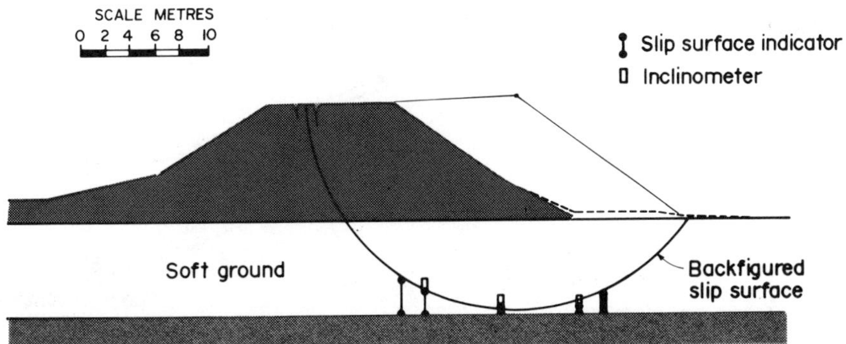

FIGURE 100. Backfigured shear slip surface, Juturnaiba trial embankment (from Coutinho, 1985).

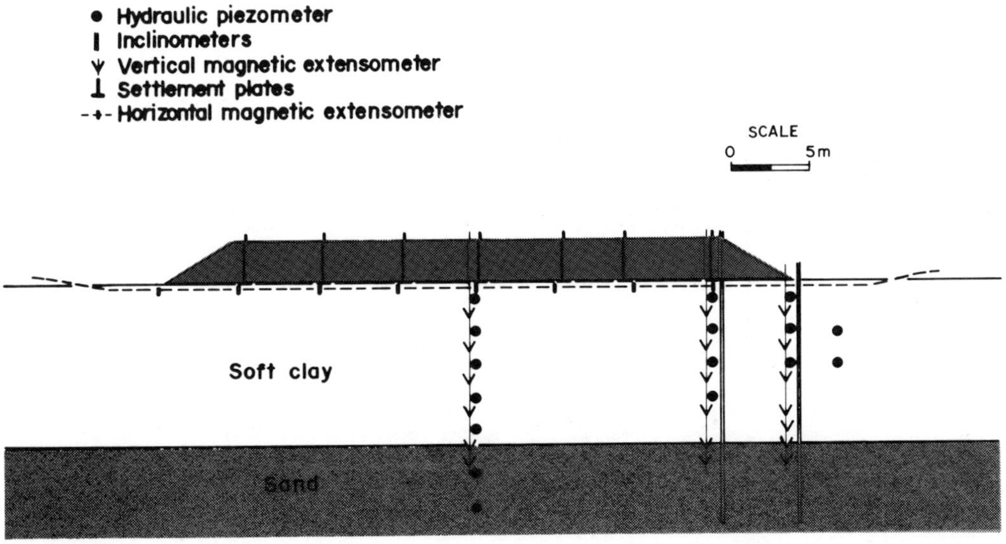

FIGURE 101. Instrumentation of the trial embankment II for consolidation observation of Rio de Janeiro clay (from Collet, 1985).

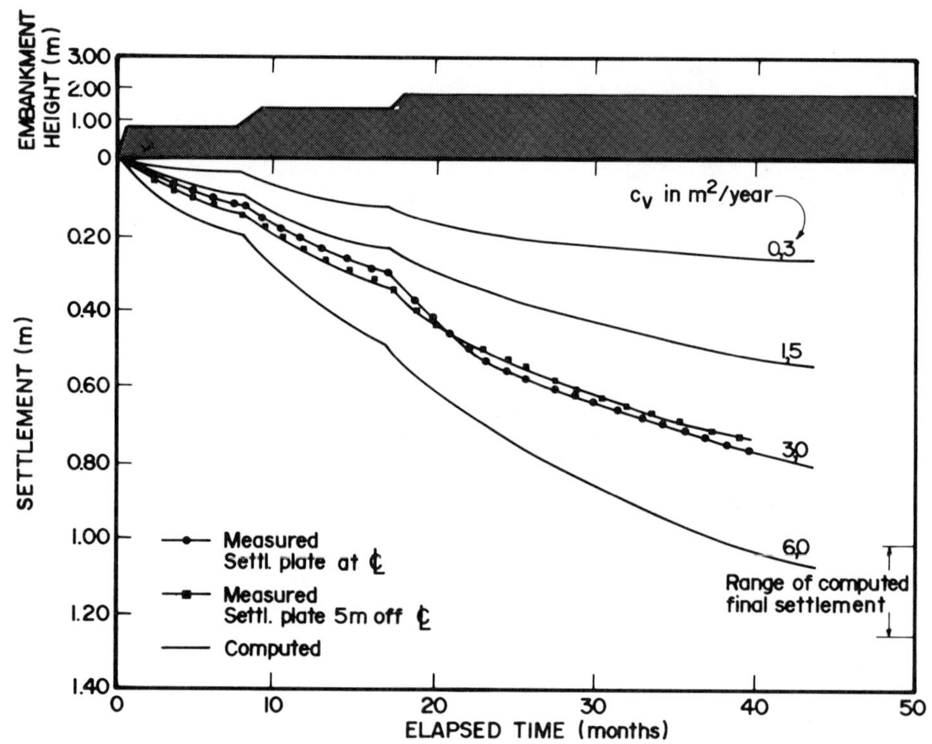

FIGURE 102. Measured and computed settlements, Trial Embankment II on Rio de Janeiro clay (section without drains) (data from Collet, 1985).

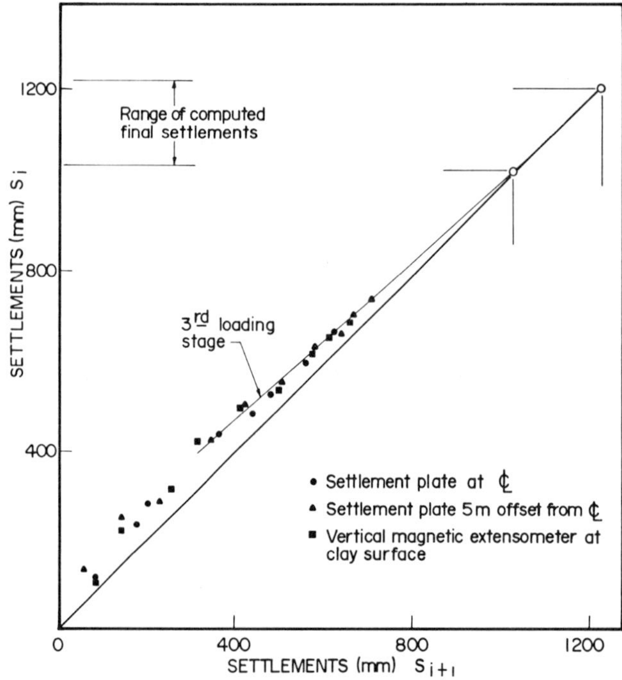

FIGURE 103. Application of Asaoka's (1978) method for final settlement computation (data from Collet, 1985).

Pore pressure observations can be used for stability control through an effective stress stability analysis. Examples of pore pressure use for this purpose can be found in references: Cole (1973), Elias and Storch (1970), Margason and Symons (1969) and Symons (1976). The use of measured pore pressures in stability analyses has been discussed in section 5.

2. *Inclinometer Observations*—Inclinometer observations at the embankment toe are shown in Figure 95. Figure 95(a) presents the variation of angle, i.e., the observed vertical inclination with depth; and the resulting displacements are shown in Figure 95(b). It is interesting to note the remarkable change of inclinations observed just before the embankment failed. This was subsequently used for suggesting a quick method for construction control through instrumentation. A plot of inclinometer observations against relative embankment height is shown in Figure 96. Maximum values of vertical inclination, rather than horizontal displacements, were employed, since these measurements can be directly related to shear strains. In fact, the shear strain, γ_{vh}, is a sum of the vertical inclination θ, plus the horizontal inclination, β, at the same point. The θ values under the embankment toe were directly measured by the inclinometer, and β values were not measured. However, at a point located near the horizontal part of a failure surface, β may be very small. Thus, the shear strain, γ_{vh}, could be considered to be equal to θ_{max}, i.e., the maximum observed vertical inclination measured at the embankment toe.

   Data in Figure 96 were obtained from three inclinometer casings. Correlating these observations with the observed embankment behaviour, a safe limit of γ_{vh} or θ_{max} equal to 3% could be defined. Beyond this safe value, severe embankment cracking was observed in the trial embankment.

3. *Horizontal displacements at the ground surface*—A typical set of observations of ground surface horizontal displacements are shown in Figure 97. Results from different instruments are plotted together, enabling an assessment of accuracy of field observations (Ortigão et al., 1983).

7.9.2 TRIAL EMBANKMENT AT JUTURNAIBA DAM SITE

The experience gained with the previously described trial embankment, enabled the same research group to conduct a more refined instrumentation programme at the Juturnaiba Dam site, in Brazil (Coutinho, 1985). Figure 98 shows a cross section of the fully instrumented trial embankment built up to failure with the purpose of yielding strength and deformation parameters for the design and construction of the main embankment dam on soft foundation. The improvements in the instrumentation adopted for this new site were:

(a) Concentration of all instruments in the main section, including those for deep settlements monitoring

(b) A profilometer was employed for continuous settlement monitoring at the base of the embankment. Results are shown in Figure 99 (Palmeira and Ortigao, 1981).

(c) Careful positioning of the inclinometer casings and slip surface indicators yielded very consistent data for the backfiguring of the slip surface, as shown in Figure 100.

7.9.3 TRIAL EMBANKMENT II FOR SETTLEMENT OBSERVATIONS ON RIO DE JANEIRO CLAY

In 1981–82 another trial embankment was constructed on Rio de Janeiro soft clay aiming at settlement observations and testing efficacy of vertical drains. Measurements are still (1985) being performed and partial results and analysis were reported by Collet (1985). A typical instrumented section is shown in Figure 101. The instruments installed under the centre and the base of the embankment are for monitoring foundation consolidation, whereas those installed close to the toe of the embankment have the additional purpose of monitoring the failure to be induced by the last loading stage, scheduled to take place in a few years.

Typical settlement observations, shown in Figure 102, are compared to computed settlements. It is concluded that the backfigured values of the coefficient of consolidation c_v are about close to those obtained from laboratory consolidation tests in the normally consolidated range.

Figure 103 shows the application of Asaoka's (1978) method, previously described, for treatment of consolidation settlement data. According to this method, final settlements are in the order of (900 ± 100) mm.

REFERENCES

Aas, G., "A Study of the Effect of Vane Shape and Rate of Strain on the Measured Values of in Situ Shear Strength of Clays," *Proc. 6th ICSMFE,* Montreal, *Vol. 1*, pp. 141–145 (1965).

Aas, G., "Vane Tests for Investigation of Anisotropy of Undrained Shear Strength of Clays," *Proc. Geot. Conf. on Shear Strength of Natural Soils and Rocks,* Oslo, *Vol. 1*, pp. 3-8 (1967).

Aboshi, H., H. Yoshikuni, and S. Maruyama, "Constant Loading Rate Consolidation Test," *Soils and Foundations, Vol. 10*(1), pp. 43–56 (1970).

Almeida, M. S. S., "Discussion on 'Embankments Failure on Clay Near Rio de Janiero'," *ASCE, Journal of Geotechnical Engineering, Vol. 111*(2), pp. 253–256 (February 1985).

Almeida, M. S. S., A. M. Britto, and R. H. G. Parry, "Numerical Analysis of Stage Constructed Embankments on Soft Clays," submitted to the *Canadian Geotechnical Journal*" (1985).

Almeida, M. S. S., M. C. R. Davies, and R. H. G. Parry, "Centrifuged Embankments on Strengthened and Unstrengthened Clay Foundations," *Geotechnique, Vol. 35*(4), p. 425–441 (December 1985).

Almeida, M. S. S. and J. A. R. Ortigao, "Performance and Finite Element Analyses of a Trial Embankment on Soft Clay," *International Symposium on Numerical Models in Geomechanics,* Zurich, pp. 548–558 (1982).

Almeida, M. S. S. and R. H. G. Parry, "Small Cone Penetrometer Test and Piezocone Tests in Laboratory Consolidated Clays," *Geotechnical Testing Journal, Vol. 8*, no. 1 (March 1985).

Andresen, A., "Exploration, Sampling and In Situ Testing of Soft Clay," chapter 3 of *Soft Clay Engineering*, edited by E. W. Brand and R. P. Brenner, Elsevier, Amsterdam, pp. 241–310 (1981).

Asaoka, A., "Observational Procedure of Settlement Prediction," *Soils and Foundations, Vol. 18*(4), pp. 87–101 (1978).

Asaoka, A. and M. Matsuo, "An Inverse Problem Approach to Settlement Prediction," *Soils and Foundations*, 20(4): 53–66 (1980).

ASCE, "Soil Improvement, History, Capabilities and Outlook," Report by the Committee on Placement and Improvement of Soils of the Geotechnical Engineering Division, ASCE, 182 pp. (February 1978).

ASTM-D2573, "Standard Method for Field Vane Shear Test in Cohesive Soil" (1972).

ASTM-D3441, "Standard Method for Deep Quasi-Static Cone and Friction-Cone Penetration Test of Soil" (1979).

Atkinson, J. H. and P. L. Bransby, *The Mechanics of Soils. An Introduction to Critical State Soil Mechanics*, McGraw-Hill, 375pp (1978).

Azzouz, A. M., M. M. Baligh, and C. C. Ladd, "Corrected Field Vane Strength for Embankment Design," ASCE *Journal of Geotechnical Engineering, Vol. 109*, no. 5, pp. 730–733 (May 1983).

B.G.S. (British Geotechnical Society), *Symposium on Field Instrumentation in Geotechnical Engineering*, Butterworths, London (1973).

Baguelin, F., J. F. Jezequel, E. Lemee, and A. Le Mehaute, "Expansion of Cylindrical Probes in Cohesive Soils," ASCE *Journal of Soil Mechanics and Foundation Division, Vol. 98*, SM 11, pp. 1129–1142 (November 1972).

Baguelin, F., J. F. Jezequel, and D. H. Shields, *The Pressuremeter and Foundation Engineering*, Transtech Publications, 617 p. (1978).

Balasubramanian, A. S. and R. P. Brenner, "Consolidation and Settlement of Soft Clay," in *Soft Clay Engineering*, Edited by E. W. Brand and R. P. Brenner, Elsevier, pp. 367–557 (1981).

Baligh, M. M. and J. N. Levadoux, "Pore Pressure Dissipation After Cone Penetration," Publication R80-11, MIT (1980).

Baligh, M. M., V. Vivatrat, and C. C. Ladd, "Cone Penetration in Soil Profiling," ASCE, *JGED, Vol. 106*(GT4), pp. 447–461 (1980).

Begemann, H. K. S. P., "The Friction Jacket Cone as an Aid in Determining the Soil Profile," *Proc. 6th ICSMFE*, Montreal, *Vol. 1*, pp. 17–20 (1965).

Biot, M. M., "General Theory of Three-dimensional Consolidation," *J. Appl. Phys.*, 12, pp. 155–164 (1941).

Bishop, A. W. and P. A. Green, "The Development and Use of Trial Embankments," *Proc. Symp. on Field Instrumentation in Geotechnical Engineering*, Butterworths (publishers), London, pp. 13–37 (1973).

Bjerrum, L., "Embankments on Soft Ground," State-of-the-art report, *ASCE Conf. on Performance of Earth Supported Structures, Lafayette, Vol. 2*, pp. 1–54 (1972).

Bjerrum, L., "Problems of Soil Mechanics and Construction on Soft Clays and Structurally Unstable Soils," *Proc. 8th ICSMFE, Vol. 3*, pp. 111–159, Moskow (1973).

Bourges, F. and C. Mieussens, "Deplacements Lateraux a Proximité des Remblais sur Sols Compressibles, Methode de Prevision," Bulletin des Laboratoire des Ponts et Chaussees, Paris, no. 101, pp. 73–100 (1979).

Broms, B. B., "Problems and Solutions to Construction in Soft Clay," *Proc. 6th Asian Regional Conf. on SM & FE, Vol. 2*, pp. 3–38 (1979).

BS 1377, "Methods of Test for Soils for Civil Engineering Purposes," British Standards Institution (1975).

Burghignoli, A., "An Experimental Study of the Structural Viscosity of Soft Clays by Means of Continuous Consolidation Tests," *VII ECSMFE Brighton, Vol. 2*, pp. 23–28 (1979).

Burland, J. B., "A Method of Estimating the Pore Pressures and Displacements Beneath Embankments on Soft, Natural Clay Deposits," *Proc. Roscoe Memorial Symp.*, Cambridge: G. T. Foulis, pp. 503–536 (1971).

Cadling, L. and S. Odenstad, "The Vane Borer: an Apparatus for Determining the Shear Strength of Soils Directly in the Ground," *Proc. Swedish Geotechnical Institute*, no. 2, 87pp (1950).

Campanella, R. G., P. K. Robertson, and D. Gillespie, "Cone Penetration Testing in Deltaic Soils," *Canadian Geotechnical Journal, Vol. 20*(1), pp. 23–35 (February 1983).

Carrier, W. D., "Consolidation Parameters Derived from Index Tests," *Geotechnique, Vol. 35*(2), pp. 211–213 (June 1985).

Carrier, W. D. and J. F. Beckman, "Correlations between Index Tests and the Properties of Remoulded Clays," *Geotechnique, Vol. 34*(2), pp. 211–228 (1984).

Ching, R. K. H. and D. G. Fredlung, "Some Difficulties Associated with the Limit Equilibrium Method of Slices," *Canadian Geotechnical Journal, Vol. 20*(4), pp. 661–672 (1983).

Clayton, C. R. I., N. E. Symons, and M. C. Matthews, "Site Investigation," Granada Publ., London, 424 pp. (1982).

Cole, K. W., Discussion, *Proc. Symp. on Field Instrumentation in Geotechnical Engineering*, Butterworths, London, pp. 623–632 (1973).

Collet, H. B., "Field Vane Tests in Clay Deposits Near Rio de Janeiro," MSc thesis, Federal University of Rio de Janeiro (1978).

Collet, H. B., "The Embankment II on Rio de Janeiro Clay to Assess the Efficacy of Vertical Drain Types," Progress Report, Brazilian Road Research Institute (in Portuguese), 104 pp. (1985).

Cortes, H. V. M., "Some Projects of Embankments on Soft Clays in Brazil," Lecture Notes, Brazilian Society of Soil Mechanics, Brasilia (November 1981).

Costa-Filho, L. M., D. Gercovich, L. A. Bressani, and J. E. Thomaz, "Discussion on Embankment Failure on Clay Near Rio de Janeiro," ASCE, *Jour. Geot. Eng. Div., Vol. 111*(2), pp. 259–262 (February 1985).

Costa-Filho, L. M., M. L. G. Werneck, and H. B. Collet, "The Undrained Shear Strength of a Very Soft Clay," *Proc. 9th ICSMFE*, Tokyo, *Vol. 1*, pp. 79–82 (1977).

Coutinho, R. Q., "Field Measurements at Juturnaiba Dam," DSc thesis, Federal University of Rio de Janeiro, Brazil (1985).

D'Appolonia, D. J., H. G. Poulos, and C. C. Ladd, "Initial Settlement of Structures on Clay," ASCE, *JSMFD,* SM10, pp. 1359-1357 (1971).

Desai, C. S. and J. F. Abel, "Introduction to the Finite Element Method," Van Nostrand Reinhold, New York (1972).

DiBiaggio, E. and F. Myrvoll, "Field Instrumentation for Soft Clay," in *Soft Clay Engineering,* E. Brand and Brenner, eds. Amsterdam:Elsevier (1981).

DIN 4096 "Vane Testing," German Standards (May 1980).

de Ruiter, J., "Current Penetrometer Practice," state-of-the-art Report, ASCE Symposium on Cone Penetration Testing and Experience, St. Louis (1981).

Douglas, B. J. and R. S. Olsen, "Soil Classification Using Electric Cone Penetrometer," ASCE Symposium on Cone Penetration Testing and Experience, St. Louis, pp. 209-227 (1981).

Duncan, J. M. and H. B. Seed, "Strength Variations Along Failure Surfaces in Clay," *Jour. Soil Mech. Found. Eng., Vol. 92,* SM6, pp. 81-104 (1966).

Duncan, J. M. and H. G. Poulos, "Modern Techniques of Analysis of Engineering Problems on Soft Clay," Chapter 5 of *Soft Clay Engineering,* edited by E. W. Brand and R. P. Brenner, Elsevier, Amsterdam (1981).

Dunn, C. S. and S. S. Razouki, "Two-Dimensional Consolidation under Embankments," *Jour. Inst. of Highway Eng., Vol. 21*(10), pp. 12-24 (1974).

Dunnicliff, C. J. "Equipment for Field Deformation Measurements," *Proc. 4th Pan Am Conf. on SMFE,* Puerto Rico, *Vol. 2,* pp. 319-332 (1971).

Elias, V. and H. Storch, "Control and Performance During Construction of a Highway Embankment on Weak Soil," Highway Research Record, no. 323, pp. 60-70 (1970).

Fredlung, D. G. and J. Krahn, "Comparison of Slope Stability Methods of Analysis," *Can. Geot. Jour.,* 14(3), pp. 429-439 (1977).

Gibson, R. E., K. Knight, and P. W. Taylor, "A Critical Experiment to Examine Theories of Three-dimensional Consolidation," *Proc. Europ. Conf. on Soil Mech. and Found. Eng.,* Wiesbaden, *Vol. 1,* pp. 69-76 (1963).

Giroud, J. P., *Tables pour le Calcul des Foundations, Vol. 2,* Ed. Dunod, Paris, pp. 434-449 (1973).

Gorman, C. T., T. C. Hopkins, R. C. Deen, and V. P. Dinevich, "Constant Rate of Strain and Controlled Gradient Consolidation Testing," ASTM *Geotech Testing J., vol. 1,* no. 1, pp. 3-15 (March 1978).

Gould, J. P. and C. J. Dunnicliff, "Accuracy of Field Deformation Measurements," *Proc. 4th Pan Am Conf. on SMFE,* Puerto Rico, *Vol. 1,* pp. 313-366 (1971).

Gunn, M. J. and A. M. Britto, *CRISP—User's and Programmers' Manual,* Cambridge University Engineering Department (1981).

Hanna, T., *Foundation Instrumentation,* Transtech Publications, 371 pp. (1973).

Henkel, D. J., "The Shear Strength of Saturated Remoulded Clays," ASCE, *Proc. Conf. on Shear Strength of Cohesive Soil,* Boulder, Colorado, pp. 533-544 (1960).

Head, K. H., "Continuous Consolidation Test," *Ground Engineering, Vol. 16*(3), pp. 24-25 (April 1983).

Herrero, O. R., "Universal Compression Index Equation," ASCE, *Journal of Geotechnical Engineering, Vol. 106*(GT11), pp. 1179-1200 (November 1980).

Hoare, S. D., "Consolidation with Flow Restrictor: An Improved Laboratory Test," University of Oxford Department of Engineering Science, Report no. SM014/ELE/80 (October 1980).

Hoeg, K., O. B. Andersland, and E. N. Rolfsen, "Undrained Behaviour of Quick Clay Under Load Test at Asrum," *Geotechnique,* 19(1), pp. 101-115 (1969).

Hvorslev, J. M., "Subsurface Exploration and Sampling of Soils for Civil Engineering Purposes," Research Report, Waterways Experimental Station, Reprinted by Engineering Foundation, 521 pp. (1949) (1965).

Hvorslev, M. J., "Time Lag and Soil Permeability in Groundwater Observations," Bulletin no. 36, Waterways Experimental Station, Corps of Engineers, U. S. Army, 50 pp. (1951).

ISSMFE, "International Society for Soil Mechanics and Foundation Engineering," Report to the Subcommittee on Standardization of Penetration Testing in Europe, *Proceedings 9th International Conference on Soil Mechanics and Foundation Engineering,* Tokyo, *Vol. 3,* Appendix 5, pp. 95-152 (1977).

Jamiolkowski, M., C. C. Ladd, J. T. Germaine, and R. Lancellotta, "New Developments in Field and Laboratory Testing of Soils," State-of-the Art, XI International Conf. on Soil Mech. and Found. Eng., San Francisco (August 1985).

Janbu, N., "Earth Pressure and Bearing Capacity by Generalized Procedure of Slices," *Proc. 4th Int. Conf. on Soil Mech. and Found. Engineering,* London, *Vol. 2,* pp. 207-212 (1957).

Janbu, N., O. Tokhein, and K. Senneset, "Consolidation Tests with Continuous Loading," X ICSMFE Stockholm, *Vol. 1,* pp. 645-654 (1981).

Kirkpatrick, W. M. and A. J. Khan, "The Reaction of Clays to Sampling Stress Relief," *Geotechnique,* no. 1 (1984).

Kovari, K. (editor), *Proc. Symp. on Field Instrumentation in Geomechanics,* Zurich, 2 Vols., 1453 pp. (1983).

Kovari, K. and Ch. Amstad, "Fundamentals of Deformation Measurements," *Proc. Symp. on Field Measurements in Geomechanics,* edited by K. Kovari, Zurich: A. Balkema, pp. 219-239 (1983).

Lacasse, S. M., C. C. Ladd, and M. M. Baligh, "Evaluation of Field Vane, Dutch Cone Penetrometer and Piezometer Testing Devices," Res. Report, Dept. of Civil Eng., MIT (1980).

Ladd, C. C., "Estimating Settlements of Structures Supported on Cohesive Soils," MIT Soils Publication, no. 272, 99 pp. (1973).

Ladd, C. C. and R. Foot, "New Design Procedure for Stability of Soft Clays," ASCE, *JGED, Vol. 100*(7), pp. 763-786.

Ladd, C. C., R. Foot, K. Ishihara, F. Schlosser, and H. G. Poulos, "Stress-Deformation and Strength Characteristics," *Proc. 9th ICSMFE,* Tokyo, *Vol. 2,* pp. 421-494 (1977).

Lambe, T. W., "Methods of Estimating Settlement," *Jour. Soil Mech. Found. Div.,* ASCE, 90(SM5), pp. 43-67 (1964).

Lambe, T. W. and R. V. Whitman, *Soil Mechanics,* John Wiley & Sons, New York, 553 pp. (1969).

Lemasson, H., "Ensemble Carottier a Piston Stationnaire, Scissometre," Remblais sur Sols Compressibles, Special T, Laboratoire de Ponts et Chaussees, pp. 276–281 (1973).

Leroueil, S., F. Tavenas, P. Rochelle, and M. Roy, "Behaviour of Destructured Natural Clays," ASCE, *JGED, Vol. 105*(GT6), pp. 759–778 (1979).

Leroueil, S., F. Tavenas, C. Mieussens, and M. Peignaud, "Construction Pore Pressures in Clay Foundations Under Embankments. Part II: Generalized Behaviour," *Canadian Geotechnical Journal,* 15, pp. 66–82 (1978).

Levadoux, Y. N. and M. M. Baligh, "Pore Pressure During Cone Penetration in Clays," MIT Research Report, Report no. R80-15, Cambridge, Mass. (1980).

Lewis, W. A., R. T. Murray, and I. F. Symons, "Settlement and Stability of Embankments Constructed on Soft Alluvial Soils," *Proc. of the Institution of Civil Engineers,* Part 2, *Vol. 59,* pp. 571–593 (December 1975).

Lowe, J., III, E. Jones, and V. Obrician, "Controlled Gradient Consolidation Test," ASCE, *Jnl SMF div., Vol. 95,* SM!, pp. 77–97 (1981).

Lowe, J., III and P. F. Zacchev, "Subsurface Explorations and Sampling," chapter one, *Foundation Engineering Handbook,* edited by H. F. Winterborn and H. Y. Fang, Van Nostrand Reinhold, New York, pp. 1–65 (1975).

LPC, "Essay au Scissometre de Chantier LPC," Mode Operatoire, Laboratoires de Ponts et Chaussees, Paris (1980).

Magnan, J. P. *Théorie et Pratique de Drains Verticaux.* Paris:Technique et Documentation Lavoisier, 335 p. (1983).

Magnan, J. P. and J. M. Deroy, "Analyse Graphique des Tassements Observés sous les Ouvrages," Bull. Liais. Lab. Ponts Chauss., No. 109, pp. 45–52 (1980).

Magnan, J. P. and C. Mieussens, "Les Remblais d'essai: Un Outil Efficace Pour Ameliorer les Projets d'ouvrages sur Sols Compressibles," Bulletin des Liaison des Laboratoires des Ponts et Chaussees, Paris, ref 2445, pp. 79–96 (March–April 1980).

Magnan, J. P. and C. Mieussens, "Remblais sur Sols Compressibles – Instrumentation et Exploitation des Mesures," Bulletin de Liaison des Laboratoires de Ponts et Chaussees, Paris, ref 2608, pp. 61–70 (September-October 1981).

Margason, G. and I. F. Symons, "Use of Pore Pressure Measurements to Control Embankment Construction," *Proc. 7th ICSMFE,* Mexico, *Vol. 2,* pp. 307–315 (1969).

Mayne, P. W., "Cam-clay Predictions of Undrained Strength," ASCE, *Journal of Geotechnical Engineering, Vol. 106*(GT11), 1219–1242 (November 1980)

Mayne, P. W. and F. H. Kulhawy, "K_0-OCR Relationship in Soils," ASCE, *Jour. Geot. Engineering, Vol. 108*(GT6), pp. 851–872 (June 1982).

Mesri, G., "Discussion on New Design Procedure for Stability of Soft Clays," ASCE, *JGED,* 101(GT4), pp. 409–412 (1975).

Mesri, G. and Y. K. Choi, "Settlement Analysis of Embankments on Soft Clays," ASCE, *Journal of Geotechnical Engineering,* 111(4), pp. 441–464 (1985).

Milovic, D. M., "Effect of Sampling on Some Soil Characteristics," *Sampling of Soil and Rock,* ASTM STP 483, American Society of Testing and Materials, pp. 164–179 (1971).

Mitchell, J. K. and R. K. Katti, "Soil Improvement State-of-the-art Report," *Proc. of the 10th ICSMFE,* Stockholm, *Vol. 4* (1981).

Morgenstern, N. R. and V. E. Price, "The Analysis of the Stability of General Slip Surfaces," *Geotechnique,* 15(1), pp. 70–93 (1965).

Murray, R. T., "Two and Three-Dimensional Consolidation Theory," Chapter 4 of *Development of Soil Mechanics,* Ed. C. R. Scott, pp. 103–147 (1978).

OCDE, "Constrution des Routes sur Sols Compressibles," Rapport du Groupe de Recherche Routiere CM1, Organization de Cooperation et de Developpement Economique, Paris, 160 pp. (1979).

Olson, R. E., "Consolidation under Time Dependent Loading," ASCE, *JGED, Vol. 103*(GT1), pp. 55–60 (1977).

Olson, R. E. and C. C. Ladd, "One-Dimensional Consolidation Problems," ASCE, *JGED, Vol. 105*(GT1), pp. 11–30.

Ortigão, J. A. R. and E. M. Palmeira, "Geotextile Performance at an Access Road on Soft Ground Near Rio de Janeiro," *Proc. 2nd Int. Conf. on Geotextiles,* Las Vegas, *Vol. 1,* pp. 353–358 (1982).

Ortigão, J. A. R., M. L. G. Werneck, and W. A. Lacerda, "Embankment Failure on Clay Near Rio de Janeiro," ASCE, *Journal of Geotechnical Engineering, Vol. 109,* II, pp. 1460–1479 (1983).

Ortigão, J. A. R., M. L. G. Werneck, and W. A. Lacerda, closure to "Embankment Failure on Clay Near Rio de Janeiro," ASCE *Journal of Geotechnical Engineering, Vol. 3,* no. 2, pp. 262–264 (February 1985).

Ortigão, J. A. R., W. A. Lacerda, and M. L. G. Werneck, "The Behaviour of the Instrumentation of an Embankment on Clay," *Int. Symp. on Field Measurements in Geomechanics,* Zurich, *Vol. 1,* pp. 703–718 (1983).

Palmeira, E. M. and J. A. R. Ortigao, "Construction and Performance of a Full Profile Settlement Gauge: Profilometer for Embankments," *Solos e Rochas, Vol. 4*(2) (August 1981).

Parry, R. H. G. and J. H. McLeod, "Investigation of Slip Failure in Flood Levee at Lauceston, Australia," 5th Australian-New Zealand Conf. Soil Mech. and Found. Eng., Aukland (1967).

Parry, R. H. G., "Stability of Low Embankments on Soft Clays," *Proc. Roscoe Mem. Symp. Stress Strain Behaviour of Soils,* Foulis Pub., pp. 643–668 (1971).

Parry, R. H. G. and C. P. Wroth, "Shear Properties of Soft Clays," Report presented at Symp. on Soft Clay, Bangkok, Thailand (1977).

Peck, R. B., "The Observational Method in Applied Soil Mechanics," Rankine Lecture, *Geotechnique, Vol. 19*(2) (June 1969).

Penman, A. D. M., "Instrumentation for Embankments Subjected to Rapid Drawdown," Building Research Station, Current Paper, CP no. 1/72, 21 pp. (1972).

Perloff, W. H., "Pressure Distribution and Settlement," in *Foundation Engineering Handbook,* Eds. H. F. Winterkorn and H. Y. Fang, Van Nostrand Reinhold, New York, pp. 148–196.

Pilot, G. and M. Moreau, *La Stabilité des Remblais sur Sols Mous, Abaques de Calcul,* Eyrolles, Paris, 151 pp. (1973).

Pilot, G., "Study of Five Embankment Failures on Soft Soils," *Proc. ASCE Specialty Conference on Earth and Earth-Supported Structures,* Purdue Univ., *Vol. 1.1,* pp. 81–99 (1972).

Pilot, G., "Methods of Improving the Engineering Properties of Soft Clay," chapter 9 of *Soft Clay Engineering,* edited by E. W. Brand and R. P. Breenner, Elsevier Publishing Co., Amsterdam, pp. 633–696 (1981).

Pilot, G., B. Track, and P. La Rochelle, "Effective Stress Analysis of The Stability of Embankments on Soft Soils," *Canadian Geotechnical Journal, Vol. 19,* no. 4, pp. 433–450 (November 1982).

Pinto, C. S., "Discussion on Slope Stability," (in Portuguese), *Proc. of the V Brazilian Conf. on Soil Mechanics,* Sao Paulo, *Vol. 4,* pp. 342–356 (1974).

Poulos, H. G., "Difficulties in Prediction of Horizontal Deformations of Foundations," ASCE, *JSMFD,* 98 (SMB), pp. 843–848.

Poulos, H. G. and E. H. Davis, *Elastic Solution for Soil and Rock Mechanics,* John Wiley and Sons, New York, 411 pp. (1974).

Robertson, P. K. and R. G. Campanella, "Interpretation of Cone Tests. Part I: Sand," *Canadian Geotechnical Journal, Vol. 20*(4), pp. 718–733 (1983a).

Robertson, P. K. and R. G. Campanella, "Interpretation of Cone Penetration Tests. Part II: Clay," *Canadian Geotechnical Journal, Vol. 20,* pp. 734–745 (1983b).

Rocha-Filho, P., "Influence of Excess Pore Pressure on Cone Measurements," *Proc. 2nd EPSOT: European Symp. on Penetration Testing,* Amsterdam (1982).

Rocha-Filho, P. and J. A. Alencar, "Piezocone Tests in the Rio de Janeiro Soft Clay Deposit," *Proc. 11th ICSMFE,* San Francisco (1985).

Roscoe, K. H. and J. B. Burland, "On the Generalized Behaviour of 'Wet' Clay," *Engineering Plasticity,* Eds. J. Heyman and F. Leckie, Cambridge University Press, pp. 535–609 (1968).

Rowe, P. W., "The Relevance of Soil Fabric to Site Investigation Practice," *Geotechnique, Vol. 22,* pp. 193–300.

Rowe, P. W., "Representative Sampling in Location, Quality and Size," *Sampling of Soil and Rock,* ASTM STP 483, American Society for Testing and Materials, pp. 77–108 (1971).

Roy, M., M. Temblay, F. Tavenas, and P. La Rochelle, "Development of Pore Pressures in Static Penetration Tests in Sensitive Clay," *Canadian Geotechnical Journal, Vol. 19*(2), pp. 124–138 (1982).

Rozsa, L. and L. Vidacs, "New and Easy Method for Measuring Settlement of Embankments," *Proc. Symp. on Field Measurements in Geomechanics,* Zurich: Balkema, *Vol. 1,* pp. 765–772 (1983).

Sallfors, G., "Preconsolidation Pressure of Soft, High-plastic Clays," Geotechnical Department, Chalmers Univ. of Technology, Gothenburg, 231 pp. (1975).

Sanglerat, G., *The Penetrometer and Soil Exploration,* Elsevier Publishing Co., Amsterdam, 464 pp. (1972).

Schofield, A. N. and C. P. Wroth, *Critical State Soil Mechanics,* McGraw-Hill, London (1968).

Scholfield, A. N., "Cambridge Geotechnical Centrifuge Operations," *Geotechnique, Vol. 30*(3), pp. 227–268 (1980).

Senneset, K., N. Janbu, and G. Svano, "Strength and Deformation Parameters from Cone Penetration Tests," *Proc. 2nd ESOPT, European Symp. on Penetration Testing,* Amsterdam, pp. 863–870 (1982).

Senneset, K. and N. Janbu, "Shear Strength Parameters Obtained from Static Cone Penetration Tests," ASTM Symposium San Diego (1984).

Shaap, L. H. J. and H. M. Zuidberg, "Mechanical and Electrical Aspects of the Electrical Cone Penetration Tip," *Proc. 2nd ESOPT: European Symp. on Penetration Testing,* Amsterdam, *Vol. 2,* pp. 841–851.

Schiffman, R. L., V. Pane, and R. E. Gibson, "The Theory of One-Dimensional Consolidation of Saturated Clays. An Overview of Nonlinear Finite Strain Sedimentation and Consolidation," *Proc. of Symp. on Sedimentation Consolidation Models,* San Francisco, pp. 1–29 (October 1984).

Simon, R. M., C. C. Ladd, and J. T. Christian, "Finite Element Program FEECON for Undrained Deformation Analyses of Granular Embankments on Soft Clay Foundations," MIT Research Report 72-9, Soils Publication, no. 294, 99 pp. (1972).

Skempton, A. W., "The Pore Pressure Coefficients A and B," *Geotechnique,* 4(4), pp. 143–147 (1954).

Smith, R. E. and H. E. Wahls, "Consolidation Under Constant Rates of Strain," ASCE, *Jour. of Soil Mech. and Found. Eng., Vol. 95*(SM2), pp. 519–539 (1969).

Soares, M. M., "The Instrumentation of a Diaphragm Wall for the Excavation for the Rio de Janeiro Underground," *Int. Symp. on Field Measurements in Geomechanics,* Zurich, *Vol. 1,* pp. 553–564 (1983).

Spencer, E., "A Method of Analysis of the Stability of Embankments Assuming Parallel Interslice Forces," *Geotechnique,* 17(1), pp. 11–26 (1967).

Symons, I. F., "Assessment and Control of Stability of Road Embankments Constructed on Soft Subsoils," Transport and Road Research Laboratory, Report no. 711 (1976).

Tavenas, F., R. Blanchet, R. Garneau, and S. Leroueil, "The Stability of Stage-Constructed Embankments on Soft Clays," *Canadian Geotechnical Journal, Vol. 15,* pp. 283–305 (1978).

Tavenas, F., M. Brucy, J. P. Magnan, P. LaRochelle, and M. Roy, "Analyse Critique de la Théorie de Consolidation Unidimensionnelle de Terzaghi," *Rev. Française de Géotechnique,* No. 7, pp. 29–43 (1979).

Tavenas, F., C. Mieussens, and F. Bourges, "Lateral Displacements in Clay Foundations under Embankments," *Canadian Geotechnical Journal, Vol. 16*(3), pp. 532–550 (1979).

Tavenas, F. and S. Leroueil, "The Behaviour of Embankments on Clay Foundations," *Can. Geot. Jour.,* 17, pp. 236–260 (1980).

Terzaghi, K. and H. B. Pech, *Soil Mechanics and Engineering Practice,* 2nd edition, John Wiley & Sons, 729 pp. (1967).

Torstensson, B. A., "The Pore Pressure Probe," Nordiske Geoteckniske Mote, Oslo, paper no. 34.1–34.15 (referenced by Robertson and Campanella, 1983b) (1977).

Trak, B., P. La Rochelle, F. Tavenas, S. Leroueil, and M. Roy, "A New Approach to the Stability Analysis of Embankments on Sensitive Clays," *Canadian Geotechnical Journal,* 17, pp. 526–544 (1980).

UFRJ, "Brazilian Symposium on Field Instrumentation in Soil Mechanics and Foundation Engineering," Federal University of Rio de Janeiro (UFRJ), Rio de Janeiro, Brazil (1975).

Umehara, Y. and K. Zen, "Constant Rate of Strain Consolidation for Very Soft Clayey Soils," *Soils and Foundations, Vol. 20,* no. 2 (June 1980).

Vaughan, P. R., "The Measurement of Pore Pressures with Piezometers," *Proc. Symp. on Field Instrumentation,* in *Geotechnical Engineering,* Butterworths, London, pp. 411–422 (1973).

Vermeiden, J., "Improved Sounding Apparatus, as Developed in Holland Since 1936," *Proc. 2nd ISCMFE,* Rotterdam, *Vol. 1*, pp. 280–287 (1948).

Whitman, R. V. and W. A. Bailey, "Use of Computers for Slope Stability Analysis," ASCE, *JSMFD,* 93(4), pp. 475–498 (1967).

Wiesel, C. E., "Some Factors Influencing Vane Test Results," *Proc. 8th ICSMFE,* Moscow, *Vol. 1.2*, pp. 475–479 (1973).

Wizza, A., J. Christian, E. Davies, and S. Heiberg, "Consolidation at Constant Rate of Strain," *ASCE Jnl. SMF Div., Vol. 97* (1971).

Wroth, C. P. and J. M. O. Hughes, "An Instrument for the In Situ Measurement of the Properties of Soft Clays," *Proc. 8th ICSMFE,* Moscow, *Vol. 1*, pp. 487–494 (1973).

Wroth, C. P., "The Interpretation of In Situ Soil Tests," *Geotechnique,* Vol. 34(4), pp. 449–489 (1984).

Zienkiewicz, O. C., *The Finite Element Method,* McGraw-Hill (1977).

CHAPTER 9

Stability of Axisymmetric Excavation in Clay

A. M. BRITTO* AND O. KUSAKABE**

ABSTRACT

The stability of an axisymmetric excavation has been given little attention in the past, even though axisymmetric excavations are common in the construction industry. Analytical upper bound solutions (using the limit analysis theorems) for the failure of unsupported and supported excavations are presented for conditions where the soil is treated as a rigid plastic material satisfying the Tresca yield criterion. For the case of the unsupported excavation the soil was considered to possess (1) the uniform undrained shear strength and (2) undrained shear strength increasing linearly with depth. A number of wall failure mechanisms and base failure mechanisms were considered for the unsupported excavations. Comparison of the upper bound solutions is made with centrifuge tests and finite element analyses. The critical wall failure mechanism is well supported by centrifuge test results. Based on this mechanism stability charts are proposed for both supported and unsupported axisymmetric excavations.

INTRODUCTION

The stability of cut slopes is a problem which has been studied extensively. Following the work of Coulomb [9], who considered the stability of an unsupported vertical face of a soil, a variety of ground and loading conditions have been considered by various research workers (Taylor [37]; Janbu [16]; Spencer [36]).

If the soil is assumed to behave as a Tresca material, with a uniform undrained shear strength C_u, the critical height of an unsupported vertical cut D_{cr} under conditions of plane strain is bounded by the solutions

$$\frac{3.63C_u}{\gamma} < D_{cr} < \frac{3.83C_u}{\gamma}$$

where γ is the bulk unit weight of the soil. These upper and lower bounds are due, respectively, to Chen [7] and Pastor [23].

A more difficult problem, which has been given little attention, is the stability of an axisymmetric (cylindrical) excavation. Axisymmetric excavations are quite common in the construction industry; man holes; inspection or access chambers, service entrances, and excavations for bored piles or piers are mainly axisymmetric. The understanding of the mechanics of the behavior of axisymmetric excavations should also throw light on the complex three-dimensional behavior of trench headings and open-pit excavations. The advancing face of a rectangular trench could be approximated as semicircular in plan, or in a further idealization—conditions of axisymmetry could be assumed. Rectangular or square open-pit excavations may be idealized as axisymmetric excavations as a first approximation.

This chapter deals with the stability of axisymmetric excavation in clay and is divided into two parts: unsupported excavations and supported excavations. The method adopted in the study of this problem is a combination of upper bound calculations and the centrifuge modelling technique. From the viewpoint of the theory of plasticity, both upper and lower bound solutions are essential, particularly in the case of a new upper bound calculation which could be unsafe. The idea of the limit theorem is to bracket the exact solution between upper and lower bound solutions within the framework of rigid plasticity. It is often the case that a good lower bound is difficult to obtain, and then it is hard to evaluate how close is an upper bound solution to the exact solution. There exist, however, an alternative approach which the engineers can rely on; that is a combination of upper bound solutions and careful laboratory experiments. Well controlled centrifuge model tests can be regarded as independent

*University Engineering Department, Cambridge University, Cambridge, England

**Department of Civil Engineering, Utsunomiya University, Utsunomiya, 1 Shii, Japan

physical events, and the results obtained from them can stand as the "exact solution" by which the degree of correctness of upper bound solutions can be evaluated.

AXISYMMETRIC MECHANISM

In this section only an immediate collapse (undrained failure) is considered. Under undrained conditions the soil may be treated as a $\phi_u = 0$ material, obeying the Tresca yield criterion. The undrained condition means that there is no volume change during plastic flow—incompressibility condition. The equation of the incompressibility condition for axisymmetric velocity fields can be written as

$$\frac{\partial U}{\partial r} + \frac{U}{r} + \frac{\partial V}{\partial z} = 0 \qquad (1)$$

where

U is the velocity component in the radial direction
V is the velocity component in the vertical direction

Any velocity in the radial direction causes a circumferential strain rate in the axisymmetric deformation, so that in general the soil cannot remain rigid during plastic flow and only the vertical component can be constant. In its simplest form of a solution of the Equation (1), the vertical velocity is constant. Thus the Equation (1) reduces to

$$\frac{\partial U}{\partial r} + \frac{U}{r} = 0 \qquad (2)$$

Therefore in the axisymmetric deformation the possible sets of simple admissible velocity fields may be either

$$U = 0, \quad V = const \qquad \text{(region A)}$$

or

$$U = \frac{c}{r}, \quad V = const \qquad \text{(region B)}$$

The region A behaves as a rigid block with no radial velocity and its movements are always parallel to the vertical z axis. The region B is referred to as a shearing zone.

UPPER BOUND CALCULATION

The work done and the energy dissipated are calculated according to the following equation for the Tresca material,

$$\int_S pdS + \int_V \gamma V dv = 2\int_V C_u|\varepsilon|_{max} dv + \int_S C_u(\Delta V)dS \qquad (3)$$

where

S denotes surface integral
v denotes volume integral
p is surface pressure
γ is unit weight of soil
$|\varepsilon|_{max}$ is the largest principal plastic strain rate

The first term on the left hand side is the work done by the external loads, the second term is the work done due to the self weight of the soil under a given velocity field. The first term on the right hand side is the energy dissipated due to shearing within shear zones and the second term is the energy dissipated due to the velocity jump along line of discontinuity. Since we have adopted the simple velocity fields each term of the Equation (3) can be integrated analytically and no numerical integration is required.

In the problems of stability the Equation (3) is often rearranged to the expression of stability number as

$$N = \frac{\gamma D}{C_u} \qquad (4)$$

where D is the representative length for a given problem.

UNSUPPORTED EXCAVATIONS

Critical Failure Mode for C_u = Constant Case

Figure 1 shows the idealization of the problem where the excavation is of depth, D, and radius, r_0. The soil is assumed to be saturated and to have uniform undrained shear strength, C_u, and to obey the Tresca yield criterion. There appears to be four ways of obtaining a stability solution for this problem: (1) Using the bearing capacity formula; (2) the limit equilibrium method; (3) finite element analyses; and (4) bound theorems of plasticity (limit analysis). The authors (Britto and Kusakabe [5]) critically examined the possible failure mechanisms of a vertical unsupported axisymmetric excavation. Figures 2 and 3 show possible failure mechanisms for wall failure and base failure respectively. These mechanisms consist of rigid blocks and shearing zones; some are based on the mechanisms used by Kobayashi [17] who analyzed the forming process under a circular punch.

In order to examine the critical failure mode the variations of the stability number [$N = (\gamma D/C_u)$] with aspect ratio D/r_0 for all mechanisms given in Figures 2 and 3 are presented in Figure 4.

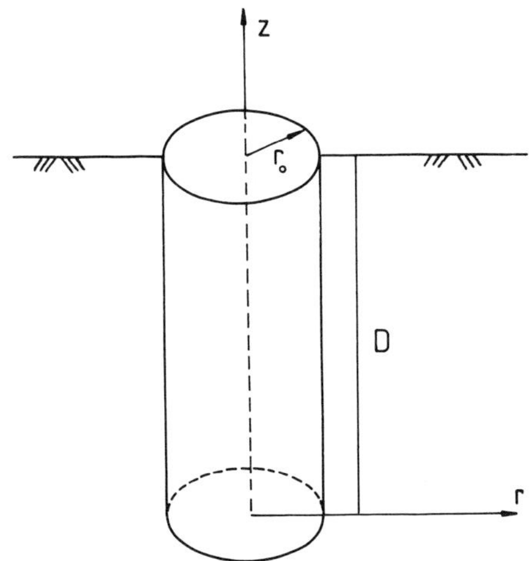

FIGURE 1. Axisymmetric excavation.

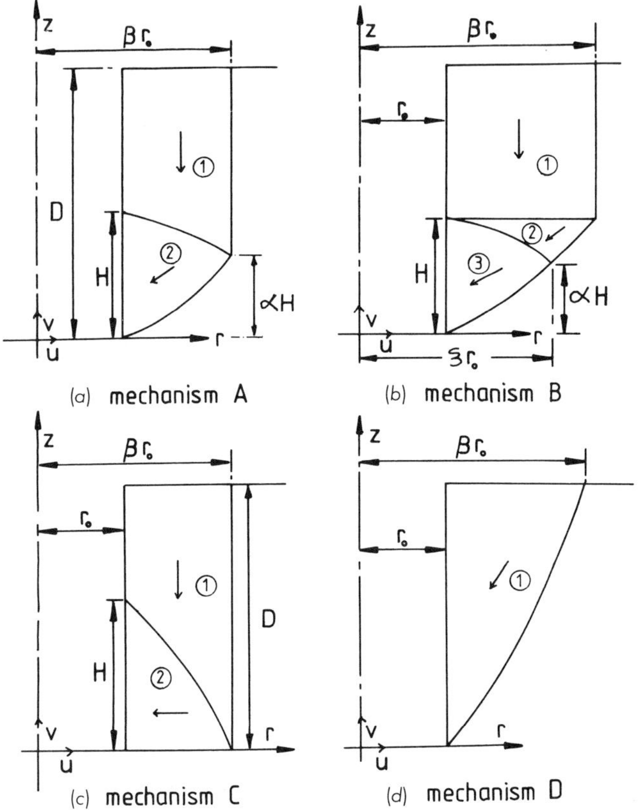

FIGURE 2. Wall failure mechanisms.

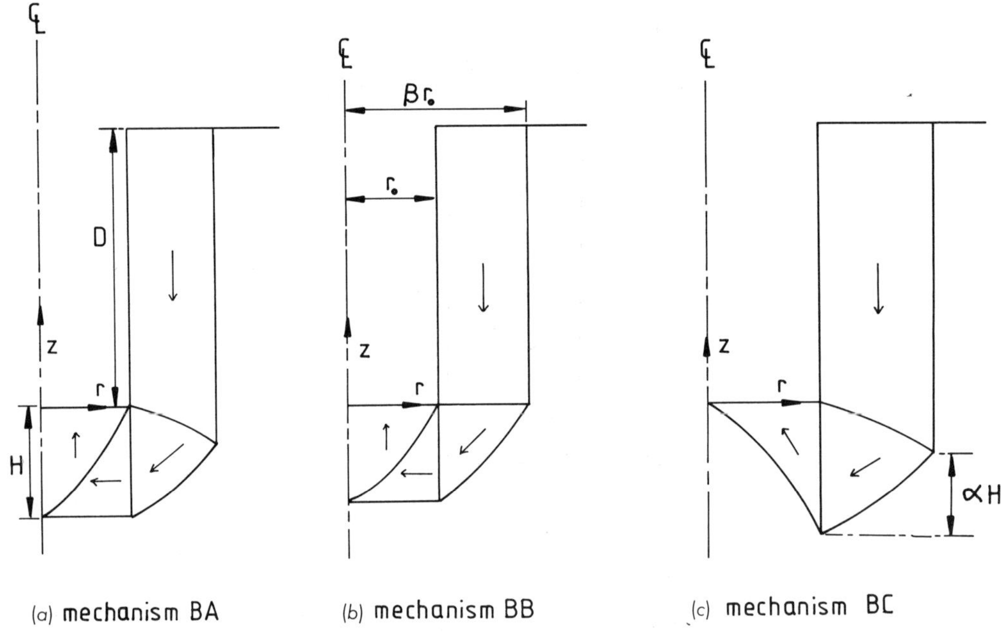

FIGURE 3. Base failure mechanisms.

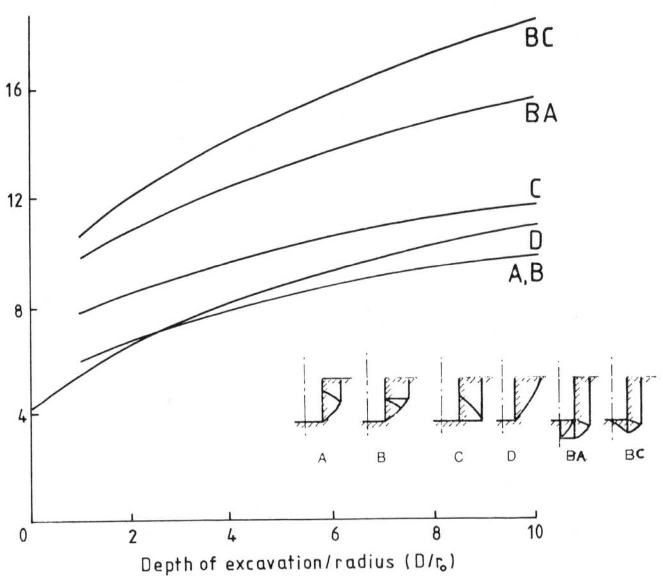

FIGURE 4. Variation of stability number N with D/r_o ratio.

The upper bound theorem states that if a work calculation is performed for a kinematically admissible failure mechanism, the loads thus deduced will be higher than or equal to those for failure. Therefore the assumed failure mechanism which yields the smallest value of N must be selected as the critical mode of failure. As is seen in the figure mechanism D gives the smallest values of N for $D/r_0 < 2.5$ and mechanisms A and B yield the smallest values of N for $D/r_0 \geq 2.5$.

From these results, it can be said that the most critical failure mode of the axisymmetric unsupported excavation is the wall failure mode, mechanism D for $D/r_0 < 2.5$, mechanisms A and B for $D/r_0 \geq 2.5$. The expressions of the stability number N for these three mechanisms are:

Mechanism A

$$N = \left[2\beta^2 \ln \beta - (\beta^2 - 1) + 2\frac{r_0}{H}(\beta^2 - 1)(\beta - 1) + \frac{4H(\beta^3 - 1)}{3r_0(\beta^2 - 1)}(2\alpha^2 - 2\alpha + 1) + 2\beta\frac{D}{r_0}\left(1 - \alpha\frac{H}{D}\right)\right] \div \left[(\beta^2 - 1)\left(1 - \frac{H}{2D}\right)\right] \tag{5}$$

Mechanism B

$$N = \left[2\beta^2 \ln \beta - (\beta^2 - 1) + \frac{2r_0}{H}\frac{(\beta^2 - \xi^2)(\beta - 1)}{(1 - \alpha)} + \frac{4H}{3r_0}\frac{\alpha(\xi^3 - 1)(\beta^2 - 1)}{(\xi^2 - 1)^2} - \frac{4}{3}\frac{H}{r_0}\frac{(1 - \alpha)(\xi^3 - 1)}{(\xi^2 - 1)} + \frac{4}{3}\frac{H}{r_0}\frac{(1 - \alpha)(\beta^3 - \xi^3)}{(\beta^2 - \xi^2)} + 2\beta\frac{D}{r_0}\left(1 - \frac{H}{D}\right)\right] \div \left[(\beta^2 - 1)\left(1 - \frac{H}{2D}\right)\right] \tag{6}$$

Mechanism D

$$N = \left[2\beta^2 \ln \beta - (\beta^2 - 1) + \frac{r_0}{D}(\beta^2 - 1)(\beta - 1) + \frac{4}{3}\frac{D}{r_0}\frac{(\beta^3 - 1)}{(\beta^2 - 1)}\right] \div \left[\frac{1}{2}(\beta^2 - 1)\right] \tag{7}$$

This result is compared with the results of various researchers in Figure 5.

Curve A is Skempton's [32] bearing capacity factor, N_c, for deep circular footings. Bjerrum and Eide [2] analyzed a number of field-strutted excavations using Skempton's bearing capacity formula. They stated that the base failure of an excavation (due to unloading) is the reverse of a failure by loading of a deep footing. Good agreement was obtained between the analyses based on the bearing capacity formula and field data including some axisymmetric cases.

Curves B and C are limit-equilibrium solutions by Prater [28] for wall failure and base failure, respectively. Extending the simple triangular failure mechanism of a vertical cut under plane strain conditions, a conical failure surface was considered for the axisymmetric situation. Assuming that the excavation wall was suupported, he derived the expression of the load on the wall. By minimizing this expression with respect to the cone angle and then equating the load on the wall to zero, he finally obtained an expression for the critical height of the excavation, D_{cr}. This result is plotted in terms of N as the curve B in Figure 5. The base failure considered by Prater is again an extension of the plane strain situation (Terzaghi [38]). Making use of Skempton's bearing capacity factor but using $N_c = 6$ to obtain the smallest value of the critical height of the excavation, the curve C was derived.

Pastor and Turgeman [25], using an upper bound calculation, obtained $N = 5.298$ for $D/r_0 = 1$. Pastor [24] also presented a lower bound solution of $N = 3.464$ for $D/r_0 = 2$. Curve D is the wall failure mechanism A shown in Figure 2 and curve E the wall failure mechanism D in Figure 2. Meyerhof [20] analyzed the stability of slurry supported axisymmetric, rectangular and square excavations using Rankine's pressure theory. For the axisymmetric excavations assigning zero value for the slurry density, an expression for the critical depth of unsupported excavation is derived as

$$D_{cr} = \frac{4C_u}{\gamma}\left[\ln\left(\frac{D}{r_0} + 1\right) + 1\right] \tag{8}$$

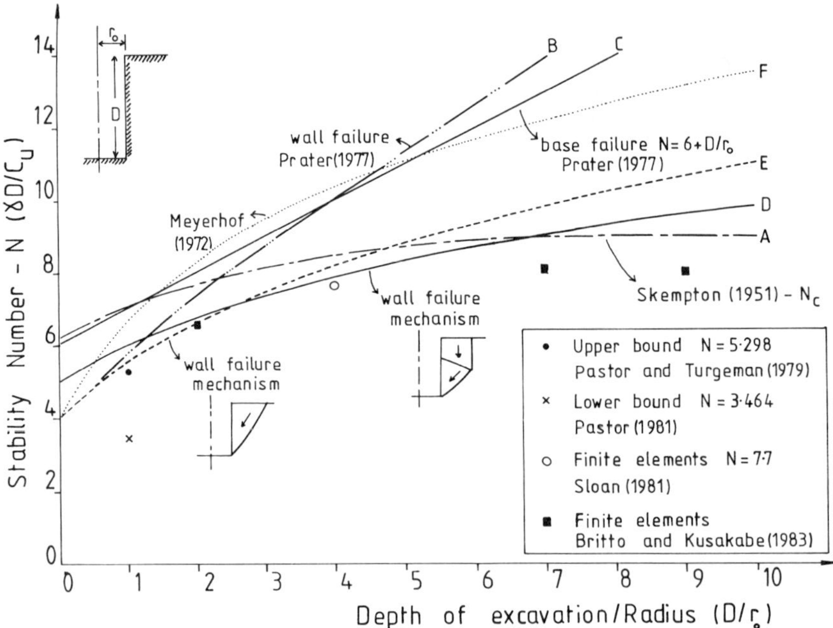

FIGURE 5. Comparison of stability number N for wall and base failure mechanisms.

and this is shown as curve F in Figure 5. Sloan [34] analyzed the problem using the finite element method and found collapse conditions at $N = 7.7$ for $D/r_0 = 4$. Finite element analysis carried out by the authors gave $N = 6.6$ for $D/r_0 = 2$ and $N = 8.12$ for $D/r_0 = 7$ and 9. In these calculations, the soil was considered to behave as a Tresca material with a ratio of Young's modulus, E_u, to the undrained shear strength, C_u, of 100 and a Poisson's ratio of 0.49. These results are close to the curves D and E of the wall failure mechanism, for $D/r_0 < 7$.

From these comparisons it appears that the most critical failure mode of the axisymmetric unsupported excavation is the wall failure mode, and the best upper bound solution to date is that of curve D for $D/r_0 \geq 2.5$ and of curve E for $D/r_0 < 2.5$. However, Skempton's N_c line gives a lower value for $D/r_0 > 7$, which implies that it is possible for unsupported deep excavations to experience a large and unacceptable base heave probably due to local failure. But this mode of base failure is only possible if the walls of the excavation are laterally supported. The stability of supported axisymmetric excavation will be discussed in Supported Excavation, below.

Experimental Verification of the Failure Mode

The previous subsection developed the notion of an axisymmetric mechanism and its application to stability of excavations. While lower bound solutions are needed, they are not available at this stage, therefore a search was made for other evidence in prior publication. Not much was available. Upper bound calculations prove fruitful and generate potentially useful design curves but how close are they to the exact solutions? The reliability of upper bound calculations depends on their being based on realistic mechanisms. In the absence of lower bounds there is a need for experimental verifications with correctly modelling prototype situations. Therefore, the centrifuge model tests for this purpose are described in this subsection.

Principles of geotechnical centrifuge modelling have been appreciated since 1930's (Bucky [6]; Pokrovsky [27]), and have been explained by several workers (Schofield [30,31], for example). Through the efforts of previous workers engaged in centrifuge testing, we have now reached the stage where test of models made of soils in geotechnical centrifuge have been accepted as a method of study of mechanics of ground deformation with more control of ground conditions than in tests of prototype scale.

The geotechnical centrifuge used in this study was Cambridge geotechnical centrifuge which has an effective radius of 4 m.

The procedure adopted in these tests was as follows. A vertical hole was excavated in a mass of kaolin clay that had been consolidated in a circular consolidometer in the laboratory. This was then lined with a rubber bag containing a heavy liquid (zinc chloride) whose specific gravity was made the same as that of the clay. After the clay mass was accelerated in the centrifuge to the predetermined gravity level (75 g, in this test series) and once equilibrium has been achieved under the centrifugal acceleration field, the heavy liquid was dumped to simulate an excavation event.

The failure mechanism was detected by a radiograph technique. Detailed explanation will be found elsewhere (Britto *et al.* [4]; Kusakabe [18]). An X-ray radiograph shown in Figure 6 gives a clear picture of the failure mechanism of the axisymmetric excavation. Two discontinuities are visible; one starts developing at the toe of the excavation, the other developing from the middle of the excavation wall. The vertical discontinuities assumed in the upper bound calculation are not detectable. However, this was anticipated because a certain magnitude of displacement will be needed for discontinuity lines to be visible. More discussions on this respect is given by Kusakabe [18]. The correctness of failure mechanism is qualitatively supported by the result.

The more detailed comparison of the failure mechanism for the axisymmetric excavation is given in Figure 7 for an excavation with a radius of 30 mm and a depth of 185 mm. Dotted lines are predicted by the upper bound calculation and the solid lines the experimental observation. A remarkable quantitative agreement between the theoretical and experimental results is seen in the figure. In particular, the discontinuity line starting from the toe of the excavation is extremely well predicted by the theory. However, the prediction for the other discontinuity line developing from the middle of the excavation wall is poor although the point of start on the wall is well predicted.

Lead shots were placed on the wall of the excavation vertically at 20 mm intervals, so that the wall movements before and after the test would be measured by X-ray radiographs. As seen in the Figure 7, both the direction and magnitude of the deformation along the excavation wall are also well predicted by the theory. The upper portion of the wall descends vertically (8.0 mm observation, 8.3 mm prediction) and the lower portion of the wall moves inwardly (maximum 13 mm observation, 15.0 mm prediction) with the angle of θ (45°–55° observation, 64° prediction).

Geometry of Critical Mechanisms

Mechanism A is the critical failure mode for aspect ratio $D/r_0 > 2.5$ and is strongly supported by experimental observation. The variation of the geometric parameters α, β and H/D is shown in Figure 8. The variations of α and H/D with D/r_0 are small; $\alpha \doteqdot 0.8$ and $H/D \doteqdot 0.5$. Figure 8 shows that β increases linearly with D/r_0 and governs the width of the zone subjected to plastic deformation. The relationship between β and D/r_0 is given by

$$\beta \doteqdot 1 + 0.4D/r_0 \tag{9}$$

i.e., the width of the plastic zone is $(\beta - 1)r_0 = 0.4D$.

Although a real soil seldom behaves as a rigid plastic material, the practical implication of this results is that the settlement zone due to the failure of an axisymmetric excavation would extend to a distance of about 0.4 times the depth of the excavation.

Figure 9 shows the surface settlement profiles of a number

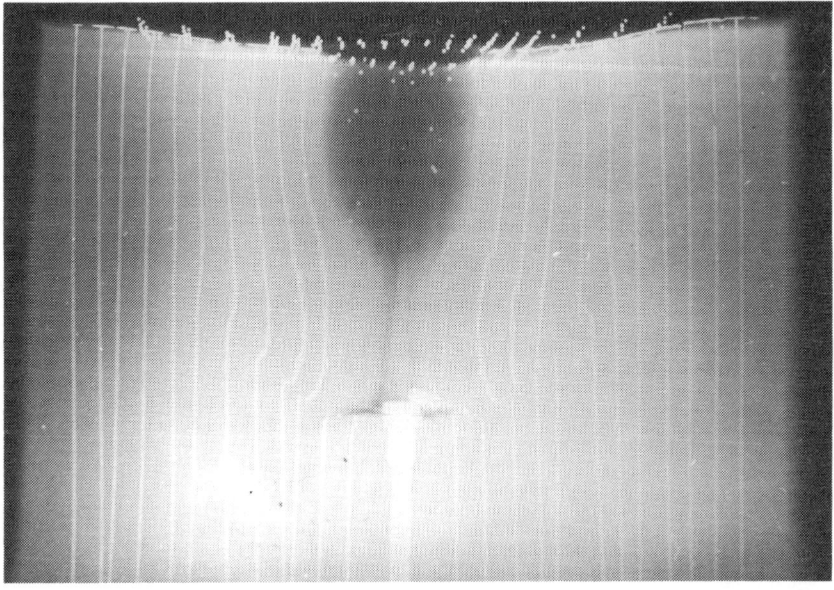

FIGURE 6. Radiograph of centrifuge test (after Kusakabe [18]).

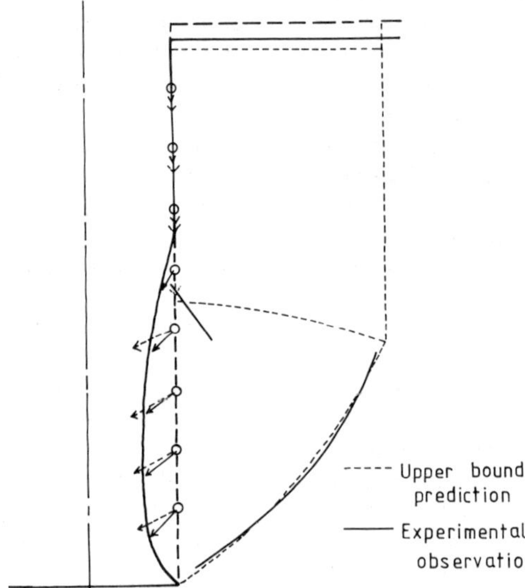

FIGURE 7. Comparison of failure mechanisms between upper bound prediction and centrifuge test.

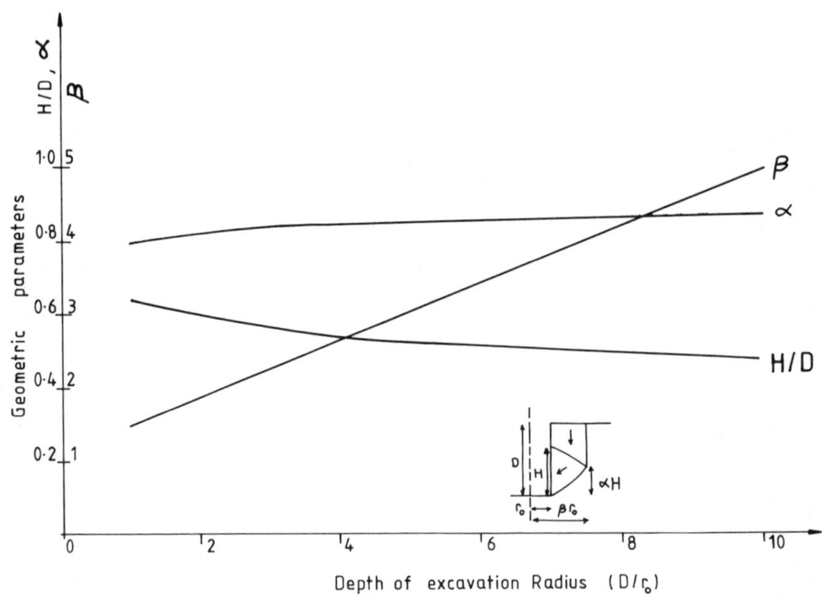

FIGURE 8. Variation of geometric parameters for wall failure mechanism A.

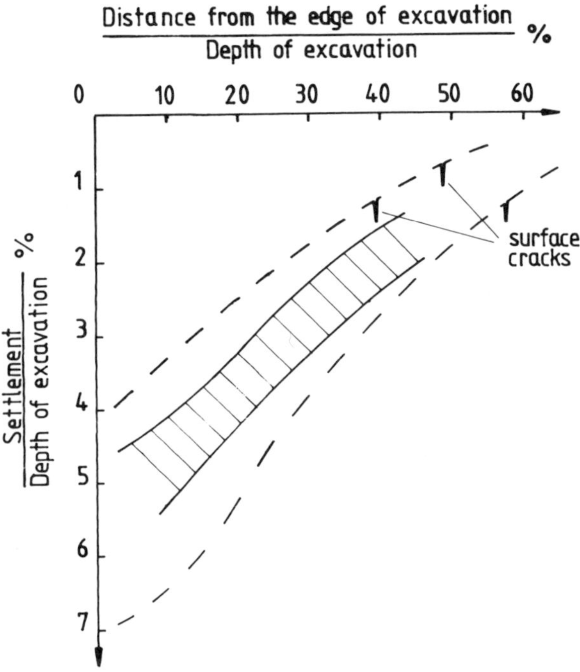

FIGURE 9. Comparison of surface settlement profiles in centrifuge tests (after Phillips [26]).

of centrifuge tests (Kusakabe [18]; Philips [26]). From this it can be seen that the settlement profile can be approximated to be linear and the width of the settlement zone is about 0.6 to 0.75D, where D is the depth of the shaft. The surface settlement profile observed in this test could be considered to extend linearly from a point of maximum settlement equal to the upper bound prediction, at the edge of the excavation to zero at a radius $\beta_e r_0$. Then by comparison of the ground loss from the upper bound prediction with that observed in the centrifuge tests, Figure 10, an empirical rule is proposed to calculate the 'equivalent width' of the surface settlement zone, $(\beta_e - 1)r_0$:

$$\pi r_0^2(\beta^2 - 1)V_0 = \frac{\pi}{3} r_0^2(\beta_e^2 + \beta_e - 2)V_0 \qquad (10)$$

therefore

$$\beta_e = \frac{\sqrt{12\beta^2 - 3} - 1}{2}$$

Defining the settlement trough to be of width to depth ratio, S_e,

$$S_e = (\beta_e - 1)r_0/D \qquad (11)$$

the value can be evaluated knowing values of β for various D/r_0 ratios from the upper bound solution. This function has been plotted in Figure 11 and it can be seen that, this compares well with the centrifuge test results.

The ratio of the radial velocity of the wall to the vertical velocity of the ground surface (U_0/V_0) is plotted against D/r_0 in Figure 12 for mechanisms A, B and D. This ratio is found to be approximately unity for $D/r_0 < 2.5$. For $D/r_0 > 2.5$ the ratio increases with increasing D/r_0 for mechanisms A and B. This shows that horizontal displacements could be more indicative than surface settlements of approaching failure, especially for excavations with higher D/r_0 ratios.

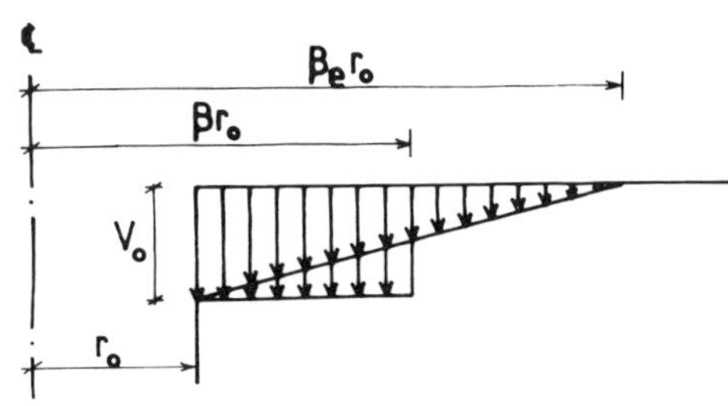

FIGURE 10. Effective surface settlement profiles.

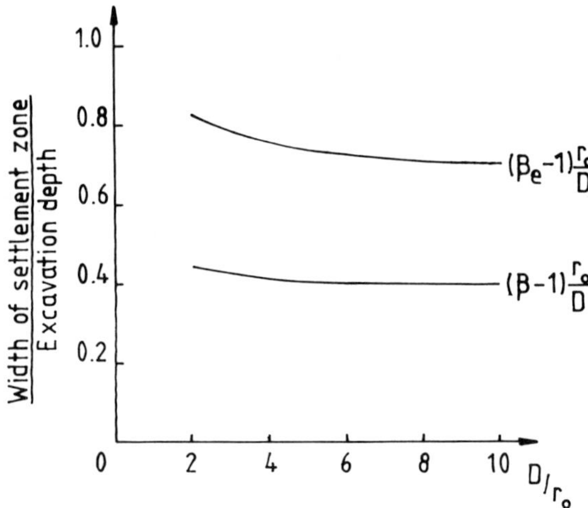

FIGURE 11. Variation of effective settlement profile with D/r_o ratio.

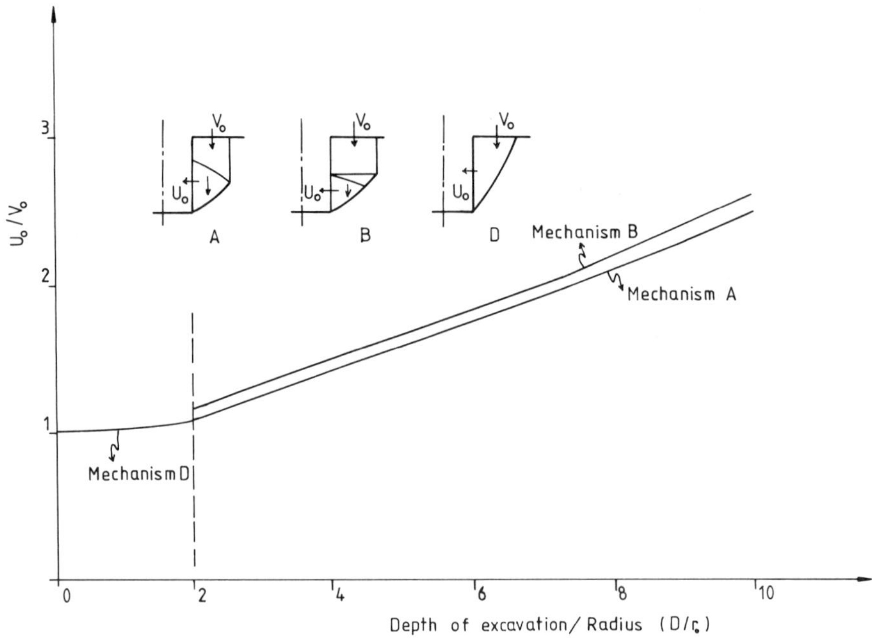

FIGURE 12. Variation of wall displacement to surface settlement ratio for wall failure mechanisms.

TABLE 1. Stability Number and Geometric Parameters for Mechanism A.

D/r_0	N	α	β	H/D
1.0	5.99	0.795	1.48	0.629
2.0	6.75	0.817	1.89	0.586
3.0	7.36	0.831	2.28	0.558
4.0	7.87	0.840	2.66	0.537
5.0	8.31	0.865	3.04	0.521
6.0	8.70	0.852	3.42	0.509
7.0	9.04	0.856	3.80	0.498
8.0	9.35	0.859	4.19	0.489
9.0	9.63	0.861	4.57	0.481
10.0	9.88	0.863	4.96	0.473

TABLE 2. Stability Number and Geometric Parameter for Mechanism B.

D/r_0	N	β	D/r_0	N	β
0.2	4.38	1.18	6.0	9.37	4.27
0.4	4.71	1.34	7.0	9.86	4.75
0.6	5.02	1.49	8.0	10.3	5.23
0.8	5.30	1.62	9.0	10.7	5.71
1.0	5.56	1.75	10.0	11.1	6.20
1.5	6.14	2.04			
2.0	6.64	2.31			
3.0	7.50	2.82			
4.0	8.22	3.31			
5.0	8.83	3.79			

Stability Charts for C_u = Constant Case

The failure mechanism for unsupported axisymmetric excavation was supported by the centrifuge model tests. This provides the basis for useful and reliable stability charts obtained based on the failure mechanism.

Tables 1 and 2 are stability numbers and geometric parameters for mechanisms A and D, respectively, which were found to be the best upper bound solutions to date. For a guide for design, the envelope of the smallest value of N should be used, which is shown in Figure 5.

Effect of Shear Strength Profile

In the preceding sections, the soil was assumed to possess a uniform undrained shear strength. It is often the case, however, that the shear strength increases with depth, and this must be taken into consideration when calculating the stability of the excavation. For the plane strain situation Gibson and Morgenstern [12] and Hunter and Schuster [15] analyzed the stability of cut slopes in clay with undrained shear strength, C_u, varying linearly with depth. Odenstad [22] has also considered the variation of shear strength with depth. Snitbhan *et al.* [35], Chen *et al.* [8], and Reddy and Srinivasan [29] have applied the upper bound theorem of limit analysis to plane strain problems for layered soils and nonhomogeneous, anisotropic $C - \phi$ soils.

The variation of undrained shear strength is shown in Figure 13. The origin of the axes is chosen as the center of the base of the excavation and therefore $C_u(z)$ is given by

$$C_u(z) = C_{u0} + k(D - z) \tag{12}$$

in which C_{u0} = the shear strength at the surface; k = the rate of increase in undrained shear strength with depth; and D = the depth of the excavation.

The proposed upper bound mechanism and the admissible velocity field is the same as the mechanism A given in Figure 2. In order to include the case of uniform shear strength described in the previous section, the stability number is defined as

$$N = \frac{\gamma D}{C_{u0} + kD} \tag{13}$$

and a new parameter, M, which describes the shear strength profile in the ground is introduced as

$$M = \frac{C_{u0}}{C_{u0} + kD} = \frac{\text{undrained shear strength at the ground surface}}{\text{undrained shear strength at the level of the excavation base}} \tag{14}$$

The critical depth, D_{cr}, is then given by $(NC_{u0}/\gamma M)$. For a

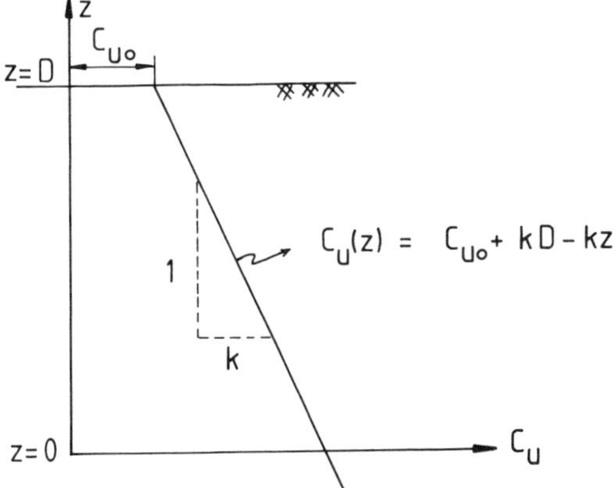

FIGURE 13. Undrained shear strength profile.

soil with uniform undrained shear strength, N, reduces to $\gamma D/C_{u0}$ and M becomes unity. When the shear strength at the surface is zero, N reduces to γ/k, and M becomes zero. Therefore, the following expression of N covers the case when the shear strength linearly increases with depth as well as the uniform shear strength case given in the previous section.

$$N = \Bigg\{\Bigg[2\beta^2 \ln\beta - (\beta^2 - 1) + 2(\beta^2 - 1) \times (\beta - 1)\frac{r_0}{H} + \frac{4}{3}\left(\frac{H}{r_0}\right)^2 \times \frac{(\beta^3 - 1)}{(\beta^2 - 1)}(2\alpha^2 - 2\alpha + 1) + 2\beta \times \left(\frac{D}{r_0} - \frac{\alpha H}{r_0}\right)\Bigg] - \frac{r_0(1 - M)}{D} \times \Bigg[\frac{H}{D}\frac{1}{D(\beta^2 - 1)}\Bigg[\beta^2(\beta^2 - 2\alpha)\ln\beta + \frac{1}{4} \times (1 - 2\alpha)(\beta^4 - 1) + (\alpha\beta^2 - \beta^2 + \alpha) \times (\beta^2 - 1)\Bigg] + (\beta^2 - \alpha)(\beta - 1) - \frac{1}{3} \times (1 - \alpha)(\beta^3 - 1) + \frac{4(1 - \alpha)^2}{(\beta^2 - 1)^2}\left(\frac{H}{r_0}\right)^2 \times \Bigg[\frac{1}{3}(\beta^2 - \alpha)(\beta^3 - 1) - \frac{1}{5} \times (\beta^5 - 1)(1 - \alpha)\Bigg] + \alpha\Bigg\{\frac{1}{3}(\beta^3 - 1) - (\beta - 1) + \frac{4\alpha^2}{(\beta^2 - 1)^2}\left(\frac{H}{r_0}\right)^2 \times \Bigg[\frac{1}{5}(\beta^5 - 1) - \frac{1}{3}(\beta^3 - 1)\Bigg]\Bigg\} + \beta\left(\frac{D^2}{r_0^2} - \frac{\alpha^2 H^2}{r_0^2}\right)\Bigg]\Bigg\} \div \left[(\beta^2 - 1)\left(1 - \frac{H}{2D}\right)\right] \tag{15}$$

The resulting values of the stability number, N, are plotted in Figure 14 for various values of M in the range 0–1. It is seen that the stability number decreases as the parameter M decreases. For a given value of shear strength at the ground level, as k increases from zero, the value of M changes from unity to zero. Correspondingly, the shear strength available in the region where the wall failure mechanism develops, decreases with the decrease in M. Therefore, the result shown in Figure 14 agrees with physical intuition. The same trend of a decrease in N, with a decrease in M was obtained by Hunter and Schuster [15] for the stability of slopes under plane strain conditions.

It is interesting to consider the physical significance of the results presented for low values of M and, in particular, for the case of $M = 0$. As shown N, then, equals γ/k and the depth of the excavation, D, does not enter into this expression. As D/r_0 tends to zero, plane strain conditions are approached, and γ/k tends to 2. Gibson [13] considered this case, and showed that $\gamma/k \leq 2$, for a vertical cut to remain stable. He also pointed out that, since this condition is independent of the depth, if a cut of given depth remains stable, then the depth can be increased at the same factor of safety. If the ground-water table is at the surface of the clay, then

$$k = (\gamma - \gamma_\omega)\frac{dC_u}{d\sigma'} \tag{16}$$

therefore

$$\frac{\gamma}{k} = \frac{1}{\dfrac{dC_u}{d\sigma'}\left(1 - \dfrac{\gamma_\omega}{\gamma}\right)} < 2$$

Skempton [33] had shown that clays normally consolidated in nature, under conditions of zero lateral strain, rarely exhibit values of $dC_u/d\sigma'$ greater than 0.5. Taking $\gamma = 18$ kN/m^3 and $\gamma_\omega = 10$ kN/m^3

$$\frac{\gamma}{k} = 3.6 > 2$$

Thus, a vertical cut is unstable. Similar arguments can be put forward for the axisymmetric case. For stability

$$\frac{\gamma}{k} = \frac{1}{\dfrac{dC_u}{d\sigma'}\left(1 - \dfrac{\gamma_\omega}{\gamma}\right)} < N \tag{17}$$

Now Figure 14 shows $N < 5$ for D/r_0 less than 10, and all known values of γ/k are greater than 5. Therefore, axisymmetric excavation ($D/r_0 < 10$) are always unstable.

Thus, the physical significance of the results presented is that excavations in normally consolidated clay with zero shear strength at the surface are inherently unstable. In other words, excavation is only possible with some form of lateral support.

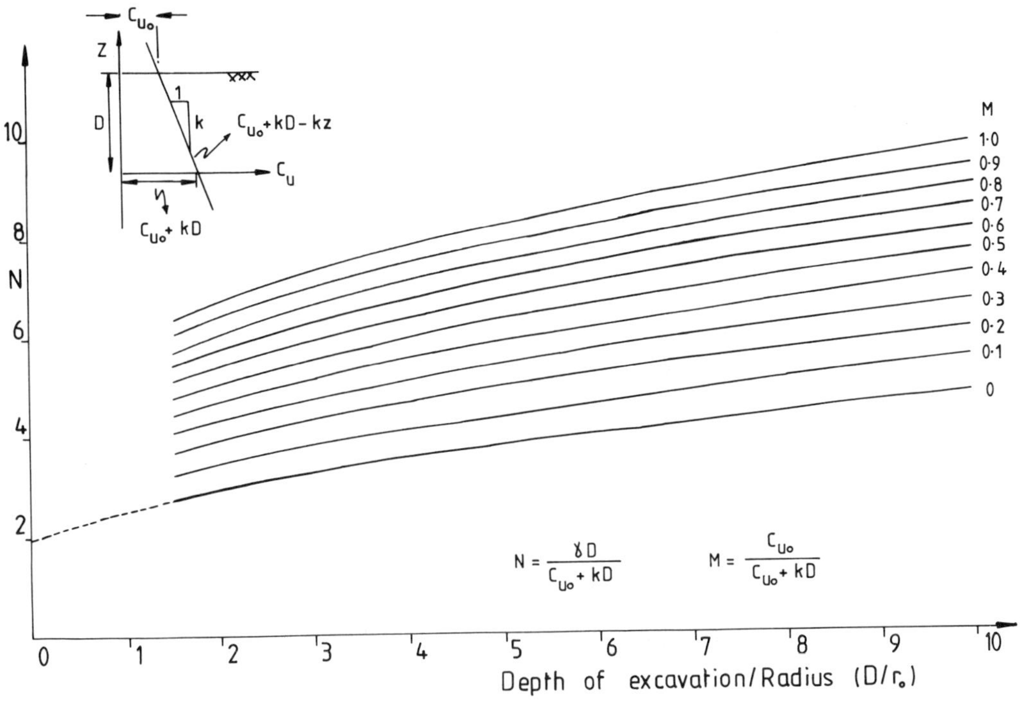

FIGURE 14. Variation of stability number N with D/r_0 and M.

The variations of the geometric parameters α and H/D with M are shown in Figures 15 and 16, respectively. It can be seen that α decreases with the increase of M and lies within 0.8 and 1 for all values of D/r_0 and M. For the case of M less than 0.1, α equals unity. This means that the discontinuity *bc* in Figure 17, becomes a horizontal line, and the mechanism *A* reduces to such that the shearing zone is overlain by a rigid and annular disc which descends vertically. The variation of the parameter H/D which governs the size of the shearing zone at the lower part of the excavation, is presented in Figure 16. It is noted that H/D decreases as M and D/r_0 increases. For M in the range of 0 and 0.1, H/D equals unity. This indicates that the shearing zone extends right to the ground surface. It was found earlier that for the previously mentioned range of M values α was equal to unity. Therefore, the mechanism becomes such that the entire plastic region consisting of a single shear zone is bounded by a parabolic discontinuity line, which is the mechanism *D* in the previous section. There is a transition mechanism for M values in the range of 0.1 and 0.2, but it is dependent on the particular D/r_0 ratio.

Figure 18 is a plot of N against M for different D/r_0 ratios. The different zones where the three different mechanisms are applicable, are also indicated. Experience shows that deformations before collapse are similar in pattern to those at collapse. Thus, an indication of the movements of buried pipes in the neighborhood can be obtained from these mechanisms. Figure 19 shows the buried pipes at three levels for the three mechanisms and also their direction of movement. The direction and magnitude of the displacement of the pipes are dependent on which zone they lie. The pipes denoted by *a*, lie within the rigid block region, and undergo only vertical settlement. Pipes denoted by *b* (which are within the shearing zone) deform both vertically and horizontally. There is a horizontal movement at the ground surface for mechanism *D* whereas for mechanism *A* there is no horizontal component. For uniform case, the value of β was approximately expressed by

$$\beta = 0.4D/r_0 + 1 \tag{18}$$

The variation of the parameter β is presented in Figure 20(a,b). It is noted that the variation of β is small for M lying in the range 0.2 and 1.0 [Figure 20(b)]. In this range the width of the plastic zone may be approximately written in a form $(\beta - 1)r_0 = 0.4D$ which is the same as that for the uniform case. For $M = 0.1$, β can be approximately given by

$$\beta = 0.4\frac{D}{r_0} + 1.55 \left(\frac{D}{r_0} > 3\right)$$

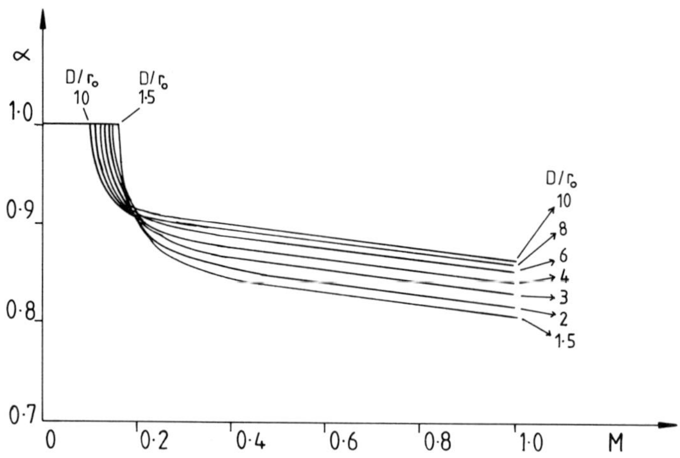

FIGURE 15. Variation of α with D/r_o and M.

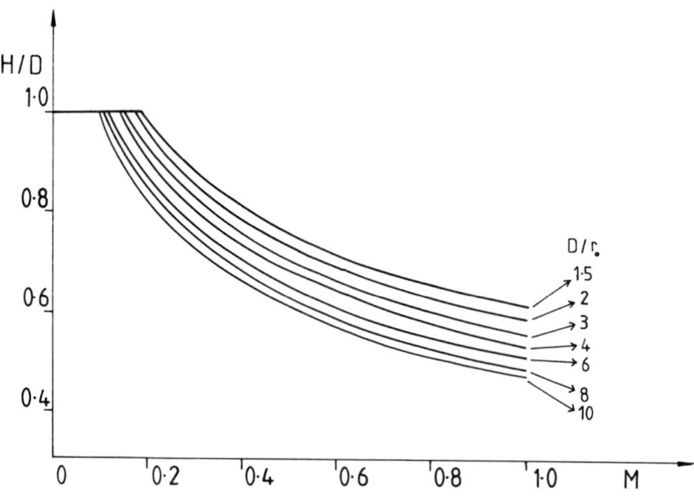

FIGURE 16. Variation of H/D with D/r_o and M.

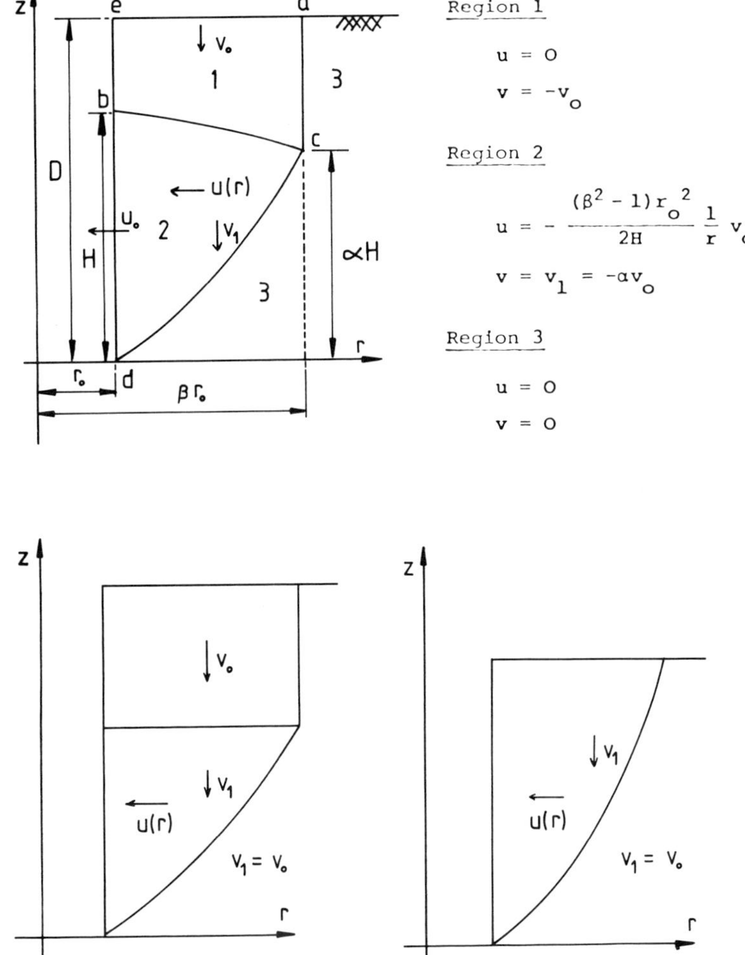

FIGURE 17. Upper bound mechanisms for wall failure.

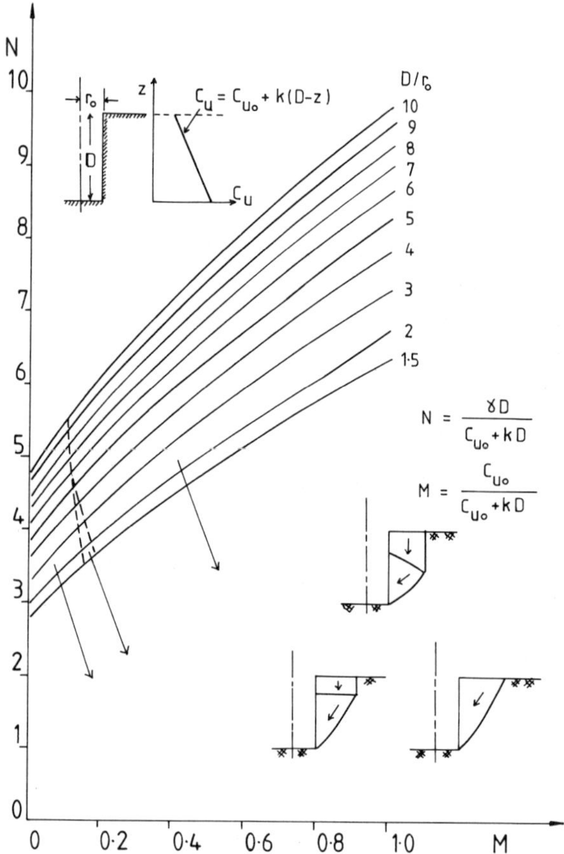

FIGURE 18. Variation of stability number N with D/r_0 and M.

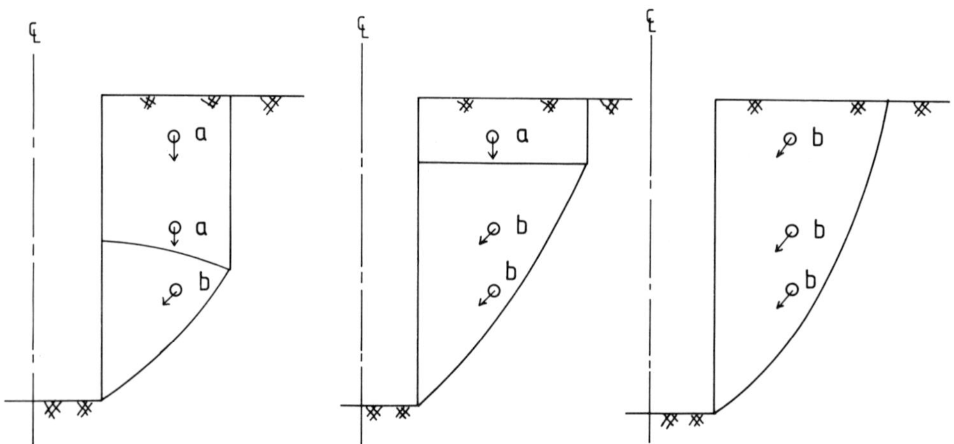

FIGURE 19. Direction of movements of pipes buried in the vicinity of the excavations.

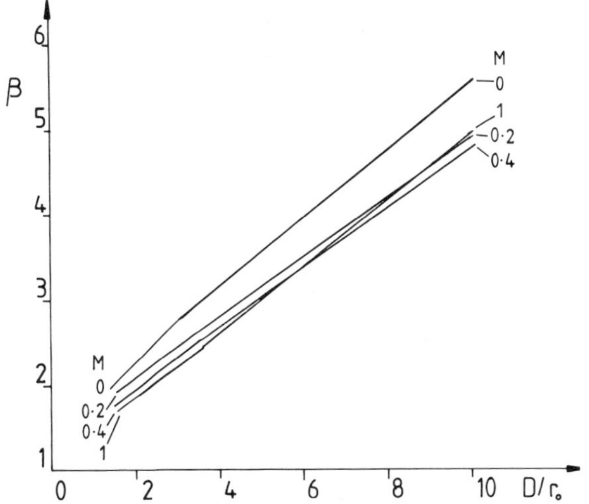

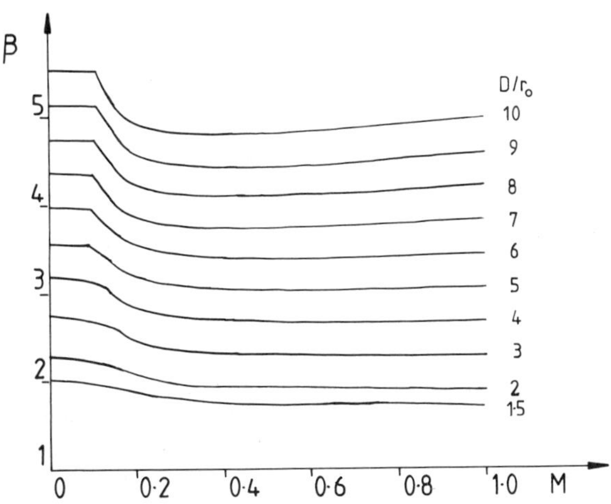

FIGURE 20. Variation of β with D/r_0 and M.

If one assumes that the surface settlement could be approximated to be linear then an equivalent width of the settlement zone can be calculated $[(\beta_e - 1)r_0]$ from the β values above (see section 4.3).

SUPPORTED EXCAVATION

Introduction

It is common practice to provide some form of support to the face of an excavation, in an attempt to increase the stability of the excavation and to reduce the ground movement associated with it. However, the question of how effective the support will be is a difficult one. This is because the effectiveness of the support is governed not only by the type of support system, the method of excavation, and soil conditions at the site, but also by how and when the support system is installed during the excavation process.

Bjerrum *et al.* [3] presented a state of the art report on supported excavations. Their main concern was the anchored or braced excavations in which the face was fully supported from the top to the bottom. It is often seen in practice, however, that the bottom part of the excavation face is left unsupported. For shallow excavations, such as trenches for gas and water mains, top timbering is quite often the only form of support. This is often for reasons of convenience, for example, workers may wish to install pipes in the bottom of the trench. In addition, top timbering is perhaps believed to be effective from the experience over the years. This also applies to support systems installed progressively during the course of the excavation; the support is extended stage by stage up to the current level of excavation and then the next stage of excavation is carried out (Terzaghi and Peck [39]).

Another practical way of supporting the excavation face and reducing the ground movement is to make use of a bentonite slurry (here after simply referred to as slurry). Recently this method has been used widely because of its advantages (Fernandes Renau [10]).

Stability of Wall

Wall failure mechanisms for supported excavations are shown in Figure 21; mechanisms E and F are for rigid lateral support and slurry support, respectively and are based on the mechanism for an unsupported axisymmetric excavation. Expressions of stability numbers are as follows:

Mechanism E

$$N = \left[2\beta^2 \ln \beta - (\beta^2 - 1) + 2(\beta^2 - 1) \times (\beta - 1)\frac{r_0}{H} + \frac{4}{3}(1 - 2\alpha + 2\alpha^2) \times \frac{(\beta^3 - 1)}{(\beta^2 - 1)}\frac{H}{r_0} + 2\beta\left(1 - \frac{\alpha H}{D}\right)\frac{D}{r_0}\right] \div \left[(\beta^2 - 1)\left(1 - \frac{1}{2}\frac{H}{D}\right)\right] \qquad (19)$$

Mechanism F

$$N = \left[2\beta^2 \ln \beta - (\beta^2 - 1) + 2(\beta^2 - 1) \times (\beta - 1)\frac{r_0}{H} + \frac{4}{3}(1 - 2\alpha + 2\alpha^2)\right.$$

$$\times \frac{(\beta^3 - 1)}{(\beta^2 - 1)} \frac{H}{r_0} + 2\beta \left(1 - \frac{\alpha H}{D}\right) \frac{D}{r_0}\Bigg] \tag{20}$$

$$\div \left[(\beta^2 - 1) \left\{\left(1 - \frac{1}{2}\frac{H}{D}\right) - \frac{\gamma_t}{\gamma}\frac{H}{D}\left(\frac{Q}{H} - \frac{1}{2}\right)\right\}\right] H \le Q$$

$$N = \Bigg[2\beta^2 \ln \beta - (\beta^2 - 1) + 2(\beta^2 - 1) \times (\beta - 1)\frac{r_0}{H} + \frac{4}{3}(1 - 2\alpha + 2\alpha^2)$$

$$\times \frac{(\beta^3 - 1)}{(\beta^2 - 1)} \frac{H}{r_0} + 2\beta \left(1 - \frac{\alpha H}{D}\right) \frac{D}{r_0}\Bigg] \tag{21}$$

$$\div \left[(\beta^2 - 1) \left\{\left(1 - \frac{1}{2}\frac{H}{D}\right) - \frac{2}{2}\frac{\gamma_t}{\gamma}\frac{D}{H} - \left(\frac{Q}{D}\right)^2\right\}\right] H \ge Q$$

For fully filled case, putting $Q = D$ in the expression of (20) we have

$$N = \Bigg[2\beta^2 \ln \beta - (\beta^2 - 1) + 2(\beta^2 - 1) \times (\beta - 1)\frac{r_0}{H} + \frac{4}{3}(1 - 2\alpha + 2\alpha^2)$$

$$\times \frac{(\beta^3 - 1)}{(\beta^2 - 1)} \frac{H}{r_0} + 2\beta \left(1 - \frac{\alpha H}{D}\right) \frac{D}{r_0}\Bigg] \tag{22}$$

$$\div \left[(\beta^2 - 1) \left\{\left(1 - \frac{1}{2}\frac{H}{D}\right) - \frac{\gamma_t}{\gamma}\frac{H}{D}\left(\frac{D}{H} - \frac{1}{2}\right)\right\}\right]$$

The stability number for mechanism E and F are given in Table 3. In order to see the effectiveness of support system for various ratios of D/r_0, it appears reasonable to compare the stability number for the supported case as a ratio of the unsupported case, i.e., in terms of μ_N, which is defined by

$$\mu_N = \frac{\text{stability number for supported excavations}}{\text{stability number for unsupported excavations}} \tag{23}$$

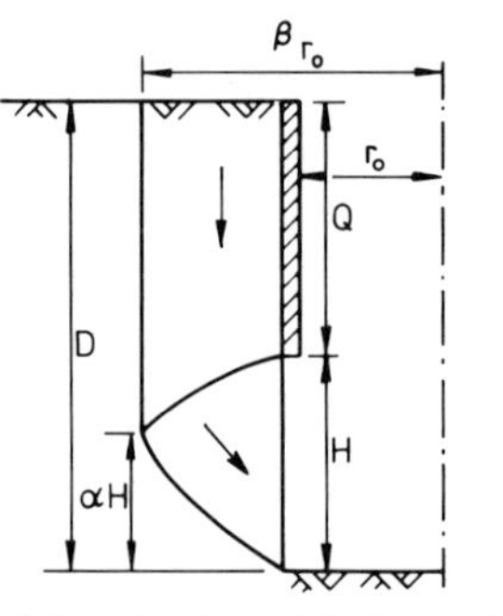

(e) mechanism E (axisymmetry)

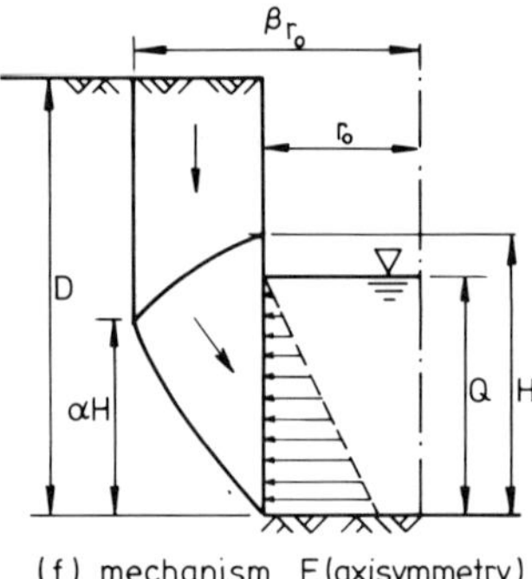

(f) mechanism F (axisymmetry)

FIGURE 21. Wall failure mechanisms for supported excavations.

The calculated results are given in Figure 22, in which μ_N is plotted against the ratio of Q/D for various values of D/r_0. The figure indicates that the rigid lateral support is less effective than the slurry support when $\gamma_t/\gamma > 0.6$. The rigid lateral support in mechanism E has no effect on the failure mechanism until the support length reaches an adequate length to prevent the horizontal movement at the bottom of the excavation wall. Therefore, the stability number does not increase for a certain range of Q/D, and it then increases rapidly with the increase in Q/D. In contrast to the rigid lateral support, the stability number of mechanism F with slurry support shows that the slurry support is useful even for small values of Q/D. It is also noted in Figure 22 that the rigid top support gives a high value of μ_N for a small value of D/r_0, while the slurry support gives a high value of μ_N for a large value of D/r_0. As D/r_0 tends to zero, the situation approaches that of plane strain. Figure 22 implies that as D/r_0 becomes small, there is a change over and as a result the slurry support might become less effective compared to the top support.

Another interesting result from the study of mechanism F is that the values of the geometric parameters, α, β and H/D for the fully filled slurry case, i.e., $Q = D$ are found to be identical to those for the unsupported excavation.

TABLE 3(a). Stability Number for Wall Supported Axisymmetric Excavation—Mechanism E.

D/r_0 \ Q/D	0	0.1	0.2	0.3	0.4	0.5	0.6	0.7	0.8	0.9
1	5.99	5.99	5.99	5.99	5.99	6.08	6.30	6.76	7.67	10.02
2	6.75	6.75	6.75	6.75	6.75	6.79	6.98	7.40	8.28	10.58
3	7.36	7.36	7.36	7.36	7.36	7.38	7.54	7.93	8.79	11.08
4	7.87	7.87	7.87	7.87	7.87	7.88	8.02	8.39	9.23	11.51
5	8.31	8.31	8.31	8.31	8.31	8.32	8.43	8.79	9.62	11.90
6	8.70	8.70	8.70	8.70	8.70	8.70	8.80	9.14	9.97	12.25
7	9.04	9.04	9.04	9.04	9.04	9.04	9.13	9.46	10.28	12.56
8	9.35	9.35	9.35	9.35	9.35	9.35	9.42	9.74	10.56	12.85
9	9.63	9.63	9.63	9.63	9.63	9.63	9.69	10.00	10.81	13.11
10	9.88	9.88	9.88	9.88	9.88	9.88	9.94	10.24	11.04	13.35

TABLE 3(b). Stability Number for Slurry Supported Excavation with γ_t/γ = 1.0.

D/r_0 \ Q/D	0	0.1	0.2	0.3	0.4	0.5	0.6	0.7	0.8	0.9
1	5.99	6.06	6.28	6.67	7.29	8.26	9.79	12.44	17.86	34.28
2	6.75	6.83	7.09	7.54	8.27	9.38	11.15	14.21	20.43	39.30
3	7.36	7.45	7.74	8.25	9.06	10.30	12.26	15.64	22.53	43.39
4	7.87	7.87	8.28	8.84	9.72	11.07	13.19	16.86	24.31	46.87
5	8.31	8.42	8.76	9.36	10.29	11.73	14.01	17.91	25.85	49.88
6	8.70	8.81	9.17	9.81	10.80	12.32	14.72	18.84	27.21	52.55
7	9.04	9.16	9.54	10.21	11.25	12.85	15.36	19.67	28.44	54.95
8	9.35	9.48	9.87	10.57	11.66	13.32	15.94	20.42	29.54	57.12
9	9.63	9.76	10.17	10.89	12.03	13.75	16.46	21.11	30.55	59.10
10	9.88	10.02	10.44	11.19	12.36	14.14	16.94	21.74	31.48	60.92

TABLE 3(c). Stability Number for Slurry Supported Excavation with γ_t/γ = 0.8.

D/r_0 \ Q/D	0	0.1	0.2	0.3	0.4	0.5	0.6	0.7	0.8	0.9	1.0
1	5.99	6.05	6.22	6.53	7.00	7.70	8.74	10.34	13.03	18.28	29.95
2	6.75	6.81	7.02	7.37	7.92	8.73	9.93	11.79	14.88	20.90	33.74
3	7.36	7.43	7.66	8.06	8.67	9.57	10.91	12.96	16.39	23.01	36.80
4	7.87	7.95	8.20	8.64	9.30	10.28	11.73	13.95	17.66	24.78	39.36
5	8.31	8.40	8.67	9.13	9.85	10.90	12.44	14.82	18.77	26.31	41.56
6	8.70	8.79	9.07	9.57	10.33	11.44	13.07	15.58	19.75	27.65	43.49
7	9.04	9.14	9.44	9.96	10.76	11.92	13.63	16.25	20.62	28.84	45.21
8	9.35	9.45	9.76	10.31	11.14	12.35	14.14	16.87	21.41	29.91	46.75
9	9.63	9.74	10.06	10.63	11.49	12.75	14.60	17.43	22.14	30.88	48.15
10	9.88	9.99	10.33	10.92	11.81	13.11	15.02	17.94	22.80	31.77	49.42

TABLE 3(d). Stability Number for Slurry Supported Excavation with γ_t/γ = 0.6.

D/r_0 \ Q/D	0	0.1	0.2	0.3	0.4	0.5	0.6	0.7	0.8	0.9	1.0
1	5.99	6.03	6.16	6.38	6.72	7.20	7.87	8.82	10.20	12.21	14.98
2	6.75	6.80	6.95	7.21	7.60	8.16	8.93	10.03	11.62	13.84	16.87
3	7.36	7.41	7.58	7.88	8.31	8.93	9.80	11.01	12.77	15.15	18.40
4	7.87	7.93	8.12	8.44	8.91	9.59	10.52	11.84	13.73	16.25	19.68
5	8.31	8.38	8.58	8.92	9.43	10.15	11.15	12.56	14.56	17.20	20.78
6	8.70	8.77	8.98	9.34	9.89	10.65	11.71	13.20	15.28	18.03	21.75
7	9.04	9.11	9.34	9.72	10.29	11.09	12.20	13.76	15.93	18.76	22.60
8	9.35	9.43	9.66	10.06	10.66	11.49	12.65	14.27	16.51	19.43	23.38
9	9.63	9.71	9.95	10.37	10.99	11.85	13.06	14.74	17.03	20.03	24.07
10	9.88	9.97	10.22	10.65	11.29	12.19	13.43	15.16	17.51	20.57	24.71

Therefore, the ratio of μ_N can be simply given by

$$\mu_N = \frac{1}{\left(1 - \dfrac{\gamma_t}{\gamma}\right)} \tag{24}$$

For example, if $\gamma_t/\gamma = 0.6$ which roughly corresponds to excavation under submerged condition then $\mu_N = 2.5$. In other words, the stability number for fully slurry filled excavations is 2.5 times that of an unsupported excavation.

In analyzing the stability of slurry-filled trenches, Meyerhof [20] also considered axisymmetric excavations and obtained a stability number of

$$N = \frac{\gamma D}{C_u} = \frac{4\left\{\ln\left(\dfrac{D}{r_0} + 1\right) + 1\right\}}{\left(1 - \dfrac{\gamma_t}{\gamma}\right)} \tag{25}$$

Comparison of the results of Meyerhof with mechanism F is made for $\gamma_t/\gamma = 0.6$ in Table 4. As is seen in the table the values given by Meyerhof are about 30% higher than those of the present analysis, which implies that the present solutions are closer to the exact solution.

In a series of centrifuge tests on unsupported axisymmetric excavations, which was described in the earlier section, in one test, with an excavation depth of 188 mm and a diameter of 77 mm, the excavation collapsed before the heavy liquid could be completely dumped. The depth of slurry was 73 mm when failure took place. Hence, this test, unintentionally, provided the experimental data in verifying the solution from mechanism F. For this test $\gamma_t/\gamma = 1.0$, $Q/D = 73/188 = 0.39$, and $D/r_0 = 188/38.5 = 4.88$.

The stability number at failure is given by

$$N = \frac{n\gamma D}{C_u}$$

where n is the centrifugal acceleration and D the depth of the model excavation.

Now $\gamma = 16.32$ kN/m^3 and from in-flight vane test data $C_u = 25$ kPa. Therefore

$$N = \frac{75 \times 16.32 \times 0.188}{25} = 9.2$$

From Table 3 for a slurry support of $Q/D = 0.39$, N is equal to 9.38. Therefore, the experimental result is within 2% of the upper bound solution presented here.

TABLE 4. Comparison of Stability Number Between Mechanism F and Meyerhof's Solution.

D/r_0	1	2	3	4	5	6	7	8	9	10
(1) Meyerhof	16.83	20.99	23.86	26.09	27.92	29.46	30.79	31.97	33.03	33.98
(2) Present Analysis	14.98	16.87	18.40	19.68	20.78	21.75	22.60	23.38	24.07	24.71
Ratio $\frac{(1)}{(2)}$	1.13	1.24	1.30	1.33	1.34	1.35	1.36	1.37	1.37	1.38

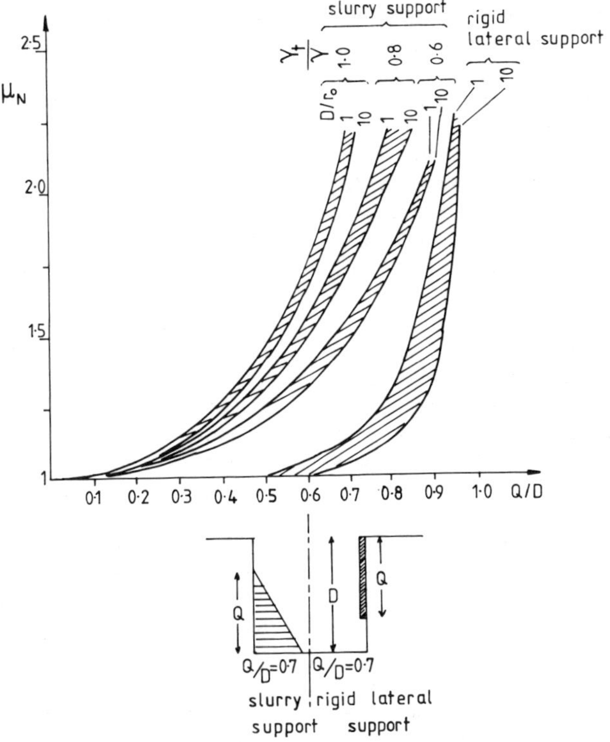

FIGURE 22. Stability number for supported excavations.

Base Failure

In the previous subsection, the stability numbers were given for a number of Q/D ratios. However, it is obvious that the base failure mode becomes more critical than the wall failure mode when the support length reaches a certain depth. Then the overall stability of the excavation is governed by base failure. It should be pointed out that the width of the excavation has to be taken into account for base failure, whereas the stability of the excavation wall is independent of the width of the excavation for the plane strain case. It is easily reasoned that the narrower the excavation, the more stable it will be for a given depth provided all other conditions remain the same. This may be due to the effect of arching. Terzaghi [38] proposed an analysis for base failure of shallow excavations under plane strain conditions. He reckoned that soil surrounding an excavation tends to act as a surcharge; thus the stability of the excavation is the reverse of the bearing capacity problem of shallow footings. This view has been shared by Skempton [32] as pointed out by Bjerrum and Eide [2].

Meyerhof [19] analyzed the bearing capacity problem of axisymmetric footing based on the failure mechanism shown in Figure 23. Gibson [11] modified the spherical cavity expansion theory by Bishop *et al.* [1] and gave the expression for N_c as

$$N_c = \frac{4}{3}\left[\ln\left(\frac{E_u}{C_u}\right) + 1\right] + 1 \tag{26}$$

where E_u is the Young's modulus.

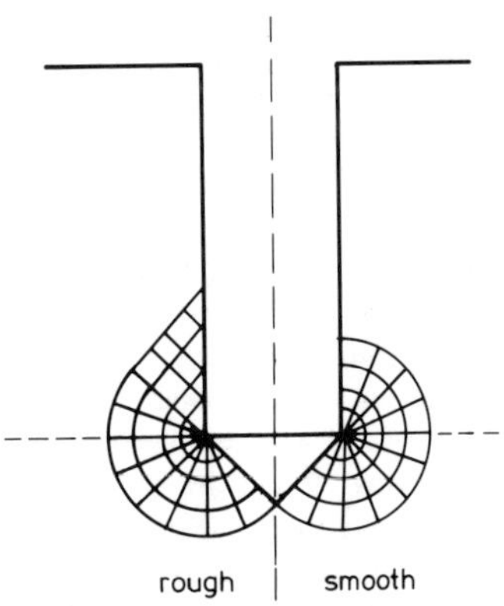

FIGURE 23. Base failure mechanisms.

Using this equation, together with field data, Skempton [32] proposed N_c values for rectangular and circular footings for various D/r_0 ratios. Bjerrum and Eide [2] analyzed the field data based on their N_c values by Skempton and obtained good agreement between field data and analytical result. Prater [28] proposed a conical shaped wall failure mechanism. Britto and Kusakabe [5] presented upper bound solutions for base failure of axisymmetric excavations as were shown in the earlier section. Compared to the value of N_c by Skempton these results gave higher values for stability numbers.

In order to check the validity of the above results of Skempton on base failure, centrifuge experiments and finite element analyses were carried out. The CRISP program (Gunn and Britto [14]) was used in the finite element analyses. This program uses an incremental approach using the tangential stiffness method. In the finite element analyses the soil was assumed to behave as a perfect elastoplastic material and to obey the Tresca yield criterion. The ratio of E_u/C_u was taken as 300 and the Poisson's ratio as 0.49. The procedure adopted in the analyses was to increase the self-weight of the soil until the base heave reached 10% of the excavation depth, when failure was deemed to have taken place.

Centrifuge tests were carried out on three different excavations with D/r_0 ratios 2, 3.5 and 5. The kaolin clay mass had an approximately uniform shear strength profile and the excavation face was fully lined with a thin metal sheet. The test procedure was to increase the centrifuge acceleration until heave at the base reached about 10% of the depth of the excavation, at which point it was assumed that failure had taken place. The results of the finite element analyses and centrifuge tests are shown in Figure 24, together with the analytical solutions. The data for C_u value in the centrifuge tests were evaluated from the results of Nadarajah [21] on the triaxial extension test which shows $C_u/\sigma'_{v0} = 0.18$.

Both the centrifuge test results and the finite element results are quite close and slightly higher than the values proposed by Skempton.

Figure 25 is a print from a radiograph of the excavation in the centrifuge test with $D/r_0 = 2$, which shows the set of lead threads placed on a vertical plane through the axis. The lead threads were injected vertically and at an angle of 45° sloping away from the axis of the excavation before the commencement of the centrifuge test. The distortion of the lead threads shows the deformations during the test. Clear discontinuity lines can be observed below the level of the base, but these do not extend towards the ground surface. These may be compared with the plastic zone developed in the finite element analyses at the stage when base heave was 10% of the excavation depth (Figure 26). It is clearly seen that the plastic zone extends below the base of the excavation but the upper regions remain elastic, which indicates that the failure is localized and the failure zone does not extend to the ground surface. This suggests that the cavity expansion type analyses (Bishop *et al.* [1]) are valid. Therefore the approach used by Bjerrum and Eide [2] is confirmed to be reasonable as a first approximation.

Stability Chart

Considering both wall failure and base failure mechanisms, the critical failure mechanism that determines the overall stability of the excavation may be given by the envelope of the smallest values of the stability number presented here. This envelope of stability number in turn may be used for generating stability charts as a design guide.

A stability chart is given for the rigid supported excavation in Figure 27. As we have seen in Figure 24, the stability numbers for base failure tend to a constant value of $N = 9.5$ around $D/r_0 = 6$. The stability numbers for base failure were determined mainly on the basis of centrifuge tests and the finite element analyses.

The slurry-supported excavations were more stable against base failure and the overall stability was governed by wall failure. Therefore the stability charts are given separately for the cases of $\gamma_l/\gamma = 0.6$, 0.8 and 1.0 in Figures 28, 29 and 30, respectively.

CONCLUDING REMARKS

Figure 5 summarizes the failure mechanisms of unsupported axisymmetric excavations and the influence of the excavation on the adjacent ground and the existing facilities. For the cases of D/r_0 less than 2.5, the failure mechanism consists of one shearing zone bounded by a parabolic curve. In this mechanism, the ground surface deforms not only vertically but also horizontally and the rate of surface settlement is the same order as the rate of radial movement of the excavation. The severe subsidence zone extends to 0.65 times the excavation depth and the damage zone may extend to a distance equal to the depth of the excavation.

In the case of D/r_0 more than 2.5, the failure mechanism consists of two regions; one is an annular ring forming a plastic zone in the lower part of the excavation, squeezing into the excavation and the other is a rigid block above it, descending vertically. In this mechanism, the surface of the ground settles only vertically, and the rate of settlement at the surface is less than the radial inward movement at the lower part of the excavation wall. The severe subsidence zone is 0.4 times the depth of the excavation, although the damage zone may extend once again as far as one depth of the excavation. These statements are approximately relevant to other shapes of locally open excavations.

Figure 31 shows a curve from the envelope of the stability numbers obtained from mechanisms *A* and *D* and is used for unsupported axisymmetric excavation. This may be used for rectangular pits, provided the aspect ratio is smaller than two.

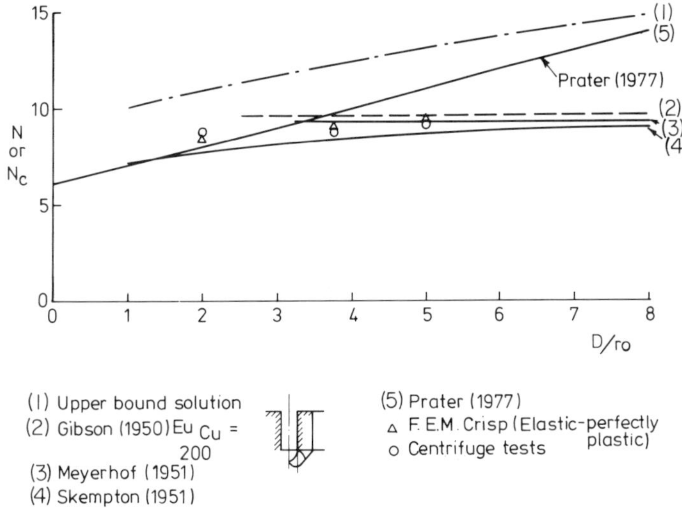

FIGURE 24. Stability number for base failure.

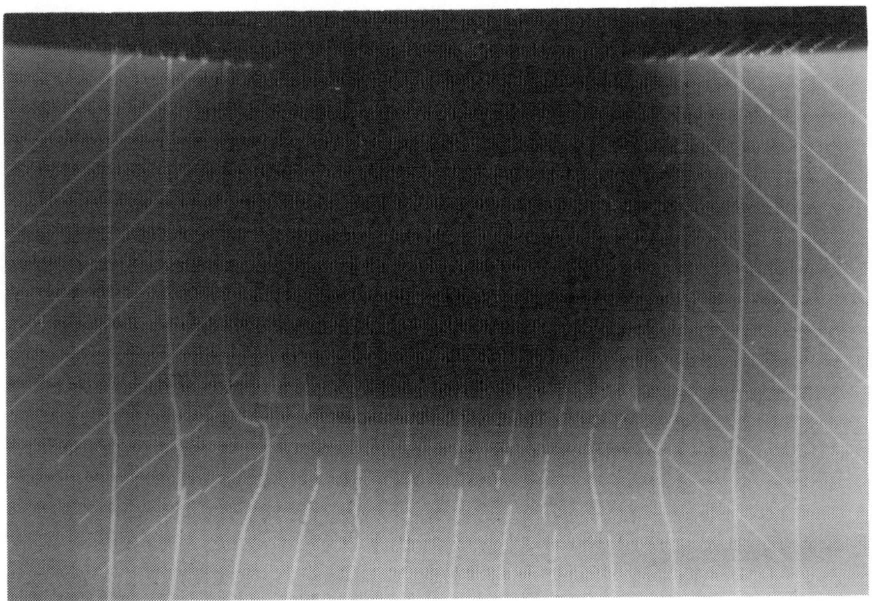

FIGURE 25. Radiograph of base failure of an axisymmetric excavation (after Kusakabe [18]).

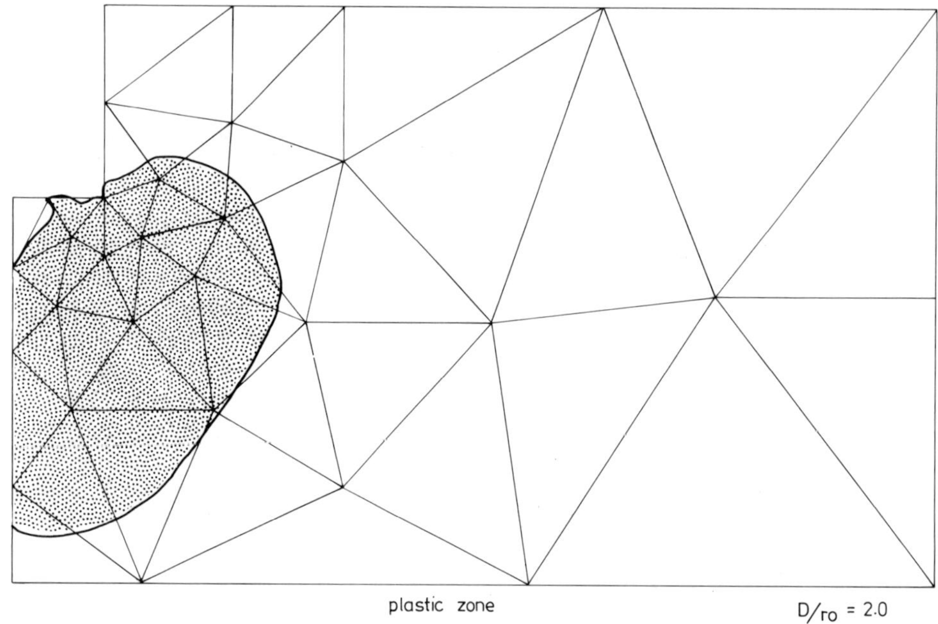

FIGURE 26. Extent of plastic zone for an axisymmetric excavation with $D/r_o = 2$.

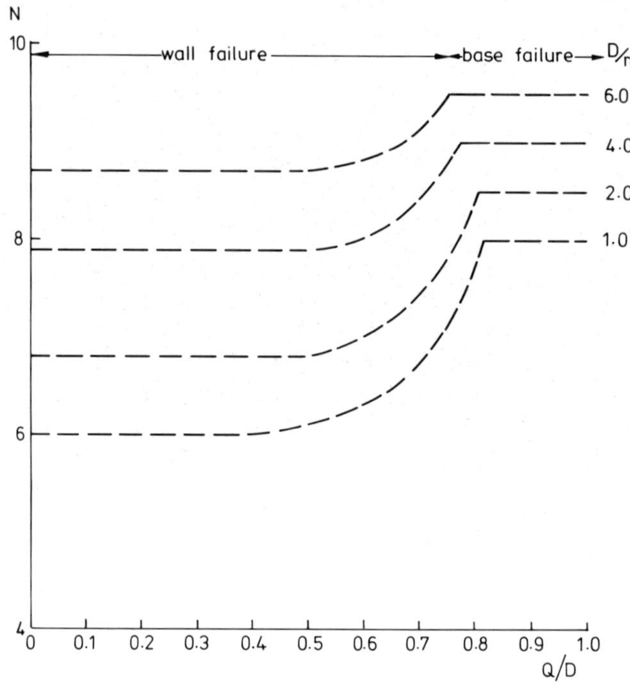

FIGURE 27. Stability chart for excavations with rigid support.

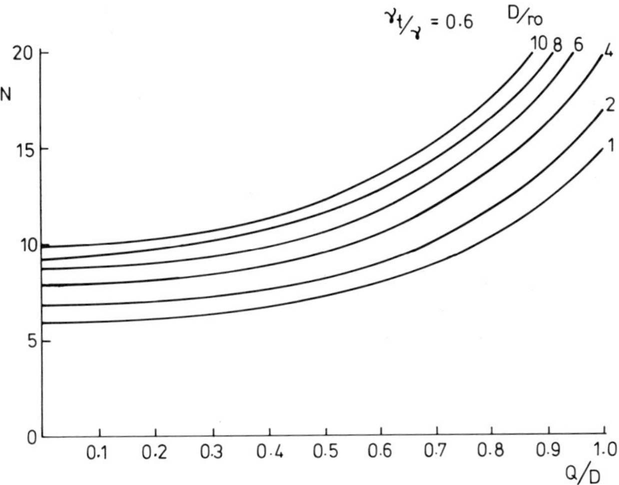

FIGURE 28. Stability chart for slurry supported excavation γ_t/γ = 0.6.

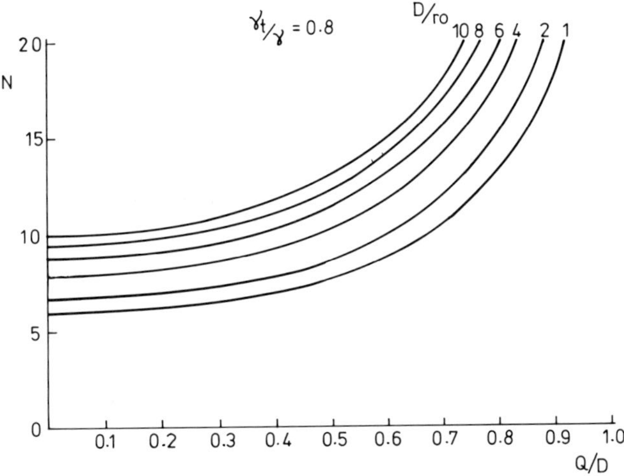

FIGURE 29. Stability chart for slurry supported excavation γ_t/γ = 0.8.

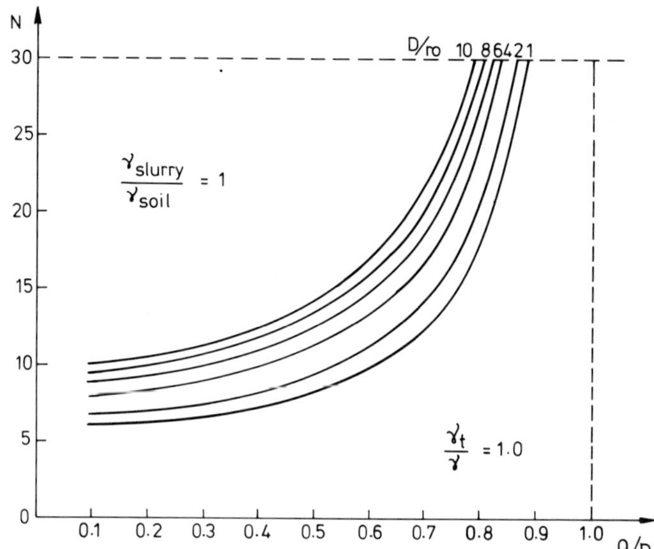

FIGURE 30. Stability chart for slurry supported excavation $\gamma_t/\gamma = 1.0$

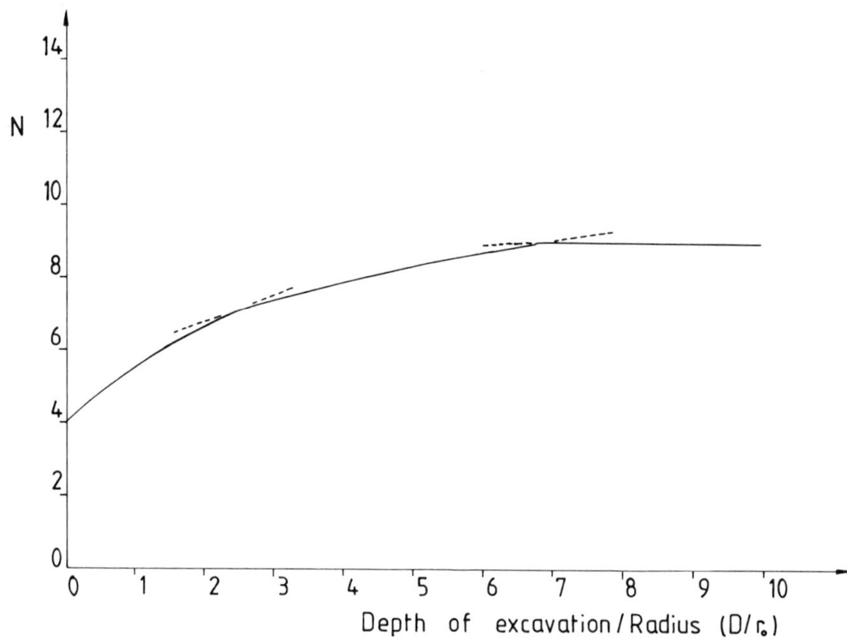

FIGURE 31. Stability chart for unsupported excavation.

Stability charts for supported excavations are obtained by considering both wall and base failure mode. For rigid wall supported excavations, Figure 27 is reconstructed from Table 3(a) for wall failure and Figure 24 for base failure. In the case of rigid support, the wall failure is the critical failure mode when the support length is small; however, the base failure mode becomes more critical and controls the overall stability of the excavation as the support length is increased. For slurry supported excavations, the wall failure is always critical irrespective of Q/D ratio.

This section presented bound solutions and the centrifuge model test results. Judged by the comparisons between the observed and predicted behavior, it has been proved that the present bound solutions can be used with confidence at least to predict the stability number and the failure mechanism for axisymmetric excavations under undrained conditions. An understanding of the behavior of the axisymmetric excavation brings us one step closer to analyzing the more complex three-dimensional problems.

However, there still remain many aspects of the excavation problems to be investigated. First of all, it is necessary to find a good lower bound solution for axisymmetric excavations. No good lower bound solution exists for this problem despite studies by Pastor and Turgeman [25] using a numerical approach. More importantly, further effort on improving upper bound solutions for three-dimensional excavations, such as rectangular pits and trench headings, is required.

The existing plasticity theory, however, does not provide any information on the behavior with time, i.e., the long term stability problem. To understand the long term stability of the excavation is in essence to trace the changes in effective stress with time. On the experimental side it needs the actual measurement of changes in both total stress and pore water pressure to determine the effective stress path. On the numerical analysis side, more detailed study on soil modelling, in particular at lower stress level, is appropriate.

ACKNOWLEDGEMENTS

The research work was carried out under the supervision of Professor A. N. Schofield and the authors are grateful to him for many stimulating discussions. Special thanks are due to Mr. M. J. Gunn and Drs. R. N. Taylor and S. W. Sloan for their constructive criticisms. The work reported here was funded by the British Gas Corporation at Killingworth, England and Transport and Road Research Laboratory, Crowthorne, England.

NOTATION

C_u = undrained shear strength of soil
D = depth of excavation
E_u = Young's modulus of soil in terms of total stresses under undrained condition
F_s = factor of safety
H = height of shear zone
N, N_s = stability number
N_c = bearing capacity factor
Q = length of support; depth of slurry
a = half-width of (plane strain) excavation
b = width of settlement zone (plane strain)
n = centrifugal acceleration
r_0 = radius of axisymmetric excavation
s = ratio of width of settlement zone to depth of excavation
α = geometric parameter governing height of shear zone
β = geometric parameter governing width of the shear zone
γ = bulk unit weight of soil
γ_t = bulk unit weight of slurry
μ_N = ratio of N for supported excavation to excavation without support

REFERENCES

1. Bishop, R. F., Hill, R., and Mott, N. F., "The Theory of Indentation and Hardness Tests," *Proceedings of the Physical Society*, 1945, Vol. 57, Part 3, 147–159.
2. Bjerrum, L. and Eide, O., "Stability of Strutted Excavations in Clay," *Geotechnique*, Vol. 6, 1956, pp. 32–47.
3. Bjerrum, L., Clausen, C. J. F., and Duncan, J. M., "Stability of Flexible Structures," *Proceedings 5th European Conference on Soil Mechanics and Foundation Engineering*, Vol. 2, 1972, pp. 169–196.
4. Britto, A. M., Kusakabe, O., and Schofield, A. N., "Trench Headings in Soft Clay," A report to the British Gas Corporation, Cambridge Univeristy, Cambridge, England, 1981.
5. Britto, A. M. and Kusakabe, O., "Stability of Unsupported Axisymmetric Excavations in Soft Clay," *Geotechnique*, Vol. 32, No. 3, 1982, pp. 261–270.
6. Bucky, P. B., "Use of Models for the Study of Mining Problems," *AIMEE Technical Publication No. 425*, 1931.
7. Chen, W. F., "Limit Analysis and Soil Plasticity," Elsevier Scientific Publishing Company, Amsterdam, 1975.
8. Chen, W. F., Snitbhan, N., and Fang, H. Y., "Stability of Slopes in Anisotropic, Nonhomogeneous Soils," *Canadian Geotechnical Journal*, Vol. 12, No. 1, 1975, pp. 146–152.
9. Coulomb, C. A., *Essai sur une Application des Regles des Maximis et Minimis a Quelques Problemes de Statique Relatifs l'Architecture*, Memoires de Mathematique et de Physique de l'Academie des Sciences, Paris, 1773. (J. Heyman, Coulomb's Memoir on Statics, Cambridge University Press, 1972.)
10. Fernandes Renau, L., Discussion on the paper by O. Eide, G. Aas and T. Josang, "Special Application of Cast-in-Place Walls for Tunnels in Soft Clay in Oslo," *5th European Conference*

on *Soil Mechanics and Foundation Engineering*, Vol. 2, 1972, pp. 366–373.

11. Gibson, R. E., Discussion on the paper by Wilson, G., "The Bearing Capacity of Screw Piles and Screw Crete Cylinder," *Journal of the Institution of Civil Engineers*, Vol. 34, 1950, pp. 382–383.
12. Gibson, R. E. and Morgenstern, N., "A Note on the Stability of Cuttings in Normally Consolidated Clays," *Geotechnique*, Vol. 12, No. 3, 1962, London, England, pp. 212–216.
13. Gibson, R. E., "The Analytical Method in Soil Mechanics," *14th Rankine Lecture, Geotechnique*, Vol. 24, No. 2, 1974, London, England, pp. 115–140.
14. Gunn, M. J. and Britto, A. M., "CRISP-Users' and Programmers' Manual," Cambridge University, Cambridge, England, 1982.
15. Hunter, J. H. and Schuster, R. L., "Stability of Simple Cuttings in Normally Consolidated Clays," *Geotechnique*, Vol. 18, No. 3, London, England, 1968, pp. 372–378.
16. Janbu, N., "Stability Analysis of Slopes with Dimensionless Parameters," *Harvard Soil Mechanics Series, No. 46*, 1954.
17. Kobayashi, S., "Upper Bound Solutions of Axisymmetric Problems," *International Journal of Engn. Industry*, Vol. 86, May 1964, pp. 122–126.
18. Kusakabe, O., "Stability of Excavations in Soft Clay," thesis presented to Cambridge University, England, in 1982, in partial fulfillment of the requirements for the degree of Doctor of Philosophy.
19. Meyerhof, G. G., "The Ultimate Bearing Capacity of Footings," *Geotechnique*, Vol. 2, 1951, pp. 301–332.
20. Meyerhof, G. G., "Stability of Slurry Trench Cuts in Saturated Clay," *Proceedings, the Specialty Conference on Performance of Earth and Earth-Supported Structures*, Vol. 1, Part 2, Purdue University, Lafayette, Ind., 1972.
21. Nadarajah, V., "Plane Strain Properties of Lightly Overconsolidated Clays," thesis presented to Cambridge University, England in 1973, in partial fulfillment of the requirements for the degree of Doctor of Philosophy.
22. Odenstad, S., Discussion on "A Note on the Stability of Cuttings in Normally Consolidated Clays," by Gibson, R. E. and Morgenstern, N., *Geotechnique*, Vol. 13, No. 2, London, England, 1963, pp. 166–170.
23. Pastor, J., "Analyse Limite: Determination Numerique de Solutions Statistiques Completes. Application au Talus Vertical," *J. Mec. Appliquee*, Vol. 2, 1978, pp. 167–196.
24. Pastor, J., "Analyse Limite et Stabilite des Fouilles," *Proceedings, 10th Int. Conf. Soil Mech.*, Stockholm, 3, 1981, pp. 505–508.
25. Pastor, J. and Turgeman, S., "Formulation Lineaire des Methodes de l'Analyse Limite en Symetrie Axiale," *Proc. 4^{e} Congr. Francais. Mec.*, Nancy, 1979.
26. Philips, R., "Trench Excavations in Clays," forthcoming Ph.D. Thesis, Cambridge University, England, 1985.
27. Pokrovsky, G. I., "Principles of Modelling of the Foundation of Structures," *Technical Physics*, Moscow, No. 2, 1934.
28. Prater, E. G., "An Examination of Some Theories of Earth Pressure on Shaft Linings," *Canadian Geotechnical Journal*, Vol. 14, 1977, pp. 91–106.
29. Reddy, A. S. and Srinivasan, R. J., "Bearing Capacity of Footings on Layered Clays," *Journal of the Soil Mechanics and Foundation Division, ASCE*, Vol. 93, No. SM2, Proc. Paper 5141, March 1967, pp. 83–99.
30. Schofield, A. N., "Cambridge Geotechnical Centrifuge Operations," *20th Rankine Lecture, Geotechnique*, Vol. 30, 1980, London, England, pp. 227–268.
31. Schofield, A. N., "Dynamic and Earthquake Geotechnical Centrifuge Modelling," State of the Art Review, *Int. Conf. on Recent Advances in Geotechnical Earthquake Engineering and Soil Dynamics*, St. Louis, 1981.
32. Skempton, A. W., "The Bearing Capacity of Clays," *Building Research Congress*, Papers presented in Division 1, 1951, pp. 180–189.
33. Skempton, A. W., "The Consolidation of Clay by Gravitational Compaction," *Quarterly Journal Geol. Soc.*, London, England, Vol. 125, 1970, pp. 373–408.
34. Sloan, S. W., "Numerical Analysis of Incompressible and Plastic Solids Using Finite Elements," thesis presented to Cambridge University, in 1981, in partial fulfillment of the requirements for the degree of Doctor of Philosophy.
35. Snitbhan, N., Chen, W. F., and Fang, H. Y., "Slope Stability Analysis of Layered Soils," *4th Asian Conference on Soil Mechanics and Foundation Eng.*, Kuala Lumpur, Malaysia, April 1975, pp. 5-26 to 5-29.
36. Spencer, E., "A Method of Analysis of the Stability of Embankments Assuming Parallel Inter-Slice Forces," *Geotechnique*, Vol. 17, No. 1, 1967, pp. 11–26.
37. Taylor, D. W., "Fundamentals of Soil Mechanics," New York, Wiley, 1948.
38. Terzaghi, K., "Theoretical Soil Mechanics," New York, Wiley, 1943.
39. Terzaghi, K. and Peck, R. B., "Soil Mechanics in Engineering Practice," 2nd Edition, John Wiley, New York, 1967.

SECTION THREE

Bearing Capacity

CHAPTER 10 Preloading: Planning and Predictions 367

CHAPTER 11 Ground Improvement by Dynamic Compaction 405

CHAPTER 12 Seasonal Temperature Effects on Foundations 423

CHAPTER 13 Centrifuge Model Tests on Bearing Capacity of Clay 439

Preloading: Planning and Predictions

ARIS C. STAMATOPOULOS* AND PANAGHIOTIS C. KOTZIAS*

GENERAL

Introduction

Preloading is the temporary loading of a construction site, for the purpose of improving subsurface soils. Preloading is sometimes called *precompression*, this term implying the resulting soil densification. Another name is *surcharge*; this term emphasizes the placement of extra load.

The oldest known application of preloading as a method of soil improvement is in the erection of the early Gothic cathedrals of western Europe, in the twelfth century. In modern times, since the 1930s, preloading has become one of the current methods of coping with poor foundation conditions.

Advantages

The greatest advantage of preloading, as compared with other methods of improving ground support, is its much lower cost. Preloading is particularly economical when the fill material with which it is effected, is subsequently used for general grading.

When preloading is performed without vertical drains, another advantage is that the construction equipment needed is the same as for simple earthmoving jobs. Such equipment can be readily mobilized and, therefore, considerable time can be saved at the initial phases of construction. The instruments required for follow-up of most preloading operations are relatively simple and inexpensive. They can be made and installed within 2–3 weeks.

Another advantage of preloading is that it allows an immediate and direct assessment of its effects. By measuring ground movements and, in some cases, also pore pressures, it is possible to follow the process of soil improvement and make predictions of future behaviour. This is not always possible in other methods of soil improvement.

Where successful, preloading secures uniformity of improvement because it eliminates local inhomogenuities. Also, by improving loose sands, it reduces considerably the danger of liquefaction by earthquake.

Special Requirements

Preloading needs space (Figure 1). The area taken up by preload operations usually extends about 10 m, or more, outside the perimeter of the planned structure. In new projects built on bare land this requirement creates no problems but in extensions to already existing installations, space can be critical.

Another requirement is the availability of fill material. Although preloading can also be effected by other means, heaping of fill is by far the most common method. Granular fill is the most desirable because it does not turn into mud when it rains, it can be readily used for embankment construction after the end of preloading, it has a high bulk density and it somewhat reduces the danger of a base failure. Clayey soils are less desirable but have been used successfully by working only when it does not rain. Ore and industrial products are, as a rule, satisfactory.

Access needs also to be considered. The transport of large quantities of earth through inhabited areas may arouse objections. Proximity to the sea, lake, or river may provide an acceptable solution to the problem of bringing in fill, especially if dredging is required for deepening the water.

In addition to the above physical requirements, preloading presupposes a good understanding between owner and engineer. The engineer should be allowed to modify the time schedule during load accumulation, introducing pauses and,

*P. Kotzias-A. Stamatopoulos Company, Ltd., Athens 11471, Greece

FIGURE 1. Preload heaps, made of sandy clay, for a tank farm in Thessaloniki, Greece (1980). Preloading lasted about six months and caused settlements of the order of 1.5 m. The results of Figure 21 are from this site (Case History 3 by Stamatopoulos and Kotzias, 1985).

FIGURE 2. Multistoried hotel in San Diego, California. This site was preloaded in 1969 by 4.5 m of earth fill for about 30 days. Preload settlements were 0.05–0.08 m.

if necessary, extending their duration. Preloading is usually carried out with a low factor of safety against a base failure and, therefore, the ability to monitor deformations with adequate instruments and to intervene at short notice, is essential.

Range of Applications as to Projects

Preloading has been used for every kind of construction, except high-rise buildings and nuclear power plants. In the following list types of projects appear, roughly, in the chronological order of first successful application:

- road embankments (1940)
- bridge abutments and box culverts
- one-storey dwellings and warehouses
- abutments of hydraulic structures
- gravity quay walls (1950)
- site developments for housing, shopping, parking, etc.
- runways
- large monumental buildings
- storage tanks for petroleum and petroleum products
- multi-storied structures (up to about 10 stories)
- light and medium industry
- canals
- heavy industry (1960)

The photographs of Figures 2 and 3 show examples of buildings constructed on preloaded ground. Interesting examples can be found also in the references by *Engineering News Record* (1953), Kyle (1953), Wilson (1953), Aldrich (1965), Darragh (1964), Johnson (1970), Rutledge (1970), and Stamatopoulos and Kotzias (1985).

Range of Applications as to Soil Types

Preloading has been used successfully on virtually every type of naturally laid or man-made soil. Natural soils have been loose sands and silts, soft silty clays, organic silts, or erratic alluvial deposits, consisting of all of these; less frequently, weak organic formations, such as peat. Man-made fills have included miscellaneous depositions, like uncompacted dredge materials, industrial wastes, like cinders, and urban dumps containing rubbish. Preloading has given satisfactory results when applied to soils lying both above and below the water table. Treated soils have had natural

FIGURE 3. Raw materials storage building in central Greece. Building has a length of 144 m, width 33 m and height 16 m. Construction (1982) was preceded by preloading with an earth fill 12 m high, over a period of about six months. Preload settlement was 0.6–0.9 m (Case History 4 by Stamatopoulos and Kotzias, 1985).

water contents 20–1000% and Atterberg limits that covered the whole range of the plasticity chart. Deposits that have given trouble are thick homogeneous layers of either plastic clay or sanitary fill (Sowers, 1964).

Methods

The most common method of preloading is by heaping fill material (Figure 1). After the end of preloading the material is removed and sometimes re-utilized in the same project either for another preload or for embankement construction.

A variant of the above method is to leave the fill in place, rather than remove it, either in part or totally. Examples of leaving part of the fill while removing the rest (the surcharge), in highway embankments, are quoted by Kleiman (1964). An example of leaving all of the fill in place, and constructing on its surface a cement plant, is given by Kotzias and Stamatopoulos (1969). The weight needed for preloading can also be applied on the ground by constructing a peripheral dike and filling its interior with water (Figure 4).

Another method is by using the final structure as a vehicle of load application. This method has been used extensively for petroleum storage tanks. The tank is constructed prior to soil improvement, then it is filled with water in increments. For each new increment, time is allowed for stabilization under the weight of the water. After the tank has been filled to the top and the time rate of settlement has diminished sufficiently, its contents are emptied and its base levelled by jacking. The method is well adapted to tanks built of flexible steel plates. It saves the cost and time involved in bringing and removing fill material. Filling with water is relatively inexpensive and can be accomplished without much delay and, in any case, it is required for testing welded joints and other parts of the construction. Preload settlements of 0.35–1.2 m for tanks 35–45 m in diameter, with differential settlements of up to 0.5 m in the tank bottom and 0.1 m in the shell, have been accommodated. Examples are described by Penman and Watson (1965) and by Darragh (1964). However, there have also been unsuccessful instances, where the soil supporting the tank could not be stabilized, and failed by shearing (base failures).

Elements of the final structure can also be utilized, as illustrated in Figure 5. Here, precast concrete blocks destined for a quay wall, are used for preloading already constructed portions of the same wall.

A different method of preloading is by lowering the water table. This is achieved either by well points, or by trenches, or by vacuum pumping in relatively deep wells (about 20 m). As the water table drops, the soil loses its buoyancy and its unit weight increases. Every one meter of water level drop produces about the same loading as half meter of fill. Sometimes, lowering of the water table is combined with heaping of fill, for increased effect. Informative examples of preloading, by causing the water table to drop, were presented by Halton et al. (1965) and by Ye et al. (1983).

When the water table is low, load can be applied, on some soils, by inundating the surface, or preponding. In this case, compression is caused partly by the weight of the water, and partly by the surface tension forces acting between touching particles. A further action of preponding is to break the precarious bonds existing in loosely deposited soils, like loess (wind-blown or aeolian deposits); as a result, the soil structure collapses and densification takes place. Preponding has been used, for example, to improve the foundation of Trenton Dam in Nebraska, U.S. (USBR, 1974), and to prepare the ground for tens of kilometers of canals.

Finally, preloading has been effected by jacking. This method has been applied mostly to individual footings, either of new buildings, or of buildings to which extra stories must be added. Jacking is also a standard method of preloading footings and piles in underpinning. An example of preloading by jacking is given by Golder (1961).

Preloading in Combination with Piling

In sites where strong support is required under concentrated heavy foundation loads while, at the same time, only moderate soil improvement is needed over extensive areas supporting light loads, preloading can be used in combination with piling. Mixed methods provide increased safety for structures that are sensitive to settlements, combined with low-cost improvement under pavements, utilities and small buildings.

Preloading near piles must be completed first, so that the down-drag forces (negative friction) on piles are minimized.

Normal Practice

The most frequent method of preloading is by embankment loading. Available experience suggests that:

- Duration of preloading, from the beginning of embankment placement to the end of load removal is, in most cases, 3–8 months. In sites where ground response is particularly fast, the duration can be reduced to 4–6 weeks, one of the controlling factors being the time needed to get reliable settlement measurements. On the other extreme there are cases, for example as reported by Jonas (1964), where preloading has taken more than 3 years.
- Height of preload heap constructed of soil fill is, in most cases, 3–8 m above original grade, with 1.5 and 18 m, probably being the minimum and maximum values, respectively.
- Settlements experienced during preloading are usually 0.3– 1.0 m with extreme values 0.05 and 2 m.

Site Preparation

Prior to heaping material for a preload operation, the site must be cleaned of surface vegetation and covered by a base layer of free draining material.

FIGURE 4. Preloading by water at the Elizabeth-Port Authority Marine Terminal on Newark Bay, NJ (Tozzoli and York, 1973). The two reservoirs shown in the photograph have a combined surface area of 26 hectares (63 acres), and were formed by 7.5 m high peripheral dikes. Reservoirs were filled with 6.4 m of water after lining with 0.3 mm PVC sheets. Consolidation lasted one year, and settlement was more than 0.6 m. (Courtesy The Port Authority of NY and NJ.)

FIGURE 5. Concrete blocks 2.3 m in height are stacked on a quay wall for preloading (east coast of Greece, 1980). They will cause a settlement of about 0.2 m in one to two months. Later they will be used in the construction of the extension of the same quay wall.

Surface vegetation must be removed to prevent future settlement by the long-term decay of wood, leaves, etc., and also to facilitate the placement of the base layer. Usually, the upper about 0.5 m of topsoil are also removed, for the same reason, by either bulldozer, or mechanical shovel, or dragline.

The base layer must have a thickness of about 0.6 m and should consist of a mixture of gravel and sand, free of clayey admixtures. It has two functions: First, to receive and discharge the water that reaches the surface of the compressible soils, as a result of the consolidation process. Secondly, to provide a working surface on which equipment can move unhindered, even under rain. In areas where sand-gravel is particularly expensive, it may prove worthwhile to consider using "geotextiles," in sheets of pervious non-woven polyester fabric, about 2 mm thick, which have a high water conductivity and considerable strength.

Conditions for Success

The decision to effect preloading must be based on a thorough investigation of subsurface conditions and a good understanding of the type of the planned structure and its sensitivity to settlements. The time schedule, availability of means and materials, access to the site, magnitude and duration of estimated settlements, must also be considered. Other factors that are significant are cost and expected reaction to earthquake or explosion.

Furthermore it should be ascertained that:

- There will be no base rupture during preloading, or during the operation of the final structure.
- The duration of preloading will be within the time allotted by the construction schedule.
- There will be no damage to adjoining structures.
- There will be no undue disturbance to nearby communities, by dust, noise, etc.
- Settlements during the operation of the final structure will be within predictions.
- The actual cost will be as estimated.

Geotechnical Problems Associated with Preloading

The foundation problems that are usually faced are those encountered in clays and loose deposits of silt and sand, under water, because these are the soils on which preloading is most likely to be undertaken.

The first problem that must be considered is that of a possible base failure under the weight of the preload embankment. The second is settlements during preloading, both in terms of magnitude and duration. The three topics of stability (possibility of a base failure), magnitude of settlement, and consolidation (duration of settlement) are reviewed later.

Site Investigations

Proper site investigations are of the utmost importance. Preloading requires higher investigation effort than other methods of coping with poor soils and, in this sense, it is more sophisticated geotechnically.

The main aims of a site investigation are to determine (a) the soil stratigraphy, (b) the composition of individual layers or formations, (c) the lower boundary of compressible soils, and (d) the elevation of the water table. Borings with continuous coring and sampling constitute the most informative type of investigation because they allow for direct visual observation. A reliable description of soils is essential from the very start and should be made by an experienced geotechnician, with particular emphasis on recording veins of silt-sand and rootholes.

The large subject of drilling, sampling and performing tests in boreholes has been covered extensively first by Hvorslev (1949) and subsequently by a number of textbooks and manuals, for example, by Lowe and Zacheo (1975).

A field measurement which is very useful in assessing the suitability of a site for preloading, is that of permeability inside borings (Cedergren, 1977 and designations E-18, E-19 of USBR, 1974). It must be borne in mind that the in situ measurement of permeability yields results that are, nearly always, higher than those of laboratory tests, sometimes by as much as 100, or even 1000 times. This great difference is due to the fact that, in the field test, the flow is mainly horizontal, or spherical, and takes place, selectively through the more pervious stratifications. On the other hand, in the laboratory test, the flow is perpendicular to the stratifications so that even one thin intercalation of impervious clay may block it.

The schematic layout of a simple permeability test that can be performed in borings is shown in Figure 6; it is assumed that soil is uniform (but not isotropic). In sites where preloading is being considered, it is recommended to perform one test every 3–4 m of boring, with L equal to about one meter. The constant head test is more convenient when the flow is relatively high, about 0.2 lit/sec or more; otherwise, the variable head is easier to manage. Common sources of important error are (a) flow through the contact between casing and surrounding soil, resulting in an overestimation and, sometimes, gross-overestimation of the k value and (b) smearing of the sides of the hole with clay, resulting in an underestimation of k. Additional sources of error, but of smaller importance, are (c) sediment accumulation at the bottom of the boring and (d) collapse of the uncased part of the hole. For calculating k, a value of the ratio of horizontal to vertical permeability, m, must be assumed; in most cases, an m value of between 10 and 100, is appropriate.

Laboratory soil tests for classification, compressibility and strength are covered in a number of books and publications such as those by Lambe (1951), and Bishop and Hankel (1957).

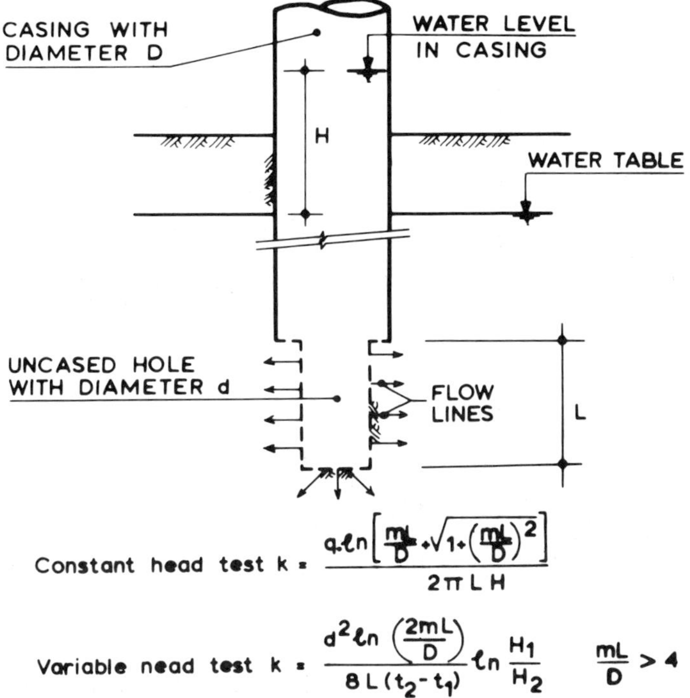

FIGURE 6. Arrangement and formulae for the measurement of field permeability inside borings. It is assumed that soil is uniform.

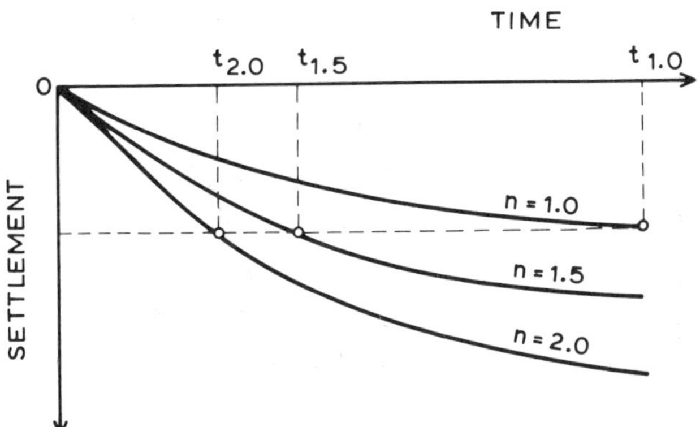

FIGURE 7. The time required for the same settlement becomes smaller with increasing n.

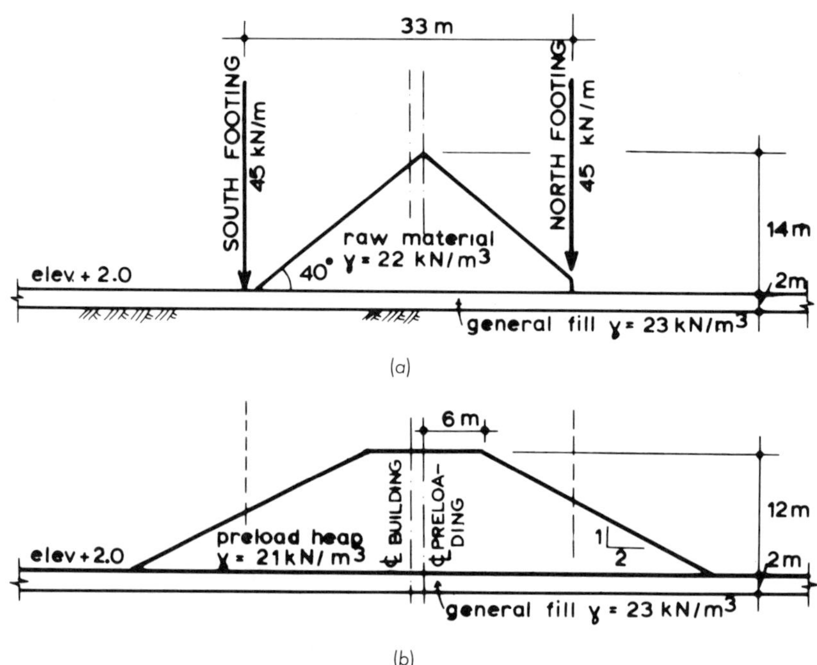

(a)

(b)

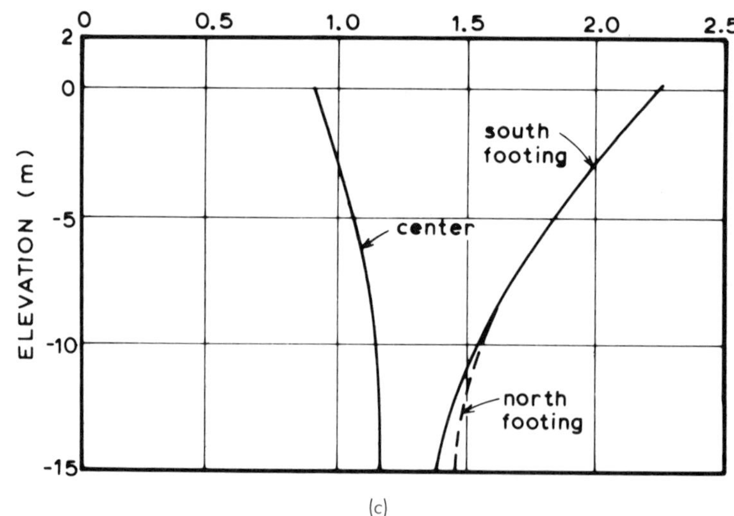

(c)

FIGURE 8. (a) Loads of the final structure; (b) Layout of preloading with n = 1.8; (c) Ratio of vertical stress vs. depth, structure centerline, and linear footings. From Case 4 of Stamatopoulos and Kotzias (1985).

Coefficient of Surcharge and Ratio of Stresses

The coefficient of surcharge (symbol n) is equal to the ratio of the weight used for preloading, to the weight of the final structure to be erected on the improved soil. The higher the value of n, the less the time required for consolidation under full preload and the higher the proven post-construction factor of safety against a base failure. But the higher the value of n, the more time and expense must be allowed for the construction of the preload heap and the greater the attention that must be given to stability. Values of n vary usually between 1.0 and 2.0, but there are exceptions to this generalization, like the case of constructing the final structure on top of the fill that effected preloading (Kotzias and Stamatopoulos, 1969, and Case History 1 of Stamatopoulos and Kotzias, 1985) where $n < 1.0$.

By using a weight of preload higher than the expected weight of the final structure, it is possible to shorten the time required for a given degree of soil improvement. This is illustrated in Figure 7, where curves of settlement versus time, for three load levels, are compared.

Although Figure 7 illustrates the effectiveness of having $n > 1.0$ in shortening the duration of preload, it is oversimplified. The illustration of Figure 7 would reflect the time relationship between preload settlement and settlement under the weight of the final structure, only if the load distribution of the preload were in direct proportion to that imposed by the final structure, at all points on the ground surface. In practice, the various parts of the subsurface are subjected to different values of the ratio R of vertical stress imposed by preload to that created by the final structure. An example of how the ratio R may differ from n is presented in Figure 8. Part (a) of this figure shows the surface loading expected of a raw materials storage building, part (b) the preload with $n = 1.8$ and part (c) the variation of R with depth, below three characteristic points of the building cross section. Here the value of R under the two linear footings varies between 2.2 at a depth of 2 m and 1.4 at a depth of 17 m, whereas under the center R varies between 0.9 and 1.2.

STABILITY-SETTLEMENT-CONSOLIDATION

Danger of a Base Failure

The most common method of preloading is by constructing an embankment. And as the supporting soil is, by the nature of the problem, weak, a base failure becomes a real possibility.

When soft clay fails under an embankment loading, it is usually pushed outward to form a linear bulge [also called a "mud-wave" (Figure 9)] parallel to the foot of the embankment. In soils that do not posess a crust, this type of base failure is more frequent than the clear-cut type, where it is possible to distinguish a single slip surface, below which the ground has not moved.

Stability Analysis for Base Failure

The stability of enbankments, that is, the possibility of a base failure occurring because of the imposed loads, is traditionally examined by trial slip surfaces. These surfaces appear on idealized sections as slip lines; the part of the section above the slip line is called the "free-body" diagram. The two types of free-bodies most frequently used are shown in Figure 10.

In a stability analysis the forces acting on a free-body are analysed and compared. Forces are of two kinds: (1) destabilizing body forces (weight, buoyancy, seismic) that tend to cause a base failure; (2) stabilizing or resisting forces (shear, passive) mobilized by virtue of the soil's strength. The ratio of the sums of the resisting to the body forces for any free-body is the factor of safety for that particular free-body. The minimum factor of safety for all free-body diagrams that can be drawn in some particular problem is the overall factor of safety for the problem, hereafter called simply the factor of safety. The slip line associated with the free body giving this factor of safety is called "critical."

There are various methods of carrying out a stability analysis and hence calculating the factor of safety for a base failure: Fellenius's, Bishop's and Sarma's are among the most notable. The general problem of stability analysis is treated extensively in textbooks such as by Lambe and Whitman (1969). The specific problem as it arises in preload embankments is addressed by Stamatopoulos and Kotzias (1985). An example of the result obtained with a preload embankment, using Bishop's method (simplified), is shown in Figure 11; the arc appearing in the same figure is the critical slip line.

In situations where accuracy of calculation is not of primary importance, either because lack of reliable shear strength data does not justify it, or because preliminary assessment of stability is needed urgently, it is possible to use the *stability chart* of Figure 12. This chart is for a simple slope, without breaks, and for natural soil and embankment having the same shear strength and unit weight, throughout. Figure 13 shows how the section of a real problem (the same as in Figure 11) must be modified so that the chart can be applied. The unit weight is taken equal to that of the embankment material, and the shear strength equal to that of the foundation. The stability chart gives the average induced shear stress along the critical arc; the factor of safety is found by dividing the average shear strength by the average induced shear stress.

Embankment–Soil Interaction

Preload embankments are usually constructed with select materials, placed and compressed in layers. By contrast, the

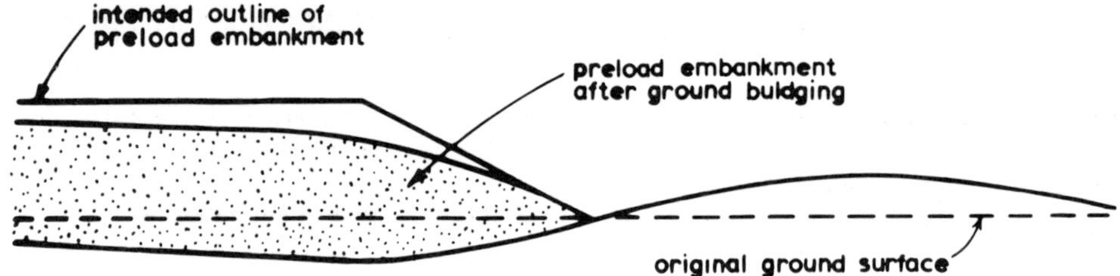

FIGURE 9. Cross section of bulging caused by an embankment on soft clay.

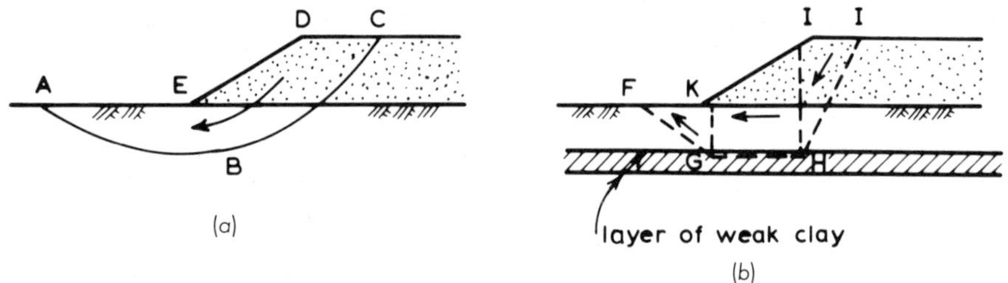

FIGURE 10. Slip failures of the circular and the sliding block types.

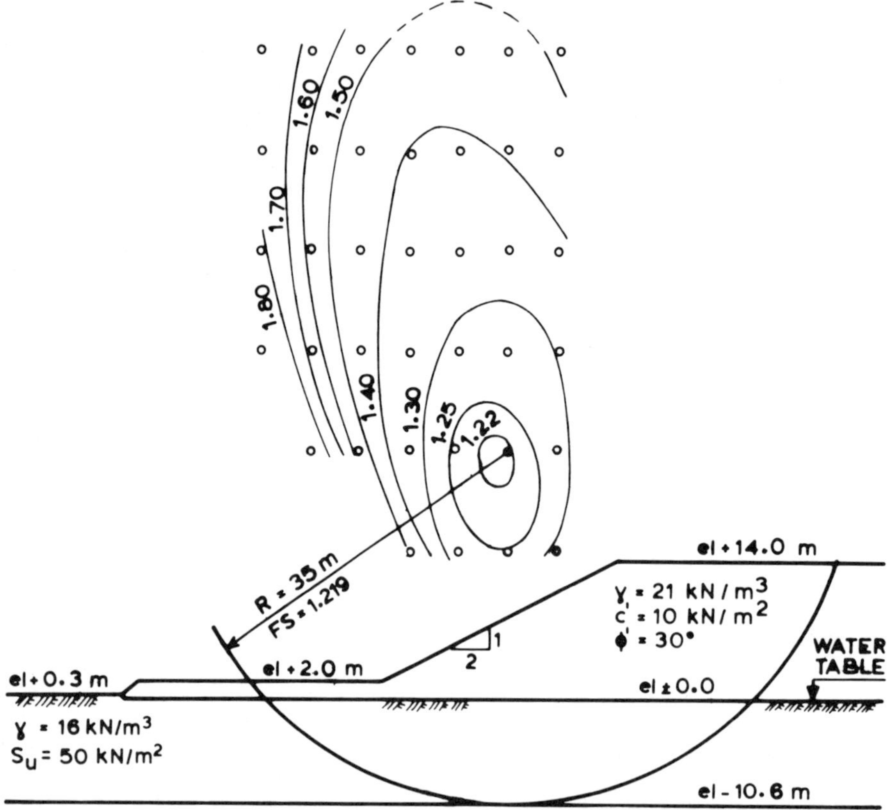

FIGURE 11. Example of a stability analysis of an embankment on soft soil, by performing repetitive calculations. Grid of points represents the centers of trial circles. The arc shown is the critical one and the factor of safety the minimum obtained from all trial circles. Calculations were made by electronic computer, using Bishop's simplified method.

underlying soil is weak and compressible. Therefore, the interaction between the embankment and the soil involves a stronger material on top of a more compressible and weaker one.

The shear stresses acting under the embankment cause a spreading-thinning out of the weak foundation soil. There results an outward movement of the sides of the embankment, away from its central part. The horizontal compressive stresses existing initially in the embankment are, consequently, reduced, sometimes to the point that the embankment develops cracks. The severity of the problem increases with the degree of disparity between the mechanical properties of the materials forming the embankment and its foundation.

If the ratio of the moduli of deformation of the two materials is large (five or more) it may be expedient to assume that the embankment has the same strength as the underlying softer material. In the light of this consideration, the assumption of the embankment having the same strength as the foundation, made in Figure 13, where the FS was calculated, roughly, by the stability chart, seems reasonable.

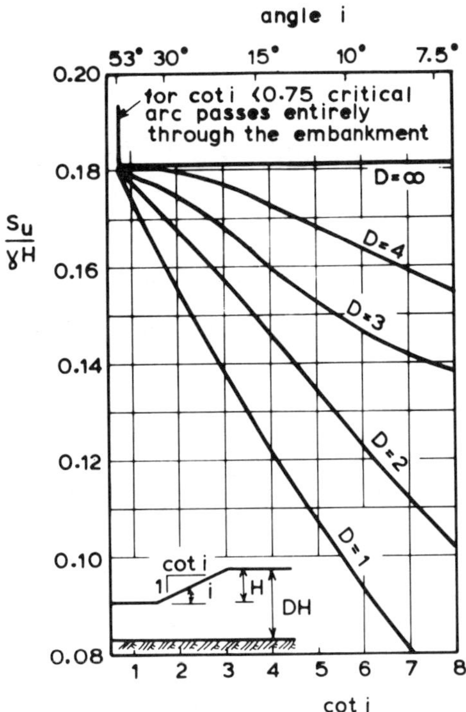

FIGURE 12. Stability chart for uniform soil and lower boundary.

Factor of Safety

A value of the factor of safety in the range of 1.1–1.3 is appropriate for preloading, as it offers reasonable safety while, at the same time, it maximizes the benefits derived from the method. The danger from a factor of safety which is too low is not only a base failure but, especially in cases where the preload embankment rests directly on plastic clay, pronounced creep.

If the calculated factor of safety is too low, it is possible to improve the stability of the embankment by taking one or more measures. One measure may be to extend the period of fill placement to allow the underlying soil to consolidate and thus gain strength. The extra time is particularly beneficial after the fill has been raised to about 60% of its final height. Extra time can be allowed either by slowing down the construction of the latter part of the embankment, or by introducing one or more pauses at some advanced stage of the construction.

The stability can also be enhanced by flattening the slope of the embankment, or by constructing a berm around it. The benefit gained by either of these measures can be determined by comparing the factors of safety calculated with, and without, the improved slope configuration. In the case of a flattened slope, the increased safety can be assessed, roughly but quickly, by the stability chart (Figure 12). In order for a berm to be really effective, it must have a height equal to about half the height of the main embankment and a width equal to about three times the depth of soft soils (Kleiman, 1964).

A further measure that may encourage progress by preventing unnecessary caution is the installation of piezometers. Piezometers allow the determination of pore water pressures needed for the calculation of the factor of safety. Piezometers should, however, be installed in the middle of the clay layers that are most suspect of developing and maintaining high excess pore water pressures. Piezometers installed in soils or at depths where excess pore pressures are smaller than maximum, may give a false indication of safety.

Test Fills for Stability

In problems involving the stability of embankments on soft soils, stability calculations are sometimes confirmed by constructing test fills. This method is particularly effective in highway embankments over poor but uniform soil conditions. Test fills should be well instrumented. Sometimes they are loaded to failure.

When preloading is contemplated over a relatively small area, it is not practical to investigate stability by a test fill. In addition to the time and expense, a test fill carried to failure endangers the integrity of foundation soils. If the test is carried to failure, it is possible to determine, by back analysis, the ultimate shear strength of the soil; if the test stops before failure, a minimum level of proven shear strength can be estimated.

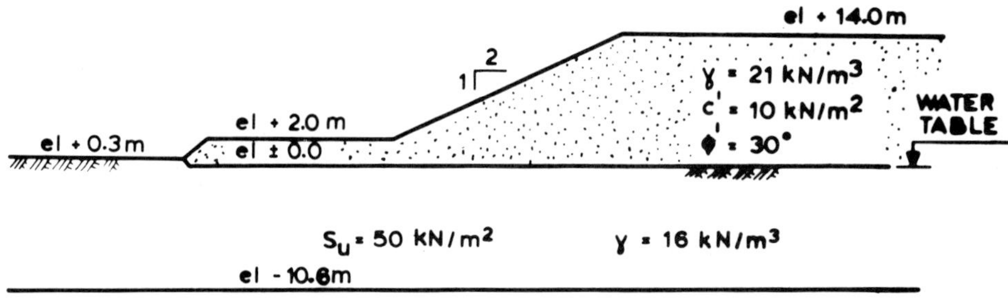

(a) SECTION IN REAL PROBLEM (Case history II)

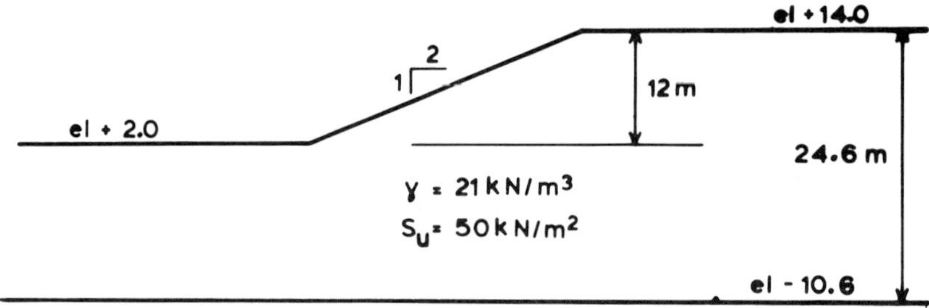

(b) SIMPLIFIED SECTION

Calculation

Enter in stability chart $D = \frac{24.6}{12} = 2.05$ $\cot i = 2$

and obtain $\frac{S_u}{\gamma H} = 0.169$. Hence induced $S_u = 0.169\,\gamma H = 0.169 \times 21 \times 12 =$

$= 42.6\ \text{kN/m}^2$

$$FS = \frac{50}{42.6} = 1.17$$

FIGURE 13. Example of using the stability chart for a simplified section.

But test fills are not useful only for clarifying stability; they also provide valuable information for the prediction of the magnitude of settlement and the duration of consolidation. A further advantage of test fills is that during their construction experience is gained in handling and compressing fill-building materials.

Prediction of the Magnitude of Preload Settlement

Predictions made at the design stage of the magnitude of settlements expected during preloading are useful for the following reasons:

- They help evaluate the progress of consolidation during preloading. For example, when the observed settlement is equal to two-thirds of the predicted value, it means that the processes of consolidation and soil improvement have progressed also by, roughly, two-thirds.
- They help predict the volume of fill that must remain in place after preloading, to compensate for ground subsidence.
- They provide the means to assess from settlement observations, the correctness of the value of compressibility assigned to the soil. A comparison between predicted and observed settlements may lead to adjustments regarding further predictions.

Predictions must be based on the results of borings and field and laboratory tests, or on the results of test fills.

When settlements are predicted by borings and tests, they are calculated as the sum of (1) the elastic initial settlement, (2) the plastic settlement due to immediate-primary compression and (3) the plastic settlement due to secondary compression. The current general methods of estimating settlements are covered adequately in a number of textbooks such as the one by Lambe and Whitman (1969). In the specific case of preload embankments the subject is treated by Stamatopoulos and Kotzias (1985). Elastic settlement is usually calculated by formulae involving the intensity and geometry of the preload embankment, the geometry of the subsurface and the soil's Young's modulus. Immediate-primary compression is calculated by first finding the increase in vertical soil stress under characteristic or critical points and second figuring the deformation of compressible soil layers. In the book by Stamatopoulos and Kotzias (1985) charts are presented, by which it is possible to calculate settlements from surface loads and soil compressibility without going through the intermediate step of finding the increase of vertical soil stress. Secondary compression is calculated from Equation (1).

When settlements are predicted by test fills, they are calculated as a multiple of the settlement caused by the test fill, providing that the size requirements stated under "Time Predictions from the Results of Test Files" are satisfied.

Examples of Settlement Records

Field records of preload settlements are important because they help in the prediction of the settlement of the permanent structure to be built on the preloaded ground. Settlement readings from real problems may give results of the type shown in Figure 14. Here, the ground is loaded gradually, at a rather slow and uneven rate, and the settlement is due to the combined effect of load applied previously and new load being added. Results of this type show that splitting the observed settlement into the three causes of the previous section (initial, immediate-primary, and secondary) is impossible. The identification of separate causes, although theoretically enlightening, is not relevant to field observations.

Real settlements during the application of load are the most difficult to analyse because they are due to a combination of causes. For each new load increment there are initial settlement and immediate compression and, at the same time, because of loads placed previously, primary compression and, also, possibly, secondary compression.

The development of field settlements in time is somewhat clarified by plotting separately the settlement after the placement of fill. Rather than differentiate among causes, it is more useful to differentiate between the settlements that occur during and after the application of load. The settlement after loading, at the central axis of the example of

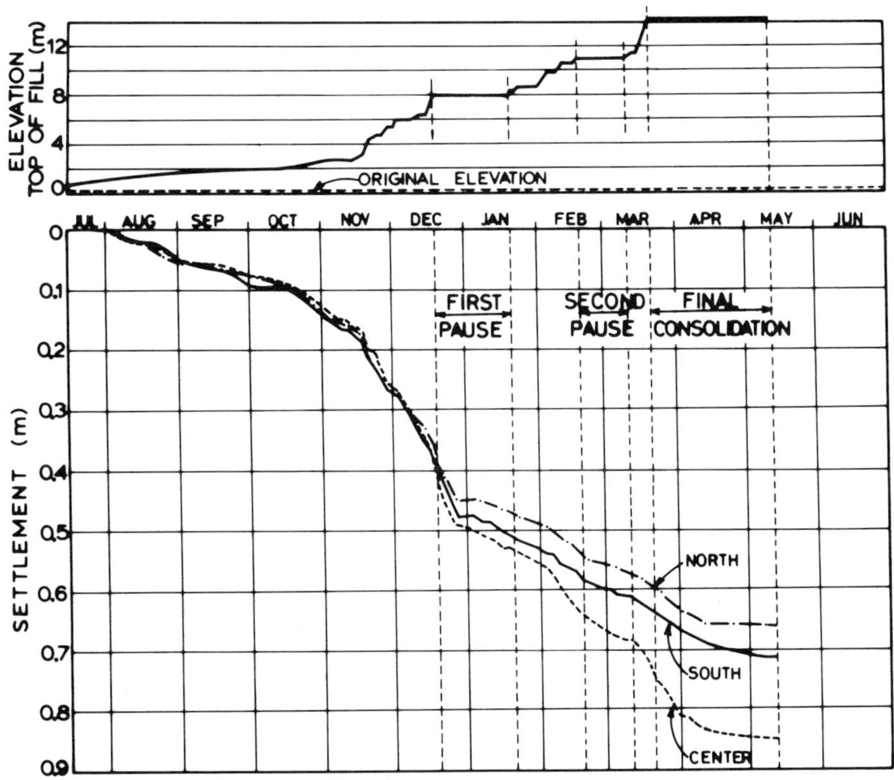

FIGURE 14. Example of loading and settlement from a real project; average settlement along axes vs. time (Stamatopoulos and Kotzias, 1985).

Figure 14, is plotted in Figure 15. The resulting graph is amenable to analysis, because it is a straight line on a logarithmic time scale.

Prediction of the Magnitude of Rebound

Removal of preloading is accompanied by rebound of the soil surface. Rebound starts during load removal and continues under the reduced constant load (Figure 16).The initial phase of rebound, taking place concurrently with the reduction in soil stress, must be attributed to elastic recovery, which is the reverse of initial settlement; the rest to the reverse of immediate-primary compression.

Initial rebound can be estimated in the same way as initial settlement (Stamatopoulos and Kotzias, 1985). The initial rebound should be expected to be of the order of one-fifth, or less, of the initial settlement.

Long-term rebound can be estimated by a modified version of the equations for immediate-primary compression. Soil parameters must be taken from the unload phase of laboratory consolidation tests (Stamatopoulos and Kotzias, 1985).

The expected rebound can be estimated as the sum of the values of the reverse initial settlement and the settlement due to the reverse immediate-primary compression.

Prediction of Magnitude of Operational Settlements

Predictions made at the design stage of the magnitude of settlements that will develop during the operation of the final structure are necessary for the assessment of the structure's future integrity and performance. In addition to maximum, also differential settlements must be foreseen.

Predictions of operational settlements, made before preloading, can be based on borings and tests, like the prediction of preload settlements. But where a coefficient of surcharge of about 1.5 or more has been used, and preload has been allowed to remain until completion of primary consolidation, calculations can be based on more favorable soil parameters. The value of Young's modulus, used in calculating initial settlement of foundations on clay, can be taken equal to about five times the value before preloading; on sands initial settlement can be omitted. Settlement due to secondary consolidation can be omitted for most soils. The settlement due to immediate-primary compression should be calculated using parameters derived from laboratory consolidation tests, where the stress alternations corresponding to preload-unload-structural reload, are reproduced.

Prediction of operational settlements can also be made from borings and tests carried out after preloading is completed. Such investigations are desirable, also, for examining the overall change of soil properties caused by preloading, such as permeability and water content, but are likely to be omitted for economic reasons.

The preceding two paragraphs concern the estimate of average settlement. Differential settlement that is not caused by differences in loading, or known variations in the thickness of compressible layers, but is due to undiscernible random inhomogenuities of the soil, is much reduced by preloading. In virgin, normally consolidated clay, differential settlements under the same loading are usually up to about 0.3 of the average (Lambe and Whitman, 1969 and Bjerrum, 1963). In sites where inhomogenuities have diminished by preloading, this figure can be assumed to be reduced, probably to 0.15 of the average.

Time Prediction

In preload problems, settlement-time predictions are of the following types:

- prediction of the behaviour of large-scale preloading, or of test fill, from the results of borings and tests
- prediction of the behaviour of large-scale preloading, from that of a test fill
- prediction of the behaviour of the permanent structure, from that of the preloading

The first of these predictions, is made by calculating, separately, the time of each of the component causes (1)–(3) mentioned previously, and then combining, to find the aggregate effect in time; this prediction is heavily dependent on theory. The other two predictions require empirical rather than theoretical methods. They are based on field data where, as we saw in a previous section, an attempt to differentiate between component parts of settlements proves usually futile; hence they concern aggregate settlements.

Time Predictions from Borings and Tests

Reverting to the causes of settlement mentioned under another section we note that time predictions of initial settlement and immediate compression are trivial, because these deformations take place shortly after load application. The prediction for secondary compression can be made from

$$\delta = C_a \log \left(\frac{t}{t_0}\right) \cdot H \tag{1}$$

This leaves us only with primary compression whose time-rate is traditionally examined by Terzaghi's Theory of Consolidation (vertical drainage).

Terzaghi's Theory of Consolidation considers a horizontal layer, or slice, of homogeneous, saturated soil, of low permeability, where the vertical stress has just been increased by σ_v (Figure 17). The upper and lower surfaces of the slice are allowed to drain freely. The differential equation governing the time rate of change of excess pore

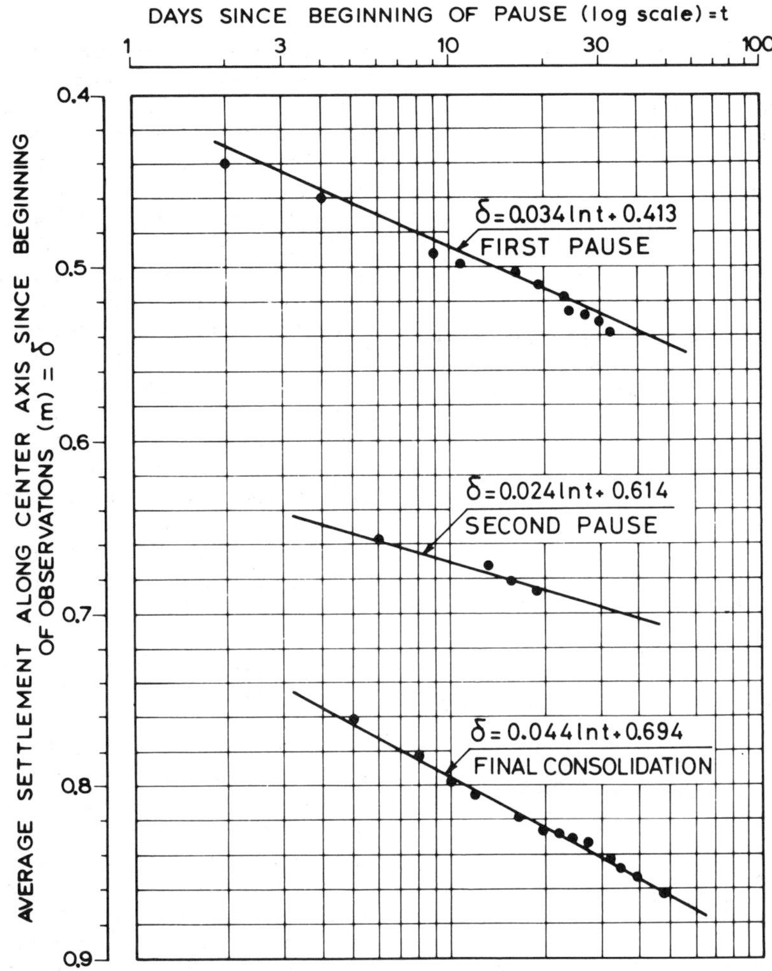

FIGURE 15. Average settlement along the center axis after the end of loading vs. log time: same project as in Figure 14.

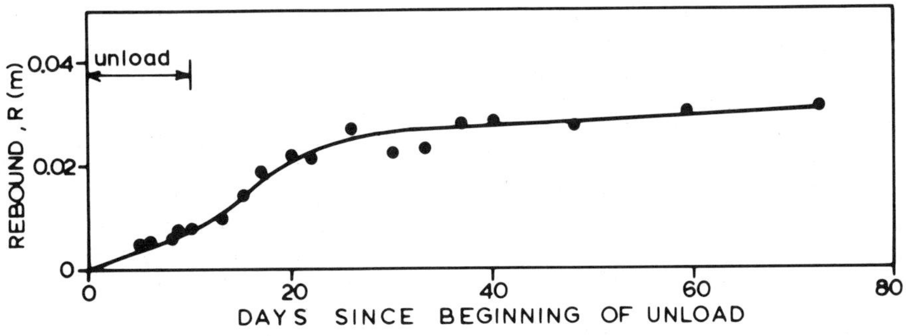

FIGURE 16. Typical rebound curve.

pressure and deformation along the z coordinate is

$$\frac{\partial u_e}{\partial t} = c_v \frac{\partial^2 u_e}{\partial z^2} \qquad (2)$$

where

$$c_v = \frac{Dk}{\gamma_w} = \text{coefficient of consolidation} \qquad (3)$$

The solution of Equation (2) is conveniently expressed by introducing the ratio U which is equal to the primary compression at time t, divided by the ultimate primary compression after infinite time:

$$U = \delta/\delta_{ult} \qquad (4)$$

then

$$U = f\left(\frac{c_u}{H^2} \cdot t\right) = f(T_v) \qquad (5)$$

where H is the maximum drainage path (Figure 17) and T_v the "time factor" (dimensionless). The variation of U with T_v is given in Figure 18.

Having obtained the relationship between U and T_v (Figure 18) we only need to know how to determine T_v. From Equations (3) and (5),

$$T_v = \frac{c_v}{H^2} \cdot t = \left(\frac{Dk}{\gamma_w}\right) \cdot \frac{1}{H^2} \cdot t \qquad (6)$$

If we could determine the value of either expression by which t is multiplied in the above equations, then for each t we could find T_v and for each T_v we could find U (Figure 18) and from it δ [Equation (4)]; H can be determined from the geometry of the subsurface. But it must be emphasized that c_v values derived from laboratory consolidation tests are, more frequently than not, at great variance with the settlement-time behaviour observed in the field; they lead to excessive estimates of settlement duration. The ratio Dk/γ_w must, therefore, be calculated by taking the value of the coefficient of permeability, k, from field tests carried out inside boreholes; D can be taken equal to the specific constrained modulus D_s (Stamatopoulos and Kotzias, 1979 and 1985) and $\gamma_w = 9.81$ KN/m^3. A comparison between predictions and a field test is given in Figure 19.

Time Predictions from the Results of Test Fills

The time rate and magnitude of settlement of large-scale preloading can be predicted reliably from a test fill. The test fill must be wide enough to exert vertical stresses even on the deepest soil formation having substantial compressibility or as wide as the large-scale preloading. Then, providing that the preload heap does not cause a base failure or excessive creep, it can be expected to cause settlements that develop in time, in like fashion to those of the test fill. Any time, figured from application of the full load, large-scale preloading can be expected to settle an amount equal to the settlement of the test fill, corresponding to the same time, multiplied by the ratio of preloading pressure to test fill pressure. The settlement-time curves of Figure 20, afford a comparison between the behaviour of a large-scale preloading and that of the test fill that preceded preloading. The test fill covered an area 40 m wide, so as to cause a substantial increase in the vertical stresses, all the way down to the lower limit of compressible soils (Case History 2 by Stamatopoulos and Kotzias, 1985). Plotted settlements and time are from application of full load, in order to focus attention on long-term behaviour. The two curves of Figure 20 are similar in shape, and the one corresponding to the full-scale loading gives about 3.5 times the settlement corresponding to the test fill; settlements are very nearly proportional to the heights of fill 11 m: 3.4 m.

Time Predictions for Permanent Structure from the Results of Preloading

Settlement-time curves plotted on a logarithmic time scale, frequently have the shape of either the letter S, or of increasing slope, or of a straight line. Examples of such settlement-time curves are presented in Figure 21. These were obtained during the preloading of 25 separate oil storage tank locations (Case History 3 by Stamatopoulos and Kotzias, 1985); they are from three individual heaps but their shapes are typical of those encountered at 2, 11 and 12 tank locations, respectively. These recurrent shapes are referred to as Case A (S-shaped), Case B (with increasing slope) and Case C (linear) (Stamatopoulos and Kotzias, 1983).

Time curve such as that of Figure 21(a) (Case A) do not arouse much concern about future settlements. However, results like those of Figure 21(b) (Case B) and Figure 21(c) (Case C) require interpretation and a method of predicting the expected settlement of the permanent structure.

ASSUMPTIONS

In order to analyze Cases B and C it must first be assumed that no further increase in the slope of the curves will take place. This assumption can not be considered to be always justified; its validity must be examined in each particular case in the light of local experience and the proportion of settlement that has developed, relative to the total expected under the applied preload.

Further assumptions are as follows:

(a) For the same subsurface conditions, settlements are directly proportional to the change of vertical stress

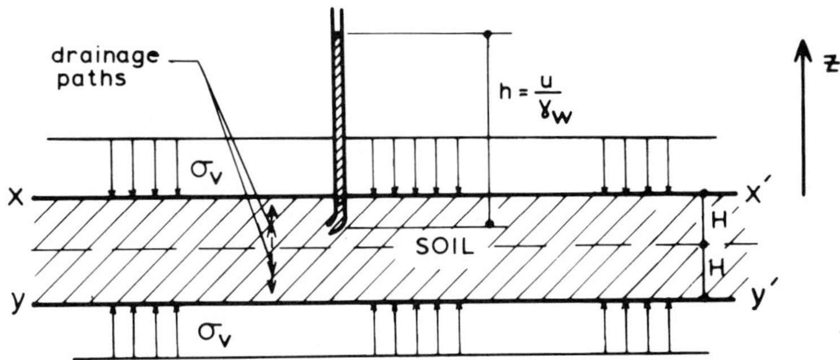

FIGURE 17. Assumptions made in deriving Terzaghi's theory of consolidation by vertical drainage. The upper and lower boundaries of a horizontal element of soil (xx' and yy') subjected to a vertical stress can drain freely. Therefore pore water pressure is zero along xx' and yy' and maximum along the horizontal mid-plane. This model represents closely the laboratory consolidation test.

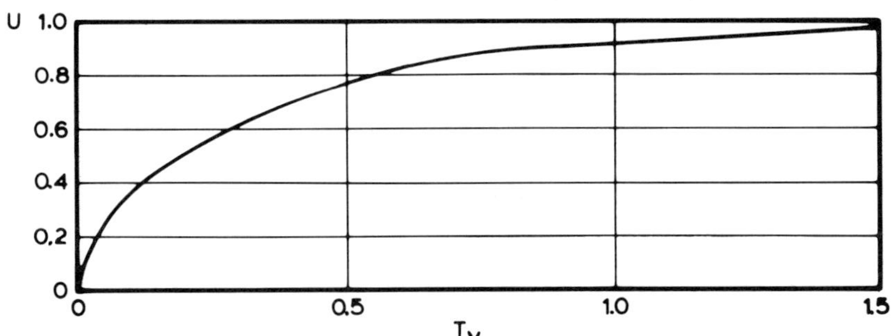

FIGURE 18. Correlation between the average degree of consolidation U and the time factor T, derived by Terzaghi's theory of consolidation; semi-log plot.

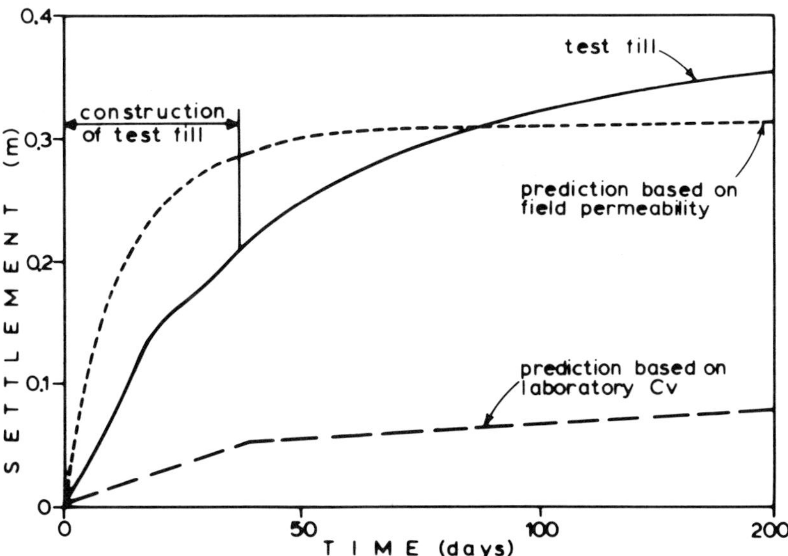

FIGURE 19. Comparison between a settlement–time curve obtained from a test fill and predictions based, first, on laboratory and, second, on field permeability. The second prediction affords a much better approximation to the field curve. From Case History 2 by Stamatopoulos and Kotzias (1985).

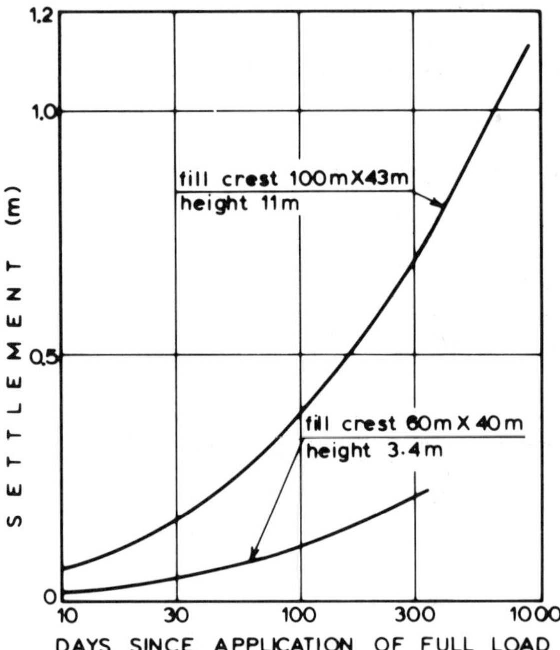

FIGURE 20. Comparison between settlement–time curves. The test fill and the large-scale preloading settled in like fashion and nearly in proportion to the applied load. From Case History 2 of Stamatopoulos and Kotzias (1985).

$\Delta\sigma_v$, induced in the compressible layer(s) by surface loading. This assumption is illustrated in Figure 22 and can be expressed as follows:

$$\frac{\delta_2}{\delta_1} = \frac{(\Delta\sigma_v)_2}{(\Delta\sigma_v)_1} \tag{7}$$

(b) After preloading for a time t_p, to a settlement δ_p, along curve II (Figure 23), let the vertical stresses change to those existing under the influence of the permanent structure. Then, the further development of settlement will be such that the settlement–time curve will converge with curve I, that would have been followed, had preloading not been applied.

PREDICTION FOR CASE B

This case is exemplified by the curve of Figure 21(b). Here it is reasonable to assume that, given enough time, the curve will show a reduction in slope so that it will become S-shaped. In the case of the generalized solution of Terzaghi's one-dimensional consolidation theory, holding for a single layer, the increase in slope takes place at approximately $T_v = 0.07 = a$, where $U = 0.28$ (Figure 24). At one-tenth this value of T_v, or $T_v = 0.007 = 0.1a$, $U = 0.08$, that is, the change in U, between $0.1a$ and a, is 0.20, or the total compression is five times the compression that occurs between T_v values $0.1a$ and a. If, instead of Terzaghi's original set-up, a modified geometrical layout had been assumed, involving more than one layer, the ratio between the total compression and the compression occurring between $0.1a$ and a would be about seven. As it is safer to accept a higher, rather than a lower ratio, it appears reasonable to accept that this ratio is seven.

Referring to Figure 25, it is assumed that a site is preloaded, for a time t_p, to a settlement value δ_p, along the preload curve II. Then the preload is removed and the permanent structure is built. Neglecting, for the time being, the rebound and recompression that will take place during preload removal and construction of the permanent structure, the δ vs. log t curve will approach curve I, corresponding to the stresses of the permanent structure. The future settlement δ_F of the permanent structure that is expected to develop at time T is $\delta_F = \delta_T - \delta_p$ and:

$$\delta_T = \frac{1}{n}\,\delta_T' \leq \frac{1}{n} \cdot 7 \cdot (\delta_a - \delta_{0.1a})$$

hence,

$$\delta_F \leq \frac{7}{n}\,(\delta_a - \delta_{0.1a}) - \delta_p \tag{8}$$

The right-hand side of Equation (8) is the probable ultimate settlement of the permanent structure. However, if the δ vs. log t curve changes slope relatively early in the process of consolidation and, after that, can be approximated by a straight line, the future settlement can be higher than given by Equation (8). In order to avoid underestimating δ_F, it may be preferable to base predictions on the final, linear part of the curve, from which the settlement at time T is

$$\delta_F = \frac{1}{n}\,\delta_T' - \delta_p = \frac{A}{n}\log T - \left(\delta_p - \frac{B}{n}\right) \tag{9}$$

where A,B are the constants of the correlation $\delta = A \log t + B$ of the final, linear part of the plot of Case B (in Figure 21(b), $A = 0.28$, $B = -0.26$). Depending on T, Equation (9) may give negative values of δ_F, in which case δ_F must be taken equal to zero.

PREDICTION FOR CASE C

This case is exemplified by the curve of Figure 21(c). It is assumed, on the safe side, that within the time T, the slope of the time curve will not be reduced, as it is expected to do sooner or later. Referring to curve II of Figure 26,

$$\delta = A \log t + B$$

for $t = 1$ day, $\delta_1 = B$

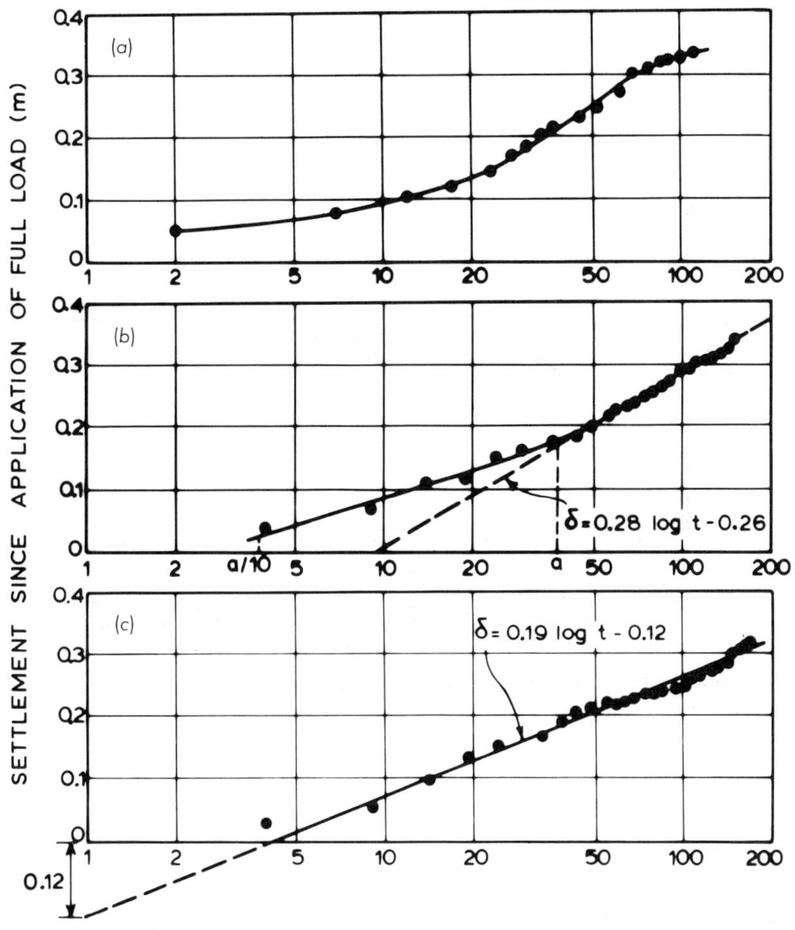

FIGURE 21. Settlement–time curves from a series of preloading. Parts (a), (b), (c), represent Cases A (S-shaped), B (increasing slope) and C (linear), respectively. From Case History 3 of Stamatopoulos and Kotzias, 1985).

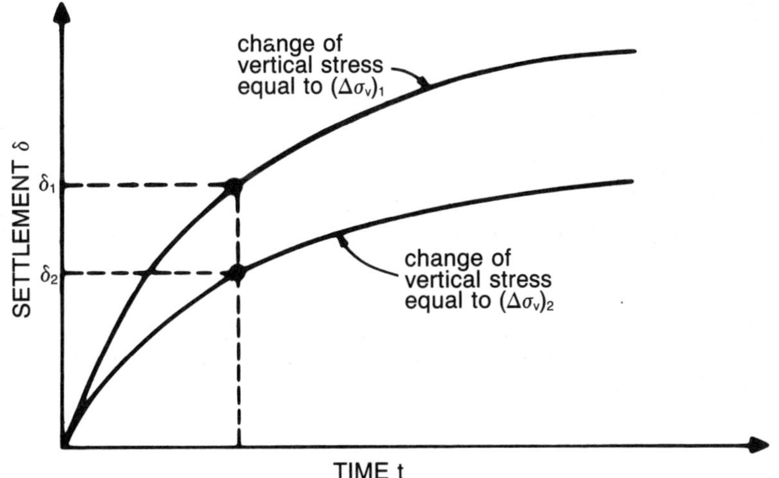

FIGURE 22. Assumption of proportionality between settlement and change of vertical stress, at all times.

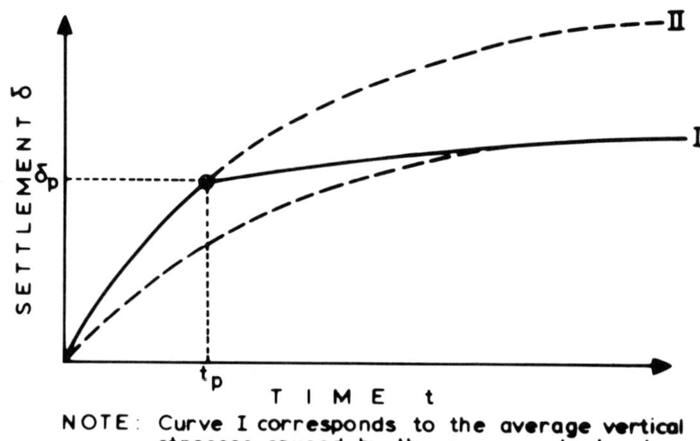

FIGURE 23. Assumption of transition from the time curve of preloading to that of structure-loading.

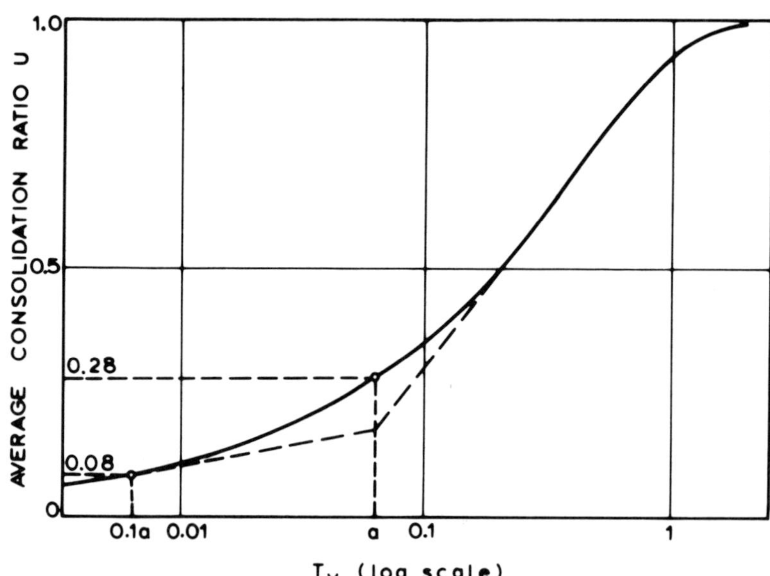

FIGURE 24. The average degree of consolidation U (Figure 18) increases by 0.20 when T changes from 0.007 = 0.1a to 0.07 = a (a is the value of T at which the straight segments intercept).

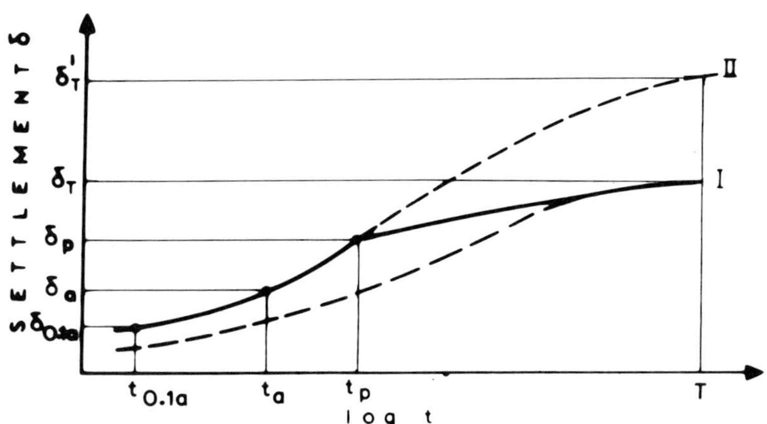

FIGURE 25. Diagram explaining the method of prediction of the final settlement for Case B.

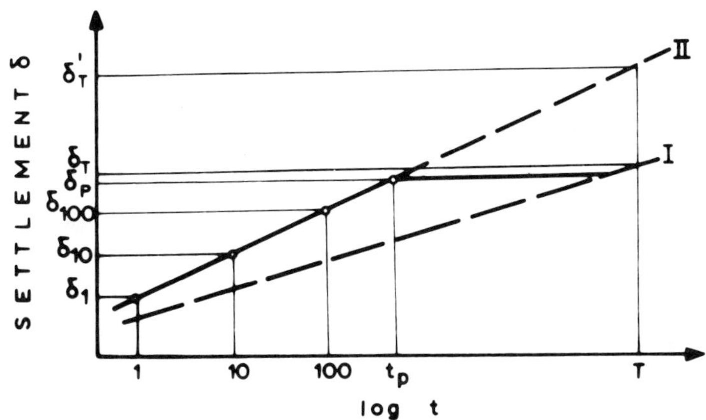

FIGURE 26. Diagram explaining the method of prediction of the final settlement for Case C; t is in days.

for t = 10 days, $\delta_{10} = A + B$, hence $A = \delta_{10} - \delta_1$
also, $\delta_F = \delta_T - \delta_p$

$$\delta_T = \frac{1}{n}\,\delta'_T = \frac{1}{n}\,(A \log T + B)$$

$$= \frac{1}{n}\,[(\delta_{10} - \delta_1) \log T + \delta_1]$$

and

$$\delta_F = \frac{1}{n}\,[(\delta_{10} - \delta_1) \log T + \delta_1] - \delta_p \qquad (10)$$

If Equation (10) gives a negative result, δ_F should be equated to zero.

The recompression due to rebound, not included in Equations (8), (9) and (10), can be estimated on the basis of readings obtained during removal of preload. At the site where the results of Figure 21 were obtained, rebound was approximately 0.03 m two weeks after load removal and then continued at a rate of about 0.13 mm/day (typical result of Figure 16). Taking the final value of rebound, just before reapplication of load, as R_f, the expected recompression should be equal to about R_f/n.

EXAMPLES OF APPROXIMATE PREDICTION

Equation (8) is used to predict, approximately, the ultimate settlement of the permanent structure, from the results of preloading of Figure 8, as follows (n = 1.65, δ_a = 0.17 m, $\delta_{0.1a}$ = 0.02 m, δ_p = 0.34 m):

$$\delta_F \leq \frac{7}{1.65}\,(0.17 - 0.02) - 0.34 = 0.30 \text{ m}$$

Equation (9) is used to predict, approximately, the settlement at time T = 30 years (= 10,950 days) from the coefficients A = 0.28 m, B = −0.26 m, found from the plot of Figure 21(b):

$$\delta_F = \frac{0.28}{1.65} \log 10{,}950 - \left(0.34 + \frac{0.26}{1.65}\right) = 0.19 \text{ m}$$

The value of R_f, that is, the value of rebound for an estimated construction period of one year, is taken equal to 0.08 m. It follows that recompression will be 0.08/1.65, or 0.05 m in addition to the values estimated above.

Equation (10) is used to predict, approximately, the future settlement of the permanent structure, from the results of Figure 21(c) as follows (n = 1.48, T = 30 yrs, δ_{10} = 0.07 m, δ_1 = − 0.12 m, δ_p = 0.32 m):

$$\delta_F = \frac{1}{1.48}\,[(0.07 + 0.12) \log 10.950 - 0.12]$$

$$- 0.32 = 0.05 \text{ m}$$

The estimated recompression is, again, an additional 0.05 m.

Time Rate of Settlement

Plots of the average daily rate of settlement versus time, during load accumulation (linear scales), provide an effective tool for monitoring the danger of a base failure. Sometimes rates are quite high; up to 80 mm/day have been observed. When the ground is settling so fast, fill placement must be discontinued to see whether or not the rate will decrease; if not, failure may be imminent and part of the load should be quickly removed.

The time rate of settlement during fill build-up is, predictably, related to the rate of accumulation of load. An example of such a correlation, obtained from the preloading of a coastal area, underlain by erratic deposits, is shown in Figure 27. The time rate of settlement during fill build-up is also related to the pore pressure induced in the soil, both quantities reflecting the intensity of the process of consolidation. An example is shown in Figure 28.

During consolidation under constant load, the plots of settlement vs. log time give, more frequently than not, linear configurations like those of the latter parts of Figures 21(b) and 21(c). Such settlement–time curves can be approximated by an equation of the form δ = A log t + B. In the case of Figure 21(b),

$$\delta = 0.28 \log t - 0.26$$

from which it follows that

$$\frac{d\delta}{dt} = \frac{0.12}{t}$$

In the case of Figure 21(c), $\delta = 0.19 \log t - 0.12$, hence,

$$\frac{d\delta}{dt} = \frac{0.082}{t}$$

In both cases the shape of the linear plot of $d\delta/dt$ versus t, is a hyperbola (providing that t is measured from an appropriate beginning).

It is noteworthy that when preloading continues until the time-rate of settlement is reduced to a final value of 0.5–1.0 mm/day, and the surcharge ratio n is 1.5 or higher, satisfactory soil improvement can be expected.

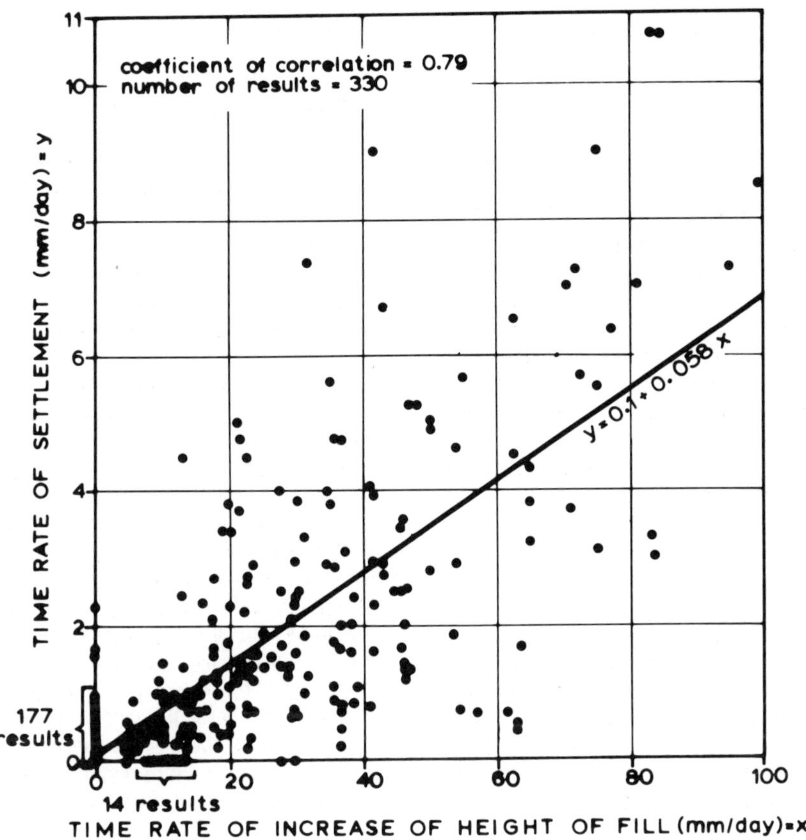

FIGURE 27. The time-rate of settlement correlates well with the time-rate of loading. Coastal area underlaid by erratic deposits (Stamatopoulos and Kotzias, 1985).

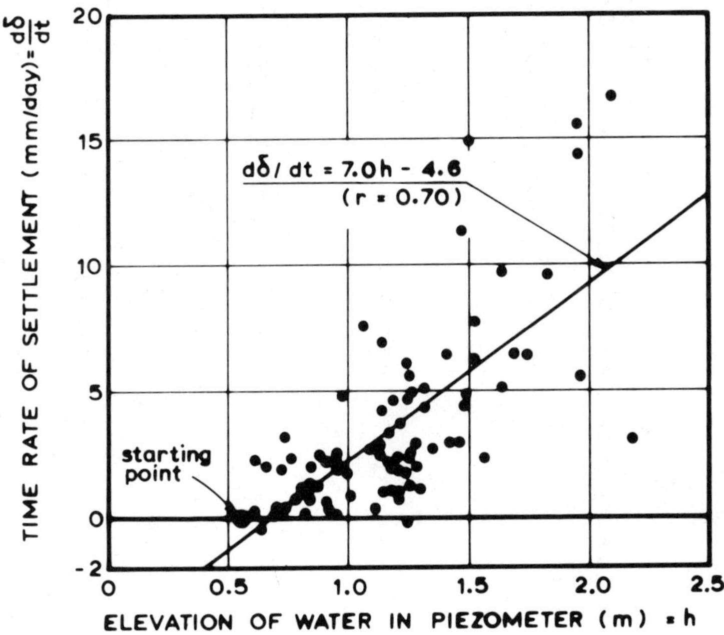

FIGURE 28. Example of correlation between the time-rate of settlement and the piezometric head (Stamatopoulos and Kotzias, 1985).

Instrumentation

Monitoring of the settlement and its time-rate, of the development of excess pore pressures, and of the progress of consolidation, such as implied by the plots of Figures 19–21 and 28, can only be achieved by instruments.

Instruments are mainly of two types: settlement plates, whose function is the measurement of the settlements of the natural ground surface, and piezometers, whose function is the measurement of pore water pressures. Less frequently, devices are installed for measuring the side movement (inclinometers), or the compression between specific elevations in the subsurface.

When deciding on the number of instruments to be installed, allowance (about 20%) should be made for damage from construction equipment.

SETTLEMENT PLATES

Settlement plates consist of a steel base plate and a string of rods at right angles to the plate (Figure 29). The plate is set horizontally on the natural soil surface, after removing the topsoil and backfilling with clean sand properly tamped. Originally, only one length of rod is attached to the plate, but as the preload fill is built, more rods are added. The plate allows the determination of the total settlement under it, by simple levelling, to the nearest about 2 mm.

Settlement plates should be given priority over other types of instruments, because the readings obtained from them are the easiest to interpret and the most conclusive. In relatively simple sites they suffice by themselves.

PIEZOMETERS

Piezometers are installed at critical or representative locations within the subsurface, in order to follow, first, the development and, later, the dissipation of excess pore water pressure.

The most common type of piezometer is the one shown in Figure 30. It is an open standpipe, installed inside a boring, with the space between standpipe and soil backfilled with impervious material, such as clay, except for a short (length about 1 m) section which is filled with a pervious medium, such as coarse sand. This section is called herein the "active part" because it allows hydraulic communication between the soil around the piezometer and the standpipe. Communication with the inside of a piezometer is secured through the "point" which is attached to, or is part of, the bottom of the standpipe. The standpipe accommodates a water head equal and opposite to the pore pressure of the soil around the active part of the piezometer [Figure 30(a)]. It should be noted, however, that the head measured is not, necessarily, equal to the maximum pressure existing next to the active part of the piezometer. In the presence of both clay and sand

FIGURE 29. Settlement plates and rods, just before installation.

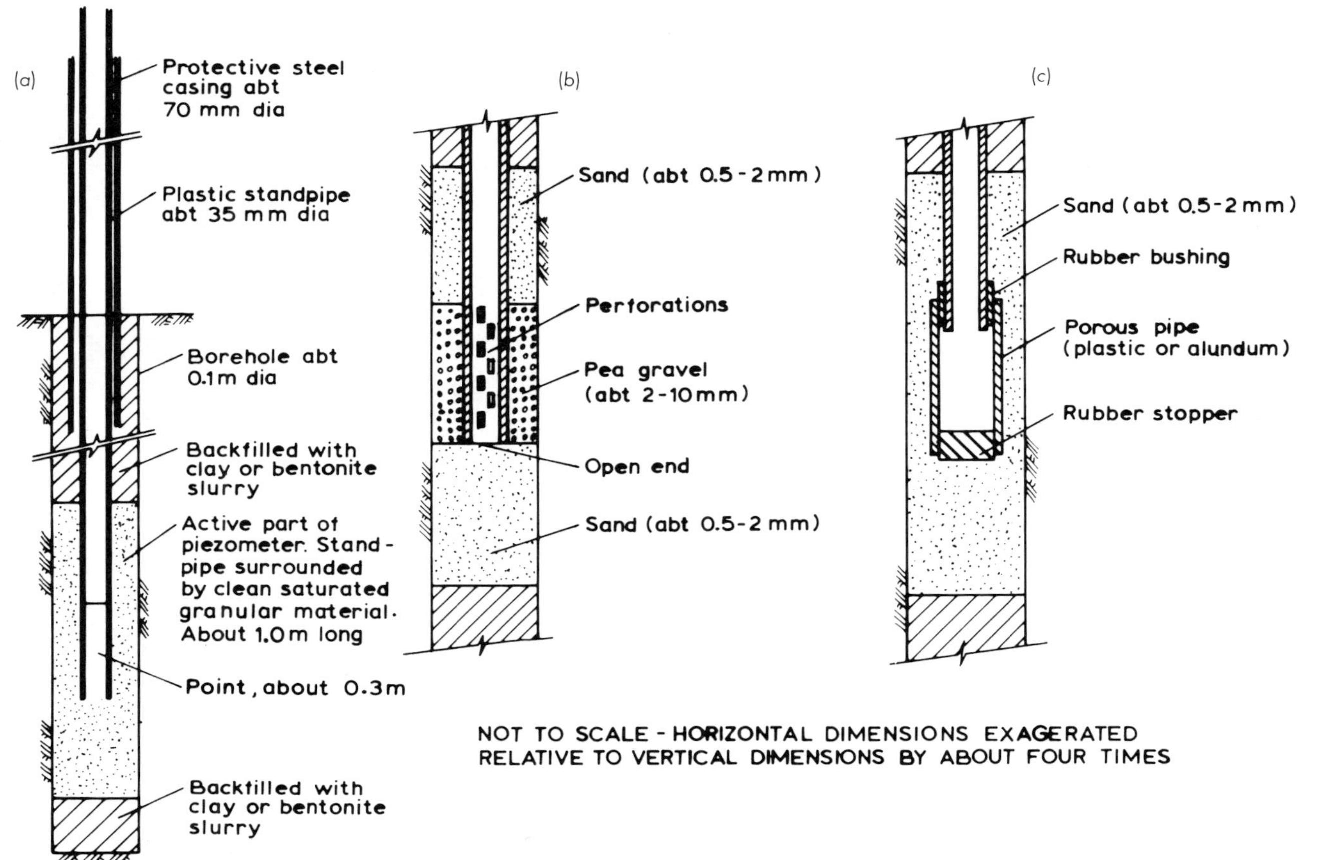

FIGURE 30. Open standpipe piezometer. (a) General layout; note that the piezometer point communicates with the soil through the "active part." (b) The piezometer point may consist of perforations covered by a plastic screen, surrounded by pea gravel. (c) The piezometric point may also consist of a porous cylinder.

layers, the measured value may be indicative of the water pressure in the sand rather than the clay.

In addition to the borehole-installed open standpoint piezometer, described above, which is suitable for most cases, there are other types of piezometer that may be useful under special conditions.

- pushed, or driven, piezometer point, with either hydraulic or electric sensing
- borehole-installed, twin-tube, closed hydraulic system
- borehole-installed, pneumatic
- borehole-installed, vibrating wire (electric)

For details the reader is referred to the designation E-27, E-28 of USBR (1974), Shannon et al. (1962), Broms (1975), Wissa et al. (1975), and manufacturers' catalogs, for example Soiltest and Geonor.

INCLINOMETERS

Inclinometers, also known as slope indicators, or borehole deflectometers, measure the average inclination along given depth intervals inside boreholes. They operate by sensing the electric resistance of a wire-wound arc, whose length is affected by the point of a pendulum, that follows the vertical direction. From the inclinations, the horizontal movement of the foundation soil, and its time rate, can be determined. If ground settlement is accompanied by important outward movement, it may be caused, to a large extent, by side displacement, rather than volume reduction; this factor must be figured in estimating the soil improvement that should be expected from preloading (Holtz and Lindskog, 1972).

Inclinometers should be used only in those cases where intensive creep under a preload heap, is a real possibility. The best source of information regarding these instruments is their manufacturers; the reader is referred, among others, to the Slope Indicator Catalog and the Terrametrics Catalog.

INSTALLATION FOR VERTICAL DIFFERENTIAL SETTLEMENT

This installation allows the determination of the vertical distance between successive telescoping couplings in borehole casing. Measurements are made by a torpedo that provides the signal of being in line with each coupling, at which time, the chain reading is taken.

Differential vertical settlements are useful only in those cases where it is essential to measure the compression of specific layers in the subsurface, independently of the total settlement. Instruments are described in manufacturers' catalogs, such as the Slope Indicator Catalog.

VERTICAL DRAINS

Introduction

Vertical drains are continuous vertical columns or trenches, made of pervious material, installed in compressible clayey soil, for the purpose of collecting and discharging the water expelled during consolidation. Vertical drains are mainly of two columnar types: (a) sand drains, made by filling cylindrical holes with sand; (b) prefabricated drains. To these two we may add vertical trenches (Cole and Garrett, 1980) which can be useful for limited depths (up to about 5 m). The two main types were first used with a difference of only a few years (Wood, 1981), the sand drain in 1934 (Porter, 1936) and the prefabricated one in 1937 (Kjellman, 1948).

Since 1980, prefabricated drains have gained most of the market because they are cheaper and faster to install (Morrison, 1982). In the technically advanced countries, and in large projects, the cost of certain prefabricated drains per unit area of treated ground is about one-third of the cost of sand drains (McGown and Hughes, 1981).

A further aspect of drains is that they reinforce the soft ground in which they are installed. This reinforcement helps in supporting the preload embankment; improvement in bearing may be of the order of 10%.

Decisions for Using Vertical Drains

Drains provide the pathway for the pore water to escape from the consolidating clay by travelling a shorter distance than would have been necessary without them and, furthermore, they allow the flow inside the soil to take place along the horizontal, which is the direction of least resistance (Figure 31). Typical spacings are 1 to 3 m. There is no doubt that there are cases where vertical drains have shortened the duration of consolidation. Still, there is reason to believe that there are cases where drains have been used without being needed (Lewis et al., 1975). The view has also been expressed that soil disturbance during drain placement may cause instability and a prolonged period of stabilization (Casagrande and Poulos, 1969).

The decision to install (or exclude) drains must be based on either local experience or field tests. It must not be based exclusively on laboratory tests and calculations applying Terzaghi's theory of consolidation, because this approach is known to overestimate (and in some instances to grossoverestimate) the time needed for consolidation.

If experience derived from nearby projects, located on the same geologic formation, indicates that consolidation takes a reasonable time (say, up to a few months) then, clearly, drains should not be considered. Alternatively, if field permeability tests lead to a time prediction (by the method of a previous section and Figure 18) which can be accommodated within the construction schedule, drains should not be considered either. In the opposite cases drains may provide the solution to the timely completion of consolidation.

Mathematical Problem

The process of consolidation, where vertical compression takes place by virtue of horizontal radial drainage, is called

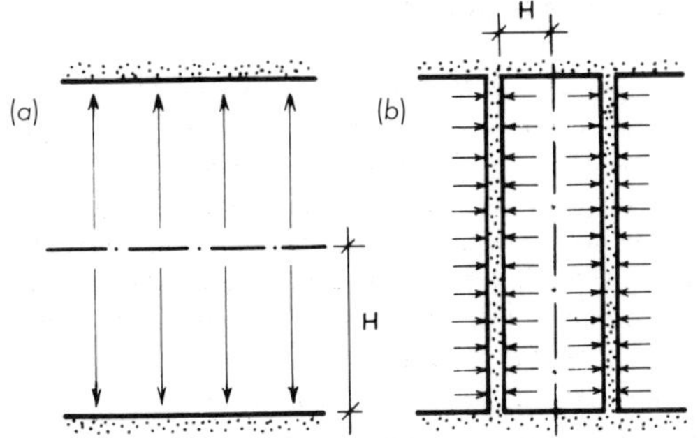

FIGURE 31. Arrows indicate direction of water flow (a) without and (b) with vertical drains. Maximum drainage path is marked H.

radial consolidation. If the soil is homogeneous, water flow will take place along the radius emanating from the center of the drain, in which case the applicable differential equation is

$$\frac{\partial u_e}{\partial t} = c_R \left[\frac{\partial^2 u_e}{\partial r^2} + \frac{1}{r} \frac{\partial u_e}{\partial r} \right] \tag{11}$$

where r is the distance from the axis of the pattern of flow lines and

$$c_R = \frac{Dk_H}{\gamma_w} = \text{coefficient of radial consolidation} \tag{12}$$

c_R is the coefficient of consolidation for radial drainage, in units of (length)²/time, assumed constant, k_H the coefficient of permeability in the horizontal direction, D the modulus of compressibility, and γ_w the unit weight of water, equal to 9.81 kN/m³.

The solution for equal vertical strains is given by:

$$U = 1 - e^{-2T_R/F(n)} \tag{13}$$

where U is the average consolidation ratio, equal to the primary compression at time t divided by the primary compression after infinite time, and e the base of natural logarithms.

$$T_R = \frac{c_R}{r_e^2} \cdot t \tag{14}$$

$$F(n) = \frac{n^2}{n^2 - 1} \ln(n) - \frac{3n^2 - 1}{4n^2} \tag{15}$$

and $n = r_e/r_w$, r_e being the radius of influence of the drain and r_w the radius of the drain. The variation of $F(n)$ with n is given in Figure 32.

When $t = 0$, $U = 0$, and as t increases, U approaches asymptomatically one.

The solution for equal vertical strains, given analytically by Equation (13), is presented graphically in Figure 33 (Leonards, 1962). The assumption of free vertical strains would have given somewhat higher U for the same T_R, but for $T_R > 0.5$ the difference is 5% or less (Richart, 1959).

A different way of presenting the solution of Equation (13) is by introducing "normalized time," that is, the ratio of physical time to the time necessary for 0.99 of primary consolidation. This solution is independent of the value of n, the ratio n entering only in the calculation of the time necessary for 99% average consolidation ($U = 0.99$) (derivation

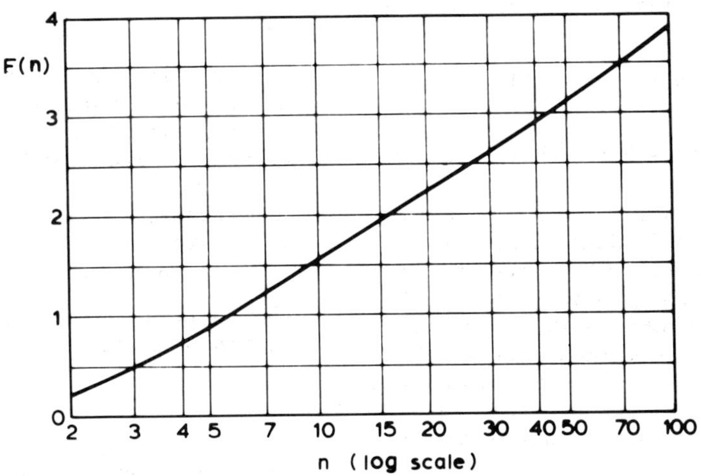

FIGURE 32. Variation of F(n) with n.

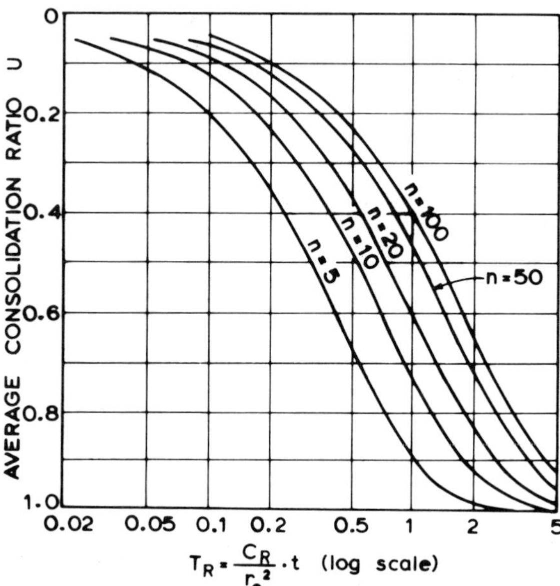

FIGURE 33. Solution of differential equation for vertical drains for various values of n.

given by Stamatopoulos and Kotzias, 1985):

$$t_{99} = \frac{2.3\ F(n)\ r_e^2}{c_R} \tag{16}$$

In any particular problem, the normalized time scale of Figure 34 can be changed into physical time, by multiplying by t_{99}.

EXAMPLE OF CALCULATION OF t_{99}

Sand drain installation with $r_e = 0.80$ m, $r_w = 0.12$ m, $c_R = 0.17 \times 10^{-6}$ m/sec

$$n = 0.80/0.12 = 6.67, \text{ hence } F(n) = 1.20$$

$$t_{99} = \frac{2.3 \times 1.20 \times 0.8^2}{0.17 \times 10^{-6}} = 10.4 \times 10^6 \text{ sec} = 120 \text{ days}$$

The solution of Figures 33 and 34 was derived without considering the effect of soil disturbance during drain placement, and, also, without checking the resistance of the drain to the flow of water that must be discharged, in order to keep the drain fully operative. These two factors were left out of the problem, temporarily, to facilitate its mathematical solution. It has been known, however, that both soil disturbance and drain resistance may affect the function of drains substantially. They are, therefore, examined directly hereunder.

Effect of Soil Disturbance

The soil is disturbed during drain placement mainly in two ways: (a) by being compressed and sheared, especially when the drains are formed by soil displacement, as when a closed-end pipe, or mandrel, is pushed into the ground, and (b) by being smeared, which causes the sand-silt veins to be covered with clay. The two actions reduce the soil permeability around the drain for different reasons but their combined effect has traditionally been called "smear." The smeared zone slows down consolidation.

In order to account arithmetically for smear it is acceptable to adjust the well radius to a value which is smaller than the actual one: the actual radius r_w is divided (or n is multiplied), by a factor of the order of 2 (Leonards, 1962). The exact value of the factor should depend on the method of placement, the type of soil, the probability of fine particles migrating from the disturbed zone to the drain, etc.

Check for Drain Resistance

In order for water to flow upwards in the drain, there must be a head differential from bottom to top, that is, the head h must be maximum at the bottom and zero at the top. But h acts as a backpressure to the excess pore pressure driving the water out of the soil, and, in this way, it retards consolidation. If the ratio $\gamma_w h/u$ is small (say, of the order of 0.02), its effect is negligible, but if it is large (say 0.2 or more), the consolidation process should be expected to slow down.

The ratio of the water pressure at the bottom of the drain, to the average excess water pressure in the soil pores at the same elevation is called "Relative Drain Resistance" and denoted by RDR (dimensionless). Its value is given by (derivation given by Stamatopoulos and Kotzias, 1985):

$$RDR = \frac{7.23\ r_e^2\ \epsilon}{t_{99}} \cdot \frac{L}{C} \cdot \frac{\gamma_w}{\sigma_{vB}} \tag{17}$$

where

r_e = the radius of influence of the drain
ϵ = the average vertical strain of the consolidating soil, under the applied preloading (dimensionless)
t_{99} = time calculated from Equation (16)
L = length of drain
γ_w = unit weight of water
σ_{vB} = vertical soil stress caused by preloading at depth L
C = drain conductivity, as explained hereunder:

For sand drains and for round prefabricated drains, filled with granular material,

$$C = \pi r_w^2 k \tag{18}$$

where k is the coefficient of permeability of filling material.

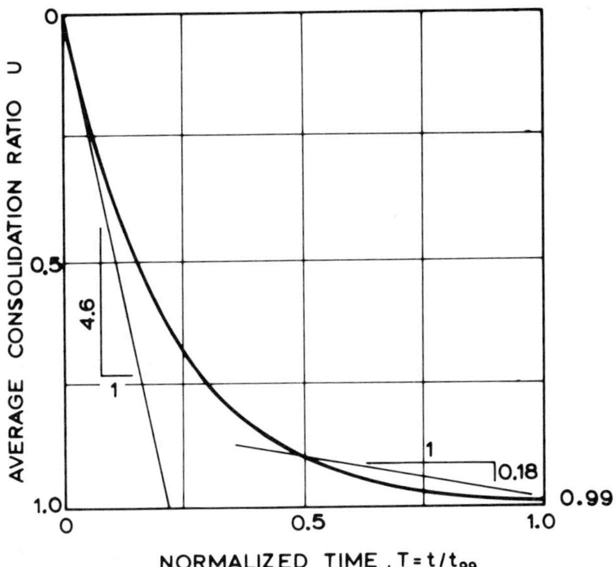

FIGURE 34. Solution of differential equation for vertical drains by using normalized time; the initial slope is 25 times greater than the slope between U = 0.90 and U = 0.99.

For prefabricated drains, consisting of sets of unfilled conduits,

$$C = \frac{\pi}{8} \cdot \frac{\gamma_w}{\mu} \cdot \Sigma\, r_c^4 \tag{19}$$

where

μ = the coefficient of viscosity of water = 0.00152 N m^{-2} sec at temperature 5°C (at 0°C it is 0.00179 and at 10°C it is 0.00131)

r_c = the radius of individual conduits (in rectangular conduits it equals the radius of a circle having the same area as the rectangular conduits)

For prefabricated drains of fibrous cross section,

$$C = a\,k \tag{20}$$

where a is the sectional area and k the coefficient of permeability of the fibrous material.

EXAMPLE OF CALCULATION OF RDR

Sand drain installation with $r_e = 0.8$ m, $\epsilon = 0.07$, $t_{99} = 120$ days, $L = 15$ m, $r_w = 0.12$ m, $k = 2 \times 10^{-5}$ m/sec, $\sigma_{vB} = 150$ kN/m^2, $\gamma_w = 9.81$ kN/m^3. From Equation (18) $C = \pi r_w^2 k = 3.14 \times 0.12^2 \times 2 \times 10^{-5} = 9.05 \times 10^{-7}$ m^3/sec and from Equation (17),

$$RDR = \frac{7.23 \times 0.8^2 \times 0.07}{120 \times 24 \times 3600} \cdot \frac{15^2}{9.05 \times 10^{-7}} \cdot \frac{9.81}{150} = 0.51$$

This value is too high and a change in the design should be considered.

Determination of c_R

The determination of c_R should be based on Equation (12), taking D from laboratory consolidation tests (Lambe, 1951 and Stamatopoulos and Kotzias, 1979), and k_H from field permeability tests.

Attempts to determine c_R from laboratory tests (Prawono, 1981, Trautwein et al., 1981), are subject to very serious errors, because the drainage of consolidation water in the field takes place through routes of higher than average permeability, the inclusion of which is unlikely in a small laboratory soil specimen.

EXAMPLE OF CALCULATION OF c_R

Let $D = 3.0$ MN/m^2, $k_H = 5 \times 10^{-8}$ m/sec

$$c_R = \frac{D\,k_H}{\gamma_w} = \frac{(3000 \text{ kN/m}^2) \times (5 \times 10^{-8}\text{m/sec})}{9.81 \text{ kN/m}^3}$$

$$= 1.5 \times 10^{-5} \text{ m}^2\text{/sec}$$

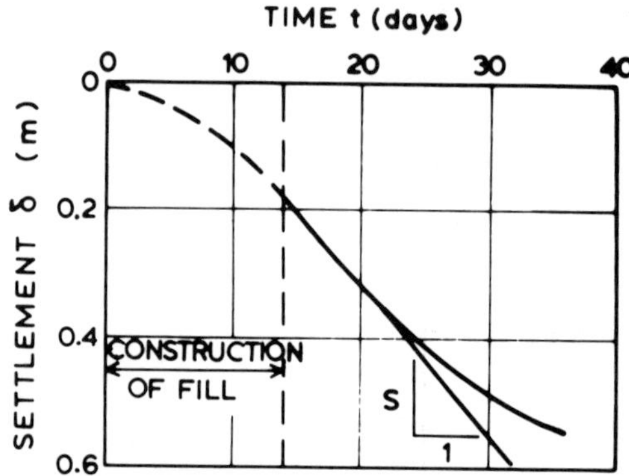

FIGURE 35. Example of settlement–time plot obtained during field test using vertical drains; the slope S of the curve, shortly after full load has been applied, can be used for estimating c_R.

If a properly instrumented field test can be performed, duplicating faithfully the type and method of installation of the vertical drain that will be used in the prototype, the value of c_R can be calculated from the plot of observed settlement versus time; an example of such a plot is shown in Figure 35. The slope S of the curve, shortly after full load has been applied, is related to the slope of the curve U vs. T of Figure 34 as follows:

$$\delta = \delta_{max}\, U \qquad t = t_{99}\, T$$

hence,

$$S = \text{initial}\ \frac{d\delta}{dt} = \frac{\delta_{max}}{t_{99}} \cdot \frac{dU}{dT} = 4.6\,\frac{\delta_{max}}{t_{99}}$$

therefore,

$$t_{99} = \frac{4.6\ \delta_{max}}{S} \tag{21}$$

and by combining Equations (16) and (21) and solving for c_R,

$$c_R = \frac{F(n)\ r_e^2}{2\ \delta_{max}} \cdot S \tag{22}$$

In Equation (22), $F(n)$ and r_e can be determined from the geometry of the drains during the test and δ_{max} can be estimated as mentioned under "Prediction of the Magnitude of Preload Settlement."

EXAMPLE OF CALCULATION OF c_R FROM FIELD TEST [EQUATION (22)]

Let $r_e = 1.3$ m, $r_w = 0.15$ m, $\delta_{max} = 0.65$ m

$S = 0.03$ m/day

$n = 1.3/0.15 = 8.7$, hence $F(n) = 1.45$

and

$$c_R = \frac{1.45 \times (1.3\ \text{m})^2}{2 \times 0.65\ \text{m}}\ (0.03\ \text{m/day})$$

$$= 6.5 \times 10^{-7}\ \text{m}^2/\text{sec}$$

Construction of Sand Drains

The holes required for the installation of sand drains in the ground have been formed by a variety of methods. The pipe or mandrel required to cast the hole can be open or closed at the bottom, and it can be sunk by jetting, driving, rotating or vibrating. The soil that initially occupied the space of the hole is either recovered by washing or augering or, alternatively, it is displaced upward and sideways.

Three of the methods of sand drain construction will be outlined. The first two are presented because they are very common, and the third because it might prove useful in an area of limited means, where a small size project might not leave room for high mobilization costs. They are (a) high pressure water jetting, (b) displacement of the natural ground and (c) wash boring. In all these methods, the formation of the hole begins by inserting a steel pipe in the ground. This pipe is later removed as the hole is filled with sand from the top. All methods are labour intensive. Hole diameter varies usually between 0.2 m and 0.5 m.

High pressure *water jetting* is accomplished by discharging water through the bottom of a steel rod, at a rate of the order of 50 lit/sec. As the water hits the bottom of the hole it causes the soil to dissociate by erosion: it then flows upwards around the jetting rod, carrying the soil fragments to the surface. As the hole progresses downwards, the steel pipe sinks by its own weight. The method has the advantage that soil disturbance is minimum, but requires large quantities of water; also the high outflow rate of water, mixed with muck, creates a disposal problem and an environmental hazard. Figure 36 shows water jetting in progress.

In the *displacement method* the lower end of the pipe is closed, for example by hinged lid bearing against the bottom; as the pipe is forced into the subsurface, the natural soil is displaced upward and sideways by shear and compression. When the pipe reaches the desired depth, sand and water are introduced at the top and the pipe is withdrawn; the lid at the bottom opens and the hole is filled with sand. This method does not need large quantities of water and a system of drainage but it creates severe soil disturbance. This disturbance (Casagrande and Poulos, 1969) may create high initial pore pressures and a zone of low permeability around the hole.

Forming the hole by *wash boring* is slower and more expensive than the above two methods but may be worthwhile in small projects because of the lower mobilization cost. In wash borings, the hole is advanced by circulating water that is introduced in the hole at rates of 1 to 2 lit/sec. The steel pipe follows the advance of the hole, either by its own weight or by light tamping. The method has the advantage of small soil disturbance, and that it can be performed by simple equipment, practically the same rig that can perform borings for site investigations.

The sand filling the holes must be sufficiently pervious to allow the unobstructed flow of water from the soil into the drain, and from the lower part of the drain to the top. The influence of the coefficient of permeability of the sand, k, on the function of the drain, is apparent from Equations (17) and (18). If k is low, so will be C, and RDR will be high, which means that the function of the drain will be hindered by a backpressure in the sand.

In addition to the requirement of satisfactory permeability, the sand must not be so coarse as to overfacilitate the migration of fine particles from the soil into the drain. Although moderate soil migration may have the beneficial effect of increasing the permeability around the drain, the same phenomenon in excess may cause clogging; for this reason, particles larger than 4 mm should be avoided, unless tests are performed to confirm that the gravel-sand mixture is compatible with the surrounding soil. A further reason for avoiding large-size particles, even in cases of well graded material, where the problem of clogging might not exist, is to avoid segregation of particles during hole filling.

Prefabricated Drains

Prefabricated drains are also known as "wickdrains" or simply "wicks"; if they consist of sand packed in a filter stocking, they are called "sandwicks." They can also be of flexible corrugated plastic pipe, wrapped inside a filter. But the most common type is band-shaped, about 100 mm wide (Figures 37 and 38); thickness is about 5 mm. The exterior of band drains is covered by filter which, in combination with the core, forms continuous vertical passages; predominant material is plastic. The main advantages of prefabricated band drains are low cost, fast installation, clean site, and small soil disturbance. The main disadvantage is that in less developed areas, the equipment needed for their installation may have to be moved from a great distance. There are more than 50 types of prefabricated drain, although a much smaller number prevails in each location. For example, according to Morrison (1982), 75% of the U.S. market is taken-up by the studded-core "alidrain," shown as the middle piece of Figure 37.

An important characteristic of prefabricated drains is their sectional permeability, that is, their ability to let water flow per unit area of their cross section; another is the perviousness of the filter around the core. The coefficient of sectional permeability claimed by manufacturers of band drains, is of the order of 10^{-4}–10^{-2} m/sec, but it depends to

FIGURE 36. Installation of sand drains by water jetting. (Courtesy Mueser-Rutledge-Johnston-Desimore.)

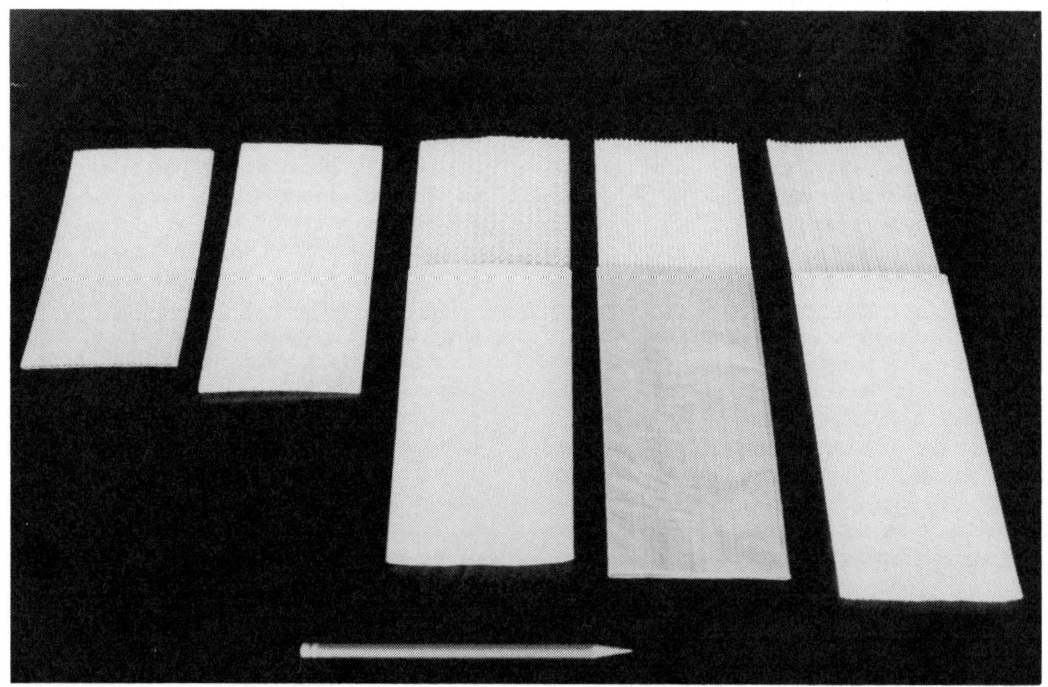

FIGURE 37. Common types of plastic band-shaped prefabricated drains. The two on the left have immovable filters; the three on the right have detachable jacketlike filters.

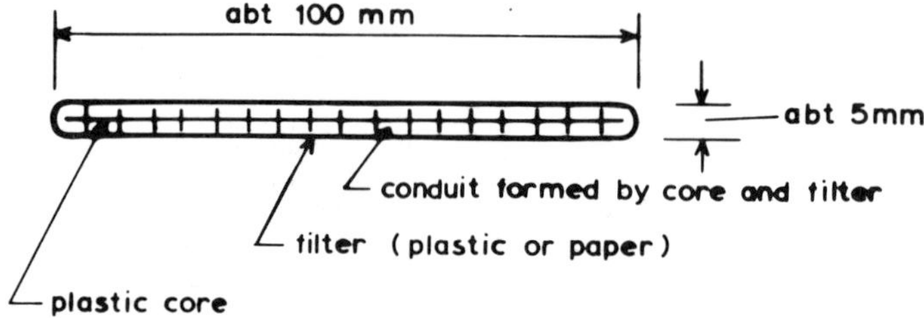

FIGURE 38. Schematic cross section of common types of band-shaped wickdrains, such as shown in Figure 37.

FIGURE 39. Equipment used in installing band-shaped drains. (Courtesy Burcan Industries Ltd.)

a certain extent on the pressure exerted laterally on the drain jacket. The drain jacket, or exterior filter, whose thickness is 0.2–0.4 mm, must have a permeability at least as high as that of the surrounding soil, so that it does not hinder the flow into the core. Laboratory apparatus and procedures for testing the hydraulic properties of band drains and band drain-soil systems are described by Hoedt (1981), Hansbo (1983), and Kremer et al. (1983). Mechanical characteristics are also significant; tensile strength is usually 0.5–2 kN (about 50–200 kgf), and elongation at yielding 6–10%.

The filter forming the outside surface of a drain should let some of the fine clay particles pass into the interior and be carried away by the discharge flow. Migration of fine particles at moderate rates is desirable because it results in a more open soil structure and high permeability around the drain. Also, the filter should be resistant to the lateral soil pressure, so that it does not bend excessively, causing the water conduits to become much narrower.

Installation equipment is as varied as the drains themselves. Common types of band drains are usually installed by a lance about 140 mm in cross section. Rigs are fitted with either open or closed mandrel; they are capable of exerting a downward force of up to more than 200 kN (20 tons), and they provide for the simultaneous driving of up to 4 wicks. Speeds of installation are of the order of 0.3–0.6 m/sec; driving depth is up to about 45 m (Nicholls and Barry, 1983). Band drains are wound around reels and as the mandrel is lowered into the ground, the band is released by unwinding. Sandwicks and flexible pipes are installed by methods and equipment similar to those used for sand drains. One type of equipment used in the installation of band drains is shown in Figure 39.

Distortion of Drains During Consolidation

Buckling or folding of sand drains or prefabricated drains caused by settlement does not, as a rule, interfere with their performance. This is a conclusion reached from observations both in the field and in the laboratory (Hansbo, 1983) and also from the prolonged service obtained by draining, even under severe settlement (Sinclair et al., 1983).

By contrast, it is well known that a base shear failure, such as indicated by the sketches of Figure 10, nullifies vertical drains by cutting off the continuity of the water path which they are meant to provide.

Effectiveness-Comparison of Vertical Drains

Although vertical drains have sometimes been used in vain, there are many instances where they have shortened the time required for soil stabilization. But in cases where they have been used and consolidation proceeded as planned, it is difficult to assert that timely completion of stabilization was due to the drains and would not have happened anyway. For this reason, the best way to assess the effectiveness of vertical drains is to compare the time-rate of settlement of the same or nearby embankments, built on ground with, and without, drains. Also, the best way of comparing the effectiveness of various types of drain is by comparing the time-rate of settlement of nearby embankments, built over ground with the different types. Published information illustrates the variety of past experience and the diversity of present opinion; also, it points to the fact that each particular case must be treated on its own merits.

Comparisons showing that vertical drains contributed can be found, for example, in Ladd et al. (1972), Tzitzas (1968), Cole and Garrett (1984), and Foott and Koutsoftas (1982). Comparisons showing that vertcial drains may have helped somewhat, but did not contribute substantially can be found in York (1976), Lewis et al. (1975), Brenner and Pbebahran (1983) and Foott and Koutsoftas (1982).

Casagrande and Poulos (1969) concluded that in some instances vertical drains have hindered rather than promoted soil improvement.

Interesting comparisons between different types of vertical drain or methods of installation are made in Nicholson and Jardine (1981), Hansbo et al. (1981), Davies and Hamphason (1981), Eriksson and Esstrom (1983) and Foott and Koutsoftas (1982). Summaries of comparisons are given by Stamatopoulos and Kotzias (1985).

CHANGE IN SOIL PROPERTIES CAUSED BY PRELOADING

Effects of Preloading

As the soft compressible soils consolidate under preloading, their density increases; hence the process is also called "soil densification." Consolidation or densification is manifested by settlement of the original ground surface. And although part of the observed settlement may be due to an outward movement of the soft foundation soils, as a rule, most of the settlement does reflect densification and the attendant soil improvement. During densification, the water content, void ratio and coefficient of permeability decrease, whereas the undrained shear strength, modulus of compressibility and penetration resistance increase.

An additional effect of preloading is that it increases the lateral stresses in the soil. As the ground surface is loaded, the soil directly underneath tends to move sideways thus causing an increase in the horizontal compressive stresses. The order of magnitude of the increase is indicated in Figure 40; this is the case of a semi-infinite embankment on elastic soil, where the increase under the full embankment is higher than one-half the maximum vertical pressure on the original soil surface. The horizontal stresses during consolidation enhance the stability of the soil structure. Also the beneficial effects of the increased lateral soil stresses remain, to a considerable degree, even after the preload is removed. This is particularly important in sands where the increased stability lessens the danger of liquefaction. In

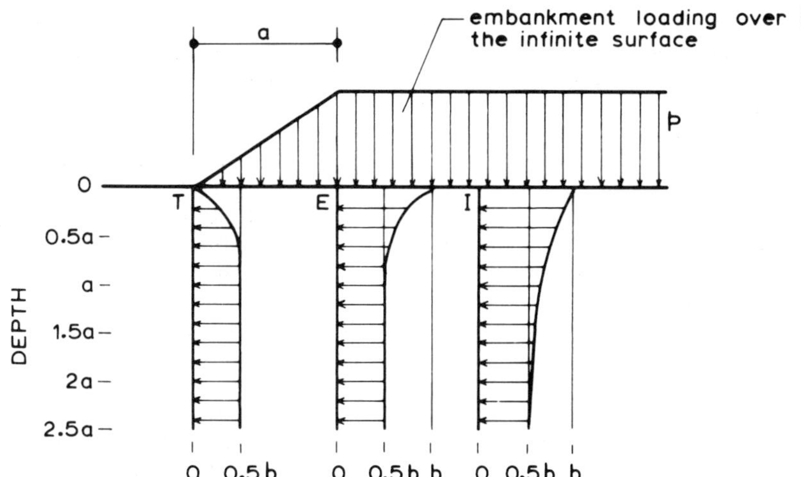

FIGURE 40. Increase in the horizontal compressive stress under the toe T, the edge E, and the interior I of an embankment (Jergensen, 1934; Poulos and Davis, 1973).

clays, higher lateral stresses should be expected to result in smaller shear stresses and, therefore, reduced secondary consolidation (Akai and Adachi, 1965).

Mathematical Expressions for Densification

The changes expected of the water content, strength and compressibility can be correlated with the preload settlement and preload pressure. If it is assumed that all of the settlement is caused by the compression of the soils undergoing improvement (side movement = 0) and that the underlying harder soils are incompressible, the following equations can be derived (derivations are given by Stamatopoulos and Kotzias, 1985).

$$\Delta w_N \cong -\left(w_N + \frac{1}{G} \right) - \frac{\delta}{h} \tag{23}$$

$$\Delta s_u \cong \frac{-G}{0.434\ C_c} \cdot s_u \cdot \Delta w_N \tag{24}$$

$$\Delta s_u \cong \frac{1 + w_N G}{0.434\ C_c} \cdot s_u \cdot \frac{\delta}{h} \tag{25}$$

$$\Delta D \cong \frac{1 + e}{0.434\ C_c} \cdot p \tag{26}$$

EXAMPLE OF CALCULATION OF Δw_N FROM EQUATION (23)

$w_N = 0.55$, $G = 2.70$, observed settlement $\delta = 0.80$ m, $h = 10.0$ m

$$\Delta w_N = -\left(0.55 + \frac{1}{2.70} \right) \frac{0.80}{10.0} = -0.074 \text{ or } -7.4\%$$

EXAMPLE OF CALCULATION OF Δs_u FROM EQUATION (25)

$w_N = 0.55$, $G = 2.70$, $\delta = 0.80$ m, $h = 10.0$ m

$C_c = 0.40$, $s_u = 20$ kN/m^2

$$\Delta s_u = \frac{1 + 0.55 \times 2.70}{0.434 \times 0.40} \times 20 \times \frac{0.80}{10.0} = 23 \text{ kN/m}^2$$

EXAMPLE OF CALCULATION OF ΔD FROM EQUATION (26)

$e = 1.5$, $C_c = 0.40$, $p = 130$ kN/m^2

$$\Delta D = \frac{1 + 1.5}{0.434 \times 0.40} 130 = 2369 \text{ kN/m}^2$$

Comments on the Derived Equations

Equations (23) and (25) are derived on the assumption that the observed settlement δ is due entirely to the densification of the compressible soils undergoing improvement. This assumption is subject to two sources of error: (a) side movement due to initial settlement and creep, and (b) the contribution to the settlement of the underlying harder soils. Whether or not there is appreciable side movement in a real problem can be checked by installing inclinometers or, more simply, by having observation points on the unloaded ground surface near the toe of the preload embankment; side movement would show as bulging, that is, there would result upward displacement. As for the possible contribution of the underlying harder soils, it can be estimated from the compressibility of these soils and the vertical stresses reaching them.

The magnitude of h appearing in Equations (23) and (25)

is in many cases obvious, but may occasionally prove elusive. When the compressible soils extend beyond the depth affected by preloading or when the transition between the compressible and the underlying harder soils is very gradual, the exact value of h is uncertain.

Increase of Standard Penetration Resistance (N)

Changes of the N value reflect degrees of soil improvement that are particularly beneficial to sand layers in earthquake-prone areas. When it comes to assessing the danger of sand liquefaction, the N-value provides a good indication of stability (Seed and Idriss, 1971, Seed et al., 1983). For example, in one case where deposits of silty sand improved by preloading to the extent that their N value increased from 6 to 16 blows/0.3 m, the approximate method of analysis of Seed et al. (1983) indicated that for a maximum ground acceleration of 0.20 g the magnitude of earthquake (Richter scale) likely to cause liquefaction increased from 5.5 to 9. As the probability of a magnitude 9 earthquake is extremely remote, it can be concluded that preloading has averted the danger of sand liquefaction for earthquakes hitting the site with ground accelerations up to 0.20 g. Another way of expressing the protection afforded by preloading would be to estimate, by the same method, that for a magnitude of 5.5 the ground acceleration necessary to cause liquefaction nearly doubled, increasing from 0.20 g to 0.39 g.

The increase of the N value induced by preloading is caused partly by densification and partly by the increased lateral compressive stresses. But regardless of the proportion in which these two factors have contributed, it is the overall improvement of the N value that reflects the advantage gained.

The amount by which the N value is likely to increase by a given preload, cannot be estimated by a relationship such as Equations (23) to (26). Yet, the available evidence points to the likelihood that a preload of the order of 200 kN/m^2 (equivalent to about 10 m of fill), properly carried out, may cause the average N value of a loose sand to increase to about 15 blows/0.3 m.

Rebound

The improved mechanical properties of the soil are first manifested by the rebound that takes place when the preload is removed. The fact that the soil bounces back to a looser state indicates that it is no longer "soft." An example of a rebound curve is given in Figure 16. The ability of the soil to rebound when unloaded is also called "spring effect."

NOTATION

A = Coefficient used in expressing settlement in terms of log t

a = Empirical factor, equal to 0.07; length of slope

B = Coefficient used in expressing settlement in terms of log t

C = Conductivity of vertical drain

C_a = Coefficient of secondary compression

C_c = Coefficient of compressibility

c' = Effective cohesion

c_R = Coefficient of consolidation by radial drainage

c_v = Coefficient of consolidation by vertical drainage

D = Modulus of compressibility; diameter of casing; depth ratio

D_s = Specific constrained modulus

d = Diameter of hole

$F(n)$ = As defined by Equation (15)

FS = Factor of safety

G = Specific gravity of soil grains

g = Acceleration of gravity

H = Height of slope; thickness of compressible soil layer; maximum drainage path in consolidating clay layer; water head

H_1, H_2 = Particular values of H

h = Thickness of layer of compressible soil; piezometric height of water column

k = Coefficient of permeability

k_H = Coefficient of permeability for horizontal flow

L = Length of uncased borehole; length of vertical drain

log = Logarithm to the base 10

m = Ratio of horizontal to vertical coefficient of permeability

N = Standard penetration resistance in number of blows/0.3 m

n = Coefficient of surcharge; r_e/r_w

p = Pressure exerted on soil surface

q = Time-rate of water flow

R = Ratio of vertical stress induced by preload to vertical stress induced by structure; radius of circle of failure; rebound

RDR = Relative drain resistance

r = Distance from the axis of vertical drain

r_c = Radius of individual conduits in wick drains

r_e = Radius of influence of vertical drains

r_w = Radius of vertical drain

S = Slope of settlement-time curve in Figure 36

s_u = Undrained shear strength

T = Time; normalized time

T_R = Time factor for consolidation by radial drainage

T_v = Time factor for consolidation by vertical drainage

t = Time

t_1, t_2 = Particular values of t

t_0 = Time at which $\Delta\delta = 0$

t_{99} = Time for $U = 0.99$

$t_a, t_{0.1a}$ = Times of preloading corresponding to $T_v = 0.07$ and $T_v = 0.007$, respectively

t_p = Time of preloading

U = Average degree of consolidation
u = Pore water pressure
u_e = Excess pore water pressure
w_N = Natural water content
z = Distance along the vertical
γ = Unit weight
γ_w = Unit weight of water
Δ = Increment, or change
δ = Settlement
δ_1 = Particular value of δ; settlement at $t = 1$ day
δ_2 = Particular value of δ
δ_{10},δ_{100} = Settlement at $t = 10$ days and $t = 100$ days
$\delta_a,\delta_{0.1a}$ = Settlement under preloading corresponding to $T_v = 0.07$ and $T_v = 0.007$, respectively
δ_{max} = Settlement under preloading for $U \cong 1.0$
δ_F = Future settlement of permanent structure
δ_p = Settlement at end of preloading
δ_T = Settlement of permanent structure at time T, without preloading
δ_T' = Settlement under preloading at time T
ϵ = Unit strain, taken positive as soil compresses
μ = Coefficient of viscosity of water
Σ = Sum, or summation
σ_v = Vertical compressive stress
σ_{vB} = Vertical stress caused by preloading at the bottom of vertical strain
φ' = Effective angle of friction

REFERENCES

Akai, K. and T. Adachi, "Study on the One-Dimensional Consolidation and the Shear Strength Characteristics of Fully Saturated Clay in Terms of Effective Stress," *Proc. 6th ICSMFE,* Montreal, no. 2/2 (1965).

Aldrich, H. P., "Precompression for Support of Shallow Foundations," *J. Soil Mech. Found. Div. Proc. of ASCE, 91* (SM2) (1965).

Bishop, A. W. and D. J. Hankel, The Measurement of Soil Properties in the Triaxial Test, Edward Arnold (1957).

Bjerrum, L., *Discussion to European Conference on Soil Mech. Found. Eng. (Weisbaden), II,* 135 (1963).

Bjerrum, L., "Generelle krav til fundamentering av forskjellige byggverk; tillatte setninger," Den Norske Ingeniorforening, Kurs i fundamentering. Oslo (1963).

Brenner, R. P. and Pbebahran, N., "Analysis of Sandwick Performance in Soft Bankok Clay," *Proc. 8th ECSMFE,* Helsinki, no. 6.6 (1983).

Broms, B. B., "Landslides," in *Foundation Engineering Handbook,* H. F. Winterkorn and H. Y. Fang, eds., Van Nostrand Reinhold, New York, p. 395 (1975).

Casagrande, L. and S. Poulos, "On the Effectiveness of Sand Drains," *Canadian Geot. J.,* 287 (1969).

Cedegren, H. R., *Seepage, Drainage, and Flow Nets,* Wiley, New York, pp. 53–76 (1977).

Cole, K. W. and C. Garret, "Two Road Embankments on Soft Alluvium," *Proc. 10th ICSMFE,* Stockholm, no. 1/12 (1981).

Darragh, R. D., "Controlled Water Tests on Preload Tank Foundation," *J. Soil Mech. Found. Div. Proc. ASCE, 90* (SM5) (1964).

Davies, J. A. and C. I. Humphason, "A Comparison Between the Performance of Two Types of Vertical Drain Beneath a Trial Embankment in Belfast," *Geotechnique, 31* (2) (1981).

"Earth Compacts Earth for Baltimore Cathedral Base," *Engineering News Record* (April 28, 1955).

Eriksson, L. and A. Esstrom, "The Efficiency of Three Different Types of Vertical Drain—Results from a Full-Scale Test," *Proc. 8th ECSMFE,* Helsinki, no. 6.10 (1983).

Foott, R. and D. Koutsoftas, "Geotechnical Engineering for the Replacement Airport at Chek Lap Kok," Symposium on Soil and Rock Improvements, Asian Institute of Technology (November 1982).

Geonor Catalog, Geonor A/S, Grini Molle, P.O. Box 99 Roa. Oslo 7.

Halton, G. R., R. W. Lou, and E. Winter, "Vacuum Stabilization of Subsoil Beneath Runway Extension at Philadelphia Airport," *Proc. 6th ICSMFE,* Montreal (1965).

Hansbo, S., "How to Evaluate the Properties of Prefabricated Drains," *Proc. 8th ECSMFE,* Helsinki, no. 6.13 (1983).

Hansbo, S., M. Jamiolkowski, and L. Kok, "Consolidation by Vertical Drains," *Geotechnique, 31* (2) (1981).

Hoedt, G. den, "Laboratory Testing of Vertical Drains," *Proc. 10th ICSMFE,* Stockholm, no. 4/22 (1981).

Holtz, R. D. and G. Lindskog, "Soil Movements Below a Test Embankment," in *Performance of Earth and Earth-Supported Structures,* ASCE Specialty Conference, Purdue University, Lafayette, Ind., vol. 1, pp. 273–384 (1972).

Hvorsley, M. J., "Subsurface Exploration and Sampling of Soils for Civil Engineering Purposes," ASCE, Waterways Experiment Station, Vicksburg, Miss. (1949).

Johnson, S. J., "Precompression for Improving Foundation Soils," *J. Soil Mech. Found. Div., Proc. of ASCE, 96* (SM1) (1970).

Jonas, E., "Subsurface Stabilization of Organic Silty Clay by Precompression," *J. Soil Mech. Found. Div. Proc. ASCE, 90* (SM5) (1964).

Jurgenson, L., "The Application of Theories of Elasticity and Plasticity to Foundation Problems," in *Contributions to Soil Mechanics 1925–1940,* Boston Society of Civil Engineers, Boston, Mass. (1934).

Kerisel, J., "Old Structures in Relation to Soil Conditions," *Geotechnique, 25* (3) (1975).

Kjellman, W., "Accelerating Consolidation of Fine-Grained Soils by Means of Cardboard Wicks," *Proc. 2nd ICSMFE,* Rotterdam (1948).

Kleiman, W. F., "Use of Surcharges in Highway Construction," *J. Soil Mech. Found. Div. Proc. ASCE, 90* (SM5) (1964).

Kotzias, P. C. and Stamatopoulos, A. C., "Preloading for Heavy Industrial Installations," *J. Soil Mech. Found. Div. Proc. ASCE, 95* (SM6) (1969).

Kremer, R. H. J., J. P. Oostveen, A. F. Van Weele, W. F. J. De Jager, and I. J. Meyvogel, "The Quality of Vertical Drainage," *Proc. 8th EWCSMFE,* Helsinki, no. 6.32 (1983).

Kyle, J. M., "The Use of Sand Drains Under Buildings at Port Newark," *Proc. 3rd ICSMFE,* Zurich, vol. 1 (1953).

Ladd, C. C., "Test Embankment on Sensitive Clay," in *Performance of Earth and Earth-Supported Structures,* ASCE Specialty Conference, Purdue University, Lafayette, Ind., vol. 1, pp. 101–128 (1972).

Lambe, T. W., *Soil Testing for Engineers,* Wiley, New York (1951).

Lambe, T. W. and R. V. Whitman, *Soil Mechanics,* Wiley, New York (1969).

Leonards, G. A., "Engineering Properties of Soils," in *Foundation Engineering,* G. A. Leonards, ed., McGraw-Hill, New York, pp. 174–175 (1962).

Lewis, W. A., R. T. Murray, and I. F. Symons, "Settlement and Stability of Embankments Constructed on Soft Alluvial Soils," *Proc. Inst. Civil Engineers* (December 1975).

Lowe, J. and P. F. Zacheo, "Subsurface Explorations and Sampling," in *Foundation Engineering Handbook,* H. F. Winterkorn and H. Y. Fang, eds., Van Nostrand Reinhold, New York (1975).

McGown, A. and F. H. Hughes, "Practical Aspects of the Design and Installation of Deep Vertical Drains," *Geotechnique, 31* (1) (1981).

Morrison, A., "The Booming Business in Wick Drains," *Civil Engnrg.* (March 1982).

Nicolls, R. A. and A. J. Barry, "Vertical Drains—A Case History," *Proc. 8th ECSMFE,* Helsinki, no. 6.21 (1983).

Nicholson, D. P. and R. J. Jardine, "Performance of Vertical Drains at Queenborough Bypass," *Geotechnique, 31* (2) (1981).

Penman, A. D. M. and G. H. Watson, "The Improvement of a Tank Foundation by the Weight of its Own Test Load," *Proc. 6th ICSMFE,* Montreal, vol. 2 (1965).

Porter, O. J., "Studies of Fill Construction over Mud Flats Including a Description of Experimental Construction Using Vertical Sand Drains to Hasten Stabilization," *Proc. 1st ICSMFE,* Cambridge, Mass., vol. 1, pp. 229–235 (1936).

Poulos, H. G. and E. H. Davis, *Elastic Solutions for Soil and Rock Mechanics,* Wiley, New York (1973).

Richart, F. E., "Review of the Theories for Sand Drains," *ASCE Trans., 124,* 709–736 (1959).

Seed, H. B. and I. M. Idriss, "Simplified Procedure for Evaluating Soil Liquefaction Potential," *J. Soil Mech. Found. Div. Proc. ASCE, 97,* (SM9) (1971).

Seed, H. B., I. M. Idriss, and I. Arango, "Evaluation of Liquefaction Potential Using Field Performance Data," *J. Geot. Engrg. Proc. ASCE, 109* (3) (1983).

Shannon, W. L., D. W. Stanley, and R. H. Meese, "Field Problems: Field Measurements," *Foundation Engineering,* G. A. Leonards, ed., McGraw-Hill, New York, pp. 1035–1041 (1962).

Sinclair, T. E. J., E. E. Rinne, and Q. Laumans, "Dike Stabilization for Long Term Stage Loading and Severe Settlement," *Proc. 8th ECSMFE,* Helsinki, no. 6.24 (1983).

Slope Indicator Catalog, Slope Indicator Co., 3668 Albion Place N., Seattle, Wash.

Soiltest Catalog, Soiltest Inc., 2205 Lee Street, Evanston, Ill., vol. 1, p. 33.

Sowers, G. F., "Fill Settlement Despite Vertical Sand Drains," *J. Soil Mech. Found. Div. Proc. of ASCE, 90* (SM5) (1964).

Stamatopoulos, A. C., and P. C. Kotzias, "Soil Compressibility as Measured in the Oedometer," *Geotechnique, 109* (6) (1980).

Stamatopoulos, A. C. and P. C. Kotzias, *Soil Improvement by Preloading,* John Wiley, New York (1985).

Terrametrics Catalog, Terrametrics Co., 16027 West 5th Ave., Gorden, Colo.

Tozzoli, A. J. and D. L. York, "Water Used to Preload Unstable Subsoils," *Civil Engngr.* (August 1973).

Tzitzas, S. S., "Prolongation of the Runway of the Airport of Corfu," *Technica Chronica,* Athens (April 1968).

USBR, Earth Manual, U.S. Department of the Interior, Bureau of Reclamation, Washington, D.C., p. 209 (1974).

Wilson, S. D., "Control of Foundation Settlement by Preloading," *J. Boston Soc. Civil Engnrs., 40* (1) (1963).

Wissa, A. E. Z., R. T. Martin, and J. E. Garlanger, "The Piezometer Probe," *Proc. ASCE Specialty Conference on In Situ Measurement of Soil Properties,* Raleigh, N.C., vol. 2, pp. 181–230 (1975).

Wood, I. R., "Preface on 'Vertical Drains,' " *Geotechnique, 31* (1) (1981).

Ye, B.-R., S.-Y. Lu, and Y.-S. Tanb, "Packed Sand Drain-Atmospheric Preloading for Strengthening Soft Foundation," *Proc. 8th ECSMFE,* Helsinki, no. 6.31 (1983).

York, D. L., "Site Preloading," Geotechnics: A Bicentennial State-of-the-Art, ASCE, Philadelphia Section, Geotechnical Group (December 1976).

Note: The meaning of the abbreviations are as follows:

ASCE: the American Society of Civil Engineering

ECSMFE: European Conference on Soil Mechanics and Foundation Engineering

ICSMFE: International Conference on Soil Mechanics and Foundation Engineering

Div.: division

Engnrg.: engineering

Found.: foundations

Geot.: geotechnical

Mech.: mechanics

Proc.: proceedings

CHAPTER 11

Ground Improvement by Dynamic Compaction

PAUL W. MAYNE*

PREFACE

Dynamic compaction is an effective and economical means of densifying loose ground in situ using the high energy impacts of a free-falling weight. Standard cranes are used to repeatedly lift and drop 2- to 20-ton steel or concrete weights onto the ground surface from heights of 10 to 100 feet in a controlled pattern. Dynamic compaction is most applicable for improvement of loose sands and granular fills, with or without groundwater conditions. Unique applications of the method include the induced consolidation of ground comprised of: sinkholes, sanitary landfills, rockfills, clayey materials, and seabeds completely underwater. The depth and degree of improvement are controlled by the levels of energy applied to the ground. Standard soil borings, cone penetration soundings, and other geotechnical tests are used to verify the effectiveness of dynamic compaction. Special concerns include the control of low-frequency vibrations on nearby structures, airborne particles, and excess pore pressures.

INTRODUCTION

As early as 1933, civil engineers have used the free-fall impact of large weights to compact loose sands [16]. The methodology developed into a popular construction technique, however, after the involvement by the French engineer Louis Menard in 1969. The process is also known by several other names, including: heavy tamping, dynamic consolidation, impact densification, and in China, the trade name of "flying goose" [6]. A detailed history of the development of dynamic compaction is given by Pearce [26].

The general purpose of dynamic compaction is to improve the strength and compressibility properties of the ground within a specified depth of improvement. The method consists of systematically dropping large weights onto the ground surface to compact the underlying ground. In addition to natural and artificial sands, dynamic compaction is very useful in treating reclaimed land and heterogeneous fill materials. Some unique applications include forming stone columns [3], underwater densification in harbors [8,13], collapsing sinkholes [20], consolidating landfills [18,35], and compacting rockfills [11,12].

Dynamic compaction has been utilized on a wide variety of soil types and conditions, primarily sandy materials and granular fills, although a limited number of cohesive soils have also been treated. Heavy tamping has been used for improving the ground conditions for many different types of civil engineering projects, including building structures, dams, bridges, highways, airports, coal facilities, dockyards, and reducing the liquefaction potential of loose soils in seismically active regions [10,13,25]. The degree of improvement resulting from dynamic compaction has been measured by a variety of field testing methods including standard penetration tests (SPT), cone penetration tests (CPT), pressuremeter tests (PMT), dilatometer soundings (DMT) and other means. Consequently, each particular project has been treated according to different criteria, depending upon the specific soil conditions, engineering purpose, and performance monitoring program employed.

Dynamic compaction is an alternative geotechnical solution for poor ground conditions. In lieu of pile foundations, excavation/replacement, vibroreplacement methods, and stone columns, spread footing foundations may be designed on dynamically-compacted soil [1,3,9,14,15,17,24,36]. The heavy tamping is usually performed using a steel weight or concrete-filled steel shell dropped from a crawler crane, as illustrated by Figure 1. For this site near Alexandria, Virginia, a large crawler crane was used to lift and drop 8-ton and 14-ton steel weights.

Most of the dynamic compaction projects to date have been completed by specialty contractors with experience in geomechanics [3,9,23]. Restrictive aspects of the method in-

*Law Engineering Testing Company, Washington, DC

(a)

(b)

FIGURE 1. Heavy tamping of a site in Fairfax County, Virginia using an 8-ton weight and 125-ton crawler crane.

clude: a limiting depth of influence, modest effect on clays, ground vibration control, groundwater at the surface, and flying debris. This chapter outlines the major factors concerning ground improvement by dynamic compaction and provides general guidelines for assessing the applicability of the method for a particular project. Additional details and information on the dynamic compaction method may be found in state-of-the-art reports by Lukas [18] and Mayne, Jones, and Dumas [20].

THEORETICAL MODELS

Several analytical theories for predicting ground response during dynamic compaction have been proposed. The repeated impacts of a large weight onto the ground is a complicated process involving both plastic and elastic deformation, energy losses, material damping, and other phenomena. Reference is made to Menard and Broise [23], Scott and Pearce [30], and Stegman [33] who discuss the major factors affecting dynamic compaction.

Currently, the design and applicability of the dynamic compaction process is determined by field test programs and previous experience. An analytical approach based on a soil behavior under transient loading is desirable and requires an understanding of dynamic stresses during impact. In the following section, a simplified theory of impact stresses and critical-state soil mechanics is used to describe the reasons why dynamic compaction works. Two generalized cases are considered: (a) sites with groundwater at considerable depth and (b) sites with loose soils generally below the groundwater level.

Dynamic Stresses During Impact

As a simple approximation, the dynamic force-time response upon impact may be assumed as a triangular impulse loading. This is a reasonable assumption as illustrated by the deceleration histories presented in Figure 2. This data was obtained at the site of a section of interstate highway north of Birmingham, Alabama [19]. Dynamic compaction was used to consolidate coal spoil left from strip mining operations with energy levels of up to 400 tonne-meters per blow. The coal spoil is comprised of gravelly sands derived from sedimentary rocks. The 20-tonne weight used at this site was constructed of two circular steel plates, connected together with steel I-sections and filled with concrete. An accelerometer was mounted at the center of the top of the weight and connected to a digital storage oscilloscope by cable. Measurements were recorded on polaroid film. The irregularities in the deceleration histories probably reflect the complex reverberation of seismic waves bouncing within the weight.

Using the conservation of momentum, the area under the

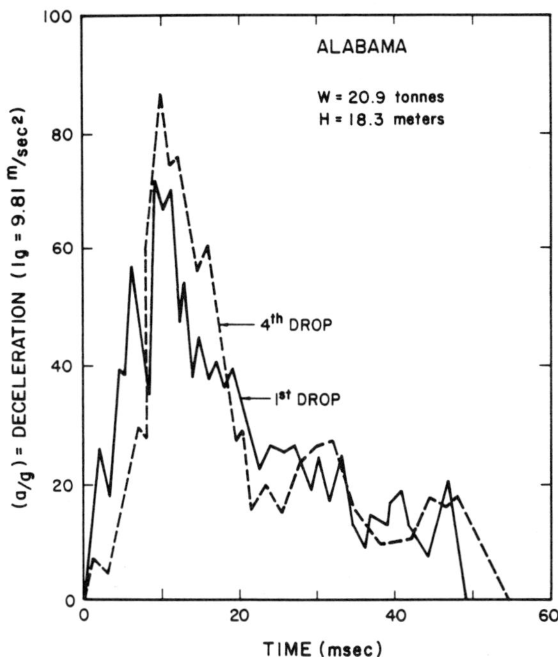

FIGURE 2. Deceleration measurements of a 23-ton steel/concrete weight during impact (reproduced by permission ASCE).

force-time curve should equal the change in momentum:

$$\frac{1}{2} F_{max} \Delta t = m \, \Delta \, v \tag{1}$$

where

F_{max} = peak dynamic force = ma_{max}
Δt = total time for deceleration
m = mass of weight = W/g
Δv = change in velocity
a_{max} = peak deceleration
g = gravitational constant = 32 ft/sec² = 9.8 m/sec²

For a free-fall system, the velocity upon impact ($v_i = \sqrt{2gH}$) will equal zero after deceleration is complete so that:

$$F_{max} = \frac{2W\sqrt{2gH}}{g\Delta t} \tag{2}$$

Actually, friction in the system inhibits a true free-fall of the weight. Assuming the natural frequency (f_n) for the system to be:

$$f_n = \frac{1}{T} = \frac{1}{2\pi}\sqrt{\frac{k}{m}} \tag{3}$$

where

$T = 2\Delta t$ = period of vibration
$k = 4\, G\, r_0/(1 - \nu)$ = vertical stiffness of the system
G = shear modulus = $E/[2\,(1 + \nu)]$
E = elastic modulus
r_o = radius of the mass
ν = Poisson's ratio

then, the maximum dynamic force becomes

$$F_{max} = \sqrt{\frac{32\; WHG\; r_o}{\pi^2\,(1 - \nu)}} \tag{4}$$

As shown by Hansbo [11,12] and Mayne [19] the shear modulus relevant to the large strains developed during dynamic compaction is about one-tenth of the low amplitude modulus as determined from geophysical tests. The actual level of shear strain at a point in the ground is measured as the ratio of peak particle velocity to shear wave velocity, which unfortunately also depends upon level of shear strain. Assuming a typical value of Poisson's ratio (0.37) and an approximate trapezoidal stress distribution, the peak dynamic stress beneath the center of impact is approximately:

$$\Delta\sigma_{z_{max}} = \frac{\sqrt{W\,H\,B\,E}}{(B + z)^2} \tag{5}$$

where

B = width of the weight = $r_o\sqrt{\pi}$
z = depth beneath impact surface

Reasonable predictions of peak dynamic vertical stress are made by this simple method when compared with measured stresses (reference Figure 3). For soils above the water table, representative values of elastic moduli E are in the range of 450 to 750 ksf (2200 to 3600 t/m²). For sites with no influence of groundwater, the underlying soil is densified by each successive impact of the dynamic compaction process and the maximum dynamic stresses will increase with number of blows. After treatment, the improved ground will exhibit a preconsolidation profile equal to the applied dynamic stress plus the in situ overburden stress, provided that sufficient energy has been imparted to the soil.

Critical-State Concepts

For sites with a high groundwater table, localized liquefaction occurs around the point of impact. This is often observed by the presence of sand boils or springs at the surface just after heavy tamping. Consequently, excess pore pressures develop, effectively reducing the soil stiffness, at least temporarily. The potential for vibration damage increases at such a site since the level of shear strain may be high. For high groundwater, a sufficient period of time for

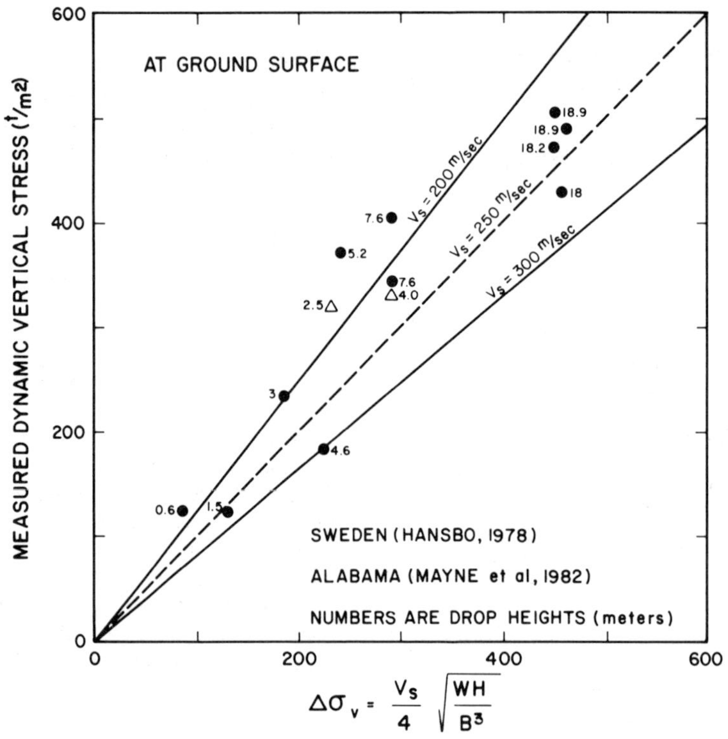

FIGURE 3. Comparison of measured and predicted dynamic stresses during heavy tamping.

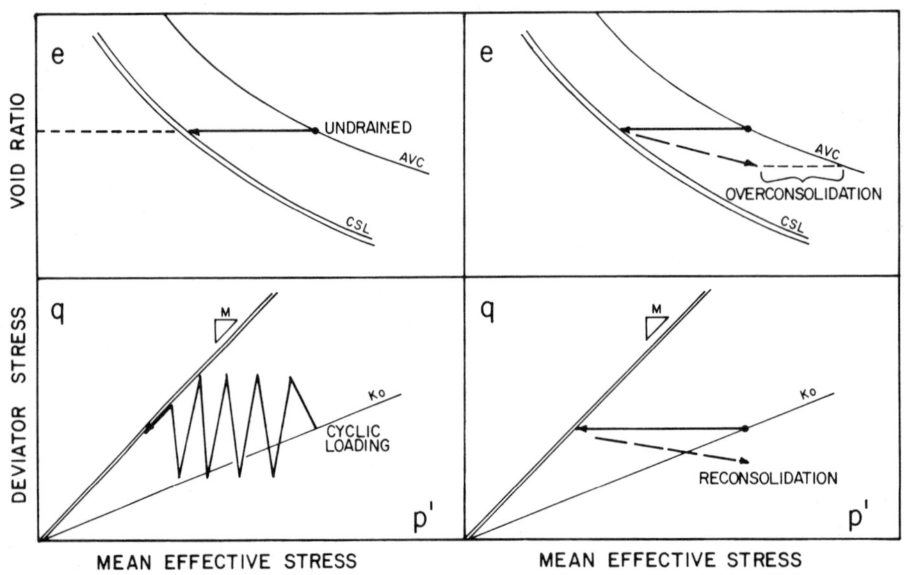

FIGURE 4. Theoretical cyclic undrained stress paths of saturated soil below the groundwater table.

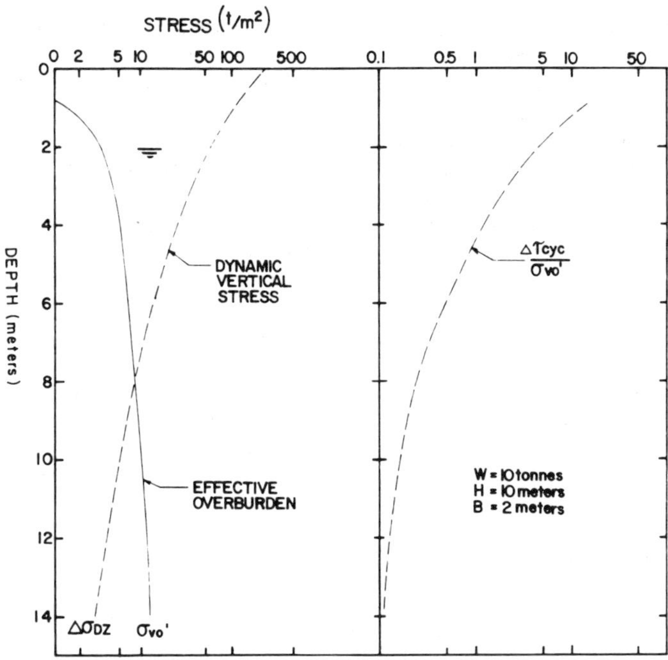

FIGURE 5. Estimated profile of induced preconsolidation with depth after dynamic compaction.

equalization of pore pressures or a recuperation period is needed between tamping passes.

The process of dynamic compaction on saturated soils below the water table can be explained by critical-state soil mechanics [38]. As shown by Figure 4, repeated impacts generate cyclic pore pressures which temporarily induce soil failure along an undrained stress path. Depending upon the subsurface drainage characteristics of the site, the excess porewater pressures will either quickly or slowly dissipate. Reconsolidation will occur along a recompression stress path back to the overburden stress, however, at a lower void ratio. The process is repeated and the stress state of the soil approaches the critical state line (CSL). The final preconsolidation profile for high groundwater conditions (see Figure 5) is thus developed in a complicated manner and depends highly on the cylic undrained behavior of the soil.

APPLIED PRACTICE

The pattern of application of dynamic compaction depends upon the desired results and depth to the groundwater level. A variety of weights and cranes have been used to perform the heavy tamping. In addition, several geotechnical testing methods are available for determining the degree of improvement. In this section, the dynamic compaction equipment, procedures, and testing methods are reviewed.

Compaction Equipment

The weights used on dynamic compaction projects have been typically constructed of steel plates, concrete-filled steel shells, and reinforced concrete. Cast iron ingots and sand-filled shells have also been used. Advantages of using bolted steel plates include adjustable weight size and ease of shipping. Steel is heavy and expensive, however, and concrete weights have the advantage of being much less costly and may be abandoned after dynamic compaction is completed.

Typical weights range in size from 2 to 20 tons (1.8 to 18 tonnes). For the expansion of the Nice Airport in France, a special tripod crane supported on 168 wheels was constructed to lift a 190-ton (172-tonne) steel weight. Base configurations are square, circular, or octagonal. The latter two are better suited for primary phases of tamping since little energy is wasted in forming the eventually circular crater shape. Square weights are better used for ironing phases where penetration is minimal. For underwater applications, special hollow shapes have been designed to increase the fall velocity through water [13].

In order to maximize the effect of dynamic compaction, cranes are utilized to lift a given weight to the highest drop height possible, considering the structural and operational limitations of the system. For typical systems with a single line pull, standard crawler cranes have essentially been restricted to maximum weights of 20 tons (18 tonnes) and drop heights of 80 feet (24 meters). For special projects, the Mega tripod lifted 44 tons (40 tonnes) to heights of 130 feet (40 m). The drop height of the Gigamachine was limited to 75 feet (23 m) because of airport safety restrictions. Recently, however, the use of double lines and quick-release mechanisms has allowed the use of standard crawler cranes to lift and drop weights of 30 to 35 tons. On very small projects, truck cranes have also been used for heavy tamping.

Heavy Tamping Procedures

Dynamic compaction is applied in a systematic controlled pattern of drops on a coordinate grid layout. The initial impacts are spaced at a distance which is dictated by the depth of the compressible layer, the depth to groundwater, and grain size distribution. Initial grid spacing is usually on the order of the thickness of the compressible layer and up to 50 drops could be used at each impact point. Typically, 5 to 15 blows per grid are applied, and often, the proximity of groundwater or crater sizes will limit the number of blows applied to each grid. Remember that the weight can become stuck in the crater. Example crater development for successive blows is presented in Figure 6 for several sites.

This first phase of the treatment is designed to improve the deeper layers. Incorrect spacing and energy at this stage could create a dense upper layer making it difficult or impossible to treat loose material below.

The initial passes are also called the "high energy phases," because the compactive energy is concentrated on distant points. The initial passes are followed at the end by a low energy pass, called "ironing," to densify the surficial layers in the upper 5 feet.

After each pass, the imprints are usually backfilled with the surrounding materials. In that case, the tamping surface is gradually lowered by an amount which is proportional to the densification achieved during each pass. In some circumstances, it may be necessary to maintain the working platform at a constant level throughout the work. For instance, in a situation of a high water table, the craters may be backfilled with imported materials. A buffer of at least five feet is required between the tamping surface and groundwater table. If the existing ground contains poor backfill materials, it may be desirable to use imported gravel or crushed stone materials to form stone columns.

In saturated fine-grained soils the process is complicated by the creation of excess pore-water pressures during compaction, a phenomenon which will reduce the effectiveness of the subsequent compactive passes. Once properly recognized, sufficient delay is planned between passes to allow these pressures to dissipate. Consequently, varying degrees of success have been reported in the dynamic compaction of saturated materials below the water table.

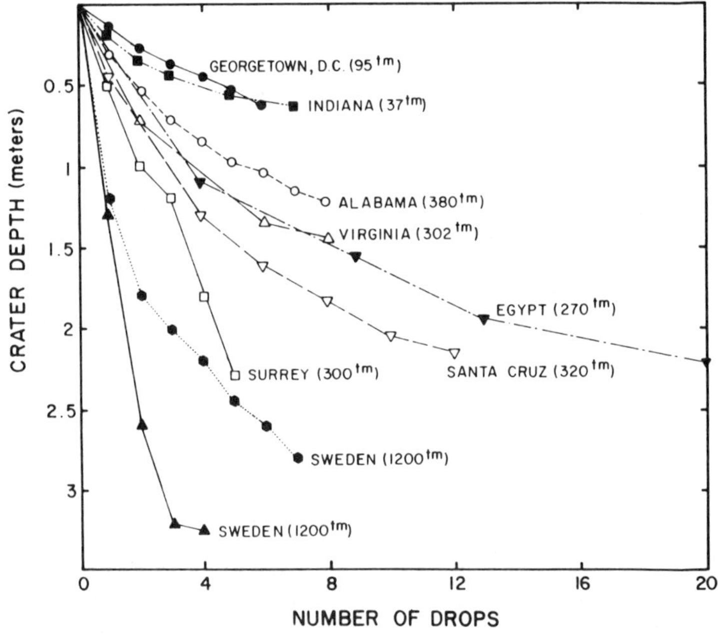

FIGURE 6. Formation of craters during successive blows at different dynamic compaction sites.

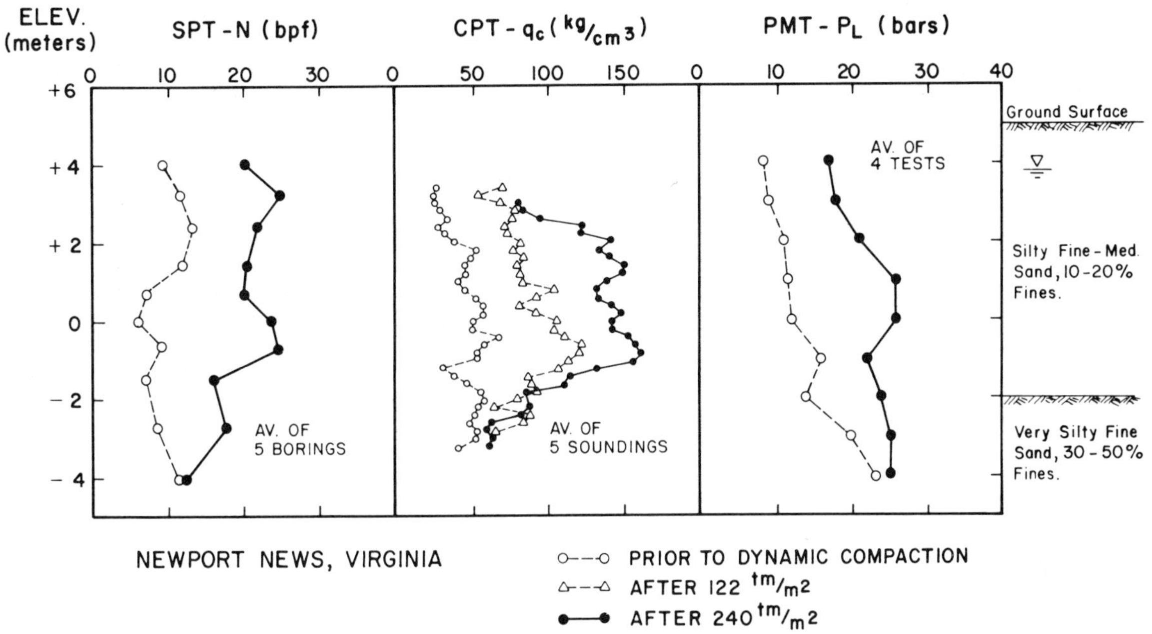

FIGURE 7. Measured improvement of silty sand by dynamic compaction at Newport News, Virginia.

Quality Control Testing

The aim of dynamic compaction is to improve the strength and compressibility characteristics of the underlying soils within some desired depth interval below the ground surface. To successfully improve ground in situ, certain quality control measures must be undertaken. Adequate ground surface coverage is verified by field surveys of the crater locations and depths. Geotechnical tests are also required to document that an adequate depth and degree of improvement has occurred.

Control testing may be divided into three types: production, environmental, and specification. Production control testing includes the documenting of imprints and elevation survey measurements. Environmental controls consist of measuring ground vibration levels and boundary surveys and may include instrumentation such as inclinometers or subsurface settlement points. Specification controls establish the minimum required goals needed to certify an allowable bearing pressure or differential settlement criterion.

The depth and degree of improvement are often evaluated by comparing field measurements before and after dynamic compaction. The most common geotechnical tests used on these projects include standard penetration tests (SPT), static cone soundings (CPT), and pressuremeter tests (PMT). Field vane shear tests, dilatometer soundings, and piezocone probes have also been utilized, as well as laboratory testing methods. In coarse gravelly soils and rockfill materials, quality control testing has included loss point soundings, becker probes, and geophysical shear wave velocity surveys.

The amount of improvement may be assessed by comparing the mean values before treatment with the measured values after improvement. For example, Figure 7 shows the average increases in SPT, CPT and PMT for a dynamic compaction site in Newport News, Virginia. The heavy tamping was applied with 16-ton and 18-ton weights falling 80 feet, and 3 crawler cranes. The apparent depth of influence is 30 feet (9 m), although most of the improvement occurred within the upper 20 feet (6 m).

GROUND RESPONSE

The application of dynamic compaction changes the ground within a limiting depth of influence. Some of the characteristics of treated ground include: increased penetration resistance, increased density, and lower elevation caused by an induced subsidence. Based on a review of data compiled from over 120 different dynamic compaction sites, generalized trends have been developed to aid the geotechnical engineer in estimating the effect of heavy tamping on a particular site.

Two measures of energy input are used to quantify heavy tamping operations: (a) the applied energy per blow (WH) and (b) the energy intensity per unit area $U = \Sigma\ (WH/S^2)$ where $S =$ spacing between grids. On typical projects, the energy per blow (WH) is between 300 and 1400 foot-tons (80 to 400 tonne-meters). Typical areal energy intensities range between 30 and 150 foot-tons per square foot (100 to 400 tm/m²).

Depth of Influence

The application of dynamic compaction at the ground surface is limited in its effect on the subsurface soils. Menard and Broise [23] suggested that the depth of influence, d_{max} is related to the square root of the energy per blow. Several investigators have modified this expression for soil type, crane efficiency, and energy level. The compilation of data from 120 different projects was reviewed to evaluate d_{max} as a function of WH [20]. The results of that study are presented in Figure 8, indicating that a reasonable estimate is approximately:

$$d_{max} = \sqrt{\frac{WH}{4n}} \tag{6}$$

in which n = units factor = 1 tonne/meter = 672 lb/ft and d_{max} is in meters or feet, respectively. A sufficient number of drops and adequate coverage of the site area must be made, of course, so that the subsurface soils "remember" the dynamic stresses imposed on them by the compaction process.

The degree of soil improvement has been observed to achieve a maximum at a critical depth, d_c, and then diminish with depth until reaching d_{max}, below which the soil properties remain unchanged. An example of the improvement in loose silty sand obtained at a coastal site for two five-story condominium buildings near Myrtle Beach, South Carolina (WH = 100 tonne-meters and U = 125 tm/m²) is shown in Figure 9. The apparent d_{max} and d_c are 5.5 and 4 meters, respectively. Above the critical depth, the mean CPT resistance increased from about 45 kg/cm² to 85 kg/cm². Usually, d_c appears to bc on the order of two-thirds d_{max}.

The observed scatter in Figure 8 is due to a variety of factors, including differences in: applied energy intensities, test methods, interpretation, groundwater conditions, etc. On

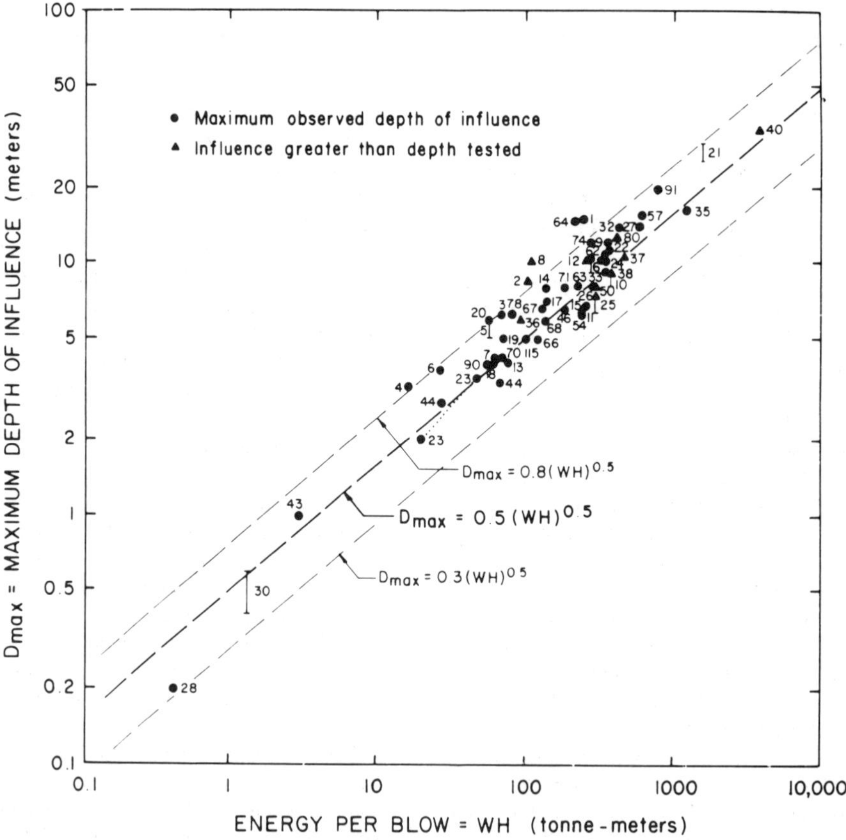

FIGURE 8. Observed depth of influence form data base of 120 sites (reproduced by permission ASCE).

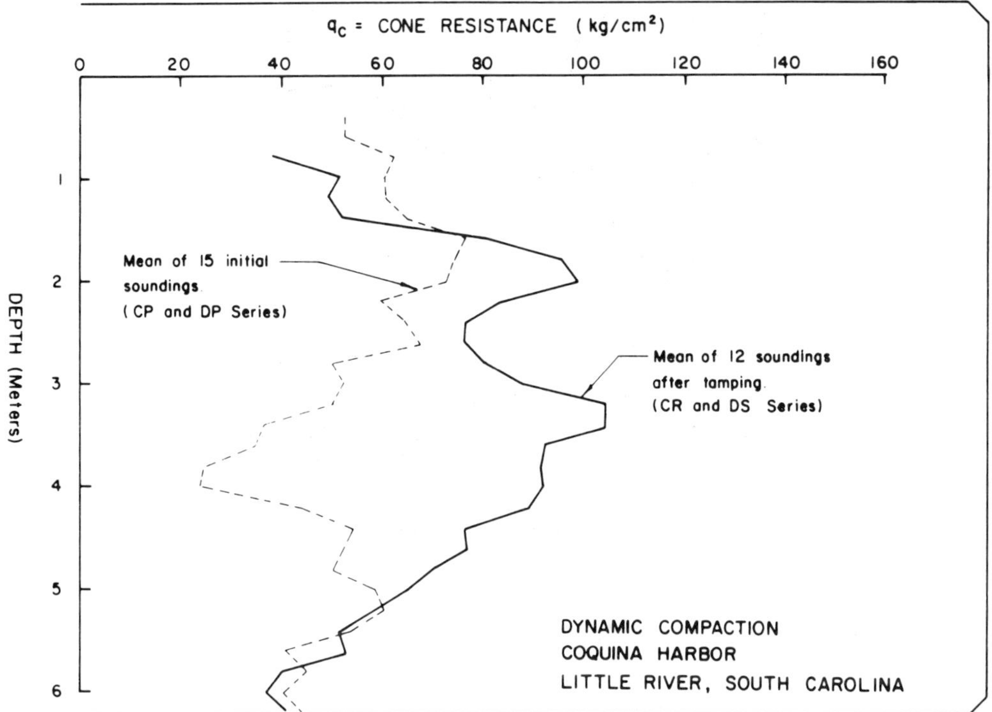

FIGURE 9. Example improvement in mean static cone resistance.

sites with favorable conditions (i.e., no groundwater, no clay or silt), a deeper depth of influence than inferred by Equation (6) may be possible. Conversely, unfavorable sites may not be as deeply treated as indicated by Equation (6). If in question, an initial test program (possibly using a truck crane to minimize costs) will be warranted to determine the applicability of dynamic compaction for a particular site.

Ground Vibrations

An undesirable side effect of dynamic compaction is the generation of ground vibrations which emanate from the point of impact. Since dynamic compaction is an attractive economical solution, its use is seen increasingly in urban and suburban communities where real estate costs are high. Ground vibrations can be potentially damaging to nearby building structures and sensitive equipment, as well as annoying to people. Consequently, careful and proper monitoring of ground vibration levels and vibration frequencies must be made in order to protect all interested parties.

Ground vibrations caused from dynamic compaction operations are unique from other types of construction activity, such as blasting, pile driving, and traffic. In this regard, vibrations from dynamic compaction are characterized by low-frequency waves which are (1) potentially more damaging than high-frequency waves and (2) below the frequency range of many commercially available vibration monitor seismographs.

Routine field measurements are taken using a vibration monitor seismograph. Usually, the seismograph package includes a triaxial component transducer or geophone, electronic signal conditioners, and a recording mechanism. A review of several leading commercial units which are available has been prepared by Stagg and Engler [32]. Most commonly the ground vibration records are written on oscillographic paper or magnetic tape, although a few units provide an electronic digital display output or ticker-tape summary of the vibration levels. Most units record the complete wave form of the measured vibration.

For most construction-related vibrations, the velocity at a point on the ground (the particle velocity) has been shown to be the best indicator of damage potential and annoyance levels [37]. For certain situations, a combination of velocity and displacement measurements may be appropriate [31].

Low frequency vibrations present a problem for those responsible for monitoring them. Many commercial seismographs cannot directly measure vibration levels when the frequency is less than 5 or 6 Hz. The restriction is primarily due to the resonant frequency and damping characteristics of the transducer. Several manufacturers provide a mag-

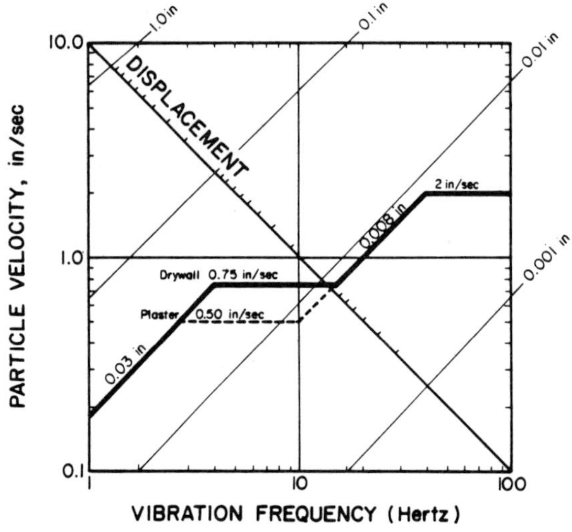

FIGURE 10. Transient vibration level criteria for probable structural damage (after U.S. Bureau of Mines, 1980).

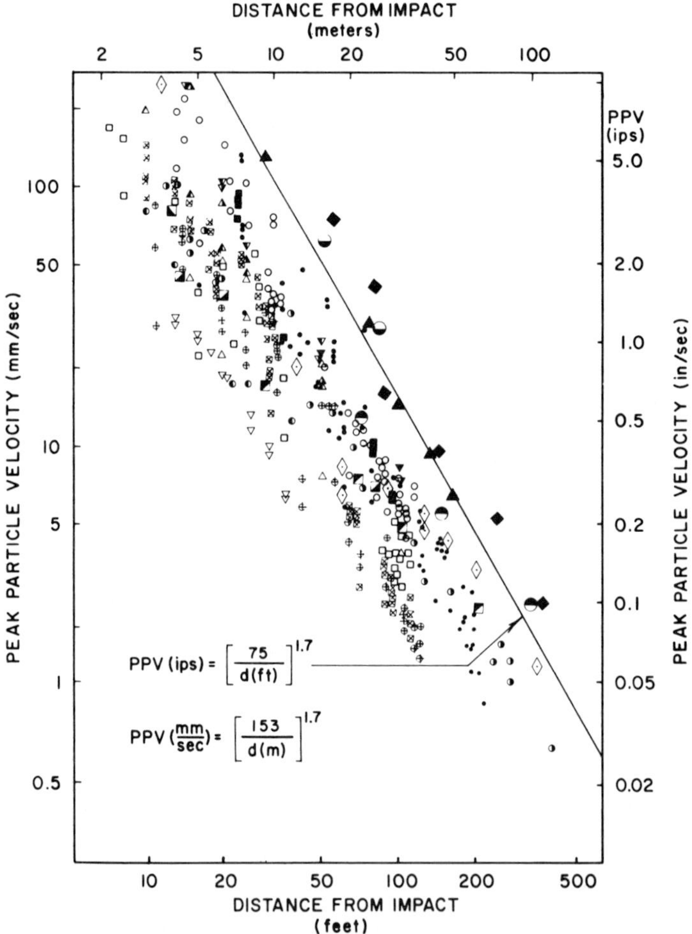

FIGURE 11. Observed attenuation of ground vibrations with distance at 12 dynamic compaction sites underlain by granular soils [21].

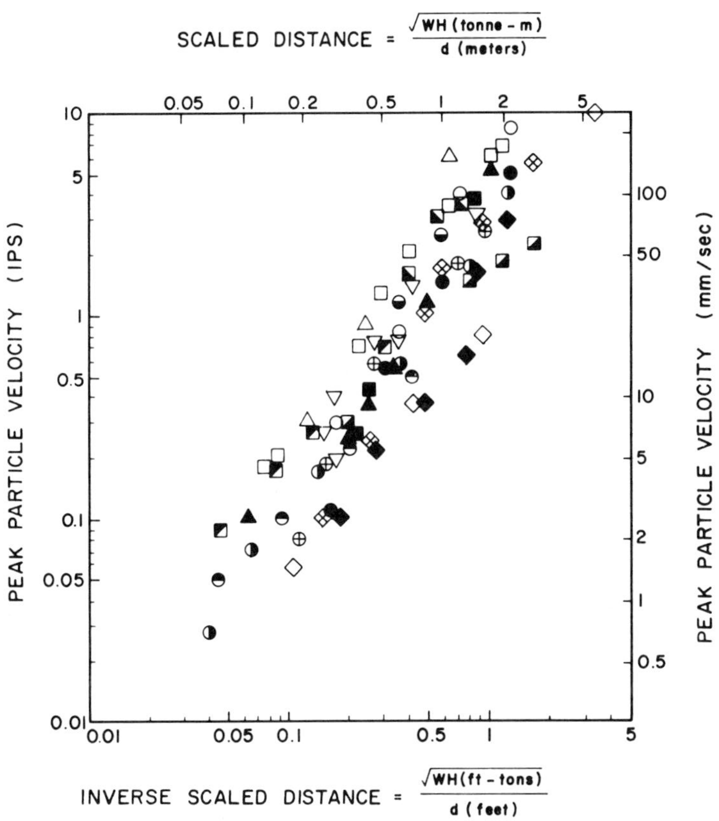

FIGURE 12. Vibration data in scaled-energy factor format.

nification factor for determining the vibration amplitude when the vibration frequency falls below the specified frequency range of the equipment. Without correction for this, the measured particle velocities may be wrong by a factor of 5 or more. At least one seismograph unit available does not measure vibration frequency at all! Readers are cautioned to check the manufacturer's specifications regarding the applicable range of the equipment used.

For many years, a limiting peak particle velocity of two inches per second (50 mm/sec) has been considered the structural vibration damage criterion for one- and two-story buildings. The primary sources of data for this basis came from blasting records from surface mining operations near residential communities. Despite the use of a 2 ips criterion, numerous litigation claims and complaints were filed in the courts. Consequently, a re-evaluation study of vibration damage was performed by the Bureau of Mines and published in 1980 [31]. Previous data and new data were analyzed using three different methodologies: (a) statistical mean and variance, (b) probability theory, and (c) observational. The extensive review culminated in particle velocity criteria dependent upon vibration frequency, as presented in Figure 10.

The amplitude of ground vibrations attenuate with distance from the point of impact. Figure 11 presents a summary of peak particle velocity data from 12 different dynamic compaction sites [21]. Based on the available data from these sites, a safe conservative upper limit (neglecting special tripod equipment) may be estimated for preliminary purposes from:

$$\text{PPV (ips)} = \left[\frac{75}{d\ \text{(feet)}}\right]^{1.7} \tag{7a}$$

$$\text{PPV (mm/sec)} = \left[\frac{153}{d\ \text{(meters)}}\right]^{1.7} \tag{7b}$$

Equation (7) does not consider the level of energy applied during dynamic compaction. The data is derived from a few sites underlain by granular materials. Extrapolation of these trends to sites underlain by clayey soils, variable fill mate-

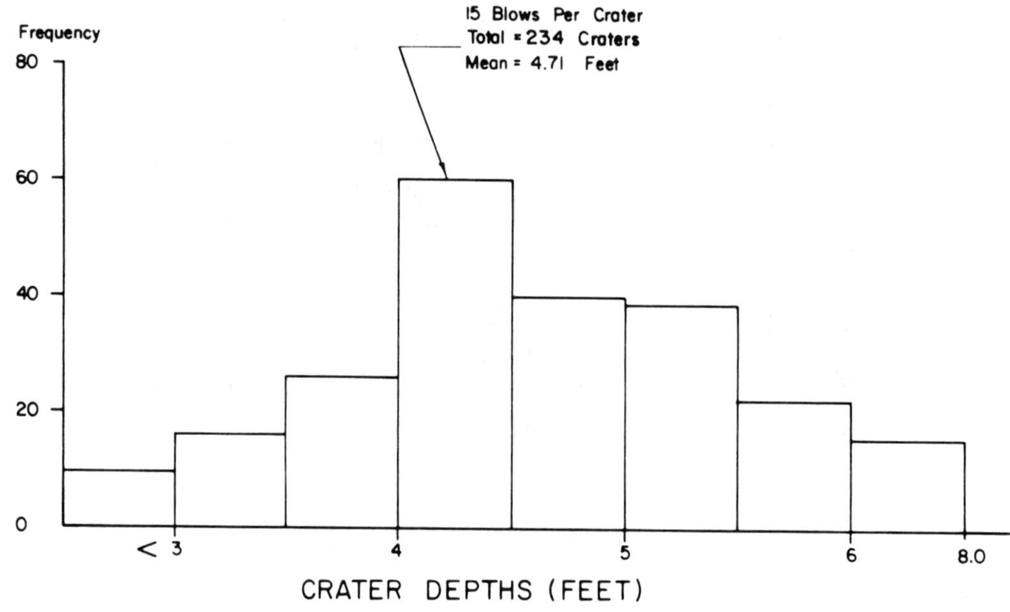

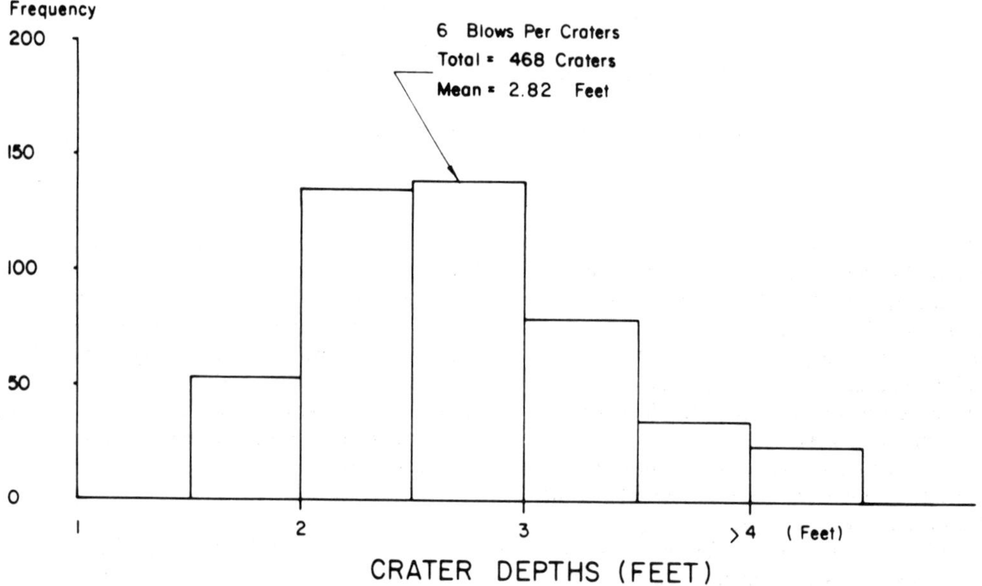

FIGURE 13. Example histograms of crater depth measurements from dynamic compaction.

rials, complex stratigraphy, shallow rock, or other dissimilarities may result in unconservative results.

Within the dynamic compaction limits, it has been observed that vibration levels increase as the treated area becomes densified. Generally, a maximum level of particle velocity is achieved after one or two passes of heavy tamping or an energy intensity of at least 150 tm/m².

In order to consider the effects of energy level, scaled distance graphs are often used to present particle velocity data. Most commonly, the scaled distance axes is defined as the ratio of the square root of the applied energy to the distance from the source. Based on the peak velocity values observed after densification, a summary of particle velocity attenuation with square-root scaled distance is presented in Figure 12. Expressions for the upper limit of the observed trend are:

$$\text{PPV (ips)} = 8\left[\frac{\sqrt{WH}}{d}\right]^{1.7} \tag{8a}$$

where d and H are in feet and W in tons and

$$\text{PPV (mm/sec)} = 92\left[\frac{\sqrt{WH}}{d}\right]^{1.7} \tag{8b}$$

where d and H are in meters and W in tonnes.

It is noted that Figures 11 and 12 have been presented for preliminary estimates of ground vibration levels and that each site should be properly monitored, as appropriate. Preconstruction surveys of existing building walls, facades, and cracks should be performed prior to tamping as a protection against frivolous claims.

Induced Subsidence

Dynamic compaction causes an induced subsidence over the area treated. In saturated materials above the groundwater table, this occurs relatively quickly. In saturated soils below the groundwater table, the subsidence occurs more slowly, as the cyclic pore pressures dissipate with time.

Since the energy is applied to specific grid points, the most obvious indications of this subsidence are the relatively large craters induced at each impact point. Typically, the craters are 3 to 6 feet deep (1 to 2 meters) and about 6 feet in diameter (2 meters), the latter primarily governed by the size of the weight. All craters are mapped and recorded to document the work done. An example of the observed range of crater depths at a site in South Carolina is presented as a histogram in Figure 13. Very large craters may indicate local zones of untreatable material such as plastic clay and possibly requiring undercutting.

After each pass of dynamic compaction, the surface of the

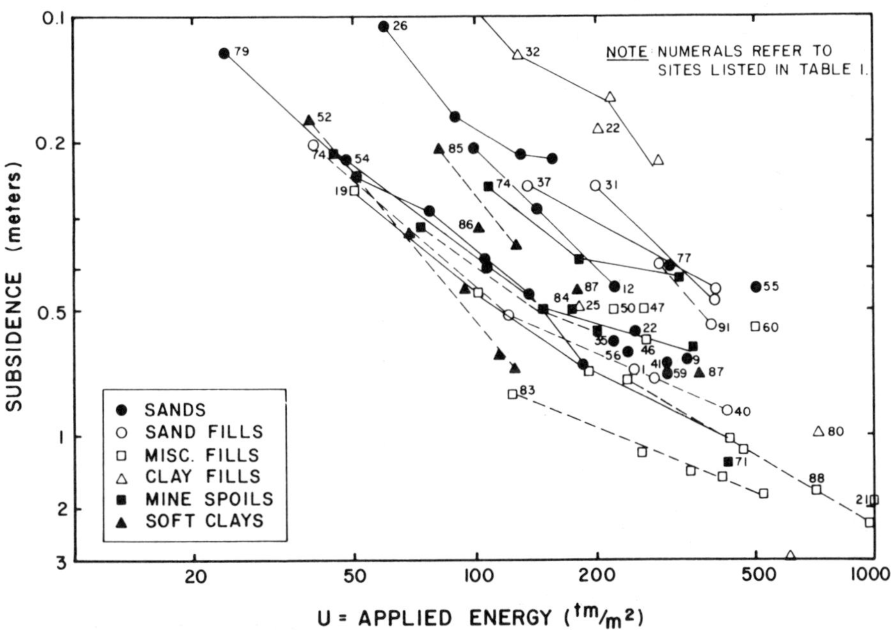

FIGURE 14. Observed subsidence at many sites as a function of energy intensity per unit area (Database from Reference 20).

site is relevelled by bulldozing surface materials into the craters. The settlement caused by each pass of compaction can then be measured by topographic surveys. Several sites are reported to have subsided as much as six feet (2 meters) or more, but typically the settlements are 1 to 2 feet. The magnitude of ground surface subsidence depends upon the applied energy per unit area [7,20]. A comparison of induced ground settlements is made in Figure 14 for different soil types and a similar trend is apparent for data obtained at several dynamic compaction sites.

Since the ground elevations will be lowered by the process, provisions may be necessary to provide new fill at the site. In addition, if imported gravel, sand, or stone is used to backfill the imprints, considerably less subsidence should be expected to occur.

VERIFICATION TESTING

Since every site has unique characteristics, it should not be presumed that dynamic compaction has been effective. Soil testing is paramount in order to verify that ground improvement has indeed occurred. Testing also allows an optimization of energy levels and applied intensities for a desired end product.

Soil variability is an inherent characteristic of most ground conditions. Consequently, a comparison of mean test values before and after tamping is often useful in evaluating the amount of improvement. As discussed previously, the depth of improvement appears to be primarily governed by the energy per blow (*WH*). Based on several studies, the degree of improvement within this zone, however, depends upon the cumulative energy intensity (*U*).

As illustrated by the results of SPT data in Figure 15, obtained at a sand fill site in Egypt, improvement was observed after 48 tm/m² [5]. After a second pass treatment, however, a cumulative intensity of 106 tm/m² resulted in an increase in mean SPT-N value from an initial 4 blows per foot (bpf) before tamping to about 12 bpf after tamping. The apparent depth of influence remained at 5.5 meters.

A general trend has been observed between the final SPT resistance after compaction and the applied intensity (see Figure 16). Separate trends are observed for sandy soils and

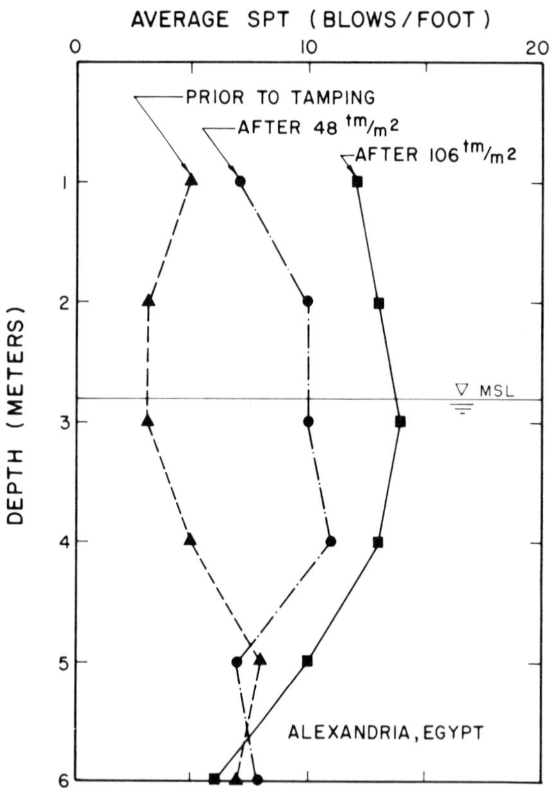

FIGURE 15. Example improvement in standard penetration test resistance showing initial, intermediate, and final results.

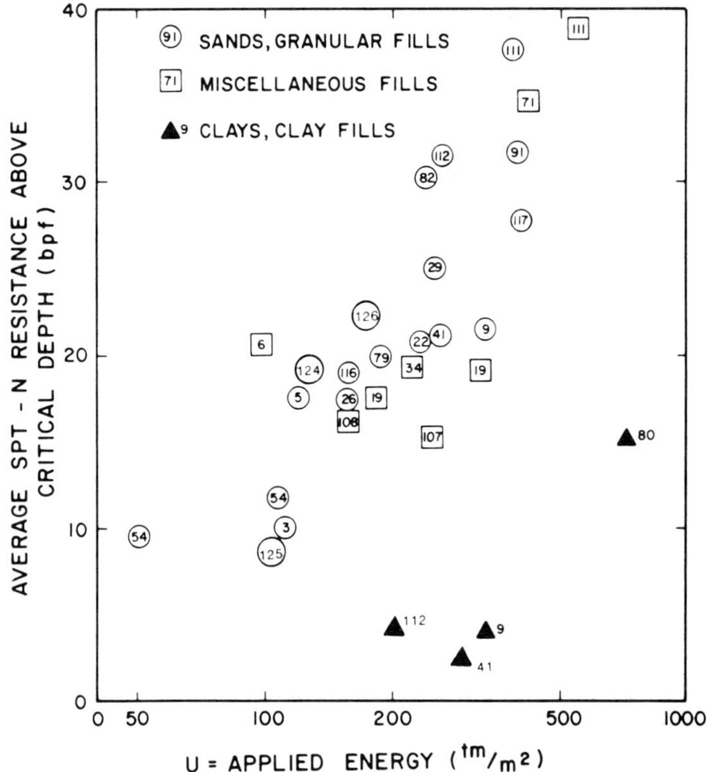

FIGURE 16. Observed trend between SPT resistance and energy intensity (reproduced by permission of ASCE). Numerals adjacent to symbols indicate specific site as listed in Reference 20.

clayey soils. For granular materials, a similar trend is found between the CPT resistance and applied intensity, as presented by Figure 17 [20]. In the case of soils below the water table, a sufficient waiting period should be allocated before testing in order for dissipation of excess pore water pressures.

The pressuremeter test (PMT) has also been used considerably for verification of dynamic compaction operations. The PMT modulus and limit pressure have also been shown to increase proportionately with the applied intensity as shown by Mayne, et al. [20].

CONCLUSIONS

Depending upon the desired level of performance, dynamic compaction is an attractive method of site improvement, especially for granular soils. Although relatively simple to apply in concept, the theoretical and practical aspects of evaluation are complicated by numerous site specific factors.

A review of data from previous dynamically compacted sites indicates the following conclusions:

1. The apparent depth of influence and ground vibration levels are controlled in part by the level of energy per blow (*WH*).
2. The induced subsidence and final penetration resistance (SPT, CPT) are partially governed by the level of energy intensity per unit area (*U*).
3. Transient low-vibration frequencies resulting from dynamic compaction require a combination velocity and frequency criterion, as recommended by two recent studies on vibration-induced damage [31,34].
4. It is extremely important to measure vibration frequencies of the waveform to determine whether these are within the range of the seismograph operating range. Many units are nonlinear below 6 Hz and thus, a magnification factor may be required.
5. Field verification testing is necessary on all projects to confirm the effectiveness of the heavy tamping program. A sufficient recuperation period should be allocated before testing if pore pressure dissipation is of concern.

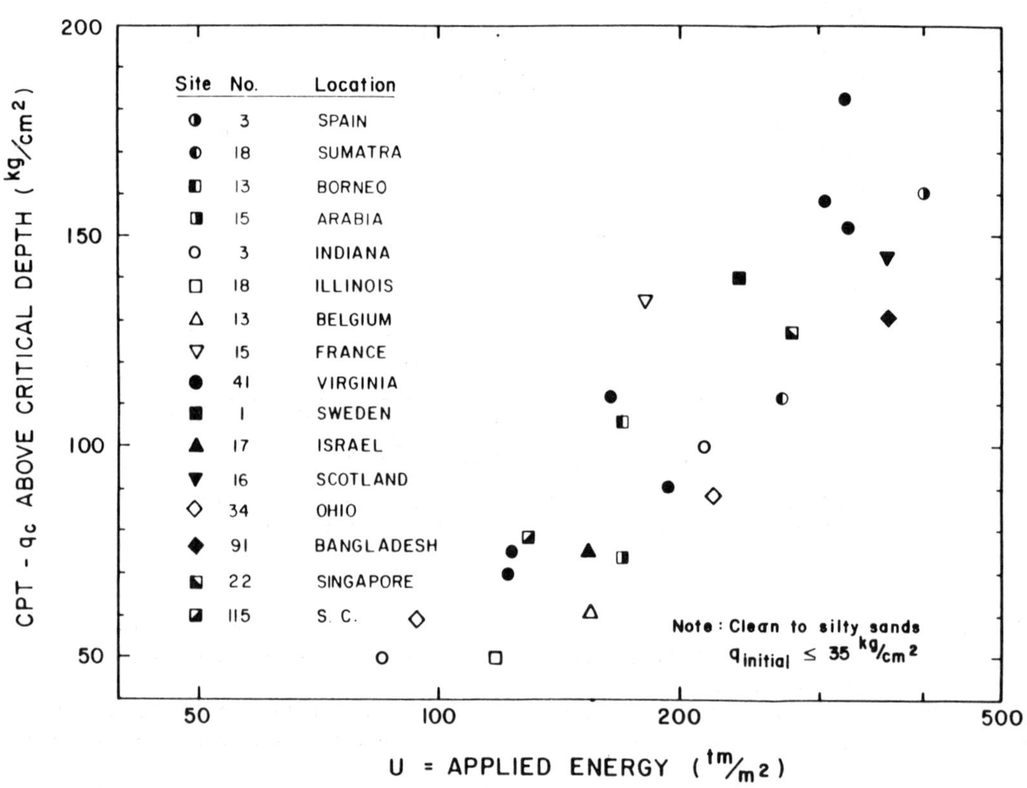

FIGURE 17. Observed trend between CPT resistance and energy intensity (reproduced by permission of ASCE).

ACKNOWLEDGEMENTS

Many thanks to Anne Bethoun and Donna Reese for their continued assistance in manuscript preparation. Messrs. Jean Dumas, John Jones, Robert Lukas, and Michel Gambin provided technical encouragement throughout the writer's study period. Appreciation is also given to the American Society of Civil Engineers for permission to reproduce selected figures and text portions from the *Journal of Geotechnical Engineering.*

REFERENCES

1. Abelev, M., "Compacting Loose Soils in the USSR," *Ground Treatment by Deep Compaction,* Institution of Civil Engineers, London, pp. 79–82 (1976).
2. Charles, J. A., "General Report—Methods of Treatment of Clay Fills," *Clay Fills*, Institution of Civil Engineers, London, pp. 318-319 (1979).
3. Dumas, J. C., "Dynamic Consolidation: Case Histories, Canadian Realizations," 1982 Reports by Geopac, Inc. 680 Birch, St.-Lambert, Quebec, J4P 2N3.
4. Dumas, J. C. and N. F. Beaton, "Limitations and Risks of Dynamic Compaction," paper presented at the ASCE National Convention, Atlanta, Georgia, Geopac, Inc., 10 pp. (May 16, 1984).
5. Edinger, P. H., "TOF Terminal, Alexandria, Egypt," Mueser-Rutledge Report No. 5534 to Black & Veatch International (March 7, 1982).
6. Fang, H. Y. and G. Ellis, "Laboratory Study of Ground Response to Dynamic Densification," Lehigh University Report No. 462.6, Fritz Laboratory (March 1983).
7. Gambin, M. P., "The Menard Dynamic Consolidation at Nice Airport," *Proceedings 8th European Conference on Soil Mechanics and Foundation Engineering*, Helsinki, 4 pp. (May 1983).
8. Gambin, M. P. and G. Bolle, "Sea Bed Soil Improvement for Lagos Dry Dock," *Proceedings 8th European Conference on Soil Mechanics and Foundation Engineering*, Helsinki, 4 pp. (May 1983).
9. Gambin, M., "Ten Years of Dynamic Compaction," *Proceedings, 8th Regional Conference for Africa on Soil Mechanics*, Harare, 363–370 (1984).
10. Gambin, M., "Consolidation Dynamique de Penitas Dam in Mexico," *Revue Francaise de Geotechnique, 1*, 59–67 (1985).
11. Hansbo, S., "Dynamic Consolidation of Rockfill," *Proceedings 9th International Conference on Soil Mechanics and Foundation Engineering, 2*, 241–246, Tokyo (1977).
12. Hansbo, S., "Dynamic Consolidation of Soil by a Falling Weight," *Ground Engineering, 11* (5), 27–36 (July 1978).
13. Hanzawa, H., "Improvement of a Quick Sand," *Proceedings 10th International Conference on Soil Mechanics and Foundation Engineering, 3*, 683–686, Stockholm (1981).
14. Hartikainen, J. and M. Valtonen, "Heavy Tamping of Ground of Aimarautio Bridge," *Proceedings 8th European Conference on Soil Mechanics, 1*, 249–252, Helsinki (1983).
15. Leonards, G. A., et al., "Dynamic Compaction of Granular Soils," *Journal of the Geotechnical Engineering Division, 106* (GT 1), 35–44 (January 1980).
16. Loos, W., "Comparative Studies for Compacting Cohesionless Soils," *Proceedings 1st International Conference on Soil Mechanics and Foundation Engineering, III*, 174–178, Harvard Univ., Cambridge, Mass. (June 1936).
17. Lukas, R. G., "Densification of Loose Deposits by Pounding," *Journal of Geotechnical Engineering, 106* (GT 4), 435–446 (April 1980).
18. Lukas, R. G., "Dynamic Compaction Manual," STS Consultants Report to Federal Highway Administration, Washington, D.C., Contract No. DTFH-61-83-C-00095 (March 1984).
19. Mayne, P. W. and J. S. Jones, "Impact Stresses During Compaction," *Journal of Geotechnical Engineering, 109* (10), 1342–1346 (Oct. 1983).
20. Mayne, P. W., J. S. Jones, and J. C. Dumas, "Ground Response to Dynamic Compaction," *Journal of Geotechnical Engineering,* 110 (6), 757–774 (June 1984).
21. Mayne, P. W., "Ground Vibrations During Dynamic Compaction," *Proceedings ASCE Detrimental Ground Movement from Man-Made Vibrations*, Detroit, 27 pp. (October 1985).
22. Mitchell, J. K., "Soil Improvement: State-of-the-Art Report," *Proceedings, 10th International Conference on Soil Mechanics and Foundation Engineering, 4*, Session 12, 509–521, Stockholm (1981).
23. Menard, L. and Broise, Y., "Theoretical and Practical Aspects of Dynamic Consolidation," *Geotechnique, 25* (1), 3–17 (March 1975).
24. Minkov, M. and Donchev, P., "Development of Heavy Tamping of Loess," *Proceedings 8th European Conference on Soil Mechanics and Foundation Engineering, 2*, 797–800 (1983).
25. Moreno, E., et al., "Dynamic Compaction of Penitas Dam Foundation," *Proceedings 7th Panamerican Conference on Soil Mechanics, 1*, 123–133 (1983).
26. Pearce, R. W., "Deep Soil Compaction by Heavy Surface Tamping," State-of-the-Art Report in Partial Fullfillment for Master of Science Degree, Imperial College, University of London, 150 pp. (July 1977).
27. Pilot, G., "Methods of Improving the Engineering Properties of Soft Clay," *Soft Clay Engineering*, Elsevier, Amsterdam, Ed. E-W. Brand, 637–696 (1981).
28. Ramaswamy, S. D., et al., "Treatment of Peaty Clay by High Energy Impact," *Journal of the Geotechnical Engineering Division, 105* (GT 8), 957–967 (Aug. 1979).
29. Schmertmann, J., "A Method for Friction Angle from DMT," *Proceedings 2nd European Conference on Penetration Testing, 2*, 853–861, Amsterdam (May 1982).
30. Scott, R. A. and R. W. Pearce, "Soil Compaction by Impact," *Ground Treatment by Deep Compaction,* Institution of Civil Engineers, Telford, Ltd., London, 19–30 (1976).
31. Siskind, D., M. Stagg, et al., "Structure Response and Damage

Produced by Ground Vibration from Surface Mine Blasting," Bureau of Mines Report RI 8507, Twin Cities, Minnesota, NTIS –PB81-157000, 74 pp. (1980).

32. Stagg, M. and A. Engler, "Measurement of Blast Induced Ground Vibrations and Seismograph Calibration," Bureau of Mines Report RI 8506, Twin Cities, Minnesota, NTIS #PB81-157828, 62 pp. (1980).

33. Stegman, B. G., "Dynamic Compaction," Thesis Submitted in Partial Requirement for Fullfillment of Degree in Master of Science, Dept. of Civil Engineering, University of Maryland, College Park, MD (March 1981).

34. Studer, J. and A. Suesstrunk, "Swiss Standard for Vibrational Damage to Buildings," *Proceedings 10th International Conference on Soil Mechanics and Foundation Engineering, 3,* 307–312, Stockholm (1981).

35. Welsh, J., "Dynamic Deep Compaction of Sanitary Landfill to Support Superhighway," *Proceedings 8th European Conference on Soil Mechanics, 1,* 319–321, Helsinki (1983).

36. West, J. M. and B. Slocombe, "Dynamic Consolidation as an Alternative Foundation," *Ground Engineering, 6* (6), 52–54 (1973).

37. Wiss, J. F., "Construction Vibrations: State-of-the-Art," *Journal of the Geotechnical Engineering Division, 107* (GT2), 167–182 (February 1981).

38. Wroth, C. P., "The Interpretation of In Situ Soil Test," *Geotechnique, 34* (4), 449–489 (December 1984).

CHAPTER 12

Seasonal Temperature Effects on Foundations

FERNANDO ASCHIERI* AND FIORE ATTILIO ULIANA*

INTRODUCTION

Temperatures in the subsoil are seldom considered by the civil engineer, except for rather special reasons, such as in solving geotechnical problems due to the presence of permanently frozen soils (permafrost), or in determining the depth of water distribution pipes.

It is suggested, however, that also the designer of big foundations should take care of subsoil temperatures: as a matter of fact, their effects (in terms of stresses and deformations in the foundations) can be sometimes as important as those due to other design actions which are normally considered (loads transmitted by the structure, differential settlements, etc.). For these foundations (only important ones, as above mentioned) thermal action should therefore be considered in the design phase, just as it is (not too often, to tell the truth) in the case of structures situated above ground level.

In the following pages some explanations will be given in order to clarify the phenomenon and also to give the designer some tools to deal with the problem, which has not been examined extensively in the past [1,2].

A number of simplifying approximations will be suggested, which are considered to be perfectly acceptable in normal cases; as it has been said, the problem of structural temperature effects is one for which "it makes (almost) no difference what you do as long as you do something" [3].

It is clear that the importance of thermal effects on a foundation mainly depends on:

- the magnitude of temperature variations in the foundation
- dimensions and form of the foundation
- stiffness of the embedding soil

*ENEL (Ente Nazionale per l'Energia Elettrica) DCO, Milano, Italy

The first item will be dealt with under Temperatures in the Subsoil, the second and the third under Static Effects. Under A Case History, at last, an example will be given concerning an existing circular foundation, having a diameter of 40.7 m (133.5 ft), where temperature variations and corresponding deformations were measured for more than two years; consequent stresses are calculated and found to be of significant importance.

TEMPERATURES IN THE SUBSOIL

Temperature Oscillations in Subsoil and in Foundations

The surface of the earth is heated periodically by the sun in its daily and annual cycle. The consequent temperature oscillations in the subsoil decrease with increasing depth, and the rate of the reduction, coeteris paribus, depends on their frequency: whereas daily temperature changes are damped down to virtually zero at a depth of a few decimeters (and are of little importance to foundations), seasonal fluctuations are appreciable to a depth which can vary from a few meters to 20 m (60 ft) and more, until a practically constant temperature layer is reached having a temperature equal to the mean annual temperature of the site. Only seasonal variations will therefore be considered here. Two typical temperature diagrams in the subsoil, measured in winter and in summer, are shown in Figure 1.

For our aims, the ground temperature field may be studied resorting to the very simplified model of an infinitely thick plate subject to sinusoidal surface temperature changes. Setting to zero the annual mean temperature, the solution of the pertinent heat flow differential equation is then:

$$T_{x,t} = \bar{T} \exp\left(-x\sqrt{\frac{\pi\rho c}{\lambda\tau}}\right) \cos\left(\frac{2\pi c}{\tau} - x\sqrt{\frac{\pi\rho c}{\lambda\tau}}\right) \quad (1)$$

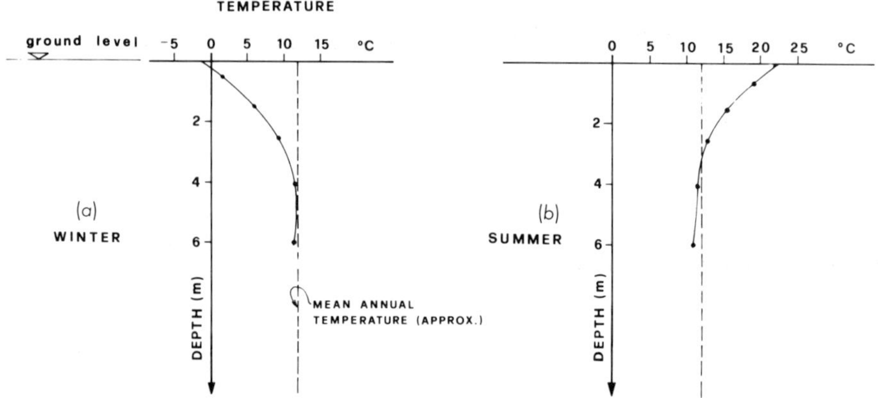

FIGURE 1. Typical winter (a) and summer (b) temperature diagrams measured in the subsoil at Porto Tolle (Northern Italy) (1 m = 3.28 ft; 1°C = 1.8°F).

in which $T_{x,t}$ = the soil temperature (which is variable with depth, x, and time, t); $\bar{T}$ = the oscillation amplitude of the surface temperature; τ = the period (1 year = 8766 h); and ϱ, c, λ = heat transfer properties of the soil medium, i.e., density, specific heat, and thermal conductivity.

The following principal features of the process may be derived [4] from Equation (1):

- Damped temperature waves move into the ground.
- The period and the sinusoidal form of the temperature oscillations do not change with depth.
- The amplitude decreases with depth according to the following e function.

$$\exp\left(-x\sqrt{\frac{\pi\rho c}{\lambda\tau}}\right)$$

- With increasing depth, the amplitude of oscillation is delayed.

For practical purposes the value of $\bar{T}$ at ground surface may be assumed to be equal to the amplitude of the seasonal fluctuations of the local atmospheric temperature; it can be derived for each location through information provided by the competent authority (for example, the Environmental Data Service in the U.S.). With good approximation $\bar{T}$ can be assumed equal to one-half of the difference between the highest and the lowest mean monthly local temperatures.

Heat transfer coefficients for different kinds of soil are discussed under Heat Transfer Coefficients in the Soil Medium.

From the above expression all needed information may be derived on the possible temperature distributions in the ground; for instance, temperature versus depth diagrams can be obtained for different values of the ratio t/τ (for example, every one-twelfth of half a cycle).

The same temperature diagrams may generally be assumed to act on a concrete foundation embedded in the ground. However, this holds true if the presence of the foundation doesn't appreciably affect heat propagation, which is certainly the case with small foundations. For "big" foundations, instead, concrete thermal properties should be used in Equation (1), as temperatures will be governed by heat flow in the foundation itself; in this sense, the term "big" means a foundation in which the least horizontal dimension is larger than, say, 2 or 3 m (7 or 10 ft).

In case of doubt, cumbersome tridimensional heat flow calculations could be avoided by repeating the above suggested unidirectional heat flow calculations with both ground and concrete properties, and selecting then the most unfavourable results. Concrete heat transfer coefficients are discussed under Heat Transfer Coefficients in Concrete.

It may be added, as a further simplification, that in the majority of cases the highest temperature action on a foundation happens at, or very near to, the moment of maximum (or minimum) surface temperature. It is then possible to trace only the temperature diagram corresponding to $t = 0$ (or $t = \tau/2$), and Equation (1) is reduced to:

$$T_x = \bar{T}\exp\left(-x\sqrt{\frac{\pi\rho c}{\lambda\tau}}\right)\cos\left(-x\sqrt{\frac{\pi\rho c}{\lambda\tau}}\right) \quad (2)$$

Only very important or special foundations would deserve a more thorough investigation, considering other values of t. If the thermal diffusivity of the medium is introduced, defined as

$$\delta = \frac{\lambda}{\rho c} \quad (3)$$

Equation (2) becomes

$$T_x = \overline{T} \exp\left(-x\sqrt{\frac{\pi}{\tau}\cdot\frac{1}{\delta}}\right)\cos\left(-x\sqrt{\frac{\pi}{\tau}\cdot\frac{1}{\delta}}\right) \quad (4)$$

Solutions of this equation are given in Figure 2 where, assuming $\overline{T}$ = 10°C (18°F), T_x can immediately be read as a function of x for different values of δ.

Of course, Figure 2 represents a summer situation; in winter, signs must be changed (as above said, $T = 0$ corresponds to the annual mean temperature).

Equation (1) is valid within a thermally homogeneous soil, but the extension is fairly easy, if really justified by the situation, to the case of a layered medium (e.g., when the foundation is covered by a layer of soil having thermal properties greatly different from concrete).

Linearization of a Temperature Diagram

To make possible simple calculations of the structural effects of temperature, the diagrams obtained by the above equations (or directly measured in the subsoil) must be linearized in the stretch of depth interested by the foundation. That is to say, an "equivalent" linear temperature distribution T_L must be found, producing the same uncracked moment and the same axial force in the section as does the non-linear temperature distribution T_{NL}.

For rectangular sections, such a diagram is obtained (Figure 3) when the following two conditions are met [5]:

- The mean temperatures T_M of the two diagrams are equal.
- The moments, about the section centerline, of the areas included by each of the temperature diagrams are equal.

The linear diagram therefore could be determined theoretically by the following expressions:

$$T_M = \frac{\int_{-d/2}^{+d/2} T_{NL}\cdot dy}{d} \quad (5)$$

$$\Delta T = \frac{12}{d^2}\cdot\int_{-d/2}^{+d/2} T_{NL}\cdot y\cdot dy \quad (6)$$

where ΔT = temperature gradient in the linear diagram; d = concrete thickness and y = distance from section centerline. It must be noted, however, that a carefully hand-

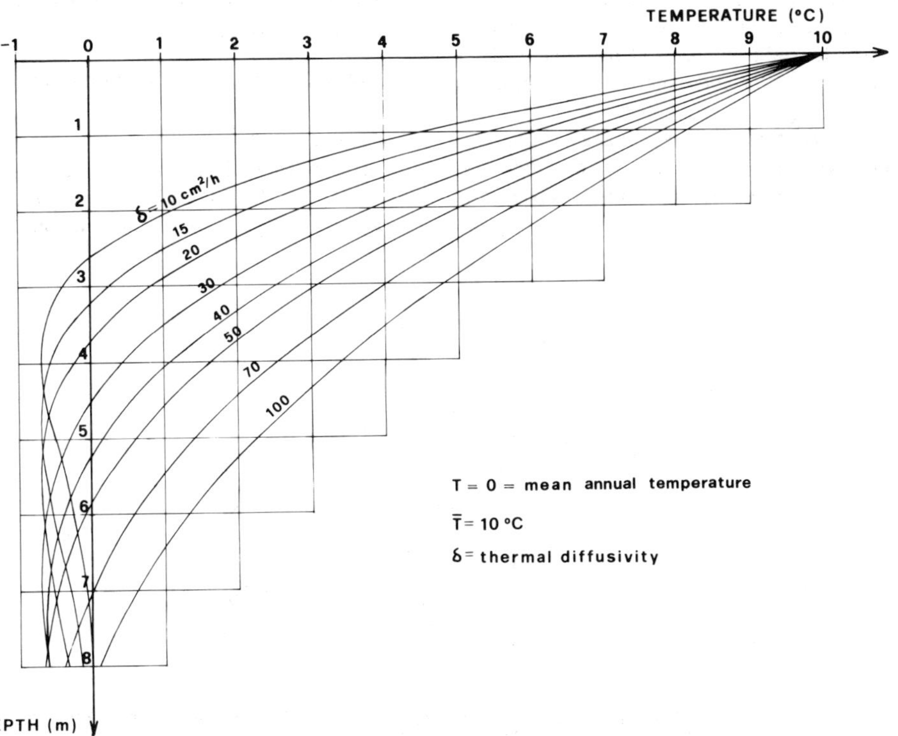

FIGURE 2. Temperature diagrams in the subsoil at t = 0 [Equation (2)], for various thermal diffusivities (1 m = 3.28 ft; 1°C = 1.8°F; 1 cm^2 · h^{-1} = 1.08 × 10^{-3} ft^2 · h^{-1}).

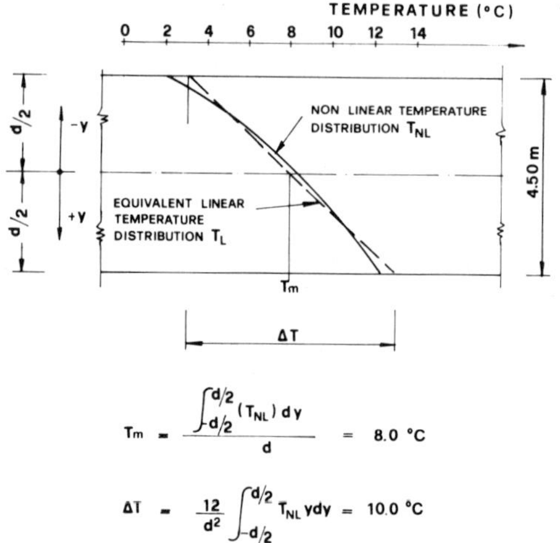

FIGURE 3. Linearization of a temperature diagram (1 m = 3.28 ft; 1°C = 1.8°F).

made linearization generally offers a good approximation and is much easier to accomplish. Neglecting the non-linearity of a temperature diagram, i.e., the differences between the original and the linearized diagrams, means to assume a stress distribution in the foundation section which is different from the actual one, but both foundation deflections and resulting flexural moments (and also axial loads, if relevant) are unchanged if the above outlined transformation is employed.

Temperature linearization in a non-rectangular concrete section is a more complicated problem (even if the basic concepts are the same) which lies beyond the scope of this chapter.

Base Temperature—Design Temperature Loadings

Base temperature is the one at which the foundation is supposed (somewhat arbitrarily) to be free of thermal stresses, and can be assumed as the temperature at which the concrete was cured; if a more precise determination is not possible, a temperature of about 20°C (70°F) can be suggested [5]. This base temperature, as well as T_M of Equation (5), will be referred here to the mean annual temperature.

Any departure from this base temperature T_B (constant through the foundation depth) gives rise to thermal strains and stresses. This means that if two design "summer" and "winter" linearized temperature diagrams are obtained (Figure 4), the thermal actions (loadings) on the foundation whose consequences should be verified are those hatched on the same figure and summarized as follows:

- in summer: a uniform temperature increase ($T_{us} = T_M - T_B$) and a negative temperature gradient ($-\Delta T$) (here the base temperature has been considered to be higher than the annual mean)
- in winter: a uniform temperature decrease ($T_{uw} = -T_M - T_B$) and a positive temperature gradient ($+\Delta T$)

As regards to the choice of the base temperature, see Preliminary Remarks where the influence of the creep phenomenon is discussed.

Heat Transfer Coefficients in the Soil Medium

SPECIFIC HEAT (c)

This quantity is expressed here in Cal · kg^{-1} · °C^{-1} (1 Cal · kg^{-1} · °C^{-1} = 1 BTU · lb^{-1} · °F^{-1} = 4.19 × 10^3 J · kg^{-1} · K^{-1}). Specific heats of common dry soils are almost

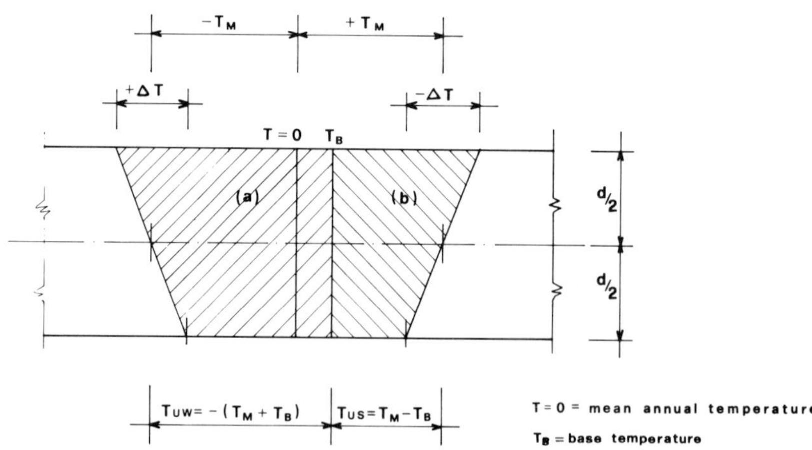

FIGURE 4. Winter (a) and summer (b) design temperature loadings.

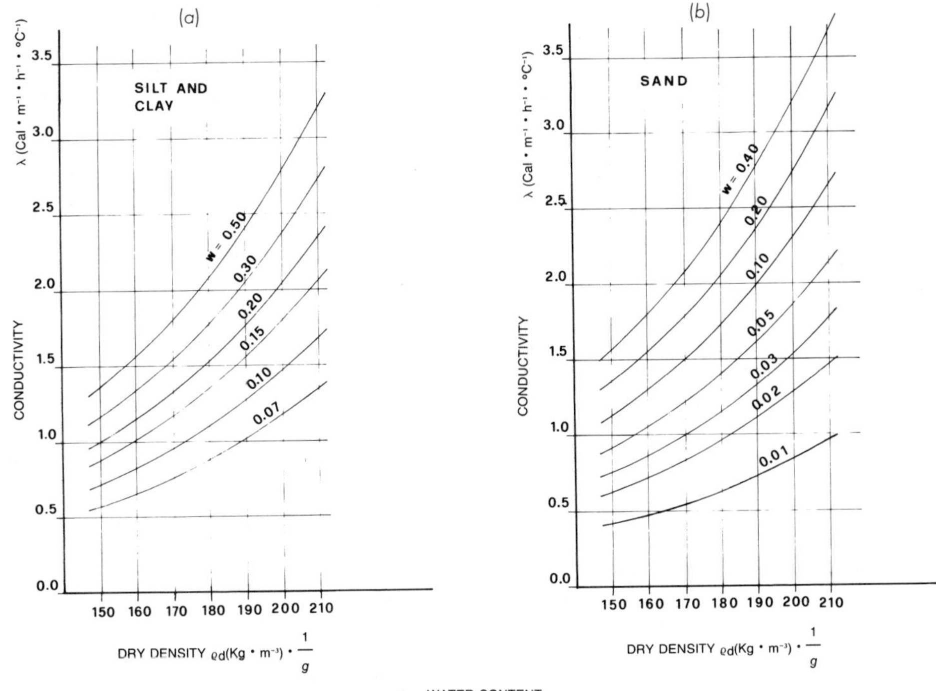

FIGURE 5. Thermal conductivity versus dry density in silt and clay soils (a) and sandy soils (b), for various moisture contents. After Kersten, (7) (1 $kg \cdot m^{-3} = 6.24 \times 10^{-2}$ $lb \cdot ft^{-3}$; 1 $Cal \cdot m^{-1} \cdot h^{-1} \cdot °C^{-1} = 0.67$ $BTU \cdot ft^{-1} \cdot h^{-1} \cdot °F^{-1}$).

equal [6] and a mean value is 0.17 $Cal \cdot kg^{-1} \cdot °C^{-1}$ at a temperature of 10°C (50°F).

Influence of moisture content in soil can be taken into account considering the specific heat of the soil-water mixture, on the basis of the relative weight of the two components, remembering that the specific heat of water is 1 $Cal \cdot kg^{-1} \cdot °C^{-1}$.

For a given water content w (ratio between the weights of water and of the dry soil in the mixture), the resulting formula is:

$$c = \frac{0.17 \times 1 + 1 \times w}{1 + w}$$

For example, if water content is 0.2, c has a value of 0.31 $Cal \cdot kg^{-1} \cdot °C^{-1}$.

THERMAL CONDUCTIVITY (λ)

This quantity is expressed here in $Cal \cdot m^{-1} \cdot h^{-1} \cdot °C^{-1}$ (1 $Cal \cdot m^{-1} \cdot h^{-1} \cdot °C^{-1} = 0.670$ $BTU \cdot ft^{-1} \cdot h^{-1} \cdot °F^{-1}$ $= 1.16$ $W \cdot m^{-1} \cdot K^{-1}$). Thermal conductivity varies with the texture of soil medium, being higher in gravel and sand than in silt or clay, and increases with increasing density and moisture content. For practical purposes, and neglecting the influence of temperature and mineral composition, two empirical formulas were given [6,7], respectively for sandy soils and for silty and clayey soils. Figure 5 was derived from these formulas, and the value of λ can be obtained as a function of both moisture content w and soil dry density ϱ_d.

DENSITY (ϱ)

This quantity is expressed here in $kg \cdot m^{-3}$ (1 $kg \cdot m^{-3} = 6.24 \times 10^{-2}$ $lb \cdot ft^{-3}$). Normal values range from 1600 to 2100 $kg \cdot m^{-3}$ for granular soils and from 1500 to 2200 $kg \cdot m^{-3}$ for silt and clay.

THERMAL DIFFUSIVITY (δ)

This quantity is expressed here in $cm^2 \cdot h^{-1}$ (1 $cm^2 \cdot h^{-1} = 1.08 \times 10^{-3}$ $ft^2 \cdot h^{-1}$). On the basis of the above mentioned soil characteristics, thermal diffusivity can be calculated, using Equation (3).

Heat Transfer Coefficients in Concrete

SPECIFIC HEAT (c)

Specific heat of concrete is almost independent from the mineralogical nature of the aggregates, while a certain influence is given by its moisture content. Normal values given in literature [8] range from 0.21 to 0.27 Cal · kg^{-1} · °C^{-1}, the higher values pertaining to concretes in a saturated condition, which would appear to be the case of massive foundations.

THERMAL CONDUCTIVITY (λ)

This quantity can assume very different values in concrete, varying from 1 Cal · m^{-1} · h^{-1} · °C^{-1} (or even less) to more than 2.7 Cal · m^{-1} · h^{-1} · °C^{-1}, depending mainly but not exclusively on the nature of the aggregates and on the moisture content [8].

A low conductivity is given for example by calcareous aggregates, a slightly higher one by basalt and trachyte, while very high values are given by gneiss and granite. As in the case of soils, high moisture content and high density increase the conductivity, due to the reduced presence of air-filled pores.

DENSITY (ϱ)

Normal weight concretes have a density varying from 2000 to 2800 kg · m^{-3}, depending on many factors, but values about 2400 kg · m^{-3} are very common.

THERMAL DIFFUSIVITY (δ)

This quantity resumes the above discussed thermal characteristics; therefore its possible variations are very large, especially due to variations of λ. Values given in literature range from about 20 to 60 cm^2 · h^{-1}. Central values from about 25 to about 45 cm^2 · h^{-1} have been derived experimentally in Italy for the concrete of several dams, so perhaps a mean value of 35 cm^2 · h^{-1} can be suggested for big foundations, unless a more precise determination is made in every single case.

As a conclusive remark, it can be added that helpfully δ appears under the square root sign in Equation (4), so a possible error in its value shouldn't affect too seriously the resulting temperatures.

STATIC EFFECTS

Preliminary Remarks

Are thermal effects important? A good answer is obtained if the problem of structural safety is evaluated assuming the limit state design approach: it can be said that the influence of thermal effects may be important, and therefore must be taken into account, when *structural serviceability* is concerned; this influence is considerably reduced, instead, when an *ultimate limit state* is reached, due to the appearance of large inelastic deformations [9]. So, in a general way, thermal effects should be considered when the limit states of deflection or cracking are checked, not when dealing with the maximum load-bearing capacity of the structure; there can be however exceptions on this point concerning not so much the structure (foundation) itself, but other structural elements connected to it; an example will be given in the next section. As regards to this point, see also [3,10].

In the analysis of thermal effects the concrete structure is frequently admitted to be uncracked; this hypothesis can often be accepted because of its simplicity, even if results may be too conservative, due to the beneficial influence of cracking in reducing thermal stresses. For a more accurate analysis of this effect, see [10].

An error is also introduced if the effect of creep in the concrete is neglected, that is if the "elastic" solution is adopted (E_c = modulus of elasticity for concrete = constant with time); creep is generally beneficial (as cracking is) in that it reduces maximum stresses, but it must be added that when the base temperature is at some distance from the annual mean temperature, creep may also cause important stresses of opposite sign with respect to the predominant stresses of the elastic solution [11]. In these cases it may be advisable, if creep is not explicitly taken into account, to consider also the effects of a second base temperature, nearer, or equal, to the annual mean. For creep effects in the concrete, see also [9,12].

Another important source of uncertainty comes from the difficulty of a realistic evaluation of the response of soil media when time-dependent consolidation and creep effects are expected: this happens with predominantly cohesive, saturated soil masses, and the presence of thermal loadings which are also time-dependent renders the analysis even more complex (of course also these effects may greatly contribute to the above-mentioned stress reversal phenomena). For some information on this problem and for an ample bibliography, see [13].

In this paragraph stresses and strains will be examined which arise in an isolated foundation (beam or plate), subject to a thermal action, as a direct consequence of restraint reactions offered by the embedding soil. Of course in more complex foundations composed by connected members, that is, in presence of statically indeterminate structures, also the effect of mutual restraint between adjacent members must be considered, resorting to one of the methods provided by the theory of structures. The effects of a constant temperature variation and of a linear temperature gradient will be examined separately.

Effects of a Uniform Temperature Variation

If the interface between a foundation and the underlying soil medium is supposed to be smooth, and the foundation is not otherwise restrained, almost stress-free horizontal movements take place, radially directed (the origin being the center of gravity of the foundation surface) and given by

$$\Delta r = r \cdot \alpha \cdot T_u$$

where r = distance from center of gravity, $\alpha = 10^{-5}$ $°C^{-1} = 5.5 \times 10^{-6} °F^{-1}$ = coefficient of thermal expansion of concrete.

As an example, consider a rectangular foundation (length L = 100 m = 328.1 ft, width b = 10 m = 32.8 ft, depth d = 2 m = 6.56 ft), with a 0.5 m (1.64 ft) thick soil cover, realized on a site where the mean annual temperature is +24°C (+75.2°F), $\bar{T}$ = 12°C (21.6°F), and cured at +15°C (59°F) (T_B = +15 −24 = −9°C = −16.2°F). If $\delta = 40\ cm^2 \cdot h^{-1}$ ($43.2 \times 10^{-3}\ ft^2 \cdot h^{-1}$) for both soil and concrete, we can see from Figure 2 that the temperature diagram in the foundation depth is almost a straight line having a value of 6°C (10.8°F) at section midheight (x = 1.5 m = 4.92 ft). The uniform annual temperature oscillation in the foundation is therefore

$$T_M = 6 \cdot \frac{12}{10} = 7.2°C\ (13.0°F)$$

(graphs in Figure 2 are set for $\bar{T}$ = 10°C = 18°F).

The two design uniform temperature variations are respectively (from Base Temperature—Design Temperature Loadings):

- in summer: $T_{us} = +T_M - T_B = +7.2 - (-9) = +16.2°C\ (+29.2°F)$
- in winter: $T_{uw} = -T_M - T_B = -7.2 - (-9) = +1.8°C\ (+3.2°F)$

This means that theoretically the foundation is almost strain-free in winter, while it expands radially from its center of gravity in summer, when maximum displacements at both ends are:

$$\Delta r = 50 \times 10^{-5} \times 16.2 = 0.0081\ m = 8.1\ mm\ (0.319\ in)$$

In the opposite limit case of a perfectly rigid restraint, relative movements will be suppressed. Horizontal compressive or tensile stresses σ will appear in the concrete, given by:

$$\sigma = \alpha \cdot T_u \cdot E_c$$

This means stresses in the order of 3 MPa (435 psi) in summer in the above considered foundation.

In real cases [13] an intermediate situation is present, but difficulties arise in the realistic estimation of the effective magnitude and distribution of tangential forces at the interface, which depend on many complex factors including, among others, soil strength, contact stresses, relative movement magnitude and soil time-dependent behaviour (the latter in particular may be of some importance in reducing tangential stresses originated by seasonal temperature variations); similar uncertainties are also present in pile-supported foundations, where the horizontal restraining action of the piling must be estimated. This means that a description of the interface condition can only be assessed on a case by case basis, with due reference to the geotechnical characteristics of the soil medium.

Here however two more realistic examples will be considered concerning the above mentioned foundation. In a first case the foundation rests directly on a dense-sand ground and shear stresses at the interface are supposed to be elastic and equal to $0.11\ MPa \cdot cm^{-1}$ ($40.5\ psi \cdot in^{-1}$) (varying linearly with the relative soil foundation movement).

In a second case, the foundation is supported by 100 reinforced concrete piles, driven in dense sand, having a diameter of 1 m (3.28 ft) and spaced approximately at 3 m (10 ft) between centers. Here again the shear force between piles and foundation is supposed to be of the elastic type, increasing linearly with the foundation movement, the reaction of each pile being equal to $740\ kN \cdot cm^{-1}$ ($422 \times 10^3\ lbf \cdot in^{-1}$).

Computational details are omitted, but principal results in the two cases are as follows:

- case 1—maximum displacement (at foundation's edge): 6.7 mm (0.264 in) maximum compressive stress (at foundation's center): 0.86 MPa (125 psi)
- case 2—maximum displacement (at foundation's edge): 7.1 mm (0.279 in) maximum compressive stress (at foundation's center): 0.61 MPa (88 psi)

In both cases these maximum values are obviously reached in summer, but care should be taken of the possible presence of tensile stresses in winter, as a consequence of creep phenomena (see Preliminary Remarks).

It appears that although rather rigid restraint conditions have been considered, in both cases the restraining effects are limited: a 10 to 20% reduction of maximum displacements in comparison to the "perfectly smooth" situation, while stresses in the concrete correspond to 20 to 30% of those due to an "infinitely rigid" restraint; stresses of this magnitude, although limited, should however be considered in a foundation's design, in that they can influence the cracking of concrete, expecially when tensile stresses are present.

A particular attention, moreover, must be paid in case 2 to the bending moments and shears induced in the piles as a consequence of foundation displacements, particularly near the foundation's edges, where movements are the highest. In fact, it could be shown that in this case such movements are too high, so that other types of piles should be considered (e.g., smaller diameters).

From the aforesaid considerations and examples concerning the effects of uniform temperature variations some conclusions can be drawn:

- Foundation movements, which are proportional to the expected temperature variation and to the dimension of the structure, can be of some relevance, especially in the external regions; connections with other structures should be checked, as well as pipes or other protrusions partly rigidly connected with the foundation and partly embedded in the ground; with good approximation movements can be simply computed under the hypothesis of smooth interface.

- For the same reasons, if the foundation is pile-supported, the effect of foundation movements should be considered in the piling design.
- Stresses in the foundation due to the soil (or piling) restraining action may influence the cracking of concrete.

Effects of a Temperature Gradient

LIMIT CASES OF INFINITELY SOFT SOIL AND INFINITELY RIGID SOIL

Here again it is useful to examine the consequences of a temperature gradient in the two hypothetical limit cases of an infinitely soft soil and infinitely rigid soil.

In the first case, the foundation is free to move. A beam is bent to a circumference and a plate (of any shape) to a spherical surface; concavity is upward if the gradient is positive (i.e., a temperature increase with depth). Curvature, constant along the beam and over the entire surface of the plate (in any direction) is given by the expression

$$\frac{1}{r} = \Delta T \cdot \alpha \cdot \frac{1}{d} \tag{7}$$

where r = radius of curvature. Expressions for edge rotations φ and deflections W (camber) in a beam are the following (see also Figure 6):

$$\varphi = \Delta T \cdot \alpha \cdot \frac{L}{2d} \tag{8}$$

$$W = \Delta T \cdot \alpha \cdot \frac{L^2}{8d} \tag{9}$$

For example, a 10°C (18°F) gradient on a beam 30 m (98.4 ft) long and 2 m (6.56 ft) thick produces a deflection of 5.00 mm (0.197 in) and edge rotation is 0.75 per thousand. The same expressions are valid in a circular plate if L is substituted by the diameter D of the plate (Figure 6).

On the other hand an infinitely rigid soil will not allow any movement, and stresses will build up in the foundation. In a rectangular beam, and under the hypothesis of uncracked concrete section, a bending moment is produced, constant along the beam, given by

$$M = -\Delta T \cdot \alpha \cdot \frac{E_c J}{d} \tag{10}$$

(where M = bending moment per unit width, positive if the upper zone of the section is compressed, and J = moment of inertia per unit width of uncracked concrete section); in fact it is evident that this moment completely neutralizes the curvature due to the gradient [Equation (7)].

In a similar way [14], in a plate of any shape a bending moment is present, constant in any point and in any direction, given by

$$M = -\Delta T \cdot \alpha \cdot \frac{E_c J}{d(1-\nu_c)} \tag{11}$$

(where ν_c = 0.16 = Poisson's ratio for concrete). Due to the plate continuity in the plane moments and stresses are increased by 20%.

It is useful for what follows to observe that the same constant distribution of bending moments is produced in a free beam or plate by an external moment M, acting at both ends of the beam, or uniformuly distributed along the edge of the plate [14]. It is clear that a positive gradient produces tensile stresses in the upper layers of the foundation and compressive stresses in the lower part; stresses are of course proportional to the distance from the neutral axis (or neutral surface) and maximum values at the opposite faces are given by

$$\sigma = \pm \frac{\Delta T \cdot \alpha \cdot E_c}{2}$$

in a rectangular beam and

$$\sigma = \pm \frac{\Delta T \cdot \alpha \cdot E_c}{2(1 - \nu_c)}$$

in a plate of any shape.

For example maximum stresses due to a 10°C (18°F) gradient are about 1.0 MPa (145 psi) in a beam and 1.2 MPa (175 psi) in a plate.

THE SOLUTION OF THE SOIL–FOUNDATION INTERACTION PROBLEM IN PRESENCE OF A THERMAL GRADIENT.

For any particular situation it may be sufficient to examine the above indicated opposite limit cases immediately showing the order of magnitude of the maximum deformations and stresses. If, on the basis of such preliminary considerations, more realistic results are needed, the soil–foundation interaction must of course be studied in more detail, and the choice must be made of an appropriate model of soil behaviour. For this, see other chapters of this book or, for example, [13].

The solution to the problem of soil–foundation interaction in the presence of a linear temperature gradient is not, to our knowledge, directly considered in the available technical literature. The problem, however, may be solved by resorting to a different though similar loading condition for which solutions are generally provided, and by modifying

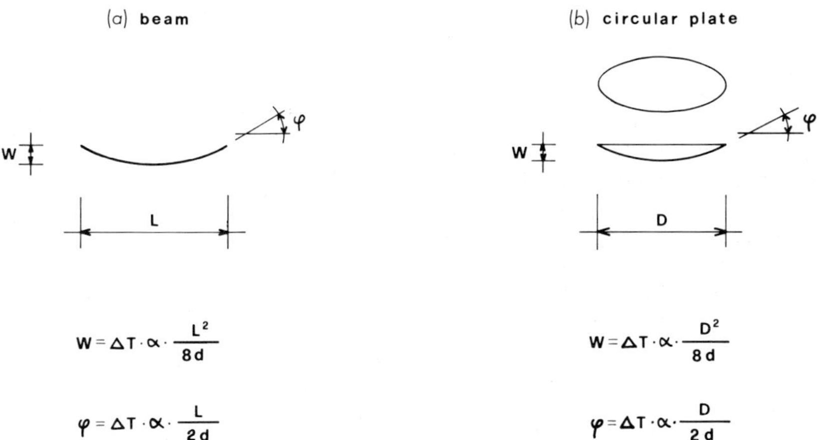

FIGURE 6. Deflections in a beam (a) and in a circular plate (b) due to a positive temperature gradient (temperature increases with depth).

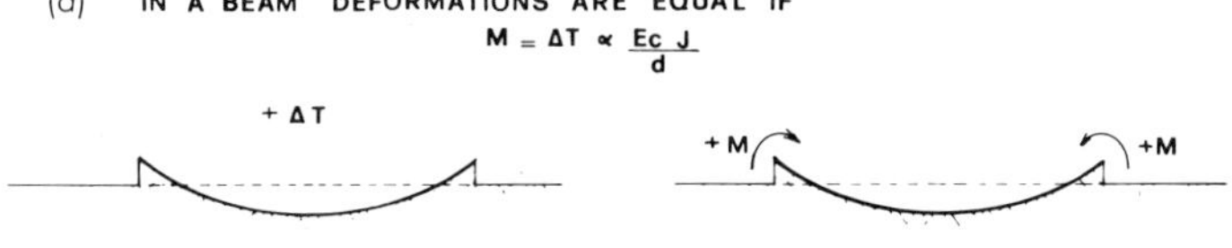

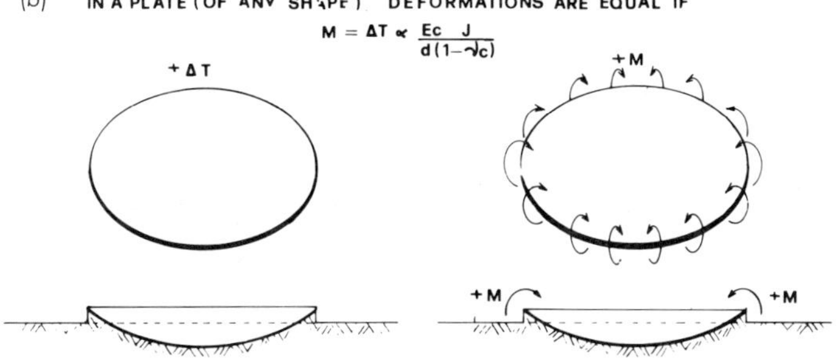

FIGURE 7. Conditions of equivalence of the deformations due to a thermal gradient and to a constant bending moment, in a beam (a) and in a circular plate (b).

some of the results thus obtained, in order to take into account the difference in the loadings.

The process can be summarized as follows:

- Step 1—An "auxiliary" loading condition is determined for the foundation, represented by a bending moment producing the same deformations (curvatures and deflections) as the thermal gradient ΔT. This moment is applied at both ends in a beam or is uniformly distributed along the edge in a plate; Figure 7 shows the relationships that must exist between M and ΔT in order to obtain this equivalence (the difference in sign with Equations (10) and (11) depends on the fact that here curvatures must be equal, not opposite). Of course these relationships keep their validity in any restraint situation, i.e., they don't depend on the mechanical response of the soil medium.
- Step 2—The analysis is performed for this auxiliary loading; reference is made for this analysis to the specialized literature on the subject. Deflections, contact stresses and internal forces can thus be evaluated.
- Step 3—Deflections, contact stresses, internal shear forces are the same for both loadings, so that results obtained in Step 2 are also valid for the thermal loading condition. Bending moments instead are different, but the difference is a constant which is equal to the external moments of the auxiliary loading condition (Figure 7); we only need to subtract from the bending moments obtained in Step 2 this constant value to find the bending moments due to ΔT.

If, for example, the already considered beam (L = 30 m = 98.4 ft, d = 2 m = 6.56 ft, ΔT = +10°C = 18°F) is supported by a Winkler type soil medium (k_s = 10 $N \cdot cm^{-3}$ = 36.8 $lbf \cdot in^{-3}$ = modulus of subgrade reaction), the external bending moment, per unit width, in the "auxiliary" loading condition is (Figure 8):

$$M = \Delta T \cdot \alpha \cdot \frac{E_c J}{d}$$
$$= +667 \text{ kN} \cdot \text{m} \cdot \text{m}^{-1} \text{ (150 lbf} \cdot \text{ft} \cdot \text{ft}^{-1}\text{)}$$

(assuming E_c = 20,000 MPa = 2.9 × 10^6 psi).

Some of the results [15] are given in Figure 8, where derivation of the bending moments corresponding to thermal action from those due to the auxiliary loading is also clearly shown. Of course, deflections (as well as contact stresses and shears) are the same for both loading conditions.

Maximum flexural stresses, in the middle of the beam, are about +0.7 MPa (100 psi) (always under the hypothesis of uncracked section) which is 70% of the stresses calculated in the limit case of a perfectly rigid restraint; an important difference is that here stresses decrease to zero at the beam ends.

Camber is equal to 2.0 + 0.8 = 2.8 mm (0.11 in) which is 55% of the value corresponding to the situation of infinitely soft soil.

A CASE HISTORY: THE PORTO TOLLE CHIMNEY FOUNDATION

Generalities

The Porto Tolle power station multiflue chimney, 250 m (820.0 ft) high, has a circular reinforced concrete foundation with a diameter of 40.7 m (133.5 ft), a thickness of 4.5 m (14.8 ft), and is supported by 415 driven closed-end steel piles, 36 m (118.0 ft) long and with a diameter of 0.50 m (19.7 in).

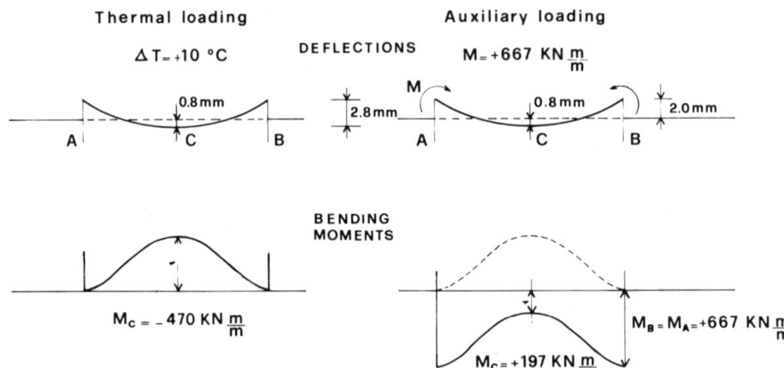

FIGURE 8. Thermal loading and corresponding "auxiliary" loading for an elastic beam resting on a Winkler soil medium and resulting deflections and bending moments (L = 30 m = 98.4 ft; d = 2 m = 6.56 ft; k_s = 10 N · cm^{-3} = 36.8 lbf · in^{-3}; 1 mm = 0.039 in; 1°C = 1.8°F; 1kN · m · m^{-1} = 224.7 lbf · ft · ft^{-1}).

Due to the poor geotechnical characteristics of the site (Figure 9), foundation settlements were regularly measured (Figures 10 and 11) during and after chimney construction as a part of the control program covering all the plant's main structures. Foundation tilting and deformation were thus periodically evaluated. The relative deflection (camber) of the foundation was obtained by calculating the difference between the center and edge of the plate (where an average of 4 equally spaced points was considered). In this report deflections will be considered positive if concavity is upward (see the sketch in the lower part of Figure 11).

Construction of the chimney's concrete shell, representing 90% of the chimney's total weight, began in February 1977 and was completed the following May. During this period, settlements were within expectation and resulting deflections were very small (Figures 10 and 11).

Little or no weight was added during the following months (Figure 10). It was therefore alarming to observe a rapid and steady increase in the deflection values from July to December of the same year (Figure 11). After a close examination it was decided that the phenomenon might be due to seasonal temperature variations and a temperature measuring program was soon prepared to check this. Thermocouples were installed at 5 different depths in the soil at the foundation's edge, at the north and south sides of the chimney (Figure 9). From January 1978 to November 1980, 40 sets of measurements were taken of both temperatures and deflections: 23 during 1978, 14 during 1979, and 3 during 1980.

Description of Phenomenon

The seasonal variation of deflections is shown in Figure 11. Apparently, the foundation concrete close to the upper surface is more influenced by seasonal atmospheric temperature than the concrete deeper down, where the temperature remains almost constant throughout the year. The corresponding thermal contractions in winter (or expansions in summer) in the concrete of the upper layers give rise to the observed variations in foundation deflections (camber) which increase in the winter (and decrease in summer).

Twelve temperature diagrams measured in the soil during 1978 are given in Figure 12. It can be seen that the temperature differential from summer to winter is about 20–25°C (36–45°F) at the surface and only a few degrees at the foundation bottom. The corresponding total amplitude of the foundation cyclic movement is about 6–8 mm (0.25–0.30 in).

Static Consequences

As already pointed out under Effects of a Temperature Gradient, stresses and deformations in a plate supported by an elastic soil and subject to a temperature gradient through its thickness depend on the relative stiffness of the plate-soil system. For a given plate and a given temperature gradient, stresses increase and deformations decrease with increasing soil rigidity.

In this study, the difficulty of a realistic determination of the supporting piling stiffness was avoided, because the relative stiffness of the system was derived through the experimental relationship found between foundation temperatures and corresponding deflections. it must be added that temperatures in the plate depth were assumed to be equal to those measured in the surrounding soil.

A preliminary processing of the experimental data consisted of transforming each measured temperature diagram into an equivalent linear distribution, as explained under Linearization of a Temperature Diagram. For some of the temperature diagrams measured during 1978 the calculated "equivalent" gradient ΔT is indicated in Figure 12. The values for the entire observation period are shown in Figure 11. The effects of non-thermal, time-dependent actions have been separated from the measured deflections, to isolate the pure, thermal-induced phenomenon. These effects are clearly visible in the general downward trend of the deflection diagram in Figure 11. They are mainly due to:

1. Primary and secondary compression in the soil strata beneath the pile tips, with corresponding settlements of the 20 m (66 ft) thick bearing stratum of sand and of the very thick underlying deposit of NC silt and silty sand
2. Increase of the chimney dead load, due to construction of the four steel liners
3. Creep in the foundation concrete
4. Differential shrinkage along the depth of the foundation, due to different exposure of the upper and the lower surfaces

These effects have been assumed to act linearly with time, which is only valid for a short period; actually, measurements made in the subsequent years indicate a marked flattening of this trend. After some trials, it was found that these effects could be represented by a straight line intersecting the deflection diagram shown in Figure 11.

The vertical distances from this line are the required thermal-induced deflections $W_{\Delta T}$. These are shown as a function of the corresponding ΔT in Figure 13. The resulting points form a narrow band which can be represented by the straight line

$$W_{\Delta T} = 0.25 \ \Delta T \qquad (12)$$

with only a slight concavity downward. The results given in Figure 13 seem a good confirmation both of the field measurements' precision and of the validity of the preceeding simplifying assumptions.

In order to interpret the statical behaviour of the soil-foundation system, it was assumed that the elastic half-space soil model could well represent, here, the restraining action of the piling. The solutions given by Gorbunov-Posadov [16]

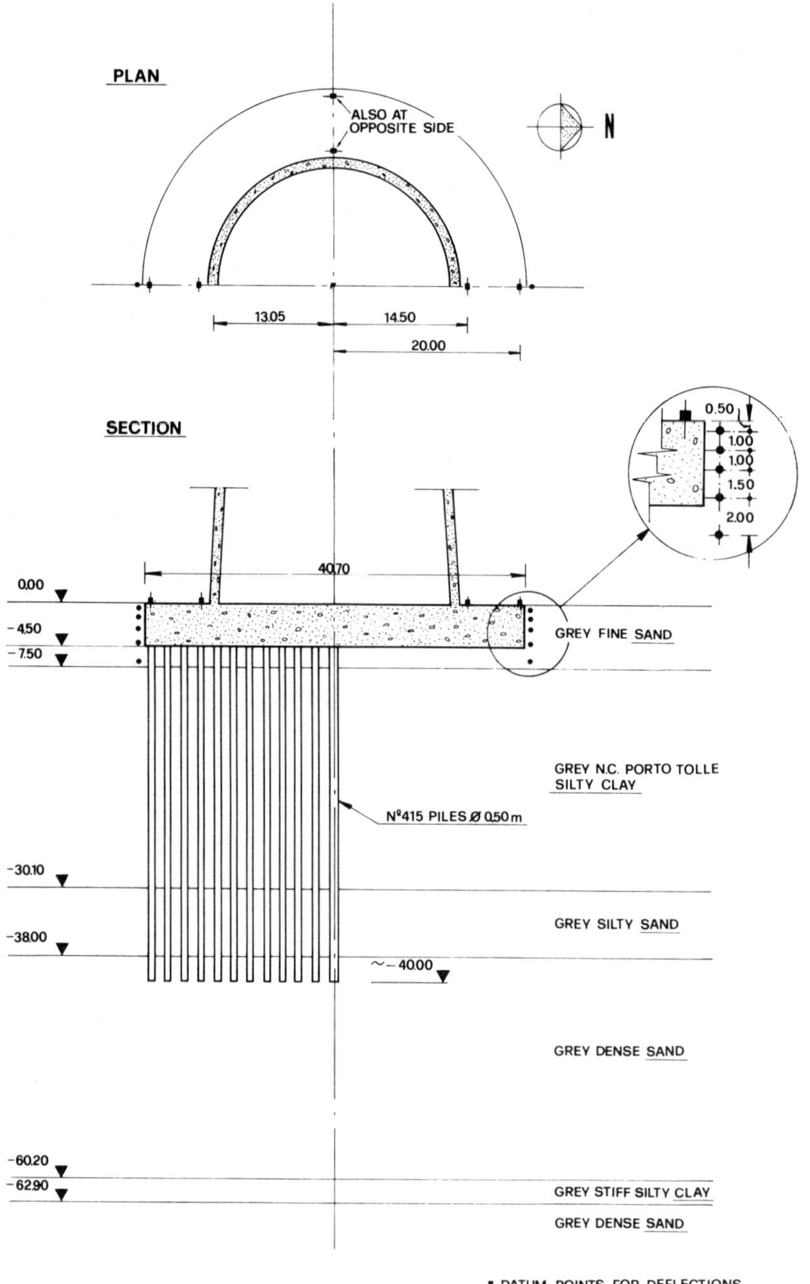

FIGURE 9. The Porto Tolle chimney foundation, soil stratigraphy and locations of deflection measurements and thermocouples (1 m = 3.28 ft).

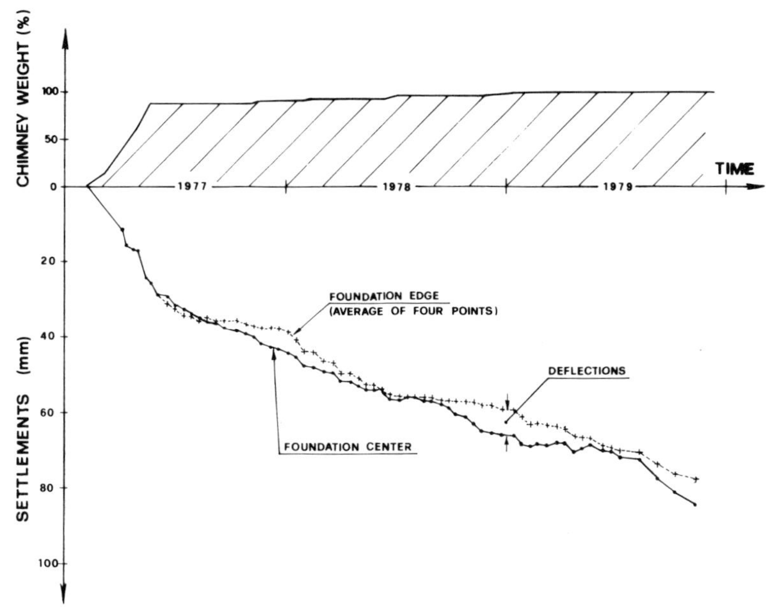

FIGURE 10. Percent of total chimney weight and foundation settlements versus time (1 mm = 0.039 in).

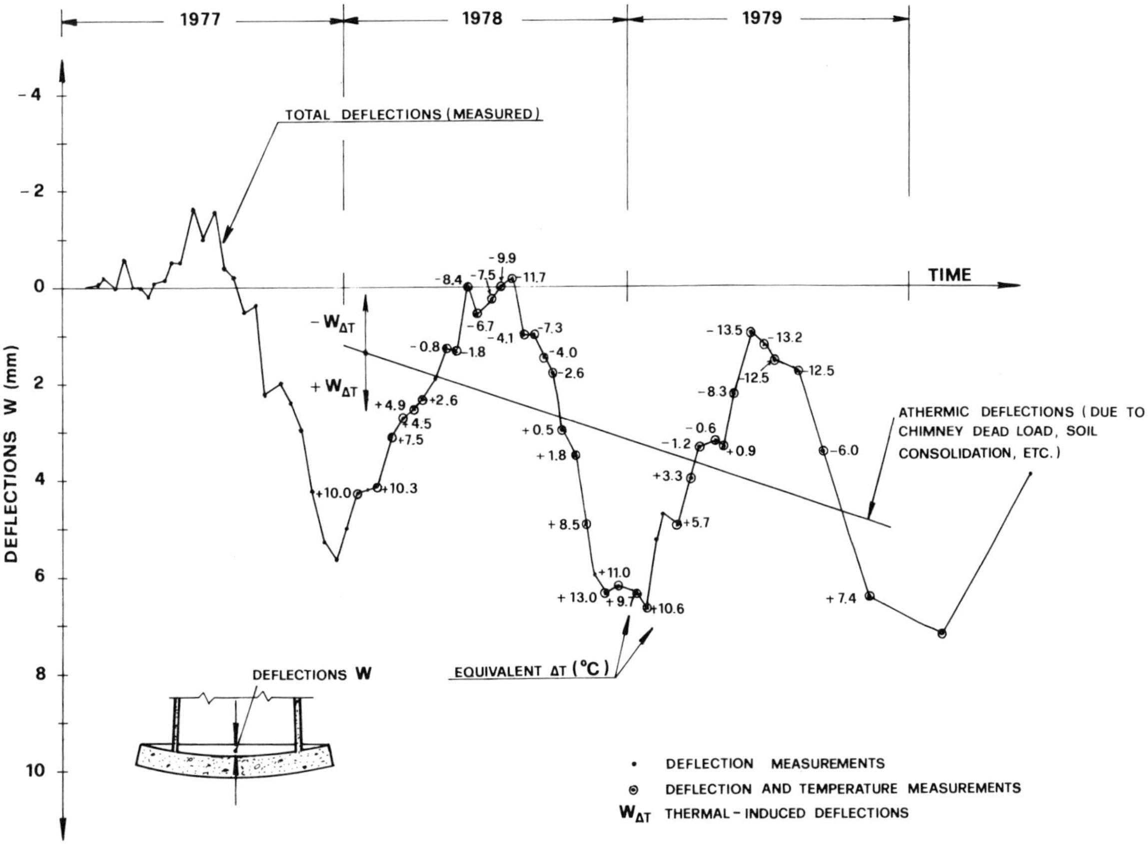

FIGURE 11. Foundation deflections (camber) versus time (1 mm = 0.039 in; 1°C = 1.8°F).

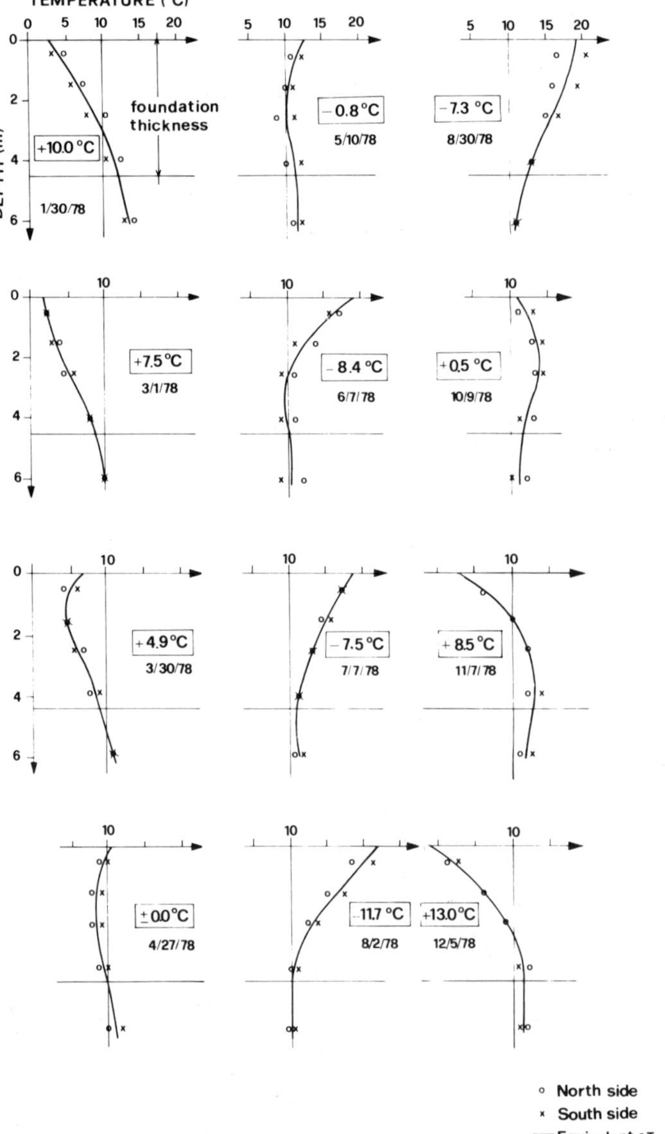

FIGURE 12. Temperature diagrams measured during 1978 and corresponding equivalent thermal gradients (1 m = 3.28 ft; 1°C = 1.8°F).

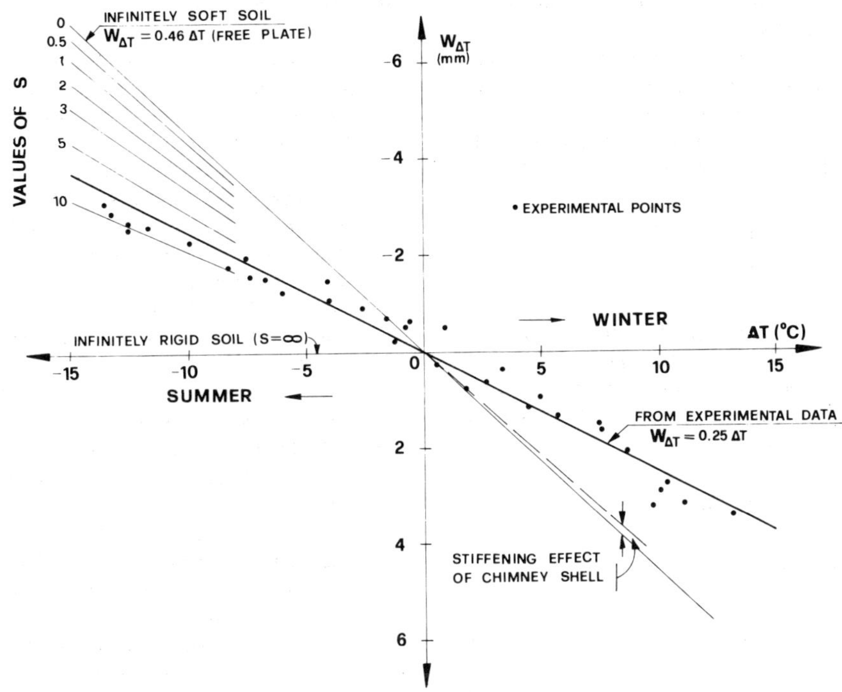

FIGURE 13. Theoretical effect of relative stiffness s on temperature-induced deflections, and experimental points (1 mm = 0.039 in; 1°C = 1.8°F).

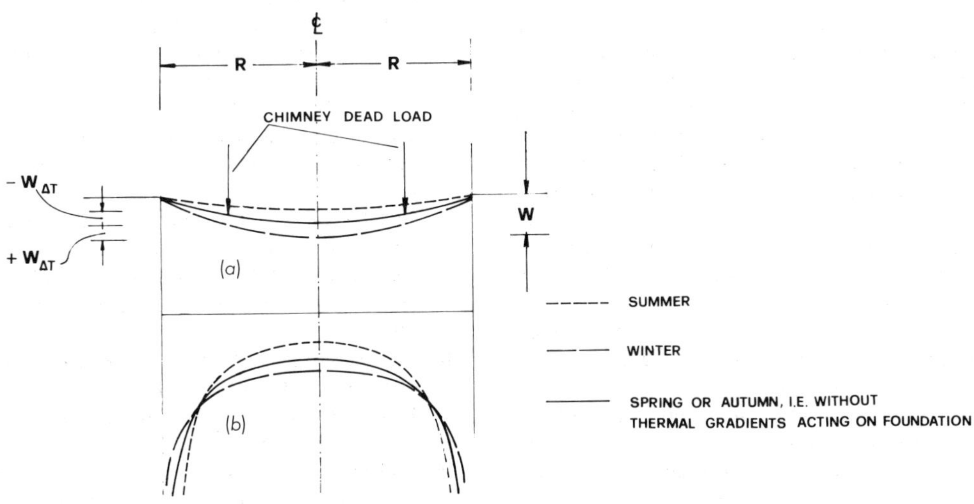

FIGURE 14. Influence of seasonal temperature oscillations on: (a) foundation deflections; and (b) contact pressure distribution (influence of temperature is exaggerated).

to the problem of elastic circular plates on an elastic, homogeneous, isotropic half-space were utilized. As these authors don't consider thermal gradients, the method outlined under Effects of a Temperature Gradient was used. Eventually the theoretical relationships between ΔT and $W_{\Delta T}$ were easily obtained for this foundation (Figure 13), for different values of the relative stiffness index, s, defined as

$$s = 3 \cdot \frac{(1 - \nu_c^2)E_t R^3}{(1 - \nu_t^2)E_c d^3} \tag{13}$$

in which E_t and ν_t = respectively, the equivalent modulus of elasticity and Poisson's ratio for the supporting medium; and R = the external radius of circular foundation. The two extreme cases of this family of lines represent the situation of infinitely soft soil ($s = 0$):

$$W_{\Delta T} = 0.46\ \Delta T \tag{14}$$

and infinitely stiff soil ($s = \infty$):

$$W_{\Delta T} = 0 \tag{15}$$

(note, comparing Equation (13) and Equation (14), that in this foundation measured deflections are about 55% of those corresponding to an infinitely soft soil).

The experimental line previously derived indicates, by comparison with the theoretical ones, that the stiffness index in this case has a value of about 7, and for this value both internal forces in the foundation and stress distribution over the contact area were determined. It was found that radial and tangential moments induced in the foundation, although not affecting safety, amount to a significant fraction of those due to dead load and wind load. Assuming $\Delta T = \pm 13°C$ (23.4°F) and E_c = 20,000 MPa (2.9×10^6 psi), and under the hypothesis of uncracked concrete section, the order of magnitude of these moments is about $\pm 3{,}000$ kN · m · m^{-1} (674×10^3 lbf · ft · ft^{-1}) at the foundation center, where the maximum value is reached; stresses are about ±0.9 MPa (130 psi) which is 60% of the values corresponding to an infinitely rigid soil.

In summer, consequently, a compression force is exerted on the upper part of the concrete section, and a tension force on the lower part; the reverse occurs in winter.

Vertical forces are also induced in the piling. In summer, the external piles are compressed and the central ones are in tension; the reverse is true in the winter (Figure 14). These additional forces however are small compared to those due to dead load and wind load.

REFERENCES

1. Smoltczyk, U., et al., "Earth Pressure Variations due to Temperature Change," *Proceedings of the 9th International Conference on Soil Mechanics and Foundation Engineering," Vol. 1,* Tokio, pp. 725–733 (1977).
2. Aschieri, F.,and F. A. Uliana, "Seasonal Temperature Effects on a Foundation,"*Journal of Geotechnical Engineering, ASCE, Vol. 110,* No. 6, pp. 796–806 (June, 1984).
3. Philleo, R. E., "Report on Symposium on Designing for Effects of Creep, Shrinkage and Temperature," *Designing for Effects of Creep, Shrinkage, Temperature in Concrete Structures,* SP-27, Americal Concrete Institute, Detroit, pp. 247–253 (1970).
4. Jacob, M., *Heat Transfer,* Vol. 1, John Wiley and Sons, Inc., New York, N.Y., p. 293 (1956).
5. ACI 349–80, *Code Requirements for Nuclear Safety Related Concrete Structures,* American Concrete Institute, Detroit (1980).
6. Winterkorn, H. F., and Hsai-Yang Fang, "Soil Technology and Engineering Properties of Soil," *Foundation Engineering Handbook,* Van Nostrand Reinhold Company, New York, N.Y., pp. 67–120 (1975).
7. Kersten, M. S., "Thermal Properties of Soil," *Bulletin No. 28,* University of Minnesota Institute of Technology, Minneapolis (1949).
8. Stucky, A., and M. H. Derron, *Problèmes Thermiques Posés par la Construction des Barrages-Réservoires,* Paul Feissly, ed., Lausanne, Switzerland (1957).
9. CEB Manuel de Calcul, "Effets Structuraux des Déformations Différées du Béton," *Bulletin No. 136,* Comité Euro-International du Béton, Paris, p. 3 (1980).
10. ACI Committee 349, "Reinforced Concrete Design for Thermal Effects on Nuclear Power Structures," *Journal of the American Concrete Institute, pp. 399–427* (Nov. Dec. 1980).
11. Chiorino, M. A., et al., "Influence of Creep on Stresses due to Temperature Variations in Concrete Structures," *Bulletin No. 154,* Comité Euro-International du Béton, Paris, pp. 219–229 (1982).
12. CEB-FIP, *Model Code for Concrete Structures,* Comité Euro-International du Béton, Bulletin No. 124/125-E, Paris (1978).
13. Selvadurai, A. P. S., *Elastic Analysis of Soil–Foundation Interaction,* Elsevier Scientific Publishing Company, Amsterdam (1979).
14. Timoshenko, S. and S. Woinowsky-Krieger, *Theory of Plates and Shells,* McGraw-Hill Book Co., New York, N.Y. (1959).
15. Hetény, M., *Beams on Elastic Foundation,* The University of Michigan Press, Ann Arbor (1946).
16. Gorbunov-Posadov, M. I., "Literature on Constructions and Architecture," *Computation of Constructions on Elastic Half-Space,* Moscow, (Russian) (1953).

CHAPTER 13

Centrifuge Model Tests on Bearing Capacity of Clay

AKIO NAKASE,* OSAMU KUSAKABE, AND JIROH TAKEMURA***

INTRODUCTION

In most cases the foundation failure takes place in saturated cohesive soils. Shear stresses in the ground increase during a loading period. However the strength of soil shows practically no change in this period, since the loading period is usually too short for the consolidation to occur. Hence the factor of safety against failure becomes minimum at or near end of loading. In such a case, the strength of soil before the loading is used for the stability analysis.

The strength of saturated soil measured in undrained shear tests is called the undrained strength C_u. This strength is independent of a confining pressure; hence the angle of shear resistance ϕu is equal to zero. Therefore, the stability analysis of cohesive soils, in which the undrained strength is considered, is called the $\phi u = 0$ analysis.

Bishop and Bjerrum (1960) collected 22 case records of failure in cohesive soils at end of construction, and stated that an accuracy of $\pm 15\%$ could be expected in the estimate of factor of safety. This statement may indicate that the $\phi u = 0$ analysis is satisfactorily accurate for practical purpose.

It is well known that in a normally consolidated clay, the undrained strength increases linearly with depth, provided that the soil density is uniform. Figure 1 shows a typical example of the relationship between measured strength and depth. The importance of considering the linear variation of the undrained strength of normally consolidated clay in stability analysis has been pointed out by many research workers.

Nakase (1966) attempted to determine the bearing capacity of a rectangular footing resting on clay of which undrained strength increases linearly with depth by assuming that the slip surface is a part of horizontal cylinder emerging from an edge of the footing. The load on the footing is considered to be evenly distributed and the effect of roughness of footing does not come into the solution. Nakase (1967) also examined three case records of failure in cohesive soils which occurred in Japan, and reported that the occurrence of failure in cohesive soils could be explained reasonably by the $\phi u = 0$ analysis, provided that the undrained shear strength–depth relationship was considered. Raymond (1967) dealt with the bearing capacity problem of a strip footing on similar clay but he did not obtain the minimum values of bearing capacity.

Davis and Booker (1973,a) calculated the bearing capacity of a strip footing on this type of clay rigorously by solving Kotter's equations for various values of a parameter kB/C_o in which k denotes the gradient of linear increase in undrained strength, B the footing width and C_0 the undrained strength at the surface of clay (Figure 2). It was found that the roughness of the footing has small effects on the bearing capacity. The ratio of the slip circle solution to this plasticity solution varies from 1.08–4.50, increasing with the values of kB/C_o. They (1973,b) pointed out that intuitively sensible approximate methods are not always reliable. This plasticity solution has later been coupled with the slip circle solution by Nakase (1981) for evaluating the bearing capacities of rectangular surface footings on normally consolidated clay.

Recently the bearing capacity of a circular footing on the surface of clay that has its undrained shear strength increasing linearly with depth, has been tackled by Houlsby and Wroth (1983), Suzuki (1985) using plasticity theorem.

There are a few recommendations where a certain undrained shear strength is taken as an equivalent strength. Skempton (1951) suggested that the average value of the undrained shear strength from the footing base down to a depth of $2B/3$ should be used as the equivalent strength provided

*Department of Civil Engineering, Tokyo Institute of Technology, Tokyo, Japan

**Department of Civil Engineering, University of Utsunomiya, Utsunomiya, Japan

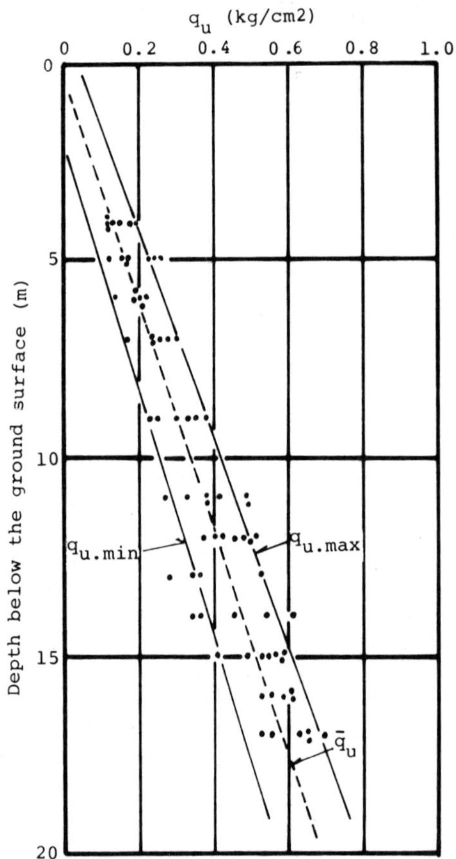

FIGURE 1. Typical relationship between measured unconfined compression strength and depth (Nakase, 1967).

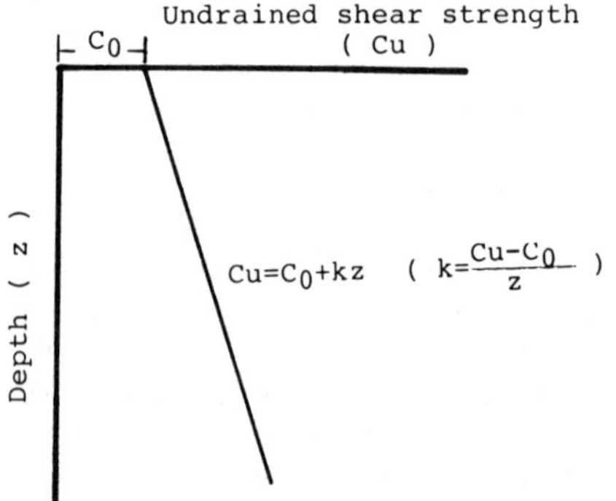

FIGURE 2. Increase of undrained shear strength with depth.

that the strength within the previously mentioned range is included into +50% of the average strength. In the case where the undrained shear strength increases linearly with depth, this limitation corresponds to the range of $kB/C_o <$ 3. Peck et al. (1953) recommended that the average value of the undrained shear strength from the footing base down to a depth of B should be used, whereas Livneh and Greenstein (1973) suggested that the strength at the depth of $0.4B$ should be used as the equivalent strength. Both recommendations were made without any restriction.

So far there have been quite a number of research workers who have studied the bearing capacities of foundations with varying strength. But as we have seen above, most of them are theoretical studies and until now there have been very few experimental studies, especially with clays, to provide any checkings or verifications on the theories.

The main reason that there had been very few experimental studies on the bearing capacities of foundations of strength increasing linearly with depth is because modeling of such a prototype is difficult under earth's gravitational field.

For small-scale models under gravitational field, only bearing capacities of small footings can be obtained. As the effects of increasing strength with depth on bearing capacities are small for small footings, these types of models serve no significant purpose. But if large-scale models are to be made, not only huge testing systems are necessary, but the consolidation time will be too long for the experiments to be practicable as well.

This difficulty in modeling, however, can be overcome by using a centrifuge. The simulation of prototype is made possible both geometrically and mechanically with the use of a centrifuge model.

Principles of centrifuge modelling have been appreciated since the 1930s (Bucky, 1931, Pokrovsky, 1934), and have been explained by several workers (for example, Schofield, 1980).

In centrifuge model tests, the weight of soil is increased and the scale of the model is reduced, both by a factor of N. The result is identical similarity at corresponding points in a model and in notational full-scale prototype of the total and effective stresses and strains in the soil of the pore-water pressures. In addition, the reduction of model scale by a factor of N means that pore-water diffustion to achieve a given time factor ($T_v = C_v t/h^2$ in Terzaghi's consolidation theory) requires times t_m in the model greatly reduced in comparison with times t_p in the prototype in the ratio $t_m/t_p = 1/N^2$. The reduction in consolidation time is one of the main merits in using centrifuge models. With models at 1/100 scale the times for dissipation of pore water pressure are reduced by 100 × 100, so that events seen in one week at the full prototype scale occur in a time of approximately 1 minute at the model scale.

This chapter shows firstly how a normally consolidated clay stratum of which undrained strength increases nearly linearly with depth can be prepared in a centrifuge. Secondly it shows the loading test results and the comparison of bearing capacities of a strip footing on the surfaces of the clay between the existing theories and the experimental results.

PREPARATION OF A NORMALLY CONSOLIDATED CLAY STRATUM IN A CENTRIFUGE

Centrifuge and Test System

Currently more than a dozen geotechnical centrifuges are in operation in various countries. There will be more new centrifuges coming in a couple of years. The centrifuge used for this particular study was Mark II of the Tokyo Institute of Technology (Kimura et al. 1982). The specifications are listed in Table 1 and the structure is illustrated in Figure 3. Various types of container, rectangular, square or cylindrical, can be mounted on most of the centrifuges currently in operation. For this particular series of tests a rectangular container shown in Photo 1 was used to carry out model tests in the plane strain condition. The main parts of the container are made of steel and the front face is made of a highly shock resistant plastic plate with the thickness of 50.8 mm, so that optical targets placed on a lateral side of clay can be viewed through in flight. It has been pointed out that the lateral strain of the soil box must be kept less than 0.1% to maintain the plane strain condition (Ko and Davidson, 1973). The stress calculation on the soil box predicts 0.08% as the lateral strain under 19.6 kN/m² of surcharge pressure and 80 g of centrifugal acceleration.

Test system is schematically shown in Figure 4. The dimensions of the package are 19.7 in (500 mm) by length, 5.19 in (150 mm) by width and 16.1 in (410 mm) by height. The height, H, of the clay block is 5.9 in (150 mm).

To reduce the action of adhesion forces between the clay and walls of the container, a grease which reported as effective by Mair (1979) was applied on the walls before the clay block was constructed. Underlying the clay block is a layer of 3.1 in (80 mm) compacted coarse sand which serves as a pervious base for reducing the consolidation time. Buried in the sand is a drainage pipe which is connected to two drainage tubes fixed at the sides of the container. The tubes are in turn connected to the upper part of the container and their ends remain open to air. This is to maintain a constant total water head during the tests by draining out any excess water produced.

Consolidation Procedures

The material used was Kawasaki clay which was dredged marine clay that has been thoroughly remolded and sifted with a 2000 sieve. Its index properties are shown in Table 2. The data of the coefficient of consolidation and the coefficient of volume compressibility obtained from a series of oedometer tests are shown in Figure 5 and Figure 6, respectively.

TABLE 1. Specifications of T.I.T. Centrifuge.

Maximum acceleration	150 g—382.6 r.p.m. at the radius of 1230 mm
Driving motor	A.C. variable speed type; 15 kW, 200 V Controllable range; 100–1350 r.p.m.
Maximum package mass	250 kg
Hydraulic slip ring	5 channels
Electrical slip ring	76 poles

The clay was remolded at the water content of about 80%, approximately equal to 1.5 times the liquid limit. Slurry was then deaired under the negative pressure of about −90 kN/m² for half an hour and poured into the container in eight layers. Preconsolidation was carried out for each layer by compressing clay with a loading plate and a bellofram cylinder. The preconsolidation pressures were 7.4 kN/m² and 9.8 kN/m² for the final surcharge pressures of 9.8 kN/m² and 19.6 kN/m² respectively. On completion of preconsolidation of each layer optical targets were placed on the surface of clay layer.

The final preconsolidation was continued until a curve obtained by plotting the settlement against the logarithm of time t had an intersection with a straight line derived by shifting the steepest part of the curve by the factor of $\log_{10} 2$ [2t method, Japan Society of Soil Mechanics and Foundation Engineering, (1975)].

TABLE 2. Properties of Kawasaki Clay.

Liquid limit LL (%)		53.5
Plastic limit PL (%)		26.6
Plasticity Index I_p		26.9
Specific gravity of soil particles G_s		2.69
Percentage of fraction of soil component (%)	sand	19.3
	silt	45.2
	clay (under 5μ)	35.5
Oedometer test	compression index C_c	0.318
	void ratio at 98 kN/m²	1.046
$(c_u/\sigma'_{vc})_{K_0}$		0.41

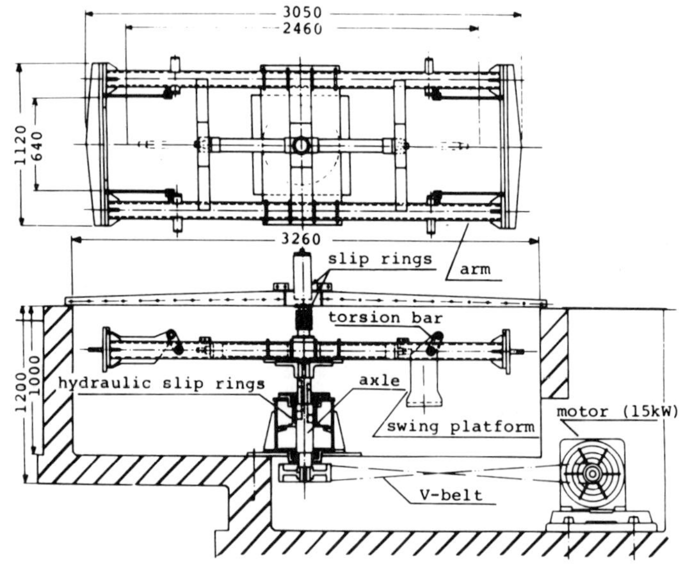

FIGURE 3. Mark II centrifuge of the Tokyo Institute of Technology.

PHOTO 1. Package for centrifuge test.

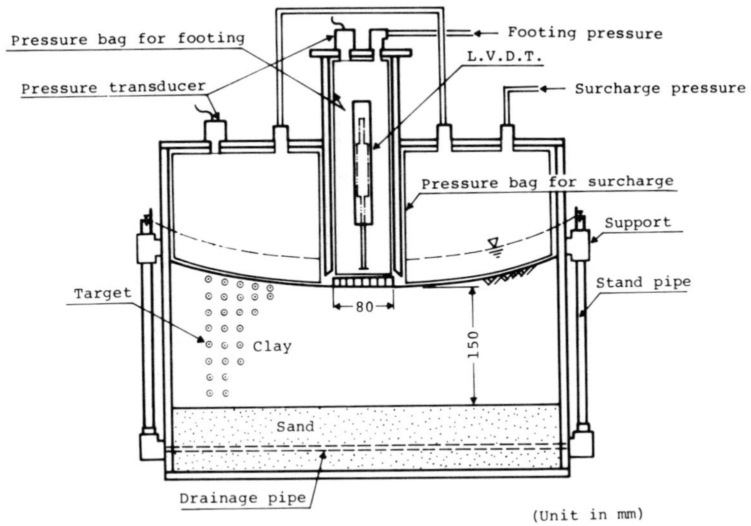

FIGURE 4. Set up of centrifuge model and location of pore pressure transducers.

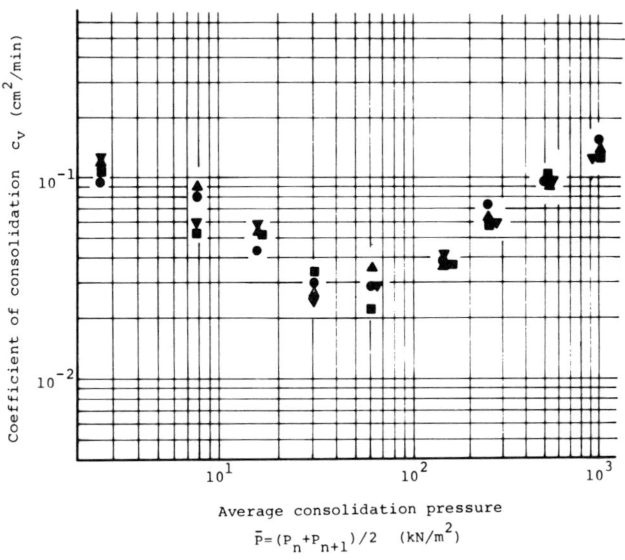

FIGURE 5. Variation of c_v with average consolidation pressure.

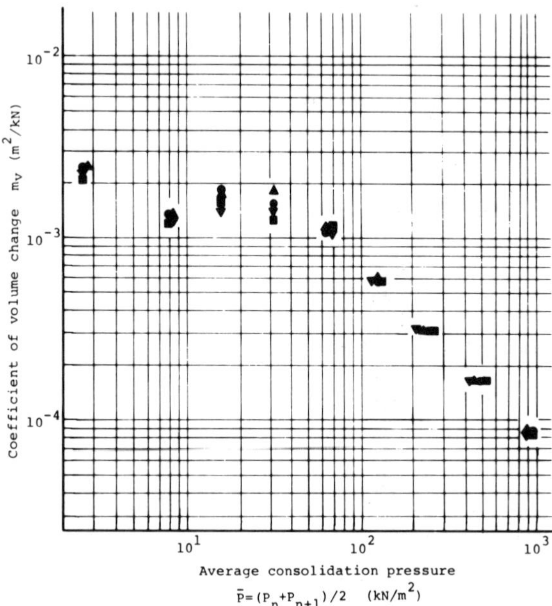

FIGURE 6. Variation of m_v with average consolidation pressure.

After preconsolidation, the pressure was removed. Six horizontal holes were drilled through clay by a small hand auger using the guide holes on the back face of the container, each having the length of 2.95 in (75 mm) and the diameter of 0.31 in (8 mm). Each hole was filled with deaired slurry and a pore pressure transducer was inserted into the hole. A final surcharge pressure of either 9.8 kN/m² or 19.6 kN/m² was then applied and the increase in preconsolidation pressure would serve to reduce the effect of disturbance due to the insertion of the transducers. The 2t method was again adopted to determine the completion of consolidation.

The final surcharge pressures on the lab floor were selected such that: (1) it can be shaped easily; (2) the footing settlement during centrifugal consolidation will not be too large to handle; and (3) the centrifugal consolidation time required to reach the desired degree of consolidation will not be too long. Consequently the change in void ratio during the centrifugal consolidation was relatively small.

After having completed consolidation on the lab floor, the upper surface of the clay block was shaped into an arc with its radius equivalent to the distance between the rotation axle of the centrifuge and the upper surface of the clay block during centrifugal operation. This was done by using a wooden frame of the same curvature and a blade for shaping. This procedure was necessary to avoid complications due to possible effect of tangential component of centrifugal force. If the surface of clay is horizontal, the strength at a point near side walls of soil box might be different from that at the centre due to the effect of tangential component even when two points are at the same depth. The bottom of the clay block can be shaped but was not shaped. This gives the tangential stress of approximately equal to 1/7 of the vertical stress to the clay at bottom corners, losing the one-dimensional consolidation condition locally. However, the error induced for not doing so is negligible in this case since the block was thick and the radius of the centrifuge is long enough. And also this part of clay plays a less important role in subsequent loading tests because failure planes are considered to be rather shallow for this type of clay as the plasticity solution suggests.

Pressure was applied to the clay by using three rubber bags installed in the container: two surcharge bags at the sides and a footing bag at the center. Inside the bags were transducers to measure the air pressures. Just underneath the footing bag, an aluminum footing of width 3.15 in (80 mm) was placed on the clay surface. In order to obtain a uniform settlement along the width of the clay block, which is necessary to satisfy the plane strain condition, and to fit the curvature of the upper surface, the model footing used was made up of eight rectangular aluminum rods bundled with adhesive tape. At each end of the footing, a layer of sponge was pasted on. This serves to reduce frictional forces and at the same time prevent clay leakage during loading tests. This footing is considered as rigid. The settlement of the centre of footing can be detected by the L.V.D.T. installed inside the footing bag.

After the preparations of the system had been completed, the system was installed in the swing of the centrifuge and centrifugal consolidation was carried out. Centrifugal acceleration was raised gradually in stages of 10 g. As the water pressure changed with centrifugal acceleration, the pressure in the bags was adjusted from time to time to maintain a net value equivalent to half of the surcharge pressure. Furthermore, the pressure in the footing bag was also corrected for the weight of the model aluminum footing. Full pressure was applied only after the centrifugal acceleration and the water level in the container had reached the proposed value. Pore pressure, air pressure, settlement and time readings together with photographs were taken.

A series of four tests with the combinations of accelerations 40 g, 80 g and surcharge pressures 1.42 psi (9.8 kPa), and 2.84 psi (19.6 kPa) were carried out. It should be noted that g is the earth's gravitational acceleration and is taken to be equal to 32.1 ft/sec² (9.8 m/s²). Two reference tests with surcharge pressures of 1.42 psi (9.8 kPa), and 2.84 psi (19.6 kPa) were performed in the earth's gravitational field. Test conditions are listed in Table 3.

The parameter kB/C_o was calculated by assuming that the undrained shear strength is strictly in proportion to the depth as

$$C_u = C_o + kz$$

where C_o denotes the undrained shear strength at the surface of clay, k the gradient of the strength increase with depth and z the depth. Then the parameter can be expressed as

$$\frac{kB}{C_o} = \frac{B}{C_o}\frac{C_u - C_o}{z} = \frac{B}{\left(\frac{C_u}{\sigma'_{vc}}\right) p_o}\,\frac{\left(\frac{C_u}{\sigma'_{vc}}\right)\gamma' Nz}{z}$$

$$= \frac{B\gamma' N}{p_o} = \frac{BN}{p_o}\frac{G_s - 1}{1 + e_o}\gamma_w$$

when B denotes the width of the footing in loading tests conducted later on, σ'_{vc} the consolidation pressure, γ' the submerged weight of the clay, N the ratio of centrifugal acceleration to the gravitational acceleration, p_o the surcharge pressure and e_o the void ratio in the model after the lab floor consolidation due to the surcharge load. Using quantities listed in Table 2, the equation for the K_o normal consolidation line is written as

$$e = 1.046 - 0.318 \log_{10} (\sigma'_{vc}/98)$$

The value of e_o can be obtained from the above equation by putting the surcharge pressure on the lab floor into σ'_{vc}.

Results of Centrifugal Consolidation

Pore pressure and footing settlement during, and water content distribution after, centrifugal consolidation were measured to ensure that the proposed model foundation of strength increasing linearly with depth was obtained. One of the pore pressure–time curve plottings during consolidation is shown in Figure 7. (One out of the six pore pressure transducers malfunctioned.) On completion of consolidation on the lab floor, the consolidation pressure was once removed, which was required to make necessary preparations for the centrifugal operation. Assuming that the effective stress is fully retained during this process, the total and effective stress and pore pressure before and immediately after the start of centrifugal consolidation is given in Figure 8. The excess pore pressure at the initial stage of consolidation is also shown. In the first extreme case where no negative excess pore pressure load even decayed, the initial distribution of excess pore pressure at the start of centrifugal consolidation would be in the shape of triangle DEFD. Or even if some negative excess pore pressure had decayed to OGCO the distribution at the start of centrifugal consolidation would be CEFOJC. As the portion of CDFOJC was in the overconsolidated area where the coefficient of consolidation, cv, is about ten times that of normally consolidated area, it might have dissipated away easily during the period between the start of centrifugal consolidation and initial measurements taken. This would again bring the initial excess pore pressure distribution to triangle DEFD.

In the second extreme case, the negative excess pore pressure created was assumed to have totally decayed before centrifugal consolidation and no significant amount of excess pore pressure formed then had ever dissipated before initial measurements taken. The initial excess pore pressure distribution would be taken in the shape of trepezium CEFOC.

These computations of excess pore pressure were made for the two extreme cases mentioned by solving the conventional Terzaghi's type of consolidation differential equation using Fourier series. Actual pore pressure–time relationship should be expected to lie in between the two curves.

It can be seen from Figure 7 that the initial portions of the curves have some irregularities. In the initial stage of centrifugal consolidation the water level was below the expected level and thus only part of the total pressure was applied to avoid the possibility of overconsolidating the clay. The full consolidation pressure was only applied after the water level had risen to the expected level. This amount of water pressure and unapplied pressure accounted for the low pore pressure detected in the initial period. Except for these irregularities, the computed curves and measured values agree reasonably well. This constitutes one of the evidences that the proposed modeling was carried out properly.

Computations were also performed to determine the settlement time relationship using the same Fourier series mentioned previously. It was found that there is extremely good agreement between computed and measured values as shown in Figure 9. This agreement forms another positive evidence.

As the effective stress varies with the time of consolidation and the depth of clay, so do coefficient of consolidation cv and coefficient of volume compressibility. Therefore equivalent constants must be selected in the consolidation computation.

Since the average values of consolidation pressures in clay

TABLE 3. Test Conditions.

Test number (1)	Surcharge (kPa) (2)	Acceleration (g) (3)	kB/C_0 (4)	Loading rate (kPa/min) (5)
TK02	19.6	1	0.03	9.8
TK01	9.8	1	0.06	9.8
TK5	19.6	40	1.19	9.8
TK7	19.6	40	1.19	39.2
TK2	9.8	40	2.29	9.8
TK6	9.8	40	2.29	39.2
TK3	19.6	80	2.38	39.2
TK4	9.8	80	4.58	39.2

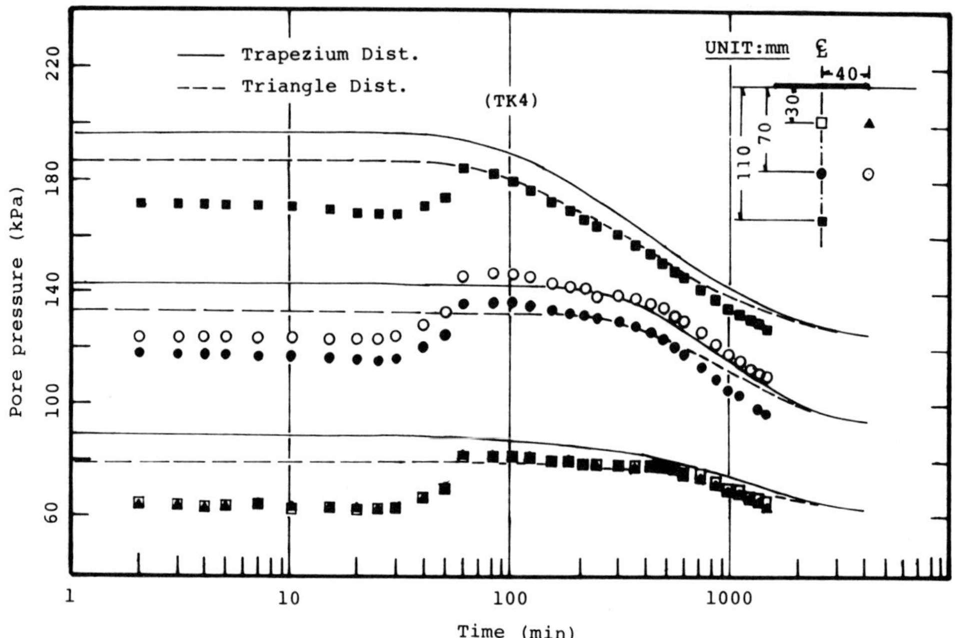

FIGURE 7. Pore pressure time behavior during consolidation.

were estimated to range from 30–60 kN/m², the value of 0.0025 cm²/min was used for c_v, based on the data shown in Figure 5.

Denoting compression index as C_c, the definition of m_v is

$$m_v = \frac{0.434\ C_c}{(1 + e)\sigma'}$$

If the effective stress stays in a narrow range, the value of m_v can be assumed to be constant. In the current experiments the change in effective stress from the initial stage of the centrifugal consolidation to the final stage is at most fivefold. Since this was considered not very large, the following approximations were employed. A representative value of effective stress $\bar{\sigma}'$ was determined as a logarithmic average value of effective stress at the surface and the bottom of clay as

$$\bar{\sigma}' = \sqrt{p_o(p_o + k_1 H)}$$

and the void ratio corresponding to the particular effective stress was obtained from the normal consolidation line. Defining the void ratio as $\bar{e}$, an approximate expression for m_v is given as

$$m_v = \frac{0.434\ C_c}{(1 + \bar{e})\sqrt{p_o(p_o + k_1 H)}}$$

This type of approximation does not hold for general cases of self-weight consolidation. The theoretical study for the self-weight consolidation is found in works by Mikasa (1963). Experimental verification of the self-weight consolidation was given by Croce et al. (1984).

The measured pore pressures and consolidation settlements are compared with the theoretical calculations in Figures 7 and 9. Computed consolidation settlements compare generally well with the observations. This shows that the one-dimensional consolidation has actually taken place in the centrifuge and that the approximation employed for the analysis is justifiable.

After the loading tests, 40 soil samples were obtained by pushing hollow aluminum tubes to the clay block at appropriate position to investigate the water content distribution in the clay model. The thickness of the aluminum foil used to make the tubes is about 0.04 in (0.1 mm). A typical one is shown in Figure 10. It should be noted here that prior to centrifugal consolidation, the water content of the model has a constant value corresponding to the surcharge pressure applied. Three computed curves are also shown. The degree of consolidation is denoted by U in the figure. It is obvious that measured values and the 90% consolidation curve (or more precisely, the curve after 1,500 min of consolidation) agree closely. The maximum discrepancy of water content was only 1.5%. This occurred at the lower portion of the clay block which is near to the pervious base and where the consolidation stress level is the highest. Therefore this can

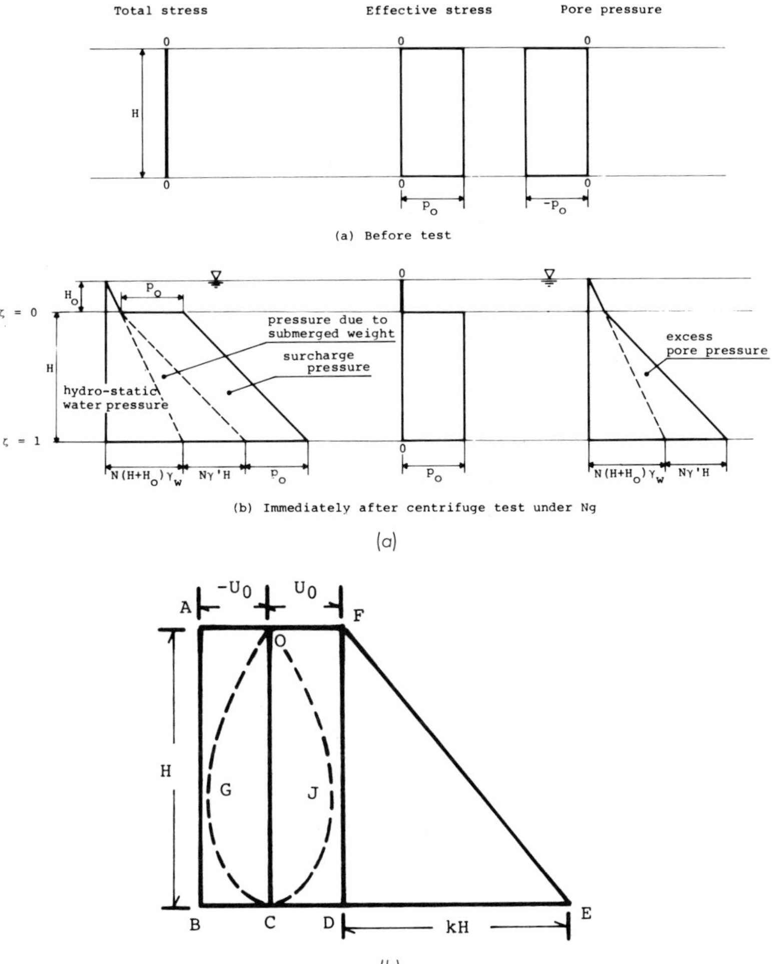

FIGURE 8. Stress change during test: (a) stress before and after centrifuge test; (b) initial distribution of excess pore pressure.

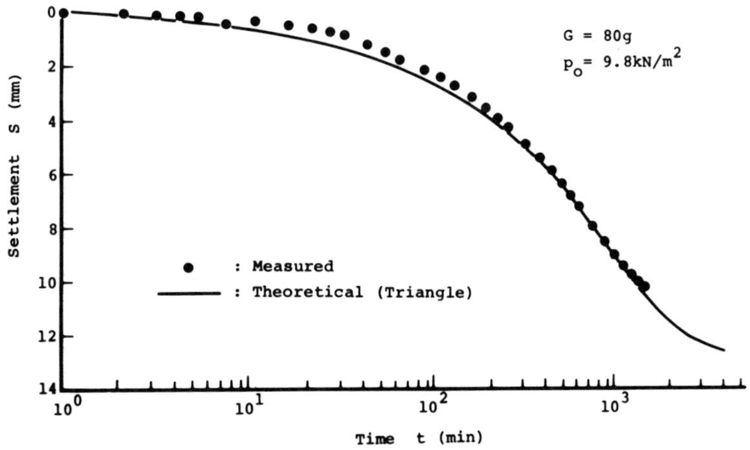

FIGURE 9. Footing settlement time relation during consolidation.

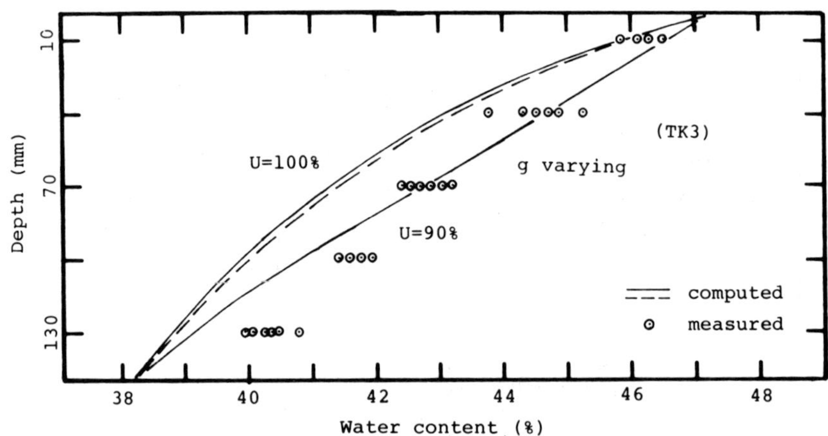

FIGURE 10. Water content distribution with depth.

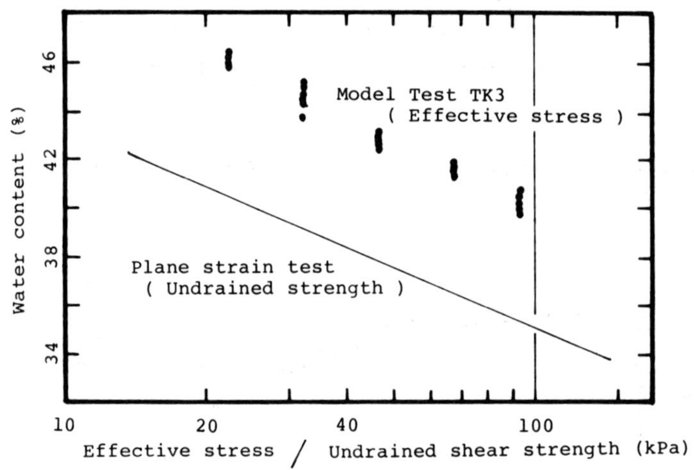

FIGURE 11. Water content–effective stress plot of TK3.

be explained by the effect of suction created upon the release of centrifugal load after the test. As the centrifugal consolidation of the test was also performed to a degree of 90%, this agreement provides the third positive evidence to the correct modeling of prototype.

As shown in Figure 10, the maximum water content difference between 100% consolidation curve and 90% curve is about 1.5%. The only way to reduce this difference is to increase the degree of consolidation by reducing the depth of the clay block or increasing the consolidation time. But to increase the degree of consolidation time from 90% to 95% in this series of tests would require more than 500 additional minutes which is about 2/5 of the time needed to bring the degree of consolidation to 90%. This can be done but will be rather meaningless in the engineering aspect. Therefore in this series when the degree of consolidation was found to reach 90% by the t fitting method, the loading tests were performed.

The centrifugal acceleration in the model varies with distance from the center of rotation axle. Computations show that the difference between the 100% consolidation water content curve with acceleration remained constant, and the variation in the direction of radius is too small to be of any importance, as seen in Figure 10. Computations on the effective stress also show that the maximum difference due to the change of acceleration is only about 3%.

In general the height of centrifugal models is limited by a requirement that the linear increase of vertical pressure due to self weight in the prototype, and the non-linear increase of radial pressure due to acceleration in the centrifuge should not differ too much. Schofield stated (1976) that models are kept within radii of $r \pm 0.1\ r$ in order to limit the errors in stress.

The water content effective stress graph of TK3 is also shown in Figure 11, together with the water content-undrained shear strength line, obtained from the plane strain compression test with no lateral strain consolidation. The effective stress was computed on the assumption that the degree of consolidation had reached 90% with the use of soil constants obtained from the oedometer tests. The Critical State Theory (Schofield and Wroth, 1969) suggests that the two lines should be parallel if the clay had been consolidated by the expected stress level and C_u/Po' is a constant, in which Po' is the effective consolidation stress. From the figure, it is obvious that they are parallel to each other. With the reasonable assumption that C_u/Po' is a constant, this shows that the clay had been consolidated by the proposed stress which increases with depth.

Some attempts have been made to measure the shear strength in centrifuge models in flight directly by using a vane apparatus developed by Davis (1981) and a penetrometer apparatus by Almeida and Parry (1983). These direct measurements can be done. But in this series of tests, no direct measurements were made with the views; (1) the size of current apparatus is not correctly modelled, (2) careful investigation on how to interpret the data is needed, and (3) more experience is required in order to use the data with confidence.

The preceding computations and measurements clearly demonstrate the correct modeling of the proposed prototype; that is to say, normally consolidated clay, of which strength increases linearly with depth, can be made by centrifuge consolidation subsequent to preconsolidation on the lab floor.

BEARING CAPACITY OF A NORMALLY CONSOLIDATED CLAY OF WHICH STRENGTH INCREASES WITH DEPTH

Modeling

The proposed prototype is a normally consolidated foundation of homogeneous and saturated clay with undrained shear strength increasing linearly with depth. The undrained shear strength at the top of the deposit has a certain value C_o. The clay stratum is considered to be deep enough so that the failure mechanism will not be influenced by any adjacent stratum. The surface strip footing under consideration is a rigid one. No special attention is given to the roughness of the footing and thus its properties are supposed to be in between that of rough and smooth ones. Computations needed to affirm the correct modeling are on: (1) the rigidity of the container and (2) the geometry of the model.

PLANE STRAIN CONDITION

Firstly, computation was whether the rigidity of container was high enough to maintain the plane strain condition. The case of surcharge pressure 2.84 psi (19.6 kPa) and acceleration 80 g which exerted the largest stress on the container was considered. The maximum value of strain in the direction of the second principal axis of this series was found to be 0.08% which is less than 0.1%, the maximum lateral strain allowable in order to maintain the plane strain condition as suggested by Ko and Davidson (1973). Therefore, the plane strain condition can be considered as satisfactorily maintained during the tests.

BOUNDARY CONDITION

Secondly, attentions were being paid to the geometry of the model such that the failure mechanism will not be affected by the boundary conditions. For a foundation of constant undrained shear strength, the slip surface will cut a depth of $0.71B$ and a distance of B from the edge of footing by Prantal failure mechanism. In the present tests where the undrained shear strength increases with depth, the range that is influenced by the failure should be smaller as the theories suggest. As the sides of the containers are $1.5B$ away from the edge of footing and the lower porous layer is more than $1.8B$ from the base of footing, they should not have any effects on the bearing capacities.

TABLE 4. Measured and Analytical Bearing Capacities.

Test number (1)	C_0 (kPa) (2)	kB/C_0 (3)	Measured values q (kPa) (4)	q/C_0 (5)	Davis & Booker (smooth) (6)	Measured/ analytical (7)	Davis & Booker (rough) (8)	Measured/ analytical (9)	Nakase (10)	Measured/ analytical (11)
TK02	8.03	0.03	55	6.84	42	1.31	42	1.31	45	1.22
TK01	4.02	0.06	24	5.97	21	1.14	21	1.14	22	1.09
TK5	8.03	1.19	52	6.47	49	1.06	54	0.96	64	0.81
TK7	8.03	1.19	64	7.96	49	1.31	54	1.19	64	1.00
TK2	4.02	2.29	41	10.2	27	1.52	31	1.32	39	1.05
TK6	4.02	2.29	48	12.0	27	1.78	31	1.55	39	1.23
TK3	8.03	2.38	59	7.34	55	1.07	64	0.92	80	0.74
TK4	4.02	4.58	49	12.2	33	1.48	39	1.26	54	0.91

Loading Test Procedures

The loading tests were conducted by increasing the air pressure in the centre rubber bag and pushing down the model footing into clay. In the test measurements of footing settlement, deformations of clay and pore pressure were made.

Loading rates were decided by considering capacity of the system, the predicted bearing capacity, and that the tests were to be carried out as far as possible under undrained condition. The loading rates decided are shown in Table 4. Two different loading rates were used in the 40 g series to investigate their effects on bearing capacities.

The longest time consumed in the loading test was about 6 minutes. By that time, no consolidation was to have had occurred below a depth of 0.5 in (13 mm), based on the assumption that pore pressures vary parabotically with depth. It is understood that partial drainage and consolidation effect have significant influence on the bearing capacities. The coefficient of permeability of material used in this series was of the order of 10^{-9}ft/sec (10^{-10}m/s). With this coefficient of permeability and corresponding loading rates of prototype mentioned, it was found by Shibata and Sekiguchi (1981) that there should not be any influence of partial drainage on the bearing capacities. Thus the tests could be considered as properly carried out under the undrained condition.

Figure 12 shows the change in the positions of the targets placed immediately below the footing during the loading test analyzed from the photographs taken. The figure shows that the footing, to a good extent, did manage to maintain its original shape during the loading test, giving evidence that it can be considered rigid.

Results of Loading Tests

DETERMINATION OF BEARING CAPACITY

The bearing capacity, q, does not seem to be determinable readily from the load settlement curves plotted in normal scale as shown in Figure 13. A method often employed in practice (De Beer, 1970), therefore, is used for determining the bearing capacity. Load intensity settlement relationships were plotted on double logarithm graph paper and the bearing capacity, q, was determined from the intersection point of two straight lines extrapolated from the initial and final portions of the curve. A typical example is shown in Figure 14. The bearing capacities determined are shown in Table 5.

COMPARISON WITH COMPUTED VALUES

The exact plasticity solution was obtained for continuous footing with rough and smooth bases, respectively by Davis and Booker (1973,a). The bearing capacity q was expressed

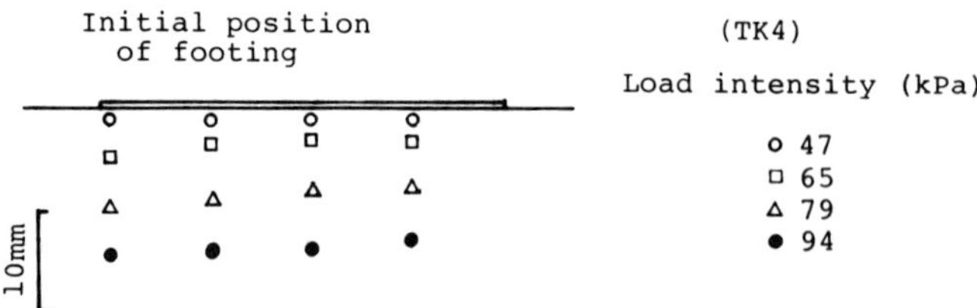

FIGURE 12. Change in positions of targets immediately below footing.

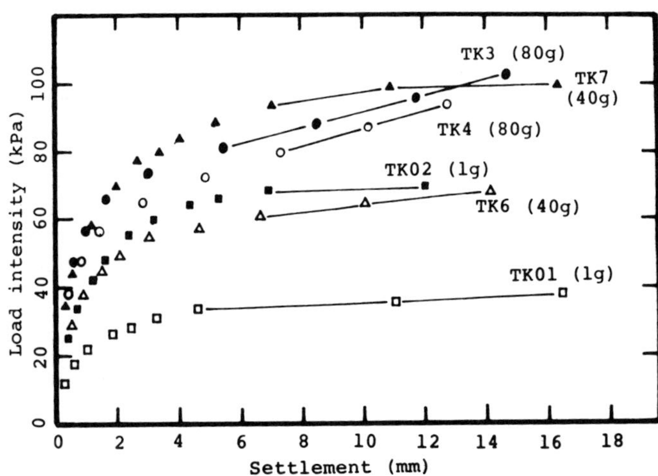

FIGURE 13. Typical load intensity settlement plots.

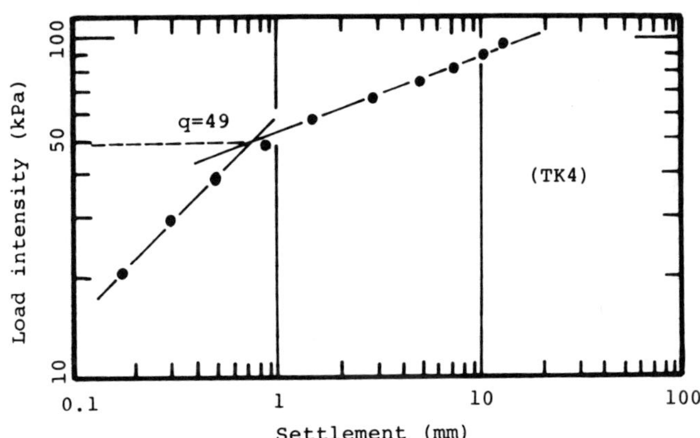

FIGURE 14. Determination of bearing capacity.

TABLE 5. Modified Values.

Test Number (1)	kB/C_0 (2)	C_0 (kPa) (3)	q/C_0 (4)	Coefficient of correlation (5)
TK5	0.95	6.85	7.59	0.87
TK7	1.12	5.57	11.5	0.88
TK2	1.25	4.80	8.54	0.84
TK6	1.74	3.45	13.9	0.87
TK3	1.61	8.10	7.28	0.93
TK4	2.30	5.29	9.26	0.95

in the form

$$q = F\left[(2 + \pi)C_o + \frac{kB}{4}\right]$$

where F is a dimensionless factor depending only on the ratio kB/C_o. The above equation can be written in the form with the bearing capacity factor

$$q = c_o\left[(2 + \pi)F + \frac{F}{4}\frac{kB}{C_o}\right] = C_oN_{co}$$

In Figure 15, the bearing capacity factor for continuous footings obtained by the slip circle method (Nakase, 1966) is shown in dashed line. Solid lines in Figure 15 show relationship between the bearing capacity factor N_c and the ratio kB/C_o, which are worked out by the F versus kB/C_u relationship by Davis and Booker. The computed values used in comparison are those from the plasticity solution and the slip circle solution.

The value of C_u/Po' used in computing C_u and k is 0.41 and is obtained from plane strain shear tests. For easier comparison, the values are plotted as dimensionless quantities of kB/C_u and $q/C_o(N_c)$ as in Figure 16, the same as Figure 15.

From the figure, the plasticity solution for smooth footings is found to be the lower limit of bearing capacities. Although the measured values are rather scattered, they do lie around the plasticity solution (rough) and the slip circle solution. It can be noted that to a certain degree this series verifies the significance of the present bearing capacity solutions of foundation that has its undrained shear strength increasing linearly with depth.

The values of k, previously mentioned, were computed with the assumption that the degree of consolidation had reached 100% before the loading tests. Therefore, it should be of interest to examine the degree of discrepancy resulting then. Figure 17 shows the measured and computed distributions of undrained shear strength with depth. The measured undrained shear strength was determined from the measured values of water content with the use of an oedometer test and plane strain compression test results. It is obvious that the observed undrained shear strength–depth relationship is closer to the 90% consolidation line than the 100% consolidation line, measuring that the actual kB/C_o values are somehow different from the computed values where consolidation was assumed to have 100% finished.

Meanwhile, analysis of the deformation characteristic

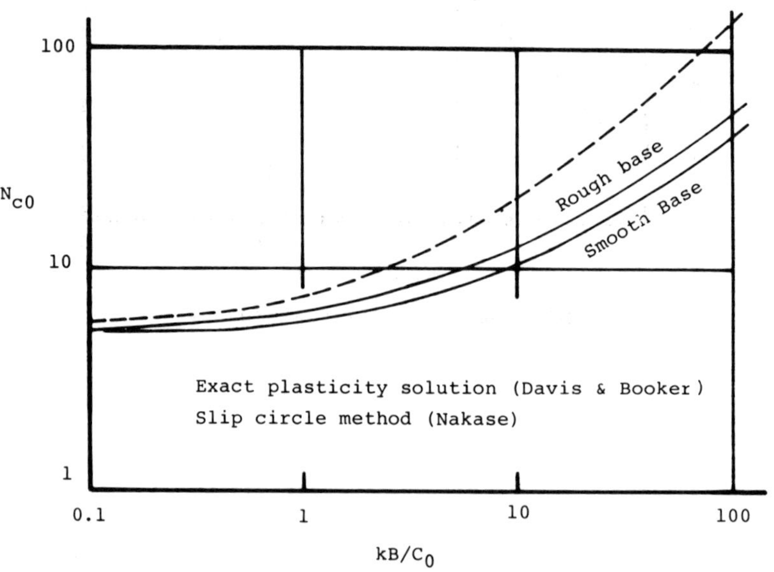

FIGURE 15. Bearing capacity factor for continuous footings.

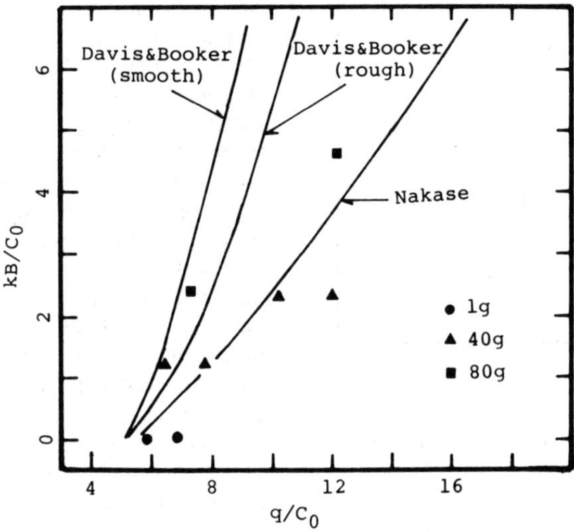

FIGURE 16. Dimensionless plot of failure criteria.

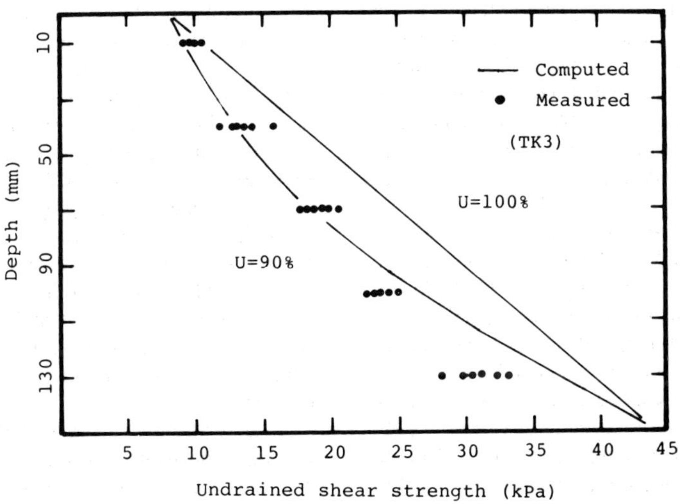

FIGURE 17. Undrained shear strength distribution of TK3.

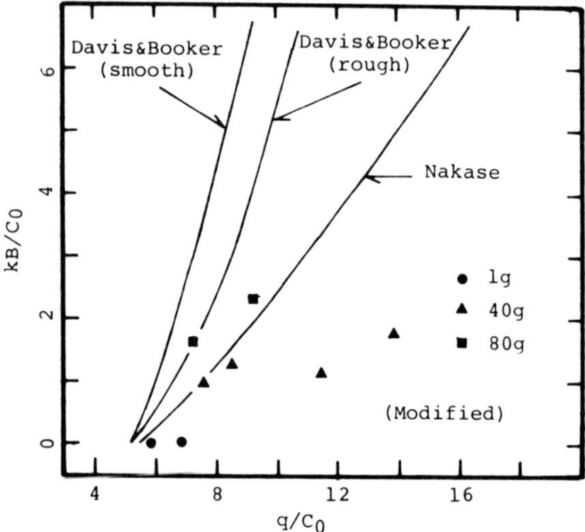

FIGURE 18. Dimensionless plot of failure criteria (modified).

based on photographs taken throughout the loading tests reveals that the clay lying below a depth of 3.94 in (100 mm) was not influenced by the failure mechanism. Furthermore, the water contents below this depth might have been altered by the suction created after the test as explained earlier. Subsequently, the measured undrained shear strength below this depth was excluded before the regression line of C_u on depth is obtained using least square method. The values of kB/C_o and C_o deduced then are presented in Table 5. All the values of kB/C_o are smaller than that of 100% consolidation.

The values of kB/C_o and q/C_o in Table 5 are then plotted in Figure 18. As compared to Figure 17, the kB/C_o values are overall reduced by the adjustment, but the previous trend of having values scattering around Nakase's and Davis-Booker's (rough) solutions remains unaltered.

In normally consolidated or lightly overconsolidated marine clays, the k value is in the range from 0.8 kN/m³ to 2.0 kN/m³, which corresponds to the submerged unit weight of clays of 4 kN/m³ to 5 kN/m³ and the C_u/Po' value of 0.2 to 0.4. As for the C_o value, 2 kN/m² to 5 kN/m² will be commonly found. If the footing width B of 1 m to 30 m is considered, therefore, the value of the ratio kB/C_o most likely encountered in practice will be in the range from 0.2 to 30. The range of kB/C_o values of the present series is 0.03–4.6. To get larger kB/C_o values in experiments, it is necessary either to increase B or k values, or to decrease C_o value. As tests with smaller C_o values will increase the degree of influence of undesirable frictional and adhesive forces on the bearing capacities, they are not recommended. Therefore the use of larger centrifuges is desirable, if possible, to cover the whole range of the ratio of kB/C_o.

Strain rate effect should be noted here. From Table 4 it is known that with the same test conditions of surcharge pressures and accelerations, the higher the loading rate, the larger the bearing capacity. In this case, the increase in bearing capacity was about 20% when the loading rate was four times larger. However it was reported that a log cycle increase in loading rate raises the shear strength by only about 3% (Parry, 1970). Therefore, this has to be explained by some other factors such as the visco-plasticity of the soil skeleton and experimental error that was contributed mainly by the adhesive forces acting on the sides of the model. Further research will be called for in this respect.

As for the bearing capacity, it is summarized that the measured values of bearing capacity are located around the slip circle solution and the plasticity solution (rough). On the other hand, the plasticity solution (smooth) is found to be the lower limit of bearing capacity. Although the measured values of bearing capacities of the present series are rather scattered, and the range of kB/C_o values is narrow as compared to the actual one usually encountered in practice, the relevance of the solutions of bearing capacity for foundation that has its undrained shear strength increasing linearly with depth is verified by the tests.

CONCLUDING REMARKS

The role of centrifuge modelling in geotechnical engineering has been discussed (Schofield, 1976). One of the roles may be to validate numerical modeling or theoretical methods by providing data in the properly scaled fashion. A series of centrifuge model tests on bearing capacity of clay, described in this section, belongs to this group.

The proposed prototype was a normally consolidated clay foundation having undrained shear strength increasing with depth. To ensure that the proposed modeling was performed accordingly, pore pressure and settlement behaviours with time were recorded and compared with computed values obtained by solving the consolidation differential equation. As a further check, water content distribution of the model was obtained after loading test. The observation gave positive evidence showing that correct modeling had been carried out successfully.

Loading tests to investigate the bearing capacity of a strip footing on normally consolidated clay were performed. The bearing capacities of the foundations were determined by plotting the load intensity settlement curve on double logarithm paper. The bearing capacities thus obtained were then compared with Davis-Booker's plasticity solution and Nakase's slip circle solution. The results reveal that Davis-Booker's solution (smooth) formed the lower limit of bearing capacity and that solutions for bearing capacity of clay, having undrained shear strength increasing with depth, is relevant.

REFERENCES

1. Almeida, M. S. S. and R. H. G. Parry, "Studies with Vane and Penetrometer Tests During Centrifuge Flight," Cambridge University, CVED/D-SOILS TR 142 (1983).
2. Bishop, A. W. and L. Bjerrum, "The Relevance of the Triaxial Test to the Solution of Stability Problems," *Proc. Research Conf. Shear Strength of Cohesive Soils, ASCE,* 462–490 (1960).
3. Bucky, P. B., "Use of Models for the Study of Mining Problems," A.I.M.M.E., Tech., Pub., No. 425, 3–28 (1931).
4. Croce, P., V. Pane, D. Znidarcic, H. Y. Ko, H. W. Olsen, and R. L. Schiffman, "Evaluation of Consolidation Theories by Centrifuge Modelling," *The Application of Centrifuge Modelling to Geotechnical Design,* W. H. Craig, ed., 380–401 (1984).
5. Davis, E. H. and J. R. Booker, "The Effect of Increasing Strength with Depth on the Bearing Capacity of Clays," Vol. 23, No. 4, 551–563 (1973).
6. Davis, E. H. and J. R. Booker, "Some Applications of Classical Plasticity Theory for Soil Stability Problems," *Proc. of Symposium on Plasticity and Soil Mechanics,* 24–41 (1973,b).
7. Davies, M. C. R., "Centrifugal Modelling of Embankments on Clay Foundation," Ph.D. Thesis, Cambridge University (1981).
8. De Beer, E. E., "Experimental Determination of the Shape Factors and the Bearing Capacity Factors of Sand," *Geotechnique, 20* (4), 387–441 (1970).
9. Houlsby, G. T. and C. P. Wroth, "Direct Solution of Plasticity Problems in Soil by Method of Characteristics," *Proc. of 4th International Conference on Numerical Methods in Geomechanics.*
10. Japanese Society of Soil Mechanics and Foundation Engineering, "On the Standardization of Unconfined Compression Tests and Triaxial Tests," (in Japanese), *20th Soil Mechanics Symposium,* 1–59 (1975).
11. Kimura, T., A. Nakase, O. Kusakabe, K. Saitoh, and A. Ohta, "Geotechnical Centrifuge Model Tests at the Tokyo Institute of Technology," Technical Report No. 30, Department of Civil Engineering, Tokyo Institute of Technology, 7–33 (1982).
12. Ko, H. Y. and L. W. Davidson, "Bearing Capacities of Footings in Plane Strain," *Proceedings, ASCE, 99,* NO.SM1, 1–23 (1973).
13. Livneh, M. and J. Greenstein, "The Bearing Capacity of Footings on Nonhomogeneous Clays," *Proceedings, 8th ICSMFE, 1,* Part 3, 151–153 (1973).
14. Mair, R. J., "Centrifugal Modelling of Tunnel Construction in Soft Clay," Thesis presented to the University of Cambridge, at Cambridge, England, in partial fulfillment of the requirements for the degree of Doctor of Engineering (1979).
15. Mikasa, M., "The Consolidation of Soft Clay," *Civil Engineering in Japan,* 21–26 (1965).
16. Nakase, A., "Bearing Capacity of Cohesive Soil Stratum," (in Japanese), Report of Port and Harbour Research Institute, Vol. 5, No. 12, 24–42 (1966).
17. Nakase, A., "Bearing Capacity of Rectangular Footings on Clays of Strength Increasing Linearly with Depth," *Soils and Foundations, 21* (4), 101–108 (1981).
18. Nakase, A., "The u = 0 Analysis of Stability and Unconfined Compression Strength," *Soils and Foundations, VII* (2), 33–49 (1967).
19. Parry, R. H. G., "Stability Analysis for Low Embankments on Soft Clays," *Proceedings, Roscoe Memorial Symposium,* Foulis, 643–668 (1972).
20. Peck, R. B., W. E. Hanson, and T. H. Thornburn, *Foundation Engineering,* John Wiley & Sons, New York, p. 252 (1953).
21. Pokrovsky, G. I., "Principles of the Modelling of the Foundations of Structures," *Technical Physics of U.S.S.R.* (2) (1934).
22. Schofield, A. N., "Cambrige Geotechnical Centrifuge Operations," *Geotechnique, 30* (3), 227–268 (1980).
23. Schofield, A. N., "The Role of Centrifuge Modelling," *Offshore Soil Mechanics,* P. George and D. Wood, eds. (1976).
24. Skempton, A. W., "The Bearing Capacity of Clays," *Proceedings, British Building Research Congress, Div. 1, Part 3,* 180–189 (1951).
25. Suzuki, H., "Bearing Capacity of a Circular Footing on Normally Consolidated Clays," Master thesis of Civil Engineering, Tokyo Institute of Technology (1985).

SECTION FOUR

Buried Structures

CHAPTER 14 Geotechnical Engineering for Pipelines and Tank Farms 459
CHAPTER 15 Cement-Mortar Lined Pipes..503
CHAPTER 16 Soil Mechanics Principles in Underground Mining 521
CHAPTER 17 Design of Buried Pipelines 555
CHAPTER 18 Corrosion of Underground Piping 575

CHAPTER 14

Geotechnical Engineering for Pipelines and Tank Farms

Syed Ahmed*

INTRODUCTION

Since the construction of the Alaskan pipeline, geotechnical engineering has come to play a vital role in the design and construction of projects which transport and store significant volumes of fluids. The specific areas of geotechnical engineering which directly relate to these projects are the subject of this section of the handbook. Many new technologies have emerged since the construction of the Alaskan pipeline. This chapter presents these new technologies along with the theoretical basis for design and construction. The material presented in this chapter should be of use to pipeline designers and contractors as well as the civil and geotechnical engineers involved in these types of projects.

PIPELINES

General

The optimization of a pipeline system must be based on an integrated effort of environmental, right-of-way, geotechnical and design functions. It is the responsibility of the design section to utilize the information from the other disciplines in the development of an optimized system for the expected service life.

The successful optimization of a pipeline system depends on a balanced integration of inputs from the different disciplines. The function of each discipline, however, can be considered separately.

Geotechnical engineering plays a crucial role from conception to completion of a pipeline project in the following areas:

1. Feasibility and Planning: Rock, soil, and swamp area delineation for route selection; effect of water crossings on project cost (directional drilling vs. dredging); geologic faults; seismic activity; groundwater conditions; assessment of method of construction (flotation canal, push method, etc.).
2. Design: Selection of appropriate foundation type for launcher/receiver facilities, mainline and mainline valves, pump and compressor stations; bearing capacity, settlement, and heave of foundations; lateral pressure analysis of anchor blocks and pipe bends; geotechnical parameters for cross country pipeline stress analysis; thermal conductivity of soil and insulation requirements; shore protection; design of levee crossings.
3. Construction: Backfill selection and compaction; stability of water crossing approaches, dewatering need and methods; methodology of road and railway crossing (cased or uncased); dredging requirements.

Feasibility & Planning

Delineation of area geotechnical conditions, i.e., determination of the character of rock, soil, swamp, and man-made fill areas, is essential in the selection of a pipeline route and designing for the type of construction that will be utilized. Figure 1 shows a generalized geologic facies map where floodplain deposits of the river are identified as late and early Holocene. A pipeline system situated in a floodplain deposit where the soils are weak and compressible present difficult problems for the pipeline designer and the contractor. Construction costs are known to have doubled due to unanticipated soil conditions. A pipeline system situated partially on soft and compressible deposits and partially on firmer deposits presents even more complicated problems of differential movements. While it may not be

*Law Engineering Testing Co., Houston, TX

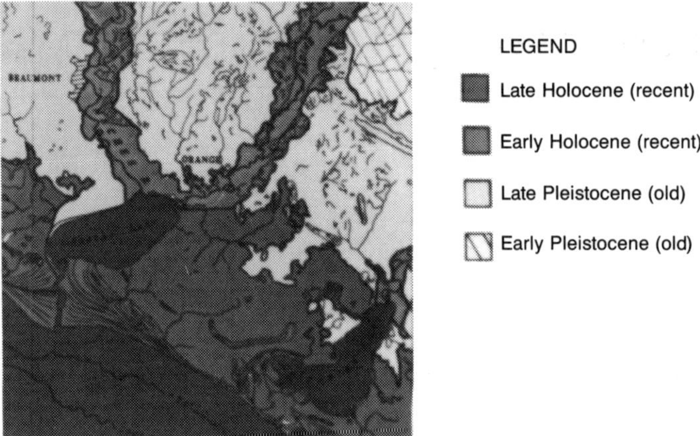

FIGURE 1. Geologic facies map.

possible to completely avoid a floodplain deposit of a large river, it is possible to design a pipeline appropriately if the boundaries of these deposits are clearly identified. Identification of geologic patterns and the presence of undesirable soil deposits does not require an extensive study. A knowledgeable geologic or geotechnical engineer can easily delineate the general boundaries of soil deposits with the aid of published maps and literature.

Many man-made waterways such as intracoastal waterways are maintained by dredging; the dredged material is typically deposited near the banks as shown in Figure 2. This material will generally consist of very fine soil particles and large amounts of water, which continues to settle and move for a long period of time. A pipeline placed on a dredged fill will therefore settle and move laterally causing stresses in the pipe. In some cases it is possible to determine the zone of probable soil movement. If this zone is shallow, pipeline could be placed below this zone; otherwise the pipe segment should be designed for the anticipated stresses due to dredge fill movement. When a pipeline must cross a dredge fill the recognition of the real problem, in most cases, is more difficult than finding a solution.

Difficult water crossings are costly in pipeline construction. Selecting a pipeline route which encounters the least number of water crossings should be an obvious first choice. It is important to recognize that large excavations are usually encountered in approaching a significant waterway (Figure 3). In those cases where a pipeline route crosses several water bodies, every attempt should be made to select the crossing at a place where water approach is easy, width is narrow, and construction modified the natural bank contours to a minimum.

The larger the excavation the more complicated is the task of restoring the river banks. Specification clauses such as, "Proper restoration of the river banks is the responsibility of the contractor" only invite uncontrolled construction bids, cost uncertainties and overruns due to change orders. An excavation of the river banks has to be stable. However, an effort to excavate long flat slopes and to make the side slopes conservatively stable complicates the problem of river bank restoration; optimization is necessary. Larger excavations accompany the problem of disposal of large volumes and the attendant environmental questions. The bank restoration is a major undertaking since the requirements of environmental regulatory agencies are stringent and the need to protect the pipelines from hazards is essential.

Various bank protection schemes have been used in the past. They include: timber and steel bulkheads, cofferdam, stabilized shell, riprap, flexible and rigid revetments, and ordinary grass turf. A survey of some rivers in south Louisiana and Texas revealed a number of these bank protection schemes in various stages of distress (Figures 4–7). This distress can be attributed to one or more of the following factors:

- improper design
- poor construction
- failure to fully realize the potential of environmental forces
- limitations of on-site soils to support the systems
- weaknesses of adjoining banks to resist washout

A proper slope protection scheme should take into consideration the behavior of on-site soils under time dependent erosion forces. Lack of this understanding can result in expensive construction, similar to that shown in Figure 5 where sheetpile bulkhead is no longer mitigating the water flow and erosion. In some other cases of inadequate bank protection, pipeline exposure (Figure 7) may invite hazards. Regardless of its elaboration, no river bank protection is maintenance free. Therefore, capitol expenditures should be carefully weighed against the life of the proposed bank protection scheme.

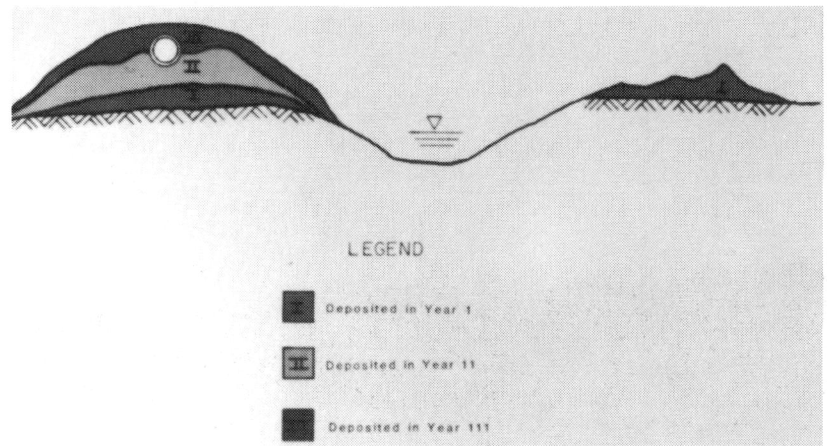

FIGURE 2. Typical dredged material deposition.

FIGURE 3. A typical approach to a waterway.

FIGURE 4. Deterioration of timber bulkhead.

FIGURE 5. Ineffective sheet piling.

FIGURE 6. Uneven settlement of rigid revetment.

FIGURE 7. Erosion of dumped shell.

A pipeline crossing a geologic fault in a seismically active area should be carefully studied. The presence of shallow sand deposits and a high water table in the seismically active area adds the problem of liquefaction which can significantly increase the total cost of the pipeline. A geologic and geotechnical literature search at the time of pipeline planning and feasibility can reveal economical choices for the pipeline route and right of way acquisition.

DESIGN CONSIDERATIONS

A preliminary geotechnical exploration should be performed before initiating the design of a pipeline system. This will aid the design process of launcher-receiver facilities (Figure 8), mainline valves (Figure 9), and pumps and compressor stations in the following manner:

1. Foundation types, i.e., shallow vs. pile, are selected for the major soil or rock types along the pipeline route and safe bearing values and pile lengths are determined to allow the sizing of the foundations after the various structural and product loads are finalized.
2. Probable soil settlement and heave problems are identified and the range of magnitudes is determined. This is vital information in designing the pipe joints and flanges in the above mentioned facilities. If the generalized range of settlement or heave is high then the foundation types and the generalized safe bearing values are determined on a more specific basis in a final geotechnical exploration and design alternatives are recommended.
3. Lateral resistance and friction parameters of the in situ soils are determined in the cross country pipeline stress analysis. These parameters are also used to assess the adequacy of the in situ soil resistance at the anchor block and pipe bend locations. If the in situ soil resistance is inadequate imported backfills are evaluated and a soil–structure interaction analysis is performed in the final geotechnical study.
4. Values of soil thermal conductivity, shear modulus, and damping are generally recommended on the basis of correlations with basic soil properties and experience. The necessary parameters are determined at specific locations in a final geotechnical study.

FIGURE 8. Launcher-receiver facilities.

The final geotechnical study is generally necessary for special problem areas. Conducting the preliminary geotechnical exploration in advance of the design allows the screening of the special problem areas and thus the design process on the bulk of the pipeline system continues without interruption while special problem areas are studied and solution formulated.

Levee crossings can be particularly notorious if the underlying soils are compressible. For these conditions a careful estimation of the rate of settlement is made in a final geotechnical study. In the writer's opinion a "hair pin" design (Figure 10) which allows the pipeline to settle with the levee without developing damaging stresses is more appropriate than the conventional design which follows the levee slope.

Shore protection of river banks may consist of a scheme as simple as dressing up of the banks or a complicated design of sheet pile or other type of wall. If it is determined that a retaining structure is required, the determination of accurate soil stratigraphy and their long-term behavior is essential.

River crossings utilizing directional drilling is becoming more popular for environmental reasons and the fact that the cost is usually competitive with conventional construction methods. Determination of subsoil stratigraphy below a riverbed is essential in designing the direction drilling scheme. If conventional dredging is used in river crossing underwater slope stability should be carefully determined in designing the dredging schemes.

FIGURE 9. Mainline valve.

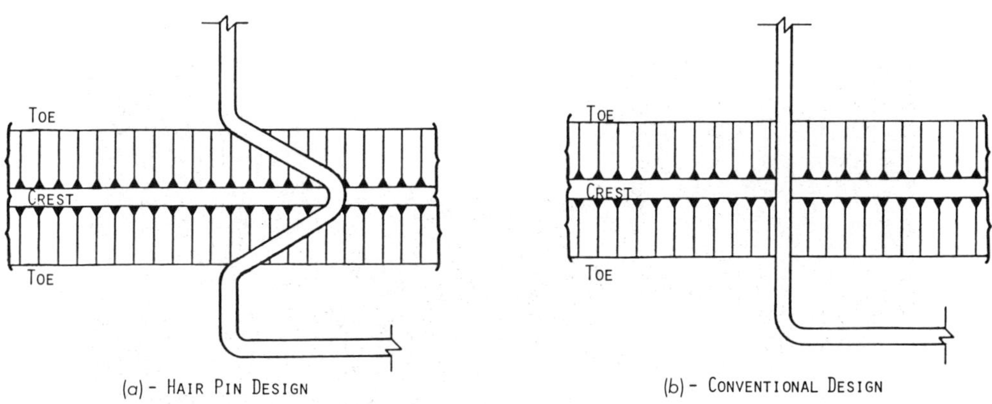

FIGURE 10. Design of levee crossings.

CONSTRUCTION CONSIDERATIONS

Regardless of the thoroughness of the design, undesirable performance can result if proper construction techniques are not used. Some construction problems arise because design documents are not compatible with the soil conditions found in the field, i.e., the absence of geotechnical exploration. A case in point is the crossing of Netches River of a crude oil pipeline. The problem associated with soil instability and upheaval resulted in more than doubling the cost and time of construction. No subsurface soil data was obtained and the visual inspection by the contractor did not reveal the abnormal conditions encountered. Other problems arise because the contractor does not recognize that changes take place in the foundation bearing soils as a result of their exposure to rain and sunshine. A soil with a calculated bearing capacity cannot yield this capacity if it is allowed to be softened by accumulated rainwater (Figure 11). Thus, detailing of proper construction sequence and the importance of protecting the integrity of the soils during construction should be clearly spelled out in the specifications.

Railroad and street crossings require special attention. Generally the pipe section under streets and railroads is cased, and in most cases sag and overbends are provided to transition from the cased section of the pipe to the remainder of the pipeline. Presence of improperly compacted backfill in the transition zones promotes undesirable differential settlement resulting in excessive stresses and damage to the cathodic protection system when the line pipe makes contact with the casing pipe. Repairs of pipe sections under railways and roads are expensive. The geotechnical specifications for construction will include the type, method and required level of compaction of the backfill. If an uncased pipeline crossing is desired then maintenance of the integrity of the in situ soils becomes an important item and geotechnical quality control of construction should be implemented to achieve a safe construction. In the long run it pays itself several times over by reducing the maintenance and pipeline shutdown cost.

The construction sequence and techniques can be used to the advantage of the entire project if proper consultation with the geotechnical engineer is maintained during construction. For example, hydrotesting of pipe sections may be used to preload and consolidate soft sediments before putting the line in service. This may eliminate the need of piling and difficult tie-in problems where the pipeline comes out of the ground and a freestanding leg connects it to a metering station.

One problem commonly encountered during construction is the extent of dewatering required. The geotechnical recommendation for this case will be based on probable water infiltration problems and will recommend the alternate dewatering methods. If water seepage can be controlled by ordinary sump pumps it shouldn't be necessary to go to the expense of a wellpoint system. The use of temporary sheeting during construction should be carefully evaluated since an improper design may be potentially dangerous during construction.

DESIGN PRINCIPLES

The sizing of straight segments of buried pipelines is based on conventional analysis methods. The line is first

FIGURE 11. Accumulated rainwater softens foundation soils

sized to react to internal pressure using the well-known relationship for hoop stresses:

$$\sigma_H = Pr/t \tag{1}$$

where σ_H = hoop stress (psi or kPa), P = internal pressure (psi or kPa), r = average pipe radius (in or mm), and t = pipe wall thickness (in or mm). If the line is subjected to a temperature differential a check must also be made to insure that longitudinal stresses and principal shear stresses do not exceed design allowables.

Segments of a buried pipeline some distance from bends in the line are fully restrained due to soil friction. The longitudinal stress in this area is:

$$\sigma_L = E\alpha\Delta T - \mu\sigma_H \tag{2}$$

where σ_L = longitudinal stress (psi or kPa), E = Young's Modulus (psi or kPa), α = coefficient of thermal expansion (in/in/°F or mm/mm/°C), ΔT = temperature differential (°F or °C), μ = Poisson's Ratio.

For designs with high temperature differentials the critical stress is usually the principal shear stress (τ max) which is obtained from the relationship:

$$\tau_{max} = \frac{(\sigma_H + \sigma_L)}{2} \tag{3}$$

At locations where the line changes direction, a thrust force is developed which may result in undesirable movement if adequate restraint is not provided. Undesirable movements of the pipe are translated into pipe distortions and associated critical stresses. The logical approach to the design of a buried pipeline, therefore, is to provide adequate restraint.

The types of restraints used to react thrust forces in addition to the inherent strength of the pipe material include anchor blocks, thrust blocks, soil uplift resistance, bearing capacity and passive resistance.

A free body diagram of a fully restrained bend segment of a pipeline is shown in Figure 12. The resultant force (F) is equal to:

$$F = PA + A_s(E\alpha\Delta T - \mu\sigma_H) \tag{4}$$

in which F is the resultant end force, P is internal pressure (psi or kPa), A is internal area of pipe (in^2 or mm^2), A_s is steel cross section area (in^2 or mm^3).

The pipe–soil interaction force, P_s, required to fully restrain a bend segment is given by:

$$P_s = F/R \tag{5}$$

where P_s is the pipe–soil interaction force (lbs/ft or N), F is the resultant end force (lbf or N), R is pipe bend radius (ft or m).

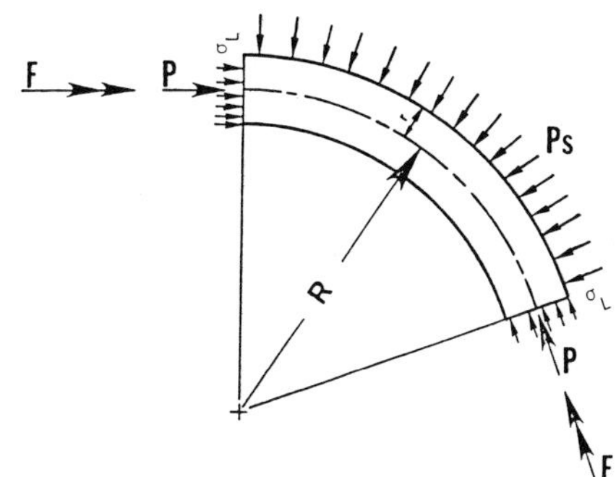

FIGURE 12. Free body diagram of a fully restrained pipe bend.

The forces resisting movement for different type bends are given below.

Type Bend	Restraining Force
Horizontal:	Soil Passive Resistance
Sagbend:	Soil bearing
Overbend:	–Weight of full pipe –Weight of soil overburden –Soil friction

The development of Equation (5) is based on the assumption of no movement at the bend segment. With the exception of the weight of full pipe and soil overburden which provide restraint without deformation, movement must occur in the soil mass to mobilize the soil restraining forces. This movement results in the development of shears and movement forces in a bend segment as shown in Figure 13. These forces develop to restrain the movement of the pipe

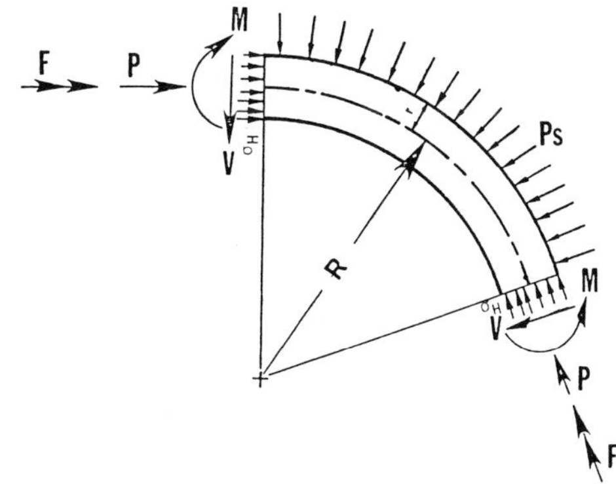

FIGURE 13. Free body diagram of a pipe bend with movement.

bends. It should be noted that as a pipe bend is displaced, the resultant end force, F, will decrease in magnitude caused by relaxation of the resistance to thermal expansion. This reduction in force is normally neglected since the movements and shears associated with movement large enough to significantly decrease the resultant end force would yield critical pipe stresses.

From Figure 13 it is apparent that the resistance to the thrust force at a bend is reacted by a combination of the structural strength of the pipe as well as the backfill soil resistance. Since the mobilization of soil resistance depends on pipe movement, insufficient lateral movement of a very rigid pipe may not take advantage of the available backfill resistance. On the other hand, excessive lateral movements may cause a failure of the backfill and the eventual separation of the pipe from the surrounding soil. Continued use of the pipeline under these circumstances may overstress the pipe bend structurally. The above formulation makes reference to horizontal bends but the conclusion applies equally to overbends and sagbends. If no contribution from soil backfill behind the bend segment is desired, buried anchor blocks can be used at either end of the bend segment to resist force F of Equation (4).

FINITE ELEMENT ANALYSIS

With the advent of high speed and powerful micro- and mini-computers a soil–pipe interaction analysis can be performed using finite element analysis procedures. The following case study describes the method and results.

Analytical Model

The computer program described by Radhakrishnan and Jones [6] was adapted to the problem in the following manner.

a) It was recognized that the actual radial force shall vary along the length of the bend in a three-dimensional manner. However, a two-dimensional analysis on a one foot long section of the bend subjected to maximum radial force per foot was considered satisfactory. The weight of steel and the weight of oil in the one foot long section of the pipe bend were combined and divided by the total volume of the one foot long pipe section to arrive at a density of a material which was the mathematical equivalent of the above two components. Other parameters were assigned to this mathematically equivalent fictitious material consistent with the nonlinear model described in the next subsection. The cross section of the pipe was idealized by a number of straight line segments since the program does not incorporate curved lines. The pipe–soil interaction force (P_s) calculated as above was distributed equally among the nodes of the finite element mesh within the pipe sections. A typical finite element mesh is shown in Figure 14; the arrows identify the nodes carrying the pipe–soil interaction force, P_s. The anchor block was modeled similarly with force F divided among the nodes representing a one foot long section of the block.
b) The soil enveloping the pipe bend was modeled as a continuum of one material by assigning parameters representative of their type determined by laboratory testing. The nonlinearity of the stress-strain relationship was modeled by hyperbolic transformation as suggested by Kondner and Zelasko [14] and defined by the following relation:

$$\frac{\varepsilon}{(\sigma_1 - \sigma_3)} = a + bE \tag{6a}$$

$$R_f = \frac{(\sigma_1 - \sigma_3)_f}{(\sigma_1 - \sigma_3)_{\text{ult.}}} \tag{6b}$$

where ϵ is strain, σ_1 and σ_3 are respectively major and minor stresses, R_f is the failure ratio of stress difference at failur to the ultimate asymptotic value of the hyperbolic stress-strain curve, a and b are respectively the intercept and slope of the straight line represented by Equation (6a). The confining pressure dependency of the stress-strain modulus was modeled in line with Janbu [13] and Duncan and Chang's [12] research as stated by the following relation:

$$E = kPa\ (\sigma_3/Pa)^n \tag{7}$$

where E is initial, unloading or reloading modulus, K is primary loading, unloading or reloading modulus number, Pa is atmospheric pressure, σ_3 is minor principal stress, and n is the modulus exponent.

All the parameters used in the analysis are summarized in Table 1.

The state of stress in the finite element mesh was studied by utilizing $\bar{p}$-$\bar{q}$ diagram. The parameters $\bar{p}$ and $\bar{q}$ are functions of effective principal stresses $\bar{\sigma}_1$, and $\bar{\sigma}_3$ as shown on Figure 15.

The boundaries of the soil continuum were established at a sufficient distance away from the pipe such that the influence of the radial force on the soil continuum was not affected by them. A relatively small grid (elements) was used in the vicinity of the pipe to closely determine the effect of movement on the backfill nearest to the source of the radial force.

The computer program generates and solves several simultaneous equations for each grid or mesh element under the selected non-linear constitutive stress–strain relationship in order to calculate stresses and displacements in the soil backfill continuum. The stresses and displacements obtained for various pipe sizes, radial forces, soil cover, and backfill types were studied to derive the conclusions presented in the following sections.

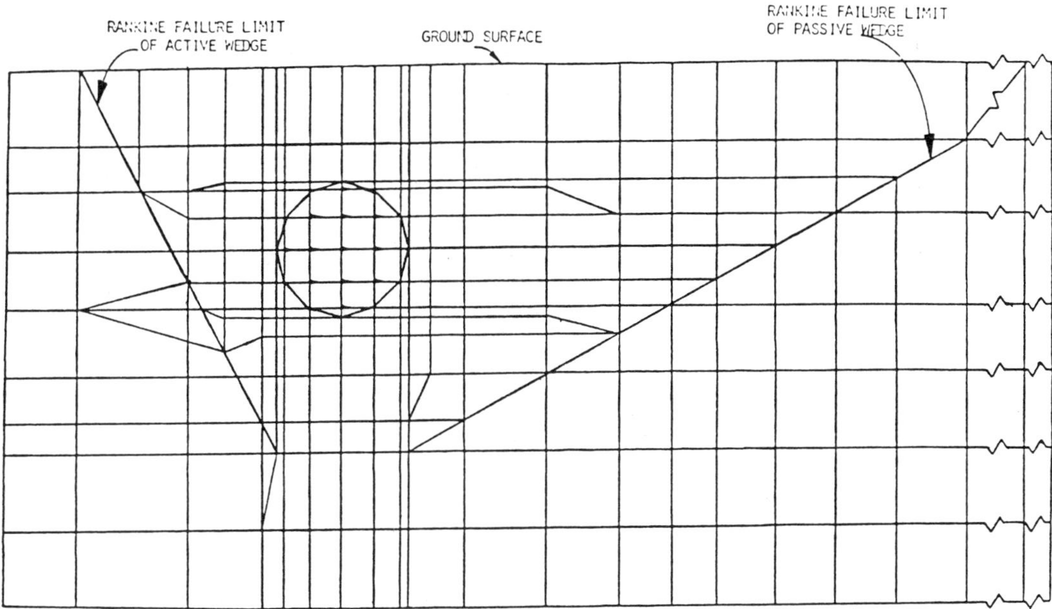

FIGURE 14. Finite element mesh of 42 in. (1067 mm) pipe bend.

Pipe Bends

The effect of backfill on the outward radial movement is illustrated by considering three states of sand denseness, namely, dense sand, medium dense sand, and loose sand. Their associated properties are given in Table 1, and their failure envelope determined by laboratory tests as depicted in Figure 15.

A 42 inch (1067 mm) diameter steel pipe bend of 120 feet (36.6 m) radius, 6 feet (1.83 m) of soil cover carrying crude oil at a design pressure of 600 psi (4134 kPa) was considered to develop an outward radial force of 8000 lb per foot (116,/21 N per m) at the bend. Two types of soil backfill were studied, namely, a dense sand and a loose sand. The resulting outward movement of the pipe bend is depicted in Figure 16 where displacement scale is magnified five times in relation with the geometric scale. The maximum radial displacements were calculated to be 0.09 inch (2.29 mm) and 0.93 inch (23.62 mm) for the dense sand and loose sand backfill continuums, respectively.

The displacements are directly related to the stresses in any soil continuum. In general, the stresses in the soil backfill elements on the inside of the bend radius will reduce (active case) from the "at rest" condition (no radial force development) while the stresses in the backfill elements on the outside of the bend will increase (passive case).

The state of stress in the finite element mesh elements exhibiting a passive case is shown in Figure 17 and Figure 18 for the dense and loose sand continuums for a single load cycle, respectively. In contrast with the dense sand continuum, which plot below the failure envelope, the state of stress in a number of elements in the region of significant passive pressure plots above the failure envelope for loose sand continuum for the same radial force, thus indicating

TABLE 1. Soil Continuum (Backfill) Parameters.

	Dense Sand	Medium Dense Sand	Loose Sand
Unit Weight	120 pcf*	100 pcf**	100 pcf
Poisson's Ratio Before Failure	0.30	0.30	0.30
Poisson's Ratio Near Failure	0.49	0.49	0.49
Coefficient of Lateral Earth Pressure at Rest	0.38	0.43	0.50
Friction Angle	38°	35°	30°
Cohesion	0	0	0
$K_{loading}$	2000	720	295
$K_{unloading}$	2120	900	1090
n	0.55	0.50	0.65
R_f	0.87	0.80	0.90

*1.92 gm per cm^3
**1.60 gm per cm^3

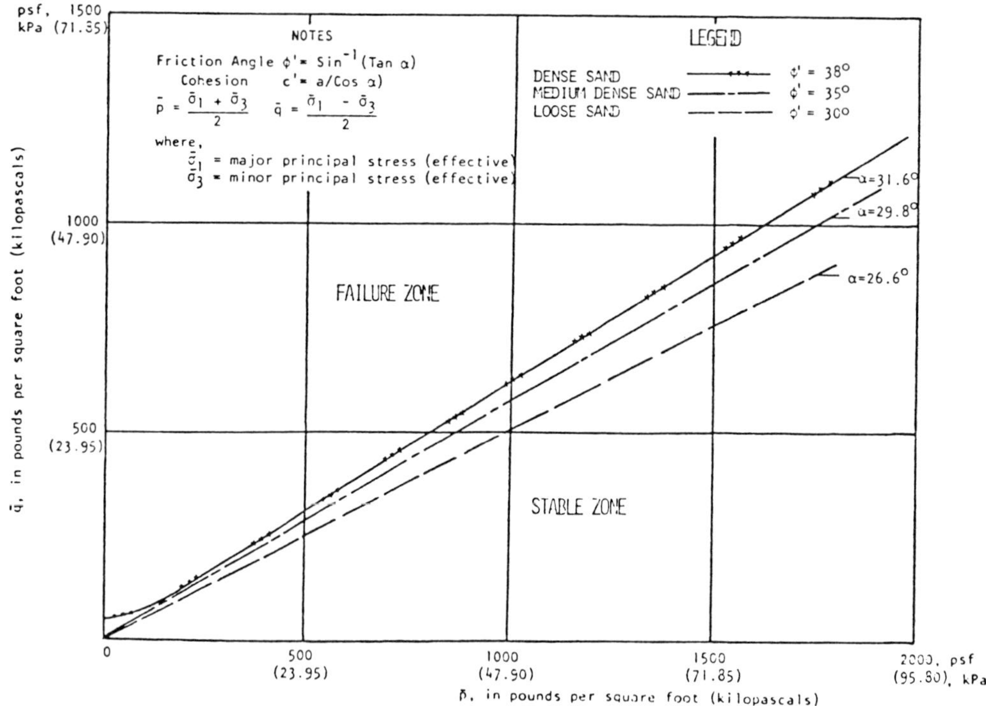

FIGURE 15. Soil backfill failure envelopes.

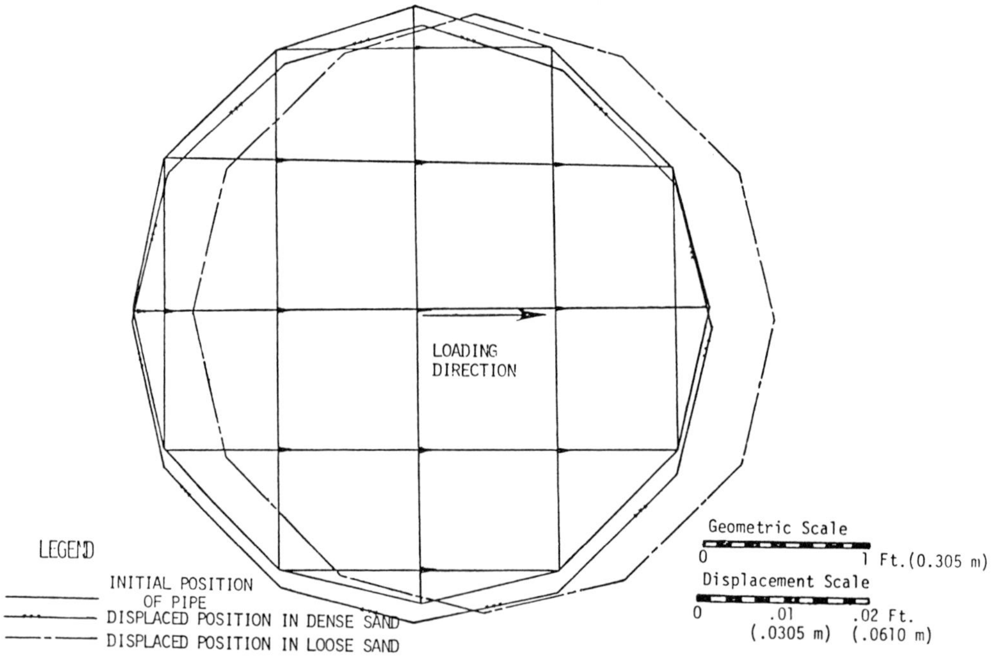

FIGURE 16. Displacement comparison of a 42 in. (1067 mm) pipe bend.

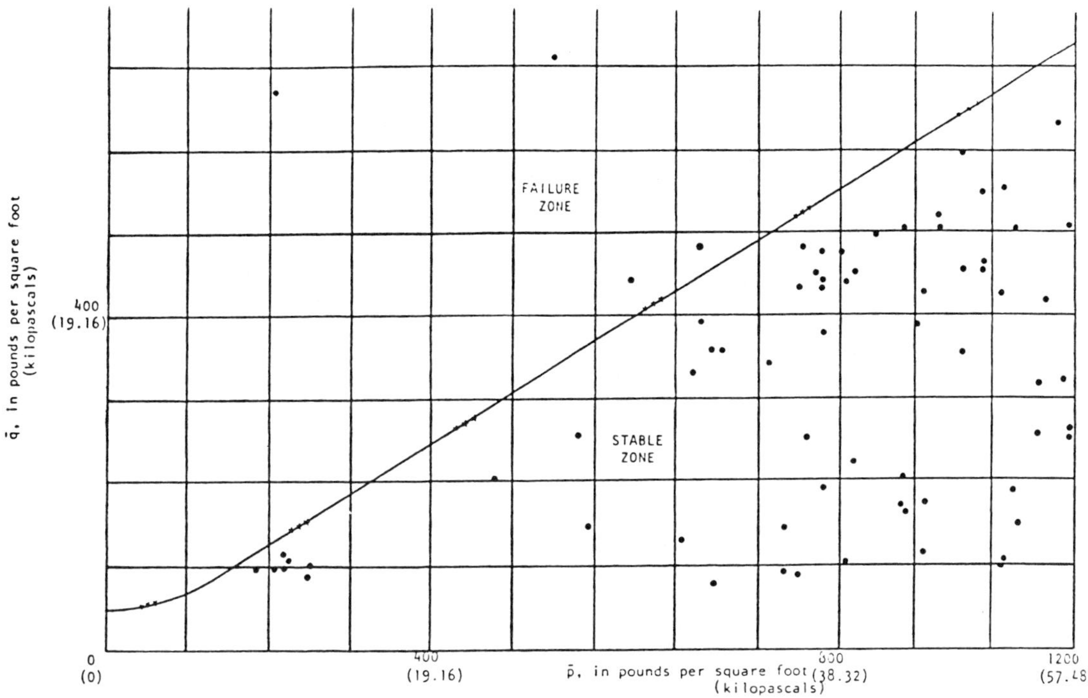

FIGURE 17. Stresses in dense sand continuum after radial load application in a 42 in. (1067 mm) pipe bend.

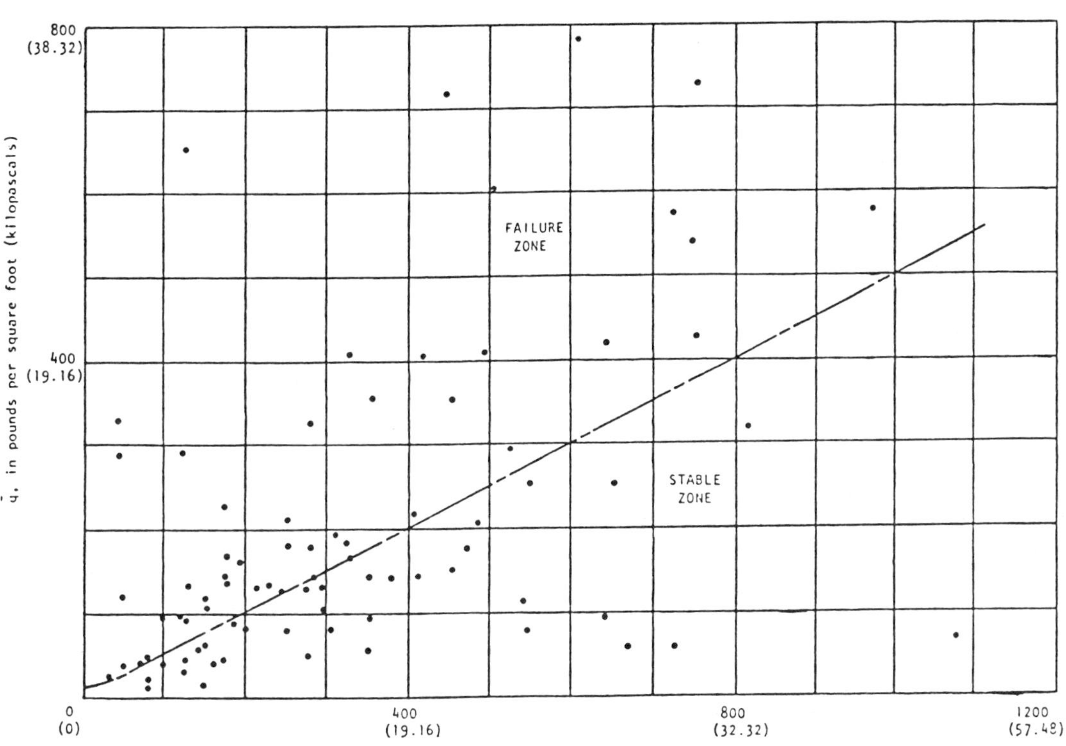

FIGURE 18. Stresses in loose sand continuum after radial load application in a 42 in. (1067 mm) pipe bend.

unstable conditions. The maximum radial displacement for loose sand was calculated to be more than ten times that of the dense sand. The conclusions about the backfill instability depend on the backfill type above and are independent of the pipe bend diameter.

It was observed that the region of significant stress increase was one to two diameters in front and below the pipe for both dense and loose sand continuums. With reference to Figure 14 where Rankine passive wedge extends one diameter below the pipe bottom and intersects the ground surface at approximately four diameters away from the pipe it is evident that the region of significant passive pressure development is overestimated when compared with finite element procedures. If the passive wedge slope were to be drawn such that the wedge intersects the bottom of the pipe, i.e., intersecting the ground at approximately two diameters away, a critical portion of the passive pressure region will be underestimated. The conventional earth pressure theories do not lend themselves to appropriately defining the region of significant stress increase. A finite element formulation on the other hand appropriately defines the critical area and will eventually result in a better and economical design.

The separation of pipe from the backfill due to high displacements and backfill failure condition represents a totally new set of boundary conditions for the stress analysis in the pipe. Under these conditions successive cycles may result in fatigue failure especially for highly stressed components. Recognition of this possibility at the time of design will alert the designer to use alternatives such as better backfill material, anchor blocks, etc.

Overburden pressure is a function of soil cover [Figure 19(a)]. On a $\bar{p}$-$\bar{q}$ diagram (Figure 15), the initial state of stress for a deep soil cover, will plot to the right of the corresponding plot for a shallow soil cover.

When the radial force is applied, a given stress path at any given point in the backfill has a shorter distance to travel before the failure envelope is reached for a shallow soil cover. The mode of backfill behavior for shallow and deep soil covers is schematically represented in Figures 19(b) and 19(c).

Analyses were performed for two cases of 16 in (406 mm) pipe with 60 ft (18.3 m) bend radius under different soil cover and density conditions. Case I: 3 ft (0.92 m) soil cover, a dense sand backfill and a radial force of 3700 lbf per ft (53,984 N per m). Case II: 10 ft (3.05 m) soil cover, a medium dense sand backfill, and a radial force of 4000 lbf per ft. (58,361 N per m). Figure 20 shows a comparison of displacements for the above two cases during the first cycle of loading; the displacement scale is magnified 17 times to illustrate the displacement mode. The zone of significant passive pressure development was found to extend approximately one to three diameters in front and under the pipe for both cases. It extended up to the surface for Case I and up to three diameters above the pipe for Case II. The results of the analysis showed that even though radial force was approximately equal, the state of stress in a relatively large number of critical elements in the dense sand continuum (which is the stronger of the two continuums) plotted above the failure envelope, indicating a widespread failure region, compared with the medium dense sand where only a few isolated zones of local failure were observed. In the writer's opinion the comparatively larger displacements and the unstable conditions in the dense sand backfill are the result of the shallow soil cover. As a corollary, this conclusion also suggests that a stronger backfill continuum is not a replacement for an adequate soil cover.

A shallow soil cover may cause pipe overbends to lift out of the ground. Determination of the optimum thickness of soil cover and the suitable backfill type is necessary to make the design cost effective.

Anchor Blocks

Figure 21 shows the results of Finite Element analysis on an anchor block 5 ft (1.52 m) by 30 ft (9.15 m) in plan dimensions and 10 ft (3.05 m) deep. The top of the block was 2 ft (0.61 m) below grade. The block was required to withstand a force of 820,000 lbf (3,649,000 N) or 2733 psf (130.91 kPa) pressure. The backfill was dense sand the properties of which are given in Table 1. The pipe was assumed to be rigidly connected to the anchor block. For the purpose of analysis, a typical 1 foot (.305 m) section of the 30 foot (9.15 m) long block was selected. The total force of 820,000 lbf (3,649,000 N) was divided by the length of the block to obtain the load on the side section shown in Figure 21. This force was equally divided among the nodes representing boundaries of the pipe. By doing the above the pipe-anchor block system was manifested into a finite element model which was similar in behavior under load application but not necessarily identical geometrically. Water table was conservatively assumed to be near the top of the block.

The solid lines in Figure 21 show the position of the anchor block before load application, while the dotted lines show its position after load application. The scale was magnified fifty times in order to make a better visual comparison. A maximum lateral movement of 0.44 inch was calculated at the top of the block; it reduces to 0.21 inch near the bottom. The block rotated by about one-thousandth of one degree on the average.

The lateral stresses resulting in the backfill adjacent to the anchor block due to application of the above force are also plotted in Figure 21. A maximum stress of 2120 psf (101.55 kPa) resulted 9 feet (2.75 m) below the top of the block. This stress diagram can be integrated over the depth of the block to obtain the total resistance of the backfill behind the one-foot (0.305 m) length of the anchor block. The centroid of the integrated stress distribution curve will be the point of application of the resultant backfill resistance. The stress distribution shown in Figure 21 takes into account the varia-

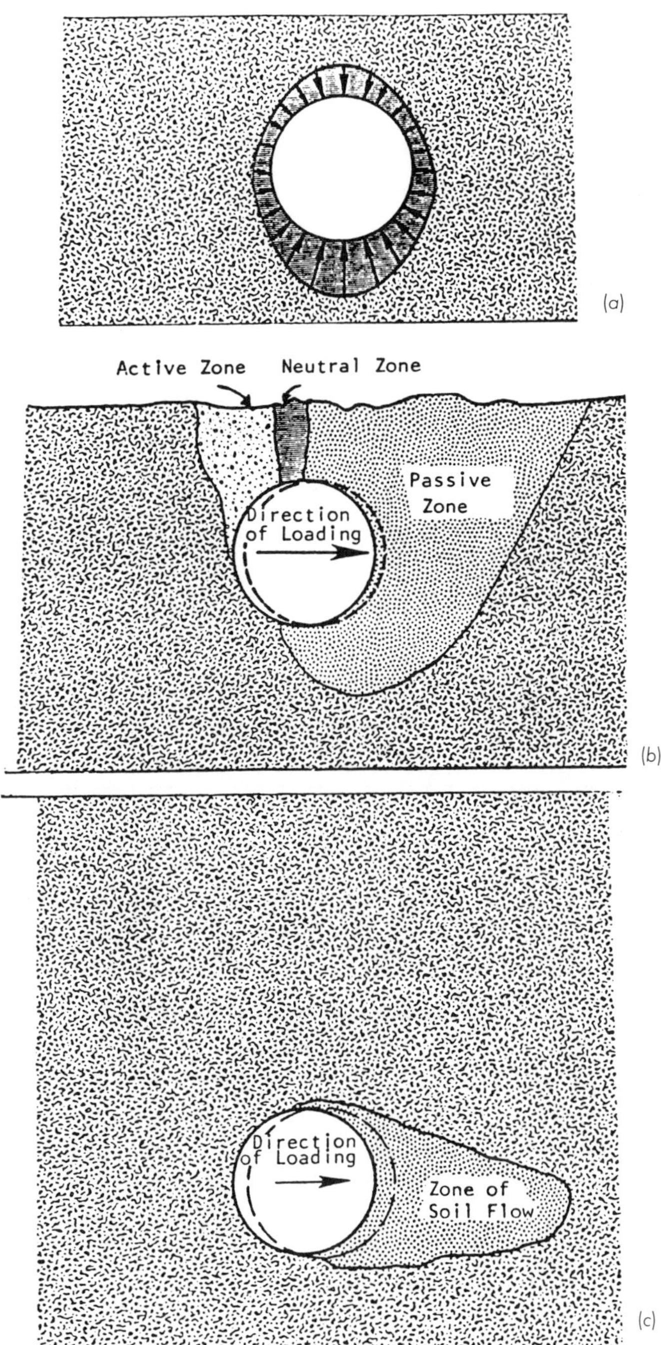

FIGURE 19. Schematic backfill behavior (a) at rest, (b) shallow soil cover, (c) deep soil cover.

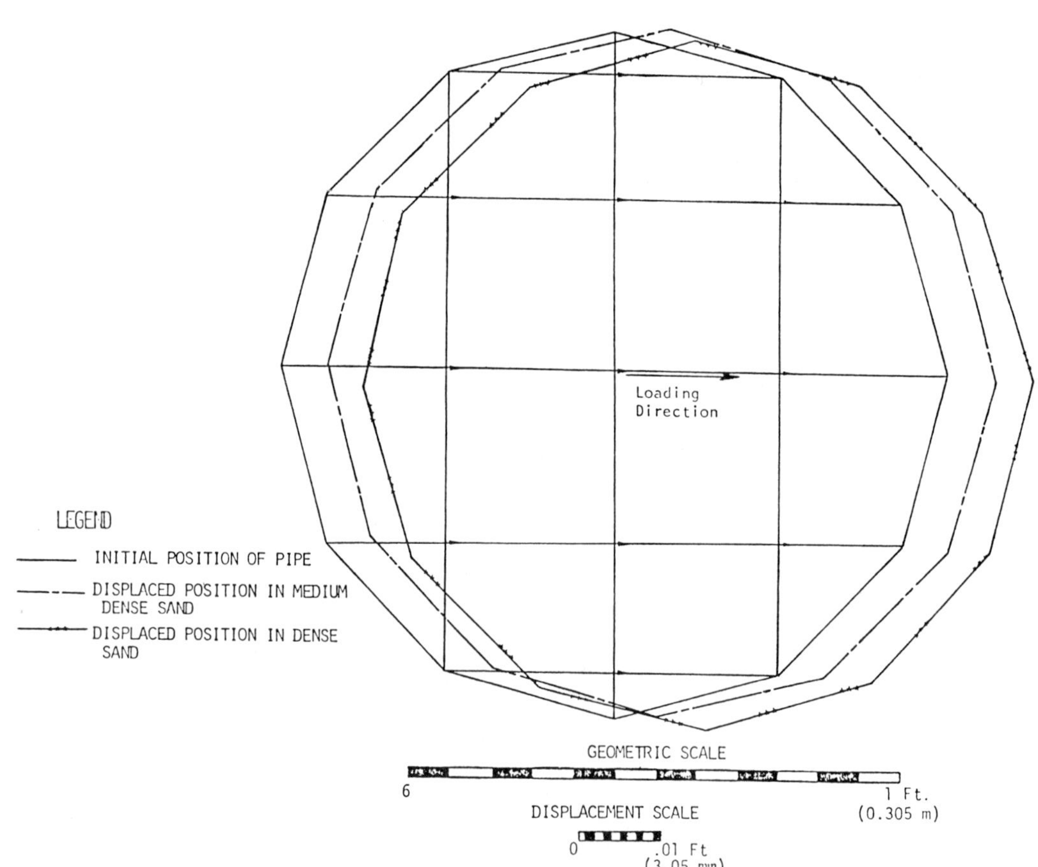

FIGURE 20. Displacement comparison of a 16 in. (406 mm) pipe bend.

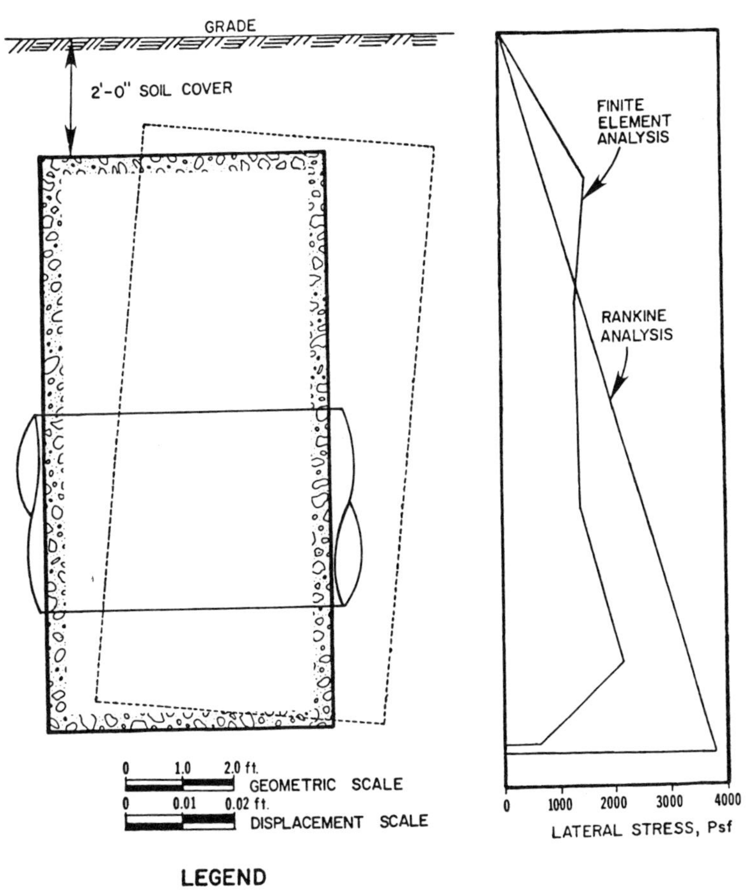

FIGURE 21. Anchor block analysis with shallow soil cover.

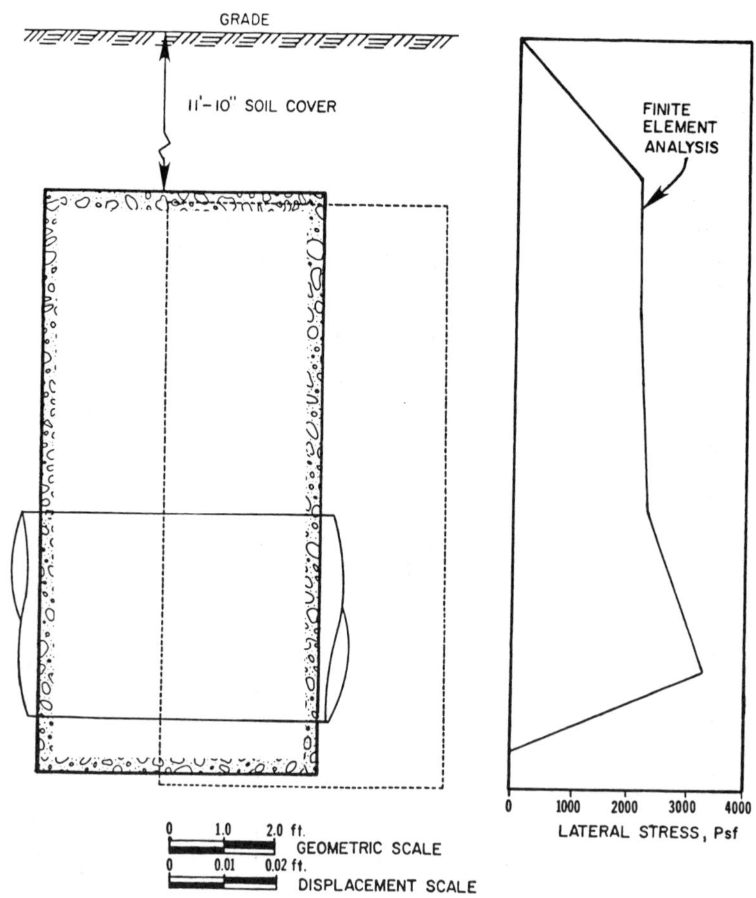

FIGURE 22. Anchor block analysis with deep soil cover.

bles of soil cover anchor block geometry, backfill nonlinearity, and the mode of anchor block displacement.

Some comparison with conventional Rankine theory can be made. The passive backfill resistance in this theory is given by the relation:

$$p = \delta h \tan^2(45 + \phi/2) \qquad (8)$$

in which p = passive backfill resistance or stress, δ = unit weight, h = depth at which p is desired, and ϕ = friction angle. According to this theory, passive resistance increases linearly from the top of the block; its maximum value near the bottom of the block will be calculated to be 3800 psf (182.51 kPa) after allowing for a suitable factor of safety and its center of pressure will be 1/3 of the block height above the base. This theory does not lend itself to account for the variables of soil cover, backfill linearity, and the displacement mode; more importantly, the displacement magnitude. Some researchers believe that several inches of displacement is required before the ultimate Rankine passive resistance is achieved. Large backfill displacements are generally intolerable for pipeline operations where anchor blocks are employed. Finite element analysis is superior in the sense that it predicts the magnitude of displacement for the require backfill resistance. Another important difference is that the area behind the block which experienced passive stress increase, delineated by Finite Element Analysis, was observed to be smaller than the Rankine passive wedge.

Figure 22 shows the results of Finite Element Analysis on an anchor block 5 ft (1.52 m) by 15 ft (4.58 m) in plan dimensions and 9 ft (2.75 m) deep. The top of the block was approximately 12 ft (3.66 m) below grade. The block was required to withstand a force of 608,000 lbf (2,705,600 N) or 4503 psf (215.69 KpA) pressure. The backfill was dense sand and the analysis was performed similar to the one described in Figure 21. The solid lines show the position of the anchor block before load application, while the dotted lines show its position after load application; the scale was magnified 200 times since the resulting deformations were very small. Lateral movement of the block varied from 0.124 inches to 0.130 inches from the top to the bottom of the block in the backfill elements immediately behind the block. Rotation of the block was negligible. In contrast with the analysis shown in Figure 21 which appeared to have rotated about the base, this block appears to translate towards the fill. The basic reason for this difference in the mode of displacement appears to be the soil cover. Deeper soil covers provide more all-around confinement, thus restricting large deformations; this was observed by the analysis shown in Figure 22 even though a higher pressure per unit area was applied in comparison with the analysis shown in Figure 21.

PRACTICAL APPLICATION

From a practical application standpoint the pipe–soil interaction force (P_s) should be computed using Equation (5) and compared with the appropriate soil resistance available for a given subsurface condition. If the available passive re-

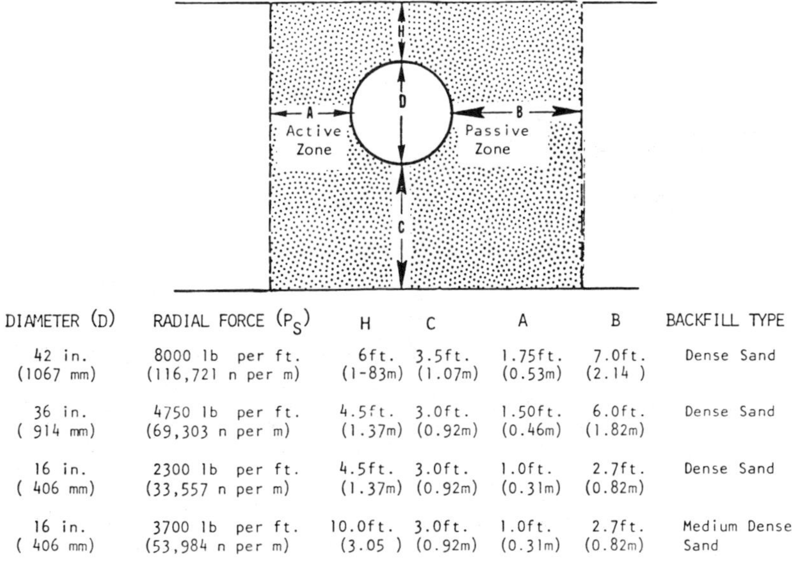

DIAMETER (D)	RADIAL FORCE (P_S)	H	C	A	B	BACKFILL TYPE
42 in. (1067 mm)	8000 lb per ft. (116,721 n per m)	6ft. (1-83m)	3.5ft. (1.07m)	1.75ft. (0.53m)	7.0ft. (2.14)	Dense Sand
36 in. (914 mm)	4750 lb per ft. (69,303 n per m)	4.5ft. (1.37m)	3.0ft. (0.92m)	1.50ft. (0.46m)	6.0ft. (1.82m)	Dense Sand
16 in. (406 mm)	2300 lb per ft. (33,557 n per m)	4.5ft. (1.37m)	3.0ft. (0.92m)	1.0ft. (0.31m)	2.7ft. (0.82m)	Dense Sand
16 in. (406 mm)	3700 lb per ft. (53,984 n per m)	10.0ft. (3.05)	3.0ft. (0.92m)	1.0ft. (0.31m)	2.7ft. (0.82m)	Medium Dense Sand

FIGURE 23. Design recommendations.

sistance is not adequate and anchor blocks are not planned, the designer should consider a more sophisticated finite element analysis to compute the pipe bend movement and the resulting stresses (Figure 13) for a proper design. Alternatively an appropriate combination of backfill type and soil cover (Figure 23) can be achieved by the finite element analysis discussed in this paper to allow for very small bend movement and design conditions approaching a fully restrained condition (Figure 12).

Based on this analysis, design recommendations similar to those shown in Figure 23, developed for this study, should be formulated which will allow tolerable bend movement and design conditions approaching a fully restrained condition (Figure 12).

It is recognized that achievement of high quality of compaction is not construction expedient especially if the bends are located in difficult access areas. Furthermore achievement of the desired compaction level closest to the pipe, which is the most critical area, is very difficult. As an alternative to compaction, we selected a sample of concrete sand, added 15 percent cement and mixed it with sufficient water to produce a pumpable slurry. The resulting slurry was left to stand for at least seven days in triaxial test molds, about 6 inches (152.4 mm) high and 3 inches (76.2 mm) in diameter, to allow the removal of excess water. Subsequently, these samples were tested in a manner similar to the testing of dense sand (Figure 15). The resulting failure envelope indicated the cement stabilized sand to be stronger than the dense sand. In our judgment a cement stabilized sand will exhibit bonding or cohesion of particles more than a compacted sand; its angle of internal friction will also be higher, thus making it a stronger material overall. Consequently, the dense sand backfill (Figure 23) can be replaced by a sand stabilized with about 15 percent cement. This alternative should be construction expedient since sand, cement and water can be combined in a conveniently located mixing plant and pumped in-place by means of common grout pumps and attenuated pipes; no compaction equipment will be required at the site of the bends. It should be pointed out here that the cement stabilized sand will set into a fairly hard mass and, if in the future pipe bends require uncovering, some breaking and chipping of the cement stabilized backfill will be required. For a given project and conditions, various sand–cement ratios should be experimented with to determine the most cost effective mixture.

TANK FARMS

General

Tank farms similar to the one shown in Figure 24 are a vital component of the petroleum product storage and distribution. In the majority of cases tank farms are located near waterways for proximity to the major transportation channels of the petroleum products. Unfortunately, soils near waterways, or in the coastal zone, are soft and compressible, and are rather undesirable from a construction standpoint. Since tank farms cover a large area of ground, the cost of a foundation can be relatively higher than the cost of the tank itself. Foundation costs of up to 200 percent of the cost of the storage tanks have been experienced. This is quite in contrast with other typical structures such as tall buildings where the cost of foundations rarely exceeds 30 percent of the cost of the building.

FIGURE 24. Tank farm.

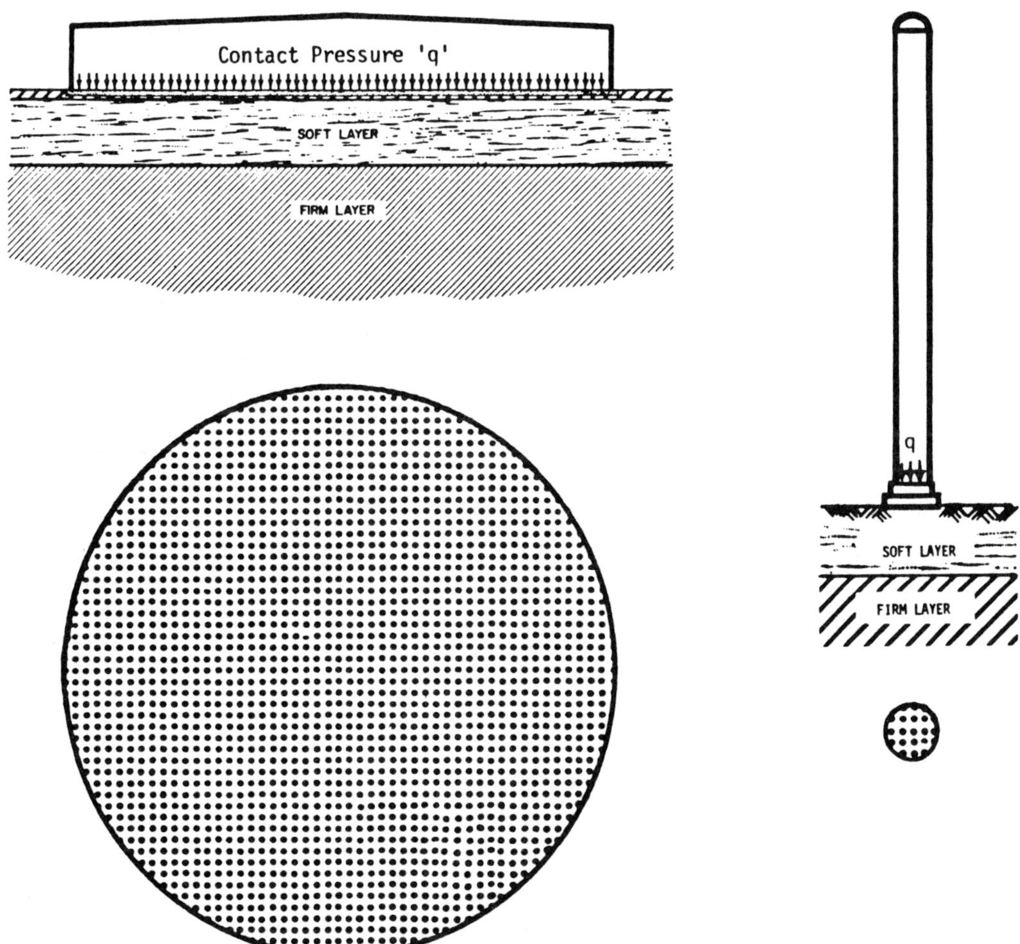

FIGURE 25. Comparison of piles required: tank vs. fractionator tower.

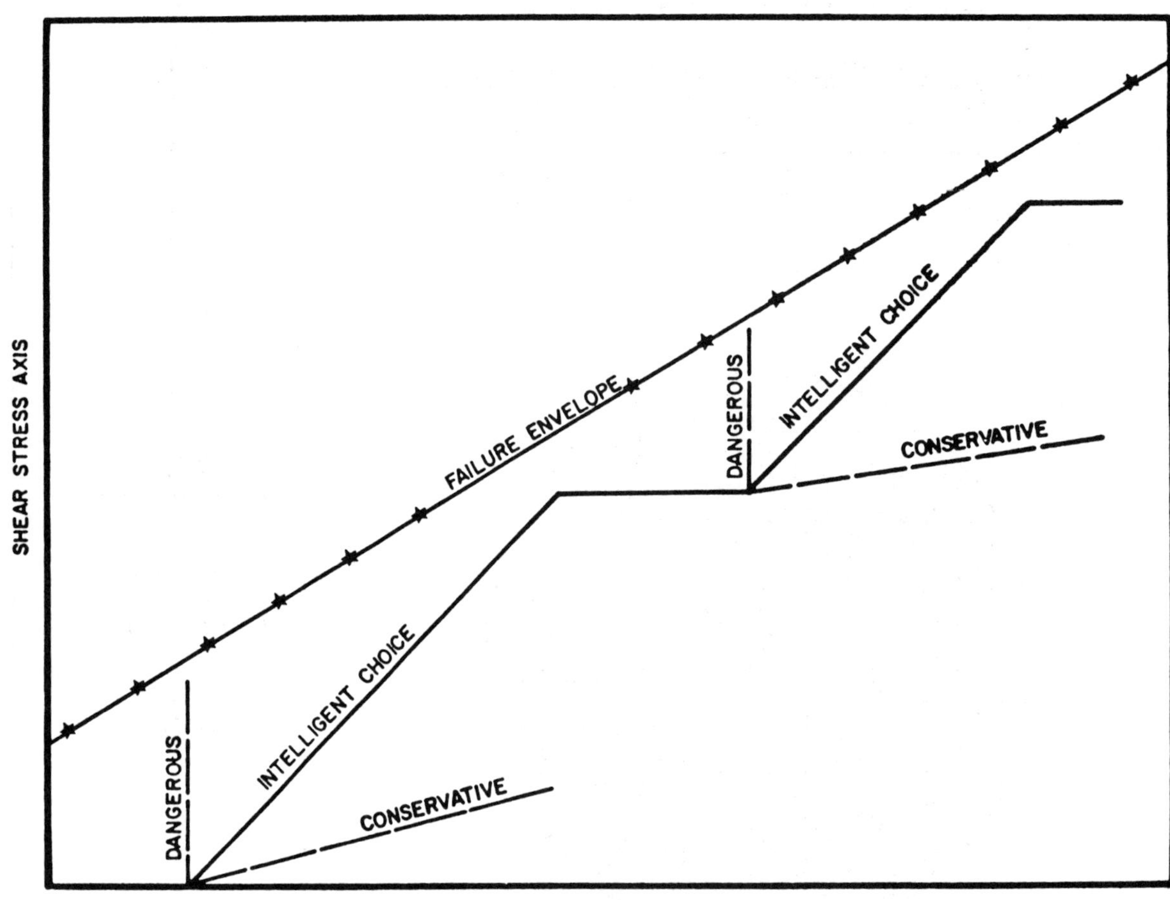

FIGURE 26. Soil improvement—conceptual.

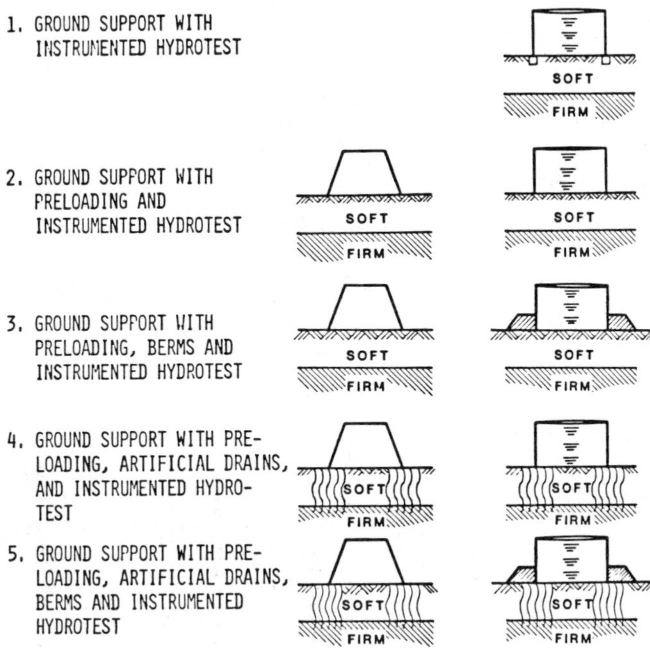

FIGURE 27. Soil improvement by static methods.

A visual comparison of foundation units required by a tank and a fractionator tower, located in poor soil conditions, is presented in Figure 25. For the same contact pressure at the base of a fractionator tower and a storage tank the number of foundation units, say piles, is considerably greater for a storage tank foundation than for the tower. Therefore, unlike small diameter isolated structures, the cost of the tank farms which cover a large surface area is a very serious item, and requires special considerations as discussed by Ahmed [9]. Such special considerations are not covered in the existing industry standards, e.g., API 650 [10].

Two types of construction are available whenever soft and compressible soils are encountered, namely:

- bypassing, using piles, to transfer the structural loads to the lower, more competent soils
- improvement of otherwise undesirable soils

Improvement of soils is based on the concept depicted in Figure 26. The soil in its natural condition has a unique relationship, called "Failure Envelope," between applied normal stress due to loading and the stress which causes the soil to shear. In any improvement scheme soil is gradually subjected to normal stresses, comparable to those which will be imposed by the tank and its products, in order to elevate the stress level which causes the soils to shear and fail. The slope of the stress path, relating the normal stress to shear stress, should be chosen such that it does not approach the failure envelope shown in Figures 27 and 28. These new approaches allow a far more economical construction than the conventional usage of piles. Each case of tank farm construction should be appropriately studied in relation to the soils present and the optimum method of soil improvement should be selected.

SOIL IMPROVEMENT

Static Methods

Figure 27 shows soil improvement methods which utilize static means to induce stresses in the soil mass.

Method 1—In this method a storage tank is placed over the ground surface in its present condition. The soil strengthening and reduction of its compressibility is achieved by inducing hydrostatic loads in several stages until the final product pressure is achieved. The circular loaded area results in high shear stresses near the edge of the tank which can promote instability. Therefore, extensive geotechnical instrumentation and monitoring is required to employ this method in very poor soil conditions. From a cost standpoint, this is the cheapest method of construction.

Method 2—In contrast with Method 1, when an area of the ground proposed to support one or more tanks is preloaded with an equivalent embankment, the shear stresses are minimized because of the sloping nature of the embank-

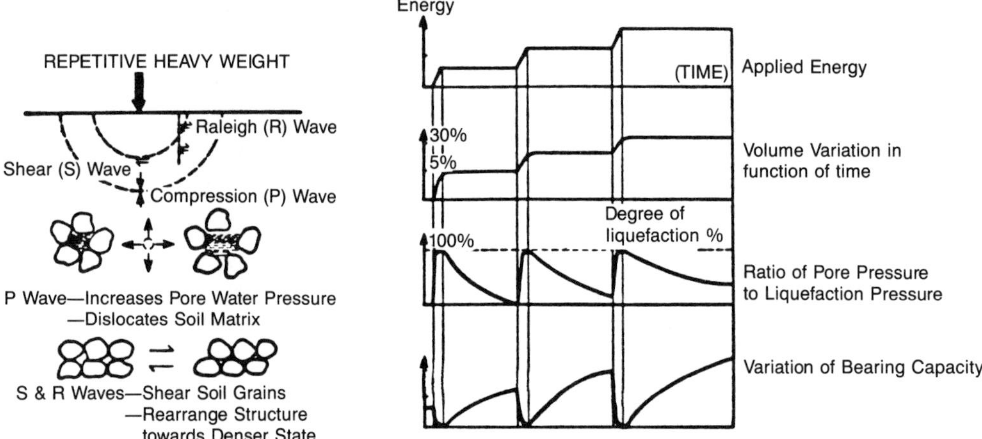

Method 6—Impact Densification

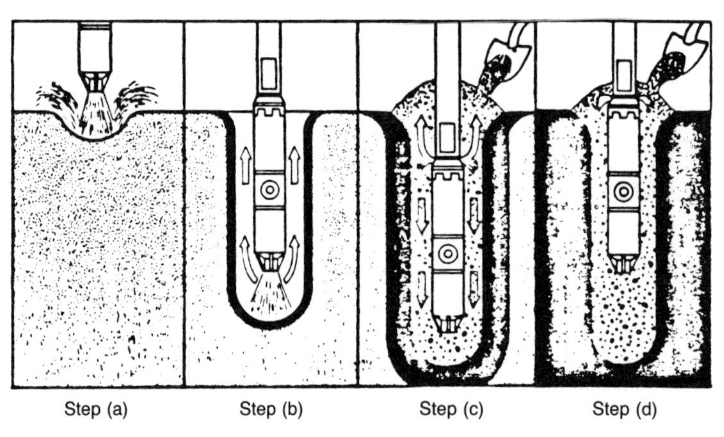

Method 7—Vibrocompaction

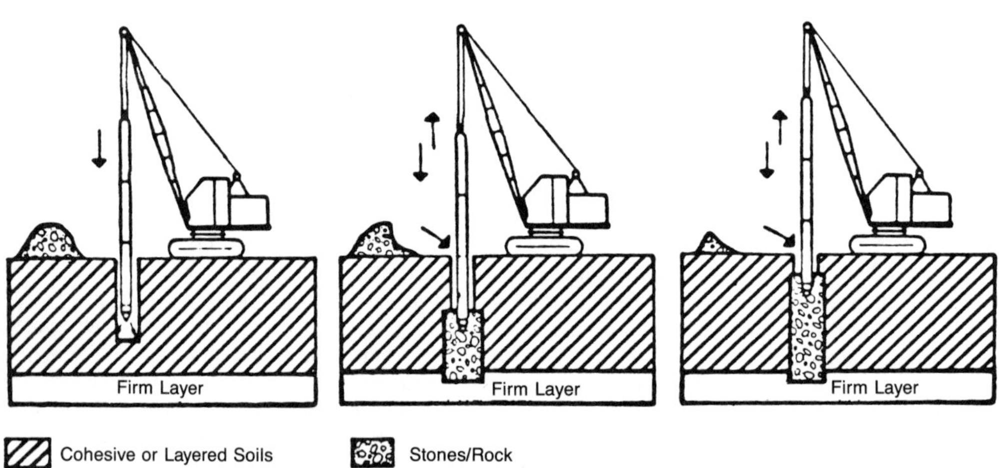

Method 8—Vibroreplacement

FIGURE 28. *Soil improvement by dynamic methods.*

ment. The improvement in the condition of soft soil is monitored by geotechnical instruments and survey measurements. Once the desired soil improvement is achieved, preload is removed and the tank is constructed. Generally tanks are proof tested by filling them with water.

Method 3—Counterbalancing berms can be used in conjunction with Method 2 to counter the "edge cutting" type of failures.

Method 4—Artificial drains can be utilized in conjunction with the above three methods to accelerate the consolidation and strengthening process in the predominantly clay, low permeability deposits. The artificial drains are band shaped consisting of a filter fabric and specially grooved plastic bands to transmit the water escaping from the soil mass as it consolidates. These drains are generally terminated in a granular layer near the surface. Installation of the drains increases the number of drainage paths, increases the overall permeability of the soil mass, and accelerates the consolidation process of the otherwise slowly consolidating clay soils. For an optimum improvement approach, the artificial drains are placed in a peripheral zone near the edge of the tank which is determined analytically. Reference is made to Morrison [21] for further details of artificial drain components.

Method 5—In some cases, artificial drains can be combined with an embankment preload, and counteracting berms in order to place the tank farm in service at an accelerated pace.

Dynamic Methods

Figure 28 shows soil improvement methods which utilize dynamic means to induce stresses in the soil mass.

Method 6—For relatively shallow deposits of loose sand, silts, debris or erratic fill, the method of impact densification is more effective than soil improvement by the above static methods. This method, however, is not suitable for clays where static procedures outlined above hold the edge. Basically, the method consists of providing large energy impacts at the ground surface by dropping a heavy weight from great heights. The shock waves cause the densification of the soil mass. For the types of soils mentioned above, this method is more economical than static surcharge methods shown in Figure 27 as pointed out by Mayne, et al. [20].

Method 7—At those sites where loose sand deposits are deep, or where a deep liquefiable stratum is confined between competent soil strata the vibrocompaction process is required. The equipment consists of a hollow cylinder in which eccentric weights are driven by a hydraulic or an electric motor. The combined effect of vibration and jetting produces a zone of densified soil as shown in Figure 28. The process is carried out throughout the site in a preselected pattern. Brown [11] has presented a good case history of the use of this method.

Method 8—Vibroreplacement approach is applicable to cohesive soils, i.e., clays and silts and many layered or mixed soils where vibrocompaction process is not applicable due to the difference in the character of granular and cohesive soils. Englehardt, et al. [13] presents an impressive case of the use of this method. The equipment used in this method is similar to that of Method 7. Following penetration of the vibrator and its withdrawal after reaching the required depth, the resulting cavity is filled with gravel or stone. A stone column is thus formed which is laterally pressured into the soil. The stone column and the in situ soil form an integrated mass which can be arranged to suit varying combinations of load, soil types and performance required. Sometimes this method is termed as "Stone Column" method.

Each of the above methods require a special and sophisticated engineering input. Subsurface soils are a complex phenomenon, each site being unique in character. Unlike new construction, the in-place soil deposits are not laid down by nature to any rigid set of specifications. Therefore, engineering solutions for the various soil conditions can neither be standardized into a set of handbook formulas nor should the solutions be generalized to include a variety of soil conditions. Competent geotechnical engineering input should be a vital component in making the optimum selection from the methods outlined above.

COST BENEFIT ANALYSIS

Before selecting a particular foundation scheme, whether it consists of bypassing the undesirable soils or one of the soil improvement approaches described above, a rational cost benefit analysis must be made. Some of the factors which should be considered in such an analysis are:

- cost of piles
- cost of pile installation
- cost of pile cap
- duration of pile foundation construction
- lead time needed to order pile material or to manufacture the piles
- cost of site improvement to support pile driving equipment
- cost of more sophisticated geotechnical studies and testing
- cost of placing and removing the fill
- cost of artificial drains
- contingencies for tank relevelling
- duration of preloading and hydrotesting
- cost of mobilization of specialty equipment
- interest on capital investment due to delay in placing the tankage facility in service
- lost revenue due to delay in placing the tankage facility in service

The obvious cost items, e.g., cost of piles, cost of installation, cost of earth moving, etc., generally do not escape the attention. However, certain other factors which impact the final cost are not properly understood. For example, the lead time needed to order or to manufacture the pile material can play a very major role in the overall cost-benefit evaluation. For a given location, pipe piles may be the easiest to drive; however, if they are not available in the vicinity, or in the quantity needed, the lead time required may play a major role in the overall cost evaluation. Similarly, the lead time required for the manufacture and transport of precast concrete piles may affect the project schedule and the cost.

Those individuals who have not used the preloading scheme as a means of ground improvement generally consider the cost of placing and removing the fill as an additional item. They frequently lose sight of the fact that fill is required to build the fire dikes, and sometimes in raising the site grade for drainage purposes. The net cost of fill placement and movement should be calculated without any particular bias. Sophisticated engineering explorations, in situ monitoring and evaluation, and such other activities which are beyond the realms of conventional engineering always cause anxiety because of the associated costs and unfamiliarity. Experience has shown that sophisticated engineering monitoring and evaluation, particularly for the soil improvement approaches, can reduce the final costs by substantial margins as pointed out by Ahmed [9] and experienced by Hunt [14].

CONSTRUCTION MANAGEMENT

A logic diagram, such as the one shown in Figure 29, is essential to a successful execution of an efficiently designed project. Figure 29 represents planning considerations of a tank farm consisting of four floating roof tanks, each of 350,000 barrels. For this tank farm, geotechnical engineering exploration concluded that a combination of preloading, artificial drains, berms and staged hydrotesting (Method 5 in Figure 4) would be appropriate.

The contract schedule of tank erection was such that the first tank could not receive any preload. Approximately 50 days were available to undertake foundation construction and installation of artificial drains prior to beginning the erection of the first tank. As shown in Figure 29, within the first 40 day period three activities were carried out concurrently; namely, ringwall foundation construction of the first tank, artificial drain preloading of the second, third and fourth tanks, and preloading of the second tank. Due to proper planning, coordination and management, no interruption or interference between these activities resulted. Since the first tank did not receive preload, it was recognized that soil improvement by hydroloading alone would be slower than the other tanks. Further, other tanks received the preload for varying periods of time, and thus required staged hydroloading to complement the preloading for varying periods of time. By properly coordinating all the activities it was possible to complete the preloading as well as staged hydroloading of all the tanks, and complete the total facility on schedule.

This example shows that a number of activities required for a soil improvement program can be carried out in what is known as "float time," the block of time when several activities are executed concurrently which do not prolong the project schedule.

For the project example shown in Figure 29, the conventional engineering and testing predicted that soil consolidation would take two to five years, and thus, the tanks could not be filled to capacity for this period of time after completion of construction. However, by using instrumentation and in situ monitoring, and making evaluations based on present-day technology, it was possible to fill the tanks to capacity in less than one year. The savings in interest on capital investment and the generation of revenue by the leasing of the tanks one year after construction, as opposed to two to five years, certainly speaks for the value of sophisticated engineering in the application of soil improvement methods.

TANK–SOIL INTERACTION

Stability and Settlement

Every soil deposit has a natural relationship, called "Failure Envelope," between the applied normal stress due to loading and the stress which causes the soil to shear. This envelope can be determined by testing soil samples under various applied normal stress and measuring the respective stress which causes the shear failure of the soil. The tank and its products impose a normal stress on the soil which produce a bulb of shear stresses within the soil mass. As long as the developed shear stress at a point in the soil mass is less than the natural shear strength of soil at that point, the conditions remain stable; otherwise failure ensues. In other words, the stress states of points in the soil mass lying below the failure envelope represent a stable soil mass, and the points above the failure envelope represent an unstable soil mass.

The soil mass is composed of three elements: solid (mineral) particles, water, and air. When a load is applied at the surface it is initially taken up by the water present in the pores, causing the pore water pressure to rise. Air is expelled rather quickly. With time the water is expelled from the pores due to the pressure gradient which exists between the loaded area and the surrounding ground. As water moves out of the pores the solid grains move closer together to fill the gap. This phenomenon is called consolidation. This chain of events results in a net reduction in volume and a more compact, stronger soil mass. Putting it differently,

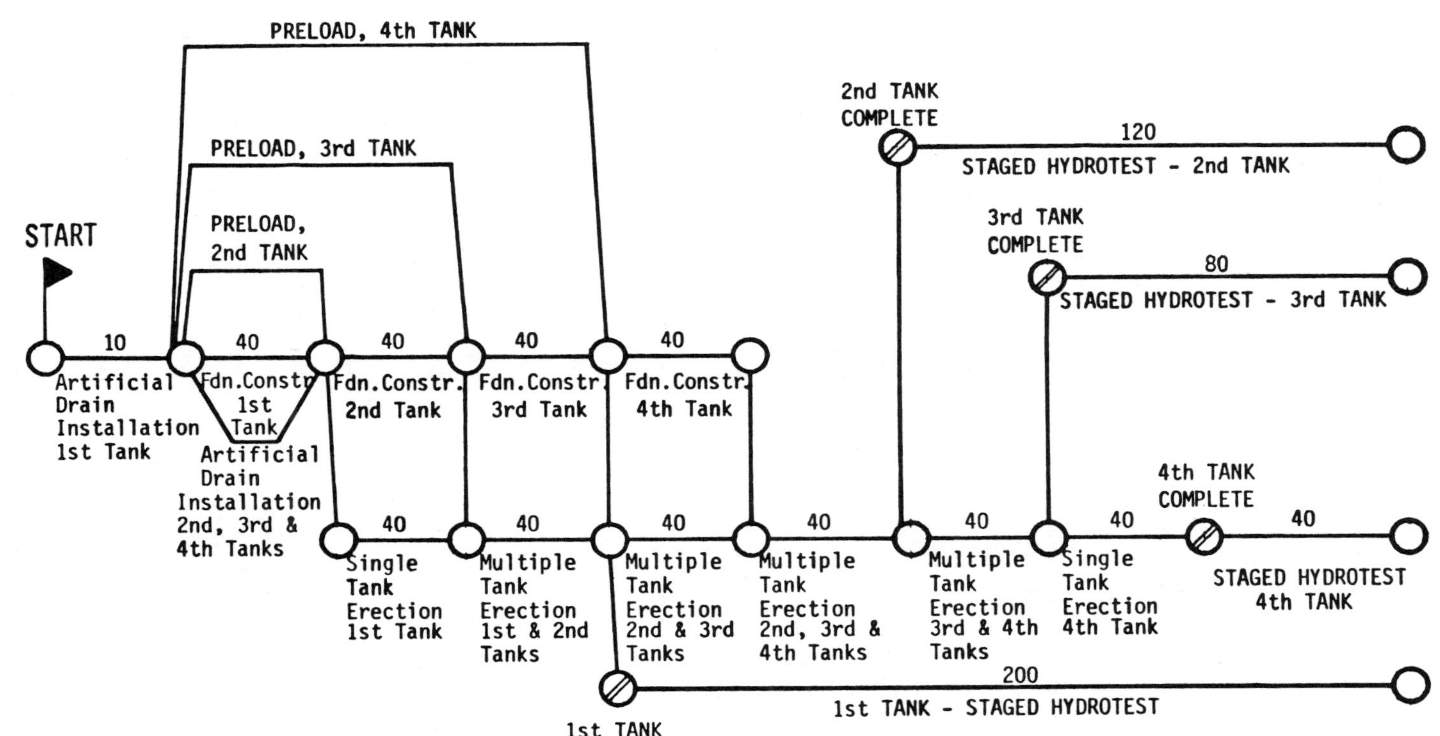

FIGURE 29. Logic diagram.

the character of soil mass changes with time due to loading. The reduction in the volume of voids (consolidation) is exhibited by the settlement of the ground surface, and the load bearing capacity of the soil mass improves due to the increase in density of solid particles per unit volume of soil mass.

It can be postulated from the above discussion that a staged loading program will result in successively higher normal stresses which will compact the soils to successively increasing levels such that shear stresses required to fail the soil will be greater. Also, consolidation, accompanied by settlement and bearing capacity, progresses to higher and higher levels with time. This process is shown schematically in Figure 30.

The gain in strength of a soil mass is related to stress distribution characteristics of the soil mass and the dissipation of pore water pressure. The present method of measurement of stresses in natural deposits is cumbersome and somewhat unreliable. Furthermore, very complex computational models are required to synthesize the stress measurement data, assuming it can be generated economically. These limitations generally preclude the measurement of stresses as a rule for everyday construction. Therefore, various analytical procedures, based on elastic or elasto-plastic theories summarized by Lambe [16] are used. The measurement of pore water pressure can be quite easily undertaken by using pneumatic or electric transducers. The progress of time dependent consolidation and strength gain can be assessed by studying the stresses determined by analytical procedures and measured pore water pressures.

Pore Water Pressure

The rise and decay of pore water pressure at selected points in the soil mass can be monitored by installing commercially available transducers at the various selected locations. Generally, several transducers are installed at increasing depths at the center of the tank, and several near the edges of the tank. In most cases, three to four transducers are sufficient to monitor the consolidating layer of clay in order that pore pressure distribution over the entire depth can be determined.

The instrument package consists of a sensing unit, called the transducer, which is installed in the ground at the selected depth, and a measuring unit which is portable and can be connected to any number of transducers to measure the pore water pressure. The typical results of pore water pressure measurements are shown in Figure 31. As seen in this figure, the pore water pressure at various stages of loading at the ground surface is highest for the transducers located at the center of the tank. This is because the stress distribution of a circular area causes the maximum normal stress at the center and minimum at the edges. With time the pore pressures dissipate, resulting in consolidation and strength gain (see Figure 7). Once a loading stage has been applied and pore pressures have dissipated to acceptable levels, their value can be subtracted from the calculated value of the stress at that point to determine the "stress path" of that particular point in the soil mass. This construction is shown in Figure 32.

The objective of staged loading is to apply only enough load such that the pore pressure resulting from it causes the stress path of the particular point in the soil mass to move on a gradual upward slope but stops short of the failure envelope. When this load is maintained for a period of time, the pore pressure dissipates (from point 1′ to point 2 in Figure 31) and now the soil mass is ready to entertain another stage of loading which results in the stress paths 2–2′ in Figure 32. Once again the loading stage is followed by a holding stage (2′–3 in Figure 9) to allow the dissipation of pore pressure. Successive stages of load are applied in a similar manner until the final product load is reached. By monitoring the pore water pressures at selected points in the soil mass and making constructions similar to those shown in Figure 32, a qualitative estimate of strength gain and overall stability of soil mass is made by the experienced geotechnical engineer.

The pore pressure data is utilized in two other ways to study the tank–soil interaction. One, this data along with basic soil properties is input in computerized algorithms to perform arc failure type of stability analysis to assess the potential of edge cutting type of failure. Second, when pore pressure at the end of the hold period of loading stages is plotted over the depth of the consolidating clay layer, it yields curves known as isochrones. The area between the successive isochrones represents the percentage of consolidation completed between these stages of loading. Since consolidation is related to settlement, a relationship between pore pressure dissipation and vertical settlement can be established.

Pore pressure data is an extremely useful tool in assessing the stability and settlement potential of the soil underneath the tank. Consequently, a careful plan should be developed to collect and analyze this data in any soil improvement program.

Vertical Settlement

As a result of loading, soil settles both vertically and laterally. This section discusses the effects of vertical settlement of soil on the tank components. From a settlement standpoint the tank–soil interaction should be studied with respect to the three components of the tank, namely shell, bottom plate, and shell-bottom plate junction.

The settlement which is uniform across the diameter of the tanks seldom affects the tank structurally. About the only thing it does is to reduce the tank elevation with respect to the surrounding ground which affects the drainage and the piping connections to the tank. When potential uniform settlement is properly estimated the surrounding area can be

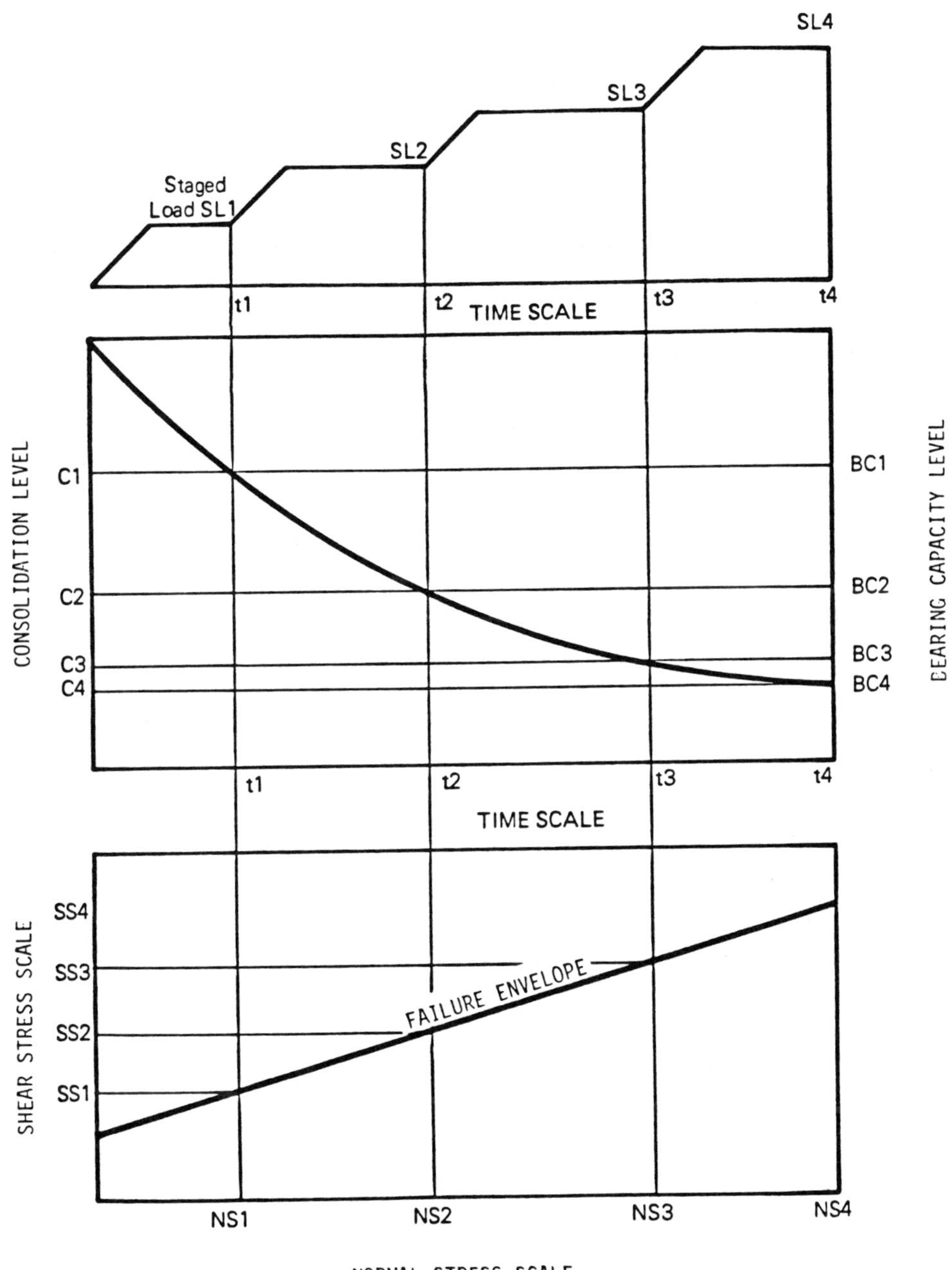

FIGURE 30. Time dependent consolidation and strength gain—conceptual.

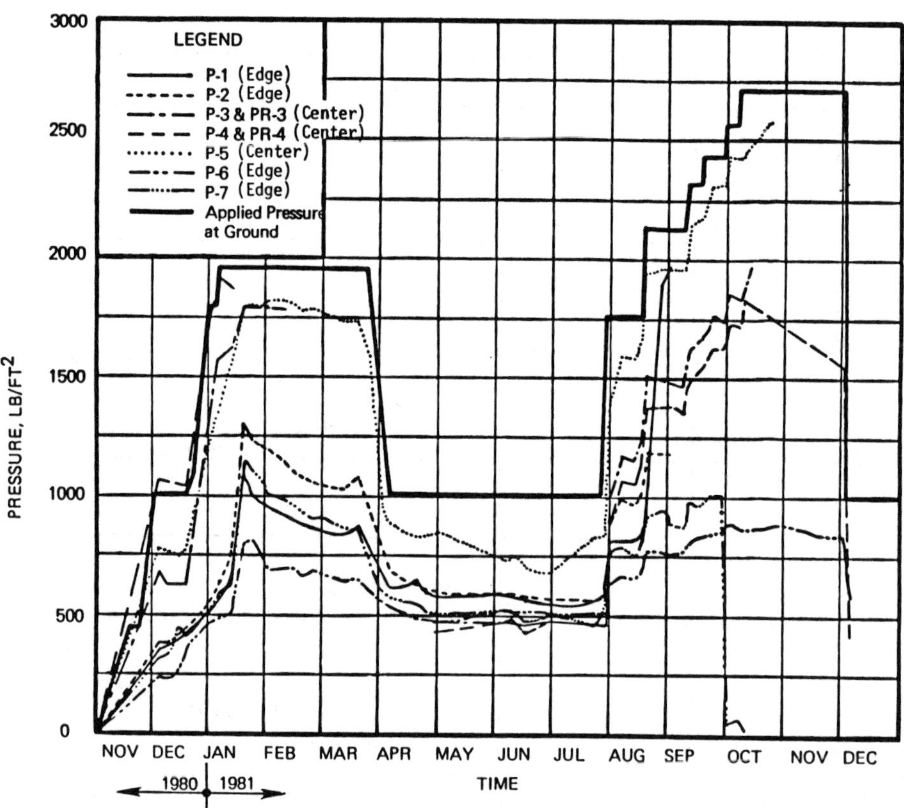

FIGURE 31. Typical pore water pressure measurements.

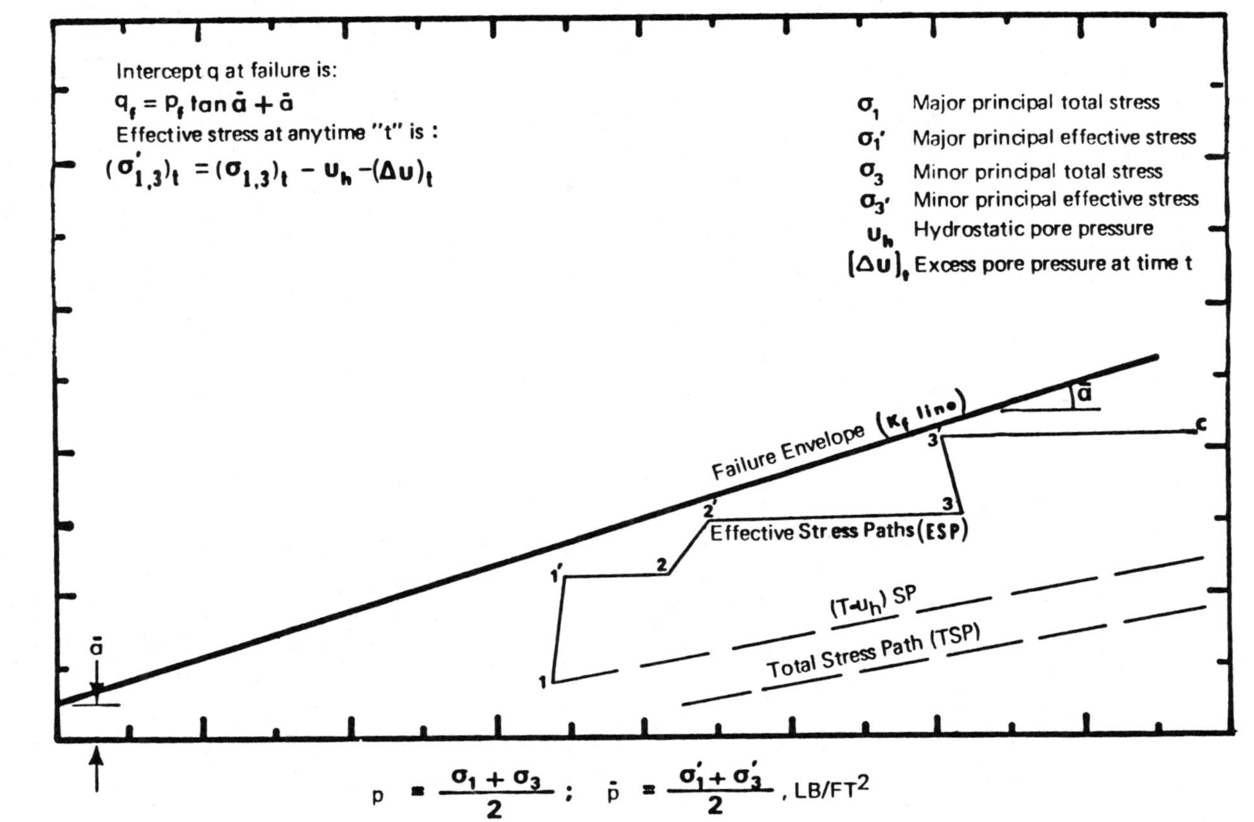

FIGURE 32. Example of stress path construction.

built higher to provide adequate site drainage over the service life of the tank farm. In addition, either flexible joints can be provided in the piping connections or a schedule of adjusting the joints can be forecasted.

When different points across the diameter of the tank settle different amounts, the result is the various modes of differential settlement. Differential settlement is the result of non-homogenous geometry or compressibility of the soil deposit, and non-uniform stress distribution across the tank diameter. Depending on the magnitude and the distribution of the differential settlement the whole tank may tilt with respect to one common plane. This mode is called "planar tilt." Otherwise the differential settlement is termed "non-planar settlement."

In order to evaluate the effect of differential settlement on the tank shell, the settlement is monitored by making survey measurements on at least 4 and as many as 32 points around the tank circumference depending on the tank diameter. The measurements are made at regular intervals during the loading process and the results are plotted on a fold-out graph. A flat curve joining all the points surveyed indicates a uniform settlement. A sinosoidal curve represents a planar tilt. However, peaks and valleys in the plot represent one or more forms of non-planar settlement. Very few problems are associated with the planar tilt. In some cases it can cause overtopping of the shell or damage to the roof seal. The non-planar settlement, on the other hand, can cause overstressing and buckling of shell, ovality, and roof binding in floating roof tanks.

The planar tilt resulting from differential settlement has insignificant detrimental effect on the tank bottom and shell-bottom plate junction. However, a non-planar settlement pattern resulting from differential settlement can cause rupture and cracking of the bottom plate and the welds. For evaluating the settlement pattern of the tank bottom, conventional survey measurements can be used for contouring. In a floating roof tank a plumb bob dropped from the roof supporting legs measures the distance between a surveyed bench mark on the top of the tank shell and the various points of the tank bottom. Thus, bottom contouring can be performed while the floating roof tank is being hydrotested. For a fixed roof tank, the contouring of the bottom has to be done after the completion of hydrotesting. The settlement profiles constructed from these measurements can be analyzed to calculate the stresses in the bottom plate. A comparison with the allowable stress values will indicate if rupture is imminent. An example of tank bottom profile measurement is shown in Figure 33. For this tank artificial drains, bearing the trade name "Alidrains" were used in a 40 foot zone around the periphery of the tank which resulted in a higher degree of consolidation near the tank edge than near its center. Therefore, the tank bottom contour near the 100 foot distance reversed itself from the "as constructed" shape; however, the resulting stresses were within tolerable limits.

Sometimes strain gauges are used to verify the calculated stresses. The bottom profile can also be measured by installing several horizontal inclinometers.

In summary, the vertical settlement is categorized into uniform and differential settlement. The differential settlement in turn can be planar or non-planar. The uniform and planar settlement are relatively easy to predict and have little detrimental effect on the tank components. The non-planar settlement, on the other hand, can cause overstressing and buckling of tank shell, rupture and cracking of tank bottom and binding of floating roof. Therefore, the innovative approaches discussed earlier should be used in conjunction with appropriate engineering evaluation, and the interaction of soil and tank should be studied at every stage in order to assess its behavior.

The existing code handbooks, e.g., API 650 [9] do not clearly recognize the different ways in which the various modes of settlement discussed above can affect the tank components. Furthermore, the criteria given in these standards are either too conservative or non-existent. More up-to-date information is available in the petroleum industry literature, e.g., Langeveld [17] and geotechnical engineering literature, e.g., Marr, et al. [18]. With the advance of the tank–soil interaction technology, the knowledge of settlement modes and acceptable magnitudes will increase and it will be reflected in the standards such as API 650. In the meantime, the existing standards should not be rigidly adhered to when evaluating the effect of vertical settlement of the structural components of the tank.

Lateral Movement

The application of the vertical load by the tank and its products is transmitted into the ground in a three-dimensional manner. In addition to the vertical settlement described above, settlement and deformation also take place in the radial or lateral direction. If the vertical load is applied too rapidly, the radial or lateral deformation may proceed at such a rate as to cause a shear failure of the soil. For this reason, the lateral movement around the tank should be measured in any scheme of soil improvement. The lateral movement profile of the consolidating soil can be developed at every stage of the loading by using the commercially available instruments called "inclinometers." The inclinometer probe is comprised of sensitive electronic accelerometers which are inserted in specially grooved plastic tubing to measure the deformation or movement profile of the subsurface. The specially grooved plastic tubes are permanently installed in boreholes around the tank periphery. The portable probe is inserted into the permanently installed plastic tubes to measure the deflection profile by reading the electronic readout of the angle of inclination of the probe. This monitoring should be done on a continuous basis. An example of the results obtained is shown in Figure 34. It can be seen in this figure that the weakest soil stratum defined as

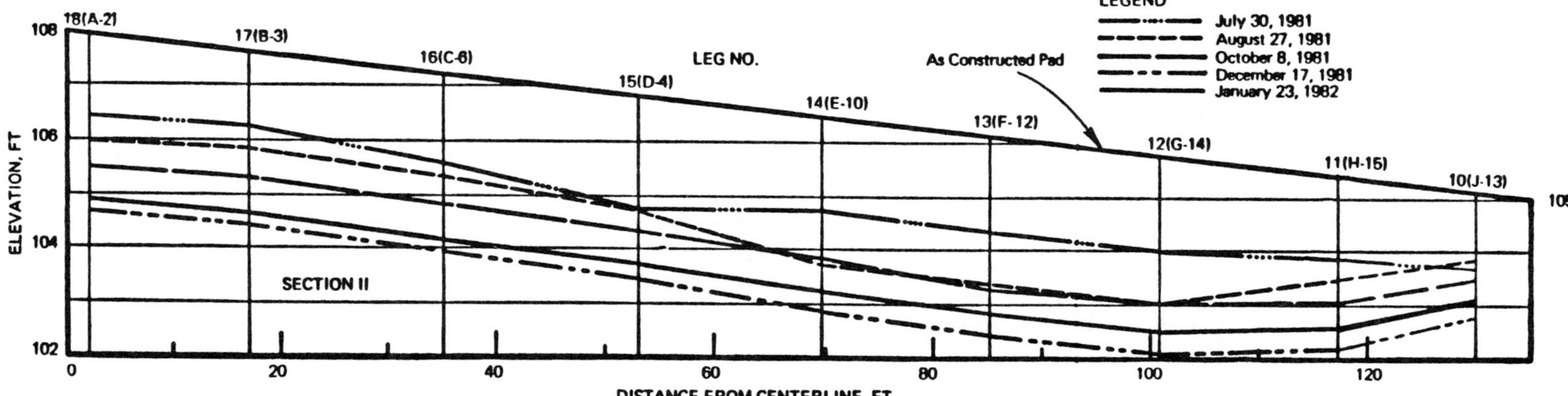

FIGURE 33. Example of tank floor settlement.

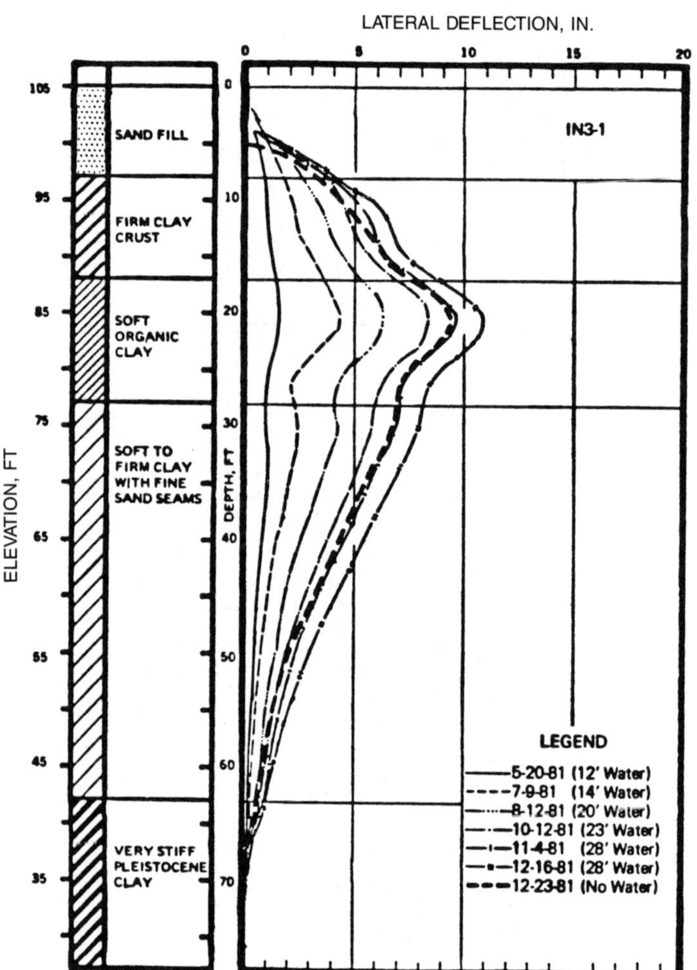

NOTE: The lateral deflection shown is the actual value under the surface load indicated.

FIGURE 34. Example of lateral movement measurement.

soft organic clay experienced the largest lateral movement. The loading stages should be applied judiciously to allow consolidation of the weakest soil stratum; otherwise an arc type of failure lying mostly within the weakest stratum can take place.

The schedule of loading will depend on the rate of lateral movement among other things. Therefore, the rate of lateral movement should be monitored very carefully by experienced and knowledgeable geotechnical engineers. A decreasing rate of lateral movement suggests a successful consolidation process. On the other hand, an increasing rate of lateral movement under constant surface load points to shear deformation which can lead to failure. In these situations, the magnitude of the surface load should be reduced and the subsoil allowed to consolidate under a lower surface load.

PERFORMANCE EVALUATION METHODS

Tank Behavior

Evaluation of the tank behavior should be done independently and concurrently for the three components of the tank, namely shell, bottom plate, and shell-bottom plate connection. Marr, et al. [18] has presented an excellent summary of performance criteria for differential settlement of tanks which is reproduced in Figure 35 and represents the state of the art.

Experience has shown that on large floating roof tanks the ovality of shell generally precedes the other modes of distress. Therefore, criterion no. III in Figure 10 is very critical. In order to estimate the non-planar settlement it is essential to subtract the uniform settlement as well as the component of settlement associated with planar tilt from the measured settlement at various points around the tank periphery. In turn, this requires the determination of the plane of tilt. Marr, et al. [12] has chosen to use only one harmonic. In the writer's opinion, higher order harmonics are necessary for complex non-planar settlement modes. The procedure is as follows.

In general one can fit a Fourier series of the form:

$$z(\alpha) = \frac{a_0}{2} + \sum_{k=1}^{m} \times (a_k \cos k\alpha + b_k \sin k\alpha)$$

to a set of n settlement points of settlement value Z_t, spaced at equal angles:

$$\alpha = \frac{2\pi t}{n}$$

$$t = 1, 2, 3, \ldots . n$$

where

$$a_k = \frac{2}{n} \sum_{t=1}^{n} z_t \cos k\alpha_t$$

$$k = 0, 1, \ldots . m$$

$$b_k = \frac{2}{n} \sum_{t=1}^{n} z_t \sin k\alpha_t$$

$$k = 1, 2, \ldots . m$$

for

$$m \leq \frac{1}{2}(n-1)$$

The value of M should be the smallest of the expression $\frac{1}{2}(n-1)$ so that the nonegative quantity

$$\varepsilon^2 = \frac{2}{n} \sum_{t=1}^{n} z_t^2 - \left\{ \frac{a_0^2}{2} + \sum_{k=1}^{m} \left(a_k^2 - b_k^2 \right) \right\}$$

is small; it will be exactly zero for $M = \frac{1}{2}(n-1)$. The maximum error for $k > M$ will be:

$$\left\{ \varepsilon^2 \frac{(n-1)}{2} - m \right\}^{1/2}$$

By selecting the appropriate value of M a series $Z(\alpha)$ is determined which represents the best fit mathematical curve of tilt plane.

An example of the use of the above procedures is shown in Figure 36 for a 280 ft diameter 32 ft high tank; also shown is the curve developed by using the first order harmonics. It is clear from the figure that the latter curve is a fictitious mathematical curve which does not relate to the physical behavior of the tank.

The determination of plane of tilt by the above procedure becomes cumbersome in those cases where uniform settlement and planar tilt are a significant percentage of the total settlement. For these cases, the writer recommends the use of the simplified method as suggested by Koczwara [15]. Regardless of the method of determination of plane of tilt, the criterion of roof binding advanced by Malik, et al. [19] and adopted by Marr, et al. [18] is valid.

At the present state of knowledge several uncertainties remain, e.g., the effect of roof girders, which does not qualify any of the above approaches to be the exact representation of physical tank behavior. Therefore, this writer recommends that physical measurements of the gap between the roof and the shell be made occasionally to verify the predictions of the analytical procedures. An example of the comparison of

COMPONENT	ASPECT OF PERFORMANCE	MODE OF FAILURE	CRITERION *	DEFINITIONS and COMMENTS
SHELL	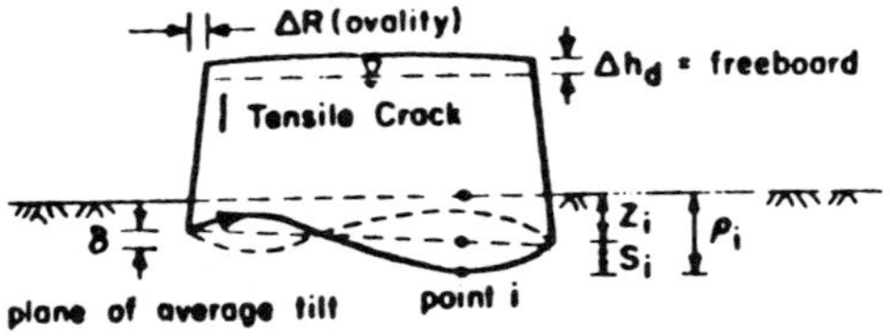 ΔR (ovality); Δh_d = freeboard; Tensile Crack; δ; Z_i; S_i; ρ_i; plane of average tilt; point i Planar Tilt: $Z_i = \bar{\rho}_i + 2(\delta/D)\cos(x_i/R + \beta)$ where $\bar{\rho}$ = average shell settlement β = orientation of plane of average tilt δ/D = angle of average planar tilt Non-Planar Settlement: $S_i = \rho_i - Z_i$; $\Delta S_i = S_i - 0.5(S_{i+1} + S_{i-1})$; $\ell = \pi D/n$	PLANAR TILT I. Overtopping of shell II. Loss of roof seal NON-PLANAR SETTLEMENT III. Binding of roof seal IV. Overstress of shell	I. $\delta \leq 2\Delta h_d$ II. $\delta \leq 2\sqrt{\Delta R_{tol} D}$ III. $\Delta S \leq \frac{\ell^2}{HD} \Delta R_{tol}$ IV. $\Delta S \leq 11 \frac{\ell^2 \sigma_y}{HE}$	Δh_d = freeboard H, D = tank dimensions ΔR_{tol} controlled by: 1. tolerance of roof seal 2. buckling of wind girder 3. distortion of cone roof σ_y = yield strength of shell E = Young's modulus of elasticity ℓ = distance between points of measured settlement n = number of peripheral survey points
BOTTOM PLATE	Non-Planar Settlement: 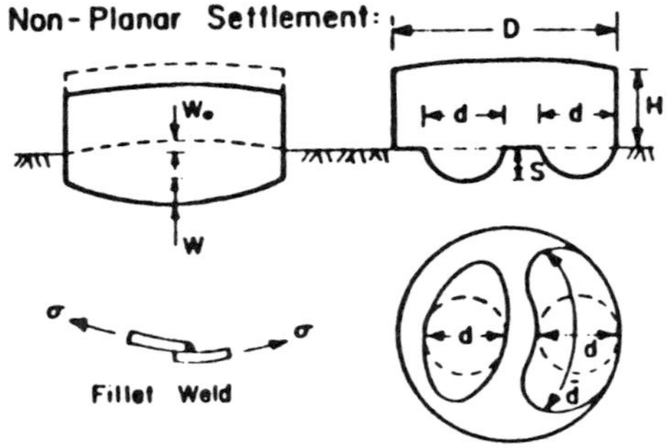 D; W_o; W; d; H; S; σ; Fillet Weld; d; $\bar{d}$	V. Rupture from dish-shaped settlement VI. Rupture from localized depressions a. remote from shell b. adjacent to shell	V. $W \leq \left[W_o^2 + \frac{0.37\sigma_f D^2}{FS \cdot E}\right]^{0.5}$ VI. a. $S \leq d\left[\frac{0.28\sigma_f}{FS \cdot E}\right]^{0.5}$ b. $S \leq d\left[\frac{2.25\sigma_f D}{d^{0.75} FS\, E\, H}\right]^{0.5}$ d and S in meters	σ_f = ultimate strength appropriate for bottom plate $FS \leq 4$ localized yield possible $FS \leq 2$ severe overstress and yield possible E = Young's modulus of elasticity W_o = initial camber $d, \bar{d}$ = dimensions of local depression, m For local depressions adjacent to shell, if $d < D/4$ and $\bar{d} \geq 2d$, use criterion VI b. Otherwise use criterion VI a
SHELL BOTTOM PLATE CONNECTION	Non-Planar Settlement: 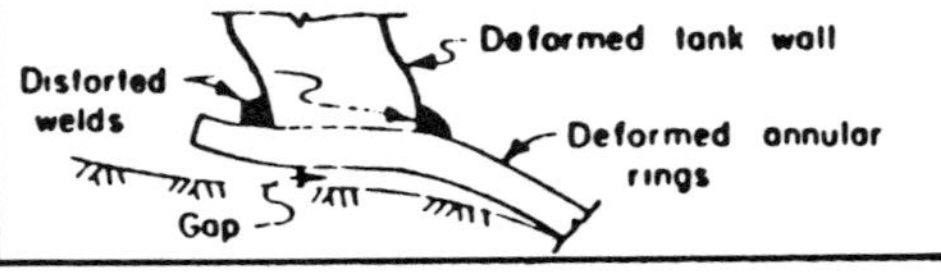 Distorted welds; Deformed tank wall; Deformed annular rings; Gap	VII. Rupture of connection as shell bridges over soft spot	VII. Surveillance and maintenance to prevent separation of shell and foundation	* Use definitions in Fig 2 and Malik et al correction for tilt to find all settlement parameters

FIGURE 35. Performance criteria for differential settlement.

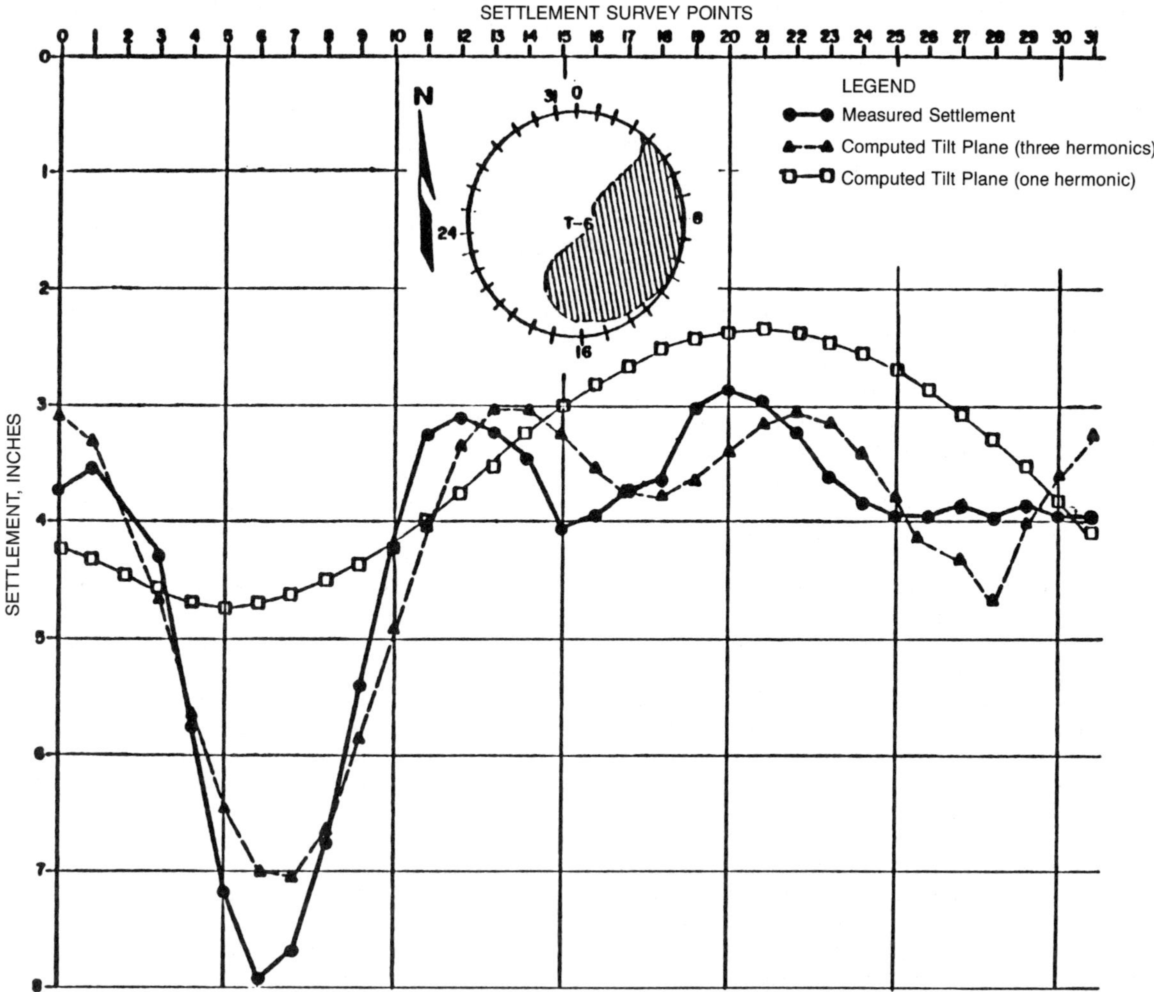

FIGURE 36. Determination of tilt plane.

the analytical procedures [15] of determining the shell ovality with the actual measurements of the gap between the roof and the shell is given in Figure 37. As can be seen in this figure, the predicted shape of shell near Point No. 15 indicated to have adequate tolerance, approximately 4 inches. However, the actual measurement showed the roof almost touching the shell at that point. This example illustrates that approximations made in developing the analytical models may not be accurate enough to predict the exact shape of the shell where seal tolerances are relatively small.

In summary, the criteria given in Figure 35 should be used as a guide in evaluating the behavior of tanks. It is strongly recommended that actual physical measurements be made to verify the various analytical criteria.

Soil Behavior

The evaluation of soil behavior using static methods of soil improvement should be done in the three main areas, namely pore water pressure, vertical settlement, and lateral movement. In using the dynamic methods, emphasis should be placed on the actual details of execution of the particular method and the end result verification. For impact densification (Method 6) and vibrocompaction (Method 7), the end result verification is done by conducting Dutch Cone probing for relative density determination and pressuremeter testing to assess the improvement in volume change due to loading. In addition, piezometers are also installed to monitor the liquefaction process in Method 6.

The following paragraphs present evaluation criteria for pore pressure, vertical settlements, and lateral movement. The pore pressure criterion applies only to soil improvement by static methods while discussion relative to vertical settlement and lateral movement is applied to both static and dynamic methods since tanks are generally hydrotested similar to static soil improvement Method No. 1 following one of the dynamic improvement methods.

The rate of loading for soil improvement using static methods should be in concert with pore pressure dissipation. In other words, pore pressures should not increase to the point that the stress path of any given point in the soil mass reaches the failure envelope. As explained earlier, the related stress distribution and stress path analysis is a complex analytical undertaking. A more simple diagnostic tool is necessary to routinely monitor the pore pressure behavior in the field. Through experience with soil improvement programs using static methods, the writer has developed a simple parameter termed, "pore pressure response ratio" which has been successfully used for monitoring pore pressures and has been verified analytically by constructing the stress paths. It is defined as the ratio of increment of pore pressure divided by the increment of applied stress. A similar parameter has been discussed by DeBeer [12].

Typical plots of pore pressure response ratios are shown in Figure 38 for a number of piezometers under a tank. In general, pore pressures would rise approaching the applied stress at the ground level near the center of the tank, whereas the pore pressures near the edge of the tank would be smaller. After studying the data of a number of pore pressure transducers installed in non-sensitive marine clays, the writer recommends a maximum pore pressure response ratio of 0.75 for transducers located near the edge of the tank. Near the center of the tank, this ratio can approach unity as long as the response ratio near the edge of the tank is smaller than 0.75. Pore pressures must dissipate when the load is held at a constant level during any stage of the soil improvement program. If, after the application of the load, pore pressures continue to increase under constant load, it indicates a yielding phenomenon. In this circumstance, the load should be reduced and pore pressure monitored carefully. The pore pressure response ratio limits recommended above should be used in conjunction with competent geotechnical input. The pore pressure response ratio criterion is only for ease of measurement and evaluation in the field, stress path construction and other analytical methods discussed earlier should be used to evaluate soil mass stability.

The vertical settlement is monitored using a fixed number of points around the tank periphery and several points within the tank bottom. The settlement of any point, when plotted on a semi-log paper, has a characteristic shape for a given load cycle.

The plot typically consists of an initial flat portion (representing initial settlement), followed by a steep portion (primary consolidation) which is in turn followed by a flat portion (undrained creep or secondary settlement). The transition from the steep portion to the last flat leg of the curve represents a substantial (50 to 60 percent) completion of the primary consolidation settlement. In the writer's opinion, the next stage of load can be applied at this point provided pore water pressure has dissipated to the level where there is no danger of shear failure. It is neither necessary nor practical to hold a stage of load until all the consolidation is completed under that stage of load. An example of vertical settlement plot is presented in Figure 39.

From the above discussion, it is clear that the evaluation of vertical settlement is based on the shape of the settlement curve rather than single numbers. Thus, a considerable amount of judgement is required. Only experienced geotechnical engineers should exercise this judgement. The rate and duration of stage loading should be adjusted during the soil improvement program based on this judgement. Generally the first tank on a farm should be designated as the "test case" and the rate and duration of loading should be monitored. This data should be tabulated for guidance in the loading program of subsequent tanks.

The lateral movement has not been demonstrated to yield a characteristic shape similar to that of vertical settlement, the reason being that dissimilar and incompatible phenomena are involved in producing the lateral movement,

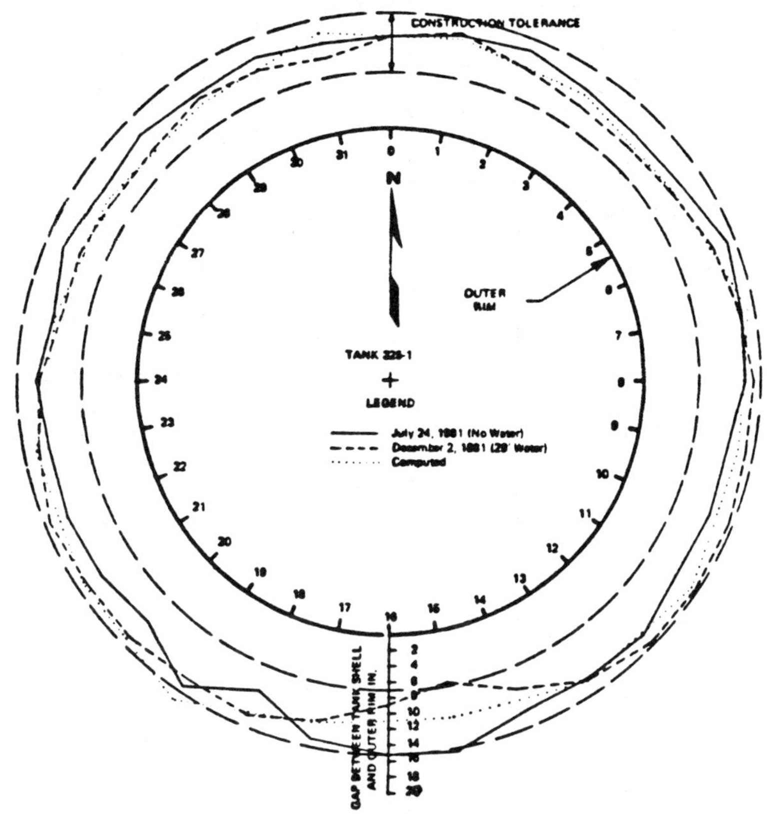

(a) Shell - Roof Gap Measurement

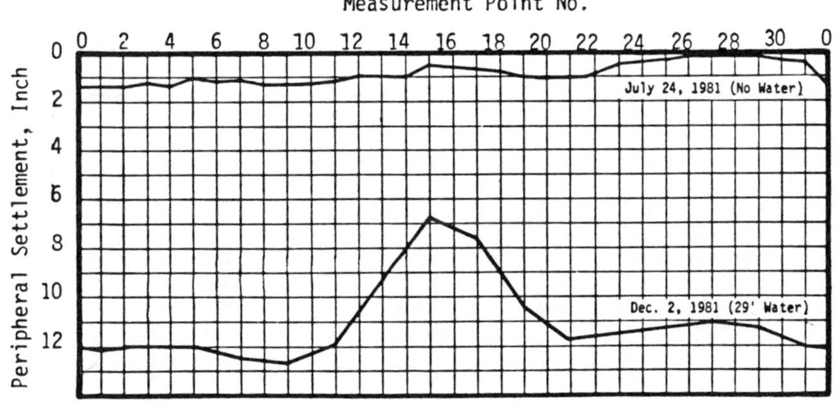

(b) Fold Out Settlement Plot

FIGURE 37. Comparison of measured and computed shell ovality.

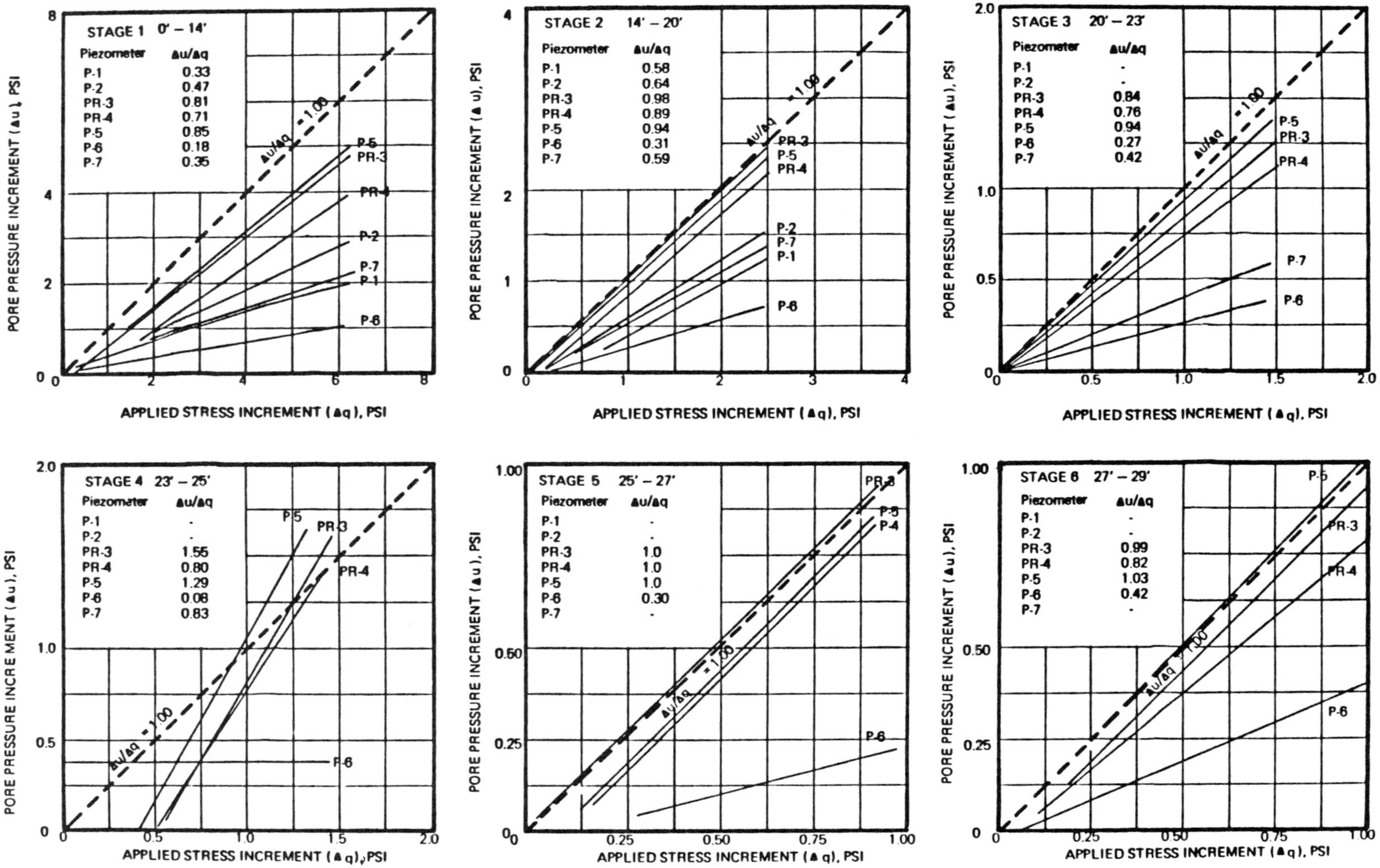
STAGE 1 0' – 14'
Piezometer Δu/Δq
P-1 0.33
P-2 0.47
PR-3 0.81
PR-4 0.71
P-5 0.85
P-6 0.18
P-7 0.35
STAGE 2 14' – 20'
Piezometer Δu/Δq
P-1 0.58
P-2 0.64
PR-3 0.98
PR-4 0.89
P-5 0.94
P-6 0.31
P-7 0.59
STAGE 3 20' – 23'
Piezometer Δu/Δq
P-1 -
P-2 -
PR-3 0.84
PR-4 0.76
P-5 0.94
P-6 0.27
P-7 0.42
STAGE 4 23' – 25'
Piezometer Δu/Δq
P-1 -
P-2 -
PR-3 1.55
PR-4 0.80
P-5 1.29
P-6 0.08
P-7 0.83
STAGE 5 25' – 27'
Piezometer Δu/Δq
P-1 -
P-2 -
PR-3 1.0
PR-4 1.0
P-5 1.0
P-6 0.30
P-7 -
STAGE 6 27' – 29'
Piezometer Δu/Δq
P-1 -
P-2 -
PR-3 0.99
PR-4 0.82
P-5 1.03
P-6 0.42
P-7 -
Δu/Δq = 1.00
PORE PRESSURE INCREMENT (Δu), PSI
APPLIED STRESS INCREMENT (Δq), PSI

Note: Piezometers P-1, P-2, P-6 and P-7 are under the tank edge.
Piezometers PR-3, PR-4 and P-5 are under the tank center.

FIGURE 38. Example of pore pressure response ratio plots.

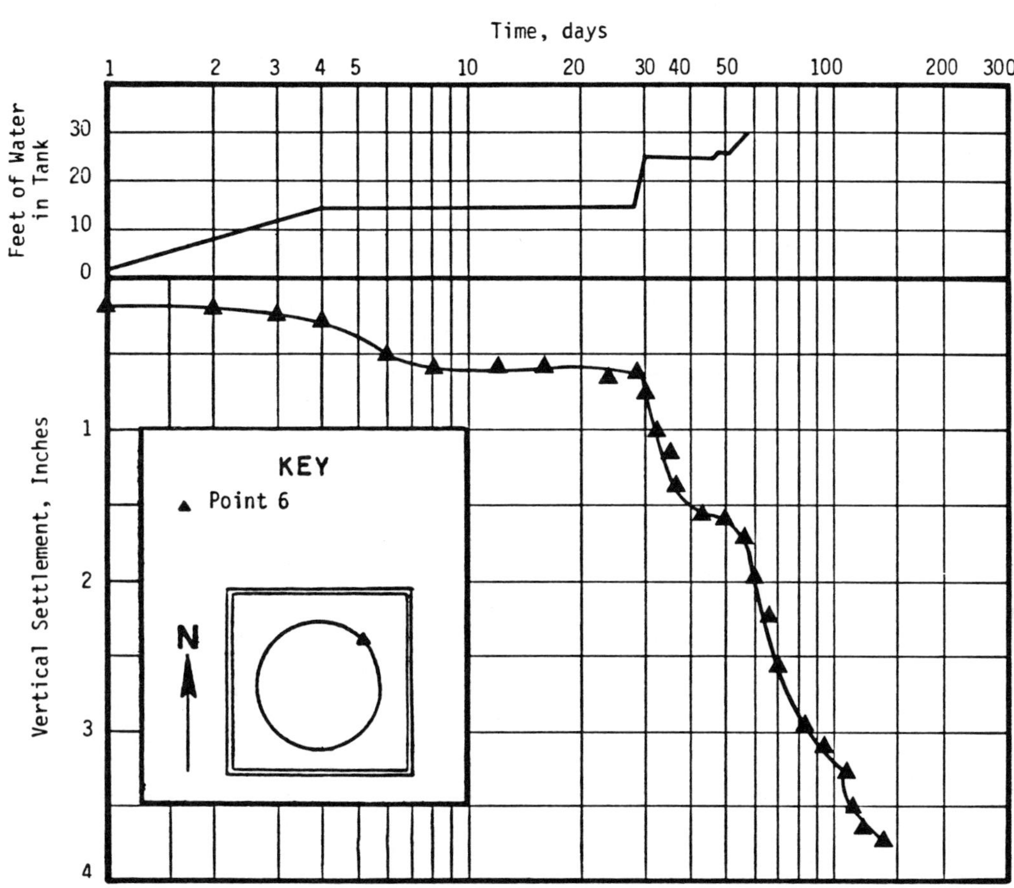

FIGURE 39. Example of vertical settlement plot.

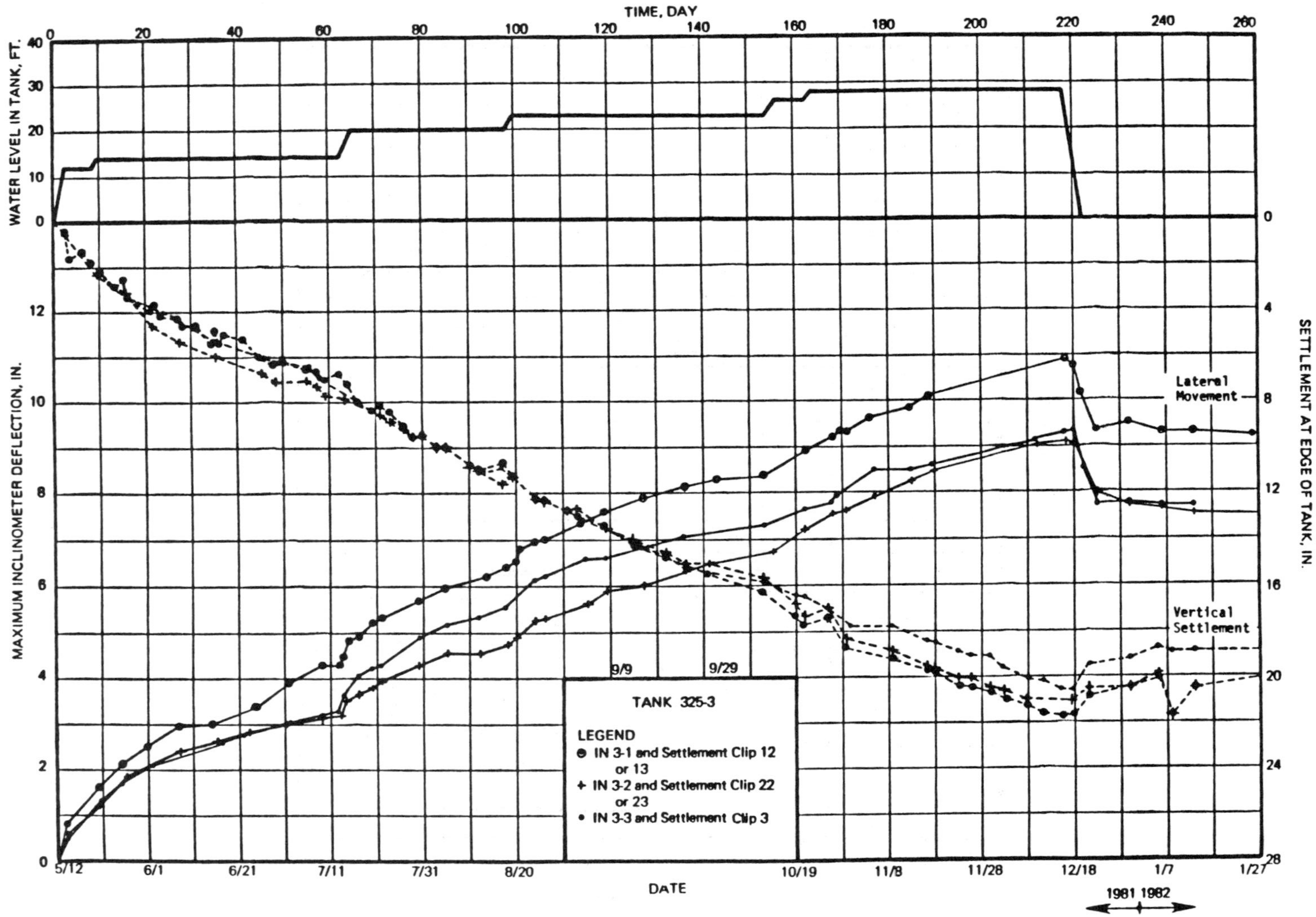

FIGURE 40. Progress of maximum lateral movement.

i.e., three-dimensional consolidation, plastic and shear deformation. In the writer's experience, a simple and practical means of the evaluation of lateral movement is to plot the maximum lateral movement magnitude measured by inclinometers on an arithmetic time plot similar to that shown in Figure 40. This figure also shows the plots of corresponding vertical settlement of points on the tank periphery near the three inclinometers installed to measure the lateral movement. The deflection of inclinometer casing rises sharply during the application of a stage of load. Subsequently, the movement levels off. When the rate of movement becomes constant the next stage of load may be applied as long as the criteria of pore pressure and vertical settlement are satisfied.

Once again, a considerable amount of judgement is required and the first tank should be used as a test case. The absolute magnitude of lateral movement is significant only from the standpoint that soil heave outside the tank edge or other bearing capacity type of failure is not suspected. After evaluating the data of the first tank, the criterion for the tolerable rate of lateral movement should be established on a site specific basis.

REFERENCES

1. Ahmed, S., et al., "Influence of Soil–Pipe Interaction on Pipeline Design," *Proceedings of ASCE Convention and Exposition*, Portland, Oregon, Paper No. 80-050 (April 14–18, 1980).
2. Audibert, J., M. E. and K. J. Numan, "Soil Restraint Against Horizontal Motion of Pipes," *Journal of the Geotechnical Engineering Division*, ASCE Vol. 103 No. GT10 (October 1977).
3. Kondner, R. L. and J. S. Zelasko, "A Hyperbolic Stress Strain Formulation For Sands," *Proceedings of Pan American Conference on Soil Mechanics and Foundations Engineering*, Brazil, Vol. I, pp 289-324 (1963).
4. Luscher, U., et al., "Pipe–Soil Interaction, Trans-Alaska Pipeline," *Proceedings of ASCE Conference on Pipelines in Adverse Environments*, New Orleans, Louisiana (January 15–17, 1979).
5. Peng, L.-C., "Stress Analysis Methods For Underground Pipe Lines," Pipe Line Industry (May 1978).
6. Radbakrishnew, N. and Jones, W. H., "Documentation For Modified U-Frame Program," Technical Report K-76-1, U. S. Army Engineer Waterways Experiment Station, Vicksburg, Mississippi (June 1976).
7. Spangler, M. G., "The Structural Design of Flexible Pipe Culverts," The Iowa State College Bulletin, Ames, Iowa, Bulletin 153 (December 24, 1941).
8. Yen, C., Bing, et al., "Soil–Pipe Interaction of Heated Oil Pipelines," *Proceedings of ASCE Convention and Exposition*, Portland, Oregon, Paper No. 80-163 (April 14–18, 1980).
9. Ahmed, S., "Economical Foundation Alternatives on Soft Ground," *Proceedings of 8th Annual Conference of Association of Muslim Scientists and Engineers*, Purdue University (September 3–5, 1982).
10. American Petroleum Institute, "Recommended Practices for Construction of Foundations for API Vertical Storage Tanks," API 650, Appendix B (1980).
11. Brown, Ralph E., "Vibroflotation Compaction of Granular Hydraulic Fills," *ASCE National Water Resources and Ocean Engineering Convention*, San Diego, California, Reprint No. 2657 (April 1976).
12. DeBeer, E. E., "Foundation Problems of Petroleum Tanks," *Annalr de l'institute Beldje de Petrol*, No. 6, pp. 25–40 (1969).
13. Engelhardt, K., W. A. Flynn and A. A. Bayuk, "Vibroreplacement—A Method to Strengthen Cohesive Soils In Situ," ASCE National Structural Engineering Meeting, Cincinnati, Ohio, Report No. 2281 (April, 1979).
14. Hunt, Roy E., "Preloading an Effective Tank Site Preparation," *Oil & Gas Journal* (December 4, 1967).
15. Koezwera, F. A., "Simple Method Calculates Tank Shell Distortion," *Hydrocarbon Processing* (August, 1980).
16. Lander, T., William and R. V. Whiteman, "Soil Mechanics," John Wiley & Sons, Inc. (1969).
17. Langveld, J. M., "Design of Large Steel Storage Tanks for Crude Oil and Liquid Natural Gas," *Proceedings of Annual Meeting of I.I.W Section: Welding of Storage Tank for Oil & Gas Industries*, pp. 35–95 (1979).
18. Marr, W., Jose A. Ramos and William T. Lambe, "Criteria for Settlement of Tanks," *Journal of Geotechnical Engineering Division*, ASCE (August 1982).
19. Malik, Z., J. Martan and C. Ruiz, "Ovalization of Cylindrical Tanks as a Result of Foundation Settlement," *Journal of Static Analysis*, Vol. 12, No. 4 (1977).
20. Mayne, P. W., J. S. Jones and R. P. Fedosick, "Predicting Subsurface Improvement Due to Impact Densification," ASCE Geotechnical Conference, Las Vegas, Nevada (April 1982).
21. Morrison, A., "The Booming Business in Wick Drains," Civil Engineering (March 1982).

CHAPTER 15

Cement-Mortar Lined Pipes

Chang-Ning Sun* and Joe Hunt*

INTRODUCTION

Carbon steel pipes (jointed or welded), cast iron pipes (jointed), and ductile iron pipes (jointed) have advantages of long laying length, tight joints, ability to withstand high internal pressures and external loadings. The disadvantages are that all three types of pipes are subject to internal corrosion and incrustation and internal linings are needed for long-term service. Cast iron and ductile iron pipes are highly resistant to corrosion in most natural soils and usually do not need external coating and/or wrapping; carbon steel pipes are subject to external corrosion and usually need some form of external protective coating and/or wrapping. This chapter only covers the application of cement-mortar lining to the internal surface of smooth pipe barrels of steel and iron pipes.

Lining the interior surface of carbon steel, cast iron, and ductile iron pipes with cement mortar for the prevention of corrosion and incrustation in new pipeline and for the rehabilitation of existing pipeline has gained broad acceptance in the United States in the last 50 years [1,3,11,14,16,17,20,26, 54,55]. A number of testing programs have been carried out [15,20,44,45,46,51]. Field application [1,3,14,21,38,54,55] of the cement-mortar lining and pipeline rehabilitation [11,14, 19,30,35,54] have been successful. Many standards and construction specifications have been issued by government agencies [21,22,32,33,35,38], professional and technical institutions [34,41,42,43], and consulting organizations and constructors [10,28,36,40,50]. Cement-mortar lining is considered to be durable [26] and the cracks, if there are any, are self-healing [12,53]. Cement-mortar linings are suited for pipelines carrying water, sanitary sewage, oil, and gas, but not for pipelines carrying aggressive liquids, certain chemicals, and certain industrial wastes which would react with the lining material [41]. Cement-mortar lining offers smooth interior surface, reduced pumping efforts, and reduced number of pipe breaks due to excessive internal pressure.

Curing of the cement-mortar lining is the most important step in either the in-place or shop-applied lining close operation. Well designed, constructed, and cured cement-mortar lining should work satisfactorily for many decades.

HISTORICAL DEVELOPMENT

The use of cement-mortar lining for water mains was reported by the French Academy of Sciences in 1836 [42]. Many towns in New England area began to use cement-mortar lining for cast-iron pipe as early as the middle 1800s [26]. In the 1870s, pipe made of 20-gage (0.95-mm) thin wrought-iron shell with a ½-in. (13-mm) thick cement-mortar lining and a ½-in. (13-mm) thick cement-mortar coating came into use in the New England and middle atlantic states; failure of this type of pipe usually occurred when the outer cement-mortar coating broke [26]. In 1922, the first recorded cement-mortar lined cast-iron pipe was installed in the water distribution system of Charleston, South Carolina [41]; this new cast-iron pipe was lined in place by means of a projectile drawn through the pipe. In 1933, a 72-in. (1.82-m) diameter steel pipe that had been in service was lined in place with cement mortar by the centrifugal process [42]. Great Britain began to use cement-mortar lining in the middle 1930s and the process was soon adopted by other countries [14].

For many years the smallest diameter of pipe that could be lined in place by the centrifugal process was 24 in. (610 mm) as a man was required to ride the lining machine through the pipe. In 1950, a remote-controlled centrifugal lining ma-

*Geology and Geotechnical Engineering Group, Tennessee Valley Authority, Knoxville, TN

TABLE 1. Size and Wall Thickness of Iron Pipes [16].

Type of Pipe	Cast Iron	Ductile Iron
Size (Nominal)	3 in. to 48 in. (76 mm to 1.21 m)	3 in. to 54 in. (76 mm to 1.37 m)
Wall Thickness	0.32 in. to 1.95 in. (8.1 mm to 49.5 mm)	0.25 in. to 1.05 in. (6.4 mm to 26.7 mm)
Number of Different Size-Thickness Selections	160	110

TABLE 2. Shop-Applied Cement-Mortar Lining Thickness for Iron Pipes [41].

Pipe Size (Nominal)	Standard Thickness*	Double Thickness*
3 in. to 12 in. (76 mm to 305 mm)	1/16 in. (1.6 mm)	1/8 in. (3.2 mm)
14 in. to 24 in. (356 mm to 610 mm)	3/32 in. (2.4 mm)	3/16 in. (4.8 mm)
30 in. to 54 in. (762 mm to 1.37 m)	1/8 in. (3.2 mm)	¼ in. (6.4 mm)

*Permitted tolerance shall be +1/8 in. (3.2 mm), no minus tolerance allowed.

TABLE 3. Shop-Applied Cement-Mortar Lining Thickness for Steel Pipes [43].

Pipe Size (Nominal)	Lining Thickness	Tolerance
4 in. to 10 in. (102 mm to 254 mm)	¼ in. (6.4 mm)	±1/32 in. (0.8 mm)
11 in. to 23 in. (279 mm to 584 mm)	5/16 in. (7.9 mm)	+1/8 in. (3.2 mm) −1/16 in. (1.6 mm)
24 in. to 36 in. (610 mm to 915 mm)	3/8 in. (9.5 mm)	
Over 36 in. (Over 915 mm)	1/2 in. (12.7 mm)	

chine was introduced making possible in-place lining of smaller diameters [42]. Presently, pipelines ranging from 4 in. (102 mm) to 20 ft (6.16 m) in diameter can be cleaned and cement-mortar lined in place [42,54]. In this process, the cement mortar is centrifugally applied by a rapidly revolving head. The cement mortar is thrown in a tangential direction, and stuck to the wall upon impact. The centrifugal lining machine can be used either in the field or in the shop. The field application is carried out after the pipe has been installed and trench backfilled; the shop application is carried out before the transporting of the pipe to the site of installation.

The spinning process [41,43] is used exclusively for shop-applied cement-mortar lining. The pipe is placed horizontally on a set of rotating wheels and a calculated amount of cement mortar is introduced into the pipe. The cement mortar is spread evenly on the interior surface of the pipe and densified by the centrifugal force resulting from spinning the pipe at high speed about its longitudinal axis. Lining as thin as 1/16 in. (1.6 mm) can be produced uniformly along the pipe. Almost all cast-iron and ductile-iron pipes installed in water system today are cement-mortar lined by the spinning process in the shop.

PIPE SIZES AND CEMENT-MORTAR LINING THICKNESSES

Iron and steel pipes with shop-applied cement-mortar lining, if manufactured in accordance with the industrial stan-

TABLE 4. Shop-Applied Cement-Mortar Lining Thickness for Oilfield Steel Pipes [34].

Pipe Size (OD)	Selected Lining Thickness* Minimum	Selected Lining Thickness* Maximum
2 3/8 in. (60 mm)	1/8 in. (3.2 mm)	1/4 in. (6.4 mm)
2 7/8 in. (73 mm)		
3 1/2 in. (89 mm)	5/32 in. (4.0 mm)	5/16 in. (7.9 mm)
4 1/2 in. (114 mm)		3/8 in. (9.5 mm)
6 5/8 in. (168 mm)	3/16 in. (4.8 mm)	7/16 in. (11.1 mm)
8 5/8 in. (219 mm)	1/4 in. (6.4 mm)	5/8 in. (15.9 mm)
10 3/4 in. (273 mm)	3/8 in. (9.5 mm)	7/8 in. (22.2 mm)
12 3/4 in. (324 mm)		
16 in. (406 mm)	7/16 in. (11.1 mm)	1 in. (25.4 mm)
20 in. (508 mm)	1/2 in. (12.7 mm)	1 1/8 in. (28.6 mm)

*Permitted tolerance shall be ±1/32 in. (±0.8 mm) for pipe sizes 6 5/8 in. (168 mm) and smaller and ±1/16 in. (±1.6 mm) for pipe sizes 8 5/8 in. (219 mm) and larger.

dards [34,41,43], should have good quality; users only have to choose the pipe size, pipe wall thickness, and cement-mortar lining thickness. In-place iron and steel pipes are cleaned and then lined; mortar mix, lining application, and curing are important quality control items.

Tables 1 and 2 show the current available sizes, wall thicknesses, and shop-applied cement-mortar lining thicknesses for the iron pipes. Tables 3 and 4 show the shop-applied cement-mortar lining thicknesses for steel and oilfield steel pipes. Table 5 shows the in-place applied cement-mortar lining thicknesses for both iron and steel pipes.

CODES AND STANDARDS

The shop application of cement-mortar lining is covered by ANSI/AWWA C104/A21.4–80 [41] for iron pipes and AWWA C205-80 [43] for steel pipes. The in-place application of cement-mortar lining is covered by AWWA C602-76 [42] for both iron and steel pipes. The shop application and field installation of cement-mortar lining for oilfield steel pipes are covered by API RP 10E-78 [34]. An item by item comparison between the standards is given in Table 6.

SPECIFICATIONS

For the procurement of pipes with shop-applied cement-mortar lining, the quality assurance of the cement-mortar lining is not an issue. The standards [34,41,43] are adequate and only thickness of the cement-mortar lining has to be specified.

The in-place installation of cement-mortar lining is not as straightforward. In addition to the AWWA and API standards [34,42], a construction specification is usually needed to specify the technical requirements and contractual responsibilities of the contractor. Many construction specifications were issued for these reasons [7,10,17,21,22,28,32, 33,35,36,38,40,50]. In many cases, the pipeline has already been constructed and the trench backfilled; the lining contractor only has to install the cement-mortar lining. A specification for in-place applied cement-mortar lining may include some of the following items:

- Materials
  - Cement
  - Sand
  - Admixture
  - Water
  - Wire reinforcement
  - Curing compound
  - Seal coat
  - Handling, delivery, and storage
  - Material certificate
- Mortar
  - Mix design
  - Slump
  - Temperature
  - Testing for compressive strength
  - Number of testing samples taken per day
- Temporary bypass pipeline
  - Service pipes
  - Fire hydrants
- Appurtenance work
  - In-line valve replacement
  - Installation of chlorine cocks
- Protection of appurtenances
  - Temporary stopper for existing outlets, branch connections, openings, air valves, and blowoffs
- Dewatering
  - System dewatering
  - Access opening dewatering
- Site preparation and clearing
  - Clearing and grading work area before work begins
  - Removing equipment and unused materials and clear site after work is finished
- Access opening
  - Excavation
  - Backfilling
  - Existing manhole or pipe cutting
- Pipe interior surface cleaning
  - Removing all rust, scales, deposits, tubercules, incrustations, debris, loose or foreign materials
- Lining application
  - Mixing equipment
  - Lining machine—lining and troweling
  - Handwork—lining and/or troweling
  - Curing
- Inspection
  - Lining thickness measurement
  - Looking for lining surface cracks, rough areas, blistered areas, sandpockets, voids, and separation of lining from steel surface
- Repairs
- Joints
  - Welded or mechanically jointed
  - Joint compound
  - Welded joint vacuum box testing
- Performance testings
  - System hydrostatic test
  - Pressure drop test for determining the C coefficient
- Safety and hazards requirements
  - OSHA standards
  - State and federal regulations.

STRUCTURAL PROPERTIES OF CEMENT-MORTAR LINING

Most shop-applied cement-mortar linings are very thin compared to the thickness of the pipe wall and, therefore, the structural integrity of the thin shop-applied lining is not an issue. No testing is needed to verify the linings' structural integrity.

TABLE 5. In-Place Applied Cement-Mortar Lining Thickness [42].

Type of Pipe	Pipe Size (Nominal)	Lining Thickness*
Old and New Iron	4 in. to 10 in. (102 mm to 254 mm)	1/8 in. (3.2 mm)
	12 in. to 36 in. (305 mm to 915 mm)	3/16 in. (4.8 mm)
	Over 36 in. (Over 915 mm)	1/4 in. (6.4 mm)
Old Steel	4 in. to 12 in. (102 mm to 305 mm)	1/4 in. (6.4 mm)
	14 in. to 22 in. (356 mm to 559 mm)	5/16 in. (7.9 mm)
	24 in. to 60 in. (610 mm to 1.52 m)	3/8 in. (9.5 mm)
	Over 60 in. (Over 1.52 m)	1/2 in. (12.7 mm)
New Steel	4 in. to 12 in. (102 mm to 305 mm)	3/16 in. (4.8 mm)
	14 in. to 36 in. (356 mm to 915 mm)	1/4 in. (6.4 mm)
	42 in. to 60 in. (1.07 m to 1.52 m)	3/8 in. (9.5 mm)
	66 in. to 90 in. (1.68 m to 2.29 m)	7/16 in. (11.1 mm)
	Over 90 in. (Over 2.29 m)	1/2 in. (12.7 mm)

*Permitted tolerance shall be + 1/8 in. (3.2 mm), no minus tolerance allowed.

TABLE 6. Comparison Between Standards.

Item	ANSI/AWWA C104/A21.4	AWWA C205	AWWA C602	API RP 10E
Inspection				
Shop		*		*
Field			*	
Material				
Cement	*	*	*	*
Sand	*	*	*	*
Admixture			*	*
Water	*	*	*	*
Wire Reinforcement		*		*
Protective Painting		*		
Curing Compound		*		
Seal Coat	*			
Surface Cleaning	*	*	*	*
Lining				
Mortar Mix	*	*	*	*
Mixing Equipment		*	*	
Lining Machine	*	*	*	*
Handwork	*	*	*	
Thickness	*	*	*	*
Curing	*	*	*	*
Defects and Repair	*	*	*	*
Field Joint		*		*
Temporary Bypass Pipeline			*	
Handling and Delivery		*		*

*Items covered in the standards.

TABLE 7. Testing of Cement-Mortar Lined Carbon Steel Pipes.

Type of Test	TVA [15,44,45,46]	City of Detroit [20]	Nippon Steel [51]
Cement-Mortar Specimen			
Compressive Strength Test	*	*	
Tensile Strength Test	*	*	
Flexural Strength Test	*		
Shear Test		*	
Stress-Strain Curve	*		
Density Test	*		
Slump Test	*	*	*
Water Absorption Test		*	
Curing Compound			
Water Retention Test	*		
Softening Point Test	*		
Saybolt Viscosity Test	*		
Flash Point Test	*		
Elcometer Adhesion Test	*		
Lining Thickness Measurement	*		*
Pipe Section			
Three-Edge-Bearing Test	*	*	*
Cyclic Loading Test	*		*
Impact Test	*		
Drop Test	*		
Torsion Test	*		
Long Pipe			
Bending Test	*		*
Impact Test	*		*
Drop Test	*		*
Pressure Test		*	
Lime Absorption Test		*	
Elbow			
Bending Testing	*		
Field Load Test			
Static	*		
Dynamic	*		

*Tests performed in the testing programs.

Large pipes are usually lined in place for convenience and cost saving. The in-place applied linings usually have thicknesses comparable to those of the pipe walls. The linings may take up to 20 percent of the total applied load [44] and their structural integrity becomes an important issue.

Miles of iron and steel pipes of sizes up to 20.5 ft (6.25 m) have been cement-mortar lined with good results in many regions of the U.S. [1,3,5,11,14,16,17,19,20,21,26,35,38,54, 55]. In recent years, carbon steel pipes with in-place applied cement-mortar linings are used extensively in power plants and safety-related systems [15,21,29,38,44,45,46]. During the 1971, San Fernando earthquake, cement-mortar linings were damaged only at places where the steel pipe suffered similar damage or separated. Cement-mortar lining which had been in place for only 12 hours was not damaged by this earthquake [55]. The ability of cement-mortar lining to resist seismic loading has also been verified by dynamic load tests [44,45,46].

Field and laboratory tests on cement-mortar lined pipes have been performed under controlled conditions. These testing programs simulated dead load (loading from a roller without vibration, three-edge-bearing test, torsion and bending tests), low frequency load (cyclic loading test), large dynamic load at 28 Hz (loading from a roller with vibration), large acceleration load with a major frequency content of 1-100 Hz (transportation vibration measurement), line load with very short duration (drop test), point load with very short duration (impact test), pressure test, and tests on mortar specimens and curing compound [15,20,44, 45,46,51]. The types of tests performed are listed in Table 7. A typical three-edge-bearing test is shown in Figure 1; the test clearly demonstrated the flexibility of the cement-mortar lining. A bending test is shown in Figure 2. The set-up for a field dynamic loading test is shown in Figure 3.

IN-PLACE APPLICATION OF CEMENT-MORTAR LINING

Centrifugal lining machines are used for the in-place application of cement-mortar linings. The cement mortar is mixed first and then pumped (see Figure 4), through rubber hose, to the lining machine (see Figure 5). The cement mortar is then centrifugally applied to the interior surface of the pipe by a rapidly revolving head and immediately troweled (see Figures 6 and 7).

Before application of cement mortar to the pipe wall, certain preparatory steps must be taken first; after application, curing must be carried out properly.

Quality Control Items

Experience and testing have shown that an economical and structural sound cement-mortar lining can be applied in place if the following rules are observed:

1. The thickness of a cement-mortar lining should not be greater than the thickness of the pipe wall. The cement-

FIGURE 1. Three-edge-bearing test.

FIGURE 2. Bending test.

FIGURE 3. Field dynamic loading test.

FIGURE 4. Mixing and pumping equipment.

FIGURE 5. Centrifugal lining machine.

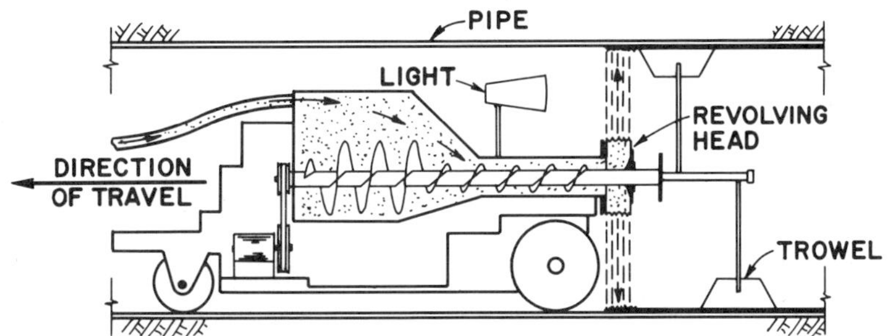

FIGURE 6. Lining machine details.

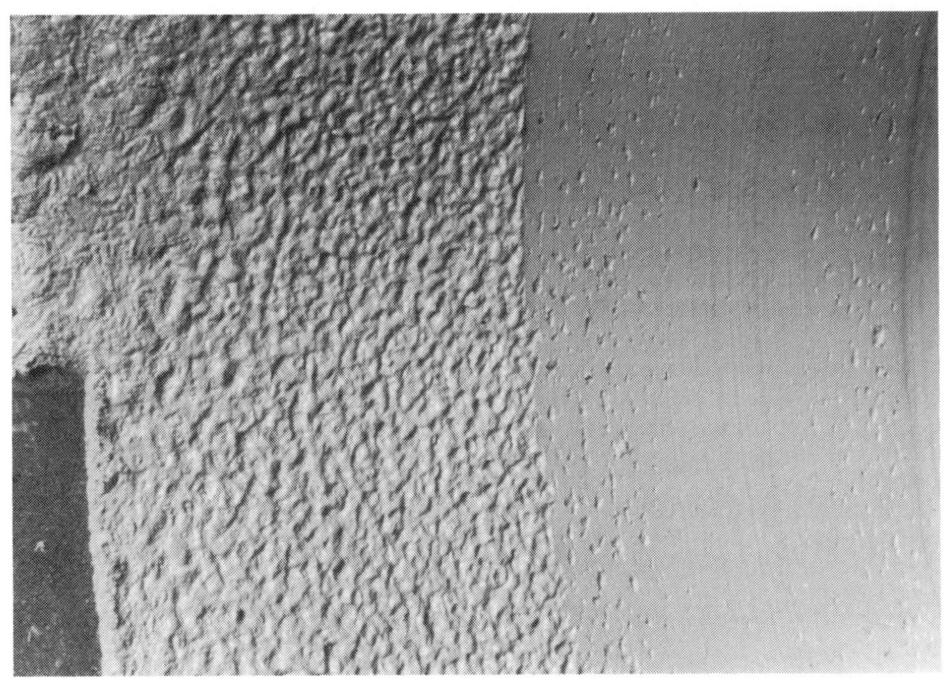

FIGURE 7. Cement-mortar lining: before and after troweling.

mortar lining and pipe wall form a composite, cylindrical shell structure, each taking a portion of the load in according to its stiffness. If the cement-mortar lining is not thicker than the pipe wall, the lining will fail only after plastic hinges are formed in the pipe. The load bearing function should be given mostly to the pipe; the cement-mortar lining's main function should be limited to providing isolation and protection for the interior surface of the pipe from the liquid it is carrying.

2. In-place applied cement-mortar lining has a practical thickness limit of 3/4 in. (19 mm); thicker lining leads to instability and failure of the upper portion of the lining. Thicker lining must be applied by layers.
3. The proportion of cement to sand should be about 1:1 either by weight or by volume. Water should be adjusted for workability.
4. During lining placement, the ambient temperature should be above 40°F (4.5°C) and the mortar temperature should be below 85°F (29.5°C).
5. The interior surface of the pipe should be cleaned but not polished to white metal condition.
6. The experience, ability, and cooperation of the workmen on the job are the most important factors in a successful in-place lining operation. They are the ones who estimate the water–cement ratio, monitor the mortar consistency, set the rate of traveling of the lining machine, set the mortar pumping pressure, set the rate of mortar application, and adjust the mechanical trowels for specified lining thickness.
7. Proper curing is essential for the production of high quality cement-mortar linings. If water curing process is used, electric gages with remote recording device should be installed inside the pipe to record the temperature and humidity inside the pipe. If mechanical gages were used, the cover at the end of the pipe would have to be removed every time the gages are read. Uncovering the pipe will lead to a sudden decrease of the humidity inside the pipe and is harmful to the cement-mortar lining being cured; inaccurate readings will also result.
8. Lining thickness should be checked by using nondestructive testing instruments such as the eddy-current thickness gage, not by drilling a hole through the cement-mortar lining.
9. Hydrostatic test should be performed after the lining is installed and cured properly.

Cement Mortar

Cement mortar for in-place lining is composed of cement, sand, pozzolanic materials, admixtures, and water. The materials should conform to the following American Society for Testing and Materials (ASTM) specifications:

C 144-81 Standard Specification for Aggregate for Masonary Mortar
C 150-84 Standard Specification for Portland Cement
C 494-82 Standard Specification for Chemical Admixtures for Concrete
C 618-84 Standard Specification for Fly Ash and Raw or Calcined Natural Pozzolan for Use as an Admixture in Portland Cement Concrete

Slump test and strength test for cement mortar should conform to the following ASTM standards:

C 143-78 Standard Test Method for Slump of Portland Cement Concrete
C 109-84 Standard Test Method for Compressive Strength of Hydraulic Cement Mortars (Using 2-in. or 50-mm Cube Specimens)

Cement mortar mix used by various organizations has a sand to cement ratio ranging from 1.1 to 3 by weight and a pozzolanic material to cement ratio ranging from 0.15 to 0.5 by weight. Water to cement ratio is in the range of 0.3 to 0.4 by weight. The amount of retarding admixture in the mortar is usually less than 0.1 percent; admixture to aid pumping efficiency is less than 0.01 percent. Table 8 shows different mix designs used for various projects.

The selection of optimum slump for cement mortar mix mainly depends on the pipe diameter and required workability. Cement mortar applied to larger pipes needs a lower slump mix design so that the centrifugally applied cement mortar will adhere to a flatter surface. In smaller pipes, mortar adhesion is not critical and higher slump is used to provide a better workability for troweling operation. The suggested limits of slump of cement mortar, using a 12 in. (305 mm) cone and conforming to ASTM C 143, are shown in Figure 8 and the relation between slump and water–cement ratio is shown in Figure 9.

Access Opening

To gain access opening for in-place lining operation, a trench of adequate dimensions is excavated, pipes are cut

TABLE 8. Cement-Mortar Mix Design.*

Cement	Sand	Pozzolanic Material
1	1.1	—
1	1.5	0.2
1	1.5	0.5
1	1.7	0.15
1	1.9	0.25
1	2	—
1	3	—

*Proportion by weight.

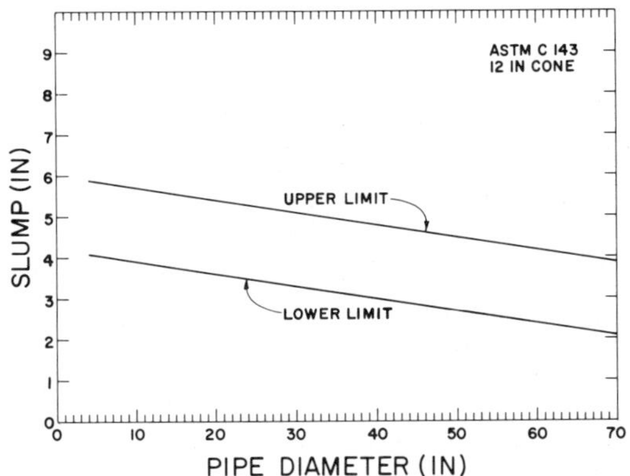

FIGURE 8. Slump limits of cement mortar.

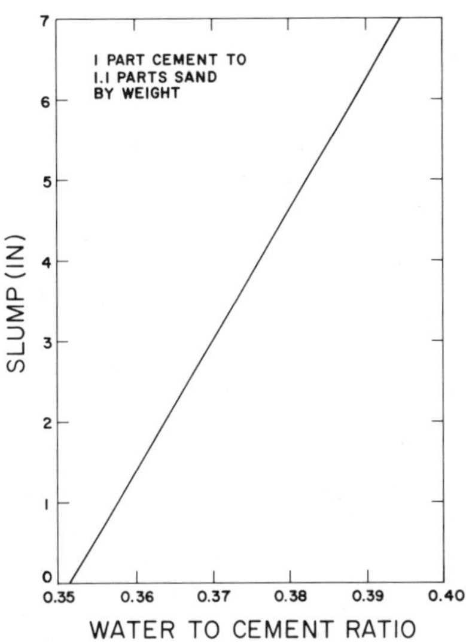

FIGURE 9. Water to cement ratio vs. cement-mortar slump.

(see Figure 10) and, after lining, the closure piece is welded back and trench backfilled. Care should be taken during excavation to avoid equipment impact on the pipe and the degree of compaction of the backfill should be approximately equal to the original condition. The distance between access openings may be over 1,000 ft (300 m) and the entire pipeline between access openings can be lined in one continuous operation. Manholes and air valves are sometimes used to feed cement mortar to the lining machine. For large diameter steel pipes, only a half shell or smaller opening is necessary.

Pipe Cleaning

Cleaning the interior surface of old iron and steel pipelines is an important step before the in-place application of cement-mortar lining. It is also often needed for relatively new iron and steel pipes. Lining operation should follow as soon as practicable after cleaning.

For pipelines of 42 in. (1.07 m) in diameter and smaller, the use of a steel drag scraper followed by a rubber "squee-gee" seems to be very effective [7]. As shown in Figure 11, a steel drag scraper consists of serrated steel blades mounted on a cylindrical chassis, in three or four rows. The drag cables are attached to the front and rear bolts. The steel drag scraper is pulled through the pipeline forwards and backwards until the incrustation is removed from the pipe wall. Up to 600-ft (182-m) length of pipeline can be cleaned each time. Figure 11 also shows the rubber "squee-gee" which is used to remove all the dislodged debris. Figures 12 and 13 show the pipe interior surface conditions before and after cleaning operation.

Other generally used cleaning methods are hydraulic cleaning, mechanical cleaning, boring machine and flail, brush cleaning, and hand cleaning [14]. Less used cleaning methods are acid cleaning [14], solvent cleaning, flame cleaning, and sand blast [34].

FIGURE 10. Access opening.

FIGURE 11. Steel drag scraper and rubber "squee-gee."

FIGURE 12. Pipe interior surface condition before cleaning.

FIGURE 13. Pipe interior surface condition after cleaning.

FIGURE 14. Cement-mortar lining after troweling.

Lining Operation

As shown in Figures 4, 5, 6, and 7, a short piece of open pipe is placed next to the cut end of the pipeline to guide the lining machine and adjust the wheels and trowels. The quality of mortar is examined at the mixer, and then the hose is connected to the lining machine. Mortar is then pumped through the hose and dumped until a desirable consistency is obtained at the lining machine. The lining machine is pulled through the pipeline to the far end. The machine travels backwards during the entire lining operation (see Figure 6) so that the rubber hose and electrical cable would not interfere with the freshly placed mortar lining.

When the mortar reaches the lining machine via hose, it is centrifugally applied to the interior surface of the pipe by a rapidly revolving head. The mortar is thrown in a tangential direction, and adheres to the wall upon impact. At this stage, the lining has an unsmooth surface similar to that of an orange peel (see Figure 7).

Immediately following the application of the mortar to the pipe wall, the troweling operation is conducted with rotating steel trowels for larger pipes and conical-shaped drag trowels for pipes smaller than 20 in. (508 mm) in diameter. After troweling, the surface of the lining looks similar to the rifling in a gun barrel (see Figure 14), but is actually quite smooth (see Figure 7).

At areas where machine lining is not practical, the lining should be applied by hand or shortcrete [50]. Wire reinforcement is needed for vertical pipes [34,43].

Closure Piece

Closure pieces are cleaned, lined, and cured separately (see Figures 15, 16, 17, 18) and then joined or welded to the main portions of the pipeline. The methods for cleaning, lining, and curing the closure pieces may be different from those used on a long pipeline. Handwork may be needed and curing compound may be used without the undesirable side effects when used in a long pipeline.

Curing

Curing is the most important factor in the entire lining operation. This is required to control shrinkage cracks in the cement rich mortar. Moist curing (water curing) is simple to conduct, with no undesirable side effects. Compound curing has limited usage since it contains volatile material and, as a result, special protective equipment may be needed for the inspector. Also, inspection of compound cured lining is difficult due to the fact that any cracks in the lining may be concealed under the layer of the curing compound. Steam curing is four times as efficient as moist curing, but seldom used in the field because the generation of steam is not a simple task in the field. Shop-applied cement-mortar lining is often cured in a steam chamber.

Immediately upon completion of the lining of a section of pipe between access openings, or of a day's run of the machine, curing should be initiated. The starting end of the pipe should be covered with plastic sheet (see Figure 10) or airtight end cap no later than 30 minutes after the beginning of the lining operation to maintain moisture in the lining. Hand troweling and other finish work also should be completed within 30 minutes.

It generally takes 30 minutes for the cement mortar to attain initial set. During the first 30 minutes, the temperature of the lining may rise considerably and excess water may appear on the surface of the lining and accumulate at the bottom of the pipe. The pipe end should not be covered during the initial set period; but, it should be covered soon after the initial set. The airtight cover should be kept in place throughout and beyond the specified curing period. When additional moisture is required to maintain a moist condition, water should be introduced inside the pipe approximately one hour after lining or 30 minutes after the initial set so that the lining is not damaged. Usually over 90 percent relative humidity can be maintained by using sandbags to pond water in the pipe.

The exterior surfaces of pipe exposed to sunlight should be covered with burlap and sprinkled with water in the daytime during the period of lining, finishing, and curing to prevent undesirably high temperature in the pipe.

The suggested curing period is 8 days [34], or 7 days [22,33,35,42]. The relative humidity should be kept above 90 percent [22] and the temperature inside the pipe should be kept above 50°F (10°C) [22].

Lining Thickness

The lining thickness is determined by pumping pressure, travel speed of the lining machine, and adjustment of trowel arms, blades, or cone. It is possible to drill a small hole through the cement-mortar lining, measure the thickness, and, later, fill the hole with cement mortar. This would be time-consuming and would damage the lining to some degree. The working environment in a pipe is congested, humid, and hot. An accurate and fast way of measuring the lining thickness for the many thousand feet of pipeline is desired. A better way is to use nondestructive testing instruments.

All thickness measurements are based on either the ultrasonic or eddy-current method. Ultrasonic gages, operating on wave reflection theory, are not suitable for underground application. No free boundary can be defined for buried pipes. Reflection and refraction of ultrasonic waves at the lining–pipe and pipe–soil interfaces will hamper an accurate measurement of the thickness of the lining. The density, water content, and even material composition of the surrounding soils are far from uniform. Also, nearby pipes and other objects may affect the results [48]. Eddy-current method for thickness determination is based on the principle

FIGURE 15. Closure piece before cleaning.

FIGURE 16. Closure piece after cleaning.

FIGURE 17. Lined closure piece with curing compound.

FIGURE 18. Lined closure piece without curing compound.

that an alternating magnetic field adjacent to a metallic body induces a measurable electric current (eddy current) within the metal. The strength of the eddy current is related to the distance from the probe to the surface of the metallic body [44,48]. It is a relatively simple operation to determine the thickness of the cement-mortar lining of a steel or iron pipe. The pipe is a good metallic object and the lining and the surrounding soils have practically no effect on the accuracy of measurement due to their low conductivity compared to that of the pipe. Another fact is that the exterior wrapping and coating of the pipe are good insulators, and they will add stability to the thickness measurement. An eddy-current thickness gage with direct digital readout on thickness is easy to use and very accurate, each measurement taking only a few seconds. The range of measurement is between 0.0025 mm and 50 mm.

The generally accepted procedure of lining thickness measurement calls for 10 measurements to be made randomly along each 100 ft (30.5 m) of pipe and 8 out of the 10 measurements should be within the specified tolerance [22,48].

Defects and Repairs

Before initial set of the mortar, only limited inspection is possible in large pipes and nearly impossible in smaller pipes. But any obvious defects such as falling of the lining should be rejected, the lining surrounding the defective area should be removed, and the pipe is relined.

After initial set, defects in lining are easier to detect. Video camera may be used in smaller pipes and visual inspection may be conducted in pipes of 24 in. (610 mm) in diameter and larger. Defects include excessive cracks, voids, sand pockets, thin spots, and loose areas [22,41,42, 43]. Defective lining should be removed to the pipe wall; the edges of the lining not removed are perpendicular or slightly undercut. The cut-out area and adjoining lining should be thoroughly wetted, and the cement mortar applied and troweled smooth with the adjoining lining [22,41]. Small cut-out areas may be patched by hand, and large areas may be relined by machine. Repaired areas should be cured properly.

Excessive cracks having a width greater than 0.02 in. (0.5 mm) should be repaired by the above procedure. Cracks having a width between 0.001 and 0.02 in. (0.025 and 0.5 mm), if concentrated in an area, should be repaired by brushing of cement paste [22]. Cracks smaller than 0.001 in. (0.025 mm) are barely detectable by the naked eye and need no repair. Circumferential cracks are not as serious as longitudinal cracks [41].

JOINTS IN CEMENT-MORTAR LINED PIPES

It is essential that joints in cement-mortar lined pipes be completely sealed inside; otherwise, the lining's isolation and protection functions would be lost. Joints are needed for:

1. Closure piece at access opening
2. Replacement of damaged pipe section
3. Connecting shop-lined pipes
4. Branch and fitting connections

These joints are either welded or mechanically jointed. Straight pipes are cut by using power pipe cutter, and branch connection openings are cut by torch.

For pipes with welded joints and of 24 in. (610 mm) in diameter and larger, the joints can be finished by hand troweling; for smaller pipes, special details such as welded sleeve joint are needed.

Mechanically jointed pipes may have screwed joints, welded slip-on flange joints, Dresser-style coupling, etc. Cement joint compound, plastic joint compound, or asbestos gasket are used with mechanical joints. Suppliers should be consulted for details.

See References [21,34,35,38,40,43] for more information on both welded joints and mechanical joints.

AUTOGENOUS HEALING OF CRACKS

It has been well established that cement-mortar will expand and contract on wetting and drying. Prolonged exposure to drying conditions can result in cracking of cement-mortar lining in iron and steel pipes. Once in service, the lining will absorb water and expand, resulting in tightening of the lining and closing of the cracks [12,53]. During the autogenous healing process, small cracks become undetectable by the naked eye and large cracks are filled with white crystalline material. Chemical analysis showed that the white crystalline material was calcium carbonate [53]. Strength test showed that most of the healed specimens were stronger than the equivalent specimens without cracks, the recovery in strength was greater for the specimens with higher cement contents, and the rate of healing decreased with age [12].

CORROSION AND INCRUSTATION

Corrosion tubercles and incrustation in an iron or steel pipe form a complex environment in which electro-chemical, and microbiological interactions are occurring concomitantly. This environment offers a good place for the growth of aerobic, anaerobic, and facultative bacteria [52]. Corrosion and incrustation are two different, but related processes.

Corrosion arises in iron or steel pipes where oxygen is ready accessible. The major forms of corrosion in a water pipe are uniform attack, grooving, pitting, and tuberculation [19]. The stages in the development of corrosion in cast iron pipes [37] and steel pipe [49] are very complicated. Additionally, corroded iron pipes do not exhibit the superficial rust appearance of steel. Iron oxides leached from the graphite-iron matrix will weaken the pipe wall [39].

Incrustation is a form of iron bacteria. Iron bacteria are microorganisms which feed on the iron and manganese in solution [27]. Chemical analysis showed that incrustation specimens are composed predominantly of iron, manganese, and silica [4,27]. Scanning electron microscope examination showed that the microorganism responsible for the incrustation is iron bacteria species such as Gallionella ferruginea or Sphaerotilus leptothrix group [27,52].

It is not easy to single out either corrosion or incrustation for treatment. Many processes such as scraping, pigging, sphering, chemicals, and interior lining were designed to treat both of them [4,8]. The chemicals, including corrosion inhibitors, oxygen scavengers, and bactericides, have been found not effective [4]. Chlorination was not only ineffective [27]; it even increased the buildup of incrustation in steel pipes [4]. Cement-mortar lining provides an alkaline environment and effectively isolates and protects iron and steel pipes from corrosion and incrustation.

CONCLUSIONS

Application of cement-mortar lining to the interior surface of iron and steel pipes is a proven way of preventing corrosion and incrustation in these pipes. The integration of cement-mortar lining with iron or steel pipe wall offers many structural, hydraulic, and economical advantages. The in-place lining procedure is flexible and suited for many situations; once clearly understood, its usage is almost limitless.

Readers are directed to the references for further information such as the occurrence of microorganism [2], using of camera for inspection [6], other lining materials [13], pipe cleaning devices [18,24], and seismic analysis of continuous buried pipelines [47].

REFERENCES

1. Adrian, G. W., "Steel Pipe Design for Second Los Angles Aqueduct," *Journal of the Pipeline Division*, Proceedings of the ASCE, Vol. 93, No. PL3, pp. 33–43 (November 1967).
2. Allen, M. J., R. H. Taylor and E. E. Geldreich, "The Occurrence of Microorganisms in Water Main Encrustations," *Journal of the American Water Works Association*, Vol. 72, No. 11, pp. 614–625 (November 1980).
3. Aroni, S. and G. L. Fletcher, "Observations on Mortar Lining of Steel Pipelines," *Transportation Engineering Journal*, ASCE, Vol. 105, No. TE6, pp. 667–681 (November 1979).
4. Bain, W. S., R. E. Taylor and C. F. Bowman, "Study of Corrosion in Carbon Steel Raw Water Piping Systems," Report No. SP304, Tennessee Valley Authority (1979).
5. Burnett, G. E. and P. W. Lewis, "New Developments in Tests of Coatings and Wrappings," *Journal of the American Water Works Association*, Vol. 48, No. 2, pp. 100–120 (February 1956).
6. Clarke, F., "Camera Pig Checks Weldments, Internal Coating Offshore," *Pipeline & Gas Journal*, pp. 24–25 (March 1981).
7. "Cleaning Procedure for Surface Preparation," RIB Specification C1905A-003, Raymond International Builders, Inc. (1982).
8. "Control of Internal Corrosion in Steel Pipelines and Piping Systems," NACE Standard RP-01-75, Recommended Practice, National Association of Corrosion Engineers (1975).
9. Cornet, I., "Protection with Mortar Coatings," *Materials Protection*, Vol. 6, No. 3, pp. 56–58 (March 1967).
10. "Curing Procedure for Applied Lining," RIB Specification C1905A-005, Raymond International Builders, Inc. (1982).
11. Dahl, R., "Cleaning and Lining Small-Diameter Pipe," *Journal of the American Water Works Association*, Vol. 66, No. 8, pp. 483–484 (1974).
12. Dhir, R. K., C. M. Sangha and J. G. L. Munday, "Strength and Deformation Properties of Autogenously Healed Mortars," *ACI Journal*, pp. 231–236 (March 1973).
13. Frye, S. C., "Epoxy Lining for Steel Water Pipe," *Journal of the American Water Works Association*, Vol. 66, No. 8, pp. 498–501 (August 1974).
14. Goulding, H., "Concrete Lined Pipes," *Proceedings, First International Conference on the Internal and External Protection of Pipes*, University of Durham, Bedford, England, pp. C3–23 to C3–36 (September 9–11, 1975).
15. Hand, F. R., C. N. Sun and J. M. Hoskins, "Full Scale Testing and Seismic Qualification of Cement Mortar Lined Carbon Steel Pipe," *Eighth World Conference on Earthquake Engineering*, San Francisco, California, pp. VII 279–286 (July 21–28, 1984).
16. "Handbook, Ductile Iron-Cast Iron Pipe," Cast Iron Pipe Research Association (1978).
17. "Handbook for Sewer System Evaluation and Rehabilitation," EPA 430/9-75-021, U.S. Environmental Protection Agency, Water Programs Operations, Government Service Administration (December 1975).
18. Hardy, W. R., "Pigs in the Water Pipe," *Journal of the American Water Works Association*, pp. 909–914 (August 1968).
19. Holler, A. C., "Corrosion of Water Pipes," *Journal of the American Water Works Association*, Vol. 66, No. 8, pp. 456–457 (August 1974).
20. Hornshaw, J. E. and L. Hart, "Tests of Centrilining by the City of Detroit," City of Detroit Internal Report (1940).
21. Hubble, J. D., W. W. Avril and C. F. Bowman, "Cement Mortar Lining of the Essential Raw Cooling Water System," Bellefonte Nuclear Plant Construction Specification No. N4M-924, Tennessee Valley Authority (1983).
22. Hubble, J. D., C. F. Bowman and W. W. Avril, "Cement Mortar Lining of Piping for Safety-Related Carbon Steel Piping," Standard Specification No. MEB-SS-3.14, Tennessee Valley Authority (1982).
23. "Internal Corrosion Control," Paragraph 462, ANSI B31, Case 108, the American Society of Mechanical Engineers, pp. 129–131 (June 1972).

24. Jones, B. G., "Pigs Clean Up Pipeline," *Water & Wastes Engineering*, pp. 51–56 (October 1978).
25. "Maintenance and Repair: Control of Mineral Deposit (Scale) Buildup in Water Pipes," Technical Note No. 77-12, Department of the Army (June 7, 1977).
26. Miller, W. T., "Durability of Cement-Mortar Lining in Cast-Iron Pipe," *Journal of the American Water Works Association*, Vol. 57, No. 6, pp. 773–782 (June 1965).
27. Moore, M. O., "Incrustation in Water Pipelines," *Journal of the Pipeline Division*, ASCE, Vol. 94, PL1, pp. 37–47 (October 1968).
28. "Mortar Lining and Placement Procedure," RIB Specification C1905A-004, Raymond International Builders, Inc. (1982).
29. O'Keefe, W., "Corrosion-Resistant Piping for Utility and Industrial Powerplants," *Power*, Vol. 125, No. 4, pp. S1-S24 (April 1981).
30. Ouellette, H. and B. J. Schrock, "Rehabilitation of Sanitary Sewer Pipelines," *Transportation Engineering Journal*, ASCE, Vol. 107, No. TE4, pp. 497–513 (July 1981).
31. Padley, T. J., "Protection of Ductile Iron Pipe by Cementitious Coatings," *Proceedings, First International Conference on the Internal and External Protection of Pipes*, University of Durham, Bedford, England, pp. C2-13 to C2-22 (September 9–11, 1975).
32. "Pipe, Steel (Cement-Mortar Lining and Reinforced Cement-Mortar Coating)," Fedearl Specification SS-P-385a (1964).
33. "Pipe, Steel (Coal Tar Enamel or Cement-Mortar Lining, and Coal Tar Enamel Coated and Wrapper)," Federal Specification WW-P-1432B (1980).
34. "Recommended Practice for Application of Cement Lining to Steel Tubular Goods, Handling, Installation, and Joining," API RP 10E, American Petroleum Institute (March 1978).
35. "Rehabilitation of Crystal Springs Pipeline No. 1 from Randolph Avenue, South San Francisco to University Mound Reservoir, San Francisco," City and County of San Francisco Technical Specification (1981).
36. "Repair Procedure for Nonconforming Lining," RIB Specification C1905A-006, Raymond International Builders, Inc. (1982).
37. Smith, D. C. and B. McEnaney, "The Influence of Dissolved Oxygen Concentration on the Corrosion of Grey Cast Iron in Water at 50°C," *Corrosion Science*, Vol. 19, No. 6, pp. 379–394 (1979).
38. Smith, L. C., C. F. Bowman and H. R. Corbett, "Cement Lining of the Essential Raw Cooling Water System," Watts Bar Nuclear Plant Construction Specification No. N3M-921, Tennessee Valley Authority (1982).
39. Speed, H. D. M. and M. J. Rouse, "Renovation of Water Mains and Sewers," *Journal of the Institution of Water Engineers and Scientists*, Vol. 34, No. 5, pp. 401–424 (September 1980).
40. "Standard Contract Specifications for In-Place Cleaning and Cement-Mortar Lining of Water Mains," Ameron Pipe Lining Division (1980).
41. "Standard for Cement-Mortar Lining for Ductile-Iron and Gray-Iron Pipe and Fittings for Water," ANSI/AWWA C104/A21.4-80, American Water Works Association (1980).
42. "Standard for Cement-Mortar Lining of Water Pipelines—4 Inches and Larger—In Place," AWWA C602-76, American Water Works Association (1976).
43. "Standard for Cement-Mortar Protective Lining and Coating for Steel Water Pipe—4 Inches and Larger—Shop Applied," AWWA C205-80, American Water Works Association (1980).
44. Sun, C. N., J. M. Hoskins and F. R. Hand, "Full Scale Testing and Qualification of Cement-Mortar Lined Carbon Steel Pipe," TVA Report No. CEB 82-8, Tennessee Valley Authority (April 1982).
45. Sun, C. N., J. M. Hoskins and R. J. Hunt, "Full Scale Testing of Cement-Mortar Lined Carbon Steel Pipe," Preprint No. 84-015, ASCE 1984 Spring Convention, Atlanta, Georgia (May 14–18, 1984).
46. Sun, C. N., J. M. Hoskins and R. J. Hunt, "Testing of Cement-Mortar Lined Carbon Steel Pipes," *Journal of Transportation Engineering*, Vol. III, No. 1, pp. 17–32 (January 1985).
47. Sun, C. N. and R. J. Hunt, "Stress Intensification Factor and Flexibility Factor in Seismic Analysis of Continuous Buried Pipelines," *International Conference on Advances in Underground Pipeline Engineering*, Madison, Wisconsin, pp. 206–215 (August 27–29, 1985).
48. Sun, C. N., "Pipe Lining Thickness and Thickness Gages," *Journal of Transportation Engineering*, ASCE, Vol. 110, No. 4, pp. 447–459 (July 1984).
49. Suss, H., "Localized Corrosion Attack on Carbon Steel—Case Histories of Service Failures," ASTM STP 576, *Galvanic & Pitting Corrosion-Field and Laboratory Studies*, American Society for Testing and Materials, pp. 117–131 (1976).
50. "Technical Specification for Cement-Mortar Lining Circulating Water Pipe," Specification No. 9645-C-062.1, Bechtel Power Corporation (1976).
51. "The Test Results of Steel Pipe Mortar Lining by Field Procession Method," Nippon Steel Corporation Report (1980).
52. Tuovinen, O. H., et al., "Bacterial, Chemical, and Mineralogical Characteristics of Turbercles in Distribution Pipelines," *Journal of the American Water Works Association*, Vol. 72, No. 11, pp. 626–635 (November 1980).
53. Wagner, E. F., "Autogenous Healing of Cracks in Cement-Mortar Linings for Gray-Iron and Ductile-Iron Water Pipe," *Journal of the American Water Works Association*, pp. 358–360 (June 1974).
54. Wolfe, J. E., "Cement Mortar Lining of Large Diameter Water Pipelines," *Journal of the Pipeline Division*, ASCE, Vol. 94, No. PL1, pp. 81–88 (October 1968).
55. Wolfe, J. E., "Cement Mortar Lining of 20-Ft Diameter Steel Pipe," *Transportation Engineering Journal*, ASCE, Vol. 98, No. TE4, pp. 835–845 (November 1972).

CHAPTER 16

Soil Mechanics Principles in Underground Mining

G. E. BLIGHT*

INTRODUCTION

As world consumption of minerals and fossil fuels continues to expand, it becomes necessary to exploit the earth's resources more efficiently and completely. This means that ore bodies are being mined today that would not have been considered mineable a few years ago. Old mines are being remined to extract ores abandoned in earlier times. Ore bodies are being mined at ever-greater depths. Problems with underground water tend to become more severe.

In addition to this, some mining practices that used to be acceptable are no longer so considered, as a result of the adverse environmental effect they have at surface.

The civil engineer has always had a place in mining, but this role is now rapidly becoming more essential. In particular, the geotechnically trained civil engineer is becoming ever more important in mining. On the surface, geotechnical engineers are heavily engaged in the engineering of mine waste disposal systems and of open cast mining systems. Their activities are directed at making these systems safe, cost effective and environmentally acceptable [1,2]. This aspect of geotechnical engineering activity is fairly well known, but less familiar are the role and capabilities of the geotechnical engineer to assist and facilitate underground mining. This chapter will illustrate, by means of examples, some of the activities for which the geotechnically trained civil engineer is uniquely suited.

Examples that will be cited include the following:

1. The design and properties of fill materials for the support of underground excavations and the prevention of surface settlement
2. Stress distributions within fills contained by underground excavations
3. The design of water retaining underground plugs
4. The design and behavior of steel-reinforced walls or pillars of fill material used to provide roof support

*Department of Civil Engineering (Construction Materials), University of the Witwatersrand, Johannesburg, 2000, Republic of South Africa

USES OF FILL IN UNDERGROUND MINING

The backfilling of mining excavations to achieve various objectives has long been practised in the mining industry. Recently, there has been an intensification of research into the use of fill in mining. Waste or tailings, placed as an underground fill:

1. Provides support to the excavation and makes the mining operation safer
2. Reduces the convergence of excavations, thus reducing the release of gravitational potential energy from the superincumbent strata
3. Can be used to provide access for mining wide or high ore-bodies
4. Provides lateral support to rock pillars thus enhancing their load-carrying capacities
5. Reduces the fire hazard associated with using conventional timber supports underground
6. Can be used to direct and channel ventilation air, thus reducing air losses and heat gained by the air from the country rock
7. Reduces surface subsidence in the case of shallow mining, and stabilizes the surface in the case of long-abandoned shallow workings.

In addition to these purely functional uses, the disposal of tailings in the form of underground fill reduces (though only slightly) the pressures on the surface environment [1,2] by

1. Releasing area at the surface for more productive use than the disposal of tailings
2. Reducing problems of air, water, and visual pollution, both in the short and the long term

PROPERTIES REQUIRED OF PUMPED TAILINGS FILL

The following are the properties required of a successful pumped tailings fill:

1. The tailings should become pumpable at the lowest possible water content. Water supply is often a major problem in mining. In addition, much of the water used to place a fill eventually drains out of it and must be handled within the mine. However, the most important reason for seeking pumpability at a low water content is that the lower the water content at which the slurry is placed, the less compressible the fill subsequently.
2. During placing, the fill should settle into as dense a state as possible and after that should be as incompressible as possible under applied stress.
3. The fill should be capable of resisting the effects of shock loading produced by blasting or seismic events.

When the relative merits of a number of possible fill materials are considered, an attempt should be made to maximize the above factors, all of which, as it happens, are compatible with one another.

FILLS UNDER STRESS

When fills are subjected to stress, their behavior is governed by the principle of effective stress, which applies to all particulate materials [3]. This principle states that changes in volume or in strength of a particulate material are governed by changes in effective stress, and that volume and strength can change only if the effective stress in the material changes. The effective stress is defined by the equation

$$\sigma' = \sigma - u \tag{1}$$

in which σ is the total or applied stress in the material, u is the pressure in the water contained by the material (the porewater pressure), and σ' is the effective stress.

Changes in the volume of a fill are governed by the equation

$$\Delta\epsilon_v = -C\Delta\sigma' \tag{2}$$

in which $\Delta\epsilon_v$ is the change of volumetric strain corresponding to a change in effective stress of $\Delta\sigma'$, and C is the compressibility of the fill material. Changes of shear strength are controlled by the equation

$$\Delta\tau = \tan\phi'\Delta\sigma' \tag{3}$$

where $\Delta\tau$ is the change in shear strength corresponding to a change in the effective stress of $\Delta\sigma'$, and $\tan\phi'$ is the angle of shearing resistance of the fill material. Integrating Equation (3) yields the shear strength equation

$$\tau = c' + \tan\phi'\sigma' \tag{3a}$$

in which c' is the shear strength when $\sigma' = 0$.

Changes of volume, shear strength, and water content are inextricably linked in that the shear strength cannot change without a change of volume, the volume cannot change unless water is expelled from the fill, and the compressibility depends on the shear strength (as compression usually involves internal shearing).

The compressibility C and the angle of shearing resistance ϕ' depend on the density at which the fill is placed and increase with increasing placed density. As the dry density γ_d of a saturated fill is related to the water content w by

$$\gamma_d = \frac{G_s\gamma_w}{1 + wG_s} \tag{4}$$

in which G_s is the particle specific mass of the fill and γ_w the density of water), low water contents are associated with high dry densities and vice versa.

An increment of total stress applied to a fill will result in the expulsion of water via a time-dependent process of consolidation, which is governed by the differential equation

$$\frac{\delta u}{\delta t} = c_v \frac{\delta^2 u}{\delta z^2} \tag{5}$$

In this equation, t is the time measured from the instant of application of a stress increment, and z is a distance related to the drainage path or distance through which water must percolate in order to escape from the fill; c_v, the coefficient of consolidation, depends on both the permeability and the compressibility of the consolidating material.

The interchange between pore pressure, total stress, and effective stress according to Equation (5) is shown schematically in Figure 1: when the increment of total stress is imposed on a laterally confined fill, it is entirely transferred to the pore water, i.e., at time $t = 0$, $\Delta u = \Delta\sigma$ and $\Delta\sigma' = 0$.

The pore water immediately starts to drain from the fill, and, as it does so, stress is transferred from the pore water to the solids in the fill, i.e., at time $t > 0$, $\Delta u + \Delta\sigma' = \Delta\sigma$.

Finally, after the elapse of a period of time that is related to the dimensions and drainage conditions of the fill (z) and

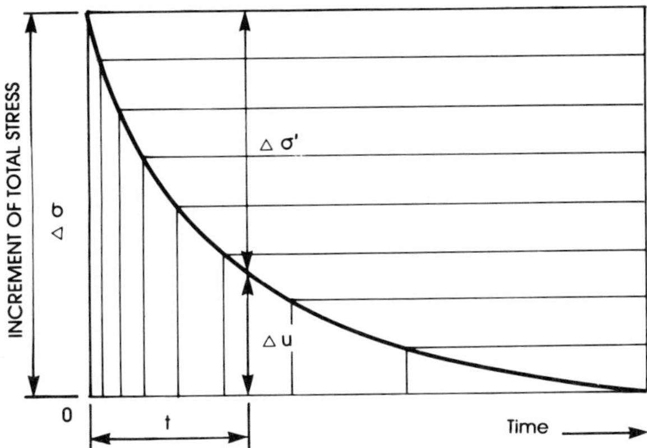

FIGURE 1. Interchange between total stress, pore water pressure, and effective stress during consolidation of a fill.

the coefficient of consolidation c_v, the entire total stress incremenent is carried by the solids in the fill, i.e., at $t \to \infty$, $\Delta\sigma' = \Delta\sigma$ and $\Delta u = 0$.

Details of the consolidation process described by Equation (5) are given in all standard soil mechanics texts (e.g., [3]) and will not be repeated here.

GRADINGS FOR MAXIMUM DENSITY

The placement density, and hence the compressibility and shear strength, of a fill greatly depend on the grading or particle-size distribution of the material. In what follows, guidance will be given as to what constitutes a satisfactory grading and what factors are most important in ensuring a high placement density.

Grading curves can be idealized to two main types: (1) uniform or Fuller gradings, for which the grading curve (the relationship between the logarithm of the particle size and the percentage finer than that size) is an inclined straight line; and (2) gap gradings, which consist of a series of inclined straight lines joined by horizontal size "gaps" in which there are no particles present.

Uniform Gradings

Most tailings materials include a relatively narrow range of particle sizes (see, for example, Figure 9) and can be approximated by a straightline grading curve. The two main variables in a straight-line grading curve are the slope (or ratio of maximum to minimum particle size) and position (size of the largest particle present).

Figure 2a shows a series of straight-line or Fuller gradings in which the ratio d_1/d_0 of the largest to the smallest particle size present ranges from 8 to 128 (gradings F0 to F4). Figure 2b shows the effect of these gradings on the settled densities of aggregates made up from two different source materials. The same compactive effort was used for each material. Figure 3a shows a series of parallel straight-line gradings F0, F5, F6, and F7, and Figure 3b compares the settled densities of aggregates having these gradings. It is

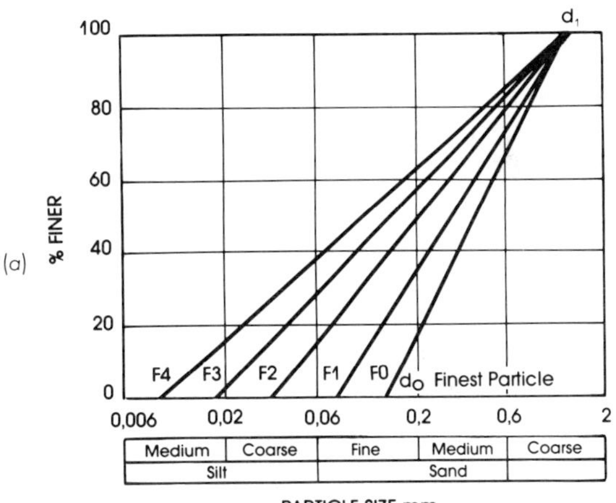

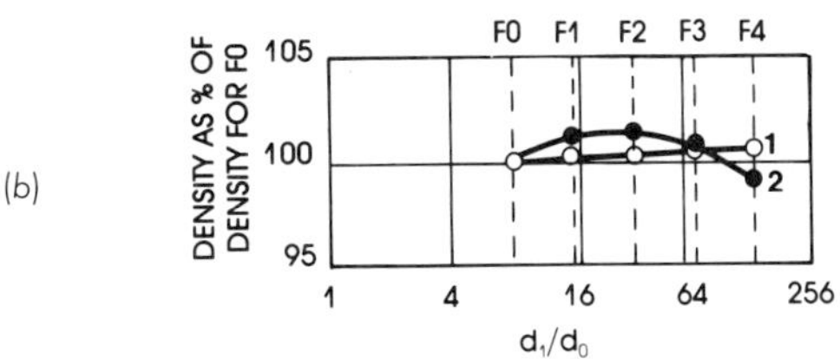

FIGURE 2. (a) Series of continuous gradings F0, F1, F2, F3, F4; (b) Settled densities of two materials made up in gradings F0 to F4.

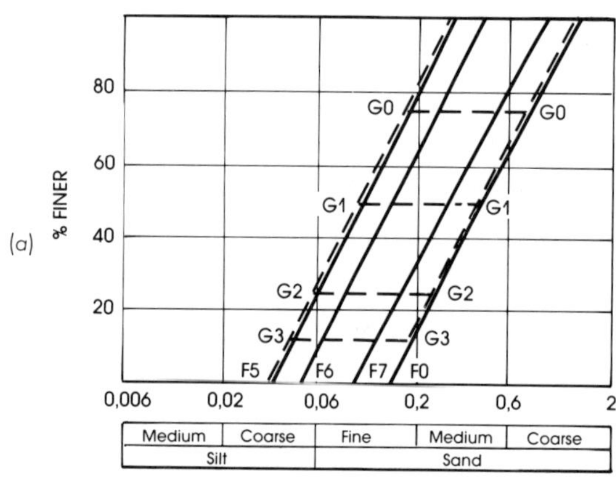

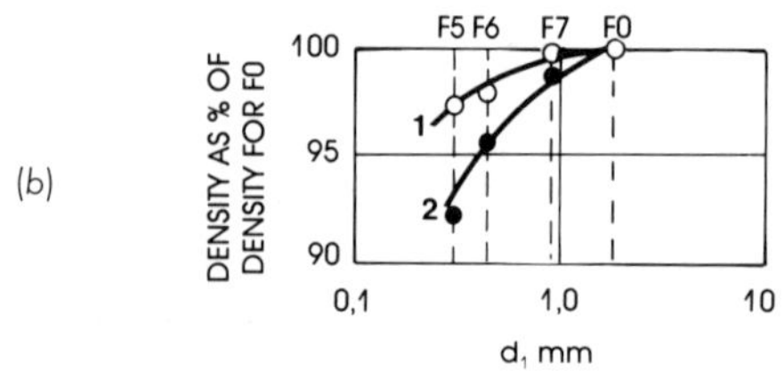

FIGURE 3. (a) Series of continuous gradings F0, F5, F6, F7 and gap gradings G0, G1, G2, G3; (b) Settled densities of two materials made up in gradings F0, F5, F6, F7.

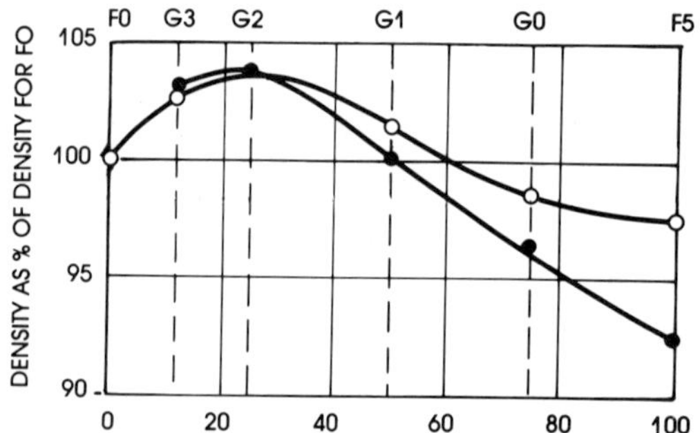

FIGURE 4. Settled densities of two materials made up in gradings F0, G0 to G3, and F5.

clear from Figures 2 and 3 that the slope of a uniform grading curve has a relatively minor effect on the settled density but that the position of the curve has a greater effect. For the same slope of grading curve, the density may decrease by up to 8 percent when the maximum particle size is decreased by a factor of 4.

Gap Gradings

Figure 3a also shows a series of gap gradings G0 to G3 in which the gap bridges between the uniform gradings F0 and F5. The settled densities corresponding to the various gap gradings are shown in Figure 4 (for the same two basic source materials). A comparison of Figure 4 with Figures 2b and 3b shows that a far greater range of densities is obtainable via a gap grading than via a uniform grading. For example, the addition of about 20 percent fines to a material with grading F0 can cause the settled density to increase by 4 percent, whereas the density of the fines alone would be 5 percent less than the density obtainable with grading F0.

The experimental results presented in Figures 2 to 4 do not provide a set of rules for producing the greatest possible settled density, but do provide a framework for experimenting with a combination of available materials to produce a satisfactory high settled density, and hence a fill of low compressibility and high strength. It can be concluded that, in order of increasing importance, the slope of a uniform grading curve, the size of the maximum particle present, and the percentage of fines in a gap grading all have an effect on the attainable settled density of a pumped fill.

WATER CONTENT FOR PUMPABILITY

One of the first requirements in any programme of testing to ascertain the properties of a potential pumped tailings fill is to estimate the minimum water content at which a slurry of the tailings will become pumpable. Once this has been established, all the laboratory specimens for further testing are prepared at that water content.

The pumpability of a slurry depends on its viscosity[1], which can be measured in a variety of ways. One suitable method is by means of a rotating-cylinder apparatus in which the slurry is contained in the annular space between a pair of co-axial cylinders. The outer cylinder is rotated at a fixed rate, and the torque transferred to the inner cylinder is measured to establish the viscosity of the slurry.

The pumpability characteristics of the slurry can then be investigated via the Hagen-Poiseuille equation, which for viscous flow in a circular pipe or a semi-circular open chan-

[1]Tailings slurries are here assumed to act as Newtonian fluids. This is not strictly true, but is a sufficiently accurate first approximation to their true behavior.

nel flowing full has the form

$$v = \frac{\gamma i D^2}{32\mu} \quad (6)$$

in which

v = the velocity of flow
γ = the unit weight of the slurry (weight per unit volume)
i = the flow gradient or the loss of head per unit length of conduit
D = the diameter of the conduit (pipe or semi-circular channel)
μ = the viscosity of the slurry

The rate of transport of dry solids can be found from

$$Q = vA\varrho_{d'} \quad (7)$$

in which

A = cross-sectional area of conduit
ϱ_d = dry specific mass of slurry (mass of dry solids per unit volume of slurry)

It appears that a viscosity of 100 P (poise) or 10 Pa.s represents the upper limit to pumpability, but lower viscosities, and hence higher water contents, may be required, depending on the specific application.

The water content at which a slurry has a viscosity of 100 P can be estimated by means of an adaptation of a simple soil mechanics test, using the liquid-limit apparatus [3]. This consists of a hemispherical cup that can be raised and dropped through a fixed height onto a hard-rubber base by means of a rotating cam. The slurry is placed in the cup, and its surface is levelled and a groove cut in it with a special tool. The water content at which the groove just closes with one blow of the cup on the base corresponds to the water content for a viscosity of 100 P. A typical relationship between viscosity measured by means of the rotating-cylinder test and the number of blows in the liquid-limit test is illustrated in Figure 5.

Alternatively, the Marsh flow cone can be used to estimate pumpability. This consists of a metal cone having an angle of 28° which discharges through a tube of 50mm length and 10mm internal diameter. A stopper is placed in the outlet tube, and with the cone in a vertical position, 1 litre of slurry is poured into it. The stopper is removed and simultaneously, a stop watch is started. The time taken for the cone to empty is the Marsh flow time. Adequate pumpability corresponds to a Marsh flow time of 12 to 15 seconds.

The grading or particle-size analysis of the tailings has a considerable influence on the water content required for the 100 P viscosity. For example, a typical silt-sized slimes requires a water content of 55 percent, whereas a coarser sand requires only 33 percent. The cement content of the slurry also influences the 100 P viscosity water content slightly,

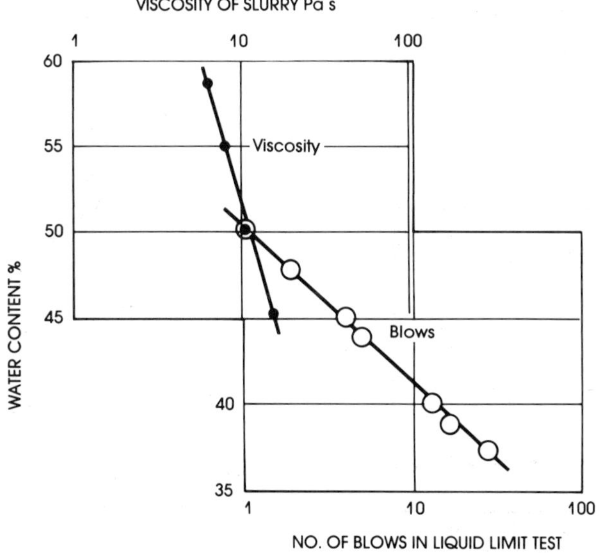

FIGURE 5. Relationship between viscosity, blow count in the liquid-limit apparatus, and water content for a typical tailings slurry.

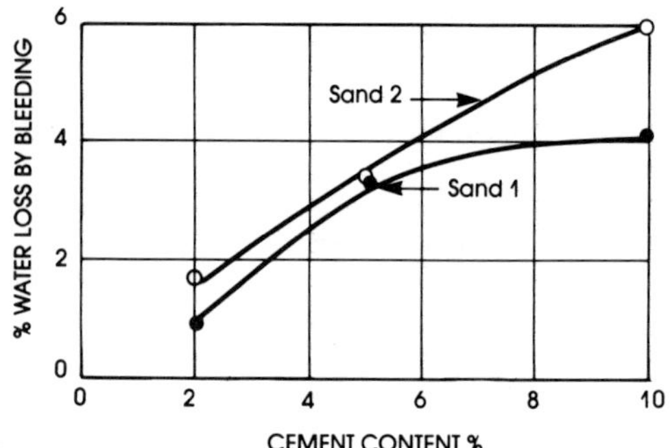

FIGURE 6. Relationship between cement content and water lost by bleeding for two cemented sand slurries.

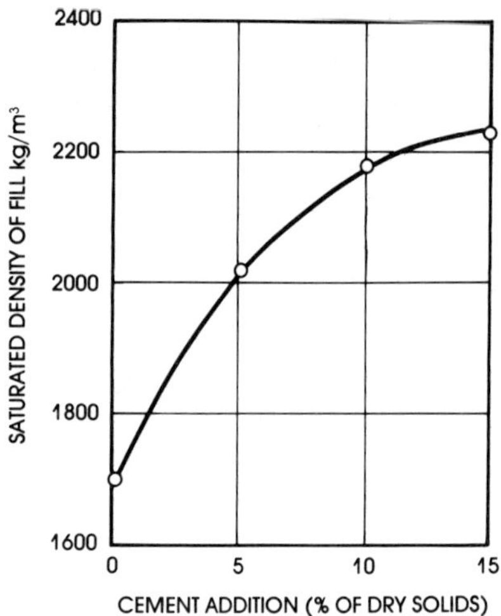

FIGURE 7. Effect of cement addition on settled density for a fill material.

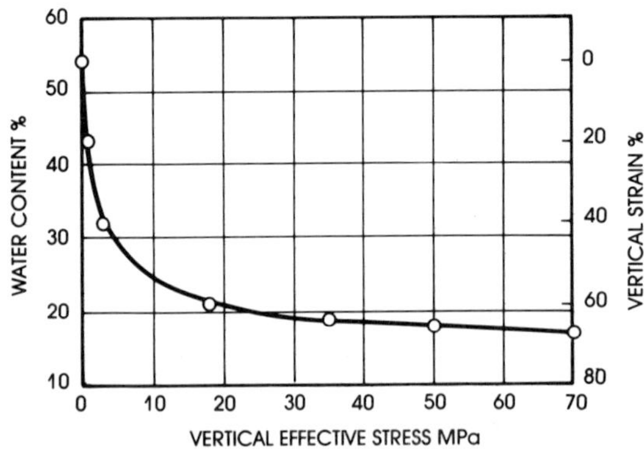

FIGURE 8. Compression curve for gold tailings containing 10 percent cement and showing that most of the compression occurs at low stresses.

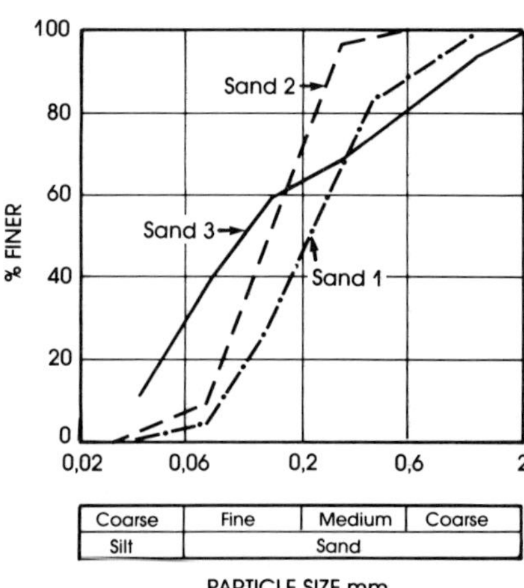

FIGURE 9. Grading curves for three typical tailing sands.

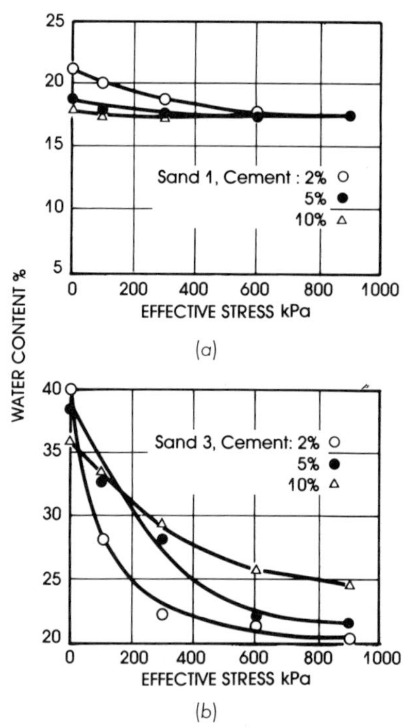

FIGURE 10. (a) Compression of sand 1 and three cement contents; (b) Compression of sand 3 and three cement contents.

e.g., for the same sand:

Cement content	100 P water content
0	33%
5	31.9%
10	31.6%

This effect appears to result from an improvement in the grading of the fill from the addition of the cement fines. The fines convert the approximate straight-line grading of the sand (sand 2 in Figure 9) to a gap grading that is similar to G3 in Figure 3a, thus improving its flow properties at a given water content.

Plasticizers, super-plasticizers and pumping aids such as sodium tripolyphosphate may be used to reduce water requirements for pumpability. As most of these additives are proprietary, the reader is referred to suppliers of these materials for further details.

SETTLEMENT OF SLURRY FILLS AFTER PLACING

A pumpable slurry always contains more water than is required to fill the voids in the settled material. If this were not so, interparticle friction and interlock (Equation 3a) would prevent the material from flowing. Once the pumped fill is in place, this surplus water drains or bleeds off and must be disposed of. Figure 6 shows the influence of cement content on two typical sand-cement slurries placed at the 100 P water content. The extent of the problem posed by the disposal of this bleeding water will clearly depend on the volume of the fill being placed, on its rate of placing, and on the configuration of the fill relative to available drainage in the mine. The more bleeding that occurs, the better will the fill ultimately perform, as bleeding is an indication of settlement into a denser state.

The amount of bleeding water is also very dependent on the grading of the fill. In Figure 6, the increase in bleeding water with increasing cement content also arises because the cement content acts as an addition of fines, thus improving the grading and allowing the particles to settle into a closer packing. Figure 7 shows the effect of cement addition on increasing the density reached by a sand fill, in this case, Sand 1, whose grading is shown in Figure 9.

COMPRESSION OF FILLS UNDER STRESS

Most of the compression undergone by a pumped fill subjected to stress occurs at low stresses. This is illustrated by Figure 8 which shows how a typical fill becomes less compressible as the stress increases. Hence, it is the compressibility at low stresses that is all-important—a fill that is incompressible at low stresses will remain incompressible at high stresses and vice versa.

It will be noted that the compression curve has been presented in terms of decreasing water content. This emphasizes the fact that a fill of saturated tailings can compress only as water is expelled from it under increasing stress.

As stated earlier, the compressibility of a fill is highly dependent on its grading, both at small and large stresses. For small stresses, this is illustrated by Figures 9 and 10: Figure 9 shows the grading curves for three sands (sands 1 and 2 have been referred to in Figure 6), while Figure 10 shows the compression curves, at low stresses, for sands 1 and 3 and the effect on compressibility of added cement fines. (All the specimens were placed at a water content corresponding to 100 P viscosity).

The water contents at zero effective stress indicate the effect of cement content on bleeding, since all the specimens were placed at approximately the same initial water content. It will be noted that the materials that bled most (lower initial water contents) are also less compressible under applied stress. This conclusion also applies to a comparison of sands 1 and 3. Sand 1 had a lower initial water content than sand 3 and proved less compressible under stress.

Apart from altering the grading of the fill, the cement content also decreases the compressibility by developing cohesive bonds between the solid particles of the fill. However, as will be seen later, at the high water contents and low cement contents used in pumped tailings fills, these bonds are relatively weak and have only a small effect on the compressibility of the fill.

To sum up, Figure 10 shows that materials that settle into a dense packing on being placed will be less compressible under subsequent loading.

DRAINAGE OF FILLS

As described earlier, the rate at which fills under compression drain is influenced primarily by the coefficient of consolidation, c_v. Fortunately, most tailings materials are relatively permeable, and values of c_v lie in the range of 10 to 50m^2 a year or more (in comparison with clays which may have c_v values as low as 0.1m^2 a year).

The value of c_v is influenced by the effective stress, and decreases with increasing effective stress as shown in Figure 11(a). However, the effect of stress on c_v, and hence on the rate of drainage, is usually relatively minor.

It may also be necessary to estimate the rate of percolation flow through a fill if the fill is being used as a water barrier. In this case, D'Arcy's law,

$$v = ki \tag{8}$$

(where v is the velocity of flow under a gradient of head i, and k is the coefficient of permeability) is applicable. The

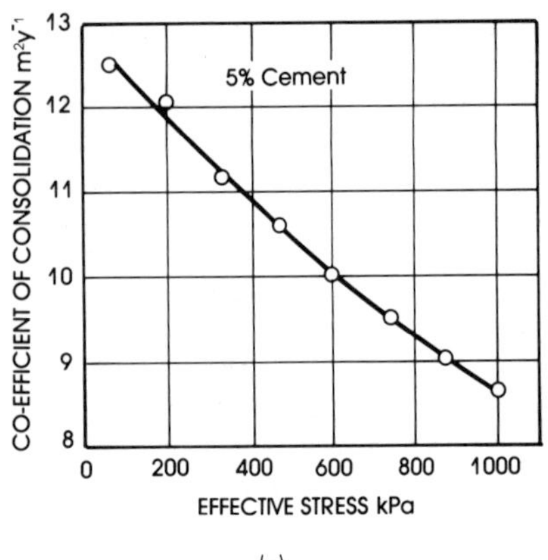

(a)

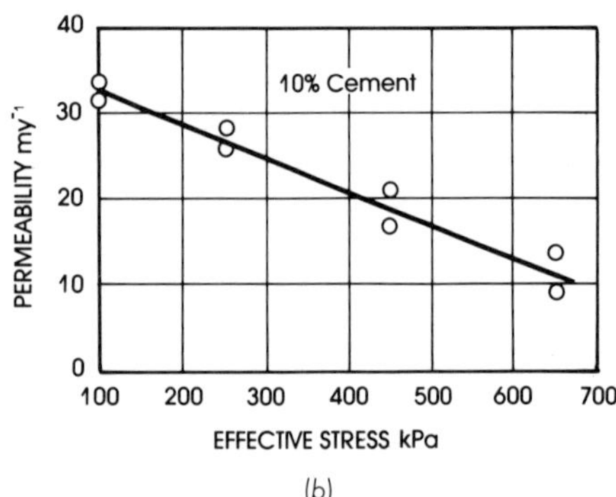

(b)

FIGURE 11. (a) Variation of coefficient of consolidation with effective stress for a typical tailings fill; (b) Variation of coefficient of permeability with effective stress for a typical tailings fill.

coefficient of permeability for tailings materials typically falls in the range 1 to 50m a year, and also decreases with increasing effective stress, typically as shown in Figure 11(b).

Although the figures quoted above are typical, both c_v and k can be drastically reduced by the presence of only a few percent of clay-sized material in the fill. It is for this reason that many fill materials are prepared by removing the fine fraction of the particle size range by means of cycloning. Figure 12, for example, shows the change in grading produced in a fine gold mine tailings by cycloning. In this case, a two-stage process was used: The tailings was cycloned and the underflow then re-cycloned. The grading curve shown in Figure 12 represents the underflow of the second cycloning. According to the Hazen formula (see Ref. 3) the coefficient of the permeability of a sand is given by

$$k = D_{10}^2 \text{ cm/sec} \tag{8a}$$

in which D_{10} is the particle size in millimetres such that 10 percent of the material is finer than D_{10}.

Applying this equation to the grading curves in Figure 12, it will be seen that cycloning would have increased the permeability a hundred-fold from 15×10^{-6}cm/sec for the run-of-mill tailings to 25×10^{-4}cm/sec for the twice-cycloned underflow.

Cycloning has the additional advantage of reducing the water requirement for pumpability.

SHEAR STRENGTH OF FILLS

Typical strength characteristics for cemented-tailings fills are shown in Figure 13. In these diagrams, the maximum

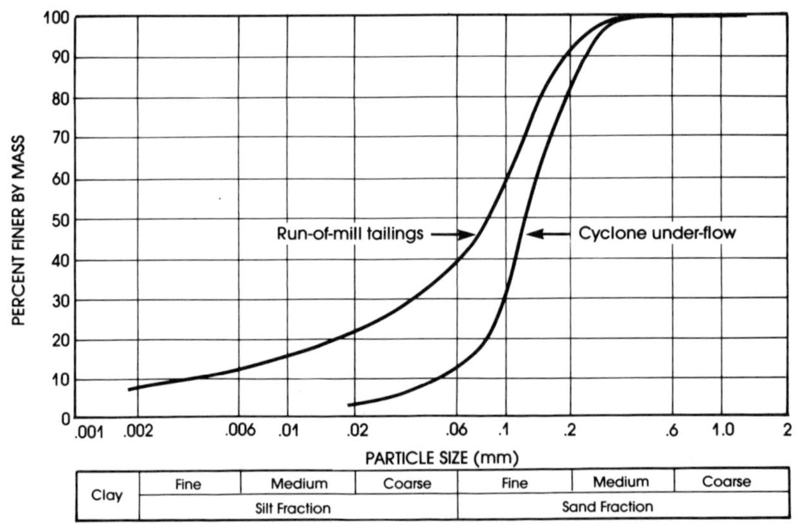

FIGURE 12. Effect of cycloning on particle size analysis of a gold tailings.

shear stress at failure ½ $(\sigma_1 - \sigma_3)_f$ has been plotted against the average principal effective stress ½$(\sigma_1' + \sigma_3)_f$. The diagrams clearly show the effect on shear strength of an increasing cement content. Increasing cement contents affect both the slope ψ of the strength lines (ψ is related to the angle of shearing resistance ϕ' by $\sin\phi' = \tan\psi$) and the cohesion c' of the material. At higher stresses, the major effect on shear strength results from the increase in ϕ'. For example, for sand 3, when ½ $(\sigma_1 + \sigma_3')_f$ was equal to 3000kPa, the strength increase as a result of an increase in the cement content from 0 to 10 percent was 550kPa, of which 110kPa resulted from the increase of cohesion and 440kPa from the increase in ϕ'. This disparity increases progressively with increasing stress.

The effect of cement content on both c' and ϕ' for sands 1, 2, and 3 is summarized in Figure 14. A comparison of Figure 14 with Figures 9 and 10 shows that the least compressible sand (1) has considerably superior strength characteristics to those of sand 3. Sand 2, which has a compressibility intermediate between that of sands 1 and 3, displays intermediate strength characteristics. One is therefore led to the conclusions that, in relation to strength, (1) the greater the initial density of the fill, the better its performance; and (2) the cementitious bonds developed by any cement that may be added to a fill play a relatively unimportant role in the strength of the material; the effect of the cement on compressibility and frictional strength via its effect on ϕ' is far more important.

This latter effect probably results from improved grading (via higher settled density) rather than any chemical bonding.

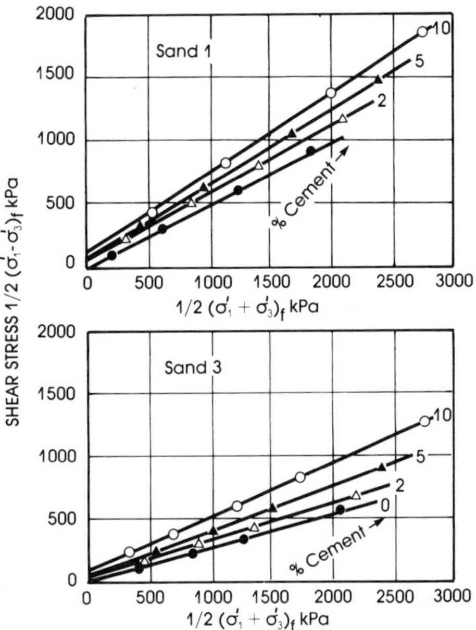

FIGURE 13. Failure stress diagrams for triaxial shear tests on two cemented sands.

ROLE OF CEMENT IN PUMPED FILLS

In view of earlier remarks, one is led to ask whether cement is necessary in a pumped fill and whether, if the major effect of the cement is to improve the settled density and compressibility of the fill by improving its grading, some other type of fine material (preferably a waste material obtainable at low cost) would not do almost as well. These are questions that should be explored in each individual case. There are certainly indications, for example, that a mixture of pulverized fuel ash (PFA) and lime added to a tailings fill can produce almost the same effect as cement [4]. PFA is produced in vast quantities as a power-station waste product, and PFA from ash dams or lagoons, can be obtained for the cost of loading and transportation. Figure 15 compares the compression curves for three slurries of gold-mine tailings containing cement and lime-PFA fines. It is evident that the use of lime-PFA mixes warrants further investigation, especially as the addition of this type of fines to a coarser material than gold-mine tailings) could be expected to have an even better effect. (The grading of the PFA was not much finer than that of the tailings).

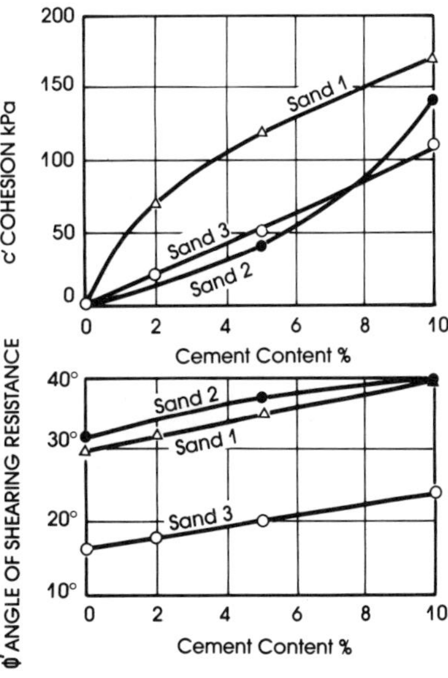

FIGURE 14. Relationship between cohesion, angle of shearing resistance, and cement content for three cemented sands.

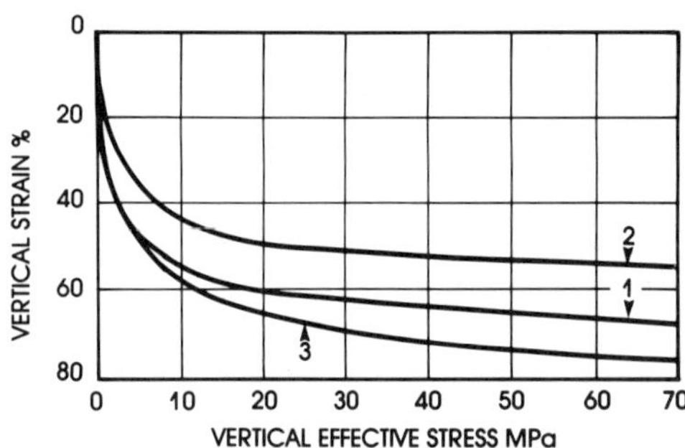

FIGURE 15. Compression of gold tailings cemented with: (1) 10 percent Portland cement; (2) 5 percent lime and 16 percent pulverised fuel ash; and (3) 6 percent lime and 12 percent pulverised fuel ash.

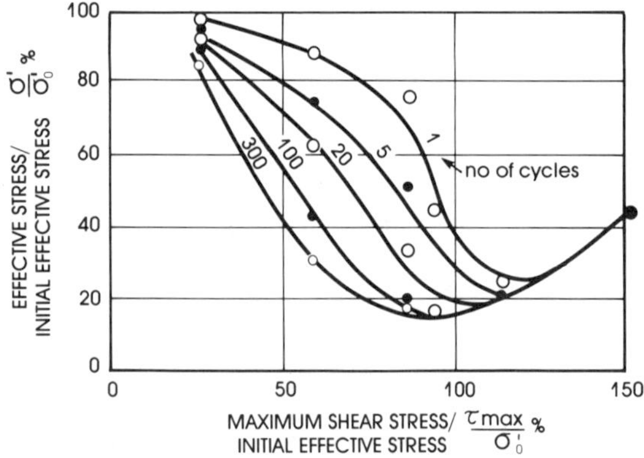

FIGURE 16. Results of undrained dynamic triaxial shear tests on a cemented gold tailings.

EFFECTS OF SHOCK OR SEISMIC LOADING

Shock loading as a result of blasting or seismic events has the effect of applying a rapidly alternating shear stress to a fill. Each application of shear stress causes an increment of pore pressure and, because the frequency of the applied shear stresses is relatively large, there is no time for drainage of the fill to allow the shear-induced pore pressure to dissipate. In terms of Equation (3),

$$\Delta\tau = \tan\phi'\Delta(\sigma - u) \tag{3b}$$

σ remains reasonably constant, but u increases with each application of shear stress, the net result being that $\Delta\tau$ is negative, i.e., the shear strength of the fill decreases. In the worst case, $\Delta\tau$ becomes equal to $-\tau$, and the fill loses all its strength and liquefies.

The effect of seismic or blast loading can be simulated in the laboratory by the application of alternating shear stresses under undrained conditions to a specimen of tailings fill in a triaxial test.

In tests of this type on specimens of both uncemented and cemented gold tailings fill, it has been found that the pore pressure first of all increases (indicating a tendency for the fill particles to move into a denser packing) and then starts to decrease (indicating a tendency for the material to increase in volume or dilate). A set of results for tests of this type is shown in Figure 16. On the vertical axis, the ratio σ'/σ_o' represents the effective stress after N cycles of repeated loading as a percentage of the initial effective stress (σ_o') in the fill. On the horizontal axis, the magnitude of the applied repeated shear stress is also expressed as a percentage of the initial effective stress. It will be seen that in this particular case, 100 cycles of a shear stress equal to σ_o' caused the effective stress in the fill to fall to 15 percent of its initial value. The effective stress (and hence the strength) did not fall to less than this, however, and the fill was in no danger of liquefying.

Figure 17 shows a set of results for similar tests presented as a series of stress paths. The material was a cyclone underflow gold tailings similar in grading to the curve shown in Figure 12. Figure 17(a) shows the results of four tests: (1) a static consolidated undrained shear test on a specimen initially consolidated to an effective stress of 10kPa; (2) dynamic undrained shear tests on two specimens, both prepared at an initial water content of 35 percent, one with flocculant added, the other without; and a similar dynamic undrained shear test on (3) a specimen sedimented out from a 100 percent water content slurry to ensure that it was in the loosest state possible.

The stress path for static shear goes directly towards and up the K_f-line for static shear, indicating the highly dilatant nature of the tailings. The K_f-line shown in Figure 17(a) was established in a higher stress range; hence specimens follow similar K_f-lines for static shear regardless of stress (up to initial consolidation stresses of at least 500kPa).

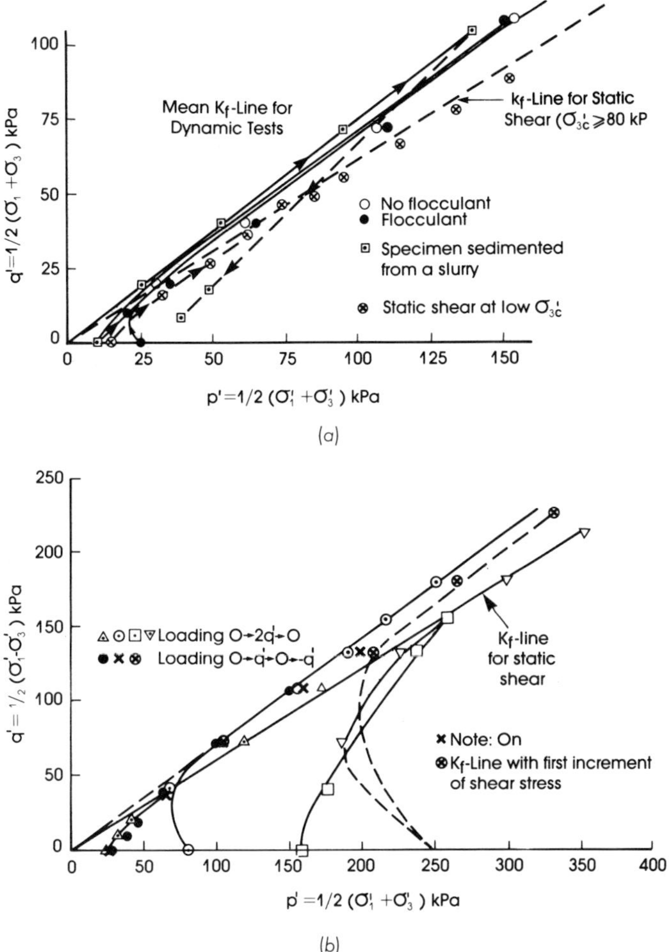

FIGURE 17. Stress paths for dynamic triaxial shear tests on uncemented cycloned gold tailings: (a) low stress range; and (b) higher stress range.

The points defining the stress paths for dynamic undrained shear represent the stresses after 25 applications of a particular shear stress q' applied in a cycle:

$$0 \text{ to } q' \text{ to } 0 \text{ to } -q' \text{ to } 0$$

After this, q' was increased and another 25 cycles applied.

As Figure 17(a) shows, the stress paths for dynamic shear are very similar to those for static shear, except that the K_f-line is steeper. All specimens dilated strongly and there was no tendency towards liquefaction. After cyclically stressing one specimen at a value of $q' = 105$kPa, q' was reduced to 9kPa. The result was that the stress state in the specimen moved to well below the K_f-line.

Figure 17(b) shows the results of further dynamic shear tests on the same material, but at a higher stress range. The figure shows that the behavior of the fill is similar at higher stresses.

The diagram shows that the type of stress cycle has little influence on the stress paths: 0 to q' to 0 to $-q'$ to 0 has a similar effect to the cycle 0 to $2q'$ to 0 to $2q'$ to 0.

It also makes no difference to the subsequent stress path if the first loading cycle places the stress state in a material on the K_f-line.

Although milled gold tailings stiffen rather than liquefying under dynamic undrained shear, it must not be forgotten that they also strain under dynamic loading. Figure 18 shows the stress path and axial strain versus minor principal effective stress relationships for a series of five replicated dynamic shear tests on a saturated gold tailings. It will be seen that the strains can be large and could result in a fill losing contact with the hanging wall of a stope as a result of

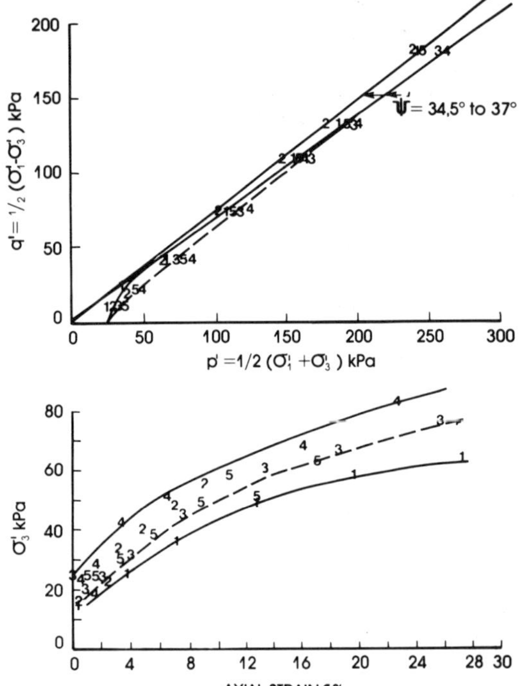

FIGURE 18. Stress paths and relationship between $\sigma_3{}'$ and axial strain for a set of replicated dynamic triaxial shear tests.

seismic or shock loading. The implications of this will be investigated below.

The conclusion to be drawn from the above study (which has been repeated with similar results on four other gold tailings materials) is that gold tailings fills do not liquefy when subjected to seismic loading. Rather, they stiffen, although undergoing strain.

A number of seismically-related failures of tailings dams have been reported in the literature, e.g., those at El Cobre, Chile [5] and Mochikoshi, Japan [6]. It is not certain how these failures occurred. One possibility is that the dynamic shaking caused the tailings to lose strength. The outer slope of the dam then failed by sliding and the impounded tailings flowed out. Another possibility is that the shaking caused the outer walls to spread and settle, i.e., to strain without losing strength. The liquid tailings in the impoundment then overtopped the dam, eroded a breach and escaped through the breached wall. In the case of the 1974 Bafokeng failure [7], the outer wall of the dam was overtopped and breached by water stored on the dam, augmented by storm water. The tailings impounded in the dam had a very low shear strength—were in fact in a liquid state—and flowed out through the eroded breach. Hence, although failures of tailings dams have been caused by seismic shaking, it is not certain if the failures resulted from liquefaction under seismic stresses.

THE USE OF FILL TO ABSORB STRAIN ENERGY IN DEEP MINES

The strain energy associated with mining excavations arises from the gravitational stresses acting on the strata above the excavation. The energy per unit of plan area available for release as a result of creating an excavation of height v_o at a depth H below the surface is given by

$$U = \gamma H v' \tag{9}$$

in which v' = the distance through which the centroid of the rock overburden will move as a result of the closure of the excavation. v' will always be less than v_o and it is clearly advantageous to keep v' as small as possible.

As the excavation is made, energy is released from the destressed zone surrounding the excavation and also as a result of closure of the excavation (i.e., a reduction of v_o). The released energy is redistributed and stored as strain energy in highly stressed zones at the perimeter of the excavation. If partial failure occurs in the rock surrounding the excavation, some of the stored energy is consumed as energy of fracture while the remainder continues to be stored in the unfailed rock. Figure 19 shows the energy balance diagrammatically. σ_v is the average virgin rock stress and V_o the volumetric convergence of the excavation ($V_o = v_o$ numerically). U_R is the energy released by the excavation and U_s that stored in the surrounding rock. If no fracture occurs around the excavation, $U_s = U_R$. If fracture does occur, the stored energy is reduced by U_F, the energy of fracture. If the excavation fails catastrophically as a so-called "rock burst" in which the excavation closes instantaneously and completely, almost the entire stored energy is converted into fracture energy and $U_F = U_R$. Thus any measure that will reduce or absorb U_R will benefit the safety of the mining operation.

If the excavation is filled with a compressible material, part of the released energy is absorbed in compressing the fill. The absorbed energy is represented by U_f in Figure 20 and for conservation of energy $U_R = U_s + U_f + U_F$. Clearly, the less compressible the fill the less the excavation will converge and the larger U_f will be as a fraction of U_R. U_R will also be decreased by the decrease in volumetric convergence and a safer excavation will result.

IN SITU MEASUREMENTS OF THE COMPRESSIBILITY OF FILLS

Very few measurements of the in situ compression-stress characteristics of fills are available, mainly because of the difficulties of gaining access to the fills after placement, as well as the difficulty of obtaining instruments capable of measuring in the high stress ranges prevailing in deep underground fills.

One such set of measurements was made in a fill used to

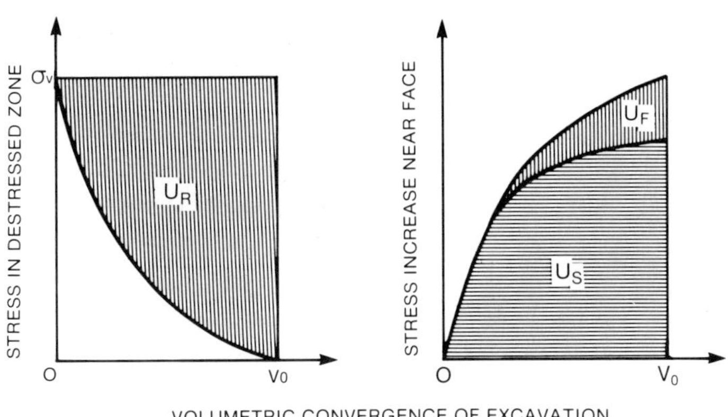

FIGURE 19. Energy balance for unfilled mining excavation.

replace a rock shaft pillar at the Stilfontein mine in South Africa. The pillar existed at a depth of 1900m below surface and was mined out and replaced by a pumped cemented tailings fill containing 10 percent of ordinary Portland cement. Figure 21 shows a plan of the area surrounding the shaft, as well as the instrumentation installed in the tailings fill. The flat jacks referred to in the figure were 200mm diameter oil-filled Freyssinet flat jacks connected by copper tubing to Bourdon pressure gauges, and pre-pressurized to 1MPa. The jacks were installed so as to measure vertical stress as well as horizontal stress in two orthogonal directions.

Figure 22 shows the field compression curve, as recorded on two of the jacks and the closure meters compared with laboratory compression curves. The figure shows that the fill performed considerably better in the field than would have been predicted from laboratory tests. The pillar extraction was, incidently, completely successfull and the shaft continued to operate without interruption, throughout.

The piezometers in the fill confirmed that compression of the fill was completely drained as none of them registered any pore water pressures.

DESIGN AND PROPERTIES OF STIFF FILLS

It follows from Figure 20 that the larger the area under the stress-compression curve of fill, i.e., the stiffer the fill, the more energy it will absorb and the more effective it will be in controlling catastrophic energy releases.

Waste usually available for preparing a stiff fill consists of rock and finely ground ore tailings or slimes. Typical parti-

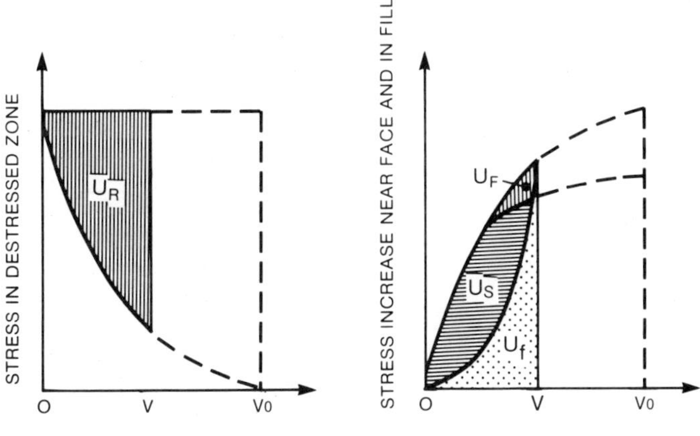

FIGURE 20. Energy balance for mining excavation filled with compressible material.

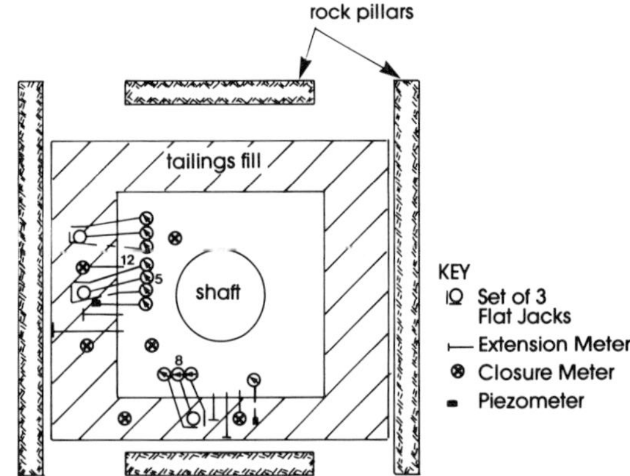

FIGURE 21. Layout of instrumented fill used to replace rock shaft pillar.

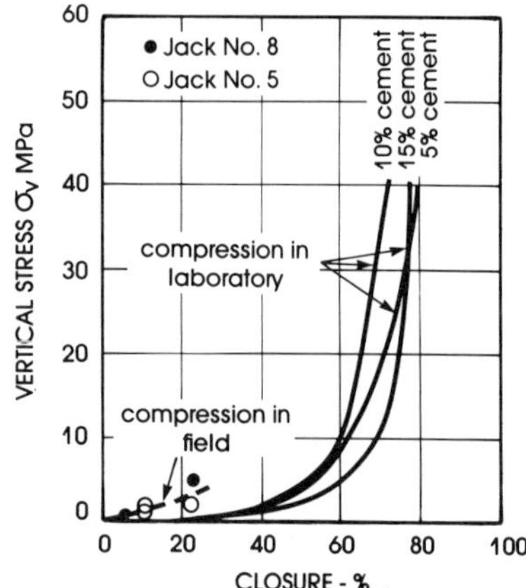

FIGURE 22. Comparison of field and laboratory compression curves for fill illustrated in Figure 21.

cle size distribution curves for these two disparate materials are shown in Figure 23. The design of the fill is based on the concept of creating a rigid skeleton of rock particles with the voids just filled with cemented tailings. As cement becomes a major cost factor when large volumes of fill are used, this approach has the added advantage of keeping cement costs to a minimum. In addition to the requirement of rigidity in situ, the fill has to be pumped into place and the fresh mix must therefore have a sufficiently low viscosity to allow of pumping.

In the case of the materials described by Figure 23, the aforementioned requirements led to a basic fill consisting of 5 parts by mass of waste rock with its voids filled by 1,6 parts of fine material. The fines consists of 1 part of tailings and 0,6 parts of either Portland cement or a Portland cement (OPC)–pulverized fuel ash (PFA) blend. The water requirement of the mix for pumpability is decided using standard concrete workability tests, and confirmed or adjusted by means of pumping tests using positive displacement concrete pumps.

Fills designed in this way have an exceptionally high total density as placed which averages over 90% of the density of the solid constituents. The maximum possible compression of the fill is thus less than 10%.

Once the fill is in situ, it is confined laterally by the sides of the excavation and is compressed vertically as the excavation starts to converge. The vertical compression takes place slowly enough to allow the fill to drain as it is compressed.

The maximum possible energy release from an excavation 1m deep at a depth of 3000m is 40MJ/m² (½ · 3,000m · 27 · 10³ N/m³ · 1m). At 10% compression the fill should be capable of absorbing 90% of this. The actual energy absorption will be less than this, however, as some convergence of the excavation will inevitably occur as the virgin rock is excavated, and before it can be replaced by fill.

PROPERTIES OF STIFF FILLS IN CONFINED COMPRESSION

Figure 24 shows the range of confined compression characteristics attainable with stiff fills. The figure also shows the range of compression characteristics for soft fills. Note the very large stresses that are necessary to simulate underground conditions in a deep mine. Figure 25 shows the range of relationships between lateral or horizontal pressure developed in stiff and soft fills during confined compression and the corresponding applied vertical stress. Stiff fills are almost self-supporting at low stresses and only start exerting appreciable lateral stress once the vertical stress exceeds about 10MPa. Soft fills, however, exert somewhat greater lateral stresses even when the applied vertical stress is low. There is a general trend for $K_o = \sigma_h'/\sigma_v'$ to increase progressively as the vertical stress increases, and the support provided by particle interlock is overcome.

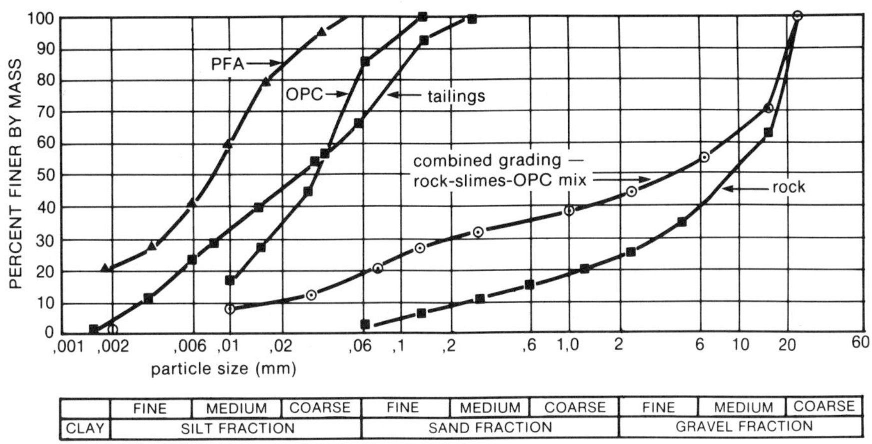

FIGURE 23. Grading curves for components of stiff fill.

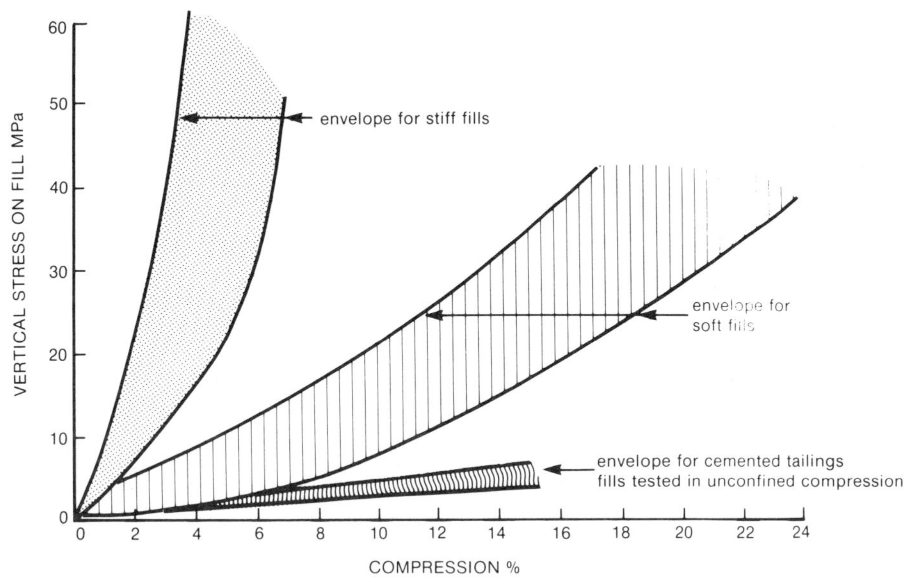

FIGURE 24. Range of confined compression characteristics obtainable with stiff and soft fills.

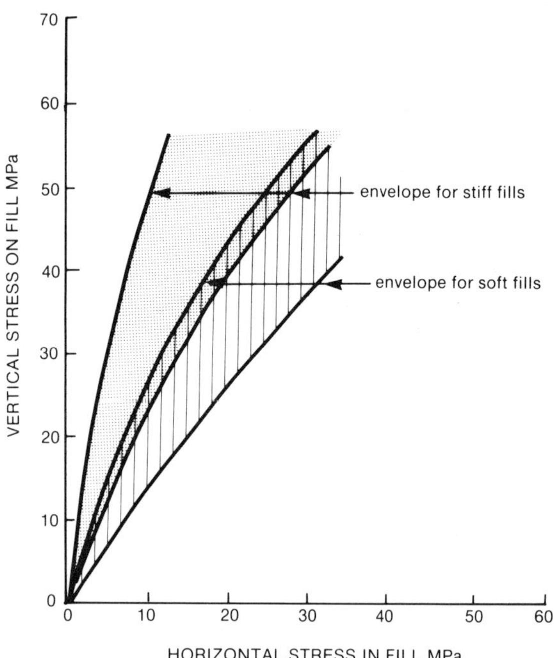

FIGURE 25. Range of relationships between vertical and horizontal stress during compression of stiff and soft fills.

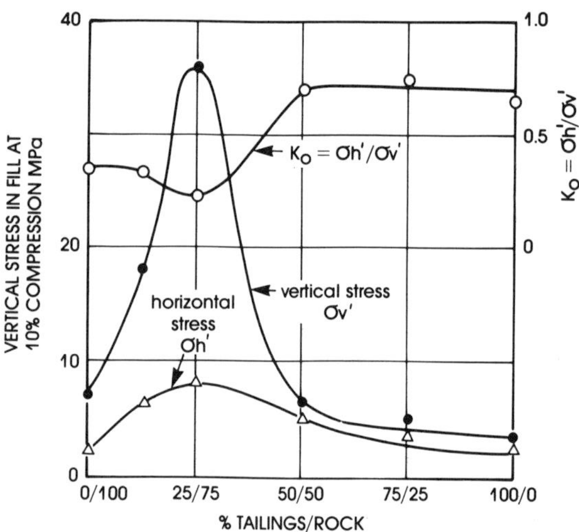

FIGURE 26. Vertical stress in fill at 10% compression for fills composed of tailings/rock mixtures.

TAILINGS–ROCK FILLS OF INTERMEDIATE STIFFNESS

Even when an extremely stiff fill such as those described above is not required, there may be advantages in increasing the stiffness of the fill to an intermediate stage. This can be achieved by adding waste rock to a basic fill of fine tailings. Figure 26 shows the result of adding various percentages of crushed rock waste to an uncemented gold tailings fill. The results have been presented as the variation of the vertical stress required to produce 10 percent compression in the fill under laterally confined conditions, versus the tailings/rock proportion. The figure indicates a very definite optimum mix of about 75 percent rock to 25 percent tailings.

It will also be seen that a reduction in the ratio of horizontal to vertical stresses, K_o occurs in the vicinity of the optimum, indicating that the stiff fill is more self-supporting than materials of lesser stiffness.

STABILITY OF FILLS

When considering the stability of a body of fill, at least two basic situations must be investigated: (1) the case of the narrow tabular fill in which the fill thickness is small in comparison with its lateral extent both on strike and dip; and (2) the case of the high fill which has a large height or thickness in comparison with its lateral dimensions. Each of these cases will now be considered in turn.

STABILITY OF NARROW TABULAR FILLS

It is convenient to consider both the static stability of fills and their stability when subjected to seismic accelerations within a single theoretical framework. This can be done by denoting the weight per unit volume of the fill as ϱa. Under static conditions, a becomes equal to g, the acceleration due to gravity. Under seismic loading, a becomes (g ± the seismic acceleration).

Figure 27 shows the basis for estimating stresses in a narrow inclined filled stope subjected only to vertical acceleration.

For equilibrium of the fill element in the z-direction,

$$w\sigma_z + \varrho aw\sin\beta dz = w(\sigma_z + d\sigma_z) + 2\tau dz$$

$$d\sigma_z = \left(\varrho a\sin\beta - \frac{2\tau}{w}\right) dz \qquad (10)$$

In Equation (10): σ_z is the total stress in the fill acting in the z-direction; ϱ is the mass per unit volume of the fill; a is the sum of gravitation and vertical seismic accelerations; τ is the average shear stress acting on the footwall and hanging of the fill element; and w and β are as defined in Figure 27.

The surface of the footwall and hanging will generally be

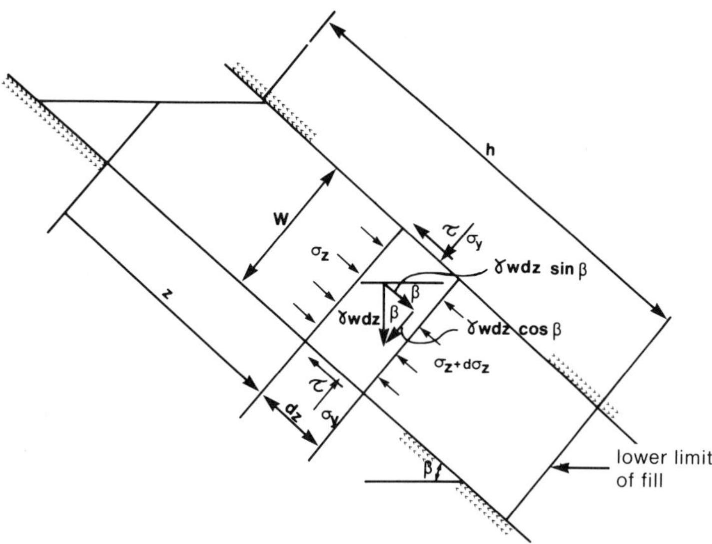

FIGURE 27. Basis for calculating the stresses in a column of fill.

sufficiently rough to ensure that if the fill moves, shearing will occur through the fill and not along the fill–rock interface. Hence τ can be written

$$\tau = (\sigma_y - u)\frac{\tan\phi'}{F} \tag{11}$$

in which $\tan\phi'$ is the angle of shearing resistance of the fill and F is the factor of safety against shearing through the fill. If the hanging and footwalls happen to be very smooth, so that shearing occurs along the interface between the rock and the fill, the angle of internal shearing resistance ϕ' must be replaced by the angle of wall friction δ' (fill on rock) to give

$$\tau = (\sigma_y - u)\frac{\tan\delta'}{F} \tag{11a}$$

$\delta\varrho$ is usually two to three degrees of arc less than ϕ'. Equation (11a) will then carry through the remainder of the analysis. Also, σ_y will be related to σ_z by

$$(\sigma_y - u) = K(\sigma_z - u) \tag{12}$$

in which K is the ratio of effective stress in the y-direction to that in the z-direction.

Hence Equation (10) becomes

$$\frac{d\sigma_z}{dz} = \varrho a \sin\beta - \frac{2K\tan\phi'}{w}\frac{(\sigma_z - u)}{F} \tag{13}$$

Provided the distribution of u with z is known, Equation (13) can be integrated to give the distribution of σ_z along the length of the stope. However, we are mainly concerned with evaluating the conditions at the lowest point of the stope that will cause the fill just to be retained. At this point, the disturbing forces must just balance the side shear forces, i.e., from Equation (13),

$$\frac{d\sigma_z}{dz} = 0$$

or

$$\varrho a \sin\beta = \frac{2K\tan\phi'}{wF}\cdot(\sigma_z - u)$$

In this condition,

$$F = \frac{2K\tan\phi'(\sigma_z - u)}{w\varrho a \sin\beta} \tag{14}$$

Putting in numerical values:

$K = 1.0$ $\quad\tan\phi' = 0.7$ $\quad w = 1\text{m}$

$\varrho = 1800\text{kgm/}^3$ $\quad\sin\beta = 0.4$

If there is no bulkhead or barrier to retain the fill $\sigma_z = 0$ and

$$F = \frac{0.0019}{a} - u$$

which shows that the fill will not be retained unless negative, i.e., capillary stresses exist in the fill. Under normal gravitational conditions ($a = g = 10\text{m/s}^2$) the value of $-u$ required for a factor of safety of 1.0 is only 5.3kPa, or 0.5m of water capillary head. If $a = 5g$, the required value of $-u$ increases to 26.3kPa or 2.6m of water head. (Typical values of seismic accelerations occurring in deep mines will be given later.)

If a bulkhead is provided, and if $u = o$, the bulkhead must be capable of resisting values of σ_z equal to the values of u calculated above, if the fill is to be retained.

The above considerations relate to a "solid" fill and the pressures required for retention are astonishingly small. If, however, the fill is in a liquid state and hence the effective stress $\sigma'_z = \sigma_z - u$ tends to be zero, $u = \sigma_z$ and from Equation (10) $d\sigma_z/dz = \varrho a\sin\beta$; $d\sigma_z/dz$ now cannot be zero, and at the lowest point of the stope ($z = h$) $\sigma_z = h\,\varrho a\sin\beta$.

For the numbers of the previous example, if $h = 50$m, $\sigma_z = 36a$ kPa; if $a = g = 10\text{m/s}^2$, $\sigma_z = 360$kPa; and if $a = 5g = 50\text{m/s}^2$, $\sigma_z = 1800$kPa.

Hence, under normal gravitational conditions, the pressure on the retaining bulkhead would increase from 5.3kPa to 360kPa, while if the acceleration is $5g$ the pressure would increase from 26.3kPa to 1800kPa.

Both of these pressures, are for practical purposes, irresistible for a temporary fill-retaining bulkhead.

A third case must also be considered: This arises if the fill does not make contact with the hanging wall. The consequence is that resistance to sliding develops only on the footwall and the bulkhead pressures and capillary pore pressures calculated earlier are doubled. This is not necessarily a serious occurrence. Even if a bulkhead should fail, allowing materials to move out of the stope, the fill will not move far as long as it does not liquefy.

A fill that does not contact the hanging is susceptible to removal by erosion if water is allowed to run over its exposed upper surface. There have been reports of certain fills that had been in place for long periods then subsequently liquefied. Erosion by water coming from higher levels is a more likely explanation for the failure of these fills than that they liquefied under seismic action. If the fill does not contact the hanging wall, Equation (10) becomes

$$d\sigma_z = \left(\varrho a\sin\beta - \frac{\tau}{w}\right) dz \qquad (10a)$$

and Equation (14) becomes

$$F = \frac{K\tan\phi'(\sigma_z - u)}{w\varrho a\sin\beta} \qquad (14a)$$

The above analysis has assumed that the fill is a frictional, cohesionless material. If the fill is cemented, it will have a cohesion and its shear strength can be expressed by

$$\tau = c' + (\sigma_y - u)\tan\phi' \qquad (11b)$$

For a cohesive fill, Equation (13) will become

$$\frac{d\sigma_z}{dz} = \varrho a\sin\beta - \frac{2[c' + k(\sigma_z - u)\tan\phi']}{wF} \qquad (13b)$$

If

$$\frac{d\sigma_z}{dz} = o$$

$$F = \frac{2[c' + K\tan\phi'(\sigma_z - u)]}{w\varrho a\sin\beta} \qquad (14b)$$

Putting in the numerical values used previously,

$$F = \frac{c' + 0.7(\sigma_z - u)}{360a}$$

If, in the absence of a bulkhead, $\sigma_z = o$ and also $u = o$, then for $a = 10\text{m/s}^2$, the value of c' required for a factor of safety of 1.0 is $c' = 3.6$kPa while if $a = 50\text{m/s}^2$, the required value is $c' = 18$kPa.

An adequate cohesion of 50kPa to 100kPa can easily be obtained by adding a small proportion of Portland cement to the fill.[2] The entire fill mass does not need to be cemented, but only a zone of limited width adjacent to the lower boundary of the stope.

SHEAR STRESSES APPLIED TO A FILL BY SEISMIC ACTION

It is convenient at this juncture to consider the shear stresses likely to be applied to a tabular fill by seismic action: Figure 28a represents an accelerogram for a typical earthquake [8], in which the vertical axis shows the recorded horizontal acceleration in a particular direction. It will be noted that the maximum acceleration does not exceed 0.2g, and hence the maximum acceleration used in an analysis such as that above, would be 1.2 g.

In contrast, Figure 28(b) shows an accelerogram for a seismic event at a depth of 3km in the ERPM mine [9] in South Africa. The peak acceleration shown in the figure is 7.7g and other evidence suggests that accelerations as high as 12g may occur in deep mines. Hence, seismic events underground may impose accelerations that are many times more severe than those imposed by natural events at the surface. However, comparing the time scales of the two accelerograms, the duration and number of high acceleration cycles

[2]See Figure 14.

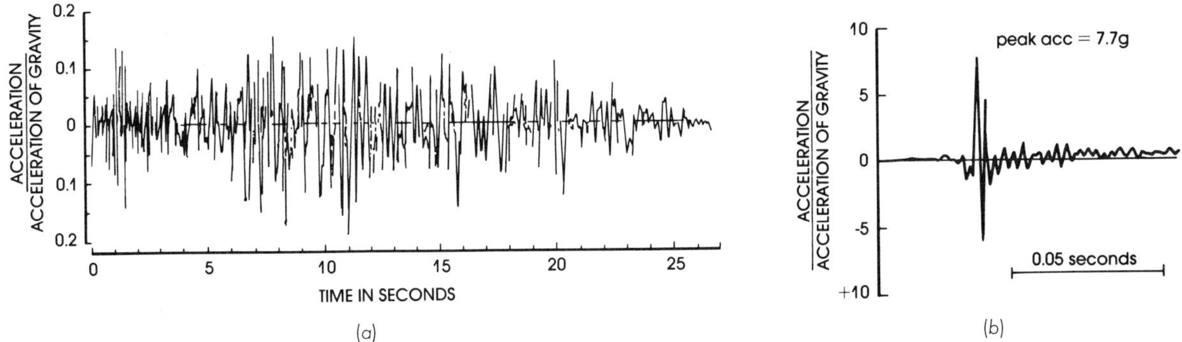

FIGURE 28. (a) Accelerogram for Olympia, Washington earthquake of April 13, 1949. Component S10E; (b) Accelerogram for horizontal component of underground seismic event.

in an underground event appear to be far less than in a natural earthquake.

In terms of the analysis in the previous section, the maximum shear stress imposed on a fill would be given by

$$\tau = \tfrac{1}{2}(\sigma_z - \sigma_y) \tag{15}$$

$$\text{if } u = o,\ \tau = \tfrac{1}{2}\sigma_z(1 - K) \tag{16}$$

Although a value of $K = 1.0$ was previously used for illustrative purposes, and is a perfectly possible value, a value of 0.5 would be more likely in a fill while a value of 0.2 would be the smallest likely value of K (corresponding to K_A for a fill having $\phi' = 42°$. This would give the biggest likely value of τ as $\tau = 0.4\sigma_z$.

From Equation (14), if $u = o$, the maximum value of σ_z is given by

$$\sigma_z = \frac{Fw\varrho a \sin\beta}{2K\tan\phi'} \tag{14c}$$

which for the numerical values used before and a factor of safety $F = 1.5$ reduces to $\sigma_z = 0.514a$ kPa with a in m/s². Hence a representative value of τ for $\beta = 25°$ would be $\tau = 0.21a$ kPa. Hence for an acceleration of 10g = 100 m/s² $\tau = 21$ kPa.

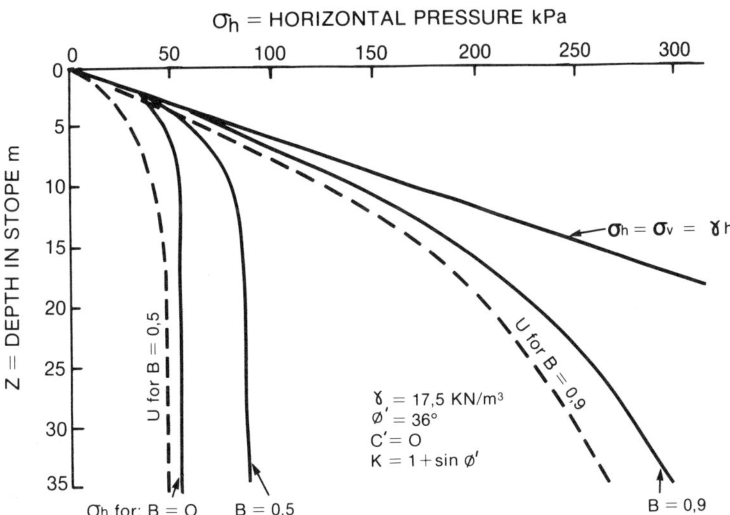

FIGURE 29. Effect of pore water pressure on distribution of horizontal stress on walls of vertical filled stope.

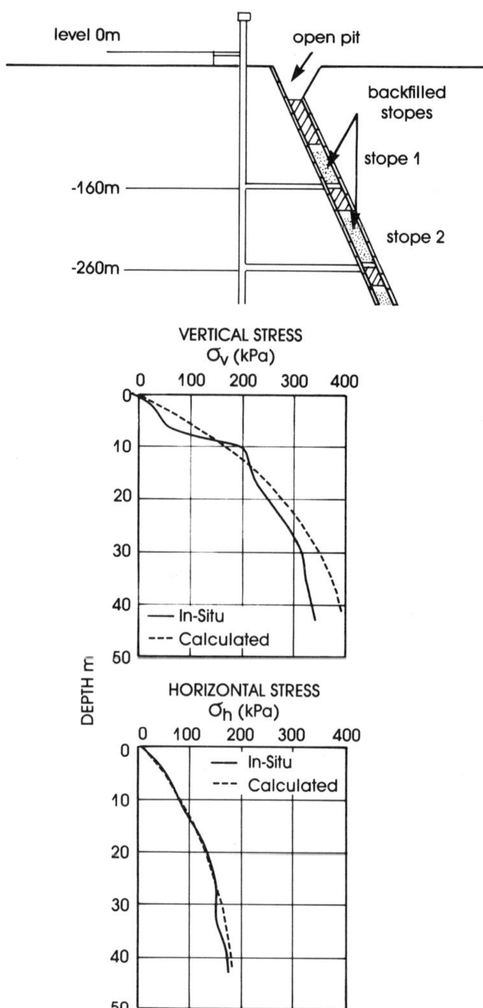

FIGURE 30. Measured vertical and horizontal stresses in a steeply inclined filled stope at Näsliden mine, Sweden.

An absolute worst case could be considered by taking $\sin\beta = 1.0$ in Equation (14c), i.e., by considering a vertical stope. In this case $\sigma_z = 1.29a$ kPa. This would increase τ to 52kPa for $a = 100$ m/s² and $w = 1$m and to 104kPa for $a = 100$ m/s² and $w = 2$m.

Hence the range of shear stress likely to be imposed on an underground fill even by a severe seismic event is unlikely to exceed 100kPa.

THE DISTRIBUTION OF VERTICAL AND LATERAL STRESS IN A COLUMN OF FILL UNDER STATIC CONDITIONS

If the complete distribution of vertical and lateral stress in a column of fill is required, Equation (13) or (13b) must be integrated.

The integration depends on the distribution of u and whether u is dependent on z, on σ_z or on both u and σ_z. It can be made formally or numerically.

For example, while the fill is being placed and while it is draining subsequent to placing, u could be represented by a linear relationship such as $u = u_o + B\sigma_z$. In this case,

$$\sigma_v = \frac{1}{A}\left\{ \varrho a \sin\beta - \frac{2}{wF}(c' - u_o K \tan\phi')\right\}\{1 - e^{-Az}\} \tag{17a}$$

$$A = \frac{2K\tan\phi'(1 - B)}{wF} \tag{17b}$$

and $\sigma_h = K(\sigma_v - u) + u$.

Under long-term conditions Equations (17a) and (17b) would still apply, with $B = 0$ and $u\ (= u_o)$ given by

$$u = \frac{-w\sec\beta\gamma_w}{2} \tag{18}$$

as an average value over the height of the stope.

Figure 29 shows the effect of pore pressure on the horizontal total stresses exerted by a cohesionless fill on the walls of a vertical stope in the case where $u = B\sigma_v$.

Figure 30 shows vertical and horizontal stresses measured in a steeply inclined stope of the Nasliden mine in Sweden by Knutsson [10]. As the figure shows, the agreement between measured and calculated pressures can be excellent.

CONSOLIDATION OF HYRAULICALLY PLACED FILL

Because the stability of a fill is very much a function of its shear strength and therefore of the effective stresses in it, it may be necessary to control the rate of filling so that pore

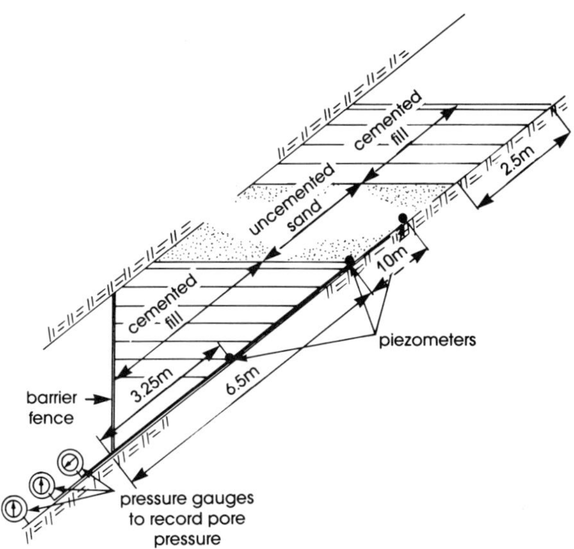

FIGURE 31. Layout of piezometers used to control rate of filling of an inclined stope.

pressures remain within acceptable limits. If the foot-wall and hanging of the stope are assumed to be completely impervious, a theory due to Gibson [11] can be used to estimate the minimum rate of filling required to ensure that pore pressures in the fill remain within acceptable limits. In practice, because of the finite permeability of the foot-wall and hanging and because drainage can also take place to the sides of the fill, greater rates of filling will be possible, but the actual rate will have to be decided on the basis of pore pressures monitored by means of piezometers.

According to Gibson's theory, the rate of filling m in meters per unit time should not exceed the following values, if the average degree of consolidation is to be at least:

U	m^2t/c_v
95%	0.25
90%	1.0
75%	4.0

In the factor m^2t/c_v, t is the time taken to completely fill the stope and c_v is the coefficient of consolidation of the fill. For example, if an average degree of consolidation of 90% is required in a fill having a coefficient of consolidation of 200 m² per year and the stope is 20m long, then the time required to fill the stope is given by:

$$t = \frac{20}{m}$$

Also:

$$1.0 = \frac{m^2t}{c_v} = \frac{m^2 \cdot 20}{m \cdot 200} = \frac{m}{10}$$

Hence, $m = 10$ m/year, and 2 years will be required to fill the stope. As mentioned earlier, actual rates will usually exceed theoretical ones, because of lateral drainage occurring into the rock on either side of the fill.

Figure 31 shows a typical layout of piezometers installed to monitor pore pressures in a fill during placing. Note that the piezometers have been placed on the foot-wall of the stope as this is the most critical position for stability should excess pore pressures exist in the fill. Also note the zone of cemented fill placed at the base of the fill to help retain it in the stope.

STABILITY OF HIGH FILLS

The treatment given in this section follows that of Mitchell [12] and his co-workers.

The high fill is used either to provide access for mining high ore bodies, or to fill high stopes in situations where the ore is mined in such a way that high inverted bottle-shaped stopes are produced, as shown in Figure 30. Mining usually proceeds in checker-board fashion and the critical condition for a high stope fill arises when one of the remnant pillars is mined out exposing one side of the fill. The face of the fill

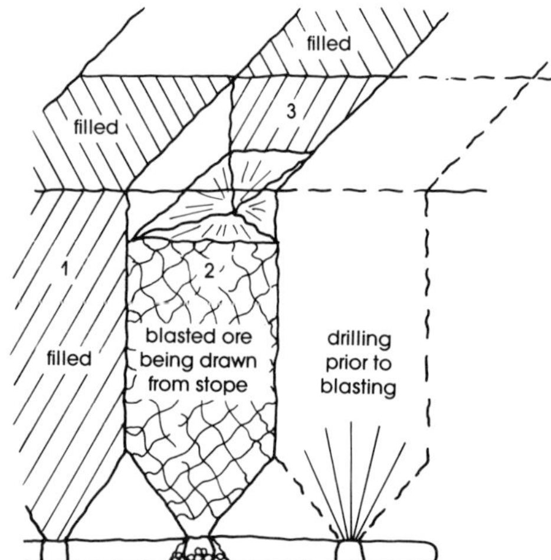

FIGURE 32. Diagram showing a possible sequence of mining and filling to form high fills.

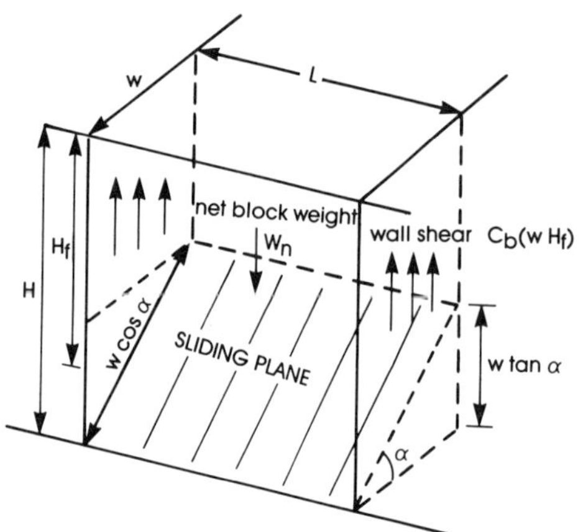

FIGURE 33. Basis for Mitchell's analysis of the stability of a high fill.

must then be capable of standing unsupported until the adjacent void has been filled. Stopes 1 and 3 in Figure 32 are approaching this critical condition. Figure 33 shows the basis for the approximate analysis developed by Mitchell [12].

It is assumed that shear stresses between the vertical sides of the fill and adjacent surfaces support some of the weight and that if the fill fails, it will do so by sliding in a plane inclined at $\alpha = 45° + \phi'/2$ to the horizontal. The net weight of the block is then calculated as

$$W_N = WH_f\,(\varrho aL - 2c') \tag{19}$$

in which

$$H_f = H - \frac{W\tan\alpha}{2}$$

The factor of safety against sliding is then given by

$$F = \frac{\tan\phi'}{\tan\alpha} + \frac{2c'L}{H_f(\varrho aL - 2c')\sin 2\alpha} \tag{20}$$

In Equation (20), a would normally be set equal to g, the acceleration due to gravity. If it is necessary to design against seismic effects, a can be given an appropriately increased value. It may also be necessary to include the effects of horizontal seismic acceleration in the analysis.

Mitchell has found good correlation between predictions based on Equation (2) and the results of model tests.

THE ANALYSIS OF FLOOD CONTROL PLUGS

Figure 34 represents a section through a cylindrical flood control plug of diameter D. The plug prevents water at pressure u_o from rising up the shaft and flooding workings higher up. The analysis proceeds along lines similar to those outlined earlier.

For equilibrium of a typical element of the plug in the z-direction:

$$\frac{\pi D^2\sigma_z}{4} = \frac{\pi D^2}{4}(\sigma_z + d\sigma_z) + \frac{\gamma\pi D^2}{4}dz\sin\beta + \pi D dz\,\frac{\tau}{F}$$

in which τ/F is the shear stress developed at the perimeter of the plug and F is the factor of safety against shear failure. Hence,

$$d\sigma_z = -4\left(\gamma\frac{D}{4}\sin\beta + \frac{\tau}{F}\right)dz \tag{21a}$$

As usual, writing

$$\tau = c' + K(\sigma_z - u) \tan\phi'$$

$$d\sigma_z = \frac{4}{D}\left\{\frac{\gamma D \sin\beta}{4} + \frac{c'}{F} + K(\sigma_z - u)\frac{\tan\phi'}{F}\right\} dz \tag{21b}$$

Seepage will occur both through the plug and through the rock walls surrounding it. Once a steady-state situation has been reached, it is likely that the seepage gradient will be approximately linear, i.e.,

$$u = u_o\left(\frac{1 - z}{H}\right) \tag{22}$$

and we have, for a linear seepage gradient:

$$d\sigma_z = -\frac{4}{D}\left\{\frac{\gamma D \sin\beta}{4} + \frac{c'}{F} + K\left[\sigma_z - u_o\left(\frac{1 - z}{H}\right)\right] \tan\frac{\phi'}{F}\right\} dz \tag{21c}$$

The distribution of stresses obtained in this case is somewhat different from that shown earlier in Figures 29 and 30, because u_o is usually so large that: (1) the selfweight of the plug is negligible (i.e., the term $\gamma D\sin\beta/4$ is negligible); and (2) the frictional component (i.e., the $\tan\phi'$ term) is small.

Equation (21b) can be integrated for any known distribution of u with z. However, if a seepage gradient somewhat less than the linear gradient given by Equation (22) is assumed, the effective stress $(\sigma_z - u)$ becomes zero and the only resistance to displacement of the plug then arises from the cohesion c'. The distribution of stress with distance along the plug then becomes linear.

Any variation of F with z can be assumed. However, the designer is usually interested in evaluating an average factor of safety for which F would be assumed constant. Figure 35 shows gradients of σ_z calculated for a plug in a vertical shaft that is retaining a water head of 900m. The shaft diameter is 10m and the plug length 33m. The value of ϕ' has been taken as 40° and the unit weight of the fill as 22.5kN/m³. Calculations have been made assuming c' as either 5MPa or 2.5MPa. The example represents the design calculations for a plug it proved necessary to place in a new vertical shaft at the Western Deep Levels mine in South Africa after the shaft was flooded during sinking. Once the plug had been concreted under water, the shaft was pumped out and the rock surrounding the plug could then be grouted to seal off the inflow of water. The plug was then mined out and shaft sinking resumed.

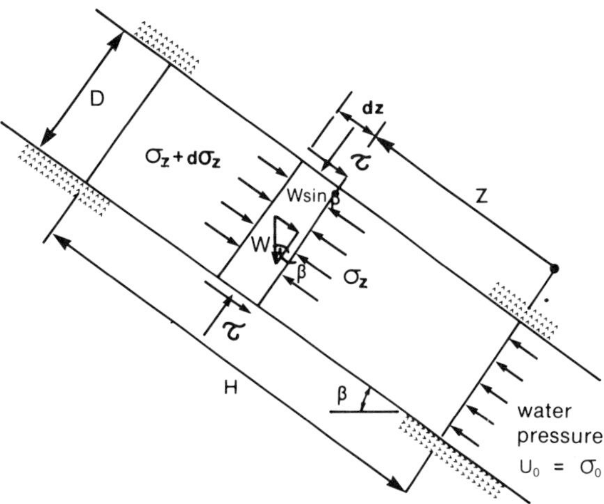

FIGURE 34. Basis for calculating stability of an underground flood control plug.

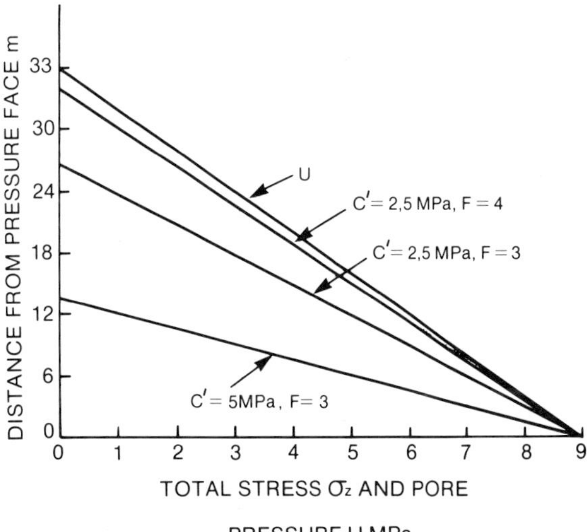

FIGURE 35. Stress gradients in a cemented fill shaft plug retaining 900m of water head.

The actual average factor of safety of the plug will be that which reduces the axial stress σ_z to zero at the free end of the plug. In this case, with $c' = 2.5\text{MPa}$, F is just over 4. The linearity of the stress gradients in this case arises from the circumstance that the frictional resistance to displacement of the plug is very small. Hence, Equation (21b) reduces approximately to

$$d\sigma_z = \frac{4c'}{DF} \cdot dz$$

or

$$\sigma_z = v_o - \frac{4c'}{DF} \cdot z \qquad (21d)$$

It is usually not necessary to calculate rates of seepage flow through such a flood control plug, as leakage through the rock will generally far exceed seepage through the plug. However, seepage rates can be calculated by applying D'Arcy's law to the plug, as follows: The seepage rate is given by

$$Q = \frac{kAu_o}{H} \qquad (23)$$

in which k is the coefficient of permeability of the plug material, A is the cross-sectional area of the plug, and u_o and H are as defined in Equation (22).

MEASURED STRESS DISTRIBUTION IN FLOOD CONTROL PLUGS

The arduous circumstances under which flood control plugs operate render it almost impossible to measure stresses in them. However, some years ago, tests were performed by Ockleston [13], on a model plug consisting of a length of steel piping (representing the opening in the rock) which was cast full of concrete. One end of the pipe was welded closed and a water pressure was applied to the concrete inside the closed end. Longitudinal strains were measured at intervals along the pipe. The results of the measurements are shown in Figure 36. Unfortunately, the results were published without any scales being shown on the axes. The trend shown is, however, very similar to that shown in Figure 35, if it is accepted that the axial strain and displacement in the pipe were directly related to the axial stress in the concrete of the plug.

COMPRESSED AIR RECEIVER PLUGS

Compressed air receivers are often constructed underground by plugging the entrance to a blind tunnel or drive.

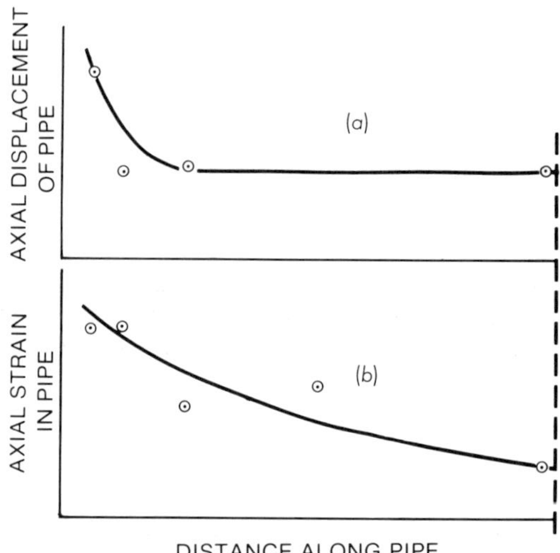

FIGURE 36. Distribution of strain and displacement in model flood control plug as measured by Ockleston [13].

The stability of the plug can be assessed in exactly the same way as for a flood control plug. The only difference is that the pore pressures that enter into the calculations are pore air pressures and not pore water pressures.

Figure 37 shows a section through such a receiver plug, while Figure 38 shows the results of a proof loading test on the plug. As the figure shows, the plug behaved almost completely elastically under the applied pressure.

LATERAL SUPPORT PROVIDED TO PILLARS BY FILL

It was mentioned earlier that fills can be used to provide lateral support to rock pillars, thus increasing their capacity to carry vertical load.

To assess the extent and effect of lateral support expected to be provided to rock pillars by a fill, a series of tests (Blight and Clarke [14]) were performed in which quartzite drill cores representing the pillars were embedded in either stiff fill or a softer material consisting of cemented fine tailings. The fill and cores were contained in steel moulds, instrumented with strain gauges in order to measure lateral stresses developed in the fill. The composite core-fill system was loaded via a rigid piston. The stress–compression relationship for the core and the relationship between compression of the core and lateral stress in the fill were recorded.

A typical set of these relationships is shown in Figure 39. The figure shows that the soft fill provided very little lateral support to the core, as the strengths of the unconfined core and the core supported by soft fill were almost identical. Relatively little lateral stress was generated in the soft fill,

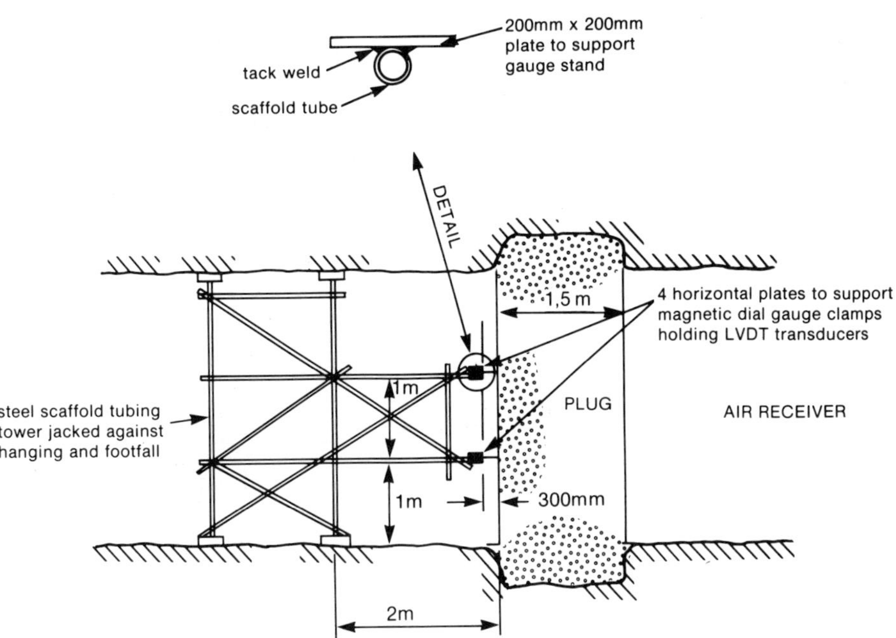

FIGURE 37. Section through a compressed air receiver plug showing arrangements to measure deflexion of plug under pressure.

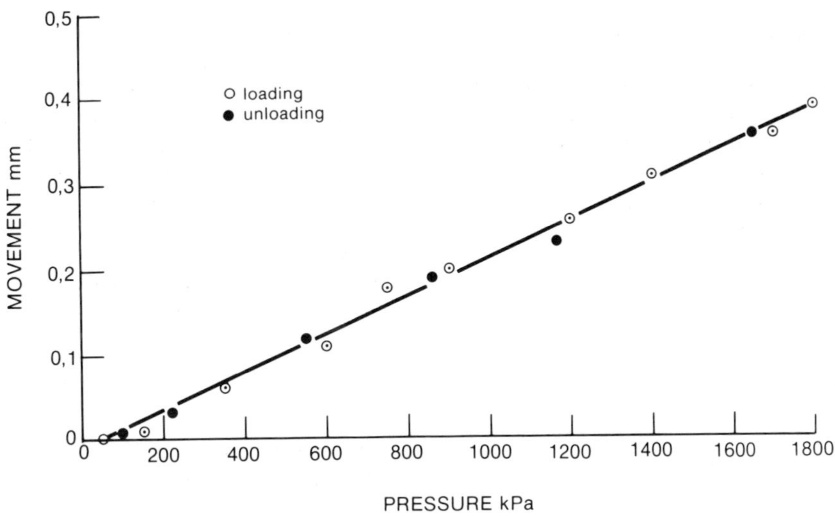

FIGURE 38. Pressure-deflexion record for air receiver plug.

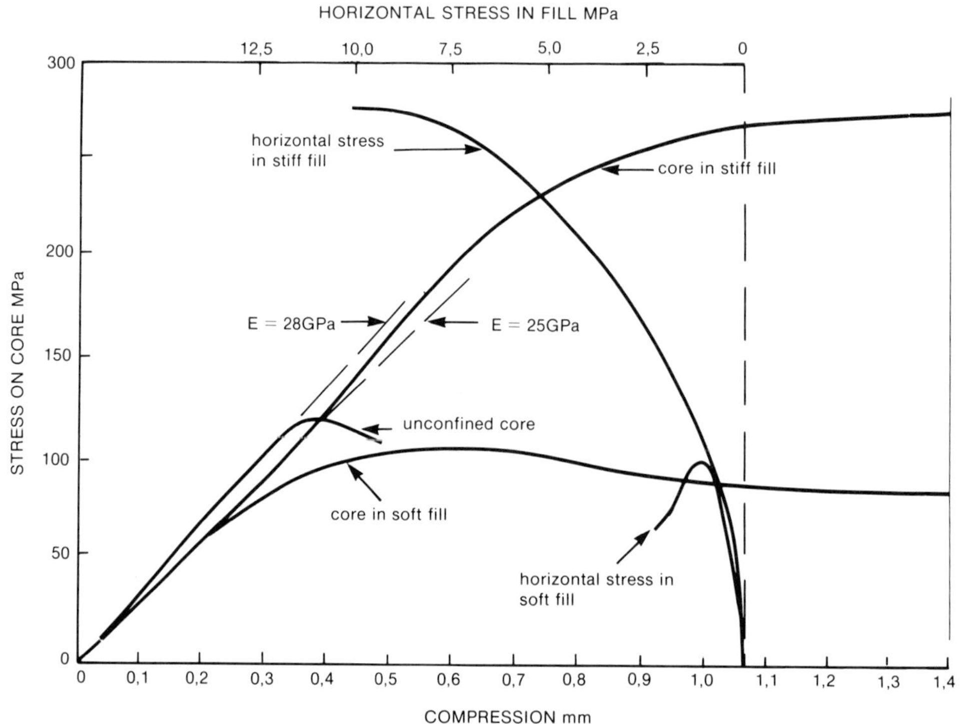

FIGURE 39. Comparison of stress-compression behaviour of unconfined drill core (representing rock pillar) with that of fill-supported cores.

but there was enough lateral support to maintain a post-failure strength of about 85% of the peak strength.

A considerable lateral stress was generated in the stiff fill and Figure 39 shows that a peak strength was not reached by the core, even at a compression of 1.4mm or 1.75%. It therefore appears that surrounding a highly stress pillar with a soft fill will not materially improve the strength of the pillar, but will provide a high level of residual resistance. A stiff fill, on the other hand, can considerably increase the strength of a pillar.

Figure 39 also shows that because the fill must be laterally compressed in order to provide lateral support, surrounding a pillar with fill has little or no effect on the pre-failure compression modulus of the pillar.

The mechanism of pillar support can be modelled as follows: Figure 40 represents a cylindrical pillar of initial height S and diameter D which is compressed at approximately constant volume by an amount v. The lateral expansion h of the pillar is then given approximately by

$$h = \mathrm{v} \cdot \frac{D}{2S} \tag{24}$$

The pressure required to expand a cylindrical pillar into a surrounding elastic fill is given by

$$\Delta\sigma_h = 2G \cdot \frac{h}{D} \quad = G \cdot \frac{V}{S} \tag{25}$$

(The process of expansion can be likened to the expansion of a cylindrical cavity in an elastic medium [15]).

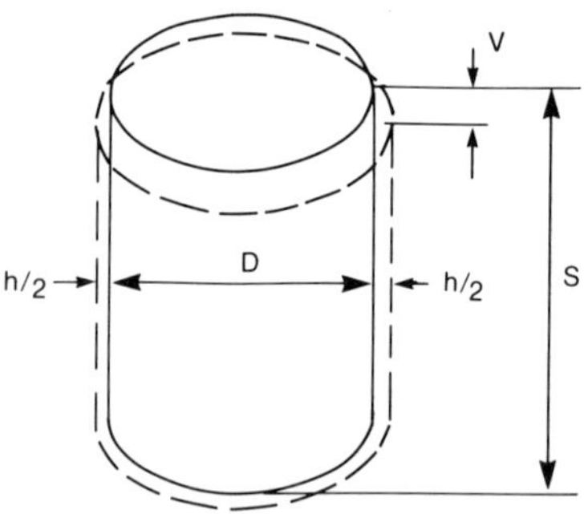

FIGURE 40. Dimensions used in analysis of pillar-fill system.

$\Delta\sigma_h$ is the expanding pressure, i.e., the supporting pressure on the pillar and G is the shear modulus of the fill.

For an elastic material, the shear modulus is related to the elastic modulus E by

$$G = \frac{E}{2(1 + \nu)} \tag{26}$$

where ν is Poisson's ratio.

ν and E for the fill can be evaluated from the results of one-dimensional compression tests such as those described by Figures 21 and 22.

It can be shown that if a material is subjected to vertical one-dimensional compression, the ratio of horizontal to vertical stresses is

$$\sigma_h/\sigma_v = K_o = \frac{\nu}{1 - \nu}$$

or

$$\nu = \frac{K_o}{1 + K_o} \tag{27}$$

Also, for such a process, the vertical strain

$$\epsilon_v = \mathrm{v}/S = \frac{\sigma_v}{E} \cdot \frac{(1 + 2K_o)}{(1 + K_o)}$$

Hence,

$$G = \frac{S}{\mathrm{v}} \cdot \sigma_v \frac{(1 - K_o)}{2} \tag{28}$$

or, from Equation (23),

$$\Delta\sigma_h = \sigma_v \frac{(1 - K_o)}{2} \tag{29}$$

In Equation (29) σ_v represents the vertical stress in the fill surrounding the pillar, while $\Delta\sigma_h$ represents the additional horizontal stress generated in the fill when the pillar and the fill have both compressed by an amount v.

The additional vertical strength of the pillar at this stage will be given by

$$\Delta\sigma_v{}^P = \Delta\sigma_h \frac{1 + \sin\phi^P}{1 - \sin\phi^P} \tag{30}$$

where ϕ^P is the angle of shearing resistance of the cohesionless fractured rock of which the pillar now consists.

For a very stiff fill at a compression of 2% (see Figures 24 and 25) $K_o = 0.2$ and $\sigma_v = 20$MPa. Hence, $\Delta\sigma_h = 8$MPa.

Reference to Figure 39 will show that in the model tests referred to earlier, $\Delta\sigma_v{}^P$ was about 100MPa when $\Delta\sigma_h$ was 7MPa; i.e., the ratio $(1 + \sin\phi^P)/(1 - \sin\phi^P)$ was 14 which corresponds to a value for ϕ^P of 60°. This is a perfectly possible value for a confined fractured rock. For a good soft fill at a compression of 2%, $K_o = 0.45$ and $\sigma_v = 5$MPa. Here $\Delta\sigma_h = 1.4$MPa which is again similar to the measured value in Figure 34.

If the failure strain for the confined pillar is taken to be the same as that for the unconfined core, $\Delta\sigma_v{}^P$ was about 12MPa which corresponds to a value for ϕ^P of 58°.

Hence there is a reasonable correspondence between the results of the model tests and the predictions of Equations (27) and (28), if suitable values are used for ϕ^P.

ARTIFICIAL PILLARS IN SHALLOW MINES

Many coal mines are relatively shallow (30 to 100m below surface) and may contain a sequence of superimposed coal seams. The top or shallowest seam of a sequence can be completely extracted if the severe surface subsidence that ensues can be tolerated. Deeper seams, however, cannot be totally extracted unless they are extracted in sequence from the shallowest to the deepest. The coal contained by different seams is often of different qualities and types and commercial demand for coal of specific qualities dictates the sequence in which the seams are extracted. As a result, most multi-seam mines are mined using the room-and-pillar or bord-and-pillar method, in which pillars of coal are left in place to support the roof. This practice results in percentages of extraction varying from 75% in very shallow seams to only 40% in deeper seams.

Although coal provides a cheap form of roof support, coal pillars represent a natural energy resource that has been sterilized and research is currently in progress to find ways and means of winning this sterilized coal, or at least, increasing the percentage extraction.

As described in the previous section, one way of achieving a greater extraction is to use back fill to strengthen pillars, thus enabling smaller pillars to be used. However, if it is important to limit vertical compressions, it must be remembered that the pillars actually have to fail and therefore, to compress appreciably before the strengthening effect of the fill will become operative.

Another technique that is currently under development (Hahn, Blight and Dison [16]), consists of replacing the coal pillars by artificial pillars or walls of horizontally reinforced granular material. In principle, any granular material can be used for this purpose. However, from environmental and cost aspects, it is preferable to use mining or other waste such as power station bottom ash. Ash is a particularly appropriate choice if the purpose of the colliery is to supply a nearby power station.

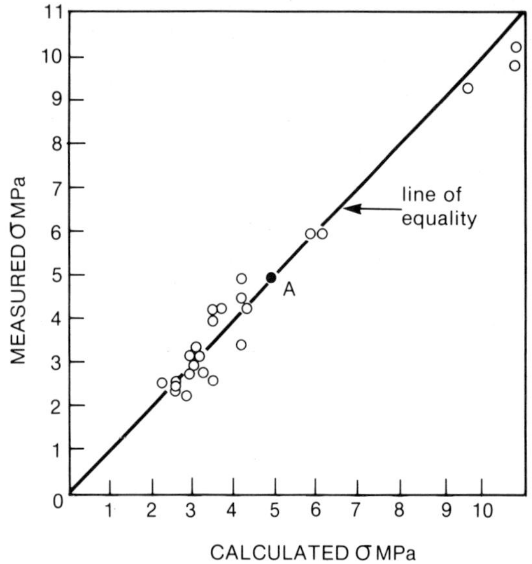

FIGURE 41. Relationship between calculated and measured strengths of walls built of horizontally reinforced granular material.

STRENGTHENING A GRANULAR MATERIAL BY HORIZONTAL REINFORCING

The effect of horizontal reinforcing on a granular material is to develop a horizontal confining stress. The horizontally reinforced granular material will tend to go into a state of failure as soon as a vertical stress is applied to it. Actual failure will not occur because of the development of tension in the horizontal reinforcing. When the horizontal reinforcement reaches its yield stress σ_{yR}, the vertical stress σ_v will be related to the horizontal stress σ_h by the relationship

$$\sigma_v = K_p \sigma_h \tag{31}$$

in which K_p is the passive pressure coefficient of the granular material.

Equating the tension in the horizontal reinforcing to the compression in the granular material $\sigma_{yR} A_R = \sigma_h (A_m - A_R)$ in which A_R is the cross-sectional area of reinforcing and A_m is the area of reinforced material, or

$$\sigma_h = \frac{\sigma_{yR} A_R}{(A_m - A_R)} \tag{32}$$

Therefore,

$$\sigma_v = \sigma_{max} = \frac{K_p \sigma_{yR} A_R}{(A_m - A_R)} \tag{33}$$

If A_R is small as compared with A_m

$$\sigma_v = K_p \cdot \sigma_{yR} \cdot p \tag{33a}$$

in which

$$p = \frac{A_R}{A_m} = \frac{A_R}{\text{vh}}$$

is the reinforcing ratio, with v the vertical and h the horizontal spacing of the reinforcing.

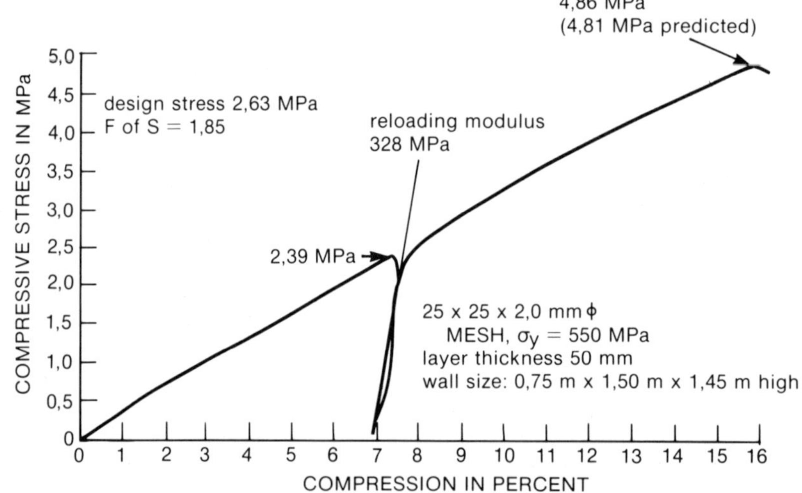

FIGURE 42. Stress-compression curve for wall of horizontally reinforced ash.

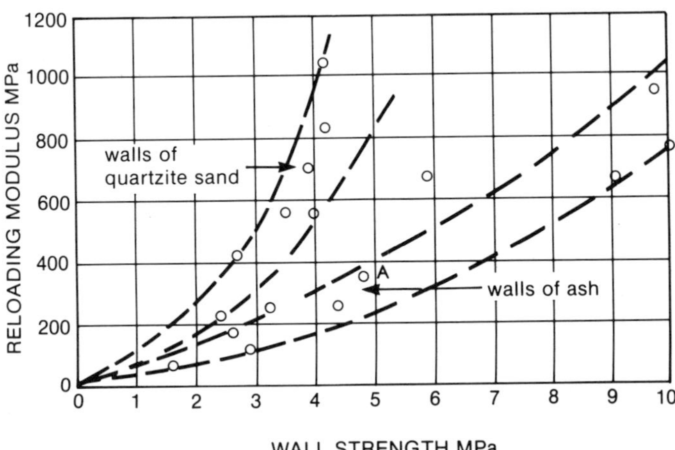

FIGURE 43. Relationship between wall strength and compression modulus on reloading for walls of horizontally reinforced granular materials.

The strength of a horizontally reinforced granular material is thus directly proportional to the reinforcing ratio p and the yield stress of the reinforcing σ_{yR}.

The properties of horizontally reinforced granular materials have been extensively investigated by means of both model and full-scale tests. Figure 41 shows a comparison of σ_v calculated from Equation (33a), with corresponding measured values. It will be seen from Figure 41 that there is an excellent correlation between calculated and measured values over a wide range of stresses.

Figure 42 (test A on Figure 41) represents the results of a typical compression test on a reinforced wall. This diagram illustrates an important feature of walls or pillars of horizontally reinforced granular material. The wall was designed to have a factor of safety of 1.6 on a design vertical stress of 2.39MPa. The wall was loaded to the design load and was then unloaded and reloaded to failure. The stress–compression curve for the first loading indicates a low compression modulus. However, the reloading curve is considerably steeper; hence the reloading modulus is much higher. Figure 42 illustrates how it is intended to use the walls of reinforced granular material in practice. As the load is subsequently transferred to the wall when the roof tends to sag, the wall will be recompressed along the recompression curve, back to the design load. The support system thus created, will be both relatively stiff and pre-tested to the design load. The stiffness is particularly important in limiting bending deflections, and therefore bending stresses in the roof.

The recompression modulus of horizontally reinforced walls is related to their strength as shown by Figure 43. The scatter of results illustrated in this figure is caused mainly by variation in the characteristics of the granular materials being used. The upper band of results corresponds to walls built of a highly frictional weathered quartzite sand. The lower band of results corresponds to walls built of power station bottom ash which, although highly frictional, is considerably more compressible than the sand.

FAILURE MECHANISM OF REINFORCED GRANULAR WALLS

As the design theory indicates, reinforced granular walls fail when the tensile strength of the horizontal reinforcing is reached. When this occurs, the reinforcing wires fracture and the granular material shears along diagonal planes inclined approximately at $(45° - \phi'/2)$ to the direction of the major principal or vertical stress.

Figure 44 shows the observed positions at which the reinforcing wires fractured in a typical test wall. The multiple fractures of the reinforcing define a system of multiple shear planes inclined at a mean angle of 26° to the direction of the major principal stress. This mean angle corresponds to the theoretical angle $(45° - \phi'/2)$ for the fill material for which $\phi = 38°$.

Theoretically, it is possible for an infinite number of shear planes to develop in the granular material. In practice, however, the development of shear planes is affected by the frictional restraint exerted on the top and bottom of the wall. The combination of end restraint and the limited height to width ratio of the walls prevents shear planes from developing fully, except in the diagonal corner-to-corner configuration.

This geometrical effect can be exploited to increase the strength of a pillar by decreasing its ratio of height to width.

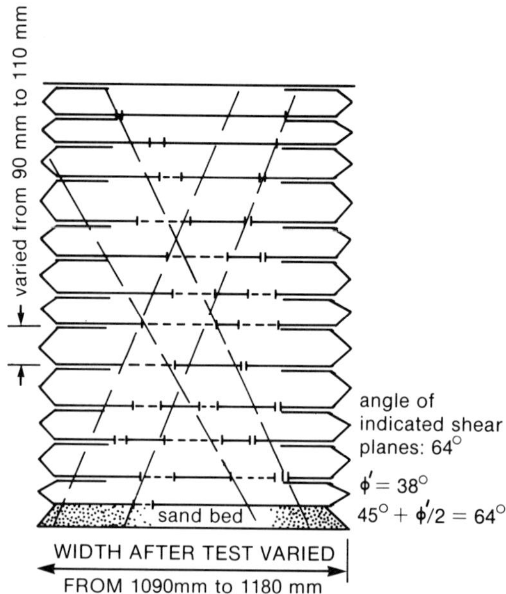

FIGURE 44. Positions of tensile failures in reinforcing of horizontally reinforced wall of granular material.

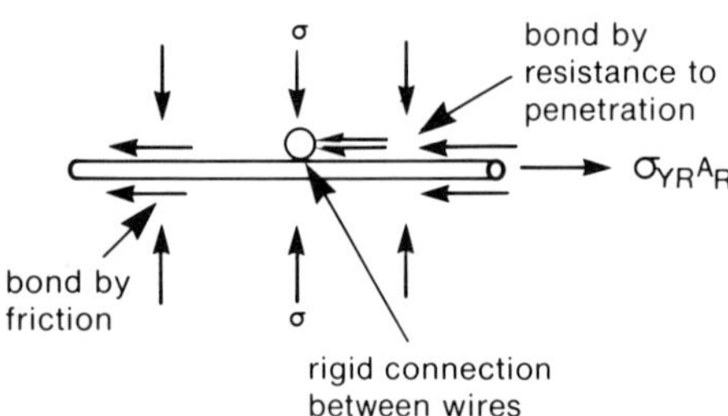

FIGURE 45. Mechanism of bond development in reinforced granular material.

BOND OF REINFORCING TO REINFORCED MATERIAL

The tension developed in the reinforcing of a horizontally reinforced granular material depends on stress transfer by bond between the reinforcing and the reinforced material. The mechanism of bond development is illustrated in Figure 45.

The average bond stress transmitted to a reinforcing wire over a length x by friction is

$$\frac{\sigma_v}{2} \cdot \left\{ 1 + \frac{1}{K_p} \right\} \tan\phi \pi D \cdot \frac{x}{2}$$

in which the diameter of the wire is D.

The most convenient form of horizontal reinforcing for a wall consists of a system of orthogonal wires rigidly bonded to each other at points where they cross. Readily available systems consist of either square or rectangular welded mesh or twisted diamond mesh. The wires that run at an angle to the direction of tension under consideration contribute considerably to the bond because they must penetrate the reinforced material before the reinforcing can slip relative to the material. If the bond wires in an orthogonal system are spaced ℓ apart, the load transferred through these wires in length x will be a minimum of:

$$\frac{x}{\ell}\left(\frac{\sigma_v}{K_p} \cdot d + 2d\sigma_v \tan\phi \right) \mathrm{h}$$

in which d is the diameter of the bond wires that are spaced h apart. Full bond resistance of the wire is therefore developed in a length x given by

$$\frac{\pi D^2}{4}\sigma_{yR} = \frac{\sigma_v}{4}\left(1 + \frac{1}{K_p} \right) \tan\phi \pi D x + \frac{\sigma_v}{4\ell}\left(\frac{4}{K_p} + 8\tan\phi \right) \mathrm{h} \cdot xd \qquad (34)$$

The value of x is astonishingly small.

For example, if

$\sigma_v = 5$Mpa, $xd = d = 2.5$mm, $\mathrm{h} = \ell = 25$mm, $\sigma_{yR} = 600$MPa, $K_p = 3.5$, $\tan\phi = 0.7$

then, $x = 98$mm.

Figure 46 illustrates the results of a series of tests on reinforced walls that was designed to investigate the validity of Equation (34). The figure shows the stress at failure in a series of tests on walls in which the number of bond wires at each side of the wall was increased progressively from zero. The spacing of the bond wires was 12.5mm and the theoretical bond length was 60mm; hence full bond could

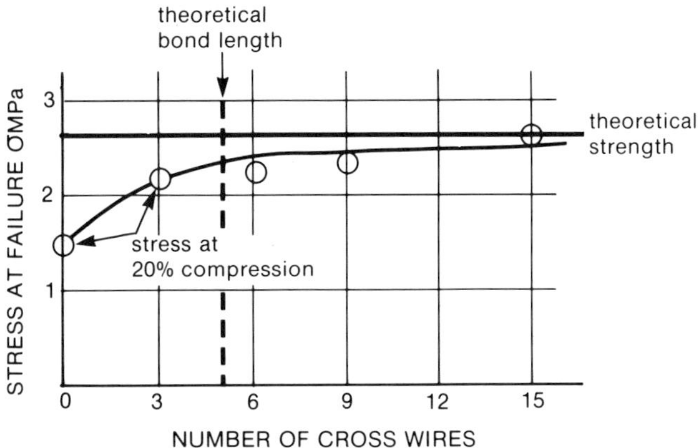

FIGURE 46. Progressive development of strength of reinforced granular walls as number of cross wires developing bond increases.

theoretically be achieved with 5 bond wires. The results in Figure 46 confirm this.

SPIRALLY OR HOOP REINFORCED COLUMNS

As an alternative to embedding reinforcing mesh in the fill, a cylindrical column of fill can be reinforced by surrounding it with a steel spiral or a series of steel hoops. A radial confining stress is then developed as shown in Figure 47.

For horizontal equilibrium of the column,

$$2Rp\sigma_h = 2T = \frac{2A_R\sigma_{yR}}{F}$$

i.e.,

$$\sigma_h = \frac{A_R\sigma_{yR}}{FRp}$$

and

$$\sigma_v = \frac{K_pA_R\sigma_{yR}}{FRp} \tag{35}$$

in which p is the pitch or spacing of the reinforcing hoops or coils.

Figure 48 shows a comparison of measured strengths of hoop and spirally reinforced columns with values predicted from Equation (35).

The agreement between theory and measurement is excellent and the comparison also shows that Equation (35) applies equally well to both spiral and hoop reinforcement.

Bond between the reinforcement and the fill plays no part in a spirally reinforced column. However, to transfer the confining stress into the fill, it is necessary to have a retaining membrane between the fill and the reinforcing. In the laboratory tests carried out so far, a woven polypropylene hessian has been used for this purpose.

The reloading moduli of spirally or hoop reinforced columns are comparable with those of mesh reinforced walls.

COMPARISON OF WALL AND COLUMN SYSTEMS

According to Equation (33a) the load carried by unit plan area of horizontally reinforced wall is

$$\sigma_v = \frac{K_p\sigma_{yR}}{F} \cdot \frac{A_R}{\text{vh}} \tag{33a}$$

A_R/vh is then the volume of reinforcement per unit volume of wall and F is the factor of safety.

Similarly, from Equation (35) the load carried per unit area of spirally reinforced column is

$$\sigma_v = \frac{K_p\sigma_{yR}}{F} \cdot \frac{A_R}{Rp} \tag{35a}$$

In this case, $2A_R/R_p$ is the volume of reinforcement per unit volume of column.

The major cost item in a reinforced granular fill support is the reinforcing steel, hence the ratios A_R/vh or $2A_R/Rp$ are, to a large extent, the key to the economic viability of the system. For the same quality of fill and reinforcing and the same factor of safety $K_p\sigma_{yR}/F$ will be the same. Hence, comparing horizontally reinforced walls with spirally reinforced columns, the two systems will be equally viable economi-

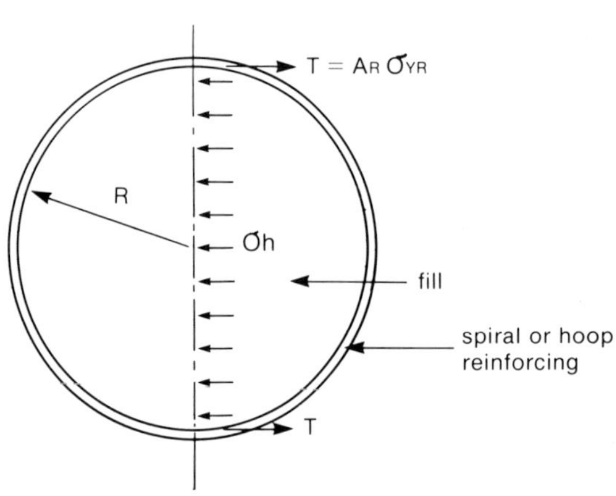

FIGURE 47. Basis for calculating the strength of a spirally or hoop reinforced granular column.

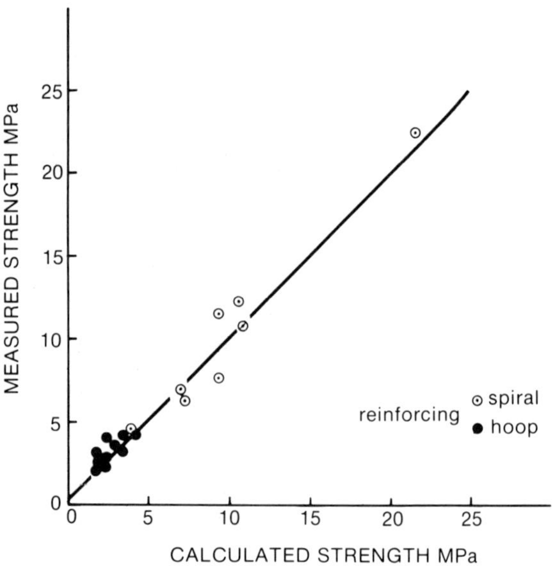

FIGURE 48. Comparison of measured and calculated strengths of spirally and hoop reinforced granular columns.

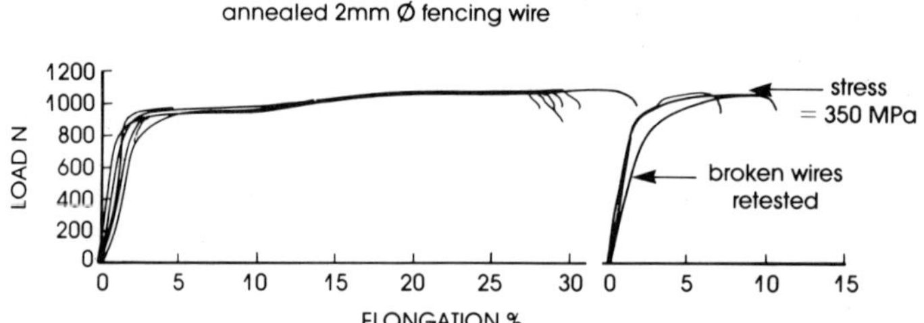

FIGURE 49. Stress-strain curves for annealed wire reinforcing.

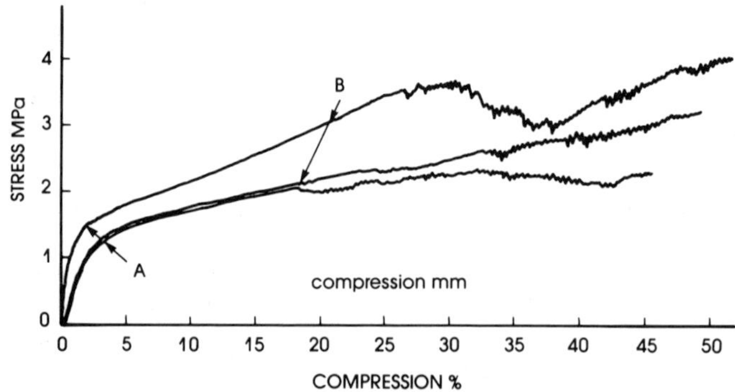

FIGURE 50. Stress-compression relationships for pillars of cemented granular material reinforced with annealed welded steel wire mesh.

cally if (1)

$$\frac{2A_R(\text{col})}{Rp} = \frac{A_R(\text{wall})}{\text{vh}} \tag{36}$$

and (2) equally strong if

$$\frac{A_R(\text{col})}{Rp} = \frac{A_R(\text{wall})}{\text{vh}} \tag{37}$$

The two requirements clearly cannot be met simultaneously. For equal strength, if A_R(col) = A_R(wall), then Rp = vh. It then follows from Equation (6) that a spirally reinforced column will require twice the steel required by the equivalent horizontally reinforced wall.

In practical terms p will be of the same order of magnitude as v (25 to 100mm). Hence, R must of necessity be of the same order of magnitude as h if A_R(col) is to be kept within realistic bounds. The situation is best illustrated by an example.

Consider a wall for which σ_v = 5MPa, K_p = 3.5, σ_{yR} = 1000 MPa, F = 1.75, h = 50mm, and A_R(wall) = 20mm^2 (equivalent to a wire diameter of 5 mm). Then

$$\text{v} = \frac{K_p \cdot \sigma_{yR} A_R}{Fh\sigma} = 160 \text{ mm}$$

Now consider an equivalent spirally reinforced column having p = 160mm.

$$R = \frac{K_p \sigma_{yR} A_R}{\sigma F p} = 50 \text{ mm}$$

which is clearly impractical.

If p is reduced to 30mm, R = 267mm, a more reasonable value. The example shows that, whereas a horizontally reinforced wall may be constructed to any plan dimensions, the radius of a spirally reinforced column is limited to a maximum of about a metre by practical considerations.

Calculating the ratio of volume of reinforcement to volume of fill for the above example, we find that for the wall A_R/vh = 0.25%, whereas for the column $2A_R/Rp$ = 0.5%, which agrees with the conclusion arrived at above.

A square or rectangular horizontally reinforced pillar will need to have the same reinforcement normal to each of its pairs of sides. Hence, the ratio of volume of reinforcement to volume of fill will be $2A_R$/vh in this case. (Because of the necessity of providing bond steel in a wall, the ratio of reinforcement volume to fill volume is actually slightly higher than A_R/vh.)

It is seen, therefore, that for isolated supports both systems require the same amount of reinforcing. The spirally reinforced column does, however, have potential advantages in that: (1) ultra-high tensile steel can be used in spiral form whereas this is not possible in mesh form because welding of the nodes locally reduces the strength; (2) the cylindrical mass of fill can be placed and compacted either before or after placing the reinforcing; and (3) the steel is open for inspection.

YIELDING PILLARS OF REINFORCED CEMENTED MATERIAL

The initial compression modulus of a reinforced granular pillar can be considerably enhanced by using a cemented material instead of an uncemented granular material.

Whereas in a shallow mine there is a need for a rigid permanent type of support, in deep mines, the need is more often for a support that has a high initial stiffness, but which will then yield while maintaining an approximately constant load. This aim can be achieved by combining cemented granular material with annealed ductile steel reinforcing.

Figure 49 shows the remarkable ductility achievable with annealed mild steel wire. The specimens tested to produce Figure 49 were cut from welded steel mesh which was annealed after welding.

Figure 50 shows the stress-compression behaviour of pillars of cemented granular material reinforced with annealed steel mesh. Points A in the figure represent the onset of failure of the cemented material which up to this point has been carrying virtually the entire load. From A to B, the load is carried by a combination of friction in the granular material and tension in the steel. Point B represents the first failure of the steel mesh reinforcing. From B onwards, the steel is failing and transferring the stress further along the reinforcing. Simultaneously, the load-bearing area of the support is increasing, and hence the stress in the composite material is decreasing.

CONCLUDING REMARKS

This chapter has illustrated a number of applications for soil mechanics principles in underground mining. There is a considerable potential for the applications described.

No doubt, there are many more potential applications as yet unthought of. As the chapter has shown, conventional soil mechanics principles are completely applicable to many problems in underground mining, even though stresses may be in a completely different range than that associated with soil mechanics on surface.

REFERENCES

1. Blight, G. E., C. E. Rea, J. A. Caldwell, and K. W. Davidson, "Environmental Protection of Abandoned Tailings Dams," *Pro-*

ceedings, 10th International Conference on Soil Mechanics and Foundation Engineering, Stockholm, Vol. 2, pp. 303–308 (1981).

2. *Guidelines for Environmental Protection, Vol. 1/83: The Design, Operation and Closure of Residue Deposits,* Chamber of Mines of South Africa (1983).
3. Lambe, T. W. and R. V. Whitman, *Soil Mechanics.* New York, Wiley (1969).
4. Avalle, D. L., "Properties of Weakly Cemented Slurries of Gold Mine Slimes." MSc(Eng) thesis, University of the Witwatersrand, Johannesburg (1976).
5. Dobry, R. and L. Alvarez, "Seismic Failures in Chilean Tailings Dams," *Journal of the Soil Mechanics and Foundation Engineering Division, ASCE, Vol. 93,* No SM6, pp. 237–260 (1976).
6. Marcuson, W. F., R. F. Ballard, and R. H. Ledbetter, "Liquefaction Failure of Tailings Dams Resulting from the Near Izu Oshima Earthquake, 14 and 15 February 1978", Pre-print, *Proceedings, 6th Pan American Conference on Soil Mechanics and Foundation Engineering,* Lima, Peru (1979).
7. Jennings, J. E., "The Failure of a Slimes Dam at Bafokeng. Mechanisms of Failure and Associated Design Considerations," *The Civil Engineer in South Africa, Vol. 21,* No 6, pp. 135–141 (1979).
8. Byrne, P. M., "The Effect of Seismic Activity on Tailings Impoundments in British Columbia," Submission to Royal Commission of Enquiry into Health and Environmental Implications of Uranium Mining (1980).
9. McGarr, A., R. W. E. Green, and S. M. Spottiswoode, "Strong Ground Motion of Mine Tremors: Some Implications for Nea Source Ground Motion Parameters," Chamber of Mines of South Africa Research Organization, Research Report no 32/80 (1980).
10. Knutsson, S., "The Näsliden Project-Stresses in the Hydraulic Backfill from Analytical Calculations and in situ Measurements," *Proceedings, Conference on Application of Rock Mechanics to Cut-and-Fill Mining,* Lulea, Sweden, Vol. 2, pp. 283–288 (1980).
11. Gibson, R. E., "The Progress of Consolidation in a Clay Layer Increasing in Thickness with Time," *Geotechnique, Vol. 8,* No 4, pp. 171–182 (1958).
12. Mitchell, R. J., *Earth Structures Engineering,* Allen and Unwin, Boston, USA (1983).
13. Ockleston, A. J., Discussion of Garret, W. S. and Campbell, Pitt, L. T., "Tests on an Experimental Underground Bulkhead for High Pressures," *Journal, South African Institute of Mining and Metallurgy,* Vol. 58, No 11, pp. 262–266 (1958).
14. Blight, G. E. and I. E. Clarke, "Design and Properties of Stiff Fill for Lateral Support of Pillars," in *Mining with Backfill, Proceedings, International Symposium on Mining with Backfill,* Lulea, Sweden, pp. 350–354 (1983).
15. Baguelin, F., J. F. Jezequel, and D. H. Shields, *The Pressuremeter and Foundation Engineering,* Trans Tech, Switzerland (1978).
16. Hahn, J. A., L. Dison, and G. E. Blight, "Supports of Reinforced Granular Fill", in *Mining with Backfill, Proceedings, International Symposium on Mining with Backfill,* Lulea, Sweden, 1983, pp. 349–354.

CHAPTER 17

Design of Buried Pipelines

H. S. Oey*

Any pipeline installed in the ground is a buried pipeline. It can be a cross country line or a municipal distribution line. It can be transporting gas, oil or water. It can be made of steel, concrete, clay or any suitable material. Regardless of the pipe material or the intended use of the buried pipe it has to be designed properly. If the pipe is to cross under a road it must be adequate to provide: (1) sufficient support for the soil, the road surface and the live loads over the pipe; (2) sufficient rigidity so that excessive flattening of the pipe does not occur; and (3) sufficient strength so that combinations of internal and external forces do not cause failure of the pipe. Cross country pipelines are designed in a similar manner excluding the road surface and corresponding live loads.

The analysis of pipeline crossings can be broken down into two main categories:

1. Determination of forces acting on the pipeline
2. Determination of the stresses and deformations in the pipe section due to combinations of the forces acting on the pipe

DETERMINATION OF FORCES ACTING ON THE PIPE

In this category the forces include the pipe's internal pressure, as well as the external pressures acting on the pipe due to the weight of soil and paving above the pipe, the wheel loads and other live loads.

Dead Load

This represents the load resulting from the soil and pavement above the pipe.

*Department of Civil Engineering, University of Texas at El Paso, El Paso, TX

MARSTON'S EARTH LOAD

Marston's theory [18] is generally accepted for soil load calculation.

Marston Rigid Pipe Method

$$W_d = C_d \gamma B_d^2 \tag{1a}$$

in which: W_d is the load on the pipe in lbs. per linear foot; C_d is a load coefficient; γ is the specific weight of soil in lbs. per cubic foot; and B_d is the width of the trench at the top of the pipe, in feet.

The soil coefficient, C_d, is a function of:

1. The ratio of the height of soil cover above top of pipe (H) to the width of trench (B_d)
2. The internal friction of the soil backfill
3. The coefficient of friction between the backfill and the sides of the trench

The coefficient C_d is obtained from the formula

$$C_d = \frac{1 - e^{-2K\mu' H/Bd}}{2K\mu'}$$

in which K = ratio of active lateral unit pressure to vertical unit pressure, $\mu = \tan \phi$ = coefficient of internal friction of fill material, $\mu' = \tan \phi'$ = coefficient of friction between fill material and sides of ditch.

The coefficient μ' may be equal to or less than μ, but cannot be greater than μ.

Values of C_d are shown in Table 1 and Nomograph 1.

Nomograph 2 can be used to find values of W_d in Kg/cm as function of H/B_d, soil type, its specific weight, and the pipe diameter.

Marston Flexible Pipe Method

Marston also determined that if the pipe is flexible and is placed with thoroughly tamped side fills having essentially

TABLE 1. Safe Working Values for the Coefficient C_d to Use in Marston's Equation for Calculating Loads on Pipes in Trenches.

Ratio H/B = height of fill above top of pipe to breadth of ditch a little below the top of the pipe (1)	Minimum possible without cohesion. These values give the loads generally imposed by granular filling materials before tamping or settling (2)	Maximum for ordinary sand. Use these values as safe for all ordinary cases of sand filing (3)	Completely saturated topsoil (4)	Ordinary maximum for clay (thoroughly wet). Use these values as safe for all ordinary cases of clay filing (5)	Extreme maximum for clay (completely saturated). Use these values only for extremely unfavorable conditions (6)
0.5	0.455	0.461	0.464	0.469	0.474
1.0	0.830	0.852	0.864	0.881	0.898
1.5	1.140	1.183	1.206	1.242	1.278
2.0	1.395	1.464	1.504	1.560	1.618
2.5	1.606	1.702	1.764	1.838	1.923
3.0	1.780	1.904	1.978	2.083	2.196
3.5	1.923	2.075	2.167	2.298	2.441
4.0	2.041	2.221	2.329	2.487	2.660
4.5	2.136	2.344	2.469	2.650	2.856
5.0	2.219	2.448	2.590	2.798	3.032
5.5	2.286	2.537	2.693	2.926	3.190
6.0	2.340	2.612	2.782	3.038	3.331
6.5	2.386	2.675	2.859	3.137	3.458
7.0	2.423	2.729	2.925	3.223	3.571
7.5	2.454	2.775	2.982	3.299	3.673
8.0	2.479	2.814	3.031	3.366	3.764
8.5	2.500	2.847	3.073	3.424	3.845
9.0	2.518	2.875	3.109	3.576	3.918
9.5	2.532	2.898	3.141	3.521	3.983
10.0	2.543	2.918	3.167	3.560	4.042
11.0	2.561	2.950	3.210	3.626	4.141
12.0	2.573	2.972	3.242	3.676	4.221
13.0	2.581	2.989	3.266	3.715	4.285
14.0	2.587	3.000	3.283	3.745	4.336
15.0	2.591	3.009	3.296	3.768	4.378
Very great	2.599	3.030	3.333	3.846	4.545

Note: Use data from column (5) unless highway subsoil is known to be material specified in columns (2), (3), (4) or (6).

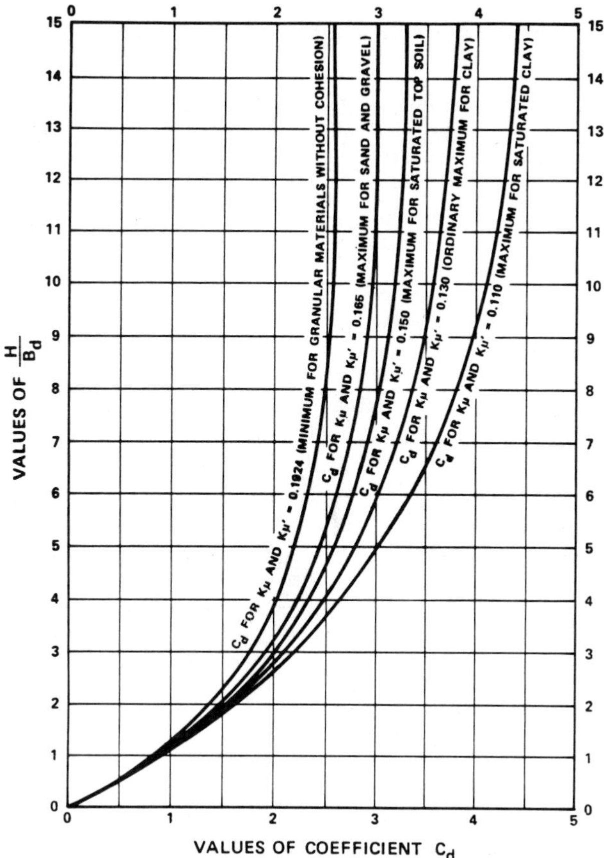

NOMOGRAPH 1. Relationship of C_d and H/B_d.

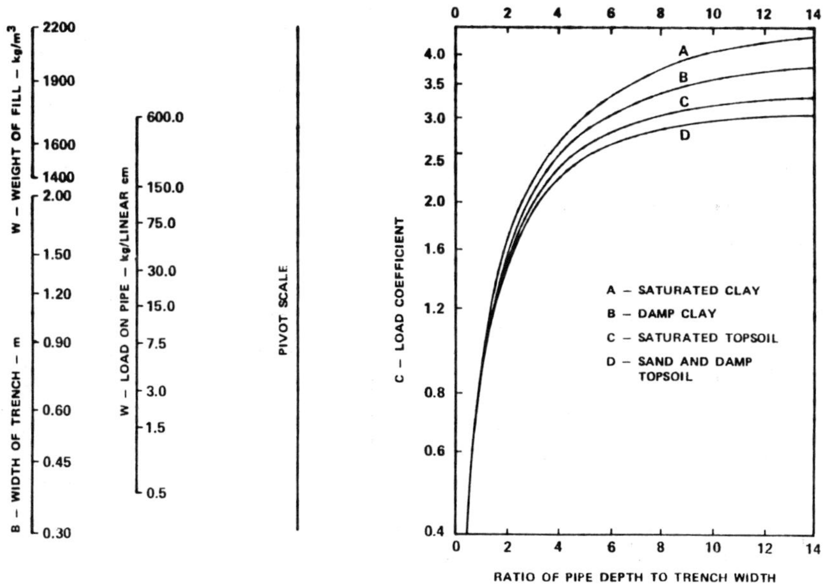

NOMOGRAPH 2. Loads on buried pipes.

the same degree of stiffness as the pipe itself, the load W_d can be reduced by a factor of D/B_d, where D equals the outside diameter of the pipe. The equation then becomes Marston's Flexible.

$$W_d = C_d \gamma D B_d \tag{1b}$$

Marston Modified Method

If the trench width is assumed to be the outside pipe diameter, the equation becomes Marston's Modified.

$$W_d = C_d \gamma D^2 \tag{1c}$$

DIRECT EARTH LOAD

Here the earth load is the weight of a unit prism of soil with a height equal to the distance from the top of the pipe of the ground surface, and a width equal to the diameter of the pipe. Thus,

$$W_d = \gamma H D \tag{2}$$

(pounds per linear foot). This method is used for the earth load computation in the American National Standard for the thickness design of ductile-iron pipe (ANSI A21.50-1976).

RELATIONSHIP

As can be seen in Figure 1, the most conservative loading would result from the use of Marston's Rigid Pipe Method. The least conservative would result from Marston's Modified Method in which the trench width is taken as the outside diameter of the pipe.

Live Loads without Slabs

This section presents the most accepted methods for computing the live load on a pipe that does not have a protective rigid slab.

SPANGLER'S SINGLE LOAD METHOD

This method applies to live or external loads on flexible pavements or no pavement, as in Boussinesq's Point Load formula discussed later.

It is assumed that the wheel load acts as a point load, IP, as shown in Figure 2. For circular conduits, the area of the section projected on horizontal plane through the top of the conduit is considered, and the load distribution is shown in the figure as σ.

The average load per linear foot (or inch), W_L, on the pipe is computed as [45]:

$$W_L = \frac{1}{L} C_T I P \tag{3}$$

in which

A is 3 ft. for long pipes

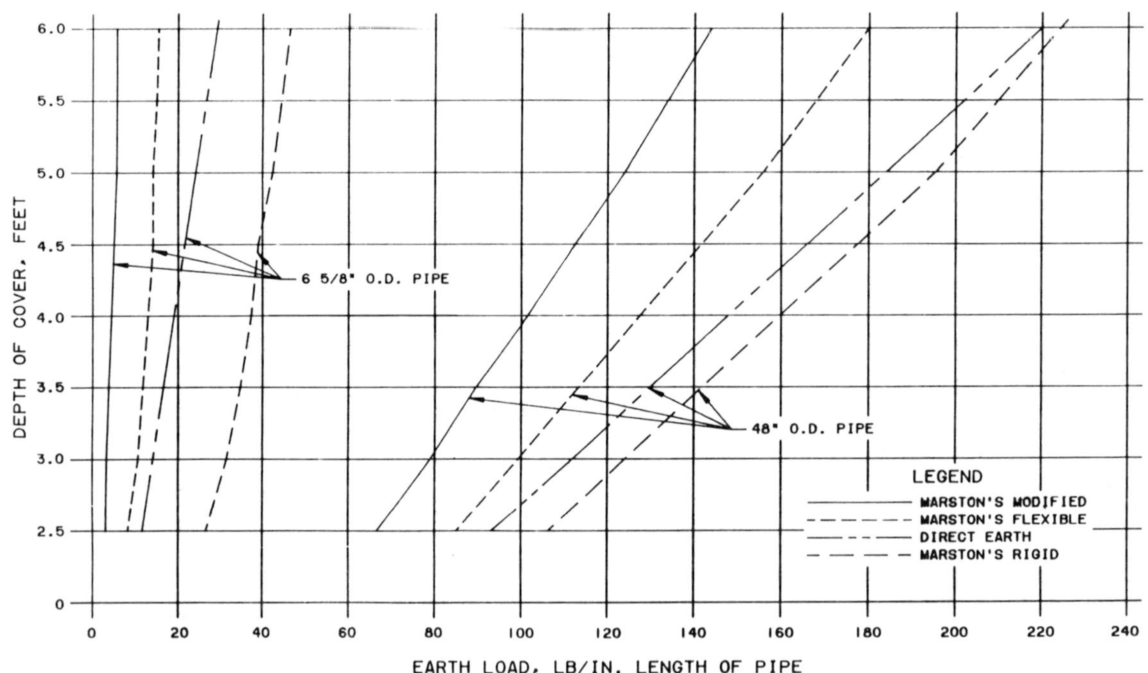

FIGURE 1. Comparison of earth loads from different formulas.

I is the impact coefficient
C_T is the Holl's influence coefficient, as shown in Table 2
P is the applied load (lbs.)

The equation is based on Boussinesq's formula for a point load applied at the surface of a semi infinite elastic medium,

$$\sigma = \frac{3P}{2\pi}\frac{z^3}{R^5}$$

in which

σ = the vertical unit pressure at a point whose coordinates are x, y, and z
P = a concentrated load applied at the surface and at the origin of coordinates
z = the depth of the point (x, y, z)
$R = \sqrt{x^2 + y^2 + z^2}$

Integration of the vertical load imposed by a concentrated surface load on a horizontal plane of dimensions a and b as shown in Figure 2 yields values of C_T tabulated in Table 2. Thus, to obtain the actual C_T for Equation (3), the value obtained from Table 2 is multiplied by 4.

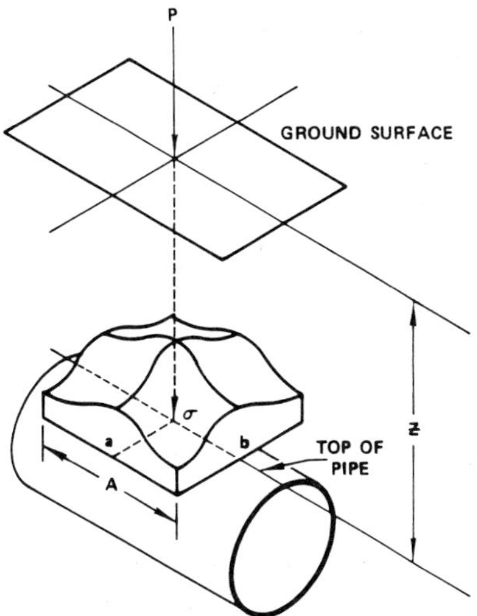

FIGURE 2. Load on a pipe due to a single point surface load according to Spangler.

TIMOSHENKO'S METHOD

This assumes that the live load is applied over a circular area as shown in Figure 3. The compressive pressure, or stress, in pounds per unit area on top of the pipe section directly under the center of the load is calculated from the formula [48]:

$$\sigma = \frac{IP}{A}\left[1 - \frac{z^3}{(r^2 + z^2)^{3/2}}\right] \tag{4}$$

This stress is used to compute W_L ($W_L = \sigma D$). It is applicable to crossings with flexible pavement or no pavement. The circular area is considered the tire print of the wheel. More realistic tire prints have been described by Sowers and Vesic [36].

The contribution of other wheels must also be considered. The assumption is made that the worst loading condition would occur when two trucks with the HS 20 loading configuration pass each other, as shown in Figure 4.

Figure 4 shows that with increasing depth, the number of wheels contributing to the live load on the pipe at a point also increases. For the loading configuration described, the maximum live load on the pipe occurs either at the centerline of the roadway (at Point 1) or underneath one of the interior wheels (at Point 2), depending on the depth of cover (z). The influence values derived from the curves in Nomograph 3 can be used to determine the contribution to vertical stress in the pipe from all the wheels in the loading configuration.

BOUSSINESQ'S POINT LOAD METHOD

This method also applies to loading on flexible pavements or no pavement. Referring to Figure 5, the normal stress in the medium at the intersection of the line of action of P and the top of the pipe section is calculated from the following formula [15]:

$$\sigma = 0.4775\frac{IP}{z^2}\ \text{lb/ft}^2 \tag{5}$$

The value is used as the uniform load on the area of the 3 ft. long section projected on a horizontal plane through the top of the conduit. Converting it to load per linear inch of the pipe, one obtains the W_L value for the load on the pipe.

RELATIONSHIP

The relationship between live load methods previously described can be seen in Figure 6. Boussinesq's Method is the most conservative and Spangler's is the least. Timoshenko's Method was derived from Boussinesq's work. The curves shown for Boussinesq and Timoshenko represent all diameter pipe where Spangler's represents that loading on a 36-inch O.D. pipe. It was found that the 36-inch O.D. pipe for Spangler's method is also representative of other diameter pipes. As depth of cover decreases from 2.5′ to zero, Spangler's and Boussinesq's Methods yield very high values.

Dead Load Plus Loads Without Slabs

Together with the load on the pipe obtained by Timoshenko's Method, the stresses due to live loads obtained in art B.2. and B.3. after conversion to W_L, load per linear foot of pipe, are selectively combined with W_d due to the earth's dead load. The worst condition is used to calculate the stresses in the pipe.

TABLE 2. Influence Coefficients for Rectangular Areas for Use in Spangler's Single Load Method.

	(n)											
m	0.1	0.2	0.3	0.4	0.5	0.6	0.7	0.8	0.9	1.0	1.2	1.4
(1)	(2)	(3)	(4)	(5)	(6)	(7)	(8)	(9)	(10)	(11)	(12)	(13)
0.1	0.00470	0.00917	0.01323	0.01678	0.01978	0.02223	0.02420	0.02576	0.02698	0.02794	0.02926	0.03007
0.2	0.00917	0.01790	0.02585	0.03280	0.03866	0.04348	0.04735	0.05042	0.05283	0.05471	0.05733	0.05894
0.3	0.01323	0.02585	0.03735	0.04742	0.05593	0.06294	0.06858	0.07308	0.07661	0.07938	0.08338	0.08561
0.4	0.01678	0.03280	0.04742	0.06024	0.07111	0.08009	0.08734	0.09314	0.09770	0.10129	0.10631	0.10941
0.5	0.01978	0.03866	0.05593	0.07111	0.08403	0.09473	0.10340	0.11035	0.11584	0.12018	0.12626	0.13003
0.6	0.02223	0.04348	0.06294	0.08009	0.09473	0.10688	0.11679	0.12474	0.13105	0.13605	0.14309	0.14749
0.7	0.02420	0.04735	0.06858	0.08734	0.10340	0.11679	0.12772	0.13653	0.14356	0.14914	0.15703	0.16199
0.8	0.02576	0.05042	0.07308	0.09314	0.11035	0.12474	0.13653	0.14607	0.15371	0.15978	0.16843	0.17389
0.9	0.02698	0.05283	0.07661	0.09770	0.11584	0.13105	0.14356	0.15371	0.16185	0.16835	0.17766	0.18357
1.0	0.02794	0.05471	0.07938	0.10129	0.12018	0.13605	0.14914	0.15978	0.16835	0.17522	0.18508	0.19139
1.2	0.02926	0.05733	0.08323	0.10631	0.12626	0.14309	0.15703	0.16843	0.17766	0.18508	0.19584	0.20278
1.4	0.03007	0.05894	0.08561	0.10941	0.13003	0.14749	0.16199	0.17389	0.18357	0.19139	0.20278	0.21020
1.6	0.03058	0.05994	0.08709	0.11135	0.13241	0.15028	0.16515	0.17739	0.18737	0.19546	0.20731	0.21510
1.8	0.03090	0.06058	0.08804	0.11260	0.13395	0.15207	0.16720	0.17967	0.18986	0.19814	0.21032	0.21836
2.0	0.03111	0.06100	0.08867	0.11342	0.13496	0.15326	0.16856	0.18119	0.19152	0.19994	0.21235	0.22058
2.5	0.03138	0.06155	0.08948	0.11450	0.13628	0.15483	0.17036	0.18321	0.19375	0.20236	0.21512	0.22364
3.0	0.03150	0.06178	0.08982	0.11495	0.13684	0.15550	0.17113	0.18407	0.19470	0.20341	0.12633	0.22494
4.0	0.03158	0.06194	0.09007	0.11527	0.13724	0.15598	0.17168	0.18469	0.19540	0.20417	0.21722	0.22600
5.0	0.03160	0.06199	0.09014	0.11537	0.13737	0.15612	0.17185	0.18488	0.19561	0.20440	0.21749	0.22632
6.0	0.03161	0.06201	0.09017	0.11541	0.13741	0.15617	0.17191	0.18496	0.19569	0.20449	0.21760	0.22644
8.0	0.03162	0.06202	0.09018	0.11543	0.13744	0.15621	0.17195	0.18500	0.19574	0.20455	0.21767	0.22652
10.0	0.03162	0.06202	0.09019	0.11544	0.13745	0.15622	0.17196	0.18502	0.19576	0.20457	0.21769	0.22654
∞	0.03162	0.06202	0.09019	0.11544	0.13745	0.15623	0.17197	0.18502	0.19577	0.20458	0.21770	0.22656

	(n)											
m	1.4	1.6	1.8	2.0	2.5	3.0	4.0	5.0	6.0	8.0	10.0	∞
0.1	0.03007	0.03058	0.03090	0.03111	0.03138	0.03150	0.03158	0.03160	0.03161	0.03162	0.03162	0.03162
0.2	0.05894	0.05994	0.06058	0.06100	0.06155	0.06178	0.06194	0.06199	0.06201	0.06202	0.06202	0.06202
0.3	0.08561	0.08709	0.08804	0.08867	0.08948	0.08982	0.09007	0.09014	0.09017	0.09018	0.09019	0.09019
0.4	0.10941	0.11135	0.11260	0.11342	0.11450	0.11495	0.11527	0.11537	0.11541	0.11543	0.11544	0.11544
0.5	0.13003	0.13241	0.13395	0.13496	0.13628	0.13084	0.13724	0.13737	0.13741	0.13744	0.13745	0.13745
0.6	0.14749	0.15028	0.15207	0.15326	0.15483	0.15550	0.15598	0.15612	0.15617	0.15621	0.15622	0.15623
0.7	0.16199	0.16515	0.16720	0.16856	0.17036	0.17113	0.17168	0.17185	0.17191	0.17195	0.17196	0.17197
0.8	0.17380	0.17739	0.17967	0.18119	0.18321	0.18407	0.18469	0.18488	0.18496	0.18500	0.18502	0.18502
0.9	0.18357	0.18737	0.18986	0.19152	0.19375	0.19470	0.19540	0.19561	0.19569	0.19574	0.19576	0.19577
1.0	0.19139	0.19546	0.19814	0.19994	0.20236	0.20341	0.20417	0.20440	0.20449	0.20455	0.20457	0.20458
1.2	0.20278	0.20731	0.21032	0.21235	0.21512	0.21633	0.21722	0.21749	0.21760	0.21767	0.21769	0.21770
1.4	0.21020	0.21510	0.21836	0.22058	0.22364	0.22499	0.22600	0.22632	0.22644	0.22652	0.22654	0.22656
1.6	0.21510	0.22025	0.22372	0.22610	0.22940	0.23088	0.23200	0.23236	0.23249	0.23258	0.23261	0.23263
1.8	0.21836	0.22372	0.22736	0.22986	0.23334	0.23495	0.23617	0.23656	0.23671	0.23681	0.23684	0.23686
2.0	0.22058	0.22610	0.22986	0.23247	0.23614	0.23782	0.23912	0.23954	0.23970	0.23981	0.23985	0.23987
2.5	0.22364	0.22940	0.23334	0.23614	0.24010	0.24196	0.24344	0.24392	0.24412	0.24425	0.24429	0.24432
3.0	0.22499	0.23088	0.23495	0.23782	0.24196	0.24394	0.24554	0.24608	0.24630	0.24646	0.24650	0.24654
4.0	0.22600	0.23200	0.23617	0.23912	0.24344	0.24554	0.24729	0.24791	0.24817	0.24836	0.24842	0.24846
5.0	0.22632	0.23236	0.23656	0.23954	0.24392	0.24608	0.24791	0.24857	0.24885	0.24907	0.24914	0.24919
6.0	0.22644	0.23249	0.23671	0.23970	0.24412	0.24630	0.24817	0.24885	0.24916	0.24939	0.24946	0.24952
8.0	0.22652	0.23258	0.23681	0.23981	0.24425	0.24646	0.24836	0.24907	0.24939	0.24964	0.24973	0.24980
10.0	0.22654	0.23261	0.23684	0.23985	0.24429	0.24650	0.24842	0.24914	0.24946	0.24973	0.24981	0.24989
∞	0.22656	0.23263	0.23686	0.23987	0.24432	0.24654	0.24846	0.24919	0.24952	0.24980	0.24989	0.25000

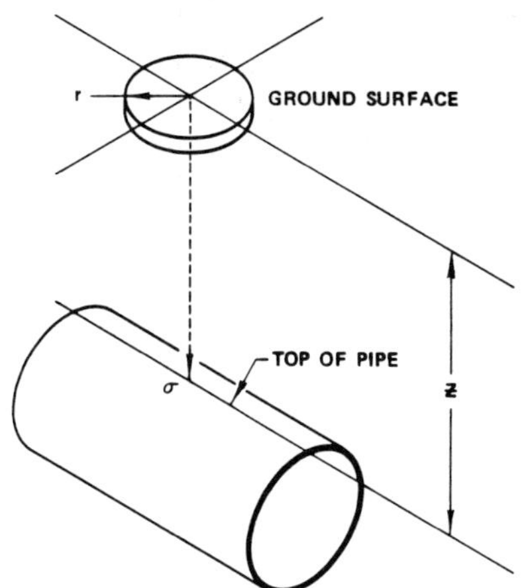

FIGURE 3. Load on a pipe due to a uniform circular surface load according to Timoshenko.

Live Loads With Protective Slabs

This section describes the most accepted methods of computing the live loads on a pipeline when a protective rigid slab is used.

PORTLAND CEMENT ASSOCIATION (PCA) METHOD

The PCA Method is for calculating vertical pressure on pipes under wheel loads on concrete pavement slabs as shown in Figure 7 [32]. This wheel loading represents two trucks passing simultaneously.

The stress from the combined wheel loads at Point 1 should be compared with that at Point 2, and the greater stress selected to obtain the maximum load per linear foot of pipe. For this loading pattern the stress at Point 2 will almost always be greater than at Point 1. Stresses are calculated from the formula:

$$\sigma = CP/R_s^2 \tag{6}$$

in which

C = load coefficient
R_s = radius of stiffness of the concrete slab

$$R_s = \sqrt[4]{\frac{Eh^3}{12(1 - \nu^2)E'}}$$

in which

E = modulus of elasticity of the concrete
h = thickness of the slab
ν = Poisson's ratio
E' = modulus of elasticity of the soil

Values of C can be found from Table 4. Values of E' are given in Table 5. For convenience, values of R_s are also tabulated for various values of h and E' in Table 3.

UNIFORM LOAD METHOD

This method applies to a live load distributed uniformly over the concrete slab on the crossing as shown in Figure 9. The unit pressure on the top of the pipe directly below the center of the uniform load is used to calculate the W_L.

Using Steinbrenner's vertical stress coefficient, C, σ is calculated as

$$\sigma = C\,\sigma_o \tag{7}$$

where $C = f(z/b,\ a/b)$. See Nomograph 4 and Figure 8 [15].

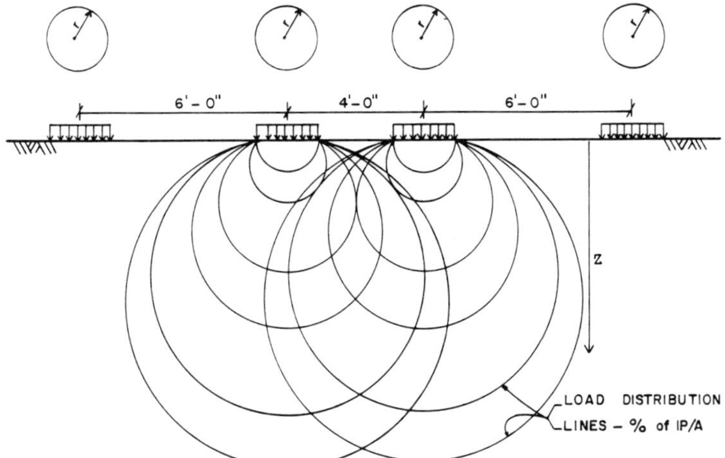

FIGURE 4. Wheel loading configuration for two trucks.

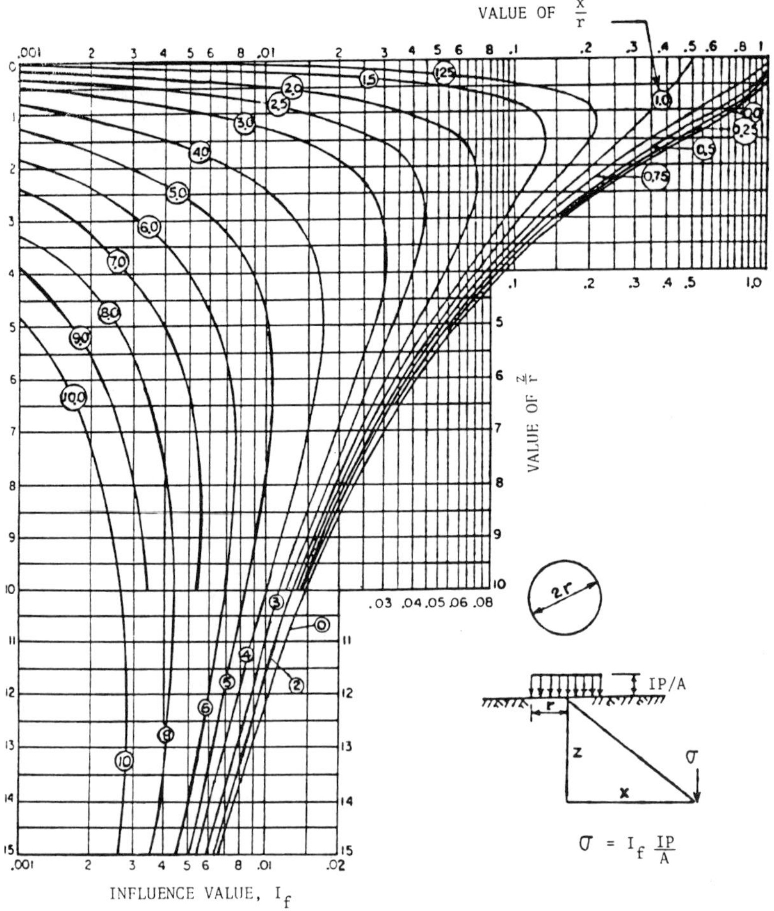

NOMOGRAPH 3. Influence values for vertical stress under uniformly loaded circular area.

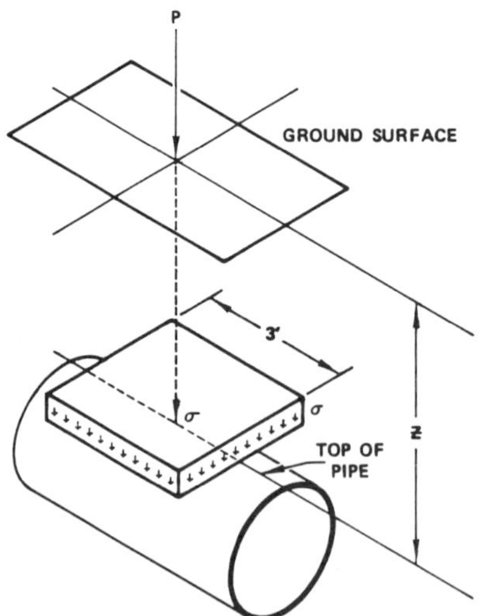

FIGURE 5. Load on a pipe due to a point surface load according to Boussinesq.

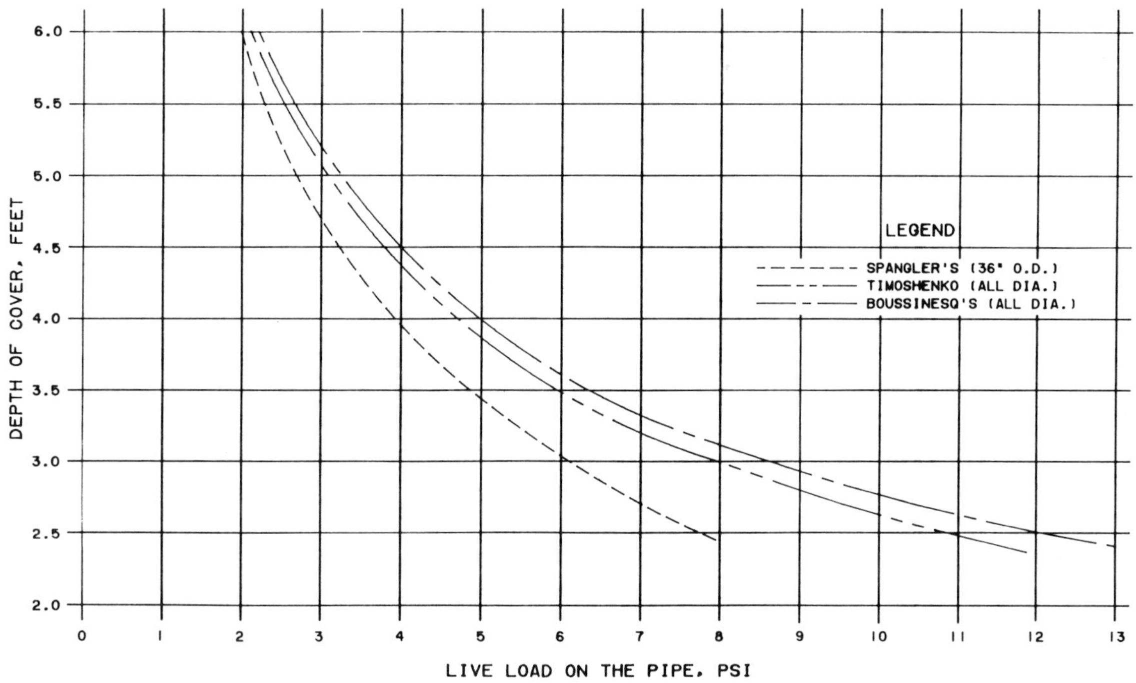

FIGURE 6. Comparison of live load effects for flexible pavements from different formulas.

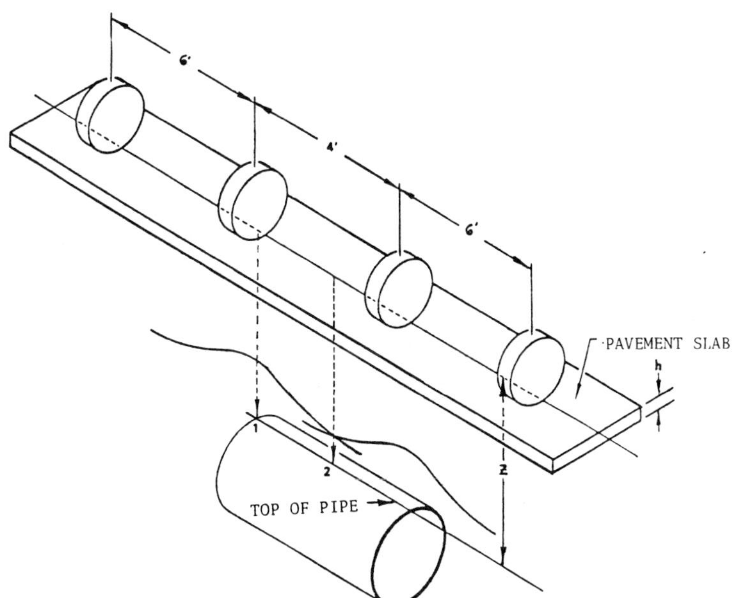

FIGURE 7. Load on a pipe due to wheel loads on a rigid pavement.

TABLE 3. Values of Radius of Stiffness R_s.

Slab h (in.)	E′ 50	100	150	200	250	300	350	400	500
6	34.84	29.30	26.47	24.63	23.30	22.26	21.42	20.72	19.59
6.5	36.99	31.11	28.11	26.16	24.74	23.64	22.74	22.00	20.80
7	39.11	32.89	29.72	27.65	26.15	24.99	24.04	23.25	21.99
7.5	41.19	34.63	31.29	29.12	27.54	26.32	25.32	24.49	23.16
8	43.23	36.35	32.85	30.57	28.91	27.62	26.58	25.70	24.31
8.5	45.24	38.04	34.37	31.99	30.25	28.91	27.81	26.90	25.44
9	47.22	39.71	35.88	33.39	31.58	30.17	29.03	28.08	26.55
9.5	49.17	41.35	37.36	34.77	32.89	31.42	30.23	29.24	27.65
10	51.10	42.97	38.83	36.14	34.17	32.65	31.42	30.39	28.74
10.5	53.01	44.57	40.28	37.48	35.45	33.87	32.59	31.52	29.81
11	54.89	46.16	41.71	38.81	36.71	35.07	33.75	32.64	30.87
11.5	56.75	47.72	43.12	40.13	37.95	36.26	34.89	33.74	31.91
12	58.59	49.27	44.52	41.43	39.18	37.44	36.02	34.84	32.95
12.5	60.41	50.80	45.90	42.72	40.40	38.60	37.14	35.92	33.97
13	62.22	52.32	47.27	43.99	41.61	39.75	38.25	36.99	34.99
13.5	64.00	53.82	48.63	45.26	42.80	40.89	39.35	38.06	35.99
14	65.77	55.31	49.98	46.51	43.98	42.02	40.44	39.11	36.99
14.5	67.53	56.78	51.31	47.75	45.16	43.15	41.51	40.15	37.97
15	69.27	58.25	52.63	48.98	46.32	44.26	42.58	41.19	38.95
15.5	70.99	59.70	53.94	50.20	47.47	45.36	43.64	42.21	39.92
16	72.70	61.13	55.24	51.41	48.62	46.45	44.70	43.23	40.88
16.5	74.40	62.56	56.53	52.61	49.75	47.54	45.74	44.24	41.84
17	76.08	63.98	57.81	53.80	50.88	48.61	46.77	45.24	42.78
17.5	77.75	65.38	59.08	54.98	52.00	49.68	47.80	46.23	43.72
18	79.41	66.78	60.35	56.16	53.11	50.74	48.82	47.22	44.66
19	82.70	69.54	62.84	58.48	55.31	52.84	50.84	49.17	46.51
20	85.95	72.27	65.30	60.77	57.47	54.92	52.84	51.10	48.33
21	89.15	74.97	67.74	63.04	59.62	56.96	54.81	53.01	50.13
22	92.31	77.63	70.14	65.28	61.73	58.98	56.75	54.89	51.91
23	95.44	80.26	72.52	67.49	63.83	60.98	58.68	56.75	53.67
24	98.54	82.86	74.87	69.68	65.90	62.96	60.58	58.59	55.41

$R_s = \sqrt[4]{\frac{Eh^3}{12(1-\nu^2)E'}}$ where $E = 4{,}000{,}000$ psi, $\nu = 0.15$ therefore $R_s = 24.1652\sqrt[4]{\frac{h^3}{E'}}$

TABLE 4. Pressure Coefficients for a Single Load.

Values of C

σ = CP/R_s^2 pounds per square foot
P = wheel load, pounds
R_s = radius of stiffness of rigid pavement slab, feet

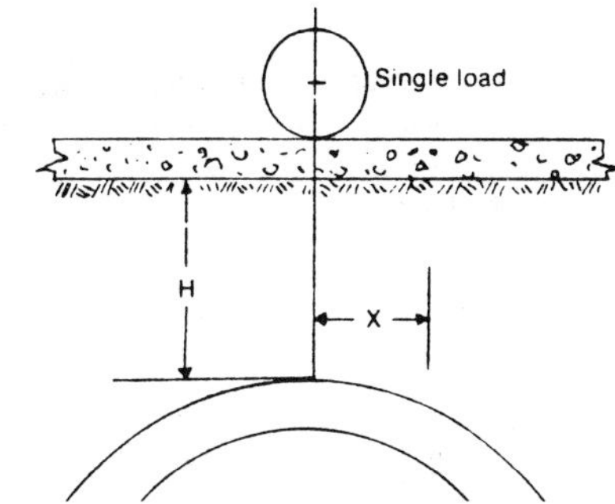

H/R_s	X/R_s 0.0	0.4	0.8	1.2	1.6	2.0	2.4	2.8	3.2	3.6	4.0
0.0	.113	.105	.089	.068	.048	.032	.020	.011	.006	.002	.000
0.4	.101	.095	.082	.065	.047	.033	.021	.011	.004	.001	.000
0.8	.089	.084	.074	.061	.045	.033	.022	.012	.005	.002	.001
1.2	.076	.072	.065	.054	.043	.032	.022	.014	.008	.005	.003
1.6	.062	.059	.054	.047	.039	.030	.022	.016	.011	.007	.005
2.0	.051	.049	.046	.042	.035	.028	.022	.016	.011	.008	.006
2.4	.043	.041	.039	.036	.030	.026	.021	.016	.011	.008	.006
2.8	.037	.036	.033	.031	.027	.023	.019	.015	.011	.009	.006
3.2	.032	.030	.029	.026	.024	.021	.018	.014	.011	.009	.007
3.6	.027	.026	.025	.023	.021	.019	.016	.014	.011	.009	.007
4.0	.024	.023	.022	.020	.019	.018	.015	.013	.011	.009	.007
4.4	.020	.020	.019	.018	.017	.015	.014	.012	.010	.009	.007
4.8	.018	.017	.017	.016	.015	.013	.012	.011	.009	.008	.007
5.2	.015	.015	.014	.014	.013	.012	.011	.010	.008	.007	.006
5.6	.014	.013	.013	.012	.011	.010	.010	.009	.008	.007	.006
6.0	.012	.012	.011	.011	.010	.009	.009	.008	.007	.007	.006
6.4	.011	.010	.010	.010	.009	.008	.008	.007	.007	.006	.005
6.8	.010	.009	.009	.009	.008	.008	.007	.007	.006	.006	.005
7.2	.009	.008	.008	.008	.008	.007	.007	.006	.006	.006	.005
7.6	.008	.008	.008	.007	.007	.007	.006	.006	.006	.005	.005
8.0	.007	.007	.007	.007	.006	.006	.006	.006	.005	.005	.005

TABLE 5. Design Values for Standard Laying Conditions.

Laying conditions (1)	Description (2)	E′ (3)	Bedding angle (4)	K_b (5)	K_a (6)
Type 1[a]	Flat-bottom trench.[b] Loose backfill.	150	30°	0.235	0.108
Type 2	Flat-bottom trench.[b] Backfill lightly consolidated to centerline of pipe	300	45°	0.210	0.105
Type 3	Pipe bedded in 4-in. minimum loose soil.[c] Backfill lightly consolidated to top of pipe.	400	60°	0.189	0.103
Type 4	Pipe bedded in sand, gravel or crushed stone to depth of 1/8 pipe diameter, 4-in. minimum. Backfill compacted to top of pipe. (Approximately 80% Standard Proctor, AASHO T-99.)	500	90°	0.157	0.096
Type 5	Pipe bedded in compacted granular material to centerline of pipe. Compacted granular or select[c] material to top of pipe. (Approximately 90% Standard Proctor, AASHO T-99.)	700	150°	0.128	0.085

[a]For 30-in. and larger pipe, consideration should be given to the use of laying conditions other than Type 1.
[b]"Flat-bottom" is defined as undisturbed earth.
[c]"Loose soil" or "select material" is defined as native soil excavated from the trench, free of rocks, foreign materials and frozen earth.

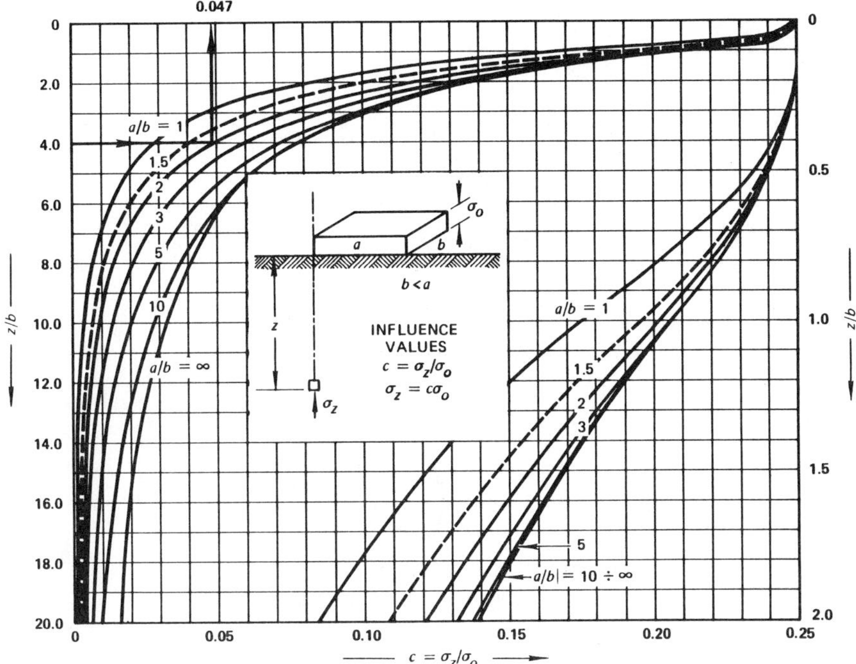

NOMOGRAPH 4. Influence values for finding vertical stress according to Steinbrenner.

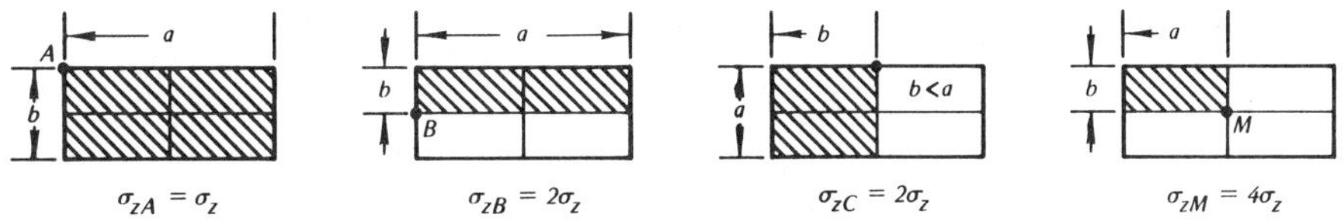

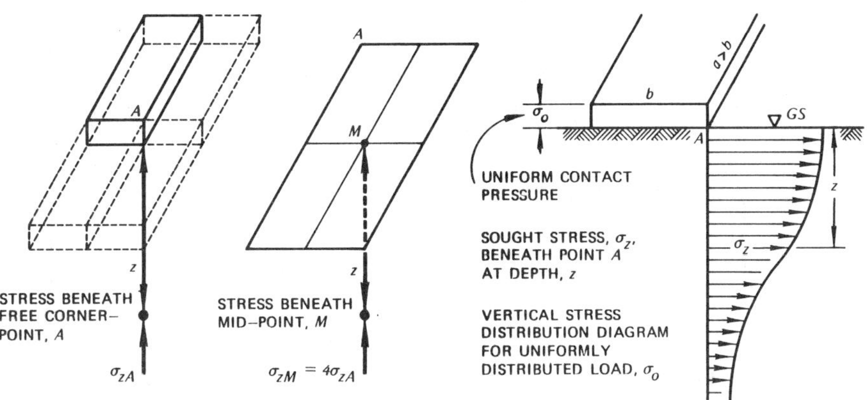

FIGURE 8. Application of Steinbrenner's nomograph.

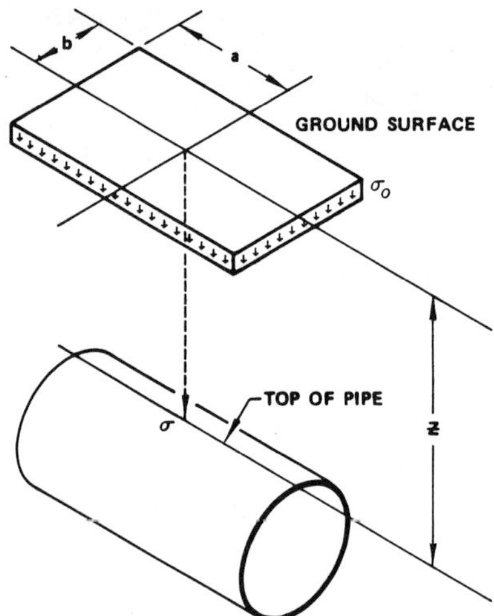

FIGURE 9. Load on a pipe due to uniform rectangular surface load (rectangular rigid pavement).

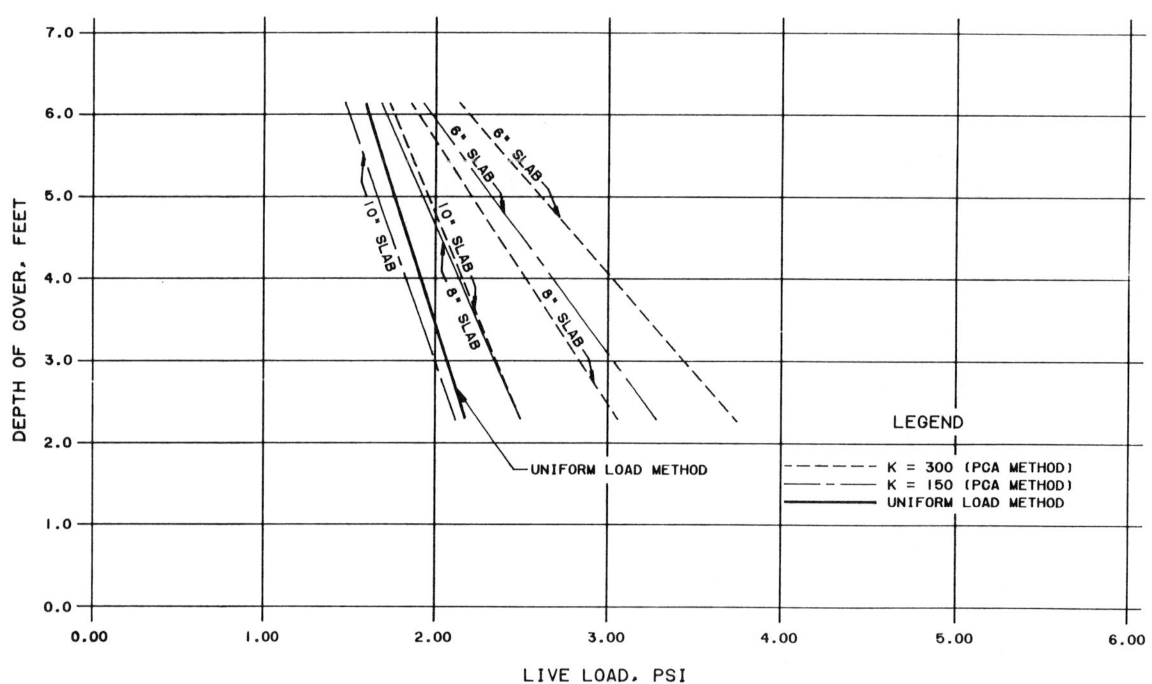

FIGURE 10. Comparison of live loads from different formulas.

Steinbrenner's diagram is prepared for uniformly loaded parallelograms based on the Boussinesq's theory.

From a given depth-to-width ratio, z/b as ordinate, and from a given length-to-width ratio, a/b, one obtains the value of $C = \sigma_z/\sigma_o$ on the absissa axis, from which the stress $\sigma_z(\sigma)$ can be calculated if the σ_o is known.

The stress obtained, σ_z, is the one acting underneath the free corner, A, at a depth z. The vertical stress, σ_{zm}, beneath the midpoint of the loaded parallelogram is $\sigma_{zm} = 4\ \sigma_z$.

RELATIONSHIP

The comparison in Figure 10 of these two methods shows that the PCA Method is more conservative. The Uniform Load Method makes no allowance for different slab thickness, and for all but a 10-inch slab with an E' value of 150 psi, the Uniform Load Method is less conservative. Therefore, the PCA Method is used in the computations of loads with a protective slab.

DETERMINATION OF THE STRESSES AND DEFORMATIONS IN THE PIPE SECTION DUE TO COMBINATIONS OF THE FORCES ACTING ON THE PIPE

Primary Hoop Stress

The internal pressure in the pipe creates hoop stress which can be computed using Barlow's Formula,

$$s_h = \frac{p}{2t}(D - 2t) \tag{8}$$

in which s_h is the hoop stress; p is the internal pressure; D is the pipe outside diameter; and t is the pipe wall thickness. This stress is also defined as the primary stress.

Secondary Stresses

Stresses in the pipe caused by external forces are defined as secondary stresses. The combined external force is usually expressed in terms of pounds per unit length of pipe, denoted by W. The combined secondary stress computed using Spangler's Stress Formula is

$$S = \frac{3K_bWEDt}{Et^3 + 3K_zpD^3} \tag{9}$$

in which

S = stress in psi
K_b = bending parameter
K_z = deflection parameter

K_b and K_z are functions of the bedding angle, and they depend on the laying condition of the pipeline as shown and described in Table 5. E is the elastic modulus of the pipe material in psi.

Deformations

Pipe deflections may be calculated with the Iowa Formula (Spangler's) [42].

$$\Delta X_1 = \frac{JK_aWr^3}{EI + .061E'r^3} \tag{10}$$

in which ΔX_1 is the maximum deflection of pipe in inches; J is the deflection lag factor (1.5 suggested by Spangler); K_a is the bedding factor (equivalent to K_z); r is the mean radius of pipe in inches; E' is the modulus of subgrade reaction in psi (Table 5) and I is the moment of inertia of pipe wall (in.^{4}/in.). For interest, Spangler [42] also suggested the use of another equation to include internal pressure, p, but not the support of the soil,

$$\Delta X_1 = \frac{0.105Wr^3}{EI + 0.21pr^3} \tag{11}$$

For type 2 laying condition (Table 5), the value of $J \times K_a$ is 0.157 as compared to 0.105. The correct value of this number is not really known. Rodabaugh [10] suggested combining the two equations qualitatively. Averaging 0.157 and 0.105 produces the following formula:

$$\Delta X_1 = \frac{0.131Wr^3}{EI + 0.061E'r^3 + 0.21pr^3} \tag{12}$$

DESIGN PROCEDURE

The proposed procedure to design vehicular crossings over pipelines is as follows.

1. Earth and other dead loads are to be calculated by the Direct Earth Load Method with Marston's Formula as a check, Equation (1).
2. Live loads should be calculated using Timoshenko's Method when a slab is not required, and PCA Method when a slab is required, Equations (4) and (6). Live load or wheel load values should be those normally used by applicable codes and specifications. The impact factor should range from 1.0 for rigid pavement to 1.5 for unpaved crossings and flexible pavements.
3. Barlow's Formula, Equation (8), is to be used for calculating hoop stress due to internal pressures. The maximum allowable value of the primary hoop stress for natural gas transmission lines is obtained from Barlow's Formula, as specified in ANSI B31.8 and DOT Part 192 (see discussion in following section), incorporating the class location factor which limits the primary stress level on the basis of

population density in the area. Other national codes or standards should be consulted for maximum values for products other than natural gas.

4. Deflection is then calculated by the Iowa Formula, Equation (10), and should be limited to 3 percent of nominal pipe diameter [45].

5. Spangler's Stress Formula, Equation (9), is used to calculate stress due to external loading. The allowable combined stresses to be used are those specified in the codes as $F_c \times$ SMYS, where F_c is the factor whose value depends on the crossing location. SMYS is the Specified Minimum Yield Strength of the pipe material.

DESIGN CODES FOR NATURAL GAS PIPELINES

Around 1926, the American Standards Association (ASA) initiated a study to develop standardized pipeline safety considerations. The code became available in 1935 as the "Tentative Standard Code for Pressure Piping." In 1951, a portion of the code was separated and in 1952 published as B31, Section 8. This B31.8 code dealt with pressure piping related to gas transmission and distribution systems. Improvements were made through the years, and the code was revised many times, the last one being in 1975 as the American National Standards Institute's ANSI B31.8-1975. The 1968 Natural Gas Pipeline Safety Act led to the development of the Department of Transportation's mandatory Minimum Federal Standards for Transportation of Natural and Other Gas by Pipeline, Code of Federal Regulation, Title 49, Part 192. Unfortunately, these regulations only define design limits for primary hoop stress to determine minimum wall thickness and flexibility requirements for unrestrained piping systems, but do not define specific secondary stress or strain limitations for buried gas pipelines.

Subpart C, Pipe Design, §192.105 to §192.113, specifies the requirements concerning the internal design pressure formula $P = (2St/D \times F \times E \times T)$ where E and T are joint and temperature factors. Values for the "design factor," F, are specified in §192.111 according to a population density class location. It is further specified that for vehicular crossings of an uncased steel pipeline the design factor should not exceed 0.60 for Class 1 locations. For crossings in Class 2 location the required design factor is 0.50. There is no mention of the crossing design requirement for Class 3 and 4 locations, but ANSI B31.8 calls for a design factor of 0.50 in Class 3 locations and 0.40 in Class 4 locations. These design factors for pipeline crossings are less than the normal 0.72 design factor for cross-country pipelines in order, possibly, to accommodate secondary stresses due to external loads, especially live loads, as well as other unforeseen but possible problems during construction, operation or maintenance.

In essence, the design factor is that fraction (or percentage) of the Specified Minimum Yield Strength (SMYS) allowed as the maximum hoop stress due to internal pressure alone. Based on the above reasoning, it is justifiable to allow the combined stress, the sum of primary and secondary stress, to reach a higher value. Furthermore, it should be noted that the live loads are not permanent in nature. It is also true that the peak flexural stresses (Secondary stresses) are only at the extreme fiber (see discussion by H. S. Oey [26]). For high pressure flexible steel conduits, failure is not usually the result of flexural or bending stresses, even when they are above the yield point. Failure is usually the result of high hoop stress wherein the stress exists through the entire thickness of the wall. The worst condition usually caused by flexural stresses alone is a permanent ring deflection.

For liquid petroleum transportation piping, ANSI B31.4-1974 specifies that the sum of combined circumferential stresses should not exceed 0.72 SMYS. However, the equivalent longitudinal tensile stress for restrained pipes is allowed to reach 90 percent of SMYS. Values of this design factor are found in various codes, each recommending its own range of applications.

Available literature suggests that a value of 0.80 SMYS is a reasonable value for the allowable combined circumferential stress at pipeline vehicular crossings for Class 1 locations. Similarly, 0.70 SMYS should be used for Class 2 and 3 locations, and 0.60 SMYS for Class 4 locations. These applications will also retain sufficient assurance for a reasonable factor of safety against sources of stress or damage other than the ones discussed in this chapter.

SAMPLE CALCULATIONS

Pipe outside diameter, $D = 30$ in. (76.2 cm)
Wall Thickness, $t = 0.324$ in. (0.823 cm)
SMYS of pipe material, 52 ksi (358.54 MPa)
MAOP of line, $P = 674$ psi (4647.23 KPa)
(MAOP = Maximum Allowable Operating Pressure)
Class Location, Class 1, $F_c = 0.80$
Depth of cover, $h = 4$ft. (1.22 m), Type 2 trench
Wheel Load, 16,000 lbs. (71.168 KN)
Unit weight of soil, $\gamma = 120$ lb./ft.^{3} (18.85 KN/m^3)
Class of soil, thoroughly wet clay
Modulus of elasticity of steel, $E = 30 \times 10^6$ psi (206,850 MPa)

Dead Load (W_d)

MARSTON EARTH LOAD METHOD

Marston Rigid Pipe Method, Equation (la)

Trench width = Outside Diameter + 12″ (30.48 cm) = B_d

$$B_d = 3.5 \text{ ft. } (1.07 \text{ m})$$

$$\frac{H}{B_d} = \frac{4}{3.5} = 1.143$$

C_d = 1.010 (Table 1)

W_d = (1.010)(120)(3.5)²/12 = 123.7 lb./linear inch of pipe (21.662 KN/m)

Marston Flexible Pipe Method, Equation (lb)
W_d (flexible) = W_d (Rigid) × D/B_d

W_d = (123.7) (2.5)/(3.5) = 88.3 lb./linear inch of pipe (15.463 KN/m)

DIRECT EARTH LOAD METHOD, EQUATION (2)

$$W_d = \gamma HD$$

$$W_d = \frac{(120)(4)(2.5)}{12} = 100 \text{ lb./linear inch of pipe (17.512 KN/m)}$$

Live Load Without a Protective Slab (W_L)

SPANGLER'S SINGLE LOAD METHOD, EQUATION (3)

An impact coefficient of 1.5 is generally acceptable for flexible pavements. Refer to Figure 2, a = 1.5 ft., b = 1.25 ft. and z = 4 ft. (0.457 m, 0.381 m, 1.219 m), m = 1.5/4 = 0.375 (Table 2), n = 1.25/4 = 0.3125 (Table 2) C_T = 0.04603 (Table 2).

$4C_T$ = (0.04603) 4 = 0.1841 because there are 4 rectangles of $a \times b$ ft.² each sharing the same corner.

W_L = 1/3 (1.5)(0.1841)(16,000) = 1473.0 lb./ft. = 122.0 lb./linear inch of pipe (21.505 KN/m).

TIMOSHENKO'S CIRCULAR LOAD METHOD, EQUATION (4)

The value of r, radius of circle of uniform load, is computed as follows: w = 80 psi = 11,520 psf = tire pressure. The circular area of contact between tire and pavement is A = (1.5)(16,000)/11,520 = 2.083 ft.² = πr^2. Thus, $r = \sqrt{A/\pi} = \sqrt{2.083/\pi}$ = 0.814 ft. (0.25 m). The worst condition is at a point beneath one of the wheels (x_1 = 0) when two trucks are simultaneously crossing the pipeline. The total load on the pipe beneath point 1 due to P_1 is σ_1, due to P_2 is σ_2, due to P_3 is σ_3, and due to P_4 is σ_4. The stress due to P_1 is calculated from Equation (4), σ_1 = 680.6 lb/ft²

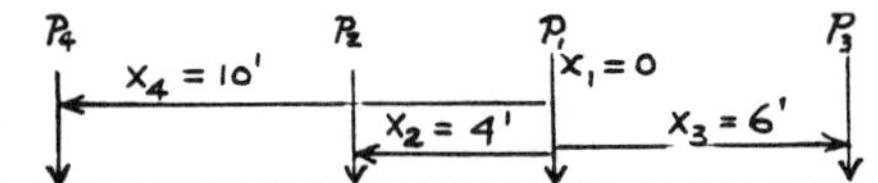

Refer to Nomograph 3.

For P_2, z/r = 4/0.814 = 4.91 and x_2/r = 4.91; therefore I_f = 0.0105. For P_3, z/r = 4.91 and x_3/r = 6/0.814 = 7.37; therefore I_f = 0.0036. For P_4, z/r = 4.91 and x_4/r = 10/0.814 = 12.28; therefore I_f = 0.001.

Adding all the σ's, $\sigma_1 + \sigma_2 + \sigma_3 + \sigma_4$, we obtain σ = 680.6 + (0.015 + 0.0036 + 0.001) 11,520 = 906 psf. Converting it to W_L = 906 (30/144) = 189 lb./inch. (31.16 KN/m).

BOUSSINESQ'S POINT LOAD METHOD

σ = 0.4775 IP/Z^2 = 0.4775 × (1.5)(16,000)/4² = 716.25 psf (34.3 KPa).

W_L = 716.25 (30/144) = 149.2 lb./linear inch of pipe (26.128 KN/m).

Maximum Combined Stress

Using the Direct Earth Load Method and Timoshenko's Live Load Method for reasons described previously $W = W_d + W_L$ = 289 lb./linear inch of pipe (48.7 KN/m). Using this value and applying Barlow's and Spangler's formula [Equations (8) and (9)], the combined stress is calculated,

$$S_t = \frac{p(D - 2t)}{2t} + \frac{3K_bWEDt}{Et^3 + 3k_zpD^3}$$

$$S_t = \frac{(674)[30-2(0.324)]}{(2)\ (0.324)}$$

$$+ \frac{(3)(0.210)(289)(30 \times 10^6)(30)(0.324)}{(30 \times 10^6)(0.324)^3 + (3)(0.105)(675)(30^3)}$$

$$S_t = 30{,}530 + 7{,}862 = 38{,}392 \text{ psi (264.7 MPa)}$$

When there is no internal pressure in the pipe, p = 0.

$$S_t = \frac{(3)(0.210)(289)(30 \times 10^6)(30)(0.324)}{(30 \times 10^6)(0.324)^3}$$

$$= 53{,}652 \text{ psi (337.13 MPa)}$$

The above calculations revealed that the crossing is in adequate because 53,652 exceeds 41,600 (0.8 SMYS) (Class 1). To improve the crossing, the use of a concrete slab can reduce the load on the buried pipe.

Live Load With a Protective Slab

Assume 10 ft. by 20 ft., 6 in. thick (3.05 m × 6.10 m × 0.15 m) concrete slab is used for rigid pavement over the crossing. For rigid pavements, there is no impact factor. Either the PCA or the Uniform Load Method can be applied. Again, results of both methods are presented for comparison.

PORTLAND CEMENT ASSOCIATION (PCA) METHOD

It is possible that four wheels from two different trucks are positioned above the pipe simultaneously with a minimum distance of 4 ft.

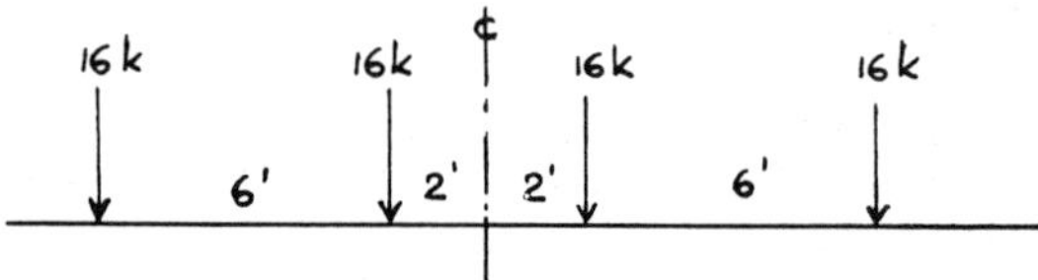

Maximum Pressure Due to Two Trucks

Assuming a value of 300 for E' (Table 5), the radius of stiffness of the slab is:

$$R_s = \sqrt[4]{\frac{Eh^3}{12(1 - \nu^2)E'}} \text{ in.} = \sqrt[4]{\frac{(4{,}000{,}000)\ 6^3}{12(1 - 0.15^2)300}}$$

At a point directly below the centerline of the roadway and at the top of the pipe section: $H/R_s = 4/1.86 = 2.16$ and $x_1/R_s = 2/1.86 = 1.08$.

From Table 4, for single load, by interpolation, the value of C is 0.043 ($= C_1$). The other wheel is at 8 ft. distance from the centerline, $H/R_s = 2.16$ and $X_2/R_s = 8/1.86 = 4.30$. The value of C is 0.006 ($= C_2$). Thus the total C is $C_t = 2(C_1 + C_2) = 2(0.043 + 0.006) = 0.098$.

The combined effect of all four wheels produced a stress of

$$\sigma = 0.098\ (16{,}000/1.86^2)\ \text{psf} = 454.6,\ (21.77\ \text{KPa})$$

$W_L = (454.6)(30/144) = 94.7$ lb./linear inch of pipe (16.583 KN/m)

Pressure on the Pipe Due to the Weight of the Concrete Slab

If the concrete slab is 10 ft. long and 20 ft. wide, using Steinbrenner's coefficient with $z/b = 4/5 = 0.8$ and $a/b = 10/5 = 2$, we obtain $C = 0.22$ and the pressure on the pipe is $4(0.22)75$ psf $= 66$ psf, and $W_d = 30/144\ (66) = 13.8$ lb./in. (2.42 KN/m).

Depending on the physical arrangement of the crossing, this load may or may not be added to the W_d due to the soil. Since the unit weight of soil was taken as 120 and of concrete 150, the difference should not alter the result too much. It is reasonably justifiable to assume that the concrete equally replaced the soil it displaced.

UNIFORM LOAD METHOD

Here the loads are assumed to be uniformly distributed over a 10×20 ft.2 area. With two trucks, the total wheel load is $4(16{,}000) = 64{,}000$ lbs. Thus, the uniform load is $64{,}000/(10 \times 20) = 320$ psf (15.33 KPa).

Using Steinbrenner's coefficient with: $z/b = 4/5 = 0.80$ and $a/b = 10/5 = 2$ we obtain $C = 0.22$ and the pressure on the pipe is $\sigma = 4(0.22)\ 320$ psf $= 281.6$ psf (13.49 KPa) and $W_L = (30/144)\ 281.6 = 58.7$ lb./linear inch of pipe.

Maximum Combined Stress

With the 6 in. protective concrete slab, the largest value for W_L is 94.7 lb./in. (16.6 KN/m).

Again, the combined stress is computed using Equations (8) and (9) with $W = 100.0 + 94.7 = 194.7$ lb./in. (34.1 KN/m).

$$S_t = \frac{pD}{2t} + \frac{3K_bWEDt}{Et^3 + 3K_zpD^3}$$

$$S_t = \frac{(674)[30-2(0.324)]}{(2)(0.324)}$$

$$+ \frac{(3)(0.210)(194.7)(30 \times 10^6)(30)(0.324)}{(30 \times 10^6)(0.324)^3 + (3)(0.105)(674)(30)^3}$$

$$S_t = 30{,}530 + 5{,}297 = 38{,}827 \text{ psi } (247.6 \text{ MPa})$$

$$= 0.69 \text{ SMYS}$$

The allowable value is 0.80 SMYS or 41,600 psi (286.8 MPa). When the internal pressure in the pipe is zero,

$$S_t = \frac{(3)(0.210)(194.7)(30 \times 10^6)(30)(0.324)}{(30 \times 10^6)(0.324)^3}$$

$$= 35{,}054 \text{ psi } (239.1 \text{ MPa})$$

Calculation of the maximum deflection in the pipe section under the zero internal pressure condition results in the following:

Taking E' (Table 5) equal 300 psi and the 6 in. concrete slab,

$$I = \frac{t^3}{12} = \frac{0.324^3}{12} = 0.0028 \text{ in.}^3(\text{in.}^4/\text{in.})(0.0459 \text{ cm}^3)$$

$$\Delta X = \frac{(0.157)(194.7)(15)^3}{(30 \times 10^6)(0.0028) + (0.061)(300)(15^3)}$$

$$= 0.70 \text{ in. } (1.78 \text{ cm})$$

The allowable deflection of 5 percent of 30 in. (76.2 cm) is 1.5 in. (3.8 cm), which is larger than the computed deflection of 0.70 in. (1.78 cm). Hence, the 30 in. (76.2 cm) diameter, 0.324 in. (0.823 cm) wall pipe is satisfactory under the protective slab.

REFERENCES

1. Allgood, J. R., Discussion of "Buckling of Soil-Surrounded Tubes," by U. Luscher, *Proc. ASCE*, Volume 93, No. SM5 (September 1967).
2. Britain, K. E., "A Report on the Redundancy of Casing in Today's Pipeline Design," unpublished (May, 1966).
3. Brody, O., "Estimating Supporting Strength of Flexible Pipes," *Journal of Transportation Division*, ASCE, Volume 105, No. TE4, pp. 473–482 (July 1979).
4. Brown, C. B., D. R. Green and S. Pawsey, "Flexible Culverts Under High Fills," *Proc., ASCE*, Volume 94, No. ST4, pp. 605–917 (April 1968).
5. Burns, J. Q. and R. M. Richard, "Attenuation of Stresses for Buried Cylinder," *Proc. Symp. on Soil-Structure Interaction*, University of Arizona, pp. 378–392 (1964).
6. Burton, L. H. and D. F. Nelson, "Study of Earth Pressures on Rigid Pipe," Report No. 3B-3, U.S. Bureau of Reclamation (March 1967).
7. *Concrete Pipe Design Manual*, American Concrete Pipe Association, Arlington, Virginia (1974).
8. Davis, R. E. and A. E. Bacher, "California Culvert Research Program—Description, Current Status, and Observed Peripheral Pressure," Highway Research Record No. 249, pp. 14–23 (1968).
9. *Design and Construction of Sanitary and Storm Sewers*, Manual of Engineering Practice No. 37, ASCE, New York (1960).
10. "Design Procedure for Uncased Pipeline Crossings," A letter from E. C. Rodabaugh, Battelle Memorial Institute, to W. F. Quinn, Research Council on Pipeline Crossings of Railroads and Highways (May 1, 1968).
11. Dorris, A. F., "Response of Horizontally Oriented Buried Cylinders to Static and Dynamic Loading," Technical Report No. 1-682, U.S. Army Engr., Waterways Experiment Station, Vicksburg, Mississippi (July 1965).
12. Drawsky, R., "An Accurate Design Method of Buried Flexible Conduit Structures," Highway Research Board Circular 34 (July 1966).
13. Forrestal, M. J. and G. Herrmann, "Buckling of a Long Cylindrical Shell Surrounded by an Elastic Medium," *Inter. J. of Solids and Structures*, Volume 1, No. 3, pp. 297–309 (1965).
14. Gabriel, L. H., "Analytical-Experimental Methods of Determining Soil Pressures Surrounding a Buried Conduit Using Principles of Soil-Structure Interaction," Final Report, Experimental Phase, Project No. 321, Sacramento State College (July 1968).
15. Jumikis, A. T., *Soil Mechanics*, Van Nostrand, Princeton, New Jersey, pp. 524–527 (1962).
16. Kolditz, L. C., "Natural Gas Pipeline Design and B31.8 Code," *Journal of Transportation Division*, ASCE (November 1978).
17. Luscher, U. and K. Hoeg, "The Beneficial Action of the Surrounding Soil on the Load Carrying Capacity of Buried Tubes," *Proc. Symp. on Soil-Structure Interaction*, University of Arizona, pp. 393–402 (1964).
18. Marston, A. and A. O. Anderson, "The Theory of Loads on Pipes in Ditches and Tests of Cement and Clay Drain Tile and Sewer Pipe," Bulletin 31, Iowa Engineering Experiment Station, Ames, Iowa (1913).
19. Marston, A., "The Theory of External Loads on Closed Conduits in the Light of the Latest Experiments," No. 96, Iowa Engineering Experiment Station, Iowa State University, Ames, Iowa (1930).
20. McClure, G. M., "Some Observations on Stresses in Large-Diameter Pipelines," *Gas*, pp. 117–121 (April 1957).
21. "Minimum Federal Safety Standards for Gas Lines," Part 192, Title 49, Code of Federal Regulations.
22. Moran, Proctor, Mueser, and Rutledge, Consulting Engineers, "Evaluation of Methods for Determining Earth Loads on Buried Concrete Pipe," American Concrete Pipe Association (December 1962).
23. Mouser, G. F. and R. H. Clark, "Loads on Buried Pipe, Water and Sewage Works," p. 260 (July 1970).
24. Mouser, G. F. "Basic Pipe Stress Evaluation Summary," *Journal of Transportation Division*, ASCE, Volume 105, No. TE4, pp. 349–359 (July 1979).
25. Oey, H. S. and D. P. Womack, "Analysis and Design of Pipelines under Road Crossings," paper presented at the Fall Meeting, ASCE Texas Section, College Station (October 5, 1979).
26. Oey, H. S., Discussion of "Basic Pipe Stress Evaluation Summary," by G. F. Mouser, *Journal of the Transportation Div.*, ASCE, Volume 105, No. TE4, Proc. Paper 15514 (July 1980).
27. Oey, H. S., V. L. Greggerson, and D. P. Womack, "Buried Gas Pipelines Under Vehicular Crossings," *ASCE Journal of Transportation Engineering*, Vol. 110, No. 2 (March 1984).
28. Peng, Liang-Chuan, "Stress Analysis Methods for Underground Pipelines," Parts 1 and 2, *Pipeline Industry* (April-May 1978).
29. Pettibone, H. C. and A. K. Howard, "Distribution of Soil Pressures on Concrete Pipe," *Journal of Pipeline Division*, ASCE, Volume 93, No. PL2, pp. 85–102 (July 1967).
30. Pierce, R. N., et al., "A Design Procedure of Uncased Natural Gas Pipeline Crossings of Roads and Highways," 1977 AGA Transmission Conference Paper 77-T-16.
31. Pierce, R. N., L. Osborn, R. Pleny and C. L. Rankin, "Design Considerations for Uncased Road Crossings," *Pipeline Industry* (May 1978).
32. Portland Cement Association, "Vertical Pressure on Culverts Under Wheel Loads on Concrete Pavement Slabs," Publication No. ST65.
33. Price, P. St. J., "Basis of Structural Design Criteria for Buried Gas Transmissions Pipelines," ASME Paper No. 78-Pet-73, Energy Conference and Exhibition, Houston (November 1978).
34. Sears, E. C., "Ductile-Iron Pipe Design," *Journal AWWA, 56*:4 (January 1964).
35. *Soil Mechanics, Foundation, and Earth Structures Design Manual*, (NAVFAC DM-7), Department of the Navy, Naval

Facilities Engineering Command, pp. (7-5-1-)-(7-5-16) (March 1971).

36. Sowers, G. F. and A. B. Vesic, "Vertical Stresses in Subgrades Beneath Statically Loaded Flexible Pavements," Highway Research Board Bulletin 342, p. 102 (1962).

37. Spangler, M. G., R. Winfrey and C. Mason, "Experimental Determination of Static and Impact Loads Transmitted to Culverts," Bulletin 79, Iowa Engineering Experiment Station, Ames, Iowa (1926).

38. Spangler, M. G., "A Preliminary Experiment on the Supporting Strength of Culvert Pipes in Actual Embankment," Bulletin 76, Iowa Engineering Experiment Station, Ames, Iowa (1926).

39. Spangler, M. G., "The Structural Design of Flexible Pipe Culverts," No. 153, Iowa Engineering Experiment Station, Iowa State University, Ames, Iowa (1941).

40. Spangler, M. G., and R. L. Hennessy, "A Method of Computing Live Loads Transmitted to Underground Conduits," *Proc. Highway Research Board*, Volume 26, p. 179 (1946).

41. Spangler, M. G., "A Theory on Lands on Negative Projecting Conduits," *Proc. Highway Research Board*, Volume 30, pp. 153–161 (1950).

42. Spangler, M. G. "Secondary Stresses in High Pressure Pipelines," *The Petroleum Engineer* (November 1954).

43. Spangler, M. G., *Soil Engineering*, Internation Textbook Company, Scranton, Pennsylvania, 2nd Edition (1960).

44. Spangler, M. G., *Culverts and Conduits*, Leonard's Foundation Engineering, McGraw-Hill Book Company, New York (1962).

45. Spangler, M. G., "Pipeline Crossing Under Railroads and Highways," American Water Works Association, Volume 56, No. 8 (August 1964).

46. "Structural Analysis and Design of Pipe Culverts," Highway Research Board Report No. 116, Northwestern University, Evanston, Illinois (1971).

47. "Thickness Design of Ductile-Iron Pipe," ANSI S21.50-1976, (AWWA C150-76).

48. Timoshenko, S. P. and J. N. Goddier, *Theory of Elasticity*, 3rd Edition, McGraw-Hill Book Company, New York, p. 405 (1970).

49. Wang, L. R. L. and R. C. Y. Fung, "Seismic Design Criteria for Buried Pipeline," *Proc. Pipeline Division, Specialty Conference*, New Orleans, pp. 130–145 (1979).

50. Watkins, R. K., "Failure Conditions of Flexible Culverts Embedded in Soil," *Proc. Highway Research Board*, Volume 39, pp. 361-371 (1960).

51. Watkins, R. K., "Structural Design of Buried Circular Conduits," Highway Research Record No. 145, pp. 1-16 (1966).

52. Watkins, R. K., M. Ghavami and G. R. Longhurst, "Minimum Cover for Buried Flexible Conduits," *Journal of Pipeline Division*, ASCE, Volume 94, PLI, pp. 155–171 (October 1968).

53. White, H. L. and J. P. Layer, "The Corrugated Metal Conduit as a Compression Ring," *Proc. Highway Research Board*, Volume 39, pp. 389–397 (1960).

CHAPTER 18

Corrosion of Underground Piping

PAUL N. CHEREMISINOFF* AND VINCENT M. PAPAROZZI**

INTRODUCTION

According to the National Bureau of Standards (Circular C450, 1945) about 500,000 miles of pipeline for gas, water and oil transmission is installed in the United States. Circular C579 (1957) reported about 988,000 miles of pipeline for the same uses.

The National Bureau of Standards has reported the annual cost due to the effects of corrosion to about 70 billion dollars. Any pipeline exposed to a wet environment or soil with a wide variety of materials is subject to corrosion. Engineers not fully understanding the corrosion rates, compensate by specifying a higher class of pipe with thicker walls to assure longer life. This approach results in undue consumption of our natural resources. A more severe problem than dollar costs of repair and replacement is the threat to public health as a result of corrosion. Corroded gas mains cause explosions, producing loss of life and property. Pressure interruptions in water mains can result in a backflow of contaminated liquid into the drinking water system. Without an adequate supply of water for fire protection, small fires can become disasters. The solution to the problem is understanding what causes corrosion and what measures are required to control it.

This chapter primarily deals with corrosion related to steel, ductile and prestressed concrete cylinder pipe. All other materials such as PVC, copper and asbestos cement are not used in large diameter piping applications. All materials are subject to corrosion, but concern over corrosion of a 1″ copper pipe for water service is not as critical as failure of a 96″ steel force main for water transmission.

*New Jersey Institute of Technology, Newark, NJ
**GHA Lock Joint, Inc., Wharton, NJ

UNDERGROUND CORROSION

Iron and other metals become coated with a thin film of oxide immediately after being exposed to air. This film furnishes some degree of protection against further oxidation or corrosion. At high temperatures, air diffuses through the film, increasing the film's thickness resulting in scaling. The scaling process exposes fresh surfaces for continued reaction. At normal temperatures, the passive film of iron oxide adds considerable protection to the underlying metal, increasing in thickness exceedingly slow.

The major cause of corrosion of iron in underground service is the result of an electrochemical reaction. In order for an electrochemical reaction to take place, a corrosion cell is to exist which contains the following four elements:

Anode—discharge current into the electrolyte and is corroding (ionization of metal).

Cathode—receives current and is protected.

Electrolyte—ionically conductive media.

Return Circuit—a metallic path through which electrons move from the anode to the cathode, usually along the wall of the metal surface.

In a corrosion cell, a potential difference exists between two points (cathode and anode) which are electrically connected and immersed in an electrolyte. A small current flows from the anodic area through the electrolyte to the cathode area and finally through the metal to complete the circuit.

An example of an electrochemical process is illustrated in Figure 1 by an electrical dry cell battery.

In Figure 1, the iron anode on the left is corroding. Electrons are flowing from the anode through the connecting wire to the cathode causing the copper cathode to become negatively charged. The electrolyte is water which contains positive ions hydrogen ($H+$) and negative ions, hydroxyl (OH^-). The positive charged ions are attracted to the copper cathode due to its negative charge, as the iron atoms within

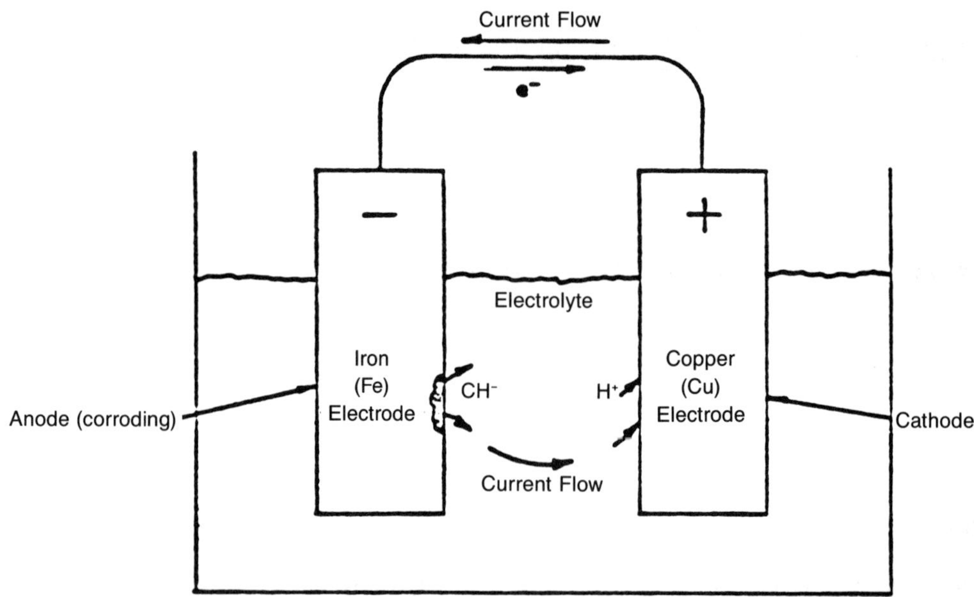

FIGURE 1. The electrochemical process illustrated by a dry cell battery system.

the anode are attracted to the negative ions in the electrolyte, causing dissolution of the metal.

RATE OF CORROSION

The rate at which a metal corrodes is affected by the following factors:

1. Polarization
2. Electrolyte resistivity
3. Voltage difference between anode and cathode
4. Anode/cathode ratio
5. Amount of energy stored in metals
6. Passivation of the metal surface

Polarization is the retardation of the corrosion process by the build-up of protective layers at the cathode or anode. The protective layer is a result of the evolution of hydrogen.

According to Ohms' Law, the greater the resistivity, the less the current flow as expressed in the equation:

$$I = \frac{E}{R}$$

I = Current Flow
E = Voltage
R = Resistance

For example salt water is more corrosive than fresh water due to its low resistivity, which allows current to flow more freely.

The greater the voltage difference between the anode and cathode, the greater the rate of corrosion. Due to polarization, the rate of corrosion may be reduced from the initial rate.

The difference in area of the anode and cathode considerably affect the rate at which the anode corrodes. If the anodic area is small in relation to the cathode, (i.e.: a steel bolt in a brass plate) the corrosion current is concentrated in a small area causing the anode to corrode rapidly. Also due to the size of the cathodic surface, polarization is not easily achieved, thus maintaining a high rate of corrosion.

By connecting a large anode to a small cathode (i.e.: brass bolt in a steel plate), corrosion is diffused over a large area. The anode corrodes more slowly because the current is not so dense. Thus, the rate of corrosion is reduced due to the rapid polarization of the cathode which stifles the current flow.

Corrosion involves the release of energy. Metals corrode at different rates because of the difference in the amount of energy stored in the metal. The rate of corrosion is expressed by Faraday's Law:

$$W = K \times I \times T$$

W = weight loss in one (1) year
K = electrochemical equivalent in pounds per ampere per year
I = corrosion current per year
T = time in years

The electrochemical equivalents for several common

metals are as follows:

iron or steel	20
lead	74
copper	45
zinc	23
aluminum	6.4
carbon	2.5

A metal exposed to air immediately forms a thin film of oxide; the ability of a metal to form a protective film affects the rate of corrosion.

CAUSES OF CORROSION

The theory of corrosion is defined as an electrochemical reaction exhibiting a voltage difference between the anode and the cathode. The causes for voltage difference can be separated into two broad categories: naturally occurring corrosion (galvanic corrosion) and stray current corrosion. Naturally occurring corrosion results from a natural reaction between the metal and its environment. As previously stated, a corrosion cell consists of an anode, cathode, electrolyte and return circuit. The following is a number of conditions in which the four elements of the corrosion cell come together to create a potential difference between the anode and cathode.

DISSIMILAR METALS

In investigating potential corrosion problems due to dissimilar metals, it is beneficial to refer to a list which defines the anodic metals. For practical purposes, corrosion engineers developed such a list known as the galvanic series (see Table 1). The metals at the top or anodic end are less noble than the metals found at the bottom or cathodic end.

Metals are anodic with respect to the metals below them. Metals close together in the series have less tendency to produce a corrosion cell than metals that are widely separated. The greatest potential difference would be found in a magnesium and platinum galvanic couple. Examples of dissimilar metal corrosion can be found on a pipe wall producing mill scale. The mill scale becomes cathodic to the pipe (see Figure 2). Cinders used in backfill for pipe installations act cathodic to steel or iron (see Figure 3).

DISSIMILAR SURFACES

Corrosion develops from differences in surface conditions. For example, scratches on a metal become anodic to other areas on the surface. Threads produce the same result. Figure 4 illustrates the path of current flow for threaded couplings. The current flows from the shiny metal area on the threaded section into the soil and completes the circuit by returning to a number of cathodic areas on the surface. Another example of a dissimilar surface cell is connecting a new metal to an old one. The new metal becomes anodic and quickly corrodes (see Figure 5).

DISSIMILAR ELECTROLYTES

A potential difference is created on a piece of metal exposed to different electrolytes. This condition is found in a long continuous pipeline that passes through different types of soils. One section of the pipeline could be installed in clay while another in sandy loam (see Figure 6).

Pipe in a trench, backfilled with sand mixed with clumps of clay cause corrosion where the clay contacts the pipe (see Figure 7). Another example of dissimilar electrolytes is soil and concrete. Pipe passing through a concrete wall experiences corrosion at the portion exposed to the soil.

OXYGEN CONCENTRATIONS

Due to crevices in metal units, the diffusion of oxygen is not uniform which results in corrosion. Oxygen content inside a crevice is very low compared to the oxygen content in the environment. Metal at the site of low oxygen concentration becomes anodic and initiates corrosion.

STRESS CORROSION

In a metal subject to different amounts of stress, the more highly stressed areas become anodic to the less stressed areas. Bolts exposed to an electrolyte are found to corrode in the center. The center of the bolt is subject to the greatest amount of stress. See Figure 8 for the flow of current.

BACTERIOLOGICAL CORROSION

Bacterial corrosion neither attacks the metal nor the coating to any sufficient degree, but does create corrosive soil conditions. The products of bacteria metabolism are usually acidic which increases the corrosiveness of the soil. Bacteria thrives in areas of low oxygen content, such as dense mucky soils. Soils high in organic content also experience bacteriological corrosion.

GRAPHITIZATION

Cast iron and ductile iron consists of graphite flakes or modules in an iron matrix. In corrosion of cast iron, the iron constituent is lost, leaving as a result the graphite and corrosion product. The structural integrity of the pipe is lost, causing the pipe to fail when subject to some sort of stress.

TABLE 1. Galvanic Series.

Electromotive Series of Metals	Galvanic Series of Metals and Alloys	
Potential	Corroded End (anodic, or least noble)	
+2.96 Lithium	Magnesium	Silver Solder
+2.93 Tubidium	Magnesium Alloys	
+2.92 Potassium		Nickel (passive)
+2.92 Strontium	Zinc	Inconel (passive)
+2.90 Barium		
+2.87 Calcium	Aluminum 2S	Chromium-Iron (passive)
+2.71 Sodium		18-8 Chromium-nickel-iron (passive)
+2.40 Magnesium	Cadmium	18-8-3 Chromium-nickel-molybdenum-iron (passive)
+1.70 Aluminum		
+1.69 Beryllium	Aluminum 17ST	Hastelloy C (passive)
+1.10 Manganese		
+0.76 Zinc	Steel or Iron	Silver
+0.56 Chromium	Cast Iron	
+0.44 Iron (Ferrous)		Graphite
+0.40 Cadmium	Chromium-Iron (active)	Gold
+0.34 Indium		Platinum
+0.33 Thallium	Ni-Resist	
+0.28 Cobalt		
+0.23 Nickel	18-8 Chromium-nickel-iron (active)	
+0.14 Tin	18-8-3 Chromium-nickel-molybdenum-iron (active)	
+0.12 Lead		
+0.04 Iron (Ferric)	Lead Tin Solders	
0.00 Hydrogen	Lead	
−0.10 Antimony	Tin	
−0.23 Bismuth		
−0.34 Copper (Cupric)	Nickel (active)	
−0.47 Copper (Cuprous)	Inconel (active)	
−0.56 Tellurium	Hastelloy C (active)	
−0.80 Silver		
−0.80 Mercury	Brasses	
−0.82 Palladium	Copper	
−0.86 Platinum	Bronzes	
−1.36 Gold (Auric)	Copper-Nickel Alloys	
−1.50 Gold (Aurous)	Monel	
	Protected End (cathodic or most notable)	

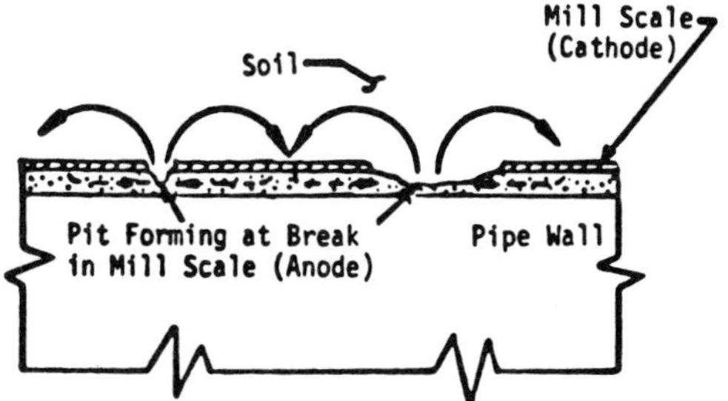

FIGURE 2. Pitting due to mill scale.

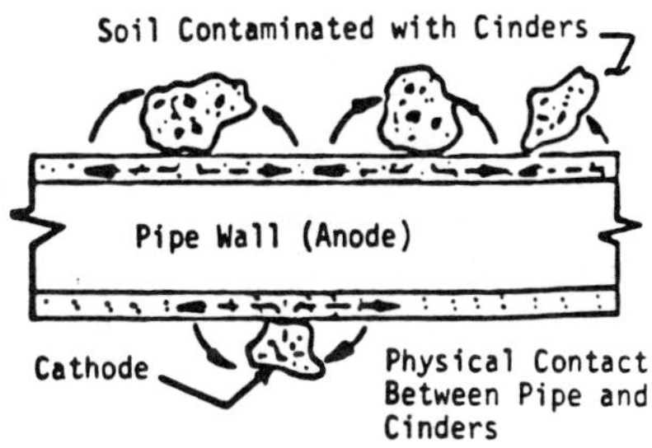

FIGURE 3. Corrosion due to cinders.

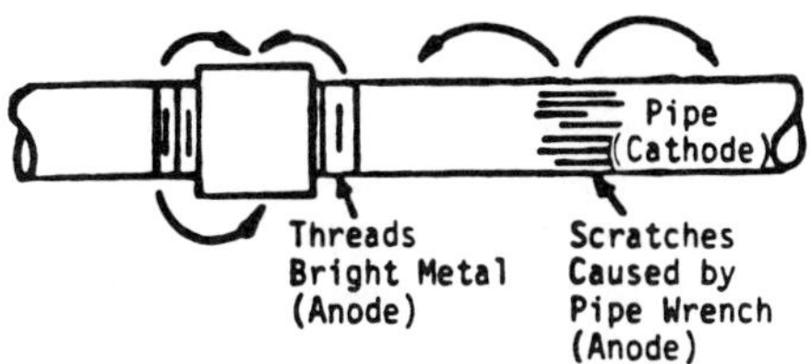

FIGURE 4. Corrosion caused by dissimilarity of surface conditions.

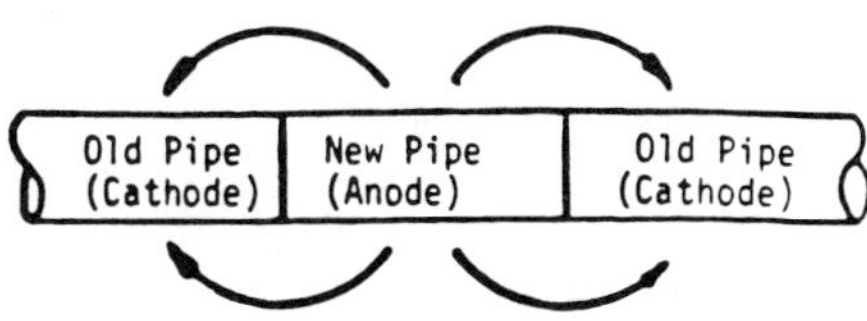

FIGURE 5. New-old pipe cell.

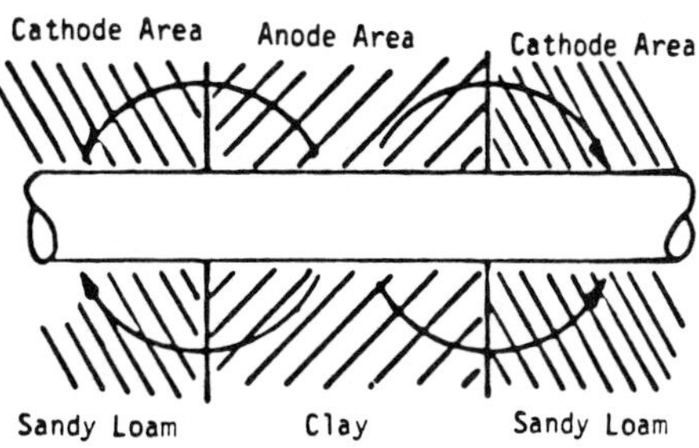

FIGURE 6. Corrosion caused by dissimilar soils.

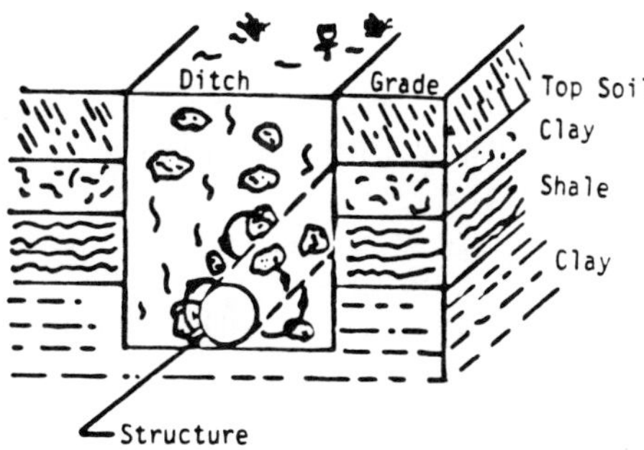

FIGURE 7. Corrosion caused by mixture of different soils.

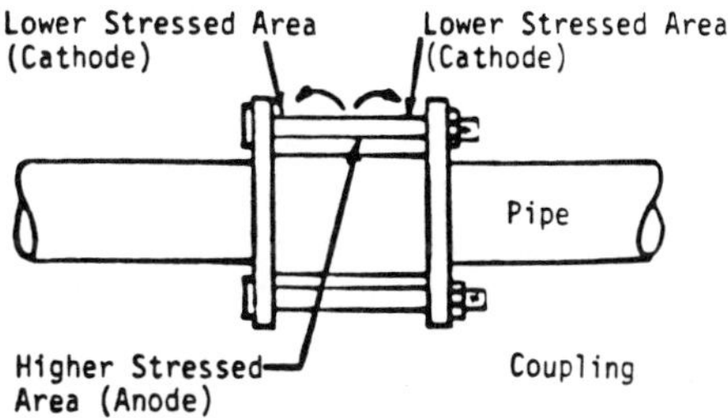

FIGURE 8. Stress corrosion.

STRAY CURRENT CORROSION

An external source furnishes the corrosion current in stray current corrosion. Corrosion from stray currents is more severe than corrosion due to dissimilar metals or any other natural occurring corrosion. According to the rate of corrosion expressed by Faraday's Law, 20 pounds of steel is lost per year for each ampere discharged from a steel structure. A galvanic cell (i.e.: dissimilar metal) possibly generates 0.01 amperes which would destroy 0.2 pounds of steel in a year.

Stray current is mainly limited to direct current. Alternating currently is not usually associated with corrosion problems.

Soil type is a major cause of galvanic corrosion. Usually, in a typical pipe installation various soils are encountered. The following is a breakdown of soils in the order of their corrosive properties as listed in the "Steel Pipe Design Manual" AWWA No M-11, page 174:

Soils grouped in order of corrosive action on steel:
Group I—Lightly Corrosive
Aeration and drainage good. Characterized by uniform color and no mottling anywhere in soil profile and by very low water table. Includes:

1. Sands or sandy loams
2. Light, textured silt loams
3. Porous loams or clay loams thoroughly oxidized to great depths.

Group II—Moderately Corrosive
Aeration and drainage fair. Characterized by slight mottling (yellowish brown and yellowish gray) in lower part of profile (depth 18–24 in.) and by low water table. Soils would be considered well drained in an agricultural sense, as no artificial drainage is necessary for crop raising. Includes:

1. Sandy loams
2. Silt loams
3. Clay loams

Group III—Badly Corrosive
Aeration and drainage poor. Characterized by heavy texture and moderate mottling close to surface (depth 6–8 in.) and with water table 2–3 ft. below surface. Soils usually occupy flat areas and would require artificial drainage for crop raising. Includes:

1. Clay loams
2. Clays

Group IV–Unusually Corrosive
Aeration and drainage very poor. Characterized by bluish-gray mottling at depths of 6–8 in. with water table at surface, or by extreme impermeability because of colloidal material contained. Includes:

1. Muck
2. Peat
3. Tidal marsh
4. Clays and organic soils
5. Adobe clay

Soil corrosion of underground cast or ductile iron pipe is dictated by the physical and chemical characteristics of the soil. One environment which serves as an adequate electrolyte for one metal, may not act as a conductive media for another. As previously mentioned, a corrosion cell consists of an anode, cathode, metallic path and an electrolyte. All metals under the right conditions can develop electrodes of a corrosion cell. The controlling factor is the type of soil (electrolyte) the pipe is exposed to, which will determine whether corrosion can be supported.

For ductile and cast iron pipe, a soil evaluation system has been developed by the Cast Iron Pipe Research Association; referred to as the 10 point system, it is used to evaluate the corrosiveness of the soil. Not all systems are applicable due to the complexity of underground corrosion.

Soil survey tests are covered by the ANSI standard A21.5, (AWWA C-105), Appendix A. The 10 point system was made public in 1968 and adopted as an appendix to the standard in 1972. The analysis is based on field experience with operating cast iron pipelines where the condition and age of the pipe are recorded along with the soil test results. Although most of the experience was with gray cast iron pipe, the soil evaluation system can be applied to ductile iron with equal accuracy. The procedure is based on the accumulation of data from five tests and observations: soil resistivity, pH, redox potential, sulfides and moisture. Each result is ranked according to its contribution to corrosivity. The points are totalled after all five observations are complete. A total of 10 or more points indicates corrosivity and a need for protective measures. Table 2 shows the soil characteristics and the points assigned to each result.

Soil resistivity is a test which measures the resistance of the soil. A low resistivity of soil indicates that the soil acts as a good electrolyte. Soil resistivity can be measured by the 4 pin system (Figure 9). The 4 pin system is not an accurate test due to the different readings that are recorded and averaged. At four locations, four different types of soil may exist from wet to dry or clean to contaminated soil. The use of a single probe (Figure 10) enables the surveyor to measure resistivity at various depths and specific locations.

Soils with a pH below 4.0 usually serve as a good electrolyte and are considered aggressive (corrosive). Neutral pH ranges from 6.5–7.5 indicates the soil is capable of supporting sulfate reducing bacteria. High pH soils (8.5–14.0) are high in dissolved salts and usually exhibit a low resistivity.

The moisture content of the soil should by measured at pipe elevation. By use of the quad-box (Figure 11), it is possible to water saturate the soil and simulate the worst possible condition.

TABLE 2. Soil Test Evaluation.*

Soil Characteristics	Points
Resistivity-ohm-cm (based on single-probe at pipe depth of water-saturated soil box)	
<700	10
700–1000	8
1000–1200	5
1200–1500	2
1500–2000	1
>2000	0
pH	
0–2	5
2–4	3
4–6.5	0
6.5–7.5	0+
7.5–8.5	0
>8.5	3
Redox Potential	
> + 100 mV	0
+50 to 100 mV	3.5
0 to + 50 mV	4
Negative	5
Sulfides	
Positive	3.5
Trace	2
Negative	0
Moisture	
Poor drainage, continuously wet	2
Fair drainage, generally moist	1
Good drainage, generally dry	0

*Ten points—corrosive to gray or ductile cast iron pipe, protection is indicated.
+If sulfides are present and low or negative redox-potential are obtained, three points shall be given for this range.

By measuring the oxidation reduction potential, the degree of aeration of the soil is determined. Low or negative results are typical of soil that is anaerobic and supports sulfate reducing bacteria. A pH meter with a platinum electrode in conjunction with the same reference electrode used for pH can be used to measure the redox potential.

Sulfide tests are qualitative and are measured by the release of hydrogen in the following reaction:

$$2N_aN_3 + I2 \rightarrow 2N_aI + 3N_2$$

A solution of 3% sodium azide in 0.1N iodine is put into a test tube with a small soil sample. Sulfides in the sample would catalyze a reaction between sodium azide and iodine, releasing nitrogen into the atmosphere. The above soil evaluation is used specifically for gray cast iron or ductile iron pipe.

PRESTRESSED CYLINDER PIPE

Concrete cylinder pipe consists primarily of a steel cylinder with reinforcement or high tensile wire (prestress wire) all embedded in a concrete or mortar encasement.

The three types of concrete cylinder pipe as designated by the American Water Works Association are:

- Reinforced Concrete Cylinder Pipe (AWWA C-300) Figure 12, consists of a steel cylinder with steel bell and spigot rings welded to each end with wire mesh reinforcement (one or two cages) placed on the outside of the cylinder. The steel assembly is encased in a concrete wall by vertical casting in steel molds.
- Prestressed Concrete Cylinder Pipe (AWWA C-301) consists of a steel cylinder with steel bell and spigot rings welded to each end with a high tensile wire wrapped either directly over the cylinder (lined cylinder pipe, Figure 13a), or wrapped around a concrete core (embedded cylinder pipe, Figure 13b). After the prestress wire is wrapped, a rich mortar coating is applied.
- Pretensioned Concrete Cylinder Pipe (AWWA C-303), Figure 14 consists of a steel cylinder with steel bell and spigot rings welded to each end with a centrifugally applied concrete lining, and pretensioned rod wrapped helically around the cylinder covered with an exterior mortar covering.

Prestressed Concrete Cylinder Pipe (PCCP) is subject to corrosion in certain environmental conditions, mainly high chloride soils, acid soils, atmospheric exposure and cathodic interference. Under most natural environments, the cement protects the steel from corrosion. The cement is chemically basic, with a pH of about 12.5. At this pH, an oxide film forms on the embedded steel surfaces passivating the steel.

The breakdown of passivation occurs in sound concrete only if certain anions exist at the steel surface. High concentrations of chloride ion and oxygen cause the iron to go into solution at the anode. For corrosion of steel in concrete to take place, anodic and cathodic reactions must take place simultaneously. Steel cannot corrode in the absence of oxygen. For example, pipe installed on ocean bottoms (outfalls) are exposed to high levels of chloride (20,000 ppm chloride) with no evidence of corrosion of embedded steel. This is due to the saturated mortar coating which limits the rate of oxygen diffusion. The concentration of chloride ion required at the steel surface to initiate corrosion is about 700 ppm. Steel in concrete will not corrode in concentrations below 700 ppm, above this level oxygen must be present to support corrosion.

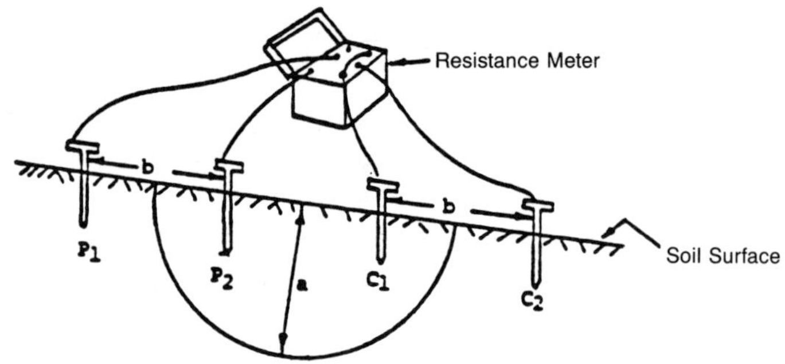

FIGURE 9. Soil resistivity will be averaged to depth a if a = b.

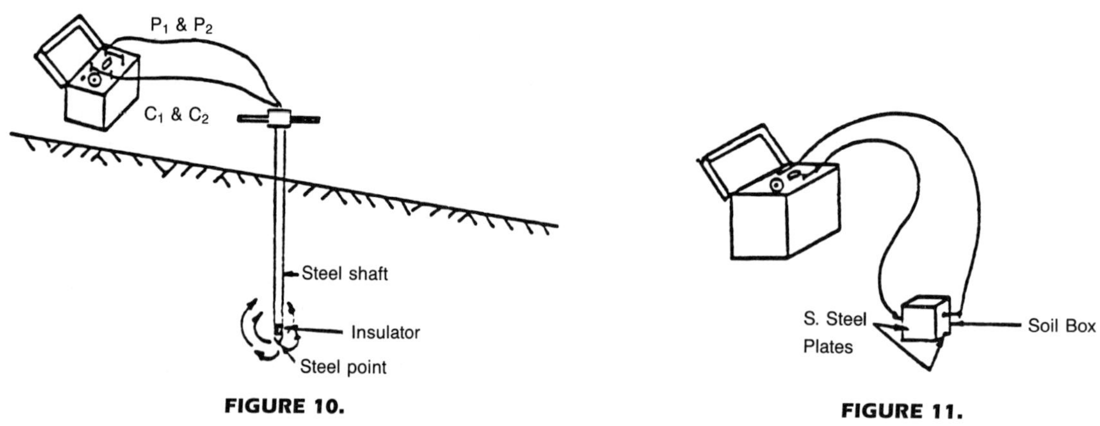

FIGURE 10.

FIGURE 11.

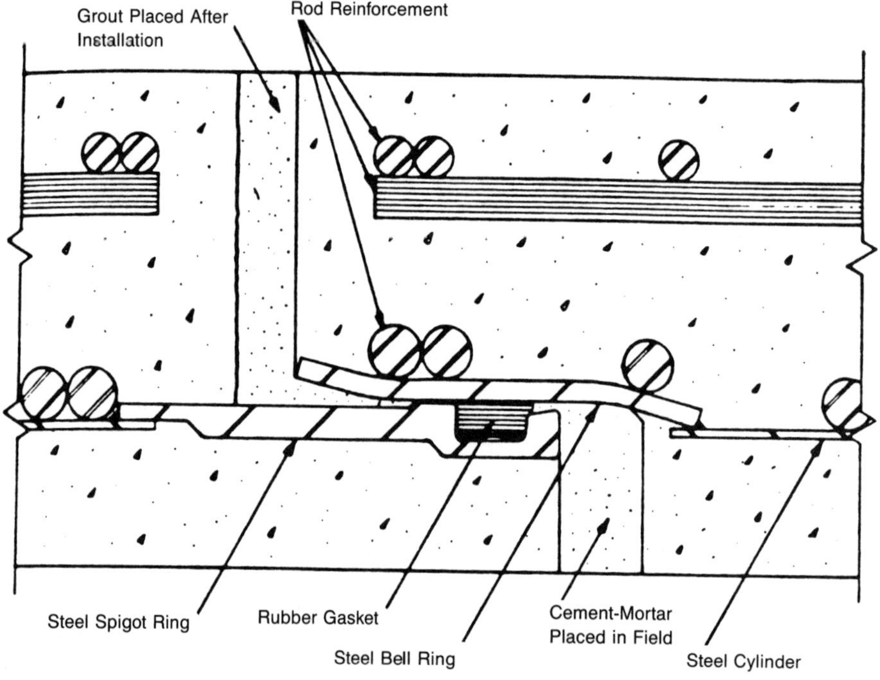

FIGURE 12. Reinforced concrete cylinder pipe.

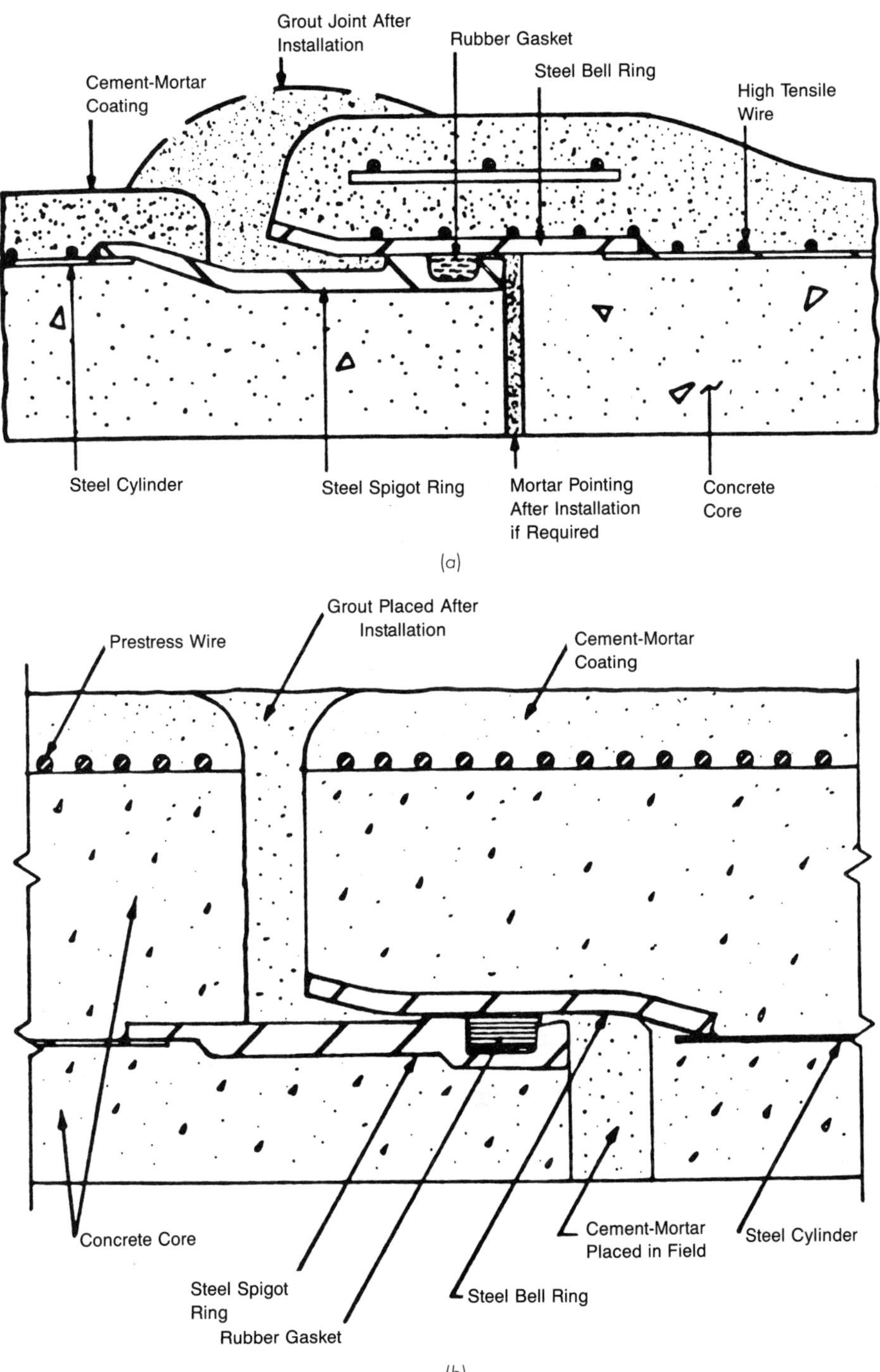

FIGURE 13. (a) Lined cylinder pipe; (b) Embedded cylinder pipe.

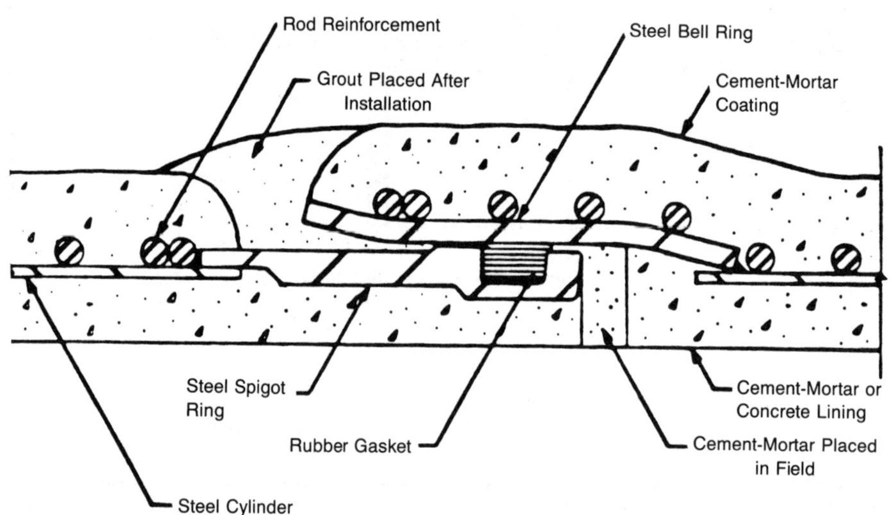

FIGURE 14. Pretensioned concrete cylinder pipe.

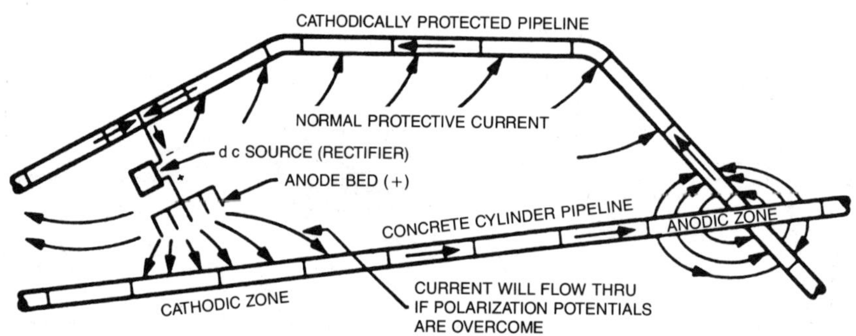

FIGURE 15. Cathodic interference.

In high chloride environments where the mortar coating is not saturated exterior coatings should be used to prevent chlorides and oxygen from reaching the steel surfaces.

Interference from impressed current protection is the most common sourcer of direct current (Figure 15). An example of this situation is the use of rectified systems to cathodically protected buried ferrous pipelines.

The current discharged from a cathodically protected pipeline will be picked up by an unprotected prestressed concrete cylinder pipeline (PCCP) if the polarization potentials of the PCCP are exceeded. Passivated steel (embedded steel) stifles external potentials of up to 1.5 volts. Corrosion is initiated at the intersection of the two pipelines, a point at which the current is discharged completing the circuit, thus causing dissolution of the metal at the anodic site. If cathodic interference with the buried concrete cylinder pipe is possible, the following precautions should be taken:

- Regulate the source of the direct current or relocate the anode bed.
- Measure the degree of interference caused by the protected pipeline. Pipe joints in close proximity to the protected structure should be bonded (+200′) and installation of test stations is also required on the concrete pipeline to measure the current.
  Note: Bonded joints refers to connecting all metal components of the pipe to facilitate electrical monitoring.
- Connect sacrificial anodes to the concrete cylinder pipe near the discharge point.

Pipe installed in acid soils (pH < 5) or soils with high sulfate content are potentially harmful to the cement mortar coating. Chemical attack of buried concrete due to acid soils is rare. Acid soils are mainly attributable to cinders, mine wastes and industrial dumps. The following precautions are recommended for pipe installed in acidic soils:

1. Completely surround the pipe with impermeable backfill (clay) to prevent acid ground water replenishment.
2. Neutralize the acid with backfill containing limestone.
3. Coat the exterior of the pipe with a barrier coating.

Soils containing high concentrations of sodium, magnesium and calcium sulfates are termed sulfates. Under certain conditions they react with hardened cement to form products which cause expansion of the cement mortar.

Sulfates are found in almost all free ground water. Evaporation of sulfate ground waters results in a high concentration of alkalis in the soil. The sulfates are dissolved in the soil by movement of surface and ground water, causing a high concentration in areas of low drainage. A change in color of the surface vegetation as compared to the surrounding areas usually indicate presence of sulfates.

The degree of corrosion of concrete in contact with ground water containing dissolved sulfates is controlled by the amount of tricalcium aluminate in the cement. The resistance of cement to sulfate attack is increased by lowering the tricalcium aluminate content in the cement.

If the soluble sulfate content for buried installations is greater than 2,000 ppm, the concrete pressure pipe industry recommends using ASTM C150 Type II cement with 5 percent (maximum) tricalcium aluminate.

The concentration of dissolved sulfates is also of importance in controlling sulfate attack. The permeability of the soil affects the sulfate build up which can be reduced by surrounding the pipe with low permeable soil (clay) during the backfill operation. By increasing the cement content in the mortar coating, the sulfate resistance is further enhanced as reported by the U.S. Bureau of Reclamation and the Portland Cement Association.

Atmospheric exposure to coated concrete pipe can present a corrosion problem due to the drastic changes in the environment. Exposed pipe is subject to alternate wetting and drying and fluctuating temperatures from day to night. During the summer months the temperature can exceed 100°F and drop below freezing (32°F) in the winter. Freezing water in concrete produces significant stresses in the coating, shortening its life.

Carbon dioxide in the atmosphere will react with the calcium hydroxide in the concrete to form calcium carbonate. The pH of the concrete coating can be reduced to a level where the passivating effect of the concrete is lost. By maintaining a minimum concrete to cement ratio of 1 to 2-1/2, the permeability is decreased.

Painting the exterior of the pipe with materials such as chlorinated rubber base paints seals the mortar coating, eliminating the attack by moisture and/or chemicals. The protective coating should be light in color to reflect the sun's damaging rays. Inspections and maintenance of the protective coating is recommended to ensure the integrity of the structure.

Steel pipelines, unlike prestressed concrete pipe requires protection from its environment. Soils corrosive to steel pipe may not have any effect on concrete embedded steel due to the alkaline environment afforded by the concrete.

Corrosion of the exterior of steel pipe is difficult to assess due to the variety of soils encountered. Soil analysis pH and moisture content are important parameters in evaluating corrosion of steel pipe. Yet, the key parameter associated with steel pipe corrosion is the measurement of resistivity of the soil. The three basic methods for controlling corrosion of underground steel pipelines are:

- Isolate and electrically insulate the pipe from its environment by coating the exterior surface.
- Impose electrical currents to counteract the currents associated with corrosion (cathodic protection).
- Create an inhibitive environment.

The object of all three methods is to inhibit the formation of rust, the first noticeable by-product of corrosion. By use of

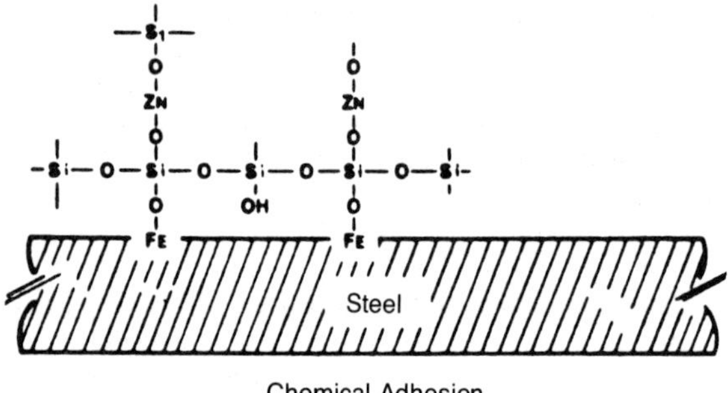

FIGURE 16. Sketch showing chemical adhesion. Inorganic zinc silicate coating chemically bonded to iron substrate.

external coatings, the steel pipe is isolated from its environment.

The effectiveness of protective coatings is only as good as its ability to stick to the surface. Surface preparation and attraction between the coating and surface is the key to good adhesion. The type of adhesion is determined by the nature of the substrate and coating. The three primary types of adhesive bonds are chemical bonds, polar bonds and mechanical bonds.

The most effective bond is the chemical bond and the most common chemical bond is zinc coating to steel (Figure 16). In the silicate matrix of the inorganic coating, an oxygen bond is attracted to an iron atom in the metal surface.

POLAR BONDS

Polar bonds or secondary valence bonds are only possible when the organic coating is close enough to the metal surface to initiate such a bond. The strength of a secondary valence attraction increases at a rate proportional to the sixth power of the intermolecular distance, and is only effective at a distance under 5A. If the surface is not cleaned, adhesion will be nullified by grains of dust, dirt or films of oil, which are thicker than 5A. Figure 17 shows secondary bonding.

MECHANICAL BONDING

Mechanical bonding is adhesion associated with surface roughness. Surface roughness or anchor pattern increases the surface area of the substrate allowing the molecules to interlock with the irregularities of the surface. If adhesion in the coating itself is poor, a deep anchor pattern is required. The high performance coatings obtain adequate adhesion with an anchor pattern of depths as low as one to two mils (25–50 μm). An additional factor associated with surface roughness is that the number of polar sites on a metal surface is directly proportional to the surface area. The adhesion is stronger due to the additional attractive forces to the polar groups on the organic coating (see Figure 18).

COATING FAILURE

Coating failure is directly related to its ability to adhere to the surface. The adhesive and cohesive strengths of the coating influence the coatings adhesion ability. The adhesion strength bonds the coating to the substrate surface. The cohesive strength is the bonding within the coating itself which holds the coating together as a single unit. If the cohesive strength is greater than the adhesive strength, adhesive failure will develop (Figure 19). A strong coating may have sufficient strength within the film to break the adhesive bond to the surface causing the coating to peel off. For best results, the adhesive strength should be greater than the cohesive strength of the coating. This failure may cause the coating to break up within itself, leaving part of the coating on the surface (Figure 20). No failure is good, but a cohesive failure will still maintain coverage over the substrate.

A third type of failure is where the substrate breaks when the coating is removed. This is not due to the adhesion of the coating but rather due to the cohesive strength of the substrate. This failure is found in concrete pipe with epoxy coatings.

Internal corrosion of cement lined pipe (steel, ductile, concrete) occurs in gravity installations by sulfur bacteria, which form H_2SO_4 (sulfuric acid) from H_2S (hydrogen sulfide) found in the sewage, causing concrete erosion. To understand how the reaction takes place the following three main nutritional requirements should be reviewed: (1) carbon source, (2) energy source, (3) electron donor.

Organisms that require organic compounds as their principal source of carbon synthesis of cell material are known as heterotrophs, while autotrophs use carbon dioxide as the principal carbon source.

Organisms obtaining energy from light are called phototrophs as opposed to chemotrophs, which oxidize organic or inorganic compounds as a source of energy. Lastly, organisms that use an organic compound as a source of electrons are organotrophs, while lithotrophs use an inorganic electron source.

In gravity installations the pipe is not flowing full, which allows for reactions by the aerobic and anaerobic bacteria to take place causing crown corrosion. Anaerobic chemoheterotrophs produce H_2S which is absorbed in condensate on the walls not in contact with the sewage. Aerobic chemolithotrophs oxidize the reduced sulfur to SO_4^{2-}-forming H_2SO_4, which attacks and erodes the concrete. Concrete pipe is subject to structural failure if this condition exists. Ductile and steel pipe do not rely on the cement lining for structural support, but deterioration of the cement lining will expose the metal to other forms of corrosion.

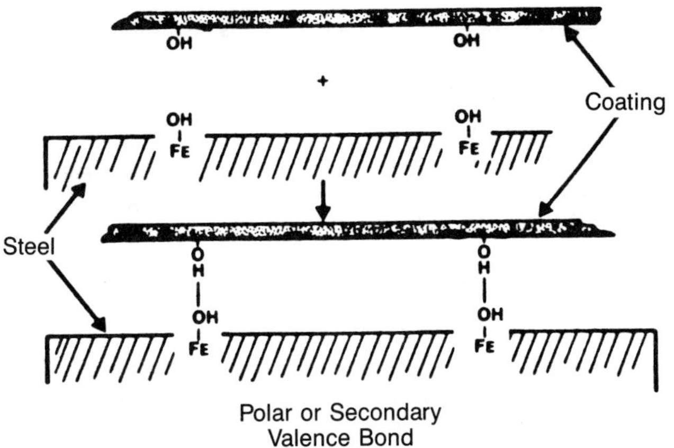

FIGURE 17. Sketch showing polar or secondary valence bond.

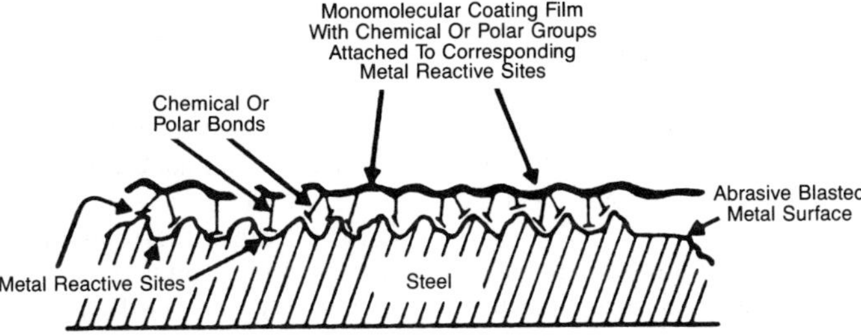

FIGURE 18. Possible association of chemical or polar bonds with metal reactive sites.

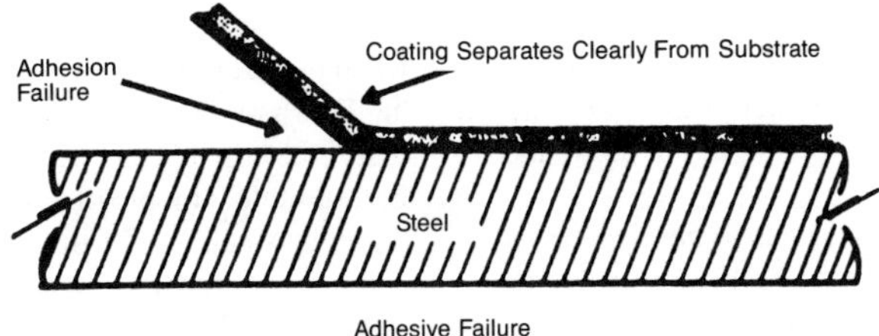

FIGURE 19. Sketch showing adhesion failure of a coating.

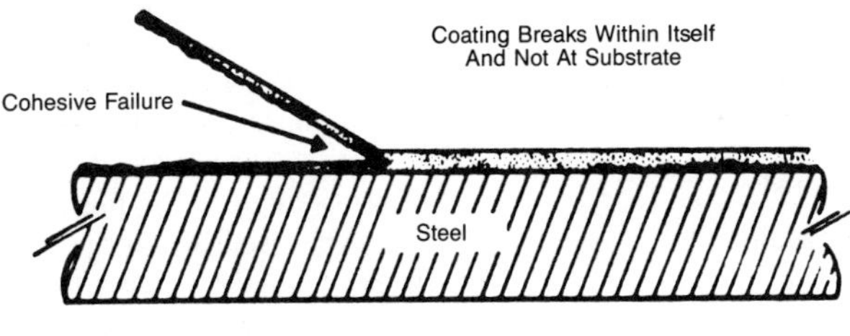

FIGURE 20. Sketch showing cohesive coating failure.

Concrete attack by sulfuric acid can be eliminated by placement of T-Lock lining. T-Lock lining is manufactured by Ameron Protective Lining Products in California, specially made for concrete pipe. This polyvinyl chloride sheeting is locked mechanically into the interior walls of concrete pipe during manufacture. The T-Lock sheeting is rolled and placed in the forms prior to concrete placement to allow for T-Lock embedment.

Some concrete sewer pipelines are designed to incorporate limestone aggregate in the concrete mix to neutralize the acidic fluid. Sewer force mains (pumped sewage pipelines) do not pose the same problem as gravity lines and usually do not require any corrosion prevention systems. A pressure line flows fill which precludes the possibility of the formation of sulfuric acid.

Coating pipe does not eliminate the possibility of corrosion. If a pinhole or holiday exists in an organic coating in a corrosive environment, a concentration cell may develop causing corrosion underneath the coating. Cathodically protecting a coated pipe may cause coating disbondment due to excessive current. Also, coatings can be easily damaged during shipping, handling, storage, or installation. Supplementing organic coating with cathodic protection is recommended in corrosive environments.

Cathodic protection is the protection of a metal structure by making it the cathode in a galvanic or electrolytic cell. One method of generating current is by use of sacrificial anodes, such as magnesium or zinc, which is used to create a galvanic cell. As noted in the galvanic series, magnesium and zinc are anodic to steel, ductile iron, or embedded steel, which causes the sacrificial anodes to corrode while the structure becomes cathodic and is protected (Figure 21). The magnesium anode is self-energized and is attached directly to the pipe. This system is used where small amounts of current is generated at a number of locations and is often used on coated steel pipelines in lightly to moderate corrosive soils.

The second method incorporates anodes energized by a direct current, commonly referred to as impressed current. Anodes, relatively inert are connected to the positive terminal of a d.c. power supply rectifier unit, with the pipe connected to the negative terminal (Figure 22).

A cathodic protection system requires adequate design. Sizing of anodes is based on a number of factors, such as current output, resistivity of the soil, life expectancy, weight of the anode and appropriate correction factors.

The advantages of cathodic protection by sacrificial anodes vs. an impressed current system are:

Sacrificial Anodes

1. No external power required
2. Minimum maintenance
3. Seldom cause interference w/foreign structures
4. Reasonable installation costs
5. Efficient use of protective current

Impressed Current

1. Larger driving voltages
2. Higher current level availability
3. Used in higher resistivity soils
4. Protects longer pipelines

The *disadvantages* of each system are:

Sacrificial Anodes

1. Limited driving voltages
2. Limited current output
3. Soil resistivity limitations

Impressed Current

1. Greater installation costs
2. Higher operation and maintenance costs
3. Danger of damage to other structures

In an impressed current system all joints are required to be electrically continuous, in order for the pipeline to act as a return current path. Any rubber gasketed joint or insulated joint will break the continuity causing the current to discharge into the soil to go around the insulation, resulting in corrosion at point of discharge (Figure 23).

Bonding of joints for prestressed concrete cylinder pipe is required when stray currents are encountered in the pipeline vicinity, or where low soil resistivities are reported due to high chloride contamination of the soil. After the line is made electrically continuous, test leads could be installed at intervals to measure the pipe to soil potential.

Pipe to soil potential readings only become significant when the pH of the coating falls below 9. As illustrated from the Pourbaix Diagram (Figure 24), with a pH of 10, no corrosion is possible no matter what the potential difference. For example, at a pH of 8, corrosion will occur if the potential is between -850 mv and -600 mv.

A pipe to soil potential of -350 mv or less (i.e., more negative) and a pH less than 9 indicates the possibility of corrosion of the prestressing wire. The reading could also indicate a bare metal structure in contact with the pipe. Such structures include connections to steel or iron pipe, valves, flanges, blowoffs, etc. The polarization shift will attenuate along the pipeline on either side of the connected bare metal. A copy of as-build drawings is required to check if any pipeline appurtenance is causing the shift in potential.

Cathode protection systems are rarely required for concrete cylinder pipe. If electrical monitoring indicates the possibility of corroding prestressing wire, a cathodic system can be installed. Sacrificial zinc anode system is preferred to eliminate the possibility of hydrogen embrittlement of the prestressing wire due to excess current transmitted by rectified systems.

A protection level of 500 mv to a CSE is adequate to prevent corrosion in a high chloride, low resistivity environ-

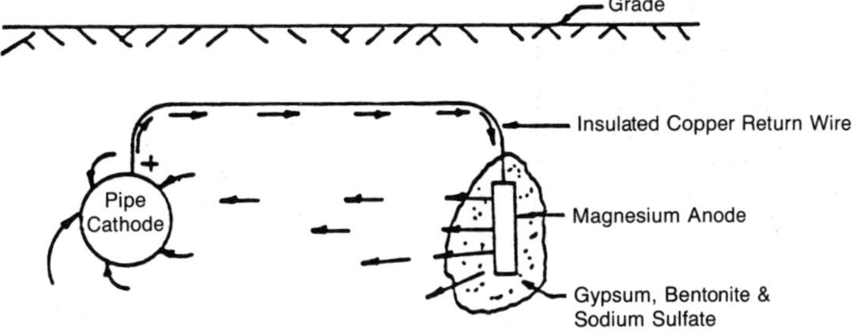

FIGURE 21.

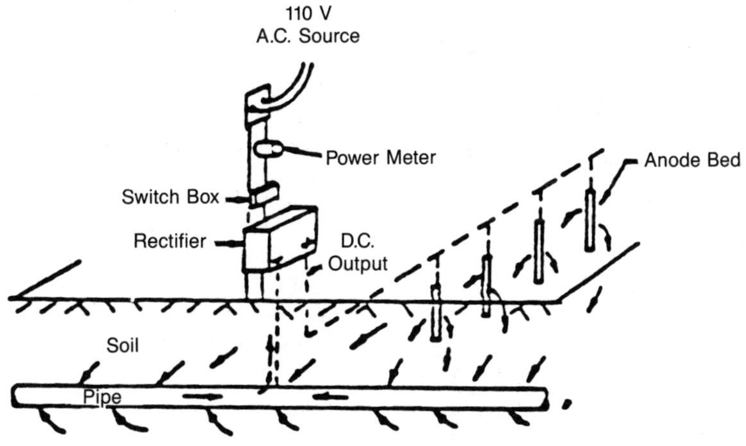

FIGURE 22.

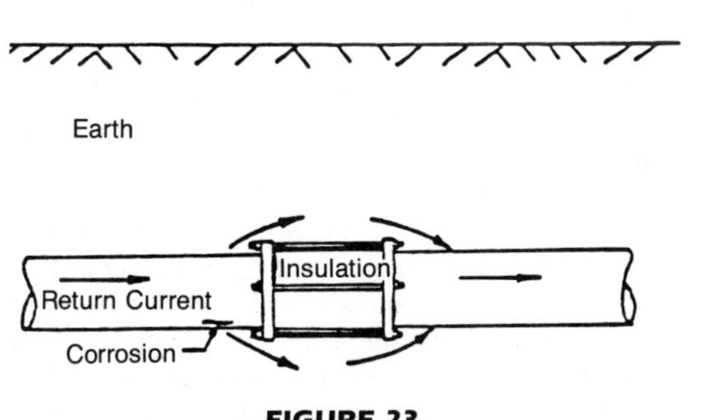

FIGURE 23.

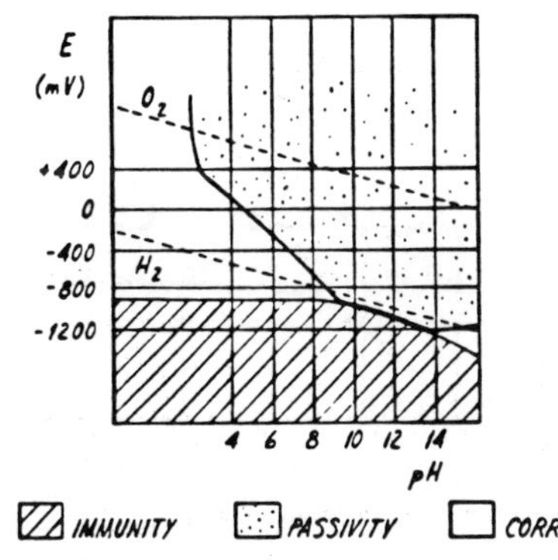

FIGURE 24. Pourbaix diagram for Fe.

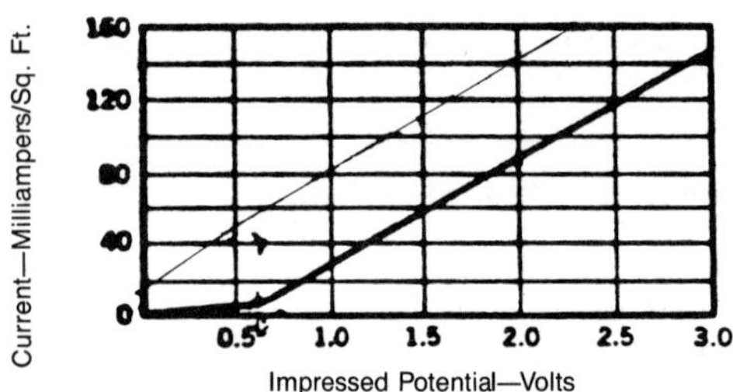

FIGURE 25. Behavior of cement coated steel cathode and bare steel anode.

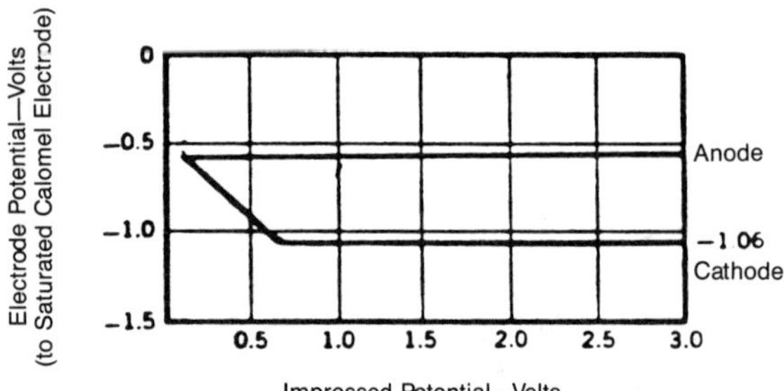

FIGURE 26. Behavior of cement coated steel cathode and bare steel anode.

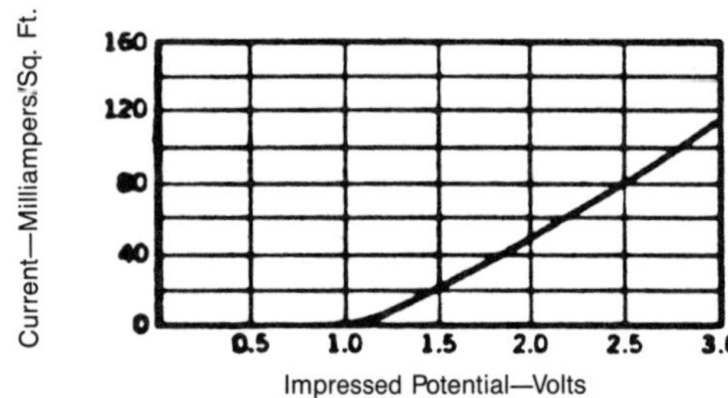

FIGURE 27. Behavior of cement coated steel anode and bare steel cathode.

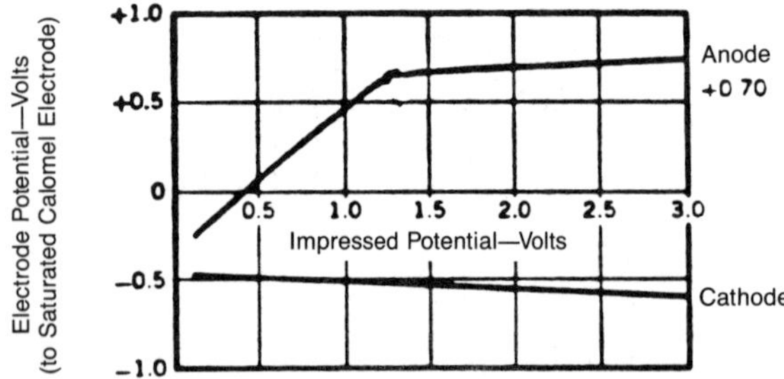

FIGURE 28. Behavior of cement coated steel anode and bare steel cathode.

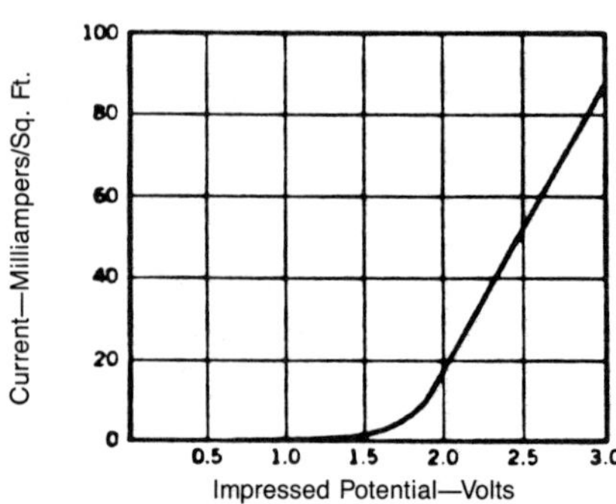

FIGURE 29. Behavior of cement coated steel electrodes.

ment. Based on experience, full cathodic protection can be provided with a current density of about 60 microamperes per square foot. The potential required to maintain 60 microamperes will vary from about −400 mv (CSE) initially to about −710 mv (CSE) after the first weeks. A potential of −850 mv to a CSE is used by corrosion engineers. Potentials more negative than approximately − 100 mv to a CSE should be avoided to prevent excessive current flow and hydrogen gas evolution. By installing galvanic systems such problems are avoided.

In 1957 the American Concrete Pressure Pipe Association directed a study which extended more than 4 years to investigate corrosion resistance of prestressed concrete pipe. The objectives of the program are as followed:

- To explain why concrete and cement mortar protects steel from corrosion
- To determine what exposure conditions lessen this protective capacity
- To ensure longevity of the protection

Cement mortar and slurry coated steel specimens were subjected to impressed cathodic voltages, and the corresponding current densities and polarization, or counter potentials were measured. Specimens that were maintained at constant driving voltages for long periods of time experienced disbondment with the mortar. This was accomplished by subjecting the concrete embedded steel electrodes to constant spring loads so that bond failure would be immediately detected.

Anodic behavior of cement mortar protected steel was also tested to measure the polarization potentials and current densities. In both cases, evolved gas was collected. The effects of environment on embedded steel was also studied. Steel was immersed in saturated calcium hydroxide solution which contained free oxygen and anions commonly found in soil and seawater. Free oxygen and anions in solution were analyzed separately and together.

The electro-chemical behavior of cathodic steel in cement mortar is illustrated in Figure 25. The current flow is negligible until the impressed voltage approaches approximately 0.60 volts.

After that point the current flow increases linearly with the voltage according to Ohm's Law, $V = IR$. The electric resistance of the circuit can be calculated. The reaction at the metal surface is described in the following equation:

$$2H_2O + 2e- = H_2 + 2OH- \ldots$$

As can be seen from the above reaction, the evolution of hydrogen gas is initiated. Under a potential of 0.6 volts, current flow is controlled by something other than circuit resistance. As shown in Figure 26 for an encased cathode, a polarization potential is developed at almost the same rate as the driving voltage is increased, up to a point at which current flows.

A maximum polarization potential of −1.06 volts to a saturated Calomel Electrode (SCE) is maintained beyond a driving voltage of 0.60 volts. The potential of the bare anode remains almost constant at the electrolytic potential.

The electro-chemical behavior of encased anodic steel is illustrated in Figure 27. With the anodic steel, there is no current flow until the driving voltage reaches about 1.1 volts. In this case, the current is less and the limiting potential is higher. Above 1.1 volts the current flow is linear to increases in potential as in accordance with Ohm's Law. With this condition oxygen instead of hydrogen is evolved at the encased electrode. The steel surface reaction is written:

$$2OH = 1/2O_2 + H_2O + 2e$$

The same principle of counter, or polarization potential for cement coated steel anode is illustrated in Figure 28. The counter potential increases proportionally with the applied potential up to a limiting value of 1.1 volts, at which point the polarization potential is about +0.70 volts (SCE). The bare cathode shifts slightly in the negative direction, as is expected under cathodic protection.

The polarization potential is drastically increased when potentials are impressed across a mortar protected anode and cathode.

According to Figure 29, current does not begin to flow until the total impressed voltage approaches 1.75 volts and is practically zero up to about 1.55 volts. It is clear that the magnitude of the potentials developed at both electrodes is equal to the potentials of the individual anode and cathode: +0.70 volts at the anode and −1.1 volts at the cathode (SCE). The overall effect is equal to the scalar sum of the individual electrode polarization potentials.

The encased cathode subjected to driving potentials as high as −1.54 volts (SCE) while under a spring load did not show signs of disbondment until nearly after 2-1/2 years; at which time the first two were then tested to destruction (by pulling of the rod) at 880 days and the third at 994 days. The measured bond strengths at these periods were the same as the control specimens. A fourth specimen under excessive protection of −2.14 volts (SCE) and current density of 400 milliamperes per square foot failed after 197 days.

Steel in cement mortar will not corrode in soil exposed to free oxygen or anions, acting independently. In the presence of some anions, chlorides, bromides, iodides and sulfides, corrosion is possible by trapping bubbles of oxygen against the steel surface. In relation to common occurrence in soils, chlorides is the most significant.

Figure 30 schematically illustrates the environment of a buried prestressed concrete cylinder pipe in a typical pipeline installation. All the elements required for electrochemical corrosion are present. The steel cylinder and joint rings comprise the electrodes, the soil and ground water provide the electrolyte for the flow of current. The soil ground water also provides the corrosive environment in the form of dissolved salts (chlorides), and free oxygen in some cases. Potential differences between various points in the pipe develop due to a number of causes such as: differences in soil characteristics, moisture content, and dissolved salt concentrations. These conditions exist along the length of a

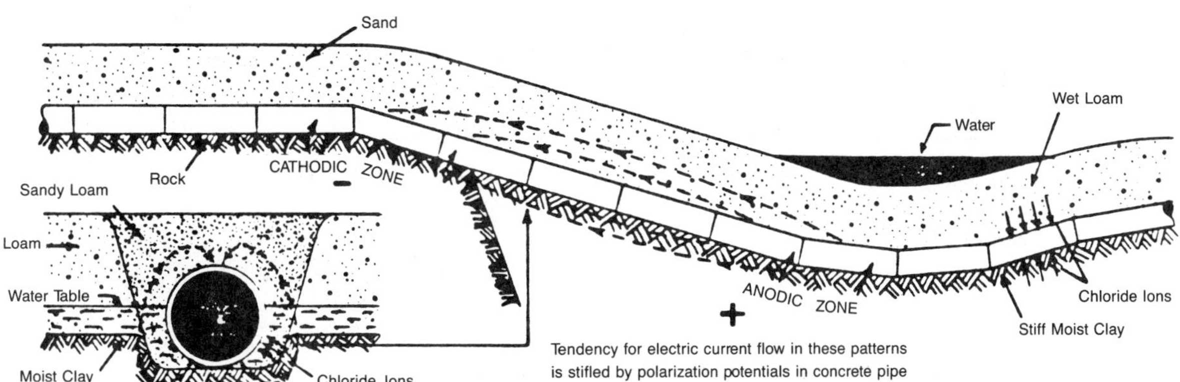

FIGURE 30. Typical soil environment of buried pipeline.

pipeline, or in large diameters, around the circumference, from top to bottom. Seasonal changes cause wetting and drying of the soil.

The net result of these differences is a cell. Each cell consists of a positive area (anode) and a negative area (cathode), an electrolyte and a return path comprising the steel cylinder. The tendency is for the current to flow from the anode through the mortar and soil to the cathode, and return via the pipe cylinder to the point of origin (anode).

With a bare steel pipe, current flow will cause the iron to go into solution resulting in corrosion at the anode. With a mortar coated pipe, the current flow is stifled. The total potential difference between the anode and cathode have to exceed 1.5 volts for the polarization potential to be overcome and for current to flow. Also the pipe to soil potential at the anode and cathode would have to exceed the individual polarization potentials.

CONCLUSION

Ductile Iron's answer to corrosion is coating. Coating in itself does not provide 100% protection because there is no such thing as a perfect coating. All coatings have pinholes or "holidays" due to the nature of the material. During installation the coating is subject to further damage. Corrosion will then be concentrated at these breaks in the coating. The surface will develop pitted corrosion as opposed to a more uniform corrosion over the surface of a bare metal pipe. Failures often occur more quickly in a coated pipe because localized corrosion penetrates the walls of a pipe faster than uniform corrosion. It is possible that graphitization of cast iron pipe at the coating breaks will stop further corrosion. To incorporate graphitization into corrosion control makes for unsound engineering practice. In stray current corrosion, the damage is more severe and occurs frequently due to the high density of the discharge current at coating faults.

Today design of ductile adds a service allowance of 0.08 in (corrosion allowance) to the pipe. Due to the extra thickness, failure is just delayed, not controlled. Steel pipe abandoned corrosion allowance in the early 1930s. The AWWA Steel Pipe Committee reported extra wall thickness is not necessary with today's methods of corrosion control.

Cathodic protection in conjunction with coating will eliminate corrosion at holidays. The current required to protect the uncoated holes can be supplied by a galvanic system of cathodic protection. In cathodic protection the magnesium or zinc anode creates a potential difference between the anode and iron pipe causing current to flow. Replacing sacrificial anodes is inexpensive as compared to repairing water main breaks.

If the system is properly designed, the anodes can last up to 30 years or more. The cost of cathodic protection and

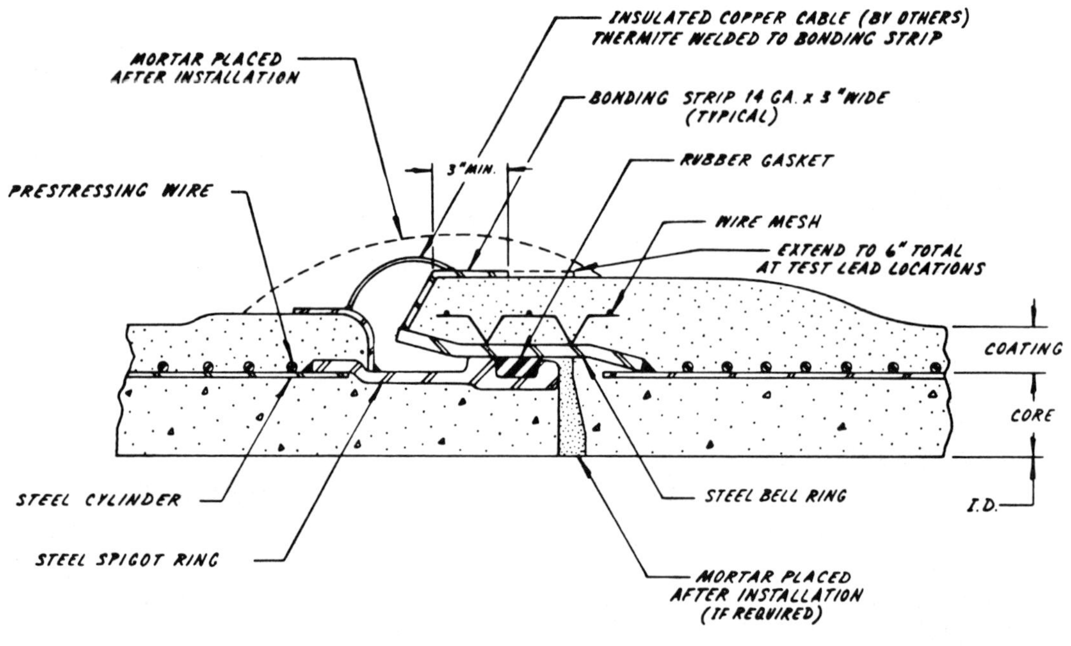

FIGURE 31. Prestressed concrete cylinder pipe (SP-5 external bonded joint detail).

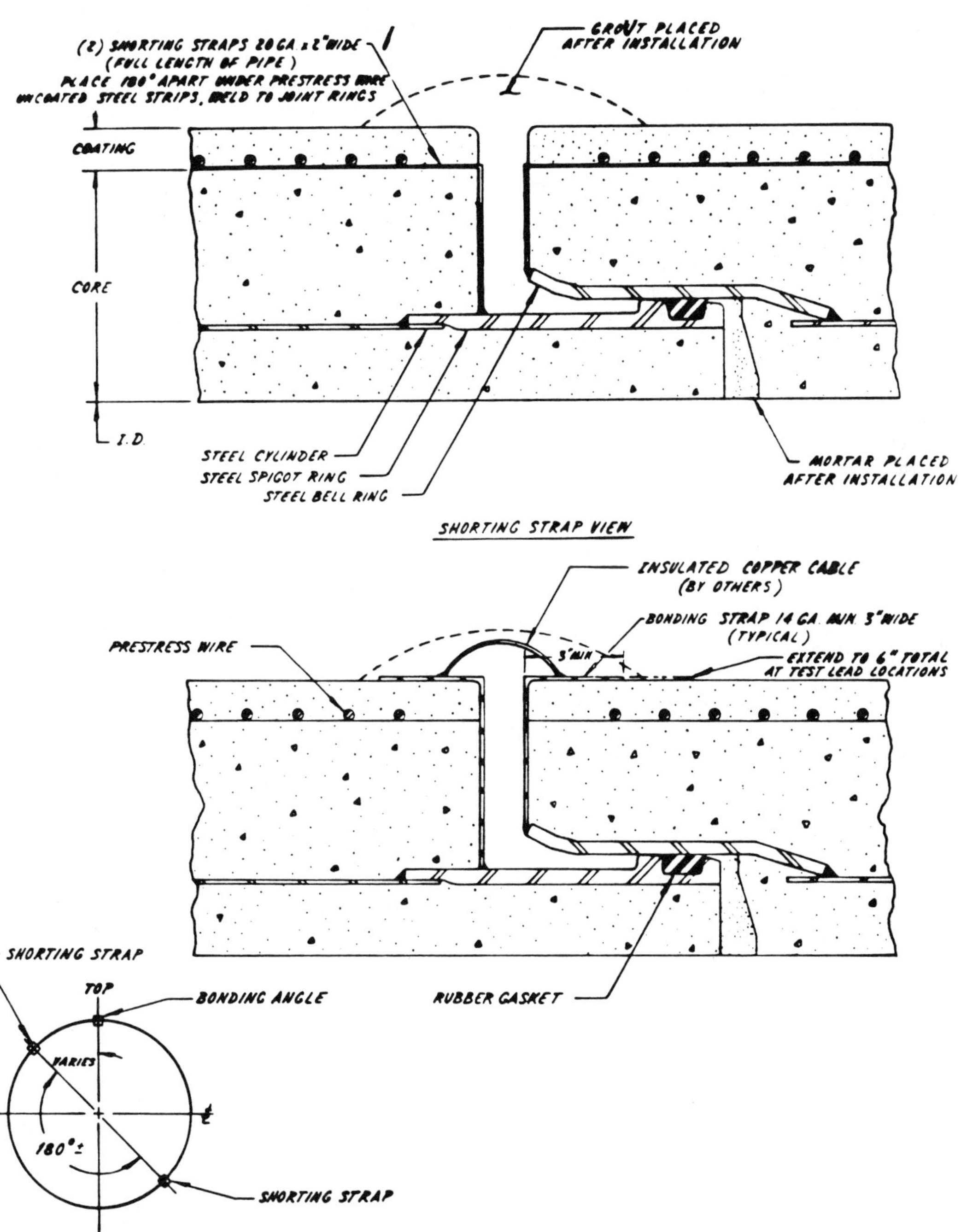

FIGURE 32. Prestressed concrete embedded cylinder pipe (SP-12 external bonded joint detail).

coating should be weighed against the cost of protecting a bare pipe.

With steel pipe, it is less expensive to coat first rather than to protect bare pipe. This is because steel pipe can be precoated in the shop or in the field. Cast iron on the other hand is not equipped to shop coat pipe on a routine basis. The result is protecting bare pipe is less expensive than coating plus protection.

For cathodic protection to be efficient, electrical continuity is required. Rubber gasketed joints do not provide sufficient continuity and as a result will not allow proper monitoring of the pipeline. Joints can be made continuous by connecting a No. 2 or 4 insulated copper cable across the joint. In stray current corrosion, bonded joints are essential to safely discharging and diffusing the current. Ductile iron's reason for unbonded joints is that individual pipe joints will experience only localized cell corrosion as opposed to subjecting the pipe to long line potentials.

The standard for protective coatings for steel pipe is found in the following specifications: AWWA C203 for coal tar enamel protective coating; AWWA C205 & AWWA C602 for cement mortar linings and coatings.

At the option of the design engineer, steel pipe can be installed with either of the protective coating systems. All coatings and linings can be applied in the plant or on the job.

Based on a study by the National Bureau of Standards between 1910 and 1955, corrosion (excluding stray current corrosion) on a steel pipe decreases with time in all but the most corrosive soils. This is largely due to the fact that corrosion products stifle corrosion. As a safeguard against corrosion extra thick wall thicknesses were specified. With today's protective coating system, wall thickness is designed in strict accordance with the internal and external pressures of that particular project without specifying extra thickness for corrosion.

Prestressed Concrete Cylinder Pipe (PCCP) is not subject to galvanic corrosion as metal pipe due to the protection of cement mortar. The fact that the steel is in an alkaline environment should not exclude the possibility of corrosion. If the mortar coating is damaged during installation, the steel will be in direct contact with the environment. If the pipe system experiences excessive surge pressures caused by the shutdown of surge relief valves, coating failure is a possibility. This is a mechanical failure which cannot be predicted and yet should be accounted for by installing bonded joints for electrical monitoring of the line. If the pipe is subject to stray currents, cathodic protection can be added at a low cost. Once the system is in place, the cost to excavate and isolate the line from stray current corrosion becomes uneconomical and in some instances the pipe is replaced. The steel components of PCCP should be electrically connected during manufacture and the joints bonded at time of installation. See Figures 31 and 32 for lined and embedded cylinder pipe respectively.

Coating of PCCP is also common practice in the case of a fluctuating water table. Wetting and drying of the mortar pipe can lead to cracking of the coating, and in the presence of high chloride content soils, the result is corrosion.

Corrosion as previously discussed is an electrical phenomenon and can be detected by electrical tests. The most common are soil resistivity and pipe-to-soil potential measurements. Pipe-to-soil potentials can be measured by making electrical contact with a continuous pipeline using test leads (insulated wire). The potential is read on a voltmeter and to a reference electrode such as a copper–copper sulfate half cell. With a continuous pipeline tests can be made at any time to check the adequacy of the pipeline.

Lastly, the most important consideration is the economics of corrosion control. The cost of coating, electrically bonding joints and/or applying cathodic protection is cheaper than replacing pipelines due to corrosion. Corrosion control should be a regular part of all new construction projects. A soil survey along the proposed route will yield valuable information. Designing a pipeline without corrosion control is like designing a building without proper foundation.

SECTION FIVE

Waves and Wave Actions

CHAPTER 19 Waves and Wave Forecasting 597
CHAPTER 20 Waves and Wave Forecasting 653
CHAPTER 21 Waves on Coastal Structures....697
CHAPTER 22 Waves on Structures 717

Waves and Wave Forecasting

EDWARD F. THOMPSON*

INTRODUCTION

Most coastal and ocean projects require an estimate of the characteristics of wind-generated gravity waves at the project site. These waves have periods of between 1 and 25 seconds; hence, they are considerably shorter than tides, storm surges, and tsunamis, which may also be important in design. Since wind waves occur continuously at exposed project sites, they can be an important concern for operations and maintenance as well as design.

Procedures for developing wave characteristics for design are presented in this chapter. The use of wave theory is covered in the following section. Subsequent sections deal with wave measurement and analysis systems and numerical and analytical wave models. The final section presents alternatives for forming statistical summaries of individual wave estimates to give wave climate information for design. References are inserted in the text as appropriate. References which are particularly comprehensive in coverage are [10,17, 38,42,45,53].

WAVE THEORY

Wave theories can be very useful for estimating certain wave characteristics such as wavelength, wave speed, crest height, wave shape, water particle accelerations, and wave forces. Theories suited for engineering use are commonly based on the following assumptions: homogeneous, incompressible fluid; no surface tension; no Coriolis effect; pressure at the surface is uniform and constant; inviscid fluid; wave does not interact with any other water motions; horizontal, fixed, impermeable bed; small wave amplitude and wave shape which does not change with time and space; long crested waves (two-dimensional). Theories can be applied to areas of gradually varying water depth by assuming the waves are adjusted to the local water depth. The theories should be used only when the bottom slope is flatter than 1 on 10.

Small-amplitude Wave Theory

The most fundamental description of a simple sinusoidal oscillatory wave is by its height H (the vertical distance between crest and trough), length L (the horizontal distance between corresponding points on two successive waves), period T (the time for two successive crests to pass a given point), and depth d (the distance from the sea bed to the stillwater level). A graphic definition of terms is given in Figure 1.

Small-amplitude wave theory has proven to be a very useful tool for many engineering applications. It is based on the additional assumption that the wave height is small in relation to the water depth and the wavelength. Expressions for various wave characteristics derived from small-amplitude theory are given in Figure 2. The approximate limits of validity for the small-amplitude assumption are shown in Figure 3.

Higher Order Wave Theories

Steep or shallow waves which exceed the range of validity for small-amplitude theory can be estimated with higher order theories. The approximate ranges of validity for higher order theories are shown in Figure 3. The higher order theories are more complex and hence more difficult to use. Tabular or graphical presentations of the solutions are generally used. Tabulated solutions from stream function theory are given by Dean [8]. Wave profiles for the 40 cases

*Coastal Oceanography Branch, U.S. Army Engineer Waterways Experiment Station, Vicksburg, MS

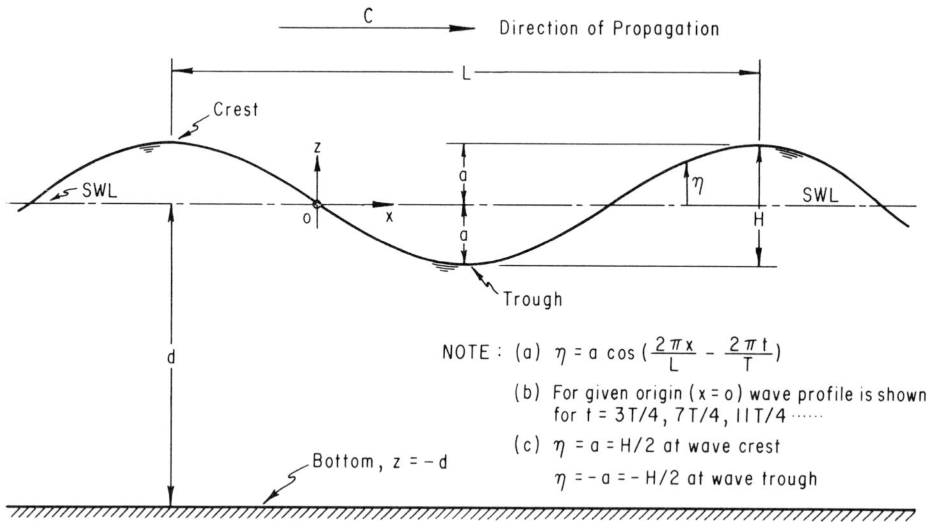

FIGURE 1. Definition of terms—elementary, sinusoidal, progressive wave [45].

RELATIVE DEPTH	SHALLOW WATER $\frac{d}{L} < \frac{1}{25}$	TRANSITIONAL WATER $\frac{1}{25} < \frac{d}{L} < \frac{1}{2}$	DEEP WATER $\frac{d}{L} > \frac{1}{2}$
1. Wave profile	Same As →	$\eta = \frac{H}{2} \cos\left[\frac{2\pi x}{L} - \frac{2\pi t}{T}\right] = \frac{H}{2} \cos \theta$	← Same As
2. Wave celerity	$C = \frac{L}{T} = \sqrt{gd}$	$C = \frac{L}{T} = \frac{gT}{2\pi} \tanh\left(\frac{2\pi d}{L}\right)$	$C = C_0 = \frac{L}{T} = \frac{gT}{2\pi}$
3. Wavelength	$L = T\sqrt{gd} = CT$	$L = \frac{gT^2}{2\pi} \tanh\left(\frac{2\pi d}{L}\right)$	$L = L_0 = \frac{gT^2}{2\pi} = C_0 T$
4. Group velocity	$C_g = C = \sqrt{gd}$	$C_g = nC = \frac{1}{2}\left[1 + \frac{4\pi d/L}{\sinh(4\pi d/L)}\right] \cdot C$	$C_g = \frac{1}{2} C = \frac{gT}{4\pi}$
5. Water Particle Velocity (a) Horizontal	$u = \frac{H}{2}\sqrt{\frac{g}{d}} \cos\theta$	$u = \frac{H}{2}\frac{gT}{L}\frac{\cosh[2\pi(z+d)/L]}{\cosh(2\pi d/L)} \cos\theta$	$u = \frac{\pi H}{T} e^{\frac{2\pi z}{L}} \cos\theta$
(b) Vertical	$w = \frac{H\pi}{T}\left(1 + \frac{z}{d}\right) \sin\theta$	$w = \frac{H}{2}\frac{gT}{L}\frac{\sinh[2\pi(z+d)/L]}{\cosh(2\pi d/L)} \sin\theta$	$w = \frac{\pi H}{T} e^{\frac{2\pi z}{L}} \sin\theta$
6. Water Particle Accelerations (a) Horizontal	$a_x = \frac{H\pi}{T}\sqrt{\frac{g}{d}} \sin\theta$	$a_x = \frac{g\pi H}{L}\frac{\cosh[2\pi(z+d)/L]}{\cosh(2\pi d/L)} \sin\theta$	$a_x = 2H\left(\frac{\pi}{T}\right)^2 e^{\frac{2\pi z}{L}} \sin\theta$
(b) Vertical	$a_z = -2H\left(\frac{\pi}{T}\right)^2\left(1 + \frac{z}{d}\right) \cos\theta$	$a_z = -\frac{g\pi H}{L}\frac{\sinh[2\pi(z+d)/L]}{\cosh(2\pi d/L)} \cos\theta$	$a_z = -2H\left(\frac{\pi}{T}\right)^2 e^{\frac{2\pi z}{L}} \cos\theta$
7. Water Particle Displacements (a) Horizontal	$\xi = -\frac{HT}{4\pi}\sqrt{\frac{g}{d}} \sin\theta$	$\xi = -\frac{H}{2}\frac{\cosh[2\pi(z+d)/L]}{\sinh(2\pi d/L)} \sin\theta$	$\xi = -\frac{H}{2} e^{\frac{2\pi z}{L}} \sin\theta$
(b) Vertical	$\zeta = \frac{H}{2}\left(1 + \frac{z}{d}\right) \cos\theta$	$\zeta = \frac{H}{2}\frac{\sinh[2\pi(z+d)/L]}{\sinh(2\pi d/L)} \cos\theta$	$\zeta = \frac{H}{2} e^{\frac{2\pi z}{L}} \cos\theta$
8. Subsurface Pressure	$p = \rho g(\eta - z)$	$p = \rho g \eta \frac{\cosh[2\pi(z+d)/L]}{\cosh(2\pi d/L)} - \rho g z$	$p = \rho g \eta e^{\frac{2\pi z}{L}} - \rho g z$

FIGURE 2. Summary of linear (Airy) wave theory—wave characteristics [45].

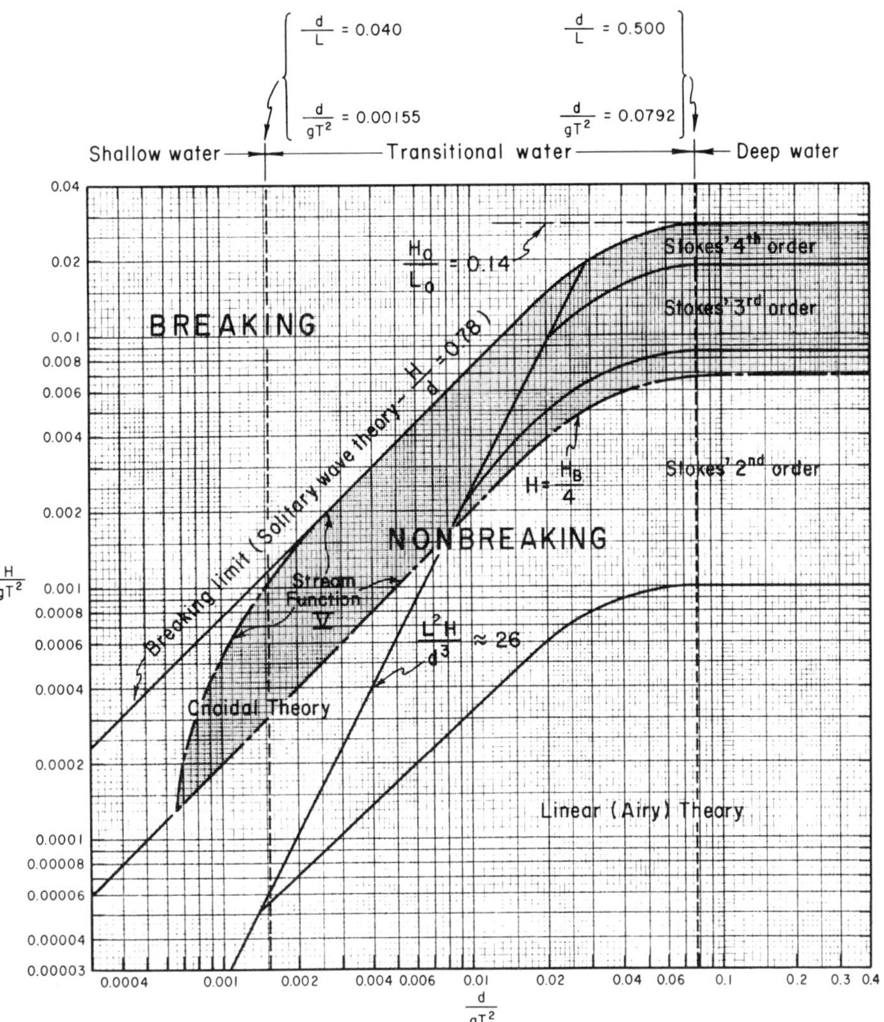

FIGURE 3. Regions of validity for various wave theories [45], after Le Mehaute [32].

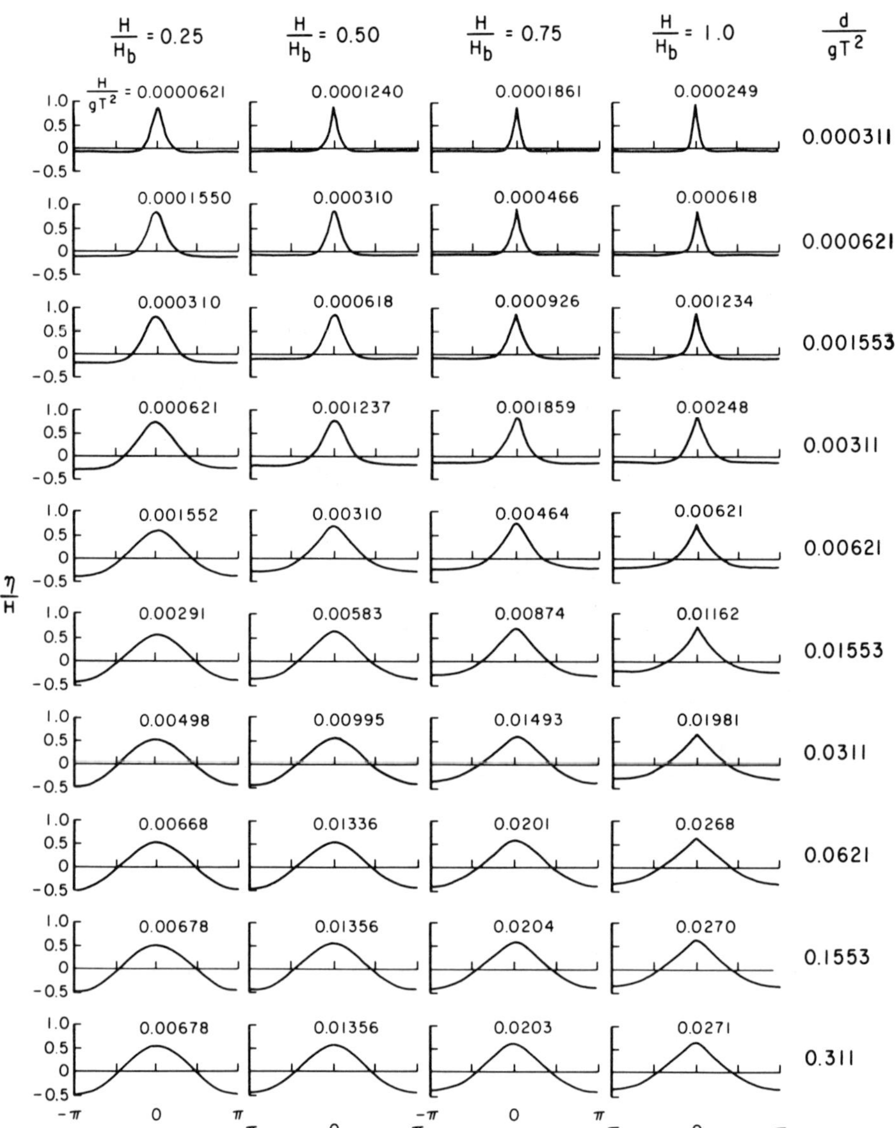

FIGURE 4. Dimensionless wave profiles for 40 cases [8]. Numbers on each plot represent the value of H/gT^2 for each case.

considered are shown in Figure 4. Many of the profiles differ considerably from the sinusoidal profiles assumed by small-amplitude theory. Additional information on higher order wave theories is available in the Shore Protection Manual [45].

WAVE OBSERVATION TECHNIQUES

The primary information from any wave observation technique is generally an estimate of significant wave height and period as defined in the following subsection. Many techniques provide additional valuable data. Commonly used techniques are described in the remainder of this section. Their particular advantages and disadvantages are given in Table 1.

Definition of Significant Wave Height and Period

The concept of a "significant wave height" and a "significant wave period" which can be used to characterize a wave field is appealingly simple. It suggests a simple transition from the experimental results in a laboratory wave tank and the theoretical results obtained with theories for uniform waves to the phenomena that occur in the real ocean.

This concept was first introduced when sailors were asked to report the height and the period of ". . . the larger, well-formed waves, and omit entirely the low and poorly formed waves" as part of the synoptic weather reports from ships. Comparisons of early wave gage records with observations led to the conclusion that the wave height H_v given by visual observers was approximately equal to the average height of the one-third highest individual waves, $H_{1/3}$. Figure 5 provides some perspective on the reliability of this approximation. The figure is based on 905 pairs of visual and instrument observations from a weather ship equipped with a shipboard wave recorder.

The parameter $H_{1/3}$ is referred to as "significant wave height" and the corresponding period is the "significant wave period." Practical techniques for estimating these parameters from wave records are presented in the following section.

Visual Observations from Shipboard

Wave observations have been collected by observers aboard ships in passage for many areas of the world over a period of many years. The observations include average height, period, and direction of sea waves (locally generated) and swell waves (generated elsewhere and propagated to the area). In modern observations, the sea direction is assumed to coincide with the wind direction.

The reliability of shipboard observations must be considered. Individual observations are highly variable. However, the observations are generally unbiased so that summaries of many observations can be useful. Height summaries provide useful estimates of the mean and distribution, as illustrated in Figure 5. Accuracy is less in the upper end of the distribution because of the small number of observations and the tendency for ships to avoid very high wave conditions. A cumulative distribution of shipboard observed wave heights should be considered reliable up to about the one percent level of occurrence or the point at

TABLE 1. Wave Observation Techniques.

Technique	Water Depth*	Advantages**	Disadvantages***
Shipboard observations	1,2	1,2	1
Shore Observations	3	1	1
Staff Gage	1,2,3	3,6	2,3
Pressure Cell (connected to shore by cable or telemetry)	2,3	4,6	3,4,5
Pressure Cell (internally recording)	2,3	4,5	3,4,5,6,7
Accelerometer Buoy	1,2	5,6	5,8,9
High Frequency Radar	1,2	5,6,7	5,9

*Water Depth: 1 = deep, 2 = intermediate, 3 = shallow.

**Advantages: 1 = inexpensive; 2 = large data set already exists; 3 = direct measurement of surface waves; 4 = relatively reliable for unattended use; 5 = relatively simple installation; 6 = adaptable to real time monitoring; 7 = spatial coverage.

***Disadvantages: 1 = low accuracy; 2 = sturdy support structure required for exposed sites; 3 = subject to fouling by marine growth; 4 = divers needed for installation and maintenance; 5 = indirect measurement of desired surface waves; 6 = prone to loss due to burial, inaccurate positioning and loss of marker buoys; 7 = frequent maintenance required for power supply and recorder; 8 = prone to loss due to collision or failed mooring; 9 = relatively expensive.

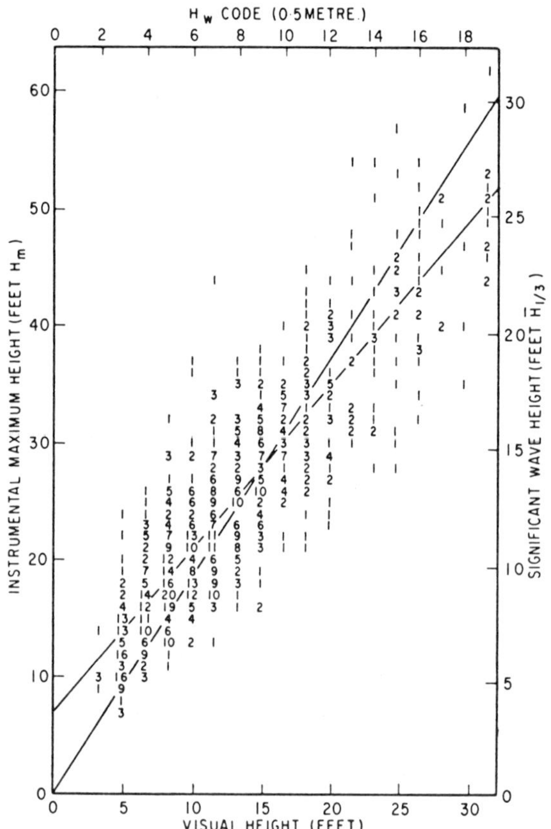

FIGURE 5. Comparison of instrument and visually observed heights [20]. For the data in the above figure $H_{1/3}$ = 1.1 H_v (1 ft = 0.305m).

which 20 observations are represented, whichever criterion is more restrictive.

Wave period is difficult to estimate aboard a moving ship, and only the overall mean period should be used. Wave directions are also somewhat difficult to estimate and should be assumed to have a resolution of 45-degrees or coarser. Shipboard observations may be obtained in the U.S. from the National Climatic Data Center, Federal Building, Asheville, North Carolina. Summaries for U.S. and other selected areas have also been published [52].

Visual Observations from Shore

Visual observations from shore can provide data on coastal phenomena at low cost. Observers typically obtain daily estimates which include breaker height, wave period, direction of wave approach, windspeed, and wind direction. Wave height and direction are visual estimates. Other parameters are often estimated with simple equipment. The skill and biases of individual observers significantly influence the validity of visual observations from shore. To be statistically descriptive of a site, observations should be taken for at least 20 days of each month for a period of at least 3 years. Additional information on an extensive program for collection of visual observations from shore is given by Berg [1] and Schneider and Weggel [43].

Wave Staff Observations

Wave staff observations are obtained with a gage which pierces the water surface. In its simplest form, the wave staff is a vertical rod with visible marks on it at measured intervals. It allows an observer to obtain better quantitative estimates of wave characteristics by watching the undulating water motions at the staff.

Most wave staffs operate as electrical sensors. They detect the instantaneous location of the water surface by using it to change the electrical properties of a circuit. Most electrical wave staffs operate as either resistive, capacitive, or wave guides. Additional information about wave staffs is available from Parker [40].

In addition to the advantages and disadvantages of the wave staff listed in Table 1, the staff may often be inexpensive relative to other instrumental observation techniques. The visually observed staff is especially economical, and it can be very useful in low-budget projects. In low energy environments, the staff gage can be mounted on a spar buoy as an alternative to a rigid mounting structure.

Pressure Cell Observations

Pressure cell observations are obtained with a pressure-sensitive gage mounted under the water surface. Pressure cells can be situated anywhere in the water column as long as they are below the elevation of the lowest expected wave trough at the lowest expected tide level. Pressure cells are often placed on a small tripod which rests on the submerged shallow bottom. The pressure cell senses dynamic pressure fluctuations created by passing surface waves. The magnitude of the fluctuations decreases exponentially with distance below the surface. The pressure fluctuations are converted to an electronic signal which can be recorded at the gage or sent to a shore station by either an armored electrical cable or a surface buoy transmitter. Additional information about pressure cells is available in Forristall [14].

Accelerometer Observations

Accelerometer buoy observations are obtained with a surface following buoy which is usually spherical in form. Buoys are routinely moored in water depths of less than 600 feet (200 meters). Buoys sense vertical acceleration which is usually integrated twice electronically to give a record of surface displacements. The observations may be recorded in

	Sensor	SLAR	SAR	Coastal Wave Imaging Radar	CODAR	ROWS	Δ k	SCR
WAVE PARAMETERS	Status	Operational	Operational	Operational	Operational	Operational	Developmental	Operational
	Height	No	(1)	No	Yes	Yes	Yes	Yes
	Length	Yes	Yes	Yes	Yes	Yes	Yes	Yes
	Direction	Yes	Yes	Yes	Yes	No	Yes	Yes
	Spectrum	Directional Wavenumber Spectrum	Directional Wavenumber Spectrum	Directional Wavenumber Spectrum	Directional Waveheight Spectrum	1-D Waveheight Spectrum	Directional Waveheight Spectrum	Directional Waveheight Spectrum
	Spatial Coverage	1 km swath	Aircraft 1 km swath	Up to 5 km radius	40 km radius	Several sq Meters	2-3 km Radius	Swath width = 1/2 aircraft altitude
	Cell Size Resolution	Depends on range	Aircraft 3 m spatial resolution	15 m range resolution	5 km spatial resolution	Several sq Meters	90 m Range Resolution	.5 -2.0 km^2 cell size
	Platform	Aircraft	Aircraft/ Satellite	Land-based tower	Land-based	Land-based tower	Land-based tower	Aircraft
	System Cost	Expensive	Very Expensive	$50 K	$150 K	$50 K	$100 K	$500 K
	Cost/ Typical Data Set	Moderate	$1000 per 5 km x 5 km patch	$100	Low	Low	Low	$2700/hr aircraft flight time
	Comments	Limited spatial resolution.	Increased resolution: more expensive than SLAR.	Inexpensive and reliable for long term operation	Large spatial coverage. Useful for getting directional spectra offshore.	High spatial resolution	Provides data at scales between ROWS and CODAR. Uses could include monitoring waves in harbors and entrances	Aircraft out of Wallops Is., VA. Expensive for remote study sites. Data acquisition quick.

(1) Theoretically possible, but no algorithm developed as yet.

NOTE: 1 km = 0.621 mile, 1m = 3.28 ft,
1 km^2 = 0.386 sq miles, 1 m^2 = 10.75 ft^2

FIGURE 6. Summary of remote sensing systems for measuring ocean waves (after Dean [9]).

the buoy or transmitted to a shore station either directly or via a satellite link.

Other Wave Recording Instruments

A variety of other wave recording instruments are available for practical use although those discussed in the preceding subsections are most widely used. The instruments are categorized as in-situ instruments and remote sensing devices.

In-situ instruments include acoustic, ultrasonic, optical (laser), and radar instruments, all of which transmit a signal toward the water surface from above or below. The reflected signal is then received and interpreted. Another approach to in-situ data recording consists of operating several instruments together at a site and analyzing the records jointly to get additional information, particularly wave direction. Typical combinations are a pressure cell or staff with two orthogonal horizontal current meters, and a spatial array of staffs or pressure cells. More information on these in-situ instruments is available [14,40,44].

A variety of remote sensing devices are available, as summarized in Figure 6. Only a few devices are suitable for routine data collection. The Coastal Imaging Radar System can be used to estimate dominant wave directions [35]. High-frequency radar, such as the Coastal Ocean Dynamics Applications Radar (CODAR) can be used to estimate coastal wave characteristics including direction [33]. Other devices are the Side Looking Airborne Radar (SLAR), Synthetic Aperature Radar (SAR), Remote Orbital Wave Spectrometer (ROWS), dual frequency radar (Δk), and Surface Contouring Radar (SCR).

WAVE ANALYSIS TECHNIQUES

Wave records are usually collected as a digital time series of surface elevation or subsurface pressure. Records should be suitably checked and edited before further analysis. Typical steps in analyzing a digital wave record are schematized in Figure 7.

The editing step should include a check for waves at short and long periods outside the range of wind wave periods, which is 1 to 30 seconds. Potential sources of waves at undesired periods are tides, water level oscillations, surf beats, electronic drift, electronic noise, and transmission interference. If any of these waves are significant in the record, they may distort estimates of wind wave characteristics. Short- and long-period contamination can be identified visually or by numerical tests. It can be removed by filtering or by considering the spectrum discussed in the following subsections.

Significant Height and Period

Significant wave height may be estimated from a digital record by direct computation of $H_{1/3}$ from the time series or by first computing a spectrum, as shown in Figure 7.

By the spectral approach, significant height is estimated as 4 times the standard deviation of the record of sea surface

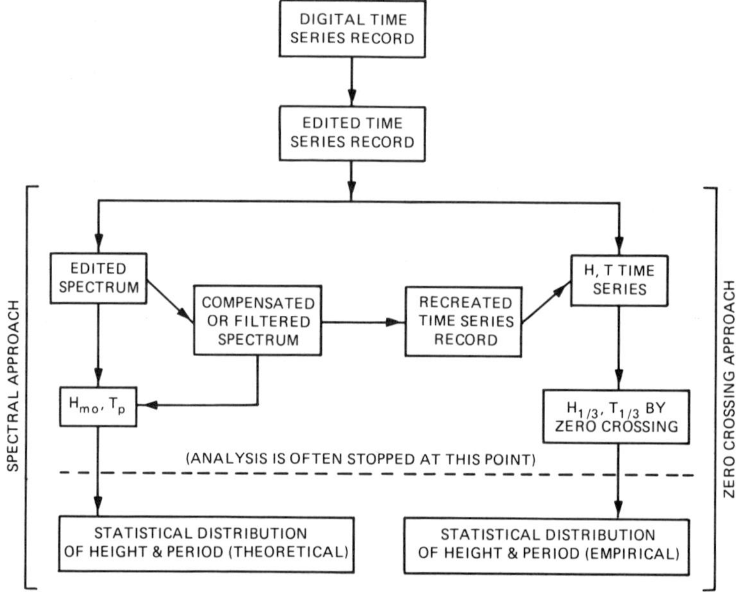

FIGURE 7. Analysis of a digital wave record.

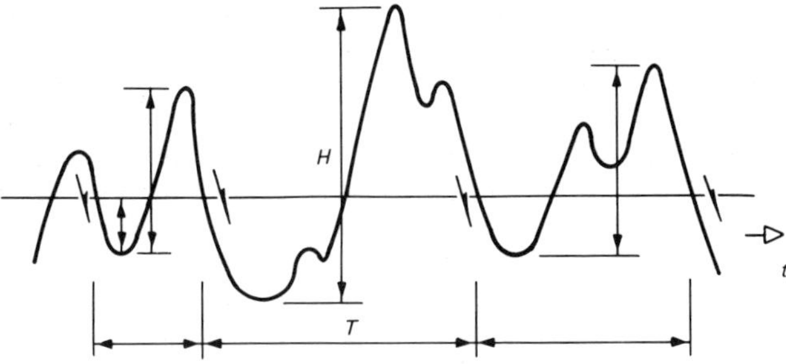

FIGURE 8. Zero downcrossing waves.

elevations. A standardized analysis package for ocean waves should be used when the spectral approach is desired. A comprehensive package has the advantages of automatically removing nonwind wave energy from the results and options for compensating pressure or acceleration signals to give estimates of surface waves. The significant height estimate is sometimes referred to as H_{mo} to clearly identify that it was obtained by a spectral approach. Wave period estimated by the spectral approach is the period corresponding to the highest energy density in the spectrum. It is called significant or peak period, T_p.

The term H_s is commonly used to designate a generalized significant wave height. When it is used, the method for estimating it and whether it represents $H_{1/3}$ or H_{mo} should be made clear.

Significant height and period may be estimated directly from the time series record by identifying all individual waves in the record. The procedure which is most widely used and most easily applied on a digital computer is the zero-crossing method. The preferred application of this method involves identifying each wave in the record as an event between two successive points at which the wave trace crosses the mean in a downward moving direction (Figure 8). Wave height is defined as the elevation difference between the highest point (crest) and lowest point (trough) of each wave. Significant wave height is computed as the average height of the highest one-third zero-crossing waves. Other wave height statistics, such as the root-mean-square wave height, are also easily computed. Significant period is computed as the average period of the one-third highest waves. Significant period may also be estimated as the mean period of all waves, although this estimate may be misleading when two or more prominent wavetrains with greatly differing average periods occur simultaneously. This procedure is called the zero downcrossing analysis procedure because each wave is defined by two downcrossings of the mean.

Both spectral analysis and zero-crossing analysis are useful in practical engineering work. Spectral analysis is more complete; but zero-crossing requires less computer capacity. Zero-crossing analysis also provides unique information when the waves are near breaking and highly nonlinear. Spectral analysis may be necessary prior to zero-crossing analysis if the time series needs to be compensated or filtered (see Figure 7).

Spectral Analysis

Spectra are becoming widely available through various field wave measurement programs, laboratory tests with programmable wave generators, and numerical wave hindcasting projects. Because of the availability and applications of spectra, practicing coastal and ocean engineers need to be familiar with spectra and their interpretation.

ENERGY SPECTRUM

A fundamental parameter for characterizing a wave field is some measure of the periodicity of the waves. For many years a significant period, which could be subjectively estimated in various ways, was used. However, the ocean surface often has waves characterized by several distinct periods occurring simultaneously. A record of the variation of sea-surface elevation with time, commonly called time series, frequently appears confusing and is difficult to interpret.

Developments in computer technology and in mathematical analysis of time series have provided a practical approach to an objective, more comprehensive analysis of periodicity in wave records. The approach is to express the time series as a sum of sine and cosine functions with different frequency and phase. Thus, the time series of sea-surface deviations from the mean surface, $\eta(t)$, is expressed by

$$\eta(t) = \sum_{j=1}^{n} a_j \cos(\omega_j t - \phi_j) \qquad (1)$$

where

a_j = amplitude
ω_j = frequency in radians
ϕ_j = phase
t = time

Frequency is often expressed in terms of hertz units where one hertz is equal to one cycle per second. One hertz is also equivalent to 2π radians per second. If the symbol f_j denotes frequency in hertz, then $2\pi f_j = \omega_j$.

The amplitudes, a_j, computed for a time series, give an indication of importance of each frequency, f_j. The sum of the squared amplitudes is related to the variance of sea-surface elevations in the original time series and hence to the potential energy contained in the wavy sea surface. Because of this relationship, the distribution of squared amplitudes as a function of frequency can be used to estimate the distribution of wave energy as a function of frequency. This distribution is called the energy spectrum and is often expressed as

$$(E_j)(\Delta f)_j = \frac{a_j^2}{2} = S_j \tag{2}$$

where $E_j = E(f_j)$ = energy density in the jth component of the energy spectrum, $(\Delta f)_j$ = frequency bandwidth in hertz (difference between successive f_j), and $S_j = S(f_j)$ is the energy in the jth component of the energy spectrum.

An energy spectrum computed from an ocean wave record is plotted in Figure 9. Frequencies associated with large values of energy density—or large values of $a^2{}_j/[2(\Delta f)_j]$; see Equation (2)—represent dominant periodicities in the original time series. Frequencies associated with small values of energy density are usually unimportant. It is common for ocean wave spectra to show two or more dominant periodicities as in Figure 9. When only one frequency is reported from a spectrum, the frequency at which the energy density is highest, f_p, is usually used. The dominant wave period, or peak period, is given as the reciprocal of f_p.

The appearance of a spectrum can be noticeably influenced by the methods used for calculation and display, neither of which is standardized in coastal and ocean engineering activities at present. The most important difference among commonly used methods is whether the spectrum is summarized as energy density at equal frequency intervals or approximately equal period intervals.

Spectral analysis procedures such as cross spectral analysis are available to extract more information when concurrent time series from several gages are obtained. The most important additional information is typically wave direction. Procedures for a triangular array are discussed in Esteva [12] and for an arbitrary gage arrangement in Borgman [2].

SPECTRAL PARAMETERS

The complete energy spectrum is too cumbersome for forming statistical summaries of wave conditions at a site. Thus simple parameters of the spectrum are very useful. The most commonly used spectral parameters are the significant wave height, H_{mo}, and the peak period, T_p. Several additional parameters are also widely used to better characterize the shape of the spectrum and the importance of nonlinearities in the time series from which the spectrum was computed. One useful parameter is the number of major peaks in the spectrum. This parameter is indicative of

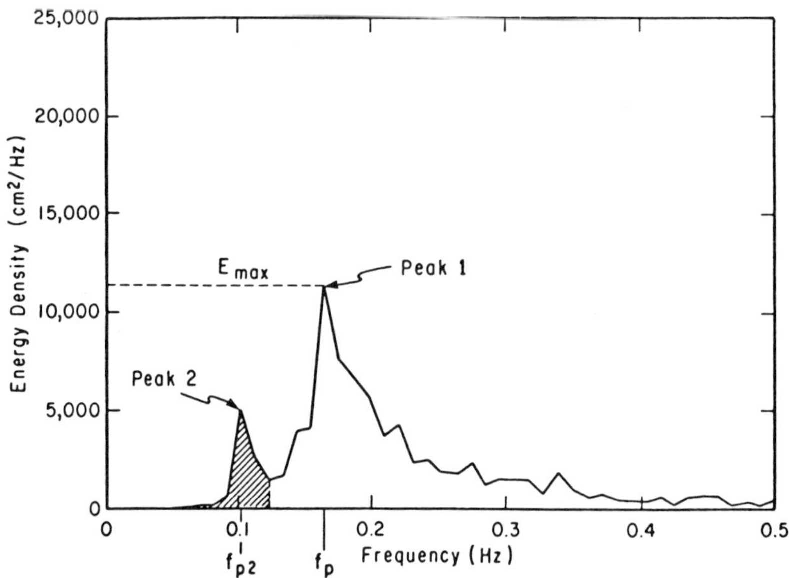

FIGURE 9. Spectrum for Wrightsville Beach, North Carolina, 0700 e.s.t., 12 February 1972; H_s = 4.2 ft (128 centimeters), Δf = 0.01074 hertz, and depth = 17.7 ft (5.4 meters).

the number of independent wave trains present, except when the measurements were taken in or near the surf zone [48].

The spectral peakedness parameter [15] is indicative of the sharpness of the spectral peak. The spectral peakedness parameter, Q_p, is computed by

$$Q_p = 2 \sum_{i=1}^{N} f(\Delta f)_i S_i^2 \left[\sum_{i=1}^{N} (\Delta f)_i S_i \right]^{-2} \tag{3}$$

The usefulness of Q_p is illustrated in Figure 10 which shows two spectra with nearly the same significant height and peak period but different values of the peakedness parameter.

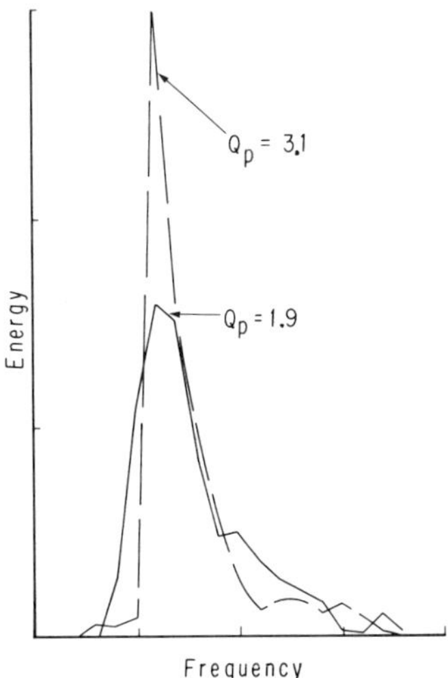

FIGURE 10. Comparison of two measured spectra from the North Atlantic Ocean with H_s = 10.8 ft (3.3 m), T_p = 10.5 sec.

PARAMETERS OF THE DISTRIBUTION OF SURFACE ELEVATIONS

Statistical moments of the distribution of sea surface elevations provide additional information. Moments are computed by

$$q_n = \sum_{i=1}^{N} \eta_i^n p(\eta_i) \tag{4}$$

where

q_n = nth moment of the distribution function of sea surface elevations
N = number of intervals in the distribution function
η_i = sea surface elevation associated with the ith interval in the distribution function
$p(\eta_i)$ = probability associated with η_i

The zeroth and first moments, q_o and q_1, are equivalent to the mean and variance of the distribution function. The third moment, or skewness, of the distribution is a very useful indicator of the extent of nonlinear deformation of wave profiles in the time series. The deformation can be significant for breaking and near breaking waves in shallow water. Since the shape of the spectrum is affected by nonlinear deformation of wave profiles (Figure 11), the skewness is helpful in interpreting spectral shape.

CERC Method for Strip Chart Records

In addition to digital records, pen and ink strip chart records are sometimes available. Although strip chart wave records are rarely collected with modern systems, they can be easily analyzed by the method in Appendix I. The method is set up for a seven-minute record length. It can be adapted to other record lengths by using the Rayleigh distribution equation (Equation 6) to compute new entries in the tabulated "number of wave to measure."

Distribution of Individual Wave Heights

The distribution of individual wave heights is well approximated by the Rayleigh distribution function. The probability density function for the Rayleigh distribution is given by

$$p(\hat{H}) = -2 \frac{\hat{H}}{H_{\text{rms}}^2} e^{-(\hat{H}/H_{\text{rms}})^2} \tag{5}$$

where $p(\hat{H})$ = probability of a given wave height, $\hat{H}$, H_{rms} = root-mean-square wave height. The cumulative form of the Rayleigh distribution function is given by

$$P(H > \hat{H}) = e^{-(\hat{H}/H_{\text{rms}})^2} \tag{6}$$

where $P(H > \hat{H})$ is the number of waves larger than $\hat{H}$, n, divided by the total number of waves in the record, N. The cumulative distribution function is plotted in Figure 12. The Rayleigh distribution was derived theoretically for the distribution of wave amplitudes in a Gaussian sea state with a narrow spectrum. However it has proved to fit empirical wave height data remarkably well, even in shallow water.

The average height for any specified fraction of the higher waves is given by the lower curve in Figure 12 in terms of H/H_{rms}. The same information is given in Table 2 for commonly used fractions of the higher waves. The table also gives the percentage of heights higher than the average.

Distribution of Individual Wave Periods

The distribution of individual wave periods is much more variable than the distribution of individual wave heights. A reasonable approximation for sea-waves is given by

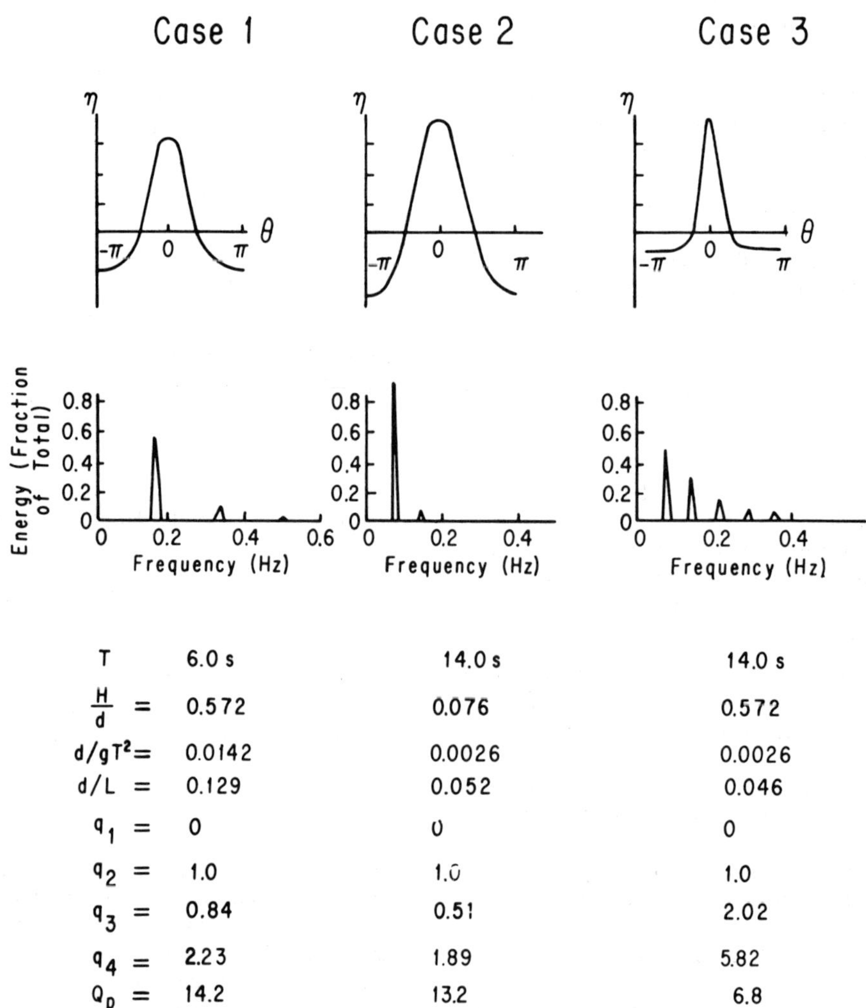

	Case 1	Case 2	Case 3
T	6.0 s	14.0 s	14.0 s
$\frac{H}{d}$ =	0.572	0.076	0.572
d/gT^2 =	0.0142	0.0026	0.0026
d/L =	0.129	0.052	0.046
q_1 =	0	0	0
q_2 =	1.0	1.0	1.0
q_3 =	0.84	0.51	2.02
q_4 =	2.23	1.89	5.82
Q_p =	14.2	13.2	6.8

FIGURE 11. Wave profiles and energy spectra for several cnoidal wave cases (record length: 512 sec; spectral bandwidth: 0.00977 hertz) [48].

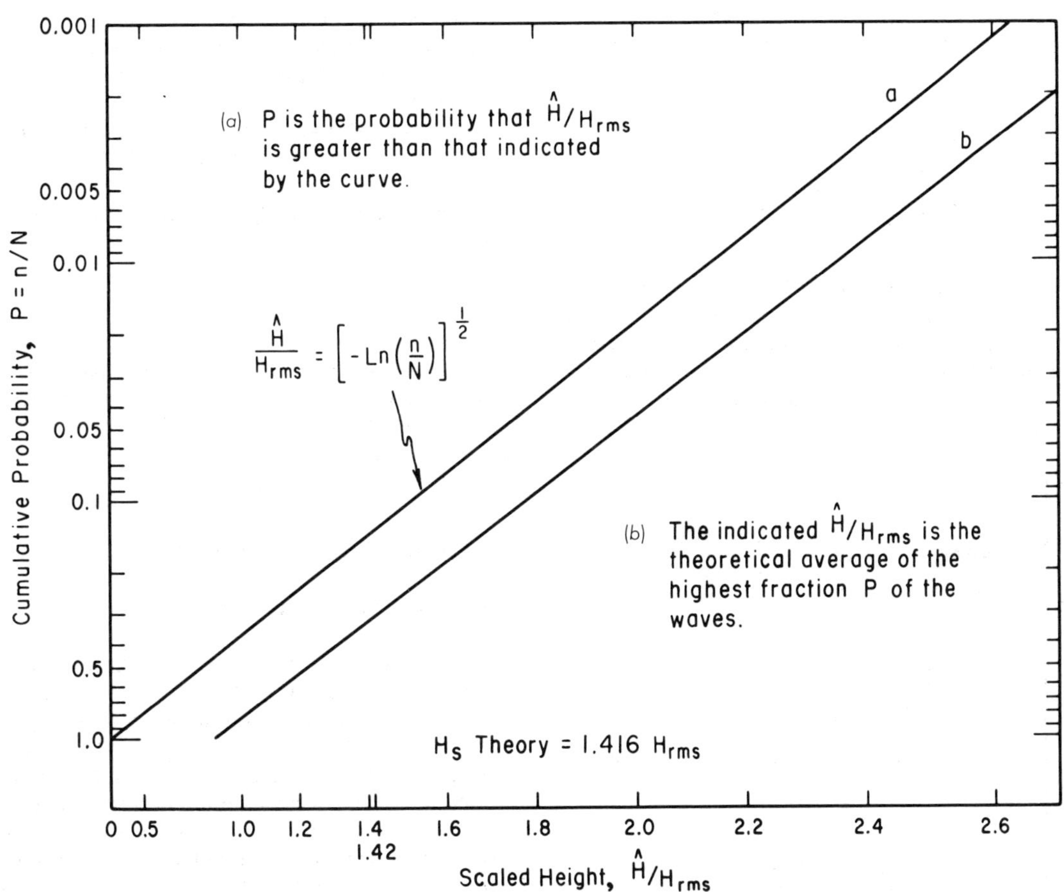

FIGURE 12. Theoretical wave height distributions [45].

TABLE 2. Wave Height Relationships Based on the Rayleigh Distribution.

Fraction of Higher Waves	Average H/H_{rms}	Percentage of Heights Higher than average H/H_{rms}
0.001	2.82	0.035
0.01	2.36	0.38
0.10	1.80	3.92
0.33	1.42	13.5
0.50	0.89	20.4

Bretschneider [5] as

$$p(\hat{T}) = 2.7\frac{\hat{T}^3}{\bar{T}^4}\exp\left[-0.675\left(\frac{\hat{T}}{\bar{T}}\right)^4\right] \tag{7}$$

where $\bar{T}$ is the mean wave period. The relationship between $\bar{T}$ and T_p depends on the particular wave condition. In general, T_p is between 0.95 and 1.5 times $\bar{T}$. This expression should not be used when two or more prominent wavetrains with widely differing periods occur simultaneously.

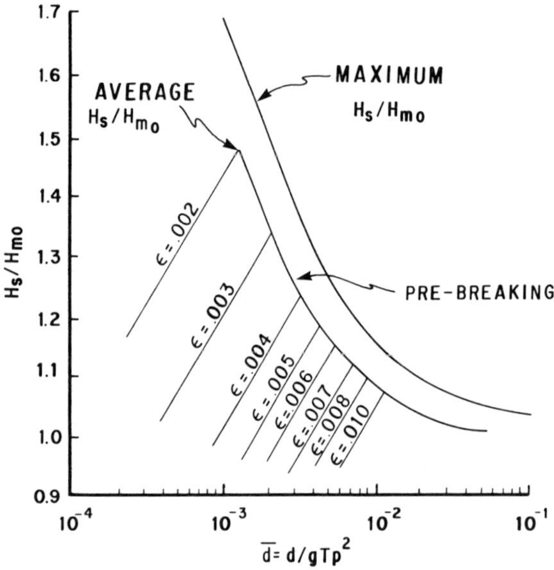

FIGURE 13. Variation of H_s/H_{mo} as a function of relative depth $\bar{d}$ and significant steepness. Both the average and maximum values of the ratio are given. The average value is read from the lower of the prebreaking line or the appropriate steepness line. Values of steepness intermediate to those plotted should be constructed as lines parallel to those plotted and spaced in between by linear interpolation. For values of steepness greater than 0.01 the ratio may be assumed to be one [50].

Intercomparisons of Analysis Procedures

The significant wave height obtained by zero crossing analysis, $H_{1/3}$, generally compares well with the estimate from spectral analysis, H_{mo}. Empirical evidence indicates $H_{1/3}$ may be 5 percent less than H_{mo} in deep water but $H_{1/3}$ can significantly exceed H_{mo} in shallow water. Differences are directly related to the change in wave profiles in shallow water such that crests become narrow and high and troughs broad and flat (Figure 4). A relationship between $H_{1/3}$ and H_{mo} is plotted in Figure 13. Wave steepness, ϵ, in the figure is defined as 0.25 H_{mo}/L_p, where L_p is the finite depth wavelength of waves at the spectral peak. A more conservative but simpler estimate of the maximum value of $H_{1/3}$ in a given shallow water depth is $H_{1/3} = 0.78\ d$ where d is the depth.

Peak spectral period, T_p, is related to zero crossing period by

$$T_p = 1.05\ T_{1/3} \tag{8}$$

In the case of multiple concurrent wavetrains, this relationship is unreliable. T_p will represent one of the wavetrains, but $T_{1/3}$ will be an average of a variety of wave periods and may not represent any one wavetrain.

COMPARISON OF GAGE RECORDS

Variability Due to Gage Type

Wave gage types used in coastal waters each have some influence on the data collected. In particular, the pressure sensitive gage provides a record in which high frequency energy is removed or severely attenuated and lower frequency energy is moderately attenuated. A simple procedure for compensating the pressure record to get an estimate of surface wave conditions is to compute significant height and period from the pressure record and to compensate for the effect of gage submersion by a factor based on linear wave theory. The factor is

$$(H_s)_{sfc} = \frac{\cosh 2\pi d/L}{\cosh 2\pi(z + d)/L}\frac{1}{\rho g}(H_s)_{pres} \tag{9}$$

where the subscripts *sfc* and *pres* are used to indicate surface conditions and underwater conditions at the gage, respectively; d is the water depth; L is the local wavelength; and z is the depth of the pressure sensor below the water surface (z is negative).

It is generally preferable to apply the compensation equation to frequency components of the energy spectrum, rather than significant wave heights. The factor for spectral

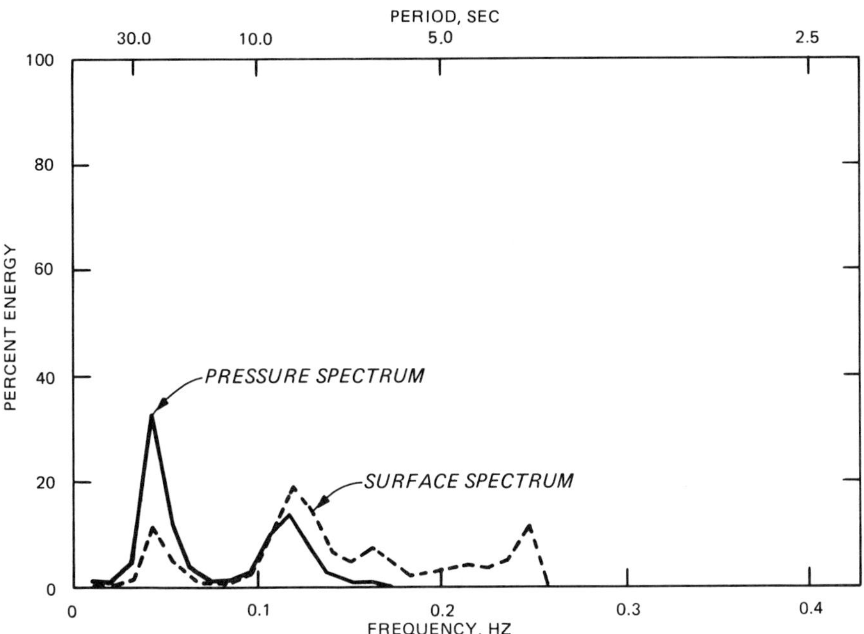

FIGURE 14. Surface spectrum computed from pressure spectrum at Pt. Mugu California. Gage was bottom-mounted in 26-ft (7.9 m) water depth [21].

application is

$$E_{sfc}(f) = \left(\frac{\cosh 2\pi d/L}{\cosh 2\pi(z + d)/L}\right)^2 \frac{E_{pres}(f)}{(\rho g)^2} \qquad (10)$$

Raw and corrected pressure spectra are illustrated in Figure 14. Significant height can be estimated from the compensated spectrum as discussed in the preceeding section. This approach can be expected to give surface wave estimates with errors less than 20 percent when no wavelengths are less than 2 times the gage depth and waves are not near the point of depth-induced breaking [13,18,19].

Another widely used gage type, the accelerometer buoy, tends to produce a spectrum with attenuated energy at the low and high frequency ends of the spectrum. Low frequency attenuation occurs because the buoy experiences very small vertical accelerations due to low frequency waves and the accelerometer does not respond well. High frequencies are attenuated because the buoy hull does not respond well to wavelengths on the order of the hull dimensions or less. These effects can sometimes be reduced by correcting the spectrum to compensate for buoy response characteristics.

Variability of Wave Spectra Due to Gage Location

Wave energy spectra are naturally variable simply because they are based on a finite length record of a wave field which varies in time and space. Spectra computed for successive records of a relatively stationary wave field are never identical and often differ noticeably. The magnitude of spectral variation in time is illustrated by spectra derived at 2-hour intervals from two pressure gages along the southern California coast (Figure 15). The significant wave height is nearly constant in the figure.

GAGES ALONG A DEPTH CONTOUR

Spatial variation of the spectrum over short alongshore distances in shallow water is also shown in Figure 15. Each spectrum in the top row of the figure can be compared to the spectrum immediately below it to see variations between spectra from two gages 80 feet (24 meters) apart. In this figure, spatial variations are smaller than temporal variations. Spatial variations would be expected to be greater if the gages were farther apart or the water depth varied between measurement points. For gages far apart, processes such as refraction, diffraction, reflection, currents, and winds can induce significant differences in spectra by differentially influencing the wave field.

Variability of spectra induced by finite length data records has been studied by Donelan and Pierson [11]. They concluded that the theory of stationary Gaussian processes provides accurate estimates of sampling variability. For 17-minute record length, the uncertainties in significant wave height and peak frequency estimates are ±12 percent and ±5 percent respectively at the 90 percent confidence level.

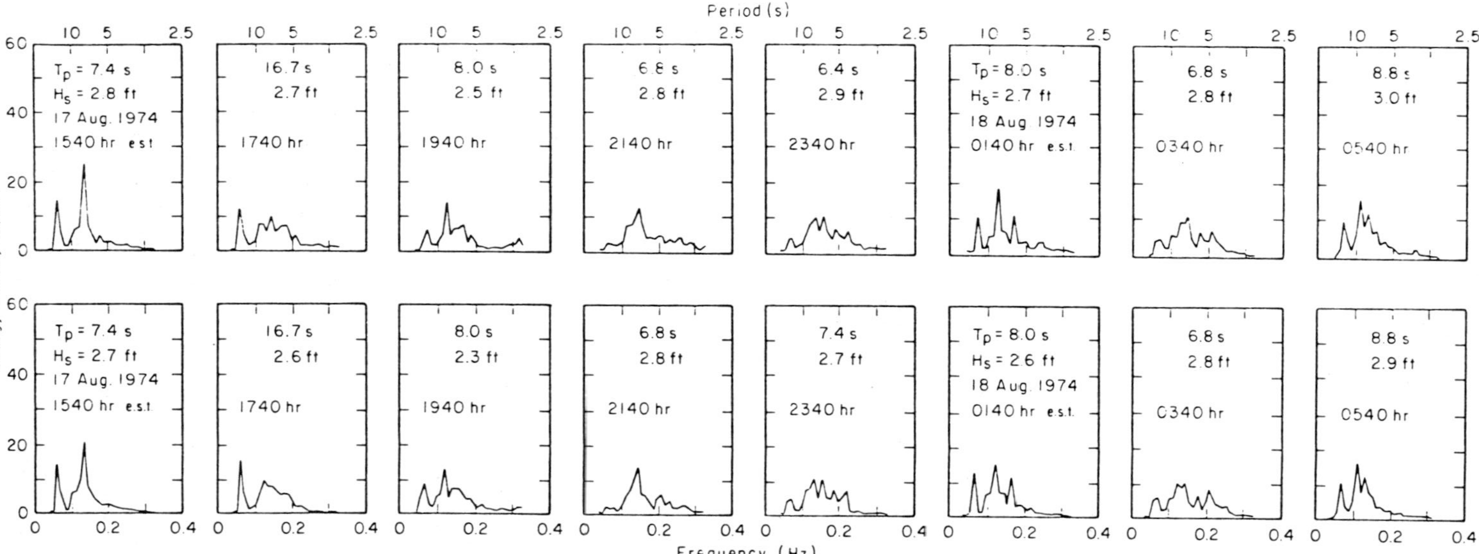

FIGURE 15. *Wave energy spectra from bottom-mounted pressure gages at Channel Islands Harbor, California.* Spectra in top row from gage 1; in bottom row from gage 2. Gages were 80 ft (24 m) apart in water depth of 20 ft (6.1 m). Spectra have been compensated for hydrodynamic attenuation due to submergence of the gage (1 ft = 0.305 m).

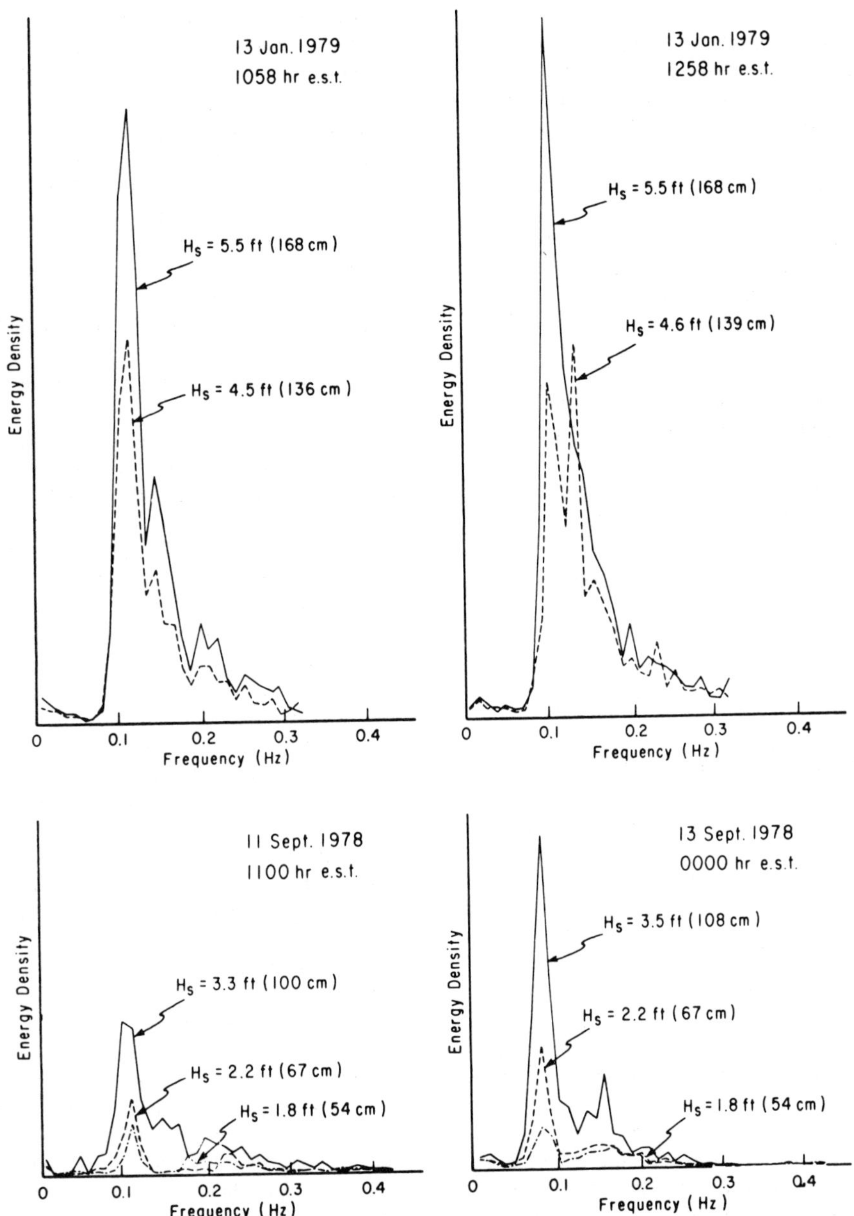

FIGURE 16. Wave energy spectra from pier-mounted continuous wire staff gages near Duck, North Carolina, showing variation along a line perpendicular to shore. Solid lines represent a gage at the seaward pier end (depth 29 ft or 8.8 m); dashlines represent a gage 480 ft (146 m) from the seaward pier end (depth 22 ft or 6.6 m); dot-dash lines represent a gage 840 ft (256 m) from the pier end (depth 17 ft or 5.1 m).

Further the height of the peak of the spectrum is generally overestimated.

GAGES ALONG A LINE NORMAL TO SHORE

Variations between spectra from gages situated along a line perpendicular to shore are shown in Figure 16. The spatial variations are more prominent in this figure than in Figure 15.

Depth-induced variations in spectra and significant heights can be large and systematic. The variations are an important consideration in interpreting data from any shallow water gage site. The effect of shallow depth on measurements can be estimated by the techniques described in the section on Simplified Wave Models.

EVALUATION OF COMMON ASSUMPTIONS ABOUT WAVES BY COMPARISON WITH OBSERVATIONS

Many widely used engineering formulas dealing with wind generated waves have been derived with assumptions about the nature of waves. When real wave conditions are not well described by the assumptions, the propriety of the formulas and designs based upon the formulas is questionable. The validity of some common assumptions is assessed in this section.

Gaussian Distribution of Water Surface Elevations

RELATIONSHIP BETWEEN SURFACE ELEVATION DISTRIBUTION AND WAVE PROFILE

The Gaussian distribution is symmetric, indicating that the same probability is associated with elevation x units above the mean and x units below the mean. Such symmetry can be expected for waves which have crest profiles which generally resemble trough profiles in width and excursion from the mean.

The distribution of measured sea surface elevations can differ noticeably from the Gaussian distribution when the measurements are taken in shallow water or when the measurements represent steep waves in relatively deep water. Shallow water waves and steep deep water waves tend to have high narrow crests and broad flat troughs (Figure 4), that lead to non-Gaussian sea surface elevation distributions. The profiles of high steep waves tend to become increasingly non-Gaussian between deep water and shallow water as illustrated by Jahns and Wheeler [29] with hurricane wave data from the Gulf of Mexico.

PARAMETERS OF SEA SURFACE ELEVATION DISTRIBUTION

Distributions of measured sea surface elevation are often normalized for convenient comparison with the Gaussian distribution. A normalized distribution has mean equal to zero and standard deviation equal to one. The third and fourth moments, λ_3 and λ_4 often called skewness and kurtosis, respectively, of a normalized distribution of sea surface elevation are defined in the section on Wave Analysis Techniques. The third and fourth moments thus defined are both equal to zero for a normalized Gaussian distribution.

The skewness for shallow water and for steep deep water waves is usually greater than zero. Skewness values up to 0.35 computed from high wave measurements during Hurricane Carla have been presented by Hudspeth and Chen [26]. Skewness values of up to about 1.5 computed from coastal shallow water wave measurements have been reported by Thompson [49]. Positive kurtosis values are also reported. Ochi et al. [39] also present extensive documentation of deviations from the Gaussian distribution in storm measurements at the CERC Field Research Facility.

Rayleigh Distribution of Wave Heights

The use of a Rayleigh distribution for wave heights is a direct consequence of the assumptions of a Gaussian distribution for sea surface elevations and a narrow band spectrum. Since the assumptions are often violated in natural wave conditions, particularly in very shallow coastal waters, the use of the Rayleigh distribution in shallow water design can only be justified by empirical evidence. Empirical data in the Shore Protection Manual [45] from several shallow water Atlantic coast gages indicate the Rayleigh distribution is a good approximation (Figure 17). As indicated, the Rayleigh distribution is increasingly conservative at cumulative probabilities less than about 0.05. Further evidence in support of the Rayleigh distribution for shallow water wave heights (including the surf zone) was presented by Thornton and Guza [51] and contrary evidence by Ochi et al. [39]. The predominant conclusion is that the Rayleigh distribution is quite satisfactory for most engineering applications. The Rayleigh distribution is less satisfactory for describing the extreme wave heights with cumulative probabilities of about 0.01 or less.

Continuity of Wave Spectra

It is often assumed that the sea surface represents a random Gaussian process and the Fourier transform of a time series of sea-surface elevations represents a continuous spectrum with an infinite number of independent frequency components. An obvious case in which spectral components are not independent is a record of steep waves with peaked crests and flat troughs. Wave profiles may be described as a summation of a wave of the fundamental frequency and waves at frequencies which are integral multiples of the fundamental, often called a Stokes wave. The spectrum has peaks at harmonics of the dominant frequency which are phase-bound to the fundamental and are clearly not independent (Figure 11).

The assumption of a continuous spectrum of independent components is adequate for most practical engineering

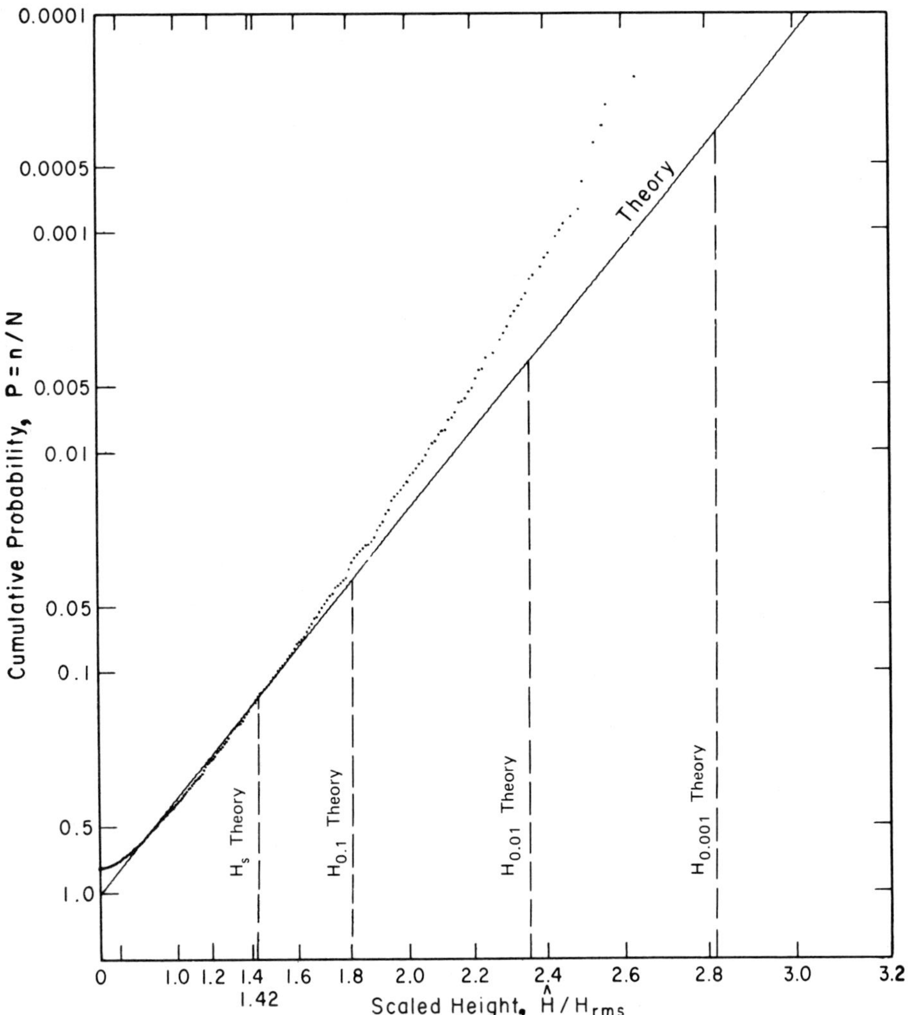

FIGURE 17. Theoretical and observed wave height distributions. (Observed shallow water waves from 72 individual 15-minute observations from several Atlantic coast wave gages are superimposed on the Rayleigh distribution curve. A total of 11,678 individual waves are represented [45].)

work. However, for very steep waves or waves in very shallow water the assumption is often incorrect. The following steepness and relative depth criteria can be used to indicate cases where the assumption of independent spectral components may be poor:

$$\text{steepness: } \frac{H_s}{gT_p^2} > 0.008$$

or (11)

$$\text{relative depth: } \frac{d}{gT_p^2} < 0.01$$

In cases where both steepness and relative depth approach the above guidelines, nonindcpcndcncc may also be a factor.

Wave Grouping

The characteristics of individual waves in a record are highly variable. This observation has led to the assumption that waves occur randomly in the record. However, there is a small but significant correlation between the heights of successive waves in a record. The tendency for high waves to occur in groups is important for several reasons. It has been demonstrated that high wave groups can be exceptionally destructive to rubble-mound structures [30]. Also, long period motion induced by quasi-periodic wave groups can generate resonant oscillation of moored floating structures such as platforms, piers, breakwaters and tethered vessels. It can also lead to oscillation in harbors and bays. Because of the significance of wave grouping, it is often necessary to include consideration of wave groups in laboratory tests of coastal phenomena. Inclusion of wave grouping effects in general criteria for coastal design is beyond the present state of art.

SIMPLIFIED WAVE MODELS

The processes of wave growth, propagation, and nearshore transformation can be predicted reasonably well by a variety of methods. The methods are based on formulas and computer programs of varying complexity. This section presents simple, self-contained methods. The following section describes more comprehensive methods available.

Estimating Wind Conditions

Winds can be estimated for wave growth models by using direct wind measurements, atmospheric pressure measurements, or a combination of both. Use of actual wind records from the site is preferred in protected areas so that local peculiarities of wind intensity such as sheltering from adjacent topography or channelization of winds along valleys is included. If wind records are not available, generalized regional wind statistics may be used if available.

WIND INFORMATION ADJUSTMENTS

Wind information must be properly adjusted for use in wave models to avoid introducing bias into the results. The following procedure provides a method for adjusting the windspeed that is reasonably quick and relatively accurate. It must be recognized that the problem of identifying the appropriate windspeed and the resultant wave estimation in irregular water bodies is complex. To achieve a simplified method, the following assumptions are made:

(a) The windfields are well organized and can be adequately described by the use of an average windspeed and direction over the entire fetch.

(b) The windspeed should be corrected to the 33-foot (10 meter) level.

(c) The windspeed should be representative of the average windspeed measured over the fetch.

(d) When the fetch length is 10 miles (16 km) or less, the wind has not fully adjusted to the frictional characteristics of the waves. In such cases, the overwater windspeed is to be estimated to be 110 percent of the overland windspeed, U_L. Thermal effects on stability of the air in this case are not applicable.

(e) When the fetch length is greater than 10 miles (16 km), thermal stability effects must be included in the windspeed transformation.

Having an observed windspeed of known direction, level above the surface, location of observation (i.e., overwater or overland), and method of windspeed description (i.e., fastest miles or a time-averaged speed), the following steps should be completed in accordance with Figure 18. The figure presents a logic diagram which leads to the adjusted windspeed required to determine the wave height and period in either deep or shallow water.

Steps

1. If the windspeed is observed at any level other than 33 feet it should be adjusted as follows:

$$U_{33} = \left(\frac{33}{Z}\right)^{1/7} U_Z = R_{33}U_Z \qquad (12)$$

where U_{33} is the windspeed at the 33 foot level and U_Z is the windspeed at distance Z above the surface. This method is valid where Z is less than 65 feet (20 meters).

2. Windspeeds are frequently described in a variety of means such as: fastest mile, 5-minute average, 10-minute average, etc. The windspeed must be averaged over the fetch or adjusted so that the average time is equal to or greater than the minimum duration, t. Figure 19 provides the means to convert the fastest mile windspeed to an equivalent duration. Figure 20 can be used to convert a windspeed of any duration to a 1-hour windspeed.

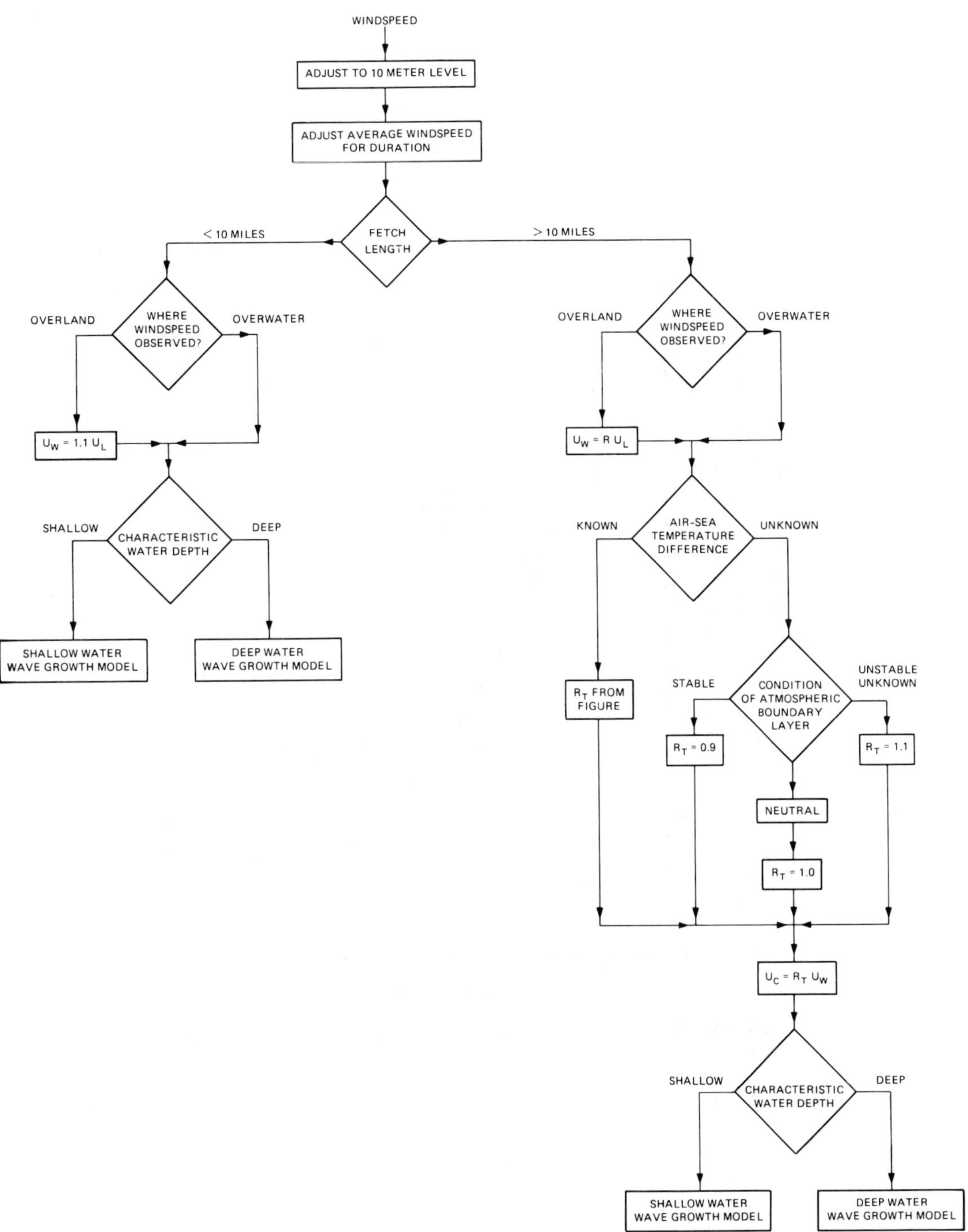

FIGURE 18. Logic diagram for determining windspeed for use in wave forecasting models.

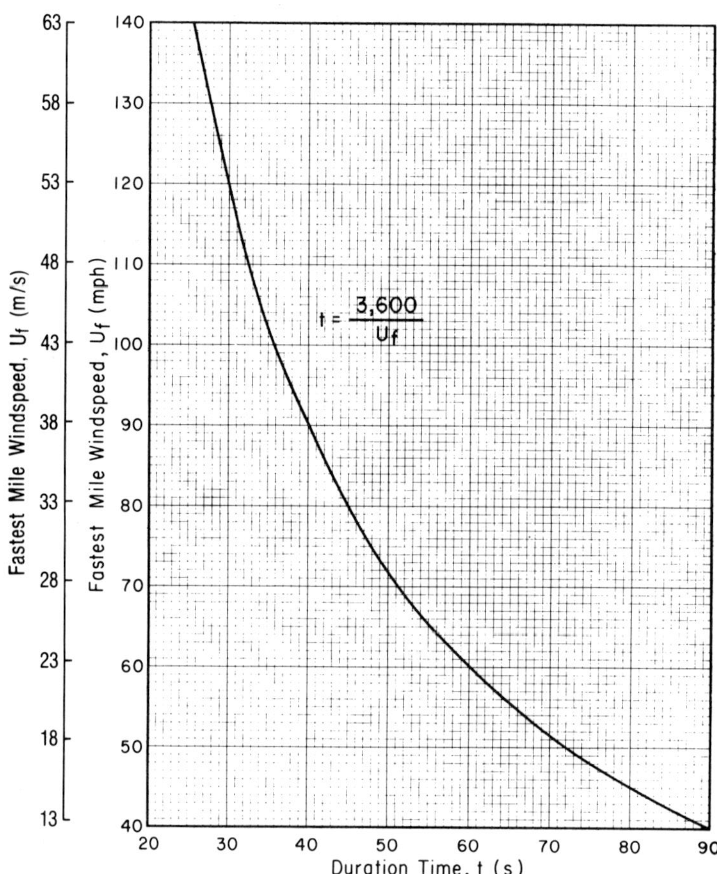

FIGURE 19. Duration of the fastest mile windspeed as a function of windspeed [45].

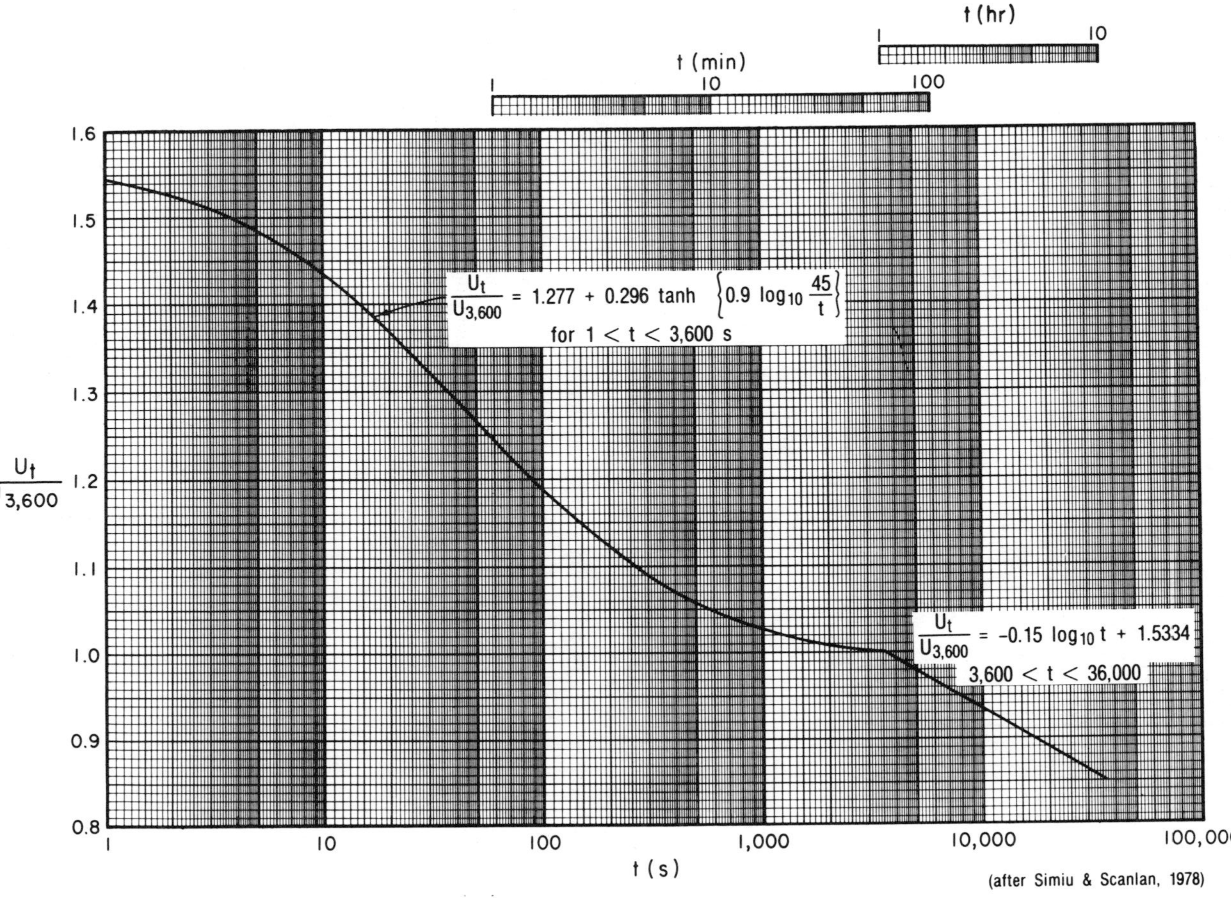

FIGURE 20. Ratio of windspeed of any duration, U_t, to the 1-hour windspeed, U_{3600} [45].

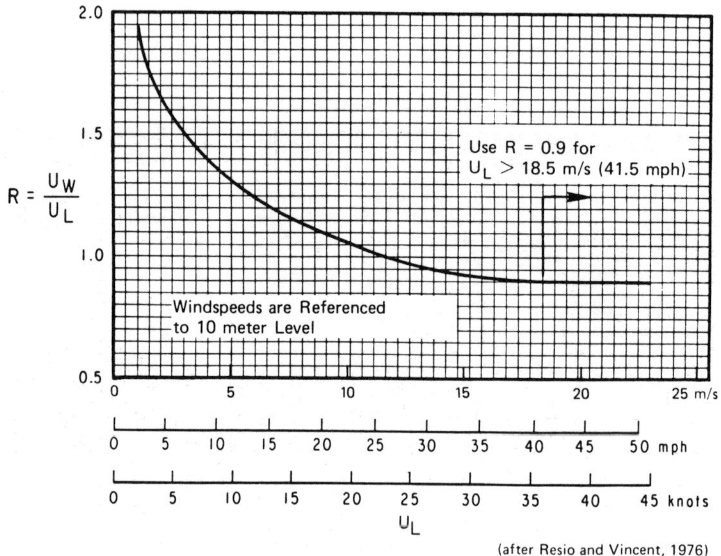

FIGURE 21. Ratio, R, of windspeed overwater, U_W, to windspeed overland, U_L, as a function of windspeed overland, U_L [45].

3. Determine if the overwater fetch distance is less than or greater than 10 miles (16 km).

4. Determine if the windspeed was observed overwater or overland. On short fetches it is assumed that the atmospheric boundary layer has not had time to fully adjust to the developing frictional characteristics of the water surface. Windspeeds observed overland, U_L, must be corrected to overwater windspeeds, U_W. For overwater fetches less than 10 miles (16km), $U_W = 1.1\ U_L$. For overwater fetches greater than 10 miles (16 km), $U_W = RU_L$, where R is determined from Figure 21. The term overland implies a measurement site that is dominantly characterized as inland. If a measurement site is directly adjacent to the water body, it may, for selected wind directions, be equivalent to overwater. Careful analysis of such a site is required.

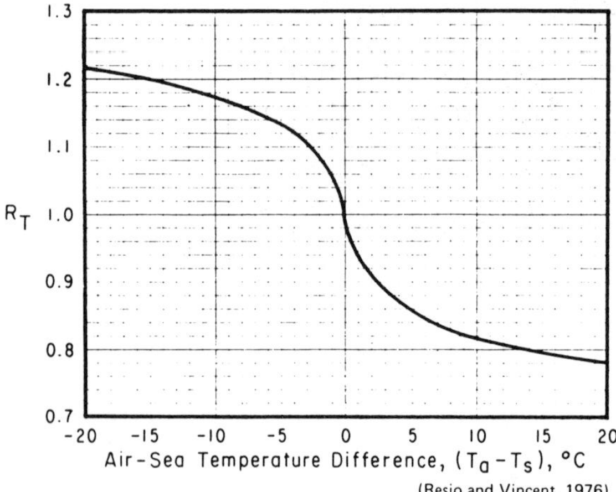

FIGURE 22. Amplification ratio, R_T, accounting for effects of air–sea temperature difference [45].

5. Determine the air–sea temperature difference. A wind stability correction is required when the air and water are different temperatures and the fetch is more than 10 miles (16 km), $U_C = R_T U_W$. If the temperature difference between the air and the sea is known, Figure 22 should be used to determine the amplification ratio, R_T. When only general knowledge of the condition of the atmospheric boundary layer is available, it should be categorized as stable, neutral, or unstable according to the following criteria:

Stable—When the air is warmer than the water, the water cools the air just above it and decreases mixing in the air column ($R_T = 0.9$).

Neutral—When the air and water have the same temperature, the water temperature does not affect the mixing in the air column ($R_T = 1.0$).

Unstable—When the air is colder than the water, the water warms the air causing the air near the water surface to rise thus increasing mixing in the air column ($R_T = 1.1$).

An unstable condition, $R_T = 1.1$, should be assumed when the boundary layer condition is unknown. Having a value for R_T, the adjusted windspeed is determined by $U_C = R_T U_W$.

Therefore, design windspeed adjustments are:

$$U_C = R_T\ R\ U_L \tag{13}$$

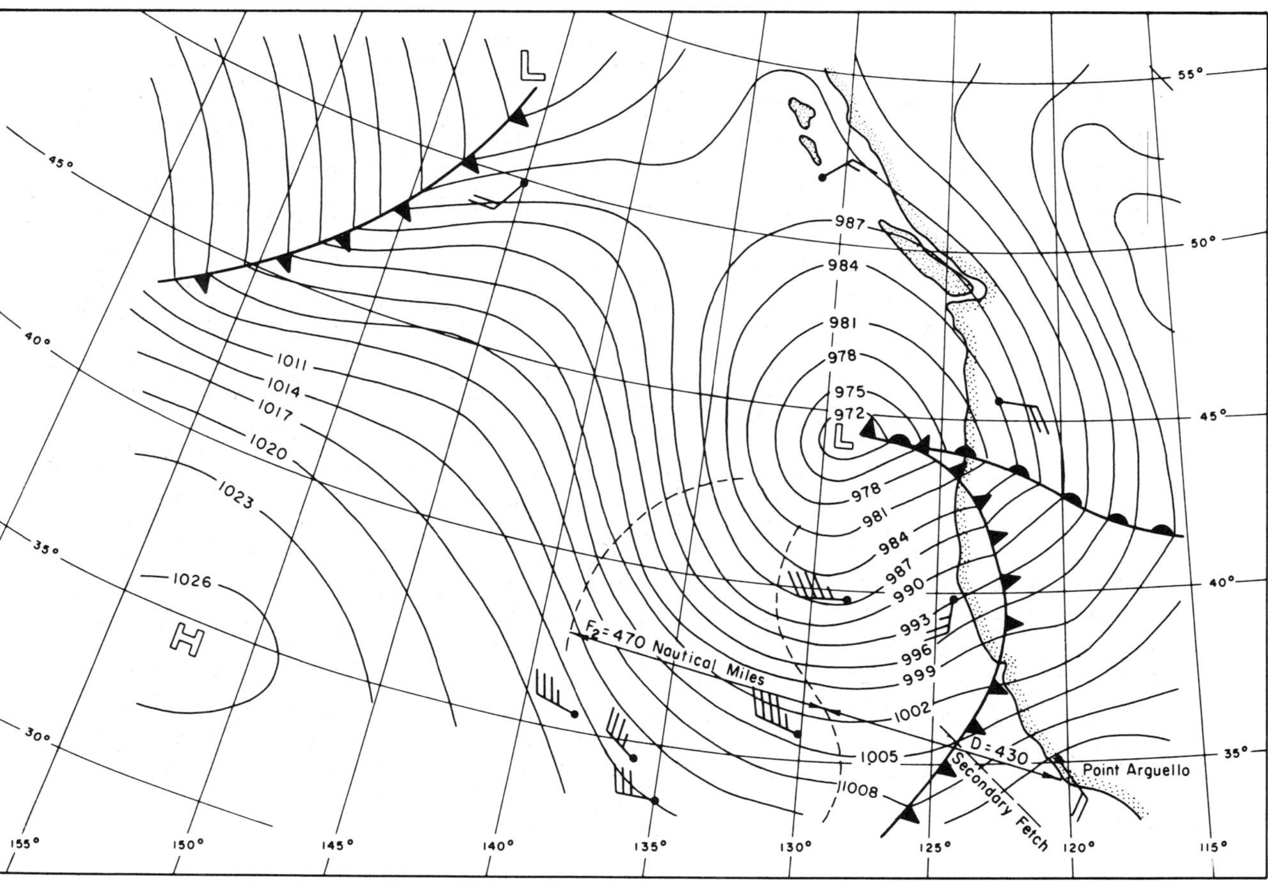

FIGURE 23. Surface synoptic chart for 0030Z, 27 October 1950 [45] (1 nautical mile = 1.85 km).

$$U_g = \frac{1}{\rho_a f} \frac{\Delta p}{\Delta n}$$

For $T = 10^\circ C$

$\Delta p = 3$ mb and 4 mb

Δn = isobar spacing measured in degrees latitude

$p = 1013.3$ mb

$\rho_a = 1.247 \times 10^{-3}$ gm/cm^3

f = Coriolis parameter $= 2\omega \sin \phi$

where

ω = angular velocity of earth, 0.2625 rad/hr

ϕ = latitude in degrees

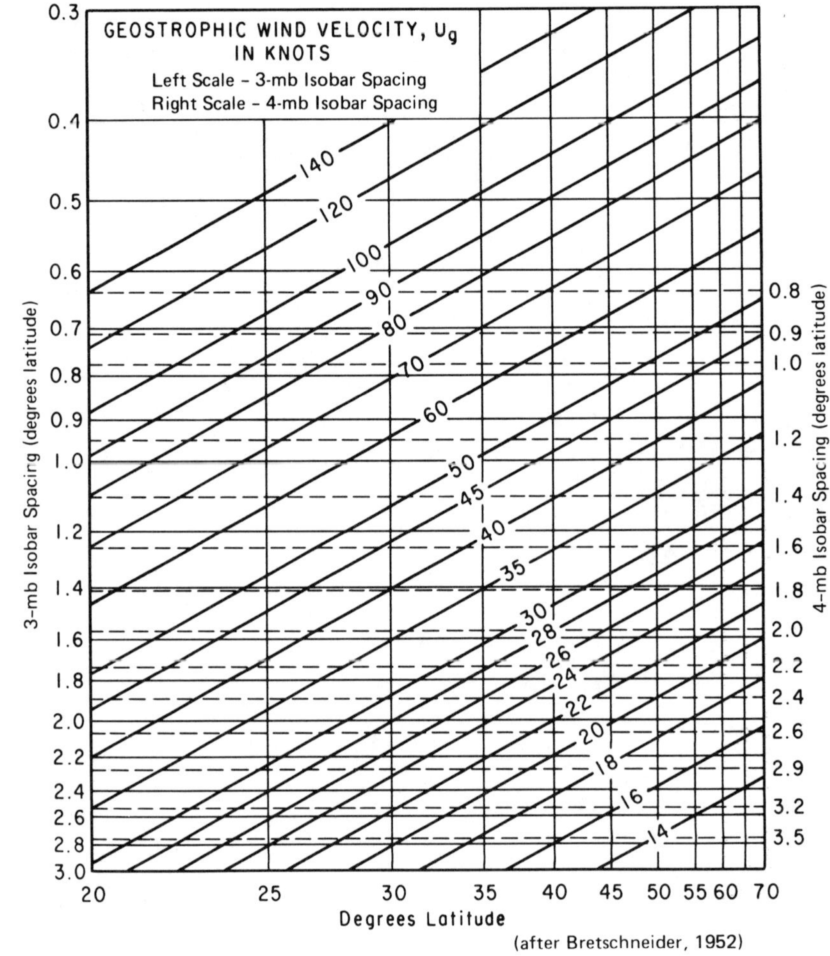

FIGURE 24. Geostrophic, or free air, wind scale [45] (1 knot = 1.15 mile/hr = 0.515 m/s).

WIND INFORMATION FROM SURFACE PRESSURE

Windspeed and direction in the open ocean is usually estimated from surface synoptic weather charts. The free air, or geostrophic, windspeed is first estimated from sea level pressure charts. Corrections to the free air wind are then made. Estimation from pressure charts should be used only for large areas and the estimated values should be compared with observations if possible to confirm their validity.

A simplified surface chart for the north Pacific Ocean is shown in Figure 23. The area labeled *L* in the right center of the chart and the area labeled *H* in the lower left corner of the chart are low- and high-pressure areas. The pressures increase moving outward from *L* (isobars 972, 975, etc.) and decrease moving outward from *H* (isobars 1026, 1023, etc.). Scattered about the chart are small arrow shafts with a varying number of feathers. The direction of a shaft shows the direction of the wind; each one-half feather represents a unit of 5 knots (2.5 meters per second) in windspeed.

Figure 24 may be used to estimate the free air windspeed. The distance between isobars on a chart is measured in degrees of latitude (an average spacing over a fetch is ordinarily used), and the latitude position of the fetch is determined. Using the spacing as ordinate and location as abscissa, the plotted or interpolated slant line at the intersection of these two values gives the geostrophic windspeed. For example, in Figure 23, a chart with 3-millibar isobar spacing, the average isobar spacing (measured normal to the isobars) over fetch F_2, located at 37 degrees N. latitude, is 0.70 degrees of latitude. Using the scales on the bottom and left side of Figure 24, a free air wind of 67 knots (34.5 m/s) is found.

After the free air wind has been estimated, the windspeed at the surface must be estimated. First the free air windspeed is converted to the 33-foot (10-meter) level speed by multiplying by R_g as given in Figure 25. R_g is a function of the free air windspeed U_g. The resulting velocity is then adjusted for stability effects by the factor R_T given in Figure 22.

WIND DURATION

Estimates of the duration of the wind are also needed for wave prediction. Synoptic weather charts are prepared at a minimum of 6-hour intervals. Thus interpolation to determine the duration may be necessary. Linear interpolation is adequate for most uses. Interpolation should not be used if short-duration phenomena, such as frontal passage or thunderstorms, are present.

FETCH

A fetch is defined as a region in which the windspeed and direction are reasonably constant. A fetch should be defined such that wind direction variations do not exceed 15 degrees and windspeed variations do not exceed 5 knots (2.5 meters

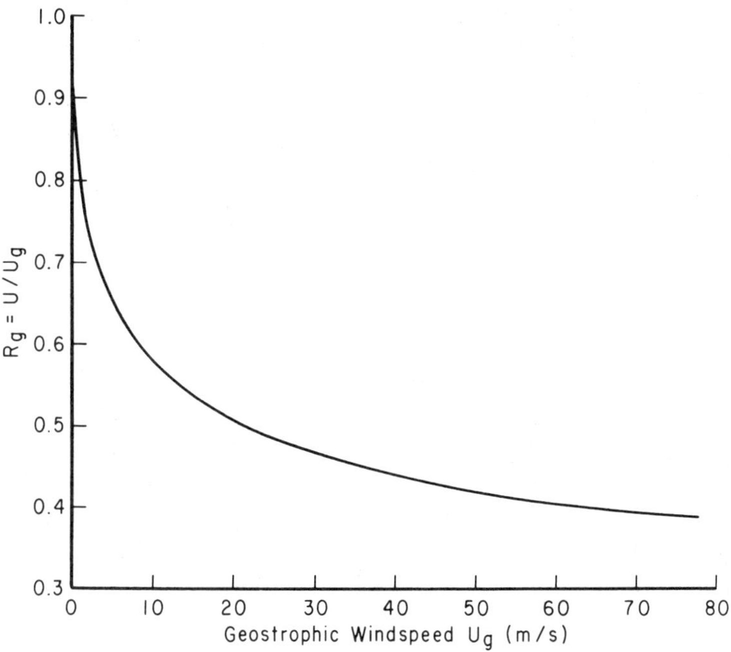

FIGURE 25. Ratio R_g of windspeed U at 33-foot (10-meter) elevation to geostrophic, or free air, windspeed U_g [45] (1 knot = 1.15 mile/hr = 0.515 m/s).

per second) from the mean. A coastline upwind from the point of interest always limits a fetch. An upwind limit to the fetch may also be provided by curvature or spreading of the isobars as indicated in Figure 26 or by a definite shift in wind direction. Frequently the discontinuity at a weather front will limit a fetch.

In inland waters (bays, rivers, lakes, and reservoirs), fetches are limited by landforms surrounding the water body. Shorelines are usually irregular. In such cases, fetch should be defined as a radial average over an arc of 24 degrees centered on the wind direction. Radials are placed at three degree intervals. Example fetch determinations are provided in Figure 27.

Wave Prediction in Deep Water

Significant wave height and peak period can be estimated from Figure 28 when the windspeed (corrected as discussed previously), duration, and fetch are known. The prediction curve is based on equations developed from the JONSWAP experiment [24,25]. The peak period is approximately 5 percent longer than significant period. In most instances they can be assumed equal. The equations in Table 3 may be used as an alternative to the figure when U.S. customary units are desired or Table 4 for SI units. Special procedures for use with hurricanes and other tropical storms are available in the Shore Protection Manual [45].

Wave Prediction in Shallow Water

If the predominant depth of water over the fetch is less than one-half the deep water wavelength, wave growth is affected by the bottom. When this occurs, the known wind-speed, duration, and fetch should be used to predict significant wave height and period from Figures 29 through 38. Wave refraction and shoaling are also a consideration as discussed in the following paragraph.

Wave Refraction and Shoaling

When waves move into shallow water, their speed decreases. This effect, referred to as shoaling, influences wave height. If the waves are moving at an angle to the bottom contours, they bend so that wave crests are more nearly parallel to the contours. This process, called refraction, also affects wave height. Refraction is generally computed on a

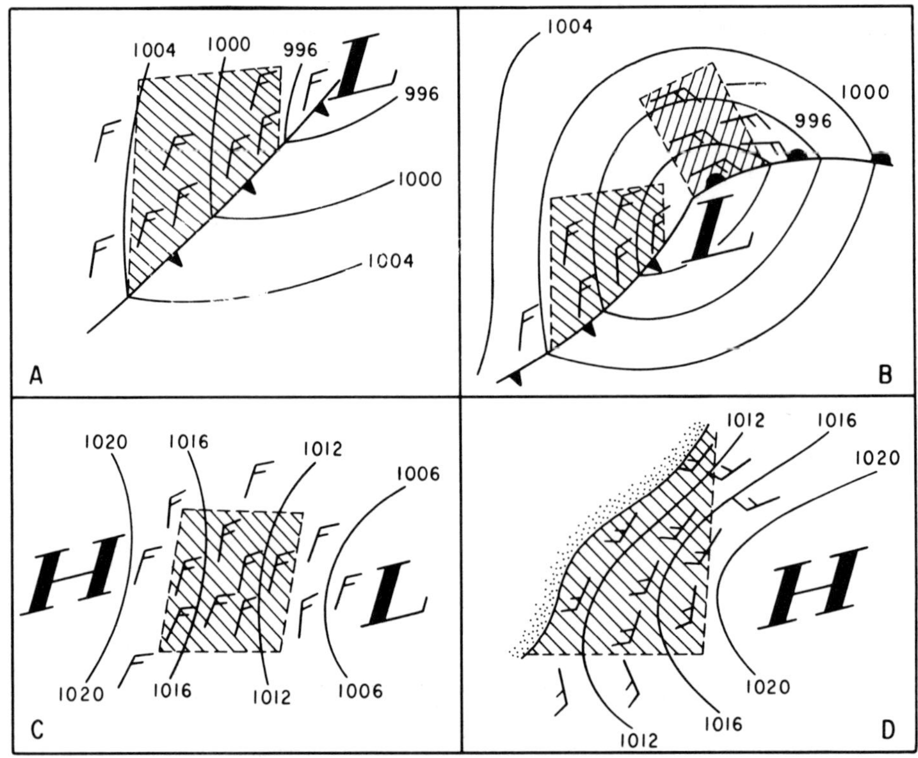

FIGURE 26. Possible fetch limitations [45].

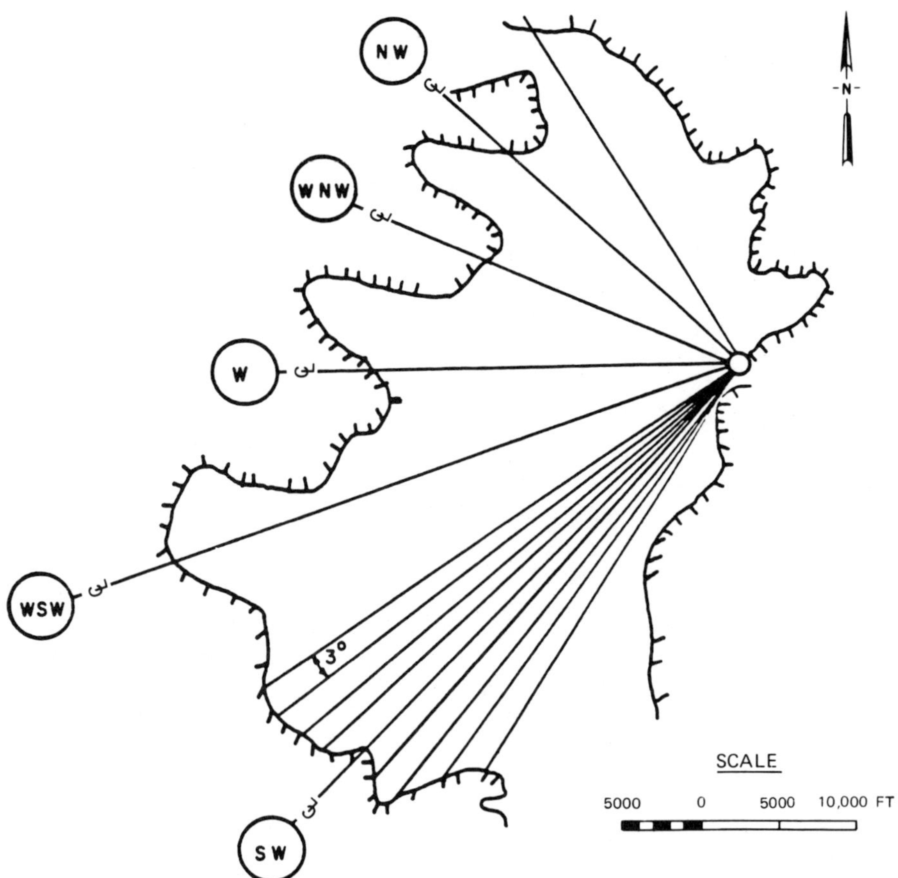

FIGURE 27. The average fetch length for each wind direction can be determined by drawing nine radials at 3 degree increments centered on the wind direction and arithmetically averaging the radial lengths as illustrated.

site-by-site basis since it depends upon the details of the bottom configuration. Computational methods are described in the following section. Rough estimates of refraction and shoaling effects on wave height and direction can be obtained from Figure 39, which is based on the assumption that bottom contours are straight and parallel.

Spectral Models

Numerous mathematical expressions for the spectral energy density function, $E(f)$, have been proposed based on theoretical considerations and analysis of field data. The expressions can be very helpful for characterizing a sea-state for modeling of wave growth, structure response to waves, and vessel response to waves. Some accepted spectral models are presented and discussed in the following paragraphs.

BRETSCHNEIDER SPECTRUM

The Bretschneider Spectrum [5] applies to deep water waves which are growing under the influence of a local wind. The spectrum is based on parameters of the wave field. Spectral energy density is given in ft^2-s by

$$E(f) = 3.36\, H_s^2 T_p (fT_p)^{-5} \exp\left[-1.25(fT_p)^{-4}\right] \quad (14)$$

JONSWAP SPECTRUM

The JONSWAP spectrum also applies to deep water waves which are being generated by a local wind. It is based on extensive wave observations collected in the North Sea as part of the Joint North Sea Wave Project [24]. Spectral energy density is given by

$$E(f) = \frac{\alpha g^2}{(2\pi)^4 f^5}\, e^{a} \gamma^{b} \quad (15)$$

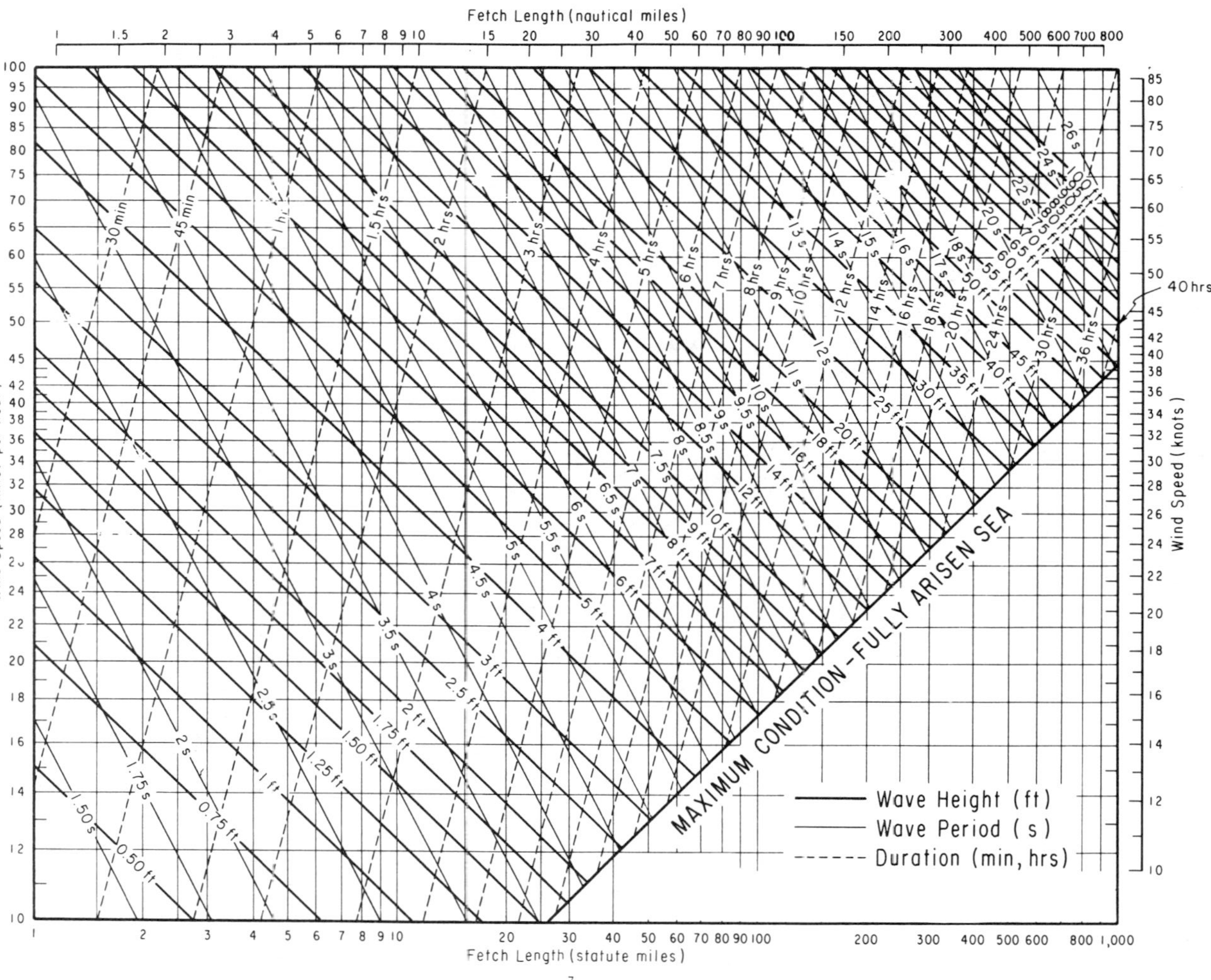

FIGURE 28. Nomograms of deepwater significant wave prediction curves as functions of windspeed, fetch length, and wind duration. Curves include the effect of nonconstant coefficient of drag on the wind (1 ft = 0.305 m, 1 mile = 1.61 km, 1 mile/hr = 0.45 m/s).

TABLE 3. Deepwater Wave Forecasting Equations[a] (U.S. Customary Units).

Units	
H_s(ft) T_p(s), U_c(miles/hr) F(miles), t(hr)	H_s(ft), T_p(s), U_c(kn) F(nmi), t(hr)
Fetch limited[b] (F, U_c)	
$H_s = 1.77X10^{-2}U_c^{1.23}F^{0.5}$	$H_s = 2.26X10^{-2}U_c^{1.23}F^{0.5}$
$T_p = 46.86X10^{-2}U_c^{0.41}F^{0.33}$	$T_p = 52.0X10^{-2}U_c^{0.41}F^{0.33}$
Duration limited[b] (U_c, t)	
$H_s = 90.79X10^{-4}U_c^{1.58}t^{0.714}$	$H_s = 1.135X10^{-2}U_c^{1.58}t^{0.714}$
$T_p = 24.16X10^{-2}U_c^{0.724}t^{0.411}$	$T_p = 26.76X10^{-2}U_c^{0.724}t^{0.411}$
Fully developed	
$H_s = 0.5634X10^{-2}U_c^{2.46}$	$H_s = 0.7963X10^{-2}U_c^{2.46}$
$T_p = 21.83X10^{-2}U_c^{1.23}$	$T_p = 25.8X10^{-2}U_c^{1.23}$
$t = 53.28X10^{-2}U_c^{1.23}$	$t = 63.2X10^{-2}U_c^{1.23}$

[a]Windspeed, U_c, in these equations must be corrected as indicated in text.
[b]It has been shown that fetch and duration are not directly interchangeable quantities for wave growth. Consequently, the duration required to reach a given fetch-limited condition cannot be obtained by the interchange of these sets of equations. The wind duration required to reach fetch-limited conditions, t_f, in hours is estimated by $t_f = 1.91F^{0.67}\ U_c^{-0.41}$ with F in miles and U_c in miles/hr.

TABLE 4. Deepwater Wave Forecasting Equations[a] (Metric Units).

Units	
H_s(m) T_p(s), U_c(m/s) F(m), t(s)	H_s(m), T_p(s), U_c(m/s) F(km), t(hr)
Fetch limited[b] (F, U_c)	
$H_s = 3.630X10^{-4}U_c^{1.23}F^{0.5}$	$H_s = 1.148X10^{-2}U_c^{1.23}F^{0.5}$
$T_p = 5.565X10^{-2}U_c^{0.41}F^{0.33}$	$T_p = 55.65X10^{-2}U_c^{0.41}F^{0.33}$
Duration limited[b] (U_c, t)	
$H_s = 28.54X10^{-6}U_c^{1.58}t^{0.714}$	$H_s = 98.76X10^{-4}U_c^{1.58}t^{0.714}$
$T_p = 1.495X10^{-2}U_c^{0.724}t^{0.411}$	$T_p = 43.28X10^{-2}U_c^{0.724}t^{0.411}$
Fully developed	
$H_s = 1.251X10^{-2}U_c^{2.46}$	$H_s = 1.251X10^{-2}U_c^{2.46}$
$T_p = 58.93X10^{-2}U_c^{1.23}$	$T_p = 58.93X10^{-2}U_c^{1.23}$
$t = 51.80X10^{-2}U_c^{1.23}$	$t = 1.439\ U_c^{1.23}$

[a]Windspeed, U_c, in these equations must be corrected as indicated in text.
[b]It has been shown that fetch and duration are not directly interchangeable quantities for wave growth. Consequently, the duration required to reach a given fetch-limited condition cannot be obtained by the interchange of these sets of equations. The wind duration required to reach fetch-limited conditions, t_f, in hours is estimated by $t_f = 1.001X10^{-2}\ F^{0.67}\ U_c^{-0.41}$ with F in m and U_c in m/s.

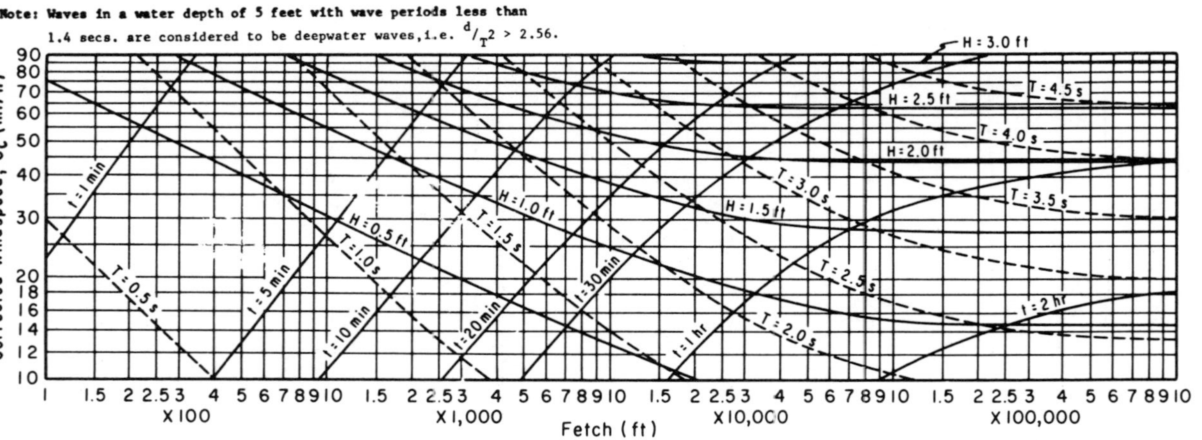

FIGURE 29. Forecasting curves for shallow water waves. Constant depth = 5 ft (1 ft = 0.305 m, 1 mile = 1.61 km, 1 mile/hr = 0.45 m/s).

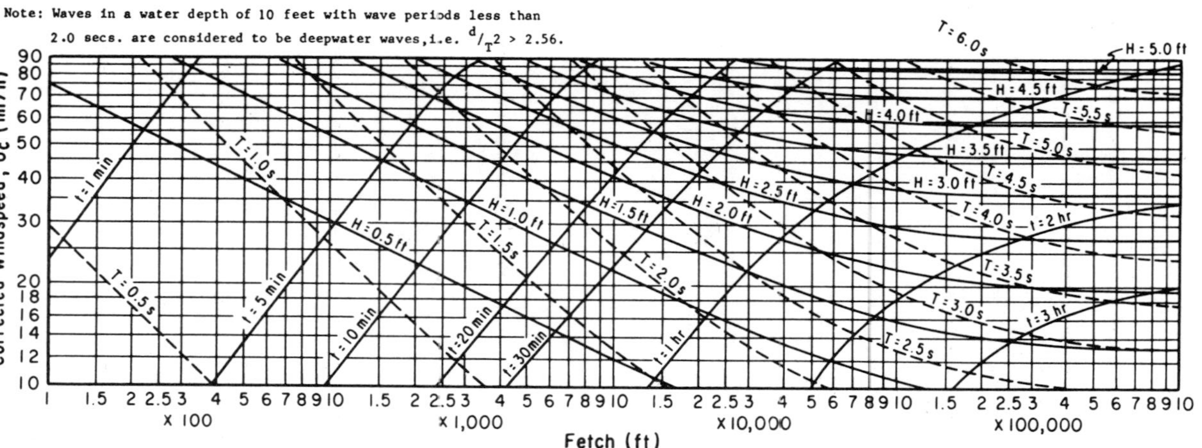

FIGURE 30. Forecasting curves for shallow water waves. Constant depth = 10 ft (1 ft = 0.305 m, 1 mile = 1.61 km, 1 mile/hr = 0.45 m/s).

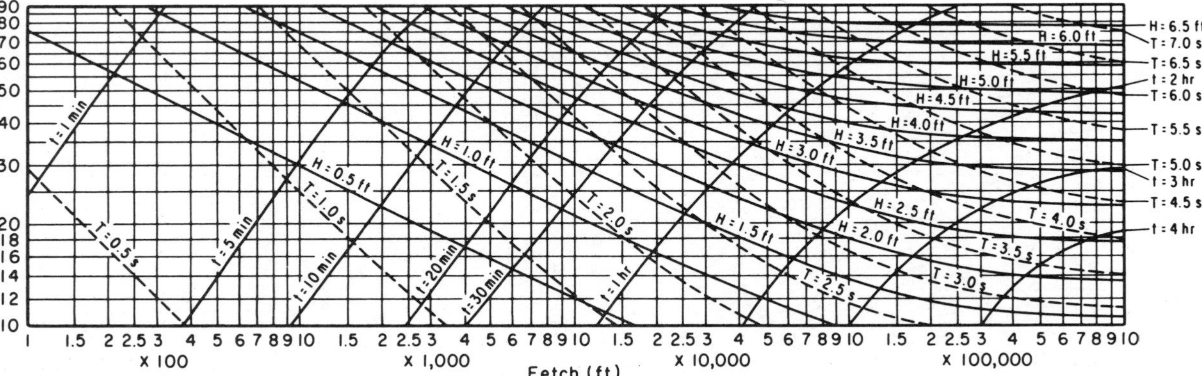

FIGURE 31. Forecasting curves for shallow water waves. Constant depth = 15 ft (1 ft = 0.305 m, 1 mile = 1.61 km, 1 mile/hr = 0.45 m/s).

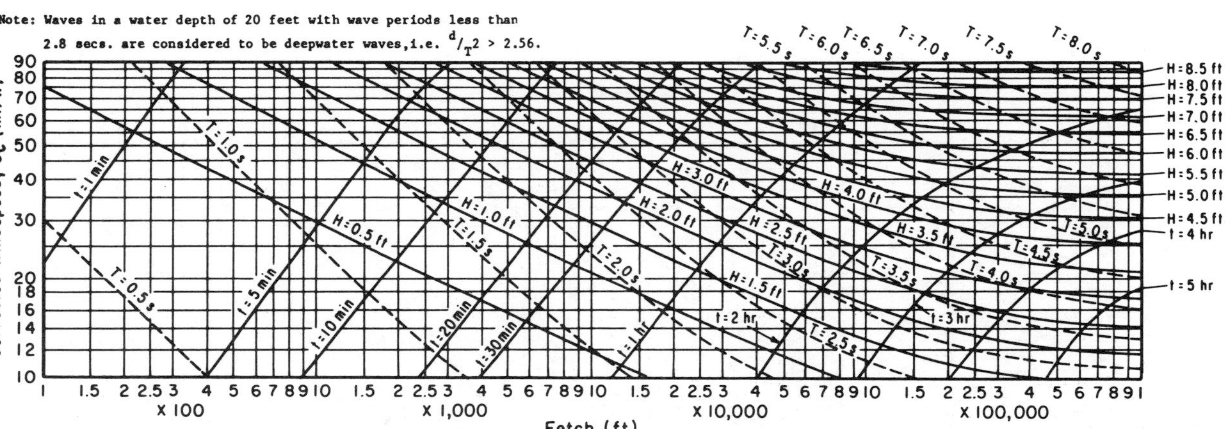

FIGURE 32. Forecasting curves for shallow water waves. Constant depth = 20 ft (1 ft = 0.305 m, 1 mile = 1.61 km, 1 mile/hr = 0.45 m/s).

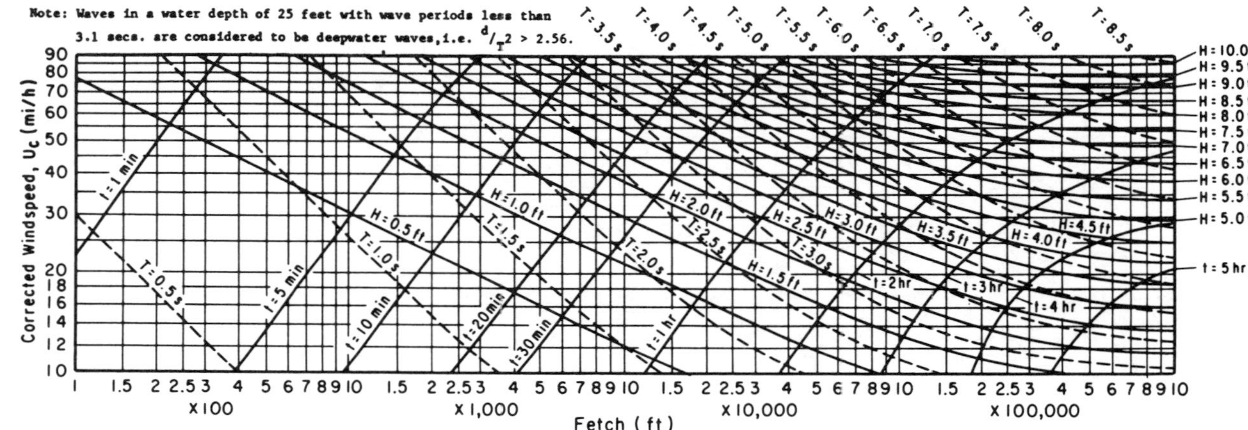

FIGURE 33. Forecasting curves for shallow wate⁻ waves. Constant depth = 25 ft (1 ᶠt = 0.305 m, 1 mile = 1.61 km, 1 mile/hr = 0.45 m/s).

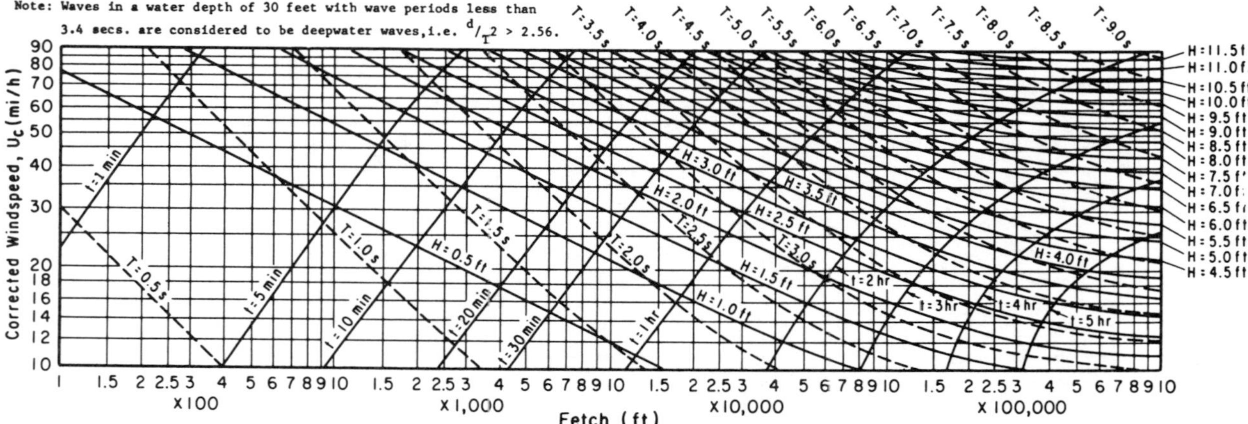

FIGURE 34. Forecasting curves for shallow water waves. Constant depth = 30 ft (1 ft = 0.305 m. 1 mile = 1.61 km, 1 mile/hr = 0.45 m/s).

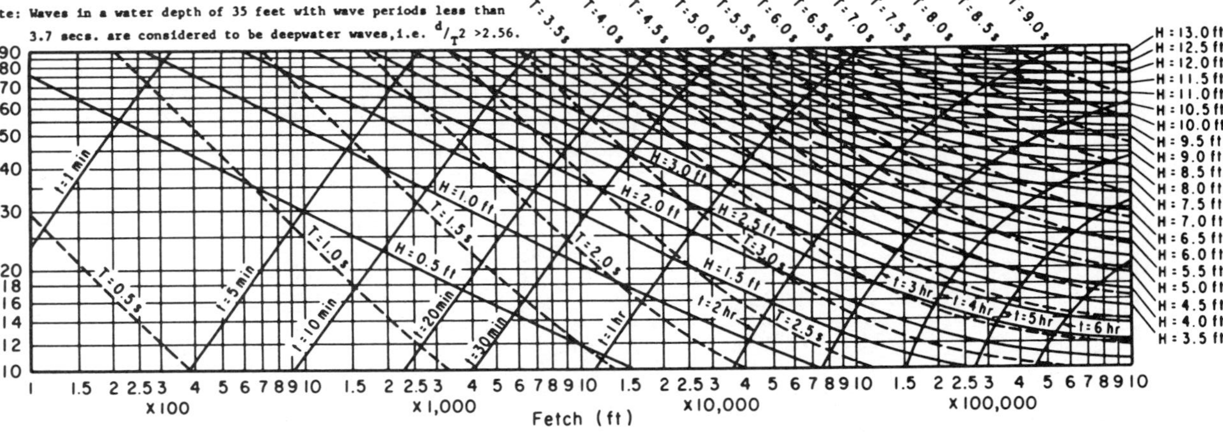

FIGURE 35. Forecasting curves for shallow water waves. Constant depth = 35 ft (1 ft = 0.305 m, 1 mile = 1.61 km, 1 mile/hr = 0.45 m/s).

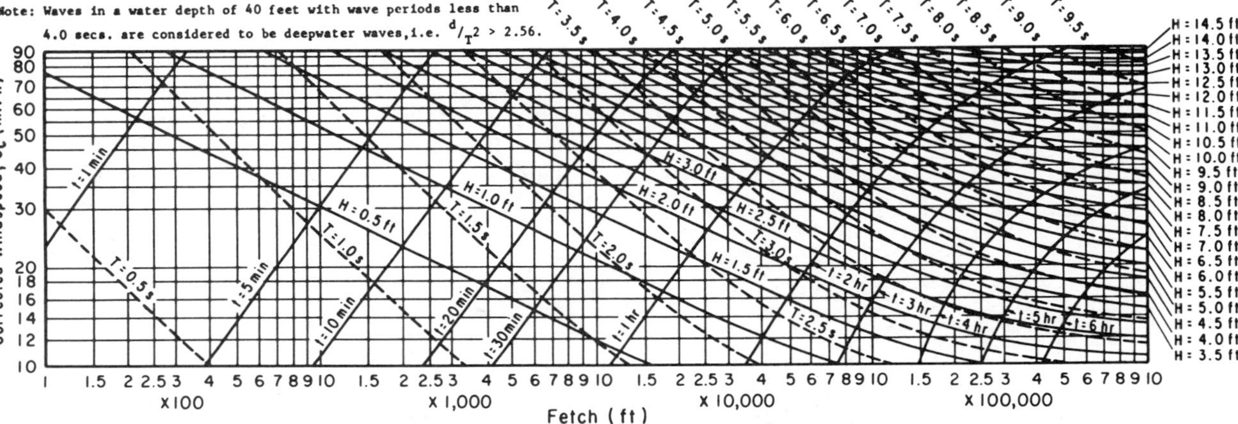

FIGURE 36. Forecasting curves for shallow water waves. Constant depth = 40 ft (1 ft = 0.305 m, 1 mile = 1.61 km, 1 mile/hr = 0.45 m/s).

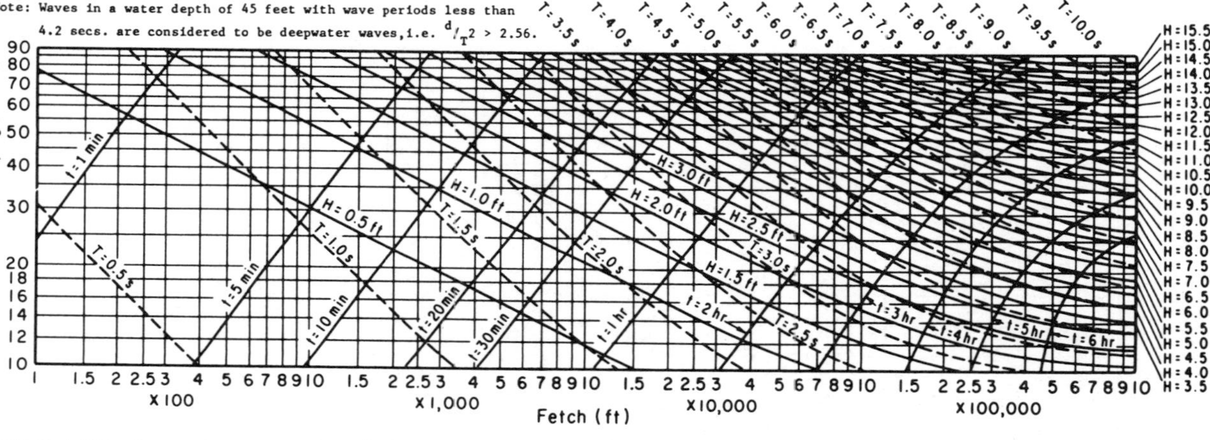

FIGURE 37. Forecasting curves for shallow water waves. Constant depth = 45 ft (1 ft = 0.305 m, 1 mile = 1.61 km, 1 mile/hr = 0.45 m/s).

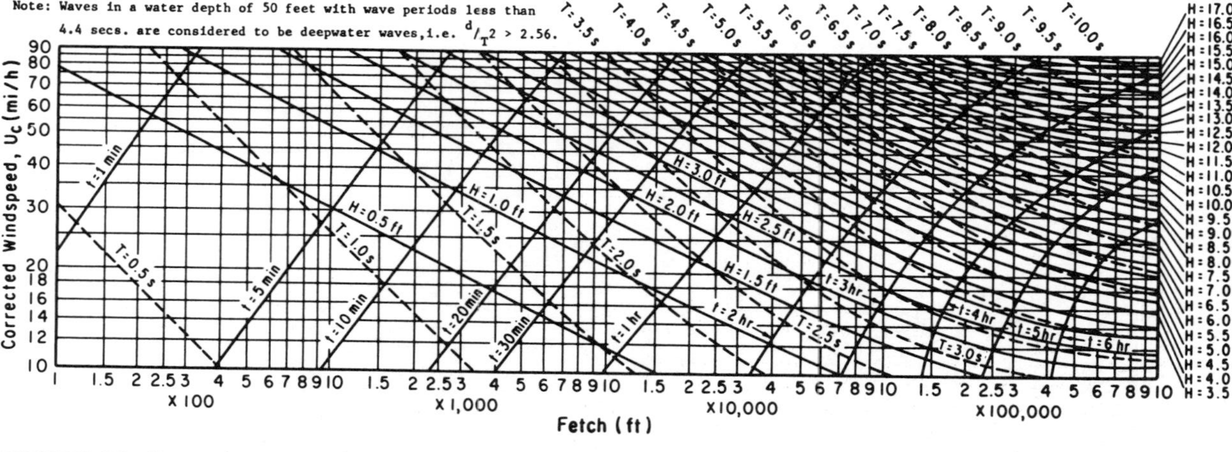

FIGURE 38. Forecasting curves for shallow water waves. Constant depth = 50 ft (1 ft = 0.305 m, 1 mile = 1.61 km, 1 mile/hr = 0.45 m/s).

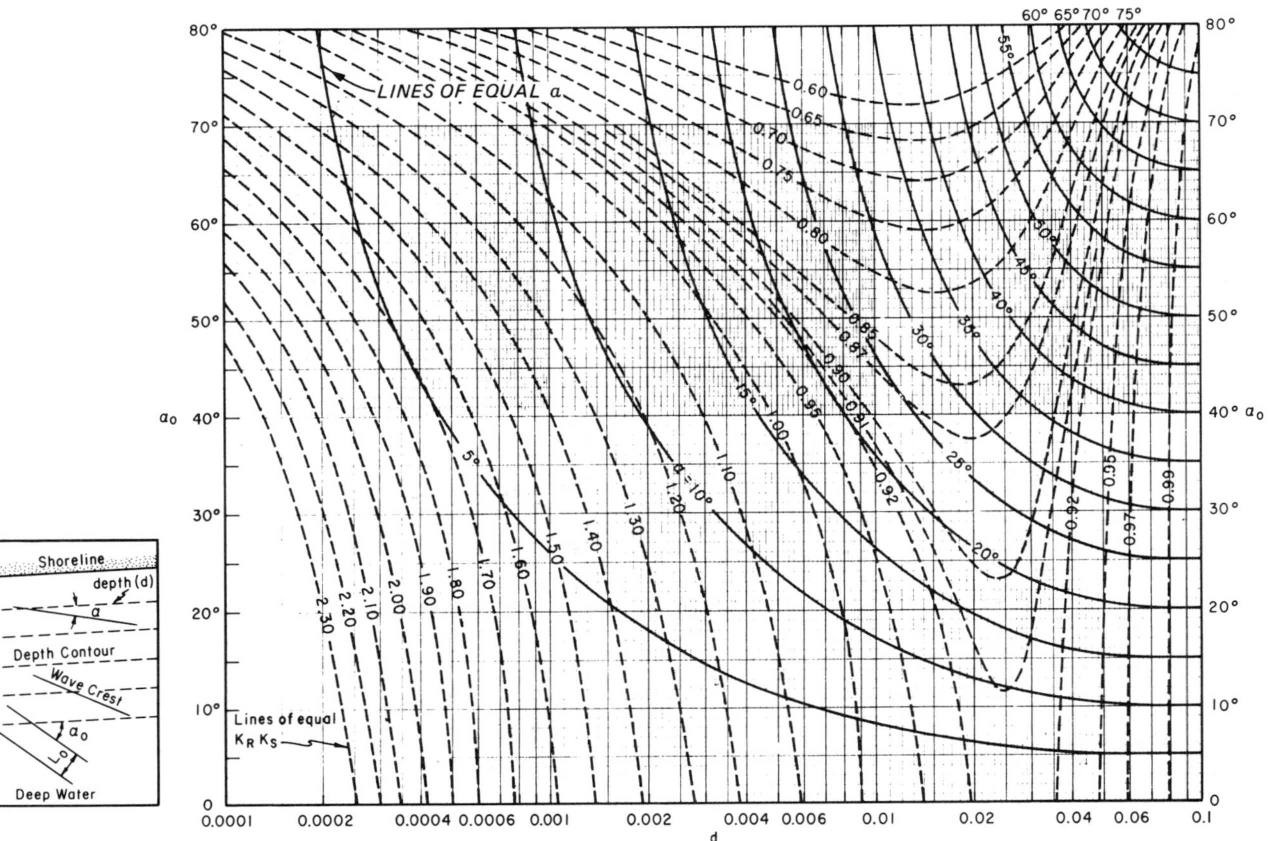

FIGURE 39. Change in wave direction and height due to refraction on slopes with straight parallel depth contours including shoaling [45].

where

$$a = -1.25(fT_p)^{-4}$$

$$b = \exp\left[\frac{-1}{2\sigma^2}(fT_p - 1)^2\right] \qquad (16)$$

The parameters α, σ, and γ may be determined either by fitting an observed spectrum or by the following expressions:

$$\sigma = \begin{cases} 0.07 \text{ for } f \leq f_p \\ 0.09 \text{ for } f \geq f_p \end{cases} \qquad (17)$$

$$\alpha = 0.0078\varkappa^{0.49} \qquad (18)$$

$$\gamma = 2.47\varkappa^{0.39} \qquad (19)$$

$$\varkappa - 2\pi U^2/(gL_p) \qquad (20)$$

where

U = windspeed at 33-foot (10-m) elevation
L_p = wavelength for waves at peak frequency

The above expressions for α and γ differ from the original JONSWAP formulation which was based on fetch and windspeed rather than wavelength and windspeed. The advantage of the above formulation is that it can easily be extended for application in shallow water as discussed in the following paragraphs [27].

The parameters α and γ may also be estimated in terms of parameters of the wave field alone (rather than windfield) by

$$\alpha = 157.9\ \epsilon^2 \qquad (21)$$

$$\gamma = 6614\ \epsilon^{1.59} \qquad (22)$$

where

$$\epsilon = H_s/(4\ L_p) \qquad (23)$$

The parameter ϵ is the significant wave steepness. The parameter γ, called the peak enhancement factor, controls the sharpness of the spectral peak. It typically ranges between 1 and 7 with a mean value of 3.3.

SHALLOW WATER SPECTRUM

The TMA spectrum characterizes waves which have been generated primarily in a local deep water area and then moved into shallow water. The spectral form is based on the assumption of complete saturation of energy at frequencies higher than f_p. No refraction effects are included. The spectral form was derived by Bouws et al. [3] from theoretical work by Kitaigorodskii et al. [31] and extensive field data from the Texel, MARSEN and ARSLOE (TMA) experiments. The TMA spectrum is given by

$$E_{TMA}(f,d) = \frac{\alpha g^2}{(2\pi)^4 f^5}\Phi(2\pi f,d)e^a\gamma^b \qquad (24)$$

The function $\Phi(2\pi f,d)$ approaches a value of one in deep water and a value of zero as depth decreases (Figure 40). It is well approximated by

$$\Phi(2\pi f,d) = \begin{cases} 0.5\omega_d^2 & \text{for } \omega_d \leq 1 \\ 1 - 0.5(2 - \omega_d)^2 & \text{for } \omega_d > 1 \end{cases} \qquad (25)$$

where $\omega_d = 2\pi f(d/g)^{1/2}$

The functions a and b are defined as with the JONSWAP spectrum [Equation (16)]. The wavelength L_p in Equations (20) and (23) is based on linear wave theory and an appropriate local water depth. Significant wave height in Equation (23) is the energy-based parameter, H_{mo}. If it can be assumed that the energy-containing frequencies are such that $\omega_d < 1$, H_{mo} is approximated by

$$H_{mo} = 0.350\ (\alpha g d)^{0.5}\ T_p \qquad (26)$$

The variation in TMA spectral shape as a function of water depth is illustrated in Figure 41. Additional details on the TMA spectrum are given by Hughes [27].

Directional Spectral Models

Ocean wave energy can be characterized by a variety of directions as well as by a variety of frequencies. Spectral representations which include both frequency distribution and angular spreading are known as directional spectral models.

Directional spectral models are based on the assumption that the spectrum may be described by the product of two functions:

$$E(f,\theta) = E(f)\ D(f,\theta) \qquad (27)$$

where

$E(f,\theta)$ = directional spectral density function
$D(f,\theta)$ = angular spreading function
θ = direction in radians

This parameterization can effectively represent the directional nature of a wavefield in the absence of complicating influences such as a large change in wind direction or the propagation of swell into a generation area.

A commonly-used form of the spreading function which is independent of frequency [34] is

$$D(\theta) = G(s)\cos^{2s}\left(\frac{\theta - \theta_o}{2}\right) \qquad (28)$$

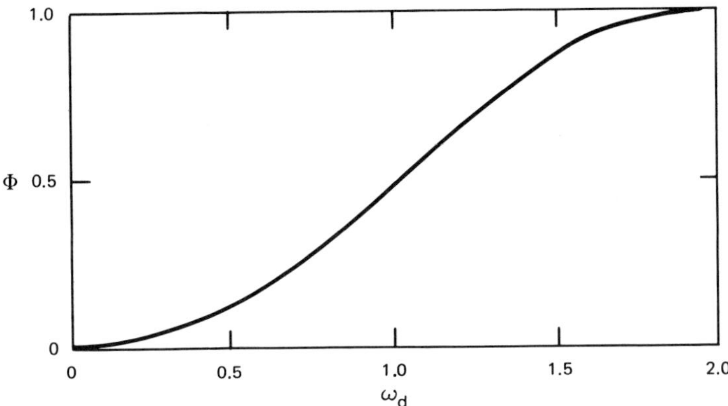

FIGURE 40. Φ as a function of ω_d [3].

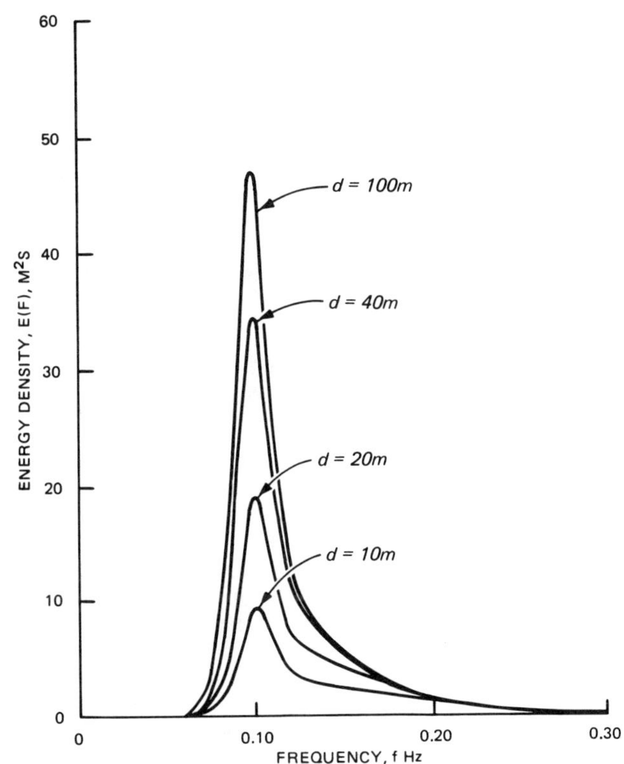

FIGURE 41. A family of wind wave spectra with identical JONSWAP parameters in frequency space. TMA spectral form [3] (1 ft = 0.305 m, 1 ft^2s = 0.093 m^2s).

where

$G(s)$ = function tabulated in Table 5
θ_o = mean wind direction
s = constant-valued spreading parameter

The parameter s controls the magnitude of directional spread as illustrated in Figure 42. Increasing the value of s causes a narrowing of the directional spread. Swell is typically represented by narrow spreads and seas by broad spreads.

More complex formulations for the spreading parameter which include a dependence on wind speed and peak spectral frequency have been proposed based on field data in deep water [23,36].

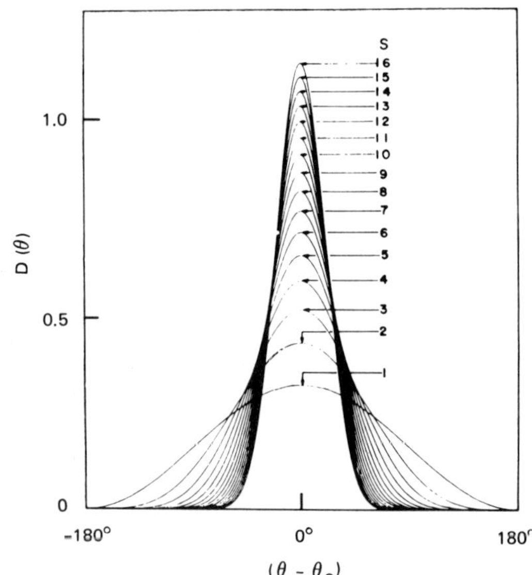

FIGURE 42. Idealized angular distribution $D(\theta)$ [36].

NUMERICAL WAVE MODELS

Because of the continually increasing capabilities and availability of digital computers, numerical wave models are becoming essential tools for practical engineering work. Models vary greatly in complexity. The simplest wave growth model is a computer program to solve the equations in Table 3. The most complex models are comprehensive spectral models which operate on a grid and simulate natural wave processes including growth, dissipation, propagation, and wave–wave interaction. The models can be used for both wave hindcasting (wind input is based on data from past events) and wave forecasting (wind input is based on predictions for the future).

Numerical modeling definitions and characteristics are discussed in the following paragraphs. Examples of considerations in selecting an appropriate model for an engineering project are given. Also a major wave hindcasting effort in the U.S., and the statistics which it has produced are discussed.

TABLE 5. Values of G(s) in the Directional Spreading Function.

s	G(s)
1	0.3183
2	0.4244
3	0.5093
4	0.5821
5	0.6467
6	0.7055
7	0.7598
8	0.8104
9	0.8581
10	0.9033
11	0.9463
12	0.9874
13	1.0269
14	1.0650
15	1.1017
16	1.1372

Numerical Modeling Definitions

Numerical model characteristics are described in the following terms:

1. Wave concept

Significant wave—model uses monochromatic wave or simple parameters of a spectrum of wave energy (Figure 43a and b).

Spectral wave—model uses a spectrum of wave energy composed of many different frequency bands (Figure 43c).

2. Time dependence

Steady state—model input does not vary with time.

Time dependent—model input changes with time.

3. Spatial configuration

Gridded—model simulates processes on a grid. Finite difference and finite element are alternative approaches for performing numerical calculations. Typically finite difference grid cells are rectangular and finite element cells are triangular (Figure 44a).

Nongridded—model simulates processes directly over entire area affecting the point of interest (Figure 44b).

4. Basic formulation

Energy equation—model solves energy equation.

Momentum equation—model solves momentum equation.

5. Wave growth

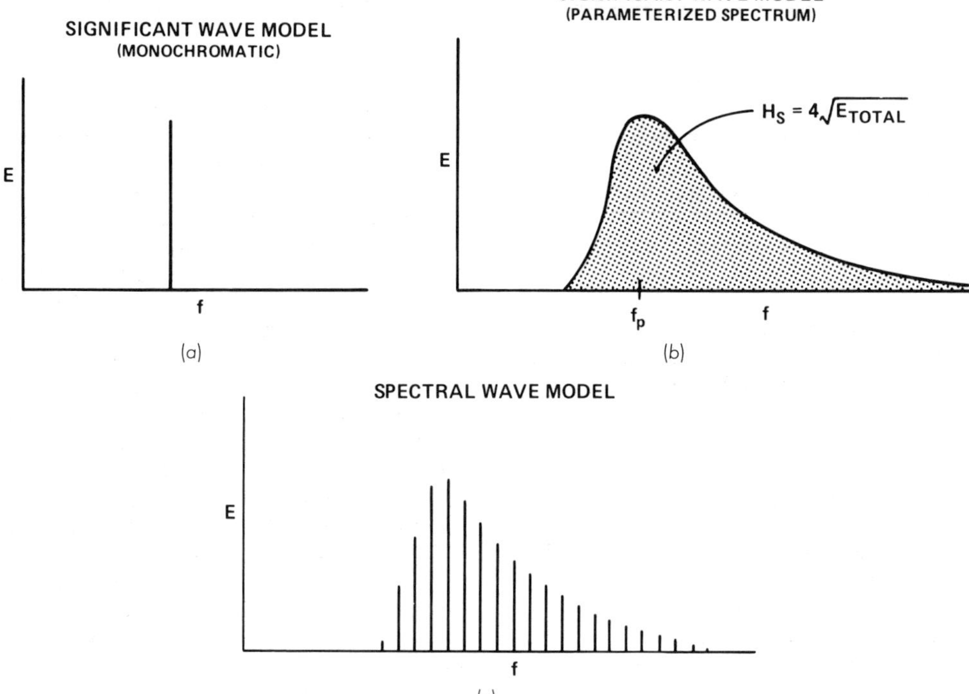

FIGURE 43. Wave model concepts.

SPATIAL CONFIGURATION

GRIDDED: PROCESSES ARE MODELED AT GRID POINTS COVERING THE WATER BODY

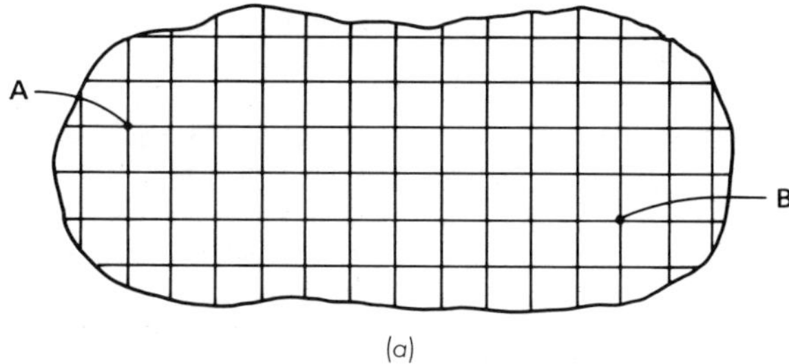

(a)

NON-GRIDDED: PROCESSES ARE APPROXIMATED OVER THE ENTIRE DISTANCE BETWEEN POINTS OF INTEREST

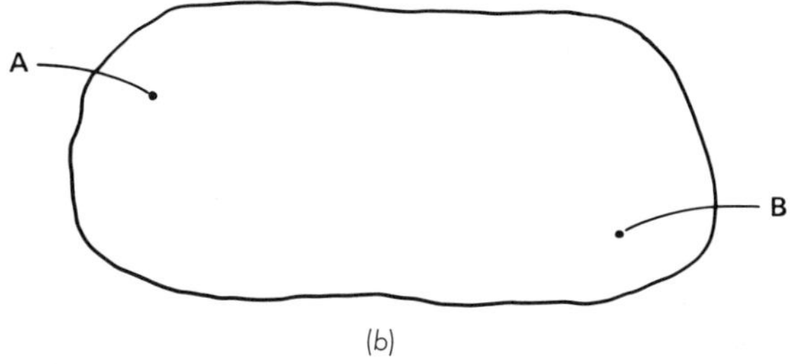

(b)

FIGURE 44. Spatial configuration of wave models.

Deep water—model can simulate wave growth in deep water.

Shallow water—model can simulate wave growth in shallow water.

Hurricanes—model can simulate wave growth due to hurricanes.

6. Currents—model can include the effect of currents.

7. Propagation—model can propagate waves in space (Figure 45).

8. Transformation in shallow water

Refraction and shoaling—modeling simulates wave refraction and shoaling. Two approaches are illustrated in Figure 46.

Bottom friction—model includes a bottom friction mechanism for energy dissipation.

Percolation—model includes a percolation mechanism for energy dissipation.

Wave breaking—model includes a wave breaking mechanism for energy dissipation.

Nonlinear interaction—model includes a mechanism for nonlinear transfer of energy between frequencies.

Diffraction (bottom induced)—model includes lateral energy transfer induced by irregular bottom.

Diffraction (structure-induced)—model includes capability for simulating diffraction around surface-piercing structures.

Growth during transformation—model includes capability for simulating additional wave growth by wind during the shallow water transformation process.

Blockage by floating or bottom-resting objects—model can simulate effects of structures or other objects floating on the surface or resting on the bottom.

9. Model basis (Figure 47)

Energy—model simulates processes in terms of modifications to wave energy.

Wave height—model simulates processes in terms of modifications to wave height.

Numerical Model Characteristics

To illustrate the range of capabilities of numerical models, a sample of 12 different models is tabulated in terms of the characteristics discussed previously (Figure 48). Model 1 in the table represents the manual methods in the Shore Protection Manual [45]. Models 2 through 12 are computerized models which generally increase in complexity with increasing model number.

Selection of Numerical Models

Some considerations in selecting a numerical model are illustrated in the following case studies.

CASE 1

Application—Wave estimates are needed at Spit A, (Figure 49) for estimating long-term sediment transport.

Appraisal—The spit is exposed to waves generated in Area B, but it is sheltered from waves generated in Areas C and D. Hence wave generation is considered only in Area B where fetches are up to 80 miles long.

Climatological wind measurements are available only at the spit. Synoptic weather maps are available. They are of little value in this study because winds are strongly channeled by topographic features (mountain ranges) and bear little relationship to synoptic pressure fields. Also wind fields over Area B are known to often be nonuniform so that the available wind measurements may not be a good representation of winds over the fetch.

The water depth in Area B is relatively great. Waves are in deep water until they approach very near to the spit.

Bottom contours near the spit are somewhat contorted including a large shoal offshore from the spit.

Currents in the shallow water area near the spit are believed to be weak, but tidal currents passing the tip of the spit are very strong.

Candidate models: Because the shallow water area is only a small part of the fetch, the candidate models can be considered in two separate categories as deep water growth models and shallow wave transformation models.

Candidate deep water growth models: 1, 9, 10, 11.

Candidate shallow transformation models: 1, 2, 3, 4, 5, 6, 7, 8, 9, 10, and 11.

Model selection—Since the wind information for Area B is very sparce in relation to the complexity of the wind fields, it is appropriate to use a simple deep water wave growth model. Also the geometry of Area B is relatively simple. Thus the models which operate on a grid appear to be unnecessary for this case. The model SPM 84 (model 1) emerges as the most efficient choice for this application.

Since considerations for the deep water growth model have led to the choice of a simple model, the shallow transformation model should also be relatively simple. It should not depend upon highly accurate input along the seaward boundary. Wave growth in shallow water is negligible. These considerations lead to elimination of the time dependent models 5, 8, and 11. Since the shallow water bathymetry is somewhat irregular and good estimates of nearshore wave direction are needed for sediment transport estimates, it is decided that the actual bathymetry must be represented in the model. Thus the nongridded models 1, 2, 3, and 4 are rejected. Model 6 is a ray-calculation routine which often leads to crossing rays in complicated nearshore areas. It is not recommended for this application. The remaining models are 7, 10, and 12. The limited accuracy of the deep water input suggests that model 10, a spectral model, is not cost-effective in comparison to 7 and 12, which are significant wave models. The nearshore complexity does not ap-

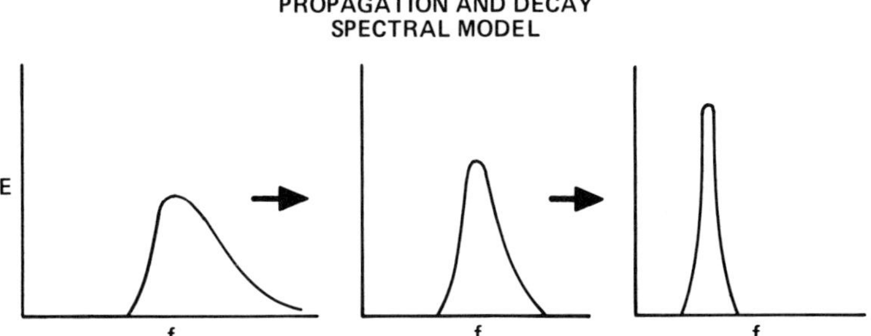

FIGURE 45. Propagation and decay for spectral model.

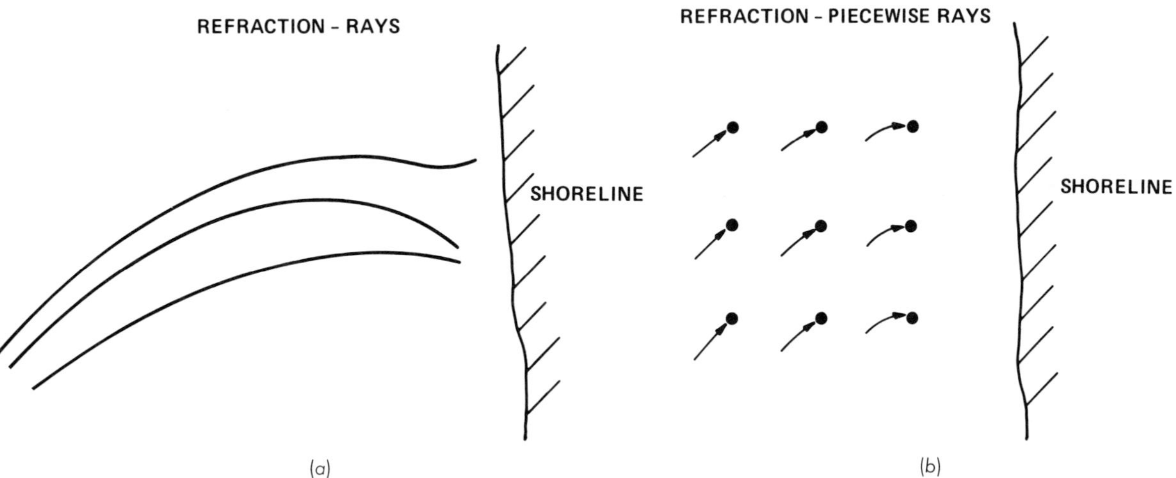

FIGURE 46. Approaches to modeling wave refraction.

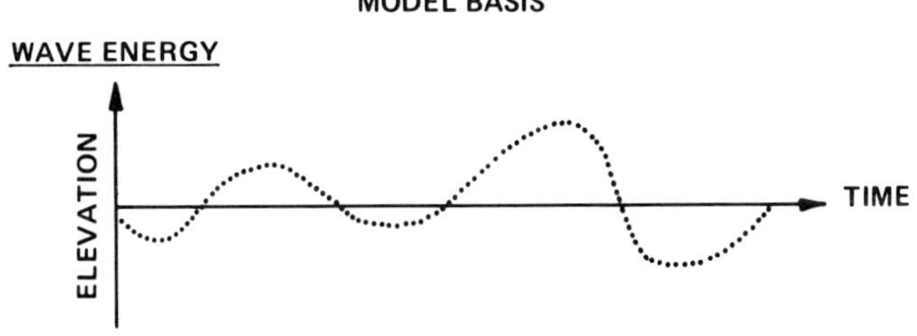

H_S = 4 X (STANDARD DEVIATION)
= 4 X (TOTAL SPECTRAL ENERGY)$^{1/2}$

WAVE HEIGHT

H_S = FUNCTION OF WAVE HEIGHT

FIGURE 47. Model basis.

Characteristics	Model SPM84 1	2	3	4	5	6	7	8	9	10	11	12
Wave concept												
Significant wave												
Monochromatic	x					x	x					x
Parameterized spectrum		x	x		x							
Spectral wave				x				x	x	x	x	
Time dependence												
Steady state	x	x	x	x		x	x			x		x
Time dependent					x			x	x		x	
Spatial configuration												
Ungridded	x	x	x	x	x	x		x				
Gridded												
Finite difference							x		x	x	x	
Finite element												x
Basic formulation												
Energy equation		x		x	x	x	x	x	x	x	x	
Momentum equation												x
Wave growth												
Deep water	x				x				x	x	x	
Shallow water	x				x					x	x	
Hurricanes	x											
Currents										x	x	
Propagation						x	x		x	x	x	x
Transformation in shallow water												
Refraction & shoaling	x			x*		x	x	x		x	x	x
Bottom friction					x					x	x	x
Percolation										x	x	
Wave breaking			x		x		x	x		x		
Nonlinear interaction					x			x		x	x	
Diffraction (bottom-induced)							x					x
Diffraction (structure-induced)	x											x
Growth during transformation					x					x	x	
Blockage by floating or bottom-resting objects												x
Model basis												
Energy		x		x	x			x	x	x	x	
Wave height	x		x			x	x					x

* **Assumes straight parallel bottom contours.**

FIGURE 48. Overview of numerical wave models.

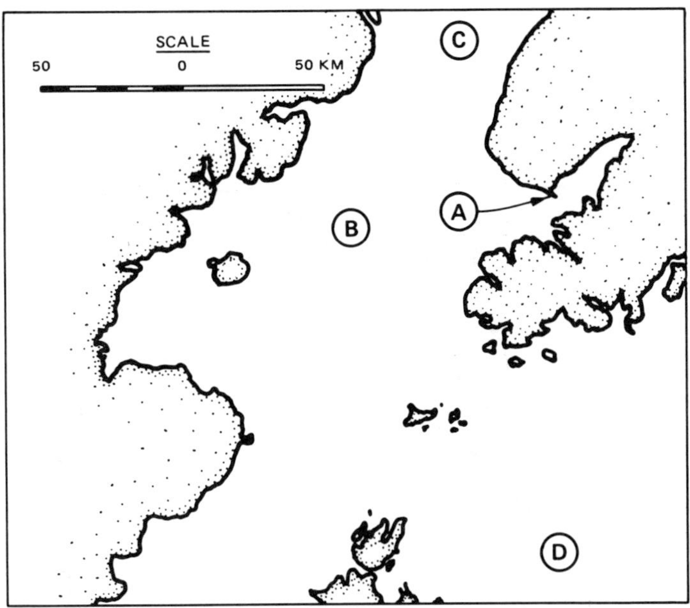

FIGURE 49. Location map for Case Study 1.

pear to be sufficient to warrant the use of model 12. Therefore the final selection is model 7 for the shallow water transformation model.

CASE 2

Application—Wave estimates are needed at Point A (Figure 50) for design of a seawall which will reduce flooding and wave overtopping to acceptable levels.

Appraisal—Point A is exposed to waves generated in the Ocean Area B through the entrance to Bay C. An additional consideration for the north side of Point A is wave energy generated in the northerly reaches of Bay C which propagates toward the project site.

Synoptic meteorological data are available for Ocean Area B. Sufficient wind measurements are available in the vicinity of Bay C to establish estimates of the winds over the bay.

The entire bay area represents shallow water for all ocean waves of interest in design. Local wave growth in the bay north of the project site may also occur to an appreciable extent in water that is shallow relative to the waves.

Bottom contours are irregular.

Currents are not expected to be a significant factor in the study.

Wave estimates from a climatological wave hindcasting program are available just seaward of the entrance to Bay C.

Candidate models—6, 7, 10, 11, and 12.

Model selection—Accurate overtopping rates are very important in this design project, so a spectral model is favored over a significant wave model. Either 10 or 11 would be an acceptable choice. Model 11 is expected to have the advantage of better representing additional wave growth in shallow water but the disadvantage of higher costs.

CASE 3

Application—Wave measurements are being collected in deep water along a coast. Just after a major storm it is desired to transfer the deep water waves to shore for a quick comparison with wave heights reported by coastal residents.

Appraisal—Wave growth between the measurement site and shore is assumed to be inconsequential.

It is usually necessary to assume that the bottom contours are straight and parallel in order to make quick estimates. This assumption is justifiable for this type of application for many coastal areas.

Candidate models—1, 2, 3 and 4.

Model selection—Model 1, for straight parallel bottom contours, is a nomogram derived for monochromatic waves (Figure 39). The model is useful if very quick estimates are needed and access to the computerized candidate models is not available. In most instances 2, 3, or 4 would be preferable to a monochromatic model.

A wave direction in deep water should be estimated. If significant refraction is expected, model 4 would be a good choice. If refraction appears to be of minor importance, 2 or 3 would also be suitable. In the case of 2 and 4, which are energy-based models, it must be remembered that the computed shallow water wave height will be somewhat lower than the crest to trough wave height seen by an observer. A procedure for estimating crest to trough height from an

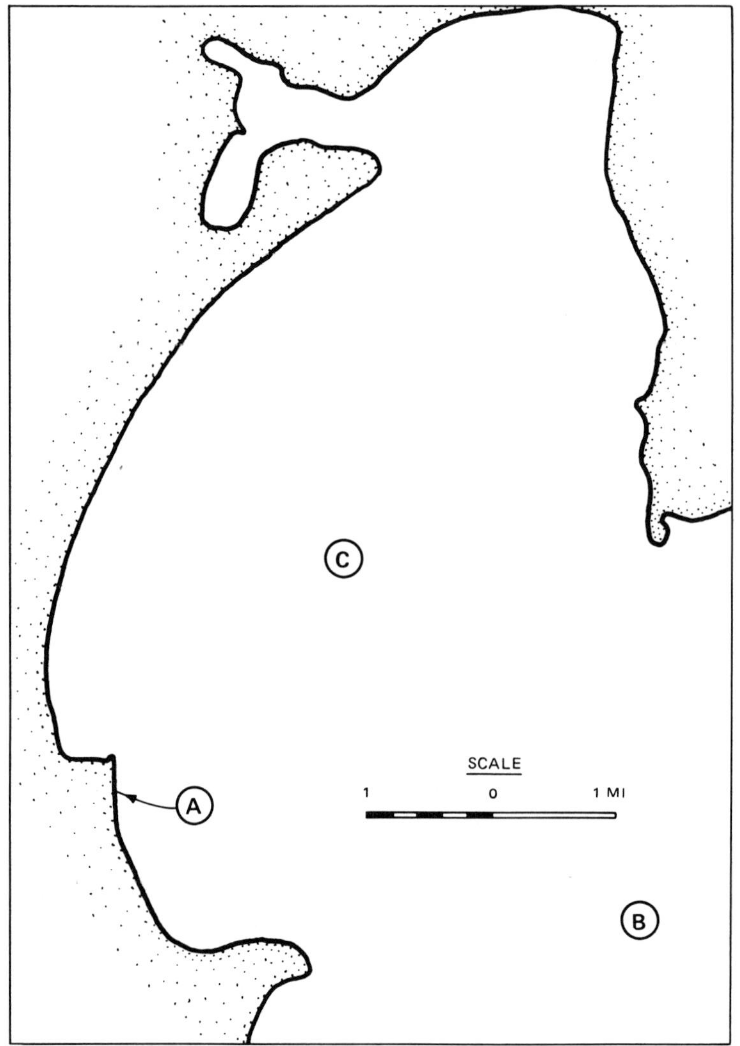

FIGURE 50. Location map for Case Study 2.

energy-based height is given in Figure 13. Since model 3 is based directly on wave height, it provides a direct estimate of shallow water height.

Wave Information Study

The Wave Information Study (WIS) is a U.S. Army Corps of Engineers project to hindcast wave climate over a 20-year period for the Atlantic, Pacific, Gulf of Mexico, and Great Lakes coasts of the United States. The hindcast period is from 1956 through 1975.

The WIS project is being executed in three phases as follows (Figure 51):

Phase I—Hindcast of deep water wave data from past meteorological data; model operates on the scale of the ocean basin.

Phase II—Hindcast at a finer scale than Phase I to better resolve sheltering effects of continental geometry; model operates on the scale of the continental shelf. Phase I data serve as boundary conditions for the seaward edge of the Phase II area.

Phase III—Transformation of Phase II wave data into nearshore region and inclusion of long waves.

The WIS uses a discrete spectral model based on an energy balance equation. The model includes wave growth in deep water and spectral wave propagation. The model is time dependent and, in the case of Phases I and II, operates on a spatial grid. Nearshore effects of sheltering, refraction, shoaling, and nonlinear interaction between different spectral components are included. Refraction in Phase III is based on the assumption of straight parallel bottom contours and uniformity of wave conditions along 10-mile stretches of coast.

The WIS results can be used to get high quality data for project sites typically by using Phase II results as input to a gridded shallow water transformation model with high quality bathymetric data. Alternatively, the existing WIS data base of shallow water wave information for simplified bathymetry and coastal configuration can be used for some stages of project planning and execution. More information on the availability and access to WIS data and programs is given by Corson [7].

STATISTICAL SUMMARIES OF INDIVIDUAL WAVE ESTIMATES

Statistical summaries of parameters from individual wave estimates are essential for defining wave climate and for predicting extreme waves at a site. The simplest statistics which are also very useful are the monthly, seasonal and annual means. Other useful summaries are described in the following sections. The quantity and quality of data available are crucial considerations in all statistical summaries. Less than

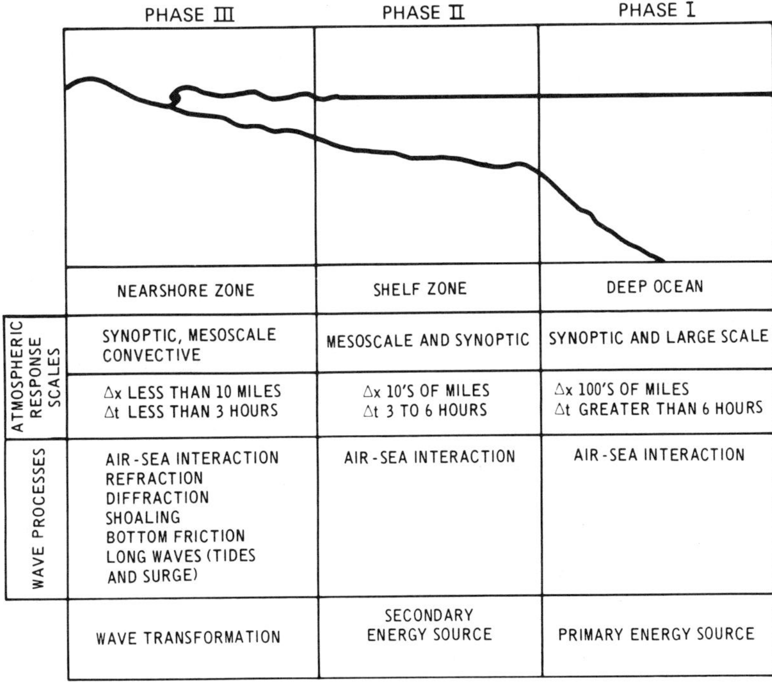

FIGURE 51. Summary of the three phases of WIS hindcasts [7] (1 mile = 1.61 km).

one year of data can be very misleading due to seasonal variations in wave climate. One complete year can give reliable estimates of routine wave conditions but not of extreme or unusual wave conditions. The possibility of biases introduced by the data collection and analysis systems must also be considered. One insidious bias on extreme values from observed or measured data is a tendency during very severe storms for missing or unreliable data in an otherwise consistent record. Good overall references on statistical summaries of wave observations are Ochi [38] and Goda [16].

Joint Distribution Tables of Wave Height and Period

Significant wave height and period statistics are often summarized as a table giving the number or percentage of occurrence of each significant height/period combination. A typical example is given in Figure 52.

Sea-State Persistence

Sea-state persistence is an estimate of how long a particular wave condition will remain. Persistence estimates are useful in planning and operational work. They are usually expressed as tables or plots of the number of consecutive hours or days the significant wave height exceeds various threshold levels. Various distribution functions have been applied to persistence data as discussed by Goda [16].

Long-Term Distributions for Wave Height

The long-term distribution for wave height is usually represented by the cumulative probability distribution of the data. It is also often fit with a model distribution function. There is no strong theoretical basis for a particular model. Several models are widely used primarily because they often provide good fits to the data.

The probability distribution functions used for long-term distribution of wave heights are given in Figure 53. These consist of the lognormal distribution and the Extremal Types I, II, and III distributions. The table includes the cumulative probability, $P(\hat{H}) = \text{Prob}\ (H \leq \hat{H})$, that is the probability that the significant wave height $\hat{H}$ is not exceeded by any randomly chosen significant height H. The general expressions for mean and variance are also given for each distribution. Expressions are in terms of the parameters α, ϵ, and θ.

In applying these distributions, data are usually plotted so that, if they follow the selected distribution, they will form a straight line. The linear ordinate scale, y, in such a plot is related to the cumulative probability and the linear abscissa scale, x, is related to $\hat{H}$ according to the relationships given in Figure 54. The slope, a and the intercept, b, for the linear relationship $y = ax + b$ are given in the table in terms of the parameters of each distribution.

OVERALL DISTRIBUTION FUNCTIONS

The Extremal Type III Distribution with lower bound, also called the Weibull Distribution, is useful for fitting the cumulative distribution of significant wave heights. A simplified form of the distribution which can be used for this purpose is given by

$$P(H_s \leq \hat{H}_s) = \begin{cases} 1 - \exp\left[-(\hat{H}_s - H_{s\,min})/\sigma_{hs}\right], & \hat{H}_s \geq H_{s\,min} \\ 0 & , \hat{H}_s < H_{s\,min} \end{cases} \tag{29}$$

where

$\hat{H}_s$ = a particular value of H_s
$H_{s\ min}$ = minimum (background) significant wave height
σ_{Hs} = standard deviation of significant wave heights

In circumstances where only the mean significant wave height, $\bar{H}_s$, is known, the distribution function can be approximated by evaluating $H_{s\ min}$ and σ_{Hs} from

$$H_{s\ min} \approx 0.38\ \bar{H}_s$$

and

$$\sigma_{Hs} = \bar{H}_s - H_{s\ min} = 0.62\ \bar{H}_s \tag{30}$$

The lognormal distribution is also used in this context. Examples of data plotted on Weibull probability paper are given in Figure 55. The same data are plotted in Figure 56 on lognormal probability paper. The trends illustrated are typical. The Weibull distribution tends to fit the moderate and high wave height ranges and the lognormal distribution fits the low and moderate wave height ranges.

EXTREME VALUE DISTRIBUTION FUNCTIONS

Extreme wave height values are a crucial ingredient in most coastal and ocean design. Often the extreme wave heights are limited by the shallow water depth as discussed earlier. For deeper water or low energy sites, extreme values are usually described in terms of significant wave height values as a function of return period. Extreme values of other height statistics such as $H_{1/10}$ can be obtained from the significant height data and a model for the distribution of individual wave heights.

The basic approaches to predicting extreme wave conditions are:

- extrapolation of long-term distribution of significant wave heights
- extreme value analysis with annual maxima
- extreme value analysis with peak significant wave heights of major storms above a certain threshold

The first approach is relatively easy to apply. However, care must be taken in regards to any statistical dependence

Wave Climatology for Atlantic City, New Jersey[a]

Period (sec)	Height (ft)											
	0-1	1-2	2-3	3-4	4-5	5-6	6-7	7-8	8-9	9-10	Total	Acc total
0.0–1.9	2.2										2.2	2.2
2.0–2.4	–											2.2
2.5–2.9	–											2.2
3.0–3.4	–	1.1									1.1	3.3
3.5–3.9	–	1.1	0.7								1.8	5.1
4.0–4.9	–	1.8	2.9	0.7							5.5	10.6
5.0–5.9	–	1.8	2.6	1.5	0.7	0.4					6.9	17.5
6.0–6.9	0.4	3.3	4.0	1.8	1.5	0.7	0.4	0.4			12.4	29.9
7.0–7.9	0.4	5.5	6.6	2.9	1.5	0.7	0.4	0.4			18.2	48.2
8.0–8.9	0.4	8.0	6.2	2.9	1.1	0.4	0.4	0.4	0.4		20.1	68.2
9.0–9.9	0.4	6.9	4.0	1.5	0.7	0.4	0.7	0.4	0.4	0.4	15.7	83.9
10.0–10.9	0.4	4.4	1.8	0.7	0.4	0.4		0.4	0.4		8.8	92.7
11.0–11.9	0.4	1.8	1.5	0.4	0.4	0.4					4.8	97.4
12.0–12.9		1.1	0.7								1.8	99.3
13.0–13.9		0.7									0.7	100.0
Total	4.4	37.6	31.0	12.4	6.2	3.3	1.8	1.8	1.1	0.4		100.0
Acc. total	4.4	42.0	73.0	85.4	91.6	94.9	96.7	98.5	99.6	100.0	100.0	

[a]Total = 12 months, January 1967 to December 1967. Distribution of height (in percent) as a function of period for 2083 observations. Each entry in the table is rounded individually, therefore, the sum and the accumulated total for each row or column may not agree with the figures as shown in the table.

FIGURE 52. Joint distribution table of significant wave height and period [22] (1 ft = 0.305 m).

Distribution (1)	Range (2)	Cumulative probability, $P(H)$ (3)	Mean (4)	Variance (5)
Lognormal	$0 < H < \infty$ $-\infty < \theta < \infty$ $0 < \alpha < \infty$	$(1/\sqrt{2\pi})\int_0^H \frac{1}{\alpha h}\exp\left[-\frac{1}{2}\left(\frac{\ln(h)-\theta}{\alpha}\right)^2\right]dh$	$\exp\left(\theta + \frac{\alpha^2}{2}\right)$	$\exp(2\theta + \alpha^2)[\exp(\alpha^2) - 1]$
Type I	$-\infty < H < \infty$ $-\infty < \epsilon < \infty$ $0 < \theta < \infty$	$\exp\left\{-\exp\left[-\left(\frac{H-\epsilon}{\theta}\right)\right]\right\}$	$\epsilon + \gamma\theta$ $(\simeq \epsilon + 0.58\theta)$	$\frac{\pi^2}{6}\theta^2$ $(\simeq 1.64\theta^2)$
Type II	$0 < H < \infty$ $0 < \theta < \infty$ $0 < \alpha < \infty$	$\exp\left[-\left(\frac{H}{\theta}\right)^{-\alpha}\right]$	$\theta\Gamma\left(1 - \frac{1}{\alpha}\right)$	$\theta^2\left[\Gamma\left(1 - \frac{2}{\alpha}\right) - \Gamma^2\left(1 - \frac{1}{\alpha}\right)\right]$
Type III$_L$ (Lower Bound)	$\epsilon < H < \infty$ $0 < \theta < \infty$ $0 < \alpha < \infty$	$1 - \exp\left[-\left(\frac{H-\epsilon}{\theta}\right)^{\alpha}\right]$	$\epsilon + \theta\Gamma\left(1 - \frac{1}{\alpha}\right)$	$\theta^2\left[\Gamma\left(1 + \frac{2}{\alpha}\right) - \Gamma^2\left(1 + \frac{1}{\alpha}\right)\right]$
Type III$_U$ (Upper Bound)	$-\infty < H < \epsilon$ $0 < \theta < \infty$ $0 < \alpha < \infty$	$\exp\left[-\left(\frac{\epsilon - H}{\theta}\right)^{\alpha}\right]$	$\epsilon - \theta\Gamma\left(1 + \frac{1}{\alpha}\right)$	$\theta^2\left[\Gamma\left(1 + \frac{2}{\alpha}\right) - \Gamma^2\left(1 + \frac{1}{\alpha}\right)\right]$

FIGURE 53. Probability distributions used to describe long-term wave heights [28].

Distribution (1)	Abscissa scale x (2)	Ordinate scale y (3)	Slope a (4)	Intercept b (5)
Lognormal	$\ln(H)$	$P(H) = \frac{1}{\sqrt{2\pi}} \int_0^y e^{-1/2t^2}\, dt$	$1/\alpha$	$-\theta/\alpha$
Type I	H	$-\ln\{-\ln [P(H)]\}$	$1/\theta$	$-\epsilon/\theta$
Type II	$\ln(H)$	$-\ln\{-\ln [P(H)]\}$	α	$-\alpha \ln\theta$
Type III$_L$	$\ln(H - \epsilon)$	$\ln\{-\ln [Q(H)]\}$	α	$-\alpha \ln\theta$
(Lower Bound)	H	$\{-\ln [Q(H)]\}^{1/\alpha}$	$1/\theta$	$-\epsilon/\theta$
Type III$_U$	$-\ln(\epsilon - H)$	$-\ln\{-\ln [P(H)]\}$	α	$\alpha \ln\theta$
(Upper Bound)	H	$-\{-\ln [P(H)]\}^{1/\alpha}$	$1/\theta$	$-\epsilon/\theta$

FIGURE 54. Scale relationships for probability papers [28].

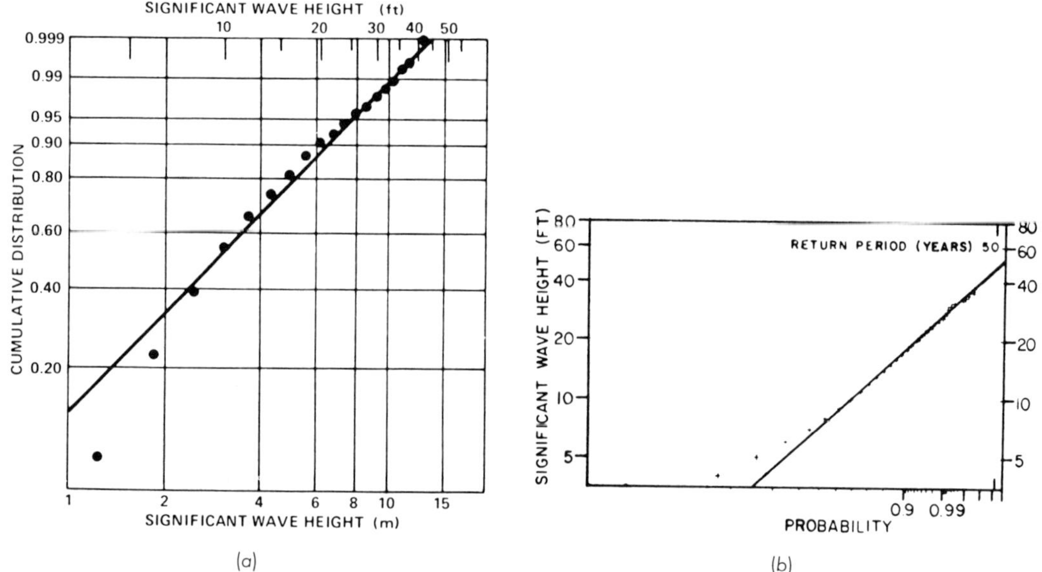

FIGURE 55. Cumulative distribution function of significant wave height plotted on Weibull probability paper: (a) from [38]; (b) from [6].

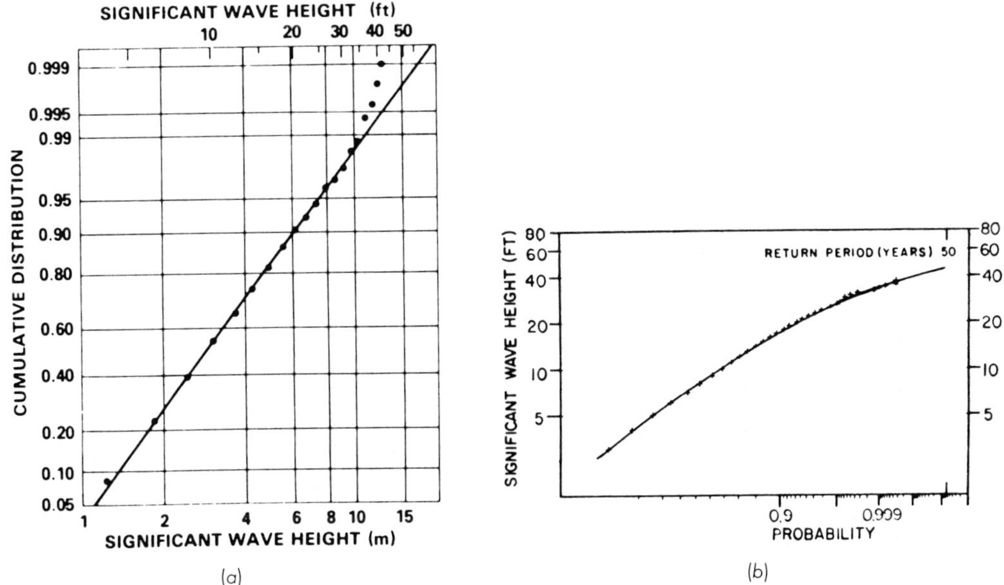

FIGURE 56. Cumulative distribution function of significant wave height plotted on log-normal probability paper for the data sets shown in Figure 55: (a) from [38]; (b) from [6].

Distribution (1)	Estimated Parameters $\hat{\alpha}$ (2)	$\hat{\theta}$ (3)	$\hat{\epsilon}$ (4)
Log-normal	$[\ln(\bar{H}^2) - 2\ln(\bar{H})]^{1/2}$	$2\ln(\bar{H}) - \frac{1}{2}\ln(\bar{H}^2)$	—
Type I	—	$\frac{\sqrt{6}}{\pi}[\bar{H}^2 - (\bar{H})^2]^{1/2}$	$\bar{H} - \gamma\hat{\theta}$
Type II	$\frac{\bar{H}^2}{(\bar{H})^2} = \frac{\Gamma(1-2/\hat{\alpha})}{\Gamma^2(1-1/\hat{\alpha})}$	$\frac{\bar{H}}{\Gamma(1-1/\hat{\alpha})}$	—

FIGURE 57. Parameters of distributions as estimated by method of moments [28].

GRAPH: Frequency of Wave Occurrence

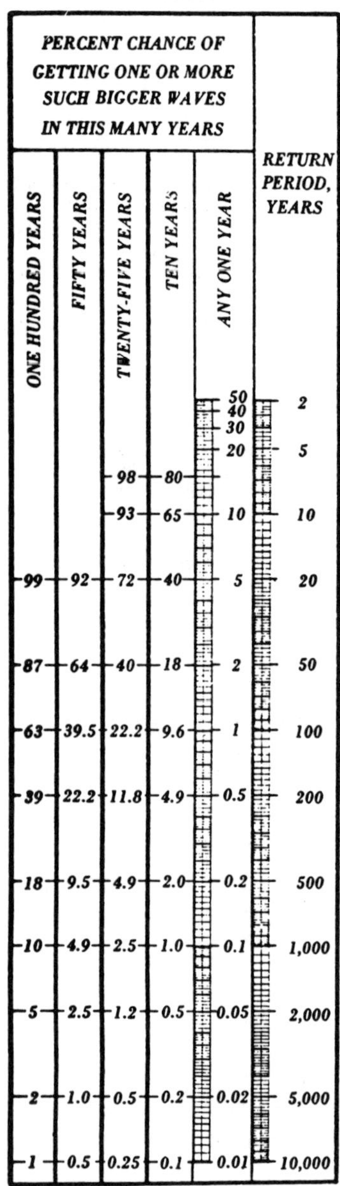

FIGURE 58. Encounter probabilities as a function of return period.

among successive observations. A method for correcting for statistical dependence was given by Nolte [37].

Possible long-term statistical variability must also be considered before observed data can be extrapolated to long return periods. A more sophisticated curve fitting and extrapolation procedure is given by Ochi [38]. The reliability of the data base is a primary concern in extreme value analysis and care should be taken to optimize it.

The steps involved in applying extreme value analysis are as follows [28]:

1. Assign a probability value to each extreme data point. The data are ordered according to wave height. The subscript m denotes the rank, with $m = 1$ for the largest wave height and $m = N$ for the smallest wave height in a sample of N wave heights. The cumulative probability is given as

$$P(H) = 1 - \frac{m}{N + 1} \tag{31}$$

2. Plot these points on an extreme value probability paper represented in Figure 54. Often it is desirable to use more than one probability paper and to select the one which gives the best fit to the data. The lognormal and Type I distributions are most often used.

3. Fit a straight line through the points to represent a trend. Often the fit is done by eye. Alternatively the best fit line may be derived by matching the mean wave height and the mean squared wave height from the data with those from the model distribution. The model parameters derived by this approach are given in Figure 57 in terms of the mean and mean squared wave heights. The Type III distribution is less amenable to this approach because it has three parameters rather than two. It has been omitted from the table.

4. Extrapolate the line to locate a design value corresponding to a chosen return period, T_r, or a chosen encounter probability, P_e. The return period is the average time interval between successive events of the design wave being equalled or exceeded. It is given by

$$T_r = r/[1 - P(\hat{H})] \tag{32}$$

where r is the time interval associated with each data point. The encounter probability, P_e, is the probability that the design wave is equalled or exceeded during a prescribed time period, L. It is given by

$$P_e = 1 - (1 - r/T_r)^{L/r} \tag{33}$$

When return period is determined for a model distribution, encounter probability may be estimated for selected time periods from Figure 58. The figure may be used, for example, to determine the percent chance of occurrence of a significant wave height with 100-year return period in time periods of 1, 10, 25, 50, and 100 years. From the figure the percent chances are 1.0, 9.6, 22.2, 39.5, and 63.0.

ACKNOWLEGEMENTS

The preparation of this chapter was supported by the U.S. Army Corps of Engineers' Coastal Engineering Research Program and Publication Program. The Technical Monitor for both programs is John H. Lockhart, Jr.

REFERENCES

1. Berg, D. W., "Systematic Collection of Beach Data," presented at the September, 1968, ASCE 11th Conference on Coastal Engineering, held at London, England.
2. Borgman, L. E., "Directional Spectra from Wave Sensors," in *Ocean Wave Climate,* Earle, M. D., and Malahoff, A., Eds., Plenum Press, New York, NY, pp. 269–300 (1979).
3. Bouws, E., H. Gunther, W. Rosenthal, and C. L. Vincent, "Similarity of the Wind Wave Spectrum in Finite Depth Water, Part I-Spectral Form," *Journal of Geophysical Research,* Vol. 90, No. C1, pp. 975–986 (1985).
4. Bretschneider, C. L., The Generation and Decay of Wind Waves in Deep Water," *Transactions of the American Geophysical Union,* Vol. 33, pp. 381–389 (1952).
5. Bretschneider, C. L., "Wave Variability and Wave Spectra for Wind-Generated Gravity Waves," *TM-118,* U. S. Army Corps of Engineers, Beach Erosion Board, Washington, D. C., Aug. (1959).
6. Carter, D. J. T. and L. Draper, "Waves at Ocean Weather Station ALPHA," *Report 69,* Institute of Oceanographic Sciences, Wormley, Godalming, England (1979).
7. Corson, W. D., "Data Processing and Management of a 20-Year Wave Climatology for U. S. Coastal Waters," presented at the May 1–3, 1985, MTS Symposium on Ocean Data: Sensor to User, held at New Orleans, La.
8. Dean, R. G., "Evaluation and Development of Water Wave Theories for Engineering Application," *Special Report No. 1,* 2 Vols., U. S. Army Engineer Waterways Experiment Station, Coastal Engineering Research Center, Vicksburg, Miss., Nov. (1974).
9. Dean, R. G., "Summary of Symposium," *Proceedings of a Symposium and Workshop on Wave-Measurement Technology,* Marine Board, U. S. National Research Council (1982).
10. Dean, R. G. and R. A. Dalrymple, *Water Wave Mechanics for Engineers and Scientists,* Prentice-Hall, Inc., Englewood Cliffs, N. J. (1984).
11. Donelan, M. and W. J. Pierson, "The Sampling Variability of Estimates of Spectra of Wind-Generated Gravity Waves," *Journal of Geophysical Research,* Vol. 88, No. C7, pp. 4381–4392, May (1983).
12. Esteva, D. C., "Evaluation of the Computation of Wave Direction with Three-Gage Arrays," *TP 77-7,* U. S. Army Engineer Waterways Experiment Station, Coastal Engineering Research Center, Vicksburg, Miss., July (1977).
13. Esteva, D. C., and D. L. Harris, "Comparison of Pressure and Staff Wave Gage Records," presented at the September 13–18, 1970, ASCE 12th Coastal Engineering Conference, held at Washington, D. C.
14. Forristall, G. Z., "Subsurface Wave-Measuring Systems," *Proceedings of a Symposium and Workshop on Wave-Measurement Technology,* Marine Board, U. S. National Research Council (1982).
15. Goda, Y., "Numerical Experiments on Wave Statistics with Spectral Simulation," *Report of the Port and Harbour Research Institute,* Vol. 9, No. 3, Japan, pp. 3–57 (1970).
16. Goda, Y., "A Review on Statistical Interpretation of Wave Data," *Report of the Port and Harbour Research Institute,* Vol. 18, No. 1, Japan, pp. 5–32, Mar. (1979).
17. Goda, Y., *Random Seas and Design of Maritime Structures,* University of Tokyo Press, Tokyo, Japan (1985).
18. Grace, R. A., "Surface Wave Heights from Pressure Records," *Coastal Engineering,* Vol. 2, No. 1, pp. 55–67 (1978).
19. Guza, R. T. and E. B. Thornton, "Local and Shoaled Comparisons of Sea Surface Elevations, Pressures, and Velocities," *Journal of Geophysical Research,* Vol. 85, No. C3, pp. 1524–1530, Mar. (1980).
20. Harris, D. L., "The Analysis of Wave Records," presented at the September 13–18, 1970, ASCE 12th Coastal Engineering Conference, held at Washington, D. C.
21. Harris, D. L., "Characteristics of Wave Records in the Coastal Zone," in *Waves on Beaches and Resulting Sediment Transport,* Academic Press, Inc., New York, N. Y., pp. 1–51 (1972).
22. Harris, D. L., "Wave Estimates for Coastal Regions," in *Shelf Sediment Transport,* Swift, Duane, and Pilkey, Eds., Dowden, Hutchinson and Ross, Inc., Stroudsburg, PA, pp. 99–125 (1972).
23. Hasslemann, D. E., M. Dunckel, and J. A. Ewing, "Directional Wave Spectra Observed During JONSWAP 1973," *Journal of Physical Oceanography,* Vol. 10, No. 8, pp. 1264–1280, Aug. (1980).
24. Hasselmann, K., T. P. Barnett, E. Bouws, H. Carlson, D. E. Cartwright, K. Enke, J. A. Ewing, H. Gienapp, D. E. Hasselmann, P. Kruseman, A. Meerburg, P. Muller, D. J. Olbers, K. Richter, W. Sell, and H. Walden, "Measurements of Windwave Growth and Swell Decay During the Joint North Sea Wave Project (JONSWAP)," *Deutsch Hydrogr. Z.,* Vol. A8 (suppl.), pp. 1–95 (1973).
25. Hasselmann, K., D. B. Ross, P. Muller, and W. Sell, A Parametric Wave Prediction Model," *Journal of Physical Oceanography,* Vol. 6, No. 2, pp. 200–228, Mar. (1976).
26. Hudspeth, R. T. and M. C. Chen, "Digital Simulation of Nonlinear Random Waves," *Journal of the Waterway, Port, Coastal and Ocean Division,* ASCE, Vol. 105, No. WW1, Proc. Paper 14376, pp. 67–85, Feb. (1979).
27. Hughes, S. A., "Summary and Applications of the TMA Shallow Water Spectrum," *Technical Report TR 84-7,* U. S. Army Engineer Waterways Experiment Station, Coastal Engineering Research Center, Vicksburg, Miss., Dec. (1984).
28. Isaacson, M. de St. Q. and N. G. MacKenzie, "Long-Term Distributions of Ocean Waves: a Review," *Journal of the Water-*

way, Port, Coastal and Ocean Division, ASCE, Vol. 107, No. WW2, Proc. Paper 16277, pp. 93–109, May (1981).

29. Jahns, H. and J. Wheeler, Long-Term Wave Probabilities Based on Hindcasting of Severe Storms," *Journal of Petroleum Technology,* pp. 473–486, Apr. (1973).
30. Johnson, R. R., E. P. D. Mansard, and J. Ploeg, "Effects of Wave Grouping on Breakwater Stability," presented at the August 27–September 3, 1978, ASCE 16th Conference on Coastal Engineering, held at Hamburg, West Germany.
31. Kitaigorodskii, S. A., V. P. Krasitskii, and M. M. Zaslavskii, "On Phillip's Theory of Equilibrium Range in the Spectra of Wind-Generated Gravity Waves," *Journal of Physical Oceanography,* Vol. 5, pp. 410–420 (1975).
32. Le Mehaute, B., "An Introduction to Hydrodynamics and Water Waves," *Water Wave Theories,* Vol. II, TR ERL 118-POL-3-2, U. S. Department of Commerce, ESSA, Washington, D. C. (1969).
33. Lipa, B. J., and D. E. Barrick, "CODAR Measurements of Ocean Surface Parameters at ARSLOE—Preliminary Results," presented at the September 20–22, 1982, IEEE Oceans '82 Conference, held at Washington, D. C.
34. Longuet-Higgins, M. S., D. E. Cartwright, and N. D. Smith, "Observations of the Directional Spectrum of Sea Waves Using the Motions of a Floating Buoy," *Ocean Wave Spectra,* Prentice-Hall, Inc., Englewood Cliffs, N. J., pp. 111–136 (1963).
35. Mattie, M. G. and D. L. Harris, "The Use of Imaging Radar in Studying Ocean Waves," presented at the August 28–September 3, 1978, ASCE 16th Coastal Engineering Conference, held at Hamburg, West Germany.
36. Mitsuyasu, H., "Observations of the Directional Spectrum of Ocean Waves Using a Cloverleaf Buoy," *Journal of Physical Oceanography,* Vol. 5, No. 4, pp. 750–760, Oct. (1975).
37. Nolte, K. G., Statistical Methods for Determining Extreme Sea States," presented at the August 27–30, 1973, 2nd International Conference on Port and Ocean Engineering Under Arctic Conditions, held at Reykjavik, Iceland.
38. Ochi, M. K., "Stochastic Analysis and Probabilistic Prediction of Random Seas," in *Advances in Hydroscience,* Chow, V. T., Ed., Vol. 13, Academic Press, New York, N. Y., pp. 218–376 (1982).
39. Ochi, M. K., S. B. Malakar, and W-C. Wang, "Statistical Analysis of Coastal Waves Observed During the ARSLOE Project," *Report UFL/COEL/TR-045,* University of Florida, Gainesville, Florida, Oct. (1982).
40. Parker, A. G., "Wave Measurements Using Surface-Mounted Instruments," *Proceedings of a Symposium and Workshop on Wave-Measurement Technology,* Marine Board, U. S. National Research Council (1982).
41. Resio, D. T. and C. L. Vincent, Estimation of Winds Over the Great Lakes, *MP H-76-12,* U.S. Army Engineering Waterways Experiment Station, Vicksburg, Miss. June (1976).
42. Sarpkaya, T., and M. Isaacson, *Mechanics of Wave Forces on Offshore Structures,* Van Nostrand Reinhold Company, New York, N. Y. (1981).
43. Schneider, C. and J. R. Weggel, "Visually Observed Wave Data at Pt. Mugu, Calif.," presented at the March 23–28, 1980, ASCE 17th International Coastal Engineering Conference, held at Sydney, Australia.
44. Seymour, R. J., "Review of Wave Measurement Using In Situ Systems," *Proceedings of a Symposium and Workshop on Wave-Measurement Technology,* Marine Board, U. S. National Research Council (1982).
45. *Shore Protection Manual,* 4th ed., 2 vols., 1984. U. S. Army Engineer Waterways Experiment Station, Coastal Engineering Research Center, U. S. Government Printing Office, Washington, D. C.
46. Simiu, E. and R. N. Scanlan, *Wind Effect on Structures: An Introduction to Wind Engineering,* Wiley, New York, N. Y., p. 62 (1978).
47. Thompson, E. F., "Wave Climate at Selected Locations Along U. S. Coasts," *TR 77-1,* U. S. Army Engineer Waterways Experiment Station, Coastal Engineering Research Center, Vicksburg, Miss., Jan. (1977).
48. Thompson, E. F., "Energy Spectra in Shallow U. S. Coastal Waters," *TP 80-2,* U. S. Army Engineer Waterways Experiment Station, Coastal Engineering Research Center, Vicksburg, Miss., Feb. (1980).
49. Thompson, E. F., "Shallow Water Surface Wave Elevation Distributions," *Journal of the Waterway, Port, Coastal and Ocean Division,* Vol. 106, No. WW2, pp. 285–289, May (1980).
50. Thompson, E. F. and C. L. Vincent, "Significant Wave Height for Shallow Water Design," *Journal of the Waterway, Port, Coastal and Ocean Division,* Vol. 111, No. 5, Sept. (1985).
51. Thornton, E. B. and R. T. Guza, "Transformation of Wave Height Distribution," *Journal of Geophysical Research,* Vol. 88, No. C10, pp. 5925–5938, July (1983).
52. U. S. Naval Weather Service Command, *Summary of Synoptic Meteorological Observations,* prepared by the National Climatic Data Center, Asheville, N. C. (1976).
53. Weigel, R. L., *Oceanographical Engineering,* Prentice-Hall, Englewood Cliffs, N. J. (1964).

APPENDIX I

Procedure for Analysis of Wave Data from 7-Minute Pen and Ink Records (Based on a Rayleigh Distribution for Wave Height [47].

1. Run the period template (Figure I-1) along the 7-minute record until a group of fairly uniform waves is found which should contain some of the highest waves. A template can be fabricated on a clear overlay such as acetate.

2. Determine the appropriate period of the waves selected in step 1 by using the template according to instructions. When the wave period on the chart falls between two of the periods shown on the template, the analyzer may approximate what is considered to be nearest to the exact period; e.g., if the period is about the same amount longer than the

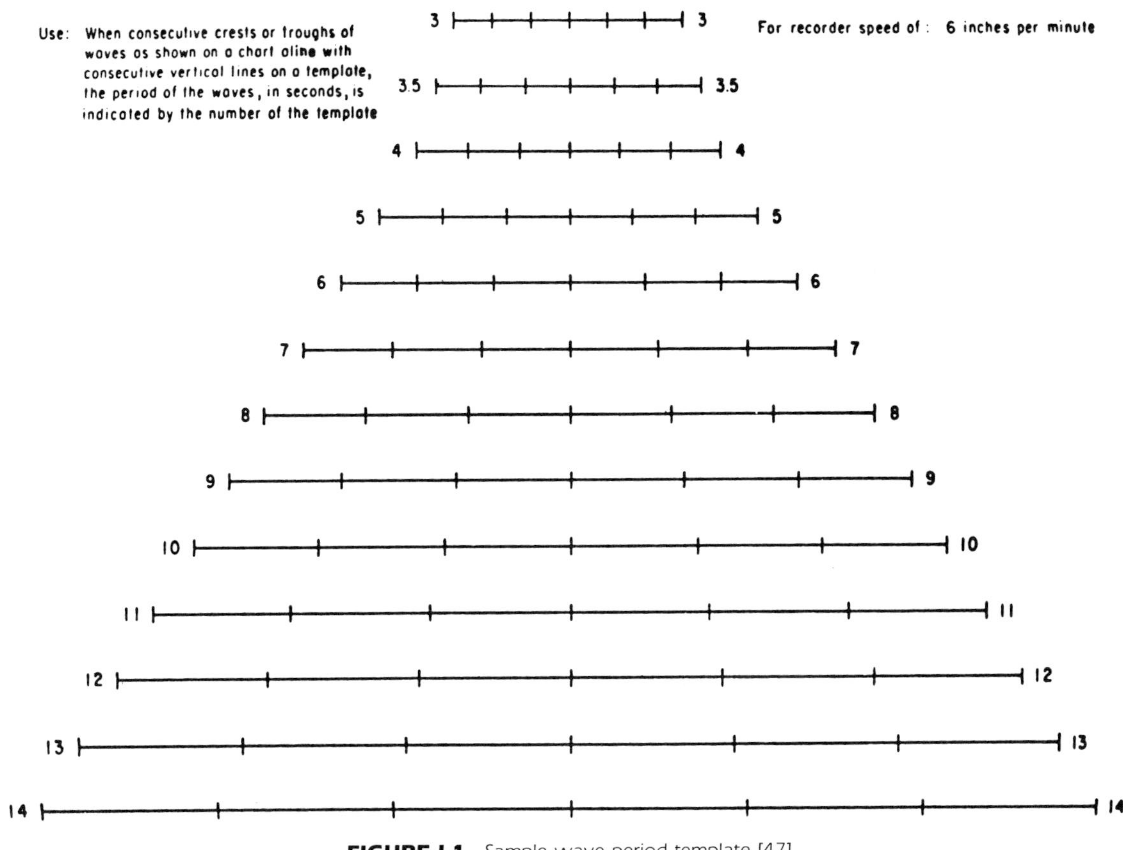

FIGURE I-1. Sample wave period template [47].

TABLE I-1. Number of Wave to Measure for Manual Analysis of 7-Minute Pen and Ink Records.

Wave period (seconds)	Number of wave to measure
3.0	19
3.5	16
4.0	14
4.5	13
5.0	11
5.5	10
6.0	9
7.0	8
8.0	7
9.0	6
10.0	6
11.0	5
12.0	5
13.0	4
14.0	4
15.0	4
16.0	4

5-second period, than it is shorter than the 6-second period, the period must be about 5.5 seconds.

3. Use Table I-1 to determine which wave should be measured in the full 7-minute record to get the approximate significant height of the waves. The wave number is determined by calling the highest wave in the full 7-minute record as wave number 1; the second highest wave is number 2, etc. Wave height is defined as the difference in elevation between a wave crest and the preceding trough.

4. Determine the height of the wave given by step 3, in terms of small divisions on the chart paper.

5. Using the appropriate relationship between chart paper divisions and actual elevations in feet or centimeters, convert the wave height determined in step 4 from chart divisions to feet or centimeters. Estimate to the nearest tenth of a foot or nearest centimeter.

CHAPTER 20

Waves and Wave Forecasting

C. T. Bishop* and M. A. Donelan*

INTRODUCTION

This section is devoted to one type of surface water waves, usually known as gravity waves or wind generated waves. Waves can be defined as undulations of the water surface. The subject of water waves covers phenomena from capillary waves, which are very small and have very short periods (of the order of 0.05 seconds), to long period waves, such as tides, tsunamis and others which can reach magnitudes of the order of 10 m (in coastal waters) and have long periods (of the order of minutes or hours). Gravity waves produced by the wind have typical periods of 0.3 to 30 seconds and wave heights mostly of the order of 1 metre, seldom exceeding 20 m. These wind waves account for a major part of the total surface wave energy and are usually the most important in coastal engineering and structural design.

Sea and Swell

The wind is the primary generating force for gravity waves. Other important factors include the fetch (open water distance over which the wind blows), the duration of the storm, and the depth of water. Wind waves can be separated into sea and swell. Waves are known as seas within the local wind generating area. When these waves travel out of the local wind generating area, so that they are no longer subject to significant wind input, they are known as swell. Seas contain waves of many different heights, periods and directions and can appear to be very complex. As waves travel out of the generating area the components of the sea disperse and attenuate at different rates. Short waves decay more quickly so that swell has a more uniform and long-crested nature than does sea. At some distance from the generating area the dispersion (long waves travelling faster than short waves) separates the components. Accordingly, a time history of swell approaching from a distant storm reveals a narrow spectrum (most waves having nearly the same frequency) in which the mean frequency slowly increases because the slower short (higher frequency) waves arrive later. At the same time, the energy of all the waves is dissipated to some extent with the result that swell has a more orderly pattern than sea, with longer wave periods, smaller wave heights and a more uniform direction.

*Research and Applications Branch, National Water Research Institute, Canada Centre for Inland Waters, Burlington, Ontario, Canada

WAVE THEORY

This section presents the simplest of wave theories, assuming monochromatic waves. This theory provides the building blocks from which actual random waves and wave processes can be analyzed.

Definitions

Figure 1 shows the basic description of a simple sinusoidal progressive wave consisting of its wavelength L (the horizontal distance between corresponding points on two successive waves), height H (the vertical distance between a crest and the preceding trough), period T (the time for two successive crests to pass a given point), and water depth d (the distance from the bed to the stillwater level SWL). If the water surface η varies sinusoidally it can be defined as:

$$\eta = a \cos \theta \tag{1}$$

where

a = wave amplitude = $H/2$ for small-amplitude theory

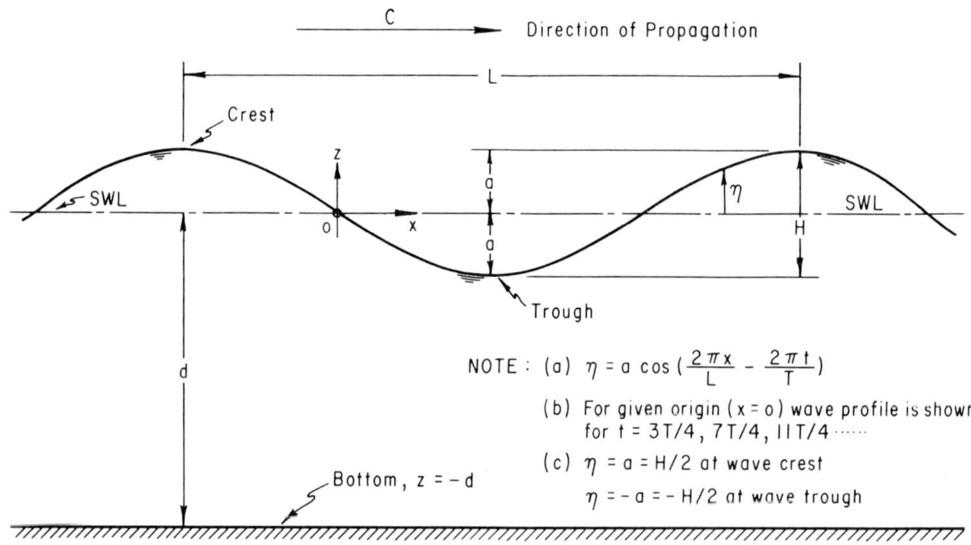

FIGURE 1. Definition sketch of simple, sinusoidal, progressive wave (*Shore Protection Manual* 1984).

$\theta = 2\pi\left(\frac{x}{L} - \frac{t}{T}\right)$ = phase angle
x = horizontal coordinate of the water surface
t = time

A wave is periodic if its motion and surface profile recur in equal intervals of time. A waveform which moves relative to a fixed point is called a progressive wave; the direction in which it moves is termed the direction of wave propagation. Water waves are considered oscillatory if the water particle motion is described by orbits that are closed for each wave period. Small-amplitude wave theory (also known as linear theory, Airy theory, first order theory) developed by Airy in 1845 describes pure oscillatory, periodic, progressive waves. In many straightforward civil engineering designs, the small-amplitude theory is adequate.

Most finite-amplitude wave theories describe nearly-oscillatory waves because the fluid is moved a small amount in the direction of wave propagation by each successive wave. This motion yields mass transport by the waves. The main finite-amplitude theories are Stokes, Stream Function, Cnoidal and Solitary. For larger wave heights (finite-amplitude waves) in constant depth the Stokes theory (Stokes 1847) or Stream Function theory (Dean 1974) can be used. In shallower and shoaling water Cnoidal wave theory can be used (Korteweg and deVries 1895). In very shallow water, as waves are about to break, Solitary wave theory (McCowan 1891) can be used. The ranges of applicability of each theory are summarized in Figure 2. The expressions given in this section are applicable to small amplitude waves (no mass transport).

Wave Celerity

The velocity of the waveform in relation to the body of fluid through which the wave propagates is termed the wave celerity C. The celerity is such that the phase function θ of Equation (1) remains constant and thus is related to wavelength and period by:

$$C = \frac{L}{T} \tag{2}$$

The small-amplitude wave theory expression for celerity versus wavelength is:

$$C = \left\{\frac{gL}{2\pi}\tanh\frac{2\pi d}{L}\right\}^{1/2} \tag{3}$$

In deep water, $d/L \geq 0.5$, the value of tanh $2\pi d/L$ approaches 1.0 (Figure 3) and the previous equation reduces to

$$C_0 = \left\{\frac{gL_0}{2\pi}\right\}^{1/2} = \frac{L_0}{T} \tag{4}$$

where the subscript "0" indicates deep water conditions. The period T remains essentially constant and independent of depth for oscillatory waves. In deep water, all the wave characteristics are virtually independent of water depth. In shallow water, $d/L \leq 0.04$, Equation (3) reduces to:

$$C = \sqrt{gd} \tag{5}$$

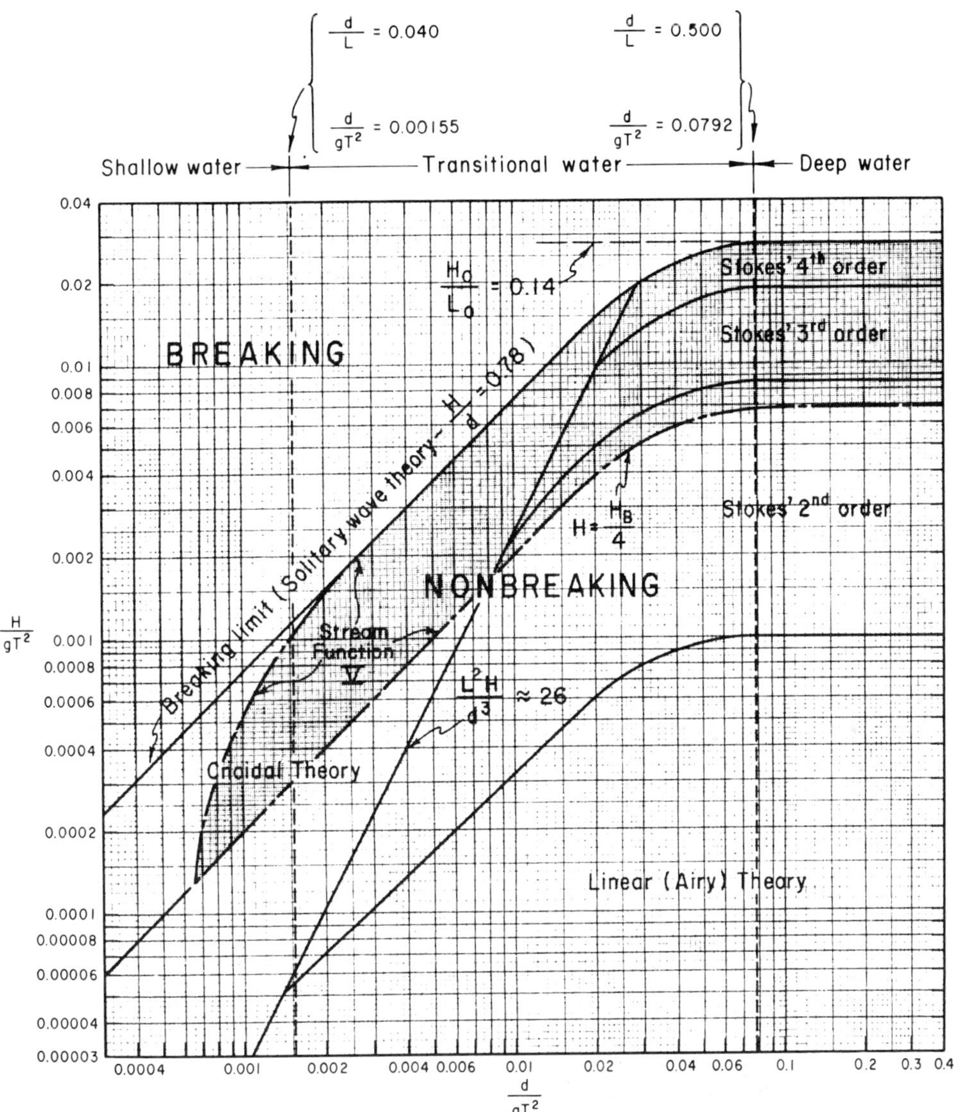

FIGURE 2. Regions of validity for various wave theories (after Le Mehaute 1969, *Shore Protection Manual* 1984).

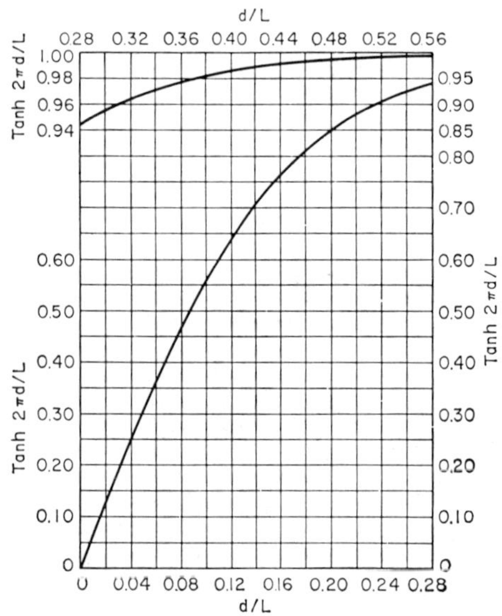

FIGURE 3. Tanh $2\pi d/L$ versus d/L (Brater and King 1976).

The water depth is considered to be transitional when $0.04 < d/L < 0.5$.

Wavelength

From Equations (2) and (3), an expression for wavelength may be obtained.

$$L = \frac{gT^2}{2\pi} \tanh \frac{2\pi d}{L} = L_0 \tanh \frac{2\pi d}{L} \tag{6}$$

Equation (6) can be solved using Figure 4 or can be rearranged to give:

$$\frac{d}{L_0} = \frac{d}{L} \tanh \frac{2\pi d}{L} \tag{7}$$

This equation can be solved most easily using published tables (Shore Protection Manual 1984; Wiegel 1964) of d/L and d/L_0, or with the help of Figures 3 and 5.

An approximate expression for Equation (6) given by Eckart (1952) (also Shore Protection Manual 1984) is:

$$L \simeq \frac{gT^2}{2\pi} \left(\tanh \left(\frac{4\pi^2}{T^2} \frac{d}{g} \right) \right)^{1/2} \tag{8}$$

The maximum error from using Equation (8) is 5 percent, occurring when $2\pi d/L \simeq 1$.

In deep water, Equation (6) reduces to:

$$L_0 = \frac{gT^2}{2\pi} \tag{9}$$

Orbital Motion

The water particles in a wave move in orbital paths as shown in Figure 6. The horizontal component u and the

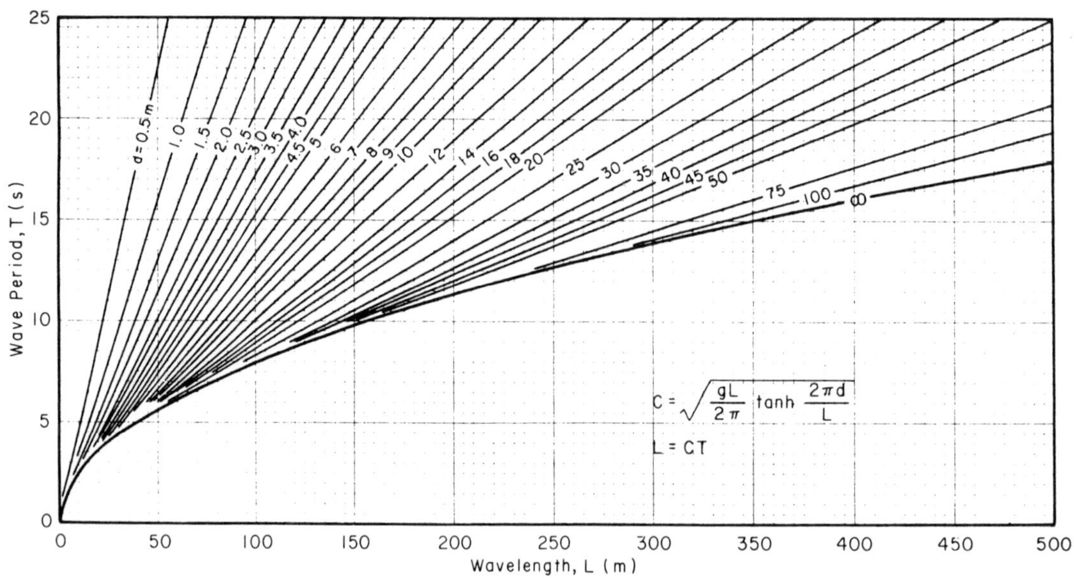

FIGURE 4. Relationship between wavelength, wave period and depth (*Shore Protection Manual* 1984).

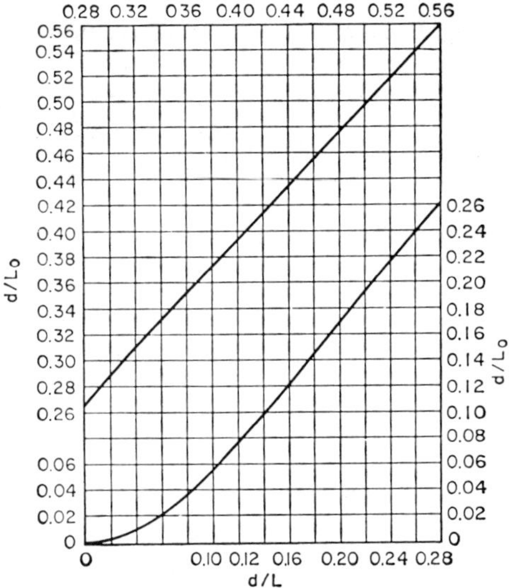

FIGURE 5. d/L versus d/L_0 (Brater and King 1976).

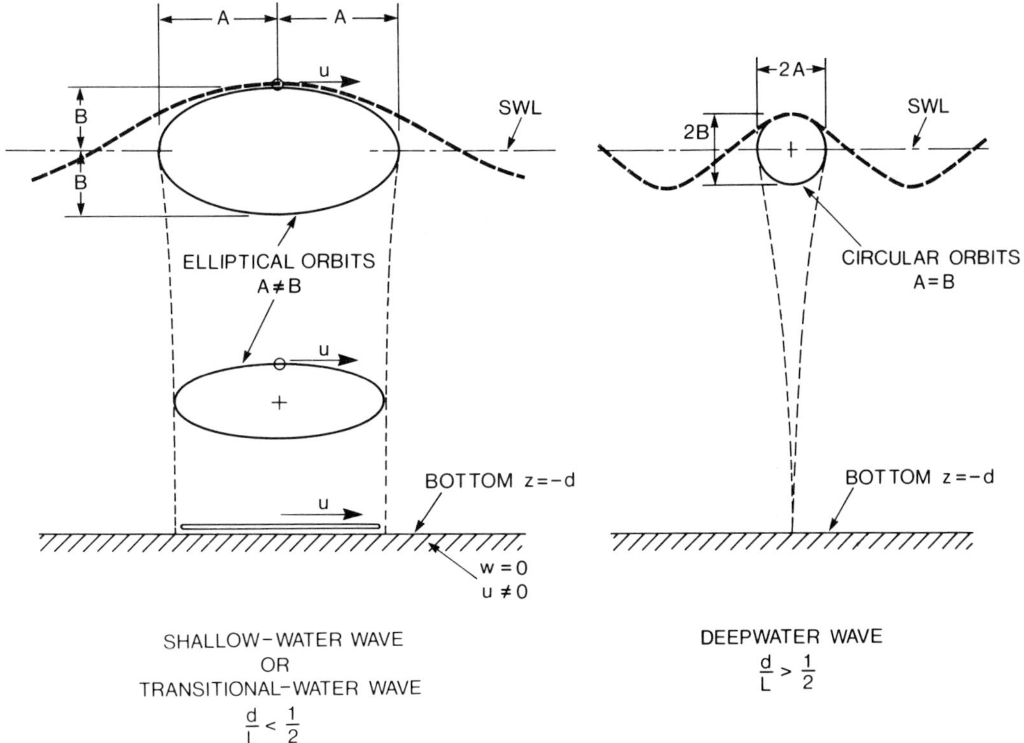

FIGURE 6. Representation of water particle motions (*Shore Protection Manual* 1984).

vertical component w of the local fluid velocity are given by:

$$u = \frac{\pi H}{T} \frac{\cosh k(z + d)}{\sinh kd} \cos \theta \tag{10}$$

and

$$w = \frac{\pi H}{T} \frac{\sinh k(z + d)}{\sinh kd} \sin \theta \tag{11}$$

where

$k = 2\pi/L$ = wavenumber
z = vertical distance measured positive upward from the mean water level

The corresponding expressions for orbital acceleration are:

$$\frac{\partial u}{\partial t} = \frac{2\pi^2 H}{T^2} \frac{\cosh k(z + d)}{\sinh kd} \sin \theta \tag{12}$$

and

$$\frac{\partial w}{\partial t} = -\frac{2\pi^2 H}{T^2} \frac{\sinh k(z + d)}{\sinh kd} \cos \theta \tag{13}$$

Water particle displacements can be characterized by A and B where

$$A = \frac{H}{2} \frac{\cosh k(z + d)}{\sinh kd} \tag{14}$$

and

$$B = \frac{H}{2} \frac{\sinh k(z + d)}{\sinh kd} \tag{15}$$

A is the major (horizontal) semiaxis of an ellipse with the minor (vertical) semiaxis equal to B. The lengths A and B are measures of the horizontal and vertical displacements of the water particles.

In deep water:

$$A_0 = B_0 = \frac{H}{2} e^{kz} \tag{16}$$

and the orbits are circular.

In shallow and transitional water the orbits are elliptical. The shallower the water, the flatter the ellipse becomes. At the bottom the water motion is entirely horizontal.

Subsurface Pressure

Subsurface pressure beneath a wave consists of a static component, ρgz, due to the hydrostatic pressure, and a dynamic component due to fluid acceleration, giving

$$p = \rho g \frac{H}{2} \frac{\cosh k(d + z)}{\cosh kd} \cos \theta - \rho gz \tag{17}$$

where p = gauge pressure = total pressure − atmospheric pressure.

The expression

$$\frac{\cosh k(d + z)}{\cosh kd} \tag{18}$$

is normally known as the pressure response factor, K_p. The dynamic pressure fluctuations are attenuated with depth as a function of wave period. The water column tends to filter out the higher frequency components of the pressure fluctuations. Equation (17) can be rearranged to give:

$$\eta = \frac{1}{K_p} \left(\frac{p}{\rho g} + z \right) \tag{19}$$

Group Velocity

The speed of a group of waves, travelling with a group velocity C_g, is generally not the same as the speed of individual waves, travelling at a wave celerity or phase velocity C. The concept of group velocity can be described by considering the interaction of two sinusoidal wave trains moving in the same direction with slightly different wavelengths and periods. Assume, for simplicity, that the heights of both components are equal. Since the wavelengths of the two component waves are assumed to be slightly different, for some values of x at a given time, the two components will be in phase and the observed wave height will be $2H$. For some other values of x, the two waves will be completely out of phase and the resultant wave height will be zero. The waves shown in Figure 7 appear to be travelling in groups.

It is the velocity of these groups that represents the group velocity, given by

$$C_g = nC \tag{20}$$

where

$$n = \frac{1}{2} \left[1 + \frac{4\pi d/L}{\sinh(4\pi d/L)} \right] \tag{21}$$

Values of n can be found in tables or from Figure 8. In deep water the group velocity is one-half the wave celerity.

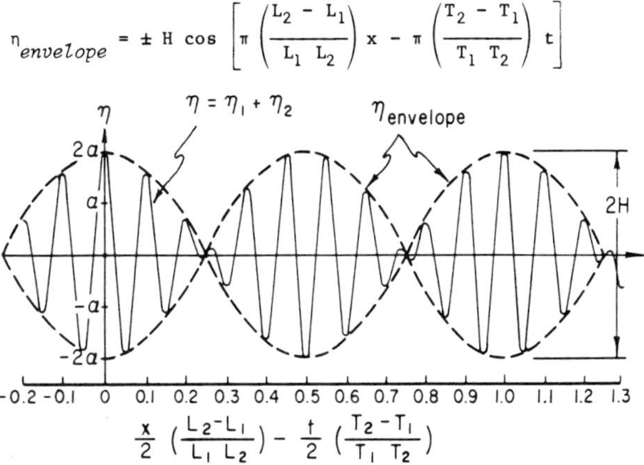

FIGURE 7. Formation of wave groups by the addition of two sinusoids having different periods (after Kinsman 1965, *Shore Protection Manual* 1984).

In shallow water, the group velocity equals the wave celerity. Wave energy is propagated at the group velocity.

Wave Energy

The total energy of a wave system is the sum of its kinetic energy and its potential energy. According to small amplitude wave theory, if the potential energy is calculated relative to the still water level and all waves are propagating in the same direction, potential and kinetic energy components are equal. Then the total wave energy in one wavelength per unit crest width is

$$E_T = E_k + E_p = \frac{\rho g H^2 L}{16} + \frac{\rho g H^2 L}{16} = \frac{\rho g H^2 L}{8} \tag{22}$$

Total average wave energy per unit surface area, termed the specific energy or energy density is

$$E = \frac{\rho g H^2}{8} \tag{23}$$

Wave energy flux per unit crest width is the rate at which energy is transmitted in the direction of wave propagation and is vectorial quantity (denoted by a ~) given by

$$\underset{\sim}{P} = E\underset{\sim}{C}_g \tag{24}$$

Random Waves

Wind waves are not monochromatic sinusoidal waves. However, the concept of a random sea made up of an infinite collection of infinitesimal sinusoids in random phase leads to a convenient spectral description that is amenable to engineering applications. Since the pioneering work of Pierson and Marks (1952), spectral analysis of wave records has found more and more frequent application in oceanography and engineering.

Consider five wave trains of different heights and frequencies (inverse periods) summed together in random phase (Figure 9). A section of the resulting wave train is shown at the bottom of the figure. The (discrete) spectrum of variance with frequency for these five wave trains is represented by Figure 10. Natural waves are represented by a continuous spectrum, i.e., an infinite set of components each having infinitesimal energy (Figure 11).

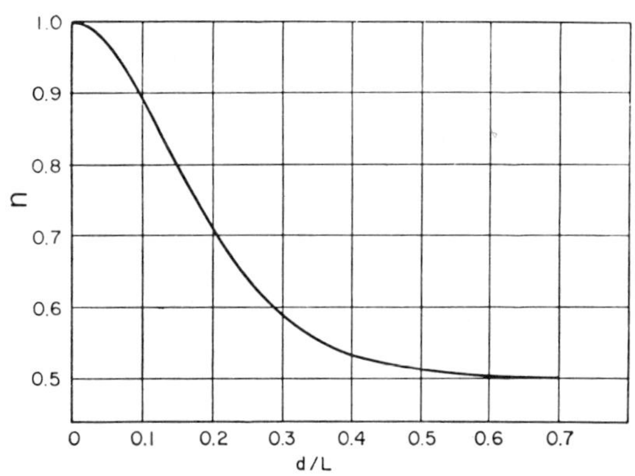

FIGURE 8. The group velocity to phase velocity ratio n versus d/L (Brater and King 1976).

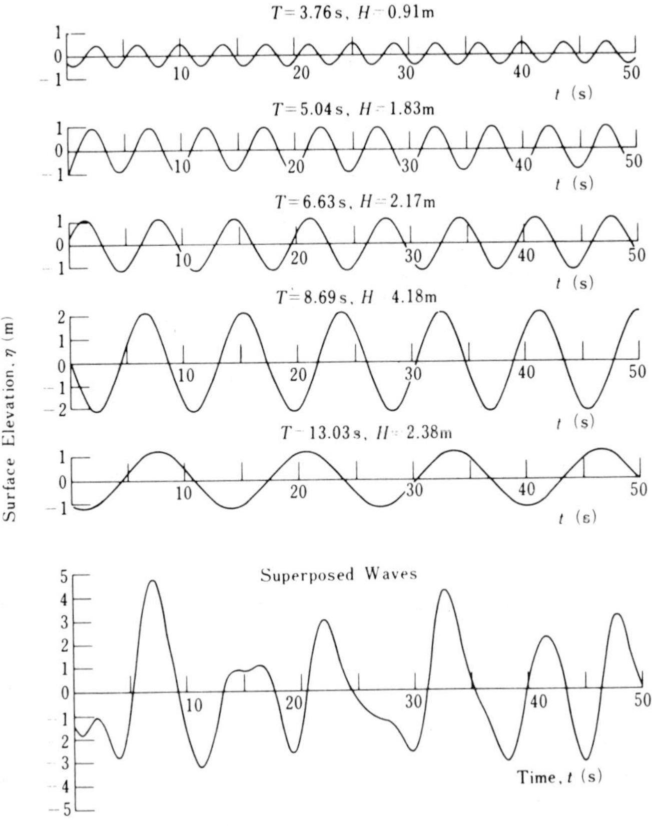

FIGURE 9. Simulation of irregular waves by superposition of five trains of sinusoidal waves (Goda 1985).

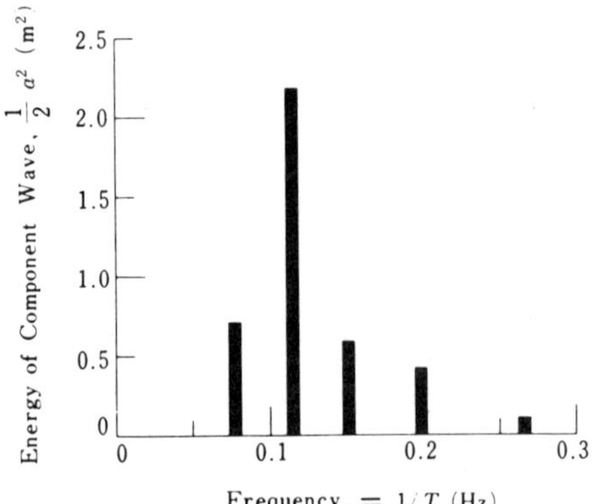

FIGURE 10. Spectral representation of waves of Figure 9 where a = H/2 (Goda 1985).

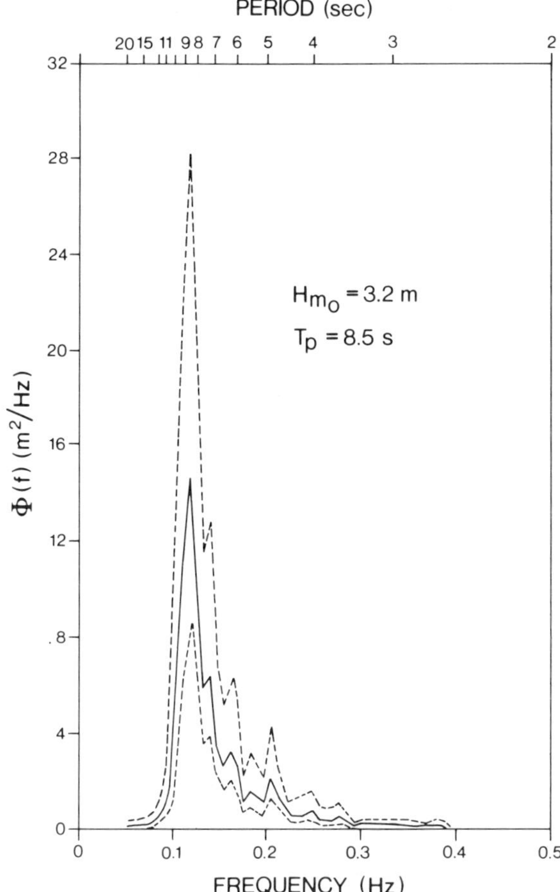

FIGURE 11. Spectrum measured by an accelerometer buoy (ARSLOE Station 910) off Duck, North Carolina, October, 1980 (dashed curves indicate 90% confidence limits).

Analysis of random seas yields a wave variance spectrum in which the wave variance spectral density $\phi(f)$ is expressed as a function of frequency f. Wave energy density is related to the variance spectrum as:

$$E = \rho g \int_0^\infty \Phi(f)df = \rho g \sigma^2 = \frac{\rho g}{16} H_{m_0}^2 \qquad (25)$$

where the area under the variance spectrum is σ^2 and a characteristic wave height is defined as:

$$H_{m_0} = 4\sigma \qquad (26)$$

It is interesting to note that the energy density in a random sea with $H_{m_0} = H$ is only one-half of the energy density of a monochromatic wave train with wave height H.

WAVE PROCESSES

Wave Refraction and Shoaling

Equation (3) shows that for $d/L < 0.5$ the wave celerity depends on the water depth. As the wave celerity decreases with depth, so does the wavelength. When waves approach the shore in such a manner that the angle between the wave crests and the bottom contours is non-zero, the depth will vary along each wave crest. The parts of the wave in shallower water will move forward more slowly than those parts in deeper water. This variation causes the wave crest to bend toward alignment with the contours. This process is known as refraction. In addition to refraction caused by variations in bathymetry, waves may be refracted by currents or any other phenomenon causing one part of a wave to travel slower or faster than another part. Refraction is an important factor in determining local wave height and direction (see, for example, Pierson 1972).

The primary assumption in refraction analysis is that wave energy flux between orthogonals remains constant; orthogonals are lines drawn perpendicular to the wave crests and extend in the direction of wave advance. In deep water the rate at which wave energy is transmitted forward across a plane between two adjacent orthogonals is

$$B_0 = \frac{1}{2} b_0 E_0 C_0 \qquad (27)$$

where b_0 is the distance between the selected orthogonals in deep water. Since the energy flux between orthogonals is taken to be constant, this energy flux may be equated to the rate at which the energy is transmitted forward between the same two orthogonals in shallow water

$$B = nbEC = B_0 \qquad (28)$$

where b is the spacing between the orthogonals in the shallower water. From Equations (23), (27), and (28), one obtains

$$\frac{H}{H_0} = \left(\frac{E}{E_0}\right)^{1/2} = \left(\frac{C_0}{2nC}\right)^{1/2} \left(\frac{b_0}{b}\right)^{1/2} \qquad (29)$$

The term $\sqrt{C_0/2nC}$ is known as the shoaling coefficient K_s. It can be found in tables of d/L_0 or d/L (Shore Protection Manual 1984) or from Figure 12. It may be seen that, neglecting refraction, the wave height tends to decrease in shoaling water until near the breaking point, when it increases again.

The term $\sqrt{b_0/b}$ is known as the refraction coefficient K_r. There is no general formula for K_r because it depends on the bathymetry and wave approach direction. Several graphical procedures for refraction analysis are available (Shore Protection Manual 1984).

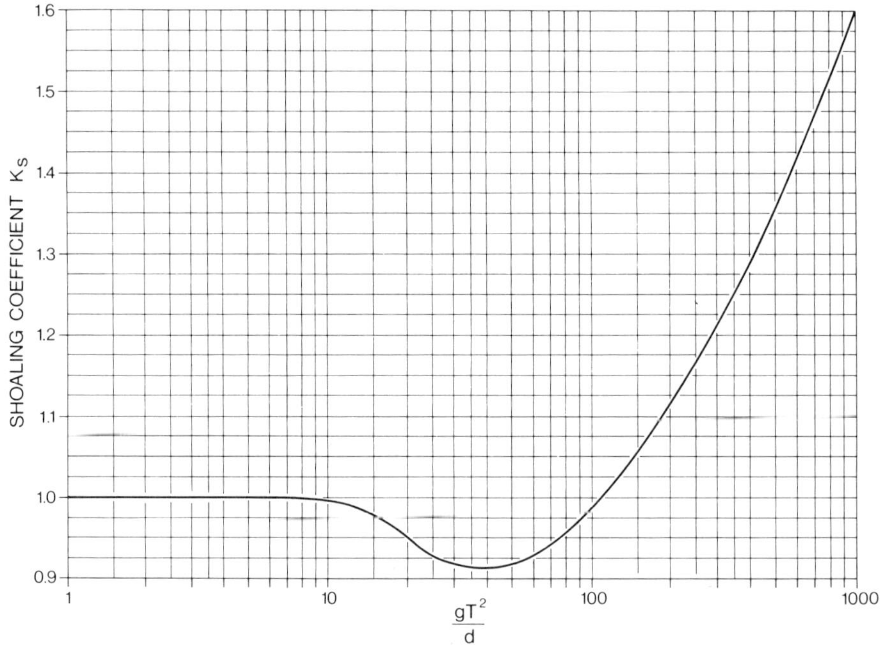

FIGURE 12. Shoaling coefficient versus gT^2/d.

The change of direction of an orthogonal as it passes over relatively simple bathymetry that can be approximated by parallel contours may be computed from Snell's law:

$$\sin \alpha_2 = \frac{C_2}{C_1} \sin \alpha_1 \tag{30}$$

where

α_1 = the angle a wave crest makes with the bottom contour over which the wave is passing

α_2 = a similar angle measured as the wave crest passes over the next selected bottom contour

C_1 = the wave celerity at the depth of the first contour

C_2 = the wave celerity at the depth of the second contour

For straight shorelines with parallel offshore contours, refraction and shoaling may be computed analytically or with the help of Figure 13. For more complex topography many computer methods are available to perform refraction analyses. These numerical methods can be very useful in determining wave refraction diagrams over a large area or when several different wave periods are of interest. Details for regular waves can be found in many references such as Wilson (1966) and Dobson (1967). In addition the Shore Protection Manual (1984) gives a detailed manual graphical procedure. A typical refraction diagram for monochromatic waves from one direction is shown in Figure 14.

The above discussion is applicable to monochromatic waves. Such conditions are approximately met by swell generated at great distance from an occanic coast. In general, random (broad-band) wind seas must be treated differently. Two approaches are available: (1) detailed calculations of wave propagation (Karlson 1969) including generation of and interaction with shore-normal and longshore currents (Birkemeier and Dalrymple 1975, Tayfun *et al.* 1976); (2) separating the spectrum into several components, refracting each separately and recombining to yield the transformed spectrum (Goda 1985). The first method involves the use of time-stepping finite difference numerical models. The second method involves a linear decomposition of the spectrum into several discrete frequency bands. The choice of how many frequency bands depends on the desired accuracy of the calculation—normally seven to ten are adequate if the bands are chosen to have equal energy (spectral variance) rather than equal spectral width. A single frequency (the centroid of this band with respect to spectral density) is associated with each band and the procedure for regular waves (outlined above) is applied to each band independently of the other spectral bands. The construction of wave rays from offshore towards the shore often leads to the occurrence of "caustics" (areas where adjacent rays cross); this can cause difficulties in interpreting the local wave height (Chao and Pierson 1972). Furthermore, a large number of rays must be constructed for various offshore approach di-

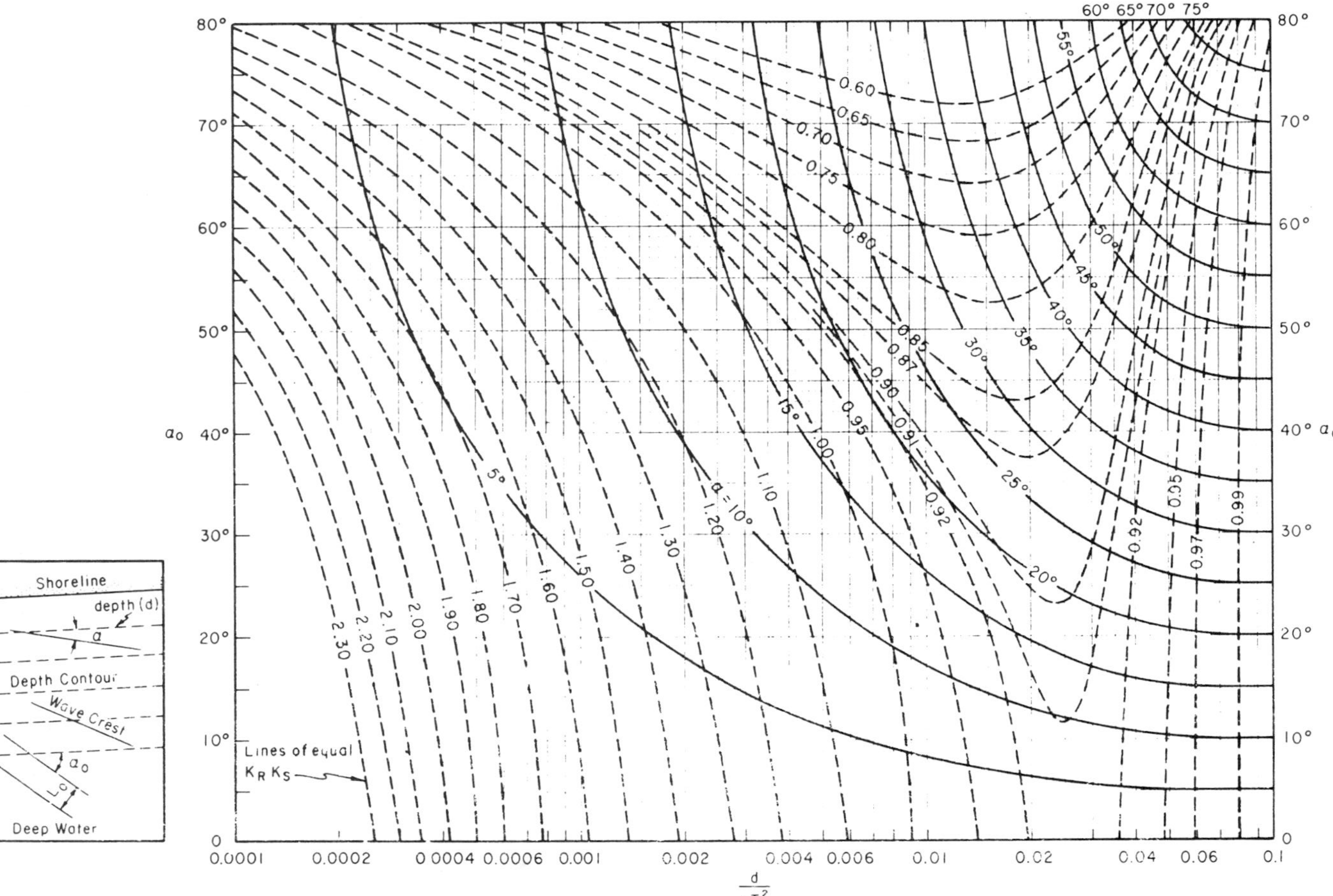

FIGURE 13. Change in wave direction and height due to refraction and shoaling on slopes with straight, parallel depth contours (*Shore Protection Manual* 1984).

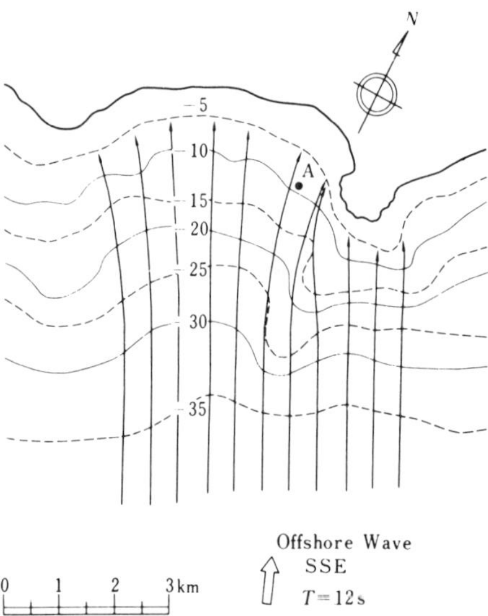

FIGURE 14. Example of wave refraction diagram (Goda 1985).

rections in order to yield wave heights at particular inshore locations. Also, the inshore conditions are highly sensitive to topographic features and to small changes in offshore conditions. In order to reduce the number of ray trajectory calculations, Dorrestein (1960) suggested constructing rays from the inshore location of interest towards the offshore. This method has been developed into a general refraction computer method for directional wave spectra by Abernethy and Gilbert (1975). The use of spectra with some directional spread, rather than long crested waves, reduces the problem of caustics and smoothes the results making them less sensitive to small topographic irregularities and to changes in offshore approach directions.

Since changes in water depth affect the refraction calculations, the results of refraction analyses depend on phase and amplitude of the tide and on storm surges, seiches and longer term water level changes such as seasonal lake level fluctuations. In general, refraction calculations should be performed for the maximum and minimum expected (over the life of the proposed structure) water levels as well as the mean. Furthermore, when refraction calculations are carried out over large distances (>50 km) the curvature of the earth must be considered (Chao 1972).

For further information on numerical refraction of wave spectra refer to Abernethy and Gilbert (1975), Brampton (1981) and Goda (1985).

Diffraction

Diffraction of water waves is the phenomenon by which energy is radiated from the point of interception of a wave train by a barrier, such as a breakwater. The effect of diffraction is shown by waves propagating into the sheltered region within the barrier's geometric shadow. Diffraction is important in determining wave height distribution within harbours or sheltered bays, or in the lee of islands.

Wave diffraction diagrams have been calculated on the basis of Sommerfeld's (1896) diffraction theory as extended or applied by Penny and Price (1944), Blue and Johnson (1949) and Wiegel (1962). Wave height reduction is given in terms of a diffraction coefficient K' defined as the ratio of local diffracted wave height H to the incident wave height H_i unaffected by diffraction. Figure 15 shows the diffraction diagram for waves approaching normal to a semi-infinite breakwater. Diagrams are available for different angles of approach at 15 degree intervals (Shore Protection Manual 1984). Overlay templates of these diagrams can be scaled to correspond to the hydrographic chart being used.

Diffraction diagrams are also available for waves passing a gap of width less than five wavelengths at normal incidence (Johnson 1952, Shore Protection Manual 1984). The diagram for a gap of width $B = 2L$ is shown in Figure 16. If the gap width is greater than $5L$, diagrams for a single breakwater may be used. An approximate determination of diffracted wave characteristics for oblique incidence may be obtained by considering the gap to be as wide as its projection in the direction of incident wave travel. More accurate results can be obtained by using diagrams in the Shore Protection Manual (1984).

Diffraction around an offshore island or structure has been dealt with by Harms (1979).

The use of these diffraction diagrams calculated for monochromatic waves should be restricted to very narrow band swell. As for refraction and shoaling, when broad-band wind seas are being considered, it is necessary to treat the spectrum as a whole rather than a "significant (regular) wave" representation of it. As before, the spectrum is partitioned into equal variance bands, each of which is diffracted separately and finally recombined. The resulting diffraction diagrams for moderately narrow band (swell) and broad band (wind sea) spectra normally incident to a semi-infinite breakwater and a breakwater gap are shown in Figures 17 and 18 respectively; in these diagrams the wavelength L corresponds to the significant wave period T_s, where T_s is defined as the average of the periods of the highest one-third of wave heights. The differences are apparent between these diffraction diagrams for random seas and the corresponding diagrams (Figures 15 and 16) for a monochromatic train of waves.

For diffraction of obliquely incident waves by a semi-infinite breakwater, rotate the axis of the breakwater in Figure 17 while keeping the wave direction and coordinate axes in their original positions. This technique produces satisfactory results when the angle between the dominant direction of wave approach and the line normal to the breakwater is less than 45 degrees. For diffraction of obliquely in-

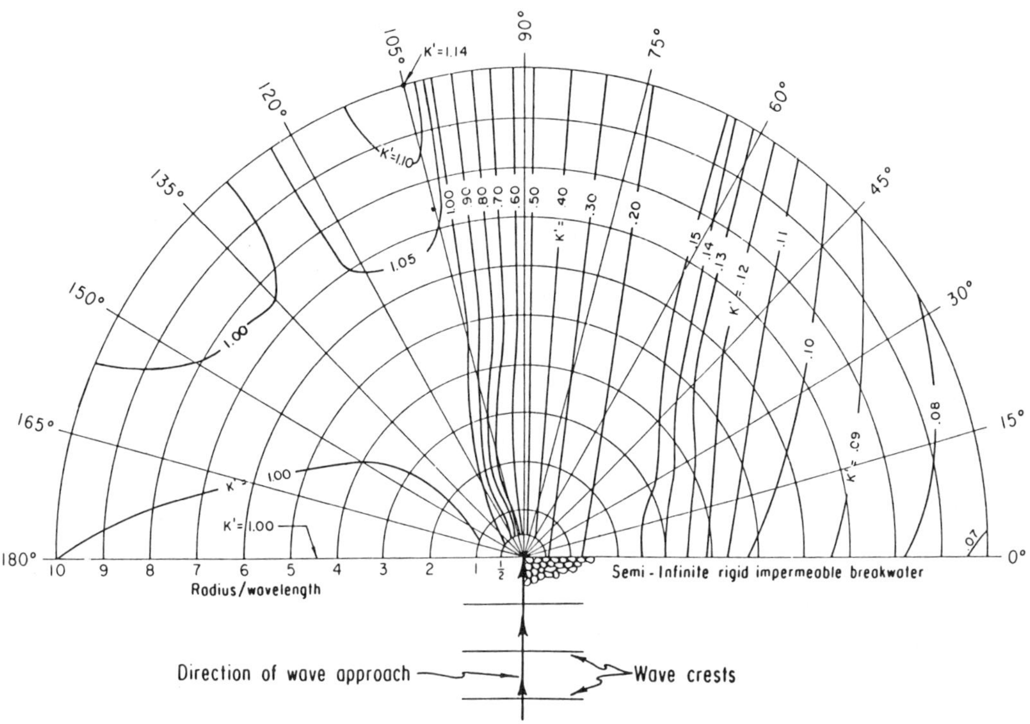

FIGURE 15. Wave diffraction diagram for monochromatic waves approaching normal to a semi-infinite rigid impermeable breakwater (after Wiegel 1962, *Shore Protection Manual* 1984).

cident waves through a breakwater gap, refer to Goda (1985).

Reflection

Water waves may be either partially or totally reflected from walls and other barriers. For incident angles (the angle between the direction of wave advance and the wall) greater than 45 degrees, waves reflect from a vertical wall in such a manner that the angle of incidence is equal to the angle of reflection. Depending on the wall's rigidity and porosity, reflections can take place with very little loss in energy. Consequently, in a harbour with insufficient wave energy absorbers, multiple wave reflections can result in a buildup of wave energy, manifested as wave agitation.

A barrier's reflection coefficient is given by the ratio of the reflected wave height H_r to the incident wave height H_i. Smooth impermeable vertical walls typically have reflection coefficients close to 1.0. Waves approaching in a direction perpendicular to such a wall can form a standing wave or clapotis, having a maximum height of $2H_i$ where the water surface is shown in Figure 19 and is given by:

$$\eta = H_i \cos \frac{2\pi x}{L} \cos \frac{2\pi t}{T} \tag{31}$$

For certain values of x, the water surface remains at the *SWL* (i.e., $\partial\eta/\partial t = 0$ for all t). These points are called nodes and are located a distance x_{node} from the vertical barrier, where:

$$x_{node} = \left(\frac{2n + 1}{4}\right) L, \quad n = 0, 1, 2, \ldots \tag{32}$$

and L is the wavelength of the incident wave. At the nodes water motion is always horizontal. On erodible beds these areas are prone to erosion.

At antinodes, located a distance $L/4$ from nodes, the water surface excursion is $2H_i$ and the water motion is always vertical.

The reflection coefficient for waves approaching at right angles to plane slopes, beaches, and rubblemound revetments and breakwaters can be estimated from Figure 20 as a function of the surf similarity parameter

$$\xi = m\sqrt{L_0/H_i} \tag{33}$$

where m is the slope the beach or structure makes with the horizontal. The curves show that the wave reflection coefficient decreases as either the wave steepness increases or as the slope decreases.

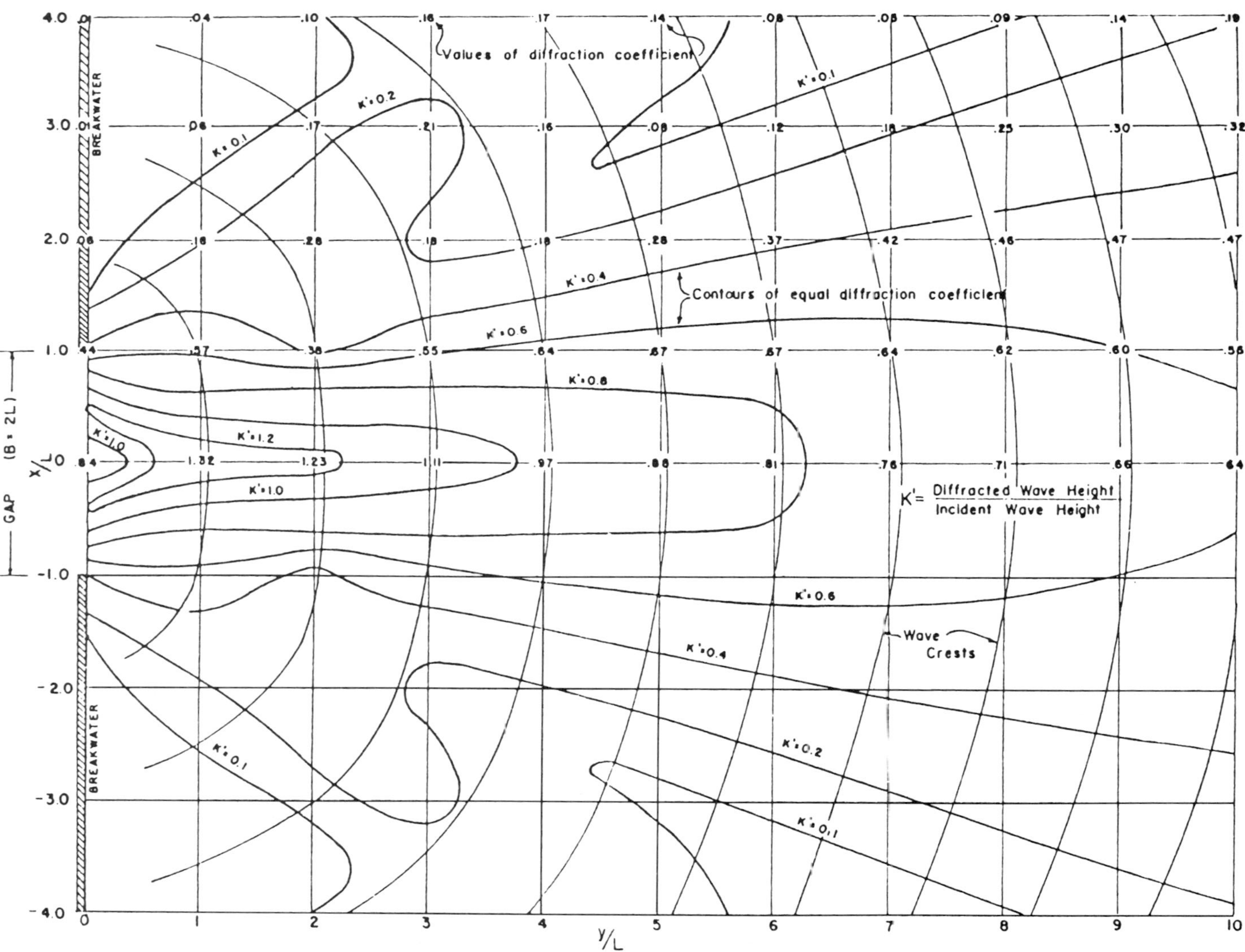

FIGURE 16. Wave diffraction diagram for monochromatic waves approaching normal to a breakwater gap of two wavelengths (*Shore Protection Manual* 1984).

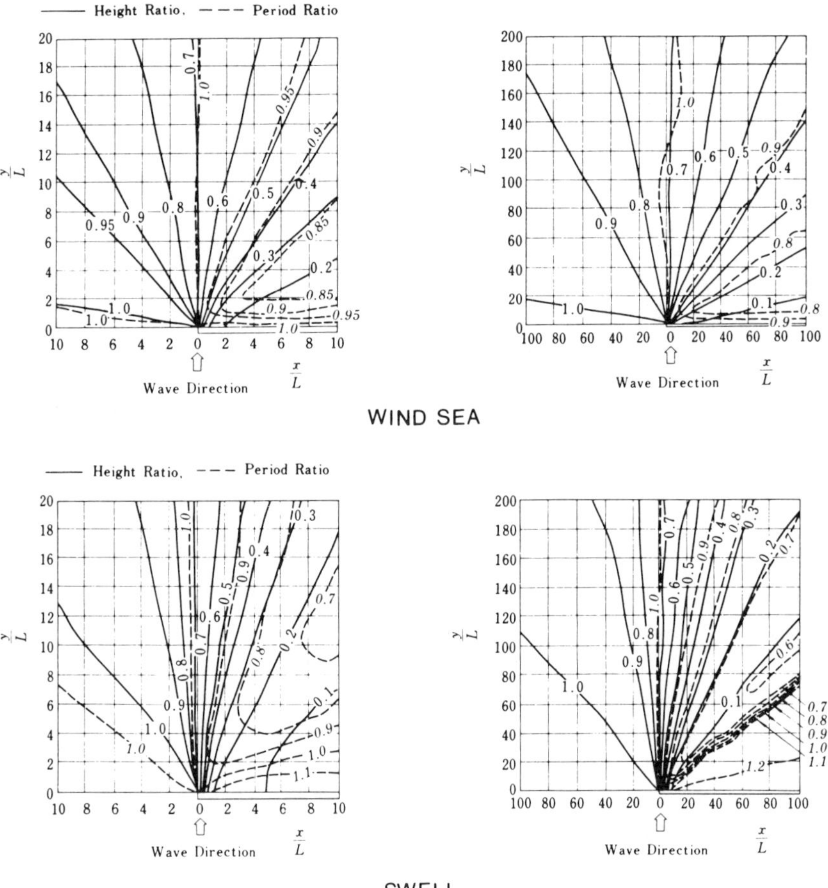

FIGURE 17. Wave diffraction diagram for irregular waves approaching normal to a semi-infinite rigid impermeable breakwater (solid lines for wave height ratio and dashed lines for wave period ratio) (Goda 1985).

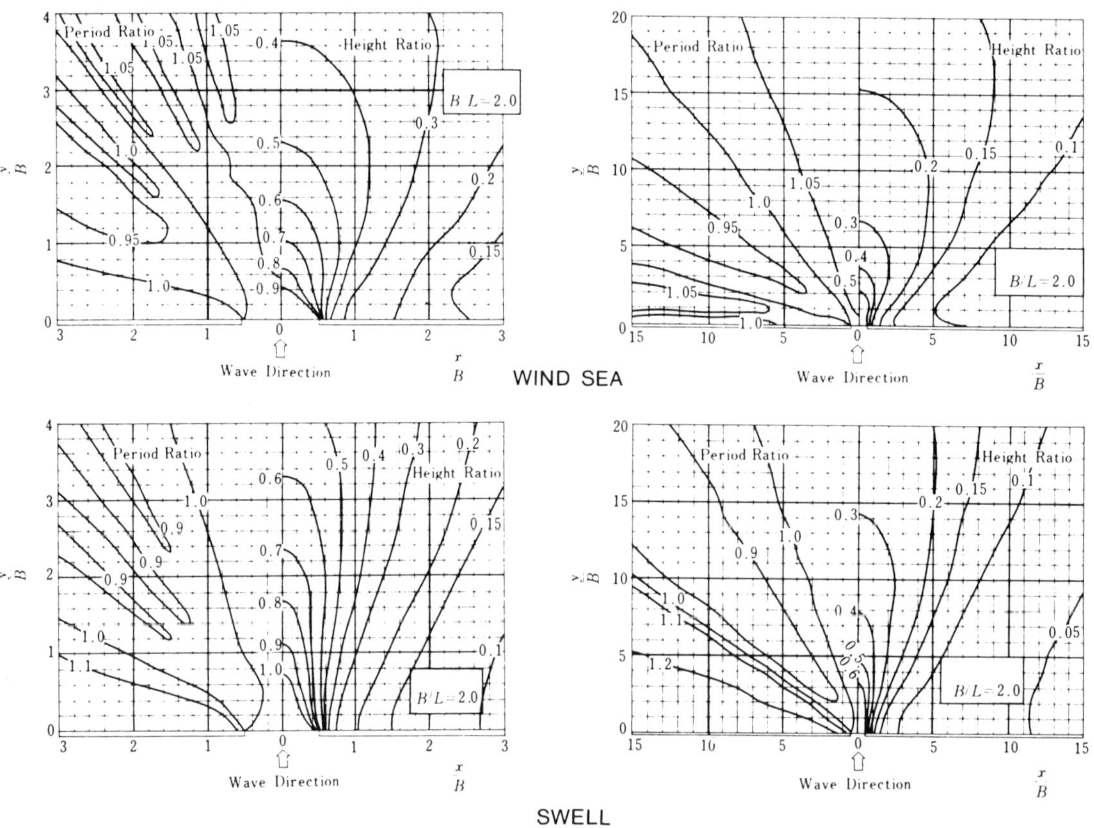

FIGURE 18. Wave diffraction diagram for irregular waves approaching normal to a breakwater gap (B) of two wavelengths (Goda 1985).

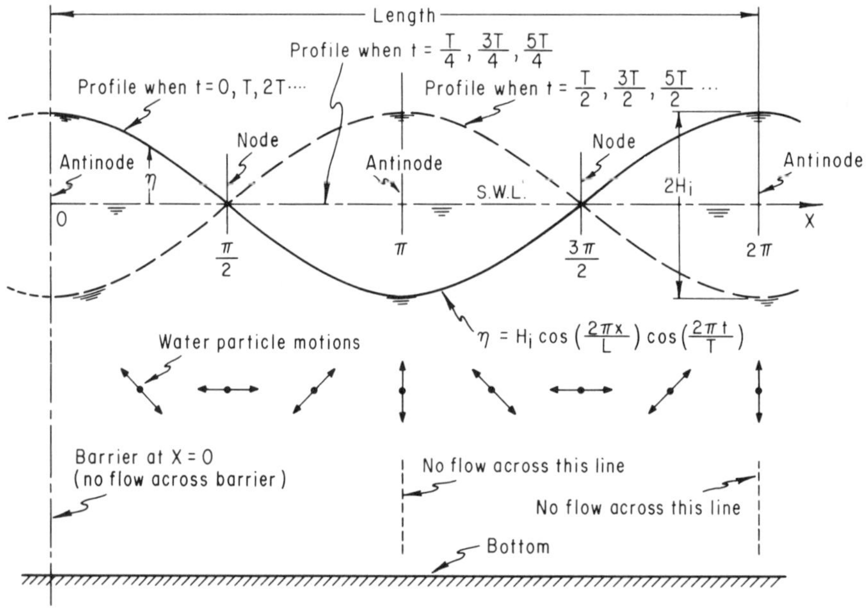

FIGURE 19. Definition sketch for a standing wave system (*Shore Protection Manual* 1984).

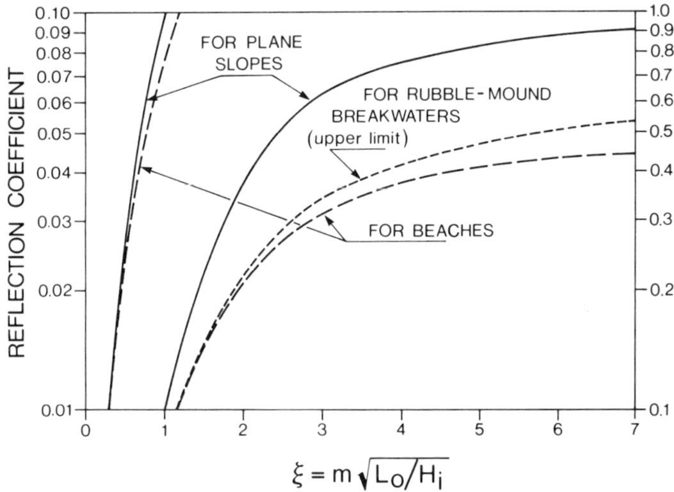

FIGURE 20. Wave reflection coefficients for slopes, beaches and rubble-mound breakwaters (after *Shore Protection Manual* 1984).

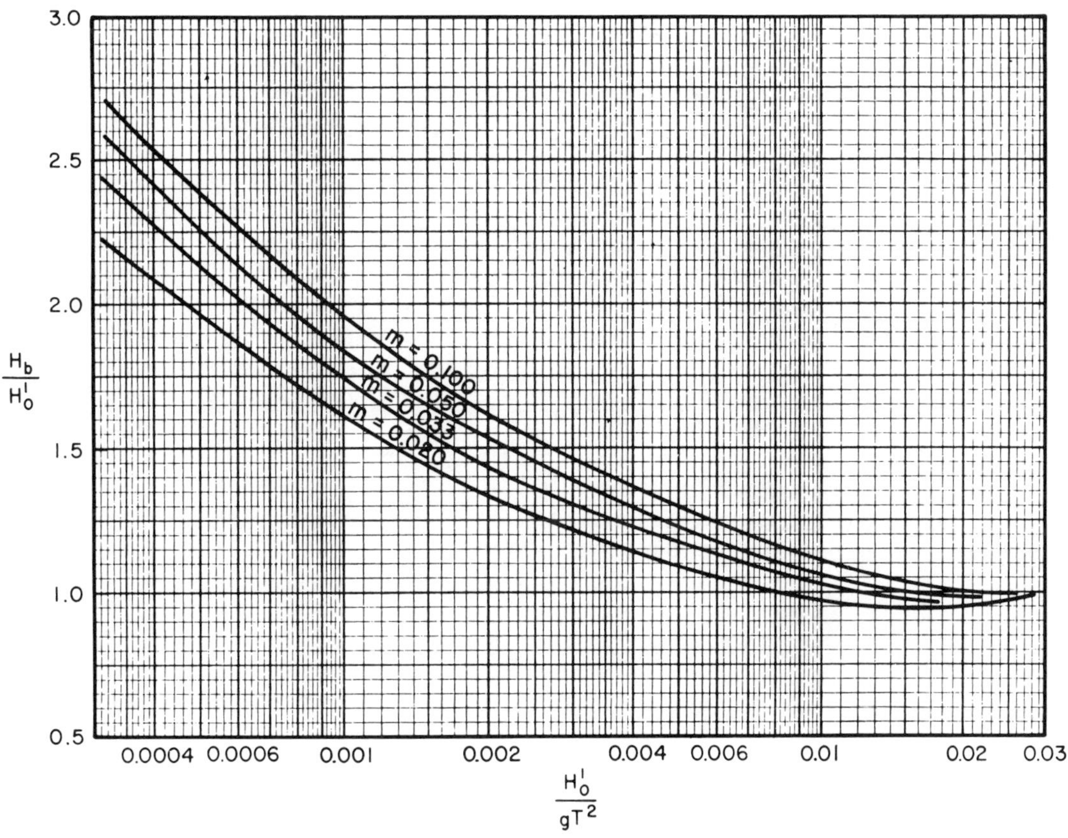

FIGURE 21. H_b/H_0' versus H_0'/gT^2 with m = beach slope (after Goda 1970, *Shore Protection Manual* 1984).

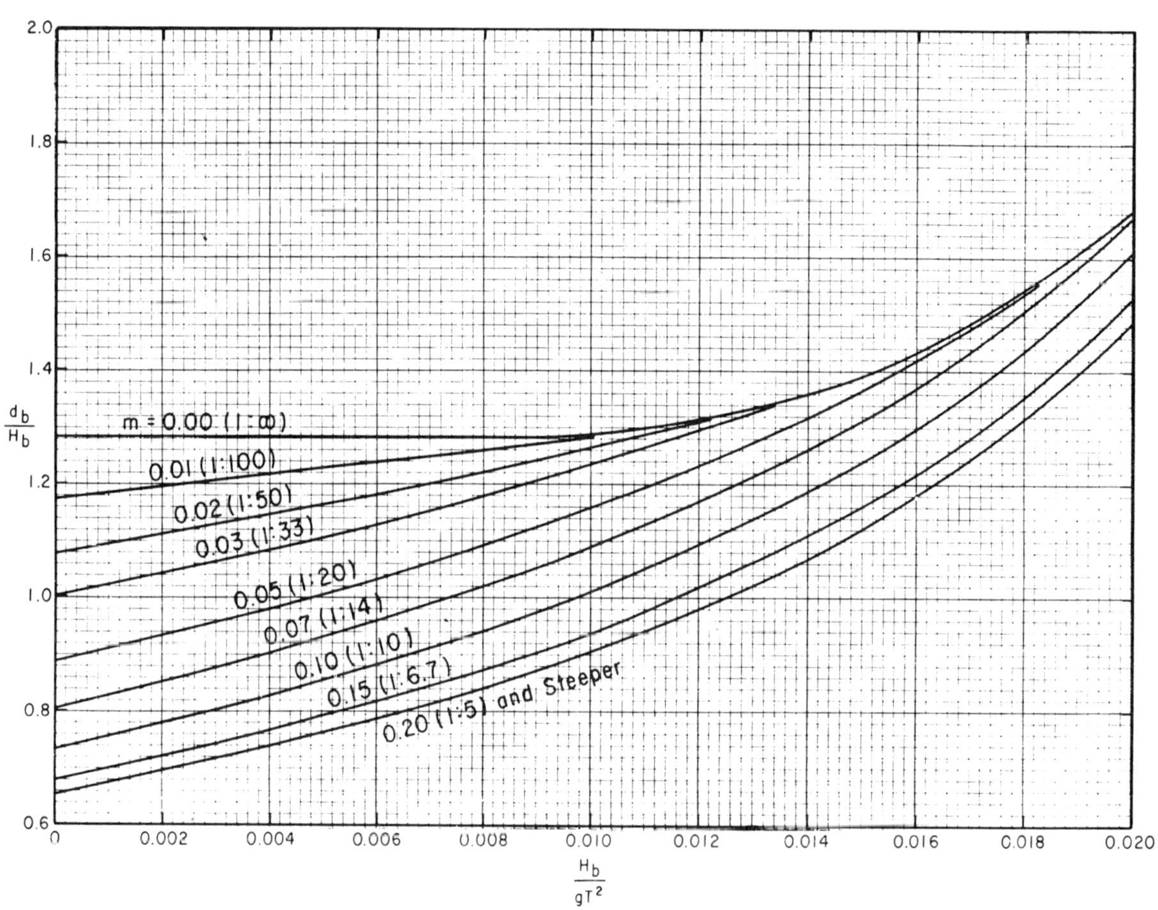

FIGURE 22. d_b/H_b versus H_b/gT^2 with m = beach slope (*Shore Protection Manual* 1984).

For incident angles between 20 and 45 degrees, so-called Mach stem reflection (Wiegel 1964) occurs in which the interaction of the wave with the wall is a combination of reflection and diffraction. The resulting wave crest travels perpendicular to the wall, increasing in height as it travels, and sometimes reaching a height several times that of the incident wave (Berger and Kohlhase 1976).

For incident angles less than 20 degrees, the wave crest bends so that it becomes perpendicular to the wall and no reflected wave appears.

Breaking

The maximum height of a wave in deep water occurs when the water particle velocity at the wave crest just equals the wave celerity. This corresponds to a limiting wave steepness (Michell 1893) of

$$\frac{H_0}{L_0} = 0.142 \simeq \frac{1}{7} \tag{34}$$

and occurs when the crest angle is 120 degrees.

As a wave propagates into shoaling water depths, the limiting steepness which it can attain decreases, and has been found to depend on the relative depth d/L and the beach slope m perpendicular to the direction of wave advance. The breaking wave height H_b is related to the unrefracted deep water wave height H_0' as shown in Figure 21. The depth of water at breaking d_b is related to H_b as shown in Figure 22 (this is sometimes simplified to $H_b/d_b = 0.78$ from Solitary Wave theory).

The preceding results are for monochromatic waves. For random waves, field measurements (Tucker *et al.* 1983, Thornton and Guza 1982, Thompson 1980) and laboratory measurements (Vincent 1985) have shown that the limiting value of H_{m0}/d_b varies from 0.55 to 0.65. For most engineering purposes, a value of 0.6 can be used (Hughes 1984). An approach to determining the statistical distribution of heights in shallow water is given later.

Limiting Wave Energy Levels in Shoaling Water Depths

Bouws *et al.* (1985, 1987) have produced a universal equilibrium spectral form for wind seas called the *TMA* spectrum valid for all water depths outside the breaker zone. A simple expression for H_{m0} has been derived from the *TMA* spectral form (Hughes 1984, Vincent 1982) which allows the prediction of the depth-limited equilibrium H_{m0} in either intermediate or shallow water.

$$H_{m_0} = \frac{L\sqrt{\alpha_1}}{\pi} \tag{35}$$

where

L = wavelength associated with f_p (from linear theory)

$$\alpha_1 = 0.0192\left(\frac{U_{10}^2}{gL}\right)^{0.49} \tag{36}$$

U_{10} = windspeed at 10 m elevation

The main assumption invoked when using the *TMA* spectrum is that the wind sea is at a steady state or equilibrium condition. Therefore, Equation (35) is not valid for fetch- or duration-limited wave conditions (if used it would provide a conservative estimate). Furthermore, use of Equation (35) assumes that the bottom topography is a gentle slope (1:100 or flatter) with smoothly varying features such that refraction and diffraction effects are negligible.

Design Approach for Irregular Waves

Goda (1976, 1985) has provided a coherent procedure for dealing with random seas approaching offshore structures or coastlines. Here we briefly summarize Goda's procedure with reference to his schematic diagram (Figure 23). The procedure starts from the generation of wind waves in deep water and follows them through dispersion, diffraction, refraction, shoaling and breaking to their final interaction with structures and coastlines. In this chapter the various natural wave modification processes outlined in Figure 23 are described. However the panels at the bottom of the figure, dealing with the interaction of waves with engineering structures, are discussed in a separate chapter. Five analysis methods are considered: (A) significant wave representation, (B) highest wave representation, (C) probability calculation, (D) irregular wave experiments, (E) spectral calculations. The types of analysis appropriate to a particular stage of the procedure are indicated by the letters A to E on the figure. Some types of analysis are general but require considerable calculation, while others are simple and generally idealized so that their range of suitability is quite restricted.

(A) *The significant wave representation* amounts to assuming the existence of a train of regular (long crested) waves having (fixed) height and period equal to the significant height and period of the random sea. The virtue of this method lies in its simplicity. The drawback is that it can lead to large errors in certain situations, e.g., the diffraction of irregular waves or in the design of structures that may be sensitive to waves larger than the significant wave.

(B) *The highest wave representation* is akin to the significant wave representation except that the height and period of the highest wave replace those of the significant wave in the hypothetical regular wave train. This method leads to conservative design and is commonly used in the design of offshore structures where personnel safety is paramount. These are the idealized methods of dealing with irregular waves. The other methods (C to E) recognize the essential random character of natural waves.

(C) *The probability calculation* method relies on observed or theoretical joint distributions of heights and periods of individual waves. This method is particularly useful in cases where the cumulative effect of small changes is of interest

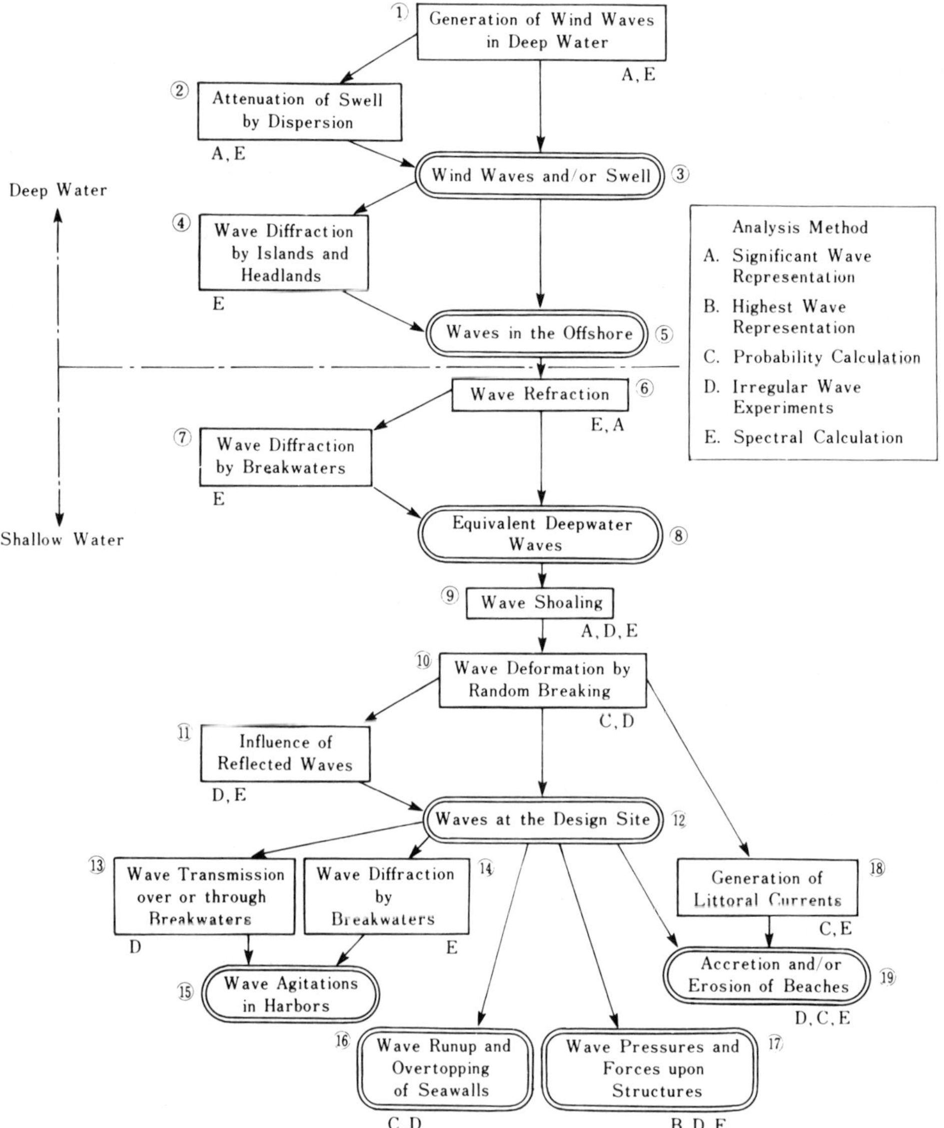

FIGURE 23. Flow of the transformations and actions of sea waves with suggested methods for their calculations (Goda 1985).

rather than catastrophic failure due to a single large wave. Such problems as wave overtopping of a seawall and the working of breakwater components are suitably handled by the probability calculation method. Pierson and Holmes (1965) have devised a method of this type for dealing directly with the probability distribution of forces on structures.

(D) *The irregular wave test* method employs physical modelling techniques to simulate (on a smaller scale) the expected irregular wave conditions at the design site. The method requires sophisticated laboratory equipment. The best design method would be to test the model with a representative series of design spectra, suitably scaled by Froude's law.

(E) *The spectral calculation* method recognizes that surface elevation changes are distributed among various frequencies (periods or wavelengths) and that decomposition into spectral components facilitates many frequency (or wavelength) dependent calculations such as refraction, diffraction, shoaling, etc. Thus instead of treating irregular waves as a hypothetical regular wave train (as in methods A and B) the essential spectral character of a random sea is preserved.

WAVE MEASUREMENT

Practical methods of acquiring information on the statistics of waves for engineering purposes depend on observing the change in elevation of the surface, or the velocity or pressure at depths smaller than the lengths of the wave components of interest.

Surface Elevation Methods

The most common methods for acquiring information on the change of surface elevation at a point are resistance, capacitance or transmission line wave gauges. In each case the sensing element or wave staff is installed vertically through the air/water interface and extending above and below it to the limits of its expected excursions. Resistance and capacitance staffs are generally taut wires or slender rods, in which the changes in the elevation of the (conductive) water affect the resistance or capacitance of the staff which is an element of a suitable electric circuit. Transmission line gauges are essentially radio wave guides consisting of a perforated pipe in which the rising and falling water alters the transmission path and so modulates the frequency of a radio frequency oscillator.

The principal advantages of resistance and capacitance staffs are low cost and good spatial resolution. The principal disadvantagaes are high susceptability to meniscus errors caused by fouling and to damage by floating debris and ice. The transmission line tube largely avoids these disadvantages at the expense of poorer spatial resolution and higher cost.

Accelerometer Buoys

In many applications wave climate data is needed where no suitable structure to support a wave staff exists or is practicable. In these circumstances a moored accelerometer buoy provides a reliable and economic method of acquiring the necessary data. These buoys contain an accelerometer suitably supported within the buoy so that it remains vertical (or nearly so) as the buoy heaves. The buoy is moored with sufficient scope so that its vertical motion approximates that of the surface. The double integral of the time history of the buoy's acceleration, corresponding to changes in surface elevation, is transmitted to shore or recorded on board.

Pressure

Many engineering applications are relatively unconcerned about the shorter (higher frequency) wind waves and instead focus attention on the larger (and long) waves near the peak of the spectrum. In these cases a record of the pressure beneath the surface may provide the most economic solution. As discussed earlier, the magnitude of the pressure disturbance associated with any particular wavelength diminishes rapidly with the ratio of wavelength to probe depth (L/z), so that an appropriate pressure response factor (K_p) must be applied to the measured spectrum in order to recover the equivalent spectrum of surface elevation. Ideally, the pressure port of the instrument should be mounted flush with the bottom but if the water depth (d) is such that $L_{min}/d < 2$ (where L_{min} is the wavelength of the shortest wave desired) then it will be necessary to mount the pressure port above the bottom to avoid burying the desired signal in measurement noise. When this is done care must be taken in the design of the pressure port to minimize contamination of the pressure signal by the flow around the port (Cavaleri 1980).

For many applications ($H/L < 0.05$) the pressure response factor given by small amplitude wave theory is adequate:

$$K_p = \frac{\cosh k(d + z)}{\cosh kd} \tag{37}$$

In deep water ($L_{max}/d < 2$) the pressure response factor simplifies to

$$K_p = e^{kz} \tag{38}$$

For monochromatic waves, a subsurface wave pressure head fluctuation, H_p, can be related to the surface wave height, H by

$$H = \frac{H_p}{K_p} \tag{39}$$

To account for observed deviations from Equation (39), an empirical correction factor, N, has been introduced by many investigators (Cavaleri 1980, Grace 1978, Tubman and Suhayda 1976, Draper 1957) as in:

$$H = \frac{NH_p}{K_p} \tag{40}$$

for regular waves, or

$$(\rho g)^2 \Phi_s(f) = \left[\frac{N(f)}{K_p(f)}\right]^2 \Phi_p(f) \tag{41}$$

for irregular waves where

$\Phi_s(f)$ = surface wave variance spectral density
$\Phi_p(f)$ = subsurface pressure variance spectral density

Other investigators (Forristall 1982, Esteva and Harris 1970, Simpson 1969) have found linear theory to be adequate ($N = 1.0$). Bishop and Donelan (1987) indicate that linear theory is adequate to compensate pressure records to give surface wave heights within five percent; reasons for greater discrepancies in previous studies are discussed.

Pressure (wave) recorders are often an appropriate wave measuring technique solution in harbours or shipping channels where buoys and wave staffs are hazardous or vulnerable. Their main disadvantage is their poor response to short waves, but this can be turned to advantage to eliminate "wind wave noise" when information is required on long shallow water waves such as seiches, tides and surges. In fact additional filtering is often applied to reduce the demands on data reduction when long period motions are of interest. Caution should be taken in using mechanical filtering for this purpose as is done in conventional tide gauges (Noye 1968, 1972).

Other Methods

The methods given above have long held a secure place in engineering practice where the emphasis is on local wave climatology, i.e., long term measurements in one or a few chosen spots. Many of the more modern methods have been devised for remote sensing and are particularly well suited to obtaining good areal coverage at a few chosen times. A short list of other methods is given here with appropriate references: radar altimetry (Barnett and Wilkerson 1967); laser altimetry (McClain *et al.* 1982, Tsai and Gardner 1982); stereo photogrammetry (Cote *et al.* 1960 and Holthuijsen 1981); synthetic aperature radar (Alpers *et al.* 1981); surface contour radar (Kenney *et al.* 1979); scanning-beam microwave radar (Jackson *et al.* 1985); shipboard radar (Young *et al.* 1985) shore based high frequency radar (Barrick 1972, Teague *et al.* 1977); optical slope distributions (Irani *et al.* 1981).

MEASUREMENT OF WAVE DIRECTION

The measurement of wave direction requires greater instrumental complexity (and hence cost) than does the measurement of wave height. For this reason a climatology of heights (and periods) is often obtained by measurement and the corresponding distributions of wave direction inferred from the wind climatology. Differences in wave and wind-sea direction can be quite large in fetch limited situations (Donelan *et al.* 1985) and the effectiveness of some structures (e.g., breakwaters) depends sensitively on wave direction. Thus the additional effort of acquiring a climatology of directions as well as heights and periods may in some instances be justified.

Arrays of Wave Staffs

Estimates of the directional properties of waves may be deduced from simultaneous recordings from several wave staffs or pressure sensors suitably arranged in the same general area. Barber (1963) has described the method of analysis and Munk *et al.* (1963) have applied the method to a sparse array of pressure recorders for examining the properties of long swell. An excellent analysis of the design and properties of such arrays is given by Davis and Regier (1977). High resolution directional spectra can be obtained from multi-element arrays (Donelan *et al.* 1985), but engineering studies generally require information on only the lowest few moments of the directional distribution (often mean direction and root-mean-square spread are sufficient) and therefore cannot justify the considerable cost and expenditure of time and effort required to maintain and analyse multi-element arrays. On the other hand, three wave staffs (or pressure sensors) arranged at three corners of a square of suitable size (depending on the wavelengths of interest) will often yield sufficient wave climatological information (spectra, direction and spread) for engineering purposes and may reap substantial rewards in tighter design and hence lower construction costs.

Pitch-Roll-Heave Buoys

Perhaps the most common method of estimating the directional spectrum is by the use of moored buoys containing sensors that respond to heave, pitch and roll (Longuet–Higgins *et al.* 1963). Such buoys are now commercially available and often come equipped with analysis software so that the desired engineering information from any episode of wave recording is at hand almost immediately from radio or satellite transmission of the analysed data or soon after recovery of onboard recordings.

The basic information sensed by the buoy includes buoy slope in two orthogonal earth-referenced directions and surface elevation (double integral of vertical acceleration). Often the mooring is designed to reduce the likelihood of capsize by providing an erecting force. This reduces the buoy's response to surface slope, but, provided the effect is the same in both orthogonal horizontal directions, the buoy's direction sensing fidelity is not affected. Thus care should be taken to ensure the symmetry of mooring, bridles, etc. The use of such moored directional buoys in areas of strong (>40 cm/s) current is not recommended.

The analysis of the information obtained from ''pitch-roll'' buoys is given by Longuet–Higgins *et al.* (1963).

Pressure-Velocity Sensors

In cases where it is inconvenient to install buoys or wave staffs at the surface a third method, using submerged sensors, often finds favour. The simplest arrangement (Nagata 1964) employs a pressure sensor and two current meters arranged to yield two horizontal orthogonal components. Apart from depth response factors, which degrade the signals from short waves, these pressure-velocity combinations yield equivalent information to pitch-roll-heave buoys.

Two advantages of this method are: (1) the sensors are submerged and hence less subject to damage due to ships, ice, etc.; (2) observed velocities that are incoherent with the pressure signal are due to currents and not waves. These currents are themselves useful in design specifications.

WAVE STATISTICS

The variability of wave heights is best regarded in a statistical sense as a two-scale process. Within a given storm or wave record wave heights vary considerably, but if conditions are quasi-steady during the storm then the statistical distribution of wave heights can be well-modelled by certain distributions that depend on average properties of the particular sea state. This is the short term wave height distribution. When designing coastal or offshore structures, one is also interested in wave conditions over the life of the structure. A much longer time scale associated with the frequency of storms affects the average properties of the wave field, the statistical description of which constitutes a wave "climate." This is the long term wave height distribution.

Clearly these aspects of wave variability depend on the wind, in particular its strength and temporal behaviour. In coastal areas the direction of the wind, topography and bathymetry are also important factors in determining the distribution of wave heights.

Short Term (Intra-Storm) Height and Period Distributions

The vast majority of observations of changes in surface elevation of a wind-driven sea are in the form of time series of elevation η (or subsurface pressure) at a point. A great deal of success in describing the statistical properties of the surface elevation starts from the assumption that these arise through the independent propagation of a spectrum of waves of various wavelengths. If the components are assumed to be statistically independent then the probability density function, $p(\eta)$ of the surface elevation various, η could be Gaussian:

$$p(\eta) = (2\pi m_0)^{-1/2} \exp\left\{-\frac{\eta^2}{2m_0}\right\} \tag{42}$$

where m_r are the moments of the frequency spectrum $\Phi(f)$ of η:

$$m_r = \int_0^\infty f^r \Phi(f) df \tag{43}$$

The zeroth moment

$$\begin{aligned} m_0 = \sigma^2 &= \int_0^\infty \Phi(f) df = \overline{\eta^2} \\ &= \lim_{T\to\infty} \frac{1}{T} \int_0^T \eta^2 dt \end{aligned} \tag{44}$$

is the variance of surface elevation.

The Gaussian assumption appears to be quite closely satisfied for large natural waves and has been applied successfully as early as 1953 (St. Denis and Pierson) to calculate the motion of ships in a confused sea. Generally speaking the Gaussian assumption is valid in the majority of engineering applications where one is interested in waves near the spectral peak and not in the smaller scale waves at higher frequencies.

The envelope of a Gaussian process has a Rayleigh distribution and if the spectrum is sufficiently narrow (in practice this is generally so) the crest to (preceeding with intervening zero crossing) trough height may be identified with the double amplitude of the envelope. Thus, the probability density function of heights $p(H)$ is:

$$p(H) = \frac{H}{4\sigma^2} \exp\left\{-\frac{H^2}{8\sigma^2}\right\} \tag{45}$$

or in terms of the normalized height

$$h = \frac{H}{m_0^{1/2}} \tag{46}$$

giving

$$p(h) = \frac{h}{4} \exp\left\{-\frac{h^2}{8}\right\} \tag{47}$$

The probability of exceedance of a given wave height H_+, the cumulative probability, is the integral of (45) from H_+ to ∞

$$P(H_+) = \int_{H_+}^\infty p(H) dH = \exp\left\{-\frac{H_+^2}{8\sigma^2}\right\} \tag{48}$$

or

$$P(h_+) = \exp\left\{-\frac{h_+^2}{8}\right\} \quad (49)$$

These probabilities are graphed in Figure 24.

If one wants to know the average height (or height squared—relevant to force calculations) of waves exceeding h_+, these may be readily computed from (47) and (49)

$$\overline{h_+^n} = \frac{\int_{h_+}^{\infty} h^n p(h)dh}{P(h_+)} \quad (50)$$

where n is the power of height whose average is desired. Functions (49) and (50) (with $n = 1$ & 2) are listed in Table 1 for several commonly used reference heights.

Calculation of wave forces on structures requires information on periods associated with wave heights as well as the heights themselves. The joint distribution of normalized heights and periods, τ (Longuet-Higgins 1983) is given by:

$$p(h, \tau) = \frac{h^2}{2\pi^{1/2}\nu[1 + (1 + \nu^2)^{-1/2}]\tau^2} \exp\left\{-\frac{h^2}{8}\left[1 + \left(1 - \frac{1}{\tau}\right)^2 \Big/ \nu^2\right]\right\} \quad (51)$$

where

$$\tau = \frac{Tm_1}{m_0} \quad (52)$$

and the spectral width ν is

$$\nu = \left(\frac{m_0 m_2}{m_1^2} - 1\right)^{1/2} \quad (53)$$

The theoretical distribution (51) is shown in Figure 25 for two values of the spectral width, ν. The spectral width may be determined directly from observed spectra or from empirical spectral parameterizations. For example, the wind sea spectra of Donelan *et al.* (1985), described in Spectral Analysis, below, have spectral width $\nu = 0.42$, almost independent of wave age (Table 2). For swell, on the other hand, the spectral width is considerably less and the value $\nu = 0.1$ may be taken as typical.

Integration of (51) over appropriate regions of the h, τ plane yields the desired average heights:

$$h_q = q^{-1}\int_{h_+}^{\infty}\int_0^{\infty} hp(h, \tau)d\tau dh \quad (54)$$

where q is the probability of exceedance of the height h_+:

$$P(h_+) = \int_{h_+}^{\infty}\int_0^{\infty} p(h, \tau)d\tau dh = q \quad (55)$$

For example, the significant height (average height of highest 1/3 of waves) is given by (54) and (55) with $q = 1/3$.

The most probable (mode) period associated with a given height, h is τ_m

$$\tau_m = 2\{1 + (1 + 32\nu^2/h^2)^{1/2}\}^{-1} \quad (56)$$

This is graphed (dashed curve) in Figure 25 and listed in Table 1.

The probability density of τ (regardless of h) is:

$$p(\tau) = \left\{\nu\tau^2[1 + (1 + \nu^2)^{-1/2}] \times \left[1 + \left(1 - \frac{1}{\tau}\right)^2 \nu^{-2}\right]^{3/2}\right\}^{-1} \quad (57)$$

and is graphed in Figure 26 for two values of the spectral width parameter, ν.

The expected height of the highest wave in a sample of N waves is given by (Longuet–Higgins 1952)

$$\hat{E}(h_{max}) = \int_0^{\infty} [1 - \{1 - P(h)\}^N]dh \quad (58)$$

where $\hat{E}$ denotes "expected value of."

For the Rayleigh distribution (narrow frequency band waves) Longuet–Higgins showed that

$$\hat{E}(h_{max}) \cong (8 \ln N)^{1/2} + \gamma_E\left(\frac{1}{2}\ln N\right)^{-1/2} \quad (59)$$

where γ_E = Euler's constant = 0.5772. The approximation of Equation (59) is within 1.5% of the exact value for $N = 20$ and quickly improves with increasing N.

Another useful estimate for engineering purposes (Ochi 1973) is the extreme value of h that has a certain probability ($q < 0.05$) of being exceeded in N waves. Since q^N is the number of waves that exceed h_q on average, then the expected value of h_q is given by:

$$\hat{E}(h_q) \cong (8 \ln q^{-1})^{1/2} + \gamma_E\left(\frac{1}{2}\ln q^{-1}\right)^{-1/2} \quad (60)$$

The statistical estmates are based on assumed knowledge of certain moments of the distribution. In fact all that are ever available are estimates of these moments, so that error bands are associated with all the above calculations. See

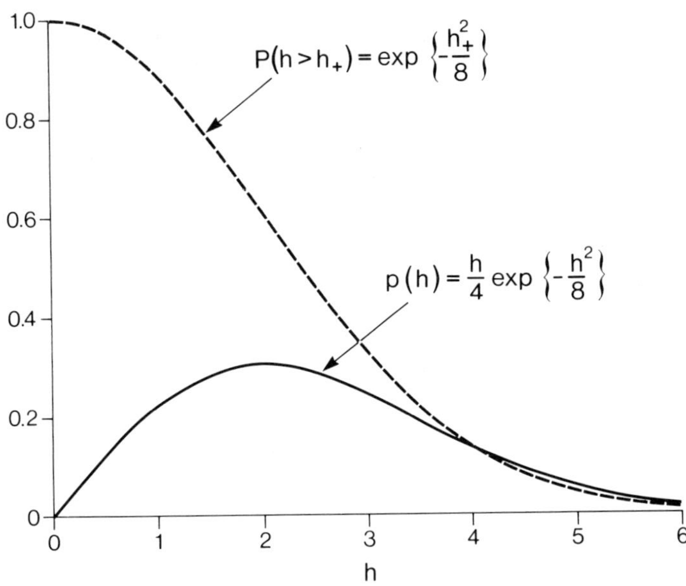

FIGURE 24. The probability density function of normalized wave heights p(h) (Rayleigh) and the cumulative distribution, P(h).

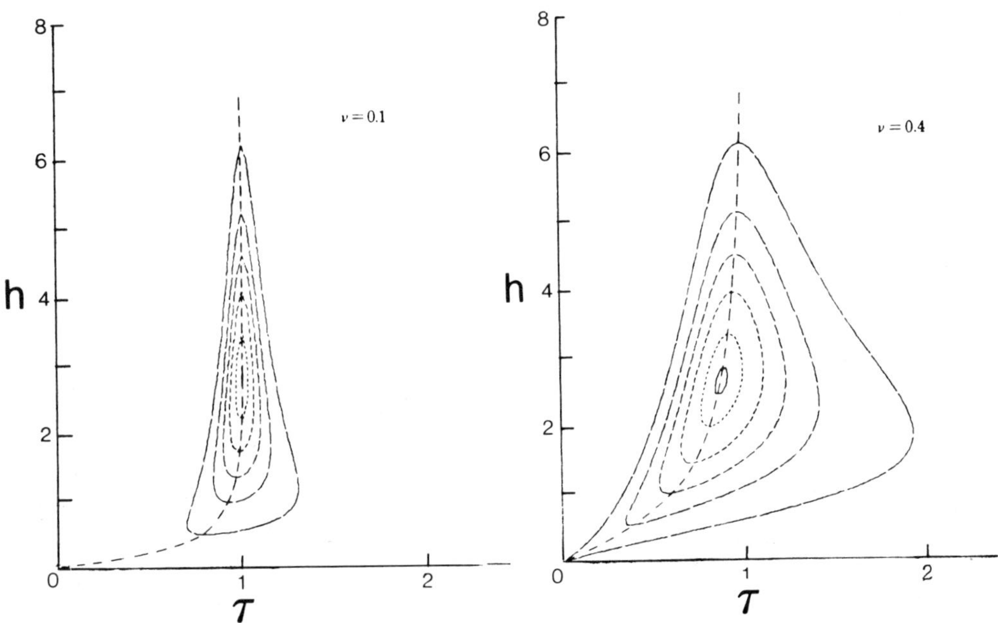

FIGURE 25. The theoretical joint distribution of normalized heights and periods $p(h,\tau)$ for two values of spectral width: $\nu = 0.1$ is appropriate for swell and $\nu = 0.4$ for a typical wind sea. The contours trace $p/p_{max} = 0.99$, 0.90, 0.70, 0.50, 0.30, 0.10 from the centre contour outwards (after Longuet-Higgins 1983).

TABLE 1. Some Wave Height and Period Statistics (Heights and Periods Are Normalized).

Symbol	Description	Normalized Height	Probability of Exceedance	Root-Mean-Square	Most Probable (Normalized) Period	
					$\nu = 0.1$	$\nu = 0.4$
$\bar{h}_{.01}$	average of highest 1% of the waves	6.68	0.004	6.70	1.00	0.97
$h_{.01}$	height, exceeded by 1% of the waves	6.07	0.010			
$\bar{h}_{0.1}$	average of highest 10% of the waves	5.10	0.039	5.14	1.00	0.96
$h_{0.1}$	height, exceeded by 10% of the waves	4.29	0.100			
$h_s = \bar{h}_{1/3}$	significant wave height (average height of the highest 1/3 of the waves)	4.01	0.134	4.10	1.00	0.93
$h_{1/3}$	height, exceeded by 1/3 of the waves	2.96	0.333			
$\bar{h} = \bar{h}_{1.0}$	average wave height	2.51	0.456	2.83	0.99	0.85
$h_{0.5}$	median wave height	2.35	0.500			
h_{mode}	most probable wave height	2.00	0.607			

Confidence Limits on Variance and Characteristic Height, below.

Probability Distribution of Run Length

Although the average statistical properties of waves conform well with a random model, there is a tendency for high waves to appear in groups (Rye 1982; Nolte and Hsu 1973). The grouping of high waves is significant in many engineering problems where structures or vessels may be excited near their resonant frequencies and in coastal problems such as the stability of breakwaters. One useful index of grouping is the probability distribution of run length or the likelihood of observing a given number of consecutive waves exceeding a chosen height. Kimura (1980) has developed a theory for the probability distribution of run length based on the Rayleigh distribution of wave height and requiring the correlation coefficient between pairs of adjacent heights in a record. Kimura's theory has also been summarized by Goda (1985).

TABLE 2. Peak Enhancement Parameter, γ Spectral Width, ν and Peakedness, Q_p of a Model Wind Sea [Equation (64)] for Various Wave Ages, W.

W	State of Development	γ	ν	Q_p
1.2	Fully developed	1.70	0.414	1.90
1.0	Increasingly fetch or	1.70	0.422	1.95
0.8	duration limited	2.28	0.419	2.18
0.6	↓	3.03	0.419	2.51
0.4		4.09	0.418	2.94
0.2		5.89	0.408	3.65

Ewing (1973) has obtained an approximate empirical relationship between mean length $\bar{\ell}$ of runs of waves above a chosen (normalized) height, h_c and Goda's (1970) spectral peakedness parameter, Q_p.

$$\bar{\ell}(h_c) = \sqrt{2}Q_p h_c^{-1} \tag{61}$$

and

$$Q_p = 2m_0^{-2}\int_0^\infty f\Phi^2(f)df \tag{62}$$

The values of Q_p computed from the wind sea spectrum [Equation (64)] for various wave ages are given in Table 2.

Wave Statistics in Shoaling Water

A great deal of theoretical and experimental work has been done on the statistics of wind waves in deep water and the results are summarized in the previous section. However, in many applications the engineer requires estimates of probability of occurrence of waves of various heights and periods in shallow or shoaling water. Various attempts to address wave statistics in shoaling water have been made (Goda 1975, 1985; Thompson and Vincent 1985).

Figure 27 from Thompson and Vincent (1985) allows the prediction of H_s/H_{m_0} as a function of d/gT_p^2 and the significant wave steepness, ε defined as

$$\varepsilon = \frac{H_{m_0}}{4L} \tag{63}$$

H_{m_0} and L can be calculated from Equations (35) and (6) [or (8)], H_s can be obtained from Figure 27 (use the

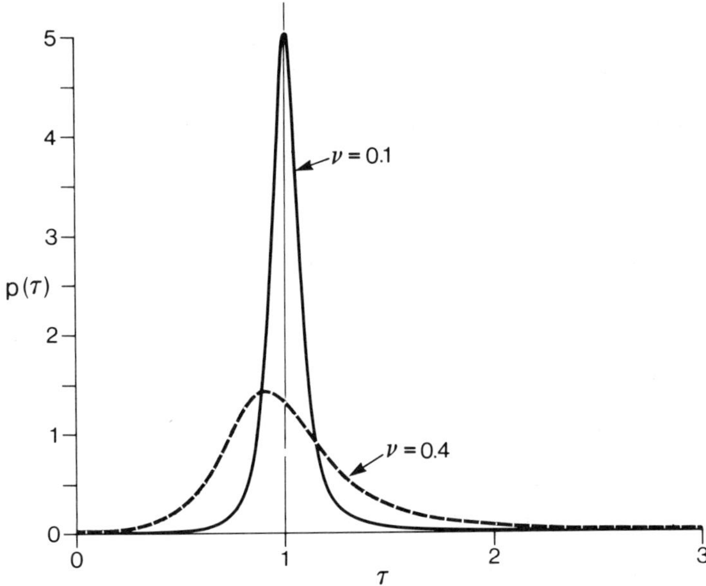

FIGURE 26. The probability density function of normalized periods p(τ) for values of the spectral width parameter $\nu = 0.1$ and 0.4 (after Longuet-Higgins 1983).

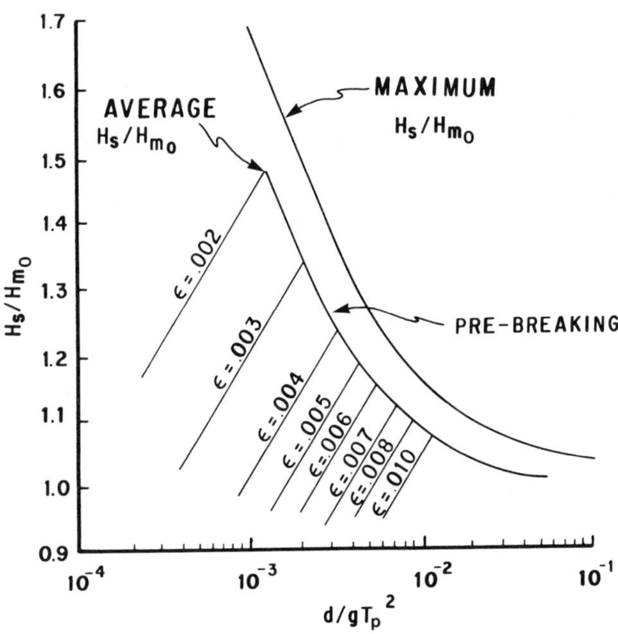

FIGURE 27. Relation between H_s and H_{mo} in shoaling water (after Thompson and Vincent 1985).

"Average" curve). This method also assumes that refraction and diffraction effects are negligible.

When wave conditions are quasi-steady and the combination of beach slope and propagation distance is sufficiently small, the wave period at the peak of the variance spectrum can be assumed to be relatively constant (change of less than 10 percent) from deep to shallow water (Vincent 1985).

Long Term Distribution of Extremes

There are two basic approaches to estimating extreme waves: (1) From records of wave statistics construct an appropriate distribution (e.g., Gumbel, Weibull) of maxima (Fisher and Tippet 1928) and from this distribution infer the required extremes. An excellent review of this method and its pitfalls is given by Muir and El-Shaarawi (1986). (2) From records of wind statistics, construct an appropriate distribution of worst storms and from this distribution infer the required extreme storms. Then apply these storms in a wind-wave hindcast model to determine the extreme sea states during the desired forecast period. The extreme sea states may then be used to deduce extreme heights and the probability of exceedance of these heights using the statistical models described in Short Term (Intra-Storm) Height and Period Distributions, above.

The first method is in common use but the second offers several advantages over the first: (a) A long history of wave climatology is generally not available in most areas, whereas long term meteorological records are. The estimation of extremes appropriate to times much longer than the available record is fraught with error. Thus a much better estimate of extremes in meteorological conditions is possible than of extremes in the wave heights deduced directly from the wave climatology. (b) Local effects such as fetch limits, bottom topography causing refraction and limiting wave heights and lengths are not included in the first method, whereas in the second a good hindcast model will incorporate all these effects and produce a sea state appropriate to the extreme storms. Thus the second method is highly recommended.

ANALYSIS OF WAVE RECORDS AND PRESENTATION OF WAVE DATA

Spectral Analysis

The systematic analysis of wind sea spectra has its origin in the work of Kitaigorodskii (1962). Since then, various parametric descriptions of observed wave spectra have been proposed. Pierson and Moskowitz (1964) proposed a spectral form on the basis of observations of "fully developed" waves in deep water and Hasselmann *et al.* (1973) extended the concept to include fetch-limited waves in earlier stages of development. Bouws *et al.* (1985) have extended this work to include depth-limited conditions. Donelan *et al.* (1985) have proposed a spectral form that includes fetch and duration limited cases in deep water in which the waves and wind need not be colinear. The spectral parameters are related to the wave age W (ratio of wave celerity at the spectral peak to the wind speed component in the direction of travel of the peak waves) and in the limit of full development ($W = 1.2$) their spectral form approaches the Pierson–Moskowitz form. The spectral representation due to Donelan *et al.* (1985) is derived from data in the wave age range $0.2 < W < 1.2$ and is given by:

$$\Phi(f) = \alpha g^2(2\pi)^{-4} f^{-4} f_p^{-1} \exp\left\{-\left(\frac{f_p}{f}\right)^4\right\} \gamma^{\exp\{-(f-f_p)^2/2\mu^2 f_p^2\}} \tag{64}$$

where

$$\alpha = 0.006W^{-0.55} \tag{65}$$

$$\mu = 0.08(1 + 4W^3) \tag{66}$$

$$\gamma = \begin{cases} 1.7 - 6.0 \log W; & 0.2 < W \le 1 \\ 1.7; & 1 < W < 1.2 \end{cases} \tag{67}$$

and f is the cyclic frequency and f_p its value at the spectral maximum. The spectral parameters (α, μ, γ) are functions of the wave age W only. Spectra from Equation (64) for a 10 m/s wind speed and various fetches are shown in Figure 28, and for a 100 km fetch and various wind speeds in Figure 29.

Wind waves are not long-crested but, rather, are distributed about the wind direction with appreciable angular spread. Empirical directional spectra have been given by several researchers including Mitsuyasu *et al.* (1975), Hasselmann *et al.* (1980) and Donelan *et al.* (1985). The directional spectrum or frequency-direction distribution as given by (Donelan *et al.* 1985) is:

$$F(f, \theta) = \frac{1}{2}\Phi(f)\beta \operatorname{sech}^2 \beta\{\theta - \bar{\theta}(f)\} \tag{68}$$

where $\bar{\theta}(f)$ is the mean wave direction and β, the spreading parameter, is given by:

$$\beta = \begin{cases} 2.61\left(\dfrac{f}{f_p}\right)^{+1.3}; & 0.56 < \dfrac{f}{f_p} \le 0.95 \\ 2.28\left(\dfrac{f}{f_p}\right)^{-1.3}; & 0.95 < \dfrac{f}{f_p} < 1.6 \\ 1.24; & \text{otherwise} \end{cases} \tag{69}$$

The (normalized) shape of the directional distribution (Donelan *et al.* 1985) is given in Figure 30.

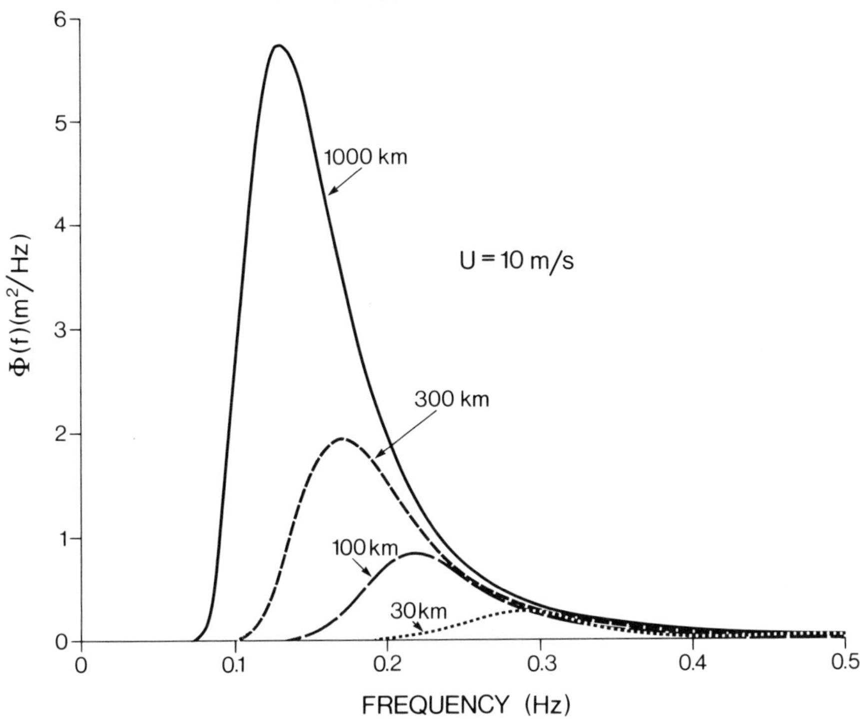

FIGURE 28. Spectra from Equation (64) for a 10 m/s wind (at 10 m height) and various fetches.

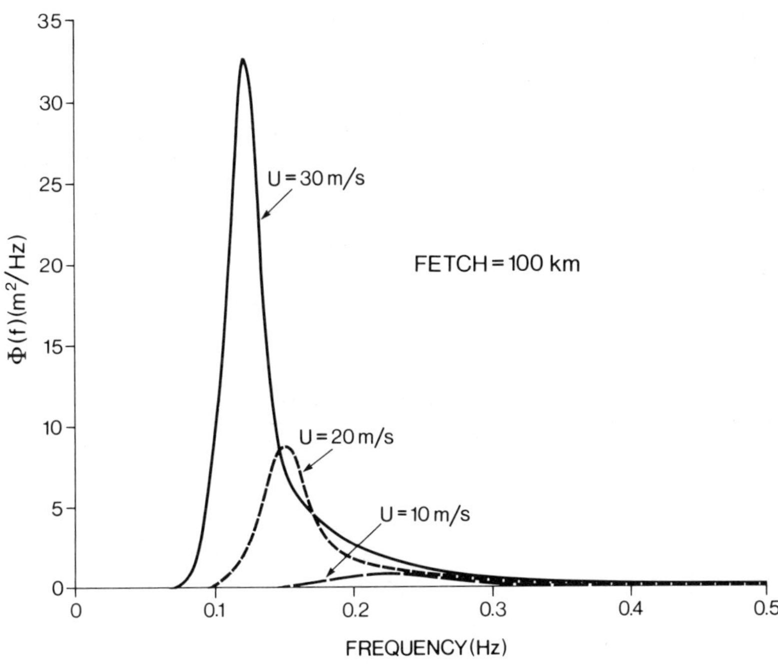

FIGURE 29. Spectra from Equation (64) at 100 km fetch and various wind speeds.

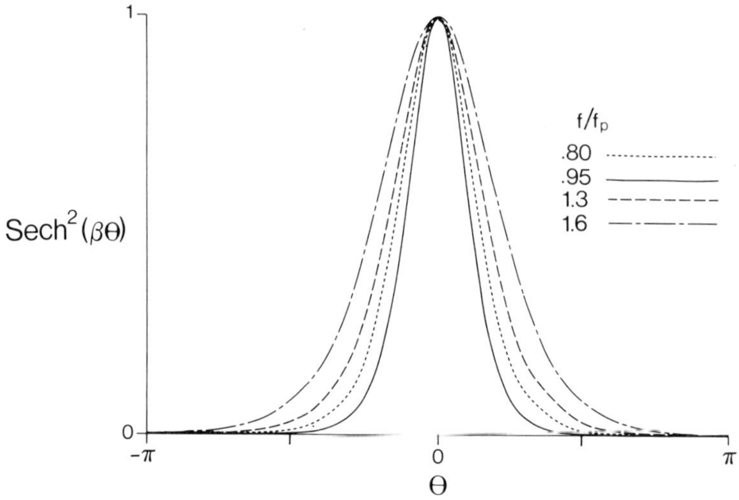

FIGURE 30. The directional spreading of wind waves for various frequencies relative to the peak frequency f_p (Donelan et al. 1985).

Sampling Variability of Spectral Estimates

Spectral analysis of a wave record of finite duration leads to a statistical *estimate* of the true spectrum of the process in the same sense that an average of a sample of random data is an estimate of the true but unknown first moment of the probability density function associated with the sample. The wave record is a (generally short) sample drawn from a much larger population and as such can yield only an estimate of the underlying spectrum of the process. The question is, how accurate is the estimate of the true but unknown spectrum? The simplest model of the statistical behaviour of waves that yields reliable answers to this question is that of a stationary Gaussian process. Although deficient in some respects (e.g., surface is free of skewness, i.e., symmetrical about the mean water level) the model permits the calculation of theoretical confidence limits on spectra which are in excellent accord with observations (Donelan and Pierson 1983).

The spectral estimates, computed from a wave record, are chi-square distributed with d_f degrees of freedom. The number of degrees of freedom, d_f associated with each estimate depends on the length of the record in seconds, t_R, and the frequency bandwidth of the estimate, Δf, where

$$d_f = 2\Delta f t_R \tag{70}$$

Normally fast Fourier transform (*FFT*) methods are used to obtain a large number of elemental spectral estimates which are then averaged in bands of equal spectral width, Δf.

Once the value of d_f has been established, chi-square tables (e.g., Abramowitz and Stegan 1965) will yield the confidence limits on the spectral estimates.

Multiplicative factors corresponding to 95 percent and 5

TABLE 3. Confidence Limits on Spectra.

Degrees of Freedom d_f or D_f	Multiply Smoothed Spectral Estimates by this Factor to Get Appropriate Confidence Limit			
	5%	10%	90%	95%
2	0.334	0.434	9.49	19.50
4	0.422	0.514	3.76	5.63
8	0.516	0.599	2.29	2.93
16	0.608	0.680	1.72	2.01
30	0.685	0.745	1.46	1.62
d_f, $D_f > 30$	$10^{-(d_f-1)^{-1/2}}$	$6.3^{-(d_f-1)^{-1/2}}$	$6.3^{+(d_f-1)^{-1/2}}$	$10^{+(d_f-1)^{-1/2}}$

percent confidence limits (enclosing a 90 percent confidence band) and to 90 percent and 10 percent confidence limits are given in Table 3 for various values of d_f. For d_f larger than 30 the approximations given by Blackman and Tukey (1958) are listed.

The confidence bands decrease quickly at first with increasing degrees of freedom and then more slowly. So that while it is good practice to obtain long enough records so that d_f is large, a value of 30 is a reasonable compromise. Much larger values generally can be obtained only at the expense of over-smoothing of the spectra or accepting non-stationary data. With 30 degrees of freedom the spectrum is known to within a factor of 2 at the 80 percent confidence level. That is, eight times out of ten the true spectrum will be within the 10 percent and 90 percent confidence limits. Increasing the number of degrees of freedom to 90, by tripling the record length or widening the smoothing window so that Equation (70) yields 90, narrows the 80 percent confidence bands to a factor of 1.48.

Confidence Limits on Variance and Characteristic Height

The estimation of the total variance of the record, σ^2 $(=m_0)$ leading to such estimates as the characteristic height, $H_{m0} = 4\sigma$ is subject to similar inaccuracies as in the previous section. Since the variance is derived from the entire spectrum, the total degrees of freedom, D_f associated with the variance is larger than the degrees of freedom, d_f associated with each spectral estimate. Donelan and Pierson (1983) describe a method to calculate D_f from a spectrum. The confidence limits on the variance for various values of D_f are given in Table 3. The limits of heights (characteristic height, mean height, etc.) are given by the square root of the multiplicative factors in Table 3.

For wind seas (with very little or no swell) Skafel and Donelan (1983) give the following approximation to the calculation of D_f:

$$D_f \approx 0.86N_p \tag{71}$$

where N_p is the number of peak periods, T_p in the record. Figure 31 shows the confidence limits corresponding to Equation (71).

Confidence Limits on Peak Frequency

Within the confidence limits of discrete spectral estimates the estimated spectral values differ from the true values. In some cases the estimated spectral value at a frequency different from the true peak may be larger than the estimated value at the true peak frequency; accordingly, the estimated peak frequency may be in error by an amount which depends on the confidence limits on the spectrum and the shape of the spectrum. Clearly for given confidence limits a narrow spectrum will be subject to less error than a broad spectrum.

The number of degrees of freedom can be increased by broadening the bandwidth Δf of the averaged bands, however this reduces the possible resolution of the peak frequency. This reduction in resolution may be offset by assigning the peak frequency to the weighted (by spectral estimate) mean of the three spectral bands at and on either side of the highest band.

In Table 4 are summarized the coefficients of variation (ratio of standard deviation to mean) and bias of measured peak frequency for various values of the spectral peakedness parameter, Q_p, bandwidth, Δf and degrees of freedom, d_f. The spectral shape [Equation (64)] corresponds to wind-seas of various wave ages. This table has been prepared by Monte Carlo simulation in which the estimated peak frequency is determined from the centroid of the three estimates at and about the estimated peak, as described above. A similar method was applied by Günther (1981) to the *JONSWAP* spectrum.

Zero Crossing Method

The spectral methods described above are recommended for engineering design. However, they require considerable machine calculation and there are occasions when a hand graphical technique is useful in providing quick preliminary answers. The so-called "zero crossing method" developed by Tucker (1963) is suitable for this purpose and is described below. The method is based on the assumption of a Rayleigh distribution of wave heights and it simplifies a wave record (such as Figure 32) so that it may be summarized by only two parameters, a wave height and a wave period. The steps in this metod are listed below. The square bracketed figures refer to the record in Figure 32.

1. Sketch the mean water level as closely as possible.
2. Count the number of crests $-$ N_c [10].
3. Count the number of zero-up-crossings $-$ N_z when the record crosses the mean water level in an upward direction [7].
4. The average crest period, $\overline{T}_c$ and the average zero-up-crossing period, $\overline{T}_z$, may be calculated as

$$\overline{T}_c = \frac{t_R}{N_c} \quad \text{and} \quad \overline{T}_z = \frac{t_R}{N_z} \tag{72}$$

   where t_R is the length of the record in seconds.
5. Determine the values of A, B, C and D, the highest crest, the second highest crest, the lowest trough and the second lowest trough respectively.

From these values two estimates of σ, the root mean square value of water surface displacement [Equation (44)] may be

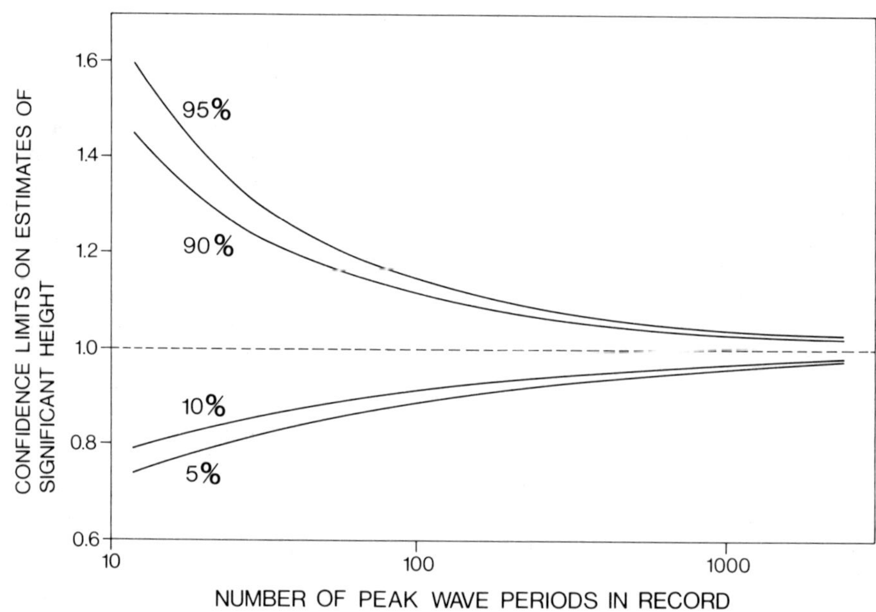

FIGURE 31. Confidence limits on (relative) significant height as a function of the number of peak wave periods in a record.

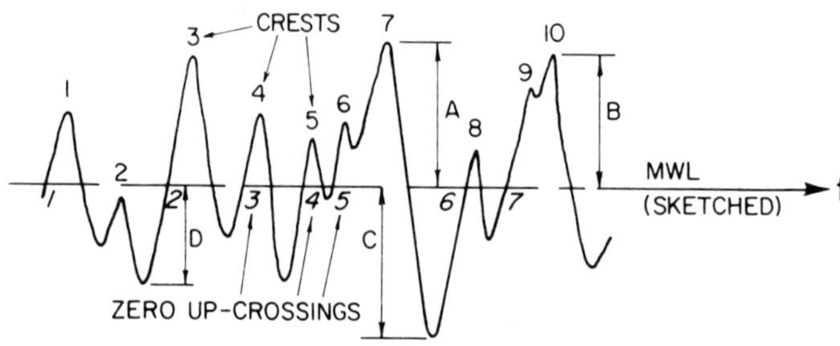

FIGURE 32. Typical wave record with notation for manual analysis.

obtained as

$$\sigma_{x,1} = (A + C) \cdot \frac{1}{2} \cdot (2\lambda)^{-1/2} \times [1 + .289\lambda^{-1} - .247\lambda^{-2}]^{-1} \quad (73)$$

$$\sigma_{x,2} = (B + D) \cdot \frac{1}{2} \cdot (2\lambda)^{-1/2} \times [1 - .211\lambda^{-1} - .103\lambda^{-2}]^{-1} \quad (74)$$

where σ_x is the value of σ estimated by the zero crossing method and

$$\lambda = \ln N_z \quad (75)$$

From these two estimates of σ the significant wave height may be estimated as

$$H_{x,s} = 2[\sigma_{x,1} + \sigma_{x,2}] \quad (76)$$

This method is only applicable to records with a narrow frequency band.

The zero crossing method yields $H_{x,s}$ and $\overline{T}_z$. The remainder of the distributions of H and T may also be obtained from the record by grouping and analysis of the individual waves. One must be careful of definitions at this point, however, and in a number of studies different wave height definitions are used. The normal definition for wave height, H, is the vertical distance between a wave crest and the preceding trough (Figure 32) where crest and trough are defined as maxima and minima of the record. The second definition sometimes used in connection with the zero crossing method is zero crossing wave height, H_z. This is defined as the vertical distance from the maximum crest between a zero up crossing and the subsequent zero down crossing to the minimum trough between the same zero up crossing and the preceding zero down crossing (Figure 32). This second definition removes all the small wavelets that do not cross the mean water level. A careful distinction must be made, for instance, between H_s and $\overline{H}_{z,s}$, $H_{0.1}$ and $\overline{H}_{z,0.1}$ where the first of each pair refers to the normal definition of wave height while the second uses the zero crossing wave heights H_z.

The estimates of heights and periods are assumed to be normally distributed about the mean, so that given the coefficient of variability (ratio of standard deviation to mean) the probability that the true height (or period) lies within a certain range about the mean is immediately available from tables of the normal distribution. The coefficient of variation of this method is about 10%. A sufficiently long ($N_z > 50$) record should be used to achieve even this limited accuracy.

TABLE 4. Variability of Estimates of Peak Frequency.

W	Q_p	$\frac{\Delta_f}{f_p}$	d_f	Bias %	Coefficient of Variation %
0.4	2.94	0.05	10	0.2	4.5
			20	0.1	3.4
			40	0.1	2.9
			80	0.0	2.4
			160	0.0	1.9
		0.1	10	0.3	4.4
			20	0.2	2.7
			40	0.2	1.7
			80	0.1	0.8
			160	0.1	0.4
		0.2	10	1.1	2.7
			20	1.0	1.5
			40	1.0	0.9
			80	1.0	0.6
			160	1.0	0.5
0.8	2.18	0.05	10	2.0	8.5
			20	1.5	7.4
			40	1.1	6.5
			80	0.8	5.4
			160	0.6	4.5
		0.1	10	2.4	9.0
			20	1.3	7.3
			40	1.2	6.5
			80	1.1	5.5
			160	1.0	4.5
		0.2	10	4.0	8.9
			20	2.8	5.8
			40	2.4	4.1
			80	2.0	2.2
			160	1.7	0.8
1.2	1.90	0.05	10	4.1	11.2
			20	2.6	9.6
			40	2.2	8.6
			80	1.8	7.3
			160	1.4	5.7
		0.1	10	4.9	12.1
			20	3.2	10.1
			40	2.6	8.6
			80	2.1	7.4
			160	1.8	6.1
		0.2	10	7.1	12.4
			20	5.5	9.1
			40	4.5	7.3
			80	3.6	5.1
			160	2.8	3.9

Directional Spectral Analysis

The most common method of obtaining directional wave information for engineering purposes has been by the use of pitch-roll buoys. The method of analysis (direct Fourier transform method) is given by Longuet–Higgins *et al.* (1963).

Where directional resolution of better than ±30° is required, the direct Fourier transform method may not be adequate. Finer directional resolution may be achieved with "Maximum Likelihood Methods," which strive to minimize the variance between the true spectrum and the estimate, subject to certain constraints. A comprehensive description of these methods is given by Capon *et al.* (1967), Capon (1969), Long and Hasselmann (1979) and Isobe *et al.* (1984). The sampling variability of estimates of wave direction is discussed by Borgman *et al.* (1982) and by Kuik and Van Vledder (1984).

WAVE PREDICTION

General

Wave forecasting refers to the prediction of wave conditions in the future; wave hindcasting is the estimation of past wave conditions from past weather records. Wave prediction is of fundamental importance to the coastal engineer because measured wave data sufficient for engineering design is seldom available at any given location. Basically, wave conditions can be considered to be a function of wind velocity, duration of wind, fetch, depth of water and wave decay rates. A special case, known as fully arisen sea (*FAS*), exists when the wave conditions depend only on the wind speed because there is no net wave decay and the other variables are sufficiently large that they have no effect; for wind speeds of engineering interest, *FAS* conditions occur primarily in the oceans.

Simplified wave prediction models yield estimates of a characteristic wave height (or energy) and period. Usually the dominant wave direction is assumed coincident with the wind. Most of these simplified models can be used with an empirical wave spectral shape to provide an estimate of the wave spectral characteristics (assuming no wave decay or incident swell). The two most widely used models are the Sverdrup-Munk-Bretschneider (*SMB*) model (Sverdrup and Munk 1947, Bretschneider 1958, 1973) and the so-called *JONSWAP* model (Hasselmann *et al.* 1973). Both use the wind speed at a 10 m height above the water level, hereafter referred to as U_{10}, as the main input parameter.

Estimating Wind Speed

The greatest source of error in wave prediction involves the estimation of wind speed. The best estimate of wind speed is an overwater wind speed measurement at the location of interest. If the wind is not measured at the 10 m elevation, the wind speed must be adjusted. The simple approximation

$$U_{10} = U_z \left(\frac{10}{z}\right)^{1/7} \tag{77}$$

can be used if the anemometer's elevation, z in metres is less than 20 m.

Most engineering calculations of waves are concerned with quite strong winds so that the atmospheric boundary layer is nearly neutrally stratified; i.e., the stabilizing or destabilizing effect of an air-sea temperature difference is nullified by the strong mechanical wind mixing. A useful index of the degree of stratification is the bulk Richardson number, R_b:

$$R_b = \frac{(T_A - T_W)zg}{T_A U_z^2} \tag{78}$$

where T_A, T_W are air (at height z) and water surface temperature in degrees Kelvin.

If $|R_b| < 0.01$ Equation (77) is valid. Otherwise suitable corrections for stability must be made in establishing UN_{10}, i.e., the equivalent neutral stability wind at the 10 m height. It is this wind speed (UN_{10}) that is appropriately applied to the wave prediction formulae to follow. Formulae for calculating UN_{10} are given in Large and Pond (1981), in which a different stability index is used. The relationship between this stability index and R_b is given by Donelan *et al.* (1974).

Overwater wind data is usually unavailable so data from nearby sites on land (airports) is often used. Winds over the water are usually stronger than those over the land. A relationship between overwater (U_w) and overland (U_L) wind speeds for the Great Lakes can be found in Resio and Vincent (1977) or the Shore Protection Manual (1984). A detailed comparison of overwater and overland wind measurements in Lake Erie can be found in Phillips and Irbe (1978). If the land station is close to the water (within 500 m) the correction factor for onshore winds can be assumed equal to unity. Typical correction factors for offshore winds of engineering significance are in the range of 1.2 to 1.5.

Estimating Fetch

Fetch is the open water distance over which the wind blows with reasonably uniform speed and direction. For the *JONSWAP* and *SMB* models, fetch is usually estimated as the straight line distance in the direction of the wind to the edge of the wave generating area (coastline, island, shoal or discontinuity in the weather system). For irregular shorelines some form of fetch-averaging can be done. The Shore Protection Manual (1984) recommends extending 9 radials, at 3 degree intervals (i.e., wind direction ±12 degrees) from the point of interest to the upwind fetch boundary and arithmetically averaging the measured lengths.

JONSWAP Model

The *JONSWAP* results (Haselmann *et al.* 1973, Shore Protection Manual 1984) relating characteristic wave height (H_{m0}), period of the spectral peak (T_p), fetch (F) and wind duration (t), can be expressed as:

$$\begin{aligned}&\text{Fetch-limited}\\ H_{m0} &= 0.0016g^{-1/2}U_{10}F^{1/2};\\ &\text{Duration-limited}\\ &8.29 \times 10^{-5}g^{-2/7}U_{10}{}^{9/7}t^{5/7}\end{aligned} \tag{79}$$

$$\begin{aligned}&\text{Fetch-limited}\\ T_p &= 0.286g^{-2/3}(U_{10}F)^{1/3};\\ &\text{Duration-limited}\\ &0.0676g^{-4/7}U_{10}{}^{4/7}t^{3/7}\end{aligned} \tag{80}$$

where duration (t) in seconds, is given by

$$t = 68.8(gU_{10})^{-1/3}F^{2/3} \tag{81}$$

These equations are valid in deep water ($d/L > 0.5$) for locally generated seas. The relations are given in a nomogram in Figure 33. The nomogram is entered with values of U, F and t. The intersection of U and F will yield one set of values for H_{m0} and T_p (fetch-limited waves); the intersection of U and t will yield another (duration-limited waves). The smaller of the two sets of values is chosen as the design set. It is essential that fetch-limited wave calculations be checked to see if they are duration-limited; likewise, duration-limited cases should be checked to see if they are fetch-limited.

The *JONSWAP* equations are presented in dimensionless form in Figures 34 and 35 with adjustments for shallow water effects (Shore Protection Manual 1984). Use of the deepwater wave curve will give the same results as Figure 33 for fetch-limited conditions. For transitional and shallow water depths, assumed to be constant over the fetch, compute the value of gd/U_{10}^2 and use the appropriate curve. The curves labelled maximum wave height (Figure 34) and maximum wave period (Figure 35) indicate full wave growth, meaning that the wave parameters to the right of these curves are not fetch-limited. When using Figures 34 or 35 the storm duration t should be calculated using Equation (81) to check whether the waves are duration-limited. More detailed nomograms for shallow water wave prediction can be found in the Shore Protection Manual (1984).

Estimating Wave Direction

Until recently, most simple wave prediction models assumed coincident wind and wave directions ($\theta = 0$ where θ is the angle between wind and wave directions). However, it has been shown that waves with frequencies near the spectral peak can, in non-stationary (Hasselmann *et al.* 1980) or fetch-limited (Donelan *et al.* 1985) conditions, travel at off-wind angles. Values of θ up to 50 degrees have been observed in Lake Ontario (Donelan 1980). In fact, if the fetch gradient about the wind direction is large, one can expect the wave direction to be biased towards the longer fetches if the reduced generating force of the lower wind component ($U \cos \theta$) is more than balanced by the longer fetch over which it acts (Donelan 1980).

A relation for the dominant wave energy direction (ψ) versus wind direction (ϕ) in deep water can be determined for any given location (Figure 36). For fetch-limited conditions Donelan *et al.* (1985) found that the ψ versus θ relation for a point with known fetch distribution F_ψ could be obtained by maximizing the expression

$$\cos \theta F_\psi^{0.426} \tag{82}$$

The ψ versus θ relation for point 0 on an elliptical lake is shown in Figure 37.

A simple manual procedure for obtaining the ψ versus θ relation for any point in deep water is given below.

1. Starting in the wind direction and working towards longer fetches, extend radials from the point of interest to the fetch boundary in the upwind direction as far as the fetch continues to increase. Radials should be at some convenient interval depending on the fetch lengths and the desired resolution. An interval of 3 to 5 degrees would suffice in many applications.
2. Measure the fetch lengths, and average them over 30°C (± 15° from each radial.
3. Compute $\cos \theta F_\psi{}^{0.426}$.
4. The maximum value of the expression $\cos \theta F_\psi^{0.426}$ for any particular wind direction gives the corresponding dominant wave direction.

This procedure can easily be computerized. Calculation of wave direction, while not warranted for all wave predictions, may be important in the design of deep water structures,the estimation of wave climates, etc. In shallower and transitional water depths, refraction effects may alter wave directions further.

Knowing the ψ versus θ relation, somewhat improved accuracy in the prediction of H_{m0} and T_p can be achieved (Bishop 1983) using the fetch-limited equations of Donelan (1980):

$$H_{m_0} = 0.00366g^{-0.62}(U_{10} \cos \theta)^{1.24}F_\psi^{0.38} \tag{83}$$

$$T_p = 0.54g^{-0.77}(U_{10} \cos \theta)^{0.54}F_\psi^{0.23} \tag{84}$$

$$t = 30.1g^{-0.23}(U_{10} \cos \theta)^{-0.54}F_\psi^{0.77} \tag{85}$$

where F_ψ is the fetch in the dominant wave direction. Unless

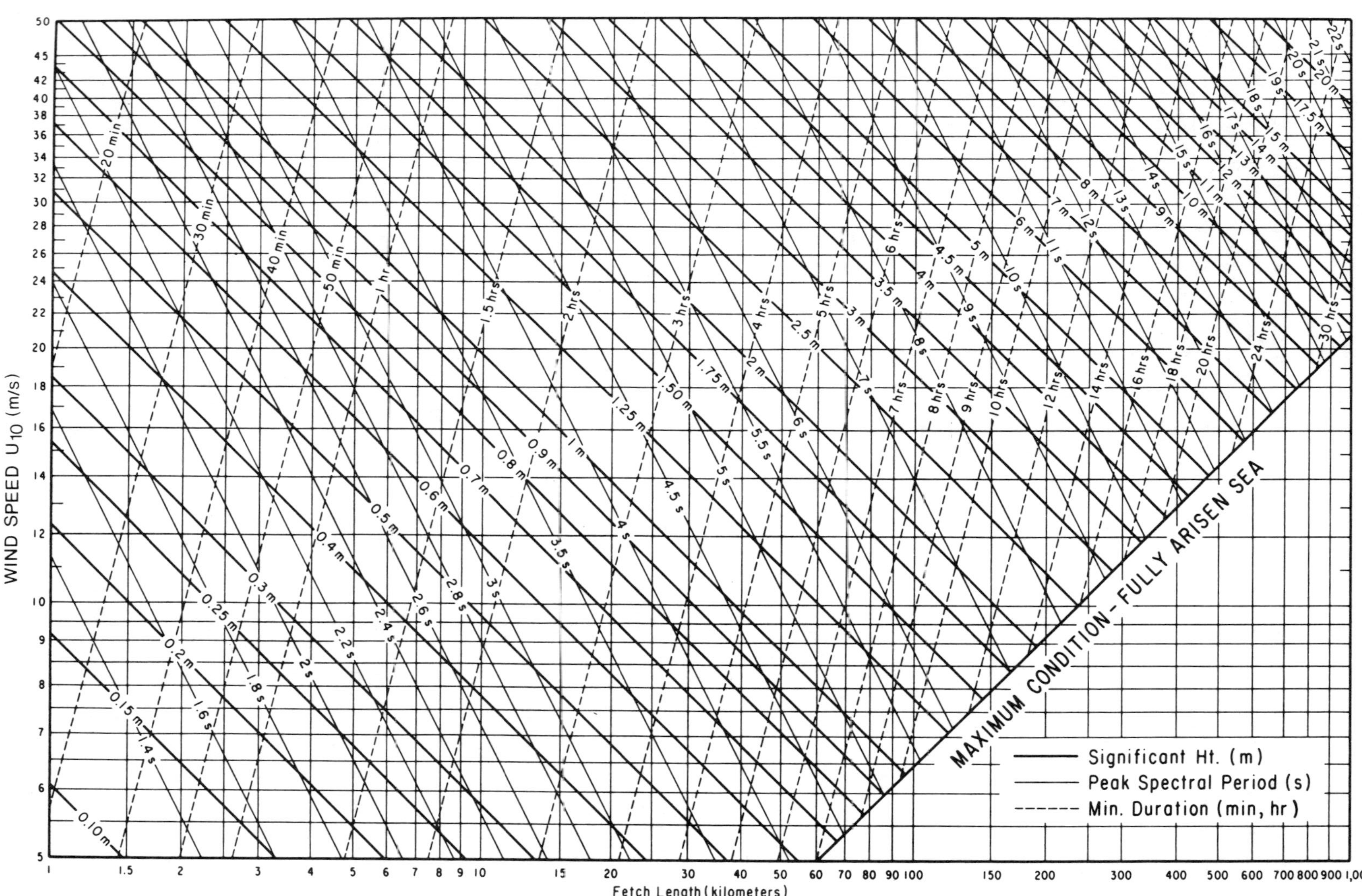

FIGURE 33. Nomogram of deepwater wave prediction curves (after *Shore Protection Manual* 1984).

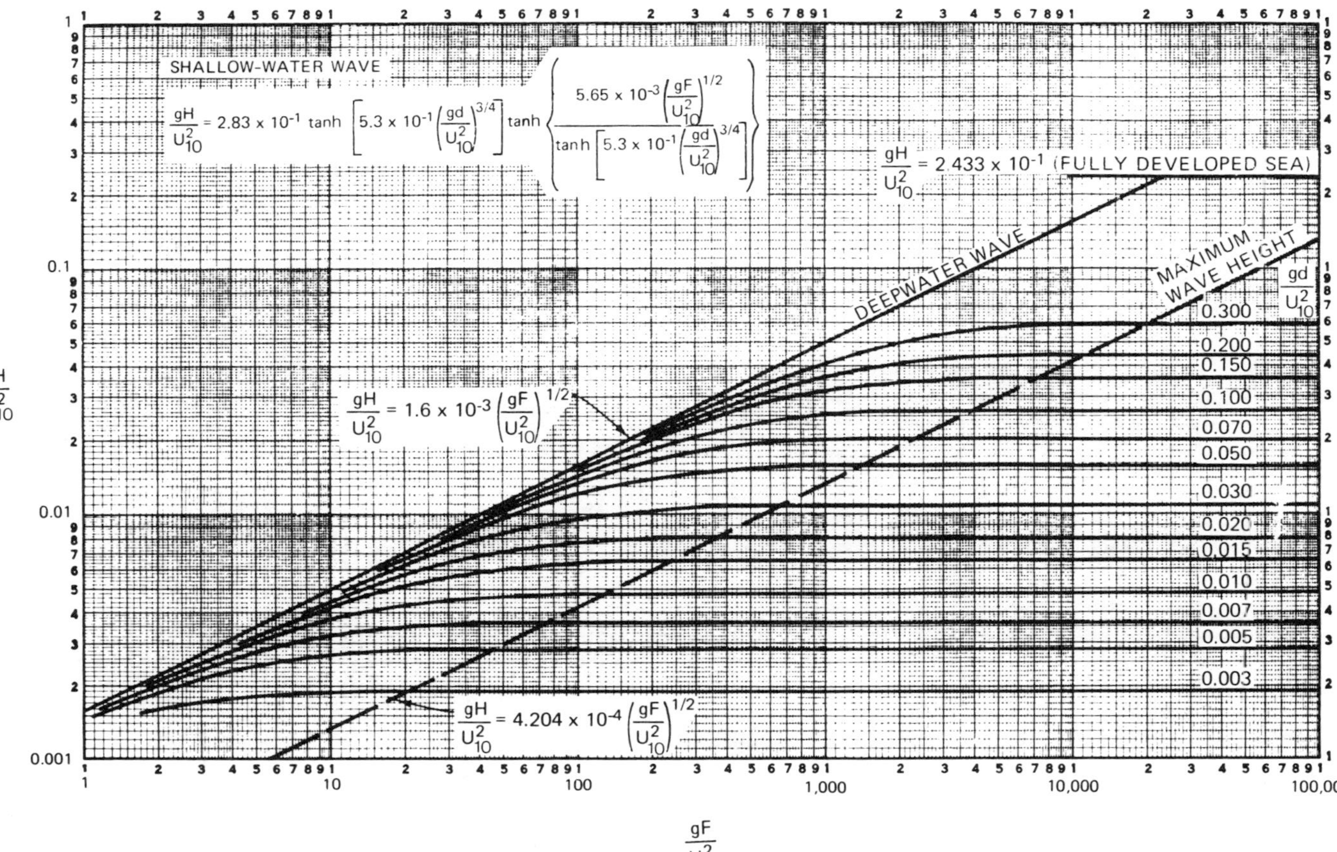

FIGURE 34. Wave height prediction curves for different values of constant water depth (after *Shore Protection Manual* 1984).

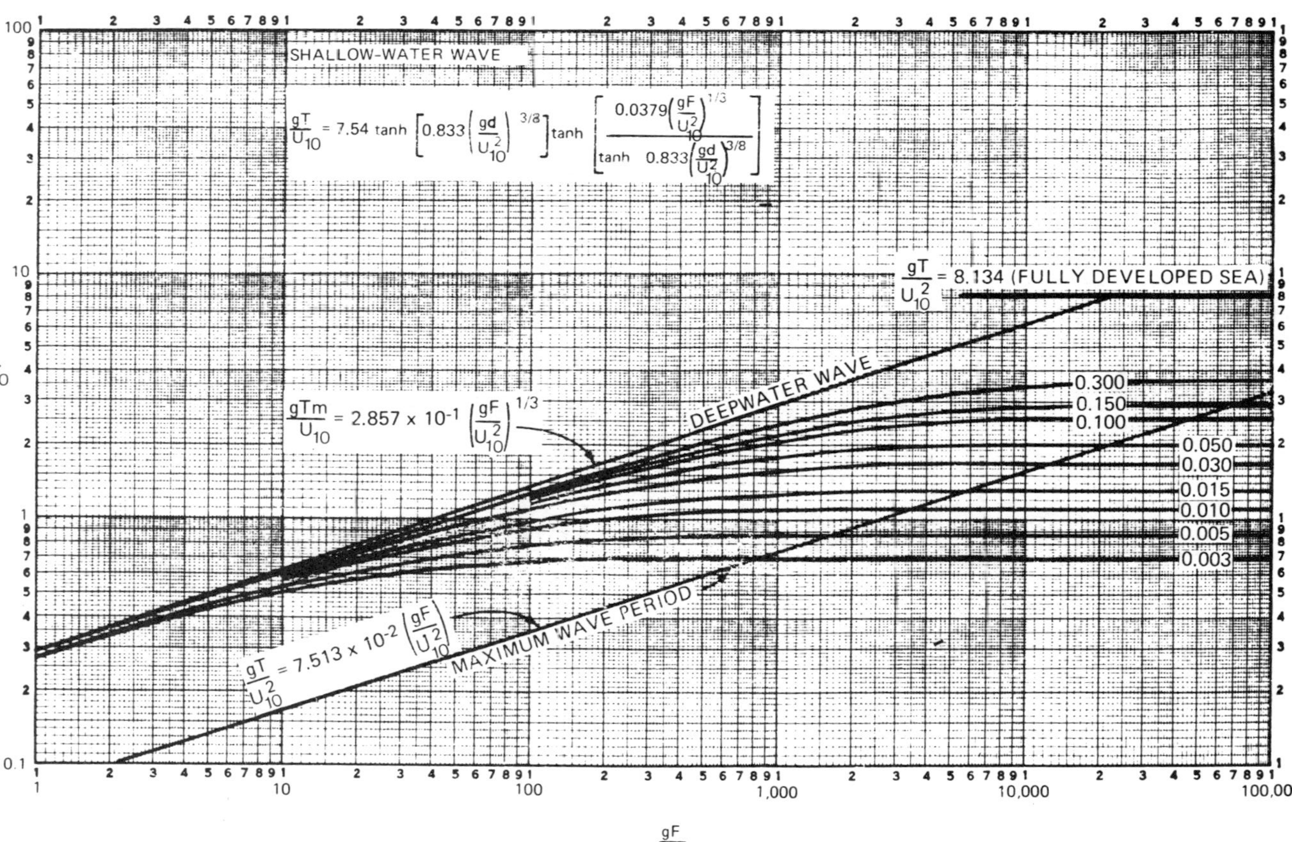

FIGURE 35. Wave period prediction curves for different values of constant water depth (after *Shore Protection Manual* 1984).

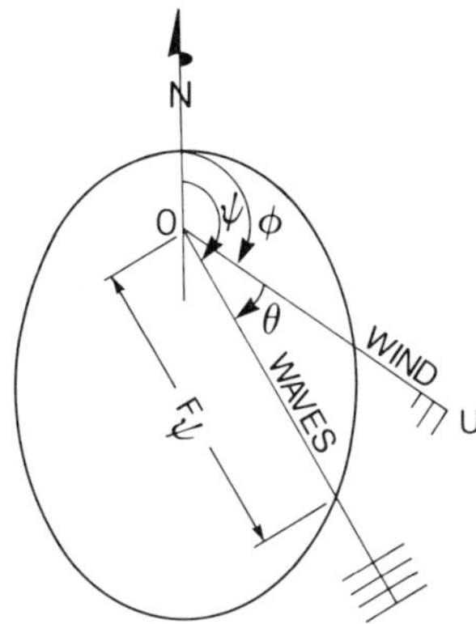

FIGURE 36. Definition sketch of wave direction versus wind direction (Donelan 1980).

θ is small, the use of F (fetch in the wind direction) rather than F_ψ in these equations will result in poorer accuracy of prediction than the use of the *JONSWAP* equations.

Wave calculations must be checked to see if the waves are duration-limited. A value of F_ψ can be calculated using Equation (85) with a known wind duration. If this value of fetch is less than the geometric fetch, then the waves are duration-limited and the lesser of fetch should be used. This in turn affects the ψ versus θ relation for the particular conditions of wind speed, direction and duration. New values for ψ and θ should be determined using duration-limited fetches, and then revised values of H_{mo} and T_p can be predicted.

To avoid overdevelopment of the calculated waves, the value of F_ψ used in Equations (83) and (84) must be subject to the criterion:

$$\frac{gF_\psi}{(U_{10}\cos\theta)^2} \leq 9.47 \times 10^4 \tag{86}$$

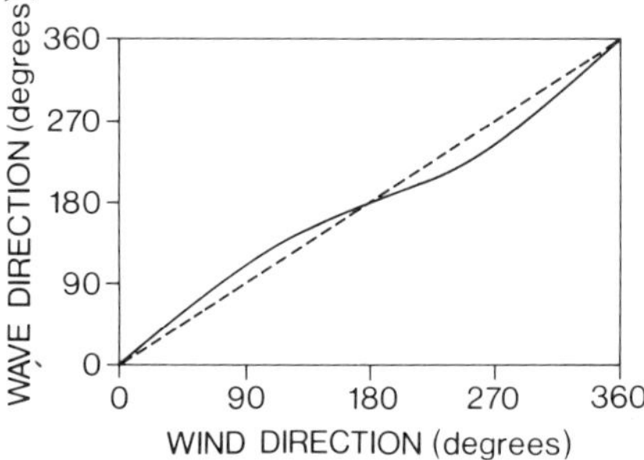

FIGURE 37. Relation for peak wave energy direction versus wind direction at a focus of the elliptical lake in Figure 36 (Donelan 1980).

For wind speeds of engineering significance this criterion is rarely exceeded on inland lakes. However, if the geometric fetch from the wind direction exceeds that given by Equation (86), the waves are fully developed and they approach from the wind direction with

$$\frac{gH_{mo}}{(U_{10})^2} = 0.285 \tag{87}$$

$$\frac{gT_p}{(U_{10})} = 7.54 \tag{88}$$

Numerical Models

The computation of wave climates, wave spectra and wave directional information is best done using numerical models. For the Great Lakes, a widely used spectral model is that of Resio and Vincent (1977), while the parametric model of Schwab *et al.* (1983) gives estimates of H_{m_0}, T_p and the dominant wave direction at each grid point. For oceanic conditions the discrete spectral ocean wave models of Pierson *et al.* (1966), Pierson (1982) and Golding (1985) and the hybrid (parametric wind sea, discrete swell) models of Günther and Rosenthal (1985) and DeVoogt *et al.* (1985) are commonly used.

REFERENCES

Abernethy, C. L. and Gilbert, G., 1975, "Refraction of Wave Spectra," *Hydraulics Research Station*, Report INT 117, Wallingford, England.

Abramowitz, M. and Stegun, I., (Editors), 1965, *Handbook of Mathematical Functions*, U.S. Dept. of Commerce, National Bureau of Standards, Applied Mathematics Series, 55, 1046 pp.

Airy, G. B., 1845, "Tides and Waves," *Encyclopedia Metropolitana*, 192, 241–396.

Alpers, W. R., Ross, D. B., and Rufenach, C. L., 1981, "On the Detectability of Ocean Surface Waves by Real and Synthetic Aperature Radar," *J. Geophys. Res.*, 86, 6481–6498.

Barber, N. F., 1963, "The Directional Resolving Power of an Array of Wave Detectors," *Ocean Wave Spectra*, Prentice-Hall, Inc., 137–150.

Barnett, T. P. and Wilkerson, J. C., 1967, "On the Generation of Ocean Wind Waves as Inferred from Airborne Radar Measurements of Fetch-Limited Spectra," *J. Mar. Res.*, 25, 292–321.

Barrick, D. E., 1972, "First-Order Theory and Analysis of *MF/HF/VHF* Scatter from the Sea," *IEEE Transactions*, Antennas Propagation, AP-20, 2–10.

Berger, U. and Kohlhase, S., 1976, "Mach-Reflection as a Diffraction Problem," *Proceedings of 15th Conference on Coastal Engineering, ASCE*, Vol. 1, 796–814.

Birkemeier, W. A. and Dalrymple, R. A., 1975, ''Nearshore Water Circulation Induced by Wind and Waves,'' *Proc. Symp. Modelling Technique, ASCE*, 1062–1081.

Bishop, C. T., 1983, ''Comparison of Manual Wave Prediction Models,'' *Journal of Waterway, Port, Coastal and Ocean Engineering, American Society of Civil Engineers*, Vol. 109, No. 1, 1–17.

Bishop, C. T. and Donelan, M. A., 1987, "Measuring Waves with Pressure Transducers," *Coastal Engineering*, 11:309-328.

Blackman, R. B. and Tukey, J. W., 1958, "The Measurement of Power Spectra from the Point of View of Communications Engineering,'' Dover Publications, Inc., 190 p.

Blue, F. L., Jr. and Johnson, J. W., 1949, ''Diffraction of Water Waves Passing Through a Breakwater Gap,'' *Trans. Amer. Geophysical Union*, Vol. 30, No. 5, 705–718.

Borgman, L. E., Hagan, R. L., and Kuik, A. J., 1982, ''Statistical Precision of Directional Spectrum Estimation with Data from a Tilt-and-Roll Buoy,'' In: *Topics in Ocean Physics*, edited by Osborne, A. R. and Rizzoli, P. M., Noord-Holland, Amsterdam, 418–438.

Bouws, E., Günther, H., Rosenthal, W., and Vincent, C. L., 1985, ''Similarity of the Wind Wave Spectrum in Finite Depth Water, 1. Spectral Form,'' *J. of Geophysical Research*, Vol. 90, No. C1, 975–986.

Bouws, E., Günther, H., Rosenthal, W. and Vincent, C. L., 1987, "Similarity of the Wind Wave Spectrum in Finite Depth Water, 2. Statistical Relations Between Shape and Growth Stage Parameters," Dt. Hydrogr. Z. 40.

Brampton, A. H., 1981, "A Computer Method for Wave Refraction,'' *Hydraulics Research Station*, Report IT 172, Wallingford, England.

Brater, E. F. and King, H. W., 1976, *Handbook of Hydraulics*, Sixth Edition, McGraw-Hill.

Bretschneider, C. L., 1958, ''Revisions in Wave Forecasting; Deep and Shallow Water,'' *Proceedings of 6th Conference on Coastal Engineering, American Society of Civil Engineers*, New York.

Bretschneider, C. L., 1973, "Prediction of Waves and Currents," *Look Laboratory*, Hawaii, vol. III, 1:1-17.

Capon, J., Greenfield, R. J., and Kolker, R. J., 1967, "Multidimensional Maximum-Likelihood Processing of a Large Aperture Seismic Array,'' *Proceedings IEEE*, Vol. 55, 192–211.

Capon, J., 1969, ''High-Resolution Frequency Wavenumber Spectrum Analysis,'' *Proc. IEEE*, 57, No. 8.

Cavaleri, L., 1980, ''Wave Measurement Using Pressure Transducer,'' *Oceanologica Acta*, Vol. 3, 3:339–345.

Chao, Y-Y., 1972, ''Refraction of Ocean Surface Waves on the Continental Shelf,'' *Offshore Technology Conference*, OTC 1616, 1965–1974.

Chao, Y-Y. and Pierson, W. J., 1972, ''Experimental Studies of the Refraction of Uniform Wave Trains and Transient Wave Groups Near a Straight Caustic,'' *J. Geophys. Res.*, 77, 4545–4554.

Cote, L. J., Davis, J. O., Marks, W., McGough, R. J., Mehr, E., Pierson, W. J., Jr., Ropek, J. F., Stephenson, G., and Vetter, R. C., 1960, ''The Directional Spectrum of Wind-Generated Sea as Determined from Data Obtained by the Stereo Wave Observation Project,'' *Meteor. Pap.*, New York University, College of Engineering, Vol. 2, No. 6, 88 pp.

Davis, R. E. and Regier, L., 1977, ''Methods for Estimating Directional Wave Spectra from Multi-Element Arrays,'' *J. Mar. Res.*, 35, 453–477.

Dean, R. G., 1974, ''Evaluation and Development of Water Wave Theories for Engineering Application,'' *Special Report No. 1*, U.S. Army Corps of Engineers, Coastal Engineering Research Center, Ft. Belvoir, Virginia.

DeVoogt, W. J. P., Komen, G. J., and Bruinsma, J., 1985, ''The KNMI Operational Wave Prediction Model GONO,'' *Proc. Symp. Wave Dynamics and Radio Probing of Ocean Surface*, Miami, 1981, Plenum Press.

Dobson, R. S., 1967, ''Some Applications of Digital Computers to Hydraulic Engineering Problems,'' TR-80, Chapter 2, Department of Civil Engineering, Stanford University, Palo Alto, California.

Donelan, M. A., 1980, ''Similarity Theory Applied to the Forecasting of Wave Heights, Periods and Directions,'' *Proc. Canadian Coastal Conf. 1980*, National Research Council, Canada, 47–61.

Donelan, M. A., Birch, K. N., and Beesley, D. C., 1974, ''Generalized Profiles of Wind Speed, Temperature and Humidity,'' *Internat. Assoc. Great Lakes Res., Conf. Proc. 17*, 369–388.

Donelan, M. A. and Pierson, W. J., 1983, ''The Sampling Variability of Estimates of Spectra of Wind-Generated Gravity Waves,'' *Journal of Geophysical Research*, Vol. 88, No. C7, 4381–4392.

Donelan, M. A., Hamilton, J., and Hui, W. H., 1985, ''Directional Spectra of Wind-Generated Waves,'' *Phil. Trans. R. Soc. Lond.*, A 315, 509–562.

Dorrestein, R., 1960, ''Simplified Method of Determining Refraction Coefficients for Sea Waves,'' *J. Geophys. Res.*, Vol. 65, No. 2, 637–642.

Draper, L. G., 1957, ''Attenuation of Sea Waves with Depth,'' *La Houille Blanche*, Vol. 12, 6:926–931.

Eckart, C., 1952, ''The Propagation of Gravity Waves from Deep to Shallow Water,'' *Gravity Waves*, Circular No. 521, National Bureau of Standards, Washington, D.C.

Esteva, D. and Harris, L., 1970, ''Comparison of Pressure and Staff Wave Gage Records,'' *Proceedings of the 12th Coastal Engineering Conference, American Society of Civil Engineers*, 1:101–116.

Ewing, J., 1973, ''Mean Length of Runs of High Waves,'' *J. Geophysical Research*, Vol. 78, No. 12, 1933–1936.

Fisher, R. A. and Tippet, L. H. C., 1928, ''Limiting Forms of the Frequency Distribution of the Largest or Smallest Number of a Sample,'' *Proc. Camb. Phl. Soc.*, 24, 180–190.

Forristall, G. Z., 1982, ''Subsurface Wave-Measuring Systems,'' In: *Measuring Ocean Waves*, National Academy Press, Washington, D.C., 194–209.

Goda, Y., 1970, ''Numerical Experiments on Wave Statistics with Spectral Simulation,'' *Report of Port and Harbour Research Institute*, Vol. 9, No. 3, 3–57.

Goda, Y., 1975, "Irregular Wave Deformation in the Surf Zone," *Coastal Engineering in Japan*, Vol. 18, 13–26.

Goda, Y., 1976, "Irregular Sea Waves for the Design of Harbour Structures," *Transactions Japan Society of Civil Engineers*, Vol. 8, 267–271.

Goda, Y., 1977, "Numerical Experiments on Statistical Variability of Ocean Waves," *Report of Port and Harbour Research Institute*, Vol. 16, No. 2, 3–26.

Goda, Y., 1985, "Random Seas and Design of Maritime Structures," University of Tokyo Press, Japan, 323 p.

Goda, Y., Takayama, T., and Suzuki, Y., 1978, "Diffraction Diagrams for Directional Random Waves," *Proceedings of the 16th Conference on Coastal Engineering, American Society of Civil Engineers*, 1:628–650.

Golding, B., 1985, "The U.K. Meteorological Office Operational Wave," *Proc. Symp. Wave Dynamics and Radio Probing of Ocean Surface*, Miami, 1981, Plenum Press.

Grace, R. A., 1978, "Surface Wave Height from Pressure Records," *Coastal Engineering*, 2:55–67.

Günther, H., 1981, "A Parametric Surface Wave Model and the Statistics of the Prediction Parameters," Edited by Geophysikalisches Institut University of Hamburg, Hamburger Geophysikalische Einzelschriften 55.

Günther, H. and Rosenthal, W., 1985, "The Hybrid Parametrical (HYPA) Wave Model," *Proc. Symp. Wave Dynamics and Radio Probing of the Ocean Surface*, Miami, 1981, Plenum Press.

Harms, V. W., 1979, "Diffraction of Water Waves by Isolated Structures," *Journal of the Waterway, Port, Coastal and Ocean Division, American Society of Civil Engineers*, Vol. 105, WW2:131–147.

Hasselmann, K., Barnett, T. P., Bouws, E., Carlson, H., Cartwright, D. E., Enke, K., Ewing, J. A., Gienapp, H., Hasselmann, D. E., Kruseman, P., Meerburg, A., Muller, P., Olbers, D. J., Richter, K., Sell, W., and Walden, H., 1973, "Measurements of Wind-Wave Growth and Swell Decay During the Joint North Sea Wave Project (*JONSWAP*), *Deut. Hydrogr. Z., Suppl. A*, 8, No. 12, 22 pp.

Hasselmann, D. E., Dunckel, M., and Ewing, J. A., 1980, "Directional Wave Spectra Observed During *JONSWAP* 1973," *J. Phys. Oceanogr.*, 10, 1264–1280.

Holthuijsen, L. H., 1981, "The Directional Energy Distribution of Wind Generated Waves as Inferred from Stereophotographic Observations of the Sea Surface," *Thesis*, Delft University of Technology, Rep. No. 81-2, 193 pp.

Hughes, S. A., 1984, "The *TMA* Shallow-Water Spectrum Description and Applications," U.S. Army Waterways Experiment Station, Coastal Engineering Research Centre, Technical Report 84-7, Vicksburg, Mississippi.

Irani, G. B., Gotwols, B. L., and Bjerkaas, A. W., 1981, "Ocean Wave Dynamics Test: Results and Interpretations," *Rep. No. STD-R-537*, The Johns Hopkins University, Applied Physics Laboratory, 202 pp.

Isobe, M., Kondo, K., and Horikawa, K., 1984, Extension of MLM for Estimating Directional Wave Spectrum," *Proceedings Symposium on Description and Modelling of Directional Seas*, Paper No. A-6, 1–15, Technical University of Denmark.

Jackson, F. C., Walton, W. T., and Baker, P. L., 1985, "Aircraft and Satellite Measurement of Ocean Wave Directional Spectra Using Scanning Beam Microwave Radars," *J. Geophys. Res.*, 90, No. C1, 987–1004.

Johnson, J. W., 1952, "Generalized Wave Diffraction Diagrams," *Proceedings of the 2nd Conference on Coastal Engineering, American Society of Civil Engineers*, New York.

Karlsson, T., 1969, "Refraction of Continuous Ocean Wave Spectra," *Proc. ASCE*, Vol. 95, No. WW4, pp. 437–448.

Kenney, J. E., Uliana, E. A., and Walsh, E. J., 1979, "The Surface Contour Radar, A Unique Remote Sensing Instrument," *IEEE Trans. Microwave Theory and Techniques*, MTT-27, No. 12, 1080–1092.

Kimura, A., 1980, "Statistical Properties of Random Wave Groups," *Proceedings 17th Coastal Engineering Conference, American Society of Civil Engineers*, Vol. 3, 2955–2973.

Kitaigorodskii, S. A., 1962, "Applications of the Theory of Similarity to the Analysis of Wind-Generated Wave Motion as a Stochastic Process," *IZv. Akad. Nank SSR*, Ser. Geofiz, 1, 73–80.

Korteweg, D. J. and De Vries, G., 1895, "On the Change of Form of Long Waves Advancing in a Rectangular Canal, and on a New Type of Long Stationary Wave," *Philosophical Magazine*, Series 5, 39:422–443.

Kuik, A. J. and van Vledder, G. Ph., 1984, "Proposed Method for the Routine Analysis of Pitch-Roll Buoy Data," *Proceedings Symposium on Description and Modelling of Directional Seas*, Paper No. A-5, 1–13, Technical University of Denmark.

Large, W. G. and Pond, S., 1981, "Open Ocean Momentum Flux Measurements in Moderate to Strong Winds," *J. Phys. Oceanog.*, 11, 324–336.

Long, R. B. and Hasselmann, K., 1979, "A Variational Technique for Extracting Directional Spectra from Multi-Component Wave Data," *Journal of Physical Oceanography*, Vol. 9, No. 2, 373–381.

Longuet-Higgins, M. S., 1952, "On the Statistical Distributions of the Heights of Sea Waves," *Journal of Marine Research*, Vol. IX, No. 3, 245–266.

Longuet-Higgins, M. W., 1983, "On the Joint Distribution of Wave Periods and Amplitudes in a Random Wave Field," *Proc. R. Soc. Lond.*, A 389, 241–258.

Longuet-Higgins, M. S., Cartwright, D. E., and Smith, N. D., 1963, "Observations of the Directional Spectrum of Sea Waves Using the Motions of a Floating Buoy," In: *Ocean Wave Spectra*, pp. 111–136, Englewood Cliffs, N.J.: Prentice Hall, Inc.

McClain, C. R., Chen, D. T., and Hart, W. D., 1982, "On the Use of Laser Profilometry for Ocean Wave Studies," *Journal of Geophysical Research*, 87, No. C12, 9509–9515.

McCowan, J., 1891, "On the Solitary Wave," *Philosophical Magazine*, Series 5, 32:45–58.

Michell, J. H., 1893, "On the Highest Waves in Water," *Philosophical Magazine*, 5th Series, Vol. 36, 430–437.

Mitsuyasu, H., Tasai, F., Suhara, T., Mizumo, S., Ohkuso, M.,

Honda, T., and Rikiishi, K., 1975, ''Observations of the Directional Spectrum of Ocean Waves Using a Cloverleaf Buoy,'' *J. Phys. Oceanogr.*, 5, 750–760.

Muir, L. R. and El-Shaarawi, A. H., 1986, "On the Calculation of Extreme Wave Heights: A Review," *Ocean Engineering*, Vol. 13, 1:93–118.

Munk, W. H., Miller, G. R., Snodgrass, F. E., and Barber, N. F., 1963, ''Directional Recording of Swell from Distant Storms,'' *Phil. Trans. R. Soc. Lond.*, A 255, 505–584.

Nagata, Y., 1964, ''The Statistical Properties of Orbital Wave Motions and Their Application for the Measurement of Directional Wave Spectra,'' *Journal of the Oceanographic Society of Japan*, Vol. 19, No. 4, 169–191.

Nolte, K. G. and Hsu, F. H., 1973, ''Statistics of Ocean Wave Groups,'' *4th Offshore Tech. Conf.*, No. 1688.

Noye, B. J., 1968, ''The Frequency Response of a Tide-Well,'' *Proc. 3rd Australian Conf. on Hydraulics and Fluid Mechanics*, Sydney, 65–71.

Noye, B. J., 1972, ''On the Differential Equations for the Conventional Tide-Well System,'' *Bull. Australian Math. Soc.*, Vol. 7, 251–267.

Ochi, M. K., 1973, ''On Prediction of Extreme Values,'' *J. Ship Research*, March 1973, 29–37.

Palmer, R. Q., 1957, ''Wave Refraction Plotter,'' U.S. Army Corps of Engineers, Beach Erosion Board, Vol. 11, Bulletin No. 1, Washington, D.C.

Penny, W. G. and Price, A. T., 1944, ''Diffraction of Sea Waves by a Breakwater,'' *Artificial Harbors, Technical History No. 26, Sec. 3-D*, Directorate of Miscellaneous Weapons Development.

Phillips, D. W. and Irbe, J. G., 1978, "Lake to Land Comparison of Wind, Temperature and Humidity on Lake Ontario During the International Field Year for the Great Lakes," Report CLI-2-77, Atmospheric Environment Service, Downsview, Ontario.

Pierson, W. J., 1972, ''The Loss of Two British Trawlers—A Study of Wave Refraction,'' *J. Navigation*, 25, 291–304.

Pierson, W. J., 1982, ''The Spectral Ocean Wave Model (SOWM),'' *A Northern Hemisphere Computer Model for Specifying and Forecasting Ocean Wave Spectra*, David W. Taylor, Naval Ship Research and Development Center, DTNSRDC-82/011.

Pierson, W. J. and Marks, W., 1952, ''The Power Spectrum Analysis of Ocean-Wave Records,'' *Trans. Amer. Geophys. Union*, 33, No. 6, 834–844.

Pierson, W. J. and Moskowitz, L., 1964, ''A Proposed Spectral Form of Fully Developed Wind Seas Based on the Similarity Law of S. A. Kitaigorodskii,'' *Journal of Geophysical Research*, Vol. 69, No. 24, 5181–5190.

Pierson, W. J. and Holmes, P., 1965, ''Irregular Wave Forces on a Pile,'' *Journal of the Waterways and Harbors Division, Proceedings of the American Society of Civil Engineers*, Vol. 91, No. WW4, 1–10.

Pierson, W. J., Tick, J., and Baer, L., 1966, ''Computer Based Procedures for Preparing Global Wave Forecasts and Wind Field Analyses Capable of Using Wave Data Obtained by a Spacecraft,'' *Proc. 6th Naval Hydrodynamics Symp.*, Publ. ACR-136, Office of Naval Research, Department of the Navy, Washington, D.C.

Resio, D. T. and Vincent, C. L., 1977, ''Estimation of Winds Over the Great Lakes,'' *Journal of the Waterway, Port, Coastal and Ocean Division, American Society of Civil Engineers*, Vol. 103, WW4, 265–283.

Rye, H., 1982, ''Ocean Wave Groups,'' Dept. Marine Tech., Norwegian Inst. Techn., Report UR-82-18, 214 pp.

Schwab, D. J., Bennett, J. R., Lui, P. C., and Donelan, M. A., 1984, ''Application of a Simple Numerical Wave Prediction Model to Lake Erie,'' *J. Geophys. Res.*, 89, No. C3, 3586–3592.

Shore Protection Manual, 1984, U.S. Army Corps of Engineers, Coastal Engineering Research Center, Vicksburg, Mississippi.

Shuto, N., 1974, ''Nonlinear Long Waves in a Channel of Variable Section,'' *Coastal Eng. in Japan*, 17, 1–12.

Simpson, J. H., 1969, ''Observations of the Directional Characteristics of Waves,'' *Geophysical Journal of the Royal Astronomical Society*, 17:92–120.

Skafel, M. G. and Donelan, M. A., 1983, ''Performance of the CCIW Wave Direction Buoy at ARSLOE,'' *IEEE J. Oceanic Eng.*, OE-8, No. 4, 221–225.

Sommerfeld, A., 1896, ''Mathematische Theorie der Diffraktion,'' *Mathematische Annalen*, 47, 317–364,

St. Denis, M. and Pierson, W. J., 1953, ''On the Motions of Ships in Confused Seas,'' *Trans. Soc. Naval Architects and Marine Engineers*, 61, 280–357.

Stokes, G. G., 1847, ''On the Theory of Oscillatory Waves,'' *Mathematical and Philisophical Papers*, Cambridge University Press, London, 1:314–326.

Sverdrup, H. U. and Munk, W. H., 1947, ''Wind, Sea and Swell: Theory of Relations for Forecasting,'' Publication No. 601, U.S. Navy Hydrograpic Office, Washington, D.C.

Tayfun, M. A., Dalrymple, R. A., and Yang, C. Y., 1976, ''Random Wave-Current Interactions in Water of Varying Depth,'' *Ocean Engineering*, 3, No. 6, 403–420.

Teague, C. C., Tyler, G. L., and Stewart, R. H., 1977, ''Studies of the Sea Using HF Radio Scatter,'' *IEEE Journal of Oceanic Engineering*, OE-2, No. 1, 12–19.

Thompson, E. F., 1980, ''Energy Spectra in Shallow U.S. Coastal Waters,'' Technical Paper 80-2, U.S. Army Corps of Engineers, Coastal Engineering Research Center, Fort Belvoir, Virginia.

Thompson, E. F. and Vincent, C. L., 1985, ''Significant Wave Height for Shallow Water Design,'' *ASCE, J. Waterway, Port, Coastal and Ocean Engineering*, Vol. 111, No. 5, 828–842.

Thornton, E. B. and Guza, R. T., 1982, ''Energy Saturation and Phase Speeds Measured on a Natural Beach,'' *Journal of Geophysical Research*, Vol. 87, C12:9499–9508.

Tsai, B. M. and Gardner, C. S., 1982, ''Remote Sensing of Sea State Using Laser Altimeters,'' *Applied Optics*, 21, No. 21, 3932–3940.

Tubman, M. W. and Suhayda, J. H., 1976, ''Wave Action and Bottom Movements in Fine Sediments,'' *Proceedings of the 15th*

Coastal Engineering Conference, American Society of Civil Engineers, 2:1168–1183.

Tucker, M. J., 1963, ''Analysis of Records of Sea Waves,'' *Proceedings of the Institute of Civil Engineers*, Vol. 26, 10:305–316.

Tucker, M. J., Carr, M. P., and Pitt, E. G., 1983, ''The Effect of an Offshore Bank in Attenuating Waves,'' *Coastal Engineering*, Vol. 7, 2:133–144.

Vincent, C. L., 1985, ''Energy Saturation of Irregular Waves During Shoaling,'' *Journal of Waterway, Port, Coastal and Ocean Engineering, ASCE*, Vol. 111.

Wiegel, R. L., 1962, ''Diffraction of Waves by a Semi-Infinite Breakwater,'' *Journal of the Hydraulics Division, American Society of Civil Engineers*, Vol. 88, HY1:27–44.

Wiegel, R. L., 1964, *Oceanographical Engineering*, Prentice-Hall, Inc., Englewood Cliffs, New Jersey.

Wilson, W. S., 1966, "A Method for Calculating and Plotting Surface Wave Rays," TM-17, U.S. Army Engineers Waterways Experiment Station, Vicksburg, Mississippi.

Young, I., Rosenthal, W., and Ziemer, F., 1985, ''A Three-Dimensional Analysis of Marine Radar Images for the Determination of Ocean Wave Directionality and Surface Currents,'' *J. Geophys. Res.*, 90, No. C1, 1061–1067.

CHAPTER 21

Waves on Coastal Structures

A. W. SAM SMITH*

INTRODUCTION

The interaction between waves and structures remains one of the most imprecise processes known to coastal engineering. The serious scientific design of coastal structures was barely commenced by the late thirties, but the generally successful results of the studies were structures constructed in comparatively shallow water and exposed to only moderate wave energy. Prior to this, coastal structures were generally the result of continual trial and error. When damaged, these structures were simply repaired after each damage episode until an effective result was attained.

Between the thirties and the sixties, a large number of comparatively successful coastal structures resulted. This was the period during which site experience could be related to model wave flume tests, using shallow water waves. In the seventies, maritime trade became dependent upon the use of very much larger vessels than previously. This lead to the construction of much larger ports on exposed coast and in much deeper water. For the design of the structures required to protect these ports from wave action, the shallow water design procedures and model testing techniques developed by the sixties, were extrapolated into the new conditions. The result was a string of major coastal disasters and failures, almost unprecedented in congruent terrestrial heavy construction.

Various factors ensured this result. Not only were the then accepted design and model testing procedures unsuitable for deep water high wave energy conditions, but the crucial factors that govern the structure/wave interaction process itself were grossly over-simplified. These were then masked by the use of arbitary co-efficients and scaling factors. The study of waves upon structures is currently therefore in much turmoil—the old design construction tools have been proven inadequate and new and more reliable techniques still remain to be evolved. This section investigates the structure/wave interaction of hydraulic forcing functions and structural response, in terms of current knowledge against future research requirements.

TYPES OF COASTAL STRUCTURE

The response of a coastal structure to wave input depends to a large extent upon the ratio of the mass of the structure to the mass of the wave. Large monolith structures resist wave forces entirely, by their own mass and weight. This class of structure usually consists of mass concrete, large keyed blocks, concrete or steel caissons or grouted rubble, with either a concrete or bitumen grout. These structures are not designed to move and thus can be described as rigid structures.

Nearly all the other coastal structures can then be described as flexible. These structures are constructed from much smaller wave resisting elements, either entirely or are at least fully faced with such units. Each unit or element can positively move, even if only to a limited extent, under wave action. The whole structure is then also free to move. If the toe section scours or the foundations settle in any way, then the individual elements can rattle-down or settle. This then ensures a minimum loss of integrity to the structure as a whole. These elements, or armour units, hold other valuable response properties, due to their mobility. Armour units tend to follow a stable weight/slope relationship against wave force, so if their wave capacity at any given slope is exceeded, then they may again rattle-down and form a flatter slope.

This class of structure is usually called a rubble mound, although their wave resisting elements (or armour) may con-

*Queensland Institute of Technology, Brisbane, Queensland, Australia

sist of concrete blocks and complex concrete "shapes," as well as natural rock rubble. Large armour units hold a much higher wave force capacity than smaller units and are thus stable at steeper slopes—and the extreme cases are the gravel and sandy beaches found in Nature. These are natural rubble structures, but the efficiency of all rubble structures depends upon their unit mobility and overall flexibility. Sandy beaches are extremely mobile and change their shape extensively in the face of variable wave attack. Gravel beaches respond in the same way, but to a lesser degree, whilst a well designed rubble mount should move very little. The rubble mound should be proportioned to avoid gross changes to its geometry, but the individual armour units must hold some individual mobility if they are to attain an optimum wave energy capacity. Rubble mound walls can be felt and heard to rattle and growl, under heavy wave attack, by simply standing on top of them.

The hydraulic efficiency that rubble armour tends to demonstrate, must be purchased at a price. The rubble must hold a suitable capacity to hold itself up against gravity, as well as resist wave energy. This factor is often overlooked. To ensure structural stability, rubble armour must retain enough interlocking capacity to remain in place, whilst still allowing the limited flexibility required of the units themselves. Rubble must therefore attain an effective balance between gravity and wave forces.

In addition to the two classes of coastal structures discussed above, some designers also evolve hybrid-structures, in an attempt to obtain some of the hydraulic benefits of both. Sometimes a rigid monolith is utilised to provide a massive core to the structure, and this is then faced on one or both sides by flexible rubble armour. Another approach is to construct a minimum volume rubble core and faces, but then top this with a rigid cap monolith and overtopping resistant wave wall. Although the hydraulic effectiveness of a hybrid structure appears readily amenable to theoretical analysis and model test confirmation, its gravity loading stress distribution effectiveness is not.

For the monolith capped rubble wall, a particular problem arises in providing a reliable support to the monolith itself. The rubble armour immediately below the monolith, particularly among the monolith edges, must accept very high applied dead load stresses, as well as support itself. The safe slope at which the rubble immediately beneath the monolith will stand, becomes governed by the applied surcharge monolith dead load. The sub-monolith rubble must therefore be laid much flatter than the angle, that would be adequate for the rubble on its own, in resisting both self gravity and wave forces. The Sines breakwater might represent an extreme example.

The interface between the rigid and flexible elements of hybrid structures may cause concern in other ways. The monolith core of a hybrid rubble structure provides a hard energy reflective "face" behind the rubble. In a full rubble structure, the input wave energy becomes progressively attenuated and dissipated, but with an abrupt reflective face, a significant wave energy load may be "bounced-back" and become added to the wave backwash forces. This response can also occur with any type of impervious core, in a rubble structure. Similar wave reflection behaviour may be induced by the impervious base of a cap monolith, but extra high impact loads are possible as "hammer-blow," when the air pockets within the mound are compressed by the wave front. Then marked "vibration" impact may become added to peak dead load stresses, applied by the monolith, to its supporting rubble.

In these ways, the monolith capped rubble mound becomes particularly sensitive to progressive failure. The flexible rubble has a very significant capacity for movement, so any particular rubble unit that is overloaded, will merely move aside to release its overstress. The rigid monolith cap, on the other hand, can not do this. Its behaviour is brittle, so as soon as even moderate volumes of its support are lost, its behaviour response will be by major fracture. Once a rigid monolith becomes fractured into pieces, its major wave resistance element (i.e. its mass) becomes greatly reduced. Then each broken section, without any surcharge or interlock, is highly prone to impact, scour and individual movement, induced by wave overtopping. Until wave/structure interaction theory can explain the prototype behaviour of hybrid structures—their use should be approached with considerable caution.

DEEPWATER WAVES

Ocean waves are one of Nature's most variable energy transfer mechanisms. Waves are manufactured by the wind, through a transfer of energy, and the waves then carry this energy until it becomes dissipated by wave attenuation over great distances or by breaking upon the shore. The forcing function of the wind is remarkably variable—its input to the waves is affected by wind velocity, duration, direction, fetch length and its own pressure gradients. Waves during manufacture are called a "sea" or a "rising sea", so the sea will reflect all these manufacture variabilities, but very particularly the wind duration. If the wind changes rapidly, then the sea may be only slightly, significantly or fully risen, when the wind changes. Even before this, the waves may have been carrying a different wave form, generated somewhere else at some previous time. The new applied wind must then "erase" this message, before it can redevelop its own energy imprint on the ocean.

This process may then become slightly, partially or wholly completed, all depending upon the ratio of the waves previous message energy and the new wind's own time dependent erasing and wave growth capacity energy. Under short wind duration conditions, this erasure/new growth type of process, may be quite incomplete—the result will be multiple wave forms and crossed seas (or crossed seas and

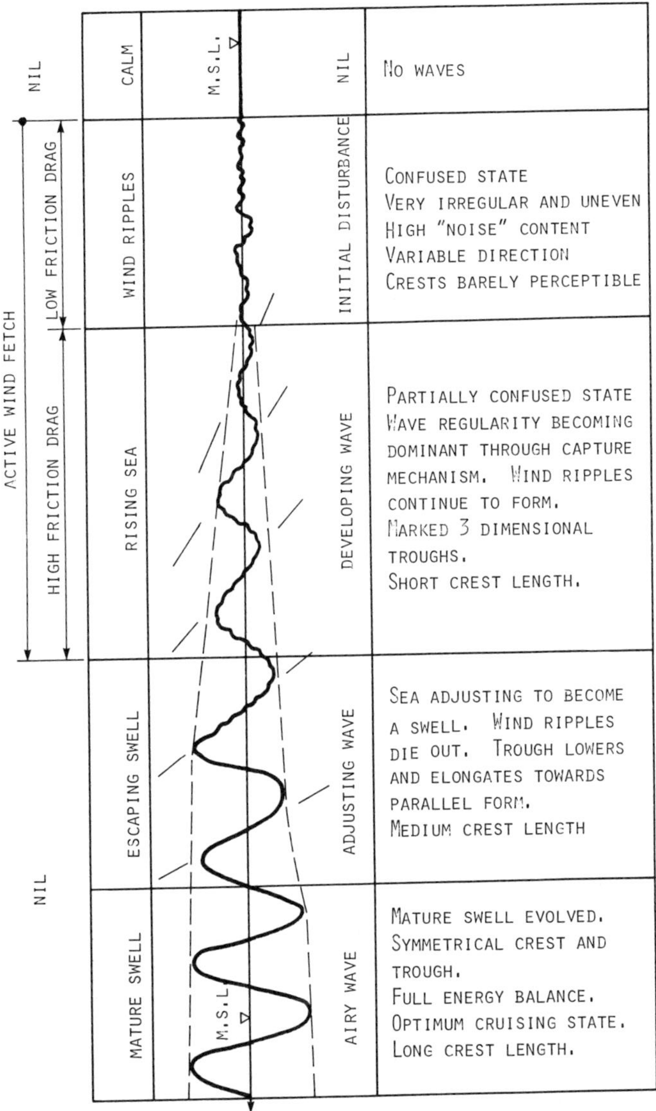

FIGURE 1. The process of the manufacture of swell waves in the ocean.

swells) of the most complex kind. Central ocean sea states may abound with this highly complex mix of confused conditions over extended periods of time.

Once the wind input dies down or the waves escape from the wind manufacturing zone, then the rising or risen sea adjusts itself to become an ocean swell. This adjustment process is quite complex, but its main features are:

(a) The waves attain a long term stability by reaching a balance, between kinetic and potential energy. This becomes equal, resulting in the development of the Airy wave form.

(b) The "escaped" waves become comparatively regular through a process of trapping and amalgamating uneven and random partially formed wave components that are not Airy.

(c) The wave crest length thus increases, and this increases the wave translation directional alignment towards parallel. Smaller more random and less well developed wave forms are absorbed until a common swell direction is attained.

This adjustment process takes time, however, an escaped sea may require dozens or hundreds of nautical miles "run" before the fully developed Airy swell becomes completely "mature." The real feature of a free mature swell is that with minimum energy loss, it may run for a thousand miles or so with only an energy density decrease due to an increase of radial attenuation along the wave crests. This attenuation may become extensive if the generating wind field, i.e. either High or Low was itself of a circular cell form. The winds generated by troughs or ridges, tend to a much more linear plan pattern.

A simplified presentation of the wave development process is shown in Figure 1, commencing with the inception of wind ripples through to a fully escaped mature swell. As can be seen, the class of deepwater waves to which any structure may be exposed depends upon where it happens to be positioned within the wave form continuum. Since Nature's weather patterns are highly variable in size, geometry pressure gradient and track velocity and course, the deepwater wave climate is almost infinitely variable. When waves reach any structure, the wave manufacture process may be existing at any stage, but the wave energy loading also varies greatly throughout the entire process. Further variability is introduced by the wave direction. In deepwater, wave fronts may be travelling in all directions. Of all waves measured in deepwater, only those approaching the structure will strike it. All waves travelling in the opposite direction will vanish out into the ocean and cannot then form part of the structures "exposure" wave climate. Then intermediate or significantly oblique waves, may affect the structure or not, depending upon how much they may be refracted during shoaling up the continental shelf or coastal contours.

A particular problem facing coastal structural designers then must be a rational assessment of the probable deepwater wave climate that can be expected to reach the structure. Most sea-state wave detectors are inertial devices that are insensitive to wave direction, and many cannot possibly differentiate even between a wave translation direction and its reciprocal. Deep ocean seas and swells, commonly "run" in varying directions, but an inertial device merely adds their traces together, as if they both formed part of a single direction wave train. The recording of deepwater wave climates, in thus an exercise beset by very high uncertainty, when the data comes to be applied to the exposure of coastal structures. Even the deepwater wave form itself at the point of measurement may be quite un-representative of the final incident deepwater wave. What may be measured as an escaping sea well offshore could well have become an adjusting or fully mature swell by the time it reaches the prototype site. Except in the fully developed "free" ocean swell form, ordinary deepwater waves are very much prone to progressive change.

DEEPWATER WAVE CLIMATE

Since it requires a finite wave and wind fetch to manufacture high energy waves, the basic wave force input to any coastal structure is generated in deepwater conditions. Whilst deepwater waves are changed during the shoaling process, the shoaling process itself remains dominated by the wave force input from deepwater. Any assessment of the wave input to a structure therefore must commence with the consideration of the incident deepwater wave climate. This remains a topic, however, where uncertainty and variability reign supreme.

The normal expectation, is that coastal structures are most prone to change from only the largest waves, but the largest waves tend to be generated during storm events. The larger storms within the weather "climate" approach random events, and although it is common to consider ocean waves as likewise following a "climate," there is much evidence to suggest that there exist three grades of wave climate. This seems to follow from the manner in which ocean storms (as random events) follow different distributions, to the occurrences of normal weather systems. The normal weather up to and including moderate storms, is generated by the comparatively large reasonably circular progression of highs and lows (with their associated troughs and ridges) that continually circle the globe. These weather systems manufacture a wave climate that follows the Log-normal distribution in both wave height and period. Large storms, as more random events, become impressed on the normal weather as an added variable, and accordingly manufacture waves that follow the more skew and extreme Weibull distribution. Extreme storms are then even more skew and random, so their waves follow the Gumbel distribution in wave height. These latter are the great hurricane typhoon and tropical cyclone classes of events.

The major problem then is that each of these frequency distributions hold unique properties at their extreme tails, so they cannot be simply added or combined. The coastal structure designer thus has to assess the exposure of each structure from three separate ocean wave climates—each of differing severity and probabilistic distribution. A Weibull distribution cannot be extrapolated from a Log-normal, nor a Gumbel from a Weibull. The Log-normal wave climate remains the on-going norm, varying by the hour and the day. The Weibull storm is much rarer—perhaps one or two per year might be expected, but the extreme Gumbel event may only appear a few times per century.

This range of probabilities of each class of these events provides the greatest problems in all wave climate forecasting exercises. Only a single year might be an adequate period to measure the range of the basic Log-normal wave climate at a particular site. However it may require a hundred years or more to detect the frequencies of the more extreme events. Reliable detailed weather records of ocean conditions of this time span simply do not exist, yet it remains that the largest waves are manufactured by the extreme events.

A further vexing complexity arises in addressing extreme event wave climates. Extreme event storms tend to be of quite finite "size," e.g. an intense hurricane may carry extreme wave manufacturing wind fields of only 100 miles diameter. The wave exposure of a coastal structure therefore depends not only upon an extreme events probability of occurrence per se, but also upon how close the event approaches the structure. The assessment of the "worst" wave exposure therefore requires not only the prediction of a storm return, but also the prediction of its probable track and speed of translation. Under current states of knowledge, these parameters are almost complete intangibles, fraught with extreme levels of uncertainty. It is not only through wave height alone that extreme events endanger coastal structures. Additional wave exposure is induced by surge and wind set-up increases in sea level. These factors are also generated by extreme storms, so their prediction requires the same classes of data on storm intensity, track and speed of translation as above.

How then may the coastal engineer begin to forecast his structure's potential wave climate? If he relies upon measurements of the normal wave climate, then he will probably vastly under-rate the extreme event wave properties. But then, if he attempts to synthesize the properties of the extreme events, his predicted wave train wind set-up and surge could be widely in error due to lack of data. For many years, coastal structures were proportioned to accept the "design wave." Recently however, the concept of the measured prototype wave spectrum, has gained greater support. As discussed above, both approaches hold significant inadequacies—but for considering both "freak" waves and extreme storms, the original concept should not be completely discarded. Wave damage to structures is the damage per wave multiplied by the number of waves.

Freak waves do exist, but their probability will almost certainly not be detected within normal weather system wave statistics. This phenomenon, together with a better understanding of the behaviour of natural wave groups, will probably lead in the future, to the use of a "design wave train" for assessing coastal structure exposure. A single wave seldom destroys a structure, but 3 or 4 "freak" waves, or groups of them, can cause the same havoc as several hundred smaller waves and damage is cumulative. Further study will almost certainly be capable of determining that freak waves, wave groups and compound simultaneous wave trains are all capable of rational description in probabilistic terms. Until this result becomes effected, the assessment of the wave exposure climate for any individual structure will remain largely within the realms of simple judgement—very much more a matter of "art" than of fact.

SHOALING WAVES

If deepwater wave form variability were not enough, then the extra variability induced on waves by the shoaling process, is even more bewildering. The changes that are forced on the incoming wave form depend not only upon its particular shape at the initiation of shoaling—but also upon the properties of the shoaling seabed. A "soft," or porous seabed, affects incoming waves very differently from a "hard" or impervious seabed. Then for a structure built within the shoaling zone, its position within the zone, i.e. the depth of water, becomes even more important than for deepwater structures. Thus the range of potential wave input to a structure approaches the infinite. Finally once a coastal structure is built within the shoaling zone, it may change the incident wave climate itself, through the agencies of wave reflection and toe scour.

The most simple case of the wave shoaling phenomenon, is the ideal sediment rich porous bed beach, as shown in Figure 2, for an incoming wave train with a mature swell geometry. Porous seabed sediments are highly mobile and they exert very high drag forces on the waves. The most obvious is the manner in which the waves become compressed, i.e. their wave length decreases, as depicted in Figure 2. The drag resistance of a porous mobile frictional seabed sediment is highly complex and multi-faceted. The porosity of the sediment or the water filled voids between the particles causes the seabed boundary layer to become not a single surface, but a three dimensional hydraulic interface zone—instead of a simple "plane." Static wave drag forces are thus induced in depth, well below the apparent interface "surface." Further drag forces are induced by the sediment's capacity for movement. As a wave passes over the sediment, some particles are washed backwards and forwards, and "work" is expended during this action. Finally the movements of the seabed sediments tend to result in the construction of surface ripples and dunes. These sediment "structures" increase the seabed roughness, which in turn again increases the equivalent surface drag.

The cumulative effect of these additive seabed drag

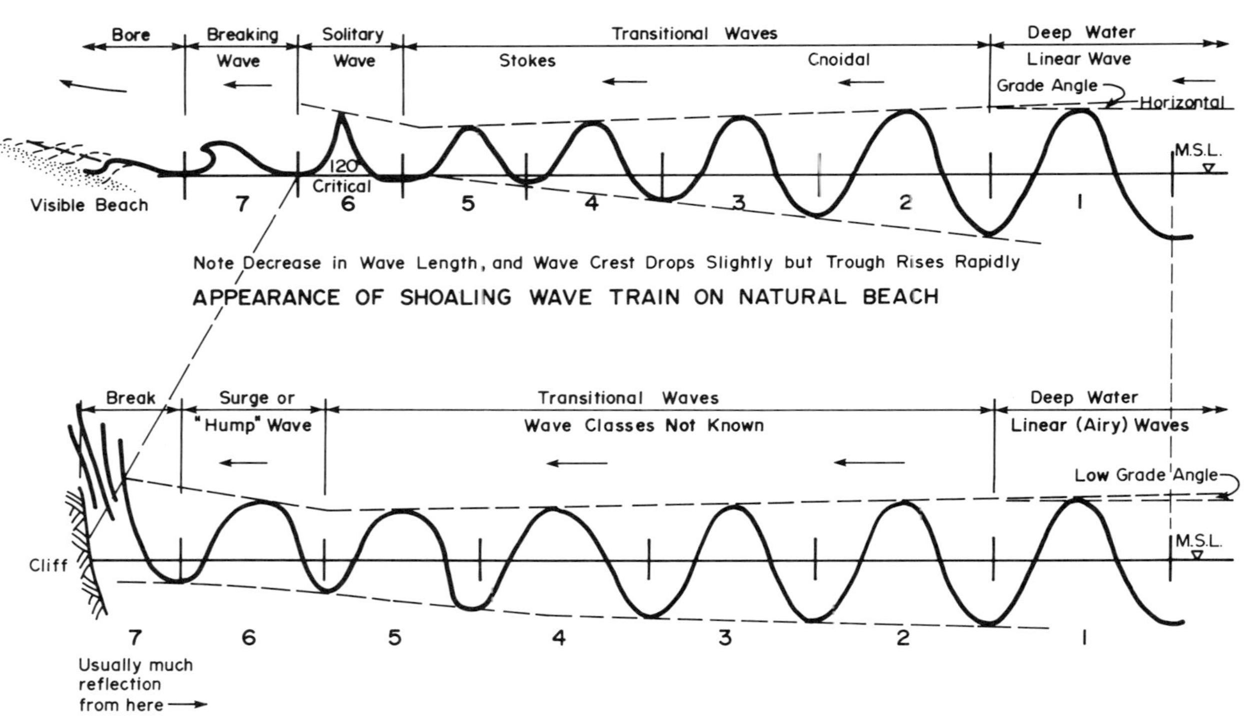

FIGURE 2. The process cf wave shoaling on beach and hard seabed coasts.

forces, induced by a porous sediment-rich seabed upon waves, may be very profound. The ultimate work capacity of the final breaking wave, as shown in Figure 2, may be only a quarter or less of the potential work capacity, of the incident deepwater swell wave, before it touches its wave "base" or initially interacts with the seabed sediments. The most striking feature of the porous bed shoaling wave is how the wave form, or "shape," changes during the shoaling process. If the beach slope is long enough and the seabed is of completely porous sediment—then the shoaling wave progressively changes from a Sine-form Airy swell to a solitary wave form at breaking. In this, the principal change in form occurs in the wave trough which rises to meet the mean sea level—the wave crest height may remain almost constant until the solitary wave cusp, or "point," rises above the average crest elevation. The visual impression is that this class of shoaling wave progressively flows through a sequence of Airy to Cnoidal, Cnoidal to Stokes and Stokes to Solitary. Then in the ideal case, the solitary wave becomes completely unstable—and breaks.

If a coastal structure is to be constructed on a natural porous seabed, the implications are obvious, but the potential wave variability becomes bewildering. The wave exposure of the structure not only depends upon the incident deepwater wave height, but also upon how far through the wave shoaling process the wave has run before it strikes the structure. The shallower the water at the structure site, the lesser the shoaling wave's work capacity, and the more unlike the Airy form, is its very shape and geometry. The over-topping capacity of a shoaling wave in its solitary state, may be much greater than in its deepwater or transitional state, but its total work capacity is then much less. The design criteria for a coastal structure on a porous bed therefore vary as the water depth increases. At all stages through the shoaling process, the wave length, crest to trough height and wave form are varying. In deeper water, face armour stability may dominate. Whilst in shallow water, over-topping may become more crucial.

Further wave exposure complexity is then generated by the depth limitation of breaking waves. On a porous seabed, natural waves tend to peak and break, in a water depth roughly equal to the wave height. This is typical of average waves on an average beach form, but during major storms another variable is introduced. Porous seabed sediments are highly mobile, and during major storms, the largest waves move inshore sediments offshore, to form a storm-bar, see Figure 3. The generation of a storm bar flattens the seabed inshore of the bar, but the sediments thus released accumulate to form a bar that its crest is *above* the mean seabed slope. The locally shallow bar then forces the largest waves to break on it. These waves then re-form to about half their previous height and continue shorewards. The storm bar then acts as a cropping filter. The deepwater largest waves are reduced, both in height and power, to become comparatively minor waves amongst a train, whose largest waves and are now those with a height, equal to the water depth, over the storm bar. That is all the waves from the deepwater train that were *not* forced to break.

Sometimes two or even three storm bars may be formed offshore, with each acting as progressive filters on the wave train. Unfortunately the depth of water over a bar is also variable. As shown in Figure 3, breaking waves generate a local wave set-up, that temporarily increases the water depth, and this increase will allow larger waves to pass through the bar shoaling zone, until such time as the wave set-up drains away. The cropping effect of a "filtering" storm bar is thus an additional significant variable on its own. The maximum unbroken wave that can cross a storm bar is dependent upon what has happened to the last few waves that preceded it. The total behaviour of multiple storm bars is

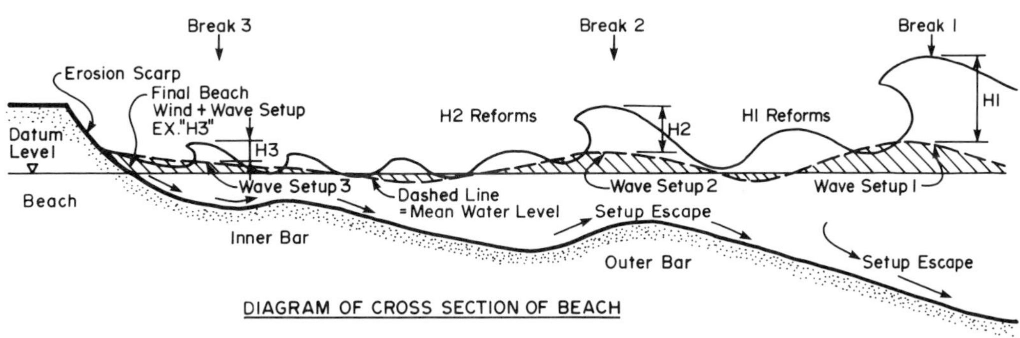

FIGURE 3. Storm wave-break behaviour on a natural beach.

then even more complex. It must also be kept in mind that until the storm bar is actually formed—then somewhat larger waves than could pass the bar may penetrate further shorewards, into shallower water, before breaking.

When a coastal structure is built on the porous seabed of the shoaling zone, the structure itself will then react with the natural wave train. Most coastal structures—even flexible rubble mounds—are quite "hard," compared with the highly mobile natural porous seabed sediments. Except under extreme waves, therefore, coastal structures tend to reflect a high proportion of the incoming "normal" classes of waves. Under comparatively mild conditions, a reflection co-efficient of one half or more is quite normal for unbroken waves. The consequences of this phenomenon is then shown in Figure 4. Due to the interaction between incident and reflected waves, the hydraulic stirring capacity, in front of the structure is enhanced. As a result of this, the reflected wave generates its own partial "beach" in front of the structure. This results in the so-called "scour-hole," but it is in fact a "negative" or offshore sloping local beach. In this, the in-coming wave reduces its shoaling energy loss, within the negative beach zone, so this ensures its higher continued reflection co-efficient. The structure/wave interaction is subtle, but the resultant is very obvious.

When a major storm arrives, however, this quasi-stable wave reflection relationship, becomes destroyed. The larger storm waves rapidly flatten the seabed in front of the structure—its "bar" effect is eliminated and a larger height of wave, is able to reach the structure itself. This larger wave then breaks directly *on* the structure and the reflection action becomes reduced almost to zero. This storm effect is depicted in Figure 5, which demonstrates the minimum design wave exposure that could be applied under these circumstances.

The above discussion concerns the ideal case where the shoaling seabed consists of an equivalent semi-infinite thickness of free porous material. On many natural beach zones, these ideal conditions do not exist. The porous seabed sediments may be quite shallow and over-lay buried layers of hard rock, impermeable clay or cohesive material. A thin layer of porous sediment may be quite capable of "training" normal moderate waves, in the form of Figure 2, but it may prove completely inadequate for coping with trains of very large waves. Then the available sediment capacity may be totally overwhelmed, and the sediment will be swept off the impermeable seabed and deposited in the offshore zone, during Nature's efforts to construct a deeper water storm bar. Then in the swept-zone, the seabed friction will be drastically reduced and much larger waves may penetrate the shallows. The relationship between porous seabed thickness and effective wave "training" capacity is currently unknown, but the resultants can be readily observed on the natural prototype. Wherever near seabed impervious reef structures (or their equivalents) occur offshore of natural beaches, it is readily observed that beach erosion is more pronounced shorewards of these zones than in shoaling areas that consist entirely of porous sediments during major storm events. On impervious seabeds, the training capacity of the shoaling zone is greatly reduced, and can be seen to be so.

The behaviour of shoaling waves traversing a completely impervious or "hard" seabed, unfortunately, has barely been studied at all. In Nature, most high energy beaches consist of porous sediments, right out to even major storm-wave wave-base depths. Totally "hard" or impervious seabeds generally occur on rocky coasts, where the shoreline is either of cliff and/or exposed reef formations. On this class of coastline, the higher work energy of the shoaling waves is very obvious. Shoaling storm waves on hard seabeds smash against the coast with remarkable fury—throwing spray, perhaps hundreds of metres into the air, quite unlike the much more subdued progressive breaking surf on a porous seabed beach. There are two major visual features that identify an almost completely impervious shoaling seabed. The first indication is that the wave shortening phenomenon is greatly reduced or almost absent. The second is the manner in which the shoaling wave "fattens" its surface profile, so that it "surges" up the coast, rather than "rolling." Both these phenomena naturally follow from the greatly reduced effective friction, exerted on the waves by the impermeable seabed. These effects are crudely depicted in Figure 2, which is applicable to a typical rocky coastline. This dia-

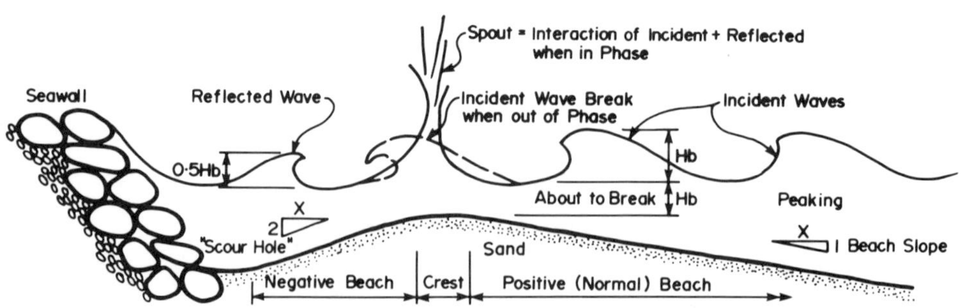

FIGURE 4. Wave reflection and toe scour on a beach in front of a hard structure.

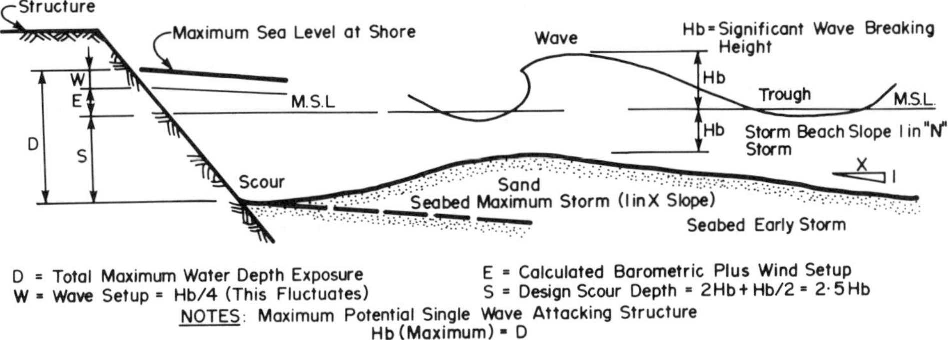

FIGURE 5. Design wave exposure for a hard structure on a beach.

gram is in scale with the same incident deepwater swell, as shown in Figure 2, strictly thus, for purposes of direct comparison.

Both cases in Figure 2, are again only for the cases where the incident deepwater wave is a mature swell. If the incident shoaling wave is either in its escaping or building stage at the inception of shoaling—then the shoaling process will diverge from the ideal conditions, shown in these diagrams. Building and adjusting waves do not carry fully developed orbit fields, so most of their energy is concentrated within their surface and near-surface forms. Any wave with an undeveloped deep mature orbit field will thus be less affected by seabed friction, so it may penetrate much further shorewards in the shoaling zone than its equivalent height "mature" wave. Again, this particular topic does not feature within coastal research reports at all—but it is of great concern to the practical coastal structure designer. These waves approaching a structure will be much less affected by shoaling, so their impact and over-topping capacities must be much more extreme.

The conclusion, therefore, must be that the prediction of the size and capacity of shoaling waves remains a subject grossly under-valued and thus comparatively unknown. It is a subject beset by almost infinite variability, in terms of the real prototype and to extrapolate the theoretical properties of pure Airy waves, as if they were unaffected by the shoaling process, must approach simple nonsense. A vast amount of on-site prototype research will be required to determine the real forms and work capacities, of all the various kinds of shoaling waves, that actually occur in Nature. At the moment our ignorance is profound—we have a very long research task ahead.

WAVES ON STRUCTURES

The manner in which waves interact with a structure is predominantly governed by the form of the structure itself. Rigid structures resist wave attack, through holding a very high mass, and thus a very high inertia. In this, they act as almost a single unit of armour that is so large that wave energy should be incapable of moving it. A rigid structure can destroy wave energy in at least three ways:

(a) The wave impacts upon the structure such that it is converted into a large volume of spray, which rises high into the air. This is an efficient destruction mechanism, but the impact forces are very high. Shock waves are induced in the structure and these can be dangerous if transmitted into a non-rigid foundation.

(b) The wave is almost totally reflected. This is a much more efficient process than impaction, since the wave forces are minimal, but full reflection is usually only economically possible for the smaller or non-extreme waves. Wave reflection however, may generate dangerous toe scour (see Figure 4) if the rigid structure is founded upon an erodable seabed, and the water depth is significant. Wave reflection is very common in Nature, on hard rocky or cliff coastlines.

(c) The wave is forced to expend its energy by uprush across a sloping rigid hard artificial "beach." The wave is forced to lift a weight of water a distance (with only partial breaking) to expend its volume of potential work. This is again an efficient process, but on a smooth rigid slope, the uprush height that must be allowed, becomes excessive. A reflective wall at the top of a well sloping rigid seawall is thus a common feature, often allied with a stepped or roughened surface on the sloping surface, to increase the uprush drag.

The degree to which any of these three processes become dominant is generally controlled by the slope of the rigid structure. Large vertical structures attract the impact mechanism, whilst gently sloping structures follow the uprush solution. All rigid structures, however, tend to hold a high reflection co-efficient and this will control the wave behaviour, for a fair percentage of the total input, of the more moderate waves. Some relatively steep rigid structures may demonstrate a combination of all three wave destruction mechanisms, simultaneously for much of the time. An unfortunate feature of rigid structures, is their sensitivity to over-load. Larger waves than the design capacity of the structure can rapidly lead to massive over-topping, even if the structure holds its structural integrity. Quite small errors in the design wave exposure can then result in widespread damage, landwards of the structure.

The wave absorption process on a flexible or rubble type structure is a very different thing. Flexible structures usually consist of a very large number of small units (called armour) which hold a very finite capacity for movement. This follows from the comparatively open packing of the units, that leaves a significant volume of voids, on the surface of the structure. Wave energy can thus be absorbed in two ways:

(a) The combination of high voids and the large surface areas of the units that thus become available exert very high friction drag forces on the waves. When this is allied with a comparatively gentle slope, the process is not only efficient, but it is also gradual. It may take several seconds for a wave to sweep a rubble mound face, but only a hundreth of a second, for a wave to impact on a vertical rigid face. Wave shock forces are thus minimised.

(b) In addition to drag forces, armour units are moved by wave forces, and extra wave "work" is expended upon doing this. If the movement is not too large, this is also an efficient process—particularly since the degree of movement increases with increasing wave power. Armour rocking or movement is often a powerful factor in the over-load capacity of a flexible structure.

The combination of these two processes, generally makes flexible structures much more efficient than their cost equivalent rigid structures, hence perhaps their growing popularity for high exposure conditions. Although the drag and movement mechanisms generally dominate flexible structure behaviour, these mounds can also be tuned to generate a significant uprush wave energy capacity through variations in the slope and armour size. During smaller unbroken wave attack, a significant wave reflection property can also be designed into the structure. In the use of flexible rubble structures, mankind is in fact closely following Nature. Sandy and gravel beaches are Nature's own rubble mounds—it is just that the armour units (sand grains and gravel stones) are much smaller than most of mankind's armour. Natural beaches follow exactly the same "rules" as man-made mounds; there exists a quite precise relationship between wave energy, beach slope and particle (i.e. mini-armour) size.

Flexible rubble mounds hold other important operational and service benefits, compared with rigid structures. The most important is probably the ease with which they can be repaired, after reasonable overloading damage. The integrity of the rubble mound depends upon its individual units and not the total unshaken mass, as is required for the rigid structure. Flexible structure armour units can simply be *added* to damaged areas—whereas the damaged rigid structures may require complete reconstruction before it can be put back in service. Toe scour is another matter of considerable concern to the coastal structure designer. Toe scour under a rigid structure can result in tipping over and almost complete destruction. On the other hand, a flexible mound will allow its comparatively "free" armour units to simply "rattle-down" to fill the scour void. Provided that it is "topped-up" therefore, a rubble mound holds a high "self-healing" capacity against toe scour and seabed erosion. Many rigid structures, in fact, have to be protected against toe scour, through the provision of a submerged rubble structure—this then is called a scour mattress—but it is still a flexible structure, discharging this very same capacity.

The manner in which the "final" wave that reaches the structure interacts with it to discharge its hydraulic energy is therefore largely controlled by the *type* of the structure itself. But the properties of the final wave are controlled by its previous shoaling history. This wave history is again pre-controlled by two other parameters, the type of seabed and the depth of water. Of these two parameters, the class of seabed is the most important. As is shown diagrammatically in Figure 2, the final shoaling wave that has shoaled over an impervious seabed (position 6) holds a cross section over twice that of the porous seabed wave—its length is greater also. The "hard" seabed wave may thus contain two or three times the work capacity of the "soft" seabed wave for the identical deepwater wave train. Then in both cases the further offshore and the deeper the water, then the higher the energy. The increase in work capacity, however, is now much more marked for the porous seabed; the wave energy increase with water depth rises at a faster *rate*, as does the wave length. The potential for a design wave exposure "error," is thus larger for a porous shoaling seabed. Then in-between there will exist the variable case for a partially porous seabed, i.e. sediment thicknesses that are inadequate for storm conditions on the existence of seabed reef structures, covering only part of the seabed.

The problem facing coastal structure designers is that the properties of real shoaling waves, whether on a hard *or* soft bottom, are almost entirely unknown. Plain observation demonstrates that shoaling waves are *not* linear or of Airy form—so until these transitional wave properties become understood, coastal designers can only expect structural failures to continue. The actual storm waves that so damaged the Sines breakwater, almost certainly would have been carrying nearly twice the work capacity that the designer might have anticipated from the small scale model tests. For the present, therefore, the designer of comparatively deep water coastal structures must be extremely conservative. If he relies upon only the consideration of the surface "shape" of deepwater waves, he will be grossly under-estimating their total power. The form of the shoaling seabed exerts influences very much greater than is normally appreciated.

THE BEHAVIOUR OF FLEXIBLE ARMOUR

The behaviour response of hydraulic armour to wave forces, is not a single simple process. Armour response is affected not only by the incident wave power, but also by the packing and disposition of the armour itself. Random placed or pell-mell armour, up to the point of failure, may react very differently from pattern placed or "regular" armour. Further variation is then imposed by the "grading" of the armour, i.e. if the armour is of constant shape and size, or if it is variable in both these parameters. The most important factor, however, remains the particular shape and weight of the armour units themselves; it is this factor that largely governs the density and packing behaviour of the armour mass.

Under wave attack, the response of any individual armour unit follows a progressive three stage behaviour mode. During the first stage, the armour unit opposes wave power, purely by its surface drag—the wave loses energy through turbulence, as it rushes around the unit. The second stage arrives when the maximum armour unit surface drag becomes exceeded. The wave then tries to wash the unit away and the unit begins to rock and lift. Extra wave power is then absorbed by the work required to lift or move the unit through a distance over a time. The final stage arrives when the waves begin to float the unit free. At this stage, the final unit reserve capacity is generated by its interlocking capacity or contact friction with its neighbouring units. When this last resort armour capacity is overwhelmed by wave power, the structure unravels and enters a total failure path. The crucial factor of the whole process remains that each stage of the armour's work capacity is not entered, until the entire capacity of the previous stage has been exhausted.

Unfortunately the particular properties of the armour units that contribute to each stage of their hydraulic response are not nearly so clear cut. During the first stage, the units drag capacity will be dominated by its surface roughness, but the total surface available to the waves will vary with the proportions of the voids between the units. The void's proportions are then, however, governed by the unit packing—this is in turn governed by the unit shape and grading. During the second stage, the unit's weight will come into play, in addition to its maximum drag capacity. For maximum stability at this stage, therefore, the unit should hold both a large rough surface *and* a large mass, but these two parameters are completely inter-related. A very large surface unit cannot hold a very high mass and vice-versa, so at this stage, the ratio of one to the other becomes predominant, in considering the volumetric efficiency of the armour.

In volumetric terms, very high surface area armour, due to its shape alone, must contain a high proportion of voids. This results in a low density of armour solids in the structure, with the inevitable consequence of a high density of attacking fluid. The reverse is then the case, for low surface area high mass armour. Uniform concrete cubes, for example, if tipped pell-mell, hold high voids, but when block stacked, they hold practically no voids at all, so there develops here, the third inter-related parameter, of regularity and pattern of placement. The potential variability between these three parameters, for both man-made and natural armour units, therefore approaches the infinite. This is, in fact, the case for many natural sand and gravel beaches.

At the third and final stage of armour unit response, the interlocking mechanism is largely one of preventing an over-loaded unit from the second stage, from washing away. By developing an interlock, the unit derives some extra capacity, by sharing its over-load with other adjoining units that are not over loaded at the same instant. This interlocking capacity is almost independent of the armour unit mass—is thus entirely a matter of unit "shape." Its most direct expression is with the slope that the units can stand, without collapse under gravity. The standard engineering expression of interlock capacity is usually expressed in the form of the unit's angle of internal friction or ∅ angle. It may be concluded, therefore, that the shape of the armour unit dominates the initial pure drag armour failure level—the inter-relation between surface and mass, the second level—and the shape again—the final level.

In all this, the significance of the "size" of the armour unit can be seen. A large rough surface unit holds a greater total area than a small unit of similar shape, so it carries a larger work capacity, due to its size alone, during the first failure level. The same applies during the second failure level because the larger unit is heavier, and even at the final level, larger units share their overload with neighbours that also have a higher individual work capacity. Unfortunately for real armour units, size and weight are not interchangeable values.

The above scenario leads to several interesting conclusions regarding the factors of safety for flexible armour structures. The armour behaviour stages set out could be regarded as representing three rather marked energy capacity thresholds, all of which must be passed before total failure eventuates. If under the worst conditions, the first (stationary drag) threshold is never exceeded, then the structure can never fail in service. It will even then hold a very high overload potential, due to its reserve movement capacity above the first threshold, but below the second. When the worst conditions bring the armour up to the limit of the second threshold, failure is then very close. The extra work capacity of the ultimate interlocking mechanism is very slight. In a strict sense, the armour crossing the interlocking threshold has already "failed." The inter-locking may merely restrain the unit against "floating-away." Interlocking is thus highly variable and dangerous to rely upon—particularly for the smallest armour units.

The degree of unit interlocking is also not merely a matter of the variation in size between individual units. The position of the armour units on the actual structure may be equally as critical. In multi-layer armour construction, the lower armour layers clearly hold a much higher interlocking capacity than can the surface layer—they are entirely surrounded by armour, whereas the surface layer (in cross section) can only interlock on three out of the four "surfaces" bounding the units. A further matter of great import is that the flatter the slope of the armour (except for full "touching" pattern placements) the less the inter-locking potential—the individual surface units are more exposed. The most vulnerable unit on any flexible armoured structure is usually the leading edge top unit—it can interlock on only *two* out of the four surfaces, bounding this unit's position. This provides at least one explanation for the high failure rate for armour units in this location on the prototype. Under high wave run-up, these units are highly prone to early dislodgement and rolling shorewards, hence the need for larger units in this, the most exposed location of any rubble mound. If the armour is designed to always remain *below* the second behaviour threshold, however, these graded interlocking capacity considerations become largely irrelevent.

It seems that most flexible armour structures are not designed upon a basis of zero damage, which would keep *all* the armour units below the second threshold, but upon some "acceptable degree of failure," expressed in terms of "percentage damage." This is generally expressed on the basis of the number of units per unit area, per layer. These damage "rates" or "quotients" (usually derived from model tests) tend to be accepted as expressing a high degree of precision, when in fact they are beset with very high levels of uncertainty. Whether any individual unit will be grossly displaced (i.e. exceed all three thresholds) depends upon the size of the unit, its exposure location in the structure, and its degree of interlocking—yet these critical parameters are never reported.

Natural rock armour is usually highly variable, in both size and shape, and when tipped or placed in the structure, the initial packing and interlocking attained is often very far from the optimum. It has long been observed that after several years of wave exposure, rubble structures often look quite different from their original as-built state. Wave attack has moved the units sufficiently to readjust the armour placement well beyond that envisaged by the designer. This phenomenon is often an unexpected "saving-grace," even moderate waves tend to rattle-down armour that is too small or dangerously exposed. Usually Nature herself has either flattened the slope or bedded in the units to attain a superior packing. This on-site bonus is, however, absent from pattern placed constant sized armour. Sometimes, therefore, a moderate amount of damage to the as-built structure may be beneficial, but often it is not. It depends upon whether a moderate shape adjustment can be safely tolerated—or whether the adjustment generates other dangerous consequences—such as decreasing the crest over-topping resistance or exposing sub-standard armour, to more direct wave attack. If the latter class of resultant occurs, then the initial damage merely becomes the "trigger" for an on-going and irreversible failure sequence. It is only very recently, that it has been realised that coastal structures must be closely monitored in service for very long time periods beyond their initial construction.

The concept of the theoretical degree of acceptable damage is therefore not a matter of simple initial design economy, but what the consequences of this damage will be on the prototype in service into the future. The consequences, for example, of a 10% damage allowance for a structure which will receive a 10% wave overload during its life will be certain destruction. In this case the safety probabilities will become almost completely additive. Where the real peak wave power that will strike the structure is in any way indeterminate and likely to exceed expectation, then any armour design for less than zero damage is an imprudent gamble. The final armour interlocking threshold remains a very fragile and minor proportion of the armour's total hydraulic resistance. The final interlocking capacity of rubble face armour remains so uncertain that it has resisted any reliable probability assessment even to the present day.

THE BEHAVIOUR OF PATTERN PLACED ARMOUR

The behaviour response of uniform pattern placed armour, is very different from the behaviour of "free" flexible armour. In pattern placing, the basic objective is usually to use light comparatively low hydraulic drag units and balance the armour resistance by maximising the interlocking capacity. If the interlocking unit friction can be built-up to a sufficiently high level, then the structure can be made to respond as if it were rigid, rather than flexible. In this, the interlocking friction "binds" or "ties" the smaller units into a coherent mass, capable of accepting much higher wave power than could the units, if they were individually "free" or not interlocked. The same philosophy is applicable to flexible rubble mounds that are later bound together with cement grout or bitumen intrusion. In this case, however, some direct adhesion is added to the full unit "contact" containment.

Normal rubber armour, due to shape and size variability, cannot be stacked or pattern placed, but there are many manufactured armour units that can. The plain concrete cube is the most simple and its textured equivalents, such as the Cob, Grobbelaar, Stolk etc., can also be very closely packed. Other regular units amenable to high friction packing are the Tribar, Tri-long, Tripod and their like, together with hexaforms such as the Seabee.

In plain pattern placed armour, the work capacity and failure mode sequence becomes reversed in comparison with free flexible armour. Initially the entire wave power is distributed by the interlocking. Then only after the interlocking capacity is exceeded, can the weight factor come into play. Finally, only after the weight factor has been exhausted, i.e. the unit has become dislodged, can the drag component become exposed. The inherent hazard with interlocking armour, is that usually the weight and drag factors, are quite small in comparison with the interlocking capacity. It may thus only require a single unit to fracture or become dislodged to then initiate a progressive and irreversible total failure in the structure, i.e. the classic "dominoe" effect.

A smooth faced coastal structure of interlocking units, is relatively inefficient in accepting wave power, because the most effective wave energy absorbtion mechanism remains the armour's turbulent friction drag component. The coastal structural designers' responses, therefore, have been to develop interlocking units with a high surface texture, e.g. deep indentations, hollow centres and the like. This stratagem ensures that significant drag and interlocking resistance factors may become exerted by the armour units simultaneously. This then results in smaller more efficient structures. On the other hand, deeply textured units are much lighter than their solid equivalents, so if any actual release of the interlocking force occurs, then the units will wash away much more easily. Textured interlocking units are thus prone to higher rates of failure, under even slight overload, than solid units, and they will unravel much more quickly.

When the texture applied to armour units approaches the extreme, another hazard appears. Comparatively thin complex shaped sections in a unit then begin to suffer stress concentrations, i.e. "notch" and Griffith's failure behaviour. The units become "brittle" and readily damaged—the dominoe failure risk becomes highly accentuated. The brittleness factor is also affected by the weight of the armour unit. For any given shape, the stress concentration is proportional to the units mass—a 5 tonne Dolos may be nearly indestructable—but a 50 tonne unit will contain dangerous stress levels, due entirely to its own weight. It is expensive to construct concrete armour formwork with smooth curves and generous fillet radii, but sharp corners are an open invitation to excessive stress concentration in any unit. This hazard is absent in small scale model testing—but it becomes very real with large units on the prototype. In structural strength, the Tetrapod, for example, is much more robust than Dolos. A very large Quadripod would be extremely fragile. Designing concrete textured armour, in terms of geometrical similitude at the expense of structural integrity, remains a dangerous exercise.

In highly textured armour, the degree of interlocking becomes very much a matter of the particular unit's shape. Cubes and Tribars, for example, can be packed in very intimate contact to attain an almost impermeable structure. On the other hand, Dolosse have a high voids percentage, their interlocking relies on the mechanical "hooking" of their flukes, and there are many classes of unit that fall in between these extremes.

A particular feature of regular patterned armour is that it usually holds two very different angles of internal friction (i.e. stability slope) depending upon how it is placed. The problem is that the careful regular placement stability slope is usually much steeper than in the "disturbed," random or pell mell condition. Cubes are readily pattern stacked to provide a vertical face, but pell mell cubes may hold a ∅ angle of less than 45°. The initial response of steep pattern placed armour to disturbance, is therefore usually a drastic reduction in its angle of repose. The loss of only a few units may be enough to totally trigger the response. Even Dolosse can be pattern stacked at 90° (in the form of a crib-wall) but its pell mell ∅ is only 65°. Natural rubble by comparison will seldom stand steeper than 35°, but this angle tends to a constant.

The crucial aspect of pattern placed interlocking armour is that to maintain its integrity, the pattern must not become disturbed. The prevention of failure is pattern armour thus depends very heavily upon the stability of the foundation at the toe of the structure. The toe foundation must not be flexible—it must be sufficiently rigid so as to prevent any movement in the armour it supports. If the toe settles differentially, then the armour will begin to lose interlock and ultimate failure of the structure will become inevitable.

Where the structure is founded on a hard seabed, a rigid toe support to the interlocking armour may be attainable, but where the seabed consists of deep deposits of soft material—silts, sands or gravels—this may become impossible. Since a breakwater itself tends to induce seabed scour at the toe in soft materials, soft seabeds provide a poor vehicle for a rigid "base." The usual approach is to provide a seabed mattress of armour seawards from the toe, but to become effective, the mattress must be massive in its own right and be capable of accepting its own toe scour. What little evidence that there is suggests that current toe mattresses are too small, i.e. of inadequate width and thickness. There seems to be no reliable method of designing toe mattresses and practically no sound records of how those installed on the prototype actually behave. A major problem would seem to follow from the fact that rubble mattresses are themselves flexible structures, so it may be quite unreasonable to expect them to behave as rigid structures in the first place.

The current conclusion therefore must be that regular interlocking armour can only be secure enough to meet its potential capacity, if it is founded upon an unassailable rigid foundation. If it is not, then even the slightest distortion of its foundation—or the most moderate hydraulic overload—then will induce irreversable unravelling behaviour, that *will* lead to catastrophic failure. The hazard of quite unexpected failure, under modest over-load is high and the factor of safety against this condition is irreversable and very narrow indeed.

THE ARMOUR BACKING

Main outer face armour is usually proportioned to resist only *incoming* incident wave, but its prototype performance is also significantly affected by its backing. The wave power absorption process of unit friction drag requires both finite time and a finite distance of wave "run" to become complete. The efficiency of the process thus depends upon not only the unit surface texture, but also the proportion of voids and how far the voids penetrate the structure. If the armour is backed by an impermeable surface within the minimum power absorption zone, then the residual wave power will become reflected, and wave energy will flow *back* through the armour into the sea. Depending upon the thickness of the armour void zone and the slope of the impermeable backing, the reflected power may be in the form of backrush or pure reflection or both. The construction of a rigid monolith cap structure on a flexible rubble structure, may also induce the same effect.

Face armour is very sensitive to reflected wave power, its mechanical stability against a seawards energy flow, is much less than against landward energy flow. In the seawards direction, the face armour has no backing or support other than its gravity interlock. The most dangerous armour zone is usually near the toe—here the backrush forces are at a maximum, and due to its submergence bouyancy, the effective armour unit weight is at a minimum. Under these conditions, it is possible for the dynamic and static pore pressures, to roll-out or even float away, armour units from the toe zone. This process it must also be noted, does not terminate at the ocean water level, backrush and reflected wave forces, can exist at significant depths, *below* the sea surface. This is particularly so, for large structures founded in deep water—the deeper the water, the more extensive is the wave's submerged orbit field component. This component is reflected just as readily, as is the wave's surface shape component.

The mechanics of reflected wave power are highly complex and largely unknown. The variability of the process is extreme. Reflection is not only affected by incoming wave height and period, but also by water depth, armour size and shape, structure slope and the degree of permeability of the armour backing. It is not even known which variables are important or how the variables interact. It is even likely the resonance effects, might be extremely important—but of these we seem to understand very little. Wave direction may also become an important parameter. A wave crest sweeping and breaking progressively along a structure, induces the highest possible pore pressure gradients. When imposed over small areas, these pressures added to backwash, can apply high hydraulic torque to toe zone armour units, and readily loosen them.

It seems to be impossible to "tune" a rubble structure towards universal wave power absorption, by only the armour's friction drug process. Under various wave climates, the same structure will affect the waves in different ways. Under modest wave input, even a very porous rubble structure, may be seen to almost entirely reflect the waves. Under larger waves, the structure can entirely absorb the wave power within the armour zone, by friction drag. But then under the very large or extreme waves, pure reflection and/or backrush, may become dominant again. The wave-structure interaction process, is certainly by no means simple—the variability can be bewildering to the extreme.

STRUCTURAL FAILURE

In dealing with flexible armour, it seems to be a very common assumption that there exists a single relationship between the wave power, armour weight and the slope of the structure, for any given unit shape factor. On the prototype structure, this is not so. The wave power applied to an armour unit also depends upon where that unit is placed in the structure. For a wave breaking at water level, the wave force applied to the armour is high in impact, but the force is nearly horizontal. In this zone the maximum friction drag on the unit is called into action, but this being nearly horizontal, the gravity component of the unit is largely inactivated. The wave forces applied to a unit near the crest of the structure however, are rather different. Here the wave force within the uprush, holds a high vertical component—so in this zone both the drag *and* gravity resistance of the unit are activated. It is this vertical component of the uprush, that usually dislodges the top leading edge unit, on a structure—this unit has no surcharging units resting *on* it—so its interlocking capacity is the lowest in the entire structure. Thus, if this particular unit is to hold an equal stability to the breaking zone units, it must be much heavier to compensate for its reduced interlocking factor.

Then, a rather different wave power climate will be applied to the armour units *below* water level. Here wave reflection and backwash forces may pre-dominate, but for horizontal elliptical wave orbits, the input force may still tend to be horizontal. In this however, the *backwash* vertical downwards component may represent the largest force on the units leading towards instability. If the breakwater is constructed in very deep water (such as Sines) however, the sub-surface wave orbits may be much more circular than elliptical, so armour units will be exposed to stress fields, that are nearly equal in all directions. Then, the upwards and downwards wave orbit pressures may become important, particularly if added to uprush or downrush static hydraulic heads—these of course, being induced upon the incoming waves—by the structures itself, all from *their* previous waves.

It might thus well be concluded, that the concept of a single simple response mode, on the part of armour units against wave attack, is a gross simplification of the most irrational kind. Real armour on the real prototype does not respond like that at all.

For rubble structures founded on a soft seabed, the most dangerous failure mode does not necessarily arise within the armour itself, but is usually the result of toe scour. In a random rubble structure, the face armour tends to repair the scour hole, by rattling down and filling the scour void. This apparently satisfactory self healing response however, will induce potentially dangerous side effects. Firstly, the armour dropping into the scour will tend to steepen the face slope, and thus reduce its wave capacity. Secondly and usually more importantly, the subsidence of the armour will reduce the freeboard of the structure, making it much more open to overtopping. Figure 6, depicts the failure mode of a simple boulder revetment, triggered by toe scour. This diagram is based on direct observation, with the final debris location effected by excavation, through the post-storm beach. Figure 7, then shows the equivalent failure of a grouted revetment. Here the grouting initially prevents any slope adjustment, but the final result is very much the same, overtopping has been delayed, but an undercutting process has replaced it. Even very modest simple log walls fail very rapidly from toe scour, see Figure 8. In each of these three prototype examples, the wave resisting elements were all founded at mean sea level. The prototype sediments liquified when excavated to below this level, and local costs precluded cofferdamming and de-watering. The peak waves that destroyed all these structures, were between 2 and 3 metres high.

These examples were for revetment walls, but double sided structures like full rubble mounds, begin to fail in the same manner. The main difference is that the background erosion, as shown in the figures, is replaced by dislodgement of the trailing edge armour units, caused by the overtopping uprush cascading down the near face. An unex-

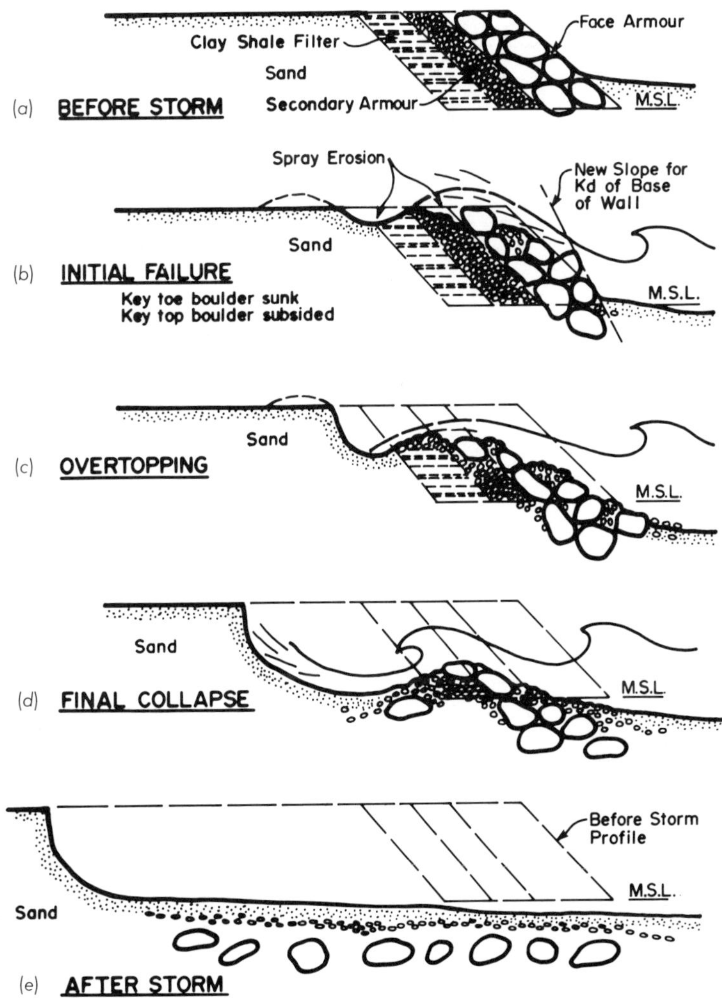

FIGURE 6. Observed failure mode of a rubble revetment on a beach.

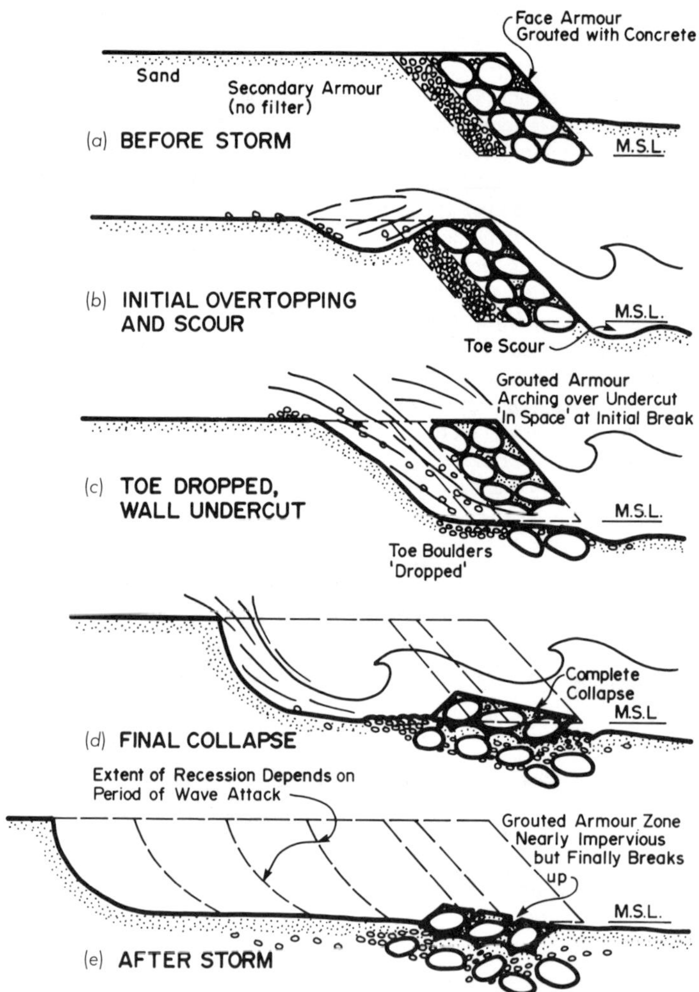

FIGURE 7. Observed failure mode of a grouted rubble revetment on a beach.

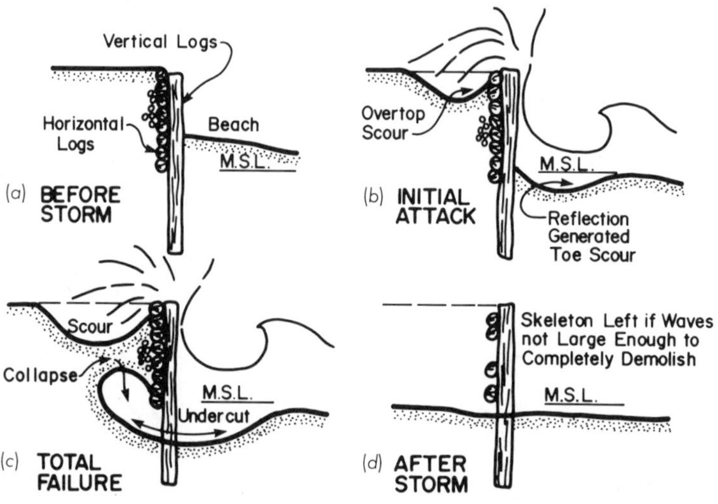

FIGURE 8. Observed failure mode of a vertical timber revetment on a beach.

pected feature of rubble mound failure, is that whilst the first damage to face armour is usually concentrated within the smallest units, toe settlement is usually initiated by the subsidence of the largest units. Prototype experience clearly demonstrates that all armour founded on a soft bed, must be supported by a graded filter backup layer, but unfortunately reliable design procedures to ensure an effective onsite behaviour mode here, remain largely unknown.

Large rigid monoliths usually fail from loss of foundation support (i.e. settlement or scour) or the monolith sections become bodily translated. The superstructure at Sines, is an example of the former—and the similar capping structure at Keelung, is an example of the latter. Coastal structures may fail in other ways, e.g. foundation overstress, degregation of the core and weathering disintegration of the armour units themselves. These other failure modes however, need not be associated with the structure/wave response process.

GRAVITY FORCES

In the design of coastal structures it is easy to proportion the armour size and geometry for wave resistance properties alone. In addition to absorbing wave power, these structures must also maintain the capacity to hold themselves up; clearly any structure which is at its limit under body or gravity forces alone, cannot be expected to hold any useful reserve to accept additional hydraulic forces. The stability of most flexible structures is derived from the interlocking capacity of the indiviudal units. Accordingly, the interlocking armour capacity available to resist wave attack, will be only that left-over from the gravity force demand. For rubble structures built near to their natural angle of repose, the interlocking component will be almost entirely required by their self support. The toe zone at Sines, for example, would have had a factor of safety of almost unity, had it been built in air—let alone in water, and then become exposed to large waves as well.

Plain gravity forces may also induce high local stress concentrations in coastal structures. These must be expected in zones where the structure profile is changed, and at the interface of zones containing significantly different types of armour units. A particularly dangerous zone occurs along the edges of large rigid monoliths, founded on a flexible bed. Large flexible rubble mounds founded on deep gravel or sand seabeds, must also be expected to induce some tensile principal shear stress field at the seabed toe zone—it is even likely that this phenomenon may induce or at least accellerate toe scour, and damage toe mattresses that are inadequately thick.

These considerations apply to all structures whether rigid or flexible and pattern placed or random placed, with respect to their armour units. It must be quite dangerous therefore, to blindly extrapolate the interlocking behaviour of small models, onto very heavy structures in the prototype. The total structure in the field, may not be nearly as "strong" under wave attack, as might have been predicted from small scale model wave-flume tests. In steep large heavy rubble mounds therefore, it is possible for a structural failure to become initiated, as soon as the armour unit gravity component of wave resistance, is activated in the highest body force stress zones.

WAVE RESISTANCE DESIGN

The design of massive monolith and totally pattern placed interlocking armour structures, must be dominated by basic structural concepts of simple stability. In this, the crucial aspect must be the attainment of a foundation, sufficiently rigid to ensure a complete avoidance of settlement and distortion, anywhere in the structure. Only after this provision has been met, can the proportion and properties of the individual monolith components or armour units, be addressed. Then the design enters the realms of dynamic hydraulic behaviour, but unit friction will remain the principal stability criteria.

In the design of flexible rubble structures, this basic design sequence becomes reversed. The individual armour units dynamic resistance must be considered first, and then the gross structure checked for static gravity forces. A third step then becomes necessary, where both unit stability factors must be combined. Gravity forces may usually be derived by classic soil and structural design principles—the prime design uncertainty, always remains the dynamic behaviour of the individual wave resisting elements. Although the subject of a comparatively vast volume of research and study, subsequent to the pioneering work of Iribarren in 1938, no reliable method of predicting the behaviour of prototype hydraulic armour, has yet been attained. This is now very obvious from the number of major coastal structures, that have been failing since the seventies.

In the definitive study of rubble mound behaviour Per Bruun (1985) lists at least 16 different formulae that have been proposed for the design of flexible armour units. Unfortunately most of the stability coefficients that have been derived, are based on small scale model tests, but there are many grounds for suspecting that Reynolds Number, does not scale model flume waves onto the prototype. A typical and perhaps the most widely used formula, is that after Hudson in the C.E.R.C. Shore Protection Manual (1984). This is of the form:

$$W = \frac{w_r H^3}{K_D(S_r - 1)^3 \cot\theta}$$

where

W = weight in pounds of an individual armour unit in the primary cover layer. (When the cover layer is two quarry stones in thickness, the stones comprising the primary cover layer can range from about 0.75 W to 1.25 W with about 75 percent of the individual stones weighing more than W. The maximum weight of individual stones depends on the size or shape of the unit. The unit should not be of such a size as to extend an appreciable distance above the average level of the slope.)

w_r = unit weight (saturated surface dry) of armor unit, lbs./ft.^{3},

H = design wave height at the structure site in feet,

S_r = specific gravity of armor unit, relative to the water at the structure, ($S_r = w_r/w_\omega$),

w_ω = unit weight of water, fresh water = 62.4 lbs./ft.^{3}, sea water = 64.0 lbs./ft.^{3},

θ = angle of structure slope measured from horizontal in degrees,

and

K_D = stability coefficient that varies primarily with the shape of the armor units, roughness of the armor unit surface, sharpness of edges and degree of interlocking obtained in placement.

Although apparently comparatively simple, this formula (as with most others) is still surrounded by significant restrictions. Initially the equation was derived only for natural rubble armour, but it has since been extended (see Table 1) for other man-made concrete interlocking armour units, even when their progressive total hydraulic behaviour modes, are quite different (see above). The formula is also restricted to conditions where the crest of the structure, is high enough to prevent major overtopping. This means therefore, that the equation cannot be applied to crest or top armour—but these units are often the most dangerous and vulnerable, within the entire structure. By analogy it must also be in-applicable in predicting toe unit behaviour, yet the toe armour is often the crucial trigger of ultimate failure. The Hudson formula must therefore, be restricted to the probable behaviour of the central face armour, so its predictive value is extremely limited.

Table 1 from S.P.M. (1977) then sets out the various stability coefficient (K_D) values, that are generally applied with the Hudson formula. The K_D value is applied to describe the variations on armour unit shape, roughness, sharpness and degree of interlocking—but as a single value coefficient for any given shape, the K_D does not hold a linear relationship with the armour unit weight. This should be obvious. The armour unit hydraulic drag "strength" component, is proportional to the unit's surface area, which depends upon size, shape and surface texture. The armour unit's hydraulic weight "strength" component, is then proportional to the unit's mass and closeness of packing. But the armour unit's interlocking "strength" component is proportional to the angle of internal friction, yet this also varies with unit weight, alignment, packing and voids. Any attempt to try and describe all these variables with a single coefficient, is bound to be inadequate. Then as discussed previously above, the order in which each of these progressive armour "strength" components are called into play, is a variable as well—depending upon the type of unit, its packing and its foundations.

What is perhaps an even greater hazard in this type of stability formula, is that it is based on model tests for "zero damage" although this actually represents a 5% damage in terms of the number of units that have been moved or displaced by the waves. If the case of a flexible natural rubble armour structure, this means that 5% of the units have exceeded their final (interlocking) strength threshold. If 5% have done this, then perhaps 50% of the units have already exceeded their second (weight factor) strength threshold and probably every unit, its first (or friction drag) strength threshold. In these terms, a 5% displacement rate already represents a gross hydraulic overload and a total factor of safety well below unity. Every unit displaced, exposes several more units to a higher level of wave power. This will result in a greater number of other units exceeding their second (gravity weight) threshold, due to the resultant loss in the original *total* unit interlock component.

Since most stability coefficients have been derived from model tests, most of the values usually adopted are only for nonbreaking waves. In this, the K_D values only simulate some of the armour units on the prototype, for some of the time—usually those very close to the ambient water level. Plunging waves, have a much higher impact force and a much higher local water velocity than do nonbreaking waves, but these conditions tend to apply to the armour units higher up the structure's face. In this location, the armour interlocking surcharge is less and its gross combined "strength" is lower as well, but the friction drag demand is much higher.

The final interlocking strength capacity of rubble armour, depends not upon the angle at which it is stacked, but the difference between this angle and the armour's natural angle of repose—and the former must be the lesser, if a plain gravity failure is to be avoided. It must also be noted, that the interlocking strength of a face unit is rather different, under backwash and core reflected waves, than from incident waves.

It is clear from the K_D values of Table 1, that manufactured high texture units, such as the Tribar and Dolos, held much higher maximum values than natural stone. In random placement, the gross armour unit weight per unit volume of structure, is generally much lower. The extra apparent strength of the textured units, cannot come from their individual weights—or even their surface areas—but from their much higher interlock strength components. This remains however, a dangerous expedient. The interlocking strength component for a unit with high friction drag and weight strength components, represents a final moderate

TABLE 1. Suggested K_D Values for Use in Determining Armor Unit Weight.

			No-Damage Criteria and Minor Overtopping				
			Structure Truck		Structure Head		Slope
			K_D §		K_D		
Armor Units	n*	Placement	Breaking Wave	Nonbreaking Wave	Breaking Wave	Nonbreaking Wave	cot θ
Quarrystone							
Smooth rounded	2	random	2.1	2.4	1.7	1.9	1.5 to 3.0
Smooth rounded	>3	random	2.8	3.2	2.1	2.3	‖
Rough angular	1	random†	†	2.9	†	2.3	‖
Rough angular	2	random	3.5	4.0	2.9 2.5 2.0	3.2 2.8 2.3	1.5 2.0 3.0
Rough angular	>3	random	3.9	4.5	3.7	4.2	‖
Rough angular	2	special‡	4.8	5.5	3.5	4.5	‖
Tetrapod and Quadripod	2	random	7.2	8.3	5.9 5.5 4.0	6.6 6.1 4.4	1.5 2.0 3.0
Tribar	2	random	9.0	10.4	8.3 7.8 7.0	9.0 8.5 7.7	1.5 2.0 3.0
Dolos	2	random	22.0	25.0	15.0 13.5	16.5 15.0	2.0¶ 3.0
Modified Cube	2	random	6.8	7.8	—	5.0	‖
Hexapod	2	random	8.2	9.5	5.0	7.0	‖
Tribar	1	uniform	12.0	15.0	7.5	9.5	‖
Quarrystone (K_{RR})							
Graded angular	—	random	2.2	2.5			

*n is the number of units comprising the thickness of the armor layer.
†The use of single layer of quarrystone armor units subject to breaking waves is not recommended, and only under special conditions for nonbreaking waves. When it is used, the stone should be carefully placed.
‡Special placement with long axis of stone placed perpendicular to structure face.
§Applicable to slopes ranging from 1 on 1.5 to 1 on 5.
‖Until more information is availabe on the variation of K_D should be limited to slopes ranging from 1 on 1.5 to 1 on 3. Some armor units tested on a structure head indicate a K_D-slope dependence.
¶Stability of dolosse on slopes steeper than 1 on 2 should be substantiated by site specific model tests.

overload capacity. A unit that relies almost entirely upon its final interlocking friction however, is always closer to dislodgement, so its total progressive factor of safety is clearly much lower. A Dolos under breaking waves, might have an ultimate K_D value of 22, with a factor of safety of one or less against total failure, but its safe K_D value at the friction drag/weight factor threshold, where a reliable factor of safety exists, might be only between 5 and 10. The textured armour stability coefficients of Table 1, all contain very different ultimate factors of safety. Very angular quarry stone in comparison, might have an ultimate interlocking failure K_D of 4, but a friction/weight factor threshold K_D of over 3.75—a very different ratio.

Any armour unit displacement, is an incipient or actual failure and a highly variable and unreliable interlocking, may represent a hazardous resisting force, that relies upon a narrow margin for success. The higher the percentages of unit friction drag and weight, the greater the security against total displacement of the armour. Full interlocking pattern placed semirigid armour, is probably even more prone to very low overload capacity and a more catastrophic failure. A particular feature of the Hudson formula is that the armour weight is proportional to the wave height cubed. A very small underestimate of wave height, will readily lead to a gross error in the minimum weight of armour required.

In general, the overall conclusion must be that the real hydraulic behaviour of armour units exposed to wave power, remains almost completely unknown. In avoiding this issue, it seems therefore, that the general research effort has been applied to merely manipulating overall coefficients that are in fact variables and not constants. The problem has been simplified to such a state, that the reliability of the armour design process, has become dangerously low. Of all the major researchers in the field, only Hedar (see Per Bruun—1985 and I.C.E., Breakwaters, Design & Construction—1984) has considered unit friction, core permeability and the differences between the armour properties of drag, weight and interlock. In this, he apparently stands alone, but at least a rational approach has now finally appeared.

CONCLUSION

An attempt has been made herein, to effect an overall review of current knowledge, concerning the wave/structure interaction process. The conclusion is that the understanding of the process is dangerously inadequate—so much so that the recent spate of major coastal structural failures, can almost certainly be directly traced to this inadequacy. Wave/structure interactions, are not only complex, but also nearly infinitely variable in both three dimensional space and in time. It seems almost certain that a vast research effort, both in theory and on-site prototype instrumentation, will be required before any positive progress can be attained. This discussion has tranversed many of the aspects of wave/structure interaction, where it is believed that further research is most necessary. These particular aspects have been selected largely as the result of observation on the full sized prototype, so it is likely that other equally important topics, may have been overlooked.

REFERENCES

A review of this kind, which is largely concerned with future progress, has little call for an extended list of references. In this sense, it is the principles that are important, not the details. In addition, the very large volume of recent coastal structures that have failed, often in dramatic form, have rendered a large volume of previous coastal engineering concepts, well obsolete. In these terms, there are probably only three up to date reference works currently available, that are now of general application to the topic. For those who wish to read further, these are:

1. "Shore Protection Manual," C.E.R.C. Dept. of the Army U.S. Corps. of Engineers, 2 Vols. (1984).
2. "Design and Construction of Mounds for Breakwaters and Coastal Protection," Per Bruun (Editor). Elsevier (1985).
3. "Breakwater, Design & Construction," I.C.E. Conf. Proceedings. Thos. Telford (1984).

CHAPTER 22

Waves on Structures

Shi Leng Xie*

INTRODUCTION

The action of ocean waves is one of the most important aspects in the design and construction of marine structures, such as breakwaters, jetties, sea walls, open-sea piers, offshore platforms, etc. Wave forces exerted on various types of structures are discussed primarily in this Chapter. For the rubble mound structures, the stability of the individual armor rock or artificial block under the action of the waves is considered instead of the wave forces on it. Owing to that local scour in front of the structures is related to the stability of the foundation of the structures, the scouring patterns of the sea bed induced by the waves are also described. The design criteria for the various parts of the structures are to be mentioned in the appropriate paragraphs.

STANDING WAVE FORCES ON A VERTICAL WALL

When progressive waves attack a vertical impermeable wall situated in water of sufficient depth with their crests parallel to the wall axis, they are almost totally reflected. Clapotis, or standing waves are formed by superimposing incident and reflected waves. The wave height of the clapotis is twice that of the incoming wave.

In 1928, Sainflou [1] presented a formula for the pressure of a standing wave based on the elliptically trochoidal wave theory. This formula has been in general use for many years. The mid-elevation of the clapotis above still-water level is given by

$$h_0 = \frac{\pi H^2}{L} \coth \frac{2\pi d}{L} \tag{1}$$

*The First Design Institute of Navigation Engineering, The Ministry of Communications, Tianjin, China

in which H is the incident wave height; L is the wave length; and d is the water depth. The crest of the clapotis above still-water level is $H + h_0$ (Figure 1a). The wave pressure at the bottom is derived as

$$p_d = \frac{\gamma H}{\cosh \dfrac{2\pi d}{L}} \tag{2}$$

in which γ is the specific weight of water.

For the simplified Sainflou formula, straight line is used to substitute the derived pressure curve. Then, for the wave at crest position, the wave pressure at the still-water level is obtained as

$$p_0 = (p_d + \gamma d)\frac{H + h_0}{H + h_0 + d} \tag{3}$$

The wave pressure diagram for the wave at trough position is shown in Figure 1(b). In Figure 1, P_c and P_t denote the resulting horizontal wave forces when wave crest and wave trough at wall respectively. The direction of P_t is opposite to the direction of the incoming waves. P_u represents the vertical wave force, which is to be discussed later.

Some experiments conducted in laboratories proved that Sainflou's method was an oversimplification. This method underestimates the wave forces when d/L is less than about 0.1 [2]. Nagai [3] proposed to consider three regions for calculating the maximum simultaneous pressures of clapotis.

(a) Region of Shallow Water Waves, i.e., $0.135 \leqq d/L < 0.35$ [Figure 2(a)]:

The equation of maximum wave pressure at position z below still-water level on a vertical wall surface is

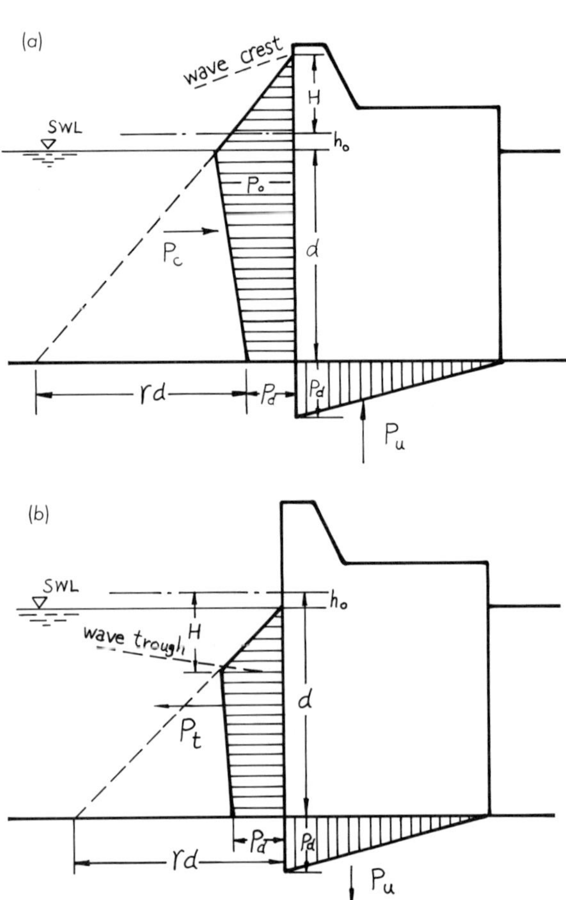

FIGURE 1. Sainflou's simplified method for standing wave pressures: (a) wave crest at wall; (b) wave trough at wall.

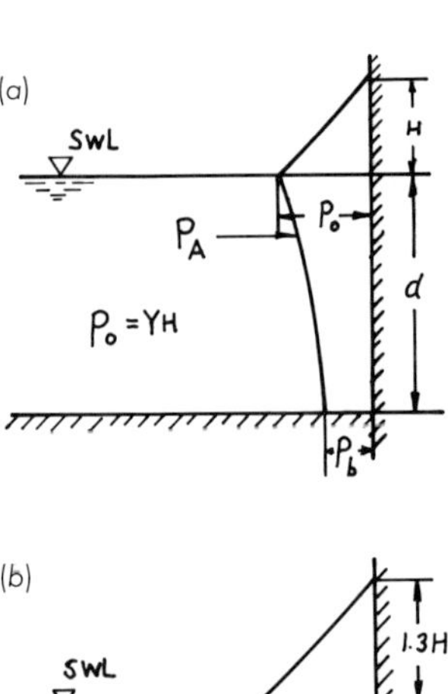

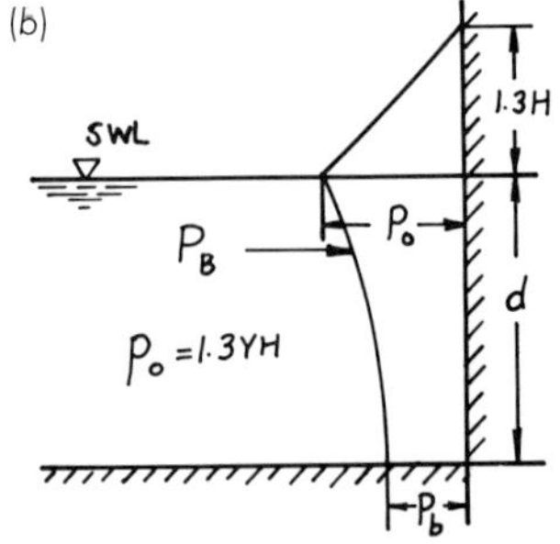

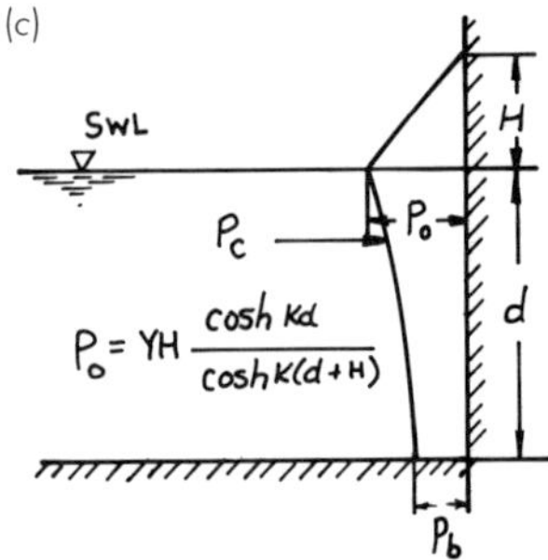

FIGURE 2. Nagai's standing wave pressure diagram: (a) for $0.135 \leq d/L < 0.35$; (b) for $d/L < 0.135$ and $H/L < 0.04$; (c) for $d/L \geq 0.35$.

given as

$$p = \gamma H \frac{\cosh k(d + z)}{\cosh kd} \tag{4}$$

in which $k = 2\pi/L$. The vertical coordinate z measured from the still-water level is positive upward. Equation (4) can be derived by the small amplitude wave theory. It reduces to Equation (2) when $z = -d$.

The vertical distribution of the wave pressures above still-water level is assumed triangular as $p = 0$ at $z = H$, and $p = \gamma H$ at $z = 0$. The resultant wave force can be obtained by

$$P_A = \gamma \left(\frac{H^2}{2} + \frac{H}{k} \tanh kd \right) \tag{5}$$

(b) Region of Very Shallow Water Waves, i.e., $d/L < 0.135$ and $H/L < 0.04$ [Figure 2(b)]:

The wave pressures above still-water level are assumed to have a triangular distribution as $p = 0$ at $z = 1.3H$ and $p = 1.3\gamma H$ at $z = 0$. The distribution of the wave pressures below still-water level is given by

$$p = \gamma H \left[\frac{\cosh k(d + z)}{\cosh kd} + \frac{0.3(d + z)}{d} \right] \tag{6}$$

The resultant wave force is

$$P_B = \gamma \left[\frac{(1.3H)^2}{2} + 0.15Hd + \frac{H}{k} \tanh kd \right] \tag{7}$$

(c) Region of Deep Water Waves, i.e., $d/L \geqq 0.35$ [Figure 2(c)]:

When $z \leqq 0$, the equation of wave pressure is expressed as

$$p = \gamma H \frac{\cosh k(d + z)}{\cosh k(d + H)} \tag{8}$$

And when $z > 0$, the equation is

$$p = \gamma \left[H \frac{\cosh k(d + z)}{\cosh k(d + H)} - z \right] \tag{9}$$

The resultant wave force becomes

$$P_c = \gamma \left[\frac{H}{k} \tanh k(d + H) - \frac{H^2}{2} \right] \tag{10}$$

Although other equations derived from high order wave theories are available, they do not seem to give better results on the whole than those formulae described above.

When dealing with the stability of breakwater or similar structures, the uplift of the standing wave should also be taken into account. The horizontal distribution of the intensity of the uplift acting on the bottom of the wall is assumed triangular in general. The intensity is equal to p_d obtained from Equation (2) at the front toe of the wall, and zero at the rear toe, as shown in Figure 1(a). When Equation (6) or (8) is utilized to calculate the lateral wave pressures, however, the appropriate pressure at the bottom should be used for the intensity of uplift at the front toe of the wall.

The vertical wave force should be added to the buoyancy of the portion of the structure below still-water level in design calculation. Then, the direction of the vertical wave pressure on the bottom of the wall in a wave trough should be taken as downward [Figure 1(b)].

From engineering point of view, the limiting depth of the clapotis can be taken as $2H$.

All the equations cited above are obtained from theories and experiments of regular waves. In nature, however, sea waves are irregular in both height and period. Then, it is important to define a representative wave in the wave train for designing.

The design wave height generally adopted for vertical wall is $H_{1/100}$, although $H_{1\%}$ or H_{max} may sometimes be used according to the different design standards.

$$H_{1/100} = 1.66H_s \tag{11}$$

$$H_{1\%} = 1.51H_s \tag{12}$$

$$H_{max} = \sqrt{\frac{\ln N}{2}} H_s \tag{13}$$

in which H_s is the significant wave height, i.e., the average height of the highest one third of the waves in a wave train; $H_{1/100}$ is the average height of the highest 1/100 of the waves; $H_{1\%}$ is the wave height with the probability of exceedance of 1% in a wave train; H_{max} is the most probable maximum wave height in a train of N waves, for instance, $H_{max} = 1.86H_s$ when $N = 1{,}000$. Equations (11), (12) and (13) are all derived from Rayleigh Distribution. It has been proved by observations that this theoretical distribution coincides with the wave height distribution in deep water in particular. The ratios of $H_{1/100}$, $H_{1\%}$ and H_{max} to H_s may be a little smaller than those given by the equations cited above in relatively shallow water.

The design wave period generally adopted is T_s or $\overline{T}$. T_s is the period of the significant wave, and $\overline{T}$ is the average wave period. $T_s \cong 1.1\overline{T}$.

The wave length is obtained by

$$L = \frac{gT^2}{2\pi} \tanh \frac{2\pi d}{L} \tag{14}$$

in which g is the acceleration of gravity.

The return period of the design wave can be taken as 50 to 100 years depending on the importance of the structure as well as on the reliability of the wave data.

BREAKING WAVE FORCES ON A VERTICAL WALL

Shock pressures of very high intensity exerted by breaking waves on a vertical wall, for instance, as high as $110\gamma H$ [4], have been recorded in some laboratory tests. But their area of action is quite limited and the duration is very short, usually less than 0.01 second. Moreover, the conditions for occurring such high shock pressures may scarcely exist in reality. Therefore, such extreme shock wave pressures are not used in practice for the statical calculation of the breakwaters. The ordinary shock pressures of the breaking waves will be introduced in the following.

The breaking conditions of waves in front of a vertical wall are different from that on beach, because of the effect of the reflection of waves from the wall. According to the results of their comprehensive model tests, the researchers of Dalian Institute of Technology of China suggested to distinguish the breaking waves in front of a vertical wall with rubble mound foundation into two types and presented their relevant wave pressure formulae [5].

(a) Breaking Waves in Front of a Wall with Low Rubble Mound Foundation (Type 1):

When the height of the rubble mound foundation above the sea bed is small, the influence of the mound on the waves may be neglected. In such a case, the waves are almost totally reflected from the wall. The preceding wave reflects back and meets its successor at a distance $x = L/2$ from the wall. Thus, the breaking point of the waves is always situated at that antinode. The conditions for occurring such kind of breaking waves can be determined as $d_1/d > 2/3$, $d < 2H$ and $i \leqq 0.1$, d_1 being the water depth above the top of the rubble mound, and i being the slope of the sea bed (Figure 3).

The wave pressure diagram can be constructed via the wave pressures at four elevations. The wave pressure at the still-water level:

$$p_0 = k_1 k_2 \gamma H \tag{15}$$

$$k_1 = 1 + 3.2(i)^{0.55} \tag{16}$$

$$k_2 = -0.1 + 0.1\frac{L}{H} - 0.0015\left(\frac{L}{H}\right)^2, \quad \text{for } 30 \geqq L/H \geqq 14 \tag{17}$$

The wave pressure at the sea bed:

$$p_d = \begin{cases} 0.6p_0, & \text{when } d/H \leqq 1.7 \\ 0.5p_0, & \text{when } d/H > 1.7 \end{cases} \tag{18}$$

The wave pressure at $H/2$ below the still-water level equals 0.7 p_0. And the wave pressure at H above the still-water level equals zero.

The lifting force on the bottom of the wall is given by

$$P_u = \mu \frac{p_{d1} b}{2} \tag{19}$$

in which p_{d1} is the wave pressure at the bottom of the wall (Figure 3); b is the width of the wall; and μ is a coefficient which equals 0.7.

(b) Breaking Waves in Front of a Wall with High Rubble

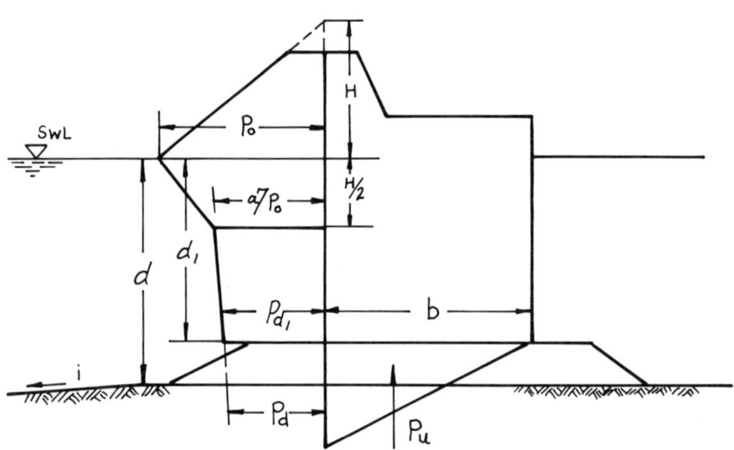

FIGURE 3. Pressure diagram for breaking wave type 1.

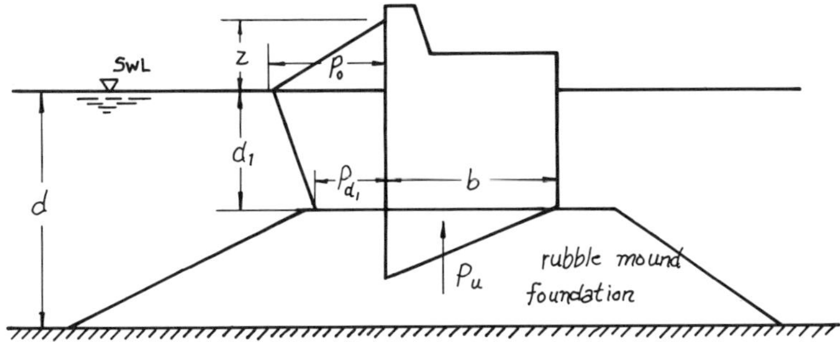

FIGURE 4. Pressure diagram for breaking wave type 2.

Mound Foundation (Type 2):

When the height of the rubble mound foundation above the sea bed is large, the size of the mound and its permeability will have a great influence on the transformation and breaking of waves. Since the water depth decreases rapidly as the wave propagating on the mound, the wave transforms severely and finally breaks on the mound at a distance less than $L/2$ from the wall. The conditions for occurring this kind of breaking waves are determined as $2/3 \geqq d_1/d > 1/3$ and $d_1 < 1.8H$, or $d_1/d \leqq 1/3$ and $d_1 < 1.5H$.

The wave pressure diagram is constructed through the wave pressures at three elevations. The zero pressure point is located at a height z above the still-water level.

$$z = \left(0.27 + 0.53 \frac{d_1}{H}\right) H \qquad (20)$$

The wave pressure at still-water level is given by

$$p_0 = 1.25\gamma H \left(1.8 \frac{H}{d_1} - 0.16\right)\left(1 - 0.13 \frac{H}{d_1}\right) \qquad (21)$$

$$\text{when } 2/3 \geqq d_1/d > 1/3$$

or

$$p_0 = 1.25\gamma H \left[\left(13.9 - 36.4 \frac{d_1}{d}\right) \times \left(\frac{H}{d_1} - 0.67\right) + 1.03\right]\left(1 - 0.13 \frac{H}{d_1}\right) \qquad (22)$$

$$\text{when } 1/3 \geqq d_1/d \geqq 1/4$$

The wave pressure at the bottom of the wall:

$$p_{d1} = 0.6 p_0 \qquad (23)$$

The lifting force on the bottom of the wall can be calculated by Equation (19).

Equations (20) to (23) should be used within a limitation of $d_1 \geqq 0.6H$.

The wave force formulae for the two types of breaking waves based on the model tests have been checked by some prototype examples with good results.

For the broken wave condition, that is the progressive waves breaking somewhere seaside of the place of the breakwater before it is built, it is shown by the tests that Equations (15) to (19) can still be used to predict the wave force as long as $i \leqq 0.02$. In such a case, the design wave height can be taken as $H = 0.78d$.

Goda [6] has proposed a new method according to the result of model tests and the examination of prototype breakwaters. This method deals with both the standing and breaking wave forces in a unified formula. On the whole, the results given by the Goda's formula are close to that given by the formulae cited in this paragraph [5].

SCOURING IN FRONT OF A VERTICAL WALL

For the overall stability of breakwater or similar structures, the scouring of sea bed in front of the structure by waves should be taken into consideration. The present author has conducted model tests with regular waves as well as with irregular waves to study the scouring patterns in front of a vertical wall at Delft University of Technology, The Netherlands [7,8].

It was found in the tests that there were two basic scouring patterns in front of a vertical wall under the action of standing waves (Figure 5). Type I occurred when the material of the sand bed was finer. The fine material was transported largely in suspension, which was scour at the nodes and deposited near the antinodes of a standing wave. Type II occurred when the material of the sand bed was coarser. The coarse material moved mainly as bed load, which was scour half-way between the nodes and the antinodes and deposited at the nodes.

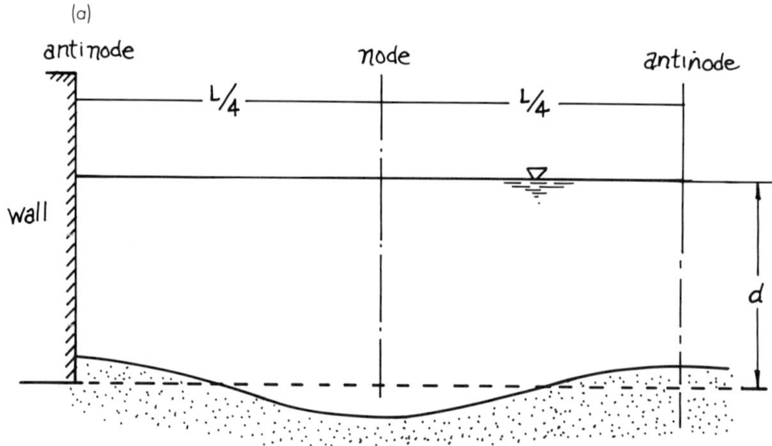

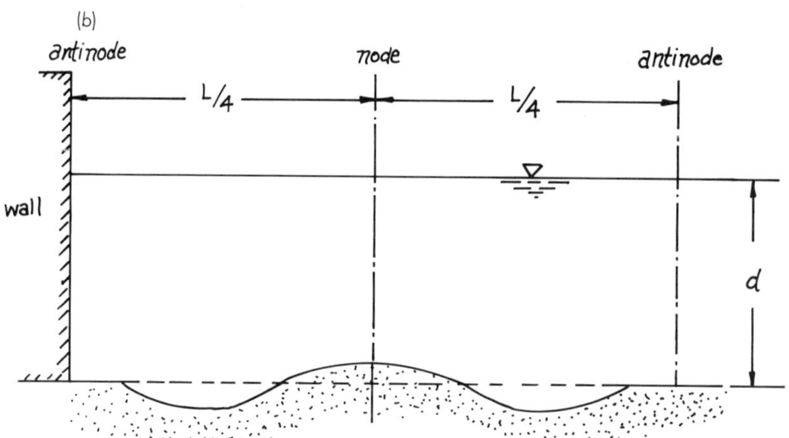

FIGURE 5. Two basic scouring patterns in front of a vertical breakwater: (a) for relatively fine sand; (b) for relatively coarse sand.

Analysis of the test results shows that a dimensionless parameter $(u_{max} - u_{crit})/w$ can be used to distinguish between the two scouring patterns. It is scouring Type I for relatively fine material when $(u_{max} - u_{crit})/w \geqq 16.5$. It is scouring Type II for relatively coarse material when that parameter is less than 16.5. u_{max} is the maximum horizontal orbital velocity of the standing waves at the bottom of the node,

$$u_{max} = \frac{2\pi H}{T \sinh kd} \tag{24}$$

u_{crit} is the critical velocity for sand particles. Comparing with the test results shows that Bagnold's formula presented in 1946 [9] is superior to many newly established formula for estimating the critical velocity in case of the standing waves. Bagnold formula can be expressed as

$$u_{crit} = 2.40\Delta^{2/3}D_{50}^{0.433}T^{1/3} \tag{25}$$

in which $\Delta = (\rho_s - \rho)/\rho$ is the relative mass density of sand, and ρ_s and ρ are the mass densities of sand and water respectively. D_{50} is the medium diameter of the sand particles. It should be pointed out that the unit of length in Equation (25) is in meter only, since the coefficient of which is dimensional. w is the fall velocity of the sand particle.

For the relatively fine material, a trochoid is adopted to predict the equilibrium scouring profile as an approximation. Then, the following equations should be used:

$$x_t = \frac{L}{4\pi}\theta + R\sin\theta \tag{26}$$

$$z_t = -R\cos\theta \tag{27}$$

in which x_t and z_t are horizontal and vertical coordinates of the curve respectively. x_t is measured from the node, and z_t from a height z_0 above the original flat sand bed, positive upward.

$$R = \frac{1 - \sqrt{1 - \dfrac{8\pi}{L} Z_{sm}}}{\dfrac{4\pi}{L}} \tag{28}$$

$$z_0 = R - Z_{sm} \tag{29}$$

θ is from 0 to 2π. Z_{sm} is the maximum depth of the trough of the scouring profile when the equilibrium condition is reached. The test results indicate that Z_{sm} can be well expressed by

$$Z_{sm} = \frac{0.4H}{(\sinh kd)^{0.135}} \tag{30}$$

According to the comparison of the test results of the irregular waves with those of the regular waves, it reveals preliminarily that if H_s is taken as the equivalent wave height, the maximum scouring depth is on the conservative side.

When there is a rubble protective layer on the sea bed in front of a breakwater, the scouring bottom profile will differ from that without a protective layer (Figure 6).

The influence of a protective layer on the sand bed is limited mainly to a distance of $L/2$ from the wall. If Z'_{sm} refers to the ultimate maximum depth of the first scouring trough from the wall when there is a protective layer on the sand bed, Z'_{sm} decreases with the increase of the width of the protective layer l, and so does the width of the scouring trough. The distance from the wall to the first scouring trough increases with l. There is essentially no scouring over a distance of $L/2$ from the wall when $l = 3L/8$.

When the scouring profile in front of a breakwater is

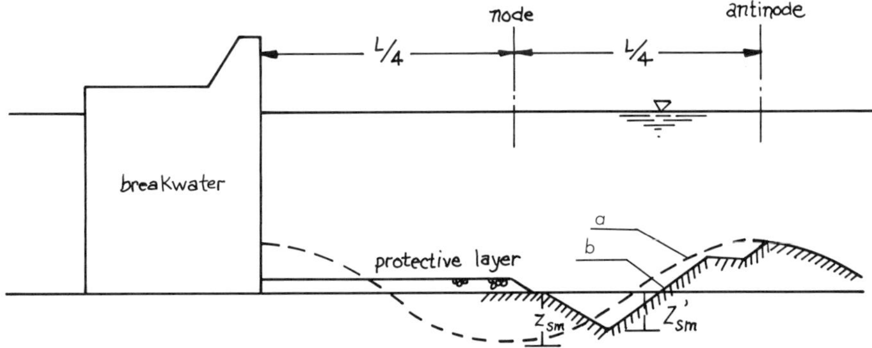

FIGURE 6. Influence of a protective layer: (a) bottom profile without protective layer; (b) bottom profile with protective layer.

known, it is possible to examine its influence on the stability of the foundation of the breakwater, and then to decide whether a protective layer is necessary and what its dimension should be. The influence of the scouring patterns on the stability of the foundation of the vertical breakwaters has been examined with some typical examples by utilizing the slip circle method in References [7] and [8]. It was found that a relatively wide protective layer on the sea bed in front of the breakwater, for instance, $l = L/4$ as generally recommended in literature, is not always necessary.

WAVE FORCES ON VERTICAL PILES

Waves acting on piles exert forces which are the result of drag and inertial forces and are generally calculated by Morison's formula [10]. The horizontal force per unit length of a vertical cylindrical pile (Figure 7) is expressed as

$$f = f_D + f_I = C_D \frac{1}{2}\frac{\gamma}{g} Du|u| + C_M \frac{\gamma}{g}\frac{\pi D^2}{4}\frac{du}{dt} \tag{31}$$

in which f_D is the drag force per unit length of pile; f_I is the inertial force per unit length of pile; D is the diameter of the pile; u and du/dt are the horizontal velocity and acceleration of water particle at the axis of the pile respectively, both calculated as if the pile were non-existent; C_D is the drag coefficient; and C_M is the mass or inertia coefficient.

Different wave theories, such as small amplitude wave, Stokes wave, stream function wave, cnoidal wave, solitary wave, etc., may be used to calculate u and du/dt according to their relative validity in different regions (d/L and H/L). In the following, small amplitude wave theory is to be used to obtain the wave forces on a pile at first. And diagrams of correction coefficients for force and moment will be provided, thus making it possible to consider the non-linear effect of the wave in shallow water. When the equations of u and du/dt for small amplitude wave are substituted into Equation (31), then

$$f = f_{Dm} \cos \omega t|\cos \omega t| - f_{Im} \sin \omega t \tag{32}$$

in which f_{Dm} and f_{Im} are the maximum values of f_D and f_I respectively; $\omega = 2\pi/T$; and t is time.

$$f_{Dm} = C_D \frac{\pi^2}{2}\frac{\gamma}{g}\frac{DH^2}{T^2}\left[\frac{\cosh k(d + z)}{\sinh kd}\right]^2 \quad \text{when } \omega t = 0° \tag{33}$$

$$f_{Im} = C_M \frac{\pi^3}{2}\frac{\gamma}{g}\frac{D^2H}{T^2}\frac{\cosh k(d + z)}{\sinh kd} \quad \text{when } \omega t = 270° \tag{34}$$

The diagram of vertical distribution of f can be constructed by substituting different z into above equations.

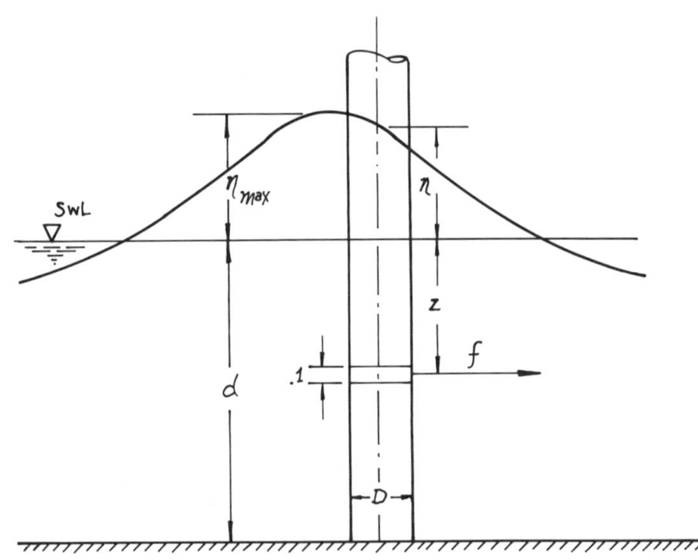

FIGURE 7. Wave forces on a vertical pile.

The resulting horizontal force acting on the pile F can be obtained by integration of Equation (32) from $z = -d$ to $z = \eta$, η being the elevation of wave profile at the pile. Then

$$F = F_{Dm} \cos \omega t |\cos \omega t| - F_{Im} \sin \omega t \qquad (35)$$

in which F_{Dm} and F_{Im} are the maximum resulting drag force and the maximum resulting inertial force on the pile respectively.

$$F_{Dm} = C_D \frac{\gamma D H^2}{2} K_1 \qquad (36)$$

$$F_{Im} = C_m \frac{\gamma \pi D^2 H}{8} K_2 \qquad (37)$$

where

$$K_1 = \frac{2kS_1 + \sinh 2kS_1}{8 \sinh 2kd} \qquad (38)$$

$$K_2 = \frac{\sinh kS_2}{\cosh kd} \qquad (39)$$

in which S_1 is the elevation of wave crest measured from $z = -d$, $S_1 = d + \eta_{max}$; and S_2 is the elevation of wave profile at the pile when the inertial force reaches its maximum, which is also measured from the mudline, $S_2 = d + \eta_{max} - (H/2)$. η_{max} may be evaluated from Figure 8, which is obtained mainly by experiments [11], and is a correction for the small amplitude wave theory.

The maximum resulting force F_m can be derived from $dF/d(\omega t) = 0$, then

(a) When $F_{Dm} \leqq 0.5F_{Im}$,

$$F_m = F_{Im} \qquad (40)$$

which occurs at $\omega t = 270°$.

(b) When $F_{Dm} > 0.5F_{Im}$

$$F_m = F_{Dm} \left[1 + 0.25 \left(\frac{F_{Im}}{F_{Dm}} \right)^2 \right] \qquad (41)$$

the phase of which is

$$\sin \omega t = -0.5 \frac{F_{Im}}{F_{Dm}} \qquad (42)$$

The moments of F_{Dm}, F_{Im} and F_m about the mudline are expressed as follow:

$$M_{Dm} = C_D \frac{\gamma D H^2 L}{2\pi} K_3 \qquad (43)$$

$$M_{Im} = C_M \frac{\gamma D^2 H L}{16} K_4 \qquad (44)$$

where

$$K_3 = \frac{1}{\sinh 2kd} \left[\frac{k^2 S_1^2}{16} + \frac{kS_1}{16} \sinh 2kS_1 - \frac{1}{32} (\cosh 2kS_1 - 1) \right] \qquad (45)$$

$$K_4 = \frac{1}{\cosh 2kd} [kS_2 \sinh kS_2 - \cosh kS_2 + 1] \qquad (46)$$

(a) When $M_{Dm} \leqq 0.5M_{Im}$

$$M_m = M_{Im} \qquad (47)$$

(b) When $M_{Dm} > 0.5M_{Im}$

$$M_m = M_{Dm} \left[1 + 0.25 \left(\frac{M_{Im}}{M_{Dm}} \right)^2 \right] \qquad (48)$$

The phase of the maximum moment M_m is the same as that of F_m.

There are many books and articles, for example, [12], dealing with the selection of C_D and C_M. From engineering point of view, the values of C_D and C_M may be suggested as $C_D = 1.2$ when $Re \leqq 5 \times 10^5$; $C_D = 0.7$ when $Re > 5 \times 10^5$; and $C_M = 2.0$. The Reynolds Number Re is expressed by

$$Re = \frac{u_{m0} D}{\nu} \qquad (49)$$

in which ν is the kinematic viscosity of the fluid; and u_{m0}

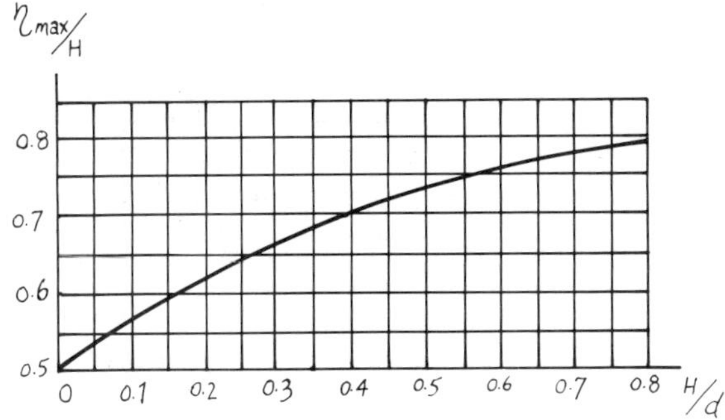

FIGURE 8. η_{max}/H versus H/d.

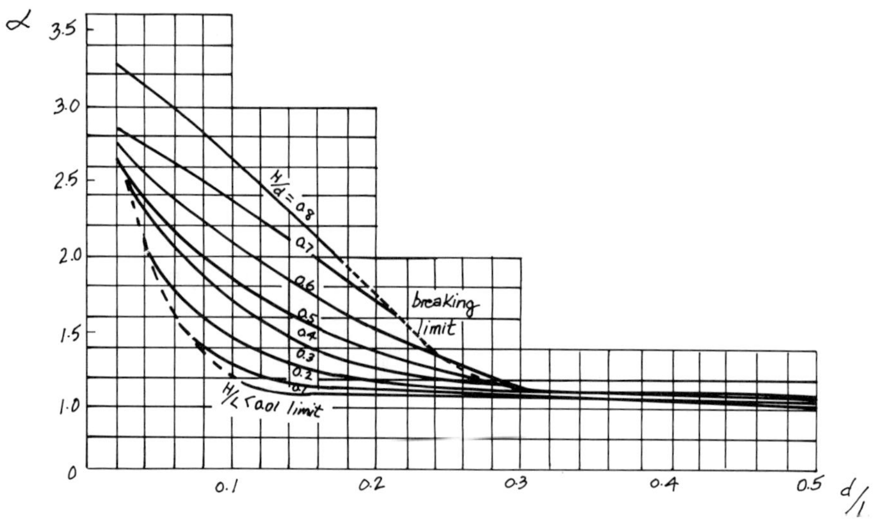

FIGURE 9. Correction coefficient α.

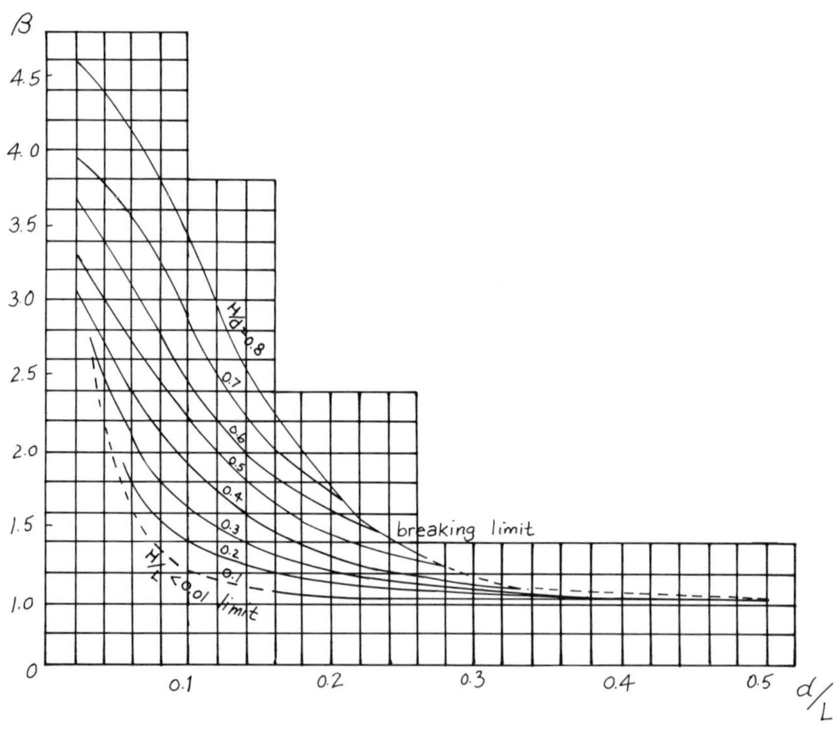

FIGURE 10. Correction coefficient β.

is the maximum horizontal velocity at $z = 0$, which is given by

$$u_{m0} = \frac{\pi H}{T} \coth kd \tag{50}$$

Comparison with the test results, however, has shown that the above equations derived from small amplitude wave theory for the drag force are only valid for relatively small wave height and relatively deep water ($H/d \leqq 0.2$ and $d/L \geqq 0.2$). Otherwise, the correction coefficients should be applied to F_{Dm} and M_{Dm} [11]. These coefficients are derived from Stoke second order and cnoidal wave theories. Then

$$F_{Dm} = C_D \frac{\gamma D H^2}{2} K_1 \alpha \tag{51}$$

$$M_{Dm} = C_D \frac{\gamma D H^2 L}{2\pi} K_3 \beta \tag{52}$$

The correction coefficients α and β are given in Figures 9 and 10.

For a pile-supported structure, the maximum total force on a series of piles should be determined. The resulting force on a single pile at position x can be calculated approximately by

$$F = F_{Dm} \cos kx|\cos kx| + F_{Im} \sin kx \tag{53}$$

in which $x = 0$ at wave crest, and the direction of x is coincided with that of wave propagation.

The total force on a group of N piles is

$$F_{total} = \sum_{n=1}^{N} F_n \tag{54}$$

In order to obtain the maximum total force on group piles, a series of different positions of pile relative to the wave crest should be chosen.

The maximum total moment on group piles can be estimated with the same procedure. The resulting moment acting on a pile at x is

$$M = M_{Dm} \cos kx|\cos kx| + M_{Im} \sin kx \tag{55}$$

The total moment on N piles is

$$M_{total} = \sum_{n=1}^{N} M_n \tag{56}$$

The design wave height generally adopted for piles is $H_{1/100}$ or H_{max}, and the design wave period is T_s. The return period of the design wave for piles is the same as that for a vertical wall.

WAVE FORCES ON A LARGE CYLINDRICAL STRUCTURE

The Morison equation described in the above paragraph is valid for the ratio D/L is small. When the diameter of the cylindrical structure, such as pier or dolphin, reaches a significant fraction of a wave length ($D/L > 0.2$), the incident waves will undergo significant scattering or diffraction. In such a case, the diffraction theory presented by MacCamy and Fuchs [13] is generally used.

When only the maximum values of f, F and M are concerned about, those equations can be expressed similar to Equations (34), (37) and (44).

$$f_m = C_M \frac{\pi^3}{2} \frac{\gamma}{g} \frac{D^2 H}{T^2} \frac{\cosh k(d + z)}{\sinh kd} \tag{57}$$

$$F_m = C_M \frac{\gamma \pi D^2 H}{8} \tanh kd \tag{58}$$

$$M_m = C_m \frac{\gamma D^2 H L}{16} \frac{(kd \sinh kd - \cosh kd + 1)}{\cosh 2kd} \tag{59}$$

The coefficient C_M derived from diffraction theory should be used in Equations (57), (58) and (59), which is given in Figure 11. It should be pointed out that the phase angle is not the same as that indicated in Equation (34), but dependent on D/L.

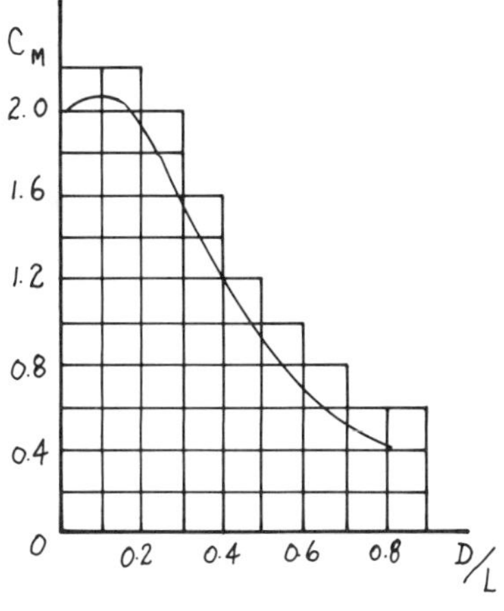

FIGURE 11. Coefficient C_M for large cylindrical structure.

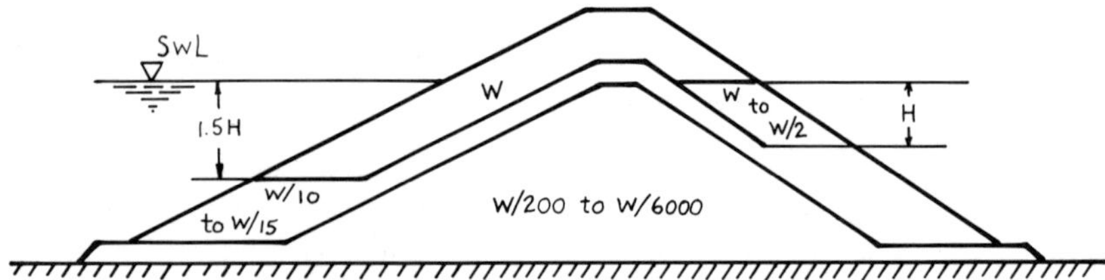

FIGURE 12. A rubble mound breakwater.

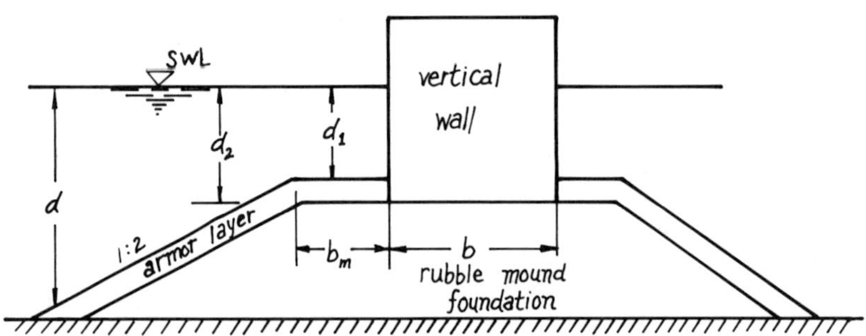

FIGURE 13. Cross section of a composite breakwater.

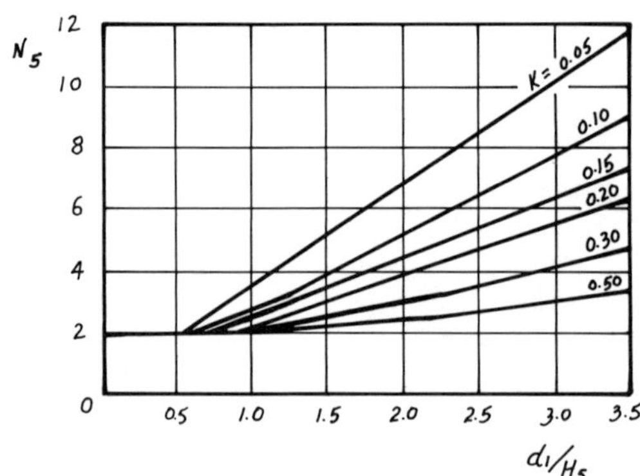

FIGURE 14. Stability number of quarry stones.

TABLE 1. Suggested K_d Values.

Armor Units	n	Placement	Structure Trunk K_D* Breaking Wave	Structure Trunk K_D* Nonbreaking Wave	Structure Head K_D Breaking Wave	Structure Head K_D Nonbreaking Wave	Structure Head Slope cot α
Stone, rounded	2	random	1.2	2.4	1.1	1.9	1.5 to 3.0
Stone, angular	2	random	2.0	4.0	1.9 1.6 1.3	3.2 2.8 2.3	1.5 2.0 3.0
Stone, angular	2	special**	5.8	7.0	5.3	6.4	1.5 to 3.0
Tetrapod	2	random	7.0	8.0	5.0 4.5 3.5	6.0 5.5 4.0	1.5 2.0 3.0
Tribar	2	random	9.0	10.0	8.3 7.8 6.0	9.0 8.5 6.5	1.5 2.0 3.0
Tribar	1	uniform	12.0	15.0	7.5	9.5	1.5 to 3.0
Dolos	2	random	15.8†	31.8†	8.0 7.0	16.0 14.0	2.0 3.0

*Applicable to slopes ranging from 1 on 1.5 to 1 on 5.
**Special placement with long axis of stone placed perpendicular to structure face.
†If no rocking (<2%) instead of no-damage criteria (<5% displacement, rocking, etc.) is desired, reduce K_D 50%.

BREAKING WAVE FORCES ON A CYLINDRICAL PILE

The forces exerted by breaking waves on a cylindrical pile consist of two components, a slowly varying force and a much larger but very short duration quasi-impact force [14]. Only a few measurements are available of forces induced by breaking waves on piles located in the surf zone. The approximate method proposed in Shore Protection Manual [15] published by U.S. Army Coastal Engineering Research Center is cited in the following.

The maximum horizontal force on a pile is

$$F_m = 1.5\gamma DH_b^2 \tag{60}$$

when $Re \leqq 5 \times 10^5$; or

$$F_m = 0.88\gamma DH_b^2 \tag{61}$$

when $Re > 5 \times 10^5$. The maximum moment about the mudline is

$$M_m = 1.11 d_b F_m \tag{62}$$

The breaking wave height H_b and the breaking depth d_b should be determined according to the deep water wave steepness H_0/L_0 and the bottom slope i.

STABILITY OF RUBBLE MOUND STRUCTURES

A rubble mound structure is composed of primary cover layer, one or two underlayers and core (Figure 12). Quarry stones or specially shaped concrete blocks can be used as armor units to form the primary cover layer.

Among many empirical and semi-empirical formulae dealing with the stability of armor units on rubble mound structures, the formula developed by Hudson [16] is used most extensively. The Hudson formula is expressed as

$$W = \frac{\gamma_r H^3}{K_D(S_r - 1)^3 \cot \alpha} \tag{63}$$

in which W is the weight of an individual armor unit; γ_r is the unit weight of armor unit; S_r is the specific gravity of armor unit relative to the water at the structure, $S_r = \gamma_r/\gamma$; α is the angle of structure slope measured from horizontal; and K_D is the stability coefficient which is dependent primarily on the type of armor, the part of structure (trunk or head) and the type of wave attacking structure (breaking or nonbreaking). The suggested K_D values for different types of armor are listed in Table 1 [15]. These K_D values correspond to no-damage criteria and minor overtopping. In Table 1, n is the number of units comprising the thickness of the armor layer.

The design wave height H in Equation (63) is generally taken as H_s. However, laboratory studies with irregular waves have shown that the equivalent wave height for the wave train is depended on the shape of the spectrum [17], the ratio of H_{max} to H_s [18], the presence of wave grouping [19], etc. In relatively deep water where the large waves in the spectrum can reach the structure without being reduced by prior breaking [20] or when the wave grouping is predominant at the site, the $H_{1/10}$ is recommended to use in Equation (63). $H_{1/10}$ is the average height of the highest one tenth of the waves in a wave train, $H_{1/10} = 1.27H_s$ according to the Rayleigh Distribution.

The return period of the design wave can be taken as 25 to 100 years.

Besides to consider the hydraulic stability, it is necessary to consider the structural stability for thc slender concrete armor unit like dolos. At present, it is very difficult to calculate the wave forces acting on the armor units. Therefore, some empirical data have to be used in designing. For instance, a dolos without reinforcement can be considered strong enough to withstand stresses induced by normal handling and the action of design wave when its weight is not greater than 20 tons [21].

STABILITY OF RUBBLE MOUND FOUNDATION OF A VERTICAL WALL

K. Tanimoto *et al.* [22] of the Port and Harbor Research Institute, Ministry of Transport of Japan carried out tests with irregular waves to study the stability of armor units for the rubble mound foundation of a vertical wall. A formula is developed as

$$W = \frac{\gamma_r H_s^3}{N_s^3(S_r - 1)^3} \tag{64}$$

in which N_s is the stability number, which is dependent on d_2/H_s, K as well as on the type of the armor. d_2 is the depth at the crest of rubble mound foundation excluding the armor layer (Figure 13). The parameter K represents the combined effects of the relative water depth and the relative distance from the vertical wall on the maximum horizontal velocity at the bottom, which is expressed as

$$K = \frac{\dfrac{4\pi d_2}{L_2}}{\sinh \dfrac{4\pi d_2}{L_2}} \sin^2\left(\frac{2\pi b_m}{L_2}\right) \tag{65}$$

in which L_2 is the wave length corresponding to the depth of d_2, which is calculated by Equation (14) through T_s; and b_m is the berm width of the rubble mound foundation.

According to the test results, the stability number N_s is expressed as function of d_2/H_s and K in Figure 14 for two-layer quarry stone armor with a slope of 1 on 2.

The significant wave height is adopted in Equation (64), but it is by no means to indicate that H_s is the equivalent wave height in a wave train. Comparing the test results of irregular waves with that of regular waves shows that the equivalent wave height would be about $1.37H_s$ for the stability of armor units on rubble mound foundation.

REFERENCES

1. Sainflou, G., "Essai sur les Digues Maritimes Verticales," *Annales des Ponts et Chaussees,* Vol. 98, No. 4, 1928.
2. Hudson, R. Y., "Wave Forces on Breakwaters, Engineering Aspects of Water Waves: A Symposium," *Trans. ASCE,* Vol. 118, 1953, 653–674.
3. Nagai, S., "Pressures of Standing Waves on Vertical Wall," *Proc. ASCE,* Vol. 95, WW1, 1969, 53–76.
4. Denny, D. F., "Further Experiments on Wave Pressures," *J. Inst. Civil Engrs.*, Feb. 1951, 330–345.
5. Li, Y. C., Y. X. Yu, and M. T. Xoy, "Investigation of Wave Pressure on Vertical Wall," *Proc. Intern. Conf. on Coastal and Port Engi. in Developing Countries,* Vol. 1, 1983, 755–776.
6. Goda, Y., "New Wave Pressure Formula for Composite Breakwaters," *Proc. 14th Coastal Engi. Conf.*, Vol. 3, 174, 1702–1720.
7. Xie, S. L., "Scouring Patterns in Front of Vertical Breakwaters and Their Influences on the Stability of the Foundations of the Breakwaters," Dept. of Civil Engi., Delft University of Tcchnology, The Netherlands, 1981, 1–61.
8. Xie, S. L., "Scouring Patterns in Front of Vertical Breakwaters," *Acta Oceanologica Sinica*, Vol. 4, No. 1, 1985, 153–164.
9. Bagnold, R. A., "Motion of Waves in Shallow Water: Interaction between Waves and Sand Bottoms," *Proc. Royal Society of London*, Series A, Vol. 187, 1946, 1–15.
10. Morison, J. R., "The Design of Piling," *Proc. 1st Conf. Coastal Engineering*, 1951, 254–258.
11. *Ministry of Communications*, PRC, Harbor Hydrology and Hydraulics, Technical Specifications for Harbor Engineering (in Chinese), 1978.
12. Sarpkaya, T. and M. Isaacson, "Mechanics of Wave Forces on Offshore Structures," Van Nostrand Reinhold, 1981.
13. MacCamy, R. C. and R. A. Fuchs, "Wave Forces on Piles: A Diffraction Theory," U.S. Army Corps of Engineers, Beach Erosion Board, Tech. Memo. No. 69, 1954, 1–17.
14. Wiegel, R. L., "Forces Induced by Breakers on Piles," *Proc. 18th Coastal Engineering Conf.*, Vol. 2, 1982, 1699–1715.
15. U.S. Army, Coastal Engineering Research Center, *Shore Protection Manual,* Vol. II, 1984.
16. Hudson, R. Y., "Laboratory Investigations of Rubble-Mound Breakwaters," *Proc. ASCE*, Vol. 85, No. WW3, 1959, 93–121.

17. Carstens, T., A. Torum, and A. Tratteberg, "The Stability of Rubble Mound Breakwaters against Irregular Waves," *Proc. 10th Conf. Coastal Engineering*, Vol. 2, 1966, 958–971.
18. Ouellet, Y., "Effects of Irregular Wave Trains on Rubble-Mound Breakwaters," *Proc. ASCE*, Vol. 98, No. WW1, 1972, 1–14.
19. Johnson, R. R., E. P. D. Mansard, and J. Ploeg, "Effects of Wave Grouping on Breakwater Stability," *Proc. 16th Coastal Engineering Conf.*, Vol. 3, 1978, 2228–2243.
20. Zwamborn, J. A., "Analysis of Causes of Damage to Sines Breakwater," *Coastal Structures '79*, Vol. I, 1979, 422–441.
21. Scholtz, D. J. P., J. A. Zwamborn, and M. van Niekerk, "Dolos Stability: Effect of Block Density and Waist Thickness," *Proc. 18th Coastal Engineering Conf.*, Vol. 3, 1982, 2026–2046.
22. Tanimoto, K., T. Yagyu, and Y. Goda, "Irregular Wave Tests for Composite Breakwater Foundations," *Proc. 18th Coastal Engineering Conf.*, Vol. 3, 1982, 2144–2163.

SECTION SIX

Coastal Structures

CHAPTER 23 Coastal Structure 735
CHAPTER 24 Offshore Structures 781
CHAPTER 25 Ports and Harbors 829

Coastal Structure

KAZUMASA MIZUMURA*

COASTAL STRUCTURES

Introduction

There are seawalls, bulkheads, revetments, groins, jetties and breakwaters as the coastal structures. They are principally classified by their objections and functions. For example, seawalls, bulkheads and revetments are the structures which protect lands from waves, surges and currents. Groins protect against land erosion due to waves and currents; jetties maintain flow direction and depth in river mouths and seas and prevent shoaling in navigational waterways. Breakwaters keep calmness in ports and harbors, protect shore from erosion and develop beaches.

Seawalls, Bulkheads and Revetments

DEFINITION AND FUNCTION

They are the coastal structures which are built parallel to shoreline and separate land and sea regions as shown in Photos 1 to 3. The purpose of the coastal structures is to protect lands from disaster due to waves. Especially, the determination of their extent is very important and the shore protection works around them must be done. Because these structures protect land from erosion immediately behind it and in result minor land erosion is occasionally expected at both ends of them. The height of the coastal structures are designed by the conditions of wave runup, overtopping or transmission. Since scours often occur at the toe of the coastal structures, the toe must be protected by rubble stones or concrete blocks as shown in Photos 4 to 5.

*Department of Civil Engineering, Kanazawa Institute of Technology, Ishikawa 921, Japan

TYPES

Seawalls are the most massive coastal structures resisting wave action. Bulkheads are secondly massive coastal structures to retain fill and to protect land from exposure to waves. Revetments are the lightest coastal structures which defend shore from erosion due to weak waves or currents. The types of seawalls are curved-face seawalls, stepped seawalls and rubble-mound seawalls. The types of bulkheads are concrete, steel and timber and the types of revetments are rigid, cast-in-place concrete, flexible or articulated armor assist.

GROUND ELEVATION IN FRONT OF A STRUCTURE

After the construction of coastal structures scours occur at the toe of the structures in short time. The scouring makes toe stones sink and finally attains equilibrium condition. Therefore, for designing them the cross-section is overbuilt or excess stones or concrete blocks are prepared at the toe. As the experience to predict the maximum depth of a scour the following is given [43]: the maximum depth of a scour trough below the natural bed is about equal to the height of the maximum unbroken wave that can be supported by the original depth of water at the toe of the structure. For long term effects erosion without the construction of the coastal structure must be considered.

Groins

DEFINITION AND FUNCTION

The groins (Figure 1) are the coastal structures which are built perpendicular to the shoreline. They trap sand to form wide beaches, stabilize beaches against waves and currents, decrease littoral drift and movement of sand on beaches, and control sand accretion on the downdrift side. Therefore, erosion on the downdrift side must be noticed after the construction. In general, to obtain effective function the groins are constructed between the wave breaker zone and

PHOTO 1. Seawall. (Courtesy of JSCE Slide Library.)

PHOTO 2. Seawall. (Courtesy of JSCE Slide Library.)

PHOTO 3. Seawall. (Courtesy of JSCE Slide Library.)

PHOTO 4. Seawall and wave breaking work.

PHOTO 5. Seawall and wave breaking work. (Courtesy of JSCE Slide Library.)

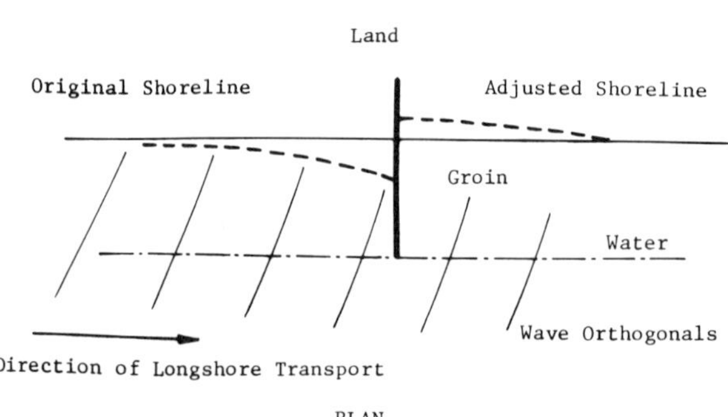

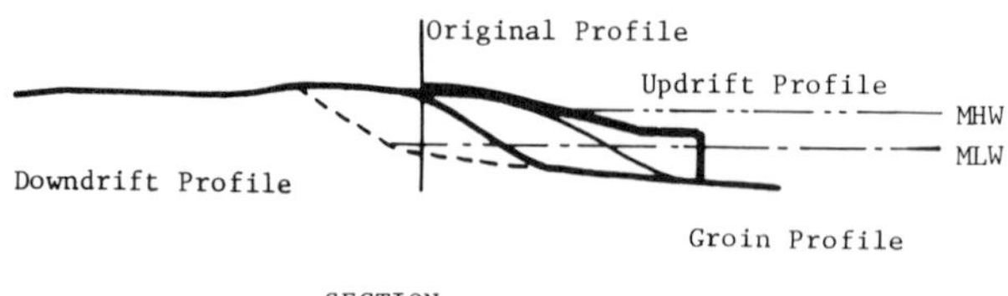

FIGURE 1. Illustration of a typical groin (Reference 43).

the shore. They are made up by timber, steel, stone or concrete. Photo 6 shows the shoreline change in one year after the construction of a groin. Photo 7 gives the view of the port after 5 years from the previous photo. Photo 8 is the result of shoreline estimates by the hydraulic model test [27]. To protect long range of shoreline a series of groins are used and they are called a groin system or groin field (Photo 9 and Figure 2). The effect of a groin system with T shape is also viewed in Photos 10 and 11.

TYPES

The types of groins are classified into permeable groins, impermeable groins (high and low groins) and adjustable groins. Permeable groins reduce fast change of the shoreline after the construction of groins and are used in rich-drift areas to widen or prevent recession of specific beach areas and to reduce scalloping of the shoreline. Impermeable groins are used to retain fill placed to restore or widen a beach. If it is necessary to maintain a sand supply downdrift, the groin may be built low enough to allow overtopping. Adjustable groins consist of removable panels between piles. These panels can be added or removed to maintain the groin at a specific height above the beach level.

DIMENSIONS OF GROINS

The dimensions of groins are dependent upon incident wave forces, types and materials. Especially, length, shapes, wave directions and littoral drift are important factors to determine the dimensions of groins. The length is determined by littoral drift to be trapped or shore state on the down drift side and the groin in the direction of length is divided into the following three parts as shown in Figure 3:

(a) Horizontal shore section
(b) Intermediate sloped section
(c) Outer section

The horizontal shore section is the part of the groin located on the shore. The intermediate sloped section is the part of the groin connecting the horizontal shore section and the outer section and is built on the beach in the same slope as the original beach slope. The outer section is almost horizontal and its length depends on the design slope of the updrift beach. To determine the groin length it is necessary to predict the final shape of the shoreline and the beach and to obtain the position of the breaker zone to normal waves. The details of the method is represented in [43]. The spacing of groins must be determined by considering that turbulence is induced and erosion of shore occurs in the case of narrow spacing and effect of groin system dies out in the case of wide spacing. As a guide to the spacing of groins, the following general rule is suggested [43]: The spacing between groins should equal two to three times the groin length from the berm crest to the seaward end.

LIMITATION AND ECONOMICS

To prevent shore erosion, groins or groin systems are very often used. But the following must be considered before designing them:

1. Natural sand is supplied and groins will be effective
2. Scope of erosion on downdrift side and stability in the structure of groins
3. Economic determination of a groin system must be compared with nourishment

Jetties

DEFINITION AND FUNCTION

Jetties are the structures to direct river flows or tidal currents, or to prevent shoaling from littoral drift in waterways. Still more, they reduce cross currents or wave action at the river mouths and at the entrance of ports and harbors and stabilize the position of inlets. Photos 12 and 13 represent the blockade of a river mouth and jetties, respectively.

TYPES

The principal construction materials are stone, concrete, steel and timber. Asphalt has occasionally been used as a binder. The types of jetties are rubble-mound jetties and sheet-pile jetties.

Breakwaters

DEFINITION AND FUNCTION

Breakwaters are classified into shore-connected and offshore types. The shore-connected type protects shore area, harbor, anchorage, or basin from wave action. Breakwaters for navigation purposes are constructed to create calm water in a harbor area, to provide protection for safe mooring and to protect for harbor facilities (Photos 14 to 18). The offshore type is constructed to protect shore, to trap littoral drift and to provide navigation safety around ports and harbors (Photos 19 to 23). Still more, in the lee of a shore-connected breakwater typical erosion and accretion region occur. Figure 4 shows the direction of incident waves and ENE direction is dominant in the case study. Figures 5, 6, and 7 represent equi-mean diameter curves, littoral currents and change of sea bottom topography, respectively. These give the typical patterns of bed material, flow and sea bottom topography near the shore-connected breakwater [23].

TYPES

The types of shore-connected breakwaters are rubble mound, composite, concrete-caisson, sheet-piling cell, crib, or mobile. The type of offshore breakwaters is principally rubble mound.

PHOTO 6. Groin and shoreline change, August 1978. (Courtesy of JSCE Slide Library.)

PHOTO 7. Groin and shoreline change, October 1983. (Courtesy of JSCE Slide Library.)

PHOTO 8. Groin and shoreline estimate. (Courtesy of ASCE.)

PHOTO 9. Groin system. (Courtesy of JSCE Slide Library.)

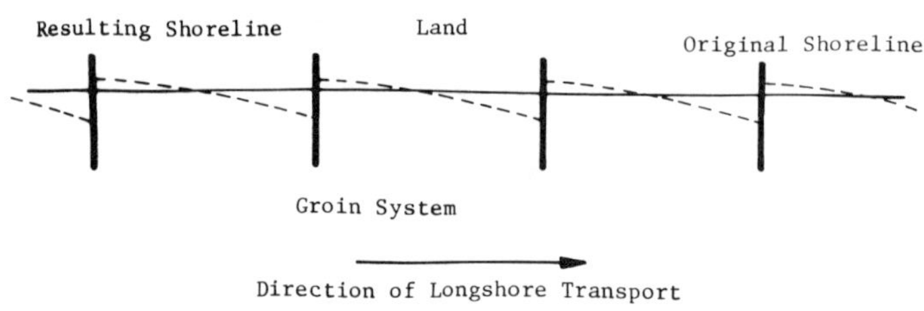

FIGURE 2. General shoreline configuration for two or more groins (Reference 43).

PHOTO 10. T-shaped groin system. (Courtesy of Dr. Tamai (Kouchi University.)

PHOTO 11. T-shaped groin system in hydraulic model test.

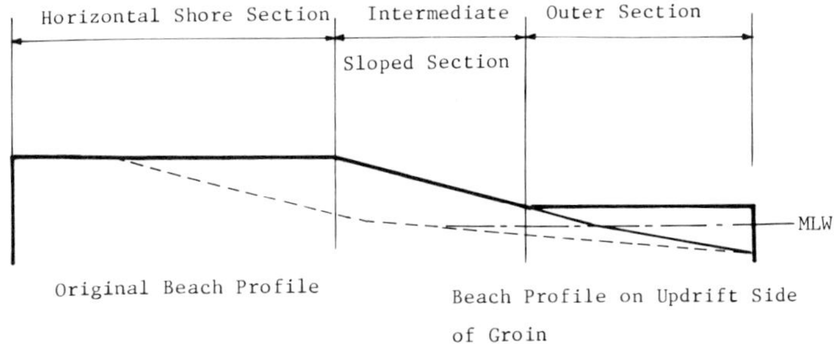

FIGURE 3. Representation of intermediate sloped groin section designed perpendicular to the beach (Reference 43).

PHOTO 12. River blockade. (Courtesy of JSCE Slide Library.)

PHOTO 13. River mouth and jetties. (Courtesy of JSCE Slide Library.)

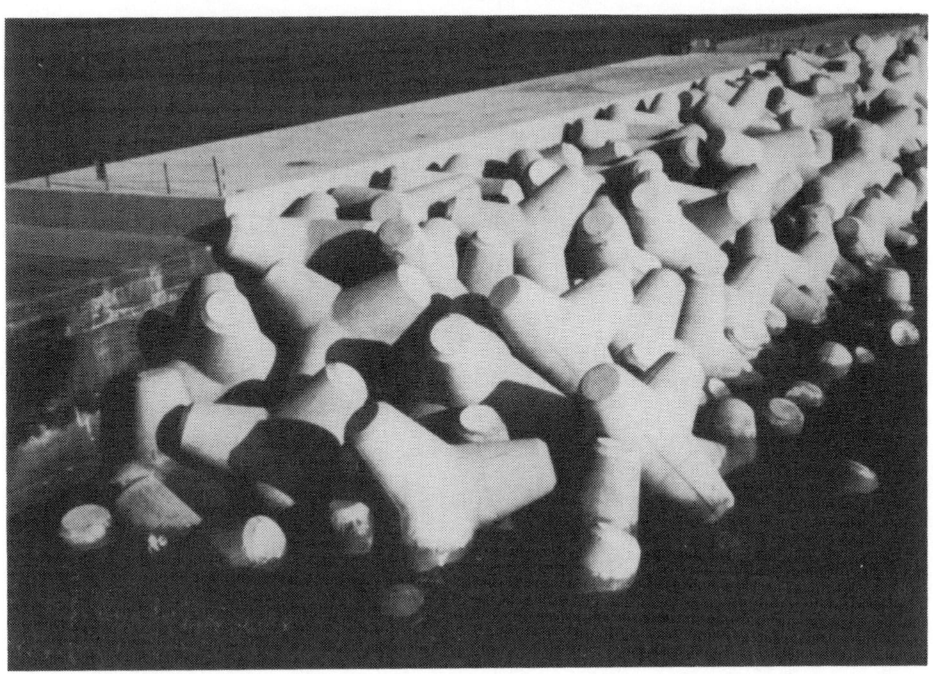

PHOTO 14. Breakwater and concrete blocks. (Courtesy of JSCE Slide Library.)

PHOTO 15. Shore-connected breakwater and wave breaking work. (Courtesy of JSCE Slide Library.)

PHOTO 16. Shore-connected breakwater and concrete blocks.

PHOTO 17. Rubble-mound breakwater. (Courtesy of JSCE Slide Library.)

PHOTO 18. Breakwater, against waves and littoral drift of sand. (Courtesy of JSCE Slide Library.)

PHOTO 19. Offshore breakwater, against waves and littoral drift of sand. (Courtesy of JSCE Slide Library.)

PHOTO 20. Offshore breakwater.

PHOTO 21. Offshore breakwaters and shoreline change. (Courtesy of JSCE Slide Library.)

PHOTO 22. Offshore breakwaters. (Courtesy of JSCE Slide Library.)

PHOTO 23. Offshore breakwaters. (Courtesy of JSCE Slide Library.)

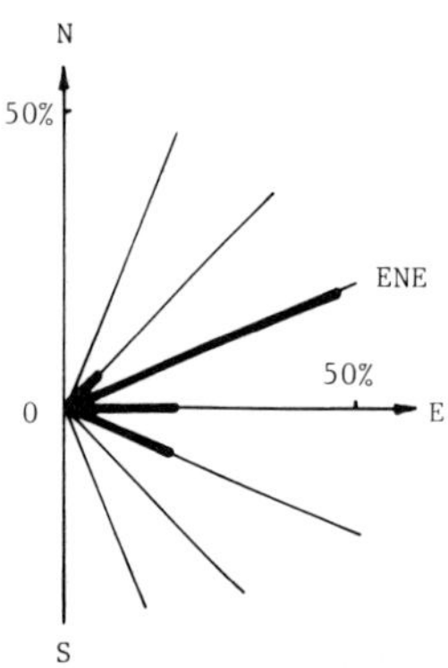

FIGURE 4. Distribution of wave directions (References 2, 25, 27, 42).

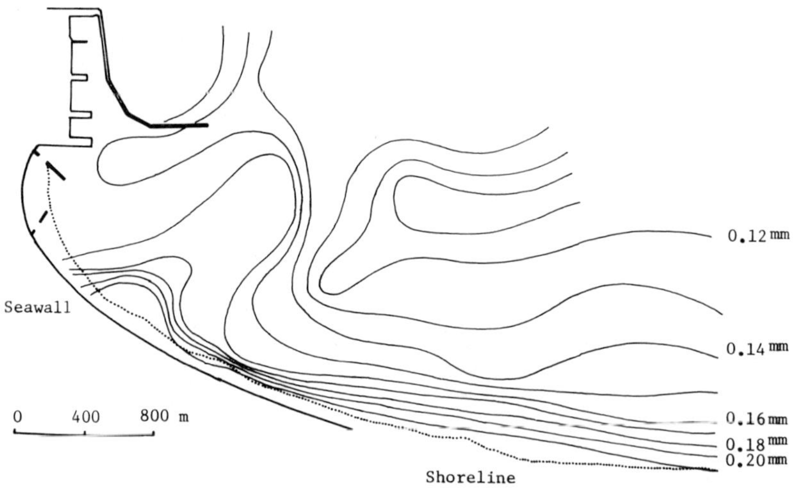

FIGURE 5. Curves of Equi-Mean Diameters (References 2, 25, 27).

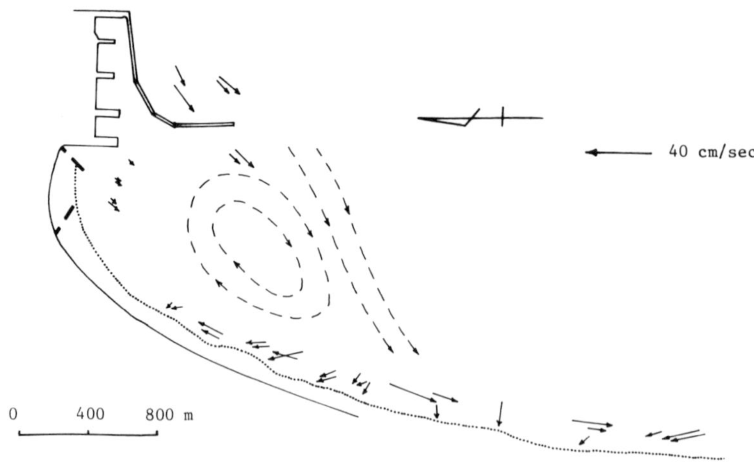

FIGURE 6. Measured nearshore currents (References 2, 25, 27, 42).

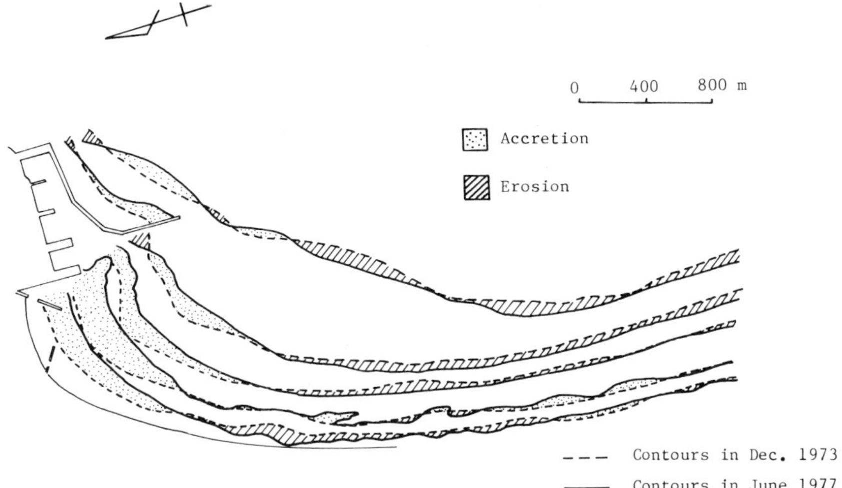

FIGURE 7. Comparison of past sea bottom topography (References 2, 25, 27, 42).

SITING

The arrangements of shore-connected breakwaters and offshore breakwaters must be designed to keep a calm water in harbors by considering refraction, diffraction and reflection of waves. Simultaneously the arrangements also must be considered to prevent sand accumulation or shoaling in ports and harbors and accompanying erosion or scouring [26]. For shore protection offshore breakwaters are the most effective means, but the several points for siting them must be considered beforehand. The offshore breakwater is discontinuous in length by considering economical condition. But if the spacing or opening of the discontinuity is large enough, the shore facing to the discontinuity will be eroded by incident wave and the region of the opening between the breakwaters will be scoured by the induced intense rip current. It is economically for us to select the siting place near the shore. Namely, the offshore breakwaters may be sited on the onshore side of the wave breakwater zone during storm weather. But the beach is often lost if the breakwaters are built near the shore, since the sea bottom on the offshore side of the breakwaters is scoured and the slope becomes steeper. Figure 8 shows the change of transmitted wave height through the permeable offshore breakwater [10]. The flow velocity and direction in an idealized permeable offshore breakwater is represented in Figure 9 (a) and (b). The water depth, the wave height and period are 8 m, 1.5 m and 3 sec, respectively [28]. The relationship between the ratio of wave transmission and the movement of the shoreline is generally not found. Figures 10 and 11 plot the relationship between the length of offshore breakwaters and the distance from the shorelines and the relationship between the length of offshore breakwaters and the water depths. Figure 12 shows the relationship between the distance from the shorelines and the ones of the discontinuity between offshore breakwaters. Figure 13 represents the relationship between the total heights of offshore breakwaters and the low water levels. Large circles explain that the tops of the offshore breakwaters are under the water surface. In Figure 14 the relationship between the ratio of the lengths of offshore breakwaters to the distances from the shorelines and that of the heights of offshore breakwaters to the water depths. Figures 15 and 16 describe the system of offshore breakwaters and groins and their effect on the shoreline changes in hydraulic model test [26]. After the construction of offshore breakwaters intense flow is induced by incident wave in the gaps among stones or concrete blocks, the foundation of the offshore breakwaters is scoured and they

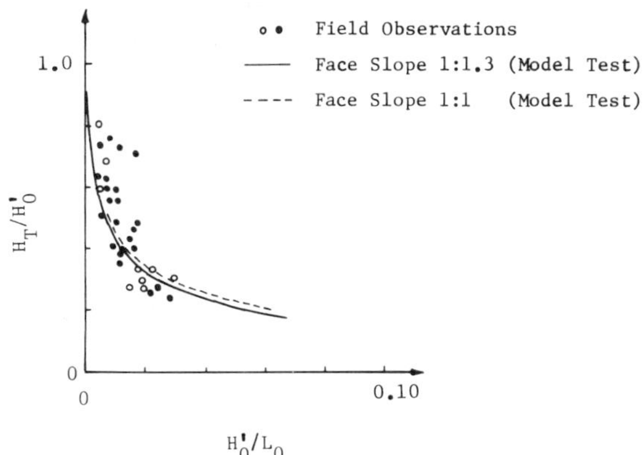

FIGURE 8. Wave transmission without overtopping (Reference 10).

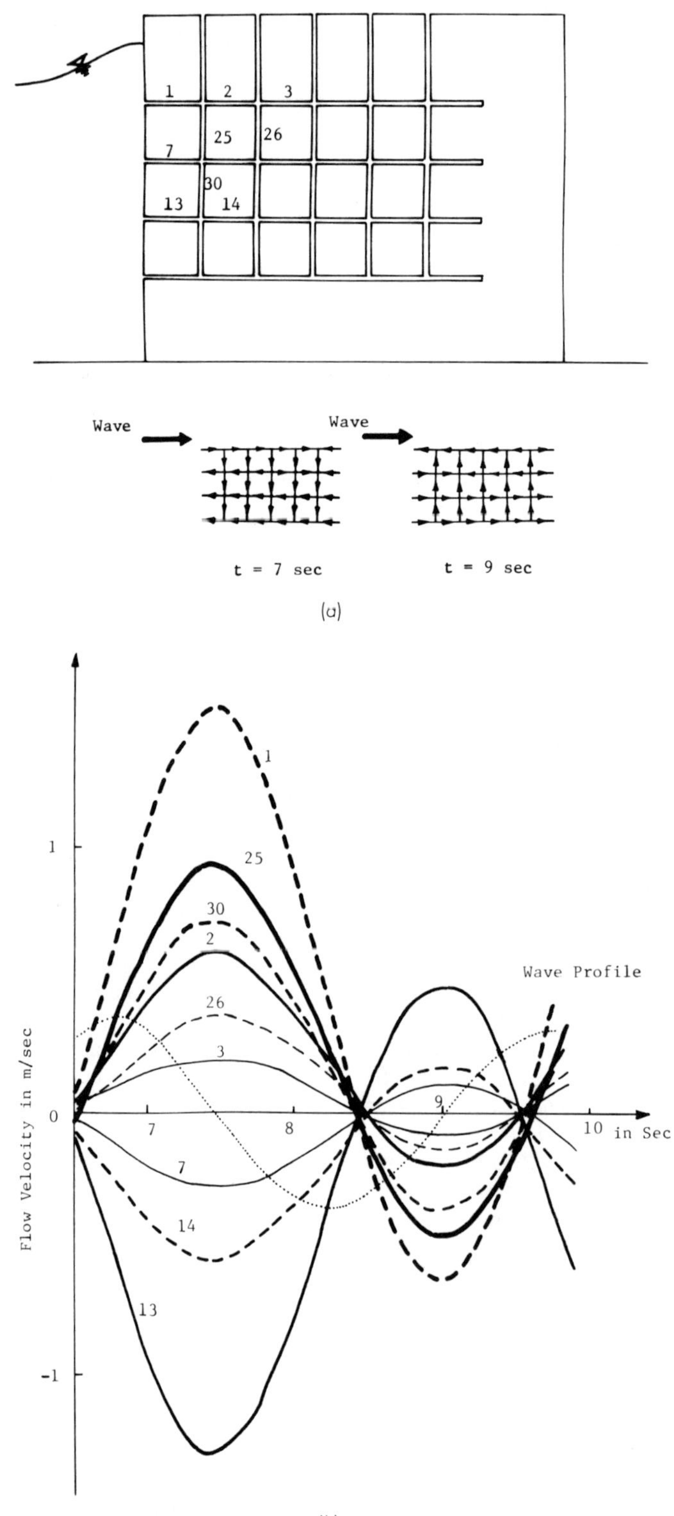

FIGURE 9. (a) An idealized rubble-mound breakwater and flow pattern (Reference 28); (b) flow velocities in an idealized rubble-mound breakwater.

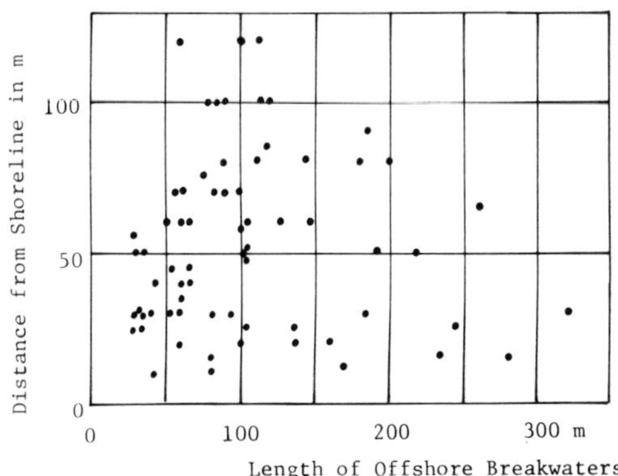

FIGURE 10. Relationship between length of offshore breakwaters and distance from shoreline (Reference 49).

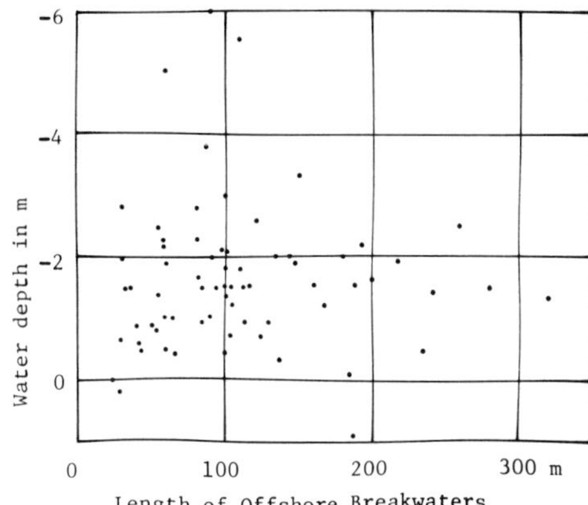

FIGURE 11. Relationship between length of offshore breakwaters and water depth at their siting (Reference 49).

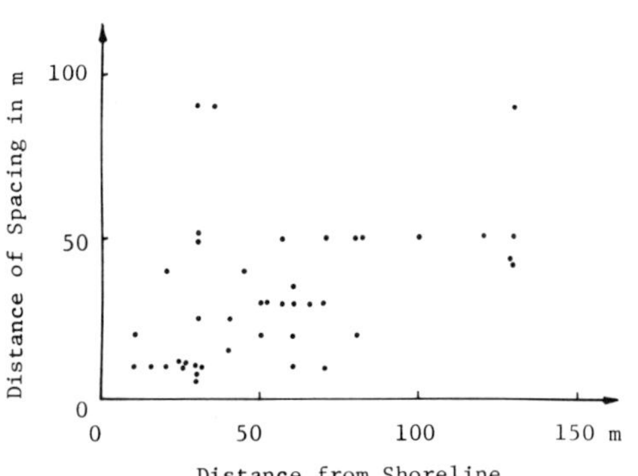

FIGURE 12. Relationship between distance of offshore breakwaters from shoreline and their spacing (Reference 49).

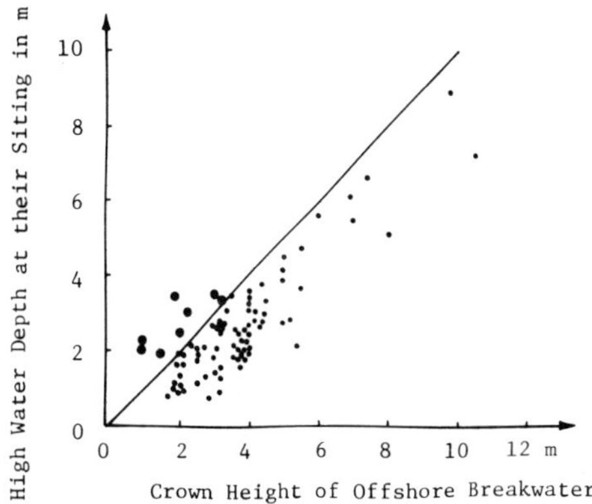

FIGURE 13. Relationship between crown height of offshore breakwaters and high water depth at their siting (Reference 49).

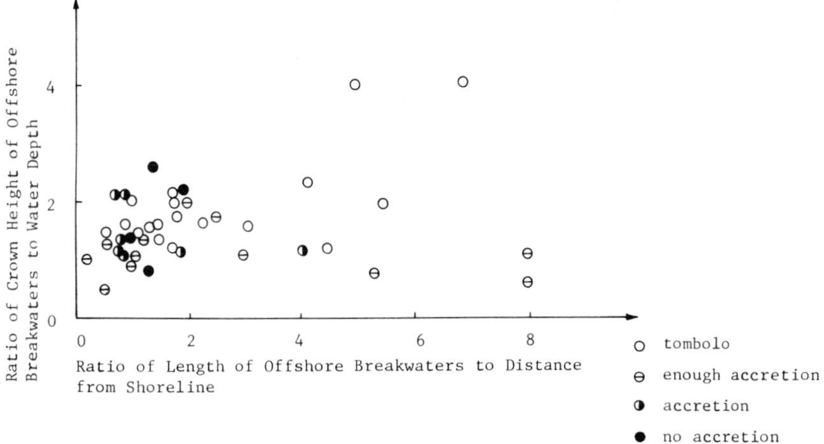

FIGURE 14. Relationship between ratio of length of offshore breakwaters to distance from shoreline and ratio of crown height of offshore breakwaters to water depth (Reference 49).

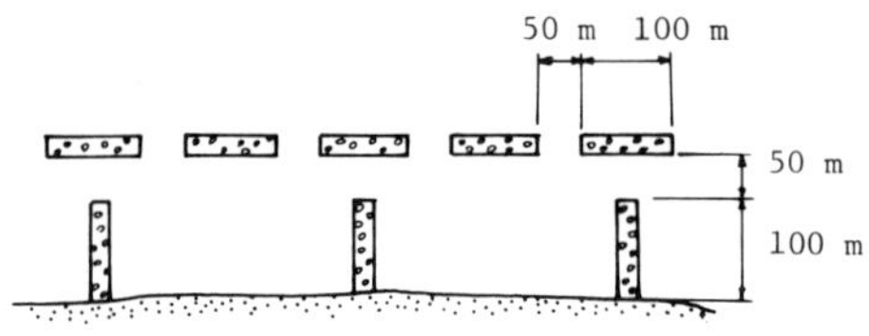

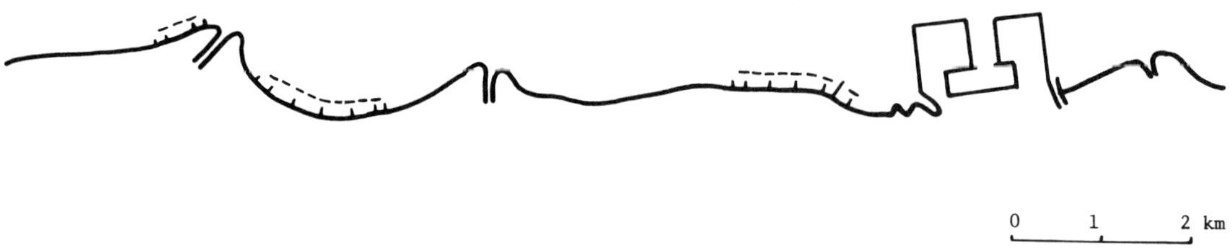

FIGURE 15. Shore protection method (Reference 26).

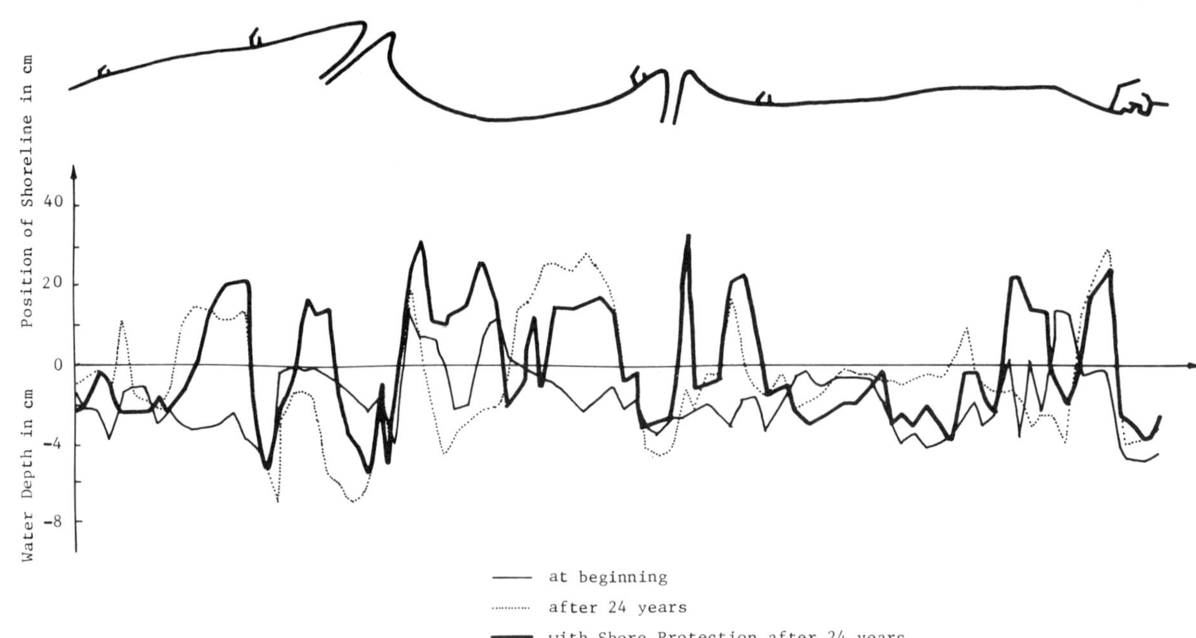

FIGURE 16. Effect of shore protection method (Reference 26).

sink. In result, mean diameters of sand under the offshore breakwater become large, because of sawing effect as wave breaking. To prevent the settlement of offshore breakwaters the foundation filter layer under them must be fixed. The foundation filter layer is made by steel form and rubble stones are packed in it.

WAVE FORCES

Introduction

It is very important to compute accurate wave forces for designing of coastal structures. The wave force is defined to be equal to the integrated resultant wave pressure over the surface of the coastal structure. Fixed coastal structures are generally classified into three types: (a) pile-supported structures such as piers and offshore platforms, (b) wall-type structures such as seawalls, bulkheads, revetments and some breakwaters, and (c) rubble structures such as many groins, revetments, jetties and breakwaters. Individual structures consist of some of the above three types. The types of waves acting on these structures are nonbreaking, breaking, or broken waves. The wave pressure over the vertical wall or the circular pile has been intensively studying to protect coastal structures from wall action. When a breakwater is built on the porous foundation such as rubble stones, uplift force acts under the breakwater due to wave action. For rubble structure design differentiation among all three types of wave action is not necessary to consider.

Wave Force on the Vertical Wall

When the small amplitude wave attacks the vertical wall, the maximum wave pressure occurs during the passage of the wave crest. It is called a standing wave pressure [Figure 17(a)]. The increase of the wave height forms a depression on the crest of the wave pressure curve as shown in Figure 17(b). As the wave height becomes large and exceeds the wave breaking limit, an antecedent wave crest grows faster than a rear one. It is called a breaking wave pressure [Figure 17(c)]. As the wave height becomes larger and the incident wave breaks, the wave pressure curve forms shock wave pressure part and the flowing flat one as shown in Figure 17(d).

STANDING WAVE FORCE

When the small amplitude wave attacks the vertical wall of a coastal structure which is located at the deeper water depth than the breaking zone, the incident and the reflected wave form the standing wave. The form of the standing wave is

$$\eta = H_I K_R \cos kx \cos \sigma t + \frac{H_I}{2}(1 - K_R)\cos(kx - \sigma t) \tag{1}$$

in which H_I = the incident wave height, K_R = the reflection coefficient, k = wave number and σ = angular frequency. Then, the velocity potential is

$$\phi = \frac{H_I c}{2}\frac{\cosh k(h+z)}{\sinh kh} \times \{\sin(kx - \sigma t) - K_R \sin(kx + \sigma t)\} \tag{2}$$

in which c = celerity, h = water depth, x and z = rectangular Cartesian coordinate system as shown in Figure 18. The substitution of the velocity potential into the pressure equation gives as:

$$\frac{p}{\rho g} = -z + \frac{H_I}{2}\frac{\cosh k(h+z)}{\cosh kh} \times \{2K_R \cos kx \cos \sigma t + (1 - K_R)\cos(kx - \sigma t)\} \tag{3}$$

in which ρ = water density and p = wave pressure. The wave pressure distribution on the vertical wall is given in Figure 19 when the wave crest and trough attain there. In the case of the finite amplitude wave the standing wave is calculated by the Stokes wave or the trochoidal wave theories. Let us consider the Sainflou's formula in which the trochoidal wave is employed. This is derived from the Sainflou's exact solution by approximating the exact distribution of the wave pressure with usage of linear function. The wave pressure distribution is represented in Figure 20. In the figure,

$$p_1 = (p_2 + wh)\frac{H + \delta_0}{h + H + \delta_0} \tag{4}$$

$$p_1' = w(1 - \delta_0) \tag{5}$$

$$p_2 = p_2' = \frac{wH}{\cosh kH} \tag{6}$$

in which $\delta_0 = \pi H^2/L$, w = specific weight of water, p_1 = the wave pressure at the normal free surface when the wave crest attains on the wall, p_1' = the wave pressure at $H - \delta_0$ when the wave trough attains on the wall, p_2 and p_2' = the wave pressure at the bottom of the wall when the wave crest and trough attains the wall, respectively and H = the significant wave height at the coastal structure before the construction of the structure. The physical condition in which each formula of the wave pressure distribution is applied is described in Figure 21 [9]. Figure 22 represents the comparison between the Sainflou's formula and the experimental values [34].

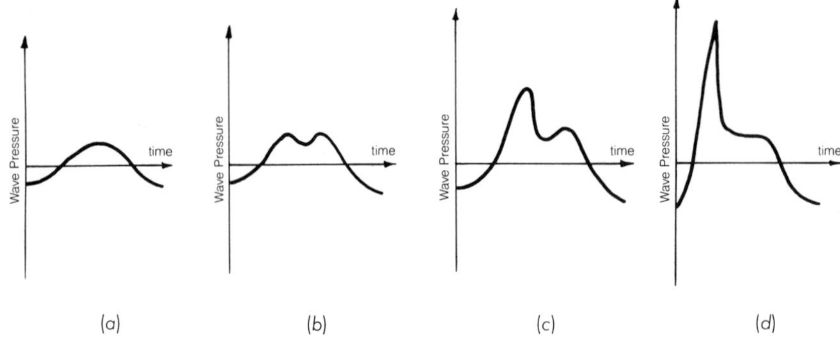

FIGURE 17. Wave pressure (Reference 17).

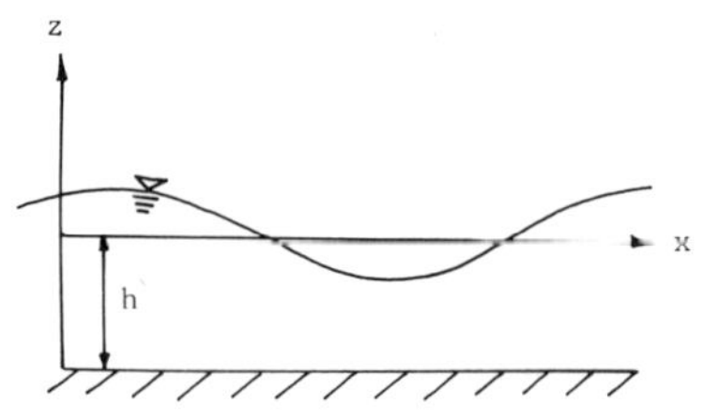

FIGURE 18. Coordinate system.

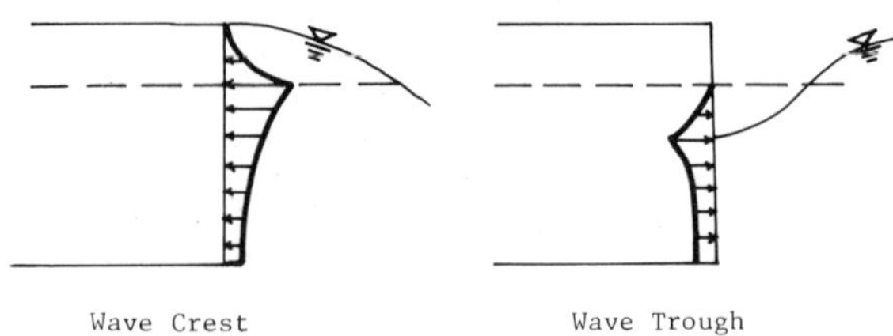

FIGURE 19. Standing wave pressure.

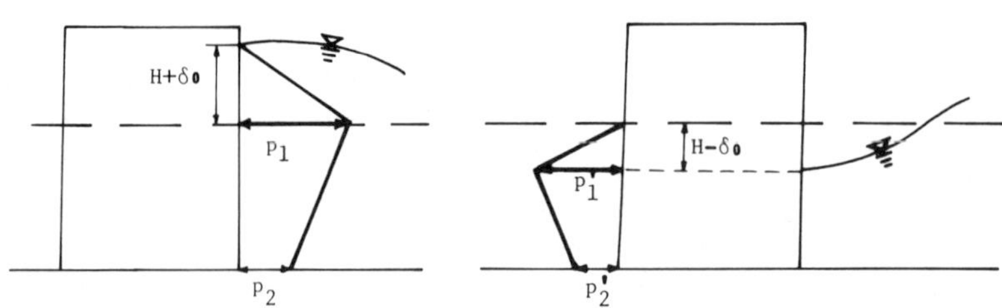

FIGURE 20. Wave pressure due to finite amplitude wave.

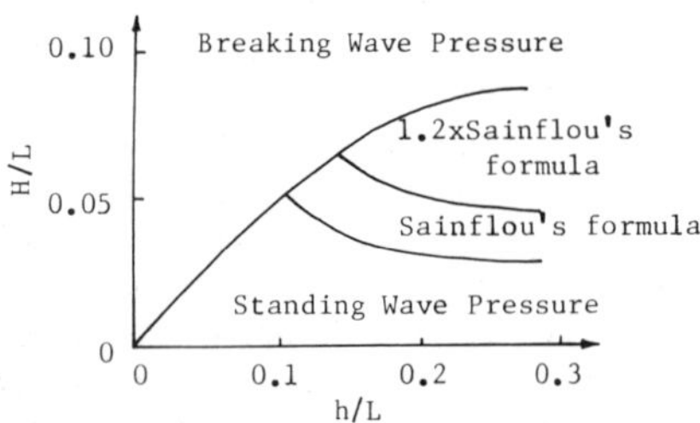

FIGURE 21. Comparison of each wave pressure formula (Reference 9).

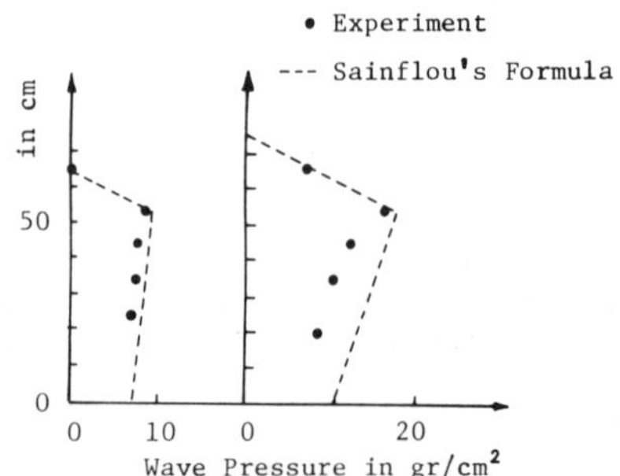

FIGURE 22. Comparison of experimental value with Sainflou's formula (Reference 34).

WAVE FORCE DUE TO BREAKING WAVE

The maximum wave pressure acts on the wall when the wave breaks in front of the wall. In this case extremely strong wave pressure may often occur if the slope of sea bottom is steep or if the wave condition is peculiar. It is called a shock pressure due to breaking wave. It is caused by the momentum transmission of moving water mass in the incident wave to the wall of the coastal structure for very short time and by the compression of air which is caught in the water mass by the breaking wave. By assuming that the uniformly distributed wave pressure due to breaking wave acts on the vertical wall from the sea bottom to the free surface, Hiroi (1919) derived the following experimental formula [17]:

$$p = 1.5wH \tag{7}$$

in which H = the incident wave height. The wave pressure acts on the vertical wall from the wall bottom to the lower one between $1.25H$ above the still water level and the top of the wall as shown in Figure 23. The computed wave pressure does not coincide with the observed local wave pressure, but the computed wave force on the wall is almost the same as the observed one. In the application of the above formula it must be noticed that the wave pressure is larger on the free surface and becomes small on the sea bottom. When the strong shock wave pressure is considered on the free surface, the following Minikin's formula is employed. Minikin obtained it by using observations in Europe and Bagnold's results [43]. In this formula he considered that the shock wave pressure is caused by catching air into the breaking wave and by collision of the wave on the wall. The definition sketch for Minikin's formula is plotted in Figure 24 [43]. The wave pressure is written as follows:

$$p_{max} = 102.4wd\left(1 + \frac{d}{h}\right)\frac{H}{L} \tag{8}$$

$$p_z = p_m\left(\frac{H - 2|z|}{H}\right)^2 \tag{9}$$

in which z = coordinate system measured vertically upwards from the still water level. The static pressures above and under the still water level are given as follows:

$$p_s = w\left(\frac{H}{2} - z\right) \tag{10}$$

$$p_s' = w\frac{H}{2} \tag{11}$$

The total wave pressure during wave breaking is the summation of the shock wave pressure and the static one.

WAVE FORCE AFTER BREAKING WAVE (BROKEN WAVE)

When the bottom slope where the coastal structure is sited as milder than 1/50, the wave pressure is calculated by applying Hiroi's formula with 90% of the water depth instead of the breaking wave height. When the coastal structure is sited on the onshore side of the shoreline, the total wave pressure is given by the summation of the dynamic and the static wave pressure as shown in Figure 25. The dynamic wave pressure p_m uniformly acts from the bottom of the coastal structure to the position of h' above the bottom. The static wave pressure is linearly distributed from the bottom to the position of h' above the bottom. They are

$$\begin{aligned} p_m &= 0.5wh_b\left(1 - \frac{X_1}{X_2}\right)^2 \\ p_1 &= wh' \\ h' &= 0.7H_b\left(1 - \frac{X_1}{X_2}\right) \end{aligned} \tag{12}$$

in which X_1 = the horizontal distance from the shoreline to the wall in meters, $X_2 = 2H_b \cot\theta$: the horizontal distance from the shoreline to the assumed position of wave runup, $\tan\theta$ = the sea bottom slope, H_b = the breaking wave height in meters and h_b = the water depth at the breaking point.

WAVE FORCE DUE TO IRREGULAR WAVE

The different design formula between the standing and the breaking wave is employed. The design wave pressure obtained by Hiroi's formula and Sainflou's formula becomes discontinuous at the boundary between the application regions of two formulas. Still more, it is not clear which wave in irregular waves must be used in the formula of the wave pressure. To solve this problem, Ito [16] showed the concept of the expected sliding distance, based on the consideration that the stability of the breakwater must be examined by including the behavior of the breakwater under the abnormal wave condition. As the formula to estimate the sliding limit, he proposed the following formula in which the wave pressure is continuous in the region between the standing and the breaking wave.

$$p = \begin{cases} 0.7wH \\ \left[0.7 + 0.55\left(\frac{H}{d} - 1\right)\right]wH \end{cases} \tag{13}$$

in which d = water depth on the rubble-mound breakwater and H = the maximum wave height. The computational method of the wave pressure in this case is represented in Figure 26. In the figure, the values of p_1, p_2 and p_3 are

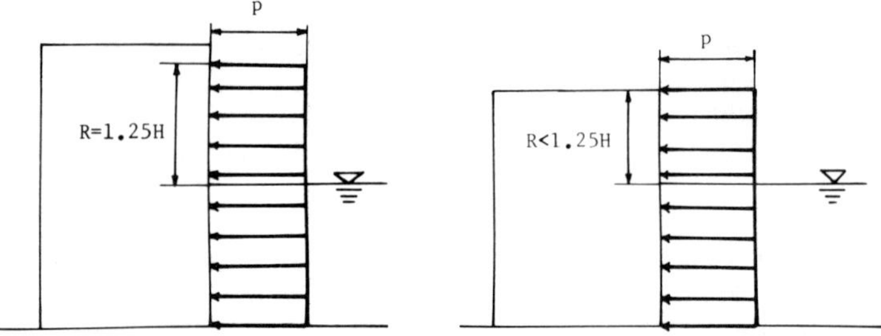

FIGURE 23. Definition sketch for Hiroi's formula.

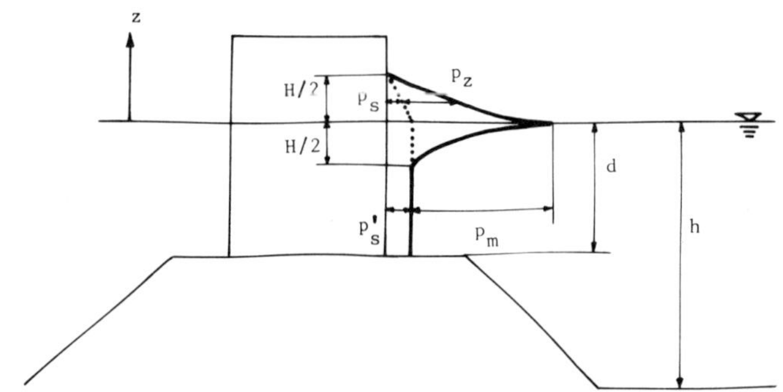

FIGURE 24. Shock wave pressure.

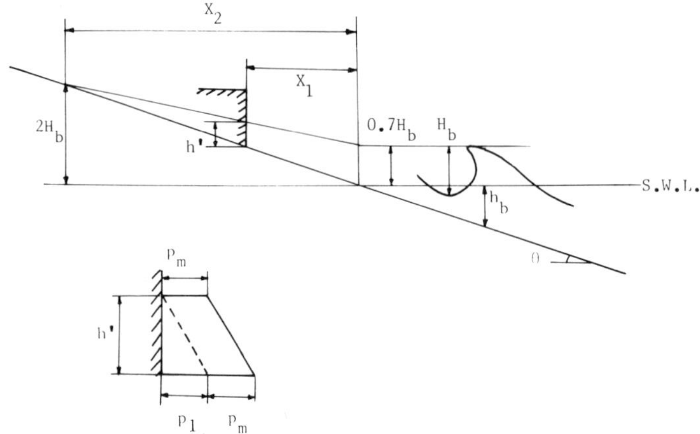

FIGURE 25. Wave pressure acted on structure on shore.

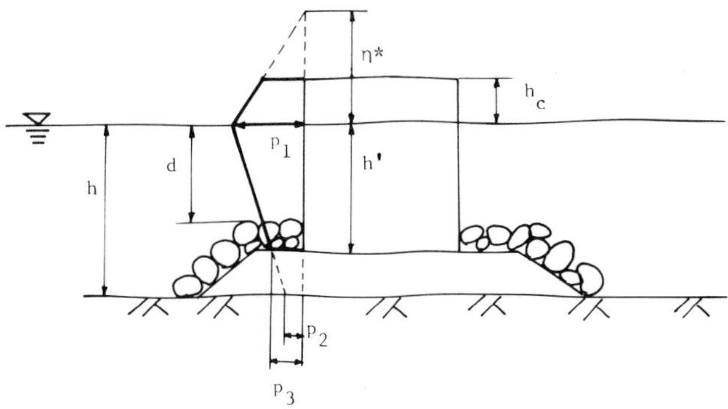

FIGURE 26. Wave pressure due to irregular waves.

given as follows:

$$p_1 = \frac{1}{2}(1 + \cos\alpha)(\beta_1 + \beta_2\cos^2\alpha)wH_{max} \quad (14)$$

$$p_2 = \frac{p_1}{\cosh\frac{2\pi h}{L}} \quad (15)$$

$$p_3 = \beta_3 p_1 \quad (16)$$

in which

$$\beta_1 = 0.6 + \frac{1}{2}\left[\frac{\frac{4\pi h}{L}}{\sinh\frac{4\pi h}{L}}\right]^2$$

$$\beta_2 = \min\left\{\frac{h_0 - d}{3h_0}\left(\frac{H_{max}}{d}\right)^2, \frac{2d}{H_{max}}\right\}$$

$$\beta_3 = 1 - \frac{h'}{h}\left[1 - \frac{1}{\cosh\frac{2\pi h}{L}}\right]$$

d = the water depth above the mound, h' = the vertical distance between the still water level and the bottom of the caisson, h_0 = the water depth at the distance $5H_{1/3}$ on the offshore side from the wall, $\eta^* = 0.75\ (1 + \cos\alpha)\ H_{max}$, α = the direction of the incident wave, $H_{max} = 1.8H_{1/3}$ and $T_{max} = T_{1/3}$.

EFFECT OF WAVE BREAKING WORKS

To reduce the shock wave pressure on the wall, wave breaking works such as rubble stones (ripraps) or concrete blocks are sited in front of the wall. But enough top width and height of wave breaking works give the reduction of the shock wave pressure. In the case of which the siting water depth of the coastal structure is less than $0.5H$ and the top position of the wave breaking works is greater than the design high water level plus $0.5H$, the averaged wave pressure after the siting of the wave breaking works is obtained by

$$p_b' = wH_{1/3} \quad (17)$$

in which $H_{1/3}$ = the significant wave height before the construction of the coastal structure. The wave pressure is assumed to act uniformly on the wall from the structure bottom to the height H above the still water level. However, the above Equation (17) is not appropriate to compute wave pressure due to standing waves, because it is only used to compute the wave pressure due to breaking waves. Tanimoto *et al.* [46] proposed the following equation to evaluate the wave pressure in the cases of the standing and the breaking wave:

$$\eta^* = 0.75(1 + \cos\alpha)\lambda H_{max}$$

$$p_1 = \frac{1}{2}(1 + \cos\alpha)\lambda\beta_1 wH_{max} \quad (18)$$

$$p_2 = \frac{p_1}{\cosh\frac{2\pi h}{L}}$$

in which η^*, p_1 and p_2 are described in Figure 26 and λ = the reduction coefficient of the wave pressure, 0.8.

Wave Forces on Circular Piles

When the nonbreaking wave acts on the circular structure which is vertically built in water, normal $p(\theta)$ and shear stress $\tau(\theta)$ on the surface of the structure appear. θ is an angle which is counterclockwisely measured from the x-

axis. The hydrodynamic forces acted on the circular pile of vertical length dz are written by integrating the x- and y-component of $p(\theta)$ and $\tau(\theta)$ over the pile surface. Namely,

$$dF_T = \left[-\int_0^{2\pi} a p(\theta) \cos\theta d\theta - \int_0^{2\pi} a\tau(\theta) \sin\theta d\theta \right] dz \tag{19}$$

$$dF_L = \left[-\int_0^{2\pi} a p(\theta) \sin\theta d\theta + \int_0^{2\pi} a\tau(\theta) \cos\theta d\theta \right] dz \tag{20}$$

in which dF_T and dF_L = the hydrodynamic forces in x- and y-direction and a = the radius of the circular pile. Therefore, the hydrodynamic forces can be computed, if $p(\theta)$ and $\tau(\theta)$ are known. The pressure distribution $p(\theta)$ on the surface of the circular pile is dependent upon vortices shed behind the pile. The generation of shed vortices are strongly influenced by the Reynolds number (Re) and Keulegan–Carpenter number ($K.C.$) [18]. They are

$$Re = \frac{u_m D}{\nu} \qquad N_{K.C.} - \frac{u_m T}{D} \tag{21}$$

in which u_m = the maximum velocity of water particle induced by waves, D = the diameter of the circular pile and T = the wave period. The pressure distribution $p(\theta)$ induced on the surface of the circular pile in the oscillatory flow is given by the following pressure equation:

$$p(\theta) = \left[-\rho \frac{\partial \phi}{\partial t} - \frac{1}{2}\rho(u^2 + v^2) - \rho g z \right]_{r=a} \tag{22}$$

in which u and v = the flow velocity in x and y direction. To compute $p(\theta)$ the term $\partial\phi/\partial t$ must be transformed in order to include the effect of the shed vortices. When the flow in the boundary layer formed along the surface of the circular pile is laminar, the shear stress τ is represented by

$$\tau = \rho\nu \left(\frac{\partial u}{\partial r} \right)_{r=a} \tag{23}$$

in which ν = the dynamic viscosity and r = the radial coordinate. In the case of which the effect of the shedding vortices is neglected, the shear stress distribution on the surface of the circular pile is expressed by using the water particle velocity in the boundary layer as follows [41]:

$$\tau(\theta) = -2\rho u_m \sqrt{\nu\sigma} \left[\sin\theta \cos\left(\sigma t + \frac{\pi}{4}\right) - \frac{\sin 2\theta}{2\pi} \left(\frac{u_{max} T}{D} \right) \left\{ (2 - \sqrt{2}) \times \cos\left(2\sigma t - \frac{\pi}{4}\right) + \frac{\sqrt{2}}{2} \right\} \right] \tag{24}$$

in which

$$\tau(\theta) = \begin{cases} \tau(\theta) & \text{if } |\theta_s| \le |\theta| \le \pi \\ 0 & \text{otherwise} \end{cases}$$

and θ_s = the separation point of the boundary layer.

LONGITUDINAL FORCE

Morison *et al.* [31] assumed that the hydrodynamic force due to waves is the summation of the force dF_D induced by the shed vortices behind the circular pile and the force dF_I produced by the acceleration of a fluid under the existence of the circular pile. They are called Morison's formula and given as follows:

$$dF_T = dF_D + dF_I \qquad dF_D = C_D \rho \frac{|u|u}{2} D dz \qquad dF_I = C_M \frac{\pi D^2}{4} \rho \frac{\partial u}{\partial t} dz \tag{25}$$

in which dF_D = the drag force produced by the steady flow, dF_I = the inertial force, C_D = the drag coefficient and C_M = the mass coefficient. In the case of the circular pile the mass coefficient is obtained to be 2.0 by the potential theory. In Morison's formula the evaluation of the drag coefficient C_D and the mass coefficient C_M is very important. The values of C_D and C_M are obtained by using the phase difference between the drag and the inertial force. When the wave form is sinusoidal, the longitudinal wave force is given by

$$dF_T = dF_D + dF_I = \left\{ \frac{C_D \rho}{2} D u_m^2 |\sin\sigma t| \sin\sigma t + C_M \rho \frac{\pi D^2}{4} \times \left(\frac{\partial u}{\partial t} \right)_{max} \cos\sigma t \right\} dz \tag{26}$$

When the velocity of water particle is maximum or minimum, $dF_I = 0$ and $dF_T = dF_D$ in the above equation. By measuring dF_T, C_D is computed as:

$$C_D = \frac{dF_T}{\rho \dfrac{|u|u}{2} Ddz}$$

When the acceleration of water particle is maximum, $dF_D = 0$ and $dF_T = dF_I$ in the above equation. From the measurement of dF_T, C_M is calculated as:

$$C_M = \frac{dF_T}{\dfrac{\pi D^2}{4} \rho \dfrac{\partial u}{\partial t} dz}$$

C_D value is the function of the Reynolds number and the shape of the body immersed in water and it becomes almost constant for a large value of Re number ($Re > 10^4$). C_D value of the circular cylinder is 1.17. Keulegan and Carpenter showed that C_D and C_M value variate with respect to $K.C.$ number [18]. Sarpkaya defined $\beta = D/(\pi\delta)$ as another parameter to control C_D and C_M value. δ indicates the thickness of the boundary layer, $\delta = \sqrt{2\nu/\sigma}$. As β value becomes large, C_D value decreases and C_M values increases. For the smooth circular pile C_D value decreases with respect to the increase of Re number. For the rough circular pile C_D value first decreases and then increases with respect to increase of Re number. Let us consider the predominance of F_D or F_I in F_T. By assuming small-amplitude approximation and by setting $(F_D)_{max} = (F_I)_{max}$, the following equation is obtained [14]:

$$\frac{H}{D} = \frac{C_M}{C_D\left(\dfrac{h}{L}\right)} \frac{\sinh^2 \dfrac{2\pi h}{L}}{1 + \dfrac{\sinh(4\pi h/L)}{(4\pi h/L)}} \qquad (27)$$

Equation (27) is plotted in Figure 27 for $C_M = 2.0$ and $C_D = 1.0$. For large h/L and small H/D the inertial force is predominant and for small h/L and large H/D the drag force excels.

BREAKING WAVE FORCE ON VERTICAL CIRCULAR PILES

When the circular pile is built in the breaking zone, the shock wave force acts on it as if it acts on the vertical wall. To evaluate the force, Goda [6] proposed the following equation:

$$F_T = \int_{-h}^{\eta} \frac{\rho}{2} C_D Du|u| dz + \int_{-h}^{\eta} \rho C_m \frac{\pi}{4} D^2 \frac{\partial u}{\partial t} dz + \int_{(1-2)\eta_0}^{\eta_0} dF_I \qquad (28)$$

in which η_0 and λ are represented in Figure 28. The third term in the above equation represents the shock wave pressure and dF_I is expressed as follows [6]:

$$dF_I = \frac{\rho}{2} \pi D u^2 \left(1 - \frac{2ut}{D}\right) \qquad (29)$$

If the impact velocity of the breaking wave to the pile is assumed to be uniform C_b from $(1 - \lambda)\,\eta_0$ to η_0 as,

$$F_I = \rho g D H_b^2 \lambda \frac{\pi C_b^2 \eta_0^2}{2gH_b^2} \left(1 - \frac{2tC_b}{D}\right) \qquad (30)$$

in which λ is given in Figure 29.

TRANSVERSE FORCES (LIFT FORCES) ON VERTICAL CYLINDRICAL PILES

The magnitude of the transverse forces is equal to or greater than that of the hydrodynamic forces and the design of the structure must take the transverse forces into consideration. The transverse force dF_L is dependent upon the unsymmetrical flow pattern behind the circular pile. Therefore, the transverse force is negligible in comparison with the hydrodynamic force during small $K.C.$ number. But for a large $K.C.$ number unsymmetrical vortices are generated and shed and in result irregular transverse force is induced. That is, the irregularity of the transverse force is based on the generation of vortices. The ratio of the tenth maximum transverse force to the total hydrodynamic force is given to be dependent upon $K.C.$ number as shown in Figure 30 [15].

WAVE FORCES ON A LARGE COASTAL STRUCTURE

Recently, large coastal structures such as oil tanks are constructed in the sea. Especially, in the case of which the diameter of the circular cylinder D is larger than one fifth of the wave length ($D/L > 0.2$), there exists the theoretical solution. Let us consider the theoretical solution in the case of Figure 31 [17]. If the fluid is inviscid and the motion is irrotational, the velocity potential ϕ satisfies the following Laplace equation:

$$\frac{\partial^2 \phi}{\partial r^2} + \frac{1}{r}\frac{\partial \phi}{\partial r} + \frac{1}{r^2}\frac{\partial^2 \phi}{\partial \theta^2} + \frac{\partial^2 \phi}{\partial z^2} = 0 \qquad (31)$$

The above equation is solved by the boundary conditions on the surface of the circular cylinder and on the free surface and by the radiation condition at infinity. Then, the solution is shown as follows [17]:

$$\phi = \frac{gH}{2\sigma} e^{-i\sigma t} \frac{\cosh k(h+z)}{\cosh kh} \left[J_0(kr) - \frac{J_0'(ka)}{H_0^{(2)\prime}(ka)} H_0^{(2)}(kr) + 2\sum_{n=1}^{\infty} i^n \left\{ J_n(kr) - \frac{J_n'(ka)}{H_n^{(2)\prime}(ka)} H_n^{(2)}(kr) \right\} \cos n\theta \right] \qquad (32)$$

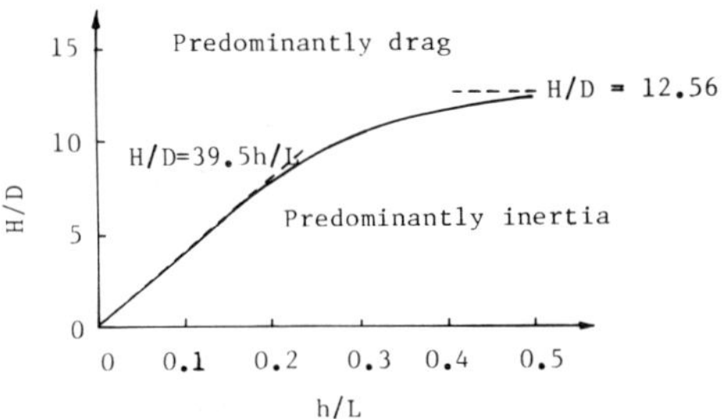

FIGURE 27. H/D such that $(F_D)_{max}/(F_I)_{max} = 1$, based on small-amplitude wave theory and $C_D = 1$, $C_M = 2$ (Reference 14).

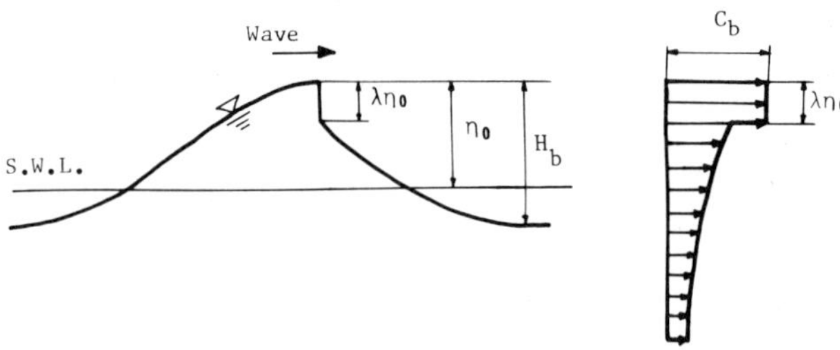

FIGURE 28. Definition sketch for breaking wave pressure.

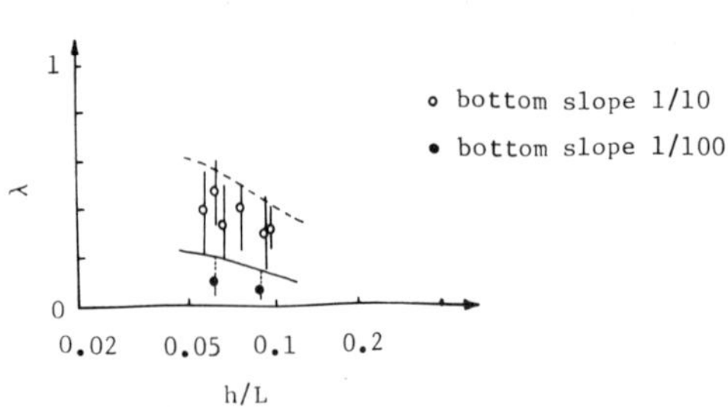

FIGURE 29. Experimental values for λ (Reference 6).

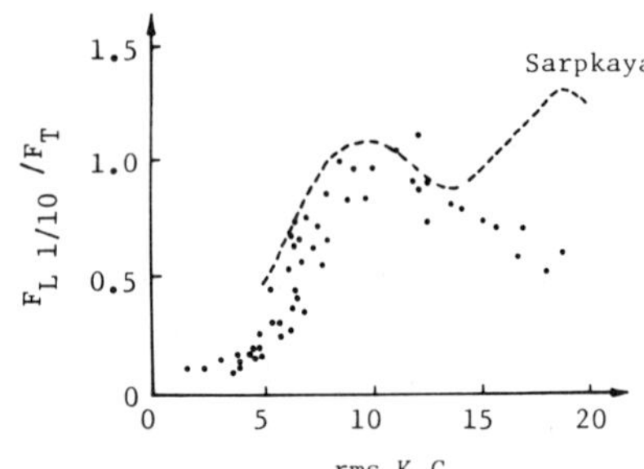

FIGURE 30. Magnitude of lift force (Reference 40).

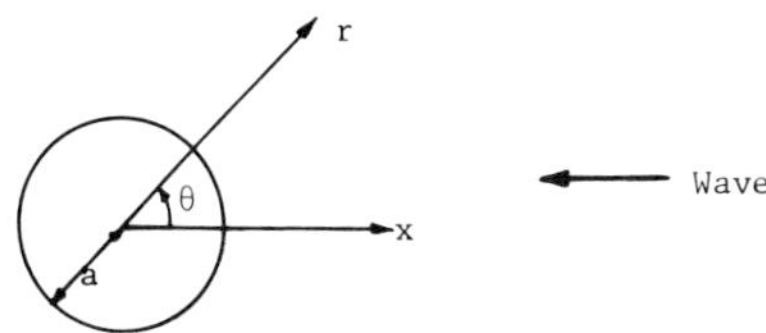

FIGURE 31. Coordinate system.

in which $a = D/2$, $i = \sqrt{-1}$, J_0 and J_n = the Bessel function of the first kind of order zero and order n, $H_0^{(2)}$ and $H_n^{(2)}$ = the Hankel function of the second kind of order zero and order n and " ′ " denotes differentiation with respect to r. The pressure p is obtained by the pressure equation as:

$$\frac{p}{\rho g} = \frac{1}{2}\left\{\frac{\cosh k(h+z)}{\cosh kh}\right\} e^{-i\sigma t} \sum_{n=0}^{\infty} \delta_n i^n \times \{J_n(kr) + \alpha_n H_n^{(2)}(kr)\}\cos\theta - \frac{z}{H} \tag{33}$$

in which

$$\alpha_0 = -\frac{J_1(ka)}{H_1^{(2)}(ka)}$$

$$\alpha_n = \frac{-nJ_n(ka) + kaJ_{n+1}(ka)}{nH_n^{(2)}(ka) - kaH_{n+1}^{(2)}(ka)}$$

for $n \geq 1$ and δ_n is the Neumann number ($\delta_0 = 1$ for $n = 0$ and $\delta_n = 2$ for $n \geq 1$). Then, the wave pressure on the surface of the circular pile p_a is

$$\frac{p_a}{\rho g} = -\frac{\text{iH}}{\pi ka}\frac{\cosh k(h+z)}{\cosh kh} e^{-i\sigma t} \times \sum_{n=0}^{\infty}\frac{\delta_n i^n(\cos n\theta)}{H_n^{(2)\prime}(ka)} - z \tag{34}$$

The hydrodynamic force dF_T is obtained by integrating p_a over the surface. It is

$$dF_T = \frac{2\rho gH}{k}\frac{\cosh k(h+z)}{\cosh kh} \times \mathcal{H}(ka)\cos(\sigma t - \beta)dz \tag{35}$$

in which $\mathcal{H}(ka) = \{J_1'^2(ka) + Y_1'^2(ka)\}^{-1/2}$, $\beta = \tan^{-1}\{J_1'(ka)/Y'(ka)\}$ and $Y_1(ka)$ = the Bessel function of the second kind of order one. The above solution explains the experimental data very well for large values of h/H, in which h = the water depth and H = the wave height. But for small values of h/H it does not, because the nonlinearity of the wave motion becomes remarkable or the flow separates behind the pile. For arbitrary shapes of the coastal structure the finite element method or the boundary element method are developing by considering the nonlinearity of the motion [29,30].

Uplift Forces

In general upward forces act vertically under the caisson of the breakwater. When the caisson is located in the sea of water depth h, the static pressure of triangular shape works on the side wall of the caisson and the upward force wh acts vertically on the caisson bottom as shown in Figure 32. Let us consider the uplift force induced by the standing wave with small and finite amplitude. When the crest and the trough of the wave act on the caisson, the uplift force of trapezoidal shape distributes as represented in Figure 33. The uplift pressure p_u due to only the wave action is given by

$$p_u = \frac{wh}{\cosh kh} \tag{36}$$

in which w = the specific weight of water, H = wave height, k = wave number and h = water depth. The assumption of linear distribution of the uplift pressure between both sides of the caisson bottom shows that the distribution shape is triangular. If the top height of the caisson is less than $H + \delta_0$ and wave overtopping occurs, the buoyancy force is assumed to act on under the whole caisson bottom and the resultant uplift force is supposed to be included in this buoyancy force. When the top of the caisson (breakwater) is higher than $1.25H$ and the overtopping does not occur, the uplift pressure p_u at the toe of the breakwater is expressed by

$$p_u = 1.25wH \tag{37}$$

Then, the distribution shape of the uplift pressure is triangular. If the top of the caisson is lower than $1.25H$, the buoyancy force is assumed to act on under the whole part of the caisson and to include the uplift force. For irregular waves the uplift pressure of the triangular distribution (Figure 34) acts on the base of the breakwater caisson with or without overtopping. The maximum uplift pressure is given by

$$p_u = (1 + \cos\alpha)\beta_1\beta_2 wH_{max}/2 \tag{38}$$

in which α = the direction of incident wave. When concrete blocks or ripraps are placed along the base of the caissons

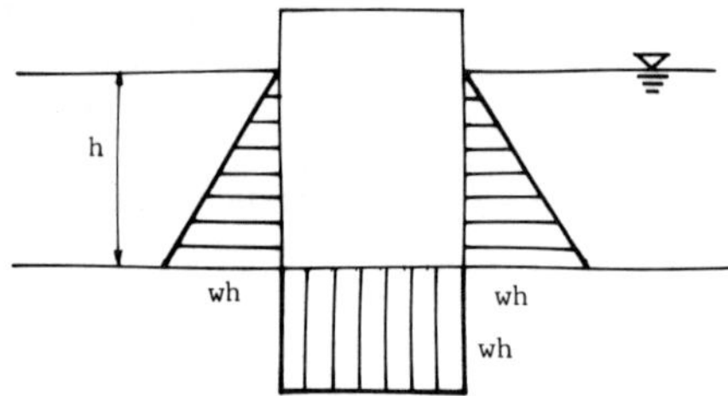

FIGURE 32. Buoyancy of caisson breakwater.

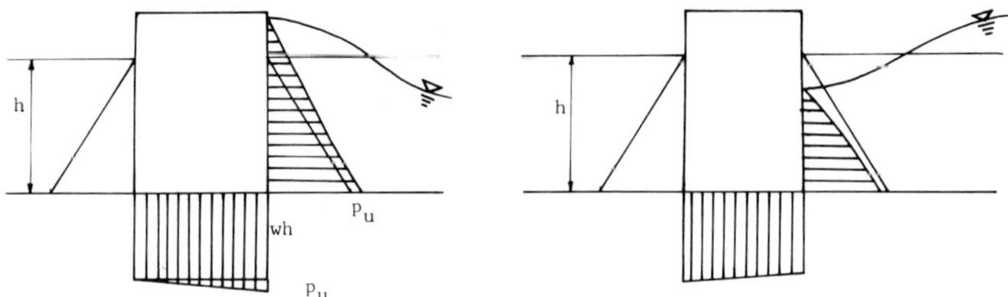

FIGURE 33. Uplift force

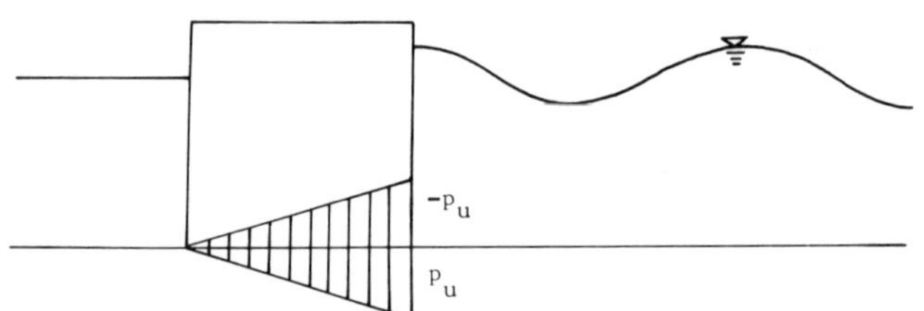

FIGURE 34. Uplift force due to irregular wave.

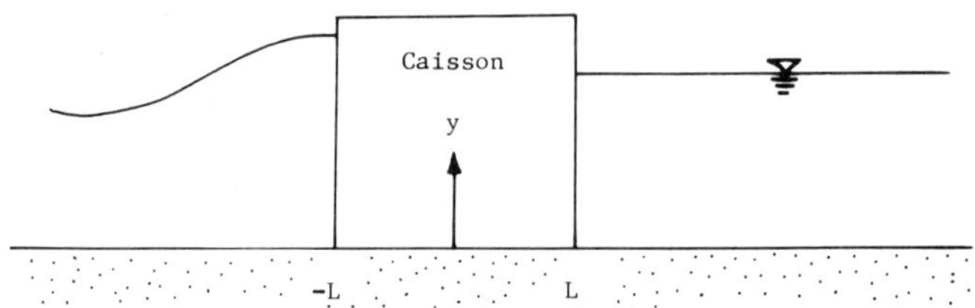

FIGURE 35. Definition sketch.

for protection, the maximum uplift pressure is also expressed by

$$p_u = (1 + \cos \alpha)\lambda\beta_1\beta_3 w H_{max} \quad (39)$$

When a rigid caisson is located on the sea bed of isotropic poro-elastic half-space saturated with water as shown in Figure 35, the dynamics of the two phases are governed by Biot's equation [32]. By using an approximate scheme, an analytic solution is obtained and the pore pressure in sea bed is given in Figure 36.

Stability of Rubble Structures

A rubble structure is composed of several layers of natural stones or concrete blocks as armor units. Armor units are placed in order to obtain good wedging or interlocking action between individual unit. To check the stability of armor units on the slope of the structure, Iribarren's or Hudson's formula is employed. These formulas are derived from the equilibrium condition between the external and friction force acting on a rubble stone or a concrete block. Iribarren expressed the force F acting on the armor unit on the slope as shown in Figure 37.

$$F = \left(1 - \frac{\rho}{\rho_d}\right) W\{f \cos\theta - \sin\theta\} \quad (40)$$

in which W = the weight of the armor unit in the air, θ = the angle between the slope and the horizon, ρ_d = the density of the armor unit, ρ = the water density and f = friction coefficient. When the wave force P acting on the armor unit is assumed to be proportional to the square of the wave-induced velocity v_r, the cross-sectional area of the armor unit is also proportional to $(W/\rho_d g)^{2/3}$. Therefore, the wave force P is written by

$$P = m'\rho g \left(\frac{W}{\rho_d g}\right)^{2/3} v_r^2 \quad (41)$$

in which m' = a constant value. If v_r in the above equation is supposed to be proportional to the velocity of the water particle in the breaking zone,

$$v_r \propto gh_b \quad (42)$$

in which h_b = the water depth at the breaking zone. After the wave breaking the wave height is almost in proportion to the water depth. Therefore,

$$v_r \propto gH \quad (43)$$

in which H = the wave height after the wave breaking. The substitution of the above equation into P gives the following equation by using the stability condition of rubble structures such as $F \geq P$:

$$W \geq \frac{m'\rho g \left(\frac{W}{\rho_d g}\right)^{2/3} H}{\left(1 - \frac{\rho}{\rho_d}\right)(f\cos\theta - \sin\theta)} = \frac{K\rho_d g H^3}{\left(\frac{\rho_d}{\rho} - 1\right)^3 (f\cos\theta - \sin\theta)^3} \quad (44)$$

in which K = a constant value. Next, let us consider the stability condition of the armor units when the armor units form layers in Figure 38. Then, the existence of other armor units induces lift forces and the weight of the armor units therefore decreases. The lift force P_L is assumed to be proportional to the cross-sectional area $(W/\varrho_d g)^{2/3}$ and the wave height. Then, the lift force is written by

$$P_L = m\rho g H \left(\frac{W}{\rho_d g}\right)^{2/3} \quad (45)$$

in which m = a constant parameter. With the usage of the friction coefficient among armor units f', the stability condition for armor units is given as follows:

$$W \geq \frac{K'f'\rho_d H^3}{\left(\frac{\rho_d}{\rho} - 1\right)^3 (f'\cos\theta - \sin\theta)^3} \quad (46)$$

in which K' = a constant value. Hence, Hudson proposed the following revised equation from the above equation:

$$W \geq \frac{\rho_d H^3}{K_D \left(\frac{\rho_d}{\rho} - 1\right)^3 \cot\theta} \quad (47)$$

in which K_D = a stability number. The above equation is called "Hudson's formula." The K_D value is represented on Table 1. The problem in the case of using Hudson's formula is that it does not include the effect of the wave period.

WAVE RUNUP, OVERTOPPING, AND TRANSMISSION

Introduction

The coastal structures to protect the shore generate wave reflections, wave runup and overtopping. Wave transmission

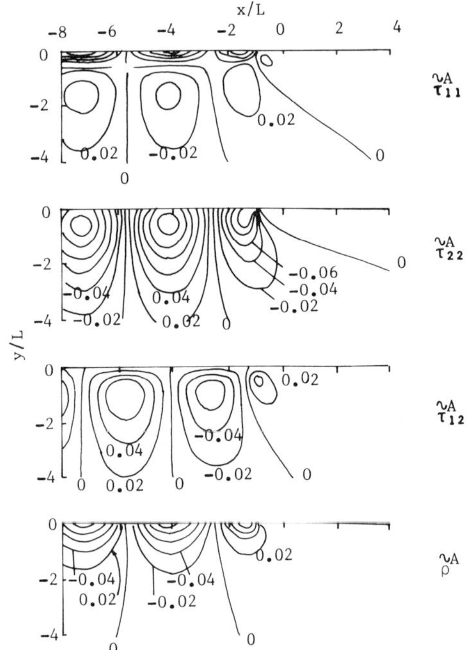

FIGURE 36. Contours of stresses and pore pressure; wave number = 1/L, m = stiffness parameters, and ratio of shear modulus of solid matrix of bulk modulus of fluid with air bubbles = 1 (Reference 32).

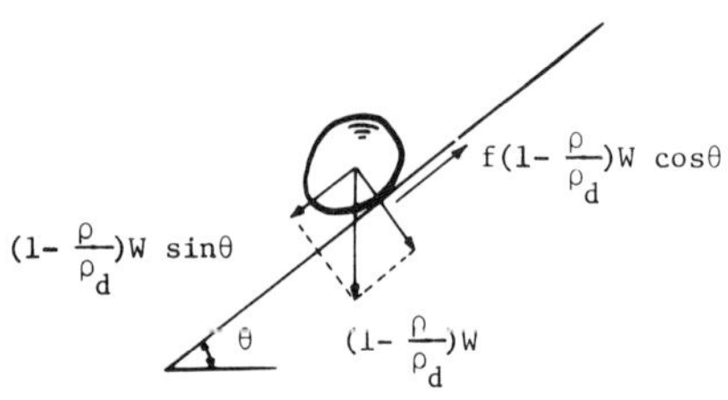

FIGURE 37. Forces on an armor unit (Reference 17).

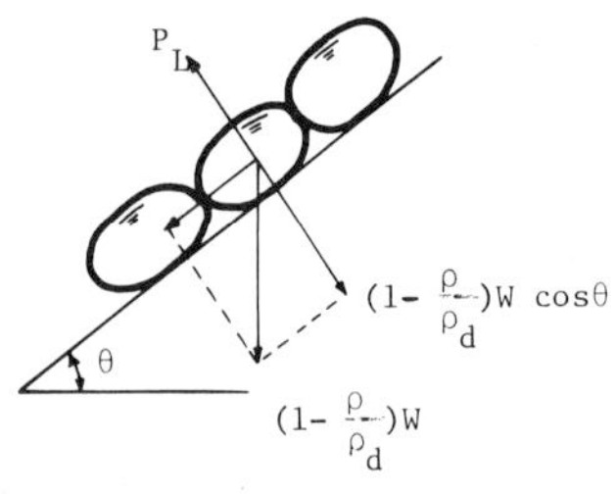

FIGURE 38. Forces on armor units (Reference 17).

phenomena occur in the cases of submerged offshore breakwaters or permeable breakwaters.

Wave Runup

The wave runup height R is defined as the vertical distance above the still water level to the location in which wave will runup the face of a structure as shown in Figure 39. The wave runup height is one of important factors to design the required height of the structure if wave overtopping is not permitted. The phenomenon of the wave runup (Photo 24) is influenced by many factors such as wave structure shapes, water depth at the toe of the structure, sea bottom topography *et al.* On the occasion of the computation of the wave runup height, the following three cases must be considered:

1. The structure is located on the offshore side of the breaking zone
2. The structure is located on the onshore side of the breaking zone
3. The structure is located on the shore.

COMPUTATION OF WAVE RUNUP HEIGHT

First let us consider the case of which wave does not break on the face of the structure which is located on the offshore side of the breaking zone. When the angle between the structure face and the horizon is equal to $\pi/2$, the standing wave is formed. If the angle θ is less than $\pi/2$ and the wave steepness is less than the following equation:

$$\frac{H}{L} = \sqrt{\frac{2\theta_c}{\pi}} \frac{\sin^2 \theta_c}{\pi} \tag{48}$$

the incident wave is completely reflected [21]. In the above equation θ_c is a limit angle such as wave does not break on the slope and H and L mean wave height and length, respectively. Miche [21] analytically derived the wave runup height R for small wave steepness as follows:

$$\frac{R}{H} = \sqrt{\frac{\pi}{2\theta}} \tag{49}$$

But the large value of wave steepness increases the wave runup height on the face of the vertical wall [35]. Therefore, the effect of the wave steepness on the wave runup height on the slope must be considered. Based on aforementioned idea, Takada [45] proposed the following equation:

$$\frac{R}{H'} = \left[\sqrt{\frac{\pi}{2\theta}} + \left(\frac{\eta_s}{H_1} - 1\right)\right] \tag{50}$$

in which H_0' = equivalent deep water wave height, η_s =

TABLE 1. K_D Values.

Armor Units	n	Placement	Structure Trunk		Structure Head	
			K_D		K_D	
			Breaking Wave	Nonbreaking Wave	Breaking Wave	Nonbreaking Wave
Quarrystone						
Smooth Rounded	2	random	2.5	2.6	2.0	2.4
Smooth Rounded	>3	random	3.0	3.2	—	2.9
Rough Angular	1	random	2.3	2.9	2.0	2.3
Rough Angular	2	random	3.0	3.5	2.7	2.9
Rough Angular	>3	random	4.0	4.3	—	3.8
Modified Cube	2	random	7.0	7.5	—	5.0
Tetrapod	2	random	8.3	10.2	—	—
Tribar	2	random	9.0	10.4		
Dolos	2	random	22.0	25.0		

n = the number of units comprising the thickness of the armor layer.

the wave runup height on the vertical wall, K_s = the shoaling coefficient, and H_1 = the wave height on the toe of the slope. As η_s/H_1, he used the following equation which Miche derived by the standing wave theory.

$$\frac{\eta_s}{H} = 1 + \frac{\pi H_1}{L} \coth kh \times \left(1 + \frac{3}{4 \sinh^2 kh} - \frac{1}{4 \cosh^2 kh}\right) \tag{51}$$

When the wave breaks on the slope in this case, Hunt [12] obtained the following equation by using experimental data:

$$\frac{R}{H} = \frac{1.01 \tan \theta}{(H/L_0)^{1/2}} \tag{52}$$

Takada [45] proposed the wave runup height as:

$$\frac{R}{H} = \left[\sqrt{\frac{\pi}{2\theta}} + \left(\frac{\eta_s}{H_1} - 1\right)\right] K_s \left(\frac{\cot \theta_c}{\cot \theta}\right)^{2/3} \tag{53}$$

In the case of which the structure is located on the onshore side of the breaking zone, the wave runup height is gained by using the result of the previous case or the wave runup diagram (Figure 40). Since the phenomena of wave breaking are not solved, the wave runup diagram is employed to compute the wave runup height in field. When the slope of the sea bottom topography is 0.1, the experimental result for smooth slopes and different wave steepness is obtained by Saville [39]. It is represented in Figure 41. On the other hand, when the slope of the structure face is less than 0.1, the wave runup height is gained by Savage [36] and plotted in Figure 42. This is drawn based on the experimental result conducted in the uniform water depth. The wave runup height first increases and then decreases as the slope of the structure face decreases.

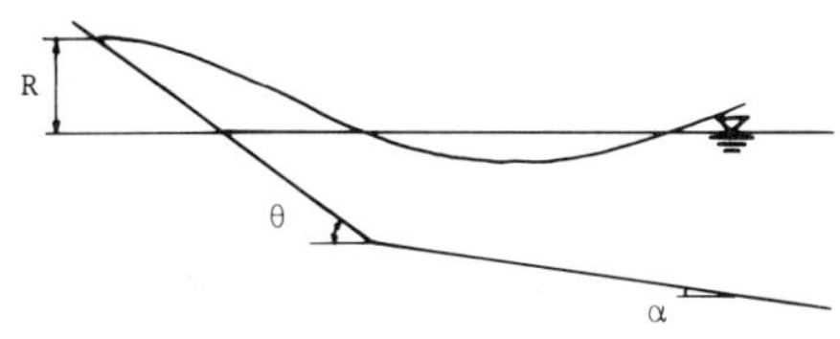

FIGURE 39. Definition sketch for wave runup.

COMPOSITE SLOPE

Next let us consider the wave runup height on the composite slope as shown in Figure 43. To obtain the wave runup height Saville [38] assumed that a composite slope can be replaced by a hypothetical, uniform slope running from the bottom, at the point where the incident wave breaks, up to the point of maximum runup on the structure. The wave runup height on the hypothetical uniform slope can be obtained by the diagram of wave runup (Figure 44). The iteration process is continued until the assumed wave runup height becomes equal to the obtained one by Figure 44. Ripraps or concrete blocks as armor units on the slope (Figure 45) are employed to reduce the wave runup height as shown in Figures 46 (a,b) and 47.

ROUGHNESS AND POROSITY

To provide the necessary design guidance, Battjes [4], Ahrens [1] and Stoa [44] have suggested the use of a roughness and porosity correction factor that allows the use of

PHOTO 24. Wave runup. (Courtesy of JSCE Slide Library.)

various smooth-slope design curves for application to other structure slope characteristics. This roughness and porosity correction factor r is defined as the ratio of runup on a rough permeable or other nonsmooth slope to the runup on a smooth impermeable slope. This is expected by the following equation:

$$r = \frac{R(\text{runup on a rough slope})}{R(\text{runup on a smooth slope})} \tag{54}$$

This roughness and porosity correction factor r is given on Table 2 [43]. Figure 48 compares the wave runup on roughened and permeable slopes, R_1, with the runup on smooth slopes, R_2 for comparable slope and (H_0'/T^2)-values. In the figure H_0' = deep water wave height, in feet, d = median diameter of roughness material, in feet, and k = permeability of permeable materials, in square feet. Madsen *et al.* [19] also analytically obtained the wave runup height by using the linearized equation of long waves. They used the water channel as shown in Figure 49 and plotted the result in Figure 50.

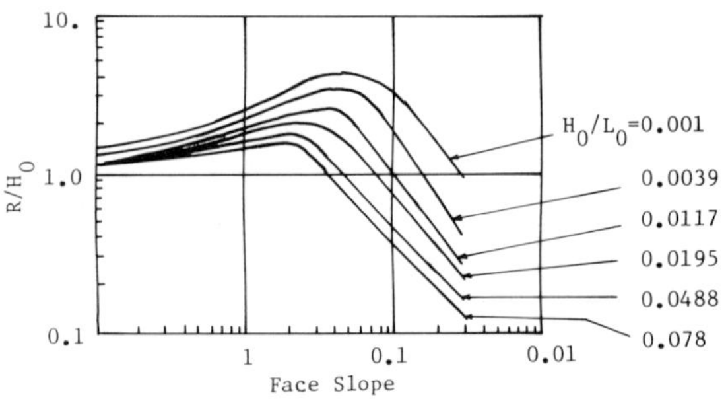

FIGURE 40 Diagram for wave runup (Reference 36).

IRREGULAR WAVE

To apply the result of the wave runup height on the slope of the structure for regular waves to irregular waves, Battjes [3] derived probability density functions of wave height and period by assuming that Hunt's Equation (52) is applied to each component of irregular waves. The probability distribution function of occurrence $F(R)$ and the probability density function $f(R)$ for the wave runup height R are written as follows:

$$F(R') = \int_0^\infty dH' \int_0^{R'^2/H'} f(H', L_0')dL_0' \tag{55}$$

$$f(R') = \frac{dF(R')}{dR'} = 2R' \int_0^\infty \frac{1}{H'} f\left(H'; \frac{R'^2}{H'}\right) dH' \tag{56}$$

in which $\overline{H}$ and $\overline{L}_0$ are averaged wave height and period in deep water, respectively, $H' = H/\overline{H}$, $L_0' = L_0/\overline{L}_0$ and $R' = R/(\sqrt{\overline{H}\overline{L}_0}\tan\theta)$. Application of Rayleigh distribution to the joint probability density function of H and L_0 with unit correlation coefficient gives the following equation:

$$F(R') = 1 - \exp\left\{-\left(\frac{\pi}{4}\right) R'^2\right\} \tag{57}$$

This shows the Rayleigh distribution of the wave runup height R. Figure 51 gives the comparison between the above equation and the observed data. n and R_{50} show the excess

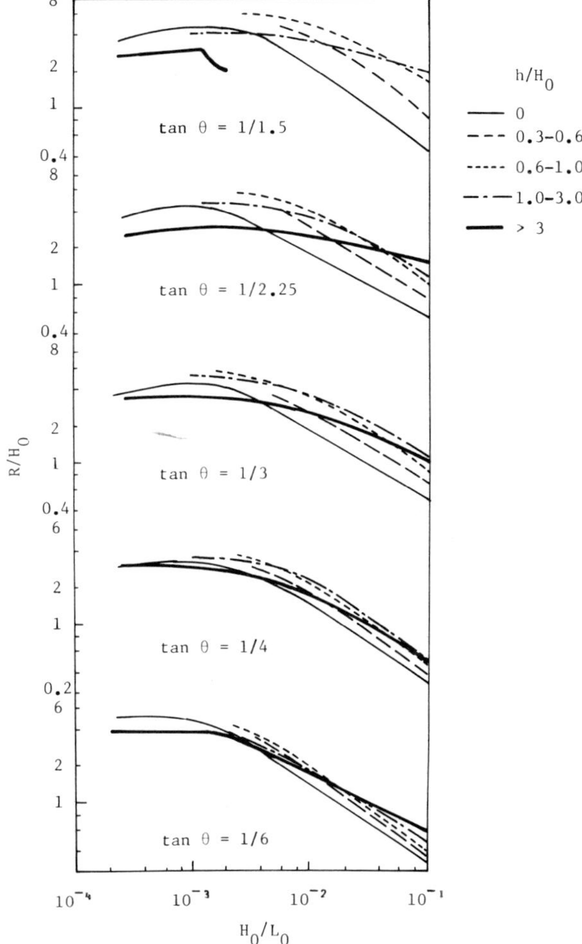

FIGURE 41. Runup height on mild slope (Reference 39).

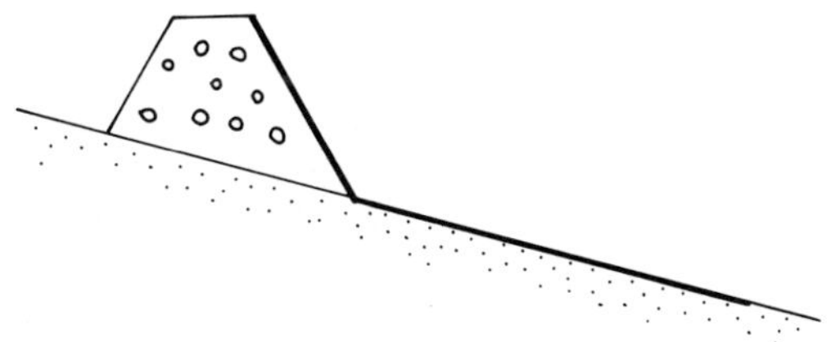

FIGURE 42. Sketch for wave runup on structure on shore (Reference 37).

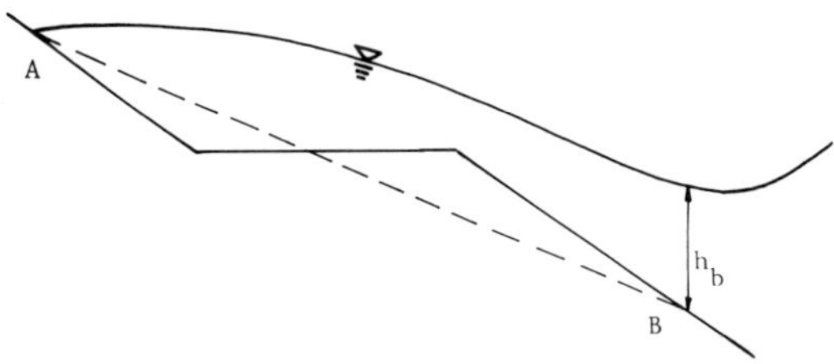

FIGURE 43. Explanation for composite slope.

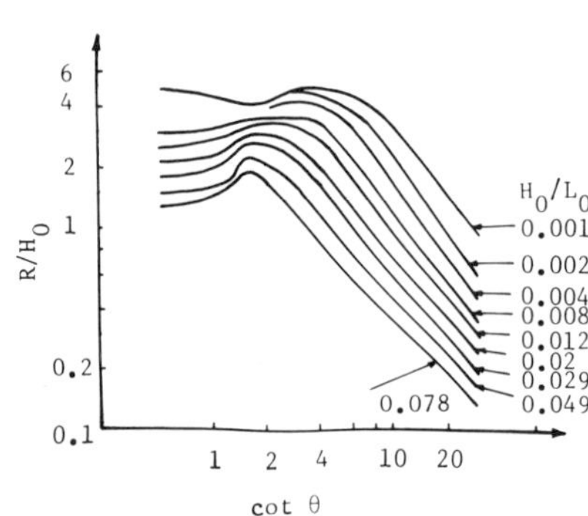

FIGURE 44. Wave runup diagram (Reference 38).

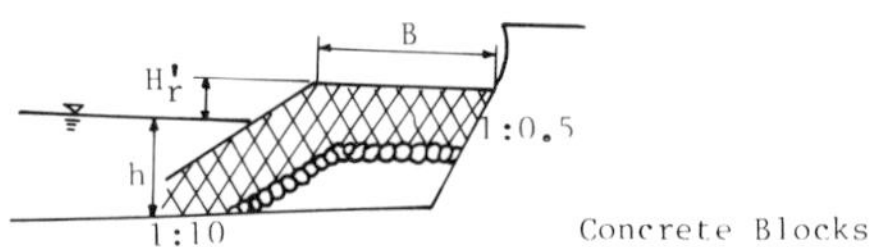

FIGURE 45. Definition sketch for wave breaking work (Reference 48).

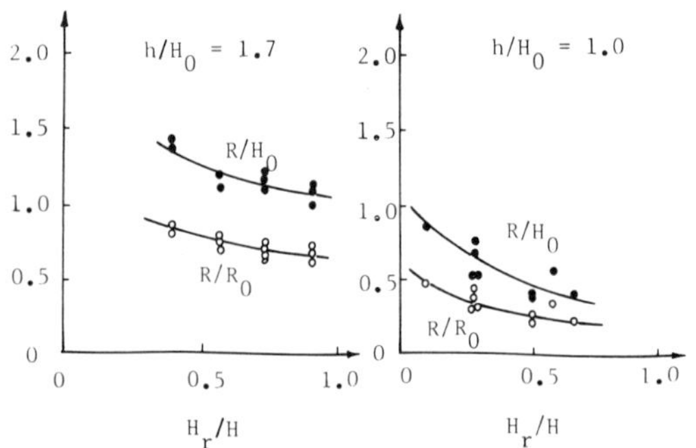

FIGURE 46. Effect of concrete blocks on wave runup (Reference 48).

TABLE 2. Value of r for Various Slope Surface Characteristics [4].

Slope Surface Characteristics	Placement	*r*
Smooth, impermeable	—	1.0
Concrete blocks	Fitted	0.90
Basalt blocks	Fitted	0.85 to 0.90
Gobi blocks	Fitted	0.85 to 0.90
Grass	—	0.85 to 0.90
One layer of quarrystone (impermeable foundation)	Randum	0.80
Quarrystone	Fitted	0.75 to 0.80
Rounded quarrystone	Randum	0.60 to 0.65
Three layer of quarrystone (impermeable foundation)	Randum	0.60 to 0.65
Quarrystone	Randum	0.50 to 0.55
Concrete armor units (50% void ratio)	Randum	0.45 to 0.50

probability and the wave runup height of $n = 50\%$, respectively.

Overtopping

OVERTOPPING RATE

In designing of coastal structures it may be too costly to build them to preclude the overtopping by the largest waves (Photo 25). Therefore, if the overtopping occurs to some extent, the rate of wave overtopping must be estimated beforehand. Because the required capacity of pumping facilities to dewater a shoreward area will depend on the rate of wave overtopping. To obtain the overtopping rate, Takada [45] expressed it from the wave form in Figure 52 when the wave runs up the slope as follows:

$$Q = a\,\frac{(R - H_c)^2}{2}\left(\frac{x_0}{R} - \cot\theta\right) \tag{58}$$

in which a = a constant value and other variables are explained in Figure 53. It is not easy to apply the above equation to get the rate of the wave overtopping, because the values of x_0 and a are dependent upon the slope of the structure face. A reanalysis of Saville's data indicates that the overtopping rate per unit length of structure can be expressed by [43],

$$Q = (gQ_0^* H_0'^3)^{1/1} \exp\left[-\left\{\frac{0.217}{\alpha}\tanh^{-2}\frac{H_c}{R}\right\}\right] \tag{59}$$

in which $0 \le H_c/R < 1.0$, g = the gravitational acceleration, H_0' = the equivalent deepwater wave height, H_c = the height of the structure crest above the still water level, R = the runup on the structure that would occur if the

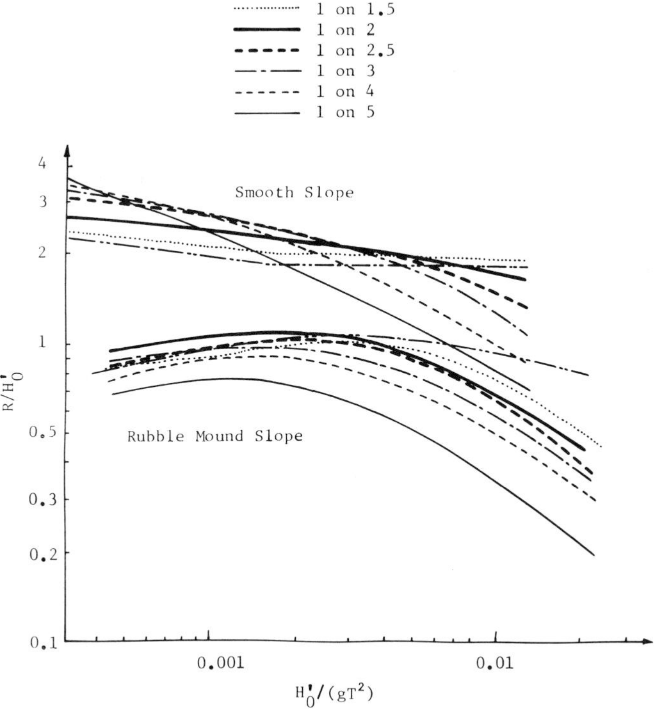

FIGURE 47. Comparison of wave runup on smooth slopes with runup on permeable rubble slopes (data for $h/H_0' > 3$) (Reference 43).

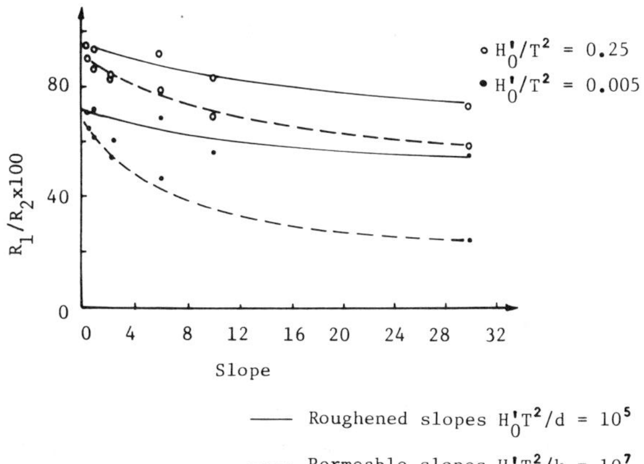

FIGURE 48. Runup on roughened and permeable slopes compared with smooth slopes (Reference 36).

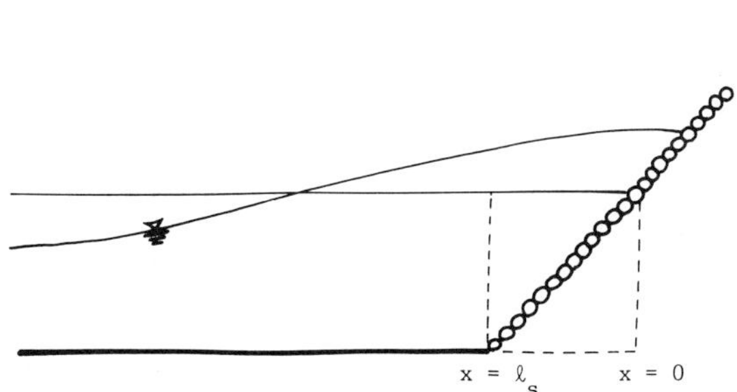

FIGURE 49. Definition sketch for wave runup (Reference 19).

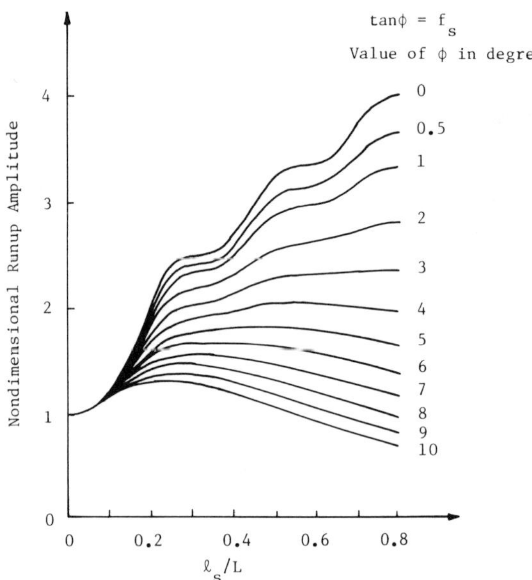

FIGURE 50. Theoretical prediction of runup on rough slope (Reference 19).

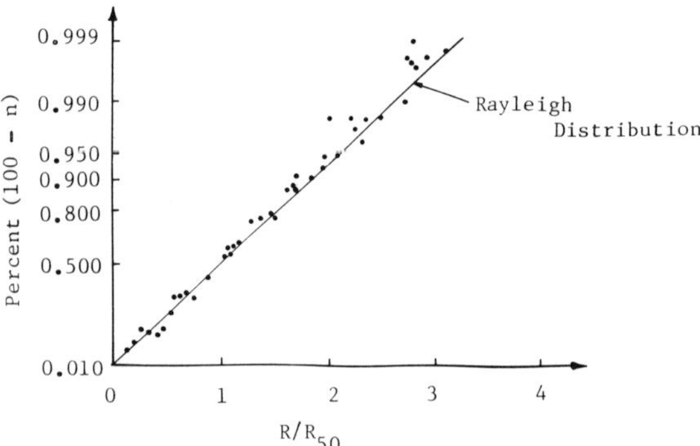

FIGURE 51. Wave runup distribution (Reference 3).

structure were high enough to prevent overtopping corrected for scale effects, and α and Q_0^* = empirically determined coefficients that depend on incident wave characteristics and structure geometry. The diagram for one case is shown in Figure 54 [43]. By conducting experimental investigation, Tominaga [17] obtained the following relationship between the difference of the wave runup height and the height of the structure crown above the still water level and the overtopping rate for different slopes of the structure face when the sea bottom slope is 1/30:

$$Q = K_1(R - H_c)^b \tag{60}$$

For standing waves the water depth at the structure toe and the wave steepness strongly influence the overtopping rate. Therefore, he suggested as:

$$Q = aA \tag{61}$$

Takada [45] proposed the following equation as A:

$$A = \frac{1}{2}\frac{1 + \cot^2\theta}{\cot\gamma - \cot\theta} \times (R - H_c)^2 + 0.15H(R - H_c) \tag{62}$$

in which

$$\cot\gamma = \begin{cases} 67\left(\dfrac{H}{L}\right)(\cot\theta)^{1.6} & \text{for } \cot\theta \geq 1 \\ \left\{n + \dfrac{n(n-1)}{2}\cot^2\theta\right\}^{1/2}\cot\theta & \text{for } \cot\theta < 1 \end{cases}$$

$$n = -3.224\log_{10}\left\{\frac{1}{1+\left(\dfrac{67H}{L}\right)^2}\right\}$$

$$a = \begin{cases} 7.6(\cot\theta)^{0.73}\left(\dfrac{H_0}{L_0}\right)^{0.83} & \text{for sloped wall} \\ 9.3\left(\dfrac{R-H_c}{H}\right)^{1/2}\left(\dfrac{H_0}{L_0}\right) & \text{for vertical wall} \end{cases}$$

FACTORS TO INFLUENCE OVERTOPPING

Next let us consider the factors to influence the overtopping rate such as the face slope of the structure, the water depth at the structure toe and wind effect. The overtopping rate increases as the face slope of the structure increases and it attains the maximum rate when the slope is 1:2 or 1:3. The water depth at the structure toe prescribes whether the structure is located in the breaking wave zone or the standing wave zone. The water depth to the maximum

PHOTO 25. Overtopping. (Courtesy of JSCE Slide Library.)

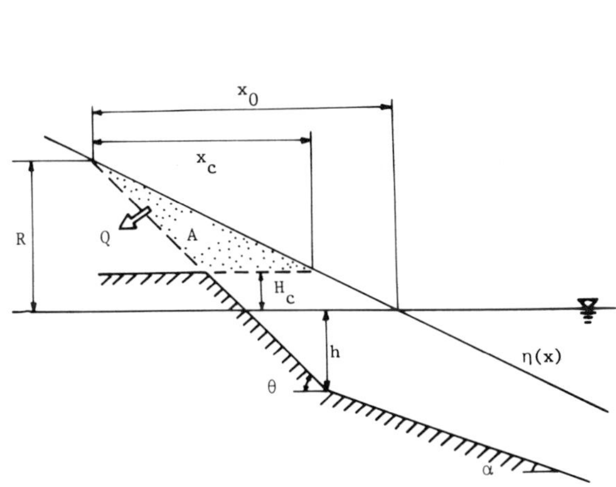

FIGURE 52. Definition sketch for overtopping (References 45, 46).

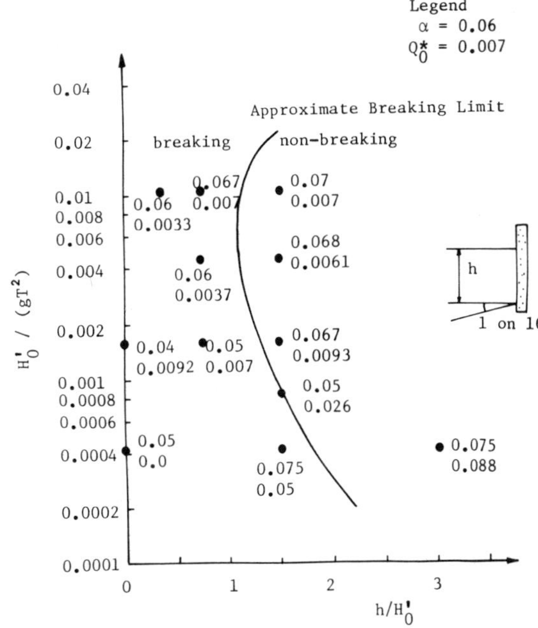

FIGURE 53. Overtopping parameters, α and Q_0^* (smooth vertical wall on a 1:10 nearshore slope) (Reference 43).

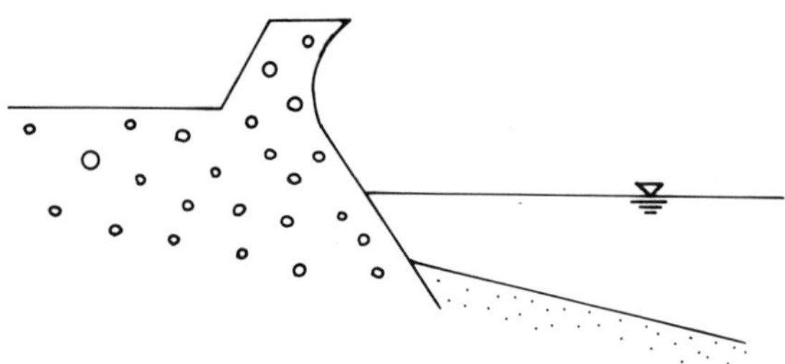

FIGURE 54. Parapet.

overtopping rate is expressed by Takada [45] as follows:

$$\frac{H_0}{L_0} = (1 - d_1 \cot\theta)(0.6 \tan\alpha + 0.12) \times \left[\frac{\sinh\frac{4\pi h_R}{L_R} + \frac{4\pi h_R}{L_R}}{\sinh\frac{4\pi h_R}{L_R}}\right]^{1/2} \tanh\frac{2\pi h_R}{L_R} \quad (63)$$

in which $d_1 \cong 0.09$ for $\cot\theta \leq 3$ and L_R = the wave length at the water depth at the structure toe h_R. The wind effect on the overtopping rate is also important, because strong wind induces storm waves. Takada [46] showed the overtopping rate with the use of the correction factor β when the wind was blowing as follows:

$$Q(v) = \beta Q(0) \quad (64)$$

in which $Q(0)$ = the overtopping rate in the windless condition. The β-value is given by the following equations:

$\overline{\beta} = \exp(0.112v/c_0)$ for deeper zone than the wave breaking point

$\overline{\beta} = \exp(0.088v/c_0)$ for the zone between the structure toe and the wave breaking point

$\overline{\beta} = \exp(0.231v/c_0)$ for shallower zone than the water depth of the structure toe

in which $\overline{\beta}$ = the average value of β, v = wind velocity and c_0 = celerity in the deep water.

IRREGULAR WAVE

In general the incident wave to the coastal structure is irregular and the irregularity effect on the overtopping rate must be considered. When the overtopping rate q for given wave height and period is known, the expected overtopping rate is expressed by

$$q_{\exp} = \int_0^\infty q_0(H|T_{1/3})p(H)dH \quad (65)$$

in which $q_0(H|T_{1/3})$ = the overtopping rate for given wave height and significant wave period without correlation between wave height and period and $p(H)$ = the probability density function of deepwater height (Rayleigh Distribution). Goda [35] obtained the overtopping rate for irregular waves by using flow discharge formula over a weir. Then, he obtained as:

$$\frac{q_{\exp}}{\sqrt{2g(H_{1/3})_0^3}} = \int_0^\infty \frac{q(\eta)}{\sqrt{2g(H_{1/3})_0^3}} p(\eta)d\eta \quad (66)$$

in which $(H_{1/3})_0$ = deepwater significant wave height. Based on the above equation Goda made the diagram to calculate the overtopping rate on the vertical wall and the wave absorbing revetment when the irregular wave attacked them. These diagrams [35] are used for different wave steepness and sea bottom slope. To reduce the overtopping rate a parapet (Figure 34 and Photo 26) or the wave breaking works are employed.

Wave Transmission

Let us consider wave transmission through a step, a submerged breakwater and a permeable breakwater. The wave transmission through complicated breakwater is studied by using *FEM* or *BEM* [5].

WAVE TRANSMISSION ON A STEP

Let us consider the wave transmission when the wave propagates on a step as shown in Figure 55. This phenomenon is analytically solved by Neuman [33], Miles [22], Ijima *et al.* [13]. Assuming that a fluid is inviscid and incompressible and the motion is irrotational and two-dimensional, the velocity potential ϕ satisfies the following Laplace equation:

$$\frac{\partial^2\phi}{\partial x^2} + \frac{\partial^2\phi}{\partial z^2} = 0 \quad (67)$$

The boundary conditions on the free surface and on the sea bottom are expressed by

$$\frac{\partial\phi}{\partial z} = \frac{\sigma^2\phi}{g} \quad \text{at } z = 0 \quad (68)$$

$$\frac{\partial\phi}{\partial z} = 0 \quad \text{at } z = -h \quad (69)$$

The value of ϕ is finite at infinity. The solution of the Laplace equation to satisfy the boundary condition on the

PHOTO 26. Parapet. (Courtesy of JSCE Slide Library.)

sea bottom is

$$\phi = (Ae^{ikx} + Be^{-ikx})\frac{\cosh k(z+h)}{\cosh kh} + \sum_{n=1}^{\infty}(C_n e^{-ik_n x} + D_n e^{ik_n x})\frac{\cos k_n(z+h)}{\cos k_n h} \tag{70}$$

in which n = an integer, k and k_n satisfy the following equations derived from the boundary condition on the free surface [Equation (68)].

$$\frac{\sigma^2 h}{g} = kh \tanh kh = -k_n h \tanh k_n h \tag{71}$$

for $n = 1, 2, 3, \ldots$. In Figure 55 the velocity potential ϕ_I and ϕ_{II} are defined in the zone I and II, respectively. They

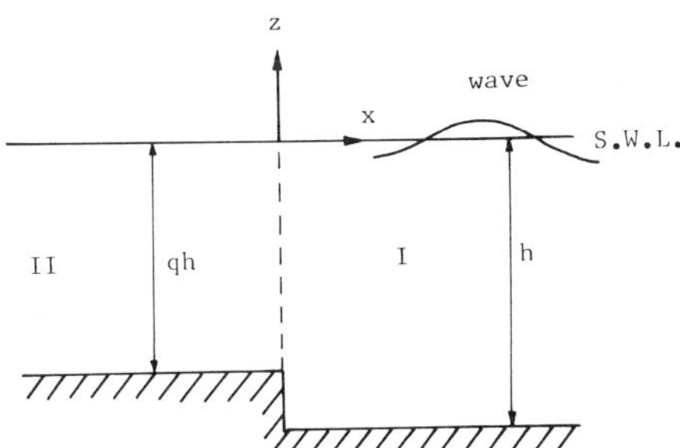

FIGURE 55. Definition sketch for wave transmission over a step.

are expressed by

$$\phi_{\mathrm{I}} = (Ae^{ikx} + Be^{-ikx})\frac{\cosh k(z+h)}{\cosh kh} + \sum_{m=1}^{\infty} C_m e^{-k_m x}\frac{\cos k_m(z+h)}{\cos k_m h} \tag{72}$$

$$\phi_{\mathrm{II}} = He^{ik'x}\frac{\cosh k'(z+qh)}{\cosh k'qh} + \sum_{n=1}^{\infty} I_n e^{ik_n' x}\frac{\cos k_n'(z+qh)}{\cos k_n' qh} \tag{73}$$

The boundary conditions at $x = 0$ are

$$\phi_{\mathrm{I}} = \phi_{\mathrm{II}}, \qquad \frac{\partial\phi_{\mathrm{I}}}{\partial x} = \frac{\partial\phi_{\mathrm{II}}}{\partial x} \quad \text{for } 0 > z \geq -qh, \qquad \frac{\partial\phi_{\mathrm{I}}}{\partial x} = 0 \quad \text{for } -qh > z \geq -h \tag{74}$$

By substituting of Equations (72) and (73) into Equation (74) and by integrating the resultant equation with respect to z from $-h$ to 0 and $-qh$ to 0 in zones I and II, respectively, the reflection coefficient K_R and the transmission coefficient K_T are obtained in Figure 56. In the above calculation the wave breaking phenomena are not included. The wave transmission from the shallow water to the deep water is also solved by using the same method. The com-

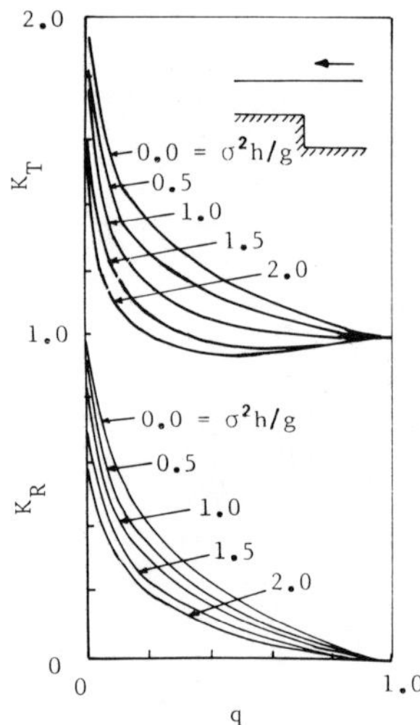

FIGURE 56. Coefficients of wave reflection and transmission when wave transmits from deep water to shallow water (Reference 13).

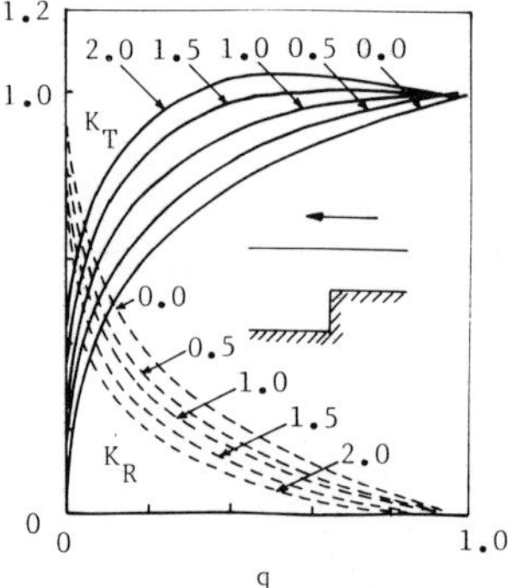

FIGURE 57. Coefficients of wave reflection and transmission when wave transmits from shallow water to deep water (Reference 13).

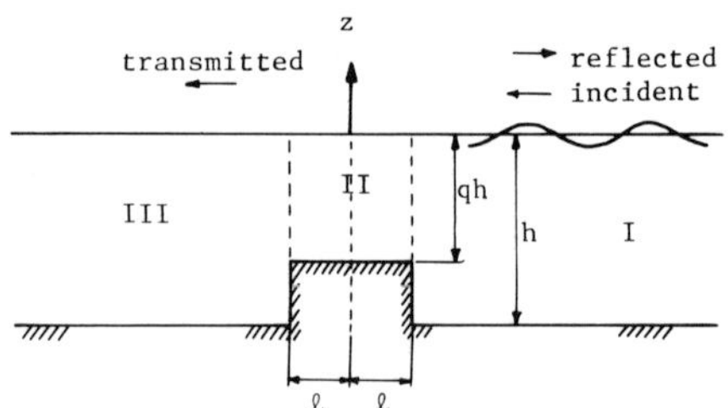

FIGURE 58. Definition sketch for submerged breakwater.

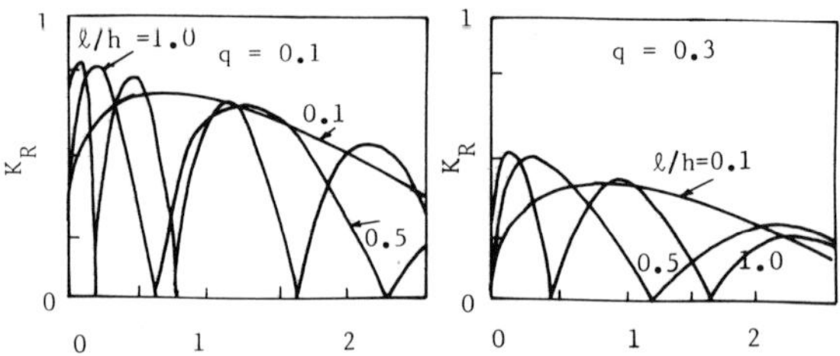

FIGURE 59. Wave reflection coefficient (Reference 13).

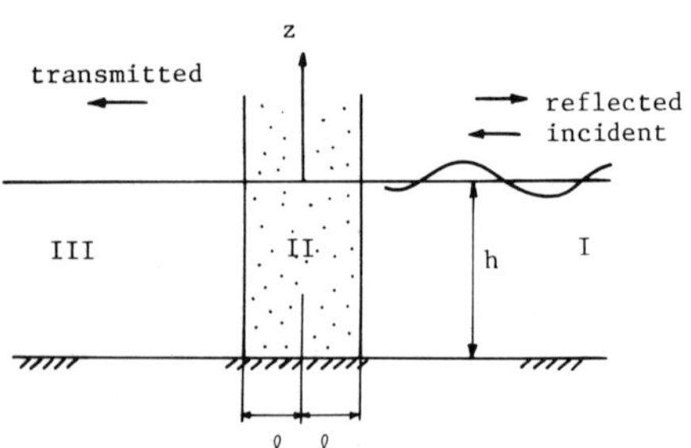

FIGURE 60. Definition sketch for permeable breakwater.

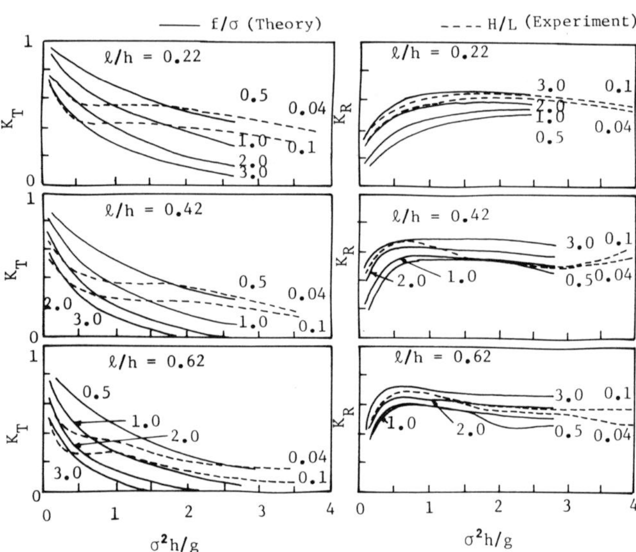

FIGURE 61. Coefficients of wave reflection and transmission through permeable breakwater (Reference 13).

puted reflection and transmission coefficient are plotted in Figure 57.

WAVE TRANSMISSION OVER A SUBMERGED BREAKWATER

Let us consider the wave transmission over a submerged breakwater as shown in Figure 58 [13,20]. Assuming that the width of the submerged breakwater is $2l$ and the whole zone is divided into three zones by $x = \pm l$ (Figure 55), the velocity potential ϕ_{I}, ϕ_{II} and ϕ_{III} on each zone are given by

$$\begin{aligned}
\phi_{\mathrm{I}} &= (Ae^{ik(x-l)} + Be^{-ik(x-l)})\frac{\cosh k(z+h)}{\cosh kh} + \sum_{m=1}^{\infty} C_m e^{-ik_m(x-l)}\frac{\cos k_m(z+h)}{\cos k_m h} \\
\phi_{\mathrm{II}} &= \left(D\frac{\cos k'x}{\cos k'l} + E\frac{\sin k'x}{\sin k'l}\right)\frac{\cosh k'(z+qh)}{\cosh k'qh} + \sum_{n=1}^{\infty}\left(F_n\frac{\cosh k_n'x}{\cosh k_n'l} + G_{T_n}\frac{\sinh k_n'x}{\sinh k_n'l}\right) \times \frac{\cos k_n'(z+qh)}{\cos k_n'qh} \\
\phi_{\mathrm{III}} &= He^{ik(x+l)}\frac{\cosh k(z+h)}{\cosh kh} + \sum_{m=1}^{\infty} I_m e^{k_m(x+l)}\frac{\cos k_m(z+h)}{\cos k_m h}
\end{aligned} \tag{75}$$

in which k and k_n are determined by the following equation:

$$\frac{\sigma^2 qh}{g} = kqh\tanh kqh = -k_n qh \tan k_n qh \tag{76}$$

for $n = 1, 2, 3, \cdots$. The boundary conditions at $x = \pm l$ are the same as that at $x = 0$ in the case of the wave transmission on a step. From these results the reflection and transmission coefficient for $q = 0.1$ and 0.3 are plotted in Figure 59. For the long waves the reflection K_R and transmission coefficient K_T are expressed by

$$\begin{aligned}
K_R &= \frac{1-q}{\sqrt{(1-q)^2 + q(\tan k'l + \cot k'l)}} \\
K_T &= \frac{\sqrt{q}(\tan k'l + \cot k'l)}{\sqrt{(1-q)^2 + q(\tan k'l + \cot k'l)}}
\end{aligned} \tag{77}$$

WAVE TRANSMISSION THROUGH A PERMEABLE BREAKWATER

Let us consider the wave transmission through a permeable breakwater as shown in Figure 60. Assuming that the averaged velocities in the permeable breakwater in x and z direction are $\bar{u}$ and $\bar{w}$ and the flow resistance is defined to be in proportional to the flow velocity (the flow is laminar), the velocity potential $\bar{\phi}$ satisfies the following Laplace equation:

$$\frac{\partial^2\bar{\phi}}{\partial x^2} + \frac{\partial^2\bar{\phi}}{\partial z^2} = 0 \tag{78}$$

The boundary condition on the free surface is given by

$$\frac{\partial\bar{\phi}}{\partial z} = \left(1 - \frac{if}{\sigma}\right)\frac{\sigma^2\bar{\phi}}{g} \tag{79}$$

in which f = resistance coefficient and $i = \sqrt{-1}$. The solution is

$$\bar{\phi}(x, z) = \sum_{r=0}^{\infty}(D_r\cos\bar{k}_r x + E_r\sin\bar{k}_r x) \times \frac{\cosh\bar{k}_r(z+h)}{\cosh\bar{k}_r h} \tag{80}$$

in which D_r and E_r are constants to be determined from the boundary conditions and $\bar{k}_r$ is explained by

$$\bar{k}_r h\tanh\bar{k}_r h = \left(2 - \frac{if}{\sigma}\right)\frac{\sigma^2 h}{g} \tag{81}$$

for $r = 0, 1, 2, \cdots$. In the case of which the breakwater width is $2l$, the velocity potential ϕ on each zone is written by

$$\begin{aligned}
\phi_{\mathrm{I}} &= \{Ae^{ik(x-l)} + Be^{-ik(x-l)}\}\frac{\cosh k(z+h)}{\cosh kh} + \sum_{n=0}^{\infty} C_n e^{-k_n(x-l)}\frac{\cos\bar{k}_n(z+h)}{\cos\bar{k}_n h} \\
\phi_{\mathrm{II}} &= \sum_{r=0}^{\infty}\left(D_r\frac{\cos\bar{k}_r x}{\cos\bar{k}_r l} + E_r\frac{\sin\bar{k}_r x}{\sin\bar{k}_r l}\right) \times \frac{\cosh\bar{k}_r(z+h)}{\cosh\bar{k}_r h} \\
\phi_{\mathrm{III}} &= Fe^{ik(x+l)}\frac{\cosh k(z+h)}{\cosh kh} + \sum_{n=1}^{\infty} G_{T_n} e^{k_n(x+l)}\frac{\cos k_n(z+h)}{\cos k_n h}
\end{aligned} \tag{82}$$

The boundary conditions at $x = \pm l$ are shown as follows:

$$\frac{\partial \phi_{\mathrm{I}}}{\partial x} = \frac{\partial \overline{\phi}}{\partial x}, \ \phi_{\mathrm{I}} = \beta \phi_{\mathrm{II}}$$
$$\frac{\partial \phi_{\mathrm{III}}}{\partial x} = \frac{\partial \overline{\phi}}{\partial x}, \ \phi_{\mathrm{III}} = \beta \phi_{\mathrm{II}} \quad (83)$$

in which λ = void ratio and $\beta = (1 - if/\sigma)/\lambda$. Figure 61 represents the coefficients of the reflection K_R and the transmission K_T. The calculated reflection coefficient explains the experimental one very well but the calculated transmission coefficient does not represent the experimental one. Because the flow resistance in the calculation is assumed to be in proportion to the velocity instead of the square of the velocity in turbulent flow.

ACKNOWLEDGEMENTS

For permission of reprinting figures the author wishes to express his sincere appreciation to Dr. Iwagaki (Kyoto Univ.), Dr. Sawaragi (Osaka Univ.), Dr. Toyoshima (Toukai Univ.), American Society of Civil Engineers, Elsevier Scientific Pub. Company, Japan Society of Civil Engineers, McGraw-Hill Book Company Inc., the Institution of Civil Engineers, and U.S. Army Coastal Engineering Research Center. He also deeply thanks Japan Society of Civil Engineers for allowance to use many photographs from Slide Library. Finally, he expresses his gratitude to the editors who gave him a chance to join in the project.

REFERENCES

1. Ahrens, J., "Prediction of Irregular Wave Runup," CETA 77-2, U.S. Army, Corps of Engineers, Coastal Engineering Research Center, Fort Belvoir, Va., July 1977.
2. Arakida, T., Tsurukawa, Y., and Mizumura, K., "Flow Pattern and Shoreline Change in the Lee of a Breakwater, Part 1," *25th Japanese Conf. on Coastal Engrg., JSCE*, 1978, pp. 199–203 (in Japanese).
3. Battjes, J. A., "Run-up Distribution of Waves Breaking on Slope," *Jour. of the Waterways, Port, Coastal Engineering Div., ASCE*, Vol. 97, No. WW1, 1971, pp. 91–114.
4. Battjes, J. A., "Wave Runup and Overtopping," Technical Advisory Committee on Protection Against Inundation, Rijkswaterstaat, The Hague, Netherlands, 1974.
5. Bird, H. W. K. and Shepherd, R., "Wave Interaction with Large Submerged Structures," *Jour. of the Waterways, Port, Coastal and Ocean Div., ASCE*, Vol. 108, No. WW2, May 1982, pp. 146–162.
6. Goda, Y., "Wave Forces on Structures," *Summer Lecture Note on Hydraulic Engineering, B, JSCE*, 1967 (in Japanese).
7. Goda, Y., "Study on Designing Wave Pressure of Breakwater," *Report of the Port and Harbor Research Institute*, Vol. 12, Part 3, 1973 (in Japanese).
8. Goda, Y. and Kakizaki, S., "Study on Standing Waves of Finite Amplitude and Amplitude and its Pressure," *Report of the Port and Harbor Research Institute*, Vol. 12, Part 3, 1972 (in Japanese).
9. *Handbook of Hydraulic Formula*, Japan Society of Civil Engineers, 1973 (in Japanese).
10. Hattori, S., "Development of Coastal Zone and Wave Controls," *Summer Lecture Note on Hydraulic Engineering, B, JSCE*, 1975 (in Japanese).
11. Hudson, R. Y., "Laboratory Investigation of Rubble-Mound Breakwaters," *Proceedings of the American Society of Civil Engineers, ASCE*, Waterways and Harbors Div., Vol. 85, No. WW3, 1959, pp. 93–121.
12. Hung, I. A., "Design of Seawalls and Breakwaters," *Proceedings of the American Society of Civil Engineers, ASCE*, Waterways and Harbors Div., Vol. 85, No. WW3, 1959, pp. 122–152.
13. Ijima, T., "Solution of Boundary Value Problems on Wave Theory and its Applications," *Summer Lecture Note on Hydraulic Engineering, B, JSCE*, 1971 (in Japanese).
14. Ippen, A. T. (ed.), *Estuary and Coastline Hydrodynamics*, McGraw-Hill Book Co., Inc., New York, 1966.
15. Issacson, S. Q. I. and Maull, D. J., "Transverse Forces on Vertical Cylinders," *Jour. of the Waterway, Port, Coastal and Ocean Div., ASCE*, Vol. 102, No. WW1, Feb. 1976, pp. 49–60.
16. Ito, Y., Fujishima, M., and Kitatani, T., "Study on the Stability of Breakwaters," *Report of the Port and Harbor Research Institute*, Vol. 5, Part 4, 1966 (in Japanese).
17. Iwagaki, Y. and Sawaragi, T., *Coastal Engineering*, 1st ed., Kyoritsu Shuppan Co. Ltd., Tokyo, Japan, 1979 (in Japanese).
18. Keulegan, G. H. and Carpenter, L. H., "Forces on Cylinders and Plates in an Oscillating Fluid," *NBS Report No. 4821*, Natural Bureau of Standards, Washington, D.C., 1956.
19. Madson, O. S. and White, S. M., "Energy Dissipation on a Rough Slope," *Jour. of the Waterway, Port, Coastal and Ocean Div., ASCE*, Vol. 102, No. WW1, Feb. 1976, pp. 31–48.
20. Mei, C. C. and Black, J. L., "Scattering of Surface Waves by Rectangular Obstacles in Waters of Finite Depth," *Jour. of Fluid Mech.*, Vol. 38, Part 3, pp. 499–511.
21. Miche, R., "Mouvements Ondulatores de la mer in Profondeur Constante ou Decroissante," *Annals des Ponts et Chaussees*, Paris, Vol. 114, 1944.
22. Miles, J. W., "Surface-Wave Scattering Matrix for a Shelf," *Jour. of Fluid Mech.*, Vol. 28, Part 4, 1967, pp. 755–767.
23. Mizumura, K., "Littoral Currents and Characteristics of Beaches Around Breakwaters," *ASCE Special Conf., Coastal Structure*, Mar. 1979, pp. 778–791.
24. Mizumura, K., "Flow Pattern and Shoreline Change in the Lee of a Breakwater, Part 2," *26th Japanese Conf. on Coastal Engrg., JSCE*, 1979, pp. 53–57 (in Japanese).

25. Mizumura, K., "Littoral Drift of Sand Near Port of Oarai," *Proc. of 17th Int. Conf. on Coastal Engrg., ASCE*, 1980, pp. 2159–2173.
26. Mizumura, K. and Shiraishi, N., "Laboratory Study on the Shoreline Changes Behind a Coastal Structure," *Coastal Engineering*, Elsevier Scientific Pub. Co., Vol. 5, 1981, pp. 51–81.
27. Mizumura, K., "Shore Line Change Estimates Near Oarai, Japan," *Jour. of the Waterway, Port, Coastal and Ocean Div., ASCE*, Vol. 108, No. WW1, Feb. 1982, pp. 65–80.
28. Mizumura, K., "Flow Analysis in Indearized Rubble-Mound Breakwater," *Jour. of the Waterway, Port, Coastal and Ocean Div., ASCE*, Vol. 110, No. WW3, Aug. 1984, pp. 344–355.
29. Mizumura, K., "Nonlinear Water Waves Developed by an Accelerated Circular Cylinder," *Proc. of 7th Int. Conf. on BEM in Engrg.*, 1985 (to be published).
30. Mizumura, K., "Nonlinear Water Waves over Wavy Bed," *Proc. of 7th Int. Conf. on BEM in Engrg.*, 1985 (to be published).
31. Morrison, J. R. *et al.*, "The Force Exerted by Surface Waves on Piles," *Petroleum Transactions*, 189, TP 2846, 1950.
32. Mynett, A. E. and Mei, C. C., "Wave-Induced Stresses in a Saturated Poro-Elastic Sea Bed Beneath a Rectangular Caisson," *Geotechnique*, 32, No. 3, 1982, pp. 235–247.
33. Newman, J. N., "Propagation of Water Waves Over an Infinite Step," *Jour. of Fluid Mech.*, Vol. 23, 1965, pp. 399–415.
34. Nagai, S., "Theory of Wave Pressure," *Summer Lecture Note on Hydraulic Engineering, B, JSCE*, 1966 (in Japanese).
35. Sato, S. and Goda, Y., *Coasts, Ports and Harbors*, 2nd ed., Shokokusha, Tokyo, Japan, 1981 (in Japanese).
36. Savage, R. P., "Laboratory Data on Wave Run-up on Roughed and Permeable Slopes," *Jour. of the Waterways and Harbor Div., ASCE*, Vol. 84, No. WW3, 1958, pp. 852–870.
37. Saville, T., Jr., "Wave Run-up on Shore Structure," *Jour. of the Waterways and Harbor Div., ASCE*, Vol. 82, No. WW2, 1956.
38. Saville, T., Jr., "Wave Run-up on Composite Slopes," *Proc. of 6th Int. Coastal Engineering Conf., ASCE*, 1958.
39. Saville, T., Jr., "Discussion: Laboratory Investigation of Rubble Mound Breakwater by R. Y. Hudson," *Jour. of the Waterways and Harbor Div., ASCE*, Vol. 86, No. WW3, 1960, p. 151.
40. Sawaragi, T., Nakamura, T., and Kida, H., "On Lift Force on a Circular Pile," *22nd Japanese Conf. on Coastal Engrg., JSCE*, 1975 (in Japanese).
41. Schlichting, H., *Boundary Layer Theory*, 4th ed., English ed. (by Kestin, J.), McGraw-Hill Book Company, Inc., New York, 1960.
42. Shiraishi, N., Arakida, T., Endo, T., and Mizumura, K., "Shoreline Changes in the Lee of a Breakwater," *ASCE Special Conf. Coastal Zone*, Mar. 1978, pp. 1401–1418.
43. *Shore Protection Manual*, Vol. II, U.S. Army, Coastal Engineering Research Center, 1977.
44. Stoa, P. N., "Reanalysis of Wave Runup on Structures and Beaches," U.S. Army, Corps of Engineers, Coastal Engineering Research Center, 1977.
45. Takada, A., "Wave Runup, Overtopping and Reflection," *Proc. of JSCE*, Vol. 182, 1970, pp. 19–30 (in Japanese).
46. Takada, A., "Wave Runup and Overtopping," *Summer Lecture Note on Hydraulic Engineering, B, JSCE*, 1977 (in Japanese).
47. Tanimoto, K., Moto, K., Ishizuka, S., and Goda, Y., "Investigation on Design Formula of Wave Force on a Breakwater," *23rd Japanese Conf. on Coastal Engineering, JSCE*, 1976, pp. 11–16.
48. Toyoshima, O., *Coastal Engineering in Field, Storm Surge*, 1st ed., Morikita Shuppan Co., Ltd., Tokyo, Japan, 1969 (in Japanese).
49. Toyoshima, O., *Coastal Engineering in Field, Coastal Erosion*, 1st ed., Morikita Shuppan Co., Ltd., Tokyo, Japan, 1972 (in Japanese).

CHAPTER 24

Offshore Structures

SUBRATA K. CHAKRABARTI*

INTRODUCTION

The offshore engineering is a relatively young field that is still experiencing childhood in the hands of its researchers. Since its inception it has come a long way in achieving great progress in a very short time. The first offshore oil platform was built 30 years ago in the Gulf of Mexico in 20 ft. water depth. Today more than 2000 major platforms exist worldwide. Many innovative structures have found their way into the coastal and deeper waters of the oceans. Because of the hostile nature of the environment, the problem is challenging and becoming more so as new oil and gas fields open up in oceans from the Arctic to the Antarctic and from the Pacific to the Atlantic and all the bodies of water in between. Numerous problems are faced by the engineers in the design, installation and operation of such offshore structures.

One of these problems constantly needing attention is the determination of waves and wave actions on such a structure. Much has been done in developing analytical theories to determine wave forces on submerged or semisubmerged structures and the responses of these structures to the waves. However, much remains to be done in understanding the physics of fluids on the structures fully.

The design of an offshore structure involves the study in the following areas:

- environmental conditions at the site including the wind, waves & current
- wave action on structures that are fixed or moving
- motions of structures that are moored, or being towed to site
- soil study and foundation design
- structural and fatigue analysis

*CBI Industries Inc., Plainfield, IL

This chapter covers the first three subjects. It discusses the choice of design waves and wave spectra. The interaction of small and large fixed structures with ocean waves is discussed. Finally, the response of floating moored structures in waves is investigated. The results are presented collectively in a form such that they may be applied in the design work.

Because of the compl.. .nature of the ocean waves and complicated geometries of the structures, certain simplifications are warranted in attempting to compute the reaction of the waves on the structures. Where analytical study is not possible or unreliable, experimental data from the laboratory and the field have been resorted to. Wave action on submerged structures can be divided into two parts: 1) wave action on small structures and 2) wave interaction with large structures. In the respective areas, various analytical techniques have been demonstrated to be worthwhile design tools.

The study of compliant structures in waves includes: 1) moored floating structures, e.g. ships, buoys, floating storage units, etc., 2) articulated towers, 3) tension leg platform, and 4) guyed towers. Different devices are used in mooring a structure, e.g. catenaries, lines and fenders, mooring arms, tension legs, etc. Numerical solutions are available to determine the added mass and damping coefficients of oscillating bodies in water. For restricted degrees of oscillations of a structure, certain approximate closed form solutions for motion are possible. General purpose numerical programs on a finite difference scheme have been developed to study the motions of a moored floating object.

In designing a structure, there are two basic approaches to translate a given sea state into hydrodynamic loads and the structural response.

Deterministic Approach

The deterministic method of structural analysis can be pseudo-dynamic and dynamic. In the pseudo-dynamic analysis, the design sea state is given in terms of a specified wave height and wave period (and direction). The loads due to these waves are computed which are used in the analysis of the static strength of the structure.

Bea (1979) analyzed the design of a steel, tubular membered, template-type structure in waves for three different areas in the Eastern Gulf of Alaska (GOA), Southern California (SC) and Santa Barbara Channel (SBC). The design was based on expected maximum wave height as the primary index. The maximum wave height probabilities for these areas, along with Baltimore Canyon (BC) and the Gulf of Mexico (GOM) are shown in Figure 1 based on oceanographic hindcasting model. These values compare with the 1979 API recommended values.

In the deterministic method, the structural response may be calculated on the basis of a single wave (as described) or a wave energy spectrum. In this case, the response is obtained through time independent formulas. In the dynamic analysis, a time history of the sea surface is obtained either from recorded data or from the simulation of a sea spectrum. The response is obtained numerically based on this time series. The structural analysis is made as shown in Figure 2.

British Standard Institute (1979) recommends single wave method with a given height and period and a suitable wave

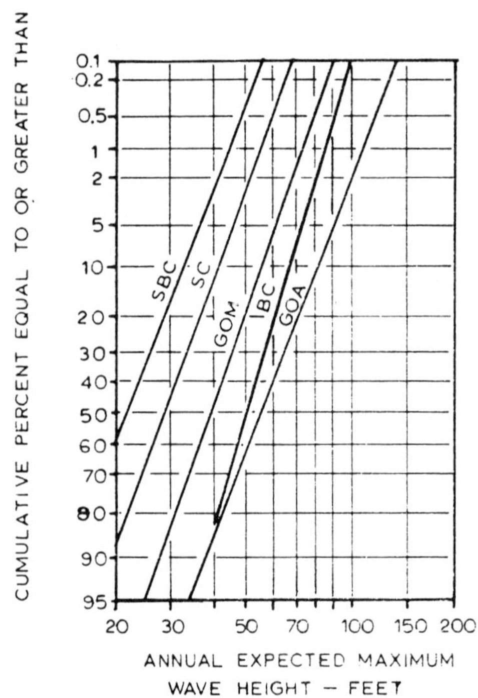

FIGURE 1. Projected storm conditions in the North American continental waters [from Bea (1979)].

theory for wave load computation. Linear wave theory for swell, Stokes fifth-order theory for storm waves and stream function theory for shallow water are utilized. For a structure whose natural frequencies approach the frequencies of the predominant wave energy components, API recommends a dynamic analysis.

Stochastic Approach

Ocean waves are random in nature and are mathematically described by statistical means in terms of its energy spectrum. In this case, the wave loads are determined by spectral analysis. In this approach, all calculations are performed in the frequency domain as opposed to the time domain of the deterministic analysis. The steps in this approach are also outlined in Figure 2. The responses are obtained in a probabilistic sense.

According to the Det Norske Veritas (1977) rules, the wave determination could be deterministic or stochastic. The wave period range should be between $\sqrt{6.5H} < T < \sqrt{15H}$ for deterministic and $\sqrt{13H_s} < T < \sqrt{30H_s}$ for stochastic approach in which the single wave height, H and the significant wave height, H_s are in meters. A range of wave periods may be selected in the deterministic approach. P-M spectrum will normally apply in open seas. Similar recommendations have been provided by the Norwegian Petroleum Directorate (1977) which gives a period range of 12 to 20 sec. For the North Sea application, JONSWAP spectrum is recommended.

The above two approaches will be developed for a fixed structure in the determination of the wave forces. The same methods will be applied, namely pseudo-dynamic and stochastic, in the analysis of a moving rigid structure.

DESIGN WAVE ENVIRONMENT

There are two basic approaches considered in the design of offshore structures. One uses a single wave method in which the design wave is represented by a wave period and a wave height. One of the reasons for using this approach is the simplicity in the design analysis and easy determination of the extreme wave conditions (as opposed to design sea state). If this method is used then it is recommended that several possible single design waves of varying period be analyzed and the structure design be based on the worst loads experienced for any of these design waves.

It should be kept in mind that the highest waves do not necessarily produce the largest responses. This is particularly true for the compliant structures where the natural frequency of the structure plays an important role in determining the responses of the structure in waves.

The other method is the wave spectrum approach. In this case a suitable wave spectrum model is chosen representing the proper energy density distribution of the sea waves at the

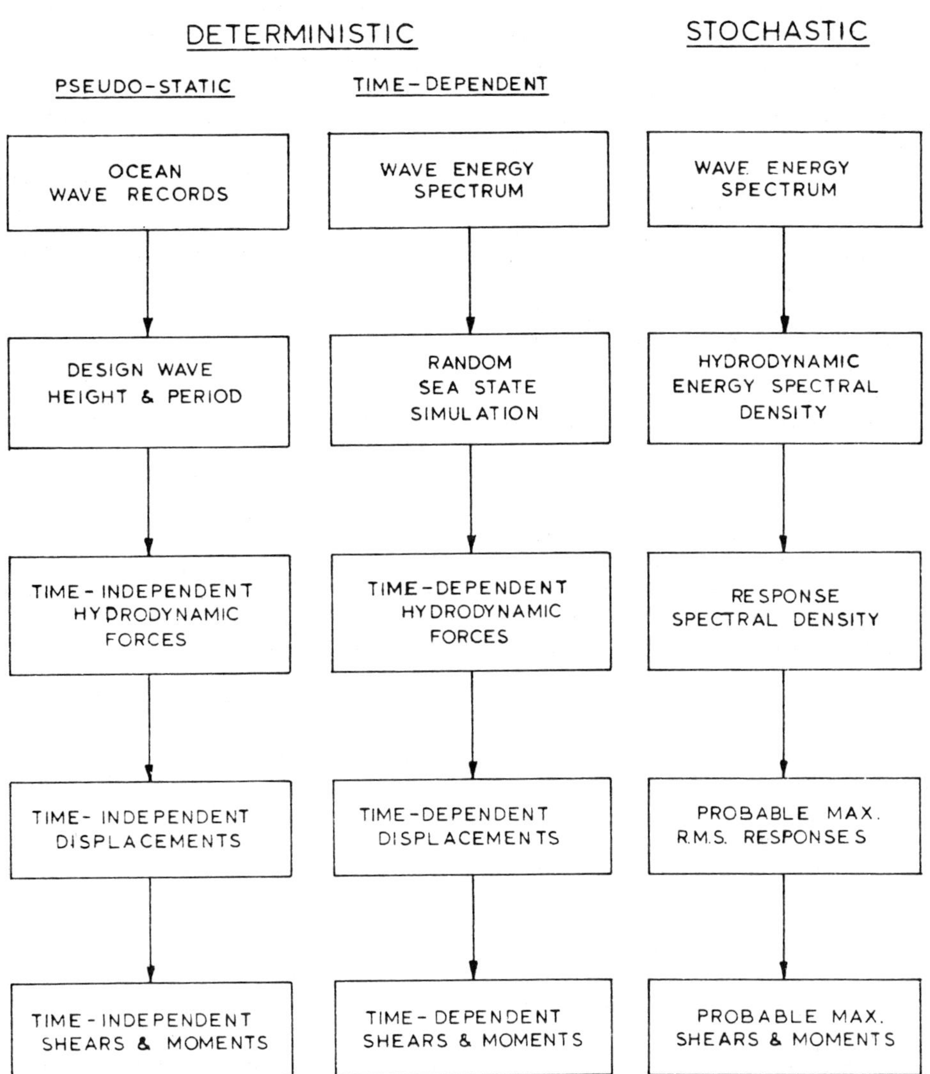

FIGURE 2. Chart showing different methods of solution.

site under consideration. The most suitable spectrum is a measured design spectrum at the site although this spectrum is seldom available. As an alternative one chooses one of the theoretical spectrum models available based on the fetch, wind and other meteorological conditions of the site.

Waves in the open ocean are most often generated by winds. Waves travelling out of a distant generating area are called swells while waves generated locally are known as sea. The ultimate growth of the sea depends primarily on wind speed, duration and fetch. Attempts have been made by various researchers to formulate the spectrum of wind-wave or a swell, known as forecasting.

There is no theoretical set of forecasting curves; all the "theoretical," semi-empirical, and empirical procedures rely on physical measurements, and these measurements in the open ocean are few in number, largely unreliable for a number of reasons, usually for varying winds and fetches, and often for moving fetches. However, better and better forecasting methods are made available which are based on more reliable and accurate field data.

The mathematical spectrum models are generally based on significant wave height, wave period and shape factors. The most common one-parameter spectrum is the Pierson Moskowitz (1964) model based on the significant wave height or wind speed. There are several two parameter spectra available. Some of the commonly used ones are modified P-M, Bretschneider (1961), Scott (1965), ISSC (1964), and ITTC (1966). JONSWAP spectrum (Hasselman, 1973) is a five parameter spectrum, but usually three of the parameters are held constant.

Pierson-Moskowitz Spectrum

In 1964 Pierson and Moskowitz proposed a formula for an energy spectrum distribution of a wind generated sea state. This spectrum commonly known as P-M model has since then been extensively used by ocean engineers as one of the most representative for waters all over the world. It has found many applications in the design of offshore structures.

The P-M spectral model describes a fully-developed sea determined by one parameter, namely, the wind speed in which the fetch and duration are considered infinite. For the applicability of such a model, the wind has to blow over a large area at a nearly constant speed for many hours prior to the time when the wave record is obtained and the wind should not change its direction more than a certain specified small amount. In spite of these limitations, the P-M model has been found to be useful in representing a severe storm wave in offshore structure design.

The P-M spectrum model is written as

$$S(\omega) = \alpha g^2 \omega^{-5} \exp\left[-0.74\left(\omega\frac{U_w}{g}\right)^{-4}\right] \quad (1)$$

where $\alpha = 0.0081$, ω = wave frequency, U_w = wind speed, and g = gravitational acceleration. Alternatively,

$$S(\omega) = \alpha g^2 \omega^{-5} \exp\left[-1.25\left(\frac{\omega_0}{\omega}\right)^4\right] \quad (2)$$

in which ω_0 = frequency at the spectral peak. Note that the relationship between ω_0 and U_w may be established from Equations (1) and (2).

An equivalent expression for the P-M spectrum in terms of the cyclic frequency, $f(=\omega/2\pi)$ may be written as

$$S(f) = \frac{\alpha g^2}{(2\pi)^4} f^{-5} \exp\left[-1.25\left(\frac{f_0}{f}\right)^4\right] \quad (3)$$

where $f_0 = \omega_0/2\pi$. Note the relationship between $S(\omega)$ and $S(f)$

$$S(f) = 2\pi\, S(\omega) \quad (4)$$

A modified form of P-M spectrum in terms of the significant wave height, H_s, and peak period, ω_0, is written as

$$S(\omega) = \frac{5}{16} H_s^2 \frac{\omega_0^4}{\omega^5} \exp\left[-1.25\left(\frac{\omega_0}{\omega}\right)^4\right] \quad (5)$$

This is the two parameter form of the P-M spectrum modified. For the one-parameter P-M spectrum, the above formula is applicable if the peak frequency is related to the significant wave height H_s by

$$\omega_0^2 = 0.161\, g/H_s \quad (6)$$

A plot of the one-parameter P-M model for an H_s = 20 ft. is shown in Figure 3. The relationship between $S(f)$ and $S(\omega)$ given in Equation (4) can be found from Figure 3. Also, the peak frequency can be checked from the significant height and Equation (6).

Bretschneider's Spectrum

On the basis of the assumption that the spectrum is narrow-banded and the individual wave height and wave period follow the Rayleigh distribution, Bretschneider (1961) derived the following form of the spectral model

$$S(\omega) = 0.1687 H_s^2 \frac{\omega_s^4}{\omega^5} \exp[-0.675(\omega_s/\omega)^4] \quad (7)$$

where $\omega_s = 2\pi/T_s$ and T_s = significant wave period defined as the average period of the significant waves.

ISSC Spectrum

The International Ship Structures Congress (1964) suggested slight modification in the form of the Bretschneider

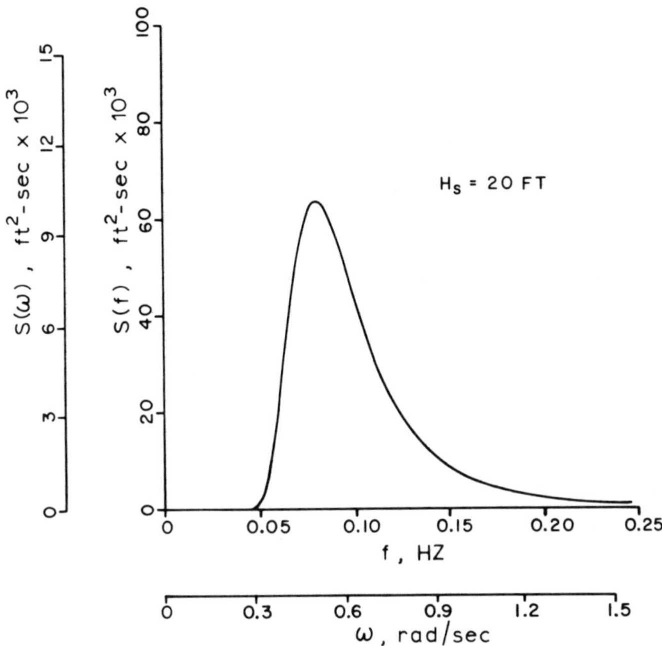

FIGURE 3. One parameter Pierson-Moskowitz energy density spectral model.

spectrum,

$$S(\omega) = 0.1107 H_s^2 \frac{\bar{\omega}^4}{\omega^5} \exp[-0.4427(\bar{\omega}/\omega)^4] \quad (8)$$

in which $\bar{\omega} = 2\pi/\bar{T}$ and $\bar{T}$ = mean period.

JONSWAP Spectrum

The JONSWAP spectrum was developed during a joint North Sea wave project (1973) and hence the name. The formula for the JONSWAP spectrum can be written by modifying the P-M formulation as follows:

$$S(\omega) = \alpha g^2 \omega^{-5} \exp[-1.25(\omega_0/\omega)^4] \gamma^{\exp[-(\omega-\omega_0)^2/2\tau^2\omega_0^2]} \quad (9)$$

in which γ = peakedness parameter, τ = shape parameter (τ_a for $\omega \le \omega_0$ and τ_b for $\omega > \omega_0$). The quantities α, τ_a and τ_b are taken as constants with values of $\alpha = 0.0081$, $\tau_a = 0.07$ and $\tau_b = 0.09$. The quantity γ may vary from 1 to 7 with a mean value of 3.3 found from the North Sea wave data. The P-M and JONSWAP spectra are compared in Figure 4. Note that the JONSWAP spectrum has a sharper peak than the P-M spectrum.

The JONSWAP spectrum is usually considered a two parameter spectrum in terms of γ, and ω_0. However, in a design case, usually the significant height, H_s and average zero-crossing period, T_z of a random wave are specified. Unfortunately, these quantities for the JONSWAP spectrum may not be obtained simply in a closed form and the values of γ and T_0 are calculated numerically by trial and error from Equation (9) using H_s and T_z. A detailed analysis of the relationship among these four parameters showed that H_s and T_z may be related to T_0 and γ by the following two polynomial equations:

$$H_s = (0.11661 + 0.01581\,\gamma - 0.00065\,\gamma^2)\,T_0^2 \quad (10)$$

and

$$T_0 = (1.49 - 0.102\,\gamma + 0.0142\,\gamma^2 - 0.00079\,\gamma^3)\,T_z \quad (11)$$

From the above equations, for $\gamma = 1$

$$H_s = 0.1317\,T_0^2 \quad (12)$$

which has an error of less than 1% for P-M spectrum [Equation (6)] and

$$T_0 = 1.4014\,T_z \quad (13)$$

with an error of about 0.1%. Note that if H_s and T_0 are

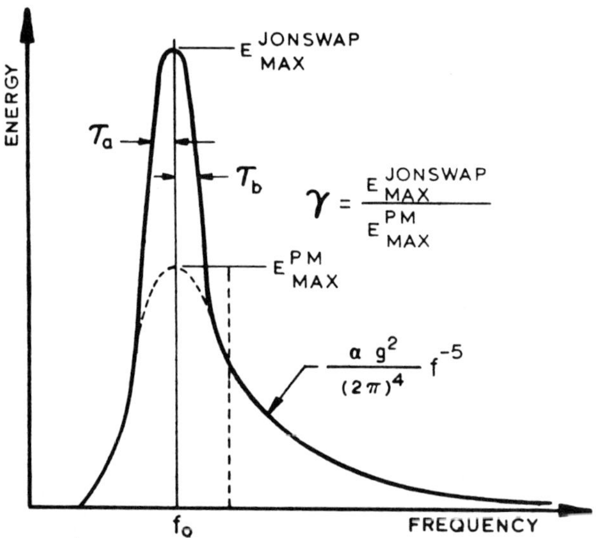

FIGURE 4. Definition sketch of JONSWAP spectrum (after Hasselmann, 1973).

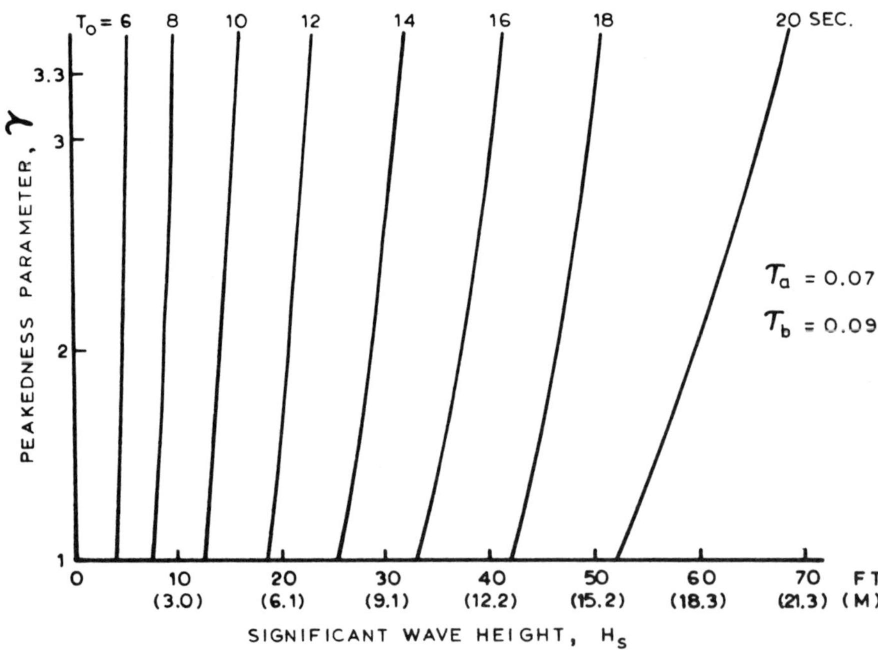

FIGURE 5. Significant wave heights, H_s as a function of overshoot parameter, γ and peak period, T_o for the JONSWAP spectrum.

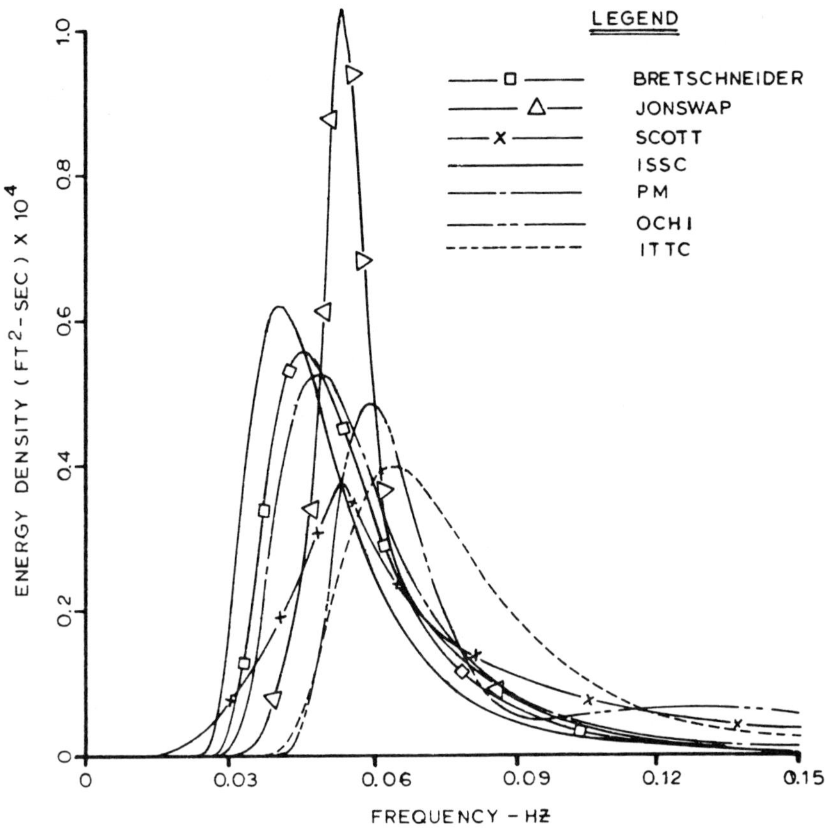

FIGURE 6. Example of variation of energy distribution of different spectral models.

specified only the first equation is needed. This relationship is shown in Figure 5.

Goda (1979) derived an approximate expression for the JONSWAP spectrum in terms of H_s and ω_0 as follows:

$$S(\omega) = \alpha^* H_s^2 \frac{\omega_0^4}{\omega^5} \exp \times \left[-1.25 \left(\frac{\omega_0}{\omega} \right)^4 \right] \gamma^{\exp[-(\omega-\omega_0)^2/(2\tau^2\omega_0^2)]} \quad (14)$$

where

$$\alpha^* = \frac{0.0624}{0.230 + 0.336\gamma - 0.185(1.9 + \gamma)^{-1}} \quad (15)$$

Note that for $\gamma = 1$, $\alpha^* = 0.312$ which reduces to the P-M spectrum [Equation (5)].

The various spectra described here are compared in Figure 6. The significant wave height chosen for this comparison of the spectral distribution is taken as $H_s = 53.8$ ft. (corresponding to a maximum wave height of 100 ft.). The JONSWAP spectrum uses $\gamma = 3.3$ so that from Equation (10), $T_0 = 18.2$ sec. For the P-M model Equation (6) gives $T_0 = 20.2$ sec. The JONSWAP model is most sharply peaked and probably most suitable for the extreme storm waves such as used in the example.

FIXED RIGID STRUCTURES

Wave forces on an offshore structure are generally calculated in three different ways. One of the approaches known as the Morison theory considers the force to be composed of inertia and drag forces added together. The components involve an inertia (or mass) coefficient and drag coefficient which must be determined experimentally. This method is usually applicable to smaller objects (compared to wave length) when drag force is significant. When the drag force is small and the inertia force predominates, but the object is still relatively small, a second approach is to utilize pressure-area method on the surface of the object called Froude-Krylov (F-K) hypothesis. For certain objects, e.g., a sphere, a horizontal cylinder, etc., the expressions of F-K force and inertia part of the Morison force are equivalent; however, for other objects they are different. A force coefficient is again needed in F-K theory to include the effect of added mass and wave diffraction. In this method forces on a typical symmetric object may be obtained in a closed form. However, values of force coefficients, likewise, are determined from experimental data. A two-dimensional strip theory is sometimes used for long slender structures to evaluate these coefficients. A third method applicable to large objects considers a numerical technique, is general with respect to the shape of the object and needs a large size computer. In a few instances, a closed form solution is possible for basic structural shapes. It is generally known as linear diffraction theory. It solves the Laplace's equation along with appropriate linear boundary conditions in terms of a total potential which is the sum of an incident and a scattered potential. This method is mathematically more pleasing and desirable. However, the first two methods are more readily used.

This part of the chapter deals with these three methods of analysis. The basic problem with the first two methods has been the selection of the appropriate coefficients. The applicability of these methods is illustrated below with the help of force calculation on a vertical cylindrical structure by various methods due to Airy waves. While the chart is strictly applicable to a vertical cylinder in deep water it should help in establishing the regions of wave force formulations for other similar structures.

The areas of drag and inertia force predominance and the regions of applicability of the various wave force formulas for a vertical cylindrical structure are shown in Figure 7. The limits are stated in terms of the Keulegan-Carpenter number ($KC = u_0D/T$) and the diffraction parameter, $\pi D/L$, and are applicable at least in the deep water region. Note that D = cylinder diameter, T = wave period, u_0 = maximum horizontal water particle velocity and L = wave length. In Figure 7, the Region I represents no-drag all-inertia situation where diffraction effect is negligible so that only the inertia part of Morison's equation is needed for the determination of the wave force. Region II constitutes the diffraction region and total wave forces should be calculated based on linear wave diffraction theory. The drag effect is small and can still be ignored in Region III. Both drag and diffraction phenomena may be important in Region IV. In Region V, drag effect becomes significant and complete Morison formula is required for total forces. In Region VI, drag force is more predominant than the inertia force. F-K theory can be applied in Regions I, II and III as long as appropriate force coefficients are available.

Morison Equation

The semi-empirical force model developed by Morison, et al. (1950) has been the most widely used method in determining forces on small diameter vertical cylindrical members in an offshore structure. This approach depends on a knowledge of the water particle kinematics and experimentally determined coefficients. The wave force per unit length of a vertical cylindrical pile is written as

$$\begin{aligned} f &= f_I + f_D \\ &= \varrho C_M \frac{\pi}{4} D^2 \dot{u} + \frac{1}{2} \varrho C_D D |u| u \end{aligned} \tag{16}$$

in which f_I = inertia force, f_D = drag force, ϱ = mass density of water, u = horizontal water particle velocity and $\dot{u}$ = horizontal water particle acceleration.

The inertia coefficient, C_M, and drag coefficient, C_D, are determined experimentally. The inertia force, f_I may be interpreted as that due to a pressure gradient caused by the acceleration of the ambient fluid. The drag force, f_D, is due to a flow separation and associated wake formation around the structure.

Linear wave theory, also known as Airy theory, is most commonly used in determining u and $\dot{u}$. According to the linear theory, the horizontal water particle velocity is

$$u = \frac{\pi H}{T} \frac{\cosh ks}{\sinh kd} \cos(kx - \omega t) \tag{17}$$

and the vertical water particle velocity is

$$v = \frac{\pi H}{T} \frac{\sinh ks}{\sinh kd} \sin(kx - \omega t) \tag{18}$$

in which H = wave height, T = wave period, d = water depth, s = vertical distance from the flat ocean floor, k = wave number ($= 2\pi/L$), $\omega = 2\pi/T$, x = horizontal distance and t = time. These quantities are defined in Figure 8.

The water particle accelerations in the horizontal and vertical directions respectively are given by

$$\frac{\partial u}{\partial t} = \frac{2\pi^2 H}{T^2} \frac{\cosh ks}{\sinh kd} \sin(kx - \omega t) \tag{19}$$

$$\frac{\partial v}{\partial t} = -\frac{2\pi^2 H}{T^2} \frac{\sinh ks}{\sinh kd} \cos(kx - \omega t) \tag{20}$$

Note from the expressions of the horizontal and vertical velocities that the horizontal velocity of a water particle is maximum (or minimum) when the vertical velocity is zero and vice versa. Since the amplitudes of the two velocities are generally different it may be inferred from this that a water particle describes an elliptical orbit about its mean position in a complete wave cycle. In deep water, however, the two amplitudes approach each other and the orbit is circular.

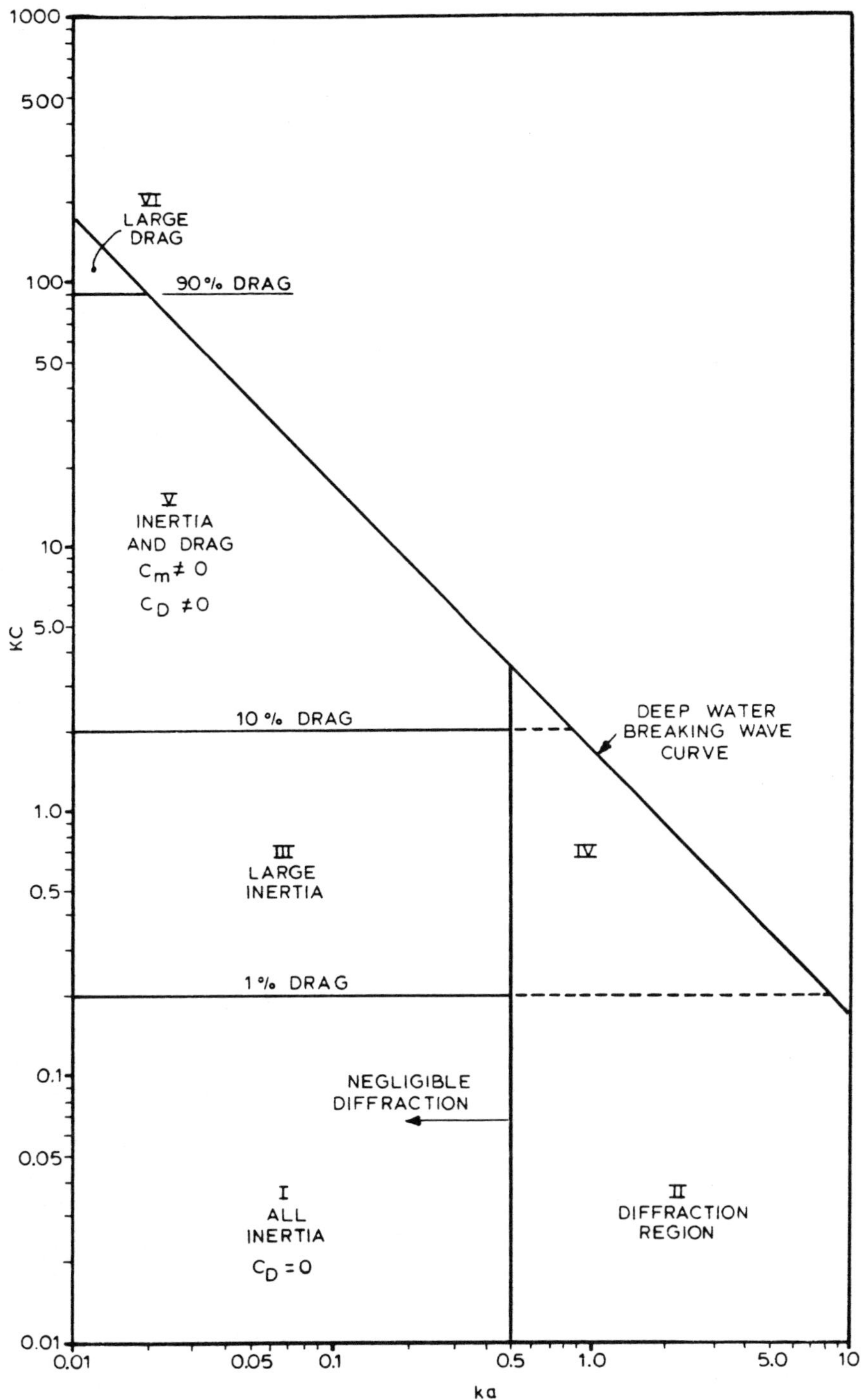

FIGURE 7. Regions of application of wave force formulas.

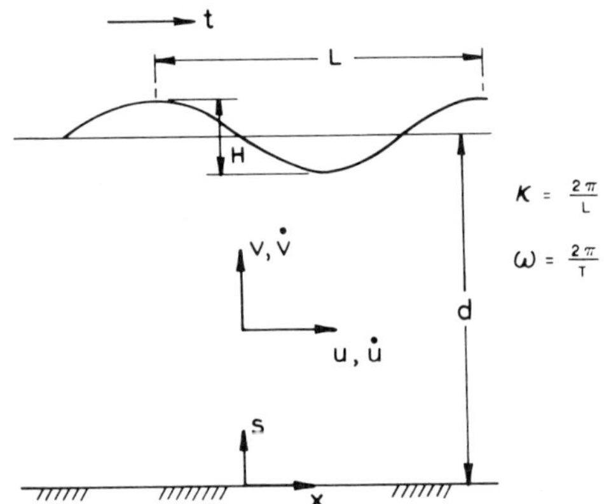

FIGURE 8. Definition sketch for wave in finite water depth.

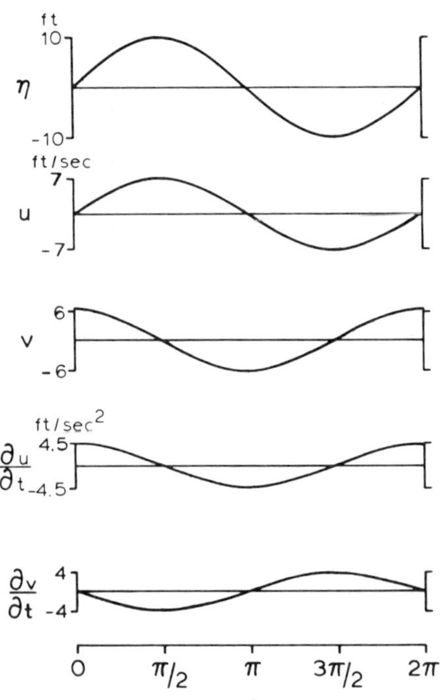

FIGURE 9. Example profiles of wave and kinematic properties by Linear theory.

An example of the water particle velocities and accelerations for a wave of height 20 ft. and period 10 sec. in a water depth of 100 ft. is given in Figure 9. The quantities are computed at the SWL at a station given by $x = 0$. Note that u, and $\delta v/\delta t$ are in phase with the wave profile while v and $\delta u/\delta t$ are out of phase by 90°. The wave length is obtained from the formula

$$L = \frac{gT^2}{2\pi} \tanh kd \tag{21}$$

From Equation (21) it is seen that the wave length, L, is dependent on T and d. Since $k = 2\pi/L$, the wave length cannot be computed directly, and an iterative technique may be used starting with the deep water value for the wave length. However, simpler approximate formula for the wave length may be used as follows

Approximation	Criteria	Wave Length Formula
Deep water	$d/L \geq 1/2$	$L_0 = gT^2/2\pi$
Intermediate water	$1/2 > d/L > 1/20$	$L = L_0[\tanh(2\pi d/L_0)]^{1/2}$
Shallow water	$d/L \leq 1/20$	$L = T\sqrt{gd}$

Based on the linear theory, the maximum inertia force and the maximum drag force may be computed from Equation (16) for given values of C_M and C_D. The most common values of these coefficients are $C_M = 2$ and $C_D = 1$. The ratio of the maximum drag to the maximum inertia and the associated phase angle of the maximum in-line force from Equation (16) and above values of the hydrodynamic coefficients are plotted in Figure 10. This figure shows the importance of the relative contribution of drag and inertia as a function of the quantities, u_0T/D also defined as the Keulegan-Carpenter number (KC). For example drag effect is negligible for $KC < 4$ while the inertia contribution is equal to drag for $KC \approx 16$.

Nonlinear wave theories, e.g., Stokes fifth-order theory [Skjelbreia and Hendricksen (1961)], stream function theory [Dean (1961)] and Cnoidal theory [LeMéhauté (1976)] are also used to compute forces in the Morison formula.

In addition to the in-line force, the oscillatory flow in the drag regime shown in Figure 7 also produces a transverse or lift force. The lift force is generated due to the unsymmetric wake (or pressure distribution around the structure) due to alternate eddy formation (and shedding) on one side of the structure or the other. The lift force is expressed analogous to the drag force as

$$f_L = \frac{1}{2} \varrho D\, C_L\, u^2 \tag{22}$$

where C_L is the lift coefficient. The lift force is irregular in nature and the lift coefficient is given in terms of an rms or a maximum value. Sometimes the right-hand side of Equa-

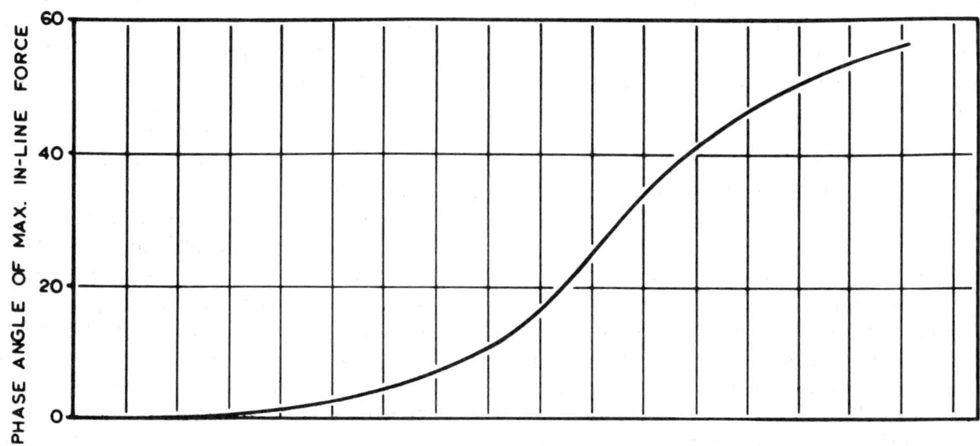

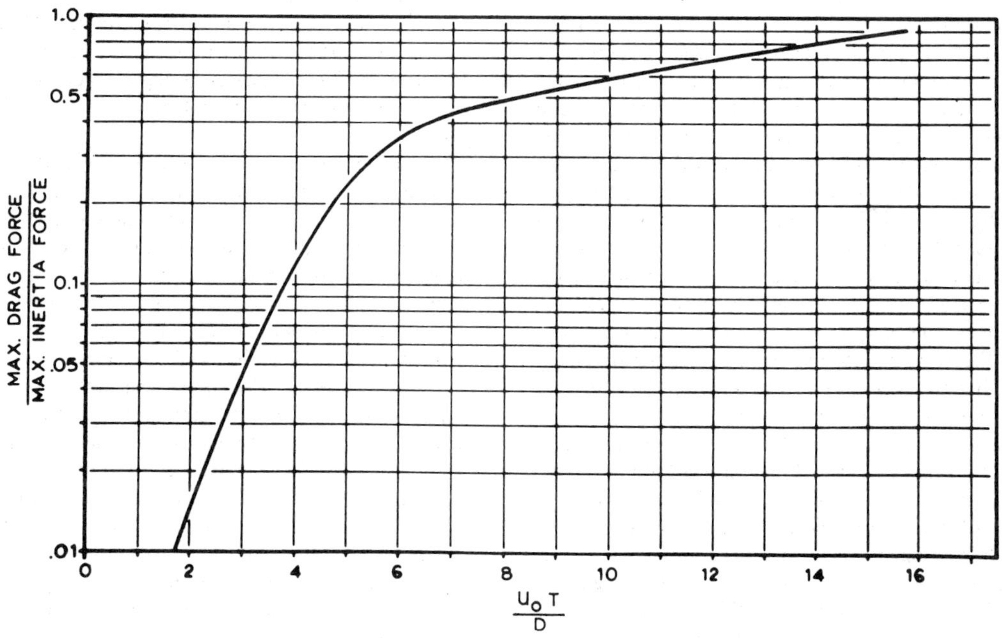

FIGURE 10. Ratio of maximum drag to maximum inertia and phase angle of maximum in-line force.

tion (22) is expressed in a Fourier series and the Fourier components of the lift coefficients are obtained from experiment.

It is clear from the above formulas for forces in Equations (16) and (22) that the hydrodynamic coefficients are either obtained from experiments in the laboratory with oscillatory flows past cylinders and cylinders placed in waves or derived from the ocean tests.

Laboratory Tests with Oscillating Cylinders and in Waves

Many tests have been performed in the past with cylinders. Sarpkaya (1976, 1977) tested horizontal cylinders in a sinusoidally oscillating water tunnel. These results clearly show the dependence of C_M, C_D and C_L on Reynolds number, Keulegan-Carpenter number and the surface roughness parameter. These tests have produced very promising results in establishing functional relationships of the hydrodynamic coefficients with small scatter. Results for C_M and C_D versus KC from Sarpkaya's experiments (1976) for different values of Re and $\beta = Re/KC$ for smooth circular cylinders are shown in Figures 11 and 12. Note that as C_M increases the value of C_D decreases and vice versa. This trend is typical of all experiments. Also, the most change in their values takes place for KC between about 8 and 15. This is the area where the fluid flow around the cylinder is least understood. In this area, C_M experiences an "inertia crisis" with a minima and C_D correspondingly exhibits a maxima. A clear picture of this phenomenon is evident in the plots of these coefficients vs. KC. The values of the lift force coefficient, C_L from these tests are shown in Figure 13. The lift coefficient corresponds to its maximum value in that the maximum value of force and velocity from the measured traces were used in computing C_L from Equation (22). Also shown in Figure 13 are the ratio of the lift force frequency to the oscillating water column frequency, f_r. Thus, $f_r = 2$ represents a lift force profile having twice the frequency of the oscillating frequency. Since lift force is generally irregular the frequency corresponds to the predominant force component. For smooth cylinders the value of C_D tends to approach 0.65 and C_M approaches 1.8 at higher values of Re. Similarly, the asymptotic value of C_L is approximately 0.25. For rough cylinders, these values experience almost a parallel shift. Results from Sarpkaya's (1977) tests are shown in Figure 14. The cylinders were artificially roughened with uniform sand grains of mean particle size K. The corresponding smooth cylinder values are also shown for various KC values. Note that the roughness parameter is defined as the ratio of mean particle size and cylinder diameter, K/D and varied in the test from 0.005 to 0.02.

In tests with horizontal cylinders, e.g., Sarpkaya's, the total force on the cylinder may be measured and it does not pose any problem in reducing this information into the coefficient values. However, in a test with a vertical cylinder, if only the total forces on the cylinder are measured, then only average values of the coefficients may be calculated and presented in terms of the average Re and KC as was done by many investigators [Isaacson and Maull (1976), Gaston and Ohmart (1979)]. The variation of the coefficient values over the length of the cylinder in this case remains unknown. While it is still possible to obtain qualitative information, application to an actual situation becomes difficult.

In a series of experiments with smooth vertical cylinder, Chakrabarti (1980) measured forces on small sections of the cylinder in waves. C_M and C_D were considered invariant over a wave cycle and obtained by a least squares technique using Morison's equation. The mean values of C_M and C_D are shown in Figure 15 as a function of KC over a range of Re between 2×10^4 and 3×10^4. Sarpkaya's results for the same range of KC and Re are superimposed here and show a general agreement albeit limited, except for C_M near $KC = 10$ and at higher KC values where Chakrabarti's data are sparse and need further verification. The data from wave tests, in general, had more scatter around the mean values. The KC varied between 0 and 50. Thus, the drag predominance is covered at the upper range of KC. Note that KC values cover the range experienced in a prototype situation. However, the large values of Re experienced by a cylinder in the ocean environment cannot be achieved in a laboratory test.

Gaston and Ohmart (1979) tested a one foot diameter vertical cylinder in regular waves for an Re range of $2\text{–}3 \times 10^5$. The specimen was tested as smooth and rough for three different roughnesses. The average value of C_D was found to double due to roughness while the C_M value increased only about 10% in the range of the test.

Chakrabarti (1982) used silica uniformly glued on cylinders to roughen the surfaces. Instrumented sections were used as before. The inertia and drag coefficients were computed using stream function theory for wave kinematics and a least squares technique. The cylinder roughness, K/D was varied from 0.002 to 0.02. The results (Figure 16) show that the drag coefficient increases in value with roughness while the inertia coefficient is relatively insensitive to the roughness of the cylinder.

From the experimental results of Sawaragi, et al. (1976) for the values of maximum (1/10th significant) lift coefficients versus rms KC values, simple empirical formulas for C_L were found as follows.

$$C_L = 0.245\,(KC + 1) \qquad KC \leq 9 \tag{23}$$

and

$$C_L = -0.155\,KC + 3.85 \qquad KC > 9 \tag{24}$$

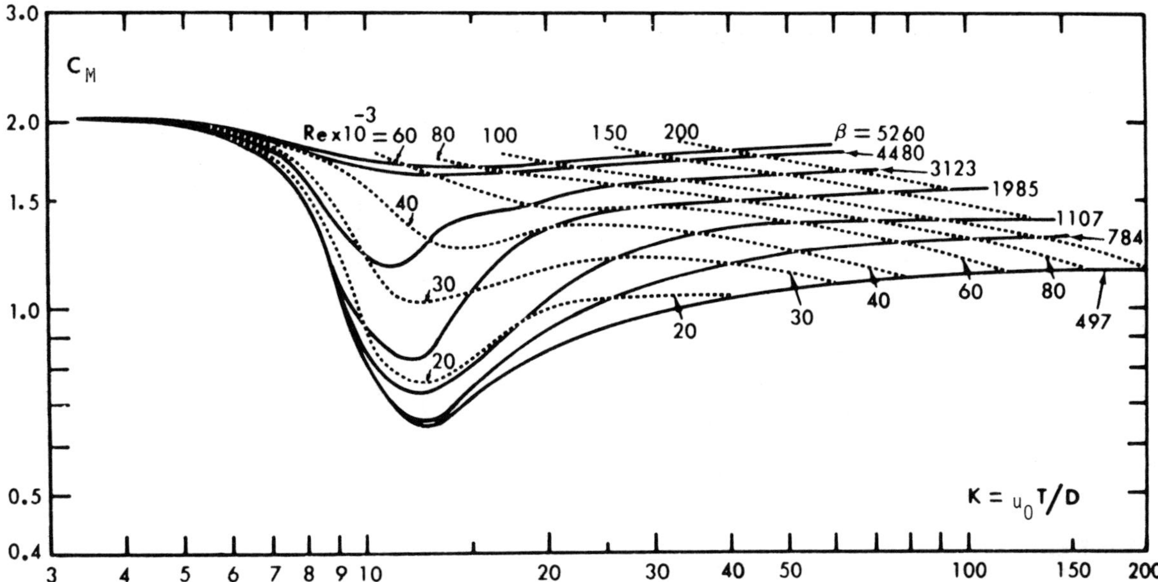

FIGURE 11. C_M versus KC number for various values of Re and β = Re/KC for smooth circular cylinder [Sarpkaya (1976)].

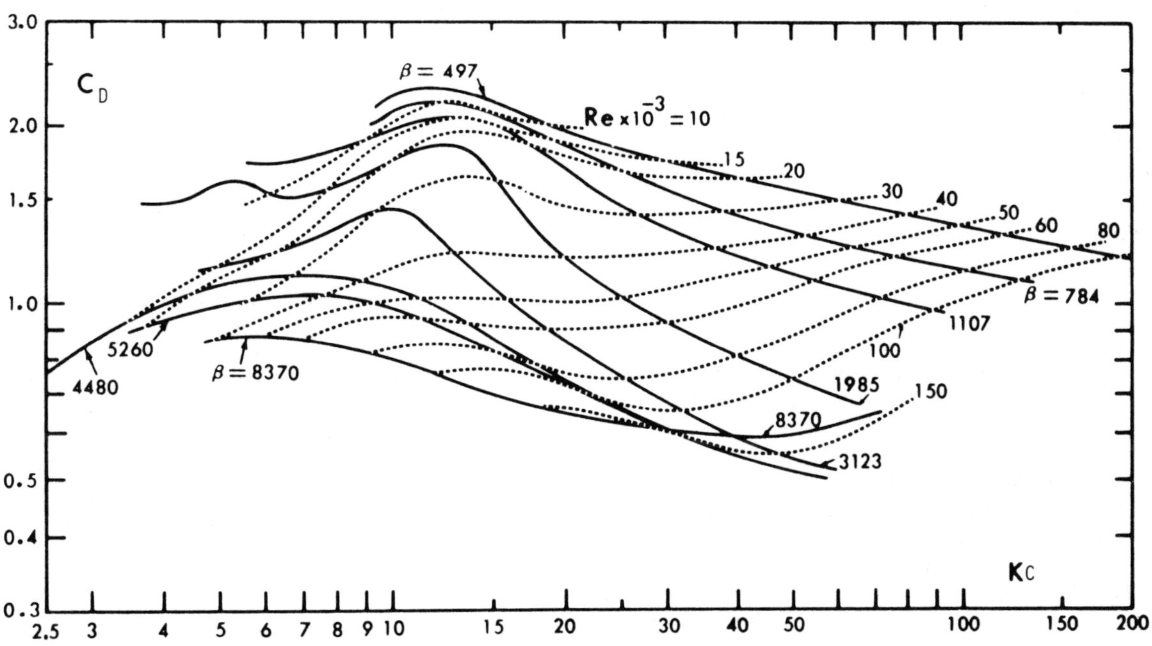

FIGURE 12. C_D versus KC number for various values of Re and β = Re/KC for smooth circular cylinder [Sarpkaya (1976)].

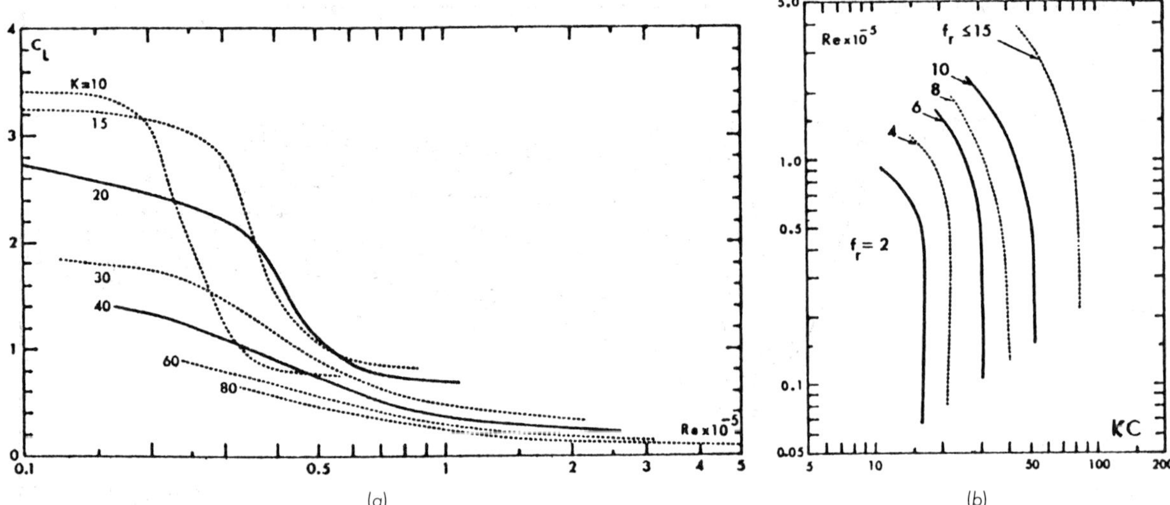

FIGURE 13. (a) Lift force coefficient, C_L versus Re and KC; (b) Ratio of lift force frequency and oscillating frequency, f_r versus Re and KC for a horizontal cylinder in an oscillatory flow [from Sarpkaya (1976)].

The phase angle of the maximum lift force was found to be quite random. But the phase difference between the in-line and transverse forces was not too large. For a conservative design, these two forces were suggested to be in-phase at maxima.

The mean maximum lift coefficients (based on the maximum value of the measured lift force) for the roughened cylinder obtained by Chakrabarti (1982) for three different roughnesses are shown in Figure 16. The lift coefficients did not change much with roughness. The scatter of the data about these mean values, however, was quite large. More consistent and extensive results on the maximum lift coefficients have been provided by Sarpkaya (1977) from his U-tube tests.

The time varying lift force being irregular in nature cannot be constructed from these values. Therefore, a predominant lift frequency may be assigned to the lift force profile (assuming regular). If the lift force profile is analyzed by a Fourier series or an FFT method, then the frequency at which the highest amount of energy is present is termed predominant lift frequency. It has been found that the ratio of the predominant lift frequency to the wave frequency is a function of *KC* and *Re* and can be related to the number of eddies formed and shed. The ratio for a smooth cylinder is near an integer number and increases with *KC* (Figure 13). There is some dependence on *Re* at the higher end of *Re* ($\geq 10^5$). The corresponding results from rough cylinders (Figure 17) are similar.

Alternatively, the lift force profile may be represented by a Fourier series and the Fourier components of the lift coefficients may be computed. The first five Fourier components of the lift coefficients have been presented (Figure 18) by Chakrabarti (1976). These values allow us to reconstruct the lift force profile if arbitrary phase angles, ϕ_n are assigned to each component.

$$f_L = \left(\frac{1}{2}\varrho D u_0^2\right) \sum_{n=1}^{5} C_{L_n} \cos(n\omega t + \phi_n) \tag{25}$$

The lift coefficient presented by Sarpkaya corresponds to the maximum lift force. It does not provide the lift force profile or its phase relationships with the in-line force. Therefore, the resultant force profile, e.g., Equation (25), cannot be constructed from this information. For this computation the Fourier components of lift force and their phase angles similar to the ones presented by Isaacson (1976) and Chakrabarti (1976) should be known. If a single lift coefficient is known, then the predominant lift frequency corresponding to this value may be obtained from Figure 13 and an approximate lift force profile may be constructed for this single frequency and C_L value from Equation (25).

If the lift force and the in-line force are vectorially combined then a resultant normal force may be computed:

$$f_R = \sqrt{f^2 + f_L^2} \tag{26}$$

The ratio of the maximum values of f_R and f was found to be about 1.4 by Sawaragi, et al. (1976) compared to 1.6 by Chakrabarti at a $KC \approx 15$. Thus, the lift force may substantially increase the maximum total force on a structural member and should be included in an offshore structure design.

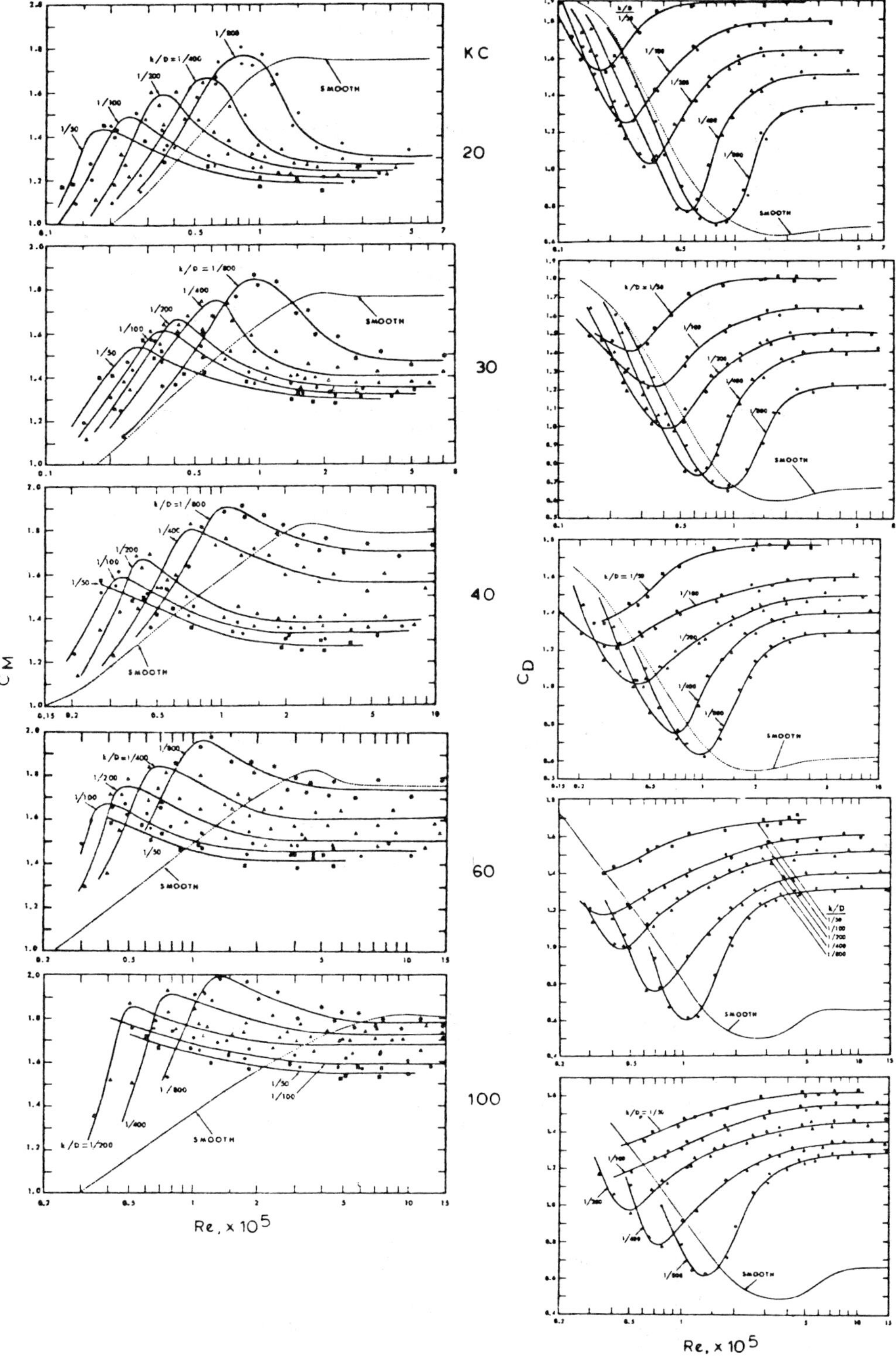

FIGURE 14. Values of C_M and C_D as functions of Re, KC, and roughness coefficient (K/D) [from Sarpkaya (1977)].

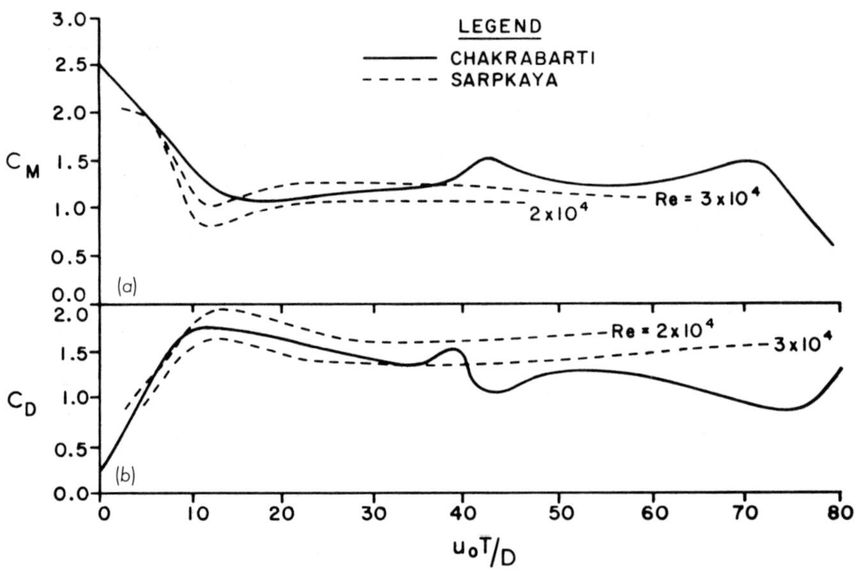

FIGURE 15. Comparison of wave tank test results with Sarpkaya's two-dimensional flow test results.

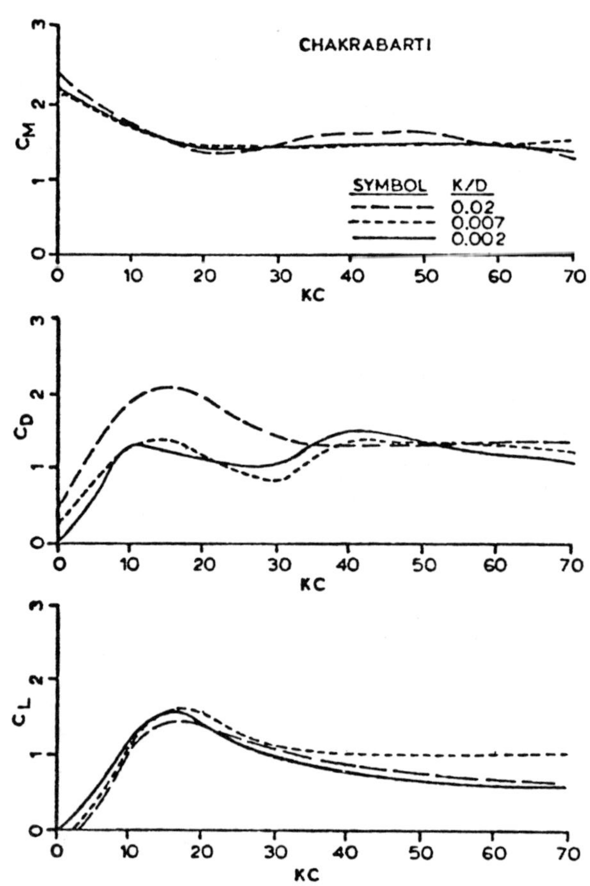

FIGURE 16. C_M, C_D and C_L values vs. KC from instrumented sections of a rough vertical cylinder.

Wave Forces on an Inclined Cylinder

The concept of the Morison equation can be extended to an inclined cylinder in waves. The force expression is written in terms of normal velocity, w and acceleration, $\dot{w}$:

$$\underset{\sim}{f} = C_M \varrho \frac{\pi}{4} D^2 \dot{\underset{\sim}{w}} + \frac{1}{2} \varrho C_D D |\underset{\sim}{w}| \underset{\sim}{w} \quad (27)$$

The force per unit length, $\underset{\sim}{f}$, is a vector quantity since $\underset{\sim}{w}$ and $\dot{\underset{\sim}{w}}$ are not necessarily in-line for all cylinder orientations. The horizontal and vertical forces on the cylinder are obtained by resolving $\underset{\sim}{f}$ in the horizontal and vertical directions when the velocity vector is written as

$$\underset{\sim}{w} = u_x \underset{\sim}{i} + u_y \underset{\sim}{j} + u_z \underset{\sim}{k} \quad (28)$$

and

$$\dot{\underset{\sim}{w}} = \dot{u}_x \underset{\sim}{i} + \dot{u}_y \underset{\sim}{j} + \dot{u}_z \underset{\sim}{k} \quad (29)$$

The normal velocity w is written as

$$\underset{\sim}{w} = \underset{\sim}{C} \times (u \underset{\sim}{i} + v \underset{\sim}{j}) \times \underset{\sim}{C} \quad (30)$$

where $\underset{\sim}{C}$ is the unit vector along the cylinder axis directed up or down so that

$$C = C_x \underset{\sim}{i} + C_y \underset{\sim}{j} + C_z \underset{\sim}{k} \quad (31)$$

In the spherical coordinate system,

$$C_x = \sin\phi\cos\psi, \; C_y = \cos\phi, \; C_z = \sin\phi\sin\psi \quad (32)$$

in which ϕ = angle of the cylinder axis to the vertical axis; ψ = angle of the cylinder projection to the x-axis. If the cross-products in Equation (30) are carried out, then the velocity components along x, y and z are

$$u_x = u - C_x(C_x u + C_y v) \quad (33)$$

$$u_y = v - C_y(C_x u + C_y v) \quad (34)$$

$$u_z = -C_z(C_x u + C_y v) \quad (35)$$

Therefore, the forces per unit length on a randomly oriented cylinder are calculated in the x (in-line), y (vertical) and z (transverse) directions respectively from the following expression:

$$f_x = \varrho C_M \frac{\pi}{4} D^2 \dot{u}_x + \frac{1}{2} \varrho C_D D |\underset{\sim}{w}| u_x \quad (36)$$

$$f_y = \varrho C_M \frac{\pi}{4} D^2 \dot{u}_y + \frac{1}{2} \varrho C_D D |\underset{\sim}{w}| u_y \quad (37)$$

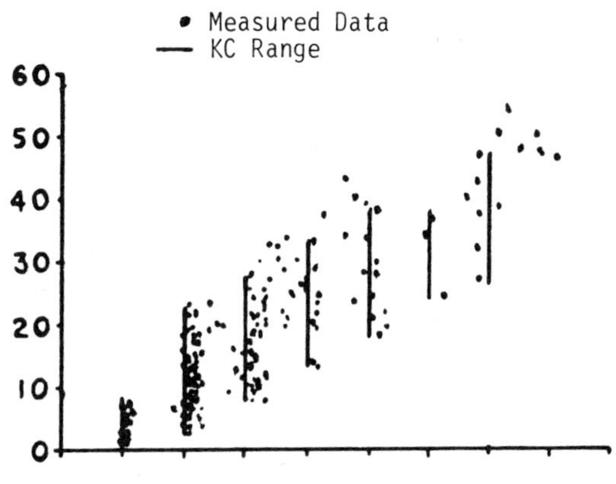

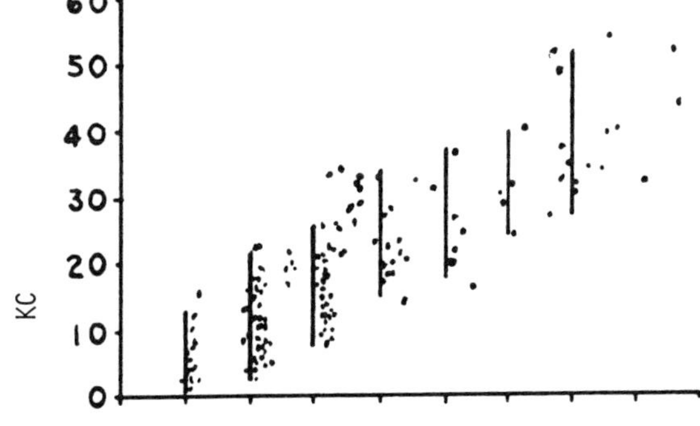

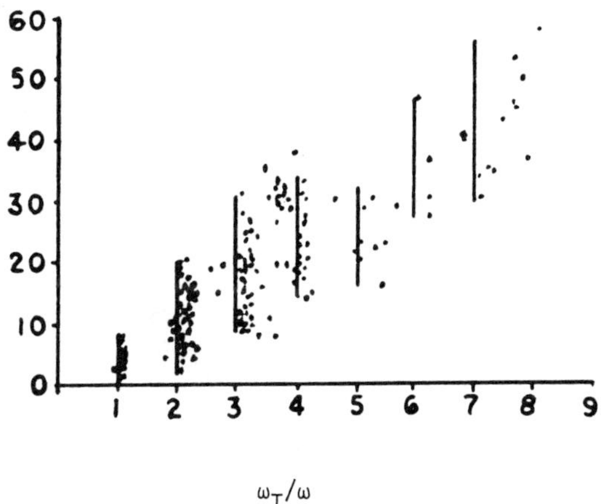

FIGURE 17. Ratio of predominant lift frequency to wave frequency vs. KC for rough cylinders in waves.

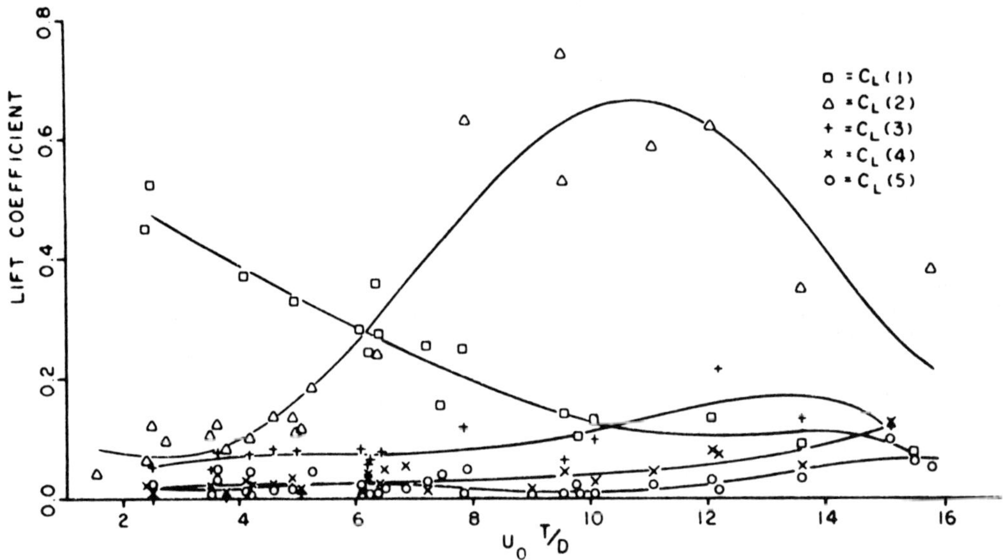

FIGURE 18. Components of C_L vs. KC from instrumented sections of vertical cylinder.

and

$$f_z = \varrho C_M \frac{\pi}{4} D^2 \dot{u}_z + \frac{1}{2} \varrho C_D D |\underset{\sim}{w}| u_z \quad (38)$$

Data on the C_M and C_D values on the inclined cylinders due to wave motions are limited. A few experiments were conducted on inclined members. Tests done in a wave tank at the writer's institution (1977) provided the values of C_M and C_D versus KC values for inclined members at various orientations as shown in Figure 19. Because of the limited and narrow range of Reynolds number, the dependence of C_M and C_D on Re could not be shown. The figure demonstrates that as long as normal components of velocity and accelerations are considered on an inclined cylinder the values of C_M and C_D are insensitive to the angle of inclination of the cylinder to the waves.

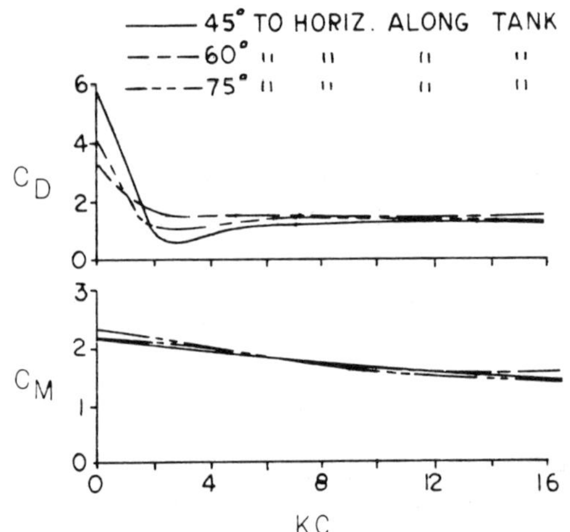

FIGURE 19. C_M and C_D values vs. KC from instrumented sections of an inclined cylinder.

During these tests the transverse forces on the cylinders were also measured. Plots of the two-component force profiles on these sections showed considerable increase in the resultant force. The ratio of the resultant and in-line forces is plotted in Figure 20 as a function of KC for all cylinder orientations tested. The plot indicates an increase of as much as 80% in the resultant force due to the presence of the transverse force.

Multiple Cylinders

Sarpkaya (1979) tested a group of cylinders in a circular array in his oscillating water tunnel. He found that C_M and C_D values did not depend on Re and presented them as functions of KC. As in his other tests, he found small scatter in the data. Others have tested oscillating cylinder arrays in still water; but the results are somewhat limited.

Chakrabarti (1979–1981) ran a series of tests with vertical cylinders in a linear array in waves. Two to five cylinders at different spacings, S/D (S = cylinder spacing, D = cylinder diameter), at 0° and 90° to the waves were tested. Results on C_M, C_D and C_L for the 3 and 5 cylinder arrays at different spacings are given in Figures 21 and 22. It is found that C_M and C_D substantially increase in value when the

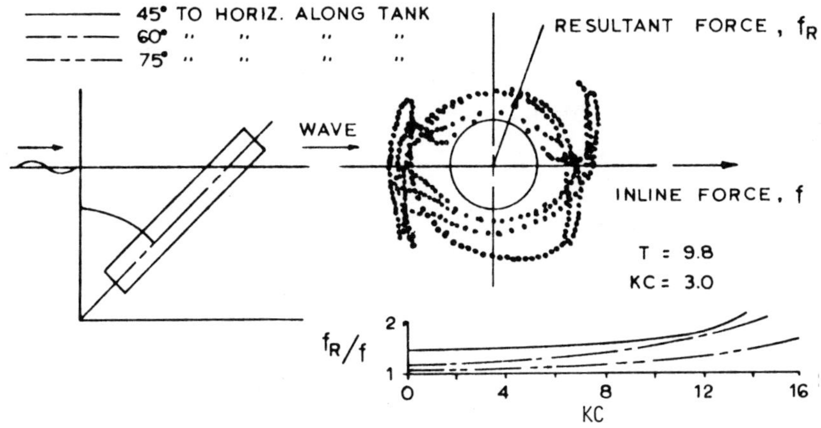

FIGURE 20. Ratio of resultant force to in-line force vs. KC for inclined cylinders.

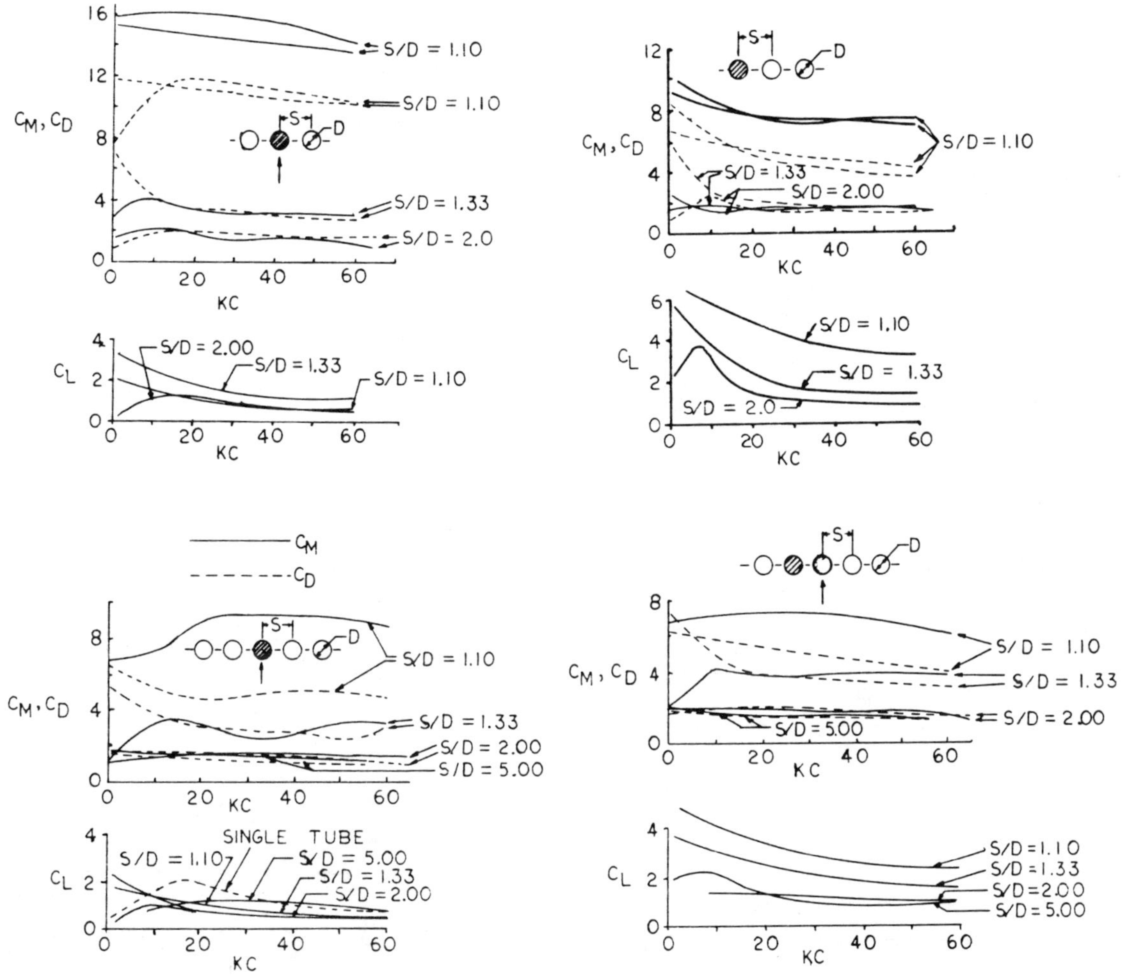

FIGURE 21. C_M, C_D and C_L values vs. KC and spacing parameters for vertical cylinder array normal to flow in a wave tank. (Note: Coefficients apply to shaded tubes.)

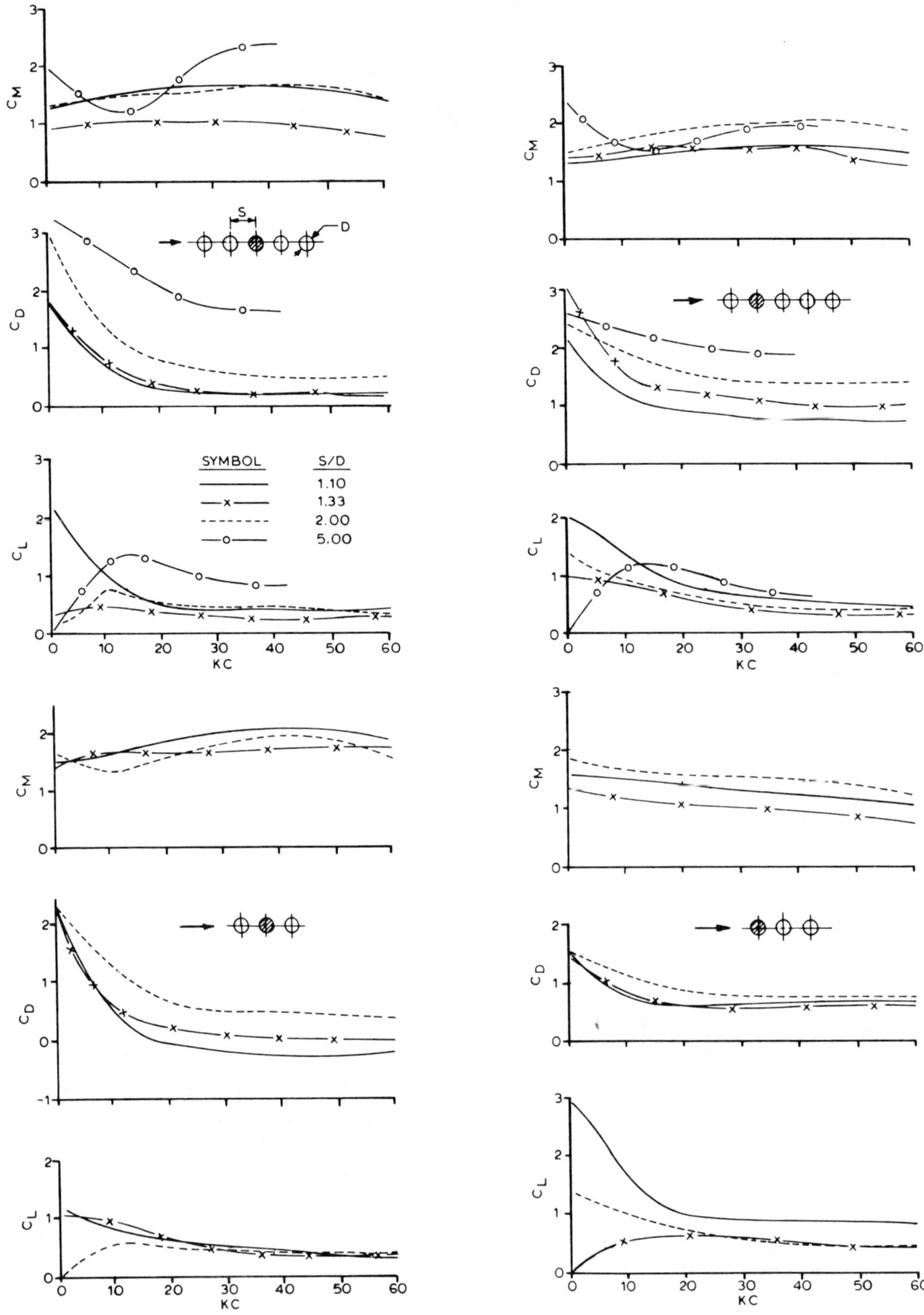

FIGURE 22. C_M, C_D and C_L values vs. KC and spacing parameters for vertical cylinder array in tandem in a wave tank. (Note: Coefficients apply to shaded tubes.)

cylinders are at 90° to the waves as the spacing among the cylinders is decreased. The reverse is found to be generally true for the lift coefficient. At 0°, the trend is quite the opposite. C_L values increase with the decrease in S/D while C_M and C_D values show a corresponding reduction in their values. As before, no dependence on *Re* could be established. However, if the coefficients are independent of *Re* as found by Sarpkaya for riser array, then these results are directly applicable to the prototype case.

Ocean Tests in Waves

It is clear from previous discussion that the laboratory tests while providing useful data fail to duplicate the prototype situation principally due to the limited range of Reynolds number. To circumvent this problem, from early 1950s many tests have been carried out in the field to measure the ocean wave forces on vertical piles (Table 1). The earlier tests did not make any attempts to measure the water particle kinematics because of lack of reliable instrumentations. In one such ocean test, forces on circular cylindrical piles from about 0.5 ft. to 5 ft. in diameter in about 50 ft. water depth were measured. The analysis of Wiegel, et al. (1957) found C_D to have a large scatter when plotted versus *Re*. An average value of C_M was found to be 2.5.

In September, 1952, wave forces on a 30-inch-pile were measured in the Gulf of Mexico in a 38 ft. water depth. Total reactions on the pile were measured. In the presence of a steady current of 0.55 ft/sec., Wilson (1965) obtained best-fit weighted mean C_M, C_D values in all waves of 1.53 and 1.79, respectively.

Wave Projects I and II were carried out in the Gulf of Mexico. Project I by Standard Oil Company of California used four instrumented piles 1 ft. to 4 ft. in diameter in 33 ft. water depth. During Project II, data from hurricane Carla (September 1961) on a 3.7-ft.-pile in 99 ft. water depth were recorded. The wave data from these projects have been analyzed by several investigators, e.g., by Dean and Aagaard (1970), Evans (1969) and Wheeler (1969). No measurements of the water particle kinematics were made. Different wave theories and analysis techniques produced different results. The data were re-analyzed by Dean, et al. (1974) under a joint industry project. The final results on C_M and C_D vs. *Re* are shown in Figure 23.

Ohmart and Gratz (1979) analyzed wave forces on a platform in the Gulf of Mexico in Hurricane Edith. The average values of C_M and C_D were 1.06 and 0.70. The best fit of peak forces yielded $C_M = 1.7$ and $C_D = 0.70$.

During a Christchurch Bay Project in the U.K. extensive testing was carried out with one instrumented large column and one wave staff with force sleeves. The water particle velocities were also measured simultaneously. Average values of C_M and C_D derived by Bishop (1979–1980) for a large and small instrumented column are shown versus *KC* in Figure 24. The reported variability of data about the average is attributed partly to the spatial separation of force and velocity measurements and to the vortex shedding.

A joint industry test on an Ocean Test Structure (OTS) was carried out in the Gulf of Mexico in which the local forces and water particle velocities were measured. The wave force transducers consist of circular member 16 in. O.D. and 32 in. long at a depth of 15 ft. from SWL. Heidemen, et al. (1979) obtained inertia and drag coeffi-

TABLE 1. Summary of Ocean Wave Force Measurements.

Year	Project	Location	KC	Range of RE $\times 10^{-5}$	Remarks
1952	Texas A&M University	Gulf of Mexico			Linear filter technique was used in data anlaysis; presence of current was estimated from data.
1953	Signal Oil & Gas Co.	off Davenport CA		0.3–9	Linear theory was used for kinematics; C_M & C_D computed when at crest, trough or SWL.
1953–1963	Wave Force Projects I & II	Gulf of Mexico		0.2–60	Stream function theory was used and two different directions (in-line and resultant) are analyzed.
1971–1974	CAGC Eugene Island	Gulf of Mexico		3–30	Water particle velocities were measured. A least squared method was used in the analysis assuming that the coefficients are independent of Re or KC.
1976	Christchurch Bay Tower	North Sea U.K.	2–40	2–40	Force records were analyzed by various means; more results are expected from this study in the future.
1977–1978	Ocean Test Structure	Gulf of Mexico	5–50	2–8	Methods of least square error over half cycle and short segments of f_I and f_D dominance were used for C_M and C_D calculation.

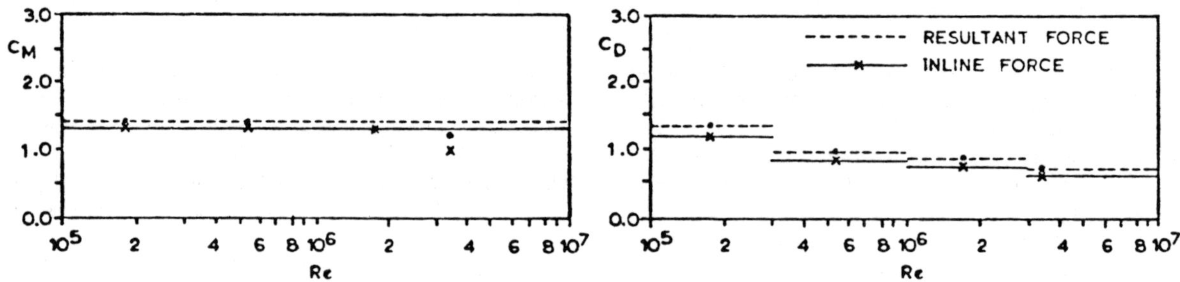

FIGURE 23. C_M and C_D values vs. Re from ocean wave projects [Dean, et al. (1974)].

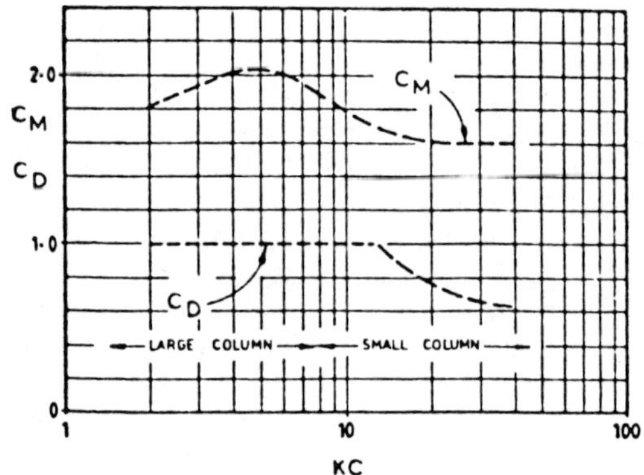

FIGURE 24. C_M and C_D values by mean square analysis vs. KC from Christchurch Bay Project [Bishop, et al. (1980)].

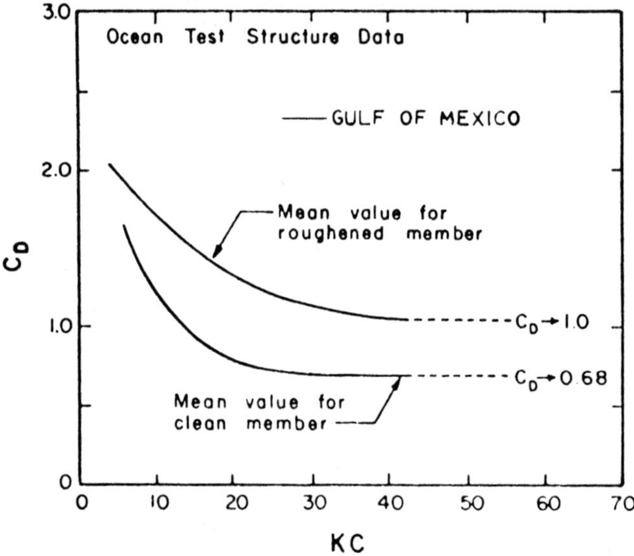

FIGURE 25. Mean C_D values vs. KC from OTS data [Heideman, et al. (1979)].

cients from these measurements for both clean and fouled specimens as functions of KC number. The results show an increase in average C_D with relative roughness and a corresponding decrease in C_M. The variation of C_D is shown in Figure 25 while average C_M values are given in Table 2.

Discussion on Hydrodynamic Coefficients

The values of the inertia, drag and lift coefficients on a vertical cylinder in the laboratory as well as field tests have been presented. In a controlled environment in the laboratory it has been possible to produce data with small scatter. On the other hand, the ocean tests have produced considerable scatter because of the random nature of waves, wave directionality, coupling with current, and measurement difficulty. The wave tank tests are limited in scope and cannot duplicate the ocean environment mainly because of the scaling problem. However, they help understand the physical processes of flow effects around a cylinder or a cylinder group in waves and the important parameters affecting the forces on it. Thus, the results from laboratory tests coupled with the ocean tests will provide the appropriate values of the coefficients for the forces on a member of an offshore structure required for its design. The coefficients obtained from measured forces on a small segment rather than averaged over the length of the cylinder are easier to interpret and apply. Many different methods of analysis for the computation of the coefficients are available. Most of them assume a constant (time-invariant) coefficient over a wave cycle. The least squares method of analysis is the most popular and, generally, preferable.

No set values of C_M, C_D and C_L for a prototype case can be prescribed with confidence with the present knowledge on them. Designers should be extremely careful when applying test results to actual design problems. Different companies and certifying agencies have their own rules in establishing these values for a certain case. Generally, for vertical cylindrical piles, the value of C_M is varied between 1.5–2.0 while C_D is considered between 0.5–1.0.

The API guideline (1979) recommends use of Morison formula for performing the wave force computation on a small member. Typical values of C_M and C_D are given as 1.5 to 2.0 and 0.6 to 1.0, respectively. The British Standard Institution (1979) guidelines are as follows. Morison's equation is used for wave loading with appropriate values of the coefficients. For Stokes fifth-order theory, $C_D = 0.8$ in the splash zone and $= 0.6$ elsewhere and $C_M = 2.0$ unless found lower by diffraction analysis are used. If Airy wave theory is used, $C_D = 1.0$. In case of large cones and pads in the transition area of a steel structure, a diffraction analysis of the complete leg or bottle is recommended for the C_M value. For low-taper members, $C_M = 2.0$. Vertical forces may be computed by the Froude-Krylov pressure. Vortex shedding effects on risers or conductor pipes should be investigated. Wave slam on a horizontal member near the MWL is determined as a drag force using $C_D = 3.5$.

TABLE 2. Inertia Coefficients from OTS Data [Heideman, et al. (1979)].

		Analysis Method			
		I		II	
Specimen Condition	K/D	C_M	St. Dev.	C_M	St. Dev.
Clean (smooth)	0.00	1.51	0.31	1.65	0.28
Fouled (rough)	0.03	1.25	0.34	1.43	0.35

The DNV guidelines (1977) recommend that the wave loads be computed by Morison equation for $ka < 0.2\pi$ where a is the radius (characteristic dimension) of the body. The values of C_M for two- and three-dimensional bodies have been given from potential theory. The C_D values for smooth and rough cylinders in steady flow are plotted. In oscillating flow, $C_D \geq 0.7$. The coefficients are considered functions of Re, KC and roughness in waves. The slamming coefficient should be at least 3.0 for horizontal circular cylinders in the splash zone. The vortex shedding frequency should be determined from Strouhal number (St) which is given as a function of Re. Formulas for the vortex shedding induced forces and their dynamic amplification due to resonance have been given.

The Norwegian Petroleum Directorate (1977) makes similar recommendations. In computing wave loads, appropriate linear or nonlinear wave theories should be used. For large structure, e.g. having cross-sections larger than 15% of the wave length, diffraction theory should be applicable. For smaller members, the Morison equation based on relative velocity including current and structure motion should be employed. The values of C_D, C_M and C_L are functions of Re, KC, St, roughness and proximity to other members or boundaries. The minimum value of C_D for smooth cylinder is 0.9 and of C_L for $Re < 3 \times 10^5$ and $KC < 60$ is 0.2. For high Re and KC numbers, steady current values for C_D and C_L are recommended.

Froude-Krylov Theory

Limited information is available for C_M and C_D values for shapes other than a circular cylinder. For these objects whose sizes are large enough that the drag force is small compared to inertia, while the diffraction effect is not significant the F-K theory is applicable.

The Froude-Krylov theory is based on the pressure-area method. The pressure is obtained from an incident wave theory. Based on linear wave theory, the expression of dynamic wave pressure is given by

$$p = \varrho g \frac{H}{2} \frac{\cosh ks}{\cosh kd} \cos(kx - \omega t) \tag{39}$$

For a submerged structure this pressure acts normal to the

TABLE 3. Summary of Wave Forces on Basic Structures.

Number	Object	Wave Force: Horizontal	Wave Force: Vertical
1	Hemisphre	$C_H \varrho V [\dot{u}_0 + S_1(ka)\omega v_0]$	$C_v \varrho V[\dot{v}_0 + S_2(ka)\omega u_0]$
2	Sphere	$C_H \varrho V \dot{u}_0$	$C_v \varrho V \dot{v}_0$
3	Horizontal half-cylinder	$C_H \varrho V[\dot{u}_0 + S_3(ka)\omega v_0]$	$C_v \varrho V[\dot{v}_0 + S_4(ka)\omega u_0]$
4	Horizontal cylinder	$C_H \varrho V \dot{u}_0$	$C_v \varrho V \dot{v}_0$
5	Rectangular block	$C_H \varrho V H\left(\frac{kl_1}{2}\right) E\left(\frac{kl_3}{2}\right) \dot{u}_0$	$C_v \varrho V H\left(\frac{kl_1}{2}\right) E\left(\frac{kl_3}{2}\right) \dot{v}_0$
6	Vertical cylinder	$C_H \varrho V B(ka) F\left(\frac{kl}{2}\right) \dot{u}_0$	0
7	Horizontal circular plate[1]	0	$C_v \varrho \frac{\pi R^2}{k} B(ka)\omega u_0$

$$H(\alpha) = \frac{\sin \alpha}{\alpha}$$

$$E(\alpha) = \frac{\sinh \alpha}{\alpha}$$

$$B(\alpha) = \frac{2J_1(\alpha)}{\alpha}$$

$$S_1(\alpha) = 3 \sum_{n=0}^{\infty} \frac{2^n n!}{(2n)!} \alpha^{n-1} J_{n+2}(\alpha)$$

$$S_2(\alpha) = 3 \sum_{n=0}^{\infty} \frac{2^n n!}{(2n)!} \alpha^{n-2} J_{n+1}(\alpha)$$

$$S_3(\alpha) = \frac{2}{\pi}\left[\frac{\cos \alpha}{\alpha} - \frac{\sin \alpha}{\alpha^2} + Si(\alpha)\right]$$

$$S_4(\alpha) = \frac{2}{\pi}\left[\frac{\cos \alpha}{\alpha} - \frac{\sin \alpha}{\alpha^2} + Si(\alpha)\right]$$

$$Si(\alpha) = \int_0^{\alpha} (\sin \beta/\beta)d\beta$$

V = structure volume

l_1 = block length

l_3 = block height

a = structure radius

ω = wave frequency $(2\pi/T)$

u,v = horizontal and vertical velocity

[1]Assumed very thin and the vertical force is experienced on one side.

TABLE 4. Inertia Force Coefficients for Basic Structures.

Basic Structure	Force Coefficients Horizontal	Force Coefficients Vertical	Range of $\pi D/L$
Hemisphere	1.50	1.10	0–0.8
Sphere	1.50	1.10	0–1.75
Horizontal Halfcylinder	2.00	1.10	0–1.0
Horizontal Cylinder	2.00	2.00	0–1.0
Rectangular Block	1.50	6.00	0–5.0

D = characteristic dimension; L = wave length

surface. The total force on the structure in a particular direction is obtained by integrating the component of this pressure in that direction over the submerged portion of the structure. The expressions for the horizontal and vertical force components in the x and y directions are written in integral form as

$$F_x = C_H \int\int_s p dA_x \tag{40}$$

and

$$F_y = C_v \int\int_s p dA_y \tag{41}$$

in which C_H and C_V = the horizontal and vertical force coefficients, and dA_x and dA_y = the projected areas in the x and y directions, respectively, of an elemental surface area of the submerged structure. Closed form expressions for a few basic structures are given in Table 3. Note that only the values of the kinematics by the linear wave theory and the structure volume are needed to compute the forces. The recommended horizontal and vertical (inertia) force coefficients for them are included in Table 4. If the range of $\pi D/L$ values is much larger than these ranges then these constant values are not applicable since the diffraction effect will alter these values depending on the size of the object. In these cases, the complete linear diffraction theory should be applied to arrive at a total force on these objects.

The results for subsurface spheres are scarce. The ideal value of C_M for sphere in potential flow is 1.5. The only tests on small spheres in waves were carried out by Grace (1978). Recently, he extended his earlier work with spheres and recommended values of inertia and drag coefficients for a sphere at low Reynolds number as follows: $C_M = 1.21$ and $C_D = 0.4$. At high Reynolds number, the steady state drag coefficient for a sphere is $C_D = 0.5$ for $Re < 5 \times 10^6$ and $C_D = 0.1$ beyond.

Free Surface Effects

When the wave height is large relative to the water depth, the effect of the changing free surface at the cylinder near the SWL on the total wave forces becomes significant. Due to the flow around the cylinder, there is usually a run-up at the "front" of the cylinder and a corresponding drawdown "behind" the cylinder.

For a large diameter cylinder, this surface effect can be considered as follows. Since the linear theory predicts the pressure up to the still water level and the pressure at the free surface is atmospheric, a simple method is to assume a linear decay of pressure from the still water level to the free surface as shown in Figure 26. For the drawdown region (below SWL), however, the pressure is calculated by the linear diffraction theory up to the free surface on the assumption that the pressure quickly goes to zero from a point close to free surface as it reaches the free surface (Figure 26).

Where the diffraction effect is small and drag effect is present, Hogben (1974) suggested extending the linear wave theory up to the free surface beyond the still water level in a hyperbolic fashion. Tests in a wave basin with fixed gravity structures have indicated that this method overestimates the total forces and particularly the overturning moment. Extending the water particle kinematics exponentially to the free surface is extremely conservative.

To obtain values up to the free surface of a linear wave, the kinematics, e.g., the horizontal water particle velocity, are sometimes extrapolated from the mean water level to the free surface with the same value as the mean water level. The other extrapolation of kinematics to the free surface consists of linearly extending the kinematics from the mean water level at the same slope as obtained at the mean water level. These two methods of extrapolation should be applied with caution as they could lead to conservative results. This is particularly true for the overturning moment calculation. Moreover, for random seas, the extrapolation method gives even higher values of forces since the kinematics of the wave components of the random wave are added together.

Therefore, a stretching method for the kinematics is recommended if they require extension up to the free surface. If a stretched approach is used to compute the kinematics, the kinematics at the free surface is considered identical to those originally calculated for the mean water level. The remaining kinematics between the free surface and the

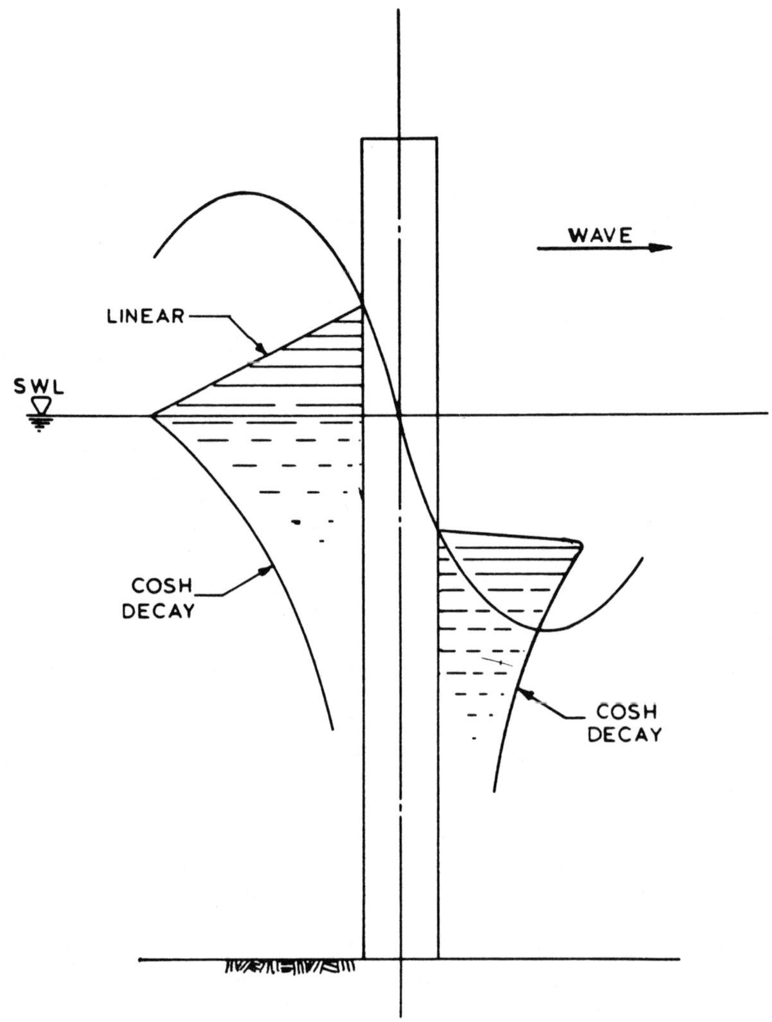

FIGURE 26. Approximation of pressure profiles up to free surface in a diffraction regime.

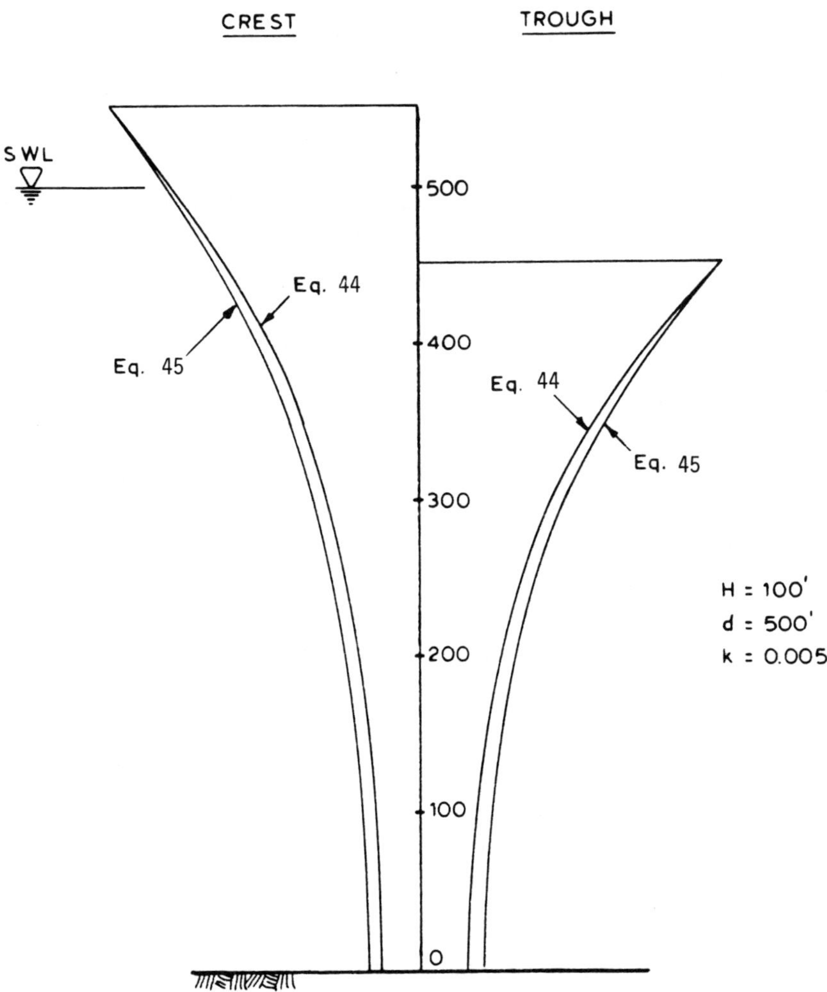

FIGURE 27. Vertical distribution of the stretched horizontal water particle velocity up to the free surface at the crest and trough.

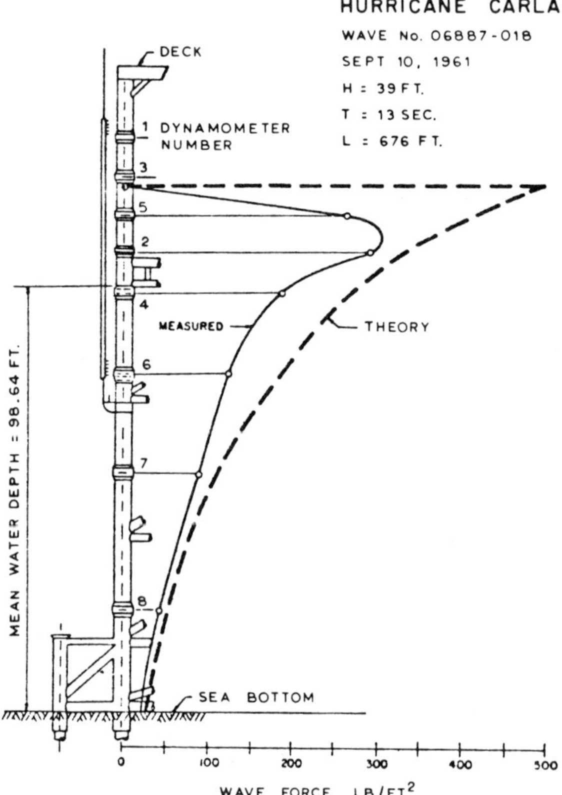

FIGURE 28. Comparison between measured and computed (by Stokes 5th order theory) dynamic pressures on a pile up to free surface [from Bea (1978)].

seafloor assume a hyperbolic cosine function so as to maintain the total velocity and acceleration over the depth of the water column. Different modifications have been suggested for Airy theory. For example:

$$u = \frac{gkH}{2\omega} \frac{\cosh ks}{\cosh k\,(d + \eta)} \cos\,(kx - \omega t) \tag{42}$$

where η is the instantaneous profile of the wave given by

$$\eta = \frac{H}{2} \cos(kx - \omega t) \tag{43}$$

Another stretching form for u is

$$u = \frac{gkH}{2\omega} \frac{\cosh ks \left(\dfrac{d}{d + \eta}\right)}{\cosh kd} \cos(kx - \omega t) \tag{44}$$

In these expressions, the quantity $(d + \eta)$ is assumed constant when deriving the expressions for the water particle accelerations. These two forms in Equations (43) and (44) have quite similar magnitude. An example of the horizontal velocities under the crest and trough by these formulas is shown in Figure 27. For nonlinear waves, e.g., Stokes fifth-order theory, the forces calculated up to the free surface as suggested by the theory are found to be conservative. Data obtained from the measured pressures on a structure in the Gulf of Mexico verifies this trend (Figure 28).

Diffraction Theory

When the structure is large, compared to the wave length, Morison's formula is no longer applicable. Figure 7 demonstrates (for a vertical cylinder) in what region the formula fails. In this case the diffraction of waves from the surface of the object should be considered. For a bottom-mounted, surface-piercing vertical cylinder, the expressions for the wave forces are known in a closed form. For a ship-shaped structure or long horizontal cylinders, a two-dimensional strip theory method is commonly applied, e.g., in naval hydrodynamic problems. For an arbitrarily-shaped structure, a three-dimensional boundary integral method or a finite element method is used.

VERTICAL CYLINDER

For a surface piercing vertical circular cylinder, the net force in the x-direction (direction of wave propagation) per unit axial length is obtained on integration of the pressure around the cylinder.

$$f = \frac{2\varrho gH}{k} \frac{\cosh ks}{\cosh kd} A(ka)\cos(\omega t - \alpha) \tag{45}$$

where a = cylinder radius (= $D/2$) and

$$A(ka) = \frac{1}{[J_1'^2(ka) + Y_1'^2(ka)]^{1/2}} \tag{46}$$

and

$$\alpha = \tan^{-1} \frac{J_1'(ka)}{Y_1'(ka)} \tag{47}$$

The horizontal force per unit length may be, equivalently, written as the inertia part of Morison's equation:

$$f = C_M \varrho \frac{\pi}{4} D^2 \dot{u}_\alpha \tag{48}$$

in which

$$\dot{u}_\alpha = \frac{gHk}{2} \frac{\cosh ks}{\cosh kd} \cos(\omega t - \alpha) \tag{49}$$

water particle acceleration at an elevation s from the bottom at a phase lag of α; and C_m = the effective inertia coefficient given by

$$C_M = \frac{4A(ka)}{\pi(ka)^2} \tag{50}$$

Note that unlike Morison equation or F-K force a phase shift in the force is caused by the diffraction of waves from the surface of the cylinder. The values of C_M are plotted against ka in Figure 29. Note that for small cylinder radius compared to the wave length ($ka \leq 0.1$), the value of C_M equals 2.0. The value of C_M increases to above 2.1 for up to about $ka = 0.3$ before it continuously decreases reaching a value of 0.2 near $ka = 4.0$. Also shown in the same plot are the values of the diffraction coefficient, C_H. This quantity is evaluated by taking the ratio of the maximum diffraction force given by Equation (45) and the maximum Froude-Krylov force (Table 3) which assumes no diffraction of waves in the presence of the object. Thus, it may be shown that

$$C_H = \frac{2A(ka)}{(\pi ka)J_1(ka)} \tag{51}$$

A few interesting points may be noted from Figure 29. The values of C_M and C_H are about the same (near 2.0) for small values of ka. This is because, for small bodies the amount of scattering of waves is small and thus the diffraction coefficient approaches the steady potential flow value. For the values of ka between 1.7 and 2.6 the value of C_H is nearly equal to 1.0. Thus, in this range of ka values, the diffraction of waves from the body is such that the net force approaches the Froude-Krilov force even for a relatively large body. In fact, the total force for part of this region is slightly lower than the Froude-Krilov force.

The values of the phase angle α from Equation (47) are plotted against ka in Figure 30. The quantity, α is a measure of the phase shift of the maximum horizontal force on the cylinder, Equation (45) with respect to the zero cross-over of the wave profile at the cylinder center line.

The total horizontal force on the cylinder is obtained by adding the force per unit length at each elevation over the entire length of the cylinder under water.

$$F = \varrho g H a^2 \frac{2A(ka)}{(ka)^2} \tanh kd \cos(\omega t - \alpha) \tag{52}$$

Similarly the overturning moment on the cylinder about its bottom center is derived by integrating the moment per unit length over the water depth.

$$M = \varrho g H a^3 \frac{2A(ka)}{(ka)^3} [kd \tanh kd - 1 + \operatorname{sech} kd] \cos(\omega t - \alpha) \tag{53}$$

SUBMERGED HALF CYLINDER

Based on an approximate deep submergence analysis, the net horizontal force (in the x direction) for a bottom seated halfcylinder of length ℓ is given by

$$F_x = 2\varrho V\dot{u} \tag{54}$$

where $V = (\pi/2)a^2\ell$ is the displaced volume (a = radius, ℓ = length of halfcylinder) and $\dot{u}$ is the water particle acceleration at the centerline (origin) of the halfcylinder given by

$$\dot{u} = \frac{gHk}{2\cosh kd}\cos\omega t \tag{55}$$

This expression for F_x is equivalent to Morison's inertia term, the effective inertia coefficient being 2.

The net vertical force (in the y direction) is given by

$$F_y = \varrho a\ell \frac{gH}{\cosh kd} C_1(ka)\cos\omega\tau \tag{56}$$

in which

$$C_1(ka) = \cos ka + \frac{\sin ka}{ka} + kaSi(ka) - 1 \tag{57}$$

and $Si(ka)$ denotes the sine integral,

$$\int_0^{ka} \frac{\sin\beta}{\beta} d\beta.$$

BOTTOM SEATED HEMISPHERE

Under the same assumption of deep submergence as before, the horizontal force (in the x direction) based on the diffraction theory is given by

$$F_x = 1.5\,\varrho\,V\,\dot{u} \tag{58}$$

where $V = 2\pi a^3/3$, the displaced volume of the hemisphere (a = radius of hemisphere) and $\dot{u}$ is the water particle acceleration at the center of the hemisphere given by Equation (55). This equation is equivalent to Morison's inertia term with an effective inertia coefficient of 1.5. The net vertical force (in the y direction) is

$$F_y = \varrho\pi a^2 \frac{gH}{2\cosh kd} C_2(ka)\cos\omega t \tag{59}$$

in which $C_2(ka)$ can be written in terms of the Bessel functions as

$$C_2(ka) = J_0(ka) + \left(ka - \frac{1}{ka}\right)\int_0^{ka} J_0(\varepsilon)d\varepsilon - \left(ka - \frac{2}{ka}\right)J_1(ka) \tag{60}$$

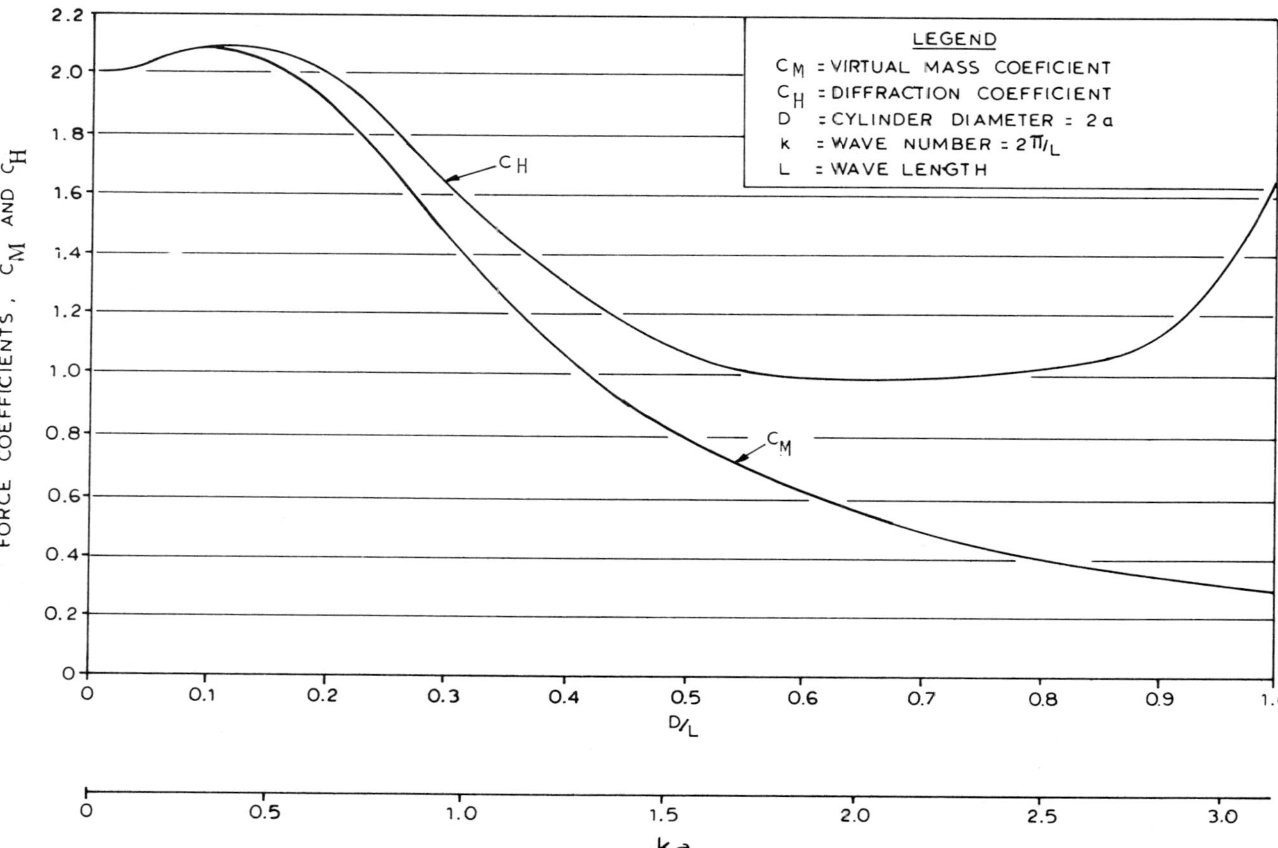

FIGURE 29. Effective inertia coefficient for a large vertical cylinder.

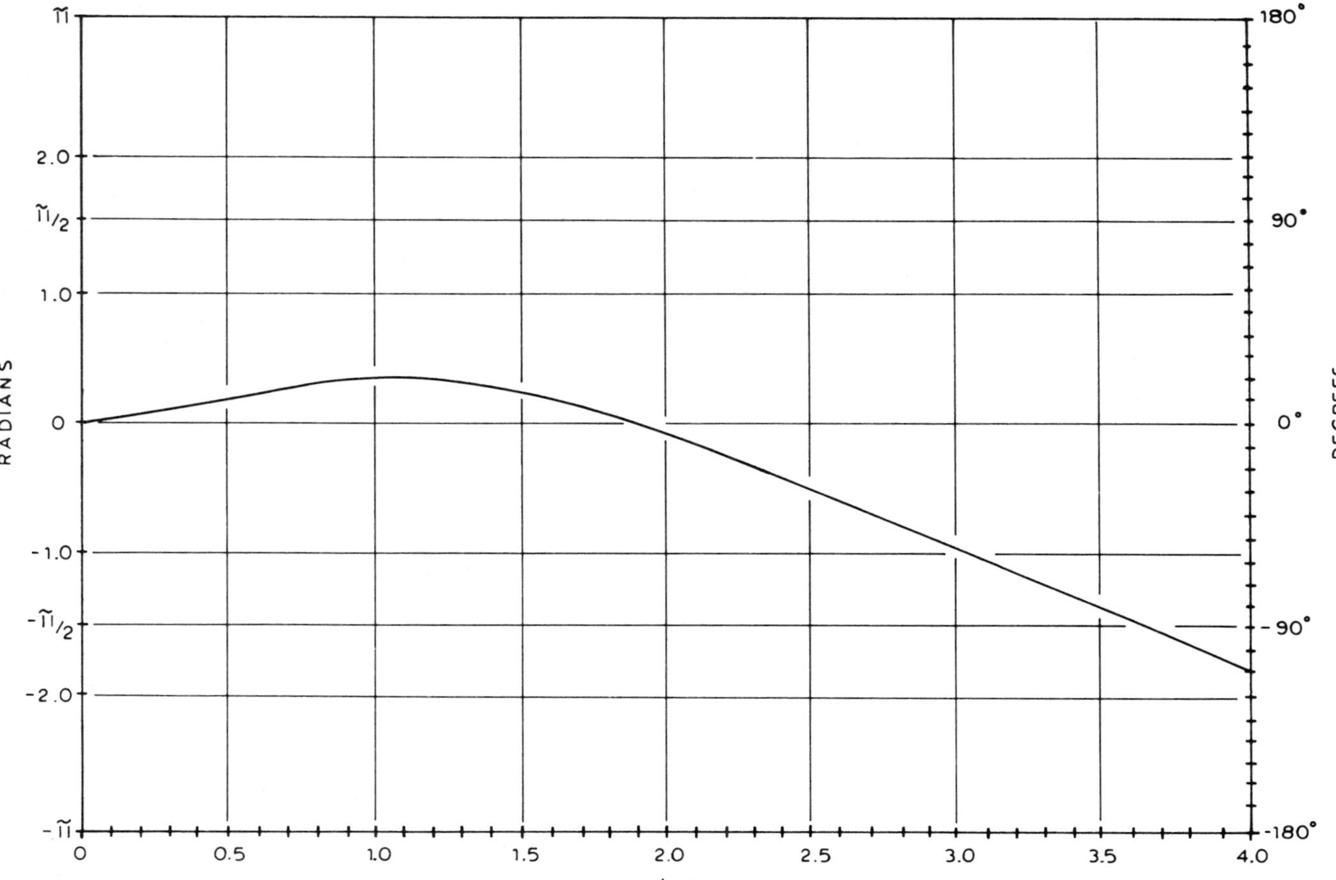

FIGURE 30. Phase shift of maximum force by wave diffraction on a vertical cylinder.

TABLE 5. Values of C_1 and C_2 versus ka.

ka	C_1	C_2
0.1	1.003	1.002
0.2	1.013	1.008
0.3	1.030	1.019
0.4	1.053	1.033
0.5	1.083	1.052
0.6	1.119	1.075
0.7	1.162	1.101
0.8	1.211	1.132
0.9	1.266	1.167
1.0	1.328	1.205
1.1	1.395	1.248
1.2	1.469	1.294
1.3	1.548	1.343
1.4	1.633	1.397
1.5	1.723	1.453
1.6	1.818	1.514
1.7	1.919	1.577
1.8	2.024	1.644
1.9	2.135	1.713
2.0	2.249	1.786
2.5	2.885	2.189
3.0	3.603	2.649
4.0	5.190	3.677
5.0	6.842	4.763

Values of C_1 and C_2 are tabulated against ka in Table 5. Note that for $ka > 3$ the variation of both $C_1(ka)$ and $C_2(ka)$ is nearly linear.

Theoretically, the asymptotic formulas apply as d/a approaches infinity. The accuracy of the results for finite values of d/a can be ascertained by a comparison with the results of the complete linear diffraction problem. This comparison for the hemisphere for the horizontal and vertical forces for $d/a = 2$ and 3 is shown in Figures 31 and 32. It shows that for $d/a \geq 2$ the approximate expressions give results with better than 5% accuracy. Similarly, for the horizontal halfcylinder, the approximate expressions are valid for $d/a \geq 4$.

COMPLETE 3-D NUMERICAL SOLUTION

The solution for the common geometries and the approximate expression are not suitable for offshore structures of composite shapes. In this case the complete boundary value problem is solved numerically. There are many numerical techniques available. The basic flow in all cases of the complete boundary value problem assumes oscillatory, incompressible and irrotational flow. Thus, the fluid velocity is represented as the gradient of a scalar potential. The total potential is represented as the sum of an incident and a scattered potential. The incident potential is known from the linear wave theory. The problem is posed to solve for the scattered potential on the surface of the structure. Generally, the structure or the flow field including the structure is divided into numerous segments.

The computation requires a large, high-speed computer with adequate core capacity. The numerical results from this theory have been tested by many laboratory and field experiments. They provide reliable results except in cases of objects with sharp edges, e.g., a square caisson, etc. Examples of the correlation on forces on fixed bodies are shown in Figure 33.

Often a small gap exists between the offshore structure and the ocean bottom. For large open-bottom structures, the linear diffraction theory shows a reduction in the total horizontal and vertical forces. The reduction in the horizontal force is small and can be neglected in a design. The vertical forces, in particular, are an order of magnitude smaller over the practical range of wave periods. The inside pressures of a slightly open structure can be represented by the mean bottom pressure of the corresponding sealed structure. In this case, the vertical force is reduced by this mean pressure times the horizontal projection of the bottom of the structure.

One of these pressures is the standard mean obtained by integrating the pressure around the bottom of the structure and dividing by the circumference. Another mean pressure can be obtained by the simple arithmetic mean of the pressures at the four stagnation points which occur fore and aft and on the sides of the body. Thus, the results from the sealed case may be used to compute the forces on the open case.

LARGE GRAVITY STRUCTURES

A typical gravity production platform consists of large volume caissons near the base of the structure and relatively smaller circular cylindrical or conical sections at various planes near the water surface. On the large lower members the wave diffraction is important whereas the upper members are subject to both the inertia and drag effects. In deep waters, high waves are nonlinear, such as in the North Sea, and fifth-order wave theory is more appropriate. The large members being near the bottom are influenced little by the higher order wave components so that the linear wave diffraction theory is still applicable using the first-order component of the nonlinear incident wave. The discrepancy of not using the larger wave length from the nonlinear waves in the linear diffraction is not significant being a decrease in the horizontal force by about 5%. For the upper members, the Stokes fifth-order wave theory is used for the water particle kinematics and wave forces are then computed by the Morison equation. In this way, the nonlinear effects of the waves are included in the design in a practical way.

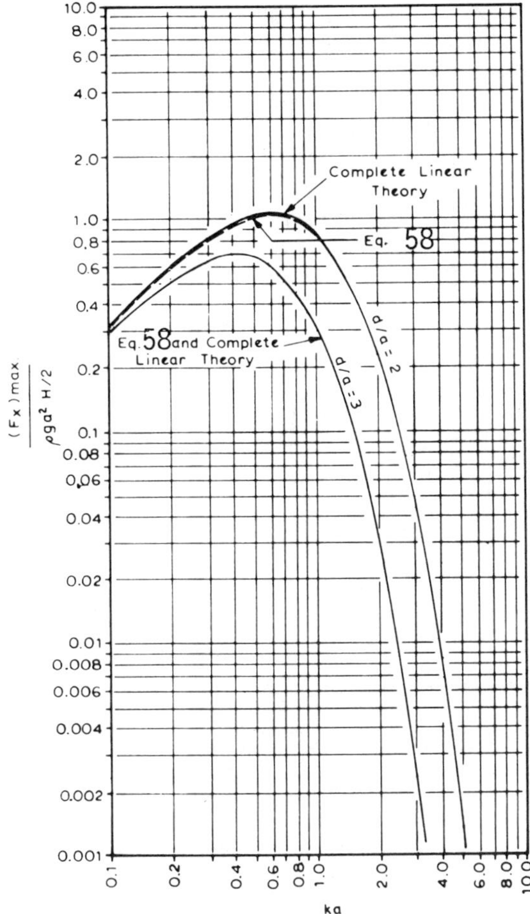

FIGURE 31. Normalized horizontal force on a hemisphere correlated with numerical solution.

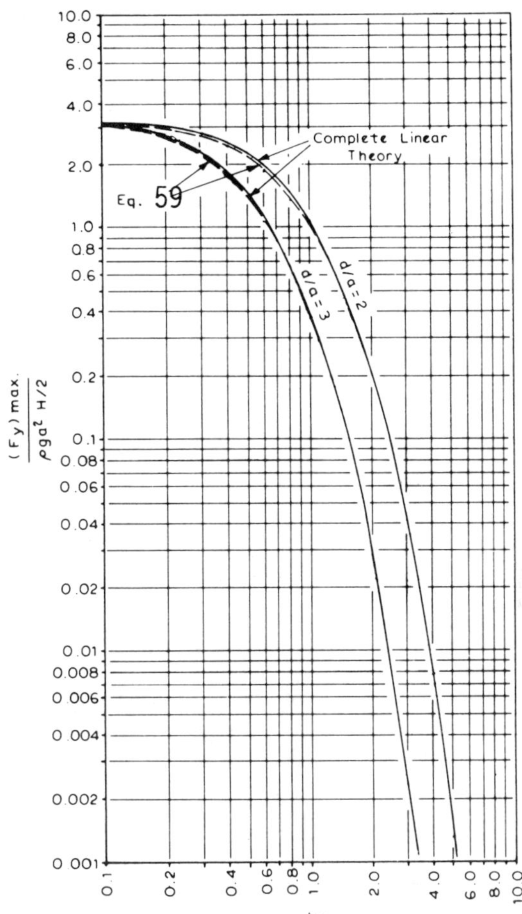

FIGURE 32. Normalized vertical force on a hemisphere correlated with numerical solution.

MULTILEGGED STRUCTURE

The proximity effects from closely-spaced members in a structure may have significant influence on the force coefficients. These include members in a lattice structure, conductor groups, or a relatively small bracing member in the neighborhood of a large member of a structure.

For members in-line to the flow, the critical relative spacing (spacing/diameter) for drag forces is about 4.5. For members lined up perpendicular to the flow, the critical relative spacing is about 2. In the case of inertial forces on large objects, the interference effect of neighboring elements of a structure can be accounted for applying a diffraction analysis. For vertical cylinders, an extension of the MacCamy-Fuchs theory which takes into account the multiple scattering can be made. Chakrabarti (1978) gives extensive results on 3- and 4-post structures. A critical relative spacing of about 5 can be considered for the proximity effects on inertial forces.

MOVING RIGID STRUCTURES

A moving rigid structure in the open sea is connected to the seafloor by some means. The moored structure is allowed to move during the passage of the wave. In order to design such a structure the motions of the structure should be known in addition to the wave forces on it. This requires solving equations of motion in various degrees of freedom. In most cases, these equations are coupled. Because of the presence of the relative velocity drag force and a nonlinear mooring line response, the equations are generally nonlinear. These equations are solved by means of a numer-

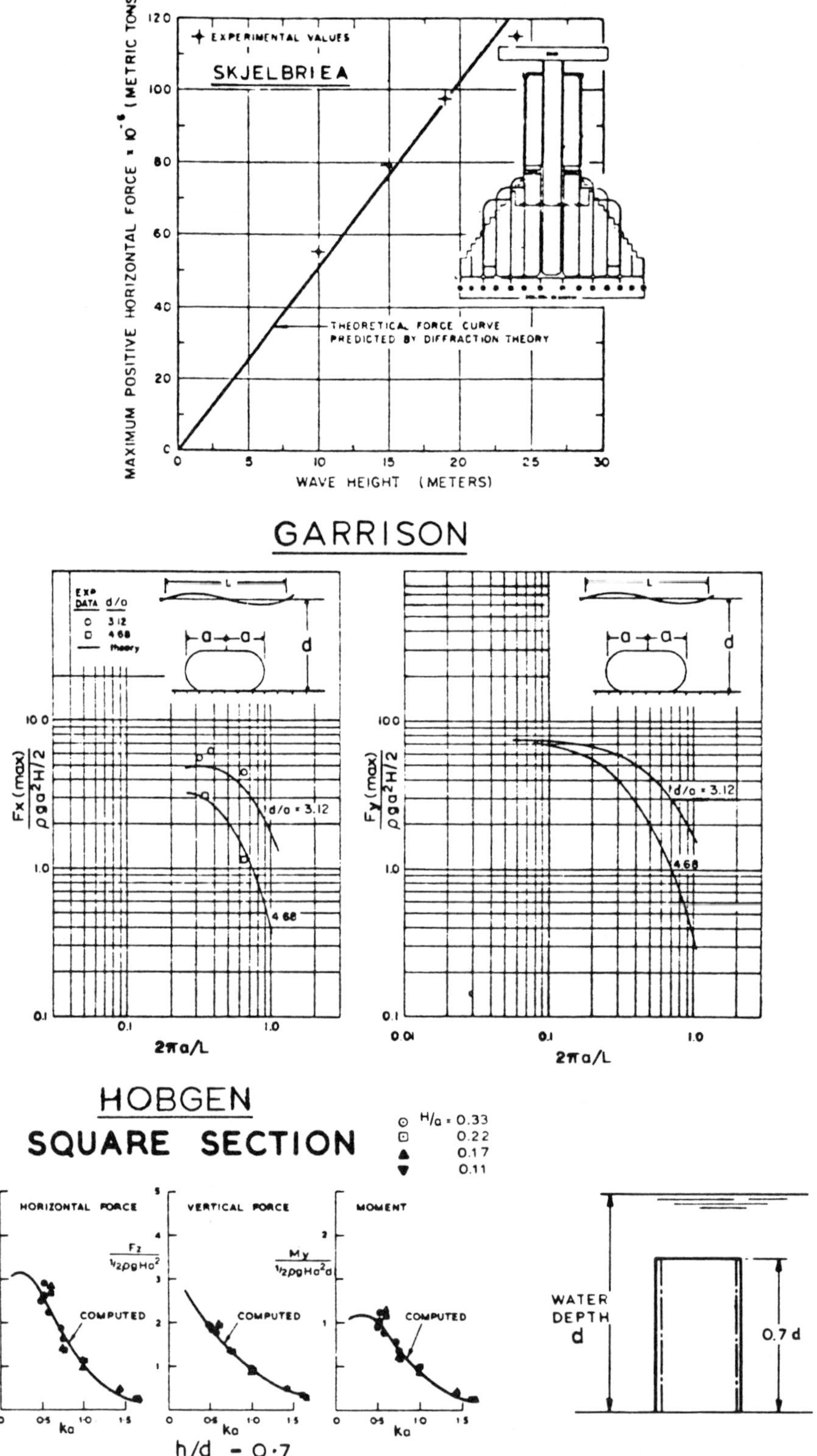

FIGURE 33. Correlation of measured wave exciting forces on offshore structures with linear diffraction theory.

ical solution, e.g., a finite difference scheme [Olsen, et al. (1978)].

In some cases, however, a simpler approach may be taken. On certain simplifications, the problem can be solved in a semi-closed form. The equations of motion for a few such systems are given here and the technique of solution is discussed. The examples included here consist of moored floating structure and articulated towers alone or moored to a tanker by a single mooring arm or hawser.

Moored Floating Structures

The motions of a large floating structure are obtained with the help of a linear potential flow theory in which the drag effect is assumed negligible. It computes the Froude-Krylov force, the diffraction force and the radiation force due to body motion. The last components provide the hydrodynamic coefficients of the body in its six degrees of motion in terms of the added mass and damping coefficients. The coupled six degrees of freedom motions of the floating rigid body are computed from the six linear, coupled differential equations as follows:

$$\sum_{k=1}^{6} [(m_{jk} + M_{jk})\ddot{x}_k + N_{jk}\dot{x}_k + (K_{jk} + C_{jk})x_k] = F_j e^{i(\omega t + \alpha_j)}, \quad j = 1, 2, \ldots 6 \tag{61}$$

in which m_{jk} is the 6×6 mass matrix representing the appropriate mass of the structure and its moments and products of inertia. The hydrostatic restoring forces are given by C_{jk} while K_{jk} are the mooring line spring constant matrix. The added mass and damping coefficients in six degrees are given by the matrices M_{jk} and N_{jk}. The total external forces and moments and the corresponding phase angles are F_j and α_j, respectively. Under the assumption of small amplitude of wave and body motion, all quantities are linear to first-order.

The mooring line reactions depend on the shape and configurations, weight, and material of mooring lines and hydrodynamic loading on them. The stress-strain characteristics of the lines are often nonlinear. It is, however, permissible sometimes to approximate the load-elongation curve by a straight line within the range of its application.

For a linear system, a pseudo-dynamic solution may be sought for the set of dynamic equation given by Equation (61). Substituting

$$x_k = x_{ko}\, e^{i(\omega t + \epsilon_k)} \tag{62}$$

where x_{ko} is the amplitude of motion and ϵ_k, its phase angle, one gets

$$[-\omega^2(m_{jk} + M_{jk}) + i\omega N_{jk} + (K_{jk} + C_{jk})]x_{ko}\, e^{i\epsilon_k} = F_j\, e^{i\alpha_j} \qquad j = 1, 2, \ldots 6 \tag{63}$$

which may be solved for x_{ko} and ϵ_k by a matrix inversion routine.

An example of such an analysis for a floating disc buoy in heave and pitch is shown in Figure 34. The correlation with the experimental results is quite good.

Articulated Towers

Articulated buoyant towers are primarily used for mooring tankers for loading oils in open waters. They are also used for permanently mooring storage tankers, flaring of gas, etc. The dynamic response of such a tower in wind, waves and currents must be known in the design of the tower. The wind and waves are generally in the same direction especially under severe environmental condition.

In one analytical approach, the steady loads on the articulated tower are computed separately. Then the steady angle of tilt caused by the steady loads, e.g., wind and current, are computed by a formula similar to

$$C \sin\phi = M_c \tag{64}$$

where C = net righting moment about the pivot point due to the buoyancy and weight of the tower including its ballast and platform weight, ϕ = azimuthal angle from vertical

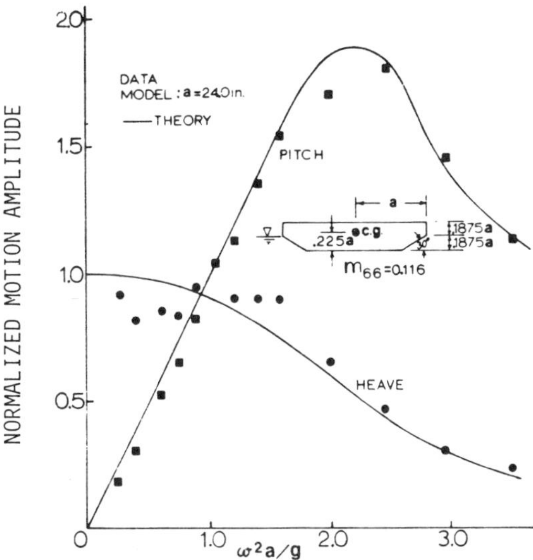

FIGURE 34. Response of a disk buoy in deep water [Garrison (1978)].

and M_c = moment about the pivot due to the steady loads. The wind and current forces in this case are computed in terms of the square of the wind or current velocity and a wind or current drag coefficient based on Reynolds number. A prescribed wind and current profile may be considered in such computations. In most numerical analyses a relative velocity between the current velocity and water particle velocity due to wave is used to compute the drag force and the drag coefficient is chosen on the basis of the relative velocity. Since the current may come from a different direction, its components in the direction of the waves and normal to them are used in the numerical computation.

The relative velocity method was used by Kirk and Jain (1977) in their analysis of a free articulated tower. The free tower has two degrees of freedom. In terms of the angle, ϕ and the angle in the horizontal plane, ψ the equations of motion of the free tower in waves are

$$I\ddot{\phi} - \dot{\psi}^2 \sin \phi \cos \phi + C \sin \phi = M_\phi \qquad (65)$$

$$I\ddot{\psi} \sin^2\phi + 2I\,\dot{\psi}\,\dot{\phi} \sin \phi \cos \phi = M_\psi \qquad (66)$$

in which I is the moment of inertia of the tower including the added mass effect, C is the righting moment of tower about the pivot and M_ϕ and M_ψ are the moments due to the wave forces on the tower about the respective axes. The wave forces are computed by the Morison equation considering the relation motion between the fluid particle and the oscillating tower. An approximate frequency domain analysis was made by Chakrabarti and Cotter (1979, 1980) for an articulated tower independently in the in-line and transverse directions and correlated with model test results.

If a tanker is coupled to the tower, then the equations for the tower, Equations (65) and (66), should include a mooring line force as a force vector at the point of attachment of the line to the tower. For the tanker, equations of motion similar to Equation (61) for a moored floating body are applicable. The coupled equations are then solved by one of many numerical schemes.

In a single point mooring system if the wind, waves and current are all in one direction, a simplified, yet useful solution may be obtained in a semiclosed form by the pseudo-dynamic method.

The static displacements of the tower-tanker system are kept separate from the dynamic oscillations and superimposed linearly. Thus, an equilibrium point is reached by the steady wind, current and wave drift load and the ship assumes a head sea position. Any oscillation is assumed to take place about this position. In this position, the surge and pitch of the ship are most important. The heave of the ship is generally small compared to pitch and has little influence on the coupled motion of the tower. Therefore, the problem is considered as a 3 degrees of freedom system—the oscillation of the tower in the plane of the wave motion, and the surge and pitch of the ship. These motions are coupled through the mooring line. Then the governing equations become

(1) Tower Oscillation

$$I_c \ddot{\psi} + B_1 (w,\dot{\psi}) + D_1 \dot{\psi} + C_c \psi + F_r \ell \cos (\theta + \phi) = M_c e^{1(\alpha 1 - \omega t)} \qquad (67)$$

(2) Ship Surge

$$M\ddot{x} + B_2 (u,\dot{x}) + D_2 \dot{x} - F_r \cos \theta = F_s e^{1(\alpha 2 - \omega t)} \qquad (68)$$

(3) Ship Pitch

$$I_s\ddot{\mu} + B_3 (w,\dot{\mu}) + D_3 \dot{\mu} + C_s \mu - F_r \times \left(H_s \cos \theta + \frac{L}{2} \sin \theta \right) = M_s e^{i(\alpha 3 - \omega t)} \qquad (69)$$

The quantities ψ, x and μ are the tower oscillation angle, ship surge and ship pitch, respectively, due to waves, I_c = virtual moment of inertia of tower, M = virtual mass of ship in surge, I_s = virtual moment of inertia of ship pitch, D = linear damping term, C_c = righting moment of tower, C_s = righting moment of ship in pitch, α = phase angles of forcing function, and θ = angle of the mooring line to the horizontal. Other quantities are defined in Figure 35. The first term in Equations (67–69) is the inertia term. The second term is a nonlinear drag term based on the relative velocity between the object and the water particle. The third term in the equations is a linear damping term. The fourth term is a righting moment term. The final term on left-hand side of Equations (67–69) is the response of the mooring system to the tower oscillation, ship surge and ship pitch, respectively. Assuming a linear spring constant for the mooring arm, the dynamic force on the arm is given in terms of the three displacements by

$$F_r = K \left[\ell \psi \cos(\phi + \theta) - x \cos \theta - \left(H_s \cos \theta + \frac{L}{2} \sin \theta \right) \mu \right] \qquad (70)$$

This term, F_r couples the three equations of motion.

The quantities on the right-hand side of the equations, M_c, F_s and M_s are respectively the moment on the tower, force on the ship in surge and moment on the ship in pitch due to waves. These forcing functions represent forces by a linear diffraction/radiation theory. In order that the equations of motion can be handled in a semi-closed form, the solutions are assumed of the following form:

$$\psi = \psi_0 e^{i(\varepsilon_1 - \omega t)} \qquad (71)$$

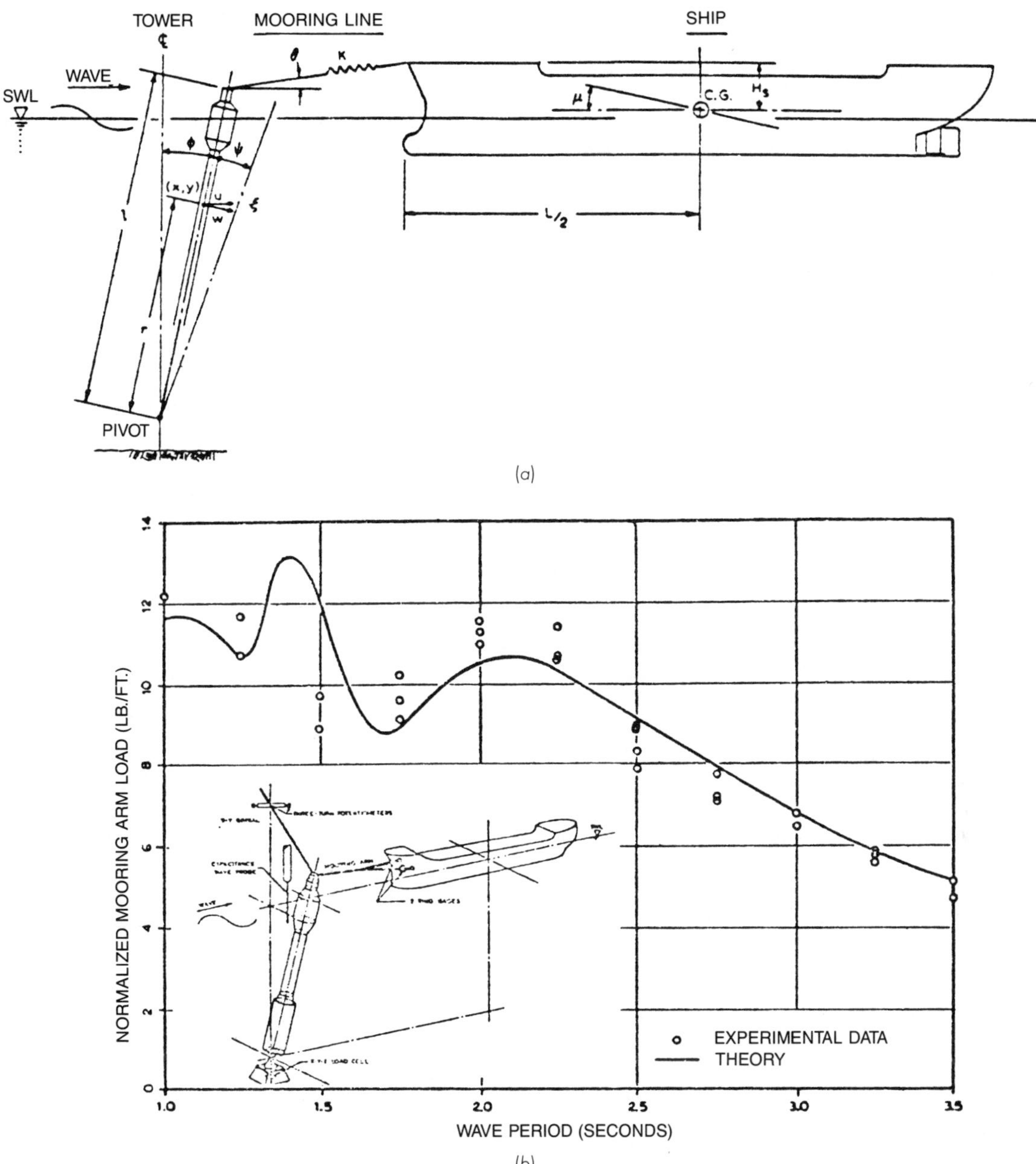

FIGURE 35. (a) Definition sketch for a simplified tower-tanker system analysis; (b) Correlation of measured mooring arm load with theory.

$$x = x_0 e^{i(\epsilon_2 - \omega t)} \tag{72}$$

$$\mu = \mu_0 e^{i(\epsilon_3 - \omega t)} \tag{73}$$

in which ψ_0, x_0 and μ_0 are the amplitudes of the tower oscillation, ship surge and ship pitch, respectively, while ϵ_1–ϵ_3 are the corresponding phase angles. This necessitates linearizing (with respect to time) the nonlinear drag terms. This is accomplished by expanding the terms in a Fourier series and retaining only the first term. For example,

$$|\dot{x} - u|(\dot{x} - u) \simeq \frac{8}{3\pi} |i\omega x_0 e^{i\epsilon_2} + u_0 e^{ikx}|(\dot{x} - u) \tag{74}$$

Approximation similar to Equation (74) is introduced in the nonlinear drag terms. Then, on substitution of the expression from Equations (71–73) for ψ, x and μ in Equations (67–69) and equating the coefficients of $e^{-i\omega t}$, the following matrix equation is obtained:

$$A_{ij} x_j = B_i \qquad i,j = 1,2,3 \tag{75}$$

where A_{ij} and B_i are functions of x_j. The solutions are obtained by a complex matrix inversion routine:

$$x_j = A_{ji}^{-1} B_i \tag{76}$$

and an iterative technique. An example of the mooring arm load for a tower-tanker system tested in a wave tank in Figure 35 shows satisfactory correlation.

This analysis is valid for a tower-tanker system that is connected by a rigid mooring arm and a tower that is large so that the forces on it are inertia-dominated. If the tower is latticed and the mooring line consists of a hawser with nonlinear spring characteristics then this formulation is not applicable. In this case, the equations of motion are highly nonlinear and should be solved numerically, e.g., Olsen, et al. (1978).

FREQUENCY DOMAIN ANALYSIS

The stochastic spectral analysis technique based on a frequency domain approach has often been applied to study platform dynamic and fatigue response. The first step in the analysis is the choice of a wave energy spectrum (e.g., Pierson-Moskowitz, JONSWAP, etc.) representing wave conditions in the design area. The second step is to determine the transfer function under wave action or a Response Amplitude Operator (*RAO*). Then the next step is to compute the corresponding response spectra. We have already discussed methods of computing force or motion responses of offshore structures. These responses provide the *RAO*. The method of computing response spectra from these *RAO*s is outlined here.

Response Amplitude Operator

The maximum response per unit wave amplitude (or height) is defined as the *RAO*. It is represented as a function of wave period or frequency. The *RAO*s may be linear or nonlinear. The linearity is determined with respect to the wave amplitude. If the responses are linear with wave amplitude, it is called a linear system. Generally, inertial systems are linear and drag systems are nonlinear. For a linear system then the response function can be written as

$$\text{Response } (t) = (RAO)\eta(t) \tag{77}$$

As an example, the dynamic pressure due to linear theory is given by

$$p(t) = \varrho g \frac{H}{2} \frac{\cosh ks}{\cosh kd} \cos (kx - \omega t) \tag{78}$$

which may be written as

$$p(t) = (RAO)_p \, \eta(t) \tag{79}$$

where

$$(RAO)_p = \varrho g \frac{\cosh ks}{\cosh kd} \tag{80}$$

Response Spectra

The response spectrum is defined as the response energy density of a structure due to the input wave energy density spectrum. The wave spectrum elevation (at a particular frequency) is multiplied by the square of the *RAO* (at the same frequency) to obtain the response spectrum at that frequency. Symbolically,

$$S_R(\omega) = (RAO)^2 S(\omega) \tag{81}$$

where S_R is the response spectrum.

The response spectra for pressure in the preceding example become

$$S_p(\omega) = \left[\varrho g \frac{\cosh ks}{\cosh kd} \right]^2 S(\omega) \tag{82}$$

Thus, the pressure response spectrum is obtained from the wave spectrum by multiplying by the square of the pressure response amplitude operator. It is seen that the RAO_p is nothing but the maximum dynamic pressure per unit wave amplitude.

Similarly, for the inertia part of Morison's equation

$$S_{fI}(\omega) = [RAO]^2_{fI} S(\omega) \tag{83}$$

where

$$[RAO]_{f_I} = C_M \rho \frac{\pi}{4} D^2 \left[gk \frac{\cosh ks}{\cosh kd} \right] \tag{84}$$

When the response function has a nonlinear relationship with the wave height (or amplitude) the *RAO* is called nonlinear. Thus, when drag is present in a system, the system *RAO* is nonlinear. For a nonlinear *RAO* the conversion of wave spectrum to response spectrum is not straightforward. In this case, an approximate method of handling the nonlinear terms is generally used.

Since the drag force is proportional to the square of the velocity, a linear approximation is written as

$$|u|u \approx C_l u \tag{85}$$

Assuming that u is normally distributed with a mean zero and standard deviation, σ_u, the most accurate linear estimate gives

$$C_l = \sqrt{\frac{8}{\pi}} \sigma_u \tag{86}$$

so that the drag force becomes

$$f_D \approx \frac{1}{2} \rho C_D D \sqrt{\frac{8}{\pi}} \sigma_u u \tag{87}$$

So, in this case the drag term may be treated as linear in constructing an *RAO* and calculating the response spectrum.

An example of the correlation of the force spectra for ocean waves is given below. During Hurricane Carla in 1961, a consortium of oil companies obtained measurements on a vertical pile supporting a Chevron oil drilling platform in the Gulf of Mexico in a water depth of about 100 ft. Data was recorded in the form of surface wave profile and wave pressures at various elevation of the pile. From these profiles the wave energy density spectrum and pressure spectra are computed by the Fourier transform technique. In order to evaluate S_F using the computed wave energy spectrum an appropriate set of values for C_M and C_D should be known. For the purpose of force spectrum correlation here, Dean and Aagaard's (1970) data are chosen as $C_M = 1.33$ and $C_D = 0.5$.

The wave energy spectrum for September 9, 1961 is utilized in conjunction with *RAO* to calculate the response spectrum and the correlation of measured and theoretical force (per unit length) spectra at dynamometer location 8 is shown in Figure 36. Note that the unit of force is given as lb/ft. (kg/m). The correlation with the measured spectra is generally good except at the low frequency end.

Another example of the calculation of the stress response spectrum from the stress *RAO* and a JONSWAP type spectrum is given in Figure 37. Note that because of two areas of high values in the stress, the response spectrum is double-peaked even though the wave spectrum has only a single peak.

Nonlinear Coupled Current and Wave Drag Force Spectrum

The drag force per unit length of a vertical cylinder in the presence of unidirectional current is given by

$$f_D(t) = \frac{1}{2} \varrho C_D D |u(t) - U|(u(t) - U) \tag{88}$$

The modified wave spectrum in deep waters due to the uniform current is written as

$$S^*(\omega) = \frac{S(\omega)}{\mu(\omega)[1 + \mu(\omega)]^2} \tag{89}$$

where

$$\mu(\omega) = (1 + U\omega/g)^{1/2} \tag{90}$$

For opposing current, U is negative in these expressions. Under the deepwater assumption, an expression for the drag force spectrum due to relative velocity is obtained. Note that the water particle velocity spectrum in deepwater are related to the wave energy density spectrum as follows:

$$S^*_u(\omega) = \omega^2 S^*(\omega) \tag{91}$$

The asterisk indicates that the spectrum has been modified by current. To the first order of approximation, the spectrum of the drag force may be shown [Borgman (1967)] to have the form

$$S_{fD}^*(\omega) = 16\, c^2\, \sigma_u^2\, [Z(\gamma) + |\gamma|\, P(\gamma)]^2 S_u^*(\omega) \tag{92}$$

where $c = \frac{1}{2} \varrho C_D D$, σ_u^2 is the variance of the modified velocity spectrum, $\gamma = U/\sigma_u$ is a parameter measuring the strength of the current,

$$Z(\gamma) = \frac{1}{\sqrt{2\pi}} e^{-(\gamma^2/2)} \tag{93}$$

and

$$P(\gamma) = \int_0^\gamma Z(x) dx \tag{94}$$

Inertia Force Spectrum Coupled with Current

In the case of the drag force spectrum, the influence of current is measured jointly by the parameter, γ, and the

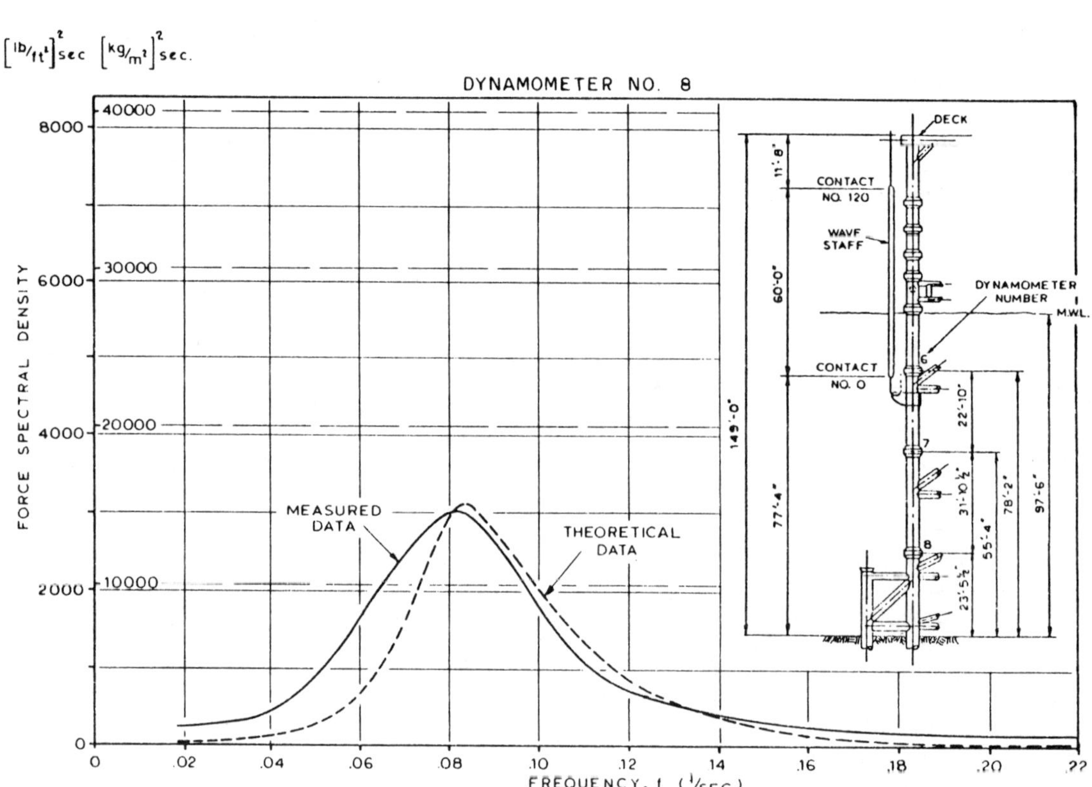

FIGURE 36. Correlation of theoretical force spectrum with measured force spectrum from field data on a vertical pile.

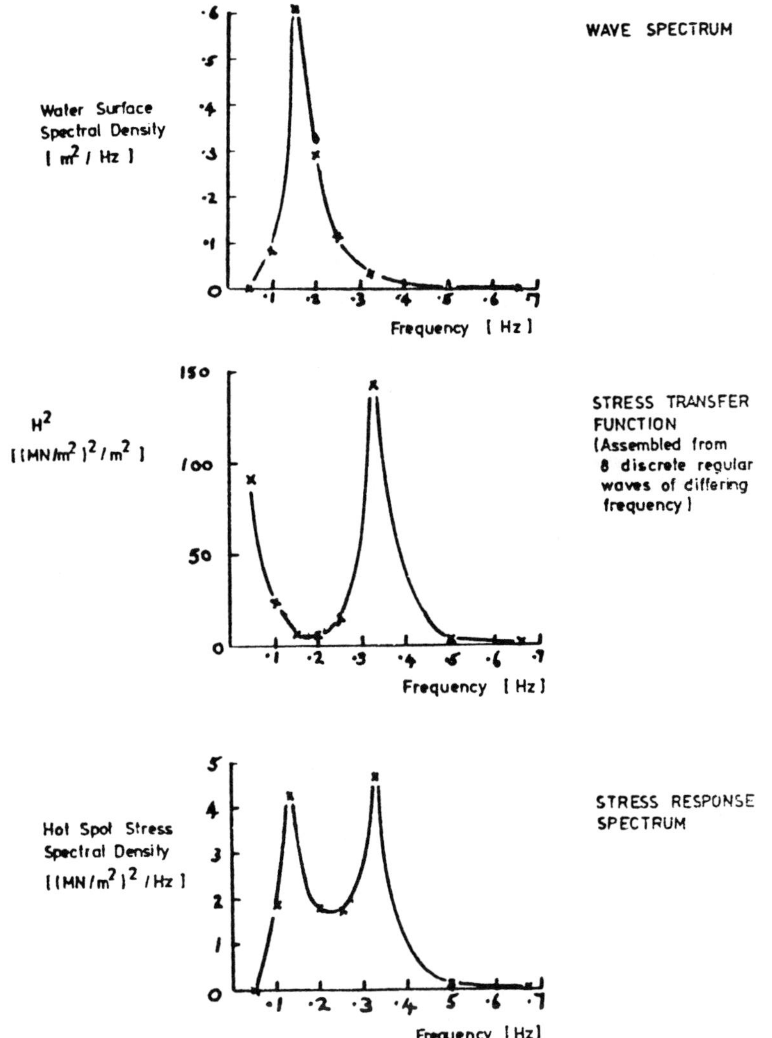

FIGURE 37. Computation of stress response spectrum [Hallam, et al., Offshore Structures Engineering (1979)].

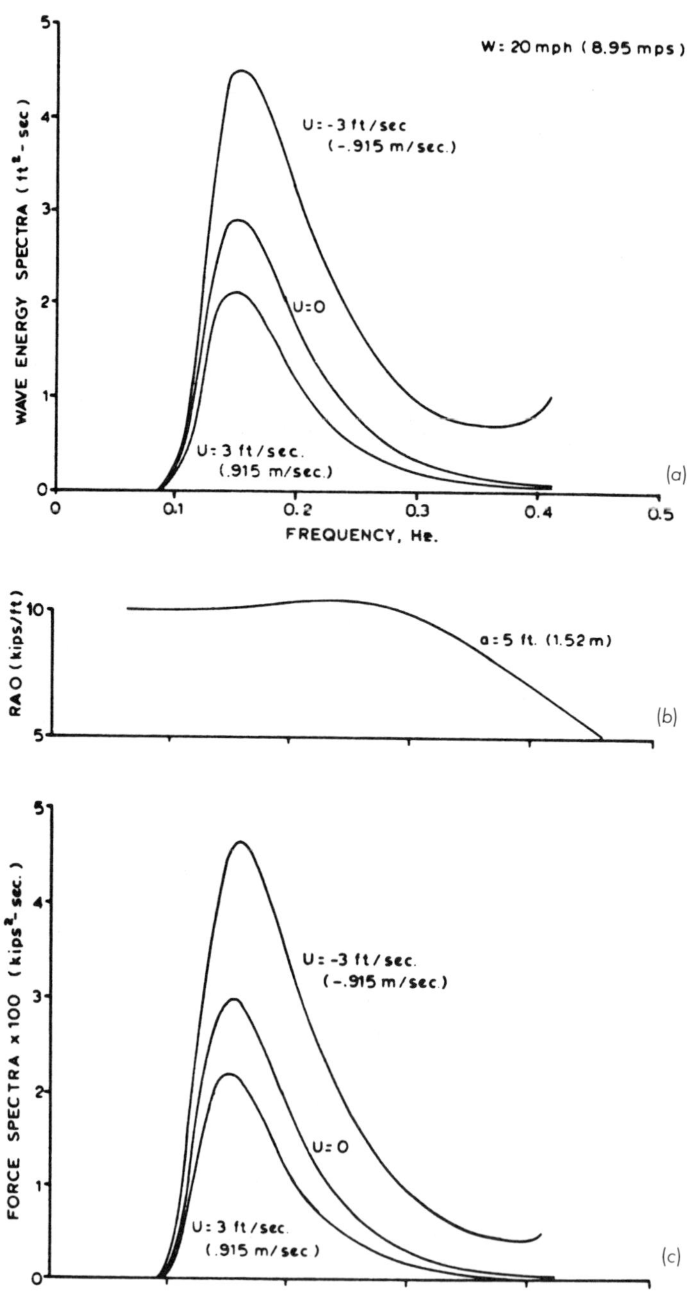

FIGURE 38. Response spectra on forces on a large vertical cylinder in the presence of current.

modification of the velocity spectrum as well as its variance due to current. For inertia force, the force spectrum is affected by the modified acceleration spectrum only. The water particle acceleration spectrum is obtained in deep water from

$$S_{\dot{u}}^*(\omega) = \omega^4 S^*(\omega) \tag{95}$$

so that the inertia force spectrum on a vertical cylinder of unit length is

$$S_{f_I}^*(\omega) = \left(C_M \rho \frac{\pi}{4} D^2\right)^2 S_{\dot{u}}^*(\omega) \tag{96}$$

For large cylinders where drag effects are negligible compared to inertia forces and diffraction effects are important, a similar analysis can be made. The total force on the cylinder in deep waters is given by

$$F(t) = \varrho g \frac{H}{2} a^2 \frac{4A(ka)}{(ka)^2} \cos(\omega t - \alpha) \tag{97}$$

The total force spectrum in this case becomes

$$S_F^*(\omega) = \left[\varrho a^3 \frac{4A(ka)}{(ka)^3}\right]^2 S_{\dot{u}}^*(\omega) \tag{98}$$

Thus, for a given wave spectrum and a steady current, the total force spectrum on a cylinder may be determined. This is illustrated in the following deep water example. The wave spectrum is assumed to be given by the P-M model. The following quantities are considered for the example: $U_w = 20$ mph (8.95 m/s), $U = \pm 3$ fps (0.915 m/s) where the minus sign indicates that the current opposes the waves and $a = 5$ ft. (1.52 m).

The three cases of positive, negative and no current are presented in Figure 38. In Figure 38(a), the P-M wave energy spectra with and without current are shown, in Figure 38(b) the *RAO* for the cylinder is drawn as a function of frequency, *f*. Knowing these two functions, the force spectra are calculated with the help of Equation (98) where in the case of current $S(\omega)$ is replaced by $S^*(\omega)$ from Equation (89). Note that while opposing current increases the force, the current flowing in the direction of the waves reduces it. Similar results have been shown by Tung and Huang (1976) using the linearized Morison equation.

REFERENCES AND BIBLIOGRAPHY

General Review Papers

1. Blanc, J. P. and M. Monro, Editors, "Design and Construction of Offshore Structures," *Proceedings of the Conference, ICE,* London, October (1976).
2. Hogben, N., "Fluid Loading on Offshore Structure, a State-of-the-Art Appraisal: Wave Loads," *Maritime Technology Monograph No. 1,* Royal Institute of Naval Architects, London, England (1974).
3. Lighthill, J., "Wave and Hydrodynamic Loading," Proceedings of the Second International Conference on the Behavior of Offshore Structures, *BHRA Fluid Engineering,* Cranfield, Bedford, England, Vol. 1, pp. 1–40 (August 1979).
4. Milgram, J. H., "Wave and Wave Forces," *Proceedings on the Behavior of Offshore Structures,* Trondheim, Norway, pp. 11–38 (1976). Also, *Sea Grant Program,* Massachusetts Institute of Technology, Report No. 76-19, 28 pp. (November 1976).
5. Shaw, T. L. (Editor), *Mechanics of Wave-Induced Forces on Cylinders,* Pitman Advanced Publishing Program, San Francisco, California (1976).
6. Tickell, R. G. and P. Holmes, "Approaches to Fluid Loading, Probabilistic and Deterministic Analyses," *Numerical Methods in Offshore Engineering,* John Wiley and Sons, Great Britain, pp. 43–86 (1978).

Watt, B. J., "Basic Structural Systems—A Review of Their Design and Analysis Requirements," *Numerical Methods in Offshore Engineering,* John Wiley and Sons, Great Britain, pp. 1–42 (1978).

Wave Environment

8. Bea, R. G., "Earthquake and Wave Design Criteria for Offshore Platforms," *Journal of the Structural Division, ASCE, Vol. 105,* ST2, pp. 401–419 (February 1979).
9. Bretschneider, C. L., "A One-Dimensional Gravity Wave Spectrum," *Ocean Wave Spectra,* Prentice-Hall, Inc., Englewood Cliffs, New Jersey, pp. 41–56 (1961).
10. Bretschneider, C. L., "Operational Sea State and Design Wave Criteria: State-of-the-Art of Available Data for U.S.A. Coasts and the Equatorial Latitudes," *Proceedings of the Fourth Conference on Ocean Thermal Energy Conversion,* New Orleans, Louisiana, George E. Loup, Ed., University of New Orleans, pp. IV-61–73 (March 1977).
11. Goda, Y., "A Review on Statistical Interpretation of Wave Data," *Report of the Port and Harbour Research Institute, Vol. 18,* No. 1 (March 1979).
12. Hasselman, K., et al., "Measurements of Wind-Wave Growth and Swell Decay During the Joint North Sea Wave Project (JONSWAP)," *Deutsche Hydrographische Zeitschrift, Erganzungsheft, Vol. 13,* NO. A (1973).
13. Hogben, N. and F. E. Lumb, *Ocean Wave Statistics,* London, Her Majesty's Stationery Office (1967).
14. Ochi, M. K., "Wave Statistics for the Design of Ships and Ocean Structures," *Transactions on the Society of Naval Architects and Marine Engineers, Vol. 86,* pp. 47–76 (1978).
15. Ochi, M. K., "A Series of JONSWAP Wave Spectra for Offshore Structure Design," *Proceedings on Second International Conference on the Behavior of Offshore Structures, BHRA Fluid Engineering,* Cranfield, Bedford, England, Vol. 1, Paper No. 4, pp. 75–86 (August, 1979).

16. Ochi, M. K. and E. N. Hubble, "Six-Parameter Wave Spectra," *Proceedings on the Fifteenth Coastal Engineering Conference,* Honolulu, Hawaii, ASCE, pp. 301–328 (1976).
17. Pierson, W. J. and L. Moskowitz, "A Proposed Spectral Form for Fully Developed Wind Seas Based on the Similarity Theory of S. A. Kitaigorodskii," *Journal of Geophysical Research, Vol. 69,* No. 24, pp. 5181–5203 (December 1964).
18. Proceedings of the Second International Ship Structures Congress, Delft, Netherlands (1964).
19. Scott, J. R., "A Sea Spectrum for Model Tests and Long-Term Ship Prediction," *Journal of Ship Research, Vol. 9,* pp. 145–152 (December 1965).

Wave Theories

20. Dean, R. G., "Relative Validities of Water Wave Theories," *Journal of Waterways and Harbor Div.*, ASCE, pp. 105–119 (February 1970).
21. Dean, R. G., "Stream Function Representation of Nonlinear Ocean Waves," *Journal of Geophysical Research, Vol. 70,* No. 18, pp. 4561–4572 (September 1965).
22. Kinsman, B., *Wind Waves,* Prentice-Hall, Inc., Englewood Cliffs, New Jersey (1965).
23. LeMéhauté B., *An Introduction to Hydrodynamics and Water Waves,* Springer-Verlag, New York (1976).
24. Skjelbreia, L. and J. A. Hendricksen, "Fifth-Order Gravity Wave Theory," *Proceedings on Seventh Conference on Coastal Engineering,* Ch. 10, pp. 184–196 (1961).

Wave Loads

25. Bea, R. G. and N. W. Lai, "Hydrodynamic Loadings on Offshore Platforms," *Proceedings of the Tenth Offshore Technology Conference*, OTC 3064, pp. 155–168 (May 1978).
26. Bea, R. G., N. Lai, and A. Niedoroda, "Assessment of the Morison Equation," Civil Engineering Laboratory, Naval Construction Battalion Center, Port Hueneme, California, Report No. CR80.022 (July 1980).
27. Garrison, C. J., "A Review of Drag and Inertia Forces on Circular Cylinders," *Proceedings on the Twelfth Offshore Technology Conference,* OTC 3760, pp. 205–218 (May 1980).
28. Hogben, N., "Wave Loads on Structures," *Proceedings of Behavior of Offshore Structures,* Trondheim, Norway, pp. 187–217 (1976).
29. Hogben, N., B. L. Miller, J. W. Searle, and G. Ward, "Estimation of Fluid Loading on Offshore Structures," *Proceedings of the Institution of Civil Engineers, Part 2, Vol. 63,* pp. 515–562 (September 1977).
30. Lundgren, H., O. Brink-Kjaer, S. E. Sand, and V. Jacobsen, "Improved Physical Basis of Wave Forces," *Proceedings on Civil Engineering in the Oceans IV,* ASCE, San Francisco, pp. 1–16 (September 1979).
31. Morison, J. R., M. D. O'Brien, J. W. Johnson, and S. A. Schaaf, "The Forces Exerted by Surface Waves on Piles," *Transactions AIME, Vol. 189,* pp. 149–154 (1950).
32. Ochi, M. K. and S. Wang, "Prediction of Extreme Wave-Induced Loads on Ocean Structures," *Proceedings on Behavior of Offshore Structures,* Trondheim, Norway, pp. 170–186 (1976).
33. Pearcey, H. H. and J. R. Bishop, "Wave Loading in the Drag and Inertia Regimes: Routes to Design Data," *Proceedings on the Behavior of Offshore Structures, BHRA Fluid Engineering,* Cranfield, Bedford, England (August 1979).
34. Ramberg, S. E. and J. M. Niedzwecki, "Some Uncertainties and Errors in Wave Force Computations," *Proceedings of Eleventh Offshore Technology Conference,* Houston, Texas, OTC 3597, pp. 2091–2011 (May 1979).

Hydrodynamic Coefficients

35. Bearman, P. W. and J. M. R. Graham, "Hydrodynamic Forces on Cylindrical Bodies in Oscillatory Flow," *Proceedings of the Second International Conference on Behavior of Offshore Structures,* Cranfield, Bedford, England, pp. 309–322 (August 1979).
36. British Ship Research Association, "A Critical Evaluation of the Data on Wave Force Coefficients," Report (August 1976).
37. Burton, W. J. and R. M. Sorensen, "The Effects of Surface Roughness on the Wave Forces on a Circular Cylindrical Pile," *Sea Grant Publication No. 211, Coastal and Ocean Engineering Division,* (*COE Report No. 121*), Texas A & M University (March 1970).
38. Chakrabarti, S. K., "Wave Forces on Submerged Objects of Symmetry," *Journal of the Waterways, Harbors and Coastal Engineering Division, ASCE, Vol. 99,* No. WW2 (May 1973).
39. Chakrabarti, S. K., "In-Line Forces on Fixed Vertical Cylinder in Waves," *Journal of Waterway, Port, Coastal and Ocean Division, ASCE, Vol. 106* (May 1980).
40. Chakrabarti, S. K., A. L. Wolbert, and W. A. Tam, "Wave Forces on Vertical Circular Cylinder," *Journal of the Waterways, Harbors, and Coastal Engineering Division, ASCE, Vol. 102,* No. WW2 (May 1976).
41. Chakrabarti, S. K., A. L. Wolbert, and W. A. Tam, "Wave Forces on Inclined Tubes," *Coastal Engineering, The Netherlands, Vol. 1* (1977).
42. Garrison, C. J., J. B. Field, and M. D. May, "Drag and Inertia Forces on a Cylinder in Periodic Flow," *Journal of the Waterway, Port, Coastal and Ocean Division, ASCE, Vol. 103,* No. WW2, pp. 193–204 (May 1977).
43. Gaston, J. D. and R. D. Ohmart, "Effects of Surface Roughness and Drag Coefficients," *Proceedings on the Civil Engineering in the Oceans IV, ASCE, Vol. II,* pp. 611–621 (1979).
44. Grace, R. A. and G. T. Y. Zee, "Further Tests on Ocean Wave Forces on Sphere," *Journal of the Waterway, Port, Coastal and Ocean Division, ASCE, Vol. 104,* No. WW1, pp. 83–88 (February 1978).
45. Isaacson, M. de St. Q. and D. J. Maull, "Transverse Forces on Vertical Cylinders in Waves," *Journal of Waterways, Harbors, and Coastal Engineering Division, ASCE, Vol. 102,* No. WW1, pp. 49–60 (February 1976).
46. Matten, R. B., "The Influence of Surface Roughness on the

Drag of Circular Cylinders in Waves," *Proceedings on Offshore Technology Conference,* Houston, Texas, OTC 2902, pp. 185–192 (1977).

47. Miller, B. L., "The Hydrodynamic Drag of Roughened Circular Cylinders," Royal Institute of Naval Architects, Spring Meeting (1976).
48. Sarpkaya, T., "In-Line and Transverse Forces on Smooth and Sand-Roughened Cylinders in Oscillatory Flow at High Reynolds Numbers," *Proceedings on the Eighth Offshore Technology Conference,* Houston, Texas, OTC 2533, pp. 95–108 (1976).
49. Sarpkaya, T., N. J. Collins, and S. R. Evans, "Wave Forces on Rough-Walled Cylinders at High Reynolds Numbers," *Proceedings on the Ninth Offshore Technology Conference,* Houston, Texas, OTC 2901, pp. 175–184 (1977).
50. Sawaragi, T., T. Nakamura, and H. Miki, "Characteristics of Lift Forces on Circular Piles in Waves," *Coastal Engineering in Japan, JSCE, Vol. 19,* pp. 59–71 (1976).

Ocean Wave Force Coefficients

51. Bishop, J. R., "RMS Force Coefficients Derived from Christchurch Bay," *Mechanics of Wave-Induced Forces on Cylinders* (Editor: T. L. Shaw), Pitman Publishing Ltd., London, G. B., pp. 334–345 (1979).
52. Bishop, J. R., R. G. Tickel, and K. A. Gallagher, "The U.K. Christchurch Bay Project: A Review of Results," *Proceedings of the Twelfth Offshore Technology Conference,* OTC 3796, pp. 9–23 (May 1980).
53. Dean, R. G. and P. M. Aagaard, "Wave Forces: Data Analysis and Engineering Calculation Method," *Journal of Petroleum Technology* (March 1970).
54. Dean, R. G., R. A. Dalrymple, and R. T. Hudspeth, "Analysis of Wave Project I and Wave Project II Data," Coastal and Oceanographic Engineering Laboratory, University of Florida, Gainesville, Florida, *Subject Report No. 4* (August 1974).
55. Evans, D. J., "Analysis of Wave Force Data," *Proceedings of First Offshore Technology Conference,* Houston, Texas, OTC 1005 (1969).
56. Heideman, C., O. A. Olsen, and P. I. Johhansson, "Local Wave Force Coefficients," *Proceedings on the Civil Engineering in the Oceans IV, ASCE, Vol. II,* pp. 684–699 (1979).
57. Ohmart, R. D. and R. L. Gratz, "Drag Coefficients from Hurricane Wave Data," *Proceedings on the Civil Engineering in the Oceans IV, ASCE, Vol. I,* pp. 260–272 (1979).
58. Starsmore, N., "Constant Drag and Added Mass Coefficients from Full-Scale Data," *Proceedings of the Thirteenth Offshore Technology Conference,* Houston, Texas, OTC 3990, pp. 357–367 (May 1981).
59. Wheeler, J. D., "Method for Calculating Forces Produced by Irregular Waves," *Proceedings of First Offshore Technology Conference,* Houston, Texas, OTC 1006 (1969).
60. Wiegel, R. L., K. E. Beebe, and J. Moon, "Ocean Wave Forces on Circular Cylindrical Piles," *Journal of the Hydraulic Division, ASCE,* (April 1957).
61. Wilson, B. W., "Analysis of Wave Forces on a 30-inch Diameter Pile under Confused Sea Conditions," U.S. Army, Coastal Engineering Research Center, *Technical Memo No. 15* (October 1965).

Multiple Cylinders

62. Chakrabarti, S. K., "Wave Forces on Multiple Vertical Cylinders," *Journal of the Waterway, Port, Coastal and Ocean Division, ASCE, Vol. 104,* pp. 147–161 (May 1978).
63. Chakrabarti, S. K., "Wave Forces on Vertical Array of Tubes," *Proceedings of Civil Engineering in the Ocean,* San Francisco, ASCE, pp. 241–259 (September 1979).
64. Chakrabarti, S. K., "Hydrodynamic Coefficients for a Vertical Tube in an Array," *Applied Ocean Research, Vol. 2,* pp. 2–12 (1980).
65. Chakrabarti, S. K., "Wave Interaction with Multiple Horizontal Cylinders," *Applied Ocean Research, Vol. 1* (1970).
66. Chakrabarti, S. K., "In-Line and Transverse Forces on a Tube Array in Tandem with Waves," *Applied Ocean Research, Vol. 3* (1981).
67. Chakrabarti, S. K., "Wave Force Coefficients for Rough Vertical Cylinders," *Journal of Waterway, Port, Coastal and Ocean Division, ASCE, Vol. 108* (November 1982).
68. Chakrabarti, S. K., "Transverse Forces on a Vertical Tube Array in Waves," *Journal of Waterway, Port, Coastal and Ocean Division, ASCE, Vol. 108* (February 1982).
69. Laird, A. D. K. and R. P. Warren, "Groups of Vertical Cylinders Oscillating in Water," *Journal of the Engineering Mechanics Division, ASCE, Vol. 80,* No. EM1, pp. 25–35 (February 1963).
70. Loken, A. E., O. P. Torset, S. Mathiassen, and T. Arnesen, "Aspects of Hydrodynamic Loading in Design of Production Risers," *Proceedings on the Eleventh Offshore Technology Conference,* Houston, Texas, OTC 3538, pp. 1591–1601 (1979).
71. Rains, C. P. and S. K. Chakrabarti, "Mechanical Excitation of Offshore Tower Model," *Journal of the Waterways, Harbors and Coastal Engineering Division, ASCE, Vol. 98,* No. WW1, pp. 35–47 (February 1972).
72. Sarpkaya, T., "Hydrodynamic Forces on Various Multiple Tube Riser Configurations," *Proceedings of the Eleventh Offshore Technology Conference,* Houston, Texas, OTC 3539, pp. 1603–1606 (1979).

Large Structures

73. Chakrabarti, S. K. and W. A. Tam, "Gross and Local Wave Loads on a Large Vertical Cylinder–Theory and Experiment," *Proceedings on the Offshore Technology Conference,* Houston, Texas, Paper No. OTC 1818 (May 1973).
74. Chakrabarti, S. K. and R. A. Naftzer, "Wave Interaction with a Submerged Open-Bottom Structure," *Proceedings of the Eighth Offshore Technology Conference,* Houston, Texas, OTC 2495 (May 1976).
75. Garrison, C. J., "Hydrodynamic Loading of Large Offshore

Structures, Three-Dimensional Source Distribution Methods," *Numerical Methods in Offshore Engineering*, John Wiley and Sons, Great Britain, pp. 87–140 (1978).

76. Hogben, N. and R. G. Standing, "Wave Loads on Large Bodies," *Proceedings on International Symposium on Dynamics of Marine Vehicles and Structures in Waves*, University College, London, Institute of Mechanical Engineering, pp. 258–277 (1975).
77. Mei, C. C., "Numerical Methods in Water-Wave Diffraction and Radiation," *Annual Review of Fluid Mechanics, Vol. 10,* pp. 393–416 (1978).
78. van Oortmerssen, G., "Some Aspects of Very Large Offshore Structures," *Ninth Naval Hydrodynamics Symposium*, Office of Naval Research, Washington, D.C. (1972).

Lift Force and Vortex Study

79. Isaacson, M. and D. Maull, "Transverse Forces on Vertical Cylinders in Waves," *Journal of the Waterways, Harbors and Coastal Engineering Division, ASCE, Vol. 102,* No. WW1, pp. 46–60 (February 1976).
80. King, R., "Review of Vortex Shedding Research and Its Application," *Ocean Engineering, Vol. 4,* Pergamon Press, pp. 141–172 (1977).
81. Sarpkaya, T., "Vortex-Induced Oscillations—A Selective Review," *Journal of Applied Mechanics, Transactions of the ASME, Vol. 46,* pp. 241–258 (June 1979).
82. Zdravkovich, M. M., "Review of Flow Interference Between Two Circular Cylinders in Various Arrangements," *Journal of Fluids Engineering, Transctions ASME, Vol. 99, Series 1,* No. 4, pp. 613–633 (December 1977).

Moored Floating Objects

83. Chao, J. C., "Dynamic Responses of Floating Structures," *Journal of Waterway, Port, Coastal and Ocean Division, ASCE, Vol. 104,* WW2, pp. 105–118 (May 1978).
84. Chung, J. S., "Motion of a Floating Structure in Water of Uniform Depth," *Journal of Hydronautics, Vol. 10,* No. 3, pp. 65–73 (July 1976)
85. Faltinsen, O. M. and F. C. Michelsen, "Motions of Large Structures in Waves at Zero Froude Numbers," *Proceedings on International Symposium on Dynamics of Marine Vehicles and Structures in Waves,* University College, London, Institute of Mechanical Engineering, pp. 91–106 (1975).
86. Mes, M. J., "Wave Dynamic Platform Analysis," *Petroleum Engineer International*, pp. 56–68 (July 1978).
87. Niedzwecki, J. M. and M. J. Casarella, "On the Design of Mooring Lines for Deep Water Applications," *Journal of Engineering for Industry, Transactions of the ASME,* pp. 514–522 (May 1976).
88. Schmitke, R. T., "Ship Sway, Roll, and Yaw Motions in Oblique Seas," *Transactions, Society of Naval Architects and Marine Engineers, Vol. 86,* pp. 26–46 (1978).
89. Faltinsen, O. M., "Theoretical Seakeeping. A State-of-the-Art Survey," *International Symposium on Advances In Marine Technology,* pp. 229–277 (1979).

Articulated Towers

90. Chakrabarti, S. K. and D. C. Cotter, "Analysis of a Tower-Tanker System," *Proceedings on the Tenth Offshore Technology Conference,* OTC 3202, pp. 1301–1310 (1978).
91. Chakrabarti, S. K. and D. C. Cotter, "Motion Analysis of Articulated Tower," *Journal of Waterway, Port, Coastal and Ocean Division, ASCE, Vol. 105,* No. WW3, pp. 281–292 (August 1979).
92. Chakrabarti, S. K. and D. C. Cotter, "Transverse Motion of Articulated Tower," *Journal of Waterway, Port, Coastal and Ocean Division, ASCE, Vol. 106,* No. WW1, pp. 65–78 (February 1980).
93. Kirk, C. L. and R. K. Jain, "Response of Articulated Towers to Waves and Currents," *Proceedings on the Ninth Offshore Technology Conference,* Houston, Texas, OTC 2798, pp. 545–552 (May 1977).
94. Olsen, O. A., et al., "Slow and High Frequency Motions and Loads of Articulated Single Point Mooring Systems for Large Tankers," *Norwegian Maritime Research,* No. 2, pp. 14–28 (1978).
95. Flory, J. F., et al., "Guidelines for Deepwater Port Single Point Mooring Design," *Report No. CG-D-49-77,* U.S. Department of Transportation, U.S. Coast Guard, Washington, D.C. (September 1977).

Frequency Domain Analysis

96. Bendat, J. S. and A. G. Piersol, *Random Data: Analysis and Measurement Procedures,* Second Edition, Wiley, New York (1971).
97. Borgman, L. E., "Spectral Analysis of Ocean Wave Forces on Piling," *Journal of Waterways and Harbours Division, ASCE, Vol. 93,* No. WW2, pp. 129–156 (1967).
98. Borgman, L. E., "Statistical Models for Ocean Waves and Wave Forces," *Advances in Hydroscience, Vol. 8,* pp. 123–156 (1972).
99. Tung, C. C. and N. E. Huang, "Interaction between Waves and Currents and Their Influence on Fluid Forces," *Proceedings on Behavior of Offshore Structures,* Trondheim, Norway, pp. 129–143 (1976).

Certifying Agency Rules and Regulations

100. American Petroleum Institute, "Recommended Practice for Planning, Designing and Constructing Fixed Offshore Platforms," *API RP2A,* Washington, D.C. (March 1979).
101. British Standard Institution, "Fixed Offshore Structures," *Draft for Development,* U.K. (1979).
102. Department of Navy, "Design Manual—Harbor and Coastal Facilities," *WAVFAC DM-26, Naval Facilities Engineering Command,* Washington, D.C. (July 1968).

103. "Codes for Offshore Structures, Design Criteria and Safety Requirements," *Offshore Structures Engineering,* Carneiro, F. L. L. B., A. J. Ferrante, and C. A. Brebbia, (Editors), Gulf Publishing Co., Houston, Texas, pp. 333–354 (1979).
104. *Det Norske Veritas,* "Rules for the Design, Construction and Inspection of Offshore Structures," Oslo, Norway (1977).
105. *Norwegian Petroleum Directorate,* "Environmental Loads—Non-mandatory Supplement to Regulations for the Structural Design of Fixed Structures on the Norwegian Continental Shelf," Stavanger, Norway (September 1977).
106. U.K. Department of Energy, "Guidance on the Design and Construction of Offshore Installation," Second Edition, Petroleum Engineering Division, HMSO, London (1977).
107. U.S. Geological Survey, "Requirements for Verifying the Structural Integrity of OCS Platforms," Conservation Division, OCS Platform Verification Program (October 1979).

CHAPTER 25

Ports and Harbors

Michihiko Noritake*

INTRODUCTION

Classes of Harbors and Ports

A harbor is a partially enclosed area of water which is protected against high waves and strong currents so as to furnish a safe and suitable anchorage for ships seeking refuge, supplies, refueling, repairs, and the loading and/or unloading of cargo.

Harbors may be classified into one of the following broad categories according to the functions they perform and the protection they offer.

Natural harbors are inlets of water where protection from storms and waves is provided by the natural indentations in the topography of the land. They are found in bays, tidal estuaries, and river mouths. Natural harbors insure safe and tranquil waters with suitable entrance to facilitate navigation of ships using them.

Seminatural harbors are enclosed on two sides by headlands and require artificial protection only at the entrance.

Artificial harbors are formed by constructing breakwaters and jetties to provide sufficient protection from waves.

Harbors of refuge are used as a haven for ships in storms, essentially located along established sea routes and dangerous coasts, serving as good anchorage for ships.

Military harbors are naval bases which have the purpose of accommodating naval vessels and serving as supply depots.

Commercial harbors provide protection for ports engaged in foreign or coastwise trade. The docks are provided with necessary facilities for loading and discharging cargo. Municipal- or government-controlled harbors are operated and managed by port authorities existing in many countries, while many commercial harbors are privately owned and operated by companies representing various trade industries.

A port is a harbor where terminal facilities are provided, and it serves as a base for commercial activities. Terminal facilities consist of piers or wharves at which ships berth, transit sheds and warehouses where goods discharged from vessels and awaiting distribution may be stored.

A port of entry is a designated place where foreign goods and foreign citizens may be cleared through a customhouse.

A free port or zone is an isolated, enclosed, and policed area in or adjacent to a port of entry where foreign goods may be discharged for immediate transshipment or stored, repacked, sorted, mixed, or otherwise manipulated without being subject to import duties.

Classes of Cargo

Materials transported by oceangoing ships may be classified into the following categories, depending on the type of handling equipment required.

General cargo refers to a wide variety of goods shipped as units, like automobiles, and items in any sort of package, like bales, bags, boxes, or barrels. Certain care is required in handling general cargo, to prevent damage.

Bulk cargo, on the other hand, includes loose, unpackaged material that is usually loaded and discharged by pouring or pumping. This category comprises dry bulk cargoes, like grain, iron ore, and coal, and also liquid or slurrified commodities, like crude oil, and refined petroleum.

Containers are large sealed boxes containing certain types of freight, usually transported over water on special ships, called containerships, and over land on truck trailers or by railroad. The use of containers aims to speed the handling of freight and decrease water transportation costs by making it unnecessary to unpack the freight at ports.

*Department of Civil Engineering, Kansai University, Suita, Osaka, Japan

TABLE 1. Dimensions of Ships.

Type of Ship	Deadweight (metric tons)	Overall Length (m)	Breadth (m)	Depth (m)	Draft Loaded (m)
Tanker	50,000	226	32.1	16.5	12.5
	100,000	270	39.0	19.2	14.6
	200,000	325	47.2	24.5	19.0
Ore carrier	50,000	222	32.6	16.8	11.9
	100,000	275	42.0	23.0	16.1
	150,000	313	44.5	24.7	18.0
General cargo	10,000	144	19.4	11.2	8.2
	20,000	177	23.4	13.8	10.0
	30,000	199	26.1	15.7	11.0
Containership	30,000	253	30.0	17.0	11.7
	40,000	280	31.9	19.0	12.5
	50,000	285	33.5	20.0	12.8

Characteristics of Ships

The length, beam (breadth), and draft of vessels anticipated to use a port are bound to influence the planning and design for improvements or for the construction of harbor entrances, approach channels, turning basins, and port or terminal facilities. Further, the design of port and terminal facilities may be dictated by the types and sizes of vessels. The relevant characteristics for representative ships of principal types are shown in Table 1.

The size of ships is generally expressed in terms of their displacement tonnage, deadweight tonnage, gross register tonnage, or net register tonnage. These are defined as follows.

Displacement tonnage is the actual weight of the ship, or the weight of water displaced by a floating ship. Displacement loaded is the weight, given in metric tons, of the loaded vessel, when fully loaded with cargo to its Plimsoll mark, or load line (1 ton = 2205 lb). Displacement light is the weight, given in metric tons, of the ship without cargo, fuel, water, and other supplies.

Deadweight tonnage is the total carrying capacity of a ship in metric tons, or the difference between a ship's displacement light and her displacement loaded to the Plimsoll mark. In other words, it is the weight of the load (cargo, fuel, water, and other supplies) that a vessel carries when loaded to the Plimsoll mark. Deadweight tonnage is measured in metric tons, and gross tonnage by volume measurement.

Ships are registered with gross or net tonnage measured in units of 100 ft^3 (2.83 m^3). Gross tonnage is the total volume of the ship including its superstructure and hatches, and net tonnage is the gross tonnage less the volume of the crew's quarters, machinery for navigation, engine room, ballast tanks, and fuel.

Selection of Port Site

The purpose of the selection of a port site is to determine a favorable location that offers refuge, convenience, and ease of operation. Refuge may be appraised on the basis of protection against ocean wave action and the security of moored ships within the harbor. Convenience includes the possibility for littoral, regional or demographic development, technical feasibility of harbor construction, and the availability of inland communications by both land and water.

The choice of a particular location for a new port highly depends upon the exposure to waves, currents, and sediment transport at the planned site. The wave exposure may be determined through available oceanographical and meteorological data, wave periods, wave heights, direction and frequency of occurrence of waves, and wind speed and direction. In order to obtain an accurate statistical measure of wave exposure, sufficient long-time wave records are necessary.

Geophysical and topographical conditions at the proposed site are of great importance in the selection of a port site. Natural protective features such as offshore inlets, reefs, submarine canyons, or outstanding capes may provide considerable protection of the harbor against wave systems.

Another major factor is depth of water. The minimum harbor depth required depends principally on the full load draft of the ships using the port. The maximum water depth is subjected to the restrictions on the construction cost of protective breakwaters. Taking these two limits into account, and with the aid of hydrographic charts of a projected site, the adequate harbor area without prohibitive amount of dredging must be determined. The area so determined may be extended shoreward by dredging, or seaward by further breakwater construction.

Soil conditions along the bottom are also very important factors in the selection of port site. Underwater excavation of rock should be shunned because it is very costly. Generally, hard cohesive materials provide good anchorage, while bottoms consisting of a deep bed of soft material, such as mud, silt, or clay are poor anchorage areas. Poor bottom soil conditions would make construction of breakwaters and wharves very costly.

Another factor for long-range planning is future area requirements. Area requirements are dependent upon the character of the port and the corresponding need for land for transportation, storage, and industry. This factor has become very important in recent years because of the modern containerized shipping which requires extensive land area. A particular port may be superior at an earlier stage of development. However, if the possibility for future expansion is limited, its future development may be difficult. The use of ports for industry is often a principal factor in planning.

COASTAL ENVIRONMENT

Waves and Wave Forces

Water waves may be caused by earthquakes, tides, or certain man-made disturbances such as explosions and moving ships; however, the most interesting waves in the design of harbors and coastal structures are those produced by winds.

WAVE FORM AND MOTION

The friction between wind and the still body of water causes the water surface to undulate. As the wind continues to blow, the wave will grow.

Depth of water has a significant effect on the characteristics of the wave. In deep water, the wave surface is approximately trochoidal in form. Deep water waves are defined as those that occur in water having a depth d greater than one half the wave length L ($d > L/2$). The trochoidal theory was developed by Gerstner [1]. This theory can predict wave profiles quite accurately. According to the trochoidal theory, the movement of waves is described by assuming that each particle of water on the wave surface rotates in a vertical plane about a horizontal axis.

Figure 1 shows the surface of a wave in deep water as a trochoid produced by the rotating particles of water. The trochoid is drawn by a point on a circle that rotates and rolls in a larger concentric circle. The center of rotation moves along a line that is elevated above the still water level. The difference h_0 that this line lies above the still water level is dependent on the wave steepness. The diameter of the smaller circle equals the wave height or amplitude H and the circumference of the larger circle is equal to the wave length L. The line thus drawn represents the waveform at the surface of the water. Particles under the wave surface also describe a circle, but the radii decrease quickly with depth.

The speed of propagation of a waveform, or wave celerity, in feet per second, is

$$C = \frac{L}{T} \tag{1}$$

where

C = wave celerity, ft/s
L = wavelength, or distance between consecutive wave crests, ft
T = wave period, or time for wave to travel one wave length L ft, s

The speed of the waveform in deep water is approximately

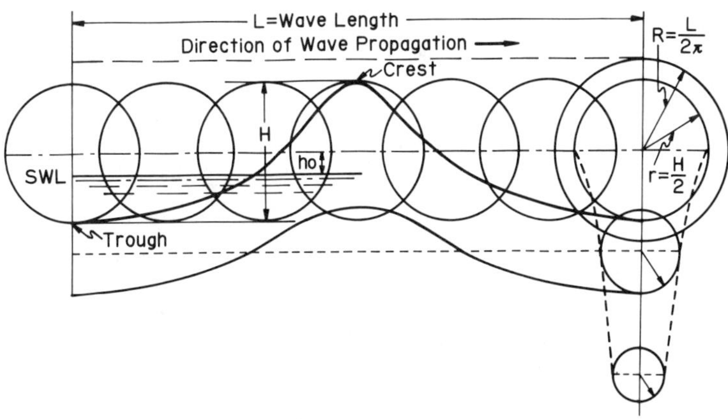

FIGURE 1. Deep water wave profiles.

equal to the velocity obtained by a body falling freely through a height equal to one half the radius of a circle, whose circumference equals the wavelength. Thus:

$$C = \sqrt{2g\frac{1}{2}\frac{L}{2\pi}} = \sqrt{\frac{gL}{2\pi}} \tag{2}$$

where g = acceleration due to gravity, 32.2 ft/s^2.

Shallow water waves are those occurring where the depth is less than one half the wave length ($d<L/2$). In shallow and transitional water, the wave motion of water particles is affected by the frictional forces of the sea bottom. This causes the form of orbital motion to change from circular to nearly elliptical with the major axis being horizontal.

The speed of wave propagation C in shallow water is a function of the water depth. The wave celerity in depths less than 1/25 of the wavelength is represented by the following equation:

$$C = \sqrt{gd} \tag{3}$$

where d = water depth, ft.

The following equation has been proposed by G. B. Airy [2] to estimate wave celerities in shallow water (applicable where $1/25 < d/L < 1/2$):

$$C = \sqrt{\frac{gL}{2\pi}\tanh\frac{2\pi d}{L}} \tag{4}$$

It should be noted that, when d becomes greater than L, the wave celerity is substantially that of a deep water wave.

As a comparison of shallow- and deep-water equations indicates, the variation in wave celerity occurs along the crest of a wave moving at an angle to underwater contours because the part of the wave in deeper water is moving faster than the part in shallower water. Consequently, when waves approach the shore at an oblique angle, the part of the wave nearest the shore slows down with the result that the wave swings around and tends to become parallel to the shoreline. At the same time, the wavelengths decrease as the wave period remains constant. This bending effect is called wave refraction.

When waves move into shallower depths, as along the shore, the orbits of the water particles become distorted owing to the friction exerted by the bottom. This causes the major axis of the elliptical path to tilt shoreward from the horizontal, and the wave gradually transforms from a completely oscillatory wave to a wave of translation. This process is shown in Figure 2. It is at this point that waves exert great forces against bulkheads, breakwaters, and other coastal structures.

CHARACTERISTICS OF OCEAN WAVES

Actual wind-generated waves vary in height, wavelength, period, and direction of propagation. For design of harbors and marine structures, it has been usual to choose a design wave height which is potentially most damaging. For nonbreaking waves, the design wave commonly is supposed to be the most extreme wave predicted to attack the structures at some infrequent intervals, say 50 or 100 years. For breaking waves, the design wave customarily is assumed to be the maximum wave breaking directly against the structures. To select the design wave conditions, therefore, statistical analysis of waves at the site must be made.

Definition of Representative Waves

It is necessary to define the wave height and period from a wave record such as that shown in Figure 3. The zero-upcrossing method is the one customarily accepted as a standard technique for defining waves. This method utilizes the time when the surface wave profile crosses the zero (still water) level upward. An individual wave height (H_1, H_2, $\cdots$) is defined by the vertical distance between the highest and lowest points between the two adjacent zero-upcrossing points, and the corresponding wave period (T_1, T_2, $\cdots$) is defined by the interval of the two crossing points. Based on the height and period data obtained by the above procedure, the following four types of representative waves are usually defined:

1. The highest wave (H_{max}, T_{max}), which refers to the wave having the maximum height in a given wave record;
2. The one-tenth highest wave ($H_{1/10}$, $T_{1/10}$), which corresponds to the average of the heights and periods of the highest one-tenth waves of a given wave record;
3. The significant wave ($H_{1/3}$, $T_{1/3}$), which corresponds to the average of the heights and periods of the highest one-third waves in a given wave record. The height $H_{1/3}$ is often called the significant wave height, and the period $T_{1/3}$ the significant wave period;
4. The mean wave ($\overline{H}$, $\overline{T}$), which corresponds to the mean wave height and period of a given wave record.

Among the above statistically representative waves, the significant wave is most frequently used as a basis for design of marine and coastal structures. The concept of the significant wave was originally introduced by Munk [3] because it represented fairly well what was estimated as "the average height of the waves" by an experienced observer.

Wave Height Variability

When the heights of individual waves on a deepwater wave record are ranked from the highest to lowest, the frequency of occurrence of waves above any given value is closely approximated by the cumulative form of the Rayleigh distribution. This fact was demonstrated by Longuet-Higgins [4]. Therefore, the probability density function $p(H)$

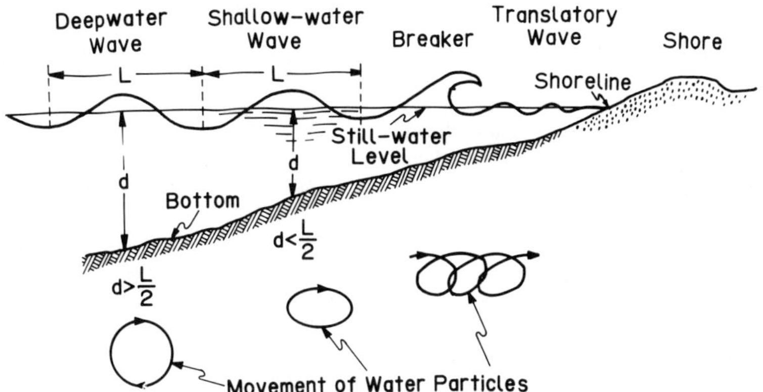

FIGURE 2. Wave transformation and water particle movement.

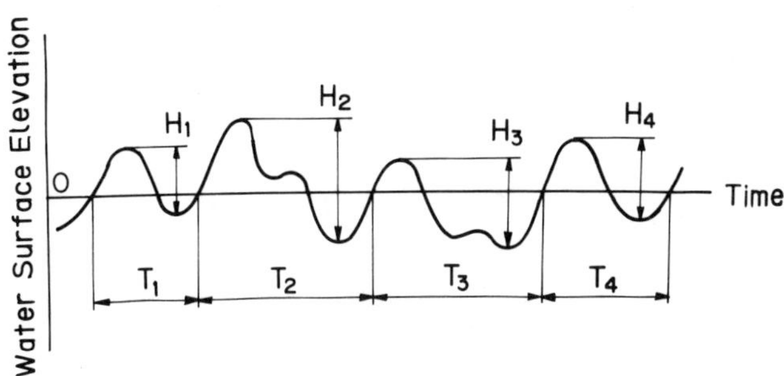

FIGURE 3. Profile of wave record.

for the distribution of individual wave heights H can be given by

$$p(H)dH = \frac{\pi}{2}\frac{H}{\overline{H}^2}\exp\left[-\frac{\pi}{4}\left(\frac{H}{\overline{H}}\right)^2\right]dH \quad (5)$$

and is shown in Figure 4, where $\overline{H}$ is the mean wave height. The probability of a prescribed value H being exceeded is given by

$$P(H) = 1 - \int_0^H p(H)dH = \exp\left[-\frac{\pi}{4}\left(\frac{H}{\overline{H}}\right)^2\right] \quad (6)$$

Assuming the Rayleigh wave height distribution stated above, the relations between the representative wave height parameters can be obtained as follows:

$$\begin{aligned} H_{1/3} &\fallingdotseq 1.60\overline{H} \\ H_{1/10} &\fallingdotseq 2.03\overline{H} \fallingdotseq 1.27H_{1/3} \end{aligned} \quad (7)$$

The maximum wave height, H_{max}, cannot be determined as a definite value, but the prediction commonly used falls within the range

$$H_{max} = (1.6 \sim 2.0)H_{1/3} \quad (8)$$

where the specific value is selected by considering the reliability of data used to estimate design storm waves, types and importance of the structure, degree of structural damage tolerable and associated maintenance and repair costs, and other factors. It should be noted that the effect of such a single wave that is almost twice as high as the significant wave may often be critical to the structure considered.

Wave Period Variability

The periods of individual waves in natural wave trains in deep water show a distribution narrower than that of wave heights. It has been empirically found that the representative period parameters are interrelated, and the following results are given from the observation data [5]:

$$T_{max} \fallingdotseq T_{1/10} \fallingdotseq T_{1/3} \fallingdotseq (1.1 \sim 1.3)\overline{T} \quad (9)$$

WAVE FORCES

There is a considerable difference in wave pressure on coastal and harbor structures depending on whether or not the waves break, or on the geometry of the face of the structure, and on its roughness and permeability. For example, forces produced by nonbreaking waves are fundamentally hydrostatic, but breaking waves exert an additional force caused by the hydrodynamic effects of turbulent water and the compression of entrapped air pockets. Hydrodynamic forces may be much greater than hydrostatic forces; therefore, structures exposed to breaking waves must be designed for greater forces than those exposed only to nonbreaking waves.

Forces Due to Nonbreaking Waves

In protected regions, or where the fetch (the horizontal distance along open water available for wave growth) is limited by landforms surrounding the body of water, and

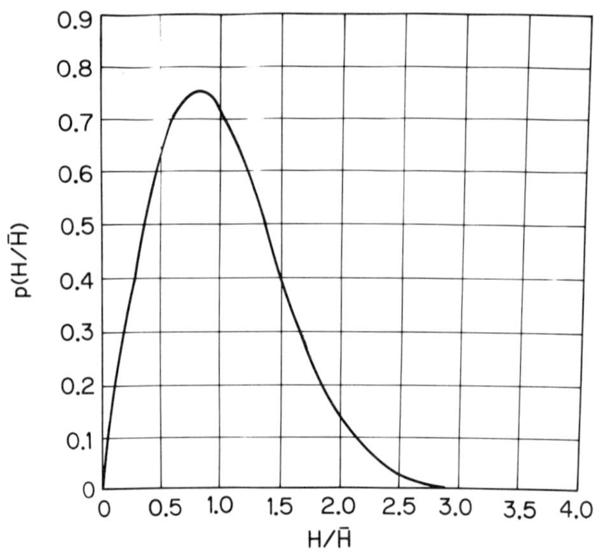

FIGURE 4. Rayleigh density distribution of wave heights.

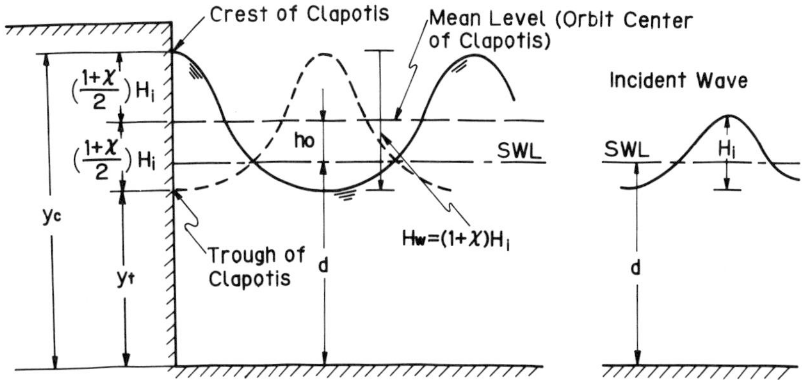

FIGURE 5. Wave conditions for nonbreaking wave forces (Source: Reference 9).

when water depth at the structure is greater than about 1.5 times the maximum expected wave height, nonbreaking waves may occur.

A method for determining the pressure due to nonbreaking waves was proposed by Sainflou [6] and his theory gives good results for long waves of low steepness, while the higher order theory by Miche [7], as modified by Rundgren [8], to consider the wave reflection of the structure, appears to give a satisfactory estimate of wave forces on vertical walls for steep waves.

Wave conditions are shown in Figure 5. The original free wave height that would exist at the structure if the structure were not present is the incident wave height H_i. The wave height that actually exists at the structure is the sum of H_i and the height of the wave reflected by the vertical face of a wall H_r. The wave reflection coefficient χ is equal to H_r/H_i. Wave height at the wall H_w is given by

$$H_w = H_i + H_r = (1 + \chi)H_i \tag{10}$$

If reflection is complete, then $\chi = 1$ and the wave height of the clapotis or standing wave at the structure will be $2H_i$. The height of the clapotis crest above the bottom y_c is given as

$$y_c = d + h_0 + \frac{1 + \chi}{2} H_i \tag{11}$$

where d is the depth from the stillwater level (SWL) and h_0 the height of the orbit center of the clapotis above the stillwater level. The height of the clapotis trough above the bottom y_t is given as

$$y_t = d + h_0 - \frac{1 + \chi}{2} H_i \tag{12}$$

The reflection coefficient depends on the slope, shape, and roughness of the reflecting wall face and possibly on wave steepness and the "wave height-to-water depth" ratio.

Pressure distributions of the crest and trough of a clapotis at a vertical wall are depicted in Figure 6. When the wave is in the crest position, pressure increases from zero at the water surface to $wd + p_1$ at the bottom, where p_1 is approximated by

$$p_1 = \left(\frac{1 + \chi}{2}\right) \frac{wH_i}{\cosh(2\pi d/L)} \tag{13}$$

where w is the unit weight of water and L the wavelength. When the wave is in the trough position, pressure increases from zero at the free water surface to $wd - p_1$ at the bottom.

Forces Due to Breaking Waves

Seawalls, bulkheads, and the inshore end of vertical wall breakwaters and jetties are commonly subjected to the force of breaking waves. Waves breaking directly against vertical wall faces exert great, short duration, dynamic pressures that act near the region where the wave crests hit the structure.

The method for determining these pressures was developed by Minikin [10,11] based partly on the results of Bagnold's study [12]. Minikin's method can predict wave forces that are excessive, as much as 15 to 18 times those calculated for nonbreaking waves.

According to Minikin, the maximum pressure occurs at the stillwater level and is given as

$$p_m = 101w \frac{H_b}{L_D} \frac{d_s}{D} (D + d_s) \tag{14}$$

where p_m is the maximum dynamic pressure, w the unit weight of water, H_b the breaker height, d_s the depth at the toe of the vertical wall, D the water depth measured one wavelength seaward from the wall, and L_D the wavelength

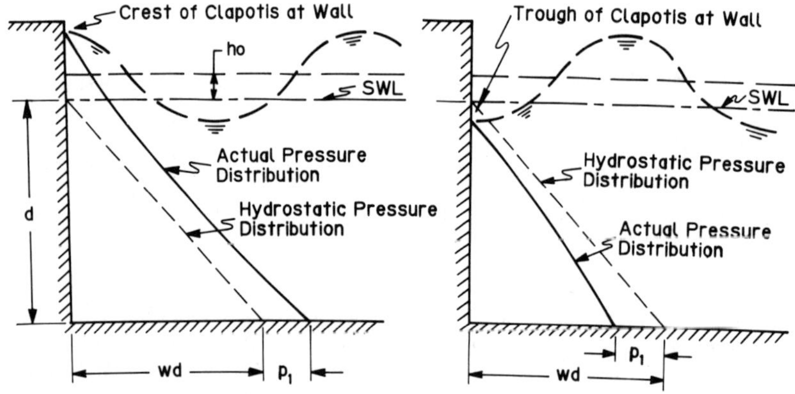

FIGURE 6. Pressure distributions for nonbreaking waves (Source: Reference 9).

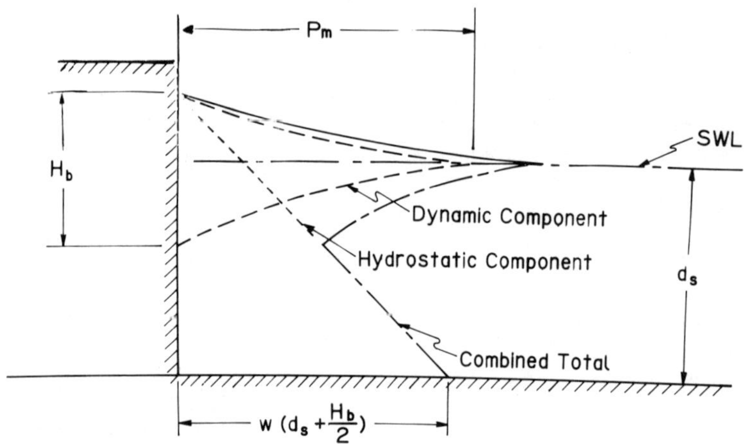

FIGURE 7. Minikin wave pressure distributions (Source: Reference 9).

in water of depth D. The dynamic pressure distribution is shown in Figure 7. The pressure diminishes parabolically from p_m at the stillwater level to zero at a distance of $H_b/2$ above and below the stillwater level. The breaking wave force represented by the area of the dynamic pressure component diagram is approximated as

$$R_m = p_m H_b/3 \tag{15}$$

and the overturning moment about the toe is given by

$$M_m = R_m d_s = p_m H_b d_s/3 \tag{16}$$

In order to determine the total force and overturning moment, the hydrostatic contribution to the force and overturning moment must be added to the results computed from Equations (15) and (16).

Formula Covering Both Breaking and Nonbreaking Wave Pressures

In the case where coastal and harbor structures are planned to be built over long stretches from the shoreline to fairly deep water, the wave pressure formula for the design of these structures must be changed from that pertaining to breaking waves to that of nonbreaking waves. At the boundary point of the applicability of the two types of pressure formulas, the predicted wave pressure suddenly alters. Therefore, the design section of a structure has, in principle, to be changed. To resolve such a singular situation, a set of wave pressure formulas for upright sections of vertical breakwaters was developed by Goda [5].

The formulas of wave pressure assume the occurrence of a trapezoidal pressure distribution when the wave is in the crest position at a vertical wall, as shown in Figure 8, regardless of whether the waves are breaking or nonbreaking. In Figure 8, h is the water depth in front of the breakwater (m), d the water depth above the armor layer of the rubble foundation (m), h' the distance from the stillwater level to the bottom of the upright section (m), h_c the crest elevation of the breakwater above the stillwater level (m), η^* the height above stillwater level to which the wave pressure is exerted (m), p_1 and p_2 are the wave pressures at the stillwater level and bottom (t/m^2), respectively. The wave height for the pressure calculation and other factors are specified as follows.

1. Design Wave. The highest wave is to be used because a breakwater should be designed to be safe against the single wave with the greatest pressure among storm waves. Its height is recommended as $H_{max} = 1.8H_{1/3}$ seaward of the surf zone, whereas within the surf zone the height is recommended as the highest of random breaking waves H_{max} at a distance $5H_{1/3}$ seaward of the wall. The period of the highest wave is usually taken as that of the significant wave, as in Equation (9), i.e., $T_{max} = T_{1/3}$.
2. Elevation to Which the Wave Pressure is Exerted.

$$\eta^* = 0.75(1 + \cos\beta)H_{max} \tag{17}$$

where β is the angle between the direction of incident wave approach and a line normal to the breakwater (deg). In order to provide safety against uncertainty in wave direction, the wave direction should be rotated by an amount of up to 15 degrees toward the line normal to

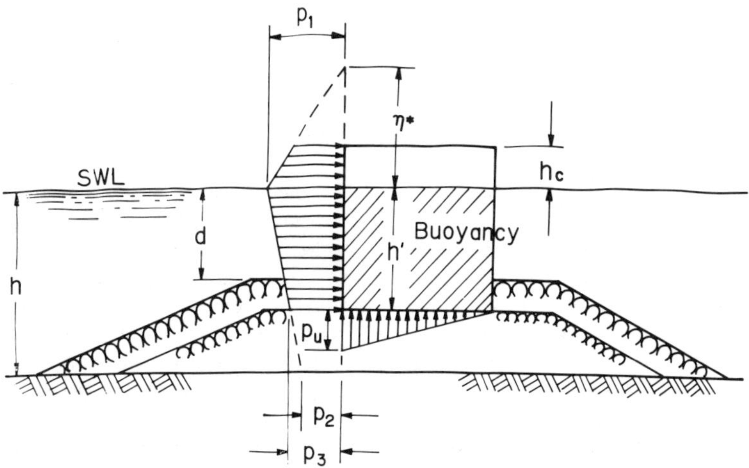

FIGURE 8. Wave pressure distribution for vertical breakwater (Source: Reference 5).

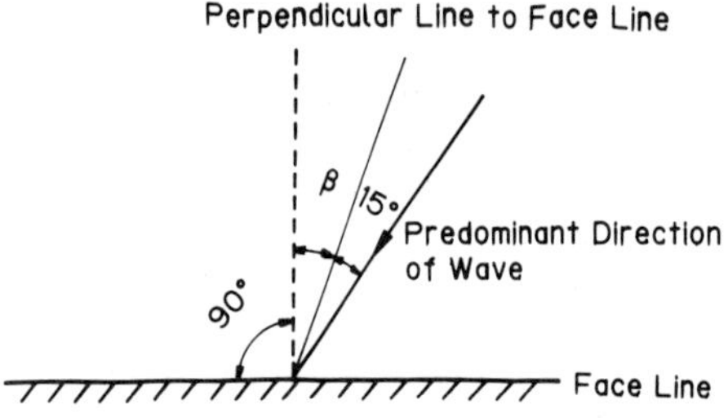

FIGURE 9. Way of taking incident wave angle.

the wall from the principal wave direction (see Figure 9).

3. Wave Pressure on the Front of a Vertical Wall.

$$p_1 = \frac{1}{2}(1 + \cos\beta)(\alpha_1 + \alpha_2\cos^2\beta)w_0H_{max} \quad (18)$$

$$p_2 = \frac{p_1}{\cosh(2\pi h/L)} \quad (19)$$

$$p_3 = \alpha_3 p_1 \quad (20)$$

where

$$\alpha_1 = 0.6 + \frac{1}{2}\left[\frac{4\pi h/L}{\sinh(4\pi h/L)}\right]^2 \quad (21)$$

$$\alpha_2 = \min\left\{\frac{h_b - d}{3h_b}\left(\frac{H_{max}}{d}\right)^2, \frac{2d}{H_{max}}\right\} \quad (22)$$

$$\alpha_3 = 1 - \frac{h'}{h}\left[1 - \frac{1}{\cosh(2\pi h/L)}\right] \quad (23)$$

and

w_0 = unit weight of sea water (t/m³)
L = wavelength at the water depth h (m)
min $\{a, b\}$ = smaller of a and b
h_b = water depth at a distance $5H_{1/3}$ seaward from the front of the wall (m)

Figure 10 represents the value of the coefficient α_1, and Figure 11 the value of $1/\cosh(2\pi h/L)$ for α_3. The symbol L_0(m) in both figures denotes the wavelength corresponding to the significant wave period in deep water.

4. Buoyancy and Uplift Pressure. The buoyancy is to be calculated for the displacement volume of the upright section in still water below the stillwater level, and the uplift pressure acting on the bottom of the upright section is assumed to have a triangular distribution with toe pressure p_u(t/m²) given by Equation (24) below, and with a heel pressure of zero (see Figure 8).

$$p_u = \frac{1}{2}(1 + \cos\beta)\alpha_1\alpha_3 w_0 H_{max} \quad (24)$$

With the above formulas for the wave pressure, the total wave force P(t/m) and its moment M_p(t-m/m) around the bottom of an upright section (see Figure 12) are given by

$$P = \frac{1}{2}(p_1 + p_3)h' + \frac{1}{2}(p_1 + p_4)h_c^* \quad (25)$$

$$M_P = \frac{1}{6}(2p_1 + p_3)h'^2 + \frac{1}{2}(p_1 + p_4)h'h_c^* + \frac{1}{6}(p_1 + 2p_4)h_c^{*2} \quad (26)$$

where

$$p_4 = \begin{cases} p_1\left(1 - \dfrac{h_c}{\eta^*}\right) & :\eta^* > h_c \\ 0 & :\eta^* \leq h_c \end{cases} \quad (27)$$

$$h_c^* = \min\{\eta^*, h_c\} \quad (28)$$

The total uplift force U(t/m) and its moment M_U(t-m/m) around the heel of the upright section (see Figure 12) are

$$U = \frac{1}{2}p_u B \quad (29)$$

$$M_U = \frac{2}{3}UB \quad (30)$$

where B is the width of bottom of the upright section (m).

Tides

The tide is the periodic rising and falling of the surface of the oceans, gulfs, bays, estuaries and other water bodies connected with the oceans caused by the gravitational attraction of the moon and sun. The moon exerts a greater influence on the tides than the sun. This influence varies directly as the mass and inversely as the cube of the distance; therefore, the ratio is nearly 7:3. Generally, the tide ebbs and flows twice in each lunar day (24 hours and 50 minutes).

The highest tides, which occur at intervals of half a lunar month, are called spring tides (see Figure 13). They occur at about the new moon or full moon, i.e., when the sun, moon, and earth fall in line, and the moon and sun act in combination. On the other hand, when the lines connecting

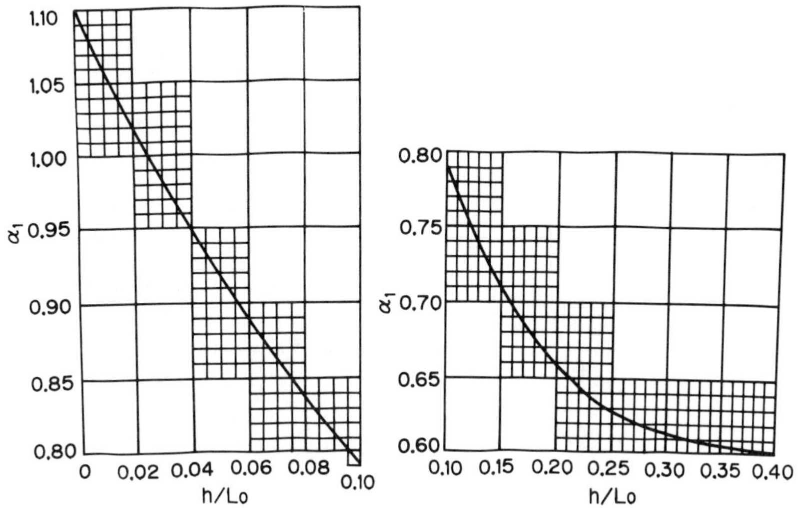

FIGURE 10. Calculation diagrams for parameter α_1 (Source: Reference 5).

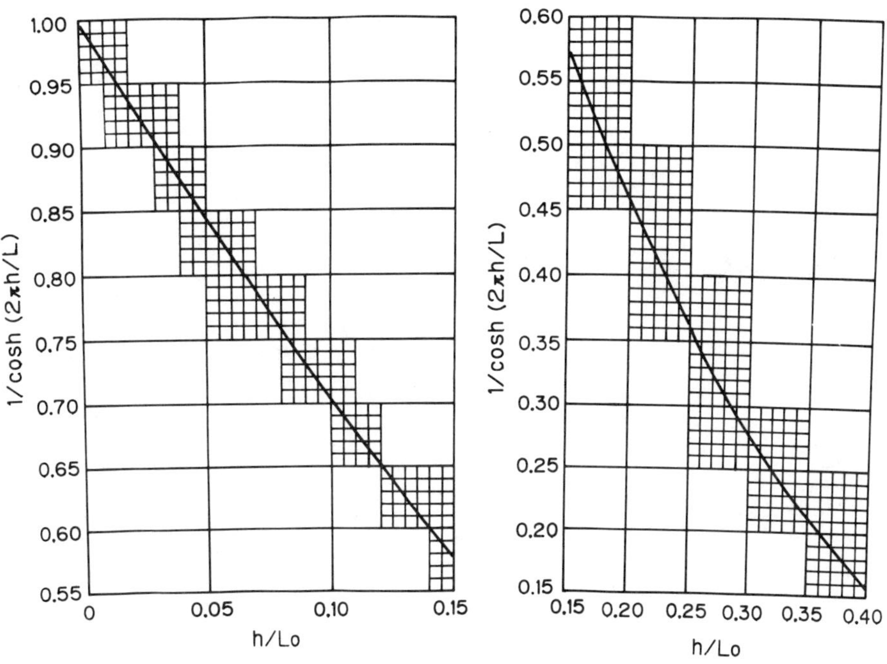

FIGURE 11. Calculation diagrams for factor of $1/\cosh(2\pi h/L)$ (Source: Reference 5).

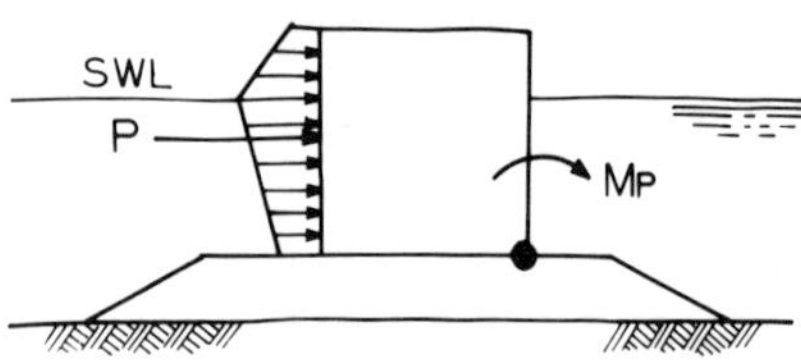

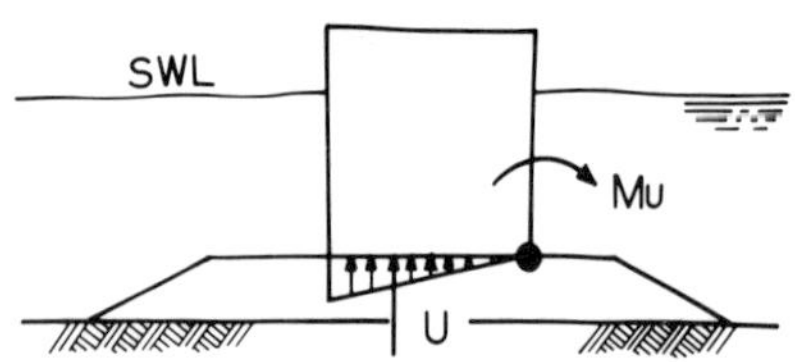

FIGURE 12. Total wave force and total uplift force as well as their moments.

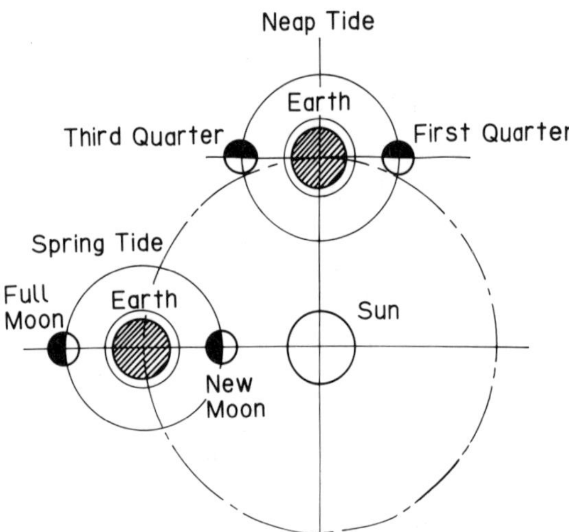

FIGURE 13. Relations between moon phase and spring tide as well as neap tide.

the earth with the sun and the moon form a right angle, i.e., when the moon is at about the first or third quarter, then the tide-generating forces of the moon and sun are subtractive, and the lowest tides of the month, called the neap tides, occur. Actually, the spring and neap tides for each location appear one to three days after the corresponding moon.

When the high (low) tide appears twice each lunar day, the tide is called semidiurnal, while if the high (low) tide occurs only once a day, the tide is called diurnal. The mean periods of the diurnal and semidiurnal tides are 24 hr 50 min and 12 hr 25 min, respectively.

In addition to the effects of the moon and sun, the magnitude of a tide at a given location and time is influenced by:

1. Geographical features
2. Physical character of the shorelines
3. Barometric pressure
4. Currents

At certain inland and confined seas, such as the Mediterranean, the Baltic, and the Gulf of Mexico, the tides are practically negligible. At other places, such as the Bay of Fundy, tides as high as 100 ft occur.

Tidal charts and tables are published for major ports and harbors of the world by the U.S. Coast and Geodetic Survey, and other organizations.

Currents

Since water is viscous, the rotating principles that compose waves do not return to their original position but rather proceed toward the wave movement. This flow is termed a current. All vertical displacements of the water level are, for reasons of continuity, related to currents. In addition, wind shear on the water surface causes currents, and in estuaries fresh water flow generates currents. Except in rare and extreme cases, current forces are normally too small to be of great significance to port and harbor structures. The currents play an important role through their effects on vessels, whether moving or moored, and through their sediment transporting capability, resulting in erosion or accumulation of materials.

Tidal currents are induced by the alternating horizontal movement of water connected with the rise and fall of the tide. For this reason, the motion of these currents has periods which correspond to the tidal constituents.

In the ocean, there exist ocean currents that move continuously in a certain direction. They are generated principally by wind and are classified as drift currents, gradient currents, density currents, and compensation currents according to the generation mechanisms.

The nearshore current system is caused mainly by wave action in and near the breaker zone, and is composed of four parts: the shoreward mass transport of water; longshore cur-

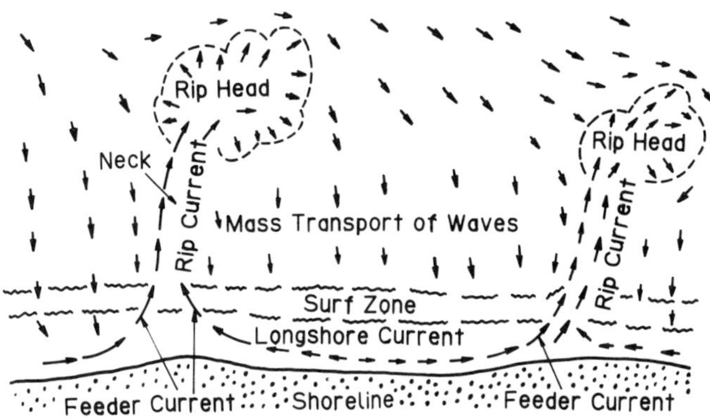

FIGURE 14. Nearshore current system.

rents; seaward return flow, including rip currents; and the longshore movement of the expanding heads of rip currents. They are shown in Figure 14. The net motions of nearshore currents usually have low velocities, but as they transport whatever sedimentary materials are moved by the water motions, they are important in investigating littoral transport.

Longshore current is the littoral current in the breaker zone flowing essentially parallel to the shoreline. Longshore currents are normally generated by waves that obliquely approach the shoreline.

Most remarkable of the exchange mechanisms between offshore and the surf zone are rip currents. Rip currents are surface jets that flow seaward from the shore. Rip current commonly appears as a visible band of agitated water and is the return movement of water piled up on the shore by incoming waves and wind. With the seaward movement converged in a narrow band, its velocity is increased. A rip is composed of three parts: the feeder currents moving parallel to the shoreline inside the breakers; the neck, where the feeder currents concentrate and flow through the breaker zone in a limited band; and the head, where the current broadens and relaxes outside the breaker line.

PLANNING AND DESIGN OF HARBORS

Harbor and coastal structures are the measures by which protection from waves, tides, currents, and winds is provided and the erosion of beaches and shorelines is controlled. Protective works such as breakwaters, jetties, groins, seawalls, revetments, and bulkheads can protect harbors, their entrances, and shores, and shelter an anchorage being used by ships to escape the storms.

Breakwaters

Breakwaters are offshore structures constructed generally parallel to the shoreline to protect a harbor, anchorage, or basin from the force of waves, thereby providing safe accommodation for shipping.

CLASSES OF BREAKWATERS

There are two main types of breakwaters: mound type which may be built of natural rock, concrete block, or a combination of rock/concrete/asphalt and wall type which may be built of natural rock, masonry, concrete block, concrete caisson, rock-filled timber crib, and concrete or steel sheetpile. The amount of rocks required for rubble mound breakwaters is more than that for vertical wall construction. Vertical wall breakwaters are less massive and, therefore, provide more advantageous harbor area. On the other hand, rubble mound breakwaters can be built on bases that would not be appropriate for the support of vertical wall breakwaters. Waves on the slopes of rubble mound breakwaters are inclined to break and be scattered, and consequently the heights of mound breakwaters need not be as high as those required for vertical wall breakwaters. The stability of vertical wall breakwaters is mainly dependent on the ability of the total structure to remain stable under the severe storm waves, while the stability of rubble mound breakwaters depends on the steadiness of the individual pieces of armor rock.

A composite breakwater is often built that is a combined structure consisting of a vertical wall such as caisson placed on top of a rubble mound foundation.

The choice of breakwater type may depend on its purpose, character, and depth of the bottom material, wave forces, availability of materials and construction equipment, and cost.

MOUND BREAKWATERS

The most popular type of breakwater is the rubble mound breakwater. Commonly, this type of breakwater consists of three distinct parts: the armor, the first underlayer, and the second underlayer or core. Typically, this type of breakwater is constructed of rocks ranging in weight from 500 lb up to 20 tons each. The smaller rocks are used as underlayer or

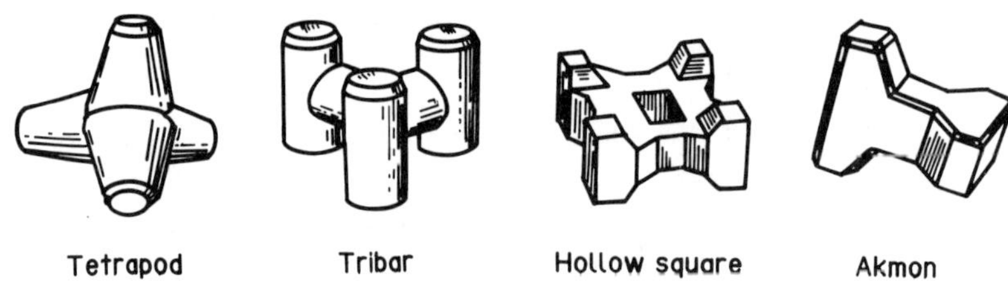

FIGURE 15. Views of concrete armor blocks.

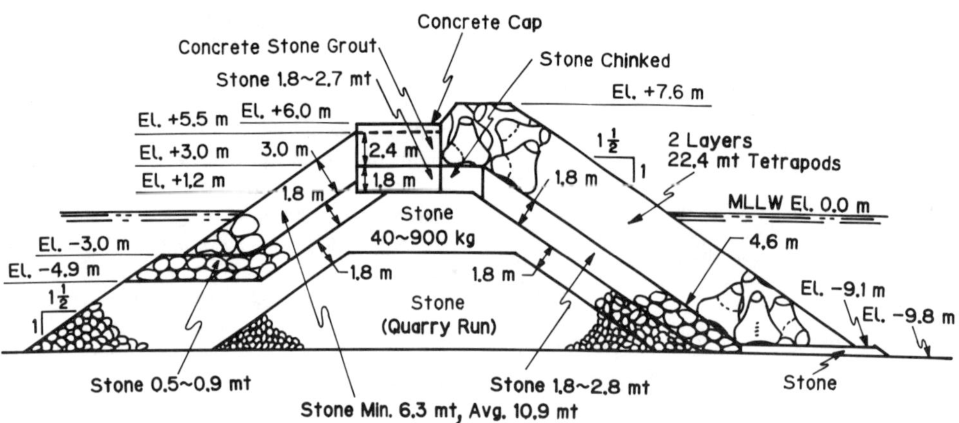

FIGURE 16. Tetrapod and rubble mound breakwater (Source: Reference 9).

core, while the largest sizes, being most resistant to wave forces serve as armor rocks that comprise the outer protective covering of the mound. Usually, the largest armor rocks are employed on the seaward side of the breakwater. Stone sizes of the material available must be great enough to withstand the wave forces so that the mound is not flattened during heavy storms.

When sufficient quantities of durable rock are not available in large enough size for armoring the breakwater, concrete blocks prefabricated in irregular shapes may be used. Various shapes of these blocks have been employed, including tetrapods, tribars, hollow squares, and akmons (see Figure 15). Figure 16 shows a rubble mound breakwater using armor units of tetrapods. The main advantages with these blocks are that they provide better interlocking as well as high permeability and a marked rough surface.

A comparatively impermeable material is often used in the core of the breakwater. The core must prevent the waves as well as sediments from penetrating into the region behind the mound. When a sand-clay or shale is employed in the inner part of the core, the breakwater is classified as a solid fill structure. The use of a fine-grained material in the core may be favorable economy-wise or so as to secure good compactness. The voids of the upper portion of the core may also be filled with asphaltic concrete mix or portland cement concrete for the purpose of improving the stability and imperviousness of the breakwater.

STABILITY OF RUBBLE MOUND BREAKWATERS

The stability of rubble mound breakwaters is dependent mainly on the ability of the individual pieces of rock that comprise the armor layer to withstand attack by storm waves. Since the wave action on a mound breakwater is complicated, it is necessary to rely on empirical formulas that are based on the results of extensive model tests.

The following stability formula developed by the U.S. Army Corps of Engineers [9] has been accepted for determining the weight of each armor rock:

$$W = \frac{w_r H^3}{K_D (S_r - 1)^3 \cot \theta} \tag{31}$$

where

W = weight of an individual armor unit in the primary cover layer, N or lb
w_r = unit weight (saturated surface dry) of armor unit, N/m^3 or lb/ft^3
H = design wave height, at the structure site, m or ft
S_r = specific gravity of armor unit, relative to the water at the structure ($S_r = w_r/w_w$)
w_w = unit weight of water, fresh water = 9,800 N/m^3 (62.4 lb/ft^3), sea water = 10,047 N/m^3 (64.0 lb/ft^3)
θ = angle of structure slope measured from horizontal, deg
K_D = stability coefficient

Table 2 gives proposed values of K_D for use in Equation (31).

The dimensionless stability coefficient, K_D, is used to account for the effect of the following major factors:

1. Shape of armor units
2. Number of units comprising the thickness of armor layer
3. Manner of placing armor units
4. Surface roughness and sharpness of edges of armor units (degree of interlocking of armor units)
5. Type of waves which attack the breakwater (breaking or nonbreaking)
6. Part of the breakwater covered by armor unit (trunk or head)

It is well-proven that the head of a breakwater or jetty normally suffers more extensive and frequent damage than the trunk of the structure. In order to allow for this fact, different values of K_D are provided for the head and trunk.

It is noted that the values in Table 2 provide little or no safety factor.

VERTICAL WALL BREAKWATERS

Vertical wall breakwaters are second major breakwaters. Vertical wall breakwaters consist of the following types:

1. Timber or precast concrete cribs filled with rock
2. Concrete caissons filled with rock or sand
3. Sheetpile breakwaters

Breakwaters built of rock-filled timber or precast concrete cribs consist of large boxlike compartments which are divided by transverse and longitudinal walls. The cribs are founded on a prepared base, then filled with rock, and sunk end to end along the line of the breakwater. Concrete caissons are bulky watertight boxes which are floated into correct position, normally settled on a prepared foundation of rubble, filled with rock or sand for stability, and capped with concrete. Concrete caissons have the advantage of diminishing installation time on water.

Cellular steel sheetpile breakwaters are difficult to build in exposed locations where storm waves are too severe, because each cell must be stable by itself. Steel and concrete sheetpile breakwaters are sometimes employed where the sea bottom is composed of soft material that extends to a great depth. The sheeting should extend to an adequate depth below the bottom to prevent bottom scour in front of the breakwater. It is usual to place riprap against the toe of sheeting to defend the bottom from scour. Such a breakwater is shown in Figure 17.

Jetties

A jetty is a structure extending into a sea to direct and confine the stream or tidal flow to a selected channel and to prevent the shoaling of the channel by littoral material.

TABLE 2. Suggested K_D Values for Use in Determining Armor Unit Weight (No-Damage Criteria and Minor Overtopping).

Armor Units	n[b]	Placement	Structure Trunk K_D[a] Breaking Wave	Structure Trunk K_D[a] Nonbreaking Wave	Structure Head K_D Breaking Wave	Structure Head K_D Nonbreaking Wave	Structure Head Slope Cot θ
Quarrystone							
Smooth rounded	2	Random	1.2	2.4	1.1	1.9	1.5 to 3.0
Smooth rounded	3	Random	1.6	3.2	1.4	2.3	[d]
Rough angular	1	Random[c]		2.9		2.3	[d]
Rough angular	2	Random	2.0	4.0	1.9	3.2	1.5
					1.6	2.8	2.0
					1.3	2.3	3.0
Rough angular	3	Random	2.2	4.5	2.1	4.2	[d]
Rough angular	2	Special[e]	5.8	7.0	5.3	6.4	[d]
Tetrapod and Quardripod	2	Random	7.0	8.0	5.0	6.0	1.5
					4.5	5.5	2.0
					3.5	4.0	3.0
Tribar	2	Random	9.0	10.0	8.3	9.0	1.5
					7.8	8.5	2.0
					6.0	6.5	3.0
Modified cube	2	Random	6.5	7.5	—	5.0	[d]
Tribar	1	Uniform	12.0	15.0	7.5	9.5	[d]

[a]Applicable to slopes ranging from 1 on 1.5 to 1 on 5.
[b]n is the number of units comprising the thickness of the armor layer.
[c]The use of single layer of quarrystone armor units is not recommended for structures subject to breaking waves, and only under special conditions for structures subject to nonbreaking waves. When it is used, the stone should be carefully placed.
[d]The use of K_D should be limited to slopes ranging from 1 on 1.5 to 1 on 3. Some armor units tested on a structure head indicate a K_D-slope dependence.
[e]Special placement with long axis of stone placed perpendicular to structure face. (*Source*: Reference 9).

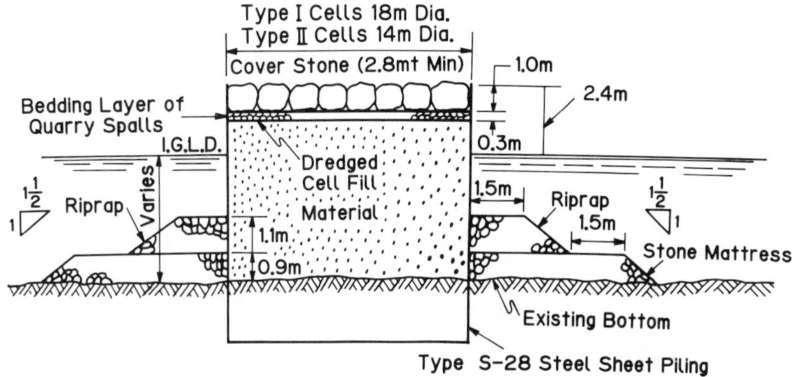

FIGURE 17. Cellular steel sheetpile breakwater (Source: Reference 9).

Jetties are built to help deepen and stabilize an entrance channel and thus facilitate navigation. Normally constructed in pairs with one on each side of a river mouth or bay entrance, the jetty serves to protect the entrance channel from wave action and excessive or otherwise undesirable currents or tides. Structurally, jetties and breakwaters are similar; however, the design standards for jetties may be somewhat lower than are those for breakwaters, because jetties are not exposed to direct wave attack to as great an extent as are breakwaters.

On account of the possibility of erosion and sedimentation caused by changes in the velocity and direction of channel currents, it is important to research the magnitude and direction of existing tidal currents and the effect that the construction of jetties might have on these currents. This is decisive in determining the spacing between jetties. The design will also depend on the existing topographical features of the area and the channel dimensions required for navigation. Engineering judgment and model studies are needed when analytical tools are insufficient.

Groins

The erosion of beach areas results from the effects of breaking waves, in particular when the waves approach the shoreline at an oblique angle. Waves approaching a shoreline obliquely produce a current that commonly parallels the shoreline. This current, named a longshore current, sweeps along the sandy particles that compose the beach bottom. The material that moves in the littoral coastal zone under the influence of waves and the longshore current is termed littoral drift.

The most common coastal protective measure is to construct a groin or a system of groins. A groin is a rigid structure that is built nearly perpendicular to the shoreline in order to retard shore erosion or to establish or maintain the beach by trapping a part of the littoral drift. The groin functions as a partial dam that causes material to deposit on the updrift side beaches. The decrease of material supply to the downdrift side induces the downdrift beach to be eroded. The shoreline changes that follow the construction of a groin system are shown in Figure 18.

There are two broad types of groins: permeable and impermeable. Permeable groins allow the passing of appreciable amount of littoral drift through the structure. Impermeable groins, the most popular type, function as a barrier to the passing of littoral drift.

A variety of groins have been built, using timber, steel, stone, and concrete. A typical design for an impermeable timber-steel sheet-pile groin is shown in Figure 19. Cellular steel sheet-pile groins also have been used successfully, as have prestressed-concrete sheet-pile groins and rubble mound groins. The latter normally is built with a core of fine quarry run material to make it sandtight.

The choice of groin type is based on the following factors:

1. Availability of materials
2. Foundation conditions
3. Topography of the shore

The hydraulic behavior of a system of groins is extremely complicated and its performance may be affected by:

1. The specific weight, size, and shape of the material that composes the littoral drift
2. The height, period, and angle of attack of approaching waves
3. The range of the tide and the magnitude and direction of tidal currents
4. The design characteristics of the groin system, including the groin orientation, length, spacing, and crown elevation

Groins normally are constructed in a straight line perpendicular to the shoreline. The use of groins of the T or L head types seems to provide little advantage. These types

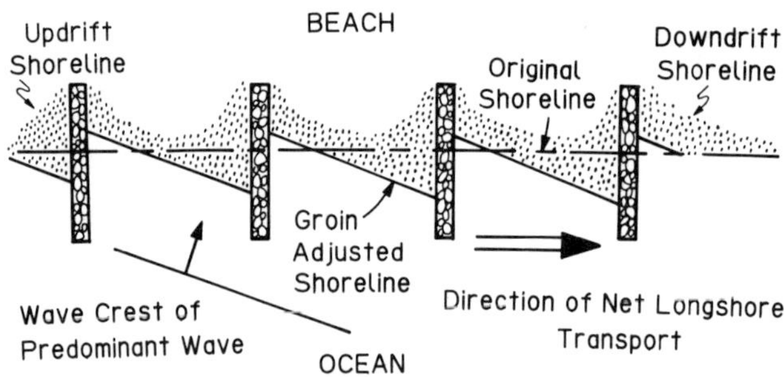

FIGURE 18. Shoreline configuration for a system of groins.

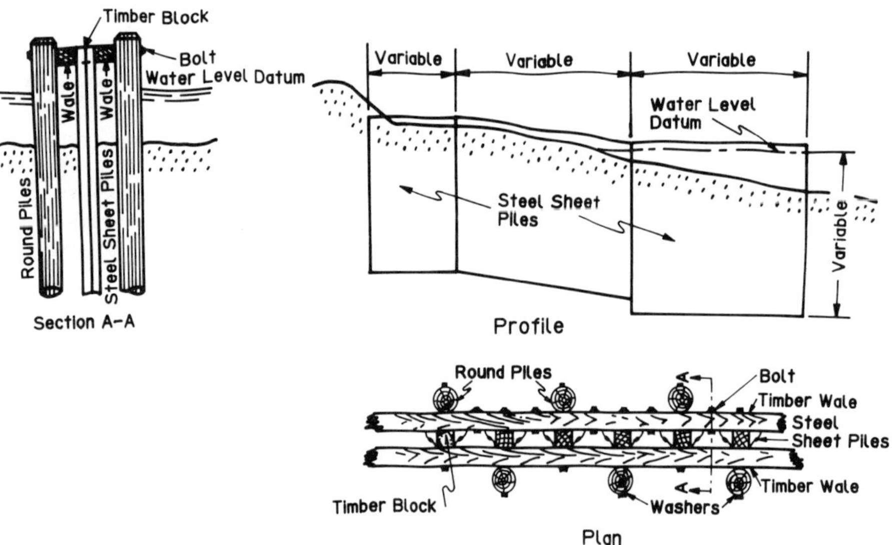

FIGURE 19. Typical timber-steel sheet-pile groin (Source: Reference 9).

may be more costly to construct and will be subjected to more scour at the end of the structure than will be suffered with the straight groin.

The length of a groin will depend on the character and extent of the prevailing erosion and the planned shape and location of the stabilized shoreline. The total length commonly is from 100 to 150 ft.

The spacing between groins depends on the groin length and the desired alignment and location of the stabilized shoreline. The U.S. Army Corps of Engineers [9] recommends that the spacing of adjoining groins should be two or three times the groin length from the berm crest to the seaward end. In reality, it is necessary to adjust the groin length and spacing according to the conditions at the locality.

The elevation of the crest of a groin determines to a certain degree the quantity of material trapped by the groin. If it is required to maintain a supply of material on the downdrift beach, the groin may be erected to a low height, permitting certain waves to overtop the structure. In case no passing of material beyond the groin is desired, the elevation of the crest should be such that heavy waves will not overtop the structure.

Seawalls, Revetments, and Bulkheads

Seawalls, revetments, and bulkheads are structures that are placed parallel to the shoreline to separate a land area from a water area. Although these structures have this same common purpose, there are great differences in specific function and design. Seawalls are rather massive structures normally placed along otherwise unprotected shores to withstand the full force of waves. Seawalls suffer the forces of waves on the seaward side and active earth pressure on the shoreward side.

A revetment is also employed to protect the shorelines against erosion by currents or light wave action. It is essentially a protective pavement supported by an earth slope. A bulkhead is not intended to withstand severe wave action, but merely to serve as a retaining wall to prevent landsliding.

Typical structural types of seawalls are shown in Figure 20. These include a sloping wall, stepped wall, and a curved wall. The curved face wall may be either a non-reentrant type, that may be considered as vertical, or a reentrant type, that turns the wave back upon itself.

Sloping or vertical face seawalls are less effective for reducing wave overtopping than the concave-curved and reentrant face structures. The volume of wave overtopping can be reduced greatly by the use of an armor block facing. The stepped seawall is employed under sound waves. It may also suffer undesirable wave overtopping when subjected to heavy waves and high winds.

Under the most severe wave conditions, massive curved-face seawalls are built most frequently. For this type of seawall, the use of a sheetpile cutoff wall at the toe of the structure is recommended to prevent or reduce the scouring and undermining of the foundation due to wave action. As a further means to prevent scour, large armor rocks may be piled at the toe of the structure. Both of these features are shown in Figure 21.

There are two broad types of revetments, rigid and flexible. The rigid type of revetment is composed of a series of cast-in-place concrete slabs. Flexible or articulated armor unit type revetments are built of riprap or interlocking con-

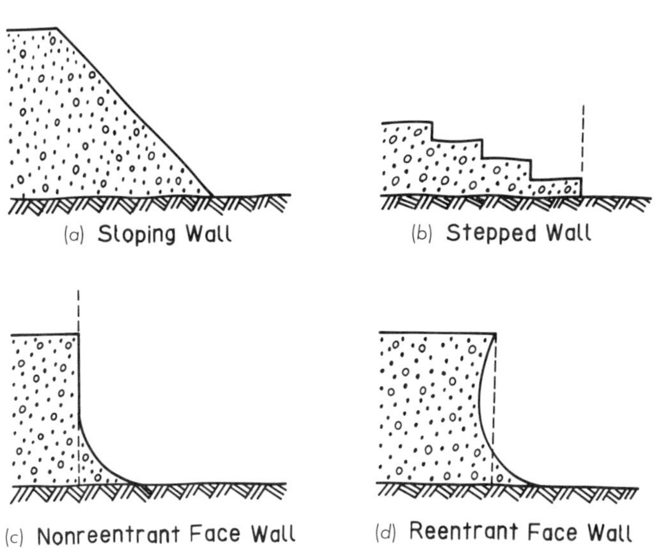

FIGURE 20. Typical shapes of seawalls (Source: Reference 9).

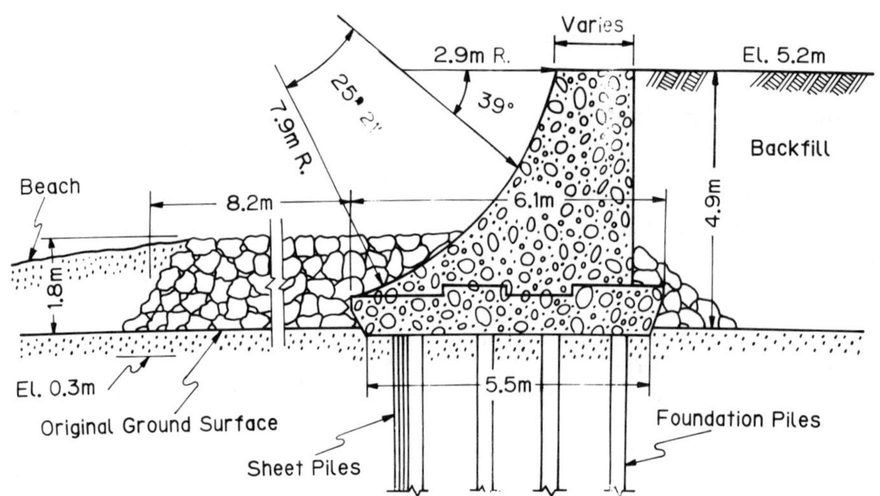

FIGURE 21. Concrete curved-face seawall (Source: Reference 9).

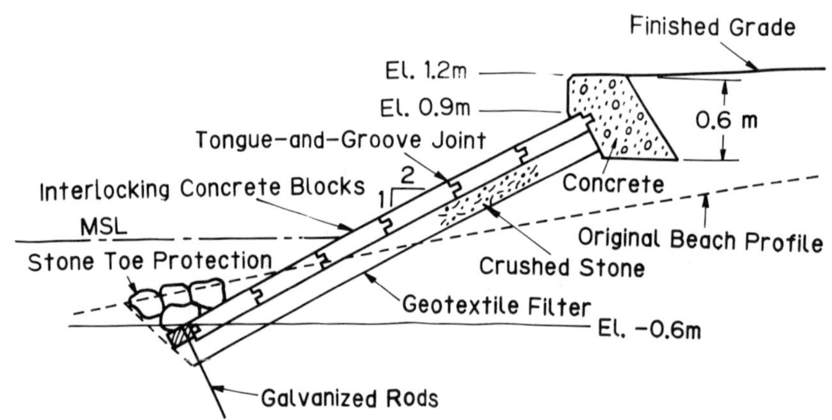

FIGURE 22. Interlocking concrete block revetment (Source: Reference 9).

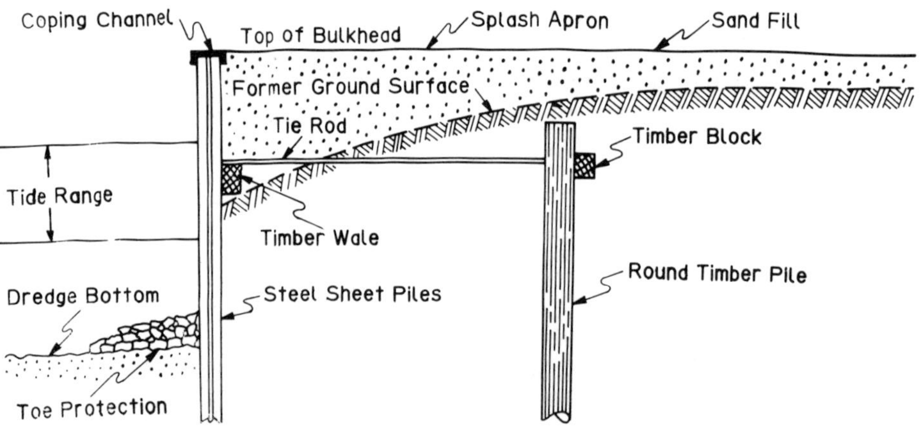

FIGURE 23. Steel sheet-pile bulkhead (Source: Reference 9).

crete blocks that cover the shore slope. Both the articulated block structure and the quarrystone or riprap structure allow for the relief of hydrostatic uplift pressure produced by wave action. The underlying geotextile filter and gravel or a crushed-stone filter and bedding layer relieve the pressure over the entire foundation area rather than through specially made weep holes. Typical interlocking concrete block revetment is shown in Figure 22.

A common type of bulkhead is shown in Figure 23. A bulkhead is supported in a cantilever manner by the soil into which it is driven. Additional support for a bulkhead may be provided by tie rods connected to anchor piles, that are driven some distance shoreward. Cellular-steel sheet-pile bulkheads are employed where rock is near the surface and sufficient penetration is impossible for the anchored sheet-pile bulkhead such as shown in Figure 23. Concrete slab bulkheads and timber sheet-pile structures may also be used.

The selections of the location and length of seawalls, revetments, and bulkheads are influenced by local circumstances. The location of the structures with respect to the shoreline usually corresponds to the line of protection against further erosion of the sea. The length of these structures depends on how much shoreline is to be protected against the sea.

An important factor in the design of seawalls, revetments, and bulkheads is the determination of the height of the structure, and it may be too expensive to build structures to preclude overtopping by the largest waves. This determination will depend on the relation between the amount of water that will pass over the top of the structure under the most severe wave conditions and the cost to install suitable pumping facilities to remove the overtopped water in a shoreward area.

Wave Runup and Overtopping

In designing a coastal structure to prevent wave overtopping, an estimation of the magnitude of wave runup is needed. Wave runup is defined as the vertical height above the stillwater level to which the rush of water from an incident wave will rise on the face of the structure. Thus the runup, when added to the stillwater elevation, determines the minimum height of the crest of the structure.

Runup depends mainly on the structure shape and roughness, water depth at the structure toe, incident wave characteristics, and bottom slope in front of a structure. The U.S. Army Corps of Engineers [9] has presented a series of empirical curves exemplified by Figure 24, by which wave runup on smooth, impermeable slopes can be estimated. Similar curves have been provided for runup on slopes covered with riprap, rubble, and concrete armor units. The curves are in dimensionless form for the relative runup R/H_0' as a function of the deepwater wave steepness, H_0'/gT^2, and structure slope, cot θ. Here, R is the runup height measured vertically from the stillwater level, H_0' the unrefracted deepwater wave height, T the wave period, and g

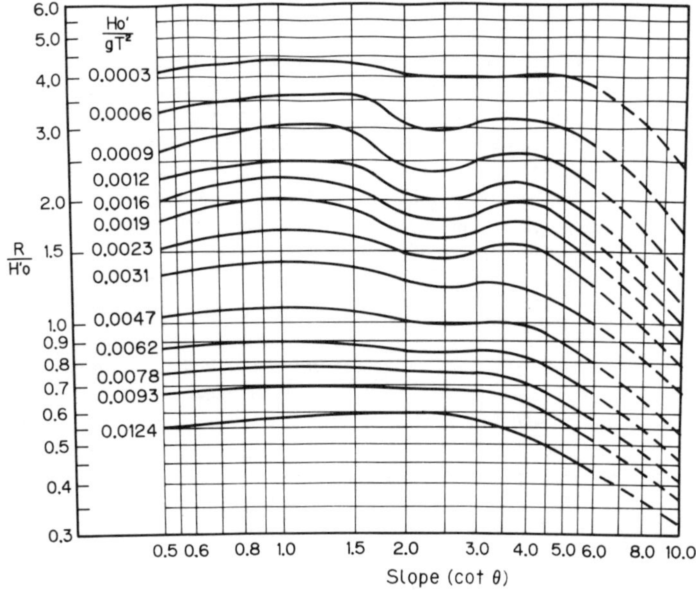

FIGURE 24. Wave runup on smooth, impermeable slopes when $D_s/H_0' = 0$ (structures fronted by a 1:10 slope) (Source: Reference 9).

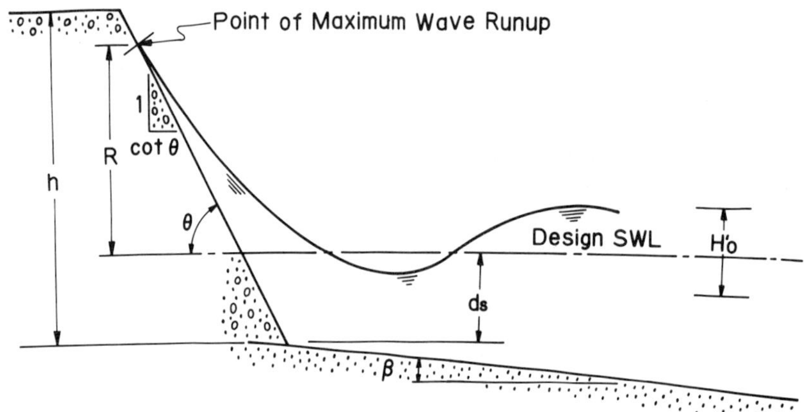

FIGURE 25. Definition sketch for wave runup (Source: Reference 9).

the acceleration of gravity. Definitions including these variables are shown in Figure 25.

There is a scale effect whereby the predicted values of wave runup from small-scale laboratory tests are smaller than those observed on prototype structures, however the predicted runup values can be adjusted by using the figure provided by the Corps of Engineers [9].

PLANNING AND DESIGN OF PORT FACILITIES

A port or a marine terminal aims to provide facilities for transshipment of ships' cargo transported to and from inland areas by rail, road, inland waterway, and pipeline. The needs of increased world trade result in increases in the number and size of the merchant fleet, which creates a growing need for the construction of new ports and the enlargement of existing port facilities.

In planning and designing port facilities, due consideration is needed to make proper decisions concerning the port layout, substructure designs, number of required berths, dimensions for berth and slip, space for apron and transit sheds, and other special facilities for the handling and storage of many types of materials.

Wharves

CLASSES OF WHARVES

A wharf is a general term for a marine structure built on the shore of navigable waters so that ships may moor and load/unload cargo or embark/disembark passengers. A marginal wharf or quay is a wharf that parallels the shoreline. A finger pier, or simply a pier, is a wharf that projects at any angle into the water. A slip is the ship's berthing and maneuvering space between two adjacent piers. A pier built in combination with a breakwater is termed a breakwater pier. A pier has berths on two sides and may also be used for accommodating vessels at the end. Piers that are more or less parallel to the shore and connected to land by a trestle and/or causeway are referred to as T- or L-shaped piers. Wharves, piers, and quays are often called docks. Marginal wharves are superior to piers on account of its easier berthing of vessels. However, piers can furnish more berths per unit length of waterfront.

SOLID FILL AND OPEN CONSTRUCTION FOR WHARVES

Basically, wharf structures fall into two principal classifications: the solid fill type, such as bulkheads, caissons, sheetpile cells, cribs, and gravity (quay) walls; and the open type. The sheetpile bulkhead, shown in Figure 26, consists of a vertical wall which is backfilled by earth supporting a paved deck. The wall may be a cantilevered, anchored steel sheetpile bulkhead, a gravity structure made of concrete or timber, or another vertical structure. The concrete caissons may have open wells and cutting edges, or they may have a closed bottom (see Figure 27). When they have a closed bottom, they are lowered onto a prepared foundation, ordinarily consisting of a gravel or crushed-stone bed or leveling layer. The solid fill type of sharf structure has the advantages of resistance to the impact of mooring vessels, stability, and strength to resist currents and tides.

In the open type wharf construction, the deck is supported by a series of piles, which may be timber, steel or reinforced concrete. Longitudinal beams may be provided to support heavy concentrated loads. A high-level deck normally has a solid deck slab (see Figure 28). A variation of the open-type wharf substructure utilizes a relieving platform on which fill is superimposed, capped by a paved deck (see Figure 29). This type of design offers the advantages of high resistance to impact. The open type of wharf is more economical in deep water locations and where tall superstructures are needed. However, the open wharf construction supported by timber piles is subject to decay by marine borers.

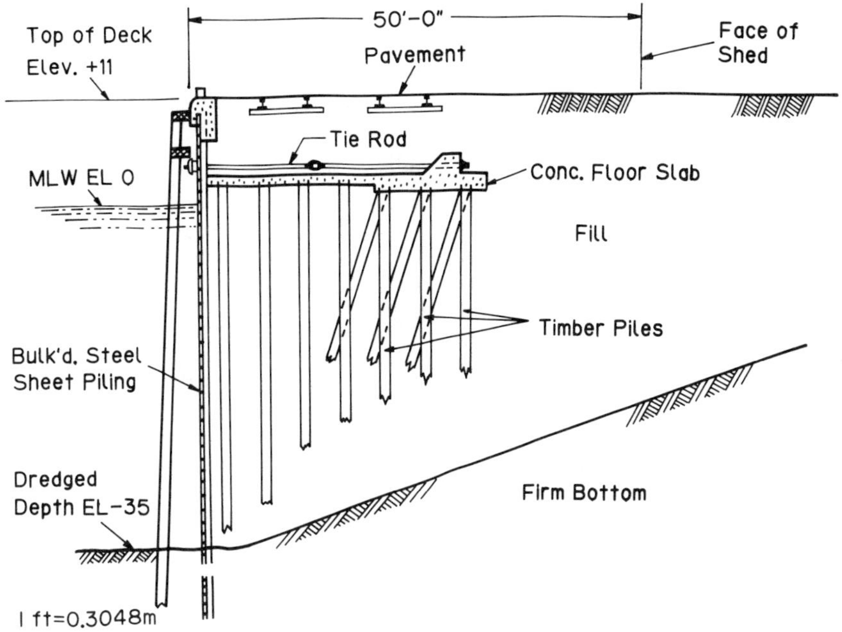

FIGURE 26. Solid-type bulkhead wharf (Source: Reference 13).

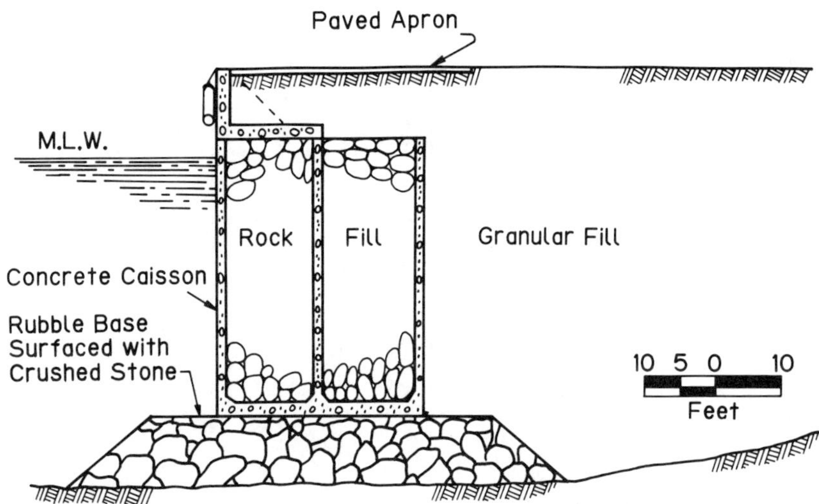

FIGURE 27. Solid-type bulkhead wharf using concrete caissons with closed bottom (Source: *Design and Construction of Ports and Marine Structures*, Second Edition, by Quinn, A. D. Copyright © 1972, McGraw-Hill. Reproduced with permission of McGraw-Hill Book Co.)

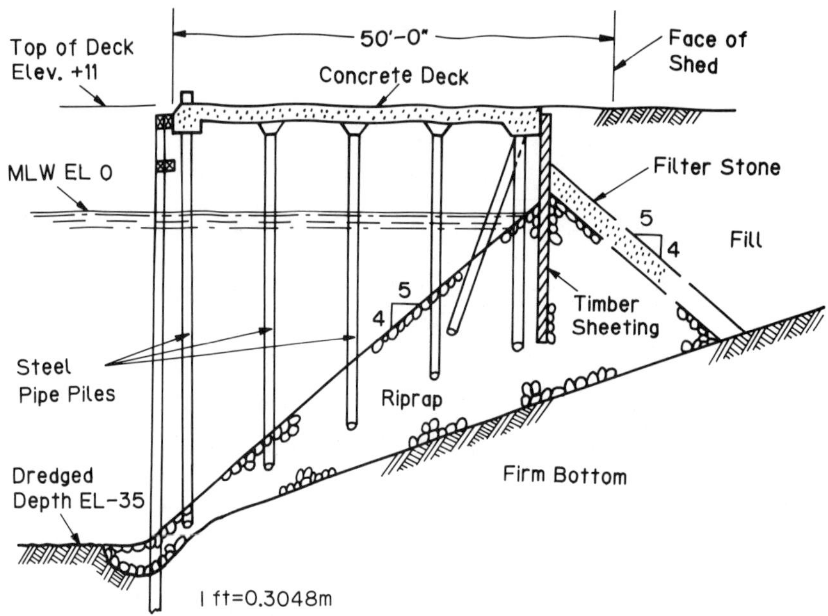

FIGURE 28. High-level open-type wharf (Source: Reference 13).

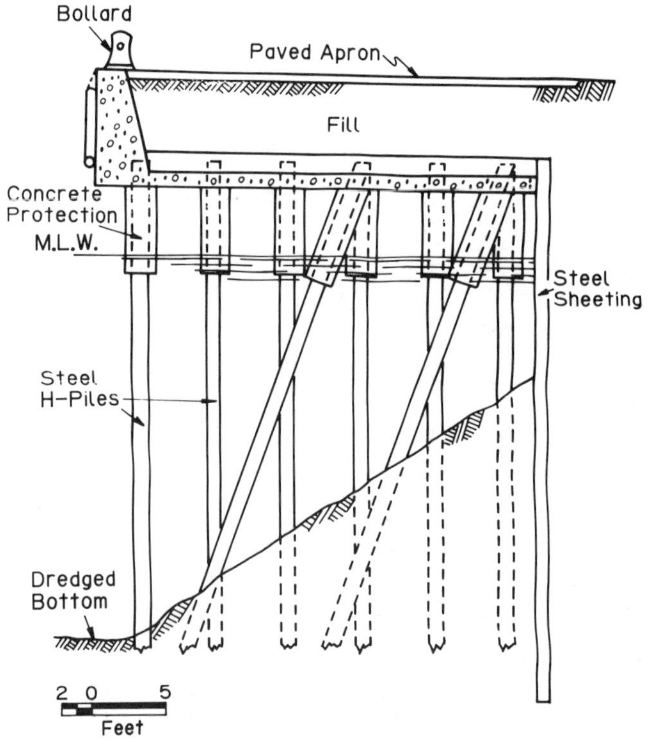

FIGURE 29. Relieving-platform type wharf (Source: *Design and Construction of Ports and Marine Structures*, Second Edition, by Quinn, A. D. Copyright © 1972, McGraw-Hill. Reproduced with permission of McGraw-Hill Book Co.).

Estimation of Optimum Number of Berths

The planning of the size of a port starts with an evaluation of the amount of present and future commerce and types of shipping. This information permits the port planner to estimate the number, type, and sizes of the ships to be accommodated.

The volume of commerce to be shipped through a port will depend on such variables as the individual national economic program, the location and trading custom of the port, and the character and size of its tributary area, or hinterland. The hinterland is defined as that region within which the overall cost of commodity flows through the port in question is less than the corresponding costs via other competing ports, based on existing rates and charges [15].

Three functions that are essential to ports are as follows [16]:

1. Unload and load ships' cargo with efficiency and dispatch.
2. Furnish sufficient temporary and long-term storage for incoming and outgoing cargo.
3. Install highway, rail, and marine transport connections for movement of cargo into and out of the port region.

Although the practical capacity of a port may be restricted by any one of these three functions, the first is often the major controlling factor. Thus, the operating capacity of berths is essentially the product of the cargo handling rate (tons/day/occupied berth) and the number and degree of utilization of berths.

The rate of loading and unloading of ship's cargo is dependent on the following factors.

1. Types of cargo
2. Vessel type and size
3. Availability and size of stevedore gangs
4. Degree of mechanization and methods of cargo handling operation

At a typical general cargo berth, the average loading or discharging rate is about 200 to 240 tons/hatch/8-hr day. For berth planning purposes, an average of 40 stevedore gang-hours per working day may be used [15]. On this basis, the maximum cargo handling rate for an occupied berth is 1000 to 1200 tons/berth/day.

The estimation of the optimum number of berths to be provided must be determined in the face of fluctuations in demand. Findings of independent studies by Fratar, *et al.* [17], Plumlee [18], Mettam [19], Nicolaou [20], Jones and Blunden [21], and Agerschou and Korsgaard [22] have shown that the types of $M/M/S(\infty)$ and $M/E_k/S(\infty)$, with $k = 2$ or 3, of queuing theory models are most suitable in order to explain the ship's movement in port. The notation $A/B/m(K)$ denotes an m-servers (berths) queuing system, where A, B, and K describe the interarrival time distribution, the service time distribution, and the size of the customer (ship) population [23]. A and B, for example, take on meanings from the following symbols whose interpretation is given in terms of distributions within parentheses: M (exponential); E_k (k-phase Erlangian). It is noted that when the value of k is one, the Erlangian distribution coincides with the exponential distribution (i.e., $E_1 = M$). Further, when the size of the customer population is supposed to be infinite, K takes the value of infinity (∞).

According to the queuing theory [24],

$$(\bar{n}_S)_k = (\bar{n}_{w,S})_k + \bar{n}_{b,S} = \frac{a^{S+1}}{(S-1)!(S-a)^2} \times \left[\sum_{n=0}^{S-1}\left(\frac{a^n}{n!}\right) + \frac{a^S}{(S-1)!(S-a)}\right]^{-1} \times \left[\frac{1+\left(\frac{1}{k}\right)}{2} + \left(1-\frac{1}{k}\right)\left(1-\frac{a}{S}\right) \times (S-1)\frac{(4+5S)^{1/2}-2}{32a}\right] + a \tag{32}$$

where

$(\bar{n}_S)_k$ = average number of ships present in a port in $M/E_k/S(\infty)$ model
$(\bar{n}_{w,S})_k$ = average number of ships waiting for berths in $M/E_k/S(\infty)$ model
$\bar{n}_{b,S}$ = average number of ships served at berths
a = traffic intensity $(=\bar{n}_{b,S})$
S = number of berths

The relationship between Q, the total cargo tonnage loaded onto and discharged from the ships during the time period of port operation considered T (usually 365 days), and the traffic intensity, a, is given by the following equation:

$$a = Q/(RT) \tag{33}$$

where R = daily rate of the cargo handling per berth.

The total cost C_S^T consumed in a port during the period T consists of two different costs: i.e., the cost of providing S berths and the cost of ships present in the port [24,25]:

$$C_S^T = c_b TS + c_s T(\bar{n}_S)_k \tag{34}$$

where

c_b = average daily cost of a berth
c_s = average daily cost of a ship

Both sides of Equation (34) are divided by $c_s T$, and the

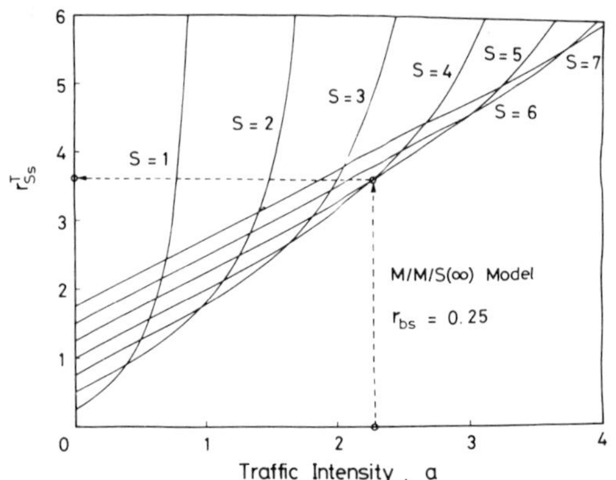

FIGURE 30. Optimum berth capacity curve (M/M/S(∞) Model) (Source: Reference 24).

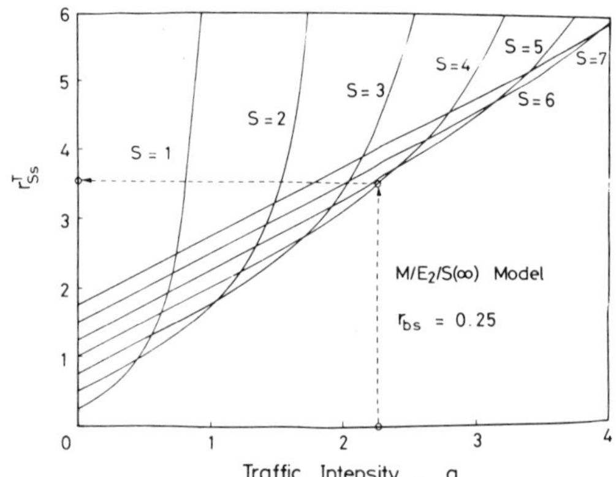

FIGURE 31. Optimum berth capacity curve ($M/E_2/S(\infty)$ Model) (Source: Reference 24).

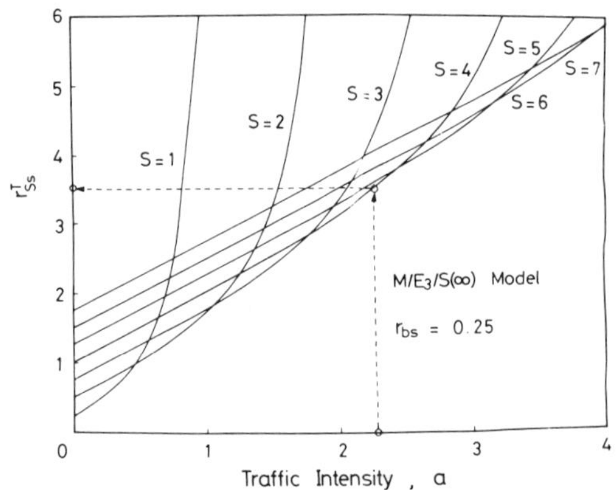

FIGURE 32. Optimum berth capacity curve ($M/E_3/S(\infty)$ Model) (Source: Reference 24).

substitution of Equation (32) yields:

$$r^T_{Ss} = r_{bs}S + \frac{a^{S+1}}{(S-1)!(S-a)^2} \times \left[\sum_{n=0}^{S-1}\left(\frac{a^n}{n!}\right) + \frac{a^S}{(S-1)!(S-a)}\right]^{-1} \times \left[\frac{1+\left(\frac{1}{k}\right)}{2} + \left(1-\frac{1}{k}\right)\left(1-\frac{a}{S}\right) \times (S-1)\frac{(4+5S)^{1/2}-2}{32a}\right] + a \tag{35}$$

where

r^T_{Ss} = ratio of the total cost for a port with S berths to the ship cost ($=C^T_S/c_sT$)

r_{bs} = berth–ship cost ratio ($=c_b/c_s$)

The relation between r^T_{Ss} and a is drawn in Figures 30, 31, and 32 for the cases of $r_{bs} = 0.25$ and $k = 1$, 2, and 3, respectively. In those figures that are called "optimum berth capacity curves," the number of berths S is regarded as a parameter.

The method to estimate the optimum number of berths by using these curves has been developed by Noritake and Kimura [24] and is exemplified as follows when given the data: $Q = 10^6$ tons; $R - 1200$ tons per day; $T = 365$ days; $c_b = \$600$ per day; and $c_s = \$2400$ per day.

1. Compute the value of berth–ship cost ratio r_{bs} from the given data of c_b and c_s. $r_{bs} = 600/2400 = 0.250$.
2. Determine the value of traffic intensity a using Equation (33). $a = 10^6/(1200)(365) = 2.28$.
3. By using Figures 30, 31, and 32, the optimum number of berths S and the value of total cost ratio r^T_{Ss} for $M/M/MS(\infty)$, $M/E_2/S(\infty)$, and $M/E_3/S(\infty)$ models are found. They are given in Columns 2 and 3, respectively, of Table 3. (The procedure to use the graphs in Figures 30, 31, and 32, is shown by the broken line in each graph.)
4. Calculate the value of total cost in a port C^T_S ($=c_sTr^T_{Ss}$) and the average number of ships present in a port $(\bar{n}_s)_k$ by the use of Equation (34). They are shown in Columns 4 and 5, respectively, of Table 3.
5. Compute the average number of ships waiting for berths $(\bar{n}_{w,S})_k$ by using Equation (32), and it is indicated in Column 6 of Table 3. By comparing this value with the actual observed data, the propriety of computation results can be verified.

The optimum berth capacity curves may provide proper planning estimates of the optimum number and capacity of berths.

Berth and Slip Dimensions

The area needed alongside a wharf for the mooring is dependent on:

1. Size and type of vessels accommodated
2. Wharf configuration
3. Berthing procedures (including the use of tugs)

Since the space requirements for bulk shipping terminals vary with the type of material and the cargo handling procedures and equipment, there are no widely admitted standards for berth and slip dimensions for these terminals. In contrast, general cargo terminals are conventional and generally accepted dimensions may be helpful in the planning of these terminals.

The necessary berth length equals the length of a ship, plus an interval between adjacent ships and room for the ships' lines. Thus, a typical berth length for general cargo ships is around 750 to 850 ft.

For a marginal wharf that accommodates general cargo ships, a minimum length of 750 ft is needed for a one-berth facility and 650 ft per berth should be allowed for multiple-berth facilities. A desirable pier length for a two-berth pier is 850 ft and for a four-berth pier, 1500 ft (see Figure 33).

When two-berth piers are used, a slip width no less than 250 ft is required. This dimension is the sum total of the beams of two ships and the length of a tug. A slip width of at least 325 ft is needed for four-berth piers. This dimension equals the beams of two ships (berthed) plus the beam of another ship (maneuvering) and the length of a tug.

The minimum and desirable pier length and slip width for a two-berth pier and a four-berth pier are summarized in Table 4.

Transit Sheds and Aprons

At a general cargo wharf, a short-term storage facility named a transit shed is provided (see Figure 33). The transit shed may also house the offices for customs and for port administration and security. The transit shed should not be used for long-term storage. When long-term storage space is needed, warehouses and open storage areas normally are placed landward of the transit shed.

For a typical dry cargo ship that carries 6250 measurement tons of cargo, an area of 85,000 to 90,000 ft^2 is required for a transit shed. This value is based on space to accommodate a single average-sized ship and includes an allowance of nearly 40% for aisle space and other nonstorage areas. The 90,000 ft^2 requirement allows for the handling of the entire shipload of an average-sized ship. With larger ships currently in service, transit sheds with space as large as 120,000 ft^2 have been located at some major ports [26]. Proportionately smaller transit sheds may be provided at small wharves where vessels are expected to unload and load only partial loads.

TABLE 3. Computation Results for Estimation of Optimum Number of Berths.

Model (1)	S (2)	r_{SS}^T (3)	C_S^T (10^6 \$) (4)	$(\bar{n}_S)_k$ (5)	$(\bar{n}_{w,S})_k$ (6)
$M/M/S(\infty)$	5	3.61	3.16	2.36	0.08
$M/E_2/S(\infty)$	4	3.54	3.10	2.54	0.26
$M/E_3/S(\infty)$	4	3.52	3.08	2.52	0.24

(*Source*: Reference 24).

The normal length of the transit shed is 500 to 550 ft. The larger value provides a sheltered storage space adjacent to the entire length of a large ship. A transit shed width of at least 165 ft is required for marginal wharves. In the case where a single shed serves two berths, one on each side of a pier, about twice this value would be needed to provide an area of 90,000 ft^2 per berth. The configuration of the transit shed may depend on local conditions [26].

A 20-ft-wide covered platform along the shoreward side of the transit shed is required to enable cargo to be directly transferred between trucks and railroad cars and the transit shed.

The area of a pier or wharf between the quay face and the transit shed is termed the apron (see Figure 33). This uncovered space is used for mooring, and for the loading, discharging, and transfer of cargo between the ship's hold and transit shed. Along the waterfront edge of the apron,

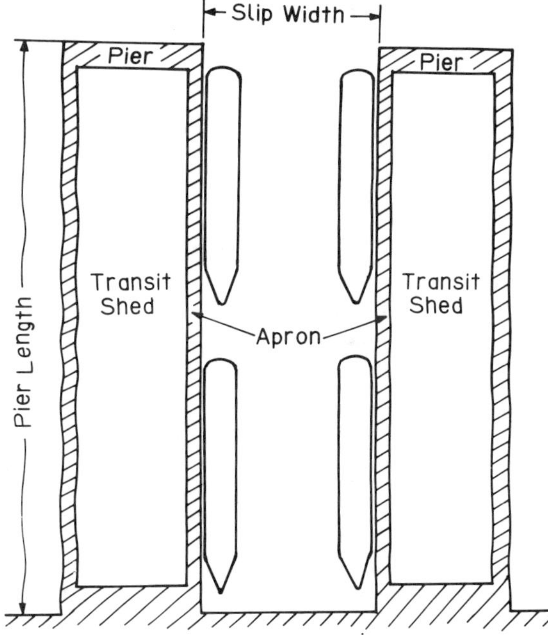

FIGURE 33. Four-berth pier and slip.

TABLE 4. Typical Dimensions for Berth and Slip.

	Pier Length (ft)		Slip Width (ft)	
	Minimum	Desirable	Minimum	Desirable
Two-berth piers	750	850	250	300
Four-berth piers	1375	1500	325	375

(*Source*: Reference 16).

space must be shared for bollards, cleats, and other mooring devices. Connections for electric power and water service must be supplied.

When direct transfer from a ship to an inland carrier is desired, railroad tracks and roadways are installed along the apron. Rail-supported gantry cranes may also be provided along the apron.

The minimum acceptable apron width will vary from 20 ft to more than 60 ft depending on the cargo-handling facilities and equipment provided. Table 5 shows principal dimensions for general cargo wharves recommended by the American Association of Port Authorities [26].

Fenders

While the ship is berthing alongside the quay wall, damage can be sustained by the dock and ship unless some kind of protective device is provided to absorb and distribute the impact energy of the mooring ship. Such a device also is needed to decrease the effect of the striking and chafing of the ship against the dock while the ship is moored. Protective installations that satisfy these requirements are termed dock fenders. Fenders may be divided into the following groups:

1. Pile fenders
2. Timber-hung fender systems
3. Rubber fenders
4. Gravity type fender systems

Original type fender systems involve a row of vertical timber piles that are driven on a slight batter and supported at the top of the quay wall. By this fender system, energy from berthing impacts is absorbed by deflection and by compression of the timber. A floating timber, termed a camel, or hard wood often is provided between the vessel and the pile fender system to distribute impact along the fender system and keep the vessel away from the quay wall. The energy absorption of timber pile fenders is influenced by the pile diameter and length and the kind of wood. The energy absorption capacities of timber fender piles diminish keenly with wear and tear.

TABLE 5. Principal Dimensions for General Cargo Terminals.

Berth lengths	
Wharves	750 ft multiples[a]
Piers	850 ft
Apron widths	
No railroad tracks	30 ft
One railroad track	30 ft
Two railroad tracks	38.5 ft
Clear stacking height—sheds	20 ft
Gross transit shed space per berth	50,000 to 120,000 ft^2
Interior column spacing	40 ft minimum

[a]1 ft = 0.3048 m. (*Source*: Reference 26).

Pile fender systems made of steel and concrete also have been used, but only rarely.

In locations where the water is tranquil and the tidal range is small, a timber-hung fender system may be employed. In this system, vertical timber members are fastened to the quay wall and end near the water level. Normally, horizontal timber members are attached between the vertical members and the quay wall. Timber-hung fender systems have low energy-absorption capabilities.

Various kinds of rubber fenders are used widely and effectively. Cylindrical or rectangular rubber blocks are used in timber fender systems to increase energy-absorption capacity (see Figure 34). These blocks, that are provided on the front side or behind or between other horizontal members or vertical piles, absorb the impact energy by compression.

The Raykin rubber fender unit is built of rubber and steel sandwiches cemented together to form an arched beam (see Figure 35). Impact energy is absorbed by this device as the resilient rubber units distort in response to the shearing forces exerted by a berthing ship. Lord fenders consist of an arcshaped rubber block bonded between two flat steel mounting plates. The lord fender absorbs the energy by the bending and compression of a rubber "buckling column."

Hollow rubber cylinder fenders may also be used. These are manufactured with outside diameters ranging from 5 to 18 inches. The inside diameter is normally one-half the outside diameter. Hollow cylindrical fenders are installed on the quay wall supported by brackets and bars or by frames suspended in heavy chains. These fenders are fitted to protect a solid and deep wall such as the face of a relieving platform-type of substructure.

The energy-absorption characteristics of rubber fenders are given in graphs available from the various manufacturers of these products.

Gravity-type fender systems are made of heavy weights such as large concrete blocks or cylinders that are suspended from the edge of the wharf deck and lifted by the berthing ship. The use of gravity fenders is not so common.

In choosing an appropriate type of fender system, the most important factors to consider are as follows:

1. Displacement tonnage of ships to be moored
2. Velocity of mooring (normal to the wharf)
3. Environmental conditions at the port

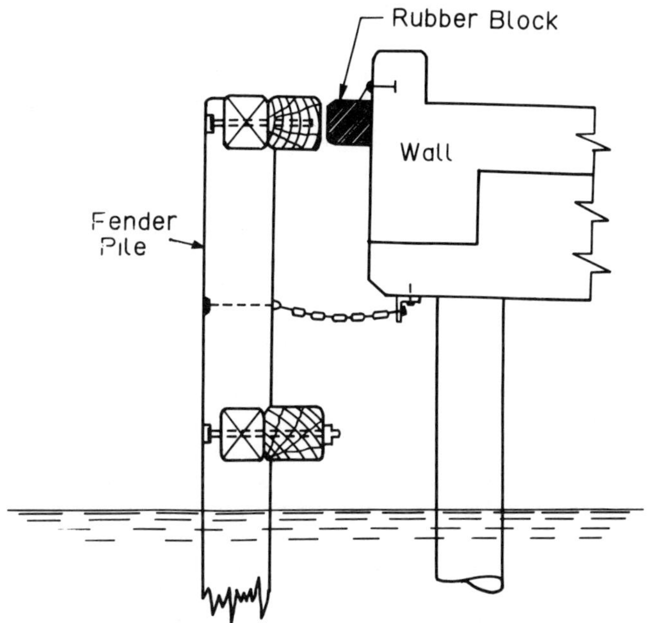

FIGURE 34. Rubber block and fender pile.

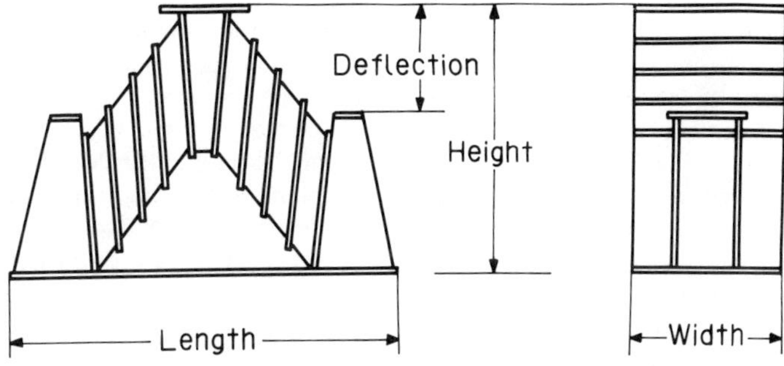

FIGURE 35. Raykin rubber fender unit.

The velocity of approach normal to the fenders adopted for design may be between about 0.1 ft/s and 1.25 ft/s, depending on such conditions as the loading condition of ship, exposure of the wharf to wind, waves, tides and currents, water depth, maneuvering space, bow thrusters, and number of tugs used. The guidance on choice of berthing velocity has been described by Lee [27].

The kinetic energy of impact is given by the equation:

$$E = \frac{1}{2} Mv^2 = \frac{1}{2} \frac{W}{g} v^2 \qquad (36)$$

where

E = energy
M = mass of ship
v = berthing velocity of ship normal to the face of wharf
W = displaced weight of ship
g = acceleration due to gravity

For vessels mooring at moderate to high velocities, the kinetic energy is increased owing to the mass of water moving alongside the vessel. To make allowance for this effect, the value of mass employed in Equation (36) should be increased by approximately 60%.

For planning purposes, one half of the kinetic energy is assumed to be absorbed by the fender system and wharf, and the share of the energy between them may depend on the type and construction of the fender system and the deflection of the wharf if it is designed as a flexible structure.

Dolphins

Dolphins are clusters of piles driven into the bottom of a harbor and bound fast together for the mooring of ships. They are usually used in combination with piers and docks to decrease the length of these structures. Dolphins are a chief part of the fixed-mooring-berth type of installation employed widely in bulk-cargo loading and discharging facilities. In addition, they are used for tying up vessels and for transferring cargo between vessels berthed along both sides of the dolphins. There are two classes of dolphins: breasting and mooring (see Figure 36).

Breasting dolphins, commonly by far the larger of the two classes, are built to support fenders that absorb berthing impact and broadside wind loads on the moored ship. Breasting dolphins may also support bollards or mooring posts to take the vessel's lines, especially spring lines for moving a vessel along the wharf or holding it against the current.

Mooring dolphins are designed for picking up the pull from the hawsers. They are located at some distance behind the berthing line, about 45° off the bow and the stern of a vessel of normal size, so that the mooring lines ordinarily will be not less than 200 ft nor more than 400 ft long. The largest vessels may need two additional dolphins, off the bow and stern. These dolphins are commonly placed so that the mooring lines will be perpendicular to the wharf, which makes them most effective for holding the vessel against a broadside wind. Mooring dolphins are furnished with bollards or mooring posts and with capstans in the case where heavy lines are to be used. The maximum pull commonly should not be greater than 50 tons on a single line, or 100 tons on a single bollard when two lines are employed.

Design of dolphins may be of the flexible or rigid type. The wood-pile clusters shown in Figure 37 are examples of the flexible type. These are driven in clusters of 3, 17, 19, etc., piles, that are bound with galvanized wire rope. The center pile of each cluster is normally allowed to stretch about 3 ft above the other piles to afford a means of fastening a vessel's mooring lines. Large steel cylinders and clusters of steel pipe piles have also been employed to construct flexible dolphins mainly for mooring. The selection of the pile dimensions depends on the loads, water depth, bottom soil conditions, and available equipment such as driving hammers or cranes on the location.

Dolphins of the flexible type have been employed for mooring small ships, less than 5000 deadweight tonnage, as an outer defense of wharves, or for breasting off somewhat larger ships from loading platforms and structures not installed to absorb the berthing impact of vessels. Bottom soil conditions must be proper for flexible type dolphins.

For larger cargo vessels and tankers of 9000 to 17,000 deadweight tonnage, a wood-platform type of rigid dolphin, using wood batter piles, may be employed for berthing and breasting. As the wood platform is relatively light, its horizontal stability highly depends on the pullout value of the wood piles. Usually, a horizontal force of around 40 to 50 tons is nearly the maximum that a dolphin of this type can withstand without becoming too bulky.

If bottom soil conditions are fitting, sheetpile cells make superior dolphins. They can be built to resist impact from the largest vessels, if provided with appropriate fenders. Heavy sheetpile cell dolphins are well-located as turning dolphins for wraping or turning the vessel around at the end of the wharf. Cellular dolphins are normally capped with a heavy concrete slab, to which the mooring post or bollard is attached. When large vessels are to be accommodated, a powered capstan should be used to draw in the heavy mooring lines.

For big vessels, dolphins may be constructed with heavy concrete platform slabs supported by vertical and batter piles, commonly of steel, although precast concrete is also employed. This type of dolphin with low-reaction-force, high-energy-absorption rubber fenders can sustain berthing impact from the largest supertankers. For this purpose, many batter piles are needed. The uplift from these piles, in turn, makes it necessary to have a large amount of deadweight because the vertical piles will withstand only a small part of the uplift. This deadweight is provided by the con-

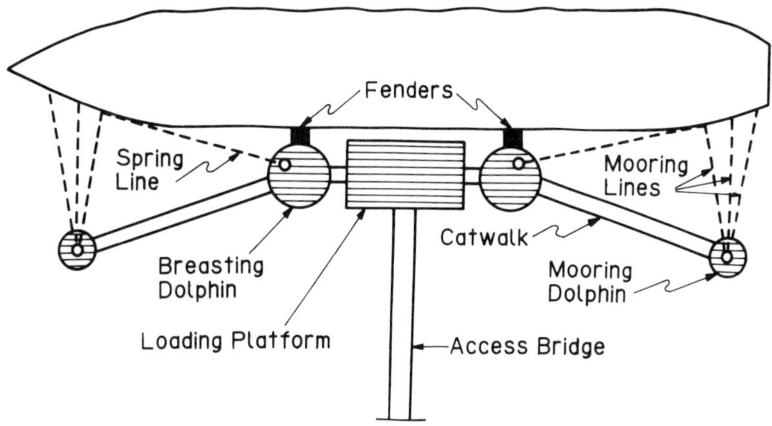

FIGURE 36. Breasting dolphins and mooring dolphins.

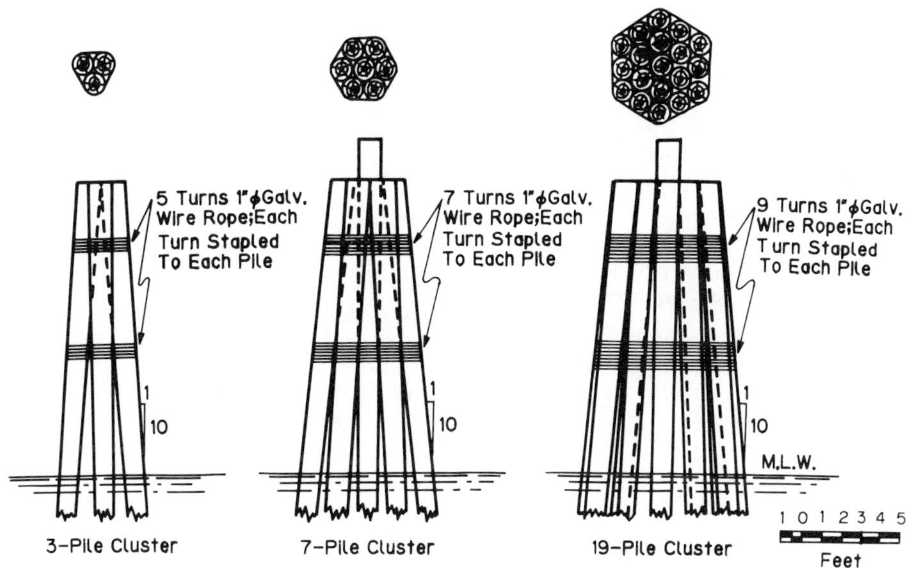

FIGURE 37. Flexible wood-pile dolphin (Source: *Design and Construction of Ports and Marine Structures*, Second Edition, by Quinn, A. D. Copyright © 1972, McGraw-Hill. Reproduced with permission of McGraw-Hill Book Co.)

crete slab, which may be 5 to 6 ft thick. An adequate number of vertical piles must be used to support this deadweight.

Container Ports

Formerly, cargo at ports was handled in small units. This package-by-package cargo handling method was time-consuming and expensive. The introduction of containers omits much of the handling of small units of cargo at a port.

In 1956, Sea-Land Service, Inc. in the United States developed a new concept in merchant shipping that came to be familiar as containerization [28]. The primary savings related to containers are those resulting from a reduction of the ship's turnaround time and in cargo handling costs at ports.

Containers are uniform modular boxes, commonly 8 by 8 ft in cross-section and 12 to 40 ft in length. Table 6 shows the standard dimensions and weights of containers proposed by the International Standards Organization. A container is charged and sealed at the place of origin and remains sealed until it arrives at its destination. However, a fraction of the containers are stuffed and stripped at the port.

Since containers are designed to be interchangeable between ship, truck, and rail transportation, it is necessary that container ports be easily accessible by highways and railroads. Container ports are, in principle, quick transit facilities at the interface between sea and land transport. Containers may arrive at a port by rail, but are more usually by trucks. Upon arrival the containers are weighed, logged in, and stored transiently in an appointed position in a marshaling yard. The containers are later transferred from the marshaling yard to shipside and lifted onto a ship by means of heavy gantry cranes. This process is reversed at the destination. A typical layout of a container port is shown in Figure 38. Advantages for a container port are as follows:

1. The throughput capacity of berth is great, often being five times as high as that of a traditional general cargo berth.
2. Overall conveyance time is less.
3. Damage to cargo is decreased.
4. Pilferage is lessened.

TABLE 6. Standard Dimensions and Weights of Containers.

Length (ft)	Volume (m³)	Weight (kg) (Empty)	Weight (kg) (Loaded)
20	31	1,850	20,320
30	46	2,600	25,400
40	62	3,200	30,580
40 (insulated)	58	4,350	30,480

(*Source*: Reference 29).

The main disadvantage of a container port is the large open space needed for the marshaling area. The requirement to furnish large and costly container cranes and other handling equipment is also a disadvantage.

The components of a typical container port are as follows:

1. Berths. The berth lengths are dependent on the size of the ships to be moored. The largest containerships in service have a length of 950 ft, a width of 110 ft, and a draft of 35 ft. The American Association of Port Authorities [26] recommends a minimum berth length of 850 ft up to a maximum of 1000 ft. Container berths of the marginal wharf type are preferable to those of the finger pier type on account of the requirement to provide an extensive land area for each berth.
2. Container Cranes. Specialized gantry cranes equipped with spreaders for container handling must be installed on shore. Usually, two or more cranes working together will load and discharge a containership. These cranes are railmounted on the wharf and commonly have a lift capacity ranging from 30 to 45 metric tons. The minimum time taken for one operational cycle of a container crane is approximately two minutes; however, this handling rate cannot be maintained over longer periods of time and averages of about twenty containers per hour are normal.
3. Marshaling Areas. One of the most important factors related to container port planning is the extent of yard area that is needed to accommodate a certain amount of container throughput. This factor is closely related to the equipment employed for transport to and from the yard, for stacking, delivery, and reception on the land side. Two types of container storage systems predominate in the marshaling area: chassis storage and stacked storage. In the chassis storage system, a container discharged from a ship is put on a semitrailer chassis and then towed by a yard tractor to an assigned spot in the yard where it is stored on the trailer until carried by a road tractor for inland transport. Similarly, arriving outbound containers are assigned positions in the marshaling yard and stay there until being pulled to the container crane by yard tractors.

   The chassis storage system affords operational efficiency, as each container is immediately available to a tractor unit. However, chassis storage needs more yard area and more chassis than do other systems.

   In the case of stacked storage, containers are stacked by straddle carriers or by a traveling bridge crane. Straddle carriers are special pieces of equipment that pick up the containers and move them between apron and storage yard or onto trucks or railroad cars. Straddle carriers can stack the containers two or three high and remove them from the stacks. Traveling bridge cranes, normally combined with tractor-trailer transport between apron and yard, can stack the containers three or four high on the

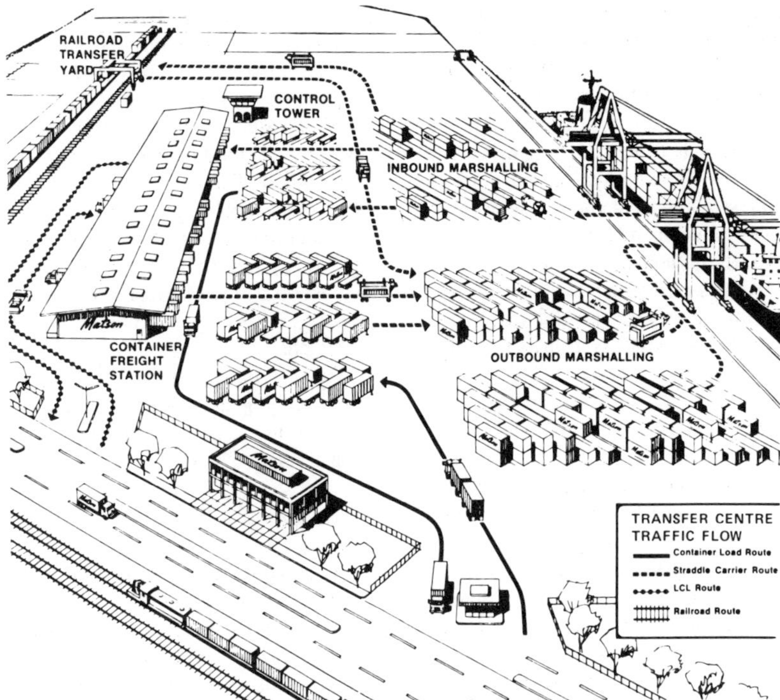

FIGURE 38. Typical layout of container terminal (Source: Reference 26).

ground. Stacked storage needs far less space than does chassis storage, but is not as efficient because rehandling is necessary on land.

4. Container Freight Station. A shed, called a container freight station, for packing and unpacking containers should be located behind the marshaling area. The area of a freight station varies widely. Its configuration is similar to a truck terminal, permitting trucks to arrive at one side of the shed and the cargo to be transferred from these trucks directly into containers on the opposite side. Thus, the configuration is preferred to be long and narrow with stress on the number of truck and container doors necessary.
5. Entry Facilities. The truck entrance to the container terminal is normally composed of two to six lanes in each direction. Each lane is furnished with a truck scale to weigh the passing containers. A receiving and delivery office is placed at the entry-exit position to keep the records and assign storage locations to incoming containers.
6. Garage and Inspection Building. A small building for the physical inspection of entering or departing containers is placed near the entrance and adjacent to the marshaling yard. In addition, a garage may be supplied to maintain the stevedoring equipment used to handle the containers in the marshaling yard.

Roll-on Roll-off Terminals

On account of the high investment costs of container ports with conventional lift-on/lift-off (*LO/LO*) facilities, a large number of berths for roll-on/roll-off (*RO/RO*) cargo handling have been constructed. These systems are often combined with conventional cargo handling systems.

Roll-on roll-off service can be supplied by three basic handling methods [29]:

1. The tractor and trailer unit drives on, stays on the ship during its voyage and drives off at the port of destination. This procedure is proper only for short sea trips because the tractor unit occupies valuable space in the ship and remains idle during the sea journey.
2. The tractor unit tows the container from the storage area onto the vessel and drives off leaving only the trailer and container for the sea trip. Another tractor unit is needed to tow the trailer off the vessel at the destination port. The towing unit may be a normal road unit or a smaller unit only for use in the port area.
3. A tractor unit tows the trailer and container onto the ship where a straddle carrier or ship crane transfers the container aboard and stacks it on the ship. Only the container remains on the ship during the sea journey. The procedure is reversed at the port of destination.

The method of handling containers, the design of berths, and the layout of terminal area are closely connected with the design of the vessel. Where roll-on/roll-off operations are employed, special mooring facilities may be needed to hold the ship in place. Shore ramps may be equipped in the bow, stern, or side of the vessel. It may become necessary to provide special adjustable shore ramps on the dock to adapt variations in ship design and in the tide. Where cargo is handled through the side of a ship, additional wharf apron area will be required to accommodate the traffic entering and leaving the ramp. As roll-on/roll-off loading and discharging can be executed in a short time, a sufficiently large space will be needed for the processing and transient storage of the cargo.

Lighter-Aboard-Ship Terminals

A new unitized cargo idea developed to maximize utilization of oceangoing vessels and port facilities is the lighter-aboard-ship (*LASH*) system. With this system, cargo is loaded onto large barges (lighters), then the barges and cargo are floated (towed) to and lifted aboard a specially designed mother vessel for the sea trip. The lighters are lifted aboard the mother vessel by a large deck straddling crane, or are loaded by a stern elevator onto the vessel after being floated to the stern. To and from the crane or elevator, the lighters are moved transversely on several decks. In the port of destination, the lighters and cargo are discharged and shifted to terminal facilities where the cargo is unloaded from the lighters (see Figure 39).

The main advantage of the *LASH* system is that it may not require special deep-water port facilities. Vessel size is not restricted by water depths at quay walls in any port, and costly port facilities such as wharves, piers, jetties, transit sheds, etc. are unnecessary. However, some of the lighter-carrying vessels are required to move both lighters and containers to a container terminal. The reduction of turnaround time is also an advantage of the *LASH* system.

Special installations of lighter-aboard-ship terminals include berths equipped with cranes and other cargo-handling equipment, anchorage areas for tugs shelterd from heavy waves and current actions, and a storage basin for the vacant and loaded lighters.

Dry Bulk Cargo Terminals

A large variety of goods can be carried unpackaged without the use of pallets or containers. Such goods, called dry bulk cargo, include grain, coal, sand, gravel, cement, salt, sugar, scrap metal, and ores. Dry bulk cargo is usually handled by shore-based facilities or ship-mounted equipment. These specialized high-capacity transfer devices are used to achieve quick turnaround of the vessels and preferably are separated from general cargo and container ports.

Dry bulk cargo vessels are inclined to be larger than general cargo vessels and may need deeper channels and docks, stronger fender systems, and larger storage areas. The storage type may be an open yard, shed, silo, or slurry pond, depending on the type of commodity.

There are a wide variety of efficient dry bulk handling systems, depending on the characteristics of commodities to be handled. They include belt conveyors, cranes or derricks with clam shell or grab buckets, pipelines that transfer light commodities pneumatically or in a hydraulic suspension, and many others. Mechanical systems normally need less power and require great investment costs, whereas slurry systems need large amounts of power and require lower investment.

At a dry bulk cargo terminal, special facilities are needed for weighing the commodity, for protection from hazards such as fires and explosions, and for controlling dust and contamination [26].

Petroleum Terminals

The major cargo category in world trade is crude oil and petroleum products in bulk, and the recent worldwide growth in demand for these products has created a need for larger tankers. The size of tankers has increased remarkably; the largest tankers are now 0.6 million dead weight tons, over 1300 ft long and more than 200 ft wide. From 1967 to 1977, the average draft of tankers in the world increased from 31 to 37 ft. The drafts of the newest tankers approach 100 ft [30].

The great draft of these tankers has created a need for many special deep-water mooring facilities. Petroleum terminals use wharves, piers, and offshore moorings depending on the water depth, bottom conditions, and the rate of discharging. While many ports have dredged deeper channels, others have constructed petroleum terminals thousands of feet off-shore, connected to the land-storage facilities by means of large submarine pipelines for loading and unloading oil.

Two fundamental types of off-shore berthing facilities have been developed: buoy moorings and fixed platforms. These facilities are becoming more and more popular due to growing tanker sizes.

A conventional and simple device for the off-shore handling of oil is called buoy moorings. It consists of proper anchorage facilities adjacent to a system of submarine pipelines equipped with hoses lying on the seabed and attached to marker buoys on the surface. Usually, a vessel ties up at a fixed position by means of multiple chain-anchor moorings. These moorings are so arranged that the vessel keeps a fixed position and orientation, facing in the direction of the wind and waves. The vessel's hoisting tackle is employed to bring the hoses to the surface and to facilitate their connection to the vessel.

Buoy moorings are apt to become impractical where more than two 12-in. hoses are needed. Hoses larger than 12 in.

FIGURE 39. Seabee vessel (Source: Lykes Lines).

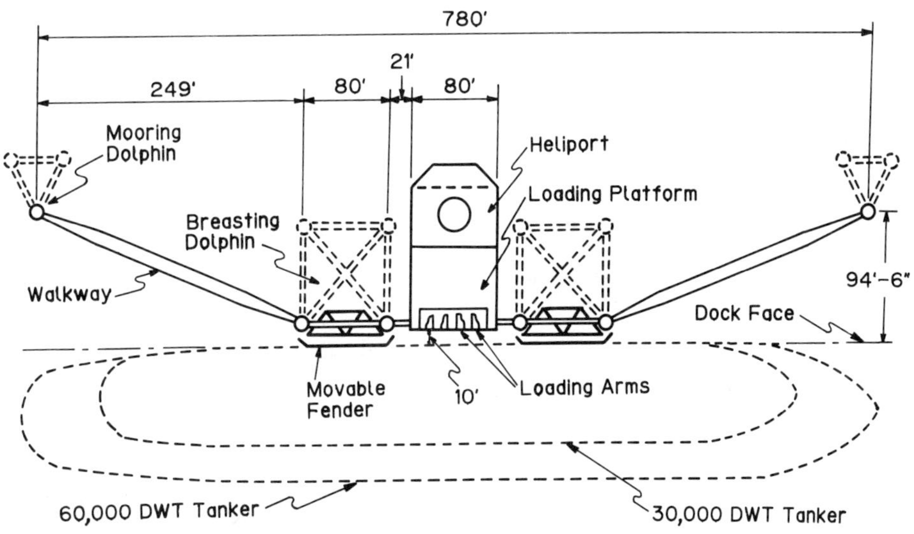

FIGURE 40. Offshore tanker berth (Source: Reference 31).

are difficult to handle with the vessel's gear. Further, high winds and unsteady seas prevent mooring of tankers and operation of the terminal. On account of the high value of the tanker's time, it has become important to construct mooring facilities capable of providing vessel turnaround in the least possible time.

Although great in capital cost, fixed platform facilities provide the most reliable and efficient tanker mooring and loading device possible without the development of an expensive deep-water terminal (see Figure 40). Normally, the fixed-platform terminal consists of a tower on which a rotating mooring boom is mounted. During loading and unloading the mooring lines are attached to the mooring boom, with the vessel heading into the wind and swells. A 500- to 600-ft loading arm stretches along the vessel to facilitate the handling of loading hoses that connect the vessel and a submarine pipeline stretching to the land. When not in use the loading arm is submerged; during the loading, it can move freely alongside the tanker to accommodate the motion of the vessel relative to the fixed-platform.

REFERENCES

1. Gerstner, F., ''Theorie die Wellen,'' *Abhandlungen der Könîgichen Bohmischen Gesellschaft der Wissenschaften*, Prague, Czechoslovakia, 1802.
2. Airy, G. B., ''On Tides and Waves,'' *Encyclopaedia Metropolitana*, 1845.
3. Munk, W. H., ''Proposed Uniform Procedure for Observing Waves and Interpreting Instrument Records,'' *Wave Project*, Scripps Institute of Oceanography, LaJolla, Calif., 1944.
4. Longuet-Higgins, M. S., "On the Statistical Distributions of the Heights of Sea Waves," *Journal of Marine Research*, Vol. *11*, No. 3, 1952, pp. 245–266.
5. Goda, Y., *Random Seas and Design of Maritime Structures*, University of Tokyo Press, Tokyo, 1985.
6. Sainflou, M., ''Treatise on Vertical Breakwaters,'' *Annals des Ponts et Chaussees*, Paris, 1928.
7. Miche, R., ''Mouvements Ondulatoires de la mer in Profondeur Constante ou Decroissante,'' *Annals des Ponts et Chaussees*, Paris, Vol. *114*, 1944.
8. Rundgren, L., ''Water Wave Forces,'' *Bulletin No. 54*, Royal Institute of Technology, Division of Hydraulics, Stockholm, Sweden, 1958.
9. *Shore Protection Manual*, 4th ed., 2 vols., U.S. Army Engineer Waterways Experiment Station, Coastal Engineering Research Center, U.S. Government Printing Office, Washington, D.C., 1984.
10. Minikin, R. R., "Breaking Waves: A Comment on the Genoa Breakwater," *Dock and Harbour Authority*, London, 1955, pp. 164–165.
11. Minikin, R. R., *Winds, Waves and Maritime Structures: Studies in Harbor Making and in the Protection of Coasts*, 2nd rev. ed., Griffin, London, 1963.
12. Bagnold, R. A., ''Interim Report on Wave Pressure Research,'' *Journal of the Institution of Civil Engineers*, Vol. *12*, London, 1939.
13. *Port Design and Construction*, American Association of Port Authorities, Washington, D.C., 1964.
14. Quinn, A. DeF., *Design and Construction of Ports and Marine Structures*, Second Edition, McGraw-Hill Book Company, New York, 1972.
15. Brant, A. E., Jr., ''The Port of Chicago,'' *Proceedings*, American Society of Civil Engineers, Vol. *84*, No. WW4, September 1958.
16. Paquette, R. J., Ashford, N. J., and Wright, P. H., *Transportation Engineering: Planning and Design*, Second Edition, John Wiley & Sons, Inc., New York, 1982.
17. Fratar, T. J., Goodman, A. S., and Brant, A. E., Jr., ''Prediction of Maximum Practical Berth Occupancy,'' *Transactions*, American Society of Civil Engineers, Vol. *126*, Part IV, 1961, pp. 632–643.
18. Plumlee, C. H., ''Optimum Size Seaport,'' *Journal of the Waterways and Harbors Division*, American Society of Civil Engineers, Vol. *92*, No. WW3, Aug. 1966, pp. 1–24.
19. Mettam, J. D., ''Forecasting Delays to Ships in Port,'' *Dock and Harbour Authority*, Vol. *47*, Apr. 1967, pp. 380–382.
20. Nicolaou, S. N., ''Berth Planning by Evaluation of Congestion and Cost,'' *Journal of the Waterways and Harbors Division*, American Society of Civil Engineers, Vol. *93*, No. WW4, Nov. 1967, pp. 107–132.
21. Jones, J. H. and Blunden, W. R., ''Ship Turn-Around Time at the Port of Bangkok,'' *Journal of the Waterways and Harbors Division*, American Society of Civil Engineers, Vol. *94*, No. WW2, May 1968, pp. 135–148.
22. Agerschou, H. and Korsgaard, J., ''Systems Analysis for Port Planning,'' *Dock and Harbour Authority*, Vol. *49*, Mar. 1969, pp. 411–415.
23. Kendall, D. G., ''Stochastic Processes Occurring in the Theory of Queues and Their Analysis by the Method of the Imbedded Markov Chain,'' *Ann. Math. Statist.*, Vol. *24*, 1953, pp. 338–354.
24. Noritake, M. and Kimura, S., ''Optimum Number and Capacity of Seaport Berths,'' *Journal of the Waterway, Port, Coastal and Ocean Engineering Division*, American Society of Civil Engineers, Vol. *109*, No. 3, Aug. 1983, pp. 323–339.
25. Wanhill, S. R. C., ''Further Analysis of Optimum Size Seaport,'' *Journal of the Waterways, Harbors, and Coastal Engineering Division*, American Society of Civil Engineers, Vol. *100*, No. WW4, Nov. 1974, pp. 377–383.
26. *Port Planning, Design and Construction*, American Association of Port Authorities, Washington, D.C., 1973.
27. Lee, T. T., ''Design Criteria Recommended for Marine Fender Systems,'' *Proceedings of Eleventh Conference on Coastal Engineering*, American Society of Civil Engineers, Sept. 1968.
28. Tozzoli, A. J., ''Containerization and Its Impact on Port De-

velopment,'' *Journal of the Waterways, Harbors and Coastal Engineering Division*, American Society of Civil Engineers, Aug. 1972.

29. Bruun, Per, *Port Engineering*, Third Edition, Gulf Publishing Company, Houston, 1981.
30. *A Statistical Analysis of the World's Merchant Fleets*, U.S. Department of Commerce, Maritime Administration, December 31, 1977.
31. Gaither, W. S. and Dalton, R. E., ''All-Weather Tanker Terminal for Cook Inlet,'' *Journal of the Waterways and Harbors Division*, American Society of Civil Engineers, Vol. *95*, No. WW2, 1969, pp. 131–148.

INDEX

accelerometer buoys 673
accelerometer observation 602
acid soils 585
anchor block analysis 475
Atterberg limits 23
axisymmetric excavation 337

backfill failure 470
backfills 469
bacteriological corrosion 577
bank protection 460
base failure mechanisms 357, 375
bearing capacity 47, 51, 175, 187, 449
bearing capacity factors 181
bearing capacity of clay 439
behavior of footings 70
behavior of shallow foundations 176
behaviour of soils 32
berth and slip dimensions 855
berths 853
Biot theories 312
borings 380
boundary condition 449
Boussinesq formulae 139
Boussinesq' point load method 559, 571
breaking wave forces 729
breaking waves 703, 757
breakwaters 739, 841
Bretschneider spectrum 625, 784
bulkheads 735, 847
buoys 674
buried pipelines 555, 591

cargo terminals 856
cathodic interference 584
cause of corrosion 575
cement-mortar 503
central limit theorem 167
centrifugal consolidation 445
centrifuge model 439
centrifuge model tests 440
characteristics of ocean waves 832
characteristics of ships 830
chassis storage 860
chimney foundation 432
classes of cargo 829
clay soil 189
clay stratum 441
closure piece 515
coal mines 547
coastal environment 831
coastal structures 697, 735
coating failure 586
codes and standards 505
coefficient of surcharge 375
cohesionless soils 193, 203
collapse load 52
column of fill 537
column systems 551
compaction equipment 409
compressibility 38
compressible soils 263
compression curves 255
compressive stress 141, 532
concrete 427
cone penetration test 285
cone penetrometer 25
confined compression 534
confining pressure 65
consistency limits 23
consolidation procedures 441
constitutive laws 65
constitutive laws of soils 65
construction expedients 225
contact pressure 88
contact pressure coefficients 114
contact pressure distribution 125
container cranes 860
container ports 860
control plug 542
corrosion 517, 575
cost benefit analysis 483
creep constants 245
cuboidal contraction 30
curing 515
current system 841
currents 840
cut slopes 337
cyclic shear behavior 16
cylindrical excavation 337
cylindrical pile 729

dead load 559
deep mines 532
deepwater wave climate 700
deepwater waves 698
dense sand 471
densification 13
depth contour 611
depth factors 181
deviatoric response 8
dhanori clay 100
diffusion analogy 141
dimensions of groins 739
dimensions of ships 830
directional spectral models 634
dissimilar electrolytes 577
dissimilar metals 577
dissipation tests 291
distortion of drains 400
distributed loads 145
dolphins 858
drain resistance 394
drainage of fills 527

drained sand 13
dredged material 460
dry bulk cargo 862
dynamic compaction 405
dynamic formulas 209
dynamic stresses 406

earthquake engineering 3
effective stress analysis 301
effects of preloading 400
elastic behavior 59
elastic continuum 143
elasto-plastic behavior 59
electrochemical process 576
electrochemical reaction 575
embankment monitoring 315
embankments 223, 226, 267
embankment-soil interaction 375
endochronic model 4
endochronic theory 3
energy spectrum 605
excavation in clay 337

factor of safety 377
failure criteria 454
failure mode 342
fetch 623, 686
field records 379
field vane test 282
fill 521
fills under stress 522, 527
finite element analyses 311
finite strains 307
fissured embankments 301
fixed rigid structures 787
flexible pavements 563
floating roof tanks 493
flood control 542
flow index 37
Fokker-Planck equation 170
footing settlement 447
footings 65
footings in clay 71, 83
footings in sand 80, 129
foundation failure 439
foundation movements 429
foundation types 464
free surface effects 805
friction parameters 464
frost depth controus 176
Froude-Krylov theory 803

galvanic series 578
gamma distribution 167
gap gradings 524
Gault clay 291
geologic facies map 460
geotechnical centrifuges 441
geotechnical engineering 459
geotechnical problems 372
Gibbs free energy 5
gradings 523
granular material 139
granular walls 549
graphitization 577
groins 735, 845
ground improvement 405
ground response 411
ground vibrations 413
groundwater table 187

harbors 829
healing of cracks 517
heave problems 464
heavy tamping 410
high fills 541
high water depth 753
hoop reinforced columns 551
hoop stress 569
horizontal extensometers 323
horizontal reinforcing 548
hydraulic armour 706
hydraulic piezometers 319
hydraulically placed fill 540
hydrodynamic coefficients 803
hysteretic loops 11

in situ measurements 532
inclination factors 182
inclinometer 324
inclinometer measurements 256, 392
incrustation 517
induced subsidence 417
inertia coefficients 803
irregular waves 671
isotropic consolidation 6

jetties 843
joints in cement-mortar 517

Kawasaki clay 441

large gravity structures 812
lateral capacity 215
lateral displacement 309
lateral load capacity 215
lateral movement 229, 490
lateral resistance 464
lateral stress 143
layered media 146
layered soils 187
laying conditions 566
levee crossings 465
limit surface 62
linear elastic formula 147
linear elasticity 139
lining machine 511
lining operation 515
lining thickness 515
liquid limit 25
liquidity index 37
live load effects 563
live loads 558, 571
loading test 450
logic diagram 485
long term settlements 305
low embankments 301

marshaling areas 860
Marston earth load method 570
Marston's Earth Load 555
material models 311
measured stresses 147
mechanical bonding 586
moored floating structures 815
Morison equation 788
mound breakwaters 841
multilegged structure 813

natural gas pipelines 570
negative skin friction 213
New York City building code 212
Nordlung's procedure 205
numerical modeling 636
numerical wave models 636

ocean tests in waves 801
oedometer tests 270
offshore breakwaters 749
offshore engineering 781
offshore structures 781
one point methods 27
orbital motion 656
oxygen concentrations 577

particle physical properties 65
particulate models 140
penetration resistance 195
penetration test 196
percussion testing 25
permeability inside borings 373
petroleum terminals 862
Pierson-Moskowitz Spectrum 784
piezocone 286

piezometer types 320
piezometers 229, 318, 390
pile foundations 175, 198
pile groups 207
pile load test 211
piles 724, 858
piles driven to rock 207
piling 370
pillars by fill 544
pillar-fill system 546
pipe bends 469
pipe cleaning 513
pipelines 459, 555
pipes 503
piston sampler 269
plane strain condition 449
plastic limit 29
plasticity 4
plasticity index 36
platic behavior 45
pneumatic piezometer 233
polar bonds 586
pore pressure measurement 277, 318
pore pressures 315, 446
pore water pressure 486
porewater pressure gauges 229
pore-water 19
port facilities 850
Portland Cement Association (PCA)
  Method 561
ports 829
precompression 367
prefabricated drains 397
preload settlement 378
preloading 367, 400
pressure settlement 91, 100, 129, 291
pressure-settlement curve 83, 91, 114
prestressed concrete 592
presumptive bearing capacity 196
probability 165
probability density function 166
protective slabs 561, 571
pumpability 525
pumped tailings 522
pumps fills 529
punching shear 193

random variable 165
random waves 659
rate of settlements 311
rebound 380
receiver plugs 544
reinforced earch 45
reinforced material 550
reinforced subsoil 49
remote sensing systems 603
response of sand 18
response spectra 818
revetments 847
rheologic models 67
rigid soil 430
rigid structures 813
rigidity 91
rigid-plastic model 45
Rio de Janeiro clay 313
river banks 460
road construction 223
roughness 100
rubber armour 708
rubble mound 729
rubble structures 765

sacrificial anodes 588
sampler quality 268
samples design 268
sand drains 396
sand response 3
scouring 721
sea and swell 653
sea bottom topography 751
sea surface elevation 614
seasonal variation 433
seawalls 847
sea-state persistance 644
seismic action 538
seismic loading 530
selection of port site 830
settlement 261, 308
settlement analysis 304
settlement gauges 227
settlement plates 390
settlement records 370
settlement-inclinometer casing 254
shallow foundations 175
shallow mines 547
shallow soil cover 475
shallow water spectrum 634
shape factors 181
shear bearing capacity 179
shear failure 176, 179
shear response 11
shear strength profile 347
shear strengths 38, 528
shoaling 624, 661
shoaling coefficient 662
shoaling water 678
shoaling waves 701
shrinkage limit 30
shrinking 30
single piles 200
site investigation 267, 372
site preparation 370
slip failures 376
slip surface indicators 323, 324
slope protection 460
slump limits 513
slurry fills 527
soft clay 267, 300
soft soil 223, 267, 430
soil 3, 23
soil behavior 496
soil classification 289
soil consistency 23
soil disturbance 394
soil environment 591
soil identification chart 289
soil improvement 479
soil mechanics 521
soil mechanics parameters 147
soil parameters 281
soil properties 400
soil property values 245, 400
soil settlement 464
soil surface 380
soil test evaluation 581
soil thermal conductivity 464
soil types 369
soil-foundation interaction 430
soil-water interaction 32
Spangler's single load method 558, 571
spectral analysis 605, 680
spectral models 625
spline function 69
stability analysis 301, 375
stability and settlement 484
stability of embankments 375
stability of fills 536
stability of wall 353
stability-settlement-consolidation 375
standing wave forces 717
states at microlevel 33
static analysis 200
static effects 428
steel electrodes 590
stiff fills 533
stochastic processes 168
storm bars 703
storm wave-break 703
Stotel embankment 241
strain energy 532
stray current corrosion 580
strength of clay 297
stress 140
stress components 144, 147
stress corrosion 577
stress distributions 139
strip footing 83
structural failure 710
subsoil 423
supported excavation 353
surface loading 141
surface pressure 623

surface settlement profiles 345
swell waves 699
swelling 30

---

tabular fills 536
tailings fill 522
tank behavior 493
tank farms 459
tank floor settlements 491
tanker berth 863
tank-soil interaction 484
temperature effects 423
temperature gradient 430
temperature loadings 426
temperature oscillations 423
temperature variation 428
tensile stresses 141
test fills 377, 382
thermal conductivity 427, 428
thermal gradient 430
tides 838
tilt plane 499
time dependent loading 306
time rate of settlement 388
Timoshenko's circular load method 571
toughness index 37
transit sheds 855
triaxial test 16, 273

---

ultimate baring capacity 83, 97
ultimate bearing pressure 123, 132
underground corrosion 575
underground piping 575
undrained shear strength 244
undrained strength 439
uniform gradings 523
uniform load method 561, 572
unsupported excavations 338
uplift forces 763

---

vane border 284
velocity field 52
verification testing 418
vertical capacity 200, 208
vertical deformations 260
vertical displacements 313
vertical drains 392
vertical settlement 486
vibration 415

---

water crossings 460
water particle movement 833
water surface elevations 614
waterways 460
wave analysis 604
wave celerity 654
wave diffraction 667
wave direction 674, 687
wave energy 659
wave energy levels 671
wave environment 782
wave equation 211
wave estimates 643
wave exposure 703
wave force formulas 789
wave force measurement 801
wave forces 724, 727, 755, 834
wave forecasting 597
wave grouping 616
wave height 601
wave height variability 832
wave measurement 673
wave models 616
wave observation 601
wave period variability 834
wave prediction 624, 868
wave pressure distribution 837
wave pressures 837
wave processes 661
wave recording 604
wave refraction 624, 661
wave runup 766, 849
wave spectra 611
wave statistics 675
wave theories 724
wave theory 597, 653
wave transmission 774
waves 597, 653, 697
waves and structures 697, 717
wharves 850
wheel loading 561
wickdrains 397
wicks 397
wind conditions 616
wind duration 623
wind speed 686

---

yielding pillars 553

---

zero crossing method 683